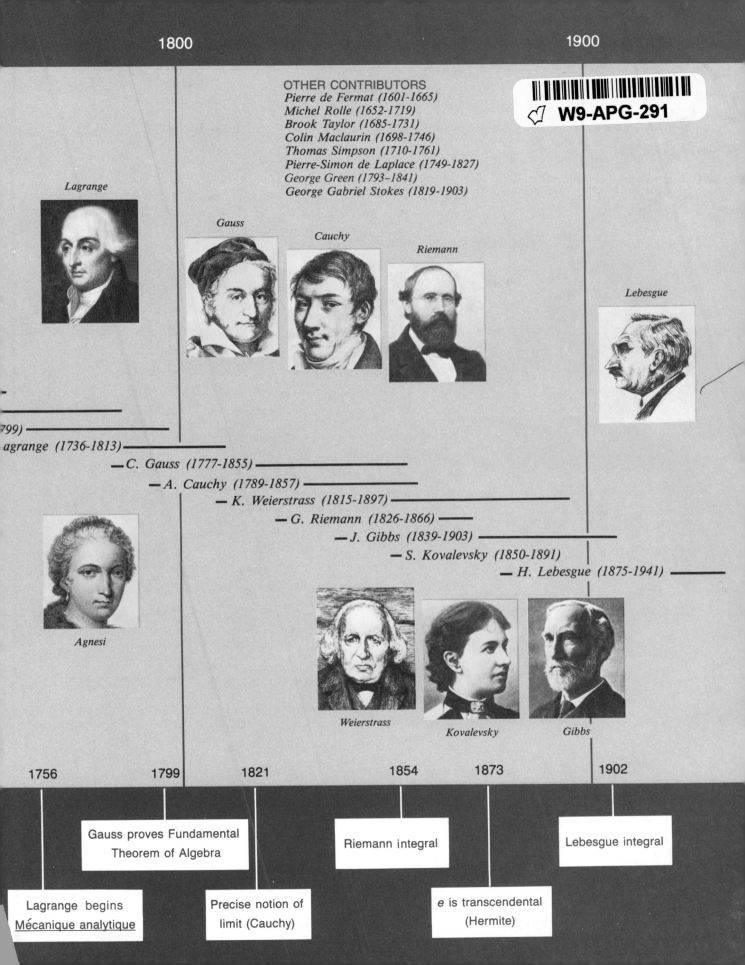

1800

1900

OTHER CONTRIBUTORS
*Pierre de Fermat (1601-1665)*
*Michel Rolle (1652-1719)*
*Brook Taylor (1685-1731)*
*Colin Maclaurin (1698-1746)*
*Thomas Simpson (1710-1761)*
*Pierre-Simon de Laplace (1749-1827)*
*George Green (1793–1841)*
*George Gabriel Stokes (1819-1903)*

*Lagrange*

*Gauss*

*Cauchy*

*Riemann*

*Lebesgue*

799) ————

agrange (1736-1813) ————

— C. Gauss (1777-1855) ————

— A. Cauchy (1789-1857) ————

— K. Weierstrass (1815-1897) ————

— G. Riemann (1826-1866) ——

— J. Gibbs (1839-1903) ————

— S. Kovalevsky (1850-1891)

— H. Lebesgue (1875-1941) ————

*Agnesi*

*Weierstrass*

*Kovalevsky*

*Gibbs*

| 1756 | 1799 | 1821 | 1854 | 1873 | 1902 |

Gauss proves Fundamental
Theorem of Algebra

Riemann integral

Lebesgue integral

Lagrange begins
Mécanique analytique

Precise notion of
limit (Cauchy)

*e* is transcendental
(Hermite)

# FORMULAS FROM GEOMETRY

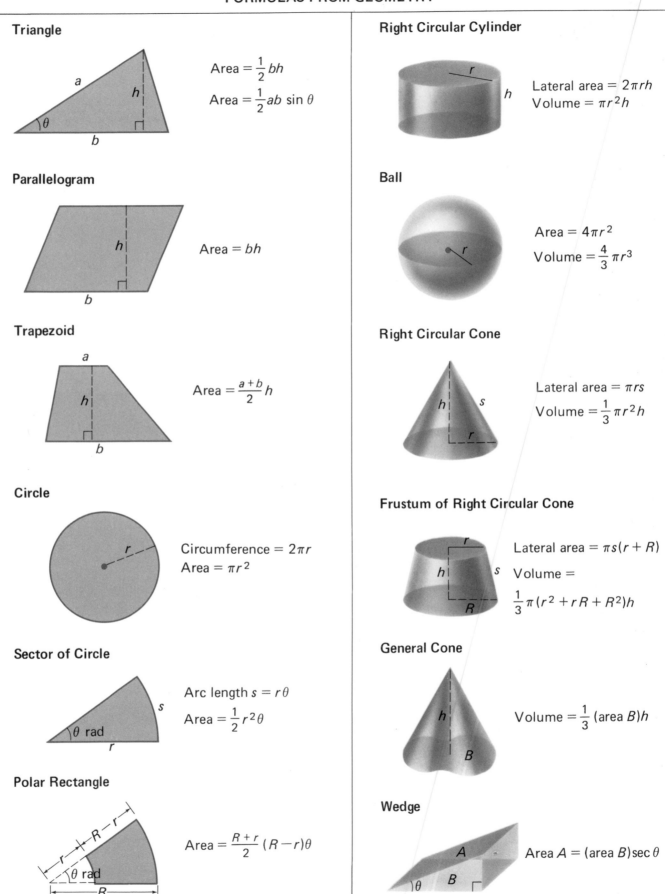

## Triangle

$$\text{Area} = \frac{1}{2}bh$$

$$\text{Area} = \frac{1}{2}ab\sin\theta$$

## Parallelogram

$$\text{Area} = bh$$

## Trapezoid

$$\text{Area} = \frac{a+b}{2}h$$

## Circle

$$\text{Circumference} = 2\pi r$$

$$\text{Area} = \pi r^2$$

## Sector of Circle

$$\text{Arc length } s = r\theta$$

$$\text{Area} = \frac{1}{2}r^2\theta$$

## Polar Rectangle

$$\text{Area} = \frac{R+r}{2}(R-r)\theta$$

## Right Circular Cylinder

$$\text{Lateral area} = 2\pi rh$$

$$\text{Volume} = \pi r^2 h$$

## Ball

$$\text{Area} = 4\pi r^2$$

$$\text{Volume} = \frac{4}{3}\pi r^3$$

## Right Circular Cone

$$\text{Lateral area} = \pi rs$$

$$\text{Volume} = \frac{1}{3}\pi r^2 h$$

## Frustum of Right Circular Cone

$$\text{Lateral area} = \pi s(r+R)$$

$$\text{Volume} = \frac{1}{3}\pi(r^2 + rR + R^2)h$$

## General Cone

$$\text{Volume} = \frac{1}{3}(\text{area }B)h$$

## Wedge

$$\text{Area }A = (\text{area }B)\sec\theta$$

# CALCULUS

Seventh Edition

# CALCULUS

## Dale Varberg

Hamline University

## Edwin J. Purcell

University of Arizona

Prentice Hall, Upper Saddle River, New Jersey 07458

**Library of Congress Cataloging-in-Publication Data**

Varberg, Dale E.
   Calculus / Dale Varberg, Edwin J. Purcell. — 7th ed.
         p.       cm.
   Includes bibliographical references and index.
   ISBN 0-13-518911-X
   1. Calculus.   2. Geometry, Analytic.   I. Purcell, Edwin J. (Edwin Joseph).
   II. Title.
   QA303.P99   1997
   515'.15—dc20                                                96-15556
                                                               CIP

Editorial Director: **TIM BOZIK**
Editor-in-Chief: **JEROME GRANT**
Assistant Vice-President of Production
   and Manufacturing: **DAVID W. RICCARDI**
Acquisitions Editor: **GEORGE LOBELL**
Production Editor: **RICHARD DELORENZO**
Managing Editor: **LINDA BEHRENS**
Prepress/Manufacturing Buyer: **ALAN FISCHER**
Manufacturing Manager: **TRUDY PISCIOTTI**
Director of Marketing: **JOHN TWEEDDALE**
Marketing Assistant: **DIANA PENHA**

Creative Director: **PAULA MAYLAHN**
Art Director: **AMY ROSEN**
Assistant to the Art Director:
   **ROD HERNANDEZ**
Interior Design: **JUDITH A. MATZ-CONIGLIO**
Cover Designer: **BRUCE KENSELAAR**
Editorial Assistant: **GALE EPPS**
Supplements Editor: **AUDRA WALSH**
Proofreader: **MARTHA WILLIAMS**

Photo credits for black and white portraits on front endpapers, New York Public Library Picture Collection.

Credit for color illustrations on pages: **251**, Garret Kallberg; **244**, NASA; **375**, Wide World; **378**, Ray Morsch/The Stock Market; **459**, "Universale" Hans Memling, Pomorskie Museum, Gdansk, Poland.

Credit for quotations in marginal boxes on pages: **4**, from G. H. Hardy, *A Mathematician's Apology* (New York: Cambridge University Press, 1941), p. 34; **143**, from Galileo Galilei, *Saggiatore*, Opere VI, p. 232, as translated on the cover of George Polya, *Mathematical Methods in Science* (Mathematical Association of America; vol. 26 in the MAA New Mathematical Library Series); **361**, from François Le Lionnais (ed.) *Great Currents of Mathematical Thought*, vol. 1 (New York: Dover Publications), pp. 68–69; **499**, from Donald E. Knuth, "Computer Science and Its Relation to Mathematics" (*Mathematical Monthly* 81, 1974, p. 323).

Credit for quotation on front endpapers: From the Foreword by Richard Courant to Carl B. Boyer, *The History of the Calculus and Its Conceptual Development* (New York: Dover Publications, 1949).

Printed in the United States of America
10  9  8  7  6  5  4  3  2

**ISBN 0-13-518911-X**

Prentice-Hall International (UK) Limited, *London*
Prentice-Hall of Australia Pty. Limited, *Sydney*
Prentice-Hall Canada Inc., *Toronto*
Prentice-Hall Hispanoamericana, S.A., *Mexico*
Prentice-Hall of India Private Limited, *New Delhi*
Prentice-Hall of Japan, Inc., *Tokyo*
Simon & Schuster-Asia Pte. Ltd., *Singapore*
Editora Prentice-Hall do Brasil, Ltda., *Rio de Janeiro*

To Jarod, Benjamin,
Aaron, Caleb, and Lauren

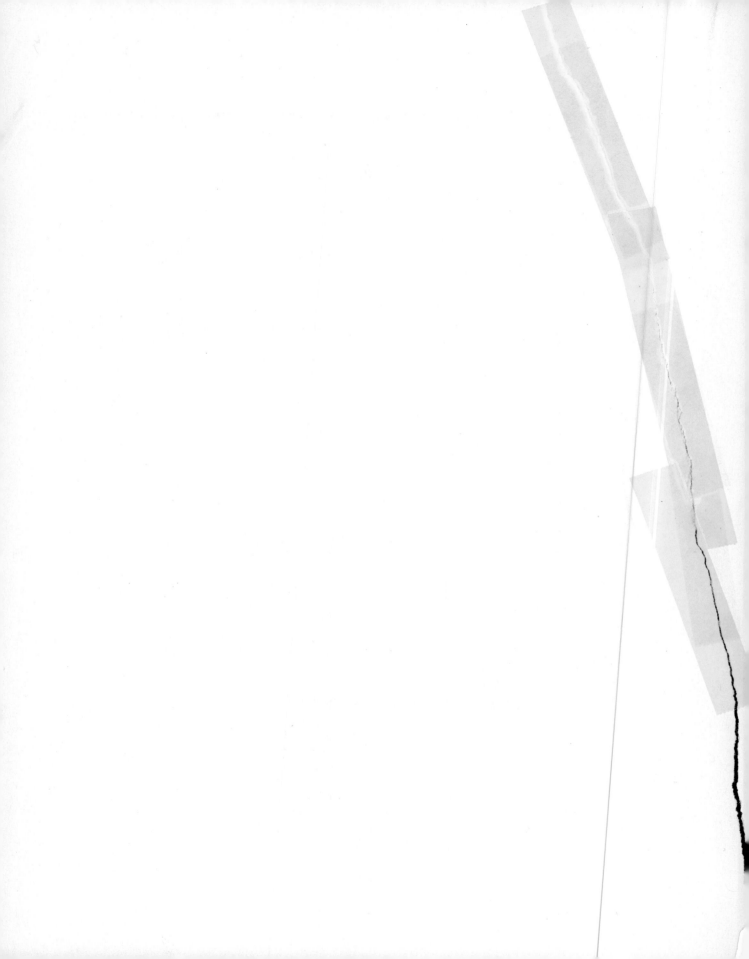

# Contents

# 15   The Derivative in *n*-Space   713

# 16   The Integral in *n*-Space   771

# 17   Vector Calculus   821

# Preface

This new edition of *Calculus* is a modest revision because users have reported great success with the previous edition and we have no desire to let this text suffer from the standard ailment of older texts, called "revisionitis".

**A Brief Text.** The seventh edition continues to be the briefest of all the successful, mainstream calculus texts. We have tried hard to prevent the text ballooning upward with new topics and alternative approaches. Students have developed some bad math habits in the last decade or so. They prefer not to read. They desire to find the appropriate worked out example so it can be matched to their homework problem. Our goal with this text continues to keep calculus as a course focused on some few basic ideas centered around words, formulas, and graphs. Solving problem sets, while crucial to developing math skills, should be subservient to this goal.

To encourage students to read the text and to reinforce our *conceptual emphasis*, we begin every problem set with our fill-in-the-blank items. These test mastery of the basic vocabulary, understanding of the theorems, and ability to apply the concepts in the simplest settings. A student who has read the lesson should be able to fill in these blanks quickly. We think students should respond to these items before proceeding to the later problems. We encourage this by giving immediate feedback; the correct answers are given at the end of the problem set.

Number sense distinguishes the mature mathematics student from the neophyte. All calculus students make numerical mistakes in solving problems, but the one with the number sense recognizes an absurd answer and reworks the problem. To encourage and develop this important ability, we have emphasized a process we call *estimation* (introduced in Section 1.2). We suggest how to make mental estimates, how to arrive at ballpark numerical answers to questions. We do this ourselves in the text in many places, and we propose that students do this, especially in problems marked with the symbol $\boxed{\approx}$.

**Use of Technology.** A significant new feature in this text is the expansion of optional computer/graphing calculator problems and projects. Slightly more than half the problem sets have some problems marked with the symbol [PC] for personal computer or with the symbol [C] to indicate that almost any type of calculator will be useful in solving these problems. Our philosophy continues to embrace *optional* use of technology.

At the end of each chapter, you will now find a full double-column page of challenging computer/graphing calculator problems.

More importantly, there is now available with this text five interchangeable technology manuals: one each for Maple, Mathematica, MATLAB, the Texas Instruments Graphing Calculators, and the Hewlett-Packard Graphing Calculators. The technology projects and problems are identical in each manual. What differs is the specific keystroke or syntax instruction that is given chapter by chapter. Thus, any given school could have one instructor teaching calculus with a TI calculator while another instructor uses, say, Maple, while another uses no technology. Each manual can be wrapped with the text for a small additional fee. An added value of using these manuals is that an instructor doesn't have to instruct the student in how to use technology; the manual will do this. The instructor need only teach math. These manuals were all written by Frank Hagin and Jack Cohen of the Colorado School of Mines after spending several years developing computer projects for students to use in a calculus lab.

**Pedagogical Concern.** Our 35 years of teaching calculus suggests *pacing* is very important. Our goal was to prepare sections of about equal length (one day's lesson). But to help students over difficult hurdles, we have spread out certain concepts. For example, the introduction of the derivative and the Chain Rule are each stretched into two lessons and vectors are first introduced in two dimensions and later in three dimensions. This concern for pacing is evident in carefully constructed problem sets that gradually lead the student from routine exercises to challenging applied problems.

Our conceptual emphasis means that definitions should be given in a *consistent* way. This implies that concepts for one-variable calculus should generalize naturally to the many variable case. Note how we achieve this for the concept of limit (Sections 2.5, 13.4, and 15.3), derivative (Sections 3.2, 13.4, and 15.4), and definite integral (Sections 5.5, 16.1, 17.2, and 17.5).

Since linear algebra is now a standard course for scientists and engineers, we think the terminology of calculus should be consistent with that subject. Thus in our book, linearity is emphasized as an important idea and vectors are written as *n*-tuples as well as in *ijk*-form.

**Supplements.**

For the Instructor:

- *Instructor's Resource Manual.* Contains worked out solutions to all exercises in the text, as well as the printed test bank. (ISBN 0-13-518929-2)
- *PH Custom Test for Windows.* Fully editable test generator with algorithmic capabilities, which provides an instructor's grade book and allows on-line testing. (ISBN 0-13-518952-7)

For the Student:

- *Student Solutions Manual.* Contains worked out solutions for all odd numbered exercises in the text. (ISBN 0-13-518937-3)
- *How to Study Calculus Booklet.* Contains strategies, suggestions, and hints for learning and achieving success in calculus. (ISBN 0-13-435116-9)
- *Technology Manuals.* Platform-specific manuals which offer additional technology problem sets and projects, as well as keystroke instructions. Written by Frank Hagin and Jack Cohen of the Colorado School of Mines.
- *Calculus with the Texas Instruments Graphing Calculator.* Hagin/Cohen (ISBN 0-13-518978-0)
- *Calculus with the Hewlett Packard Graphing Center.* Cohen/Hagin (ISBN 0-13-520339-2)
- *Calculus with Maple.* Hagin/Cohen (ISBN 0-13-518994-2)
- *Calculus with Mathematica.* Cohen/Hagin (ISBN 0-13-520347-3)
- *Calculus with MATLAB.* Hagin/Cohen (ISBN 0-13-520354-6)

**Acknowledgments.** We give special thanks to Neil Berger at the University of Illinois, Chicago for several insightful comments and for providing many new problem sets. Frank Hagin and Jack Cohen at the Colorado School of Mines helped in preparing new technology problem sets. The following faculty offered useful comments in preparing the recent editions:

Barbara Blake Bath, Colorado School of Mines;
Phillip W. Bean, Mercer University;
John D. Driscoll, University of Hartford;
Thomas Farmer, Miami University of Ohio;
William P. Francis, Michigan Technological University;
Louis F. Hoelzle, Bucks County Community College;
Paul Isihara, Wheaton College;
Calvin Jongsma, Dordt College;
William J. Keane, Boston College;
Cecilia Knoll, Florida Institute of Technology;
Joseph Johnson, Rutgers University;
Daniel J. Madden, University of Arizona;
Alexander Stone, University of New Mexico;
Bruno J. Wichnoski, University of North Carolina

# Preliminaries

Calculus is based on the real number system and its properties. But what are the real numbers and what are their properties? To answer, we start with some simpler number systems.

**The Integers and the Rational Numbers** The simplest numbers of all are the **natural numbers**,

$$1, 2, 3, 4, 5, 6, \ldots$$

With them we can *count*: our books, our friends, and our money. If we adjoin their negatives and zero, we obtain the **integers**:

$$\ldots, -3, -2, -1, 0, 1, 2, 3, \ldots$$

When we try to *measure* length, weight, or voltage, the integers are inadequate. They are spaced too far apart to give sufficient precision. We are led to consider quotients (ratios) of integers (Figure 1), numbers such as

$$\frac{3}{4}, \frac{-7}{8}, \frac{21}{5}, \frac{19}{-2}, \frac{16}{2}, \text{ and } \frac{-17}{1}$$

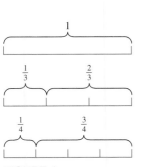

**FIGURE 1**

**FIGURE 2**

Note that we included $\frac{16}{2}$ and $\frac{-17}{1}$, though we would normally write them as 8 and $-17$, since they are equal to the latter by the ordinary meaning of division. We did not include $\frac{5}{0}$ or $\frac{-9}{0}$, since it is impossible to make sense out of these symbols (see Problem 36). In fact, let us agree once and for all to banish division by zero from this book (Figure 2). Numbers that can be written in the form $m/n$, where $m$ and $n$ are integers with $n \neq 0$, are called **rational numbers**.

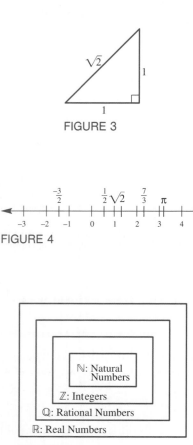

FIGURE 3

FIGURE 4

FIGURE 5

Do the rational numbers serve to measure all lengths? No. This surprising fact was discovered by the ancient Greeks several centuries before Christ. They showed that while $\sqrt{2}$ measures the hypotenuse of a right triangle with legs of length 1 (Figure 3), it cannot be written as a quotient of two integers (see Problem 43). Thus, $\sqrt{2}$ is an irrational (not rational) number. So are $\sqrt{3}$, $\sqrt{5}$, $\sqrt[3]{7}$, $\pi$, and a host of other numbers.

**The Real Numbers** Consider the set of all numbers (rational and irrational) that can measure lengths, together with their negatives and zero. We call these numbers the **real numbers**.

The real numbers may be viewed as labels for points along a horizontal line. There they measure the distance to the right or left (the **directed distance**) from a fixed point called the **origin** and labeled 0 (Figure 4). Though we cannot possibly show all the labels, each point does have a unique real number label. This number is called the **coordinate** of the point. And the resulting coordinate line is referred to as the **real line**.

There are standard symbols to identify the classes of numbers so far discussed. From now on, $\mathbb{N}$ will denote the set of natural numbers (positive integers), $\mathbb{Z}$ (from the German *Zahlen)* will denote the set of integers, $\mathbb{Q}$ (quotients of integers) the set of rational numbers, and $\mathbb{R}$ the set of real numbers. As suggested by Figure 5,

$$\mathbb{N} \subset \mathbb{Z} \subset \mathbb{Q} \subset \mathbb{R}$$

Here $\subset$ is the set inclusion symbol; it is read "is a subset of."

You may remember that the number system can be enlarged still more—to the **complex numbers**. These are numbers of the form $a + b\sqrt{-1}$, where $a$ and $b$ are real numbers. Complex numbers will rarely be used in this book. In fact, if we say *number* without any qualifying adjective, you can assume we mean real number. The real numbers are the principal characters in calculus.

**The Four Arithmetic Operations** Given two real numbers $x$ and $y$, we may add or multiply them to obtain two new real numbers $x + y$ and $x \cdot y$ (also written simply as $xy$). Addition and multiplication have the following familiar properties. We call them the field properties.

---

**THE FIELD PROPERTIES**

1. **Commutative laws.** $x + y = y + x$ and $xy = yx$.
2. **Associative laws.** $x + (y + z) = (x + y) + z$ and $x(yz) = (xy)z$.
3. **Distributive law.** $x(y + z) = xy + xz$.
4. **Identity elements.** There are two distinct numbers 0 and 1 satisfying $x + 0 = x$ and $x \cdot 1 = x$.
5. **Inverses.** Each number $x$ has an *additive inverse* (also called a *negative*), $-x$, satisfying $x + (-x) = 0$. Also, each number $x$ except 0 has a *multiplicative inverse* (also called a *reciprocal*), $x^{-1}$, satisfying $x \cdot x^{-1} = 1$.

Subtraction and division are defined by

$$x - y = x + (-y)$$

and

$$\frac{x}{y} = x \cdot y^{-1}$$

From these basic facts, many others follow. In fact, most of algebra ultimately rests on the five field properties and the definition of subtraction and division.

**Order** The nonzero real numbers separate nicely into two disjoint sets—the positive real numbers and the negative real numbers. This fact allows us to introduce the order relation < (read "is less than") by

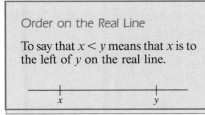

*Order on the Real Line*

To say that $x < y$ means that $x$ is to the left of $y$ on the real line.

$$\boxed{x < y \quad \Leftrightarrow \quad y - x \text{ is positive}}$$

Here the double-arrow symbol $\Leftrightarrow$ is the conjunction of $\Rightarrow$ (implies) and $\Leftarrow$ (is implied by). Thus, $\Leftrightarrow$ may be read "is equivalent to" or as "if and only if." We agree that $x < y$ and $y > x$ shall mean the same thing. Thus $3 < 4$, $4 > 3$, $-3 < -2$, and $-2 > -3$. Note the geometric interpretation of < shown in the box in the margin.

**THE ORDER PROPERTIES**

**1. Trichotomy.** If $x$ and $y$ are numbers, exactly one of the following holds:

$$x < y \text{ or } x = y \text{ or } x > y$$

**2. Transitivity.** $x < y$ and $y < z \quad \Rightarrow \quad x < z$.
**3. Addition.** $x < y \quad \Leftrightarrow \quad x + z < y + z$.
**4. Multiplication.** When $z$ is positive, $x < y \quad \Leftrightarrow \quad xz < yz$. When $z$ is negative, $x < y \quad \Leftrightarrow \quad xz > yz$.

The order relation $\leq$ (read "is less than or equal to") is a first cousin of <. It is defined by

$$\boxed{x \leq y \quad \Leftrightarrow \quad y - x \text{ is positive or zero}}$$

Order properties 2, 3, and 4 hold with the symbols < and > replaced by $\leq$ and $\geq$.

**A Bit of Logic**    Important results in mathematics are called **theorems;** you will find many theorems in this book. The most important ones occur in the text with the label *Theorem* and are usually given names (for example, Pythagorean Theorem). Others occur in the problem sets and are introduced with the words *show that* or *prove that*. In contrast to axioms or definitions, which are taken for granted, theorems require proof.

A theorem is either stated in the form "If $P$, then $Q$" or can be restated so it has this form. We often abbreviate this by $P \Rightarrow Q$. We call $P$ the hypothesis and $Q$ the conclusion of the theorem. A proof consists of showing that $P$ necessarily implies $Q$.

Beginning students (and some mature ones) may confuse $P \Rightarrow Q$ with its converse, $Q \Rightarrow P$. Clearly, these two statements are not equivalent. "If John is a Minnesotan, then he is an American" is a true statement, but its converse, "If John is an American, then he is a Minnesotan" is clearly false. On the other hand, $\sim Q \Rightarrow \sim P$, read "not $Q$ implies not $P$" and called the **contrapositive,** is equivalent to $P \Rightarrow Q$. In our example, it is true that "If John is not an American, then he is not a Minnesotan."

Because a statement and its contrapositive are equivalent, we often choose to prove a theorem stated in the latter form. Thus, to prove $P \Rightarrow Q$, we may suppose $\sim Q$ and try to deduce $\sim P$ from it; that is, we try to contradict $P$. Here is a simple example.

> ## Theorem
>
> The sum of a rational number and an irrational number is irrational.

***Proof***    The theorem could be stated as follows: "If $x = m/n$, where $m$ and $n$ are integers, and if $y$ is an irrational number, then $x + y$ is irrational." *We suppose $x + y$ is rational*, that is, we suppose that $x + y = p/q$, where $p$ and $q$ are integers. Then

$$y = \frac{p}{q} - x = \frac{p}{q} - \frac{m}{n} = \frac{np - mq}{qn}$$

This means that $y$ is a rational number, *contrary to our assumption*. The theorem is proved.    ∎

Another way to view the preceding proof is through the *Law of the Excluded Middle*, which says: Either $R$ or $\sim R$, not both. In the theorem above, let $R$ be the statement: The sum of a rational number and an irrational number is irrational. Our proof shows that $\sim R$ is false (absurd). We conclude that $R$ must be true. Any proof that begins by assuming the conclusion of a theorem is false and proceeds to show this assumption leads to an absurdity is called a **proof by contradiction.**

Occasionally, we will need another type of proof called **Mathematical Induction.** It would take us too far afield to describe this now but we have given a complete discussion in Appendix A.1.

---

### Proof by Contradiction

Proof by contradiction also goes by the name *reductio ad absurdum*. Here is what the great mathematician G. H. Hardy had to say about it.

"Reductio ad absurdum, which Euclid loved so much, is one of a mathematician's finest weapons. It is a far finer gambit than any chess gambit; a chess player may offer the sacrifice of a pawn or even a piece, but a mathematician offers the game."

## CONCEPTS REVIEW

**1.** Numbers such as $\frac{1}{2}$ and $\frac{7}{4}$ that can be represented as the ratios of two integers are called _____ numbers.

**2.** Two familiar examples of irrational numbers are _____ and _____.

**3.** Numbers that can measure lengths (together with their negatives and zero) are called _____ numbers.

**4.** Axioms and definitions are taken for granted but _____ require proof.

## PROBLEM SET 1.1

We assume that you remember how to manipulate numbers, but it won't hurt to practice a little. In Problems 1–20, simplify as much as possible. Be sure to remove all parentheses and reduce all fractions.

**1.** $4 - 2(8 - 11) + 6$

**2.** $3[2 - 4(7 - 12)]$

**3.** $-4[5(-3 + 12 - 4) + 2(13 - 7)]$

**4.** $5[-1(7 + 12 - 16) + 4] + 2$

**5.** $\frac{5}{7} - \frac{1}{13}$

**6.** $\frac{3}{4-7} + \frac{3}{21} - \frac{1}{6}$

**7.** $\frac{1}{3}\left[\frac{1}{2}\left(\frac{1}{4} - \frac{1}{3}\right) + \frac{1}{6}\right]$

**8.** $-\frac{1}{3}\left[\frac{2}{5} - \frac{1}{2}\left(\frac{1}{3} - \frac{1}{5}\right)\right]$

**9.** $\frac{14}{21}\left(\frac{2}{5 - \frac{1}{3}}\right)^2$

**10.** $\left(\frac{2}{7} - 5\right) / \left(1 - \frac{1}{7}\right)$

**11.** $\frac{\frac{11}{7} - \frac{12}{21}}{\frac{11}{7} + \frac{12}{21}}$

**12.** $\frac{\frac{1}{2} - \frac{3}{4} + \frac{7}{8}}{\frac{1}{2} + \frac{3}{4} - \frac{7}{8}}$

**13.** $1 - \frac{1}{1 + \frac{1}{2}}$

**14.** $2 + \frac{3}{1 + \frac{5}{2}}$

**15.** $(\sqrt{5} + \sqrt{3})(\sqrt{5} - \sqrt{3})$

**16.** $(\sqrt{5} - \sqrt{3})^2$

**17.** $3\sqrt{2}(\sqrt{2} - \sqrt{8})$

**18.** $2\sqrt[3]{4}[\sqrt[3]{2} + \sqrt[3]{16}]$

**19.** $\left(\frac{7}{4} + \frac{1}{2}\right)^{-2}$

**20.** $\left(\frac{1}{\sqrt{2}} - \frac{5}{2\sqrt{2}}\right)^{-2}$

A little algebraic practice is good for any calculus student. In Problems 21–34, perform the indicated operations and simplify.

**21.** $(3x - 4)(x + 1)$

**22.** $(2x - 3)^2$

**23.** $(3x - 9)(2x + 1)$

**24.** $(4x - 11)(3x - 7)$

**25.** $(3t^2 - t + 1)^2$

**26.** $(2t + 3)^3$

**27.** $\frac{x^2 - 4}{x - 2}$

**28.** $\frac{x^2 - x - 6}{x - 3}$

**29.** $\frac{t^2 - 4t - 21}{t + 3}$

**30.** $\frac{2x - 2x^2}{x^3 - 2x^2 + x}$

**31.** $\frac{12}{x^2 + 2x} + \frac{4}{x} + \frac{2}{x + 2}$

**32.** $\frac{2}{6y - 2} + \frac{y}{9y^2 - 1} - \frac{2y + 1}{1 - 3y}$

**33.** $\frac{t^2 + t - 12}{x^2 - 1} \cdot \frac{x^2 - 6x - 7}{8t - t^2 - 15}$

**34.** $\dfrac{\dfrac{x}{x - 3} - \dfrac{2}{x^2 - 4x + 3}}{\dfrac{5}{x - 1} + \dfrac{5}{x - 3}}$

**35.** Find the value of each of the following; if undefined, say so.

(a) $0 \cdot 0$     (b) $\frac{0}{0}$     (c) $\frac{0}{17}$
(d) $\frac{3}{0}$     (e) $0^5$     (f) $17^0$

**36.** Show that division by 0 is meaningless as follows: Suppose $a \neq 0$. If $a/0 = b$, then $a = 0 \cdot b = 0$, which is a contradiction. Now find a reason why 0/0 is also meaningless.

**37.** Tell whether each of the following is true or false.

(a) $-3 < -7$     (b) $-1 > -17$
(c) $-3 < -\frac{22}{7}$     (d) $-5 > -\sqrt{26}$
(e) $\frac{6}{7} < \frac{34}{39}$     (f) $-\frac{5}{7} < -\frac{44}{59}$

**38.** Assume $a > 0$, $b > 0$. Prove each statement.

(a) $a < b \iff a^2 < b^2$     (b) $a < b \iff \frac{1}{a} > \frac{1}{b}$

**39.** Prove that the average of two numbers is between the two numbers; that is, prove that

$$a < b \implies a < \frac{a + b}{2} < b$$

**40.** Which of the following are always correct if $a \leq b$?

(a) $a^2 \leq ab$     (b) $a - 3 \leq b - 3$
(c) $a^3 \leq a^2 b$     (d) $-a \leq -b$

**41.** A prime number is a natural number (positive integer) with exactly two natural number divisors, itself and 1. The first few primes are 2, 3, 5, 7, 11, 13, 17. According to the Fundamental Theorem of Arithmetic, every natural number (other than 1) can be written as the product of a unique set of primes. For example, $45 = 3 \cdot 3 \cdot 5$. Write each of the following as a product of primes. *Note*: The product is trivial if the number is prime—that is, it has only one factor.

(a) 243        (b) 127
(c) 5100       (d) 346

42. Use the Fundamental Theorem of Arithmetic (Problem 41) to show that the square of any natural number (other than 1) can be written as the product of a unique set of primes, with each of these primes occurring an *even* number of times. For example, $(45)^2 = 3 \cdot 3 \cdot 3 \cdot 3 \cdot 5 \cdot 5$.

43. Show that $\sqrt{2}$ is irrational. *Hint*: Try a proof by contradiction. Suppose $\sqrt{2} = p/q$, where $p$ and $q$ are natural numbers (necessarily different from 1). Then $2 = p^2/q^2$, and so $2q^2 = p^2$. Now use Problem 42 to get a contradiction.

44. Show that $\sqrt{3}$ is irrational (see Problem 43).

45. Show that the sum of two rational numbers is rational.

46. Show that the product of a rational number (other than 0) and an irrational number is irrational. *Hint*: Try proof by contradiction.

47. Which of the following are rational and which are irrational?

(a) $-\sqrt{9}$        (b) 0.375
(c) $1 - \sqrt{2}$      (d) $(1 + \sqrt{3})^2$
(e) $(3\sqrt{2})(5\sqrt{2})$    (f) $5\sqrt{2}$

48. Is the sum of two irrational numbers necessarily irrational? Explain.

49. Show that if the natural number $m$ is not a perfect square, then $\sqrt{m}$ is irrational.

50. Show that $\sqrt{6} + \sqrt{3}$ is irrational.

51. Show that $\sqrt{2} - \sqrt{3} + \sqrt{6}$ is irrational.

52. Show that $\log_{10} 5$ is irrational.

---

**Answers to Concepts Review:**   1. Rational   2. $\sqrt{2}$; $\pi$
3. Real   4. Theorems

---

## 1.2
## DECIMALS, CALCULATORS, ESTIMATION

Any rational number can be written as a decimal, since by definition it can always be expressed as the quotient of two integers; if we divide the denominator into the numerator, we obtain a decimal. For example (Figure 1),

$$\frac{1}{2} = 0.5 \qquad\qquad \frac{3}{8} = 0.375$$

$$\frac{13}{11} = 1.181818\ldots \qquad \frac{3}{7} = 0.428571428571428571\ldots$$

Irrational numbers, too, can be expressed as decimals. For instance,

$$\sqrt{2} = 1.4142135623\ldots, \qquad \sqrt{3} = 1.7320508075\ldots$$

$$\pi = 3.1415926535\ldots$$

```
       .375              1.181
   8 ⟌ 3.000        11 ⟌ 13.000
       2 4               11
       ─                 ──
       60                2 0
       56                1 1
       ──                ──
       40                90
       40                88
       ──                ──
                         20
   3/8 = .375            11
                         9

              13/11 = 1.181818...
```

FIGURE 1

**Repeating and Nonrepeating Decimals**   The decimal representation of a rational number either terminates (as in $\frac{3}{8} = 0.375$) or else repeats in regular cycles forever (as in $\frac{13}{11} = 1.181818\ldots$). A little experimenting with the long division process will show you why. (Note that there can be only a finite number of different remainders.) A terminating decimal can be regarded as a repeating decimal ending in all zeros. For instance,

$$\frac{3}{8} = 0.375 = 0.3750000\ldots$$

Thus, every rational number can be written as a repeating decimal. It is a remarkable fact that the converse is also true. Every repeating decimal represents a rational number. This is obvious in the case of a terminating decimal (for instance, $3.137 = 3137/1000$) and is easy to show in the general case.

**EXAMPLE 1** **(Repeating decimals are rational).** Show that

$$x = 0.136136136 \ldots \text{ and } y = 0.27171717 \ldots$$

represent rational numbers.

*Solution* We subtract $x$ from $1000x$ and then solve for $x$.

$$1000x = 136.136136 \ldots$$
$$\underline{\phantom{1000}x = \phantom{13}0.136136 \ldots}$$
$$999x = 136$$
$$x = \frac{136}{999}$$

Similarly,

$$100y = 27.17171717 \ldots$$
$$\underline{\phantom{100}y = \phantom{2}0.27171717 \ldots}$$
$$99y = 26.9$$
$$y = \frac{26.9}{99} = \frac{269}{990}$$

The decimal representations of irrational numbers do not repeat in cycles. Conversely, a nonrepeating decimal must represent an irrational number. Thus, for example,

$$0.101001000100001 \ldots$$

must represent an irrational number (note the pattern of more and more 0s between the 1s). The diagram in Figure 2 summarizes what we have said.

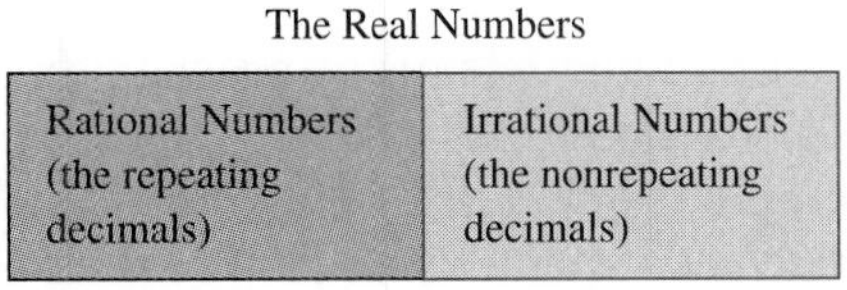

The Real Numbers

| Rational Numbers (the repeating decimals) | Irrational Numbers (the nonrepeating decimals) |
|---|---|

FIGURE 2

**Denseness** Between any two different real numbers $x$ and $y$, there is another real number. In particular, the number $z = (x + y)/2$ is a number midway between $x$ and $y$ (Figure 3). Since there is also a number $s$ between $x$ and $z$ and another number $t$ between $z$ and $y$ and since this argument can be repeated ad infinitum, we are forced to the amazing but correct conclusion that between any two different real numbers (no matter how close together), there are infinitely many other real numbers. This should destroy once and

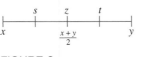

FIGURE 3

for all such notions as "the number just larger than 3." There is no such number.

Actually we can say more. Between any two distinct real numbers, there are both a rational number and an irrational number—and hence infinitely many of each variety.

**EXAMPLE 2** Find a rational number and an irrational number between $x$ and $y$ if

$$x = 0.31234158 \ldots$$
$$y = 0.31234200 \ldots$$

*Solution* Let

$$z = 0.312341600000 \ldots$$
$$w = 0.3123416010010001 \ldots$$

Then $z$ is rational (it ends in repeating 0s), whereas $w$ is irrational (note the pattern of inserting more and more 0s between the 1s). It should be clear that $x < z < w < y$. ∎

One way that mathematicians describe the situation we have been discussing is to say that both the rational numbers and the irrational numbers are **dense** along the real line (Figure 4). Every number has both rational and irrational neighbors arbitrarily close to it. The two types of numbers are inseparably intertwined and relentlessly crowded together.

One manifestation of the density property just described is that any irrational number can be approximated as closely as we please by a rational number—in fact, by a rational number with a terminating decimal representation. Take $\sqrt{2}$ as an example. The sequence of rational numbers 1, 1.4, 1.41, 1.414, 1.4142, 1.41421, 1.414213, . . . marches steadily and inexorably toward $\sqrt{2}$ (Figure 5). By going far enough along in this sequence we can get as near $\sqrt{2}$ as we wish.

FIGURE 4

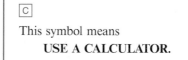

FIGURE 5

C

This symbol means
**USE A CALCULATOR.**

**Calculators** There was a time when all budding scientists and engineers walked around campuses with slide rules attached to their belts. Today they have calculators in their pockets. If you do not own one of these electronic wizards, we urge you to buy one. Be sure to get a scientific model (with sines, cosines, and logarithms) and, if you can afford it, we recommend a graphics version. You will find plenty of use for a calculator in this book, especially in problems marked with a C.

One fact that should be immediately apparent is that we cannot enter an infinite decimal into a calculator. Calculators work exclusively with decimals of some prescribed length (for example, ten digits). In fact, then, calculators handle only rational numbers with decimal expansions that terminate quickly. Thus, we must often round a number to enter it into a calculator, and usually the answer given by the calculator will also be rounded. For example, a calculator can never use the true value of $\sqrt{2}$, but must be content with an approximation such as

$$\sqrt{2} \approx 1.414213562$$

Here we have used the symbol $\approx$ to abbreviate the phrase "is approximately equal to."

Our advice is this: Do calculations that can easily be done by hand without a calculator, especially if this allows an exact answer. For example, we generally prefer the exact answer $\sqrt{3}/2$ for the sine of 60° to the calculator value 0.8660254. However, in any complicated calculation we encourage use of a calculator. You will note that our answer key at the end of the book often gives both the exact answer and a decimal approximation obtained by use of a calculator.

**Estimation**    Given a complicated arithmetic problem, a careless student will blithely press a few keys on a calculator and report the answer, not realizing that a missed parenthesis or a slip of the finger has completely spoiled the answer. A careful student with a feeling for numbers will press the same keys, immediately recognize that the answer is spurious (it's far too big or far too small) and recalculate it correctly. It is important to know how to make a mental estimate.

**EXAMPLE 3**    Calculate $(\sqrt{430} + 72 + \sqrt[3]{7.5})/2.75$.

*Solution*    A wise student approximated this as $(20 + 72 + 2)/3$ and said the answer should be in the neighborhood of 30. Thus, when her calculator gave 93.448 for an answer, she was suspicious (she had actually calculated $\sqrt{430} + 72 + \sqrt[3]{7.5}/2.75$). On recalculating, she got the correct answer: 34.434. ∎

If a man tells you that the volume of his body is 20,000 cubic inches, be suspicious. Estimate his volume this way. He is about 70 inches tall and his belt length is 30 inches, giving a radius at the waist of about 5 inches. If we approximate his volume by that of a cylinder, we find the volume to be roughly $\pi r^2 h \approx 3(5^2)70 \approx 5000$ cubic inches. He is not as big as he says he is.

Here is an example more related to calculus.

**EXAMPLE 4**    Suppose the shaded region $R$ shown in Figure 6 is revolved about the $x$-axis. Estimate the volume of the resulting solid ring $S$.

*Solution*    The region $R$ is about 3 units long and 0.9 units high. We estimate its area as $3(0.9) \approx 3$ square units. Imagine the solid ring $S$ to be slit open and laid out flat, forming a box about $2\pi r \approx 2(3)(6) = 36$ units long. The volume of a box is its cross-sectional area times its length. We estimate the volume of the box to be $3(36) = 108$ cubic units. If you calculate it to be 1000 cubic units, you need to check your work. ∎

The process of *estimation* is just organized common sense combined with reasonable number approximations. We urge you to use it all the time, especially on word problems. Before you attempt to get a precise answer,

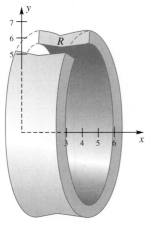

FIGURE 6

| ≈ |
| This symbol means |
| **ESTIMATE** |
| as a check of your answer. |

make an estimate. This will indicate whether or not you need to rework a problem—if your answer is not close to your estimate, you should check your work. Remember that $\pi \approx 3$, $\sqrt{2} \approx 1.4$, $2^{10} \approx 1000$, 1 foot $\approx$ 10 inches, 1 mile $\approx$ 5000 feet, and so on, and so on.

Because we think estimation is such an important skill, we will emphasize it in our exposition of calculus. Now and then, we will label a problem with the symbol shown in the margin; this symbol will suggest that it is a good idea to make a rough estimate for the answer either before you begin serious work or after you have what you think is the correct solution.

## CONCEPTS REVIEW

1. Every real number can be represented by an (unending) decimal. The decimal representation of $\frac{1}{3}$ is _____; the decimal representation of $\frac{1}{5}$ is _____; the decimal representation of $\pi$ starts out as _____.

2. The decimal representation of a number will repeat in cycles if and only if the number is _____.

3. Between any two real numbers, we can always find both _____ numbers and _____ numbers (and both in infinite supply).

4. Some rational numbers can be represented by terminating decimals (for example, $\frac{1}{2} = 0.5$, $\frac{9}{8} = 1.125$). Every _____ number can be approximated arbitrarily closely by a terminating decimal.

## PROBLEM SET 1.2

In Problems 1–6, change each rational number to a decimal by performing a long division.

1. $\frac{1}{12}$
2. $\frac{2}{7}$
3. $\frac{3}{21}$
4. $\frac{5}{17}$
5. $\frac{11}{3}$
6. $\frac{11}{13}$

In Problems 7–12, change each repeating decimal to a ratio of two integers (see Example 1).

7. $0.123123123\ldots$
8. $0.217171717\ldots$
9. $2.56565656\ldots$
10. $3.929292\ldots$
11. $0.199999\ldots$
12. $0.399999\ldots$

13. Since $0.199999\ldots = 0.200000\ldots$ and $0.399999\ldots = 0.400000\ldots$ (see Problems 11 and 12), we see that certain rational numbers have two different decimal expansions. Which rational numbers have this property?

14. Show that any rational number $p/q$, where the prime factorization of $q$ consists entirely of 2s and 5s, has a terminating decimal expansion.

15. Find a positive rational number and a positive irrational number both smaller than 0.00001.

16. What is the smallest positive integer? The smallest positive rational number? The smallest positive irrational number?

17. Find an irrational number between 3.14159 and $\pi$ (see Example 2 and note that $\pi = 3.141592\ldots$).

18. Is $\pi - \frac{22}{7}$ positive, negative, or zero?

19. Is there a number between $0.9999\ldots$ (repeating 9s) and 1?

20. Find two rational numbers between $\frac{57}{113}$ and $\frac{19}{37}$.

21. Is $0.1234567891011121314\ldots$ rational or irrational? (You should see a pattern in the given sequence of digits.)

22. Find two irrational numbers whose sum is rational.

≈ In Problems 23–32, find the best decimal approximation that your calculator allows. Begin by making a mental estimate.

23. $(\sqrt{3} + 1)^3$
24. $(\sqrt{2} - \sqrt{3})^4$
25. $\sqrt[4]{1.123} - \sqrt[3]{1.09}$
26. $(3.1415)^{-1/2}$
27. $\dfrac{\sqrt{130} - \sqrt{5}}{3^{1.2} - 3}$
28. $\dfrac{\sqrt{145} - \sqrt{35}}{(22.01)^4 - (12.1)^2}$
29. $\dfrac{(6.34 \times 10^7)(5.23 \times 10^6)}{4.21 \times 10^9}$
30. $\dfrac{(0.00121)(5.23 \times 10^{-3})}{6.16 \times 10^{-4}}$
31. $\sqrt{8.9\pi^2 + 1} - 3\pi$
32. $\sqrt[4]{(6\pi^2 - 2)\pi}$

≈ 33. Estimate the number of cubic inches in your right arm (elbow to fingertips).

≈ **34.** Estimate the number of cubic inches in your head.

≈ **35.** Estimate the length of the equator in feet. Assume the radius of the earth to be 4000 miles.

≈ **36.** About how many times has your heart beat by your twentieth birthday?

≈ **37.** The General Sherman tree in California is about 270 feet tall and averages about 16 feet in diameter. Estimate the number of board feet (1 board foot equals 1 inch by 12 inches by 12 inches) of lumber that could be made from this tree, assuming no waste and ignoring the branches.

≈ **38.** Assume the General Sherman tree (Problem 37) produces an annual growth ring of thickness 0.004 feet. Estimate the resulting increase in the volume of its trunk each year.

C **39.** Note that

$$2x^3 - 7x^2 + 11x - 2 = [(2x - 7)x + 11]x - 2$$

To calculate the expression on the right at $x = 3$, press the following keys on an algebraic logic calculator.

2 ☒ 3 ☐ 7 ☐ ☒ 3 ☐ 11 ☐ ☒ 3 ☐ 2 ☐

Use this idea to calculate the given expression in each case.

(a) $x = 2\pi$        (b) $x = 2.15$
(c) $x = 2.71828$      (d) $x = 1.1$

C **40.** Use the idea discussed in Problem 39 to evaluate $x^4 - 3x^3 + 4x^2 + 6x - 10$ at each value.

(a) $x = 1$,    (b) $x = \pi$,    (c) $x = 14.2$,    (d) $x = 1.2157$

**41.** A number $b$ is called an **upper bound** for a set $S$ of numbers if $x \leq b$ for all $x$ in $S$. For example, 5, 6.5, and 13 are upper bounds for the set $S = \{1, 2, 3, 4, 5\}$. The number 5 is the **least upper bound** for $S$ (the smallest of the upper bounds). Similarly, 1.6, 2, and 2.5 are upper bounds for the infinite set $T = \{1.4, 1.49, 1.499, 1.4999, \ldots\}$, whereas 1.5 is its least upper bound. Find the least upper bound of each of the following sets.

(a) $S = \{-10, -8, -6, -4, -2\}$
(b) $S = \{-2, -2.1, -2.11, -2.111, -2.1111, \ldots\}$
(c) $S = \{2.4, 2.44, 2.444, 2.4444, \ldots\}$
(d) $S = \{1 - \frac{1}{2}, 1 - \frac{1}{3}, 1 - \frac{1}{4}, 1 - \frac{1}{5}, \ldots\}$
(e) $S = \{x : x = (-1)^n + 1/n, n \text{ a positive integer}\}$; that is, $S$ is the set of all numbers $x$ that have the form $x = (-1)^n + 1/n$, where $n$ is a positive integer.
(f) $S = \{x : x^2 < 2, x \text{ a rational number}\}$.

**42.** The **Axiom of Completeness** for the real numbers says: *Every set of real numbers that has an upper bound has a least upper bound that is a real number.*

(a) Show that the italicized statement is false if the word *real* is replaced by *rational*.
(b) Would the italicized statement be true or false if the word *real* were replaced by *natural*?

*Note:* The real numbers $\mathbb{R}$ are the only set of numbers simultaneously having the field properties, the order properties, and the completeness property.

---

**Answers to Concepts Review:**   **1.** 0.333 . . . (3s repeat); 0.200 . . . (0s repeat); 3.14159 . . .   **2.** Rational **3.** Rational; irrational   **4.** Real

---

**1.3**
**INEQUALITIES**

Solving equations (for instance, $3x - 17 = 6$ or $x^2 - x - 6 = 0$) is one of the traditional tasks of mathematics; it will be important in this course and we assume you remember how to do it. But of almost equal significance in calculus is the notion of solving an inequality (for example, $3x - 17 < 6$ or $x^2 - x - 6 \geq 0$). To **solve** an inequality is to find the set of all real numbers that make the inequality true. In contrast to an equation, whose solution set normally consists of one number or perhaps a finite set of numbers, the solution set of an inequality usually consists of an entire interval of numbers or, in some cases, a union of such intervals.

*Intervals*   Several kinds of intervals will arise in our work and we introduce special terminology and notation for them. The double inequality $a < x < b$ describes the **open interval** consisting of all numbers between $a$ and $b$, not including the endpoints $a$ and $b$. We denote it by the symbol $(a, b)$ (Figure 1). In contrast, the inequality $a \leq x \leq b$ describes the corresponding **closed interval**, which does include the endpoints $a$ and $b$. It is

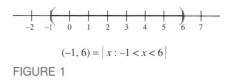

$(-1, 6) = \{x : -1 < x < 6\}$

**FIGURE 1**

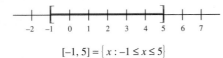

$$[-1, 5] = \{x : -1 \le x \le 5\}$$

FIGURE 2

denoted by $[a, b]$ (Figure 2). The table indicates the wide variety of possibilities and introduces our notation.

| Set Notation | Interval Notation | Graph |
|---|---|---|
| $\{x : a < x < b\}$ | $(a, b)$ |  |
| $\{x : a \le x \le b\}$ | $[a, b]$ | 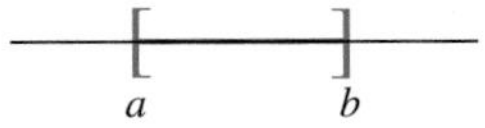 |
| $\{x : a \le x < b\}$ | $[a, b)$ |  |
| $\{x : a < x \le b\}$ | $(a, b]$ |  |
| $\{x : x \le b\}$ | $(-\infty, b]$ |  |
| $\{x : x < b\}$ | $(-\infty, b)$ |  |
| $\{x : x \ge a\}$ | $[a, \infty)$ |  |
| $\{x : x > a\}$ | $(a, \infty)$ | 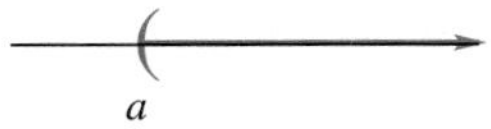 |
| $\mathbb{R}$ | $(-\infty, \infty)$ | |

**Solving Inequalities**   Just as with equations, the procedure for solving an inequality consists in transforming the inequality a step at a time until the solution set is obvious. The chief tools are the order properties from Section 1.1. They imply that we may perform certain operations on an inequality without changing its solution set. In particular:

1. *we may add the same number to both sides of an inequality;*
2. *we may multiply both sides of an inequality by a positive number;*
3. *we may multiply both sides by a negative number, but then we must reverse the direction of the inequality sign.*

EXAMPLE 1   Solve the inequality $2x - 7 < 4x - 2$ and show the graph of its solution set.

*Solution*

$$2x - 7 < 4x - 2$$
$$2x < 4x + 5 \qquad \text{(adding 7)}$$
$$-2x < 5 \qquad \text{(adding } -4x\text{)}$$
$$x > -\tfrac{5}{2} \qquad \text{(multiplying by } -\tfrac{1}{2}\text{)}$$

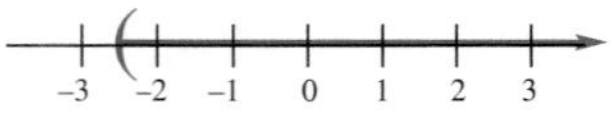

The graph appears in Figure 3.   ■

$$\left(\tfrac{-5}{2}, \infty\right) = \left\{x : x > \tfrac{-5}{2}\right\}$$

FIGURE 3

**EXAMPLE 2**   Solve $-5 \le 2x + 6 < 4$.

*Solution*

$$-5 \le 2x + 6 < 4$$
$$-11 \le 2x \quad\ \ < -2 \quad \text{(adding } -6\text{)}$$
$$-\tfrac{11}{2} \le \ \ x \quad\quad\ < -1 \quad \text{(multiplying by } \tfrac{1}{2}\text{)}$$

Figure 4 shows the corresponding graph.    ■

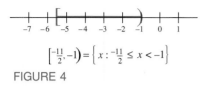

$$\left[-\tfrac{11}{2}, -1\right) = \left\{x : -\tfrac{11}{2} \le x < -1\right\}$$

FIGURE 4

Before tackling a quadratic inequality, we point out that a linear factor of the form $x - a$ is positive for $x > a$ and negative for $x < a$. It follows that a product $(x - a)(x - b)$ can change from being positive to negative, or vice versa, only at $a$ or $b$. These points, where a factor is zero, are called **split points**. They are the keys to determining the solution sets of quadratic and higher-degree inequalities.

**EXAMPLE 3**   Solve the quadratic inequality $x^2 - x < 6$.

*Solution*   As with quadratic equations, we move all nonzero terms to one side and factor.

$$x^2 - x < 6$$
$$x^2 - x - 6 < 0 \quad \text{(adding } -6\text{)}$$
$$(x - 3)(x + 2) < 0 \quad \text{(factoring)}$$

We see that $-2$ and $3$ are the split points; they divide the real line into the three intervals $(-\infty, -2)$, $(-2, 3)$, and $(3, \infty)$. On each of these intervals, $(x - 3)(x + 2)$ is of constant sign—that is, it is either always positive or always negative. To find this sign in each interval, we use the **test points** $-3$, $0$, and $5$ (any points on the three intervals would do). Our results are shown below.

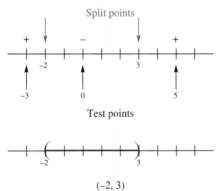

Split points

Test points

$(-2, 3)$

FIGURE 5

| Test Point | Value of $(x - 3)(x + 2)$ | Sign |
|:---:|:---:|:---:|
| $-3$ | 6 | $(+)$ |
| 0 | $-6$ | $(-)$ |
| 5 | 14 | $(+)$ |

The information we have obtained is summarized in the top half of Figure 5. We conclude that the solution set for $(x - 3)(x + 2) < 0$ is the interval $(-2, 3)$. Its graph is shown in the bottom half of Figure 5.    ■

**EXAMPLE 4**   Solve $3x^2 - x - 2 > 0$.

*Solution*   Since

$$3x^2 - x - 2 = (x - 1)(3x + 2) = 3(x - 1)(x + \tfrac{2}{3})$$

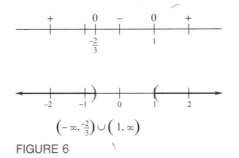

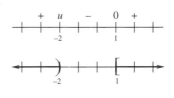

FIGURE 6

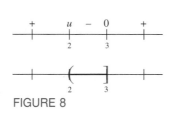

$(-\infty, -2) \cup [1, \infty)$

FIGURE 7

the split points are $-\frac{2}{3}$ and 1. These points, together with the test points $-2$, 0, and 2, establish the information shown in Figure 6. We conclude that the solution set of the inequality consists of the points in either $(-\infty, -\frac{2}{3})$ or $(1, \infty)$. In set language, the solution set is the **union** (symbolized by $\cup$) of these two intervals; that is, it is $(-\infty, -\frac{2}{3}) \cup (1, \infty)$. ∎

**EXAMPLE 5**   Solve $\dfrac{x - 1}{x + 2} \geq 0$.

*Solution*   Our inclination to multiply both sides by $x + 2$ leads to an immediate dilemma, since $x + 2$ may be either positive or negative. Should we reverse the inequality sign or leave it alone? Rather than try to untangle this problem (which would require breaking it into two cases), we observe that the quotient $(x - 1)/(x + 2)$ can change sign only at the split points of the numerator and denominator, namely, at 1 and $-2$. The test points $-3$, 0, and 2 yield the information displayed in Figure 7. The symbol $u$ indicates that the quotient is undefined at $-2$. We conclude that the solution set is $(-\infty, -2) \cup [1, \infty)$. Note that $-2$ is not in the solution set because the quotient is undefined there. On the other hand, 1 is included as the inequality is valid at 1. ∎

**EXAMPLE 6**   Solve $\dfrac{2x - 5}{x - 2} \leq 1$.

*Solution*   Rewrite the inequality successively as

$$\frac{2x - 5}{x - 2} - 1 \leq 0$$

$$\frac{2x - 5 - (x - 2)}{x - 2} \leq 0$$

$$\frac{x - 3}{x - 2} \leq 0$$

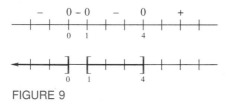

FIGURE 8

Then proceed as in Example 5. The summary shown in Figure 8 yields the solution set $(2,3]$. ∎

**EXAMPLE 7**   Solve $x^3 - 5x^2 + 4x \leq 0$.

*Solution*   The cubic $x^3 - 5x^2 + 4x$ factors as $x(x - 1)(x - 4)$, so there are 3 split points—0, 1, and 4—dividing the real line into four intervals. When we test these intervals, we obtain the information in Figure 9. The solution set is $(-\infty, 0] \cup [1,4]$. ∎

FIGURE 9

**EXAMPLE 8**   Solve $(x + 1)(x - 1)^2(x - 3) \leq 0$.

*Solution*   The split points are $-1$, 1, and 3, which divide the real line into four intervals, as shown in Figure 10. After testing these intervals, we conclude that the solution set is $[-1, 1] \cup [1, 3]$; that is, it is the interval $[-1, 3]$. ∎

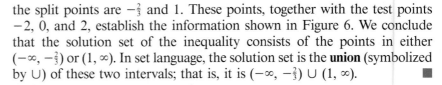

FIGURE 10

## CONCEPTS REVIEW

**1.** The set of solutions to an inequality is normally either an _____ or a union of _____ .

**2.** The set $\{x: -1 \le x < 5\}$ is written in interval notation as _____ and the set $\{x: x \le -2\}$ is written as _____ .

**3.** If $a/b < 0$, then either $a < 0$ and _____ or $a > 0$ and _____ .

**4.** The expression $(x - 3)(x + 4)/(x + 5)$ can change sign only at _____ .

## PROBLEM SET 1.3

**1.** Show each of the following intervals on the real line.

(a) $[-1, 1]$        (b) $(-4, 1]$
(c) $(-4, 1)$        (d) $[1, 4]$
(e) $[-1, \infty)$       (f) $(-\infty, 0]$

**2.** Use the notation of Problem 1 to describe the intervals below.

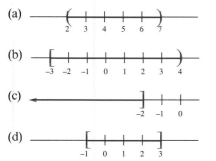

In each of Problems 3–34, express the solution set of the given inequality in interval notation and sketch its graph.

**3.** $x - 7 < 2x - 5$      **4.** $3x - 5 < 4x - 6$

**5.** $7x - 2 \le 9x + 3$     **6.** $5x - 3 > 6x - 4$

**7.** $10x + 1 > 8x + 5$    **8.** $-2x + 5 \ge 4x - 3$

**9.** $-4 < 3x + 2 < 5$    **10.** $-3 < 4x - 9 < 11$

**11.** $-3 < 1 - 6x \le 4$    **12.** $4 < 5 - 3x < 7$

**13.** $2 + 3x < 5x + 1 < 16$

**14.** $2x - 4 \le 6 - 7x \le 3x + 6$

**15.** $x^2 + 2x - 12 < 0$    **16.** $x^2 - 5x - 6 > 0$

**17.** $3x^2 - 11x - 4 \le 0$    **18.** $2x^2 + 7x - 15 \ge 0$

**19.** $2x^2 + 5x - 3 > 0$    **20.** $4x^2 - 5x - 6 < 0$

**21.** $\dfrac{x + 4}{x - 3} \le 0$       **22.** $\dfrac{3x - 2}{x - 1} \ge 0$

**23.** $\dfrac{2}{x} < 5$         **24.** $\dfrac{7}{4x} \le 7$

**25.** $\dfrac{1}{3x - 2} \le 4$     **26.** $\dfrac{3}{x + 5} > 2$

**27.** $\dfrac{x - 2}{x + 4} < 2$      **28.** $\dfrac{2x - 1}{x - 3} > 1$

**29.** $(x + 2)(x - 1)(x - 3) > 0$

**30.** $(2x + 3)(3x - 1)(x - 2) < 0$

**31.** $(2x - 3)(x - 1)^2(x - 3) \ge 0$

**32.** $(2x - 3)(x - 1)^2(x - 3) > 0$

**33.** $x^3 - 5x^2 - 6x < 0$

**34.** $x^3 - x^2 - x + 1 > 0$

**35.** Find all values of $x$ that satisfy both inequalities simultaneously.

(a) $3x + 7 > 1$ and $2x + 1 < 3$
(b) $3x + 7 > 1$ and $2x + 1 > -4$
(c) $3x + 7 > 1$ and $2x + 1 < -4$

**36.** Find all the values of $x$ that satisfy at least one of the two inequalities.

(a) $2x - 7 > 1, \{4 < x\}$ or $2x + 1 < 3$
(b) $2x - 7 \le 1, \{x \le 4\}$ or $2x + 1 < 3$
(c) $2x - 7 \le 1, \{x \le 4\}$ or $2x + 1 > 3$

**37.** Solve for $x$, expressing your answer in interval notation.

(a) $(x + 1)(x^2 + 2x - 7) \ge x^2 - 1$
(b) $x^4 - 2x^2 \ge 8$
(c) $(x^2 + 1)^2 - 7(x^2 + 1) + 10 < 0$

**38.** Solve $1 + x + x^2 + x^3 + \cdots + x^{99} \le 0$.

**39.** The formula

$$\frac{1}{R} = \frac{1}{R_1} + \frac{1}{R_2} + \frac{1}{R_3}$$

gives the total resistance $R$ in an electric circuit due to three resistances, $R_1$, $R_2$, and $R_3$, connected in parallel. If $10 \le R_1 \le 20, 20 \le R_2 \le 30$, and $30 \le R_3 \le 40$, find the range of values for $R$.

---

**Answers to Concepts Review:** **1.** Interval, intervals
**2.** $[-1, 5), (-\infty, 2]$   **3.** $b > 0, b < 0$   **4.** $3, -4, -5$

## 1.4
## ABSOLUTE VALUES, SQUARE ROOTS, SQUARES

The concept of absolute value is extremely useful in calculus and the reader should acquire skill in working with it. The **absolute value** of a real number $x$, denoted by $|x|$, is defined by

$$|x| = x \qquad \text{if } x \geq 0$$
$$|x| = -x \qquad \text{if } x < 0$$

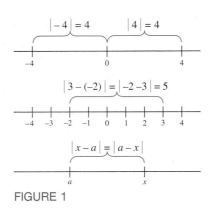

FIGURE 1

For example, $|6| = 6$, $|0| = 0$, and $|-5| = -(-5) = 5$.

This two-pronged definition merits careful study. Note that it does not say that $|-x| = x$ (try $x = -5$ to see why). It is true that $|x|$ is always nonnegative; it is also true that $|-x| = |x|$.

One of the best ways to think of absolute value is as an (undirected) distance. In particular, $|x|$ is the distance between $x$ and the origin. Similarly, $|x - a|$ is the distance between $x$ and $a$ (Figure 1).

**Properties** Absolute values behave nicely under multiplication and division, but not so well under addition and subtraction.

**PROPERTIES OF ABSOLUTE VALUE**

1. $|ab| = |a||b|$
2. $\left|\dfrac{a}{b}\right| = \dfrac{|a|}{|b|}$
3. $|a + b| \leq |a| + |b|$     (Triangle Inequality)
4. $|a - b| \geq \big||a| - |b|\big|$

$|x| < 3$

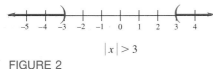

$|x| > 3$

FIGURE 2

**Inequalities Involving Absolute Values** If $|x| < 3$, then $x$ must be simultaneously less than 3 *and* greater than $-3$; that is, $-3 < x < 3$. In contrast if $|x| > 3$, then either $x < -3$ *or* $x > 3$ (Figure 2). These are special cases of the following general statements.

$$|x| < a \quad \Leftrightarrow \quad -a < x < a$$
$$|x| > a \quad \Leftrightarrow \quad x < -a \text{ or } x > a$$

We can use these facts to solve inequalities involving absolute values, since they provide a way of removing absolute value signs.

**EXAMPLE 1** Solve the inequality $|x - 4| < 1.5$ and show the solution set on the real line.

*Solution* From the first boxed statement with $x$ replaced by $x - 4$, we see that

$$|x - 4| < 1.5 \quad \Leftrightarrow \quad -1.5 < x - 4 < 1.5$$

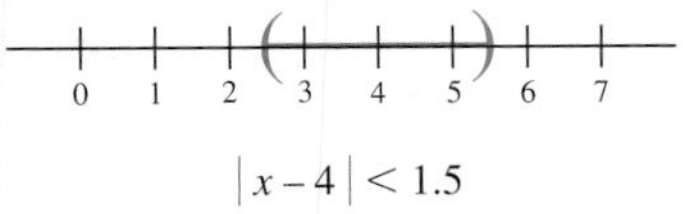

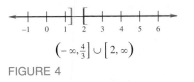

$|x - 4| < 1.5$

FIGURE 3

When we add 4 to all three members of the latter inequality, we obtain $2.5 < x < 5.5$. The graph is shown in Figure 3.

There is another way to look at this problem, and it is of equal importance. The symbol $|x - 4|$ represents the distance between $x$ and 4. Thus to say $|x - 4| < 1.5$ is equivalent to saying *the distance between x and 4 is less than* 1.5. The numbers $x$ with this property are the numbers between 2.5 and 5.5, that is, $2.5 < x < 5.5$. ∎

The boxed statements displayed just before Example 1 are valid with $<$ and $>$ replaced by $\leq$ and $\geq$, respectively. We need the second statement in this form in our next example.

**EXAMPLE 2**   Solve the inequality $|3x - 5| \geq 1$ and show its solution set on the real line.

*Solution*   The given inequality may be written successively as

$$3x - 5 \leq -1 \text{ or } 3x - 5 \geq 1$$

$$3x \leq \quad 4 \text{ or } \qquad 3x \geq 6$$

$$x \leq \quad \tfrac{4}{3} \text{ or } \qquad x \geq 2$$

$\left(-\infty, \tfrac{4}{3}\right] \cup \left[2, \infty\right)$

FIGURE 4

The solution set is the union of two intervals; it is the set $\left(-\infty, \tfrac{4}{3}\right] \cup [2, \infty)$ shown in Figure 4. ∎

When we get to the "epsilon, delta" definition of limit in Chapter 2, we will need to make the kind of manipulations illustrated by the next two examples. Delta and epsilon are the fourth and fifth letters, respectively, of the Greek alphabet and are traditionally used to stand for small positive numbers.

**EXAMPLE 3**   Let $\varepsilon$ (epsilon) be a positive number. Show that

$$|x - 2| < \frac{\varepsilon}{5} \quad \Leftrightarrow \quad |5x - 10| < \varepsilon$$

*Solution*

$$|x - 2| < \frac{\varepsilon}{5} \quad \Leftrightarrow \quad 5|x - 2| < \varepsilon \qquad \text{(multiplying by 5)}$$

$$\Leftrightarrow \quad |5||x - 2| < \varepsilon \qquad (|5| = 5)$$

$$\Leftrightarrow \quad |5(x - 2)| < \varepsilon \qquad (|a||b| = |ab|)$$

$$\Leftrightarrow \quad |5x - 10| < \varepsilon \qquad \blacksquare$$

**EXAMPLE 4**   Let $\varepsilon$ be a positive number. Find a positive number $\delta$ (delta) such that

$$|x - 3| < \delta \quad \Rightarrow \quad |6x - 18| < \varepsilon$$

Finding Delta

Note two facts about our solution to Example 4.

1. The number $\delta$ to be found must depend on $\varepsilon$. Our choice is $\delta = \varepsilon/6$.

2. Any positive $\delta$ smaller than $\varepsilon/6$ is acceptable. For example, $\delta = \varepsilon/7$ or $\delta = \varepsilon/2\pi$ are other correct choices.

*Solution*

$$|6x - 18| < \varepsilon \quad \Leftrightarrow \quad |6(x - 3)| < \varepsilon$$
$$\Leftrightarrow \quad 6|x - 3| < \varepsilon \qquad (|ab| = |a|\,|b|)$$
$$\Leftrightarrow \quad |x - 3| < \frac{\varepsilon}{6} \qquad \left(\text{multiplying by } \frac{1}{6}\right)$$

Therefore, we choose $\delta = \varepsilon/6$. Following the implications backward, we see that

$$|x - 3| < \delta \quad \Rightarrow \quad |6x - 18| < \varepsilon$$

Here is a practical problem that uses the same type of reasoning.

**EXAMPLE 5** A $\frac{1}{2}$-liter (500-cubic-centimeter) glass has an inner radius of 4 centimeters. How closely must we measure the height $h$ of water in the glass to be sure we have $\frac{1}{2}$ liter of water with an error of less than 1%, that is, an error of less than 5 cubic centimeters? See Figure 5.

*Solution* The volume $V$ of water in the glass is given by the formula $V = 16\pi h$. We want $|V - 500| < 5$, or, equivalently, $|16\pi h - 500| < 5$. Now

$$|16\pi h - 500| < 5 \quad \Leftrightarrow \quad \left|16\pi\left(h - \frac{500}{16\pi}\right)\right| < 5$$
$$\Leftrightarrow \quad 16\pi\left|h - \frac{500}{16\pi}\right| < 5$$
$$\Leftrightarrow \quad \left|h - \frac{500}{16\pi}\right| < \frac{5}{16\pi}$$
$$\Leftrightarrow \quad |h - 9.947| < 0.0947 \approx 0.1$$

Thus, we must measure the height to an accuracy of about 1 millimeter (approximately the width of the calibration marks on the glass). ■

FIGURE 5

**Square Roots** Every positive number has two square roots. For example, the two square roots of 9 are $-3$ and 3; the two square roots of 100 are $-10$ and 10. For $a \geq 0$, the symbol $\sqrt{a}$, called the **principal square root** of $a$, denotes the nonnegative square root of $a$. Thus, $\sqrt{9} = 3$ and $\sqrt{(-10)^2} = \sqrt{100} = 10$. The two square roots of 7 are $\pm\sqrt{7}$. It is incorrect to write $\sqrt{16} = \pm 4$; rather, $\sqrt{16} = 4$. Here is an important fact worth remembering.

$$\boxed{\sqrt{x^2} = |x|}$$

Most students will recall the **Quadratic Formula**. The solutions to $ax^2 + bx + c = 0$ are given by

$$\boxed{x = \frac{-b \pm \sqrt{b^2 - 4ac}}{2a}}$$

The number $d = b^2 - 4ac$ is called the **discriminant** of the quadratic equation $ax^2 + bx + c = 0$. This equation has two real solutions if $d > 0$, one real solution if $d = 0$, and no real solutions if $d < 0$.

With the Quadratic Formula, we can easily solve quadratic inequalities even if they do not factor by inspection.

**EXAMPLE 6** Solve $x^2 - 2x - 4 \leq 0$.

*Solution*  The two solutions of $x^2 - 2x - 4 = 0$ are

$$x_1 = \frac{2 - \sqrt{4 + 16}}{2} = 1 - \sqrt{5} \approx -1.24$$

and

$$x_2 = \frac{2 + \sqrt{4 + 16}}{2} = 1 + \sqrt{5} \approx 3.24$$

Thus,

$$x^2 - 2x - 4 = (x - x_1)(x - x_2) = (x - 1 + \sqrt{5})(x - 1 - \sqrt{5})$$

The split points $1 - \sqrt{5}$ and $1 + \sqrt{5}$ divide the real line into three intervals (Figure 6). When we test them with the test points $-2$, $0$, and $4$, we conclude that the solution set for $x^2 - 2x - 4 \leq 0$ is $[1 - \sqrt{5}, 1 + \sqrt{5}]$. ∎

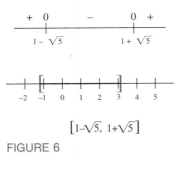

$$[1-\sqrt{5},\ 1+\sqrt{5}]$$

FIGURE 6

We mention in passing that if $n$ is even and $a \geq 0$, the symbol $\sqrt[n]{a}$ always denotes the nonnegative $n$th root of $a$. When $n$ is odd, there is only one real $n$th root of $a$, denoted by the symbol $\sqrt[n]{a}$. Thus $\sqrt[4]{16} = 2$, $\sqrt[3]{27} = 3$, and $\sqrt[3]{-8} = -2$.

**Squares**  Turning to squares, we notice that

$$\boxed{|x|^2 = x^2}$$

This follows from the property $|a|\,|b| = |ab|$.

Does the squaring operation preserve inequalities? In general, the answer is no. For instance, $-3 < 2$, but $(-3)^2 > 2^2$. On the other hand, $2 < 3$ and $2^2 < 3^2$. If we are dealing with nonnegative numbers, then $a < b \Leftrightarrow a^2 < b^2$. A useful variant of this is

$$\boxed{|x| < |y| \quad \Leftrightarrow \quad x^2 < y^2}$$

For a proof of this fact, see Problem 31.

EXAMPLE 7 Solve the inequality $|3x + 1| < 2|x - 6|$.

*Solution* This inequality is more difficult to solve than our earlier examples, because there are two sets of absolute value signs. We can remove both of them by using the last boxed result.

$$|3x + 1| < 2|x - 6| \quad \Leftrightarrow \qquad\qquad |3x + 1| < |2x - 12|$$
$$\Leftrightarrow \qquad\qquad (3x + 1)^2 < (2x - 12)^2$$
$$\Leftrightarrow \qquad 9x^2 + 6x + 1 < 4x^2 - 48x + 144$$
$$\Leftrightarrow \quad 5x^2 + 54x - 143 < 0$$
$$\Leftrightarrow \quad (5x - 11)(x + 13) < 0$$

The split points for this quadratic inequality are $-13$ and $\frac{11}{5}$; they divide the real line into the three intervals $(-\infty, -13)$, $(-13, \frac{11}{5})$, and $(\frac{11}{5}, \infty)$. When we use the test points $-14$, $0$, and $3$, we discover that only the points in $(-13, \frac{11}{5})$ satisfy the inequality. ∎

## CONCEPTS REVIEW

**1.** The inequality $|x - 2| \le 3$ is equivalent to _____ $\le x \le$ _____.

**2.** The Triangle Inequality says _____.

**3.** Which of the following are always true?

(a) $|-x| = x$,

(b) $|x|^2 = x^2$,

(c) $|xy| = |x|\,|y|$,

(d) $|x - y| \ge |x| - |y|$,

(e) $|x| \le |x - y| + |y|$,

(f) $\sqrt{x^2} = x$.

**4.** To be sure that $|5x - 20| < 0.2$, we would need $|x - 4| <$ _____.

## PROBLEM SET 1.4

In Problems 1–12, find the solution sets of the given inequalities (see Examples 1 and 2).

**1.** $|x + 2| < 1$

**2.** $|x - 2| \ge 5$

**3.** $|2x - 1| > 2$

**4.** $|4x + 5| \le 10$

**5.** $\left|\dfrac{x}{4} + 1\right| < 1$

**6.** $\left|\dfrac{2x}{7} - 5\right| \ge 7$

**7.** $|2x - 7| > 3$

**8.** $|5x - 6| > 1$

**9.** $|4x + 2| \ge 10$

**10.** $\left|\dfrac{x}{2} + 7\right| \ge 2$

**11.** $\left|2 + \dfrac{5}{x}\right| > 1$

**12.** $\left|\dfrac{1}{x} - 3\right| > 6$

In Problems 13–16, solve the given quadratic inequality using the quadratic formula (see Example 6).

**13.** $x^2 - 3x - 4 \ge 0$

**14.** $x^2 - 4x + 4 \le 0$

**15.** $3x^2 + 17x - 6 > 0$

**16.** $14x^2 + 11x - 15 \le 0$

In Problems 17–20, show that the indicated implication is true (see Example 3).

**17.** $|x - 3| < 0.5 \ \Rightarrow \ |5x - 15| < 2.5$

**18.** $|x + 2| < 0.3 \ \Rightarrow \ |4x + 8| < 1.2$

**19.** $|x - 2| < \dfrac{\varepsilon}{6} \ \Rightarrow \ |6x - 12| < \varepsilon$

**20.** $|x + 4| < \dfrac{\varepsilon}{2} \ \Rightarrow \ |2x + 8| < \varepsilon$

In Problems 21–24, find $\delta$ (depending on $\varepsilon$) so that the given implication is true (see Example 4).

**21.** $|x - 5| < \delta \ \Rightarrow \ |3x - 15| < \varepsilon$

**22.** $|x - 2| < \delta \ \Rightarrow \ |4x - 8| < \varepsilon$

**23.** $|x + 6| < \delta \ \Rightarrow \ |6x + 36| < \varepsilon$

**24.** $|x + 5| < \delta \ \Rightarrow \ |5x + 25| < \varepsilon$

**25.** On a lathe, you are to turn out a disk (thin right circular cylinder) of circumference 10 inches. This is done by continually measuring the diameter as you make the disk smaller. How closely must you measure the diameter if you can tolerate an error of at most 0.02 inches in the circumference (see Example 5)?

**26.** Fahrenheit temperatures and Celsius temperatures are related by the formula $C = \frac{5}{9}(F - 32)$. An experiment requires that a solution be kept at 50°C with an error of at most 3% (or 1.5°). You have only a Fahrenheit thermometer. What error are you allowed on it?

In Problems 27–30, solve the inequalities (see Example 7).

**27.** $|x - 1| < 2|x - 3|$      **28.** $|2x - 1| \geq |x + 1|$

**29.** $2|2x - 3| < |x + 10|$      **30.** $|3x - 1| < 2|x + 6|$

**31.** Prove $|x| < |y| \Leftrightarrow x^2 < y^2$ by giving a reason for each step below.

$$|x| < |y| \Rightarrow |x|\,|x| \leq |x|\,|y| \quad \text{and} \quad |x|\,|y| < |y|\,|y|$$
$$\Rightarrow |x|^2 < |y|^2$$
$$\Rightarrow x^2 < y^2$$

Conversely,

$$x^2 < y^2 \Rightarrow |x|^2 < |y|^2$$
$$\Rightarrow |x|^2 - |y|^2 < 0$$
$$\Rightarrow (|x| - |y|)(|x| + |y|) < 0$$
$$\Rightarrow |x| - |y| < 0$$
$$\Rightarrow |x| < |y|$$

**32.** Use the result of Problem 31 to show that

$$0 < a < b \Rightarrow \sqrt{a} < \sqrt{b}$$

**33.** Use the Properties of the absolute value to show that each of the following is true.

(a) $|a - b| \leq |a| + |b|$      (b) $|a - b| \geq |a| - |b|$
(c) $|a + b + c| \leq |a| + |b| + |c|$

**34.** Use the Triangle Inequality and the fact that $0 < |a| < |b| \Rightarrow 1/|b| < 1/|a|$ to establish the following chain of inequalities.

$$\left| \frac{1}{x^2 + 3} - \frac{1}{|x| + 2} \right| \leq \frac{1}{x^2 + 3} + \frac{1}{|x| + 2} \leq \frac{1}{3} + \frac{1}{2}$$

**35.** Show that (see Problem 34)

$$\left| \frac{x - 2}{x^2 + 9} \right| \leq \frac{|x| + 2}{9}$$

**36.** Show that

$$|x| \leq 2 \Rightarrow \left| \frac{x^2 + 2x + 7}{x^2 + 1} \right| \leq 15$$

**37.** Show that

$$|x| \leq 1 \Rightarrow \left| x^4 + \tfrac{1}{2}x^3 + \tfrac{1}{4}x^2 + \tfrac{1}{8}x + \tfrac{1}{16} \right| < 2$$

**38.** Show each of the following.

(a) $x < x^2$ for $x < 0$ or $x > 1$
(b) $x^2 < x$ for $0 < x < 1$

**39.** Show $a \neq 0 \Rightarrow a^2 + 1/a^2 \geq 2$. *Hint*: Consider $(a - 1/a)^2$.

**40.** The number $\frac{1}{2}(a + b)$ is called the average, or **arithmetic mean**, of $a$ and $b$. Show that the arithmetic mean of two numbers is between the two numbers; that is, prove that

$$a < b \Rightarrow a < \frac{a + b}{2} < b$$

**41.** The number $\sqrt{ab}$ is called the **geometric mean** of two positive numbers $a$ and $b$. Prove that

$$0 < a < b \Rightarrow a < \sqrt{ab} < b$$

**42.** For two positive numbers $a$ and $b$, prove that

$$\sqrt{ab} \leq \tfrac{1}{2}(a + b)$$

This is the simplest version of a famous inequality called the **geometric mean—arithmetic mean inequality**.

**43.** Show that, among all rectangles with given perimeter $p$, the square has largest area. *Hint*: If $a$ and $b$ denote the lengths of adjacent sides of a rectangle of perimeter $p$, then the area is $ab$, and for the square the area is $a^2 = [(a + b)/2]^2$. Now see Problem 42.

**44.** The radius of a sphere is measured to be about 10 inches. Determine a tolerance $\delta$ in this measurement that will insure an error of less than 0.01 square inches in the calculated value of the surface area of the sphere.

---

**Answers to Concepts Review:** **1.** $-1$; 5
**2.** $|a + b| \leq |a| + |b|$   **3.** b, c, d, e   **4.** 0.04

---

**1.5**
**THE RECTANGULAR COORDINATE SYSTEM**

Two Frenchmen deserve credit for the idea of a coordinate system. Pierre de Fermat was a lawyer who made mathematics his hobby. In 1629, he wrote a paper which, in effect, made use of coordinates to describe points and curves. René Descartes was a philosopher who thought mathematics could unlock the secrets of the universe. He published his *La Géométrie* in 1637. It is a

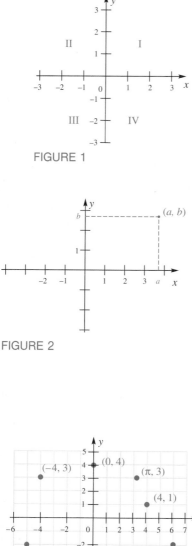

FIGURE 1

FIGURE 2

famous book and though it does emphasize the role of algebra in solving geometric problems, one finds only a hint of coordinates there. By virtue of having the idea first and more explicitly, Fermat ought to get the major credit. History can be a fickle friend; coordinates are known as Cartesian coordinates, named after René Descartes.

**Cartesian Coordinates**   In the plane, produce two copies of the real line, one horizontal and the other vertical, so that they intersect at the zero points of the two lines. The two lines are called **coordinate axes**; their intersection is labeled $O$ and is called the **origin**. By convention, the horizontal line is called the **$x$-axis** and the vertical line is called the **$y$-axis**. The positive half of the $x$-axis is to the right; the positive half of the $y$-axis is upward. The coordinate axes divide the plane into four regions, called **quadrants**, labeled I, II, III, and IV, as shown in Figure 1.

Each point $P$ in the plane can now be assigned a pair of numbers, called its **Cartesian coordinates**. If vertical and horizontal lines through $P$ intersect the $x$- and $y$-axes at $a$ and $b$, respectively, then $P$ has coordinates $(a, b)$ (see Figure 2). We call $(a, b)$ an **ordered pair** of numbers because it makes a difference which number is first. The first number $a$ is the **$x$-coordinate** (or abscissa); the second number $b$ is the **$y$-coordinate** (or ordinate).

Conversely, take any ordered pair $(a, b)$ of real numbers. The vertical line through $a$ on the $x$-axis and the horizontal line through $b$ on the $y$-axis meet at a point $P$, whose coordinates are $(a, b)$.

Think of it this way: The coordinates of a point are the address of that point. If you have found a house (or a point), you can read its address. Conversely, if you know the address of a house (or a point), you can always locate it. In Figure 3, we have shown the coordinates of several points.

FIGURE 3

**The Distance Formula**   With coordinates in hand, we can introduce a simple formula for the distance between any two points in the plane. It is based on the **Pythagorean Theorem**, which says that if $a$ and $b$ measure the two legs of a right triangle and $c$ measures its hypotenuse (Figure 4), then

$$a^2 + b^2 = c^2$$

Conversely, this relationship between the three sides of a triangle holds only for a right triangle.

Now consider any two points $P$ and $Q$, with coordinates $(x_1, y_1)$ and $(x_2, y_2)$, respectively. Together with $R$—the point with coordinates $(x_2, y_1)$—$P$ and $Q$ are vertices of a right triangle (Figure 5). The lengths of $PR$ and $RQ$ are $|x_2 - x_1|$ and $|y_2 - y_1|$, respectively. When we apply the Pythagorean Theorem and take the principal square root of both sides, we obtain the following expression for $d(P, Q)$, the (undirected) distance between $P$ and $Q$.

$$d(P, Q) = \sqrt{(x_2 - x_1)^2 + (y_2 - y_1)^2}$$

This is called the **Distance Formula**.

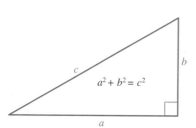

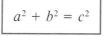
$a^2 + b^2 = c^2$

FIGURE 4

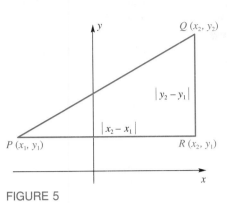

**FIGURE 5**

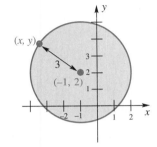

**FIGURE 6**

**EXAMPLE 1**   Find the distance between
**(a)** $P(-2, 3)$ and $Q(4, -1)$
**(b)** $P(\sqrt{2}, \sqrt{3})$ and $Q(\pi, \pi)$

*Solution*
(a) $d(P, Q) = \sqrt{(4 - (-2))^2 + (-1 - 3)^2} = \sqrt{36 + 16} = \sqrt{52} \approx 7.21$
(b) $d(P, Q) = \sqrt{(\pi - \sqrt{2})^2 + (\pi - \sqrt{3})^2} \approx \sqrt{4.971} \approx 2.23$    ■

The formula holds even if the two points lie on the same horizontal line or the same vertical line. Thus, the distance between $P(-2, 2)$ and $Q(6, 2)$ is

$$\sqrt{(-2 - 6)^2 + (2 - 2)^2} = \sqrt{64} = 8$$

**The Equation of a Circle**   It is a small step from the distance formula to the equation of a circle. A **circle** is the set of points which lie at a fixed distance (the *radius*) from a fixed point (the *center*). Consider, for example, the circle of radius 3 with center at $(-1, 2)$ (Figure 6). Let $(x, y)$ denote any point on this circle. By the Distance Formula,

$$\sqrt{(x + 1)^2 + (y - 2)^2} = 3$$

When we square both sides, we obtain

$$(x + 1)^2 + (y - 2)^2 = 9$$

which we call the equation of this circle.
More generally, the circle of radius $r$ and center $(h, k)$ has the equation

$$\boxed{(x - h)^2 + (y - k)^2 = r^2}$$

We call this the **standard equation of a circle**.

**EXAMPLE 2**   Find the equation of a circle of radius 5 and center $(1, -5)$. Also find the $y$-coordinates of the two points on this circle with $x$-coordinate 2.

*Solution*   The desired equation is

$$(x - 1)^2 + (y + 5)^2 = 25$$

To accomplish the second task, we substitute $x = 2$ in the equation and solve for $y$.

$$(2 - 1)^2 + (y + 5)^2 = 25$$
$$(y + 5)^2 = 24$$
$$y + 5 = \pm \sqrt{24}$$
$$y = -5 \pm \sqrt{24} = -5 \pm 2\sqrt{6}$$    ■

If we expand the two squares in the boxed equation and combined the constants, the equation takes the form

$$x^2 + ax + y^2 + by = c$$

---

Circle $\leftrightarrow$ Equation

To say that
$$(x + 1)^2 + (y - 2)^2 = 9$$
is the equation of the circle of radius 3 with center $(-1, 2)$ means two things:

1. If a point is on this circle, then its coordinates $(x, y)$ satisfy the equation.
2. If $x$ and $y$ are numbers that satisfy the equation, then they are the coordinates of a point on the circle.

This suggests asking whether every equation of the latter form is the equation of a circle. The answer is yes (with some obvious exceptions), as we shall see in Example 3.

In this example, we need to *complete the square,* a process important in many contexts. To complete the square of $x^2 \pm ax$, add $(a/2)^2$. Thus, adding $6^2$ to $x^2 - 12x$ and $(\frac{1}{5})^2$ to $x^2 + \frac{2}{5}x$ gives

$$x^2 - 12x + 6^2 = (x - 6)^2$$

$$x^2 + \tfrac{2}{5}x + (\tfrac{1}{5})^2 = (x + \tfrac{1}{5})^2$$

**EXAMPLE 3** Show that the equation

$$x^2 - 2x + y^2 + 6y = -6$$

represents a circle, and find its center and radius.

*Solution* We complete the squares for the expressions in both $x$ and $y$ by adding the same numbers to both sides of the equation.

$$(x^2 - 2x) + (y^2 + 6y) = -6$$

$$(x^2 - 2x + 1) + (y^2 + 6y + 9) = -6 + 1 + 9$$

$$(x - 1)^2 + (y + 3)^2 = 4$$

The last equation is in standard form. It is the equation of a circle with center $(1, -3)$ and radius 2. If, as a result of this process, we had come up with a negative number on the right side of the final equation, the equation would not have represented any curve. If we had come up with zero, the equation would have represented the single point $(1, -3)$. ■

**The Midpoint Formula** Consider two points $P(x_1, y_1)$ and $Q(x_2, y_2)$ with $x_1 \le x_2$, as in Figure 7. Note that

$$x_1 + \tfrac{1}{2}(x_2 - x_1) = x_1 + \tfrac{1}{2}x_2 - \tfrac{1}{2}x_1$$

$$= \tfrac{1}{2}x_1 + \tfrac{1}{2}x_2$$

$$= \frac{x_1 + x_2}{2}$$

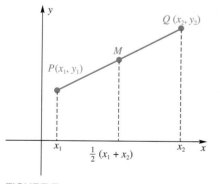

FIGURE 7

It follows that the point $(x_1 + x_2)/2$ is midway between $x_1$ and $x_2$ on the $x$-axis and, consequently, that the midpoint $M$ of the segment $PQ$ has $(x_1 + x_2)/2$ as its $x$-coordinate. In a completely similar manner, we can show that $(y_1 + y_2)/2$ is the $y$-coordinate of $M$. Thus, we have the following result.

---

The midpoint of the line segment from $P(x_1, y_1)$ to $Q(x_2, y_2)$ has coordinates

$$\left( \frac{x_1 + x_2}{2}, \frac{y_1 + y_2}{2} \right)$$

---

**EXAMPLE 4**   Find the equation of the circle having the segment from (1, 3) to (7, 11) as a diameter.

*Solution*   The center of the circle is at the midpoint of the diameter; thus, the center has coordinates $(1 + 7)/2 = 4$ and $(3 + 11)/2 = 7$. The length of the diameter, obtained from the distance formula, is

$$[(7 - 1)^2 + (11 - 3)^2]^{1/2} = [36 + 64]^{1/2} = 10$$

and so the radius of the circle is 5. The equation of the circle is

$$(x - 4)^2 + (y - 7)^2 = 25$$   ■

## CONCEPTS REVIEW

**1.** The point $(-4, 2)$ lies in quadrant _____, whereas $(2, -4)$ lies in quadrant _____.

**2.** The distance between the points $(-2, 3)$ and $(x, y)$ is _____.

**3.** The equation of the circle of radius 5 and center $(-4, 2)$ is _____.

**4.** The midpoint of the line segment joining $(-2, 3)$ and $(5, 7)$ is _____.

## PROBLEM SET 1.5

In Problems 1–6, plot the given points in the coordinate plane and then find the distance between them.

**1.** $(3, 1), (1, 1)$   **2.** $(-3, 5), (2, -2)$

**3.** $(4, 5), (5, -8)$   **4.** $(-1, 5), (6, 3)$

**5.** $(1.345, -1.234), (56.34, 89.56)$

**6.** $(\sqrt{\pi}, 3.222), (\pi, 8.145)$

**7.** Show that the triangle whose vertices are $(5, 3)$, $(-2, 4)$, and $(10, 8)$ is isosceles.

**8.** Show that the triangle whose vertices are $(2, -4)$, $(4, 0)$, and $(8, -2)$ is a right triangle.

**9.** The points $(3, -1)$ and $(3, 3)$ are two vertices of a square. Give three other pairs of possible vertices.

**10.** Find the point on the $x$-axis that is equidistant from $(3, 1)$ and $(6, 4)$.

**11.** Find the distance between $(-2, 3)$ and the midpoint of the segment joining $(-2, -2)$ and $(4, 3)$.

**12.** Find the length of the line segment joining the midpoints of the segments $AB$ and $CD$, where $A = (1, 3)$, $B = (2, 6)$, $C = (4, 7)$, and $D = (3, 4)$.

In Problems 13–18, find the equation of the circle satisfying the given conditions.

**13.** Center $(1, 1)$, radius 1

**14.** Center $(-2, 3)$, radius 4

**15.** Center $(2, -1)$, goes through $(5, 3)$.

**16.** Center $(4, 3)$, goes through $(6, 2)$.

**17.** Diameter $AB$, where $A = (1, 3)$ and $B = (3, 7)$

**18.** Center $(3, 4)$ and tangent to $x$-axis.

**19.** Find the $y$-coordinates of the two points on the circle of Problem 13 with $x$-coordinate of $\frac{1}{4}$. (see Example 2).

**20.** Find the $x$-coordinates of the two points on the circle of Problem 13 with $y$-coordinate of 1.

In Problems 21–26, find the center and radius of the circle with the given equation (see Example 3).

**21.** $x^2 + 2x + 10 + y^2 - 6y - 10 = 0$

**22.** $x^2 + y^2 - 6y = 16$

**23.** $x^2 + y^2 - 12x + 35 = 0$

**24.** $x^2 + y^2 - 10x + 10y = 0$

**25.** $4x^2 + 16x + 15 + 4y^2 + 6y = 0$

**26.** $4x^2 + 16x + \frac{105}{16} + 4y^2 + 3y = 0$

**27.** The points $(2, 3)$, $(6, 3)$, $(6, -1)$, and $(2, -1)$ are corners of a square. Find the equations of the inscribed and circumscribed circles.

≈ **28.** A belt fits tightly around the two circles, with equations $(x - 1)^2 + (y + 2)^2 = 16$ and $(x + 9)^2 + (y - 10)^2 = 16$. How long is this belt?

[C] **29.** Cities at $A$, $B$, and $C$ are vertices of a right triangle, with the right angle at vertex $B$. Also, $AB$ and $BC$ are roads of lengths 214 and 179 miles, respectively. An airplane can fly the route $AC$, which is not a road. It costs $3.71 per mile to ship a certain product by truck and $4.82 per mile by plane. Decide whether it is cheaper to ship the product from $A$ to $C$ by truck or plane and find the total cost by the cheaper method.

[C] **30.** City B is 10 miles downstream from city A and on the opposite side of a river $\frac{1}{2}$ mile wide. Mary Crane will run from city A along the river for 6 miles, then swim diagonally to city B. If she runs at 8 miles per hour and swims at 3 miles per hour, how long will it take her to get from city A to city B? Assume the rate of the current is negligible.

**31.** Show that the midpoint of the hypotenuse of any right triangle is equidistant from the three vertices.

**32.** Find the equation of the circle circumscribed about the right triangle whose vertices are (0, 0), (8, 0), and (0, 6).

**33.** Show that the two circles $x^2 + y^2 - 4x - 2y - 11 = 0$ and $x^2 + y^2 + 20x - 12y + 72 = 0$ do not intersect. *Hint*: Find the distance between their centers.

**34.** What relationship between $a$, $b$, and $c$ must hold if $x^2 + ax + y^2 + by + c = 0$ is the equation of a circle?

[C] **35.** Find the length of the crossed belt in Figure 8 that fits tightly around the circles $(x - 2)^2 + (y - 2)^2 = 9$ and $(x - 10)^2 + (y - 8)^2 = 9$. *Note*: A little trigonometry is needed to complete this problem.

FIGURE 8

**36.** Show that the set of points that are twice as far from (3, 4) as from (1, 1) form a circle. Find its center and radius.

**37.** The Pythagorean Theorem says that the areas $A$, $B$, and $C$ of the squares in Figure 9 satisfy $A + B = C$. Show that semicircles and equilateral triangles satisfy the same relation and then guess at a very general theorem.

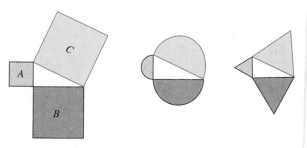

FIGURE 9

**38.** Consider a circle $C$ and a point $P$ exterior to the circle. Let line segment $PT$ be tangent to $C$ at $T$ and let the line through $P$ and the center of $C$ intersect $C$ at $M$ and $N$. Show that $(PM)(PN) = (PT)^2$.

**39.** A belt fits around the three circles $x^2 + y^2 = 4$, $(x - 8)^2 + y^2 = 4$, and $(x - 6)^2 + (y - 8)^2 = 4$, as shown in Figure 10. Find the length of this belt.

**40.** Study Problems 28 and 39. Consider a set of non-intersecting circles of radius $r$ with centers at the vertices of a convex $n$-sided polygon having sides of lengths $d_1$, $d_2$, ..., $d_n$. How long is the belt that fits around these circles (in the manner of Figure 10)?

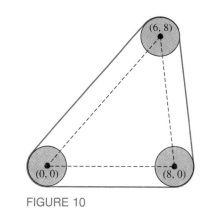

FIGURE 10

**Answers to Concepts Review:** **1.** II; IV
**2.** $\sqrt{(x + 2)^2 + (y - 3)^2}$ **3.** $(x + 4)^2 + (y - 2)^2 = 25$
**4.** (1.5, 5)

## 1.6
### THE STRAIGHT LINE

Of all curves, the straight line is in many ways the simplest. We assume that you have a good intuitive notion of this concept from looking at a taut string or sighting along the edge of a ruler. In any case, let us agree that two points—for example, $A(3, 2)$ and $B(8, 4)$ shown in Figure 1—determine a unique straight line through them. And from now on, we use the word *line* as a synonym for *straight line*.

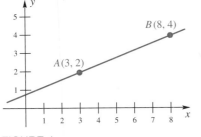

FIGURE 1

A line is a geometric object. When it is placed in a coordinate plane, it ought to have an equation, just as a circle does. How do we find the equation of a line? To answer, we will need the notion of slope.

**The Slope of a Line**    Consider the line in Figure 1. From point $A$ to point $B$, there is a **rise** (vertical change) of 2 units and a **run** (horizontal change) of 5 units. We say that the line has a slope of $\frac{2}{5}$. In general (Figure 2), for a line through $A(x_1, y_1)$ and $B(x_2, y_2)$, where $x_1 \neq x_2$, we define the **slope** $m$ of that line by

$$m = \frac{\text{rise}}{\text{run}} = \frac{y_2 - y_1}{x_2 - x_1}$$

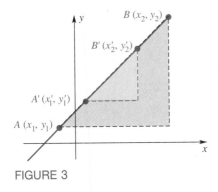

FIGURE 2

You should immediately raise a question. A line has many points. Does the value we get for the slope depend on which pair of points we use for $A$ and $B$? The similar triangles in Figure 3 show us that

$$\frac{y_2' - y_1'}{x_2' - x_1'} = \frac{y_2 - y_1}{x_2 - x_1}$$

Thus, points $A'$ and $B'$ would do just as well as $A$ and $B$. It does not even matter whether $A$ is to the left or right of $B$, since

$$\frac{y_1 - y_2}{x_1 - x_2} = \frac{y_2 - y_1}{x_2 - x_1}$$

All that matters is that we subtract the coordinates in the same order in numerator and denominator.

The slope $m$ is a measure of the steepness of a line, as Figure 4 illustrates. Notice that a horizontal line has zero slope, a line that rises to the right has

FIGURE 3

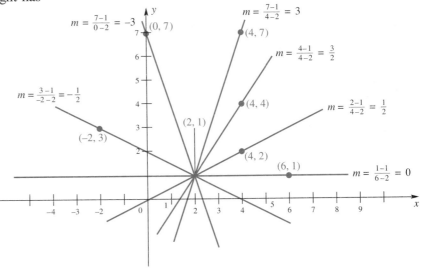

Lines of various slopes

FIGURE 4

### Grade and Pitch

The international symbol for the slope of a road (called the grade) is shown below. The grade is given as a percentage. A grade of 10% corresponds to a slope of ±0.10.

Carpenters use the term *pitch*. A 9:12 pitch corresponds to a slope of $\frac{9}{12}$.

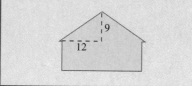

positive slope, and a line that falls to the right has negative slope. The larger the absolute value of the slope, the steeper the line. The concept of slope for a vertical line makes no sense, since it would involve division by zero. Therefore, slope for a vertical line is left undefined.

**The Point-Slope Form**    Consider again the line of our opening discussion; it is reproduced in Figure 5. We know that this line:

1. passes through (3, 2);
2. has slope $\frac{2}{5}$.

Take any other point on that line, such as one with coordinates $(x, y)$. If we use this point and the point (3, 2) to measure slope, we must get $\frac{2}{5}$—that is,

$$\frac{y - 2}{x - 3} = \frac{2}{5}$$

or, after multiplying by $x - 3$,

$$y - 2 = \tfrac{2}{5}(x - 3)$$

Notice that this last equation is satisfied by all points on the line, even by (3, 2). Moreover, none of the points not on the line can satisfy this equation.

What we have just done in an example can be done in general. The line passing through the (fixed) point $(x_1, y_1)$ with slope $m$ has equation

$$\boxed{y - y_1 = m(x - x_1)}$$

We call this the **point-slope** form of the equation of a line.

Consider once more the line of our example. That line passes through (8, 4) as well as (3, 2). If we use (8, 4) as $(x_1, y_1)$, we get the equation

$$y - 4 = \tfrac{2}{5}(x - 8)$$

which looks quite different from

$$y - 2 = \tfrac{2}{5}(x - 3)$$

However, both can be simplified to $5y - 2x = 4$; they are equivalent.

**EXAMPLE 1**    Find an equation of the line through $(-4, 2)$ and $(6, -1)$.

**Solution**    The slope $m$ is $(-1 - 2)/(6 + 4) = -\tfrac{3}{10}$. Thus, using $(-4, 2)$ as the fixed point, we obtain the equation

$$y - 2 = -\tfrac{3}{10}(x + 4)$$                   ∎

**The Slope-Intercept Form**    The equation of a line can be expressed in various forms. Suppose we are given the slope $m$ for a line and the $y$-intercept $b$ (that is, the line intersects the $y$-axis at $(0, b)$), as shown in Figure 6.

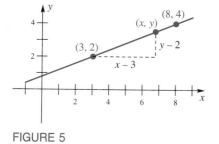

**FIGURE 5**

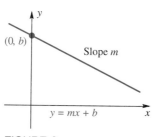

**FIGURE 6**

Choosing $(0, b)$ as $(x_1, y_1)$ and applying the point-slope form, we get

$$y - b = m(x - 0)$$

which we can rewrite as

$$\boxed{y = mx + b}$$

The latter is called the **slope-intercept** form.

What is the use of this form, you ask? Any time we see an equation written this way, we recognize it as a line and can immediately read its slope and $y$-intercept. For example, consider the equation

$$3x - 2y + 4 = 0$$

If we solve for $y$, we get

$$y = \tfrac{3}{2}x + 2$$

It is the equation of a line with slope $\tfrac{3}{2}$ and $y$-intercept 2.

**Equation of a Vertical Line**    Vertical lines do not fit within the discussion above since the concept of slope is not defined for them. But they do have equations, very simple ones. The line in Figure 7 has equation $x = \tfrac{5}{2}$, since a point is on the line if and only if it satisfies this equation. The equation of any vertical line can be put in the form

$$\boxed{x = k}$$

where $k$ is a constant. It should be noted that the equation of a horizontal line can be written in the form $y = k$.

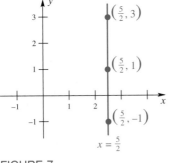

FIGURE 7

**The Form $Ax + By + C = 0$**    It would be nice to have a form that covered all lines, including vertical lines. Consider for example,

(1) $\qquad\qquad y - 2 = -4(x + 2)$

(2) $\qquad\qquad y = 5x - 3$

(3) $\qquad\qquad x = 5$

These can be rewritten (by taking everything to the left-hand side) as follows:

(1) $\qquad\qquad 4x + y + 6 = 0$

(2) $\qquad\qquad -5x + y + 3 = 0$

(3) $\qquad\qquad x + 0y - 5 = 0$

Summary: Equations of Lines

Vertical line: $x = k$
Horizontal line: $y = k$
Point-slope form:
$$y - y_1 = m(x - x_1)$$
Slope-intercept form:
$$y = mx + b$$
General linear equation:
$$Ax + By + C = 0$$

All are of the form

$$Ax + By + C = 0, \qquad A \text{ and } B \text{ not both } 0$$

which we call the **general linear equation**. It takes only a moment's thought to see that the equation of any line can be put in this form. Conversely, the graph of the general linear equation is always a line (see Problem 43).

**Parallel Lines**    If two lines have the same slope, they are parallel. Thus, $y = 2x + 2$ and $y = 2x + 5$ represent parallel lines; both have slope of 2. The second line is 3 units above the first for every value of $x$ (see Figure 8).

Similarly, the lines with equations $-2x + 3y + 12 = 0$ and $4x - 6y = 5$ are parallel. To see this, solve these equations for $y$ (that is, find the slope-intercept form); you get $y = \frac{2}{3}x - 4$ and $y = \frac{2}{3}x - \frac{5}{6}$, respectively. Both have slope $\frac{2}{3}$; they are parallel.

We may summarize by stating that *two nonvertical lines are parallel if and only if they have the same slope.*

**EXAMPLE 2**    Find the equation of the line through $(6, 8)$ which is parallel to the line with equation $3x - 5y = 11$.

*Solution*    When we solve $3x - 5y = 11$ for $y$, we obtain

$$y = \tfrac{3}{5}x - \tfrac{11}{5}$$

from which we read the slope of the line to be $\frac{3}{5}$. The equation of the desired line is

$$y - 8 = \tfrac{3}{5}(x - 6)$$

or, equivalently, $3x - 5y + 22 = 0$.    ■

**Perpendicular Lines**    Is there a simple slope condition that characterizes perpendicular lines? Yes; *two nonvertical lines are perpendicular if and only if their slopes are negative reciprocals of each other.* To see why this is true, consider two nonvertical intersecting lines $l_1$ and $l_2$. Without loss of generality, we may suppose they intersect at the origin since if not, we may translate them so they do without changing their slopes. Let $P_1(x_1, y_1)$ be a point on $l_1$ and $P_2(x_2, y_2)$ be a point on $l_2$, as shown in Figure 9. By the Pythagorean Theorem and its converse (Section 1.5), $P_1 O P_2$ is a right angle if and only if

$$[d(P_1, O)]^2 + [d(P_2, O)]^2 = [d(P_1, P_2)]^2$$

that is, if and only if

$$(x_1^2 + y_1^2) + (x_2^2 + y_2^2) = (x_1 - x_2)^2 + (y_1 - y_2)^2$$

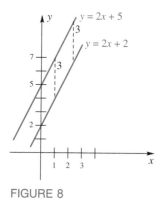

FIGURE 8

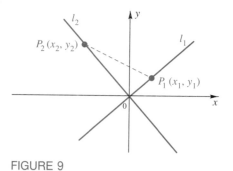

FIGURE 9

After expansion and simplification, this equation becomes $2x_1x_2 + 2y_1y_2 = 0$, or

$$\frac{y_1}{x_1} = -\frac{x_2}{y_2}$$

Now, $y_1/x_1$ is the slope of $l_1$, whereas $y_2/x_2$ is the slope of $l_2$. Thus, $P_1OP_2$ is a right angle if and only if the slopes of the two lines are negative reciprocals of each other.

The lines $y = \frac{3}{4}x$ and $y = -\frac{4}{3}x$ are perpendicular. So are $2x - 3y = 5$ and $3x + 2y = -4$, since after solving these equations for $y$, we see that the first line has slope $\frac{2}{3}$ and the second has slope $-\frac{3}{2}$.

**EXAMPLE 3**  Find the equation of the line through the point of intersection of the lines with equations $3x + 4y = 8$ and $6x - 10y = 7$ which is perpendicular to the first of these two lines (Figure 10).

*Solution*  To find the point of intersection of the two lines, we multiply the first equation by $-2$ and add it to the second equation.

$$
\begin{array}{r}
-6x - 8y = -16 \\
\underline{6x - 10y = \phantom{-1}7} \\
-18y = \phantom{-1}-9 \\
y = \tfrac{1}{2}
\end{array}
$$

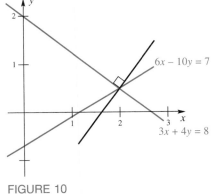

**FIGURE 10**

Substituting $y = \frac{1}{2}$ in either of the original equations yields $x = 2$. The point of intersection is $(2, \frac{1}{2})$.

When we solve the first equation for $y$ (to put it in slope-intercept form), we get $y = -\frac{3}{4}x + 2$. A line perpendicular to it has slope $\frac{4}{3}$. The equation of the required line is

$$y - \tfrac{1}{2} = \tfrac{4}{3}(x - 2) \qquad \blacksquare$$

## CONCEPTS REVIEW

**1.** The line through $(a, b)$ and $(c, d)$ has slope $m =$ _____ provided $a \neq c$.

**2.** A horizontal line has slope $m =$ _____, whereas a vertical line does not have slope.

**3.** The equation of a nonvertical line can always be written in the form $y =$ _____, whereas the equation of a vertical line can be written in the form $x =$ _____.

**4.** Every line has an equation of the form _____, called the general linear equation.

## PROBLEM SET 1.6

In Problems 1–8, find the slope of the line containing the given two points.

**1.** (1, 1) and (2, 2)

**2.** (3, 5) and (4, 7)

**3.** (2, 3) and $(-5, -6)$

**4.** $(2, -4)$ and $(0, -6)$

**5.** (3, 0) and (0, 5)

**6.** $(-6, 0)$ and (0, 6)

7. $(-1.234, 5.678)$ and $(7.654, -3.456)$

8. $(\pi, \sqrt{3})$ and $(1.642, \sqrt{2})$

In Problems 9–16, find an equation for each line. Then write your answer in the form $Ax + By + C = 0$.

9. Through $(2, 2)$ with slope $-1$.

10. Through $(3, 4)$ with slope $-1$.

11. With $y$-intercept 3 and slope 2.

12. With $y$-intercept 5 and slope 0.

13. Through $(2, 3)$ and $(4, 8)$.

14. Through $(4, 1)$ and $(8, 2)$.

15. Through $(2, -3)$ and $(2, 5)$.

16. Through $(-5, 0)$ and $(-5, 4)$.

In Problems 17–20, find the slope and $y$-intercept of each line.

17. $3y = -2x + 1$     18. $-4y = 5x - 6$

19. $6 - 2y = 10x - 2$     20. $4x + 5y = -20$

21. Write an equation for the line through $(3, -3)$ which is:

(a) parallel to the line $y = 2x + 5$;
(b) perpendicular to the line $y = 2x + 5$;
(c) parallel to the line $2x + 3y = 6$;
(d) perpendicular to the line $2x + 3y = 6$;
(e) parallel to the line through $(-1, 2)$ and $(3, -1)$;
(f) parallel to the line $x = 8$;
(g) perpendicular to the line $x = 8$.

22. Find the value of $c$ for which the line $3x + cy = 5$

(a) passes through the point $(3, 1)$;
(b) is parallel to the $y$-axis;
(c) is parallel to the line $2x + y = -1$;
(d) has equal $x$- and $y$-intercepts;
(e) is perpendicular to the line $y - 2 = 3(x + 3)$.

23. Write the equation for the line through $(-2, -1)$ which is perpendicular to the line $y + 3 = -\frac{2}{3}(x - 5)$.

24. Find the value of $k$ such that the line $kx - 3y = 10$:

(a) is parallel to the line $y = 2x + 4$;
(b) is perpendicular to the line $y = 2x + 4$;
(c) is perpendicular to the line $2x + 3y = 6$.

25. Does $(3, 9)$ lie above or below the line $y = 3x - 1$?

26. Show that the equation of the line with $x$-intercept $a \neq 0$ and $y$-intercept $b \neq 0$ can be written as

$$\frac{x}{a} + \frac{y}{b} = 1$$

In Problems 27–30, find the coordinates of the point of intersection. Then write an equation for the line through that point perpendicular to the line given first (see Example 3).

27.  $2x + 3y = 4$       28. $4x - 5y = 8$
     $-3x + y = 5$            $2x + y = -10$

29. $3x - 4y = 5$       30. $5x - 2y = 5$
    $2x + 3y = 9$            $2x + 3y = 6$

It can be shown that the distance $d$ from the point $(x_1, y_1)$ to the line $Ax + By + C = 0$ is

$$d = \frac{|Ax_1 + By_1 + C|}{\sqrt{A^2 + B^2}}$$

Use this result to find the distance from the given point to the given line.

31. $(-3, 2)$; $3x + 4y = 6$

32. $(4, -1)$; $2x - 2y + 4 = 0$

33. $(-2, -1)$; $5y = 12x + 1$

34. $(3, -1)$; $y = 2x - 5$

In Problems 35 and 36, find the (perpendicular) distance between the given parallel lines. *Hint:* First find a point on one of the lines.

35. $2x + 4y = 7$, $2x + 4y = 5$

36. $7x - 5y = 6$, $7x - 5y = -1$

37. A bulldozer costs \$120,000 and each year it depreciates 8% of its original value. Find a formula for the value $V$ of the bulldozer after $t$ years.

38. The graph of the answer to Problem 37 is a straight line. What is its slope, assuming the $t$-axis to be horizontal? Interpret the slope.

39. Past experience indicates that egg production in Matlin County is growing linearly. In 1960 it was 700,000 cases, and in 1970 it was 820,000 cases. Write a formula for the number $N$ of cases produced $n$ years after 1960 and use to predict egg production in the year 2000.

40. A piece of equipment purchased today for \$80,000 will depreciate linearly to a scrap value of \$2000 after 20 years. Write a formula for its value $V$ after $n$ years.

41. Suppose that the profit $P$ that a company realizes in selling $x$ items of a certain commodity is given by $P = 450x - 2000$ dollars.

(a) Interpret the value of $P$ when $x = 0$.
(b) Find the slope of the graph of the above equation. This slope is called the **marginal profit**. What is its economic interpretation?

**42.** The cost $C$ of producing $x$ items of a certain commodity is given by $C = 0.75x + 200$ dollars. The slope of its graph is called the **marginal cost**. Find it and give an economic interpretation.

**43.** Show that the graph of $Ax + By + C = 0$ is always a line (provided $A$ and $B$ are not both 0). *Hint*: Consider two cases: (1) $B = 0$ and (2) $B \neq 0$.

**44.** Find an equation for the line through $(2, 3)$ which has equal $x$- and $y$-intercepts. *Hint*: Use Problem 26.

**45.** Show that for each value of $k$, the equation

$$2x - y + 4 + k(x + 3y - 6) = 0$$

represents a line through the intersection of the two lines $2x - y + 4 = 0$ and $x + 3y - 6 = 0$. *Hint*: It is not necessary to find the point of intersection $(x_0, y_0)$.

**46.** Find the equation for the line which bisects the line segment from $(-2, 3)$ to $(1, -2)$ and is at right angles to this line segment.

**47.** The center of the circumscribed circle of a triangle lies on the perpendicular bisectors of the sides. Use this fact to find the center of the circle that circumscribes the triangle with vertices $(0, 4)$, $(2, 0)$, and $(4, 6)$.

**48.** Suppose that $(a, b)$ is on the circle $x^2 + y^2 = r^2$. Show that the line $ax + by = r^2$ is tangent to the circle at $(a, b)$.

**49.** Find the equations of the two tangent lines to the circle $x^2 + y^2 = 36$ that go through $(12, 0)$. *Hint*: See Problem 48.

**50.** Express the perpendicular distance between the parallel lines $y = mx + b$ and $y = mx + B$ in terms of $m$, $b$, and $B$. *Hint*: The required distance is the same as that between $y = mx$ and $y = mx + B - b$.

**51.** Show that the line through the midpoints of two sides of a triangle is parallel to the third side. *Hint*: You may assume the triangle has vertices at $(0, 0)$, $(a, 0)$, and $(b, c)$.

**52.** Show that the line segments joining the midpoints of adjacent sides of any quadrilateral (four-sided polygon) form a parallelogram.

**53.** A wheel whose rim has equation $x^2 + (y - 6)^2 = 25$ is rotating rapidly in the counterclockwise direction. A speck of dirt on the rim came loose at the point $(3, 2)$ and flew toward the wall $x = 11$. About how high up on the wall did it hit?

---

**Answers to Concepts Review:  1.** $(d - b)/(c - a)$  **2.** 0
**3.** $y = mx + b$; $x = k$  **4.** $Ax + By + C = 0$

---

## 1.7
## GRAPHS OF EQUATIONS

The use of coordinates for points in the plane allows us to describe a curve (a geometric object) by an equation (an algebraic object). We saw how this was done for circles and lines in the previous sections. Now we want to consider the reverse process: graphing an equation. The **graph of an equation** in $x$ and $y$ consists of those points in the plane whose coordinates $(x, y)$ satisfy the equation—that is, make it a true equality.

**The Graphing Procedure**  To graph an equation—for example, $y = 2x^3 - x + 19$—we follow a simple three-step procedure:

*Step 1*  Obtain the coordinates of a few points that satisfy the equation.
*Step 2*  Plot those points in the plane.
*Step 3*  Connect the points with a smooth curve.

The best way to do Step 1 is to make a table of values. Assign values to one of the variables, such as $x$, and determine the corresponding values of the other variable, listing the results in tabular form.

**EXAMPLE 1** Graph the equation $y = x^2 - 3$.

*Solution* The three-step procedure is shown in Figure 1.

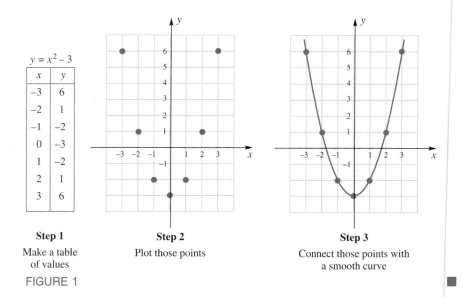

$y = x^2 - 3$

| x | y |
|----|----|
| −3 | 6 |
| −2 | 1 |
| −1 | −2 |
| 0 | −3 |
| 1 | −2 |
| 2 | 1 |
| 3 | 6 |

**Step 1**
Make a table
of values

**Step 2**
Plot those points

**Step 3**
Connect those points with
a smooth curve

FIGURE 1

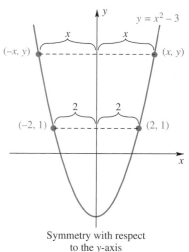

Symmetry with respect
to the y-axis

FIGURE 2

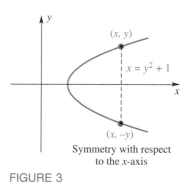

Symmetry with respect
to the x-axis

FIGURE 3

Of course, you need to use common sense and even a little faith. When you connect the points you have plotted with a smooth curve, you are assuming that the curve behaves nicely between consecutive points, which is faith. That is why you should plot enough points so the outline of the curve seems very clear; the more points you plot, the less faith you will need. Also, you should recognize that you can seldom display the whole curve. In our example, the curve has infinitely long arms, opening wider and wider. But our graph does show the essential features. This is our goal in graphing: Show enough of the graph so the essential features are visible. Later on (Section 4.7), we will use the tools of calculus to refine and improve our graphing technique.

**Symmetry of a Graph** We can save ourselves some work and also draw more accurate graphs if we can recognize certain symmetries of the graph by examining the corresponding equation. Look at the graph of $y = x^2 - 3$, drawn above and again in Figure 2. If the coordinate plane were folded along the y-axis, the two branches would coincide. For example, (3, 6) would coincide with (−3, 6), (2, 1) would coincide with (−2, 1), and, more generally, (x, y) would coincide with (−x, y). Algebraically, this corresponds to the fact that replacing x by −x in the equation $y = x^2 - 3$ results in an equivalent equation.

Consider an arbitrary graph. It is **symmetric with respect to the y-axis** if whenever (x, y) is on the graph, (−x, y) is also on the graph (Figure 2). Similarly, it is **symmetric with respect to the x-axis** if whenever (x, y) is on the graph, (x, −y) is also on the graph (Figure 3). Finally, a graph is **symmetric with respect to the origin** if whenever (x, y) is on the graph, (−x, −y) is also on the graph (see Example 2).

In terms of equations, we have three simple tests.

> The graph of an equation is:
>
> **1.** symmetric with respect to the $y$-axis if replacing $x$ by $-x$ gives an equivalent equation (for example, $y = x^2$).
>
> **2.** symmetric with respect to the $x$-axis if replacing $y$ by $-y$ gives an equivalent equation (for example, $x = 1 + y^2$).
>
> **3.** symmetric with respect to the origin if replacing $x$ by $-x$ and $y$ by $-y$ gives an equivalent equation ($y = x^3$ is a good example since $-y = (-x)^3$ is equivalent to $y = x^3$).

**EXAMPLE 2**    Sketch the graph of $y = x^3$.

*Solution*   We note, as pointed out above, that the graph will be symmetric with respect to the origin. Thus we need only get a table of values for non-negative $x$'s; we can find matching points by symmetry (Figure 4).

$y = x^3$

| $x$ | $y$ |
|-----|-----|
| 0 | 0 |
| 1 | 1 |
| 2 | 8 |
| 3 | 27 |
| 4 | 64 |

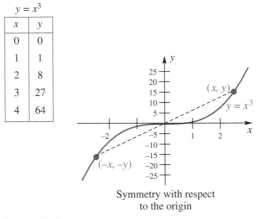

Symmetry with respect
to the origin

FIGURE 4

In graphing $y = x^3$, we used a smaller scale on the $y$-axis than on the $x$-axis. This made it possible to show a larger portion of the graph (it also distorted the graph by flattening it). We suggest that before putting scales on the two axes, you should examine your table of values. Choose scales so that all or most of your points can be plotted and still keep your graph of reasonable size.

**Intercepts**   The points where the graph of an equation crosses the two coordinate axes play a significant role in many problems. Consider, for example,

$$y = x^3 - 2x^2 - 5x + 6 = (x + 2)(x - 1)(x - 3)$$

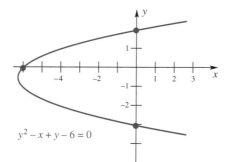

$$y^2 - x + y - 6 = 0$$

**FIGURE 5**

Notice that $y = 0$ when $x = -2, 1, 3$. The numbers $-2, 1$, and $3$ are called **x-intercepts**. Similarly, $x = 0$ when $y = 6$, and so $6$ is called the **y-intercept**.

**EXAMPLE 3**    Sketch the graph of $y^2 - x + y - 6 = 0$, showing all intercepts clearly.

**Solution**    Putting $y = 0$ in the given equation, we get $x = -6$, and so the x-intercept is $-6$. Putting $x = 0$ in the equation, we find $y^2 + y - 6 = 0$, or $(y + 3)(y - 2) = 0$; the y-intercepts are $-3$ and $2$. A check on symmetries indicates that the graph has none of the three types discussed earlier. The graph is displayed in Figure 5.    ∎

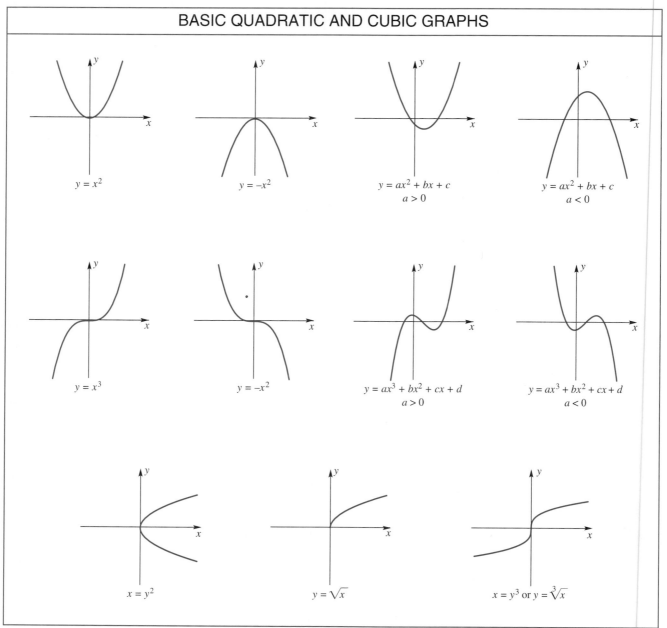

## BASIC QUADRATIC AND CUBIC GRAPHS

$y = x^2$

$y = -x^2$

$y = ax^2 + bx + c$
$a > 0$

$y = ax^2 + bx + c$
$a < 0$

$y = x^3$

$y = -x^2$

$y = ax^3 + bx^2 + cx + d$
$a > 0$

$y = ax^3 + bx^2 + cx + d$
$a < 0$

$x = y^2$

$y = \sqrt{x}$

$x = y^3$ or $y = \sqrt[3]{x}$

**FIGURE 6**

**General Quadratic and Cubic Equations**   Since quadratic and cubic equations will often be used as examples in later work, we display their typical graphs in Figure 6.

The graphs of quadratic equations are cup-shaped curves called **parabolas**. If an equation has the form $y = ax^2 + bx + c$ or $x = ay^2 + by + c$ with $a \neq 0$, its graph will always be a parabola. In the first case, the graph opens up if $a > 0$ and opens down if $a < 0$. In the second case, the graph opens right or left, according as $a > 0$ or $a < 0$. Note that the equation of Example 3 can be put in the form $x = y^2 + y - 6$.

**Intersections of Graphs**   Occasionally, we need to know the points of intersection of two graphs. These points are found by solving the two equations for the graphs simultaneously.

**EXAMPLE 4**   Find the points of intersection of the line $y = -2x + 2$ and the parabola $y = 2x^2 - 4x - 2$ and sketch both graphs on the same coordinate plane.

*Solution*   We must solve the two equations simultaneously. This is easy to do by substituting the expressions for $y$ from the first equation into the second equation and then solving the resulting equation for $x$.

$$-2x + 2 = 2x^2 - 4x - 2$$
$$0 = 2x^2 - 2x - 4$$
$$0 = 2(x - 2)(x + 1)$$
$$x = -1, \qquad x = 2$$

By substitution, we find the corresponding values of $y$ to be 4 and $-2$; the intersection points are therefore $(-1, 4)$ and $(2, -2)$.

The two graphs are shown in Figure 7.   ■

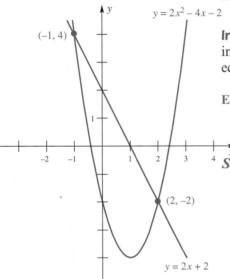

FIGURE 7

---

## CONCEPTS REVIEW

**1.** If whenever $(x, y)$ is on a graph, $(-x, y)$ is also on the graph, then the graph is symmetric with respect to the _____.

**2.** If $(-4, 2)$ is on a graph that is symmetric with respect to the origin, then _____ is also on the graph.

**3.** The graph of $y = (x + 2)(x - 1)(x - 4)$ has $y$-intercept _____ and $x$-intercepts _____.

**4.** The graph of $y = ax^2 + bx + c$ is a _____ if $a = 0$ and a _____ if $a \neq 0$.

---

## PROBLEM SET 1.7

In Problems 1–36, sketch the graph of the given equations. Begin by checking for symmetries and be sure to find all $x$- and $y$-intercepts.

**1.** $y = -x^2 + 1$

**2.** $x = -y^2 + 1$

**3.** $x = -4y^2 - 1$

**4.** $y = 4x^2 - 1$

**5.** $x^2 + y = 0$

**6.** $y = x^2 - 2x$

**7.** $7x^2 + 3y = 0$

**8.** $y = 3x^2 - 2x + 2$

**9.** $x^2 + y^2 = 4$

**10.** $3x^2 + 4y^2 = 12$

**11.** $y = -x^2 - 2x + 2$

**12.** $4x^2 + 3y^2 = 12$

**13.** $x^2 - y^2 = 4$

**14.** $x^2 + (y - 1)^2 = 9$

15. $4(x - 1)^2 + y^2 = 36$

16. $x^2 - 4x + 3y^2 = -2$

17. $x^2 + 9(y + 2)^2 = 36$

18. $x^4 + y^4 = 1$

19. $x^4 + y^4 = 16$     20. $16x^4 + 81y^4 = 144$

21. $y = x^3 - x$     22. $y = x^3 + 10x - 10$

23. $y = \dfrac{1}{x^2 + 1}$

24. $y = \dfrac{x}{x^2 + 1}$

25. $y = \dfrac{4}{x^2 + 10}$

26. $y = \dfrac{3x}{4x^2 + 7}$

27. $2x^2 - 4x + 3y^2 + 12y = -2$

28. $4(x - 5)^2 + 9(y + 2)^2 = 36$

29. $y = (x - 1)(x - 2)(x - 3)$

30. $y = (2x - 4)(3x + 9)(4x - 8)$

31. $y = x^2(x - 1)(x - 2)$

32. $y = x^2(x - 1)^2$

33. $y = x^2(x - 1)^2(x + 1)^2$

34. $y = x^4(x - 1)^4(x + 1)^4$

35. $|x| + |y| = 1$

36. $|x| + |y| = 4$

In Problems 37–46, sketch the graphs of both equations on the same coordinate plane. Be sure to find the points of intersection of the two graphs (see Example 4). You will need the quadratic formula in Problems 41–46.

37. $y = -x + 1$
    $y = (x + 1)^2$

38. $y = 2x + 3$
    $y = -(x - 1)^2$

39. $y = -2x + 3$
    $y = -2(x - 4)^2$

40. $y = -2x + 3$
    $y = 3x^2 - 3x + 12$

C   41. $y = 2.2r - 5.6$
    $y = -1.3x^2 + 4.5$

C   42. $y = 2.1x - 5.9$
    $y = 1.3x^2 - 2.2x$

C   43. $y = x$
    $x^2 + y^2 = 4$

C   44. $y = x - 1$
    $2x^2 + 3y^2 = 12$

C   45. $y - 3x = 1$
    $x^2 + 2x + y^2 = 15$

C   46. $y = 4x + 3$
    $x^2 + y^2 = 81$

47. Choose the equation which best represents Figure 8.

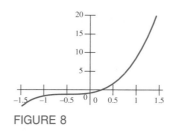

FIGURE 8

(a) $y = x^2$
(b) $y = ax^3 + bx^2 + cx + d$, with $a > 0$
(c) $y = ax^3 + bx^2 + cx + d$, with $a < 0$
(d) $y = ax^3$, with $a > 0$

48. Find the distance between the points on the circle $x^2 + y^2 = 13$ with the $x$-coordinates $-2$ and $2$. How many such distances are there?

C   49. Find the distance between the points on the circle $x^2 + 2x + y^2 - 2y = 20$ with the $x$-coordinates $-2$ and $2$. How many such distances are there?

**Answers to Concepts Review:**   **1.** $y$-axis   **2.** $(4, -2)$
**3.** $8; -2, 1, 4$   **4.** Line; parabola

# 1.8 CHAPTER REVIEW

## Concepts Test

Respond with true or false to each of the following assertions. Be prepared to justify your answer. Normally, this means that you should supply a reason if you answer true and provide a counterexample if you answer false.

**1.** Any number that can be written as a fraction $p/q$ is rational.

**2.** The difference of two rational numbers is rational.

**3.** The difference of two irrational numbers is irrational.

**4.** Between two distinct irrational numbers there is always another irrational number.

**5.** $0.999\ldots$ (repeating 9s) is less than 1.

**6.** The operation of exponentiation is commutative; that is, $(a^m)^n = (a^n)^m$.

**7.** The operation $*$ defined by $m * n = m^n$ is associative.

**8.** The inequalities $x \leq y$, $y \leq z$, and $z \leq x$ together imply that $x = y = z$.

**9.** If $|x| < \varepsilon$ for every positive number $\varepsilon$, then $x = 0$.

**10.** If $x$ and $y$ are real numbers, then $(x - y)(y - x) \leq 0$.

**11.** If $a < b < 0$, then $1/a > 1/b$.

**12.** It is possible for two closed intervals to have exactly one point in common.

**13.** If two open intervals have a point in common, then they have infinitely many points in common.

**14.** If $x < 0$, then $\sqrt{x^2} = -x$.

**15.** The value of $|x|$ is the number $x$ without its sign.

**16.** If $|x| < |y|$, then $x < y$.

**17.** If $|x| < |y|$, then $x^4 < y^4$.

**18.** If $x$ and $y$ are both negative, then $|x + y| = |x| + |y|$.

**19.** If $|r| < 1$ then $\dfrac{1}{1 + |r|} \leq \dfrac{1}{1 - r} \leq \dfrac{1}{1 - |r|}$.

**20.** If $|r| > 1$ then $\dfrac{1}{1 - |r|} \leq \dfrac{1}{1 - r} \leq \dfrac{1}{1 + |r|}$.

**21.** It is always true that $\bigl||x| - |y|\bigr| \leq |x + y|$.

**22.** It is possible to have an inequality whose solution set consists of exactly one number.

**23.** The equation $x^2 + y^2 + ax + y = 0$ represents a circle for every real number $a$.

**24.** The equation $x^2 + y^2 + ax + by = c$ represents a circle for all real numbers $a$, $b$, $c$.

**25.** If $(a, b)$ is on a line with slope $\frac{3}{4}$, then $(a + 4, b + 3)$ is also on that line.

**26.** If $(a, b)$, $(c, d)$ and $(e, f)$ are on the same line then $\dfrac{a - c}{b - d} = \dfrac{a - e}{b - f} = \dfrac{e - c}{f - d}$ provided all three points are different.

**27.** If $ab > 0$, then $(a, b)$ lies in either the first or third quadrant.

**28.** It is incorrect to say that the absolute value of a number is that number without its sign.

**29.** If $ab = 0$, then $(a, b)$ lies on either the $x$-axis or the $y$-axis.

**30.** If $\sqrt{(x_2 - x_1)^2 + (y_2 - y_1)^2} = |x_2 - x_1|$, then $(x_1, y_1)$ and $(x_2, y_2)$ lie on the same horizontal line.

**31.** The distance between $(a + b, a)$ and $(a - b, a)$ is $|2b|$.

**32.** The equation of any line can be written in point-slope form.

**33.** The equation of any line can always be written in the form $ax + by + C = 0$.

**34.** If two nonvertical lines are parallel, they have the same slope.

**35.** It is possible for two lines to have positive slopes and be mutually perpendicular.

**36.** If the $x$- and $y$-intercepts of a line are rational and nonzero, then the slope of the line is rational.

**37.** The lines $ax + y = c$ and $ax - y = c$ are perpendicular.

**38.** $(3x - 2y + 4) + m(2x + 6y - 2) = 0$ is the equation of a straight line for each real number $m$.

## Sample Test Problems

**1.** Calculate each value for $n = 1$, 2, and $-2$.

(a) $\left(n + \dfrac{1}{n}\right)^n$

(b) $(n^2 - n + 1)^2$

(c) $4^{3/n}$

(d) $\sqrt[n]{\left|\dfrac{1}{n}\right|}$

**2.** Simplify.

(a) $\left(1 + \dfrac{1}{m} + \dfrac{1}{n}\right)\left(1 - \dfrac{1}{m} + \dfrac{1}{n}\right)^{-1}$

(b) $\dfrac{\dfrac{2}{x+1} - \dfrac{x}{x^2 - x - 2}}{\dfrac{3}{x+1} - \dfrac{2}{x-2}}$

(c) $\dfrac{(t^3 - 1)}{t - 1}$

**3.** Show that the average of two rational numbers is a rational number.

**4.** Write the repeating decimal 4.1282828 . . . as a ratio of two integers.

**5.** Find an irrational number between $\frac{1}{2}$ and $\frac{13}{25}$.

[C] **6.** Calculate $\left(\sqrt[3]{8.15 \times 10^4} - 1.32\right)^2/3.24$.

[C] **7.** Calculate $\left(\pi - \sqrt{2.0}\right)^{2.5} - \sqrt[3]{2.0}$.

[C] **8.** Calculate $\sin^2(2.45) + \cos^2(2.40) - 1.00$.

In Problems 9–19, find the solution set, graph this set on the real line, and express this set in interval notation.

**9.** $1 - 3x > 0$

**10.** $6x + 3 > 2x - 5$

**11.** $3 - 2x \le 4x + 1 \le 2x + 7$

**12.** $2x^2 + 5x - 3 < 0$

**13.** $21t^2 - 44t + 12 \le -3$

**14.** $\dfrac{2x - 1}{x - 2} > 0$

**15.** $(x + 4)(2x - 1)^2(x - 3) \le 0$

**16.** $|3x - 4| < 6$

**17.** $\dfrac{3}{1 - x} \le 2$

**18.** $|12 - 3x| \ge |x|$

**19.** $|8 - 3x| \ge |2x|$

**20.** Express the given inequalities using absolute value notation

(a) $-6 \le t \le 2$

(b) $-6 \ge 2t \ge -18$

(c) $x \ge 9$ or $x \le 3$

**21.** Find a value of $x$ for which $|-x| \ne x$.

**22.** For what values of $x$ does the equation $|-x| = x$ hold?

**23.** For what values of $t$ does the equation $|t - 5| = 5 - t$ hold?

**24.** For what values of $a$ and $t$ does the equation $|t - a| = a - t$ hold?

**25.** Suppose $|x| \le 2$. Use properties of absolute values to show

$$\left|\dfrac{2x^2 + 3x + 2}{x^2 + 2}\right| \le 8$$

**26.** Write a sentence involving the word distance to express the following algebraic sentences:

(a) $|x - 5| = 3$

(b) $|x + 1| \le 2$

(c) $|x - a| > b$

**27.** Sketch the triangle with vertices $A(-2, 6)$, $B(1, 2)$, and $C(5, 5)$, and show that it is a right triangle.

**28.** Find the distance from $(3, -6)$ to the midpoint of the line segment from $(1, 2)$ to $(7, 8)$.

**29.** Find the equation of the circle with diameter $AB$ if $A = (2, 0)$ and $B = (10, 4)$.

**30.** Find the center and radius of the circle with equation $x^2 + y^2 - 8x + 6y = 0$.

**31.** Find the distance between the centers of the circles with equations

$$x^2 - 2x + y^2 + 2y = 2 \text{ and } x^2 + 6x + y^2 - 4y = -7$$

**32.** Find the equation through the indicated point which is parallel to the indicated line, and sketch both lines.

(a) $(3, 2)$: $3x + 2y = 6$

(b) $(1, -1)$: $y = \frac{2}{3}x + 1$

(c) $(5, 9)$: $y = 10$

(d) $(-3, 4)$: $x = -2$

**33.** Write the equation of the line through $(-2, 1)$ which:

(a) goes through $(7, 3)$;

(b) is parallel to $3x - 2y = 5$;

(c) is perpendicular to $3x + 4y = 9$;

(d) is perpendicular to $y = 4$;

(e) has $y$-intercept 3.

**34.** Show that $(2, -1)$, $(5, 3)$, and $(11, 11)$ are on the same line.

**35.** Figure 9 can be represented by the equation:

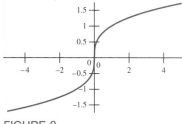

**FIGURE 9**

(a) $y = x^3$                    (b) $x = y^3$
(c) $y = x^2$                   (d) $x = y^2$

**36.** Figure 10 can be represented by the equation:

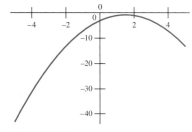

**FIGURE 10**

(a) $y = ax^2 + bx + c$, with $a > 0$, $b > 0$ and $c > 0$
(b) $y = ax^2 + bx + c$, with $a < 0$, $b > 0$ and $c > 0$
(c) $y = ax^2 + bx + c$, with $a < 0$, $b > 0$ and $c < 0$
(d) $y = ax^2 + bx + c$, with $a > 0$, $b > 0$ and $c < 0$

In Problems 37–40 sketch the graph of each equation.

    **37.** $3y - 4x = 6$           **38.** $x^2 - 2x + y^2 = 3$

    **39.** $y = \dfrac{2x}{x^2 + 2}$          **40.** $x = y^2 - 3$

**41.** Find the points of intersection of the graphs of $y = x^2 - 2x + 4$ and $y - x = 4$.

**42.** Among all lines perpendicular to $4x - y = 2$, find the equation of the one which—together with the positive $x$- and $y$-axes—forms a triangle of area 8.

### Additional Problems: Calculator Explorations

    **1.** When using a graphing calculator or computer algebra package to graph a function, you want to choose a graphing window that gives all the important details of that function. The following functions will give you some experience in choosing the correct window. In each case you should experiment with large and small windows to make sure that you can see all the details of the graph.

(a) $y = 3x^2 + 10$
(b) $y = x^3 - 9x^2 - 22x + 120$
(c) $y = (x - 1)(x - 2)(x - 3)(x - 4)$

    **2.** Consider the lines $y = mx + 4$. In the same graphing window draw these lines for $m = 0.1, 0.5, 1.0, 2.0$ and $4.0$. What similarity do you see among all these lines? Give an algebraic reason for what the graphs display.

    **3.** Consider the lines $y = -2x + b$. In the same graphing window draw these lines for $b = 1.0, 2.0, -2.0, 3.0, -3.0, 4.0$ and, $-4.0$. What similarity do you see among all these lines? Give an algebraic reason for what the graphs display.

    **4.** Graph the equations $y = x^4 + 4x^3 - 3x^2 - 2x + 1$ and $y = x^4 - 4x^3 - 3x^2 + 2x + 1$ using the same set of axis. What relationship do these two curves have to each other?

    **5.** Graph the equations $y = x^3 + bx^2 + x + 1$ for the values of $b = -5, -3, -1, 3$, and $5$. Explain qualitatively how the value of $b$ affects the shape of the curve and the number of times it crosses the x-axis.

    Most graphing utilities available on graphing calculators and personal computers permit the graphing of parametric equations. We will use this facility to explore the symmetry of graphs in problems 6–8.

    **6.** If a graph has symmetry about the y-axis, then if $(x, y)$ is on the graph, so is the point $(-x, y)$. For the specific case of the graph $y = x^2$ we can represent the whole graph parametrically by plotting two separate pieces of it. The portion to the left of the $y$-axis is given by $x_1 = -t$, $y_1 = t^2$, while the portion to the right of the axis is given by $x_2 = t$, $y_2 = t^2$. In this case a parametric plot for $0 \le t \le 5$ will give the full parabola $y = x^2$ for $0 \le x \le 5$. Set up and investigate the symmetry about the $x$-axis of the graph $x = y^2$ using a suitable parametric representation.

    **7.** Using the parametric representation $x_1 = t$, $y_1 = +\sqrt{1 - t^2}$ and $x_1 = t$, $y_1 = -\sqrt{1 - t^2}$ for $-1 \le t \le 1$ investigate the symmetries of the graph $x^2 + y^2 = 1$.

    **8.** Perform similar investigations using parametric representations for the graphs

(a) $y = x^3$
(b) $x = y^3$

# TECHNOLOGY PROJECT

In high school, you learned the solution formula for the quadratic equation $Ax^2 + Bx + C = 0$. However, you may not know any methods for solving higher-degree or transcendental equations (i.e., equations involving trig functions, logs, etc.). Oftentimes, such equations can be approximately solved by plotting to "zoom in" on the desired root.

**Exercise 1** Write an equation of the straight line $L$ that passes through $(-2, 4)$ and is perpendicular to the line with equation $x + 2y = 17$.

**Exercise 2** Solve the following generalized version of Exercise 1: Write the equation of the straight line $L$ that passes through the point $(p, q)$ and is normal to the line with equation $ax + by = c$. Check that your answer "works" with the specific numbers in part (a) of Exercise 1.

We will soon learn that:

**The slope of the tangent line to the parabola $y = x^2$ at $x = a$ is $2a$.**

Thus, the slope of this parabola at $x = 1$ is 2, the slope at $x = 3$ is 6, and so on. Please accept this as a fact for now.

**Exercise 3** Use the following hints to find the equation of the line through the point $P(2, 0)$ that is normal to the parabola $y = x^2$.

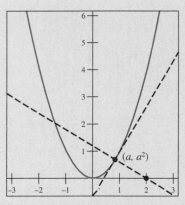

Fig. 1. Sketch for Exercise 3.

*Hints*: Draw a figure (see Figure 1) that includes a rough sketch of the normal from the given point. Use $a$ to denote the $x$-coordinate of the point where the normal cuts the parabola. Thus, the point $(a, a^2)$ is on both the parabola and the normal. The slope of the parabola at $x = a$ is $2a$, so the slope of the normal is the negative reciprocal value $-1/2a$. Show that equating the slope from $(a, a^2)$ to the given point $(2, 0)$ to the slope of the normal leads to the *cubic* equation for $a$,

$$2a^3 + a - 2 = 0$$

We can see from our sketch that $a \approx 1$, but this is *not* the exact root (try it). Instead of resorting to trial and error, plot the function near $a = 1$ and then "zoom in" on the root. The first two significant digits are 0.83; you find the next digit. Now is the time to be careful—you have a good approximation to $a$, but that is *not* quite what the exercise called for! The correct solution to the problem is to produce the requested *normal* from (2, 0) to the parabola, so state the normal line.

**Exercise 4** One problem solving maxim we've ignored in the above solution is that of trying to avoid the use of specific numbers. To maintain generality, let $(p, q)$ denote the given point (i.e., replace $(2, 0)$ by $(p, q)$). Show that the cubic equation giving the point $x = a$ where the normal from $(p, q)$ intersects the parabola $y = x^2$ is $\boxed{2a^3 - (2q - 1)a - p = 0}$. It's good practice to check that this agrees with the specific case worked out in Exercise 3, so do it!

**Exercise 5** In reference to the previous exercises, find the normal when $(p, q) = (4, 0)$. If you are uncertain how to proceed or how to present your solution, re-read Exercise 3.

**Exercise 6** In reference to the previous problems, when $(p, q) = (1, 17/4)$, show that there are *three* normals and determine them by doing three "zooms." Notice that the given point is *inside* the parabola.

**Notice how much more convenient and less error prone it is to use the *general* form for the cubic instead of setting up and simplifying the fractions for each specific point (4, 0), (1, 17/4), . . . .**

**Exercise 7**
In light of the results obtained in the lab exercises, a natural conjecture is that there are three normals for points interior to the parabola $y = x^2$ and only one for exterior points. Investigate the truth of this conjecture for the special case when the point is on the $y$-axis (that is, $p = 0$).

# 2

# Functions and Limits

## 2.1
## FUNCTIONS AND THEIR GRAPHS

Think of a function as a machine, a calculating machine (Figure 1). It takes a number (the input) and produces a result (the output). Each number put in is matched with a *single* number as output, but it could happen that several different input values give the same output value. We can state the definition more formally and introduce some notation at the same time.

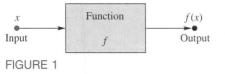

FIGURE 1

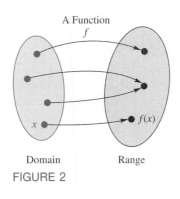

A Function

Domain          Range
FIGURE 2

> ### Definition
>
> A **function** $f$ is a rule of correspondence that associates with each object $x$ in one set, called the **domain**, a single value $f(x)$ from a second set. The set of values so obtained is called the **range** of the function. (See Figure 2.)

The definition puts no restriction on the domain and range sets. The domain might consist of the set of people in your calculus class, the range the set of grades $\{A, B, C, D, F\}$ that will be given, and the rule of correspondence the procedure your teacher uses in assigning grades.

Note that the definition does not allow more than one output to be matched with any one input. Thus, Figure 3 on the next page is not the diagram for a function.

Particularly relevant in calculus will be examples where both the domain and the range consist of sets of real numbers. For example, the function $g$ might take a real number $x$ and square it, thus producing the real number $x^2$. In this case, we have a formula that gives the rule of correspondence, namely, $g(x) = x^2$. A schematic diagram for this function is shown in Figure 4.

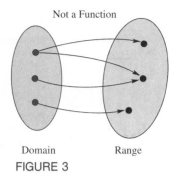

Not a Function

Domain       Range

FIGURE 3

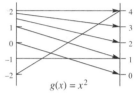

$g(x) = x^2$

Domain       Range

FIGURE 4

**Functional Notation** A single letter like $f$ (or $g$ or $F$) is used to name a function. Then $f(x)$, read "$f$ of $x$" or "$f$ at $x$," denotes the value that $f$ assigns to $x$. Thus, if $f(x) = x^3 - 4$,

$$f(2) = 2^3 - 4 = 4$$
$$f(-1) = (-1)^3 - 4 = -5$$
$$f(a) = a^3 - 4$$
$$f(a + h) = (a + h)^3 - 4 = a^3 + 3a^2h + 3ah^2 + h^3 - 4$$

A clear understanding of functional notation is critical in calculus. Study the following examples carefully. Examples such as these will play an important role in the next chapter.

**EXAMPLE 1** For $f(x) = x^2 - 2x$, find and simplify: (a) $f(4)$, (b) $f(4 + h)$, (c) $f(4 + h) - f(4)$, (d) $[f(4 + h) - f(4)]/h$

*Solution*

(a) $f(4) = 4^2 - 2 \cdot 4 = 8$

(b) $f(4 + h) = (4 + h)^2 - 2(4 + h) = 16 + 8h + h^2 - 8 - 2h$
$$= 8 + 6h + h^2$$

(c) $f(4 + h) - f(4) = 8 + 6h + h^2 - 8 = 6h + h^2$

(d) $\dfrac{f(4 + h) - f(4)}{h} = \dfrac{6h + h^2}{h} = \dfrac{h(6 + h)}{h} = 6 + h$ ∎

**EXAMPLE 2** For $g(x) = 1/x$, find and simplify $[g(a + h) - g(a)]/h$.

*Solution*

$$\frac{g(a + h) - g(a)}{h} = \frac{\dfrac{1}{a + h} - \dfrac{1}{a}}{h} = \frac{\dfrac{a - (a + h)}{(a + h)a}}{h}$$

$$= \frac{-h}{(a + h)a} \cdot \frac{1}{h} = \frac{-1}{(a + h)a} = \frac{-1}{a^2 + ah}$$ ∎

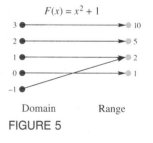

$F(x) = x^2 + 1$

Domain       Range

FIGURE 5

**Domain and Range** The rule of correspondence is the heart of a function, but a function is not completely determined until its domain is given. Recall that the *domain* is the set of elements to which the function assigns values. The *range* is the set of values so obtained. For example, if $F$ is the function with rule $F(x) = x^2 + 1$ and if the domain is specified as $\{-1, 0, 1, 2, 3\}$ (Figure 5), then the range is $\{1, 2, 5, 10\}$. The domain and the rule determine the range.

When no domain is specified for a function, we always assume that it is the largest set of real numbers for which the rule for the function makes sense and gives real number values. This is called the **natural domain**.

**EXAMPLE 3**   Find the natural domains for: (a) $f(x) = 1/(x - 3)$; (b) $g(t) = \sqrt{9 - t^2}$.

*Solution*

(a) The natural domain for $f$ is $\{x \in \mathbb{R} : x \neq 3\}$. This is read "the set of $x$'s in $\mathbb{R}$ (the real numbers) such that $x$ is not equal to 3." We excluded 3 to avoid division by 0.

(b) Here we must restrict $t$ so that $9 - t^2 \geq 0$ in order to avoid nonreal values for $\sqrt{9 - t^2}$. This is achieved by requiring that $|t| \leq 3$. Thus, the natural domain is $\{t \in \mathbb{R} : |t| \leq 3\}$. In interval notation, we could write the domain as $[-3, 3]$. ∎

When the rule for a function is given by an equation of the form $y = f(x)$ (for example, $y = x^3 + 3x - 6$), $x$ is called the **independent variable** and $y$ the **dependent variable**. Any element of the domain may be chosen as a value of the independent variable $x$, but that choice completely determines the corresponding value of the dependent variable. Thus, the value of $y$ is dependent on the chosen value of $x$.

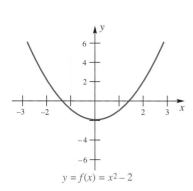

$y = f(x) = x^2 - 2$

**FIGURE 6**

**Graphs of Functions**   When both the domain and range of a function consist of real numbers, we can picture the function by drawing its graph on a coordinate plane. And the **graph of a function $f$** is simply the graph of the equation $y = f(x)$.

**EXAMPLE 4**   Sketch the graphs of: (a) $f(x) = x^2 - 2$; (b) $g(x) = x^3 - 2x$; (c) $h(x) = 2/(x - 1)$.

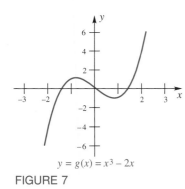

$y = g(x) = x^3 - 2x$

**FIGURE 7**

*Solution*   We use the natural domains. In the case of $f$ and $g$, this is the set $\mathbb{R}$ of all real numbers; for $h$, it is all of $\mathbb{R}$ except 1. Following the procedure described in Section 1.7 (make a table of values, plot the corresponding points, connect these points with a smooth curve), we obtain the three graphs shown in Figures 6 through 8. ∎

Pay special attention to the graph of $h$; it points to an oversimplification we have made and now need to correct. When connecting the plotted points by a smooth curve, do not do so in a mechanical way that ignores special features that may be apparent from the formula for the function. In the case of $h(x) = 2/(x - 1)$, it is clear that something dramatic must happen as $x$ nears 1. In fact, the values of $|h(x)|$ increase without bound (for example, $h(0.99) = -200$ and $h(1.001) = 2000$). We have indicated this by drawing a vertical line, called an **asymptote**, at $x = 1$. As $x$ approaches 1, the graph gets closer and closer to this line, though this line itself is not part of the graph. Rather, it is a guide line. Notice that the graph of $h$ also has a horizontal asymptote, namely, the $x$-axis.

We might ask: What is the range for each of these three functions? Our answer, obtained by looking at the graphs, is given in the following table.

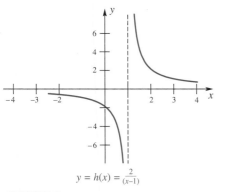

$y = h(x) = \dfrac{2}{(x-1)}$

**FIGURE 8**

| Function | Domain | Range |
|----------|--------|-------|
| $f(x) = x^2 - 2$ | $\mathbb{R}$ | $\{y \in \mathbb{R} : y \geq -2\}$ |
| $g(x) = x^3 - 2x$ | $\mathbb{R}$ | $\mathbb{R}$ |
| $h(x) = \dfrac{2}{x-1}$ | $\{x \in \mathbb{R} : x \neq 1\}$ | $\{y \in \mathbb{R} : y \neq 0\}$ |

**Even and Odd Functions**   We can often predict the symmetries of the graph of a function by inspecting the formula for the function. If $f(-x) = f(x)$, then the graph is symmetric with respect to the $y$-axis. Such a function is called an **even function**, probably because a function that specifies $f(x)$ as a sum of even powers of $x$ is even. The function $f(x) = x^2 - 2$ (graphed in Figure 6) is even; so are $f(x) = 3x^6 - 2x^4 + 11x^2 - 5, f(x) = x^2/(1 + x^4)$, and $f(x) = (x^3 - 2x)/3x$.

If $f(-x) = -f(x)$, the graph is symmetric with respect to the origin. We call such a function an **odd function**. A function that gives $f(x)$ as a sum of odd powers of $x$ is odd. Thus, $g(x) = x^3 - 2x$ (graphed in Figure 7) is odd. Note that

$$g(-x) = (-x)^3 - 2(-x) = -x^3 + 2x = -(x^3 - 2x) = -g(x)$$

Consider the third function $h(x) = 2/(x - 1)$, which we graphed in Figure 8. It is neither even nor odd. To see this, observe that $h(-x) = 2/(-x - 1)$, which is not equal to either $h(x)$ or $-h(x)$. Note that its graph is neither symmetric with respect to the $y$-axis nor the origin. (The graph of $h$ does have symmetry with respect to the point $(1, 0)$, a result that follows from the fact that $h(1 + x) = -h(1 - x)$.)

**EXAMPLE 5**   Is $f(x) = \dfrac{x^3 + 3x}{x^4 - 3x^2 + 4}$ even, odd, or neither?

**Solution**   Since

$$f(-x) = \frac{(-x)^3 + 3(-x)}{(-x)^4 - 3(-x)^2 + 4} = \frac{-(x^3 + 3x)}{x^4 - 3x^2 + 4} = -f(x)$$

$f$ is an odd function.   ∎

For more on even and odd functions, see Problems 23 and 24 of Section 2.2.

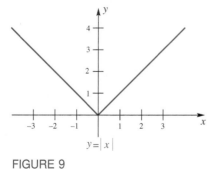

$y = |x|$

FIGURE 9

**Two Special Functions**   Among the functions that will often be used as examples are two very special ones: the absolute value function $|\ \ |$ and the greatest integer function $[\![\ \ ]\!]$. They are defined by

$$|x| = \begin{cases} x & \text{if } x \geq 0 \\ -x & \text{if } x < 0 \end{cases}$$

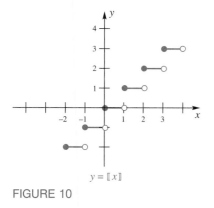

$y = [\![x]\!]$

FIGURE 10

and

$$[\![x]\!] = \text{the greatest integer less than or equal to } x$$

Thus, $|-3.1| = |3.1| = 3.1$, while $[\![-3.1]\!] = -4$ and $[\![3.1]\!] = 3$. We show the graphs of these two functions in Figures 9 and 10. The absolute value function is even, since $|-x| = |x|$. The greatest integer function is neither even nor odd, as you can see from its graph.

We will often appeal to the following special features of these graphs. The graph of $|x|$ has a sharp corner at the origin, while the graph of $[\![x]\!]$ takes a jump at each integer.

## CONCEPTS REVIEW

**1.** The set of allowable inputs for a function is called the _____ of the function; the set of outputs that are obtained is called the _____ of the function.

**2.** If $f(x) = 3x^2$, then $f(2u) =$ _____ and $f(x + h) =$ _____.

**3.** If $f(x)$ gets closer and closer to $L$ as $|x|$ gets larger and larger, then the line $y = L$ is an _____ for the graph of $f$.

**4.** If $f(-x) = f(x)$ for all $x$ in the domain of $f$, then $f$ is called an _____ function; if $f(-x) = -f(x)$ for all $x$ in the domain of $f$, then $f$ is called an _____ function. In the first case, the graph of $f$ is symmetric with respect to the _____; in the second case, it is symmetric with respect to the _____.

## PROBLEM SET 2.1

**1.** For $f(x) = 1 - x^2$ find each value.

(a) $f(1)$      (e) $f(-5)$
(b) $f(-2)$      (f) $f(\frac{1}{4})$
(c) $f(0)$      (g) $f(3t)$
(d) $f(k)$      (h) $f(2x)$
             (i) $f(\frac{1}{t})$

**2.** For $F(x) = x^3 + 3x$ find each value.

(a) $F(1)$      (d) $F(\pi)$
(b) $F(\sqrt{2})$      (e) $F(\frac{1}{t})$
(c) $F(\frac{1}{4})$      (f) $F(2.718)$

**3.** For $G(y) = 1/(y - 1)$, find each value.

(a) $G(0)$      (b) $G(0.999)$      (c) $G(1.01)$

(d) $G(y^2)$      (e) $G(-x)$      (f) $G\left(\dfrac{1}{x^2}\right)$

**4.** For $\Phi(u) = \dfrac{u + u^2}{\sqrt{u}}$, find each value.

(a) $\Phi(1)$      (b) $\Phi(-t)$
(c) $\Phi(\frac{1}{2})$      (d) $\Phi(u + 1)$
(e) $\Phi(x^2)$      (f) $\Phi(x^2 + x)$

©  **5.** For
$$f(x) = x^4 + x^3 + 3x^2 - 2x + 1$$
$$= \{[(x + 1)x + 3]x - 2\}x + 1,$$
find each value.

(a) $f(0.25)$      (b) $f(\pi)$      (c) $f(3 + \sqrt{2})$

©  **6.** For
$$g(x) = x^5 - 5x^3 + 2x^2 - \pi x + \sqrt{3}$$
$$= \{[(x^2 - 5)x + 2]x - \pi\}x + \sqrt{3},$$
find each value.

(a) $g(-1.71)$      (b) $g(3.01)$      (c) $g(\sqrt{3})$

©  **7.** For $f(x) = \sqrt{x^2 + 9}/(x - \sqrt{3})$, find each value.

(a) $f(0.79)$      (b) $f(12.26)$      (c) $f(\sqrt{3})$

©  **8.** If $G(t) = t^{\frac{3}{4}} - 3\sqrt{t}$, find each value.

(a) $G(1.23)$      (b) $G(\pi)$      (c) $G(-1.23)$

**9.** Which of the following determine a function $f$ with formula $y = f(x)$? For those that do, find $f(x)$. *Hint*: Solve for $y$ in terms of $x$ and note that the definition of function requires a single $y$ for each $x$.

(a) $x^2 + y^2 = 1$

(c) $x = \sqrt{2y + 1}$

(b) $xy + y + x = 1$

(d) $x = \dfrac{y}{y + 1}$

**10.** Which of the graphs in Figure 11 are graphs of functions of the form $y = f(x)$? (Is there a single $y$ for each $x$?)

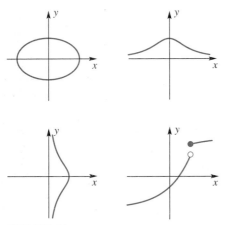

FIGURE 11

This problem suggests a rule: *For a graph to be the graph of a function $y = f(x)$, each vertical line must meet the graph in at most one point.*

**11.** For $f(x) = 2x^2 - 1$, find and simplify $[f(a + h) - f(a)]/h$. (See Examples 1 and 2.)

**12.** For $F(t) = 4t^3$, find and simplify $[F(a + h) - F(a)]/h$.

**13.** For $g(u) = 3/(u - 2)$, find and simplify $[g(x + h) - g(x)]/h$.

**14.** For $G(t) = t/(t + 4)$, find and simplify $[G(a + h) - G(a)]/h$.

**15.** Find the natural domain for each of the following. (See Example 3.)

(a) $F(z) = \sqrt{2z + 3}$

(b) $g(v) = 1/(4v - 1)$

(c) $\psi(x) = \sqrt{x^2 - 9}$

(d) $H(y) = -\sqrt{625 - y^4}$

**16.** Find the natural domain in each case.

(a) $f(x) = \dfrac{4 - x^2}{x^2 - x - 6}$

(b) $G(y) = \sqrt{(y + 1)^{-1}}$

(c) $\phi(u) = |2u + 3|$

(d) $F(t) = t^{2/3} - 4$

In Problems 17–32, specify whether the given function is even, odd, or neither, and then sketch its graph. (See Examples 4 and 5.)

**17.** $f(x) = -4$

**18.** $f(x) = 3x$

**19.** $F(x) = 2x + 1$

**20.** $F(x) = 3x - \sqrt{2}$

**21.** $g(x) = 3x^2 + 2x - 1$

**22.** $g(u) = \dfrac{u^3}{8}$

**23.** $g(x) = \dfrac{x}{x^2 - 1}$

**24.** $\phi(z) = \dfrac{2z + 1}{z - 1}$

**25.** $f(w) = \sqrt{w - 1}$

**26.** $h(x) = \sqrt{x^2 + 4}$

**27.** $f(x) = |2x|$

**28.** $F(t) = -|t + 3|$

**29.** $g(x) = \left[\!\left[\dfrac{x}{2}\right]\!\right]$

**30.** $G(x) = [\![2x - 1]\!]$

**31.** $g(t) = \begin{cases} 1 & \text{if } t \le 0 \\ t + 1 & \text{if } 0 < t < 2 \\ t^2 - 1 & \text{if } t \ge 2 \end{cases}$

**32.** $h(x) = \begin{cases} -x^2 + 4 & \text{if } x \le 1 \\ 3x & \text{if } x > 1 \end{cases}$

**33.** A plant has the capacity to produce from 0 to 100 computers per day. The daily overhead for the plant is $5,000 and the direct cost (labor and materials) of producing one computer is $805. Write a formula for $T(x)$, the total cost of producing $x$ computers in one day, and also the unit cost $u(x)$ (average cost per computer). What are the domains of these functions?

[C] **34.** It costs the ABC Company $400 + 5\sqrt{x(x - 4)}$ dollars to make $x$ toy stoves that sell for $6 each.

(a) Find a formula for $P(x)$, the total profit in making $x$ stoves.

(b) Evaluate $P(200)$ and $P(1000)$.

(c) How many stoves does ABC have to make to just break even?

[C] **35.** Find the formula for the amount $E(n)$ by which a number $n$ exceeds its square. Draw a very accurate graph of $E(n)$ for $0 \le n \le 1$. Use the graph to estimate the positive number less than or equal to 1 which exceeds its square by the maximum amount.

**36.** Let $p$ denote the perimeter of an equilateral triangle. Find a formula for $A(p)$, the area of such a triangle.

**37.** The Acme Car Rental Agency charges $24 a day for the rental of a car plus $0.40 per mile.

(a) Write a formula for the total rental expense $E(x)$ for one day where $x$ is the number of miles driven.

(b) If you rent a car for one day, how many miles can you drive for $120?

**38.** A right circular cylinder of radius $r$ is inscribed in a sphere of radius $2r$. Find a formula for $V(r)$, the volume of the cylinder, in terms of $r$.

**39.** A 1-mile track has parallel sides and equal semicircular ends. Find a formula for the area enclosed by the track, $A(d)$, in terms of the diameter $d$ of the semicircles. What is the natural domain for this function?

**40.** Let $f$ be a function with domain $\mathbb{N}$ satisfying $f(1) = 3, f(2) = 1, f(3) = 4, f(4) = 1, f(5) = 5$, and $f(6) = 9$. After finding a pattern, give a rule for $f(n)$. What is the range of this function?

**41.** What is the range of the function $f$ if $f(n)$ is the $n$th digit in the decimal expansion of $\frac{3}{13}$?

**42.** Which of the following functions satisfies $f(x + y) = f(x) + f(y)$ for all $x$ and $y$ in $\mathbb{R}$?

(a) $f(t) = 2t$ \qquad\qquad (b) $f(t) = t^2$
(c) $f(t) = 2t + 1$ \qquad (d) $f(t) = -3t$

**43.** Let $f(x + y) = f(x) + f(y)$ for all $x$ and $y$ in $\mathbb{R}$. Prove that there is a number $m$ such that $f(t) = mt$ for all rational numbers $t$. *Hint*: First decide what $m$ has to be. Then proceed in steps, starting with $f(0) = 0, f(p) = mp$ for $p$ in $\mathbb{N}$, $f(1/p) = m/p$, and so on.

**44.** A baseball diamond is a square with sides of 90 feet. A player, after hitting a home run, loped around the diamond at 10 feet per second. Let $s$ represent the player's distance from home plate after $t$ seconds.

(a) Express $s$ as a function of $t$ by means of a four-part formula.
(b) Express $s$ as a function of $t$ by means of a three-part formula.

PC To use a calculus software package efficiently, you need to discover its capabilities, its strengths, and its weaknesses. We urge you to practice graphing functions of various types using your own computer package. Problems 45–50 are designed for this purpose. The part of *True BASIC Calculus* labeled *General* will handle each of these problems with ease.

**45.** Let $f(x) = (x^3 + 3x - 5)/(x^2 + 4)$.

(a) Evaluate $f(1.38)$ and $f(4.12)$.
(b) Construct a table of values for this function corresponding to $x = -4, -3, \ldots, 3, 4$.

**46.** Follow the instructions in Problem 45 for $f(x) = (\sin^2 x - 3 \tan x)/\cos x$.

**47.** Draw the graph of $f(x) = x^3 - 5x^2 + x + 8$ on the domain $[-2, 5]$.

(a) Determine the range of $f$.
(b) Where on this domain is $f(x) \geq 0$?

**48.** Superimpose the graph of $g(x) = 2x^2 - 8x - 1$ on the graph of $f(x)$ of Problem 47 using the domain $[-2, 5]$.

(a) Estimate the $x$-values where $f(x) = g(x)$.
(b) Where on $[-2, 5]$ is $f(x) \geq g(x)$?
(c) Estimate the largest value of $|f(x) - g(x)|$ on $[-2, 5]$.

**49.** Graph $f(x) = (3x - 4)/(x^2 + x - 6)$ on the domain $[-6, 6]$.

(a) Determine the $x$- and $y$-intercepts.
(b) Determine the range of $f$ for the given domain.
(c) Determine the vertical asymptotes of the graph.
(d) Determine the horizontal asymptote for the graph when the domain is enlarged to the whole real line.

**50.** Follow the directions in Problem 49 for $g(x) = (3x^2 - 4)/(x^2 + x - 6)$

---

**Answers to Concepts Review:** **1.** domain, range **2.** $12u^2$; $3(x + h)^2 = 3x^2 + 6xh + 3h^2$ **3.** asymptote **4.** even; odd; $y$-axis; origin

---

**2.2**
**OPERATIONS ON FUNCTIONS**

Functions are not numbers. But just as two numbers $a$ and $b$ can be added to produce a new number $a + b$, so two functions $f$ and $g$ can be added to produce a new function $f + g$. This is just one of several operations on functions that we will describe in this section.

**Sums, Differences, Products, Quotients, Powers** Consider functions $f$ and $g$ with formulas

$$f(x) = \frac{x - 3}{2}, \qquad g(x) = \sqrt{x}$$

We can make a new function $f + g$ by having it assign to $x$ the value $(x - 3)/2 + \sqrt{x}$, that is,

$$(f + g)(x) = f(x) + g(x) = \frac{x - 3}{2} + \sqrt{x}$$

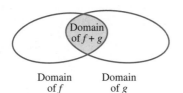

Domain     Domain
of $f$         of $g$

FIGURE 1

Of course, we must be a little careful about domains. Clearly, $x$ must be a number on which both $f$ and $g$ can work. In other words, the domain of $f + g$ is the intersection (common part) of the domains of $f$ and $g$ (Figure 1).

The functions $f - g$, $f \cdot g$, and $f/g$ are introduced in a completely analogous way. Assuming that $f$ and $g$ have their natural domains, we have the following.

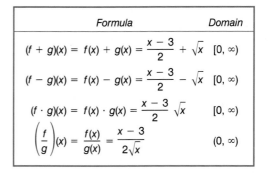

| Formula | Domain |
|---|---|
| $(f + g)(x) = f(x) + g(x) = \dfrac{x - 3}{2} + \sqrt{x}$ | $[0, \infty)$ |
| $(f - g)(x) = f(x) - g(x) = \dfrac{x - 3}{2} - \sqrt{x}$ | $[0, \infty)$ |
| $(f \cdot g)(x) = f(x) \cdot g(x) = \dfrac{x - 3}{2} \sqrt{x}$ | $[0, \infty)$ |
| $\left(\dfrac{f}{g}\right)(x) = \dfrac{f(x)}{g(x)} = \dfrac{x - 3}{2\sqrt{x}}$ | $(0, \infty)$ |

We had to exclude 0 from the domain of $f/g$ to avoid division by 0.

We may also raise a function to a power. By $f^n$, we mean the function that assigns to $x$ the value $[f(x)]^n$. Thus,

$$f^2(x) = [f(x)]^2 = \left[\frac{x - 3}{2}\right]^2 = \frac{x^2 - 6x + 9}{4}$$

and

$$g^3(x) = [g(x)]^3 = (\sqrt{x})^3 = x^{3/2}$$

There is one exception to the above agreement on exponents, namely, when $n = -1$. We reserve the symbol $f^{-1}$ for the inverse function, which will be discussed in Section 7.2. Thus, $f^{-1}$ does not mean $1/f$.

**EXAMPLE 1** Let $F(x) = \sqrt[4]{x + 1}$ and $G(x) = \sqrt{9 - x^2}$, with respective natural domains $[-1, \infty)$ and $[-3, 3]$. Find formulas for $F + G$, $F - G$, $F \cdot G$, $F/G$, and $F^5$ and give their natural domains.

*Solution*

| Formula | Domain |
|---|---|
| $(F + G)(x) = F(x) + G(x) = \sqrt[4]{x + 1} + \sqrt{9 - x^2}$ | $[-1, 3]$ |
| $(F - G)(x) = F(x) - G(x) = \sqrt[4]{x + 1} - \sqrt{9 - x^2}$ | $[-1, 3]$ |
| $(F \cdot G)(x) = F(x) \cdot G(x) = \sqrt[4]{x + 1} \sqrt{9 - x^2}$ | $[-1, 3]$ |
| $\left(\dfrac{F}{G}\right)(x) = \dfrac{F(x)}{G(x)} = \dfrac{\sqrt[4]{x + 1}}{\sqrt{9 - x^2}}$ | $[-1, 3)$ |
| $F^5(x) = [F(x)]^5 = \left(\sqrt[4]{x + 1}\right)^5 = (x + 1)^{5/4}$ | $[-1, \infty)$ |

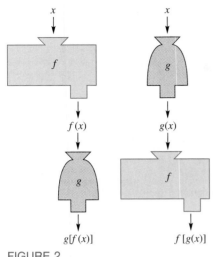

FIGURE 2

**Composition of Functions**   Earlier, we asked you to think of a function as a machine. It accepts $x$ as input, works on $x$, and produces $f(x)$ as output. Two machines may often be put together in tandem to make a more complicated machine; so may two functions $f$ and $g$ (Figure 2). If $f$ works on $x$ to produce $f(x)$ and $g$ then works on $f(x)$ to produce $g(f(x))$, we say that we have *composed* $g$ with $f$. The resulting function, called the **composite** of $g$ with $f$, is denoted by $g \circ f$. Thus,

$$(g \circ f)(x) = g(f(x))$$

Recall our earlier examples, $f(x) = (x - 3)/2$ and $g(x) = \sqrt{x}$. We may compose them in two ways.

$$(g \circ f)(x) = g(f(x)) = g\left(\frac{x - 3}{2}\right) = \sqrt{\frac{x - 3}{2}}$$

$$(f \circ g)(x) = f(g(x)) = f(\sqrt{x}) = \frac{\sqrt{x} - 3}{2}$$

We note one thing right away: Composition of functions is not commutative; $g \circ f$ and $f \circ g$ are usually different. You shouldn't be too surprised at this. If you remove your clothes and then take a bath, you are going to get a quite different result from that of doing these two operations in the opposite order.

We must also be careful in describing the domain of a composite function. The domain of $g \circ f$ is that part of the domain of $f$ (that is, those values $x$) for which $g$ can accept $f(x)$ as input. In our example, the domain of $g \circ f$ is $[3, \infty)$, since $x$ must be greater than or equal to 3 in order to give a nonnegative number ($x - 3)/2$ for $g$ to work on. The diagram in Figure 3 gives another view of these matters.

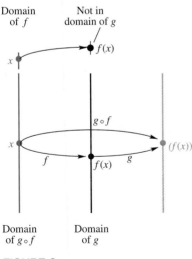

FIGURE 3

**EXAMPLE 2**   Let $f(x) = 6x/(x^2 - 9)$ and $g(x) = \sqrt{3x}$ have their natural domains. First, find $(f \circ g)(12)$; then find $(f \circ g)(x)$ and give its domain.

*Solution*

$$(f \circ g)(12) = f(g(12)) = f(\sqrt{36}) = f(6) = \frac{36}{27} = \frac{4}{3}$$

$$(f \circ g)(x) = f(g(x)) = f(\sqrt{3x})$$

$$= \frac{6\sqrt{3x}}{(\sqrt{3x})^2 - 9} = \frac{6\sqrt{3x}}{3x - 9} = \frac{2\sqrt{3x}}{x - 3}$$

The domain of $f \circ g$ is $[0, 3) \cup (3, \infty)$. (Recall that $\cup$ denotes the set operation of union.) Note that 3 is excluded from the domain to avoid division by 0. ■

In calculus, we shall often need to take a given function and decompose it, that is, break it into composite pieces. Usually, this can be done in several ways. Take $p(x) = \sqrt{x^2 + 4}$, for example. We may think of it as

$$p(x) = g(f(x)) \qquad \text{where } g(x) = \sqrt{x} \text{ and } f(x) = x^2 + 4$$

or as

$$p(x) = g(f(x)) \qquad \text{where } g(x) = \sqrt{x + 4} \text{ and } f(x) = x^2$$

**EXAMPLE 3**   Write the function $p(x) = (x + 2)^5$ as a composite function $g \circ f$.

*Solution*   The most obvious way to do it is to write

$$p(x) = g(f(x)) \qquad \text{where } g(x) = x^5 \text{ and } f(x) = x + 2 \qquad \blacksquare$$

**Translations**   Observing how a function is built up from simpler ones can be a big aid in graphing. We may ask this question: How are the graphs of

$$y = f(x) \qquad y = f(x - 3) \qquad y = f(x) + 2 \qquad y = f(x - 3) + 2$$

related to each other? Consider $f(x) = |x|$ as an example. The corresponding four graphs are displayed in Figure 4.

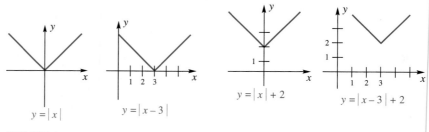

FIGURE 4

Notice that all four graphs have the same shape; the last three are just translations of the first. Replacing $x$ by $x - 3$ translates the graph 3 units to the right; adding 2 translates it upward by 2 units.

What happened with $f(x) = |x|$ is typical. Figure 5 offers an illustration for the function $f(x) = x^3 + x^2$.

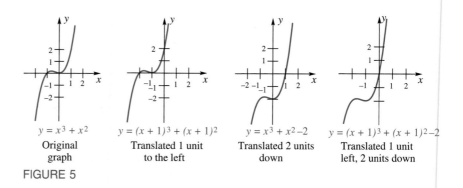

FIGURE 5

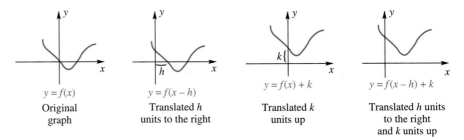

FIGURE 6

Exactly the same principles apply in the general situation. They are illustrated in Figure 6 with both $h$ and $k$ positive. If $h < 0$, the translation is to the left; if $k < 0$, the translation is downward.

**EXAMPLE 4**   Sketch the graph of $g(x) = \sqrt{x + 3} + 1$ by first graphing $f(x) = \sqrt{x}$ and then making appropriate translations.

*Solution*   The graph of $g$ (Figure 8) is obtained by translating the graph of $f$ (Figure 7) 3 units left and 1 unit up.   ∎

**Partial Catalogue of Functions**   A function of the form $f(x) = k$, where $k$ is a constant (real number) is called a **constant function**. Its graph is a horizontal line (Figure 9). The function $f(x) = x$ is called the **identity function**. Its graph is a line through the origin of slope 1 (Figure 10). From these simple functions, we can build many of the important functions of calculus.

Any function that can be obtained from the constant functions and the identity function by use of the operations of addition, subtraction, and multiplication is called a **polynomial function.** This amounts to saying that $f$ is a polynomial function if it is of the form

$$f(x) = a_n x^n + a_{n-1} x^{n-1} + \cdots + a_1 x + a_0$$

where the $a$'s are real numbers and $n$ is a nonnegative integer. If $a_n \neq 0$, $n$ is the degree of the polynomial function. In particular, $f(x) = ax + b$ is a first-degree polynomial function, or **linear function**, and $f(x) = ax^2 + bx + c$ is a second-degree polynomial function, or **quadratic function**.

Quotients of polynomial functions are called rational functions. Thus $f$ is a **rational function** if it is of the form

$$f(x) = \frac{a_n x^n + a_{n-1} x^{n-1} + \cdots + a_1 x + a_0}{b_m x^m + b_{m-1} x^{m-1} + \cdots + b_1 x + b_0}$$

An **explicit algebraic function** is one that can be obtained from the constant functions and the identity function via the five operations of addition, subtraction, multiplication, division, and root extraction. Examples are

$$f(x) = 3x^{2/5} = 3\sqrt[5]{x^2} \qquad g(x) = \frac{(x + 2)\sqrt{x}}{x^3 + \sqrt[3]{x^2 - 1}}$$

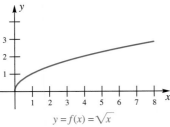

$y = f(x) = \sqrt{x}$

FIGURE 7

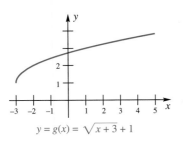

$y = g(x) = \sqrt{x + 3} + 1$

FIGURE 8

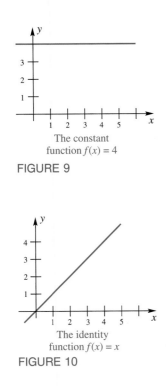

The constant function $f(x) = 4$

FIGURE 9

The identity function $f(x) = x$

FIGURE 10

The functions so far listed, together with the trigonometric, inverse trigonometric, exponential, and logarithmic functions (to be introduced later), are the basic raw material for calculus.

## CONCEPTS REVIEW

**1.** If $f(x) = x^2 + 1$, then $f^3(x) =$ _____ .

**2.** The value of the composite function $f \circ g$ at $x$ is given by $(f \circ g)(x) =$ _____ .

**3.** Compared to the graph of $y = f(x)$, the graph of $y = f(x + 2)$ is translated _____ units to the _____ .

**4.** A rational function is defined as _____ .

## PROBLEM SET 2.2

**1.** For $f(x) = x + 3$ and $g(x) = x^2$, find each value (if possible).

(a) $(f + g)(2)$          (d) $(f \circ g)(1)$
(b) $(f \cdot g)(0)$          (e) $(g \circ f)(1)$
(c) $(g/f)(3)$          (f) $(g \circ f)(-8)$

**2.** For $f(x) = x^2 + x$ and $g(x) = 2/(x + 3)$, find each value.

(a) $(f - g)(2)$     (b) $(f/g)(1)$     (c) $g^2(3)$
(d) $(f \circ g)(1)$     (e) $(g \circ f)(1)$     (f) $(g \circ g)(3)$

**3.** For $\Phi(u) = u^3 + 1$ and $\Psi(v) = 1/v$, find each value.

(a) $(\Phi + \Psi)(t)$          (d) $\Phi^3(z)$
(b) $(\Phi \circ \Psi)(r)$          (e) $(\Phi - \Psi)(5t)$
(c) $(\Psi \circ \Phi)(r)$          (f) $((\Phi - \Psi) \circ \Psi)(t)$

**4.** If $f(x) = \sqrt{x^2 - 1}$ and $g(x) = 2/x$, find formulas for the following and state their domains.

(a) $(f \cdot g)(x)$          (b) $f^4(x) + g^4(x)$
(c) $(f \circ g)(x)$          (d) $(g \circ f)(x)$

**5.** If $f(s) = \sqrt{s^2 - 4}$ and $g(w) = |1 + w|$, find formulas for $(f \circ g)(x)$ and $(g \circ f)(x)$.

**6.** If $g(x) = x^2 + 1$, find formulas for $g^3(x)$ and $(g \circ g \circ g)(x)$.

© **7.** Calculate $g(3.141)$ if $g(u) = \dfrac{\sqrt{u^3 + 2u}}{2 + u}$.

© **8.** Calculate $g(2.03)$ if $g(x) = \dfrac{(\sqrt{x} - \sqrt[3]{x})^4}{1 - x + x^2}$.

© **9.** Calculate $[g^2(\pi) - g(\pi)]^{\frac{1}{3}}$ if $g(v) = |11 - 7v|$.

© **10.** Calculate $[g^3(\pi) - g(\pi)]^{\frac{1}{3}}$ if $g(x) = 6x - 11$.

**11.** Find $f$ and $g$ so that $F = g \circ f$. (See Example 3.)

(a) $F(x) = \sqrt{x + 7}$          (b) $F(x) = (x^2 + x)^{15}$

**12.** Find $f$ and $g$ so that $p = f \circ g$.

(a) $p(x) = \dfrac{2}{(x^2 + x + 1)^3}$          (b) $p(x) = \log(x^3 + 3x)$

**13.** Write $p(x) = \log\sqrt{x^2 + 1}$ as a composite of three functions in two different ways.

**14.** Write $p(x) = \log\sqrt{x^2 + 1}$ as a composite of four functions.

**15.** Sketch the graph of $f(x) = \sqrt{x - 2} - 3$ by first sketching $g(x) = \sqrt{x}$. (See Example 4.)

**16.** Sketch the graph of $g(x) = |x + 3| - 4$ by first sketching $h(x) = |x|$ and then translating.

**17.** Sketch the graph of $f(x) = (x - 2)^2 - 4$ by making use of translations.

**18.** Sketch the graph of $g(x) = (x + 1)^3 - 3$ by making use of translations.

**19.** Sketch the graphs of $f(x) = (x - 3)/2$ and $g(x) = \sqrt{x}$ using the same coordinate axes. Then sketch $f + g$ by adding ordinates.

**20.** Follow the directions of Problem 19 for $f(x) = x$ and $g(x) = |x|$.

**21.** Sketch the graph of $F(t) = \dfrac{|t| - t}{t}$.

**22.** Sketch the graph of $G(t) = t - [\![t]\!]$.

**23.** State whether each of the following is an odd function, an even function, or neither. Prove your statements.

(a) The sum of two even functions.
(b) The sum of two odd functions.
(c) The product of two even functions.
(d) The product of two odd functions.
(e) The product of an even function and an odd function.

**24.** Let $F$ be any function whose domain contains $-x$ whenever it contains $x$. Prove each of the following.

(a) $F(x) - F(-x)$ determines an odd function.
(b) $F(x) + F(-x)$ determines an even function.
(c) $F$ can always be expressed as the sum of an odd and an even function.

**25.** Classify each of the following as a PF (polynomial function), RF (rational function but not a polynomial function), or AF (explicit algebraic function but neither PF nor RF).

(a) $f(x) = 3x^{1/2} + 1$  (b) $f(x) = 3$
(c) $f(x) = 3x^2 + 2x^{-1}$  (d) $f(x) = \pi x^3 - 3\pi$
(e) $f(x) = \dfrac{1}{x + 1}$  (f) $f(x) = \dfrac{x + 1}{\sqrt{x + 3}}$

**26.** Follow the directions of Problem 25 for each function.

(a) $g(x) = 1 - 3x + x^{52}$  (b) $g(x) = (1 + 5x)^{-2}$
(c) $g(x) = (1 + 5x)^{-1/2}$  (d) $g(x) = x^{-2} + 2x^{-1} + 3$
(e) $g(x) = x + \sqrt{x}$  (f) $g(x) = \left(\dfrac{1}{x + 2}\right)^{-2}$

 C  **27.** The relationship between the unit price $P$ (in cents) for a certain product and the demand $D$ (in thousands of units) appears to satisfy

$$P = \sqrt{29 - 3D + D^2}$$

On the other hand, the demand has risen over the $t$ years since 1970 according to $D = 2 + \sqrt{t}$.

(a) Express $P$ as a function of $t$.
(b) Evaluate $P$ when $t = 15$.

**28.** After being in business for $t$ years, a manufacturer of cars is making $120 + 2t + 3t^2$ units per year. The sales price in dollars per unit has risen according to the formula $6000 + 700t$. Write a formula for the manufacturer's yearly revenue $R(t)$ after $t$ years.

**29.** Starting at noon, plane A flies due north at 400 miles per hour. Starting 1 hour later, plane B flies due east at 300 miles per hour. Neglecting the curvature of the earth and assuming they fly at the same altitude, find a formula for $D(t)$, the distance between the two planes $t$ hours after noon. *Hint*: There will be two formulas for $D(t)$, one if $0 \le t \le 1$, the other if $t > 1$.

 C  **30.** Find the distance between the planes of Problem 29 at 2:30 P.M.

**31.** Let $f(x) = \dfrac{ax + b}{cx - a}$. Show that $f(f(x)) = x$, provided $a^2 + bc \neq 0$ and $x \neq a/c$.

**32.** Let $f(x) = \dfrac{x - 3}{x + 1}$. Show that $f(f(f(x))) = x$, provided $x \neq \pm 1$.

**33.** Let $f(x) = \dfrac{x}{x - 1}$. Find and simplify each value.

(a) $f(1/x)$  (b) $f(f(x))$  (c) $f(1/f(x))$

**34.** Let $f_1(x) = x$, $f_2(x) = 1/x$, $f_3(x) = 1 - x$, $f_4(x) = 1/(1 - x)$, $f_5(x) = (x - 1)/x$, and $f_6(x) = x/(x - 1)$. Note that $f_3(f_4(x)) = f_3(1/(1 - x)) = 1 - 1/(1 - x) = $

$x/(x - 1) = f_6(x)$, that is, $f_3 \circ f_4 = f_6$. In fact, the composition of any two of these functions is another one in the list. Fill in the composition table in Figure 11.

| $\circ$ | $f_1$ | $f_2$ | $f_3$ | $f_4$ | $f_5$ | $f_6$ |
|---|---|---|---|---|---|---|
| $f_1$ | | | | | | |
| $f_2$ | | | | | | |
| $f_3$ | | | $f_6$ | | | |
| $f_4$ | | | | | | |
| $f_5$ | | | | | | |
| $f_6$ | | | | | | |

**FIGURE 11**

Then use this table to find each of the following. You may assume the associative law holds (it does).

(a) $f_3 \circ f_3 \circ f_3 \circ f_3 \circ f_3$
(b) $f_1 \circ f_2 \circ f_3 \circ f_4 \circ f_5 \circ f_6$
(c) $F$ if $F \circ f_6 = f_1$
(d) $G$ if $G \circ f_3 \circ f_6 = f_1$
(e) $H$ if $f_2 \circ f_5 \circ H = f_5$

 PC  Use a computer in Problems 35–40.

**35.** Let $f(x) = x^2 - 3x$. Using the same axes, draw the graphs of $y = f(x)$, $y = f(x - 0.5) - 0.6$, and $y = f(1.5x)$, all on the domain $[-2, 5]$.

**36.** Let $f(x) = |x^3|$. Using the same axes, draw the graphs of $y = f(x)$, $y = f(3x)$, and $y = f(3(x - 0.8))$, all on the domain $[-3, 3]$.

**37.** Let $f(x) = 2\sqrt{x} - 2x + 0.25x^2$. Using the same axes, draw the graphs of $y = f(x)$, $y = f(1.5x)$, and $y = f(x - 1) + 0.5$, all on the domain $[0, 5]$.

**38.** Let $f(x) = 1/(x^2 + 1)$. Using the same axes, draw the graphs of $y = f(x)$, $y = f(2x)$, and $y = f(x - 2) + 0.6$, all on the domain $[-4, 4]$.

**39.** Your software package may allow the use of parameters in defining functions (as does *True BASIC Calculus*, Version 4.0). In each case, draw the graph of $y = f(x)$ for the specified values of the parameter $k$, using the same axes and $-5 \le x \le 5$.

(a) $f(x) = |kx|^{0.7}$ for $k = 1, 2, 0.5$, and $0.2$.
(b) $f(x) = |x - k|^{0.7}$ for $k = 0, 2, -0.5$, and $-3$.
(c) $f(x) = |x|^k$ for $k = 0.4, 0.7, 1$, and $1.7$.

**40.** Using the same axes, draw the graph of $f(x) = |k(x - c)|^n$ for the following choices of parameters.
(a) $c = -1, k = 1.4, n = 0.7$.  (b) $c = 2, k = 1.4, n = 1.2$.
(c) $c = 0, k = 0.9, n = 0.6$.

**Answers to Concepts Review: 1.** $(x^2 + 1)^3$  **2.** $f(g(x))$  **3.** 2, left  **4.** a quotient of two polynomial functions

We assume that you have studied trigonometry and are familiar with the definitions of trigonometric functions based on angles and right triangles. We remind you of three of these definitions in Figure 1. Do not forget them. Here, however, we are more interested in the trigonometric functions as based on the unit circle. When considered in this way, their domains are sets of numbers rather than sets of angles.

Let $C$ be the unit circle—that is, the circle $x^2 + y^2 = 1$ centered at the origin with radius 1 (Figure 2). Denote by $A$ the point $(1, 0)$ and let $t$ be any positive number. Then there is exactly one point $P(x, y)$ on $C$ such that the length of arc $AP$, measured in the *counterclockwise* direction from $A$ along the unit circle, is $t$. The circumference of $C$ is $2\pi$; so if $t > 2\pi$, it will take more than a complete circuit of the unit circle to trace the arc $AP$. If $t = 0$, $P = A$.

Similarly if $t < 0$, there is exactly one point $P(x, y)$ on the unit circle such that the length of the arc $AP$, measured *clockwise* on $C$, is $|t|$. Thus, with any real number $t$, we associate a unique point $P(x, y)$. This allows us to make the key definitions of sine and cosine (sin and cos).

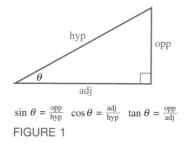

$$\sin \theta = \frac{\text{opp}}{\text{hyp}} \quad \cos \theta = \frac{\text{adj}}{\text{hyp}} \quad \tan \theta = \frac{\text{opp}}{\text{adj}}$$

FIGURE 1

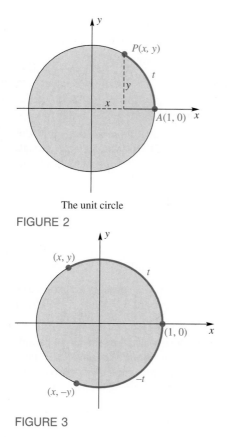

The unit circle

FIGURE 2

> **Definition**
>
> Let $t$ determine the point $P(x, y)$ as indicated above. Then
>
> $$\sin t = y \quad \cos t = x$$

**Basic Properties of Sine and Cosine**   Several facts are almost immediately apparent from the definitions just given. First, $x$ and $y$ vary between $-1$ and 1, so

$$\boxed{|\sin t| \leq 1 \qquad |\cos t| \leq 1}$$

Since $t$ and $t + 2\pi$ determine the same point $P(x, y)$,

$$\boxed{\sin(t + 2\pi) = \sin t \qquad \cos(t + 2\pi) = \cos t}$$

We say that sine and cosine are periodic with period $2\pi$. More generally, a function $f$ is **periodic** if there is a positive number $p$ such that $f(t + p) = f(t)$ for all $t$ in the domain of $f$. And the smallest such $p$ is called the **period** of $f$.

The points $P$ corresponding to $t$ and $-t$ are symmetric with respect to the $x$-axis (Figure 3), and thus their $x$-coordinates are equal and their $y$-coordinates differ in sign only. Consequently,

$$\boxed{\sin(-t) = -\sin t \qquad \cos(-t) = \cos t}$$

FIGURE 3

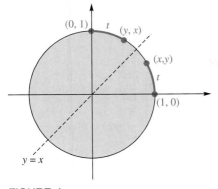

FIGURE 4

In other words, sine is an odd function and cosine is an even function.

The points $P$ corresponding to $t$ and $\pi/2 - t$ are symmetric with respect to the line $y = x$ (Figure 4), and thus have their coordinates interchanged. This means that

$$\sin\left(\frac{\pi}{2} - t\right) = \cos t \qquad \cos\left(\frac{\pi}{2} - t\right) = \sin t$$

Finally, we mention an important identity connecting the sine and cosine functions.

$$\sin^2 t + \cos^2 t = 1$$

This identity follows from the fact that on the unit circle, $y^2 + x^2 = 1$.

**Graphs of Sine and Cosine**   To graph $y = \sin t$ and $y = \cos t$, we follow our standard procedure (make a table of values, plot the corresponding points, and connect these points with a smooth curve). But how do we make a table of values? Fortunately, others have done the work for us. Extensive tables for the sine and cosine are available. One such table is Table II of the appendix; a brief table for special numbers is shown in Figure 5. With the aid of these tables or from computations on a calculator (in radian mode), we can draw the graphs in Figure 6.

| $t$ | $\sin t$ | $\cos t$ |
|-----|--------|--------|
| 0 | 0 | 1 |
| $\pi/6$ | 1/2 | $\sqrt{3}/2$ |
| $\pi/4$ | $\sqrt{2}/2$ | $\sqrt{2}/2$ |
| $\pi/3$ | $\sqrt{3}/2$ | 1/2 |
| $\pi/2$ | 1 | 0 |
| $2\pi/3$ | $\sqrt{3}/2$ | $-1/2$ |
| $3\pi/4$ | $\sqrt{2}/2$ | $-\sqrt{2}/2$ |
| $5\pi/6$ | 1/2 | $-\sqrt{3}/2$ |
| $\pi$ | 0 | $-1$ |

FIGURE 5

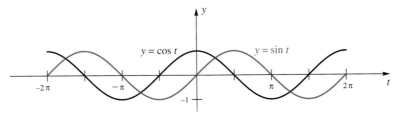

FIGURE 6

Even a casual observer might notice four things about these graphs:

1. Both $\sin t$ and $\cos t$ range from $-1$ to 1.
2. Both graphs repeat themselves on adjacent intervals of length $2\pi$.
3. The graph of $y = \sin t$ is symmetric about the origin, $y = \cos t$ is symmetric about the $y$-axis.
4. The graph of $y = \sin t$ is the same as that of $y = \cos t$ but shifted $\pi/2$ units to the right.

There are no surprises here. These are the graphical interpretations of the first four boxed formulas of the previous subsection.

**Four Other Trigonometric Functions**   We could get by with just the sine and cosine, but it is convenient to introduce four additional trigonometric functions: tangent, cotangent, secant, and cosecant.

$$\tan t = \frac{\sin t}{\cos t} \qquad \cot t = \frac{\cos t}{\sin t}$$

$$\sec t = \frac{1}{\cos t} \qquad \csc t = \frac{1}{\sin t}$$

What we know about sine and cosine will automatically give us knowledge about these four new functions.

**EXAMPLE 1** Show that tangent is an odd function.

*Solution*

$$\tan(-t) = \frac{\sin(-t)}{\cos(-t)} = \frac{-\sin t}{\cos t} = -\tan t$$

**EXAMPLE 2** Verify the following identities.

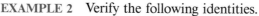

$$1 + \tan^2 t = \sec^2 t \qquad 1 + \cot^2 t = \csc^2 t$$

*Solution*

$$1 + \tan^2 t = 1 + \frac{\sin^2 t}{\cos^2 t} = \frac{\cos^2 t + \sin^2 t}{\cos^2 t} = \frac{1}{\cos^2 t} = \sec^2 t$$

$$1 + \cot^2 t = 1 + \frac{\cos^2 t}{\sin^2 t} = \frac{\sin^2 t + \cos^2 t}{\sin^2 t} = \frac{1}{\sin^2 t} = \csc^2 t$$

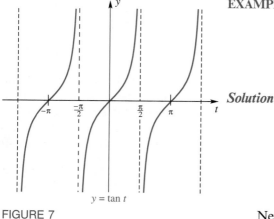

FIGURE 7

Next we graph the tangent function (Figure 7). Here we are in for two minor surprises.

First, we notice that there are vertical asymptotes at $-3\pi/2$, $-\pi/2$, $\pi/2$, $3\pi/2$, and so on. We should have anticipated this since $\cos t = 0$ at these values of $t$, which means that $(\sin t)/(\cos t)$ involves a division by zero. Second, it appears that the tangent is periodic (which we expected), but with period $\pi$ (which we might not have expected). You will see the analytic reason for this in Problem 31.

**Relation to Angle Trigonometry** Angles are commonly measured either in degrees or in radians. The angle corresponding to a complete revolution measures 360°, but only $2\pi$ radians. Equivalently, a straight angle measures 180° or $\pi$ radians, a fact worth remembering.

$$180° = \pi \text{ radians} \approx 3.1415927 \text{ radians}$$

This leads to the common conversions shown in Figure 8 and also to the following facts.

| Degrees | Radians |
|---------|---------|
| 0 | 0 |
| 30 | $\pi/6$ |
| 45 | $\pi/4$ |
| 60 | $\pi/3$ |
| 90 | $\pi/2$ |
| 120 | $2\pi/3$ |
| 135 | $3\pi/4$ |
| 150 | $5\pi/6$ |
| 180 | $\pi$ |

FIGURE 8

$$1 \text{ radian} \approx 57.29578° \qquad 1° \approx 0.0174533 \text{ radian}$$

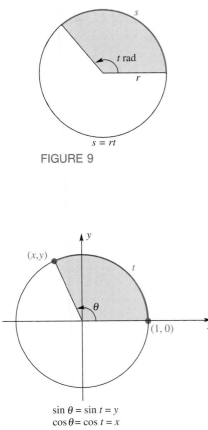

FIGURE 9

$$\sin \theta = \sin t = y$$
$$\cos \theta = \cos t = x$$

FIGURE 10

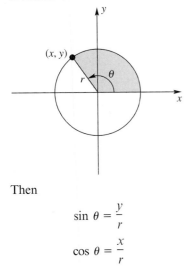

**Another View**

We have based our discussion of trigonometry on the unit circle. We could as well have used a circle of radius $r$.

Then

$$\sin \theta = \frac{y}{r}$$

$$\cos \theta = \frac{x}{r}$$

The division of a revolution into 360 parts is quite arbitrary (due to the ancient Babylonians, who liked multiples of 60). The division into $2\pi$ parts is more fundamental and lies behind the almost universal use of radian measure in calculus. Notice, in particular, that the length $s$ of the arc cut off on a circle of radius $r$ by a central angle of $t$ radians satisfies (see Figure 9)

$$\frac{s}{2\pi r} = \frac{t}{2\pi}$$

That is

$$\boxed{s = rt}$$

When $r = 1$, this gives $s = t$. In words, *the length of arc on the unit circle cut off by a central angle of $t$ radians is $t$.* This is correct even if $t$ is negative, provided we interpret length to be negative when measured in the clockwise direction.

**EXAMPLE 3**   Find the distance traveled by a bicycle with wheels of radius 30 centimeters when the wheels turn through 100 revolutions.

*Solution*   We use the boxed formula, recognizing that 100 revolutions correspond to $100 \cdot (2\pi)$ radians.

$$s = (30)(100)(2\pi) = 6000\pi$$
$$\approx 18849.6 \text{ centimeters} \qquad \blacksquare$$

Now we can make the connection between angle trigonometry and unit circle trigonometry. If $\theta$ is an angle measuring $t$ radians (Figure 10), then

$$\boxed{\sin \theta = \sin t \qquad \cos \theta = \cos t}$$

In calculus, when we meet an angle measured in degrees, we almost always change it to radians before doing any calculations. For example,

$$\sin 31.6° = \sin\left(31.6 \cdot \frac{\pi}{180} \text{ radian}\right) \approx \sin(0.552)$$

**EXAMPLE 4**   Find $\cos 51.8°$.

*Solution*   The simplest procedure is to press the right keys on a calculator. We could also use Table I of the appendix. If we want to use Table II of the appendix, we first change 51.8° to radians.

$$51.8° = 51.8\left(\frac{\pi}{180}\right) \approx 0.904 \text{ radian}$$

Thus

$$\cos(51.8°) \approx \cos(0.904) \approx 0.6184 \qquad \blacksquare$$

**List of Important Identities**    We will not take space to verify all of the following identities. We simply assert their truth and suggest that most of them will be needed somewhere in this book.

---

## TRIGONOMETRIC IDENTITIES

*Odd-even identities*

$$\sin(-x) = -\sin x$$

$$\cos(-x) = \cos x$$

$$\tan(-x) = -\tan x$$

*Cofunction identities*

$$\sin\left(\frac{\pi}{2} - x\right) = \cos x$$

$$\cos\left(\frac{\pi}{2} - x\right) = \sin x$$

$$\tan\left(\frac{\pi}{2} - x\right) = \cot x$$

*Pythagorean identities*

$$\sin^2 x + \cos^2 x = 1$$

$$1 + \cot^2 x = \csc^2 x$$

$$1 + \tan^2 x = \sec^2 x$$

*Addition identities*

$$\sin(x + y) = \sin x \cos y + \cos x \sin y$$

$$\cos(x + y) = \cos x \cos y - \sin x \sin y$$

$$\tan(x + y) = \frac{\tan x + \tan y}{1 - \tan x \tan y}$$

*Double-angle identities*

$$\sin 2x = 2 \sin x \cos x$$

$$\cos 2x = \cos^2 x - \sin^2 x = 2 \cos^2 x - 1 = 1 - 2 \sin^2 x$$

*Half-angle identities*

$$\sin\left(\frac{x}{2}\right) = \pm \sqrt{\frac{1 - \cos x}{2}}$$

$$\cos\left(\frac{x}{2}\right) = \pm \sqrt{\frac{1 + \cos x}{2}}$$

*Sum identities*

$$\sin x + \sin y = 2 \sin\left(\frac{x + y}{2}\right)\cos\left(\frac{x - y}{2}\right)$$

$$\cos x + \cos y = 2 \cos\left(\frac{x + y}{2}\right)\cos\left(\frac{x - y}{2}\right)$$

*Product identities*

$$\sin x \sin y = -\tfrac{1}{2}[\cos(x + y) - \cos(x - y)]$$

$$\cos x \cos y = \tfrac{1}{2}[\cos(x + y) + \cos(x - y)]$$

$$\sin x \cos y = \tfrac{1}{2}[\sin(x + y) + \sin(x - y)]$$

## CONCEPTS REVIEW

**1.** The natural real number domain of the sine function is _____; its range is _____.

**2.** The period of the cosine function is _____; the period of the sine function is _____; the period of the tangent function is _____.

**3.** Of the six trigonometric functions, _____ are odd functions and _____ are even functions.

**4.** If $(-4, 3)$ lies on the terminal side of an angle $\theta$ whose vertex is at the origin and initial side is along the positive $x$-axis, then $\cos \theta = $ _____.

## PROBLEM SET 2.3

**1.** Convert the following degree measures to radians (leave $\pi$ in your answer).

(a) $30°$     (b) $45°$     (c) $-60°$
(d) $240°$     (e) $-370°$     (f) $10°$
(g) $22\frac{1}{2}°$     (h) $600°$     (i) $-120°$

**2.** Convert the following radian measures to degrees.

(a) $\frac{7}{6}\pi$     (b) $\frac{3}{4}\pi$     (c) $-\frac{1}{3}\pi$
(d) $\frac{4}{3}\pi$     (e) $-\frac{35}{18}\pi$     (f) $\frac{3}{18}\pi$
(g) $\frac{9}{8}\pi$     (h) $\frac{10}{3}\pi$     (i) $-\frac{4}{3}\pi$

C   **3.** Convert the following degree measures to radians ($1° = \pi/180 = 1.7453 \times 10^{-2}$ radians).

(a) $33.3°$     (b) $46°$     (c) $-66.6°$
(d) $240.11°$     (e) $-369°$     (f) $11°$
(g) $22.5°$     (h) $359°$     (i) $-121.35°$

C   **4.** Convert the following radian measures to degrees. (1 radian = $180/\pi = 57.296$ degrees)

(a) $3.141$     (b) $6.28$     (c) $5.00$
(d) $0.001$     (e) $-0.1$     (f) $36.0$
(g) $-2.00$     (h) $1.234$     (i) $-10.0$

C   **5.** Calculate (be sure your calculator is in radian or degree mode as needed).

(a) $\dfrac{56.4 \tan 34.2°}{\sin 34.1°}$     (b) $\dfrac{5.34 \tan 21.3°}{\sin 3.1° + \cot 23.5°}$
(c) $\tan(0.452)$     (d) $\sin(-0.361)$
(e) $\cos(-0.361)$     (f) $\tan(-0.361)$

**6.** Use Table II of the appendix to find each value.

(a) $\sin(1.23)$     (b) $\cos(0.63)$
(c) $\tan(1.55)$     (d) $\sin(-1.23)$
(e) $\cos(-0.63)$     (f) $\tan(-1.55)$

C   **7.** Calculate.

(a) $\dfrac{234.1 \sin(1.56)}{\cos(0.34)}$     (b) $\sin^2(2.51) + \sqrt{\cos(0.51)}$

C   **8.** Calculate.

(a) $\dfrac{56.3 \tan 34.2°}{\sin 56.1°}$     (b) $\left(\dfrac{\sin 35°}{\sin 26° + \cos 26°}\right)^3$

**9.** Evaluate without use of a calculator.

(a) $\tan\left(\dfrac{\pi}{6}\right)$     (b) $\sec(\pi)$     (c) $\sec\left(\dfrac{3\pi}{4}\right)$
(d) $\csc\left(\dfrac{\pi}{2}\right)$     (e) $\cot\left(\dfrac{\pi}{4}\right)$     (f) $\tan\left(-\dfrac{\pi}{4}\right)$

**10.** Evaluate without use of a calculator.

(a) $\tan\left(\dfrac{\pi}{3}\right)$     (b) $\sec\left(\dfrac{\pi}{3}\right)$     (c) $\cot\left(\dfrac{\pi}{3}\right)$
(d) $\csc\left(\dfrac{\pi}{4}\right)$     (e) $\tan\left(-\dfrac{\pi}{6}\right)$     (f) $\cos\left(-\dfrac{\pi}{3}\right)$

**11.** Verify that the following are identities (see Example 2).

(a) $(1 + \sin z)(1 - \sin z) = \dfrac{1}{\sec^2 z}$
(b) $(\sec t - 1)(\sec t + 1) = \tan^2 t$
(c) $\sec t - \sin t \tan t = \cos t$
(d) $\dfrac{\sec^2 t - 1}{\sec^2 t} = \sin^2 t$
(e) $\cos t (\tan t + \cot t) = \csc t$

**12.** Verify that the following are identities (see Example 2)

(a) $\sin^2 v + \dfrac{1}{\sec^2 v} = 1$
(b) $\cos 3t = 4 \cos^3 t - 3 \cos t$
(c) $\sin 4x = 8 \sin x \cos^3 x - 4 \sin x \cos x$
(d) $(\sec \theta - 1)(\sec \theta + 1) = \tan^2 \theta$
(e) $\sin t(\tan t - \cot t) = -\sin t \dfrac{2 \cos^2 t - 1}{\cos t \sin t}$

**13.** Verify that the following are identities.

(a) $\dfrac{\sin u}{\csc u} + \dfrac{\cos u}{\sec u} = 1$
(b) $(1 - \cos^2 x)(1 + \cot^2 x) = 1$
(c) $\sin t (\csc t - \sin t) = \cos^2 t$
(d) $\dfrac{1 - \csc^2 t}{\csc^2 t} = \dfrac{-1}{\sec^2 t}$
(e) $\dfrac{1}{\sin t \cos t} - \dfrac{\cos t}{\sin t} = \tan t$

**14.** Sketch the graph of the following on $[-\pi, 2\pi]$.

(a) $y = \sin 2x$

(b) $y = 2 \sin t$

(c) $y = \cos\left(x - \dfrac{\pi}{4}\right)$

(d) $y = \sec t$

**15.** Sketch the graphs of the following on $[-\pi, 2\pi]$.

(a) $y = \csc t$

(b) $y = 2 \cos t$

(c) $y = \cos 3t$

(d) $y = \cos\left(t + \dfrac{\pi}{3}\right)$

Determine the period, amplitude, and shifts (both horizontal and vertical) and draw a complete graph over the interval $-5 \le x \le 5$ for the functions listed in Problems 16–19.

**16.** $y = 3 \cos \dfrac{x}{2}$

**17.** $y = 2 \sin 2x$

**18.** $y = 3 \cos\left(x - \dfrac{\pi}{2}\right) - 1$

**19.** $y = \tan\left(2x - \dfrac{\pi}{3}\right)$

**20.** Which of the following represent the same graph. Check your result analytically by using trigonometric identities.

(a) $y = \sin\left(x + \dfrac{\pi}{2}\right)$

(b) $y = \cos\left(x + \dfrac{\pi}{2}\right)$

(c) $y = -\sin(x + \pi)$

(d) $y = \cos(x - \pi)$

(e) $y = -\sin(\pi - x)$

(f) $y = \cos\left(x - \dfrac{\pi}{2}\right)$

(g) $y = -\cos(\pi - x)$

(h) $y = \sin\left(x - \dfrac{\pi}{2}\right)$

**21.** Find the quadrant in which the point $P(x, y)$ will lie for each $t$ below, and thereby determine the sign of $\cos t$. *Hint*: See the unit circle definition of $\cos t$.

(a) $t = 5.97$

(b) $t = 9.34$

(c) $t = -16.1$

**22.** Follow the directions of Problem 21 to determine the sign of $\tan t$.

(a) $t = 4.34$

(b) $t = -15$

(c) $t = 21.9$

**23.** Which of the following are odd functions? Even functions? Neither?

(a) $t \sin t$

(b) $\sin^2 t$

(c) $\csc t$

(d) $|\sin t|$

(e) $\sin(\cos t)$

(f) $x + \sin x$

**24.** Which of the following are odd functions? Even functions? Neither?

(a) $\cot t + \sin t$

(b) $\sin^3 t$

(c) $\sec t$

(d) $\sqrt{\sin^4 t}$

(e) $\cos(\sin t)$

(f) $x^2 + \sin x$

Use the half-angle identities to find the exact values in Problems 25–29

**25.** $\cos^2 \dfrac{\pi}{3} =$

**26.** $\sin^2 \dfrac{\pi}{6} =$

**27.** $\sin^3 \dfrac{\pi}{6} =$

**28.** $\cos^2 \dfrac{\pi}{12} =$

**29.** $\sin^2 \dfrac{\pi}{8} =$

**30.** Find identities analogous to the addition identities for each expression.

(a) $\sin(x - y)$

(b) $\cos(x - y)$

(c) $\tan(x - y)$

**31.** Use the addition identity for the tangent to show that $\tan(t + \pi) = \tan t$ for all $t$ in the domain of $\tan t$.

**32.** Show that $\cos(x - \pi) = -\cos x$ for all $x$.

**33.** Find the length of arc on a circle of radius 2.5 centimeters cut off by each central angle (see Example 3).

(a) 6 radians

(b) 225°

**34.** How far does a wheel of radius 2 feet roll along level ground in making 150 revolutions? (See Example 3.)

C **35.** Suppose that a tire on a truck has an outer radius of 2.5 feet. How many revolutions per minute does the tire make when the truck is traveling 60 miles per hour?

C **36.** A belt passes around two wheels, as shown in Figure 11. How many revolutions per second does the small wheel make when the large wheel makes 21 revolutions per second?

FIGURE 11

**37.** A 50-pound bag of corn is being dragged along the ground by a man whose arm makes an angle of $t$ radians with the ground. The force $F$ (in pounds) required is given by

$$F(t) = \frac{50\mu}{\mu \sin t + \cos t}$$

Here $\mu$ is a constant relating to the friction involved. Find $F$ in each case.

(a) $t = \dfrac{\pi}{4}$

(b) $t = 0$

C (c) $t = 1$

(d) The angle is 90°.

**38.** The **angle of inclination** $\alpha$ of a line is the smallest positive angle from the positive $x$-axis to the line ($\alpha = 0$ for a horizontal line). Show that the slope $m$ of the line is equal to $\tan \alpha$.

**39.** Find the angle of inclination of the following lines (see Problem 38).

(a) $y = \sqrt{3}x - 7$          (b) $\sqrt{3}x + 3y = 6$

**40.** Let $3_1$ and $3_2$ be two nonvertical lines with slopes $m_1$ and $m_2$, respectively. If $\theta$, the angle from $3_1$ to $3_2$, is not a right angle, then

$$\tan \theta = \frac{m_2 - m_1}{1 + m_1 m_2}$$

Show this using the fact that $\theta = \theta_2 - \theta_1$ in Figure 12.

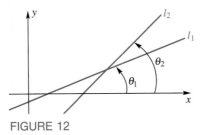

FIGURE 12

[C] **41.** Find the angle (in radians) from the first line to the second (see Problem 40).

(a) $y = 2x, y = 3x$      (b) $y = \frac{x}{2}, y = -x$

(c) $2x - 6y = 12, 2x + y = 0$

**42.** Derive the formula $A = \frac{1}{2}r^2t$ for the area of a sector of a circle. Here $r$ is the radius and $t$ is the radian measure of the vertex angle (see Figure 13).

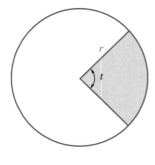

FIGURE 13

**43.** Find the area of the sector of a circle of radius 5 centimeters and vertex angle 2 radians (see Problem 42).

**44.** Suppose that the spoke of a mounted wheel of radius 2 feet rotates 10 times a second. Find the area swept out by the spoke in $\frac{1}{40}$ second; in 3 seconds.

[≈] **45.** A $33\frac{1}{3}$-rpm stereo record has a spiral groove starting 6 inches from the center and ending 3 inches from the center. If it took 18 minutes to play the record, approximately how long is the spiral groove?

**46.** A regular polygon of $n$ sides is inscribed in a circle of radius $r$. Find formulas for the perimeter, $P$, and area, $A$, of the polygon in terms of $n$ and $r$.

**47.** An isosceles triangle is topped by a semicircle, as shown in Figure 14. Find a formula for the area $A$ of the whole figure in terms of the side length $r$ and vertex angle $t$ (radians).

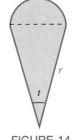

FIGURE 14

**48.** From a product identity, we obtain

$$\cos \frac{x}{2} \cos \frac{x}{4} = \frac{1}{2}\left[\cos \tfrac{3}{4}x + \cos \tfrac{1}{4}x\right]$$

Find the corresponding sum of cosines for

$$\cos\frac{x}{2} \cos\frac{x}{4} \cos\frac{x}{8} \cos\frac{x}{16}$$

Do you see a generalization?

In Chapter 11 we will have occasion to study functions of an integral variable. In preparation for this material on sequences and series we explore some aspects of functions of an integral variable in Problems 42–52.

**49.** The sum of the first $n$ integers is an example of a function of integral values of $n$. We can generate a closed form representation as a rational function of $n$ for this sum

$$S_1(n) = 1 + 2 + 3 + 4 + \ldots + n$$

by rearranging the terms in a decreasing order and then adding them to the original sum:

$$S_1(n) = 1 + 2 + 3 + \ldots + n$$
$$S_1(n) = n + (n - 1) + (n - 2) + \ldots + 1$$
$$2 \cdot S_1(n) = (n+1) + (n+1) + \ldots + (n+1) = n \cdot (n+1)$$

hence,

$$S_1(n) = \frac{n \cdot (n + 1)}{2}.$$

**50.** Another simple formula may be derived for the integer function

$$S_2(n) = 1^2 + 2^2 + 3^2 + \ldots + n^2$$

by using the formula

$$(x + 1)^3 - x^3 = 3x^2 + 3x + 1.$$

**(a)** By setting $x$ equal to the values of 0, 1, 2, 3, . . . , $n$ and adding the results, establish the formula

$$(n + 1)^3 = 3S_2(n) + 3S_1(n) + n + 1.$$

**(b)** Using the result derived above for $S_1(n)$, and solving for $S_2(n)$ in part (a) show that

$$S_2(n) = \frac{1}{6}n(n + 1)(2n + 1).$$

**51.** Find a rational function form for $S_3(n)$ using a similar technique.

**52.** Find a rational function form for $S_4(n)$ using a similar technique.

## Explorations

**1.** Circular motion can be modeled by using the parametric representations of the form $x(t) = \sin t$ and $y(t) = \cos t$. This will give the full circle for $0 \le t \le 2\pi$. If we consider a 4 foot diameter wheel rotating clockwise once every 10 seconds, show that the motion of a point on the wheel can be represented by $x(t) = 2\sin(\pi t/5)$ and $y(t) = 2 \cos(\pi t/5)$.

**(a)** Find the position of the point on the rim of the wheel when $t = 2$ seconds, 6 seconds, and 10 seconds. Where was this point when the wheel started to rotate at $t = 0$?
**(b)** How would the formulas giving the motion of the point change if the wheel was rotating *counter*-clockwise.
**(c)** At what value of $t$ is the point at (2, 0)?

**2.** The circular frequency $\nu$ of oscillation of a point is given by $\nu = \dfrac{2\pi}{period}$. What happens when you add two motions that have the same frequency or period? To investigate, we can graph the function $y(t) = 2 \sin(\pi t/5)$, $y(t) = \sin(\pi t/5) + \cos(\pi t/5)$ and look for similarities. Armed with this information, we can investigate by graphing over the interval $[-5, 5]$:

**(a)** $y(t) = 3 \sin(\pi t/5) - 5 \cos(\pi t/5) + 2 \sin((\pi t/5) - 3)$
**(b)** $y(t) = 3 \cos(\pi t/5 - 2) + \cos(\pi t/5) + \cos((\pi t/5) - 3)$

**3.** We now explore the relationship between $A \sin(\overline{\omega}t) + B \cos(\overline{\omega}t)$ and $C \sin(\overline{\omega}t + \phi)$.

**(a)** By expanding $\sin(\overline{\omega}t + \phi)$ using the sum of the angles formula show that $A = C \cos \phi$ and $B = C \sin \phi$.
**(b)** Consequently show that $A^2 + B^2 = C^2$ and that $\phi$ then satisfies the equation $\tan \phi = \dfrac{B}{A}$.

**(c)** Generalize your result to state a proposition about $A_1 \sin(\overline{\omega}t + \phi_1) + A_2 \sin(\overline{\omega}t + \phi_2) + A_3 \sin(\overline{\omega}t + \phi_3)$.
**(d)** Write an essay, in your own words, which expresses the importance of the identity between $A \sin(\overline{\omega}t) + B \cos(\overline{\omega}t)$ and $C \sin(\overline{\omega}t + \phi)$. Be sure to note that $|C| \ge max(|A|, |B|)$ and that the identity only holds when you are forming a linear combination (adding and/or subtracting multiples of single powers of) sine and cosine of the same frequency.

Trigonometric functions which have high frequencies pose special problems for graphing. We now explore how to plot such functions.

**4.** Graph the function $f(x) = \sin 50x$ using the window given by a $y$ range of $-1.5 \le y \le 1.5$ and the $x$ range given by

**(a)** $[-15, 15]$               **(b)** $[-10, 10]$
**(c)** $[-8, 8]$                 **(d)** $[-1, 1]$
**(e)** $[-0.25, 0.25]$

Indicate briefly which $x$-window gives the true behavior of the function, and discuss reasons why the other $x$-windows give results which are different.

**5.** Graph the function $f(x) = \cos x + \dfrac{1}{50} \sin 50x$ using the windows given by the following ranges of $x$ and $y$.

**(a)** $-5 \le x \le 5, -1 \le y \le 1$
**(b)** $-1 \le x \le 1, -0.5 \le y \le 0.5$
**(c)** $-0.1 \le x \le 0.1, -0.1 \le y \le 0.1$

Indicate briefly which $(x,y)$-window gives the true behavior of the function, and discuss reasons why the other $(x,y)$-windows give results which are different. In this case is it true that only one window gives the important behavior, or do we need more than one window to graphically communicate the behavior of this function?

**6.** Let $f(x) = \dfrac{3x + 2}{x^2 + 1}$ and $g(x) = \dfrac{1}{100} \cos(100x)$.

**(a)** Use functional composition to form $h(x) = (f \circ g)(x)$ as well as $j(x) = (g \circ f)(x)$
**(b)** Find the appropriate window or windows which give a clear picture of $h(x)$.
**(c)** Find the appropriate window or windows which give a clear picture of $j(x)$.

**Computer Algebra System running on a Personal Computer** PC.

**1.** Use the PC ability to factor in order to find the points where the following graphs cross the $x$-axis:

**(a)** $y = x^5 - 14x^4 - 58x^3 + 668x^2 - 135x - 3150$
**(b)** $y = x^7 + 67x^5 - 14x^6 - 168x^4 + 371x^3 - 574x^2 + 305x - 420$
**(c)** $y = x^4 - 10x^2 + 9$

2. Use the PC ability to solve simultaneous equations to find the points of intersection between the following graphs. Then use the graphics capability to check the relative correctness of your answer.

(a) $y = 3x^2 + 2x - 3$ and $y = 3x - 4$
(b) $y = x^3 - 20x^2 + 3x - 10$ and $y = 20 - 2x^2$
(c) $y = x^4 - 3x^3 + 2x^2 - x + 1$ and $y = x^2 + 5x + 1$.

3. Certain algebraic patterns prove very useful in calculus. We can use the PC to explore such patterns. In each of the following cases perform the indicated operation

(a) expand $(x + a)^2$, $(x + a)^3$, $(x + a)^5$, $(x + a)^{10}$
(b) expand $\dfrac{(x + h)^2 - x^2}{h}$, $\dfrac{(x + h)^5 - x^5}{h}$, $\dfrac{(x + h)^{10} - x^{10}}{h}$
(c) simplify $\dfrac{t^3 - 1}{t - 1}$, $\dfrac{t^5 - b^5}{t - b}$, $\dfrac{t^{11} - b^{11}}{t - b}$

4. Many computer algebra systems such as Mathematica, Maple, and Derive permit the simplification of trigono-

metric functions. In most systems simplification can go in two directions. On the one hand, the various six trigonometric functions can be replaced by powers of a single function such as the sin. On the other hand, trigonometric functions with different period arguments can be replaced by functions whose periods are $2\pi$. You will need to investigate the help files on expanding, combining, and simplifying trigonometric functions in order to perform the following operations.

Consider the function $\cot(2x) \sin^3(x)(1 + \sec(4x)) + \sec(7x) \tan(x)$

(a) Reduce the function to a rational function consisting of powers of sin and cos with period $2\pi$.
(b) Reduce the function to one containing no powers higher than the first power of all possible trigonometric functions.

---

**Answers to Concepts Review:** **1.** $(-\infty, \infty)$; $[-1, 1]$ **2.** $2\pi$; $2\pi$; $\pi$ **3.** sine, tangent, cosecant, and cotangent; cosine and secant **4.** $-4/5$

---

**2.4
INTRODUCTION TO
LIMITS**

The topics discussed so far are part of what is called *precalculus*. They provide the foundation for calculus, but they are not calculus. Now we are ready for an important new idea, the notion of *limit*. It is this idea that distinguishes calculus from other branches of mathematics. In fact, we might define calculus as *the study of limits*.

Of course, the word *limit* is used in everyday language as when one says, "I'm nearing the limit of my patience." That usage has something to do with calculus, but not very much.

**An Intuitive Understanding** Consider the function determined by the formula

$$f(x) = \frac{x^3 - 1}{x - 1}$$

Note that it is not defined at $x = 1$ since at this point $f(x)$ has the form $\frac{0}{0}$, which is meaningless. We can, however, still ask what is happening to $f(x)$ as $x$ approaches 1. More precisely, is $f(x)$ approaching some specific number as $x$ approaches 1? To get at this question, we have done three things. We have calculated some values of $f(x)$ for $x$ near 1, we have shown these values in a schematic diagram, and we have sketched the graph of $y = f(x)$ (Figure 1).

All the information we have assembled seems to point to the same conclusion: $f(x)$ approaches 3 as $x$ approaches 1. In mathematical symbols, we write

$$\lim_{x \to 1} \frac{x^3 - 1}{x - 1} = 3$$

This is read "the limit as $x$ approaches 1 of $(x^3 - 1)/(x - 1)$ is 3."

Being good algebraists (thus knowing how to factor the difference of cubes), we can provide more and better evidence.

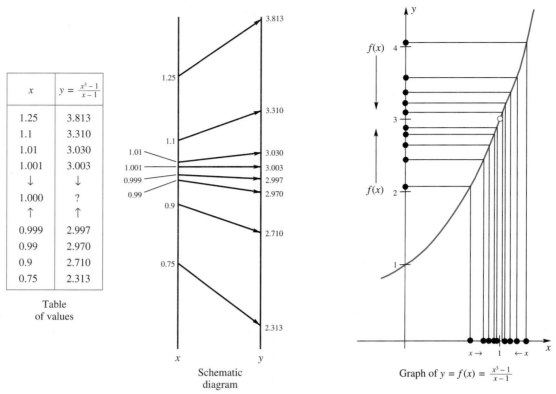

| $x$ | $y = \frac{x^3-1}{x-1}$ |
|-----|-------------------------|
| 1.25 | 3.813 |
| 1.1 | 3.310 |
| 1.01 | 3.030 |
| 1.001 | 3.003 |
| $\downarrow$ | $\downarrow$ |
| 1.000 | ? |
| $\uparrow$ | $\uparrow$ |
| 0.999 | 2.997 |
| 0.99 | 2.970 |
| 0.9 | 2.710 |
| 0.75 | 2.313 |

Table
of values

Schematic
diagram

Graph of $y = f(x) = \frac{x^3-1}{x-1}$

FIGURE 1

$$\lim_{x \to 1} \frac{x^3 - 1}{x - 1} = \lim_{x \to 1} \frac{(x - 1)(x^2 + x + 1)}{x - 1}$$

$$= \lim_{x \to 1}(x^2 + x + 1) = 1^2 + 1 + 1 = 3$$

Note that $(x - 1)/(x - 1) = 1$ as long as $x \neq 1$. This justifies the second step. The third step should seem reasonable; a rigorous justification will come later.

To be sure we are on the right track, we need to have a clearly understood meaning for the word *limit*. Here is our first attempt at a definition.

> ### Definition
>
> **(Intuitive meaning of limit).** To say that $\lim_{x \to c} f(x) = L$ means that when $x$ is near but different from $c$, then $f(x)$ is near $L$.

Notice that we do not require anything to be true right at $c$. The function $f$ need not even be defined at $c$; it was not in the example $f(x) = (x^3 - 1)/(x - 1)$ just considered. The notion of limit is associated with the behavior of a function near $c$, not at $c$.

A cautious reader is sure to object to our use of the word *near*. What does *near* mean? How near is near? For precise answers, you will have to study the next section; however, some further examples will help to clarify the idea.

**More Examples**   Our first example is almost trivial but is nonetheless important.

**EXAMPLE 1**   Find $\lim_{x \to 3}(4x - 5)$.

*Solution*   When $x$ is near 3, $4x - 5$ is near $4 \cdot 3 - 5 = 7$. We write

$$\lim_{x \to 3}(4x - 5) = 7 \qquad \blacksquare$$

**EXAMPLE 2**   Find $\lim_{x \to 3} \dfrac{x^2 - x - 6}{x - 3}$.

*Solution*   Note that $(x^2 - x - 6)/(x - 3)$ is not defined at $x = 3$, but that is all right. To get an idea of what is happening as $x$ approaches 3, we could use a calculator to evaluate the given expression, for example, at 3.1, 3.01, 3.001, and so on. But it is much better to use a little algebra to simplify the problem.

$$\lim_{x \to 3} \frac{x^2 - x - 6}{x - 3} = \lim_{x \to 3} \frac{(x - 3)(x + 2)}{x - 3} = \lim_{x \to 3}(x + 2) = 3 + 2 = 5$$

The cancellation of $x - 3$ in the second step is legitimate because the definition of limit ignores the behavior right at $x = 3$. Thus, we did not divide by 0. $\qquad \blacksquare$

**EXAMPLE 3**   Find $\lim_{x \to 1} \dfrac{x - 1}{\sqrt{x} - 1}$.

*Solution*

$$\lim_{x \to 1} \frac{x - 1}{\sqrt{x} - 1} = \lim_{x \to 1} \frac{(\sqrt{x} - 1)(\sqrt{x} + 1)}{\sqrt{x} - 1} = \lim_{x \to 1}(\sqrt{x} + 1) = \sqrt{1} + 1 = 2 \; \blacksquare$$

**EXAMPLE 4**   Find $\lim_{x \to 0} \dfrac{\sin x}{x}$.

*Solution*   No algebraic trick will simplify our task; certainly we cannot cancel the $x$'s. A calculator will help us get an idea of the limit. Use your own calculator (radian mode) to check the values in the table of Figure 2. Our conclusion—though we admit it is a bit shaky—is that

$$\lim_{x \to 0} \frac{\sin x}{x} = 1$$

We will give a rigorous demonstration in Section 3.4. $\qquad \blacksquare$

| $x$ | $\dfrac{\sin x}{x}$ |
|---|---|
| 1.0 | 0.84147 |
| 0.5 | 0.95885 |
| 0.1 | 0.99833 |
| 0.01 | 0.99998 |
| $\downarrow$ | $\downarrow$ |
| 0 | ? |
| $\uparrow$ | $\uparrow$ |
| −0.01 | 0.99998 |
| −0.1 | 0.99833 |
| −0.5 | 0.95885 |
| −1.0 | 0.84147 |

FIGURE 2

**Some Warning Flags** Things are not quite as simple as they may appear. Calculators may mislead us; so may our own intuition. The examples that follow suggest some possible pitfalls.

**EXAMPLE 5 (Your calculator may fool you).** Find $\lim_{x \to 0} \left[ x^2 - \dfrac{\cos x}{10{,}000} \right]$.

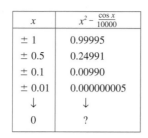

| $x$ | $x^2 - \frac{\cos x}{10000}$ |
|---|---|
| $\pm 1$ | 0.99995 |
| $\pm 0.5$ | 0.24991 |
| $\pm 0.1$ | 0.00990 |
| $\pm 0.01$ | 0.000000005 |
| $\downarrow$ | $\downarrow$ |
| 0 | ? |

FIGURE 3

*Solution* Following the procedure used earlier, we construct the table of values shown in Figure 3. The conclusion it suggests is that the desired limit is 0. But that is wrong. If we recall the graph of $y = \cos x$, we realize that $\cos x$ approaches 1 as $x$ approaches 0. Thus,

$$\lim_{x \to 0} \left[ x^2 - \frac{\cos x}{10{,}000} \right] = 0^2 - \frac{1}{10{,}000} = -\frac{1}{10{,}000} \quad \blacksquare$$

**EXAMPLE 6 (No limit at a jump).** Find $\lim_{x \to 2} [\![x]\!]$

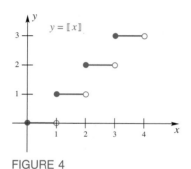

FIGURE 4

*Solution* Recall that $[\![x]\!]$ denotes the greatest integer in $x$ (see Section 2.1). The graph of $y = [\![x]\!]$ is shown in Figure 4. For all numbers $x$ less than 2 but near 2, $[\![x]\!] = 1$, but for all numbers $x$ greater than 2 but near 2, $[\![x]\!] = 2$. Is $[\![x]\!]$ near to a single number $L$ when $x$ is near 2? No. No matter what number we propose for $L$, there will be $x$'s arbitrarily close to 2 on one side or the other, where $[\![x]\!]$ differs from $L$ by at least $\frac{1}{2}$. Our conclusion is that $\lim_{x \to 2} [\![x]\!]$ does not exist. If you check back, you will see that we have not claimed that every limit we can write must exist. $\quad \blacksquare$

**EXAMPLE 7 (Too many wiggles).** Find $\lim_{x \to 0} \sin(1/x)$.

*Solution* This example poses the most subtle limit question asked yet. Since we do not want to make too big a story out of it, we ask you to do two things. First, pick a sequence of $x$-values approaching 0. Use your calculator to evaluate $\sin(1/x)$ at these $x$'s. Unless you happen on some very lucky choices, your values will oscillate wildly.

Second, consider trying to graph $y = \sin(1/x)$. No one will ever do this very well, but the table of values in Figure 5 gives us a good clue about what is happening. In any neighborhood of the origin, the graph wiggles up and down between $-1$ and 1 infinitely many times (Figure 6). Clearly $\sin(1/x)$ is

| $x$ | $\sin \frac{1}{x}$ |
|---|---|
| $2/\pi$ | 1 |
| $2/(2\pi)$ | 0 |
| $2/(3\pi)$ | $-1$ |
| $2/(4\pi)$ | 0 |
| $2/(5\pi)$ | 1 |
| $2/(6\pi)$ | 0 |
| $2/(7\pi)$ | $-1$ |
| $2/(8\pi)$ | 0 |
| $2/(9\pi)$ | 1 |
| $2/(10\pi)$ | 0 |
| $2/(11\pi)$ | $-1$ |
| $2/(12\pi)$ | 0 |
| $\downarrow$ | $\downarrow$ |
| 0 | ? |

FIGURE 5

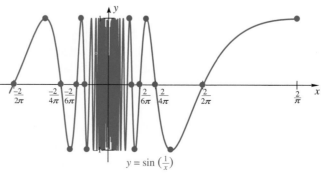

$$y = \sin\left(\tfrac{1}{x}\right)$$

FIGURE 6

not near a single number $L$ when $x$ is near 0. We conclude that $\displaystyle\lim_{x \to 0} \sin(1/x)$ does not exist. ■

**One-Sided Limits**   When a function takes a jump (as does $[\![x]\!]$ at each integer, in Example 6), then the limit does not exist at the jump points. For such functions, it is natural to introduce one-sided limits. Let the symbol $x \to c^+$ mean that $x$ approaches $c$ from the right, and let $x \to c^-$ mean that $x$ approaches $c$ from the left.

---

### Definition

**(Right- and left-hand limits).** To say that $\displaystyle\lim_{x \to c^+} f(x) = L$ means that when $x$ is near but on the right of $c$, then $f(x)$ is near $L$. Similarly, to say that $\displaystyle\lim_{x \to c^-} f(x) = L$ means that when $x$ is near but on the left of $c$, then $f(x)$ is near $L$.

---

Thus, while $\displaystyle\lim_{x \to 2} [\![x]\!]$ does not exist, it is correct to write (look at the graph in Figure 4)

$$\lim_{x \to 2^-} [\![x]\!] = 1 \quad \lim_{x \to 2^+} [\![x]\!] = 2$$

We believe that you will find the following theorem quite reasonable.

---

### Theorem A

$\displaystyle\lim_{x \to c} f(x) = L$ if and only if $\displaystyle\lim_{x \to c^-} f(x) = L$ and $\displaystyle\lim_{x \to c^+} f(x) = L$.

---

Figure 7 should give additional insight.

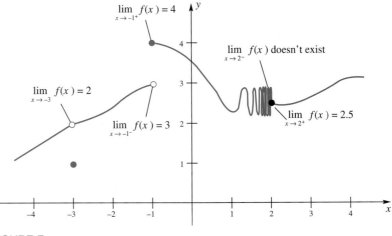

FIGURE 7

## CONCEPTS REVIEW

**1.** $\lim\limits_{x \to c} f(x) = L$ means that $f(x)$ gets near to _____ when $x$ gets sufficiently near to (but is different from _____.

**2.** Let $f(x) = (x^2 - 9)/(x - 3)$ and note that $f(3)$ is undefined. Nevertheless, $\lim\limits_{x \to 3} f(x) =$ _____.

**3.** $\lim\limits_{x \to c^+} f(x) = L$ means that $f(x)$ gets near to _____ when $x$ approaches $c$ from the _____.

**4.** If both $\lim\limits_{x \to c^-} f(x) = M$ and $\lim\limits_{x \to c^+} f(x) = M$, then _____.

## PROBLEM SET 2.4

In Problems 1–6, find the indicated limit by inspection.

**1.** $\lim\limits_{x \to 3} (x - 5)$

**2.** $\lim\limits_{t \to -1} (1 - 2t)$

**3.** $\lim\limits_{x \to 2} (x^2 + 2x - 1)$

**4.** $\lim\limits_{x \to -2} (x^2 + 2t - 1)$

**5.** $\lim\limits_{t \to -1} \dfrac{1 - 2t}{\sqrt{3t + 21}}$

**6.** $\lim\limits_{t \to -1} \dfrac{\sqrt{1 - 2t}}{(3t + 2)^3}$

In problems 7–16, find the indicated limit. In most cases, it will be wise to do some algebra first (see Examples 2 and 3).

**7.** $\lim\limits_{x \to 2} \dfrac{x^2 - 4}{x - 2}$

**8.** $\lim\limits_{t \to -7} \dfrac{t^2 + 4t - 21}{t + 7}$

**9.** $\lim\limits_{x \to -1} \dfrac{x^3 - 4x^2 + x + 6}{x + 1}$

**10.** $\lim\limits_{x \to 0} \dfrac{x^4 + 2x^3 - x^2}{x^2}$

**11.** $\lim\limits_{x \to -t} \dfrac{x^2 - t^2}{x + t}$

**12.** $\lim\limits_{x \to 3} \dfrac{x^2 - 9}{x - 3}$

**13.** $\lim\limits_{t \to 2} \dfrac{\sqrt{(t + 4)(t - 2)^4}}{(3t - 6)^2}$

**14.** $\lim\limits_{t \to 7} \dfrac{\sqrt{(t - 7)^3}}{t - 7}$

**15.** $\lim\limits_{x \to 3} \dfrac{x^4 - 18x^2 + 81}{(x - 3)^2}$

**16.** $\lim\limits_{u \to 1} \dfrac{(3u + 4)(2u - 2)^3}{(u - 1)^2}$

C   In Problems 17–26, use a calculator (or Table II) to find the indicated limit.

**17.** $\lim\limits_{x \to 0} \dfrac{\sin x}{2x}$

**18.** $\lim\limits_{t \to 0} \dfrac{1 - \cos t}{2t}$

**19.** $\lim\limits_{x \to 0} \dfrac{(x - \sin x)^2}{x^2}$

**20.** $\lim\limits_{x \to 0} \dfrac{(1 - \cos x)^2}{x^2}$

**21.** $\lim\limits_{t \to 1} \dfrac{t^2 - 1}{\sin(t - 1)}$

**22.** $\lim\limits_{x \to 3} \dfrac{x - \sin(x - 3) - 3}{x - 3}$

**23.** $\lim\limits_{x \to \pi} \dfrac{1 + \sin(x - 3\pi/2)}{x - \pi}$

**24.** $\lim\limits_{t \to 0} \dfrac{1 - \cot t}{\dfrac{1}{t}}$

**25.** $\lim\limits_{x \to \frac{\pi}{4}} \dfrac{(x - \pi/4)^2}{(\tan x - 1)^2}$

**26.** $\lim\limits_{u \to \frac{\pi}{2}} \dfrac{2 - 2\sin u}{3u}$

**27.** For the function $f$ graphed in Figure 8, find the indicated limit or function value, or state that it does not exist.

(a) $\lim\limits_{x \to -3} f(x)$

(b) $f(-3)$

(c) $f(-1)$

(d) $\lim\limits_{x \to -1} f(x)$

(e) $f(1)$

(f) $\lim\limits_{x \to 1} f(x)$

(g) $\lim\limits_{x \to 1^-} f(x)$

(h) $\lim\limits_{x \to 1^+} f(x)$

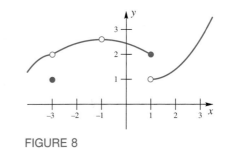

FIGURE 8

**28.** Follow the directions of Problem 27 for the function $f$ graphed in Figure 9.

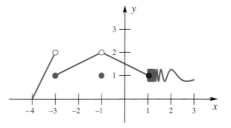

FIGURE 9

**29.** Sketch the graph of

$$f(x) = \begin{cases} -x & \text{if } x \le 0 \\ x & \text{if } 0 \le x < 1 \\ 1 + x & \text{if } x \ge 1 \end{cases}$$

Then find each of the following or state that it does not exist.

(a) $\lim\limits_{x \to 0} f(x)$

(b) $\lim\limits_{x \to 1} f(x)$

(c) $f(1)$

(d) $\lim\limits_{x \to 1^+} f(x)$

**30.** Sketch the graph of

$$g(x) = \begin{cases} -x + 1 & \text{if } x < 1 \\ x - 1 & \text{if } 1 < x < 2 \\ 5 - x^2 & \text{if } x \ge 2 \end{cases}$$

Then find each of the following or state that it does not exist.

(a) $\lim\limits_{x \to 1} g(x)$

(b) $g(1)$

(c) $\lim\limits_{x \to 2} g(x)$

(d) $\lim\limits_{x \to 2^+} g(x)$

**31.** Sketch the graph of $f(x) = x - [\![x]\!]$; then find each of the following or state that it does not exist.

(a) $f(0)$

(b) $\lim\limits_{x \to 0} f(x)$

(c) $\lim\limits_{x \to 0^-} f(x)$

(d) $\lim\limits_{x \to 1/2} f(x)$

**32.** Follow the directions of Problem 31 for $f(x) = x/|x|$.

**33.** Find $\lim\limits_{x \to 1} (x^2 - 1)/|x - 1|$ or state that it does not exist.

**34.** Evaluate $\lim\limits_{x \to 0} (\sqrt{x + 2} - \sqrt{2})/x$. *Hint:* Rationalize the numerator.

**35.** Let

$$f(x) = \begin{cases} x & \text{if } x \text{ is rational} \\ -x & \text{if } x \text{ is irrational} \end{cases}$$

Find each value, if possible.

(a) $\lim\limits_{x \to 1} f(x)$

(b) $\lim\limits_{x \to 0} f(x)$

**36.** Sketch, as best you can, the graph of a function $f$ that satisfies all the following conditions.

(a) Its domain is the interval $[0, 4]$.
(b) $f(0) = f(1) = f(2) = f(3) = f(4) = 1$
(c) $\lim\limits_{x \to 1} f(x) = 2$
(d) $\lim\limits_{x \to 2} f(x) = 1$
(e) $\lim\limits_{x \to 3^-} f(x) = 2$
(f) $\lim\limits_{x \to 3^+} f(x) = 1$

**37.** Let

$$f(x) = \begin{cases} x^2 & \text{if } x \text{ is rational} \\ x^4 & \text{if } x \text{ is irrational} \end{cases}$$

For what values of $a$ does $\lim\limits_{x \to a} f(x)$ exist?

**38.** The function $f(x) = x^2$ had been carefully graphed, but during the night a mysterious visitor changed the values

of $f$ at a million different places. Did this affect the value of $\lim\limits_{x \to a} f(x)$ at any $a$? Explain.

**39.** Find each of the following limits or state that it does not exist.

(a) $\lim\limits_{x \to 1} \dfrac{|x - 1|}{x - 1}$

(b) $\lim\limits_{x \to 1^-} \dfrac{|x - 1|}{x - 1}$

(c) $\lim\limits_{x \to 1^-} \dfrac{x^2 - |x - 1| - 1}{|x - 1|}$

(d) $\lim\limits_{x \to 1^-} \left[ \dfrac{1}{x - 1} - \dfrac{1}{|x - 1|} \right]$

**40.** Find each of the following limits or state that it does not exist.

(a) $\lim\limits_{x \to 1^+} \sqrt{x - [\![x]\!]}$

(b) $\lim\limits_{x \to 0^+} [\![1/x]\!]$

(c) $\lim\limits_{x \to 0^+} x(-1)^{[\![1/x]\!]}$

(d) $\lim\limits_{x \to 0^+} [\![x]\!](-1)^{[\![1/x]\!]}$

(e) $\lim\limits_{x \to 0^+} x[\![1/x]\!]$

(f) $\lim\limits_{x \to 0^+} x^2[\![1/x]\!]$

$\boxed{\text{PC}}$ Many calculus software packages (including *True BASIC Calculus*) have programs for calculating limits, although you should be warned that they are not infallible. To develop confidence in your program, use it to recalculate some of the limits in Problems 1–26. Then for each of the following, find the limit or state that it does not exist.

**41.** $\lim\limits_{x \to 0} \sqrt{x}$

**42.** $\lim\limits_{x \to 0^+} x^x$

**43.** $\lim\limits_{x \to 0} \sqrt{|x|}$

**44.** $\lim\limits_{x \to 0} |x|^x$

**45.** $\lim\limits_{x \to 0} (\sin 2x)/4x$

**46.** $\lim\limits_{x \to 0} (\sin 5x)/3x$

**47.** $\lim\limits_{x \to 0} \cos(1/x)$

**48.** $\lim\limits_{x \to 0} x \cos(1/x)$

**49.** $\lim\limits_{x \to 1} \dfrac{x^3 - 1}{\sqrt{2x + 2} - 2}$

**50.** $\lim\limits_{x \to 0} \dfrac{x \sin 2x}{\sin(x^2)}$

**51.** $\lim\limits_{x \to 2^-} \dfrac{x^2 - x - 2}{|x - 2|}$

**52.** $\lim\limits_{x \to 1^+} \dfrac{2}{1 + 2^{1/(x - 1)}}$

$\boxed{\text{PC}}$ **53.** Since calculus software packages find $\lim\limits_{x \to a} f(x)$ by sampling a few values of $f(x)$ for $x$ near $a$, they can be fooled. Find a function $f$ for which $\lim\limits_{x \to 0} f(x)$ fails to exist but for which your software gives a value for the limit.

---

**Answers to Concepts Review:**   **1.** $L$; $c$   **2.** 6   **3.** $L$; right   **4.** $\lim\limits_{x \to c} f(x) = M$

**2.5**
**RIGOROUS STUDY**
**OF LIMITS**

You should not believe everything you are told. It is prudent to be skeptical—not so skeptical that you will not believe anything, but skeptical enough to check a statement before you accept it. Mathematicians tend to be very skeptical people. Tell a mathematician that something is true and you will probably get the response: Prove it. But to be able to prove something requires that we be very clear about the meaning of the words we are using. This is especially true of the word *limit*, because all of calculus rests on the meaning of that word.

We gave an informal definition of limit in the previous section. Here is a slightly better, but still informal, rewording of that definition. *To say that* $\lim_{x \to c} f(x) = L$ *means that the difference between* $f(x)$ *and* $L$ *can be made arbitrarily small be requiring that* $x$ *be sufficiently close to but different from* $c$. Now let us try to pin this down.

**Making the Definition Precise**    First, we follow a long tradition in using the Greek letters $\varepsilon$ (epsilon) and $\delta$ (delta) to stand for arbitrary positive numbers. Think of $\varepsilon$ and $\delta$ as small positive numbers.

To say that $f(x)$ differs from $L$ by less than $\varepsilon$ is to say that

$$|f(x) - L| < \varepsilon$$

or, equivalently,

$$L - \varepsilon < f(x) < L + \varepsilon$$

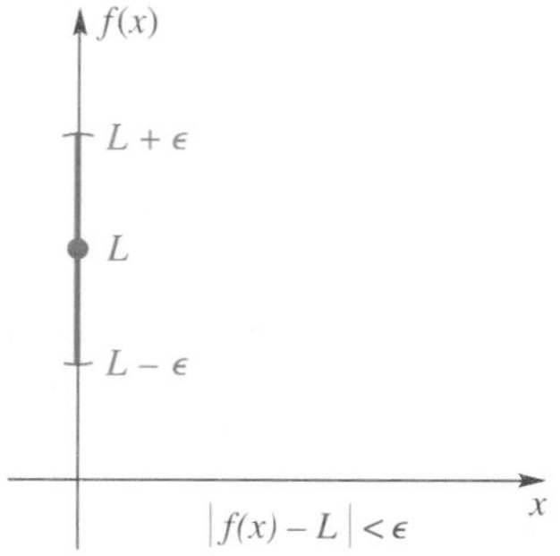

This means that $f(x)$ lies in the open interval $(L - \varepsilon, L + \varepsilon)$ shown on the graph in Figure 1.

Next, to say that $x$ is sufficiently close to but different from $c$ is to say that for some $\delta$, $x$ is in the open interval $(c - \delta, c + \delta)$ with $c$ deleted. Perhaps the best way to say this is to write

$$0 < |x - c| < \delta$$

Note that $|x - c| < \delta$ would describe the interval $c - \delta < x < c + \delta$, while $0 < |x - c|$ requires that $x = c$ be excluded. The deleted interval we are describing is shown in Figure 2.

We are ready for what some people have called the most important definition in calculus.

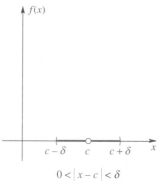

FIGURE 1

FIGURE 2

---

**Definition**

**(Precise meaning of limit).** To say that $\lim_{x \to c} f(x) = L$ means that for each given $\varepsilon > 0$ (no matter how small), there is a corresponding $\delta > 0$ such that $|f(x) - L| < \varepsilon$ provided that $0 < |x - c| < \delta$; that is,

$$0 < |x - c| < \delta \quad \Rightarrow \quad |f(x) - L| < \varepsilon$$

---

The pictures in Figure 3 may help you absorb this definition.

It is to be emphasized that the number $\varepsilon$ is given *first*; the number $\delta$ is to be produced. Suppose Doris wishes to prove to Edward that $\lim_{x \to c} f(x) = L$.

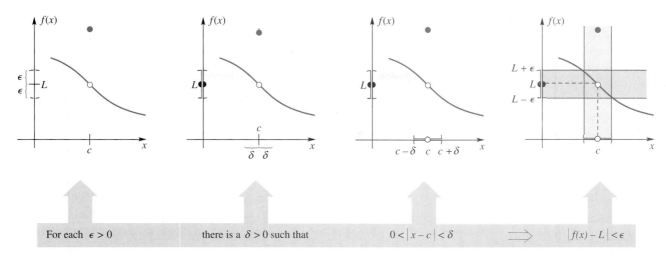

FIGURE 3

Edward can challenge Doris with any particular ε he chooses (for example, ε = 0.00001) and demand that Doris produce a corresponding δ.

Moreover, since the definition says "for each ε > 0" (not "for some ε > 0"), Doris will not have proved the limit statement if she produces a δ for one particular ε, nor even for several of them. She must be able to produce for each ε a corresponding δ. Of course, the δ she produces may depend on ε. In general, the smaller the ε Edward gives her, the smaller the δ Doris will need to return.

Is there any possible way that Doris can meet such a challenge? There is, but do not expect it to be easy.

---

**Two Different Limits?**

A natural question to ask is, Can a function have two different limits at *c*? The obvious intuitive answer is no. If a function is getting closer and closer to *L* as $x \to c$, it cannot also be getting closer and closer to a different number *M*. You are asked to show this rigorously in Problem 19.

---

**Some Limit Proofs (Optional)**  In each of the examples below, we begin with what we call a preliminary analysis. This is not part of the proof. It is the kind of work you should do on scratch paper. We include it so our proofs will not appear to float down from heaven.

**EXAMPLE 1**  Prove that $\lim_{x \to 4}(3x - 7) = 5$.

PRELIMINARY ANALYSIS  Let ε be any positive number. We must produce a δ > 0 such that

$$0 < |x - 4| < \delta \;\Rightarrow\; |(3x - 7) - 5| < \varepsilon$$

Consider the inequality on the right.

$$
\begin{aligned}
|(3x - 7) - 5| < \varepsilon \;&\Leftrightarrow\; |3x - 12| < \varepsilon \\
&\Leftrightarrow\; |3(x - 4)| < \varepsilon \\
&\Leftrightarrow\; |3||x - 4| < \varepsilon \\
&\Leftrightarrow\; |x - 4| < \frac{\varepsilon}{3}
\end{aligned}
$$

Now we see how to choose $\delta$, namely, $\delta = \varepsilon/3$. Of course, any smaller $\delta$ would work.

**FORMAL PROOF** Let $\varepsilon > 0$ be given. Choose $\delta = \varepsilon/3$. Then $0 < |x - 4| < \delta$ implies

$$|(3x - 7) - 5| = |3x - 12| = |3(x - 4)| = 3|x - 4| < 3\delta = \varepsilon$$

If you read this chain of equalities and an inequality from left to right and use the transitive properties of $=$ and $<$, you see that

$$|(3x - 7) - 5| < \varepsilon$$

If Edward were to challenge Doris with $\varepsilon = 0.01$ in this example, Doris would respond with $\delta = 0.01/3 \approx 0.0033$. Notice that $\delta = 0.0033 < \dfrac{0.01}{3}$. If Edward said $\varepsilon = 0.000003$, Doris would say $\delta = 0.000001$. If she gave an even smaller $\delta$, that would be fine.

Of course, if you think about the graph of $y = 3x - 7$ (a line with slope 3, as in Figure 4), you know that to force $3x - 7$ to be close to 5, you had better make $x$ even closer (closer by a factor of one-third) to 4. ∎

Look at Figure 5. Then convince yourself that $\delta = 2\varepsilon$ would be an appropriate choice for $\delta$ in showing that $\lim\limits_{x \to 4}(\tfrac{1}{2}x + 3) = 5$.

**EXAMPLE 2** Prove that $\lim\limits_{x \to 2} \dfrac{2x^2 - 3x - 2}{x - 2} = 5$.

**PRELIMINARY ANALYSIS** We are looking for $\delta$ such that

$$0 < |x - 2| < \delta \quad \Rightarrow \quad \left| \frac{2x^2 - 3x - 2}{x - 2} - 5 \right| < \varepsilon$$

Now for $x \neq 2$,

$$\left| \frac{2x^2 - 3x - 2}{x - 2} - 5 \right| < \varepsilon \quad \Leftrightarrow \quad \left| \frac{(2x + 1)(x - 2)}{x - 2} - 5 \right| < \varepsilon$$

$$\Leftrightarrow \quad |(2x + 1) - 5| < \varepsilon$$

$$\Leftrightarrow \quad |2(x - 2)| < \varepsilon$$

$$\Leftrightarrow \quad |2|\,|x - 2| < \varepsilon$$

$$\Leftrightarrow \quad |x - 2| < \frac{\varepsilon}{2}$$

This indicates that $\delta = \varepsilon/2$ will work (see Figure 6).

**FORMAL PROOF** Let $\varepsilon > 0$ be given. Choose $\delta = \varepsilon/2$. Then $0 < |x - 2| < \delta$ implies

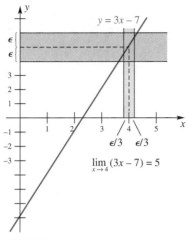

$y = 3x - 7$

$\epsilon/3 \quad \epsilon/3$

$\lim\limits_{x \to 4} (3x - 7) = 5$

**FIGURE 4**

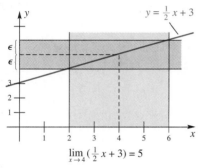

$y = \frac{1}{2}x + 3$

$\lim\limits_{x \to 4} (\frac{1}{2}x + 3) = 5$

**FIGURE 5**

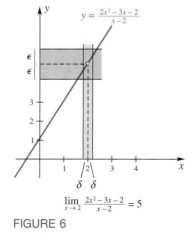

$y = \dfrac{2x^2 - 3x - 2}{x - 2}$

$\delta \quad \delta$

$\lim\limits_{x \to 2} \dfrac{2x^2 - 3x - 2}{x - 2} = 5$

**FIGURE 6**

$$\left|\frac{2x^2 - 3x - 2}{x - 2} - 5\right| = \left|\frac{(2x + 1)(x - 2)}{x - 2} - 5\right| = |2x + 1 - 5|$$
$$= |2(x - 2)| = 2|x - 2| < 2\delta = \varepsilon$$

The cancellation of the factor $x - 2$ is legitimate because $0 < |x - 2|$ implies $x \neq 2$; thus, division by 0 is avoided. ∎

**EXAMPLE 3**  Prove that $\lim_{x \to c}(mx + b) = mc + b$.

PRELIMINARY ANALYSIS   We want to find $\delta$ such that

$$0 < |x - c| < \delta \;\Rightarrow\; |(mx + b) - (mc + b)| < \varepsilon$$

Now

$$|(mx + b) - (mc + b)| = |mx - mc| = |m(x - c)| = |m|\,|x - c|$$

It appears that $\delta = \varepsilon/|m|$ should do.

FORMAL PROOF   Let $\varepsilon > 0$ be given. Choose $\delta = \varepsilon/|m|$. Then $0 < |x - c| < \delta$ implies

$$|(mx + b) - (mc + b)| = |mx - mc| = |m|\,|x - c| < |m|\delta = \varepsilon$$

There is only one problem with our choice of $\delta$. What if $m = 0$? In that case, any $\delta$ will do just fine since

$$|(0x + b) - (0c + b)| = |0| = 0$$

The latter is less than $\varepsilon$ for all $x$. ∎

**EXAMPLE 4**  Prove that if $c > 0$, $\lim_{x \to c}\sqrt{x} = \sqrt{c}$.

PRELIMINARY ANALYSIS   Refer to Figure 7. We must find $\delta$ such that

$$0 < |x - c| < \delta \;\Rightarrow\; |\sqrt{x} - \sqrt{c}| < \varepsilon$$

Now

$$|\sqrt{x} - \sqrt{c}| = \left|\frac{(\sqrt{x} - \sqrt{c})(\sqrt{x} + \sqrt{c})}{\sqrt{x} + \sqrt{c}}\right| = \left|\frac{x - c}{\sqrt{x} + \sqrt{c}}\right|$$
$$= \frac{|x - c|}{\sqrt{x} + \sqrt{c}} \leq \frac{|x - c|}{\sqrt{c}}$$

To make the latter less than $\varepsilon$ requires that we make $|x - c| < \varepsilon\sqrt{c}$.

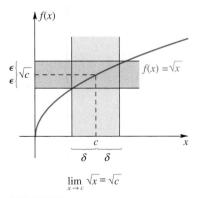

$$\lim_{x \to c}\sqrt{x} = \sqrt{c}$$

FIGURE 7

FORMAL PROOF   Let $\varepsilon > 0$ be given. Choose $\delta = \varepsilon\sqrt{c}$. Then $0 < |x - c| < \delta$ implies

$$|\sqrt{x} - \sqrt{c}| = \left|\frac{(\sqrt{x} - \sqrt{c})(\sqrt{x} + \sqrt{c})}{\sqrt{x} + \sqrt{c}}\right| = \left|\frac{x - c}{\sqrt{x} + \sqrt{c}}\right|$$

$$= \frac{|x - c|}{\sqrt{x} + \sqrt{c}} \leq \frac{|x - c|}{\sqrt{c}} < \frac{\delta}{\sqrt{c}} = \varepsilon$$

There is one further technical point. We should insist that $\delta \leq c$, for then $|x - c| < \delta$ implies $x > 0$ so that $\sqrt{x}$ is defined. Thus, for absolute rigor, choose $\delta$ to be the smaller of $c$ and $\varepsilon\sqrt{c}$. ∎

Our demonstration in Example 4 depended on rationalizing the numerator, a trick frequently useful in calculus.

**EXAMPLE 5**   Prove that $\lim_{x \to 3}(x^2 + x - 5) = 7$.

PRELIMINARY ANALYSIS   Our task is to find $\delta$ such that

$$0 < |x - 3| < \delta \quad \Rightarrow \quad |(x^2 + x - 5) - 7| < \varepsilon$$

Now

$$|(x^2 + x - 5) - 7| = |x^2 + x - 12| = |x + 4|\,|x - 3|$$

Since the second factor $|x - 3|$ can be made as small as we wish, it is enough to bound the factor $|x + 4|$. To do this, we first agree to make $\delta \leq 1$. Then $|x - 3| < \delta$ implies

$$|x + 4| = |x - 3 + 7|$$
$$\leq |x - 3| + |7| \qquad \text{(Triangle Inequality)}$$
$$< 1 + 7 = 8$$

$$\boxed{\begin{array}{l} |x - 3| < 1 \Rightarrow 2 < x < 4 \\ \qquad\qquad \Rightarrow 6 < x + 4 < 8 \\ \qquad\qquad \Rightarrow |x + 4| < 8 \end{array}}$$

**FIGURE 8**

Figure 8 offers an alternative demonstration of this fact. If we also require $\delta \leq \varepsilon/8$, the product $|x + 4|\,|x - 3|$ will be less than $\varepsilon$.

FORMAL PROOF   Let $\varepsilon > 0$ be given. Choose $\delta = \min\{1, \varepsilon/8\}$; that is, choose $\delta$ to be the smaller of 1 and $\varepsilon/8$. Then $0 < |x - 3| < \delta$ implies

$$|(x^2 + x - 5) - 7| = |x^2 + x - 12| = |x + 4|\,|x - 3| < 8 \cdot \frac{\varepsilon}{8} = \varepsilon \quad ∎$$

**EXAMPLE 6**   Prove that $\lim_{x \to c} x^2 = c^2$.

PROOF   We mimic the proof in Example 5. Let $\varepsilon > 0$ be given. Choose $\delta = \min\{1, \varepsilon/(1 + 2|c|)\}$. Then $0 < |x - c| < \delta$ implies

$$|x^2 - c^2| = |x + c|\,|x - c| = |x - c + 2c|\,|x - c|$$
$$\leq (|x - c| + 2|c|)|x - c| \qquad \text{(Triangle Inequality)}$$
$$< \frac{(1 + 2|c|) \cdot \varepsilon}{1 + 2|c|} = \varepsilon \qquad\qquad\blacksquare$$

**EXAMPLE 7**  Prove that $\displaystyle\lim_{x \to c} \frac{1}{x} = \frac{1}{c},\ c \neq 0$.

**PRELIMINARY ANALYSIS**  Study Figure 9. We must find $\delta$ such that

$$0 < |x - c| < \delta \ \Rightarrow\ \left|\frac{1}{x} - \frac{1}{c}\right| < \varepsilon$$

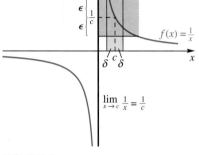

$$\lim_{x \to c} \frac{1}{x} = \frac{1}{c}$$

**FIGURE 9**

Now

$$\left|\frac{1}{x} - \frac{1}{c}\right| = \left|\frac{c - x}{xc}\right| = \frac{1}{|x|} \cdot \frac{1}{|c|} \cdot |x - c|$$

The factor $1/|x|$ is troublesome, especially if $x$ is near 0. We can bound this factor if we can keep $x$ away from 0. To that end, note that

$$|c| = |c - x + x| \leq |c - x| + |x|$$

so

$$|x| \geq |c| - |x - c|$$

Thus, if we choose $\delta \leq |c|/2$, we succeed in making $|x| \geq |c|/2$. Finally, if we also require $\delta \leq \varepsilon c^2/2$, then

$$\frac{1}{|x|} \cdot \frac{1}{|c|} \cdot |x - c| < \frac{1}{|c|/2} \cdot \frac{1}{|c|} \cdot \frac{\varepsilon c^2}{2} = \varepsilon$$

**FORMAL PROOF**  Let $\varepsilon > 0$ be given. Choose $\delta = \min\{|c|/2,\ \varepsilon c^2/2\}$. Then $0 < |x - c| < \delta$ implies

$$\left|\frac{1}{x} - \frac{1}{c}\right| = \left|\frac{c - x}{xc}\right| = \frac{1}{|x|} \cdot \frac{1}{|c|} \cdot |x - c| < \frac{1}{|c|/2} \cdot \frac{1}{|c|} \cdot \frac{\varepsilon c^2}{2} = \varepsilon \qquad\blacksquare$$

**One-Sided Limits**  It does not take much imagination to give the $\varepsilon$, $\delta$ definitions of right- and left-hand limits.

---

**Definition**

To say $\displaystyle\lim_{x \to c^+} f(x) = L$ means that for each $\varepsilon > 0$, there is a corresponding $\delta > 0$ such that

$$0 < x - c < \delta \ \Rightarrow\ |f(x) - L| < \varepsilon$$

We leave the corresponding $\varepsilon$, $\delta$ definition for the left-hand limit to the reader.

## CONCEPTS REVIEW

1. The inequality $|f(x) - L| < \varepsilon$ is equivalent to _____ $< f(x) <$ _____.

2. The precise meaning of $\lim\limits_{x \to a} f(x) = L$ is this: Given any positive number $\varepsilon$, there is a corresponding positive number $\delta$ such that _____ implies _____.

3. To be sure that $|3x - 3| < \varepsilon$, we would require that $|x - 1| <$ _____.

4. $\lim\limits_{x \to a} (mx + b) =$ _____.

## PROBLEM SET 2.5

In Problems 1–6, give the appropriate $\varepsilon$, $\delta$ definition of each statement.

1. $\lim\limits_{t \to a} f(t) = M$

2. $\lim\limits_{u \to b} g(u) = L$

3. $\lim\limits_{z \to d} h(z) = P$

4. $\lim\limits_{y \to e} \phi(y) = B$

5. $\lim\limits_{x \to c^-} f(x) = L$

6. $\lim\limits_{t \to a^+} g(t) = D$

In Problems 7–18, give an $\varepsilon$, $\delta$ proof of each limit fact (see Examples 1–5).

7. $\lim\limits_{x \to 0} (2x - 1) = -1$

8. $\lim\limits_{x \to -21} (3x - 1) = -64$

9. $\lim\limits_{x \to 5} \dfrac{x^2 - 25}{x - 5} = 10$

10. $\lim\limits_{x \to 0} \left( \dfrac{2x^2 - x}{x} \right) = -1$

11. $\lim\limits_{x \to 5} \dfrac{2x^2 - 11x + 5}{x - 5} = 9$

12. $\lim\limits_{x \to 1} \sqrt{2x} = \sqrt{2}$

13. $\lim\limits_{x \to 4} \dfrac{\sqrt{2x - 1}}{\sqrt{x - 3}} = \sqrt{7}$

14. $\lim\limits_{x \to 1} \dfrac{14x^2 - 20x + 6}{x - 1} = 8$

15. $\lim\limits_{x \to 1} \dfrac{10x^3 - 26x^2 + 22x - 6}{(x - 1)^2} = 4$

16. $\lim\limits_{x \to 1} (2x^2 + 1) = 3$

17. $\lim\limits_{x \to -1} (x^2 - 2x - 1) = 2$

18. $\lim\limits_{x \to 0} x^4 = 0$

19. Prove that if $\lim\limits_{x \to c} f(x) = L$ and $\lim\limits_{x \to c} f(x) = M$, then $L = M$.

20. Let $F$ and $G$ be functions such that $0 \leq F(x) \leq G(x)$ for all $x$ near $c$, except possibly at $c$. Prove that if $\lim\limits_{x \to c} G(x) = 0$, then $\lim\limits_{x \to c} F(x) = 0$.

21. Prove that $\lim\limits_{x \to 0} x^4 \sin^2(1/x) = 0$. *Hint*: Use Problems 18 and 20.

22. Prove that $\lim\limits_{x \to 0^+} \sqrt{x} = 0$.

23. By considering left- and right-hand limits, prove that $\lim\limits_{x \to 0} |x| = 0$.

24. Prove that if $|f(x)| < B$ for $|x - a| < 1$ and $\lim\limits_{x \to a} g(x) = 0$, then $\lim\limits_{x \to a} f(x)g(x) = 0$.

25. Suppose that $\lim\limits_{x \to a} f(x) = L$ and that $f(a)$ exists (though it may be different from $L$). Prove that $f$ is bounded on some interval containing $a$; that is, show that there is an interval $(c, d)$ with $c < a < d$ and a constant $M$ such that $|f(x)| \leq M$ for all $x$ in $(c, d)$.

26. Prove that if $f(x) \leq g(x)$ for all $x$ in some deleted interval about $a$ and if $\lim\limits_{x \to a} f(x) = L$ and $\lim\limits_{x \to a} g(x) = M$, then $L \leq M$.

27. Which of the following are equivalent to the definition of limit?

(a) For some $\varepsilon > 0$ and every $\delta > 0$, $0 < |x - c| < \delta \Rightarrow |f(x) - L| < \varepsilon$.

(b) For every $\delta > 0$, there is a corresponding $\varepsilon > 0$ such that

$$0 < |x - c| < \varepsilon \Rightarrow |f(x) - L| < \delta$$

(c) For every positive integer $N$, there is a corresponding positive integer $M$ such that $0 < |x - c| < 1/M \Rightarrow |f(x) - L| < 1/N$.

(d) For every $\varepsilon > 0$, there is a corresponding $\delta > 0$ such that $0 < |x - c| < \delta$ and $|f(x) - L| < \varepsilon$ for some $x$.

**28.** State in ε, δ language what it means to say $\lim_{x \to c} f(x) \neq L$.

**PC** **29.** Suppose we wish to give an ε, δ proof that $\lim_{x \to 3} \dfrac{x + 6}{x^4 - 4x^3 + x^2 + x + 6} = -1$. We begin by writing $\dfrac{x + 6}{x^4 - 4x^3 + x^2 + x + 6} + 1$ in the form $(x - 3)(g(x))$.

(a) Determine $g(x)$.
(b) Could we choose $\delta = \min(1, \varepsilon/n)$ for some $n$? Explain.
(c) If we choose $\delta = \min(\frac{1}{4}, \varepsilon/m)$, what is the smallest integer $m$ we could use?

---

**Answers to Concepts Review:** **1.** $L - \varepsilon$; $L + \varepsilon$
**2.** $0 < |x - a| < \delta$; $|f(x) - L| < \varepsilon$ **3.** $\varepsilon/3$ **4.** $ma + b$

---

## 2.6
## LIMIT THEOREMS

Most readers will agree that proving the existence of limits using the ε, δ definition of the preceding section is both time-consuming and difficult. That is why the theorems of this section are so welcome. Our first theorem is the big one. With it, we can handle most limit problems that we will face for quite some time.

### Theorem A

**(Main Limit Theorem).** Let $n$ be a positive integer, $k$ be a constant, and $f$ and $g$ be functions which have limits at $c$. Then

1. $\lim_{x \to c} k = k$;
2. $\lim_{x \to c} x = c$;
3. $\lim_{x \to c} kf(x) = k \lim_{x \to c} f(x)$;
4. $\lim_{x \to c}[f(x) + g(x)] = \lim_{x \to c} f(x) + \lim_{x \to c} g(x)$;
5. $\lim_{x \to c}[f(x) - g(x)] = \lim_{x \to c} f(x) - \lim_{x \to c} g(x)$;
6. $\lim_{x \to c}[f(x) \cdot g(x)] = \lim_{x \to c} f(x) \cdot \lim_{x \to c} g(x)$;
7. $\lim_{x \to c} \dfrac{f(x)}{g(x)} = \dfrac{\lim_{x \to c} f(x)}{\lim_{x \to c} g(x)}$, provided $\lim_{x \to c} g(x) \neq 0$;
8. $\lim_{x \to c}[f(x)]^n = \left[\lim_{x \to c} f(x)\right]^n$;
9. $\lim_{x \to c} \sqrt[n]{f(x)} = \sqrt[n]{\lim_{x \to c} f(x)}$, provided $\lim_{x \to c} f(x) > 0$ when $n$ is even.

These important results are remembered best if learned in words. For example, Statement 4 translates as: *The limit of a sum is the sum of the limits.*

Of course, Theorem A needs to be proved. We postpone that job till the end of the section, choosing first to show how this multipart theorem is used.

**Applications of the Main Limit Theorem** In the next examples, the circled numbers refer to the numbered statements from the list given earlier. Each equality is justified by the indicated statement.

**EXAMPLE 1**   Find $\lim\limits_{x \to 3} 2x^4$.

*Solution*

$$\lim_{x \to 3} 2x^4 \overset{\textcircled{3}}{=} 2 \lim_{x \to 3} x^4 \overset{\textcircled{8}}{=} 2 \left[ \lim_{x \to 3} x \right]^4 \overset{\textcircled{2}}{=} 2[3]^4 = 162$$

**EXAMPLE 2**   Find $\lim\limits_{x \to 4}(3x^2 - 2x)$.

*Solution*

$$\lim_{x \to 4}(3x^2 - 2x) \overset{\textcircled{5}}{=} \lim_{x \to 4} 3x^2 - \lim_{x \to 4} 2x \overset{\textcircled{3}}{=} 3 \lim_{x \to 4} x^2 - 2 \lim_{x \to 4} x$$

$$\overset{\textcircled{8}}{=} 3 \left( \lim_{x \to 4} x \right)^2 - 2 \lim_{x \to 4} x \overset{\textcircled{2}}{=} 3(4)^2 - 2(4)$$

$$= 40$$

**EXAMPLE 3**   Find $\lim\limits_{x \to 4} \dfrac{\sqrt{x^2 + 9}}{x}$.

*Solution*

$$\lim_{x \to 4} \frac{\sqrt{x^2 + 9}}{x} \overset{\textcircled{7}}{=} \frac{\lim\limits_{x \to 4} \sqrt{x^2 + 9}}{\lim\limits_{x \to 4} x} \overset{\textcircled{9}}{\underset{\textcircled{2}}{=}} \frac{\sqrt{\lim\limits_{x \to 4}(x^2 + 9)}}{4} \overset{\textcircled{4}}{=} \frac{1}{4} \sqrt{\lim_{x \to 4} x^2 + \lim_{x \to 4} 9}$$

$$\overset{\textcircled{8,1}}{=} \frac{1}{4} \sqrt{\left[ \lim_{x \to 4} x \right]^2 + 9} \overset{\textcircled{2}}{=} \frac{1}{4} \sqrt{4^2 + 9} = \frac{5}{4}$$

**EXAMPLE 4**   If $\lim\limits_{x \to 3} f(x) = 4$ and $\lim\limits_{x \to 3} g(x) = 8$, find

$$\lim_{x \to 3}[f^{2}(x) \cdot \sqrt[3]{g(x)}]$$

*Solution*

$$\lim_{x \to 3}[f^{2}(x) \cdot \sqrt[3]{g(x)}] \overset{\textcircled{6}}{=} \lim_{x \to 3} f^{2}(x) \cdot \lim_{x \to 3} \sqrt[3]{g(x)}$$

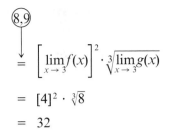

$$= \left[\lim_{x \to 3} f(x)\right]^2 \cdot \sqrt[3]{\lim_{x \to 3} g(x)}$$

$$= [4]^2 \cdot \sqrt[3]{8}$$

$$= 32 \qquad \blacksquare$$

Recall that a polynomial function $f$ has the form

$$f(x) = a_n x^n + a_{n-1} x^{n-1} + \cdots + a_1 x + a_0$$

whereas a rational function $f$ is the quotient of two polynomial functions, that is,

$$f(x) = \frac{a_n x^n + a_{n-1} x^{n-1} + \cdots + a_1 x + a_0}{b_m x^m + b_{m-1} x^{m-1} + \cdots + b_1 x + b_0}$$

### Theorem B

**(Substitution Theorem).** If $f$ is a polynomial function or a rational function, then

$$\lim_{x \to c} f(x) = f(c)$$

provided in the case of a rational function that the value of the denominator at $c$ is not zero.

The proof of Theorem B follows from repeated applications of Theorem A. Note that Theorem B allows us to find limits for polynomial and rational functions by simply substituting $c$ for $x$ throughout.

**EXAMPLE 5**   Find $\lim\limits_{x \to 2} \dfrac{7x^5 - 10x^4 - 13x + 6}{3x^2 - 6x - 8}$.

*Solution*

$$\lim_{x \to 2} \frac{7x^5 - 10x^4 - 13x + 6}{3x^2 - 6x - 8} = \frac{7(2)^5 - 10(2)^4 - 13(2) + 6}{3(2)^2 - 6(2) - 8} = -\frac{11}{2} \qquad \blacksquare$$

**EXAMPLE 6**   Find $\lim\limits_{x \to 1} \dfrac{x^3 + 3x + 7}{x^2 - 2x + 1} = \lim\limits_{x \to 1} \dfrac{x^3 + 3x + 7}{(x - 1)^2}$.

*Solution*   Neither Theorem B nor Statement 7 of Theorem A applies, since the limit of the denominator is 0. However, since the limit of the numerator is 11, we see that as $x$ nears 1, we are dividing a number near 11 by a positive number near 0. The result is a large positive number. In fact, the resulting number can be made as large as you like by letting $x$ get close enough to 1. We say that the limit does not exist. (Later in the book—see Section 4.6—we will allow ourselves to say the limit is $+\infty$.) ∎

**EXAMPLE 7** Find $\lim_{t \to 2} \dfrac{t^2 + 3t - 10}{t^2 + t - 6}$.

**Solution** Again, Theorem B does not apply. But this time, the quotient takes the meaningless form 0/0 at $t = 2$. Whenever this happens one should look for an algebraic simplification (factorization) of the quotient before trying to take the limit.

$$\lim_{t \to 2} \frac{t^2 + 3t - 10}{t^2 + t - 6} = \lim_{t \to 2} \frac{(t - 2)(t + 5)}{(t - 2)(t + 3)} = \lim_{t \to 2} \frac{t + 5}{t + 3} = \frac{7}{5}$$ ∎

**Proof of Theorem A (Optional)** You should not be too surprised when we say that the proofs of some parts of Theorem A are quite sophisticated. Because of this, we prove only the first five parts here, deferring the others to the Appendix (Section A.2, Theorem A). To get your feet wet, you might try Problems 33 and 34.

***Proof of Statements 1 and 2*** These statements result from $\lim_{x \to c}(mx + b) = mc + b$ (Example 3 of Section 2.5) using first $m = 0$ and then $m = 1$, $b = 0$. ∎

***Proof of Statement 3*** If $k = 0$, the result is trivial, so we suppose $k \neq 0$. Let $\varepsilon > 0$ be given. By hypothesis, $\lim_{x \to c} f(x)$ exists; call its value $L$. By definition of limit, there is a number $\delta$ such that

$$0 < |x - c| < \delta \;\Rightarrow\; |f(x) - L| < \frac{\varepsilon}{|k|}$$

Someone is sure to complain that we put $\varepsilon/|k|$ rather than $\varepsilon$ at the end of the inequality above. Well, isn't $\varepsilon/|k|$ a positive number? Yes. Doesn't the definition of limit require that for any positive number, there is a corresponding $\delta$? Yes.

Now with $\delta$ determined, we may assert that $0 < |x - c| < \delta$ implies

$$|kf(x) - kL| = |k|\,|f(x) - L| < |k|\frac{\varepsilon}{|k|} = \varepsilon$$

This shows that

$$\lim_{x \to c} kf(x) = kL = k \lim_{x \to c} f(x)$$ ∎

***Proof of Statement 4*** Refer to Figure 1. Let $\lim_{x \to c} f(x) = L$ and $\lim_{x \to c} g(x) = M$. If $\varepsilon$ is any given positive number, then $\varepsilon/2$ is positive. Since $\lim_{x \to c} f(x) = L$, there is a positive number $\delta_1$ such that

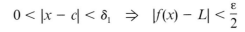

$$0 < |x - c| < \delta_1 \quad \Rightarrow \quad |f(x) - L| < \frac{\varepsilon}{2}$$

Since $\lim_{x \to c} g(x) = M$, there is a positive number $\delta_2$ such that

$$0 < |x - c| < \delta_2 \quad \Rightarrow \quad |g(x) - M| < \frac{\varepsilon}{2}$$

Choose $\delta = \min\{\delta_1, \delta_2\}$; that is, choose $\delta$ to be the smaller of $\delta_1$ and $\delta_2$. Then $0 < |x - c| < \delta$ implies

$$\begin{aligned} |f(x) + g(x) - (L + M)| &= |[f(x) - L] + [g(x) - M]| \\ &\leq |f(x) - L| + |g(x) - M| \\ &< \frac{\varepsilon}{2} + \frac{\varepsilon}{2} = \varepsilon \end{aligned}$$

In this chain, the first inequality is the Triangle Inequality (Section 1.4); the second results from the choice of $\delta$. We have just shown that

$$0 < |x - c| < \delta \quad \Rightarrow \quad |f(x) + g(x) - (L + M)| < \varepsilon$$

Thus

$$\lim_{x \to c}[f(x) + g(x)] = L + M = \lim_{x \to c} f(x) + \lim_{x \to c} g(x) \qquad \blacksquare$$

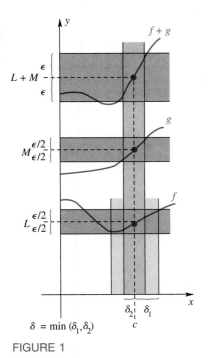

$\delta = \min(\delta_1, \delta_2)$

**FIGURE 1**

***Proof of Statement 5***

$$\begin{aligned} \lim_{x \to c}[f(x) - g(x)] &= \lim_{x \to c}[f(x) + (-1)g(x)] \\ &= \lim_{x \to c} f(x) + \lim_{x \to c}(-1)g(x) \\ &= \lim_{x \to c} f(x) + (-1)\lim_{x \to c} g(x) \\ &= \lim_{x \to c} f(x) - \lim_{x \to c} g(x) \qquad \blacksquare \end{aligned}$$

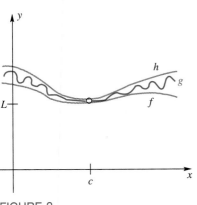

**FIGURE 2**

**The Squeeze Theorem**  Did you ever hear anyone say, "I was caught between a rock and hard place"? This is what has happened to $g$ in the following theorem (see Figure 2).

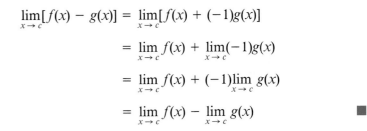

**Theorem C**

**(Squeeze Theorem).** Let $f$, $g$, and $h$ be functions satisfying $f(x) \leq g(x) \leq h(x)$ for all $x$ near $c$, except possibly at $c$. If $\lim_{x \to c} f(x) = \lim_{x \to c} h(x) = L$, then

$$\lim_{x \to c} g(x) = L.$$

***Proof*** (Optional)   Let $\varepsilon > 0$ be given. Choose $\delta_1$ such that

$$0 < |x - c| < \delta_1 \quad \Rightarrow \quad L - \varepsilon < f(x) < L + \varepsilon$$

and $\delta_2$ such that

$$0 < |x - c| < \delta_2 \quad \Rightarrow \quad L - \varepsilon < h(x) < L + \varepsilon$$

Choose $\delta_3$ so that

$$0 < |x - c| < \delta_3 \quad \Rightarrow \quad f(x) \le g(x) \le h(x)$$

Let $\delta = \min\{\delta_1, \delta_2, \delta_3\}$. Then

$$0 < |x - c| < \delta \quad \Rightarrow \quad L - \varepsilon < f(x) \le g(x) \le h(x) < L + \varepsilon$$

We conclude that $\lim_{x \to c} g(x) = L$. ∎

**EXAMPLE 8**   Assume that we have proved $1 - x^2/6 \le (\sin x)/x \le 1$ for all $x$ near but different from 0. What can we conclude from this?

***Solution***   Let $f(x) = 1 - x^2/6$, $g(x) = (\sin x)/x$, and $h(x) = 1$. It follows that $\lim_{x \to 0} f(x) = \lim_{x \to 0} h(x) = 1$, and so by Theorem C,

$$\lim_{x \to 0} \frac{\sin x}{x} = 1$$ ∎

We conclude with an important remark. *All theorems of this section are valid for right- and left-hand limits.*

## CONCEPTS REVIEW

1. If $\lim_{x \to 3} f(x) = 4$, then $\lim_{x \to 3} (x^2 + 3)f(x) = $ _____.

2. If $\lim_{x \to 2} g(x) = -2$, then $\lim_{x \to 2} \sqrt{g^2(x) + 12} = $ _____.

3. If $\lim_{x \to c} f(x) = 4$ and $\lim_{x \to c} g(x) = -2$, then $\lim_{x \to c} \dfrac{f^2(x)}{g(x)} = $ _____ and $\lim_{x \to c} [g(x)\sqrt{f(x)} + 5x] = $ _____.

4. If $\lim_{x \to c} f(x) = L$ and $\lim_{x \to c} g(x) = L$, then
$\lim_{x \to c} [f(x) - L]g(x) = $ _____, $\lim_{x \to c} \dfrac{f^2(x) + g^2(x)}{f(x) + g(x)} = $ _____, and $\lim_{x \to c} \dfrac{f^2(x) - g^2(x)}{f(x) - g(x)} = $ _____.

## PROBLEM SET 2.6

In Problems 1–12, use Theorem A to find each of the limits. Justify each step by appealing to a numbered statement, as in Examples 1–4.

1. $\lim_{x \to 1} (2x + 1)$

2. $\lim_{x \to -1} (3x^2 - 1)$

3. $\lim_{x \to 0} [(2x + 1)(x - 3)]$

4. $\lim_{x \to \sqrt{2}} [(2x^2 + 1)(7x^2 + 13)]$

5. $\lim\limits_{x \to 2} \dfrac{2x + 1}{5 - 3x}$

6. $\lim\limits_{x \to -3} \dfrac{4x^3 + 1}{7 - 2x^2}$

7. $\lim\limits_{x \to 3} \sqrt{3x - 5}$

8. $\lim\limits_{x \to -3} \sqrt{5x^2 + 2x}$

9. $\lim\limits_{t \to -2} (2t^3 + 15)^{13}$

10. $\lim\limits_{w \to -2} \sqrt{-3w^3 + 7w^2}$

11. $\lim\limits_{y \to 2} \left( \dfrac{4y^3 + 8y}{y + 4} \right)^{1/3}$

12. $\lim\limits_{w \to 5} (2w^4 - 9w^3 + 19)^{-1/2}$

In Problems 13–22, find the indicated limit or state that it does not exist. In many cases, you will want to do some algebra before trying to evaluate the limit (see Examples 5–7).

13. $\lim\limits_{x \to -1} \dfrac{x^3 - 6x^2 + 11x - 6}{x^3 + 4x^2 - 19x + 14}$

14. $\lim\limits_{x \to 4} \dfrac{x^5 + 7x^4 - 6x^3 - 104x^2 - 32x + 384}{x^5 + 7x^4 - 18x^3 - 248x^2 - 608x - 384}$

15. $\lim\limits_{x \to -1} \dfrac{x^2 + x - 2}{x^2 - 1}$

16. $\lim\limits_{x \to 2} \dfrac{x^2 + 7x + 10}{x + 2}$

17. $\lim\limits_{x \to 1} \dfrac{x^2 + x - 2}{x^2 - 1}$

18. $\lim\limits_{x \to -3} \dfrac{x^2 - 14x - 51}{x^2 - 4x - 21}$

19. $\lim\limits_{u \to -2} \dfrac{u^2 - ux + 2u - 2x}{u^2 - u - 6}$

20. $\lim\limits_{x \to 1} \dfrac{x^2 + ux - x - u}{x^2 + 2x - 3}$

21. $\lim\limits_{x \to \pi} \dfrac{2x^2 - 6x\pi + 4\pi^2}{x^2 - \pi^2}$

22. $\lim\limits_{w \to -2} \dfrac{(w + 2)(w^2 - w - 6)}{w^2 + 4w + 4}$

In Problems 23–28, find the limits if $\lim\limits_{x \to a} f(x) = 3$ and $\lim\limits_{x \to a} g(x) = -1$ (see Example 4).

23. $\lim\limits_{x \to a} \sqrt{f^2(x) + g^2(x)}$

24. $\lim\limits_{x \to a} \dfrac{2f(x) - 3g(x)}{f(x) + g(x)}$

25. $\lim\limits_{x \to a} \sqrt[3]{g(x)[f(x) + 3]}$

26. $\lim\limits_{x \to a} [f(x) - 3]^4$

27. $\lim\limits_{t \to a} [|f(t)| + |3g(t)|]$

28. $\lim\limits_{u \to a} [f(u) + 3g(u)]^3$

In Problems 29–32, find the $\lim\limits_{x \to 2} [f(x) - f(2)]/(x - 2)$ for each given function $f$.

29. $f(x) = 3x^2$

30. $f(x) = 3x^2 + 2x + 1$

31. $f(x) = \dfrac{1}{x}$

32. $f(x) = \dfrac{3}{x^2}$

33. Prove Statement 6 of Theorem A. *Hint*:

$$\begin{aligned}
|f(x)g(x) - LM| &= |f(x)g(x) - Lg(x) + Lg(x) - LM| \\
&= |g(x)[f(x) - L] + L[g(x) - M]| \\
&\leq |g(x)||f(x) - L| + |L||g(x) - M|
\end{aligned}$$

Now show that if $\lim\limits_{x \to c} g(x) = M$, there is a number $\delta_1$ such that

$$0 < |x - c| < \delta_1 \Rightarrow |g(x)| < |M| + 1$$

34. Prove Statement 7 of Theorem A by first giving an $\varepsilon$, $\delta$ proof that $\lim\limits_{x \to c} [1/g(x)] = 1 \bigg/ \left[ \lim\limits_{x \to c} g(x) \right]$ and then applying Statement 6.

35. Prove that $\lim\limits_{x \to c} f(x) = L \Leftrightarrow \lim\limits_{x \to c} [f(x) - L] = 0$.

36. Prove that $\lim\limits_{x \to c} f(x) = 0 \Leftrightarrow \lim\limits_{x \to c} |f(x)| = 0$.

37. Prove that if $\lim\limits_{x \to c} f(x) = L > 0$, then there is an interval $(c - \delta, c + \delta)$ such that $f(x) > 0$ for all $x$ in $(c - \delta, c + \delta)$, $x \neq c$.

38. Prove that $\lim\limits_{x \to c} |x| = |c|$.

39. Find examples to show that:

(a) $\lim\limits_{x \to c} [f(x) + g(x)]$ exists does not imply that either $\lim\limits_{x \to c} f(x)$ or $\lim\limits_{x \to c} g(x)$ exists;

(b) $\lim\limits_{x \to c} [f(x) \cdot g(x)]$ exists does not imply that either $\lim\limits_{x \to c} f(x)$ or $\lim\limits_{x \to c} g(x)$ exists.

40. Prove that if $\lim\limits_{x \to c} [f(x) + g(x)]$ and $\lim\limits_{x \to c} g(x)$ both exist, then $\lim\limits_{x \to c} f(x)$ must exist.

In Problems 41–48, find each of the right-hand and left-hand limits or state that they do not exist.

41. $\lim\limits_{x \to -3^+} \dfrac{\sqrt{3 + x}}{x}$

42. $\lim\limits_{x \to -\pi^+} \dfrac{\sqrt{\pi^3 + x^3}}{x}$

43. $\lim\limits_{x \to 3^+} \dfrac{x - 3}{\sqrt{x^2 - 9}}$

44. $\lim\limits_{x \to 1^-} \dfrac{\sqrt{1 + x}}{4 + 4x}$

45. $\lim\limits_{x \to 2^+} \dfrac{(x^2 + 1)[\![x]\!]}{(3x - 1)^2}$

46. $\lim\limits_{x \to 3^-} (x - [\![x]\!])$

47. $\lim\limits_{x \to 0^-} \dfrac{x}{|x|}$

48. $\lim\limits_{x \to 3^+} [\![x^2 + 2x]\!]$

49. Suppose $f(x)g(x) = 1$ for all $x$ and $\lim\limits_{x \to a} g(x) = 0$. Prove that $\lim\limits_{x \to a} f(x)$ does not exist.

**50.** Let $R$ be the rectangle joining the midpoints of the sides of the quadrilateral $Q$ having vertices $(\pm x, 0)$ and $(0, \pm 1)$. Calculate

$$\lim_{x \to 0^+} \frac{\text{perimeter of } R}{\text{perimeter of } Q}$$

**51.** Let $y = \sqrt{x}$ and consider the points $M$, $N$, $O$, and $P$ with coordinates $(1, 0)$, $(0, 1)$, $(0, 0)$, and $(x, y)$, respectively. Calculate:

(a) $\displaystyle\lim_{x \to 0^+} \frac{\text{perimeter of } \quad NOP}{\text{perimeter of } \quad MOP};$

(b) $\displaystyle\lim_{x \to 0^+} \frac{\text{area of } \quad NOP}{\text{area of } \quad MOP}.$

---

**Answers to Concepts Review:** **1.** 48 **2.** 4
**3.** $-8$; $-4 + 5c$ **4.** 0; $L$; $2L$

---

**2.7
CONTINUITY
OF FUNCTIONS**

In ordinary language, we use the word *continuous* to describe a process that goes on without abrupt changes. It is this notion as it pertains to functions that we now want to make precise. Consider the three graphs shown in Figure 1. Only the third graph exhibits continuity at $c$. Here is the formal definition.

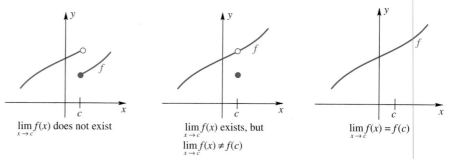

$\lim_{x \to c} f(x)$ does not exist $\qquad$ $\lim_{x \to c} f(x)$ exists, but $\qquad$ $\lim_{x \to c} f(x) = f(c)$
$\qquad\qquad\qquad\qquad\qquad\quad$ $\lim_{x \to c} f(x) \neq f(c)$

FIGURE 1

**An Analogy**

Continue to think of a function $f$ as a machine that takes an input $c$ and produces an output $f(c)$. If it is a good machine (a continous one), a small error in input will result in a small error in output. In other words, a continuous machine takes $x$ near $c$ and produces $f(x)$ near $f(c)$.

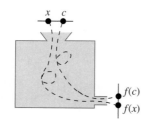

A good example of a discontinuous machine is the postage machine, which (in 1991) charged 29¢ for a 1-ounce letter but 52¢ for a letter the least little bit over 1 ounce.

> **Definition**
>
> **(Continuity at a point).** Let $f$ be defined on an open interval containing $c$. We say that $f$ is **continuous** at $c$ if
>
> $$\lim_{x \to c} f(x) = f(c)$$

We mean by this definition to require three things: (1) $\lim\limits_{x \to c} f(x)$ exists, (2) $f(c)$ exists (that is, $c$ is in the domain of $f$), and (3) $\lim\limits_{x \to c} f(x) = f(c)$. If any of these three fails, then $f$ is **discontinuous** at $c$. Thus, the functions represented by the first and second graphs above are discontinuous at $c$. They are, however, continuous at other points of their domains.

**EXAMPLE 1** Let $f(x) = \dfrac{x^2 - 4}{x - 2}$, $x \neq 2$. How should $f$ be defined at $x = 2$ in order to make it continuous there?

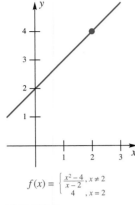

$$f(x) = \begin{cases} \frac{x^2-4}{x-2}, & x \neq 2 \\ 4, & x = 2 \end{cases}$$

FIGURE 2

*Solution*

$$\lim_{x \to 2} \frac{x^2 - 4}{x - 2} = \lim_{x \to 2} \frac{(x - 2)(x + 2)}{x - 2} = \lim_{x \to 2}(x + 2) = 4$$

Therefore, we define $f(2) = 4$. The graph of the resulting function is shown in Figure 2. In fact, we see that $f(x) = x + 2$ for all $x$. ∎

**Continuity of Familiar Functions**  Most functions that we will meet in this book are either continuous everywhere or everywhere except at a few points. In particular, Theorem 2.6B implies the following result.

> ### Theorem A
>
> A polynomial function is continuous at every real number $c$. A rational function is continuous at every real number $c$ in its domain, that is, except where its denominator is zero.

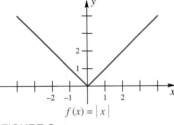

$$f(x) = |x|$$

FIGURE 3

Recall the function $f(x) = |x|$; its graph is shown in Figure 3. For $x < 0$, $f(x) = -x$, a polynomial; for $x > 0$, $f(x) = x$, another polynomial. Thus, $|x|$ is continuous at all numbers different from 0 by Theorem A. But

$$\lim_{x \to 0}|x| = 0 = |0|$$

(see Problem 23 of Section 2.5). Therefore, $|x|$ is also continuous at 0; it is continuous everywhere.

By the Main Limit Theorem (Theorem 2.6A)

$$\lim_{x \to c}\sqrt[n]{x} = \sqrt[n]{\lim_{x \to c} x} = \sqrt[n]{c}$$

provided $c > 0$ when $n$ is even. This means that $f(x) = \sqrt[n]{x}$ is continuous at each point where it makes sense to talk about continuity. In particular, $f(x) = \sqrt{x}$ is continuous at each real number $c > 0$ (Figure 4). We summarize.

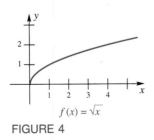

$$f(x) = \sqrt{x}$$

FIGURE 4

> ### Theorem B
>
> The absolute value function is continuous at every real number $c$. If $n$ is odd, the $n$th root function is continuous at every real number $c$; if $n$ is even, the $n$th-root function is continuous at every positive real number $c$.

**Continuity Under Functional Operations**  Do the standard functional operations preserve continuity? Yes, according to Theorem C. In it, $f$ and $g$ are functions, $k$ is a constant, and $n$ is a positive integer.

Theorem C

If $f$ and $g$ are continuous at $c$, then so are $kf$, $f + g$, $f - g$, $f \cdot g$, $f/g$ (provided $g(c) \neq 0$), $f^n$, and $\sqrt[n]{f}$ (provided $f(c) > 0$ if $n$ is even).

***Proof*** All of these results are easy consequences of the corresponding facts for limits from Theorem 2.6A. For example, that theorem, combined with the fact that $f$ and $g$ are continuous at $c$, gives

$$\lim_{x \to c} f(x)g(x) = \lim_{x \to c} f(x) \cdot \lim_{x \to c} g(x) = f(c)g(c)$$

This is precisely what it means to say that $f \cdot g$ is continuous at $c$. ■

**EXAMPLE 2** At what numbers is $F(x) = (3|x| - x^2)/(\sqrt{x} + \sqrt[3]{x})$ continuous?

***Solution*** We need not even consider nonpositive numbers, since $F$ is not defined at such numbers. For any positive number, the functions $\sqrt{x}$, $\sqrt[3]{x}$, $|x|$, and $x^2$ are all continuous (Theorems A and B). It follows from Theorem C that $3|x|$, $3|x| - x^2$, $\sqrt{x} + \sqrt[3]{x}$, and—finally

$$\frac{(3|x| - x^2)}{(\sqrt{x} + \sqrt[3]{x})}$$

are continuous at each positive number. ■

There is another functional operation that will be very important in later work, namely, composition. It, too preserves continuity.

Theorem D

**(Composite Limit Theorem).** If $\lim_{x \to c} g(x) = L$ and if $f$ is continuous at $L$, then

$$\lim_{x \to c} f(g(x)) = f\left(\lim_{x \to c} g(x)\right) = f(L)$$

In particular, if $g$ is continuous at $c$ and $f$ is continuous at $g(c)$, then the composite $f \circ g$ is continuous at $c$.

We defer the proof to the end of the section.

**EXAMPLE 3** Show that $h(x) = |x^2 - 3x + 6|$ is continuous at each real number.

***Solution*** Let $f(x) = |x|$ and $g(x) = x^2 - 3x + 6$. Both are continuous at each real number, and so their composite

$$h(x) = f(g(x)) = |x^2 - 3x + 6|$$

is also.                                                                     ∎

**EXAMPLE 4**  It will be shown later that $f(x) = \sin x$ is continuous at each real number. Conclude that

$$h(x) = \sin\left(\frac{x^4 - 3x + 1}{x^2 - x - 6}\right)$$

is continuous except at 3 and $-2$.

***Solution***  $x^2 - x - 6 = (x - 3)(x + 2)$. Thus, the rational function

$$g(x) = \frac{x^4 - 3x + 1}{x^2 - x - 6}$$

is continuous except at 3 and $-2$ (Theorem A). From Theorem D, we conclude that, since $h(x) = f(g(x))$, $h$ is also continuous except at 3 and $-2$.                                                                     ∎

**Continuity on an Interval**    So far, we have been discussing continuity at a point. We wish to discuss continuity on an interval. Continuity on an interval ought to mean continuity at each point of that interval. That is exactly what it does mean for an open interval.

When we consider a closed interval $[a, b]$, we face a problem. It might be that $f$ is not even defined to the left of $a$ (for example, $f(x) = \sqrt{x}$ has this problem at $a = 0$), so that strictly speaking $\lim_{x \to a} f(x)$ does not exist. We choose to get around this problem by calling $f$ continuous on $[a, b]$ if it is continuous at each point of $(a, b)$ and if $\lim_{x \to a^+} f(x) = f(a)$ and $\lim_{x \to b^-} f(x) = f(b)$ (called, respectively, *right continuity* at $a$ and *left continuity* at $b$). We summarize in a formal definition.

---

**Definition**

We say $f$ is **continuous on an open interval** if it is continuous at each point of that interval. It is **continuous on the closed interval** $[a, b]$ if it is continuous on $(a, b)$, right continuous at $a$, and left continuous at $b$.

---

For example, it is correct to say that $f(x) = 1/x$ is continuous on $(0, 1)$ and that $g(x) = \sqrt{x}$ is continuous on $[0,1]$.

**EXAMPLE 5**  Using the definition above, describe the continuity properties of the function whose graph is sketched in Figure 5.

***Solution***  The function is continuous on the open intervals $(-\infty, 0)$, $(0, 3)$, and $(5, \infty)$ and also on the closed interval $[3, 5]$.                                                                     ∎

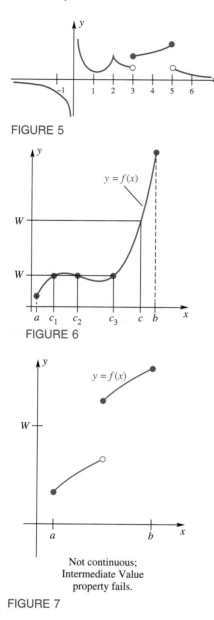

**FIGURE 5**

**FIGURE 6**

**FIGURE 7**

Not continuous;
Intermediate Value
property fails.

**FIGURE 8**

For $f$ to be continuous on $[a, b]$ means that when $x_1$ and $x_2$ are near each other and both are in $[a, b]$, then $f(x_1)$ and $f(x)_2$ are near each other. There are no jumps or abrupt changes, and so we may "draw" the graph of $f$ on $[a,b]$ without lifting our pencil from the paper. This is related to the fact that a continuous function must assume every value between any two of its values, a result we now state precisely.

> ### Theorem E
>
> **(Intermediate Value Theorem).** If $f$ is continuous on $[a, b]$ and if $W$ is a number between $f(a)$ and $f(b)$, then there is at least one number $c$ between $a$ and $b$ such that $f(c) = W$. See Figure 6.

It is clear from the graph in Figure 7 that continuity is needed for this theorem. It also seems clear that continuity is enough, and yet a formal proof of this turns out to be difficult. We leave the proof to more advanced works.

### Proof of Theorem D (Optional)

***Proof*** Let $\varepsilon > 0$ be given. Since $f$ is continuous at $L$, there is a corresponding $\delta_1 > 0$ such that

$$|t - L| < \delta_1 \quad \Rightarrow \quad |f(t) - f(L)| < \varepsilon$$

and so (see Figure 8)

$$|g(x) - L| < \delta_1 \quad \Rightarrow \quad |f(g(x)) - f(L)| < \varepsilon$$

But because $\lim_{x \to c} g(x) = L$, for a given $\delta_1 > 0$ there is a corresponding $\delta_2 > 0$ such that

$$0 < |x - c| < \delta_2 \quad \Rightarrow \quad |g(x) - L| < \delta_1$$

When we put these two facts together, we have

$$0 < |x - c| < \delta_2 \quad \Rightarrow \quad |f(g(x)) - f(L)| < \varepsilon$$

This shows that

$$\lim_{x \to c} f(g(x)) = f(L)$$

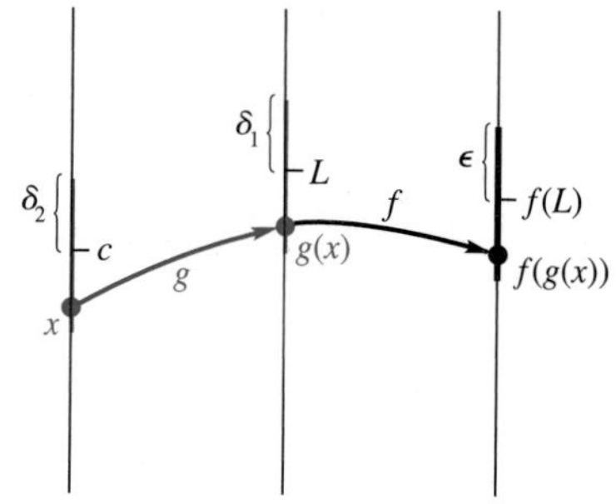

The second statement in Theorem D follows from the observation that if $g$ is continuous at $c$, then $L = g(c)$. ∎

## CONCEPTS REVIEW

**1.** A function $f$ is continuous at $c$ if _____ = $f(c)$.

**2.** The function $f(x) = [\![x]\!]$ is discontinuous at _____.

**3.** A function $f$ is said to be continuous on a closed interval $[a, b]$ if it is continuous at every point of $(a, b)$ and if _____ and _____.

**4.** The Intermediate Value Theorem says that if a function $f$ is continuous on $[a, b]$ and $W$ is a number between $f(a)$ and $f(b)$, then there is a number $c$ between _____ and _____ such that _____.

## PROBLEM SET 2.7

In Problems 1–16, state whether the indicated function is or is not continuous at 3; if not, tell why not.

**1.** $f(x) = (x - 3)(x - 4)$

**2.** $g(x) = x^2 - 9$

**3.** $h(x) = \dfrac{3}{x - 3}$

**4.** $g(t) = \sqrt{t - 4}$

**5.** $h(t) = \dfrac{|t - 3|}{t - 3}$

**6.** $h(t) = \dfrac{\left|\sqrt{(t - 3)^4}\right|}{t - 3}$

**7.** $f(t) = |t|$

**8.** $g(t) = |t - 2|$

**9.** $h(x) = \dfrac{x^2 - 9}{x - 3}$

**10.** $f(x) = \dfrac{21 - 7x}{x - 3}$

**11.** $r(t) = \begin{cases} \dfrac{t^3 - 27}{t - 3} & \text{if } t \neq 3 \\ 27 & \text{if } t = 3 \end{cases}$

**12.** $r(t) = \begin{cases} \dfrac{t^3 - 27}{t - 3} & \text{if } t \neq 3 \\ 23 & \text{if } t = 3 \end{cases}$

**13.** $f(t) = \begin{cases} t - 3 & \text{if } t \leq 3 \\ 3 - t & \text{if } t > 3 \end{cases}$

**14.** $f(t) = \begin{cases} t^2 - 9 & \text{if } t \leq 3 \\ (3 - t)^2 & \text{if } t > 3 \end{cases}$

**15.** $f(x) = \begin{cases} -3x + 4 & \text{if } x \leq 3 \\ -2 & \text{if } x > 3 \end{cases}$

**16.** From the graph of $g$, (see Figure 9) indicate the values where $g$ is discontinuous.

For each of the values state whether $g$ is continuous from the right, left, or neither.

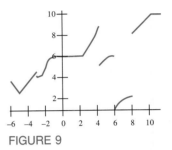

FIGURE 9

**17.** From the graph of $h$ given below (Figure 10), indicate the intervals on which $h$ is continuous.

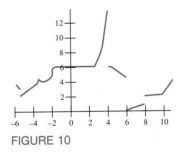

FIGURE 10

In Problems 18–23, the given function is not defined at a certain point. How should it be defined in order to make it continuous at this point? (See Example 1.)

**18.** $f(x) = \dfrac{x^2 - 49}{x - 7}$

**19.** $f(x) = \dfrac{2x^2 - 18}{3 - x}$

**20.** $g(\theta) = \dfrac{\sin(\theta)}{\theta}$

**21.** $H(t) = \dfrac{\sqrt{t} - 1}{t - 1}$

**22.** $\phi(x) = \dfrac{x^4 + 2x^2 - 3}{x + 1}$

23. $F(x) = \sin\left(\dfrac{x^2 - 1}{x + 1}\right)$

In Problems 24–35, at what points, if any, are the functions discontinuous?

24. $f(x) = \dfrac{3x + 7}{(x - 30)(x - \pi)}$

25. $f(x) = \dfrac{33 - x^2}{x\pi + 3x - 3\pi - x^2}$

26. $h(\theta) = |\sin\theta + \cos\theta|$

27. $r(\theta) = \tan\theta$

28. $f(u) = \dfrac{2u + 7}{\sqrt{u + 5}}$    29. $g(u) = \dfrac{u^2 + |u - 1|}{\sqrt[3]{u + 1}}$

30. $F(x) = \dfrac{1}{\sqrt{4 + x^2}}$    31. $G(x) = \dfrac{1}{\sqrt{4 - x^2}}$

32. $f(x) = \begin{cases} x & \text{if } x < 0 \\ x^2 & \text{if } 0 \le x \le 1 \\ 2 - x & \text{if } x > 1 \end{cases}$

33. $g(x) = \begin{cases} x^2 & \text{if } x < 0 \\ -x & \text{if } 0 \le x \le 1 \\ x & \text{if } x > 1 \end{cases}$

34. $f(t) = [\![t]\!]$

35. $g(t) = [\![t + \tfrac{1}{2}]\!]$

36. Sketch the graph of a function $f$ that satisfies all the following conditions.

(a) Its domain is $[-2, 2]$.
(b) $f(-2) = f(-1) = f(1) = f(2) = 1$.
(c) It is discontinuous at $-1$ and $1$.
(d) It is right continuous at $-1$ and left continuous at $1$.

37. Let

$$f(x) = \begin{cases} x & \text{if } x \text{ is rational} \\ -x & \text{if } x \text{ is irrational} \end{cases}$$

Sketch the graph of this function as best you can and decide where it is continuous.

38. Use the Intermediate Value Theorem to prove that $x^3 + 3x - 2 = 0$ has a real solution between 0 and 1.

39. Use the intermediate value theorem to prove that $(\cos t)t^3 + 6\sin^5 t - 3 = 0$ has a real solution between 0 and $2\pi$.

40. Show that the equation $x^5 + 4x^3 - 7x + 14 = 0$ has at least one real solution. *Hint:* Intermediate Value Theorem.

41. Prove that $f$ is continuous at $c$ if and only if $\lim_{t \to 0} f(t + c) = f(c)$.

42. Prove that if $f$ is continuous at $c$ and $f(c) > 0$, then there is an interval $(c - \delta, c + \delta)$ such that $f(x) > 0$ on this interval.

43. Prove that if $f$ is continuous on $[0, 1]$ and satisfies $0 \le f(x) \le 1$ there, then $f$ has a *fixed point*—that is, there is a number $c$ in $[0, 1]$ such that $f(c) = c$. *Hint:* Apply the Intermediate Value Theorem to $g(x) = x - f(x)$.

44. Find the values of $a$ and $b$ so that the following function is continuous everywhere.

$$f(x) = \begin{cases} x + 1 & \text{if } x < 1 \\ ax + b & \text{if } 1 \le x < 2 \\ 3x & \text{if } x \ge 2 \end{cases}$$

45. A stretched elastic string covers the interval $[0, 1]$. The ends are let go and the string contracts so that it covers the interval $[a, b]$, $a \ge 0$, $b \le 1$. Prove that this results in one point of the string (actually exactly one point) being where it was originally. See Problem 43.

46. Let $f$ be continuous on $[0, 1]$ with $f(0) = f(1) = 0$. Prove that the graph of $f$ has a chord (line segment with both ends on the graph) of length $W$, where $W$ is any number between 0 and 1.

47. Use the Intermediate Value Theorem to show that there are always two points on a circular wire ring with the same temperature. *Hint:* Put the center at the origin and let $\theta$ be the angle that a diameter makes with the $x$-axis. Define $f(\theta)$ appropriately.

48. Starting at 4 A.M., a monk slowly climbed to the top of a mountain, arriving at noon. The next day, he returned along the same path, starting at 5 A.M. and getting to the bottom at 11 A.M. Show that at some point along the path, his watch showed the same time on both days.

49. Prove that any bounded region can be circumscribed by a square. Equivalently, let $D$ be a bounded but otherwise arbitrary region in the first quadrant. Given an angle $\theta$, $0 \le \theta \le \pi/2$, $D$ can be circumscribed by a rectangle whose base makes angle $\theta$ with the $x$-axis as shown in Figure 11. Prove that at some angle, this rectangle is a square.

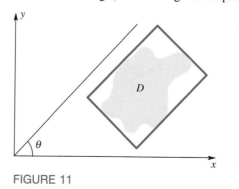

FIGURE 11

50. Show that if we know the values of a continuous function on the rational numbers, we know its values

everywhere. Equivalently, prove that if $f$ and $g$ are continuous functions and $f(x) = g(x)$ for all $x$ in $\mathbb{Q}$ (the rationals), then $f(x) = g(x)$ for all $x$ in $\mathbb{R}$.

**51.** Let $f(x + y) = f(x) + f(y)$ for all $x$ and $y$ in $\mathbb{R}$ and suppose that $f$ is continuous at $x = 0$.

(a) Prove that $f$ is continuous everywhere.
(b) Prove that there is a constant $m$ such that $f(t) = mt$ for all $t$ in $\mathbb{R}$ (see Problem 43 of Section 2.1).

In Problems 51–55 we will study linear functions. Such functions have the form $y(x) = mx + b$ where $m$ and $b$ are constants.

**52.** Show that the sum of two linear functions is also a linear function.

**53.** Show that the composition of two linear functions is also a linear function.

**54.** Show that the product of two linear functions is generally not a linear function.

**55.** Show that the quotient of two linear functions is generally not a linear function.

**56.** Prove that if $f(x)$ is a continuous function on an interval then so is the function $|f(x)| = \sqrt{(f(x))^2}$.

**57.** Show that if $g(x) = |f(x)|$ is continuous that it is not necessarily true that $f(x)$ is continuous.

**58.** Sometimes continuity of a function $f$ is said to be defined by being able to pass the $\lim\limits_{x \to c}$ "through" the function. For example if $f$ is continuous at $c$ then $\lim\limits_{x \to c} f(x) = f(\lim\limits_{x \to c} x)$. Prove or disprove this statement.

**59.** (Famous Problem). Let $f(x) = 0$ if $x$ is irrational and let $f(x) = 1/q$ if $x$ is the rational number $p/q$ in reduced form ($q > 0$).

(a) Sketch (as best you can) the graph of $f$ on $(0, 1)$.
(b) Show that $f$ is continuous at each irrational number in $(0, 1)$ but is discontinuous at each rational number in $(0, 1)$.

**60.** A thin equilateral triangular block of side length 1 unit has its face in the vertical $xy$-plane with a vertex $V$ at the origin. Under the influence of gravity, it will rotate about $V$ until a side hits the $x$-axis floor (Figure 12). Let $x$ denote the initial $x$-coordinate of the midpoint $M$ of the side opposite $V$ and let $f(x)$ denote the final $x$-coordinate of this point. Assume the block balances when $M$ is directly above $V$.

(a) Determine the domain and range of $f$.
(b) Where on this domain is $f$ discontinuous?
(c) Identify any fixed points of $f$ (see Problem 43).

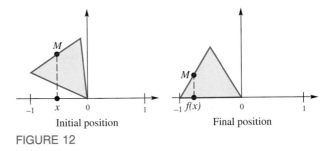

Initial position          Final position

**FIGURE 12**

**Answers to Concepts Review: 1.** $\lim\limits_{x \to c^-} f(x)$
**2.** Every integer **3.** $\lim\limits_{x \to a^+} f(x) = f(a);\ \lim\limits_{x \to b^-} f(x) = f(b)$
**4.** $a;\ b;\ f(c) = W$

---

## 2.8 CHAPTER REVIEW

### Concepts Test

Respond with true or false to each of the following assertions. Be prepared to justify your answer.

**1.** The equation $xy + x^2 = 3y$ determines a function with formula of the form $y = f(x)$.

**2.** The equation $xy^2 + x^2 = 3x$ determines a function with formula of the form $y = f(x)$.

**3.** The equation $\theta \sin \theta + t - \cos\theta = 0$ determines $t$ as a function of $\theta$.

**4.** The equation $\Phi + \Psi = |\Phi + \Psi|$ determines $\Phi$ as a function of $\Psi$.

**5.** The equation $T(\theta) = \sin(\theta)$ determines $\theta$ as a function of T.

**6.** The natural domain of
$$f(x) = \sqrt{\frac{x}{4 - x}}$$
is the interval $[0, 4)$.

**7.** The natural domain of
$$f(x) = \sqrt{-(x^2 + 4x + 3)}$$
is the interval $-3 \le x \le -1$.

8. The natural domain of $T(\theta) = \sec(\theta) + \cos(\theta)$ is all $\theta$.

9. The range of $f(x) = x^2 - 6$ is the interval $[-6, \infty)$.

10. The range of the function $f(x) = \tan x - \sec x$ is the interval $(-\infty, -1] \cup [1, \infty)$.

11. The range of the function $f(x) = \csc x - \sec x$ is the interval $(-\infty, -1] \cup [1, \infty)$.

12. The sum of two even functions (with the same domain) is an even function.

13. The sum of two odd functions (with the same domain) is an odd function.

14. The product of two odd functions (with the same domain) is an odd function.

15. The product of an even function with an odd function (with the same domain) is an odd function.

16. The composition of an even function with an odd function (assuming the final composition function is defined) is an odd function.

17. The composition of two odd functions (assuming the final composition function is defined) is an even function.

18. The function $f(x) = (2x^3 + x)/(x^2 + 1)$ is odd.

19. The function

$$f(t) = \frac{(\sin t)^2 + \cos t}{\tan t \csc t}$$

is even.

20. If the range of a function consists of just one number, then its domain also consists of just one number.

21. If the domain of a function contains at least two numbers, then the range also contains at least two numbers.

22. If $g(x) = [\![x/2]\!]$, then $g(-1.8) = -1$.

23. If $f(x) = x^2$ and $g(x) = x^3$, then $f \circ g = g \circ f$.

24. If $f(x) = x^2$ and $g(x) = x^3$ then $(f \circ g)(x) = f(x) \cdot g(x)$.

25. If $f$ and $g$ have the same domain, then $f/g$ also has that domain.

26. If the graph of $y = f(x)$ has an $x$-intercept at $x = a$, then the graph of $y = f(x + h)$ has an $x$-intercept at $x = a - h$.

27. The cotangent is an odd function.

28. The natural domain of the tangent function is the set of all real numbers.

29. If $\cos s = \cos t$, then $s = t$.

30. If $\lim_{x \to c} f(x) = L$, then $f(c) = L$.

31. If $f(c)$ is not defined, then $\lim_{x \to c} f(x)$ does not exist.

32. The coordinates of the hole in the graph of $y = \dfrac{x^2 - 25}{x - 5}$ are $(5, 10)$.

33. If $p(x)$ is a polynomial, then $\lim_{x \to c} p(x) = p(c)$.

34. If $\lim_{x \to c^-} f(x) = \lim_{x \to c^+} f(x)$, then $f$ is continuous at $x = c$.

35. If $\lim_{x \to c} f(x) = f(\lim_{x \to c} x)$, then $f$ is continuous at $x = c$.

36. The function $f(x) = [\![x/2]\!]$ is continuous at $x = 2.3$.

37. If $\lim_{x \to 2} f(x) = f(2) > 0$, then $f(x) < 1.001f(2)$ for all $x$ in some interval containing 2.

38. If $\lim_{x \to c} [f(x) + g(x)]$ exists, then $\lim_{x \to c} f(x)$ and $\lim_{x \to c} g(x)$ both exist.

39. If $0 \le f(x) \le 3x^2 + 2x^4$ for all $x$, then $\lim_{x \to 0} f(x) = 0$.

40. If $\lim_{x \to a} f(x) = L$ and $\lim_{x \to a} f(x) = M$, then $L = M$.

41. If $f(x) \ne g(x)$ for all $x$, then $\lim_{x \to c} f(x) \ne \lim_{x \to c} g(x)$.

42. If $f(x) < 10$ for all $x$ and $\lim_{x \to 2} f(x)$ exists, then $\lim_{x \to 2} f(x) < 10$.

43. If $\lim_{x \to a} f(x) = b$, then $\lim_{x \to a} |f(x)| = |b|$.

44. If $f$ is continuous and positive on $[a, b]$, then $1/f$ must assume every value between $1/f(a)$ and $1/f(b)$.

### Sample Test Problems

1. For $f(x) = 1/(x + 1) - 1/x$, find each value (if possible).

(a) $f(1)$  (b) $f\left(-\frac{1}{2}\right)$  (c) $f(-1)$

(d) $f(t - 1)$  (e) $f\left(\dfrac{1}{t}\right)$

2. For $g(x) = (x + 1)/x$, find and simplify.

(a) $g(2)$  (b) $g\left(\frac{1}{2}\right)$

(c) $g\left(\frac{1}{10}\right)$  (d) $\dfrac{g(2 + h) - g(2)}{h}$

3. Describe the natural domains of each function.

(a) $f(x) = \dfrac{x}{x^2 - 1}$      (b) $g(x) = \sqrt{4 - x^2}$

(c) $h(x) = \dfrac{\sqrt{1 + x^2}}{|2x + 3|}$

4. Which of the following functions are odd? Even? Neither even nor odd?

(a) $f(x) = \dfrac{3x}{x^2 + 1}$      (b) $g(x) = |\sin x| + \cos x$

(c) $h(x) = x^3 + \sin x$      (d) $k(x) = \dfrac{x^2 + 1}{|x| + x^4}$

5. Sketch the graphs of each of the following functions.

(a) $f(x) = x^2 - 1$      (b) $g(x) = \dfrac{x}{x^2 + 1}$

(c) $h(x) = \begin{cases} x^2 & \text{if } 0 \le x \le 2 \\ 6 - x & \text{if } x > 2 \end{cases}$

6. Suppose that $f$ is an even function satisfying $f(x) = -1 + \sqrt{x}$ for $x \ge 0$. Sketch the graph of $f$ for $-4 \le x \le 4$.

7. An open box is made by cutting squares of side $x$ inches from the four corners of a sheet of cardboard 24 inches by 32 inches and then turning up the sides. Express the volume $V(x)$ in terms of $x$. What is the domain for this function?

8. Let $f(x) = x - 1/x$ and $g(x) = x^2 + 1$. Find each value.

(a) $(f + g)(2)$      (b) $(f \cdot g)(2)$
(c) $(f \circ g)(2)$      (d) $(g \circ f)(2)$
(e) $f^3(-1)$      (f) $f^2(2) + g^2(2)$

9. Sketch the graphs of each of the following, making use of translations.

(a) $y = \frac{1}{4}x^2$      (b) $y = \frac{1}{4}(x + 2)^2$
(c) $y = -1 + \frac{1}{4}(x + 2)^2$

10. Let $f(x) = \sqrt{16 - x}$ and $g(x) = x^4$. What is the domain of each of the following?

(a) $f$      (b) $f \circ g$      (c) $g \circ f$

11. Write $F(x) = \sqrt{1 + \sin^2 x}$ as the composite of four functions, $f \circ g \circ h \circ k$.

12. Calculate each of the following without using a calculator or tables.

(a) $\sin(570°)$      (b) $\cos\left(\dfrac{9\pi}{2}\right)$

(c) $\sin^2(5) + \cos^2(5)$      (d) $\cos\left(\dfrac{-13\pi}{6}\right)$

13. If $\sin t = 0.8$ and $\cos t < 0$, find each value.

(a) $\sin(-t)$      (b) $\cos t$
(c) $\sin 2t$      (d) $\tan t$
(e) $\cos\left(\dfrac{\pi}{2} - t\right)$      (f) $\sin(\pi + t)$

14. Write $\sin 3t$ in terms of $\sin t$. *Hint*: $3t = 2t + t$.

15. A fly sits on the rim of a wheel spinning at the rate of 20 revolutions per minute. If the radius of the wheel is 9 inches, how far does the fly travel in 1 second?

In Problems 16–27, find the indicated limit or state that it doesn't exist.

16. $\lim\limits_{u \to 1} \dfrac{u^2 - 1}{u + 1}$      17. $\lim\limits_{u \to 1} \dfrac{u^2 - 1}{u - 1}$

18. $\lim\limits_{u \to 1} \dfrac{u + 1}{u^2 - 1}$      19. $\lim\limits_{x \to 2} \dfrac{1 - 2/x}{x^2 - 4}$

20. $\lim\limits_{z \to 2} \dfrac{z^2 - 4}{z^2 + z - 6}$      21. $\lim\limits_{x \to 0} \dfrac{\tan x}{\sin 2x}$

22. $\lim\limits_{y \to 1} \dfrac{y^3 - 1}{y^2 - 1}$      23. $\lim\limits_{x \to 4} \dfrac{x - 4}{\sqrt{x} - 2}$

24. $\lim\limits_{x \to 0} \dfrac{\cos x}{x}$      25. $\lim\limits_{x \to 0^-} \dfrac{|x|}{x}$

26. $\lim\limits_{x \to 1/2^+} [\![4x]\!]$      27. $\lim\limits_{t \to 2^-} ([\![t]\!] - t)$

28. Let $f(x) = \begin{cases} x^3 & \text{if } x < -1 \\ x & \text{if } -1 < x < 1 \\ 1 - x & \text{if } x \ge 1 \end{cases}$

Find each value.

(a) $f(1)$      (b) $\lim\limits_{x \to 1^+} f(x)$
(c) $\lim\limits_{x \to 1^-} f(x)$      (d) $\lim\limits_{x \to -1} f(x)$

29. Refer to $f$ of Problem 28. (a) What are the values of $x$ at which $f$ is discontinuous? (b) How should $f$ be defined at $x = -1$ to make it continuous there?

30. Give the $\varepsilon$, $\delta$ definition in each case.

(a) $\lim\limits_{u \to a} g(u) = M$      (b) $\lim\limits_{x \to a^-} f(x) = L$

31. If $\lim\limits_{x \to 3} f(x) = 3$ and $\lim\limits_{x \to 3} g(x) = -2$ and if $g$ is continuous at $x = 3$, find each value.

(a) $\lim\limits_{x \to 3} [2f(x) - 4g(x)]$      (b) $\lim\limits_{x \to 3} g(x) \dfrac{x^2 - 9}{x - 3}$
(c) $g(3)$      (d) $\lim\limits_{x \to 3} g(f(x))$
(e) $\lim\limits_{x \to 3} \sqrt{f^2(x) - 8g(x)}$      (f) $\lim\limits_{x \to 3} \dfrac{|g(x) - g(3)|}{f(x)}$

**32.** Sketch the graph of a function $f$ that satisfies all of the following conditions.

(a) Its domain is $[0, 6]$.

(b) $f(0) = f(2) = f(4) = f(6) = 2$.

(c) $f$ is continuous except at $x = 2$.

(d) $\lim_{x \to 2^-} f(x) = 1$ and $\lim_{x \to 5^+} f(x) = 3$.

**33.** Let $f(x) = \begin{cases} -1 & \text{if } x \le 0 \\ ax + b & \text{if } 0 < x < 1 \\ 1 & \text{if } x \ge 1 \end{cases}$

Determine $a$ and $b$ so that $f$ is continuous everywhere.

**34.** Use the Intermediate Value Theorem to prove that the equation $x^5 - 4x^3 - 3x + 1 = 0$ has at least one solution between $x = 2$ and $x = 3$.

## 2.9  EXPLORATIONS

The following problems are designed to enhance your understanding of functions and graphs.

**1.** Use the language of functions to describe the indicated relationship as accurately as possible. State clearly the domain and range of each function you define. Sketch the graph.

(a) How does the area of a right triangle with a hypotenuse of a fixed length depend on the acute angle.

(b) How does the area of a right triangle with a hypotenuse of a fixed length depend on the obtuse angle.

(c) How does the length of one leg of a right triangle depend on the length of the other leg, if we assume that the hypotenuse is a fixed length.

(d) How does the temperature in degrees Centigrade depend on the temperature in degrees Fahrenheit.

(e) Given a real number at most 5 units away from zero, find the maximum of the number and one minus the number.

(f) Given the angle of a piece of pie from a 10 inch pie, find the area of the piece.

**2.** Suppose a continuous function is periodic with period 1, and is linear between 0 and 0.25 and linear between $-0.75$ and 0. In addition, it has the value 1 at 0 and 2 at 0.25. Sketch the function over the domain $-1$ to 1, and give a piecewise definition of the function.

**3.** Suppose a continuous function is periodic with period 2, and is quadratic between $-0.25$ and 0.25 and linear between $-1.75$ and $-0.25$ In addition, it has the value 0 at 0 and 0.0625 at $\pm 0.25$. Sketch the function over the domain $-2$ to 2, and give a piecewise definition of the function.

**4.** Suppose a linear continuous function is even and periodic with period 1. It has the value 0 at 0 and the value 1 at 0.5. Sketch the function over the domain $-1$ to 1, and give a piecewise definition of the function.

**5.** Let $\alpha$, $\beta$, and $\gamma$ be positive constants. Characterize the changes in the graph of $y = \gamma\cos(\alpha x + \beta)$ if:

(a) $\alpha$, $\gamma$, are fixed and $\beta$ is halved.

(b) $\gamma$, $\beta$, are fixed and $\alpha$ is doubled.

(c) $\alpha$, $\beta$, and $\gamma$ are doubled.

(d) $\gamma$ is fixed and $\alpha$ and $\beta$ are doubled.

**6.** We will explore how a parabola depends on the coefficient of the linear term. We consider the specific parabola $y = P_c(x) = 2x^2 + cx - 1$ and investigate how changes in $c$ affect the graph.

(a) Using some sort of graphing device, graph $y = P_c(x)$ for positive values of $c$ on the same set of axis. Try the values $c = 1, 2, 3, 4$. How does the position of the vertex of the graph change?

(b) Graph $y = P_c(x)$ for $c = -1, -2, -3, -4$. How does the position of the vertex change?

(c) Formulate a rule which summarizes the effect that changing $c$ has on the graph of $P_c(x)$. Your rule should be valid no matter if $c$ is positive, negative or zero.

(d) Notice that all the graphs have the same $y$-intercept. Explain why that must be the case.

**7.** Consider the function $g(t) = t^3 + t$ and the function $h(t) = \alpha t$. We will consider the result of composing $g$ with $h$.

(a) Compare the graphs of $g(\alpha t) = (g \circ h)(t)$ with $\alpha g(t) = (h \circ g)(t)$ for $\alpha = 3, 2, 1/2, 1/4$.

(b) Formulate a rule which summarizes the effect that changing a positive constant $\alpha$ has on the graph of $g(\alpha t)$ and $\alpha g(t)$ compared to the graph of $g(t)$.

(c) Formulate a rule which summarizes the effect that changing a **negative** constant $\alpha$ has on the graph of $g(\alpha t)$ and $\alpha g(t)$ compared to the graph of $g(t)$.

**8.** Consider the function $\Phi(x) = \dfrac{x^3 + x}{2x - 1}$. We will consider the effect of composing this function with $\Psi(x) = |x|$.

(a) Compare the graph of $(\Phi \circ \Psi)(x) = \dfrac{|x|^3 + |x|}{2|x| - 1}$ with $\Phi(x)$.

(b) Compare the graph of $(\Psi \circ \Phi)(x) = \left|\dfrac{x^3 + x}{2x - 1}\right|$ with $\Phi(x)$.

(c) Formulate a rule which summarizes the effect of composing the absolute value function with that of another function. That is, give the relationship between the graphs of $\Phi(x)$, $|\Phi(x)|$, and $\Phi(|x|)$.

9. Consider the following statement:
"The national debt has been rising for the last two years, but less rapidly this year than last."
Sketch a possible graph for the national debt as a function of time which roughly corresponds to the above statement.

10. Consider the following statement:
"The trade deficit has been rising for the last two years, but more rapidly this year than last."
Sketch a possible graph of the trade deficit as a function of time which roughly corresponds to the above statement.

11. Consider the following statement:
"The number of school dropouts has been declining over the last five years, but in each of those years the decline has been less rapid than in the preceding year."
Sketch a possible graph of the number of school dropouts as a function of time which roughly corresponds to the above statement.

12. What are you able to deduce about the shape of a cylindrical vase based upon each of the tables below which give the measurement of the volume of the water as a function of the depth:

(a)

| Depth | 1 | 2 | 3 | 4 | 5 | 6 |
|-------|---|---|----|----|----|----|
| Volume | 4 | 8 | 11 | 14 | 20 | 28 |

(b)

| Depth | 1.5 | 2 | 3 | 4.5 | 5.5 | 6 |
|-------|-----|---|----|-----|-----|----|
| Volume | 4 | 9 | 12 | 14 | 20 | 28 |

13. Generally when a disease infects a given population various models can be used to predict the number of individuals that get sick over time. If the disease is quickly fatal, then those who catch it will not generally be a source of infection to the rest of the population. On the other hand, if we expect that all members of the population will eventually catch the disease but none will die, then the prediction will be different. It now becomes important to determine if the sick people will be a source of infection to others, or if once they are ill, they will stay home. For Figure 1 (where the x-axis is measured in months and the y-axis measures the percentage of the population infected) match a possible scenario for the epidemic. Explain your answer.

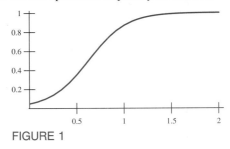

**FIGURE 1**

14. Suppose that the number of deer in a preserve are governed by the food supply which is fixed. If the number of deer becomes more than the food supply can support, then many of them die from starvation. A reasonable graph of this situation could be given by
(a) Suppose that hunters are permitted to shoot deer at a rate proportional to the number of deer in the preserve. Indicate how you would expect the graph to change. Discuss and graph various scenarios and how they would depend on the rate at which deer could be hunted.
(b) Suppose that hunters are permitted to shoot deer at a fixed rate which is independent of the number of deer in the preserve. Indicate how you would expect the graph to change. Discuss and graph various scenarios and how they would depend on the rate at which deer could be hunted.

15. At this stage in the development of calculus some limits are hard to compute exactly. Nevertheless, it is possible to make an intelligent guess as to the limit by using numerical methods. Using your calculator with the indicated values try to make an educated guess at the value of the following limits:

(a) $\lim_{x \to 0} 12 \left( \dfrac{1 - x^2/2 - \cos x}{x^4} \right)$ evaluate at $x = 0.1, 0.01,$ and $0.001$.

(b) $\lim_{x \to 0} 12 \left( \dfrac{x - x^3/6 - \sin x}{x^5} \right)$ evaluate at $x = 0.5, 0.3,$ and $0.1$.

(c) $\lim_{x \to 0} 12 \left( \dfrac{2x + \frac{8}{3}x^3 - \tan 2x}{256 x^5} \right)$ evaluate at $x = 0.5, 0.3, 0.1,$ and $0.05$.

# TECHNOLOGY PROJECT

**Exercise 1**  Consider the following piecewise defined function:

$$f(x) = \begin{cases} x^3 + 2 & \text{if } x \leq -1 \\ x^4 + 1 & \text{if } x \geq 1 \\ x^2 + x + 1 & \text{elsewhere} \end{cases}$$

(a) Find out where $f(x)$ is continuous and discontinuous and justify your conclusions.

(b) Determine values $b$ and $c$ such that the related function

$$f(x) = \begin{cases} x^3 + 2 & \text{if } x \leq -1 \\ x^4 + 1 & \text{if } x \geq 1 \\ x^2 + bx + c & \text{elsewhere} \end{cases}$$

is continuous everywhere.

**Exercise 2**  Use factoring (computer-based or by hand) to compute the following limits:

(a) $\lim\limits_{x \to 13} \dfrac{x^3 - 9x^2 - 45x - 91}{x - 13}$

(b) $\lim\limits_{x \to 13} \dfrac{x^3 - 9x^2 - 39x - 86}{x - 13}$

(c) $\lim\limits_{x \to 13}$

$\dfrac{x^4 - 26x^3 + 178x^2 - 234x + 1521}{x - 13}$

**Exercise 3**  Find

$$\lim_{x \to 0} \frac{\sqrt{25 + 3x} - \sqrt{25 - 2x}}{x}$$

by each of the following methods:

(a) Make a plot of the fraction near the point $x = 0$. This often "works" even when, as in this case, the function is not defined at the critical point.

(b) Make a table near $x = 0$.

**Exercise 4**  Evaluate the following limits as $x \to 0$ using any one of the methods introduced so far:

(a) $\dfrac{x^3 - x^2 - 4x + 4}{x - 1}$

(b) $\dfrac{\sin x}{x}$ (This is an important limit.)

(c) $\dfrac{1 - \cos x}{x}$ (So is this.)

(d) $\dfrac{\sin 5x}{x}$

(e) $(1 + x)^{1/x}$ (Another important limit.)

**Exercise 5**  Verify the assertions you made in Exercise 1 by handing in a plot for each part of the problem.

**Exercise 6**  Algebraic limits of the 0/0 form can be simplified to a form where the limit can be obtained by just evaluating the function. As we have seen, sometimes standard tools such as factoring can be used advantageously. Use the trick of multiplying top and bottom by a sum of square roots ("rationalizing the numerator") to algebraically check the result of Exercise 3 and check that your Lab results were correct.

**Exercise 7**  If, as claimed in the previous exercise, "algebraic limits of the 0/0 form can be simplified to a form where the limit can be obtained by just evaluating the function," what about

$$\lim_{x \to 8} \frac{x^{2/3} - 4}{x - 8}$$

*Hint*: Factor (i.e., divide) the quantity $x^{1/3} - 2$ from both numerator and denominator.

**Exercise 8**  You have noticed that many limits can be obtained just by evaluating the given function at the given point. State precisely a condition under which this is valid. Which of the following limits can be evaluated in this simple way? Justify your answers.

(a) $\lim\limits_{x \to 5} \dfrac{x + 5}{x^4 + x^2 + 1}$

(b) $\lim\limits_{x \to 5} \dfrac{x^2 - 25}{x - 5}$

(c) $\lim\limits_{x \to 5} \sqrt{x^6 - 5}$

**Exercise 9**  The function in Exercise 1 seems artificial. Can you think of any cases in science where a function would naturally be defined piecewise?

# 3

# The Derivative

## 3.1
## TWO PROBLEMS WITH ONE THEME

Our first problem is very old; it dates back to the great Greek scientist Archimedes (287–212 B.C.). We refer to the problem of the *tangent line.*

Our second problem is newer. It grew out of attempts by Kepler (1571–1630), Galileo (1564–1642), Newton (1642–1727), and others to describe the speed of a moving body. It is the problem of *instantaneous velocity.*

The two problems, one geometric and the other mechanical, appear to be quite unrelated. In this case, appearances are deceptive. The two problems are identical twins.

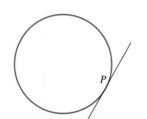

Tangent line at P

FIGURE 1

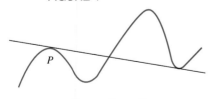

Tangent line at P

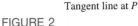
FIGURE 2

**The Tangent Line**   Euclid's notion of a tangent as a line touching a curve at just one point is all right for circles (Figure 1) but completely unsatisfactory for most other curves (Figure 2). The idea of a tangent to a curve at P as that line which best approximates the curve near P is better, but is still too vague for mathematical precision. The concept of limit provides a way of getting the best description.

Let P be a fixed point on a curve and let Q be a nearby movable point on that curve. Consider the line through P and Q, called a **secant line**. The **tangent line** at P is the limiting position (if it exists) of the secant line as Q moves toward P along the curve (Figure 3).

Suppose that the curve is the graph of the equation $y = f(x)$. Then P has coordinates $(c, f(c))$, a nearby point Q has coordinates $(c + h, f(c + h))$, and the secant line through P and Q has slope $m_{sec}$ given by (Figure 4)

$$m_{sec} = \frac{f(c + h) - f(c)}{h}$$

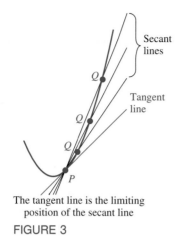

Secant lines

Tangent line

The tangent line is the limiting position of the secant line

FIGURE 3

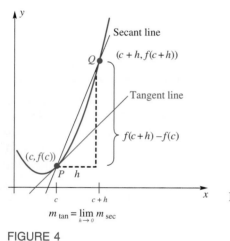

Secant line

$(c + h, f(c + h))$

Tangent line

$f(c + h) - f(c)$

$(c, f(c))$

$P$   $h$

$c$   $c + h$

$m_{tan} = \lim\limits_{h \to 0} m_{sec}$

FIGURE 4

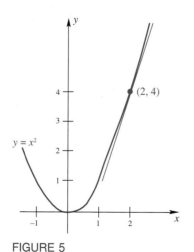

$(2, 4)$

$y = x^2$

FIGURE 5

Consequently, the **tangent line** to the curve $y = f(x)$ at the point $P(c, f(c))$— if not vertical—is that line through $P$ with slope $m_{tan}$ satisfying

$$m_{tan} = \lim_{h \to 0} m_{sec} = \lim_{h \to 0} \frac{f(c + h) - f(c)}{h}$$

**EXAMPLE 1**   Find the slope of the tangent line to the curve $y = f(x) = x^2$ at the point $(2, 4)$.

**Solution**   The line whose slope we are seeking is shown in Figure 5. Clearly it has a large positive slope.

$$m_{tan} = \lim_{h \to 0} \frac{f(2 + h) - f(2)}{h}$$
$$= \lim_{h \to 0} \frac{(2 + h)^2 - 2^2}{h}$$
$$= \lim_{h \to 0} \frac{4 + 4h + h^2 - 4}{h}$$
$$= \lim_{h \to 0} \frac{h(4 + h)}{h}$$
$$= 4 \qquad \blacksquare$$

**EXAMPLE 2**   Find the slope of the tangent line to the curve $y = f(x) = -x^2 + 2x + 2$ at the points with $x$-coordinates $-1, \frac{1}{2}, 2$, and $3$.

**Solution**   Rather than make four separate calculations, it seems wise to calculate the slope at the point with $x$-coordinate $c$ and then obtain the four desired answers by substitution.

$$m_{tan} = \lim_{h \to 0} \frac{f(c + h) - f(c)}{h}$$
$$= \lim_{h \to 0} \frac{-(c + h)^2 + 2(c + h) + 2 - (-c^2 + 2c + 2)}{h}$$
$$= \lim_{h \to 0} \frac{-c^2 - 2ch - h^2 + 2c + 2h + 2 + c^2 - 2c - 2}{h}$$
$$= \lim_{h \to 0} \frac{h(-2c - h + 2)}{h}$$
$$= -2c + 2$$

The four desired slopes (obtained by letting $c = -1, \frac{1}{2}, 2, 3$) are 4, 1, $-2$, and $-4$. These answers do appear to be consistent with the graph in Figure 6. $\blacksquare$

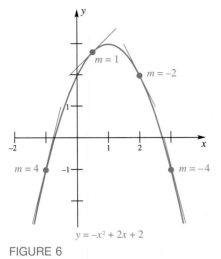

$$y = -x^2 + 2x + 2$$

FIGURE 6

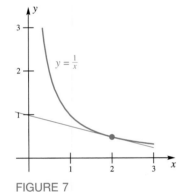

$$y = \frac{1}{x}$$

FIGURE 7

**EXAMPLE 3**   Find the equation of the tangent line to the curve $y = 1/x$ at $(2, \frac{1}{2})$ (see Figure 7).

*Solution*   Let $f(x) = 1/x$.

$$m_{\text{tan}} = \lim_{h \to 0} \frac{f(2 + h) - f(2)}{h}$$

$$= \lim_{h \to 0} \frac{\dfrac{1}{2 + h} - \dfrac{1}{2}}{h}$$

$$= \lim_{h \to 0} \frac{\dfrac{2}{2(2 + h)} - \dfrac{2 + h}{2(2 + h)}}{h}$$

$$= \lim_{h \to 0} \frac{2 - (2 + h)}{2(2 + h)h}$$

$$= \lim_{h \to 0} \frac{-h}{2(2 + h)h}$$

$$= \lim_{h \to 0} \frac{-1}{2(2 + h)} = -\frac{1}{4}$$

Knowing the slope of the line ($m = -\frac{1}{4}$) and the point $(2, \frac{1}{2})$ on it, we can easily write its equation by using the point-slope form $y - y_0 = m(x - x_0)$. The result is $y - \frac{1}{2} = -\frac{1}{4}(x - 2)$.   ∎

**Average Velocity and Instantaneous Velocity**   If we drive an automobile from one town to another 80 miles away in 2 hours, our average velocity is 40 miles per hour. That is, *average velocity* is the distance from the first position to the second position divided by the elapsed time.

But during our trip the speedometer reading was often different from 40. At the start, it registered 0; at times it rose as high as 57; at the end it fell back to 0 again. Just what does the speedometer measure? Certainly it does not indicate average velocity.

Consider the more precise example of an object $P$ falling in a vacuum. Experiment shows that if it starts from rest, $P$ falls $16t^2$ feet in $t$ seconds. Thus it falls 16 feet in the first second and 64 feet during the first 2 seconds (Figure 8); clearly it falls faster and faster as time goes on.

During the second second (that is, in the time interval from $t = 1$ to $t = 2$), $P$ fell $(64 - 16)$ feet. Its average velocity was

$$v_{\text{ave}} = \frac{64 - 16}{2 - 1} = 48 \text{ feet per second}$$

During the time interval from $t = 1$ to $t = 1.5$, it fell $16(1.5)^2 - 16 = 20$ feet. Its average velocity was

$$v_{\text{ave}} = \frac{16(1.5)^2 - 16}{1.5 - 1} = \frac{20}{0.5} = 40 \text{ feet per second}$$

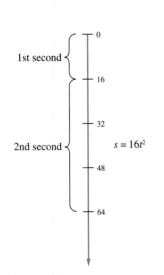

FIGURE 8

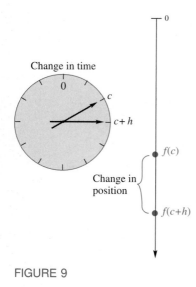

Change in time

Change in position

$f(c)$

$f(c+h)$

FIGURE 9

Similarly, on the time intervals $t = 1$ to $t = 1.1$ and $t = 1$ to $t = 1.01$, we calculate the respective average velocities to be

$$v_{ave} = \frac{16(1.1)^2 - 16}{1.1 - 1} = \frac{3.36}{0.1} = 33.6 \text{ feet per second}$$

$$v_{ave} = \frac{16(1.01)^2 - 16}{1.01 - 1} = \frac{0.3216}{0.01} = 32.16 \text{ feet per second}$$

What we have done is to calculate the average velocity over shorter and shorter time intervals, each starting at $t = 1$. The shorter the time interval, the better we should approximate the *instantaneous velocity* at the instant $t = 1$. Looking at the numbers 48, 40, 33.6, and 32.16, you might guess 32 feet per second to be the instantaneous velocity.

But let us be more precise. Suppose that an object $P$ moves along a coordinate line so that its position at time $t$ is given by $s = f(t)$. At time $c$ the object is at $f(c)$; at the nearby time $c + h$, it is at $f(c + h)$ (see Figure 9). Thus, the **average velocity** on this interval is

$$v_{ave} = \frac{f(c + h) - f(c)}{h}$$

And now we define the **instantaneous velocity** $f$ at $c$ by

$$v = \lim_{h \to 0} v_{ave} = \lim_{h \to 0} \frac{f(c + h) - f(c)}{h}$$

In the case where $f(t) = 16t^2$, the instantaneous velocity at $t = 1$ is

$$v = \lim_{h \to 0} \frac{f(1 + h) - f(1)}{h}$$

$$= \lim_{h \to 0} \frac{16(1 + h)^2 - 16}{h}$$

$$= \lim_{h \to 0} \frac{16 + 32h + 16h^2 - 16}{h}$$

$$= \lim_{h \to 0} (32 + 16h) = 32$$

This confirms our earlier guess.

Now you can see why we called the *slope of the tangent line* and *instantaneous velocity* identical twins. Look at the two boxed formulas in this section. They give different names to the same mathematical concept.

EXAMPLE 4   Find the instantaneous velocity of a falling object, starting from rest, at $t = 3.8$ seconds and at $t = 5.4$ seconds.

***Solution***   We calculate the instantaneous velocity at $t = c$ seconds. Since $f(t) = 16t^2$,

$$v = \lim_{h \to 0} \frac{f(c + h) - f(c)}{h}$$

$$= \lim_{h \to 0} \frac{16(c + h)^2 - 16c^2}{h}$$

$$= \lim_{h \to 0} \frac{16c^2 + 32ch + 16h^2 - 16c^2}{h}$$

$$= \lim_{h \to 0} (32c + 16h) = 32c$$

Thus, the instantaneous velocity at 3.8 seconds is $32(3.8) = 121.6$ feet per second; at 5.4 seconds, it is $32(5.4) = 172.8$ feet per second.    ■

**EXAMPLE 5**   How long will it take the falling object of Example 4 to reach an instantaneous velocity of 112 feet per second?

***Solution***   We learned in Example 4 that the instantaneous velocity after $c$ seconds is $32c$. Thus we must solve the equation $32c = 112$. The solution is $c = \frac{112}{32} = 3.5$ seconds.    ■

**EXAMPLE 6**   A particle moves along a coordinate line and $s$, its directed distance in centimeters from the origin at the end of $t$ seconds, is given by $s = f(t) = \sqrt{5t + 1}$. Find the instantaneous velocity of the particle at the end of 3 seconds.

***Solution***

$$v = \lim_{h \to 0} \frac{f(3 + h) - f(3)}{h}$$

$$= \lim_{h \to 0} \frac{\sqrt{5(3 + h) + 1} - \sqrt{5(3) + 1}}{h}$$

$$= \lim_{h \to 0} \frac{\sqrt{16 + 5h} - 4}{h}$$

To evaluate this limit, we rationalize the numerator (by multiplying numerator and denominator by $\sqrt{16 + 5h} + 4$). We obtain

$$v = \lim_{h \to 0} \left( \frac{\sqrt{16 + 5h} - 4}{h} \cdot \frac{\sqrt{16 + 5h} + 4}{\sqrt{16 + 5h} + 4} \right)$$

$$= \lim_{h \to 0} \frac{16 + 5h - 16}{h(\sqrt{16 + 5h} + 4)}$$

$$= \lim_{h \to 0} \frac{5}{\sqrt{16 + 5h} + 4} = \frac{5}{8}$$

We conclude that the instantaneous velocity at the end of 3 seconds is $\frac{5}{8}$ centimeter per second. ∎

### Velocity or Speed

For the time being, we will use the terms *velocity* and *speed* interchangeably. Later, in Section 3.7, we will distinguish between these two words.

**Rates of Change**  Velocity is only one of many rates of change that will be important in this course; it is the rate of change of distance with respect to time. Other rates of change that will interest us are density for a wire (the rate of change of mass with respect to distance), marginal revenue (the rate of change of revenue with respect to the number of items produced), and current (the rate of change of electrical charge with respect to time). These rates and many more are discussed in the problem set. In each case, we must distinguish between an *average* rate of change on an interval and an *instantaneous* rate of change at a point. The phrase *rate of change* without an adjective will mean instantaneous rate of change.

## CONCEPTS REVIEW

**1.** The line that most closely approximates a curve near the point $P$ is the _____ through that point.

**2.** More precisely, the tangent line to a curve at $P$ is the limiting position of the _____ line through $P$ and $Q$ as $Q$ approaches $P$ along the curve.

**3.** The slope $m_{\tan}$ of the tangent line to the curve $y = f(x)$ at $(c, f(c))$ is given by $m_{\tan} = \lim\limits_{h \to 0}$ _____.

**4.** The instantaneous velocity of a point $P$ (moving along a line) at time $c$ is the limit of the _____ on the time interval $c$ to $c + h$ as $h$ approaches zero.

## PROBLEM SET 3.1

In Problems 1–2, a tangent line to a curve is drawn. Estimate its slope (slope = rise/run). Be careful to note the difference in scales on the two axes.

**1.**

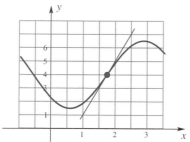

In Problems 3–6, draw the tangent line to the curve through the indicated point and estimate its slope.

**3.**

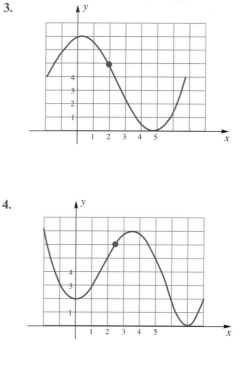

**2.**

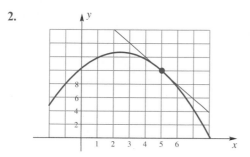

**4.**

**5.**

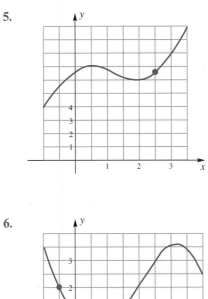

**6.**

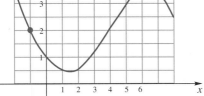

**7.** Consider $y = x^2 + 1$

(a) Sketch its graph as carefully as you can.
(b) Draw the tangent line at (1, 2).
≈ (c) Estimate the slope of this tangent line.
C (d) Calculate the slope of the secant line through (1, 2) and (1.01, $(1.01)^2 + 1.0$).
(e) Find by the limit process (see Example 1) the exact slope of the tangent line at (1, 2).

**8.** Consider $y = x^3 - 1$

(a) Sketch its graph as carefully as you can.
(b) Draw the tangent line at (2, 7).
≈ (c) Estimate the slope of this tangent line.
C (d) Calculate the slope of the secant line through (2, 7) and (2.01, $(2.01)^3 - 1.0$).
(e) Find by the limit process (see Example 1) the exact slope of the tangent line at (2, 7).

**9.** Find the slope of the tangent line to the curve $y = x^2 - 1$ at the points where $x = -2, -1, 0, 1, 2$. (see Example 2).

**10.** Find the slope of the tangent line to the curve $y = x^3 - 3x$ at the points where $x = -2, -1, 0, 1, 2$.

**11.** Sketch the graph of $y = 1/(x + 1)$ and then find the equation of the tangent line at $(1, \frac{1}{2})$ (see Example 3).

**12.** Find the equation of the tangent line to $y = 1/(x - 1)$ at $(0, -1)$.

**13.** Experience suggests that a falling body will fall approximately $16t^2$ feet in $t$ seconds.

(a) How far will it fall between $t = 0$ and $t = 1$?
(b) How far will it fall between $t = 1$ and $t = 2$?
(c) What is its average velocity on the interval $2 \leq t \leq 3$?
C (d) What is its average velocity on the interval $3 \leq t \leq 3.01$?
≈ (e) Find its instantaneous velocity at $t = 3$ (see Example 4).

**14.** An object travels along a line so that its position $s$ is $s = t^2 + 1$ meters after $t$ seconds.

(a) What is its average velocity on the interval $2 \leq t \leq 3$?
C (b) What is its average velocity on the interval $2 \leq t \leq 2.003$?
(c) What is its average velocity on the interval $2 \leq t \leq 2 + h$?
≈ (d) Find its instantaneous velocity at $t = 2$.

**15.** Suppose an object moves along a line $\sqrt{2t + 1}$ feet in $t$ seconds.

(a) Find its instantaneous velocity at $t = \alpha, \alpha > 0$.
(b) When will it reach a velocity of $\frac{1}{2}$ foot per second? (See Example 5.)

**16.** If a particle moves along a coordinate line so that its directed distance from the origin after $t$ seconds is $(-t^2 + 4t)$ feet, when did the particle come to a momentary stop (that is, when did its instantaneous velocity become zero)?

**17.** A certain bacterial culture is growing so that it has a mass of $\frac{1}{2}t^2 + 1$ grams after $t$ hours.

C (a) How much did it grow during the interval $2 \leq t \leq 2.01$?
(b) What was its average growth rate during the interval $2 \leq t \leq 2.01$?
≈ (c) What was its instantaneous growth rate at $t = 2$?

**18.** A business is prospering in such a way that its total (accumulated) profit after $t$ years is $1000t^2$ dollars.

(a) How much did the business make during the third year (that is, between $t = 2$ and $t = 3$)?
(b) What was its average rate of profit during the first half of the third year (that is, between $t = 2$ and $t = 2.5$)?
(c) What was its instantaneous rate of profit at $t = 2$?

**19.** A wire of length 8 centimeters is such that the mass between its left end and a point $x$ centimeters to the right is $x^3$ grams (Figure 10).

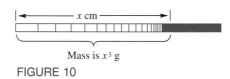

FIGURE 10

(a) What is the average density of the middle 2-centimeter segment of this wire? *Note*: Average density equals mass/length.
(b) What is the actual density at the point 3 centimeters from the left end?

**20.** Suppose that the revenue $R(n)$ in dollars from producing $n$ computers is given by $R(n) = 0.4n - 0.001n^2$. Find the instantaneous rate of change of revenue when $n = 10$ and $n = 100$. (The instantaneous rate of change of revenue with respect to the amount of product produced is called the *marginal revenue*.)

**21.** The rate of change of velocity with respect to time is called the acceleration. Suppose that the velocity at time $t$ of a particle is given by $v(t) = 2t^2$. Find the instantaneous acceleration when $t = 1$ second.

**22.** A city is hit by an Asian flu epidemic. Officials estimate that $t$ days after the beginning of the epidemic, the number of persons sick with the flu is given by $p(t) = 120t^2 - 2t^3$, when $0 \le t \le 40$. At what rate is the flu spreading at time $t = 10$; $t = 20$; $t = 40$?

**23.** The graph in Figure 11 shows the amount of water in a city water tank during one 24-hour day when no water was pumped into the tank. What was the average rate of water usage during the day? How fast was water being used at 8 o'clock?

**24.** The rate of change of electric charge with respect to time is called **current**. Suppose $\frac{1}{3}t^3 + t$ coulombs of charge flow through a wire in $t$ seconds. Find the current in amperes

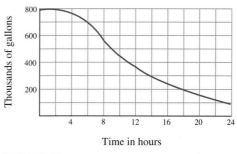

FIGURE 11

(coulombs per second) after 3 seconds. When will a 20-amp fuse in the line blow?

**25.** The radius of a circular oil spill is growing at a constant rate of 2 kilometers per day. At what rate is the area of the spill growing 3 days after it began?

**26.** Find the rate of change of the area of a square with respect to its total perimeter when its area is 1 square unit.

PC Use a computer to do Problems 27–30. In *True BASIC Calculus*, the section called *Tangent* should be helpful.

**27.** Draw the graph of $y = f(x) = x^3 - 2x^2 + 1$. Then find the slope of the tangent line at (a) $-1$, (b) $0$, (c) $1$, (d) $3.2$.

**28.** Draw the graph of $y = f(x) = \sin x \sin^2 2x$. Then find the slope of the tangent line at (a) $\pi/3$, (b) $2.8$, (c) $\pi$, (d) $4.2$.

**29.** If a point moves along a line so that its distance $s$ (in feet) from 0 is given by $s = t + t \cos^2 t$ at time $t$ seconds, find its instantaneous velocity at $t = 3$.

**30.** If a point moves along a line so that its distance $s$ (in meters) from 0 is given by $s = (t + 1)^3/(t + 2)$ at time $t$ minutes, find its instantaneous velocity at $t = 1.6$.

**Answers to Concepts Review:** **1.** Tangent line **2.** Secant **3.** $[f(c + h) - f(c)]/h$ **4.** Average velocity

---

**3.2**
**THE DERIVATIVE**

We have seen that *slope of tangent line* and *instantaneous velocity* are manifestations of the same basic idea. Rate of growth of an organism (biology), marginal profit (economics), density of a wire (physics), and dissolution rates (chemistry) are other versions of the same basic concept. Good mathematical sense suggests that we study this concept independently of these specialized vocabularies and diverse applications. We choose the neutral name *derivative*. Add it to *function* and *limit* as one of the key words in calculus.

Definition

The **derivative** of a function $f$ is another function $f'$ (read "eff prime") whose value at any number $c$ is

$$f'(c) = \lim_{h \to 0} \frac{f(c + h) - f(c)}{h}$$

provided that this limit exists.

If this limit does exist, we say that $f$ is **differentiable** at $c$. Finding a derivative is called **differentiation**; the part of calculus associated with the derivative is called **differential calculus**.

**Finding Derivatives**    We illustrate with several examples.

**EXAMPLE 1**    Let $f(x) = 13x - 6$. Find $f'(4)$.

*Solution*

$$f'(4) = \lim_{h \to 0} \frac{f(4 + h) - f(4)}{h} = \lim_{h \to 0} \frac{[13(4 + h) - 6] - [13(4) - 6]}{h}$$

$$= \lim_{h \to 0} \frac{13h}{h} = \lim_{h \to 0} 13 = 13 \qquad \blacksquare$$

**EXAMPLE 2**    If $f(x) = x^3 + 7x$, find $f'(c)$.

*Solution*

$$f'(c) = \lim_{h \to 0} \frac{f(c + h) - f(c)}{h}$$

$$= \lim_{h \to 0} \frac{[(c + h)^3 + 7(c + h)] - [c^3 + 7c]}{h}$$

$$= \lim_{h \to 0} \frac{3c^2h + 3ch^2 + h^3 + 7h}{h}$$

$$= \lim_{h \to 0} (3c^2 + 3ch + h^2 + 7)$$

$$= 3c^2 + 7 \qquad \blacksquare$$

**EXAMPLE 3**    If $f(x) = 1/x$, find $f'(x)$.

*Solution*    Note a subtle change in the way this example is worded. So far we have used the letter $c$ to denote a fixed number at which a derivative is to be evaluated. Accordingly, we have calculated $f'(c)$. To calculate $f'(x)$, we simply think of $x$ as a fixed, but arbitrary, number and proceed as before.

$$f'(x) = \lim_{h \to 0} \frac{f(x + h) - f(x)}{h} = \lim_{h \to 0} \frac{\frac{1}{x + h} - \frac{1}{x}}{h}$$

$$= \lim_{h \to 0} \left[ \frac{x - (x + h)}{(x + h)x} \cdot \frac{1}{h} \right] = \lim_{h \to 0} \left[ \frac{-h}{(x + h)x} \cdot \frac{1}{h} \right]$$

$$= \lim_{h \to 0} \frac{-1}{(x + h)x} = \frac{-1}{x^2}$$

Thus $f'$ is the function given by $f'(x) = -1/x^2$. Its domain is all real numbers except $x = 0$.  ■

Since changes in notation can cause confusion, we emphasize the formula for $f'(x)$.

$$f'(x) = \lim_{h \to 0} \frac{f(x + h) - f(x)}{h}$$

This formula says exactly the same thing as the formula in the definition at the beginning of this section.

EXAMPLE 4    Find $F'(x)$ if $F(x) = \sqrt{x}, x > 0$.

*Solution*

$$F'(x) = \lim_{h \to 0} \frac{F(x + h) - F(x)}{h}$$

$$= \lim_{h \to 0} \frac{\sqrt{x + h} - \sqrt{x}}{h}$$

By this time you will have noticed that finding a derivative always involves taking the limit of a quotient where both numerator and denominator are approaching zero. Our task is to simplify this quotient so that we can cancel a factor $h$ from numerator and denominator, thus allowing us to evaluate the limit. In the present example, this can be accomplished by rationalizing the numerator.

$$F'(x) = \lim_{h \to 0} \left[ \frac{\sqrt{x + h} - \sqrt{x}}{h} \cdot \frac{\sqrt{x + h} + \sqrt{x}}{\sqrt{x + h} + \sqrt{x}} \right]$$

$$= \lim_{h \to 0} \frac{x + h - x}{h(\sqrt{x + h} + \sqrt{x})}$$

$$= \lim_{h \to 0} \frac{h}{h(\sqrt{x + h} + \sqrt{x})}$$

$$= \lim_{h \to 0} \frac{1}{\sqrt{x + h} + \sqrt{x}}$$

$$= \frac{1}{\sqrt{x} + \sqrt{x}} = \frac{1}{2\sqrt{x}}$$

Thus, $F'$—the derivative of $F$—is given by $F'(x) = 1/2\sqrt{x}$. Its domain is $(0, \infty)$. ∎

**Equivalent Forms for the Derivative**  There is nothing sacred about use of the letter $h$ in defining $f'(c)$. Notice, for example, that

$$f'(c) = \lim_{h \to 0} \frac{f(c + h) - f(c)}{h}$$

$$= \lim_{p \to 0} \frac{f(c + p) - f(c)}{p}$$

$$= \lim_{s \to 0} \frac{f(c + s) - f(c)}{s}$$

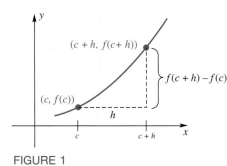

FIGURE 1

A more radical change, but still just a change of notation, may be understood by comparing Figure 1 and Figure 2. Note how $x$ takes the place of $c + h$, and so $x - c$ replaces $h$. Thus,

$$\boxed{f'(c) = \lim_{x \to c} \frac{f(x) - f(c)}{x - c}}$$

In a similar vein, we may write

$$f'(x) = \lim_{t \to x} \frac{f(t) - f(x)}{t - x} = \lim_{p \to x} \frac{f(p) - f(x)}{p - x}$$

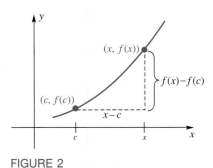

FIGURE 2

Note that in all cases, the number at which $f'$ is evaluated is held fixed during the limit operation.

**EXAMPLE 5**  Use the last boxed result to find $g'(c)$ if $g(x) = 2/(x + 3)$.

*Solution*

$$g'(c) = \lim_{x \to c} \frac{g(x) - g(c)}{x - c} = \lim_{x \to c} \frac{\dfrac{2}{x + 3} - \dfrac{2}{c + 3}}{x - c}$$

$$= \lim_{x \to c} \left[ \frac{2(c + 3) - 2(x + 3)}{(x + 3)(c + 3)} \cdot \frac{1}{x - c} \right]$$

$$= \lim_{x \to c} \left[ \frac{-2(x - c)}{(x + 3)(c + 3)} \cdot \frac{1}{x - c} \right]$$

$$= \lim_{x \to c} \frac{-2}{(x + 3)(c + 3)} = \frac{-2}{(c + 3)^2}$$

Here we manipulated the quotient until we could cancel a factor of $x - c$ from numerator and denominator. Then we could evaluate the limit. ∎

**EXAMPLE 6**    Each of the following is a derivative, but of what function and at what point?

(a) $\displaystyle\lim_{h \to 0} \frac{(4 + h)^2 - 16}{h}$     (b) $\displaystyle\lim_{x \to 3} \frac{\dfrac{2}{x} - \dfrac{2}{3}}{x - 3}$

*Solution*

(a)  This is the derivative of $f(x) = x^2$ at $x = 4$.
(b)  This is the derivative of $f(x) = 2/x$ at $x = 3$.    ∎

**Differentiability Implies Continuity**    If a curve has a tangent line at a point, then that curve cannot take a jump or wiggle too badly at the point. The precise formulation of this fact is an important theorem.

---

### Theorem A

If $f'(c)$ exists, then $f$ is continuous at $c$.

---

***Proof***    We need to show that $\displaystyle\lim_{x \to c} f(x) = f(c)$. We begin by writing $f(x)$ in a fancy way.

$$f(x) = f(c) + \frac{f(x) - f(c)}{x - c} \cdot (x - c), \qquad x \neq c$$

Therefore,

$$\lim_{x \to c} f(x) = \lim_{x \to c} \left[ f(c) + \frac{f(x) - f(c)}{x - c} \cdot (x - c) \right]$$

$$= \lim_{x \to c} f(c) + \lim_{x \to c} \frac{f(x) - f(c)}{x - c} \cdot \lim_{x \to c} (x - c)$$

$$= f(c) + f'(c) \cdot 0$$

$$= f(c) \qquad ∎$$

The converse of this theorem is false. If a function $f$ is continuous at $c$, it does not follow that $f$ has a derivative at $c$. This is easily seen by considering $f(x) = |x|$ at the origin (Figure 3). This function is certainly continuous at zero. However, it does not have a derivative there, as we now show. Note that

$$\frac{f(0 + h) - f(0)}{h} = \frac{|0 + h| - |0|}{h} = \frac{|h|}{h}$$

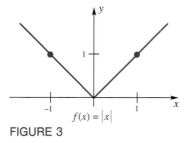

$f(x) = |x|$

**FIGURE 3**

Thus,

$$\lim_{h \to 0^+} \frac{f(0 + h) - f(0)}{h} = \lim_{h \to 0^+} \frac{|h|}{h} = \lim_{h \to 0^+} \frac{h}{h} = 1$$

### Surprise!!!

The converse to Theorem A is spectacularly false. It came as a great surprise to mathematicians when they discovered functions that were continuous everywhere but differentiable nowhere. The first three steps in the construction of such a function are shown below. Continue the process ad infinitum. In the limit, you will obtain a continuous nondifferentiable function. If this interests you, ask your teacher for more information or look at any book with the title *Real Analysis*.

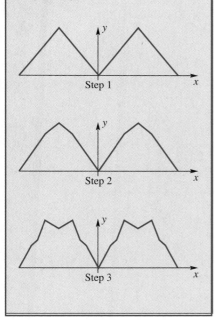

whereas

$$\lim_{h \to 0^-} \frac{f(0 + h) - f(0)}{h} = \lim_{h \to 0^-} \frac{|h|}{h} = \lim_{h \to 0^-} \frac{-h}{h} = -1$$

Since the right- and left-hand limits are different,

$$f'(0) = \lim_{h \to 0} \frac{f(0 + h) - f(0)}{h}$$

does not exist.

A similar argument shows that at any point where the graph of a function has a sharp corner, it is continuous but not differentiable. The graph in Figure 4 indicates a number of ways for a function to be nondifferentiable at a point.

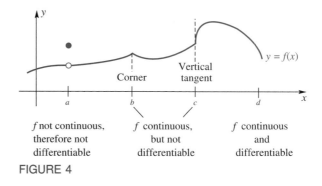

**FIGURE 4**

We claim in Figure 4 that the derivative does not exist at the point $c$ where the tangent line is vertical. This is because

$$\frac{f(c + h) - f(c)}{h}$$

increases without bound as $h \to 0$. It corresponds to the fact that the slope of a vertical line is undefined.

## CONCEPTS REVIEW

**1.** The derivative of $f$ at $c$ is given by $f'(c) =$ $\lim_{h \to 0}$ _____ . Equivalently $f'(c) = \lim_{t \to c}$ _____ .

**2.** The derivative of $f$ at $x$ is given by $f'(x) =$ $\lim_{h \to 0}$ _____ .

**3.** If $f$ is differentiable at $c$, then $f$ is _____ at $c$.

The converse is false, as is shown by the example $f(x) =$ _____ .

**4.** $\lim_{h \to 0} \dfrac{2(c + h)^2 - 2c^2}{h}$ is the derivative of $f(x) =$ _____ at $x =$ _____ .

## PROBLEM SET 3.2

In Problems 1–4, use the definition

$$f'(c) = \lim_{h \to 0} \frac{f(c + h) - f(c)}{h}$$

to find the indicated derivative (see Examples 1 and 2).

**1.** $f'(1)$ if $f(x) = x^2$

**2.** $f'(2)$ if $f(t) = (2t)^2$

**3.** $f'(3)$ if $f(t) = t^2 - t$

**4.** $f'(4)$ if $f(s) = \dfrac{1}{s - 1}$

In Problems 5–22, use $f'(x) = \lim\limits_{h \to 0} [f(x + h) - f(x)]/h$ to find the derivative at $x$ (see Examples 3 and 4).

**5.** $s(x) = 2x + 1$        **6.** $f(x) = \alpha x + \beta$

**7.** $r(x) = 3x^2 + 4$        **8.** $f(x) = x^2 + x + 1$

**9.** $f(x) = ax^2 + bx + c$    **10.** $f(x) = x^4$

**11.** $f(x) = x^3 + 2x^2 + 1$   **12.** $g(x) = x^4 + x^2$

**13.** $h(x) = \dfrac{2}{x}$        **14.** $S(x) = \dfrac{1}{x + 1}$

**15.** $F(x) = \dfrac{6}{x^2 + 1}$     **16.** $F(x) = \dfrac{x - 1}{x + 1}$

**17.** $G(x) = \dfrac{2x - 1}{x - 4}$    **18.** $G(x) = \dfrac{2x}{x^2 - x}$

**19.** $g(x) = \sqrt{3x}$        **20.** $g(x) = \dfrac{1}{\sqrt{3x}}$

**21.** $H(x) = \dfrac{3}{\sqrt{x - 2}}$    **22.** $H(x) = \sqrt{x^2 + 4}$

In Problems 23–26, use $f'(x) = \lim\limits_{t \to x} [f(t) - f(x)]/[t - x]$ to find $f'(x)$ (see Example 5).

**23.** $f(x) = x^2 - 3x$        **24.** $f(x) = x^3 + 5x$

**25.** $f(x) = \dfrac{x}{x - 5}$        **26.** $f(x) = \dfrac{x + 3}{x}$

In Problems 27–36, the given limit is a derivative, but of what function and at what point? (See Example 6.)

**27.** $\lim\limits_{h \to 0} \dfrac{2(5 + h)^3 - 2(5)^3}{h}$

**28.** $\lim\limits_{h \to 0} \dfrac{(3 + h)^2 + 2(3 + h) - 15}{h}$

**29.** $\lim\limits_{x \to 2} \dfrac{x^2 - 4}{x - 2}$        **30.** $\lim\limits_{x \to 3} \dfrac{x^3 + x - 30}{x - 3}$

**31.** $\lim\limits_{t \to x} \dfrac{t^2 - x^2}{t - x}$        **32.** $\lim\limits_{p \to x} \dfrac{p^3 - x^3}{p - x}$

**33.** $\lim\limits_{x \to t} \dfrac{\frac{2}{x} - \frac{2}{t}}{x - t}$        **34.** $\lim\limits_{x \to y} \dfrac{\sin x - \sin y}{x - y}$

**35.** $\lim\limits_{h \to 0} \dfrac{\cos(x + h) - \cos x}{h}$

**36.** $\lim\limits_{h \to 0} \dfrac{\tan(t + h) - \tan t}{h}$

**37.** From Figure 5, estimate $f'(0)$, $f'(2)$, $f'(5)$, and $f'(7)$.

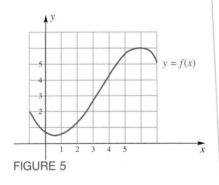

**FIGURE 5**

**38.** From Figure 6, estimate $g'(-1)$, $g'(1)$, $g'(4)$, and $g'(6)$.

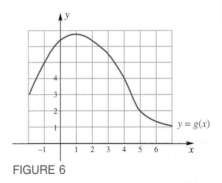

**FIGURE 6**

**39.** Sketch the graph of $y = f'(x)$ on $-1 < x < 7$ for the function $f$ of Problem 37.

**40.** Sketch the graph of $y = g'(x)$ on $-1 < x < 7$ for the function $g$ of Problem 38.

**41.** Consider the function $y = f(x)$, whose graph is sketched in Figure 7.

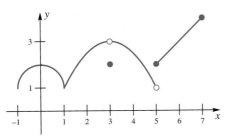

**FIGURE 7**

(a) Estimate $f(2)$, $f'(2)$, $f(0.5)$, and $f'(0.5)$.
(b) Estimate the average rate of change in $f$ on the interval $0.5 \le x \le 2.5$.
(c) Where on the interval $-1 < x < 7$ does $\lim_{u \to x} f(u)$ fail to exist?
(d) Where on the interval $-1 < x < 7$ does $f$ fail to be continuous?
(e) Where on the interval $-1 < x < 7$ does $f$ fail to have a derivative?
(f) Where on the interval $-1 < x < 7$ is $f'(x) = 0$?
(g) Where on the interval $-1 < x < 7$ is $f'(x) = 1$?

42. Suppose that $f(x + y) = f(x)f(y)$ for all $x$ and $y$. Show that if $f'(0)$ exists, then $f'(a)$ exists and $f'(a) = f(a)f'(0)$.

43. Let $f(x) = \begin{cases} mx + b & \text{if } x < 2 \\ x^2 & \text{if } x \ge 2 \end{cases}$

Determine $m$ and $b$ so that $f$ is differentiable everywhere.

44. The **symmetric derivative** $f_s(x)$ is defined by

$$f_s(x) = \lim_{h \to 0} \frac{f(x + h) - f(x - h)}{2h}$$

Show that if $f'(x)$ exists, then $f_s(x)$ exists but that the converse is false.

45. Let $f$ be differentiable and let $f'(x_0) = m$. Find $f'(-x_0)$ if (a) $f$ is an odd function; (b) $f$ is an even function.

46. Prove that the derivative of an odd function is an even function and that the derivative of an even function is an odd function.

PC Use a computer to do Problems 47–48. In *True BASIC Calculus*, the section labeled *General* will work nicely.

47. Draw the graphs of $f(x) = x^3 - 4x^2 + 3$ and its derivative $f'(x)$ on the interval $[-2, 5]$ using the same axes.
(a) Where on this interval is $f'(x) < 0$?
(b) Where on this interval is $f(x)$ decreasing as $x$ increases?
(c) Make a conjecture. Experiment with other intervals and other functions to confirm this conjecture.

48. Draw the graphs of $f(x) = \cos x - \sin(x/2)$ and its derivative $f'(x)$ on the interval $[0, 9]$ using the same axes.
(a) Where on this interval is $f'(x) > 0$?
(b) Where on this interval is $f(x)$ increasing as $x$ increases?
(c) Make a conjecture. Experiment with other intervals and other functions to confirm this conjecture.

**Answers to Concepts Review:** **1.** $[f(c + h) - f(c)]/h$, $[f(t) - f(c)]/(t - c)$   **2.** $[f(x + h) - f(x)]/h$   **3.** Continuous; $|x|$   **4.** $2x^2$; $c$

---

## 3.3
## RULES FOR FINDING DERIVATIVES

The process of finding the derivative of a function directly from the definition of the derivative, that is, by setting up the difference quotient

$$\frac{f(x + h) - f(x)}{h}$$

and evaluating its limit, can be time-consuming and tedious. We are going to develop tools that will allow us to shortcut this lengthy process—that will, in fact, allow us to find derivatives of the most complicated looking functions almost instantly.

Recall that the derivative of a function $f$ is another function $f'$. For example, if $f(x) = x^2$ is the formula for $f$, then $f'(x) = 2x$ is the formula for $f'$. Taking the derivative of $f$ (differentiating $f$) is operating on $f$ to produce $f'$. We often use the letter $D_x$ to indicate this operation (Figure 1). Thus we write $D_x f = f'$, $D_x f(x) = f'(x)$, or (in the example just mentioned) $D_x(x^2) = 2x$. All the theorems below are stated both in functional notation and in the operator $D_x$ notation.

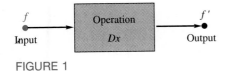

**FIGURE 1**

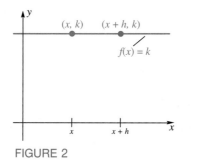

FIGURE 2

**The Constant and Power Rules** The graph of the constant function $f(x) = k$ is a horizontal line (Figure 2) which, therefore, has slope zero everywhere. This is one way to understand our first theorem.

### Theorem A

**(Constant Function Rule).** If $f(x) = k$, where $k$ is a constant, then for any $x, f'(x) = 0$—that is,

$$D_x(k) = 0$$

**Proof**

$$f'(x) = \lim_{h \to 0} \frac{f(x + h) - f(x)}{h} = \lim_{h \to 0} \frac{k - k}{h} = \lim_{h \to 0} 0 = 0 \qquad \blacksquare$$

The graph of $f(x) = x$ is a line through the origin with slope 1 (Figure 3); so we should expect the derivative of this function to be 1 for all $x$.

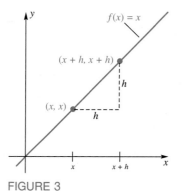

FIGURE 3

### Theorem B

**(Identity Function Rule).** If $f(x) = x$, then $f'(x) = 1$—that is,

$$D_x(x) = 1$$

**Proof**

$$f'(x) = \lim_{h \to 0} \frac{f(x + h) - f(x)}{h} = \lim_{h \to 0} \frac{x + h - x}{h} = \lim_{h \to 0} \frac{h}{h} = 1 \qquad \blacksquare$$

Before stating our next theorem, we recall something from algebra: how to raise a binomial to a power.

$$(a + b)^2 = a^2 + 2ab + b^2$$
$$(a + b)^3 = a^3 + 3a^2b + 3ab^2 + b^3$$
$$(a + b)^4 = a^4 + 4a^3b + 6a^2b^2 + 4ab^3 + b^4$$
$$\vdots$$
$$(a + b)^n = a^n + na^{n-1}b + \frac{n(n-1)}{2} a^{n-2}b^2 + \cdots + nab^{n-1} + b^n$$

### Theorem C

**(Power Rule).** If $f(x) = x^n$, where $n$ is a positive integer, then $f'(x) = nx^{n-1}$—that is,

$$D_x(x^n) = nx^{n-1}$$

**Proof**

$$f'(x) = \lim_{h \to 0} \frac{f(x + h) - f(x)}{h} = \lim_{h \to 0} \frac{(x + h)^n - x^n}{h}$$

$$= \lim_{h \to 0} \frac{x^n + nx^{n-1}h + \dfrac{n(n-1)}{2} x^{n-2}h^2 + \cdots + nxh^{n-1} + h^n - x^n}{h}$$

$$= \lim_{h \to 0} \frac{\not h \left[ nx^{n-1} + \dfrac{n(n-1)}{2} x^{n-2}h + \cdots + nxh^{n-2} + h^{n-1} \right]}{\not h}$$

Within the brackets, all terms except the first have $h$ as a factor, and so each of these terms has limit zero as $h$ approaches zero. Thus

$$f'(x) = nx^{n-1} \qquad \blacksquare$$

As illustrations of Theorem C, note that

$$D_x(x^3) = 3x^2, \qquad D_x(x^9) = 9x^8, \qquad D_x(x^{100}) = 100x^{99}$$

**$D_x$ is a Linear Operator**    The operator $D_x$ behaves very well when applied to constant multiples of functions or to sums of functions.

Theorem D

**(Constant Multiple Rule).** If $k$ is a constant and $f$ is a differentiable function, then $(kf)'(x) = k \cdot f'(x)$—that is,

$$D_x[k \cdot f(x)] = k \cdot D_x f(x)$$

In words, this says that *a constant multiplier $k$ can be passed across the operator $D_x$.*

**Proof**    Let $F(x) = k \cdot f(x)$. Then

$$F'(x) = \lim_{h \to 0} \frac{F(x + h) - F(x)}{h} = \lim_{h \to 0} \frac{k \cdot f(x + h) - k \cdot f(x)}{h}$$

$$= \lim_{h \to 0} k \cdot \frac{f(x + h) - f(x)}{h} = k \cdot \lim_{h \to 0} \frac{f(x + h) - f(x)}{h}$$

$$= k \cdot f'(x)$$

The next-to-the-last step was the critical one. We could shift $k$ past the limit sign because of the Main Limit Theorem (Part 3).    $\blacksquare$

Examples that illustrate this result are

$$D_x(-7x^3) = -7D_x(x^3) = -7 \cdot 3x^2 = -21x^2$$

and

$$D_x(\tfrac{4}{3}x^9) = \tfrac{4}{3}D_x(x^9) = \tfrac{4}{3} \cdot 9x^8 = 12x^8$$

### Theorem E

**(Sum Rule).** If $f$ and $g$ are differentiable functions, then $(f + g)'(x) = f'(x) + g'(x)$—that is,

$$D_x[f(x) + g(x)] = D_x f(x) + D_x g(x)$$

In words, this says that *the derivative of a sum is the sum of the derivatives.*

***Proof***   Let $F(x) = f(x) + g(x)$. Then

$$F'(x) = \lim_{h \to 0} \frac{[f(x + h) + g(x + h)] - [f(x) + g(x)]}{h}$$

$$= \lim_{h \to 0} \left[ \frac{f(x + h) - f(x)}{h} + \frac{g(x + h) - g(x)}{h} \right]$$

$$= \lim_{h \to 0} \frac{f(x + h) - f(x)}{h} + \lim_{h \to 0} \frac{g(x + h) - g(x)}{h}$$

$$= f'(x) + g'(x)$$

Again, the next-to-the-last step was the critical one. It is justified by appealing to the Main Limit Theorem (Part 4).   ∎

Any operator $L$ with the properties stated in Theorems D and E is called *linear*; that is, $L$ is a **linear operator** if:

1. $L(kf) = kL(f)$, $k$ a constant;
2. $L(f + g) = L(f) + L(g)$.

Linear operators will appear again and again in this book; $D_x$ is a particularly important example. A linear operator always satisfies the difference rule $L(f - g) = L(f) - L(g)$ stated next for $D$.

### Linear Operator

The fundamental meaning of the word *linear* as used in mathematics is that given in this section. An operator $L$ is linear if it satisfies the two key conditions:

- $L(ku) = kL(u)$
- $L(u + v) = L(u) + L(v)$

Linear operators play the central role in the *linear algebra* course that many readers of this book will eventually take.

Unfortunately, functions of the form $f(x) = mx + b$ are called linear functions (because of their connection with lines) even though they are not linear in the operator sense. To see this, note that

$$f(kx) = mkx + b$$

whereas

$$kf(x) = k(mx + b)$$

Thus, $f(kx) \neq kf(x)$ unless $b$ happens to be zero.

### Theorem F

**(Difference Rule).** If $f$ and $g$ are differentiable functions, then $(f - g)'(x) = f'(x) - g'(x)$—that is,

$$D_x[f(x) - g(x)] = D_x f(x) - D_x g(x)$$

*Proof*

$$D_x[f(x) - g(x)] = D_x[f(x) + (-1)g(x)]$$
$$= D_x\, f(x) + D_x[(-1)g(x)] \qquad \text{(Theorem E)}$$
$$= D_x\, f(x) + (-1)D_x g(x) \qquad \text{(Theorem D)}$$
$$= D_x\, f(x) - D_x\, g(x)$$

■

**EXAMPLE 1** Find the derivatives of $5x^2 + 7x - 6$ and $4x^6 - 3x^5 - 10x^2 + 5x + 16$.

*Solution*

$$D_x(5x^2 + 7x - 6) = D_x(5x^2 + 7x) - D_x(6) \qquad \text{(Theorem F)}$$
$$= D_x(5x^2) + D_x(7x) - D_x(6) \qquad \text{(Theorem E)}$$
$$= 5D_x(x^2) + 7D_x(x) - D_x(6) \qquad \text{(Theorem D)}$$
$$= 5 \cdot 2x + 7 \cdot 1 + 0 \qquad \text{(Theorems C, B, A)}$$
$$= 10x + 7$$

To find the next derivative, we note that the theorems on sums and differences extend to any finite number of terms. Thus,

$$D_x(4x^6 - 3x^5 - 10x^2 + 5x + 16)$$
$$= D_x(4x^6) - D_x(3x^5) - D_x(10x^2) + D_x(5x) + D_x(16)$$
$$= 4D_x(x^6) - 3D_x(x^5) - 10D_x(x^2) + 5D_x(x) + D_x(16)$$
$$= 4(6x^5) - 3(5x^4) - 10(2x) + 5(1) + 0$$
$$= 24x^5 - 15x^4 - 20x + 5$$

■

The method of Example 1 allows us to find the derivative of any polynomial. If you know the Power Rule and do what comes naturally, you are almost sure to get the right result. If you can write the answer without any intermediate steps, that is fine.

**Product and Quotient Rules**  Now we are in for a surprise. The derivative of a product of functions is *not* equal to the product of the derivatives of the functions.

Theorem G

**(Product Rule).** If $f$ and $g$ are differentiable functions, then $(f \cdot g)'(x) = f(x)g'(x) + g(x)f'(x)$—that is,

$$D_x[f(x)g(x)] = f(x)D_x g(x) + g(x)D_x f(x)$$

This should be memorized in words as follows: *The derivative of a product of two functions is the first times the derivative of the second plus the second times the derivative of the first.*

**Proof**  Let $F(x) = f(x)g(x)$. Then

$$F'(x) = \lim_{h \to 0} \frac{F(x + h) - F(x)}{h}$$

$$= \lim_{h \to 0} \frac{f(x + h)g(x + h) - f(x)g(x)}{h}$$

$$= \lim_{h \to 0} \frac{f(x + h)g(x + h) - f(x + h)g(x) + f(x + h)g(x) - f(x)g(x)}{h}$$

$$= \lim_{h \to 0} \left[ f(x + h) \cdot \frac{g(x + h) - g(x)}{h} + g(x) \cdot \frac{f(x + h) - f(x)}{h} \right]$$

$$= \lim_{h \to 0} f(x + h) \cdot \lim_{h \to 0} \frac{g(x + h) - g(x)}{h} + g(x) \cdot \lim_{h \to 0} \frac{f(x + h) - f(x)}{h}$$

$$= f(x)g'(x) + g(x)f'(x)$$

The derivation just given relied first on the trick of adding and subtracting the same thing, namely, $f(x + h)g(x)$. Second, at the very end, we used the fact that

$$\lim_{h \to 0} f(x + h) = f(x)$$

This is just an application of Theorem 3.2A, which says that differentiability at a point implies continuity there.  ∎

**EXAMPLE 2**  Find the derivative of $(3x^2 - 5)(2x^4 - x)$ by use of the Product Rule. Check the answer by doing the problem a different way.

*Solution*

$$D_x[(3x^2 - 5)(2x^4 - x)] = (3x^2 - 5)D_x(2x^4 - x) + (2x^4 - x)D_x(3x^2 - 5)$$

$$= (3x^2 - 5)(8x^3 - 1) + (2x^4 - x)(6x)$$

$$= 24x^5 - 3x^2 - 40x^3 + 5 + 12x^5 - 6x^2$$

$$= 36x^5 - 40x^3 - 9x^2 + 5$$

To check, we first multiply and then take the derivative.

$$(3x^2 - 5)(2x^4 - x) = 6x^6 - 10x^4 - 3x^3 + 5x$$

Thus,

$$D_x[(3x^2 - 5)(2x^4 - x)] = D_x(6x^6) - D_x(10x^4) - D_x(3x^3) + D_x(5x)$$
$$= 36x^5 - 40x^3 - 9x^2 + 5 \qquad \blacksquare$$

---

**Theorem H**

**(Quotient Rule).** Let $f$ and $g$ be differentiable functions with $g(x) \neq 0$. Then

$$\left(\frac{f}{g}\right)'(x) = \frac{g(x)f'(x) - f(x)g'(x)}{g^2(x)}$$

That is,

$$D_x\left(\frac{f(x)}{g(x)}\right) = \frac{g(x)D_x f(x) - f(x)D_x g(x)}{g^2(x)}$$

---

We strongly urge you to memorize this in words, as follows: *The derivative of a quotient is equal to the denominator times the derivative of the numerator minus the numerator times the derivative of the denominator, all divided by the square of the denominator.*

**_Proof_**   Let $F(x) = f(x)/g(x)$. Then

$$F'(x) = \lim_{h \to 0} \frac{F(x + h) - F(x)}{h}$$

$$= \lim_{h \to 0} \frac{\dfrac{f(x + h)}{g(x + h)} - \dfrac{f(x)}{g(x)}}{h}$$

$$= \lim_{h \to 0} \frac{g(x)f(x + h) - f(x)g(x + h)}{h} \cdot \frac{1}{g(x)g(x + h)}$$

$$= \lim_{h \to 0} \left[\frac{g(x)f(x + h) - g(x)f(x) + f(x)g(x) - f(x)g(x + h)}{h} \cdot \frac{1}{g(x)g(x + h)}\right]$$

$$= \lim_{h \to 0} \left\{\left[g(x)\frac{f(x + h) - f(x)}{h} - f(x)\frac{g(x + h) - g(x)}{h}\right] \frac{1}{g(x)g(x + h)}\right\}$$

$$= [g(x)f'(x) - f(x)g'(x)] \frac{1}{g(x)g(x)} \qquad \blacksquare$$

**EXAMPLE 3** Find the derivative of $\dfrac{(3x - 5)}{(x^2 + 7)}$.

*Solution*

$$D_x\left[\frac{3x - 5}{x^2 + 7}\right] = \frac{(x^2 + 7)D_x(3x - 5) - (3x - 5)D_x(x^2 + 7)}{(x^2 + 7)^2}$$

$$= \frac{(x^2 + 7)(3) - (3x - 5)(2x)}{(x^2 + 7)^2}$$

$$= \frac{-3x^2 + 10x + 21}{(x^2 + 7)^2}$$  ■

**EXAMPLE 4** Find $D_x y$ if $y = \dfrac{2}{x^4 + 1} + \dfrac{3}{x}$.

*Solution*

$$D_x y = D_x\left(\frac{2}{x^4 + 1}\right) + D_x\left(\frac{3}{x}\right)$$

$$= \frac{(x^4 + 1)D_x(2) - 2D_x(x^4 + 1)}{(x^4 + 1)^2} + \frac{xD_x(3) - 3D_x(x)}{x^2}$$

$$= \frac{(x^4 + 1)(0) - (2)(4x^3)}{(x^4 + 1)^2} + \frac{(x)(0) - (3)(1)}{x^2}$$

$$= \frac{-8x^3}{(x^4 + 1)^2} - \frac{3}{x^2}$$  ■

**EXAMPLE 5** Show that the Power Rule holds for negative integral exponents; that is,

$$\boxed{D_x(x^{-n}) = -nx^{-n-1}}$$

*Solution*

$$D_x(x^{-n}) = D_x\left(\frac{1}{x^n}\right) = \frac{x^n \cdot 0 - 1 \cdot nx^{n-1}}{x^{2n}} = \frac{-nx^{n-1}}{x^{2n}} = -nx^{-n-1}$$  ■

We saw as part of Example 4 that $D_x(3/x) = -3/x^2$. Now we have another way to see the same thing.

$$D_x\left(\frac{3}{x}\right) = D_x(3x^{-1}) = 3D_x(x^{-1}) = 3(-1)x^{-2} = -\frac{3}{x^2}$$

## CONCEPTS REVIEW

**1.** The derivative of a product of two functions is the first times _____ plus the _____ times the derivative of the first. In symbols, $D_x[f(x)g(x)]$ = _____.

**2.** The derivative of a quotient is the _____ times the derivative of the numerator minus the numerator times the derivative of the _____ all divided by the _____. In symbols, $D_x[f(x)g(x)]$ = _____.

**3.** The second term (the term involving $h$) in the expansion of $(x + h)^n$ is _____. It is this fact that leads to the formula $D_x[x^n]$ = _____.

**4.** $L$ is called a linear operator if $L(kf)$ = _____ and $L(f + g)$ = _____. The derivative operator denoted by _____ is such an operator.

## PROBLEM SET 3.3

In Problems 1–44, find $D_x y$ using the rules of this section.

**1.** $y = 2x^2$

**2.** $y = 3x^3$

**3.** $y = \pi x$

**4.** $y = \pi x^3$

**5.** $y = 2x^{-2}$

**6.** $y = -3x^{-4}$

**7.** $y = \dfrac{\pi}{x}$

**8.** $y = \dfrac{\alpha}{x^3}$

**9.** $y = \dfrac{100}{x^5}$

**10.** $y = \dfrac{3\alpha}{4x^5}$

**11.** $y = x^2 + 2x$

**12.** $y = 3x^4 + x^3$

**13.** $y = x^4 + x^3 + x^2 + x + 1$

**14.** $y = 3x^4 - 2x^3 - 5x^2 + \pi x + \pi^2$

**15.** $y = \pi x^7 - 2x^5 - 5x^{-2}$

**16.** $y = x^{12} + 5x^{-2} - \pi x^{-10}$

**17.** $y = \dfrac{3}{x^3} + x^{-4}$

**18.** $y = 2x^{-6} + x^{-1}$

**19.** $y = \dfrac{2}{x} - \dfrac{1}{x^2}$

**20.** $y = \dfrac{3}{x^3} - \dfrac{1}{x^4}$

**21.** $y = \dfrac{1}{2x} + 2x$

**22.** $y = \dfrac{2}{3x} - \dfrac{2}{3}$

**23.** $y = x(x^2 + 1)$

**24.** $y = 3x(x^3 - 1)$

**25.** $y = (2x + 1)^2$

**26.** $y = (-3x + 2)^2$

**27.** $y = (x^2 + 2)(x^3 + 1)$

**28.** $y = (x^4 - 1)(x^2 + 1)$

**29.** $y = (x^2 + 17)(x^3 - 3x + 1)$

**30.** $y = (x^4 + 2x)(x^3 + 2x^2 + 1)$

**31.** $y = (5x^2 - 7)(3x^2 - 2x + 1)$

**32.** $y = (3x^2 + 2x)(x^4 - 3x + 1)$

**33.** $y = \dfrac{1}{3x^2 + 1}$

**34.** $y = \dfrac{2}{5x^2 - 1}$

**35.** $y = \dfrac{1}{4x^2 - 3x + 9}$

**36.** $y = \dfrac{4}{2x^3 - 3x}$

**37.** $y = \dfrac{x - 1}{x + 1}$

**38.** $y = \dfrac{2x - 1}{x - 1}$

**39.** $y = \dfrac{2x^2 - 1}{3x + 5}$

**40.** $y = \dfrac{5x - 4}{3x^2 + 1}$

**41.** $y = \dfrac{2x^2 - 3x + 1}{2x + 1}$

**42.** $y = \dfrac{5x^2 + 2x - 6}{3x - 1}$

**43.** $y = \dfrac{x^2 - x + 1}{x^2 + 1}$

**44.** $y = \dfrac{x^2 - 2x + 5}{x^2 + 2x - 3}$

**45.** If $f(0) = 4, f'(0) = -1, g(0) = -3$, and $g'(0) = 5$, find (a) $(f \cdot g)'(0)$; (b) $(f + g)'(0)$; (c) $(f/g)'(0)$.

**46.** If $f(3) = 7, f'(3) = 2, g(3) = 6$, and $g'(3) = -10$, find (a) $(f - g)'(3)$; (b) $(f \cdot g)'(3)$; (c) $(g/f)'(3)$.

**47.** Use the Product Rule to show that $D_x[f(x)]^2 = 2 \cdot f(x) \cdot D_x f(x)$.

**48.** Develop a rule for $D_x[f(x)g(x)h(x)]$.

**49.** Find the equation of the tangent line to $y = x^2 - 2x + 2$ at the point $(1, 1)$.

**50.** Find the equation of the tangent line to $y = 1/(x^2 + 4)$ at the point $(1, 1/5)$.

**51.** Find all points on the graph of $y = x^3 - x^2$ where the tangent line is horizontal.

**52.** Find all points on the graph of $y = \frac{1}{3}x^3 + x^2 - x$ where the tangent line has slope 1.

**53.** The height $s$ in feet of a ball above the ground at $t$ seconds is given by $s = -16t^2 + 40t + 100$.

(a) What is its instantaneous velocity at $t = 2$?
(b) When is its instantaneous velocity 0?

**54.** A ball rolls down a long inclined plane so that its distance $s$ from its starting point after $t$ seconds is $s = 4.5t^2 + 2t$ feet. When will its instantaneous velocity be 30 feet per second?

**55.** There are two tangent lines to the curve $y = 4x - x^2$ that go through $(2, 5)$. Find the equations of both of them. *Hint*: Let $(x_0, y_0)$ be the point of tangency. Find two conditions that $(x_0, y_0)$ must satisfy. See Figure 4.

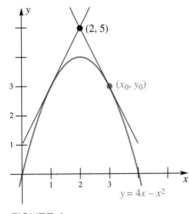

FIGURE 4

**56.** A space traveler is moving from left to right along the curve $y = x^2$. When she shuts off the engines, she will go off along the tangent line at the point where she is at that time. At what point should she shut off the engines in order to reach the point $(4, 15)$?

**57.** A fly is crawling from left to right along the top of the curve $y = 7 - x^2$ (Figure 5). A spider waits at the point $(4, 0)$. Find the distance between the two insects when they first see each other.

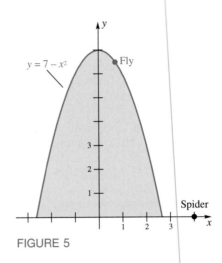

FIGURE 5

**58.** Let $P(a, b)$ be a point on the first quadrant portion of the curve $y = 1/x$ and let the tangent line at $P$ intersect the $x$-axis at $A$. Show that triangle $AOP$ is isosceles and determine its area.

**59.** The radius of a spherical watermelon is growing at a constant rate of 2 centimeters per week. The thickness of the rind is always one-tenth of the radius. How fast is the volume of the rind growing at the end of the fifth week? Assume the radius is initially 0.

PC **60.** Redo a sample of Problems 1–44 on a computer and compare your answers with those you got by hand.

---

**Answers to Concepts Review:** **1.** The derivative of the second; second; $f(x)g'(x) + g(x)f'(x)$   **2.** Denominator; denominator, square of the denominator; $[g(x)f'(x) - f(x)g'(x)]/g^2(x)$   **3.** $nx^{n-1}h$; $nx^{n-1}$   **4.** $kL(f)$; $L(f) + L(g)$; $D_x$

---

**3.4**
**DERIVATIVES OF SINES AND COSINES**

Our modern world runs on wheels. Questions about rotating wheels and velocities of points on them lead inevitably to the study of sines and cosines and their derivatives. To prepare for this study it would be well to review Section 2.3. Figure 1 reminds us of the definition of the sine and cosine functions. In what follows, $t$ should be thought of as a number measuring the length of an arc on the unit circle or, equivalently, as the number of radians in the corresponding angle. Thus, $f(t) = \sin t$ and $g(t) = \cos t$ are functions for which both domain and range are sets of real numbers. We may consider the problem of finding their derivatives.

**The Derivative Formulas**   We choose to use $x$ rather than $t$ as our basic variable. To find $D_x(\sin x)$, we appeal to the definition of derivative and use the addition identity for $\sin(x + h)$.

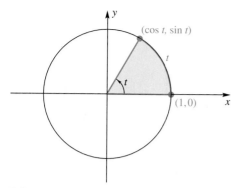

FIGURE 1

$$D_x(\sin x) = \lim_{h \to 0} \frac{\sin(x + h) - \sin x}{h}$$

$$= \lim_{h \to 0} \frac{\sin x \cos h + \cos x \sin h - \sin x}{h}$$

$$= \lim_{h \to 0} \left( -\sin x \frac{1 - \cos h}{h} + \cos x \frac{\sin h}{h} \right)$$

$$= (-\sin x) \left[ \lim_{h \to 0} \frac{1 - \cos h}{h} \right] + (\cos x) \left[ \lim_{h \to 0} \frac{\sin h}{h} \right]$$

To finish our derivation, we have two limits to evaluate. A calculator provided the table in Figure 2. It suggests and, later in this section, we will prove that

$$\lim_{h \to 0} \frac{1 - \cos h}{h} = 0 \qquad \lim_{h \to 0} \frac{\sin h}{h} = 1$$

| $h$ | $\frac{1 - \cos h}{h}$ | $\frac{\sin h}{h}$ |
|---|---|---|
| 1.0 | 0.45970 | 0.84147 |
| 0.5 | 0.24483 | 0.95885 |
| 0.1 | 0.04996 | 0.99833 |
| 0.01 | 0.00500 | 0.99998 |
| ↓ | ↓ | ↓ |
| 0 | ? | ? |
| ↑ | ↑ | ↑ |
| −0.01 | −0.00500 | 0.99998 |
| −0.1 | −0.04996 | 0.99833 |
| −0.5 | −0.24483 | 0.95885 |
| −1.0 | −0.45970 | 0.84147 |

FIGURE 2

Thus,

$$D_x(\sin x) = (-\sin x) \cdot 0 + (\cos x) \cdot 1 = \cos x$$

Similarly,

$$D_x(\cos x) = \lim_{h \to 0} \frac{\cos(x + h) - \cos x}{h}$$

$$= \lim_{h \to 0} \frac{\cos x \cos h - \sin x \sin h - \cos x}{h}$$

$$= \lim_{h \to 0} \left( -\cos x \frac{1 - \cos h}{h} - \sin x \frac{\sin h}{h} \right)$$

$$= (-\cos x) \cdot 0 - (\sin x) \cdot 1$$

$$= -\sin x$$

We summarize these results in an important theorem.

## Could You Have Guessed?

The blue curve below is the graph of $y = \sin x$. Note that the slope is 1 at 0, 0 at $\pi/2$, −1 at $\pi$, and so on. When we graph the slope function (the derivative), we obtain the red curve. Could you have guessed that $D_x \sin x = \cos x$?

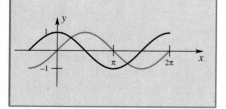

### Theorem A

The functions $f(x) = \sin x$ and $g(x) = \cos x$ are both differentiable. In fact,

$$D_x(\sin x) = \cos x \qquad D_x(\cos x) = -\sin x$$

**EXAMPLE 1**  Find $D_x(3 \sin x - 2 \cos x)$.

***Solution***

$$D_x(3 \sin x - 2 \cos x) = 3 \, D_x(\sin x) - 2 \, D_x(\cos x)$$

$$= 3 \cos x + 2 \sin x \qquad \blacksquare$$

**EXAMPLE 2**  Find $D_x(\tan x)$.

**Solution**

$$D_x(\tan x) = D_x\left(\frac{\sin x}{\cos x}\right)$$

$$= \frac{\cos x\, D_x(\sin x) - \sin x\, D_x(\cos x)}{\cos^2 x}$$

$$= \frac{\cos x \cos x + \sin x \sin x}{\cos^2 x}$$

$$= \frac{1}{\cos^2 x} = \sec^2 x$$  ∎

**EXAMPLE 3**  Find the equation of the tangent line to the graph of $y = 3 \sin 2x$ at the point $(\pi/2, 0)$ (see Figure 3).

**Solution**  We need the derivative of $\sin 2x$; unfortunately, at this point we know how to find only the derivative of $\sin x$. However, $\sin 2x = 2 \sin x \cos x$. Thus,

$$D_x(3 \sin 2x) = D_x(6 \sin x \cos x)$$

$$= 6\, D_x(\sin x \cos x)$$

$$= 6[\sin x\, D_x(\cos x) + \cos x\, D_x(\sin x)]$$

$$= 6[(\sin x)(-\sin x) + \cos x \cos x]$$

$$= 6[\cos^2 x - \sin^2 x]$$

$$= 6 \cos 2x$$

At $x = \pi/2$, this derivative has the value $-6$, which is therefore the slope of the desired tangent line. The equation of this line is

$$y - 0 = -6\left(x - \frac{\pi}{2}\right)$$  ∎

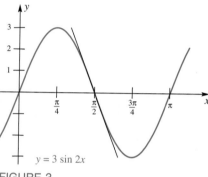

$y = 3 \sin 2x$

FIGURE 3

**EXAMPLE 4**  Consider a ferris wheel of radius 30 feet, which is rotating counterclockwise with an angular velocity of 2 radians per second. How fast is a seat on the rim rising (in the vertical direction) when it is 15 feet above the horizontal line through the center of the wheel?

**Solution**  We may suppose that the wheel is centered at the origin and that the seat $P$ was at $(30, 0)$ at time $t = 0$ (Figure 4). Thus at time $t$, $P$ has moved through an angle of $2t$ radians and so has coordinates $(30 \cos 2t, 30 \sin 2t)$. The rate at which $P$ is rising is just the derivative of the vertical

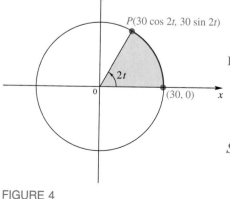

$P(30 \cos 2t, 30 \sin 2t)$

$2t$

$0$

$(30, 0)$

$x$

FIGURE 4

coordinate 30 sin 2*t* measured at an appropriate *t*-value. By Example 3,

$$D_x(30 \sin 2t) = 60 \cos 2t$$

The appropriate *t* at which to evaluate this derivative is $t = \pi/12$, since $30 \sin(2 \cdot \pi/12) = 15$. We conclude that at $t = \pi/12$, the seat *P* is rising at

$$60 \cos \left(2 \cdot \frac{\pi}{12}\right) = 60 \sqrt{3}/2 \approx 51.96 \text{ feet per second} \qquad \blacksquare$$

**Proof of Two Limit Statements**   Everything we have done in this section depends on the two limit statements.

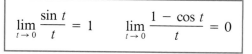

$$\lim_{t \to 0} \frac{\sin t}{t} = 1 \qquad \lim_{t \to 0} \frac{1 - \cos t}{t} = 0$$

They require proofs.

**Proof**  Consider the (by now familiar) diagram in Figure 5. Note that as $t \to 0$, the point *P* moves toward (1, 0) and, therefore, that

$$\lim_{t \to 0} \cos t = 1 \qquad \lim_{t \to 0} \sin t = 0$$

Next, for $-\pi/2 < t < \pi/2, t \neq 0$, draw the vertical line segment *BP* and the circular arc *BC*, as shown in Figure 6. (If $t < 0$, the shaded region will be reflected across the *x*-axis.) Clearly,

$$\text{Area(sector } OBC) \leq \text{area}(\triangle OBP) \leq \text{area(sector } OAP)$$

From the formulas $\frac{1}{2}bh$ (area of a triangle) and $\frac{1}{2}r^2|t|$ (area of a circular sector; see Problem 30 of Section 2.3), we obtain

$$\tfrac{1}{2}(\cos t)^2|t| \leq \tfrac{1}{2} \cos t|\sin t| \leq \tfrac{1}{2}(1)^2|t|$$

or, after multiplying by 2 and dividing by the positive number $|t| \cos t$ and noting that (sin *t*)/*t* is positive,

$$\cos t \leq \frac{\sin t}{t} \leq \frac{1}{\cos t}$$

This double inequality simply begs for the Squeeze Theorem. When we apply it, we get

$$\lim_{t \to 0} \frac{\sin t}{t} = 1$$

the first of our claimed results.

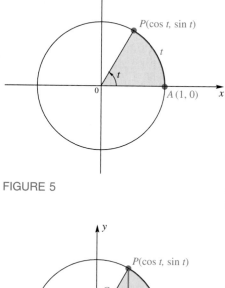

FIGURE 6

The second result follows easily from the first.

$$\lim_{t \to 0} \frac{1 - \cos t}{t} = \lim_{t \to 0} \frac{1 - \cos t}{t} \frac{1 + \cos t}{1 + \cos t} = \lim_{t \to 0} \frac{1 - \cos^2 t}{t(1 + \cos t)}$$

$$= \lim_{t \to 0} \frac{\sin^2 t}{t(1 + \cos t)} = \lim_{t \to 0} \frac{\sin t}{t} \cdot \frac{\displaystyle\lim_{t \to 0} \sin t}{\displaystyle\lim_{t \to 0}(1 + \cos t)}$$

$$= 1 \cdot \frac{0}{2} = 0 \qquad \blacksquare$$

Although the main use of these two limit statements was to prove the derivative formulas, we also can use them to evaluate other limits.

**EXAMPLE 5**  Find the following limits.

(a) $\displaystyle\lim_{t \to 0} \frac{1 - \cos t}{\sin t}$    (b) $\displaystyle\lim_{x \to 0} \frac{\sin 5x}{3x}$

*Solution*

(a) $\displaystyle\lim_{t \to 0} \frac{1 - \cos t}{\sin t} = \lim_{t \to 0} \frac{\dfrac{1 - \cos t}{t}}{\dfrac{\sin t}{t}} = \frac{0}{1} = 0$

(b) $\displaystyle\lim_{x \to 0} \frac{\sin 5x}{3x} = \lim_{x \to 0} \frac{5}{3} \cdot \frac{\sin 5x}{5x} = \frac{5}{3} \lim_{x \to 0} \frac{\sin 5x}{5x} = \frac{5}{3} \cdot 1 = \frac{5}{3}$

At the next-to-the-last step, we used the fact that when $x \to 0$, $5x \to 0$. Thus,

$$\lim_{x \to 0} \frac{\sin 5x}{5x} = \lim_{5x \to 0} \frac{\sin 5x}{5x} = \lim_{t \to 0} \frac{\sin t}{t} = 1 \qquad \blacksquare$$

## CONCEPTS REVIEW

**1.** By definition, $D_x(\sin x) = \displaystyle\lim_{h \to 0}$ _____.

**2.** To evaluate the limit in Question 1, we first use the addition formula for the sine function and then do a little algebra to obtain

$$D_x(\sin x) = (-\sin x)\left(\lim_{h \to 0} \frac{1 - \cos h}{h}\right) + (\cos x)\left(\lim_{h \to 0} \frac{\sin h}{h}\right)$$

The two displayed limits have the values _____ and _____, respectively.

**3.** The result of the calculation in Question 2 is the important derivative formula $D_x(\sin x) =$ _____. The corresponding derivative formula $D_x(\cos x) =$ _____ is obtained in a similar manner.

**4.** At $x = \pi/3$, $D_x(\sin x)$ has the value _____. Thus, the equation of the tangent line to $y = \sin x$ at $x = \pi/3$ is _____.

## PROBLEM SET 3.4

In Problems 1–12, find $D_x y$.

1. $y = 2\sin x + 3\cos x$

2. $y = \sin^2 x$

3. $y = \sin^2 x + \cos^2 x$

4. $y = 1 - \cos^2 x$

5. $y = \sec x = 1/\cos x$

6. $y = \csc x = 1/\sin x$

7. $y = \tan x = \dfrac{\sin x}{\cos x}$

8. $y = \cot x = \dfrac{\cos x}{\sin x}$

9. $y = \dfrac{\sin x + \cos x}{\cos x}$

10. $y = \dfrac{\sin x + \cos x}{\tan x}$

11. $y = x^2 \cos x$

12. $y = \dfrac{x \cos x + \sin x}{x^2 + 1}$

[C] 13. Find the equation of the tangent line to $y = \cos x$ at $x = 1$.

14. Find the equation of the tangent line to $y = \cot x$ at $x = \dfrac{\pi}{4}$.

15. Consider the ferris wheel of Example 4. At what rate is the seat on the rim moving horizontally when $t = \pi/4$ seconds (that is, when the seat reaches the very top of the wheel)?

16. A ferris wheel of radius 20 feet is rotating counter-clockwise at an angular velocity of 1 radian per second. One seat on the rim is at $(20, 0)$ at $t = 0$.

(a) What are its coordinates at $t = \pi/6$?
(b) How fast is it rising (vertically) at $t = \pi/6$?
(c) How fast is it rising (vertically) when it is rising at the fastest rate?

In Problems 17–22, follow the procedure of Example 5 to find each of the limits.

17. $\lim\limits_{x \to 0} \dfrac{\sin x}{2x}$

18. $\lim\limits_{\theta \to 0} \dfrac{\sin 3\theta}{2\theta}$

19. $\lim\limits_{\theta \to 0} \dfrac{\sin 3\theta}{\tan \theta}$

20. $\lim\limits_{\theta \to 0} \dfrac{\tan 5\theta}{\sin 2\theta}$

21. $\lim\limits_{\theta \to 0} \dfrac{\cot \pi\theta \sin \theta}{2 \sec \theta}$

22. $\lim\limits_{t \to 0} \dfrac{(\sin 3t)^2}{2t}$

23. Show that the curves $y = \sqrt{2} \sin x$ and $y = \sqrt{2} \cos x$ intersect at right angles at a certain point with $0 < x < \pi/2$.

24. At time $t$ seconds, the center of a bobbing cork is $2 \sin t$ centimeters above (or below) water level. What is the velocity of the cork at $t = 0, \pi/2, \pi$?

25. Use the definition of derivative to show that $D_x(\sin x^2) = 2x \cos x^2$.

26. Use the definition of derivative to show that $D_x(\sin 5x) = 5 \cos 5x$.

27. Let $x_0$ be the smallest positive value of $x$ at which the curves $y = \sin x$ and $y = \sin 2x$ intersect. Find $x_0$ and also the acute angle at which the two curves intersect at $x_0$ (see Problem 40 of Section 2.3).

28. From area($OBP$) ≤ area(sector $OAP$) ≤ area($OBP$) + area($ABPQ$) in Figure 7, show that

$$\cos t \le \frac{t}{\sin t} \le 2 - \cos t$$

and thus obtain another proof that $\lim\limits_{t \to 0}(\sin t)/t = 1$.

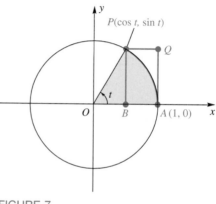

FIGURE 7

29. In Figure 8, let $D$ be the area of triangle $ABP$ and $E$ the area of the shaded region.

(a) Guess the value of $\lim\limits_{t \to 0^+} (D/E)$ by looking at the figure.
(b) Find a formula for $D/E$ in terms of $t$.
[C] (c) Use a calculator to get an accurate estimate of $\lim\limits_{t \to 0^+} (D/E)$.

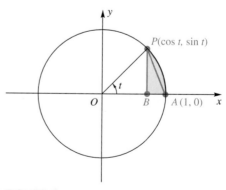

FIGURE 8

*Note*: The exact answer to (c) will be found in Problem 27 of Section 9.1.

30. An isosceles triangle is topped by a semicircle, as shown in Figure 9. Let $D$ be the area of triangle $AOB$ and $E$ be the area of the shaded region. Find a formula for $D/E$ in terms of $t$ and then calculate

$$\lim_{t \to 0^+} \frac{D}{E}$$

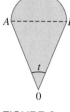

FIGURE 9

PC Problems 31–34 are computer exercises.

31. Redo Problems 7–12 on your computer and thereby check the answers.

32. Redo Problems 17–22 on your computer.

33. Let $f(x) = x \sin x$.

(a) Draw the graphs of $f(x)$ and $f'(x)$ on $[\pi, 6\pi]$.
(b) How many solutions does $f(x) = 0$ have on $[\pi, 6\pi]$? How many solutions does $f'(x) = 0$ have on this interval?
(c) What is wrong with the following conjecture? If $f$ and $f'$ are both continuous and differentiable on $[a, b]$, if $f(a) = f(b) = 0$, and if $f(x) = 0$ has exactly $n$ solutions on $[a, b]$, then $f'(x) = 0$ has exactly $n - 1$ solutions on $[a, b]$.
(d) Determine the maximum value of $|f(x) - f'(x)|$ on $[\pi, 6\pi]$.

34. Let $f(x) = \cos^3 x - 1.25 \cos^2 x + 0.225$. Find $f'(x_0)$ at that point $x_0$ in $[\pi/2, \pi]$ where $f(x_0) = 0$.

**Answers to Concepts Review:** **1.** $[\sin(x + h) - \sin x]/h$ **2.** 0; 1 **3.** $\cos x$; $-\sin x$ **4.** $\frac{1}{2}$; $y - \sqrt{3}/2 = \frac{1}{2}(x - \pi/3)$

---

## 3.5
## THE CHAIN RULE

Imagine trying to find the derivative of

$$F(x) = (2x^2 - 4x + 1)^{60}$$

You would first have to multiply together the 60 quadratic factors of $2x^2 - 4x + 1$ and then differentiate the resulting polynomial of degree 120.

Fortunately, there is a better way. After you have learned the *Chain Rule*, you will be able to write the answer

$$F'(x) = 60(2x^2 - 4x + 1)^{59}(4x - 4)$$

as fast as you can move your pencil. In fact, the *Chain Rule* is so important that you will seldom again differentiate any function without using it. But in order to state the rule properly, we need to emphasize the significance of $x$ in our $D_x$-notation.

**The $D_x$-Notation** The symbol $D_x y$ should be read "the derivative of $y$ with respect to $x$"; it measures how fast $y$ is changing with respect to $x$. The subscript $x$ indicates that $x$ is being treated as the basic variable. Thus, if $y = s^2 x^3$, we may write

$$D_x y = 3s^2 x^2 \qquad \text{and} \qquad D_s y = 2s x^3$$

In the first case, $s$ is treated as a constant and $x$ is the basic variable; in the second case, $x$ is constant and $s$ is the basic variable.

More important is the following example. Suppose $y = u^{60}$ and $u = 2x^2 - 4x + 1$. Then $D_u y = 60u^{59}$ and $D_x u = 4x - 4$. But notice that when we substitute $u = 2x^2 - 4x + 1$ in $y = u^{60}$, we obtain

$$y = (2x^2 - 4x + 1)^{60}$$

so it makes sense to ask for $D_x y$. What is $D_x y$ and how is it related to $D_u y$ and $D_x u$? More generally, how do you differentiate a composite function?

**Differentiating a Composite Function**   If David can type twice as fast as Mary and Mary can type three times as fast as Jack, then David can type $2 \cdot 3 = 6$ times as fast as Jack. The two rates are multiplied.

Suppose that

$$y = f(u) \quad \text{and} \quad u = g(x)$$

determine the composite function $y = f(g(x))$. Since a derivative indicates a rate of change, we can say that

$$y \text{ changes } D_u y \text{ times as fast as } u$$

$$u \text{ changes } D_x u \text{ times as fast as } x$$

It seems reasonable to conclude that

$$y \text{ changes } D_u y \cdot D_x u \text{ times as fast as } x$$

This is in fact true and we will suggest a formal proof in the next section. The result is called the **Chain Rule**.

---

**Theorem A**

**(Chain Rule).** Let $y = f(u)$ and $u = g(x)$ determine the composite function $y = f(g(x)) = (f \circ g)(x)$. If $g$ is differentiable at $x$ and $f$ is differentiable at $u = g(x)$, then $f \circ g$ is differentiable at $x$ and

$$(f \circ g)'(x) = f'(g(x))g'(x)$$

That is,

$$D_x y = D_u y D_x u$$

---

Maybe it will help you to remember it this way.

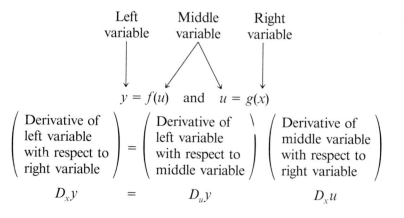

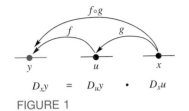

$D_x y = D_u y \cdot D_x u$

FIGURE 1

See Figure 1 for another view. Thought of this way, you should have no trouble seeing that if

$$w = f(s) \quad \text{and} \quad s = g(t)$$

then

$$D_t w = D_s w D_t s$$

**Applications of the Chain Rule** We begin with the example $(2x^2 - 4x + 1)^{60}$ introduced at the beginning of this section.

**EXAMPLE 1** If $y = (2x^2 - 4x + 1)^{60}$, find $D_x y$.

**Solution** We think of this as

$$y = u^{60} \quad \text{and} \quad u = 2x^2 - 4x + 1$$

Thus,

$$
\begin{aligned}
D_x y &= D_u y \cdot D_x u \\
&= (60u^{59})(4x - 4) \\
&= 60(2x^2 - 4x + 1)^{59}(4x - 4)
\end{aligned}
$$
∎

**EXAMPLE 2** If $y = 1/(2x^5 - 7)^3$, find $D_x y$.

**Solution** Think of it this way.

$$y = \frac{1}{u^3} = u^{-3} \quad \text{and} \quad u = 2x^5 - 7$$

Thus,

$$
\begin{aligned}
D_x y &= D_u y \cdot D_x u \\
&= (-3u^{-4})(10x^4) \\
&= \frac{-3}{u^4} \cdot 10x^4 \\
&= \frac{-30x^4}{(2x^5 - 7)^4}
\end{aligned}
$$
∎

**Notation**

The Chain Rule says that
$$D_x \sin x^3 = \cos x^3 \cdot 3x^2$$
but the answer written this way is ambiguous. Does $3x^2$ multiply $x^3$ or $\cos x^3$? We mean, of course, the latter. Therefore, we should either introduce parentheses, as in
$$(\cos x^3)(3x^2)$$
or, better, write the answer as
$$3x^2 \cos x^3$$
Note that we give our final answer in Example 3 this way. It is a practice we normally follow.
Here is another important notational point. When we write $\cos x^3$, we mean $\cos(x^3)$. If we intend $(\cos x)^3$, we write $\cos^3 x$.

**EXAMPLE 3** If $y = \sin(x^3 - 3x)$, find $D_x y$.

**Solution** We may write

$$y = \sin u \qquad \text{and} \qquad u = x^3 - 3x$$

Hence,

$$D_xy = D_uy \cdot D_xu$$
$$= (\cos u) \cdot (3x^2 - 3)$$
$$= [\cos(x^3 - 3x)] \cdot (3x^2 - 3)$$
$$= (3x^2 - 3)\cos(x^3 - 3x) \qquad \blacksquare$$

**EXAMPLE 4**  Find $D_t\left(\dfrac{t^3 - 2t + 1}{t^4 + 3}\right)^{13}.$

*Solution*  Think of this as finding $D_ty$, where

$$y = u^{13} \quad \text{and} \quad u = \frac{t^3 - 2t + 1}{t^4 + 3}$$

Then, the Chain Rule followed by the Quotient Rule gives

$$D_ty = D_uy \cdot D_tu$$
$$= 13u^{12}\,\frac{(t^4 + 3)(3t^2 - 2) - (t^3 - 2t + 1)(4t^3)}{(t^4 + 3)^2}$$
$$= 13\left(\frac{t^3 - 2t + 1}{t^4 + 3}\right)^{12} \cdot \frac{-t^6 + 6t^4 - 4t^3 + 9t^2 - 6}{(t^4 + 3)^2} \qquad \blacksquare$$

Soon you will learn to make a mental introduction of the middle variable without actually writing it. Thus, an expert immediately writes:

$$D_x(\cos 3x) = (-\sin 3x) \cdot 3 = -3\sin 3x$$
$$D_x(x^3 + \sin x)^6 = 6(x^3 + \sin x)^5(3x^2 + \cos x)$$
$$D_t\left(\frac{t}{\cos 3t}\right)^4 = 4\left(\frac{t}{\cos 3t}\right)^3 \frac{\cos 3t - t(-\sin 3t)3}{\cos^2 3t}$$
$$= \frac{4t^3(\cos 3t + 3t\sin 3t)}{\cos^5 3t}$$

**The Compound Chain Rule**   Suppose

$$y = f(u) \quad \text{and} \quad u = g(v) \quad \text{and} \quad v = h(x)$$

Then

$$D_xy = D_uyD_vuD_xv$$

**EXAMPLE 5**  Find $D_x[\sin^3(4x)]$.

*Solution*  Think of this as finding $D_xy$, where

$$y = u^3 \quad \text{and} \quad u = \sin v \quad \text{and} \quad v = 4x$$

---

**The Last First**

Here is an informal rule that may help you in using the derivative rules.

*The last step in calculation corresponds to the first step in differentiation.*

For instance, in calculating $\sin^3(4x)$ for a particular value of $x$, the last step is to cube $\sin(4x)$, so the first rule to use in differentiating $\sin^3(4x)$ is that for $u^3$. Similarly, the last step in calculating

$$\frac{x^2 - 1}{x^2 + 1}$$

is to take the quotient, so the first rule to use in differentiating is the Quotient Rule.

Then,

$$D_x y = D_u y \cdot D_v u \cdot D_x v$$
$$= 3u^2 \cdot \cos v \cdot 4$$
$$= 3 \sin^2(4x) \cdot \cos(4x) \cdot 4$$
$$= 12 \sin^2(4x)\cos(4x) \qquad \blacksquare$$

Here, too, you will soon make these substitutions in your head and write the answer immediately. It may help if you notice that, in differentiating compound composite functions, you work from the outer parentheses inward, like peeling an onion.

Let's do Example 5 again, making explicit what we have just said.

$$D_x[\sin(4x)]^3 = 3 \sin^2(4x)D_x\sin(4x)$$
$$= 3 \sin^2(4x)\cos(4x)D_x(4x)$$
$$= 3 \sin^2(4x)\cos(4x) \cdot 4$$
$$= 12 \sin^2(4x)\cos(4x)$$

**EXAMPLE 6**   Find $D_x\{\sin[\cos(x^2)]\}$.

*Solution*

$$D_x\{\sin[\cos(x^2)]\} = \cos[\cos(x^2)] \cdot [-\sin(x^2)] \cdot 2x \qquad \blacksquare$$

## CONCEPTS REVIEW

**1.** If $y = f(u)$, where $u = g(t)$, then $D_t y = D_u y \cdot$ _____. In functional notation, $(f \circ g)'(t) =$ _____ _____.

**2.** If $w = G(v)$, where $v = H(s)$, then $D_s w =$ _____ $D_s v$. In functional notation, $(G \circ H)'(s) =$ _____ _____.

**3.** If $p = f(r)$, where $r = g(s)$ and where $s = h(t)$, then $D_t p =$ _____ _____ _____.

**4.** If $y = \sin(x^2)[(2x + 1)^3]$, then $D_x y = \sin(x^2)$ _____ $+ (2x + 1)^3$ _____.

## PROBLEM SET 3.5

In Problems 1–26, find $D_x y$.

1. $y = (1 + x)^{15}$

2. $y = (7 + x)^5$

3. $y = (3 - 2x)^5$

4. $y = (4 + 2x^2)^7$

5. $y = (x^3 - 2x^2 + 3x + 1)^{11}$

6. $y = (x^5 - 5x^3 + \pi x + 1)^{101}$

7. $y = (x^3 - 2x^2 + 3x + 1)^{111}$

8. $y = (x^2 - x + 1)^{-7}$

9. $y = \dfrac{1}{(x + 3)^5}$

10. $y = \dfrac{1}{(3x^2 + x - 3)^9}$

11. $y = \sin(x^2 + x)$

12. $y = \cos(3x^2 - 2x)$

13. $y = \cos^3 x$

14. $y = \sin^4(3x^2)$

15. $y = \left(\dfrac{x + 1}{x - 1}\right)^3$

16. $y = \left(\dfrac{x - 2}{x - \pi}\right)^{-3}$

17. $y = \left(\dfrac{3x^2 + x}{1 - 2x^2 - \pi}\right)^3$

18. $y = \cos\left(\dfrac{3x^2}{x + 2}\right)$

19. $y = \cos^3\left(\dfrac{x^2}{1 - x}\right)$

20. $y = \sec^3\left(\dfrac{x}{1 - x^2}\right)$

21. $y = (3x - 2)^2(3 - x^2)^2$

22. $y = (2 - 3x^2)^4(x^7 + 3)^3$

23. $y = \dfrac{(x + 1)^2}{3x - 4}$

24. $y = \dfrac{2x - 3}{(x^2 + 4)^2}$

25. $y = \dfrac{(3x^2 + 2)^2}{2x^2 - 5}$

26. $y = \dfrac{(x^2 - 1)^3}{(4x^3 - 5)^2}$

In Problems 27–34, find the indicated derivative.

27. $D_t \left( \dfrac{3t - 2}{t + 5} \right)^3$

28. $D_s \left( \dfrac{s^2 - 9}{s + 4} \right)$

29. $D_t \left( \dfrac{(3t - 2)^3}{t + 5} \right)$

30. $D_s \left( \dfrac{s^2 - 9}{s + 4} \right)^3$

31. $D_\theta(\sin^3 \theta)$

32. $D_\theta(\cos^4 \theta)$

33. $D_x \left( \dfrac{\sin x}{\cos 2x} \right)^3$

34. $D_t[\sin t \tan(t^2 + 1)]$

In Problems 35–38, evaluate the indicated derivative.

35. $f'(3)$ if $f(x) = \left( \dfrac{x^2 + 1}{x + 2} \right)^3$

36. $G'(1)$ if $G(t) = (t^2 + 9)^3(t^2 - 2)^4$

C  37. $F'(1)$ if $F(t) = \sin(t^2 + 3t + 1)$

38. $g'\left( \tfrac{1}{2} \right)$ if $g(s) = \cos \pi s \sin^2 \pi s$

In Problems 39–46, use the Compound Chain Rule (Example 5) to find the indicated derivative.

39. $D_x[\sin^4(x^2 + 3x)]$

40. $D_t[\cos^5(4t - 19)]$

41. $D_t[\sin^3(\cos t)]$

42. $D_u\left[ \cos^4\left( \dfrac{u + 1}{u - 1} \right) \right]$

43. $D_\theta[\cos^4(\sin \theta^2)]$

44. $D_x[x \sin^2(2x)]$

45. $D_x\{\sin[\cos(\sin 2x)]\}$

46. $D_t\{\cos^2[\cos(\cos t)]\}$

47. Find the equation of the tangent line to $y = (x^2 + 1)^3(x^4 + 1)^2$ at $(1, 32)$.

48. A point $P$ is moving in the plane so that its coordinates after $t$ seconds are $(4 \cos 2t, 7 \sin 2t)$, measured in feet.

(a) Show that $P$ is following an elliptical path. *Hint:* Show $(x/4)^2 + (y/7)^2 = 1$, which is an equation of an ellipse.
(b) Obtain an expression for $L$, its distance from the origin at time $t$.
(c) How fast is $P$ moving away from the origin at $t = \pi/8$? You will need the fact that $D_u(\sqrt{u}) = 1/(2\sqrt{u})$ (see Example 4 of Section 3.2).

49. A wheel centered at the origin and of radius 10 centimeters is rotating counterclockwise at a rate of 4 revolutions per second. A point $P$ on the rim is at $(10, 0)$ at $t = 0$.

(a) What are the coordinates of $P$ at time $t$ seconds?
(b) At what rate is $P$ rising (or falling) at time $t = 1$?

50. Consider the wheel-piston device in Figure 2. The wheel has radius 1 foot and rotates counterclockwise at 2 radians per second. The connecting rod is 5 feet long. The point $P$ is at $(1, 0)$ at time $t = 0$.

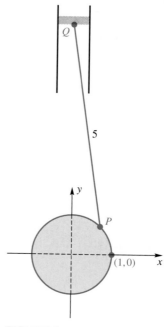

FIGURE 2

(a) Find the coordinates of $P$ at time $t$.
(b) Find the $y$-coordinate of $Q$ at time $t$ (the $x$-coordinate is always zero).
(c) Find the velocity of $Q$ at time $t$. You will need the fact that $D_u(\sqrt{u}) = 1/(2\sqrt{u})$.

51. Do Problem 50 assuming the wheel is rotating at 60 revolutions per minute and $t$ is measured in seconds.

52. Show that $D_x|x| = |x|/x, x \neq 0$. *Hint:* Write $|x| = \sqrt{x^2}$ and use the Chain Rule with $u = x^2$.

53. Apply the result in Problem 52 to find each derivative.

(a) $D_x|x^2 - 1|$

(b) $D_x|\sin x|$

54. Later in the book (Chapter 7), we will study a function $L$ satisfying $L'(x) = 1/x$. Find each of the following derivatives.

(a) $D_x L\left( \dfrac{x}{x + 1} \right)$

(b) $D_x L(\cos^4 x)$

55. In each of the following, find $f'(x)$ and write your answer in simplest form in terms of $\sin 2x$.

(a) $f(x) = -\cos 2x + \tfrac{1}{3} \cos^3 2x$
(b) $f(x) = \tfrac{3}{8}x - \tfrac{3}{32} \sin 4x - \tfrac{1}{8} \sin^3 2x \cos 2x$

**56.** Show that if a polynomial $p(x)$ is divisible by $(ax + b)^2$, then $p'(x)$ is divisible by $ax + b$.

**57.** Let $f(0) = 0$ and $f'(0) = 2$. Find the derivative of $f(f(f(f(x))))$ at $x = 0$.

**58.** Use the Chain Rule to show that the derivative of an odd function is even and the derivative of an even function is odd.

PC **59.** Let $f(x) = \sin(\sin(\sin(\sin x)))$ on $[-3\pi, 3\pi]$.

(a) Draw its graph and use it to guess whether $f$ is even, odd, or neither. Now justify your guess algebraically.

(b) Draw the graph of $f'$ and use it to guess whether $f'$ is even, odd, or neither. Justify your guess (see Problem 58).
(c) Estimate the largest value of $f(x)$.
(d) Estimate the largest value of $|f'(x)|$.

PC **60.** Follow the instructions of Problem 59 for $f(t) = \cos(t^3 - 3t)$ on $[-2, 2]$.

**Answers to Concepts Review:  1.** $D_t u$; $f'(g(t))g'(t)$
**2.** $D_y w$; $G'(H(s))H'(s)$  **3.** $D_r p D_s r D_t s$
**4.** $6(2x + 1)^2$; $2x \cos(x^2)$

---

### 3.6
### LEIBNIZ NOTATION

Gottfried Wilhelm Leibniz was one of the two principal founders of the calculus (the other was Isaac Newton). Leibniz's notation for the derivative is still widely used, especially in applied fields such as physics, chemistry, and economics. Its attraction lies in its form, a form that often suggests true results and sometimes suggests how to prove them. After we have mastered the Leibniz notation, we will use it to restate the Chain Rule, and then actually prove that rule.

**Increments**   If the value of a variable $x$ changes from $x_1$ to $x_2$, then $x_2 - x_1$, the change in $x$, is called an **increment** of $x$ and is commonly denoted by $\Delta x$ (read "delta $x$"). Note immediately that $\Delta x$ does *not* mean $\Delta$ times $x$. If $x_1 = 4.1$ and $x_2 = 5.7$, then

$$\Delta x = x_2 - x_1 = 5.7 - 4.1 = 1.6$$

If $x_1 = c$ and $x_2 = c + h$, then

$$\Delta x = x_2 - x_1 = c + h - c = h$$

Suppose next that $y = f(x)$ determines a function. If $x$ changes from $x_1$ to $x_2$, then $y$ changes from $y_1 = f(x_1)$ to $y_2 = f(x_2)$. Thus, corresponding to the increment $\Delta x = x_2 - x_1$ in $x$, there is an increment in $y$ given by

$$\Delta y = y_2 - y_1 = f(x_2) - f(x_1)$$

**EXAMPLE 1**   Let $y = f(x) = 2 - x^2$. Find $\Delta y$ when $x$ changes from 0.4 to 1.3 (see Figure 1).

*Solution*

$$\Delta y = f(1.3) - f(0.4) = [2 - (1.3)^2] - [2 - (0.4)^2] = -1.53 \qquad \blacksquare$$

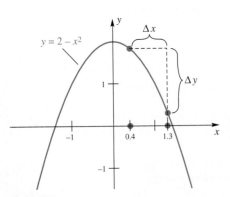

FIGURE 1

**The *dy/dx* Symbol for the Derivative**    Suppose now that the independent variable changes from $x$ to $x + \Delta x$. The corresponding change in the dependent variable, $y$, will be

$$\Delta y = f(x + \Delta x) - f(x)$$

and the ratio

$$\frac{\Delta y}{\Delta x} = \frac{f(x + \Delta x) - f(x)}{\Delta x}$$

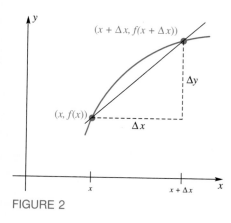

FIGURE 2

represents the slope of a secant line through $(x, f(x))$, as shown in Figure 2. As $\Delta x \to 0$, the slope of this secant line approaches that of the tangent line, and for this latter slope Leibniz used the symbol $dy/dx$. Thus,

$$\boxed{\frac{dy}{dx} = \lim_{\Delta x \to 0} \frac{\Delta y}{\Delta x} = \lim_{\Delta x \to 0} \frac{f(x + \Delta x) - f(x)}{\Delta x} = f'(x)}$$

Leibniz called $dy/dx$ a quotient of two infinitesimals. The meaning of the word *infinitesimal* is vague, and we will not use it. However, $dy/dx$ is a standard symbol for the derivative; we will use it frequently from now on. For the present, think of $d/dx$ as an operator symbol with the same meaning as $D_x$, and read it "the derivative with respect to $x$." All our past theorems on derivatives continue to apply; only our notation is different.

**EXAMPLE 2**    Find $dy/dx$ if $y = x^3 - 3x^2 + 7x$.

*Solution*

$$\frac{dy}{dx} = \frac{d}{dx}(x^3 - 3x^2 + 7x)$$

$$= \frac{d(x^3)}{dx} - 3\frac{d(x^2)}{dx} + 7\frac{d(x)}{dx}$$

$$= 3x^2 - 3(2x) + 7(1)$$

$$= 3x^2 - 6x + 7 \qquad \blacksquare$$

**EXAMPLE 3**    Find $\dfrac{d}{dt}\left(\dfrac{3t}{t^2 + 1}\right)$.

*Solution*    By the Quotient Rule,

$$\frac{d}{dt}\left(\frac{3t}{t^2 + 1}\right) = \frac{(t^2 + 1)(3) - (3t)(2t)}{(t^2 + 1)^2} = \frac{-3t^2 + 3}{(t^2 + 1)^2} \qquad \blacksquare$$

**The Chain Rule Again**    Suppose that $y = f(u)$ and $u = g(x)$. In Leibniz notation, the Chain Rule takes the particularly elegant form

$$\frac{dy}{dx} = \frac{dy}{du}\frac{du}{dx}$$

It is elegant because it is easy to remember. Just cancel the *du's* on the right side and you have the left side. Do not try to make mathematical sense out of this cancellation, but use it as a memory aid if it helps.

**EXAMPLE 4**    Find $dy/dx$ if $y = (x^3 - 2x)^{12}$.

*Solution*    Think of $x^3 - 2x$ as $u$. Then $y = u^{12}$ and

$$\frac{dy}{dx} = \frac{dy}{du}\frac{du}{dx}$$
$$= (12u^{11})(3x^2 - 2)$$
$$= 12(x^3 - 2x)^{11}(3x^2 - 2) \qquad\blacksquare$$

If $y = f(u)$, $u = g(v)$, and $v = h(x)$, then

$$\frac{dy}{dx} = \frac{dy}{du}\frac{du}{dv}\frac{dv}{dx}$$

**EXAMPLE 5**    Find $dy/dx$ if $y = \cos^3(x^2 + 1)$.

*Solution*    We can think of this as $y = u^3$, $u = \cos v$, and $v = x^2 + 1$.

$$\frac{dy}{dx} = \frac{dy}{du}\frac{du}{dv}\frac{dv}{dx}$$
$$= (3u^2)(-\sin v)(2x)$$
$$= (3\cos^2 v)[-\sin(x^2 + 1)](2x)$$
$$= -6x\cos^2(x^2 + 1)\sin(x^2 + 1) \qquad\blacksquare$$

**Partial Proof of the Chain Rule**

***Proof***    We suppose that $y = f(u)$ and $u = g(x)$, that $g$ is differentiable at $x$, and that $f$ is differentiable at $u = g(x)$. When $x$ is given an increment $\Delta x$, there are corresponding increments in $u$ and $y$ given by

$$\Delta u = g(x + \Delta x) - g(x)$$
$$\Delta y = f(g(x + \Delta x)) - f(g(x))$$
$$= f(u + \Delta u) - f(u)$$

Thus,

$$\frac{dy}{dx} = \lim_{\Delta x \to 0} \frac{\Delta y}{\Delta x} = \lim_{\Delta x \to 0} \frac{\Delta y}{\Delta u} \frac{\Delta u}{\Delta x}$$

$$= \lim_{\Delta x \to 0} \frac{\Delta y}{\Delta u} \cdot \lim_{\Delta x \to 0} \frac{\Delta u}{\Delta x}$$

Since $g$ is differentiable at $x$, it is continuous there (Theorem 3.2A), and so $\Delta x \to 0$ forces $\Delta u \to 0$. Hence,

$$\frac{dy}{dx} = \lim_{\Delta u \to 0} \frac{\Delta y}{\Delta u} \cdot \lim_{\Delta x \to 0} \frac{\Delta u}{\Delta x} = \frac{dy}{du} \cdot \frac{du}{dx}$$

This proof was very slick, but unfortunately it contains a subtle flaw. There are functions $u = g(x)$ that have the property that $\Delta u = 0$ for some points in every neighborhood of $x$ (the constant function $g(x) = k$ is a good example). This means the division by $\Delta u$ at our first step might not be legal. There is no simple way to get around this difficulty, though the Chain Rule is valid even in this case. We give a complete proof of the Chain Rule in the appendix (Section A.2, Theorem B). ∎

## CONCEPTS REVIEW

**1.** The symbol $\Delta w$ denotes an _____ (or small change) in the variable $w$. In terms of this symbol, we may define the derivative of $y$ with respect to $x$ as $\displaystyle\lim_{\Delta x \to 0}$ _____, which led Leibniz to denote this derivative by the symbol _____.

**2.** If $y = f(x)$, we may use three different symbols for the derivative of $y$ with respect to $x$. They are _____, _____, and _____.

**3.** Let $y = f(u)$, where $u = g(x)$. In Leibniz notation, the Chain Rule says $\dfrac{dy}{dx} =$ _____.

**4.** Let $w = f(t)$, where $t = g(s)$ and where $s = h(r)$. In Leibniz notation, the Compound Chain Rule says that $\dfrac{dw}{dr} =$ _____.

## PROBLEM SET 3.6

In Problems 1–4, find $\Delta y$ for the given values of $x_1$ and $x_2$ (see Example 1).

**1.** $y = 3x + 2$, $x_1 = 1$, $x_2 = 1.5$

**2.** $y = 3x^2 + 2x + 1$, $x_1 = 0.0$, $x_2 = 0.1$

[C] **3.** $y = \dfrac{3}{x + 1}$, $x_1 = 2.34$, $x_2 = 2.31$

[C] **4.** $y = \cos 2x$, $x_1 = 0.571$, $x_2 = 0.573$

In Problems 5–8, first find and simplify

$$\frac{\Delta y}{\Delta x} = \frac{f(x + \Delta x) - f(x)}{\Delta x}$$

Then find $dy/dx$ by taking the limit of your answer as $\Delta x \to 0$.

**5.** $y = x^2$

**6.** $y = x^3 - 3x^2$

**7.** $y = \dfrac{1}{x + 1}$

**8.** $y = \dfrac{x^3 + x^2}{x^3}$

In Problems 9–20, use the Chain Rule to find $dy/dx$.

**9.** $y = u^2$ and $u = \sin x$

**10.** $y = \cos u$ and $u = \dfrac{1}{x + 1}$

**11.** $y = \tan(x^2)$

**12.** $y = \tan^2 x$

13. $y = \left(\dfrac{x^2 + 1}{\cos x}\right)^4$

14. $y = [(x^2 + 1)\sin x]^3$

15. $y = \cos(x^2)\sin^2 x$

16. $y = \dfrac{(x^3 + 2x)^4}{x^4 + 1}$

17. $y = \sin^4(x^2 + 3)$ (See Example 5.)

18. $y = \sin[(x^2 + 3)^4]$

19. $y = \cos^2\left(\dfrac{x^2 + 2}{x^2 - 2}\right)$

20. $y = \sin^2[\cos^2(x^2)]$

In Problems 21–26, find the indicated derivatives.

21. $\dfrac{d}{dt}(\sin^3 t + \cos^3 t)$

22. $\dfrac{d}{ds}[(s^2 + 3)^3 - (s^2 + 3)^{-3}]$

23. $D_r[\pi(r + 3)^2 - 3\pi r(r + 2)^2]$

24. $D_t[u^3 + 3u]$ if $u = t^2$

25. $f'(2)$ if $f(x) = \left(x + \dfrac{1}{x}\right)^4$

26. $F'(0)$ if $F(t) = \cos(t^2)\sin 3t$

27. Suppose that $f(3) = 2$, $f'(3) = -1$, $g(3) = 3$, and $g'(3) = -4$. Calculate each value.

(a) $(f + g)'(3)$       (b) $(f \cdot g)'(3)$
(c) $(f/g)'(3)$       (d) $(f \circ g)'(3)$

28. If $f(2) = 4$, $f'(4) = 6$, and $f'(2) = -2$, calculate each value.

(a) $\dfrac{d}{dx}[f(x)]^3$ at $x = 2$       (b) $\dfrac{d}{dx}\left[\dfrac{3}{f(x)}\right]$ at $x = 2$
(c) $(f \circ f)'(2)$

Problems 29 and 30 refer to the graphs in Figures 3 and 4.

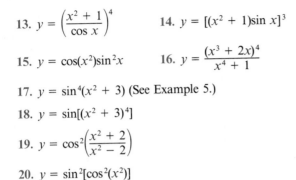

FIGURE 3

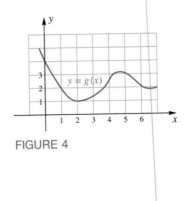

FIGURE 4

29. Find each value approximately.

(a) $(f + g)'(4)$       (b) $(f \circ g)'(6)$

30. Find each value approximately.

(a) $(f/g)'(2)$       (b) $(g \circ f)'(3)$

31. Each edge of a cube is increasing at the constant rate of 16 centimeters per minute.

(a) Find the rate at which the volume of the cube is increasing at the instant when the edge is 20 centimeters.
(b) Find the rate at which the total surface area of the cube is increasing at the instant when the edge is 15 centimeters.

32. Ships A and B start from the origin at the same time. Ship A travels due east at a rate of 20 miles per hour and ship B travels due north at the rate of 12 miles per hour. How fast are they separating after 3 hours? After 6 hours?

33. Where does the tangent line to the curve $y = x^2 \sin^2(x^2)$ at $x = \sqrt{\dfrac{\pi}{2}}$ intersect the $x$-axis?

34. The dial of a standard clock has a 10-centimeter radius. The two ends of an elastic string are attached to the rim at 12 and to the tip of the 10-centimeter minute hand. At what rate is the string stretching at 12:15 (assuming the clock is not slowed down by this stretching)?

35. Suppose $f$ is differentiable and that there are points $x_1$ and $x_2$ such that $f(x_1) = x_2$ and $f(x_2) = x_1$. Let $g(x) = f(f(f(f(x))))$. Show that $g'(x_1) = g'(x_2)$.

36. Let $f(x) = \begin{cases} x^2 \sin \dfrac{1}{x} & \text{if } x \neq 0 \\ 0 & \text{if } x = 0 \end{cases}$

(a) Find $f'(x)$ for $x \neq 0$ by using derivative rules.
(b) Find $f'(0)$ from the definition of the derivative.
(c) Show that $f'(x)$ is discontinuous at $x = 0$.

37. The hour and minute hands of a clock are 6 and 8 inches long, respectively. How fast are the tips of the hands separating at 12:20 (see Figure 5). *Hint*: Law of Cosines.

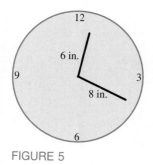

FIGURE 5

PC **38.** Find the approximate time between 12:00 and 1:00 when the distance $s$ between the tips of the hands of the clock of Figure 5 is increasing most rapidly—that is, when the derivative $ds/dt$ is largest.

---

**Answers to Concepts Review:**   **1.** Increment; $\Delta y/\Delta x$; $dy/dx$
**2.** $f'(x)$; $D_x y$; $dy/dx$   **3.** $\dfrac{dy}{du}\dfrac{du}{dx}$   **4.** $\dfrac{dw}{dt}\dfrac{dt}{ds}\dfrac{ds}{dr}$

---

## 3.7
## HIGHER-ORDER DERIVATIVES

The operation of differentiation takes a function $f$ and produces a new function $f'$. If we now differentiate $f'$, we produce still another function, denoted by $f''$ (read "eff double prime") and called the **second derivative** of $f$. It, in turn, may be differentiated, thereby producing $f'''$, which is called the **third derivative** of $f$, and so on. Thus, for example, let

$$f(x) = 2x^3 - 4x^2 + 7x - 8$$

Then

$$f'(x) = 6x^2 - 8x + 7$$
$$f''(x) = 12x - 8$$
$$f'''(x) = 12$$
$$f''''(x) = 0$$

Since the derivative of the zero function is zero, all higher-order derivatives of $f$ will be zero.

We have introduced three notations for the derivative (now also called the *first derivative*) of $y = f(x)$. They are

$$f'(x) \qquad D_x y \qquad \frac{dy}{dx}$$

called, respectively, the *prime notation*, the *dee notation*, and the *Leibniz notation*. There is a variation of the prime notation—namely, $y'$—that we will also use occasionally. All of these notations have extensions for higher-order derivatives, as shown in the chart on the next page. Note especially the Leibniz notation, which—though complicated—seemed most appropriate to Leibniz. What, thought he, is more natural than to write

$$\frac{d}{dx}\left(\frac{dy}{dx}\right) \text{ as } \frac{d^2y}{dx^2}$$

Notations for Derivatives of $y = f(x)$

| Derivative | $f'$ Notation | $y'$ Notation | $D$ Notation | Leibniz Notation |
|---|---|---|---|---|
| First | $f'(x)$ | $y'$ | $D_x y$ | $\dfrac{dy}{dx}$ |
| Second | $f''(x)$ | $y''$ | $D_x^2 y$ | $\dfrac{d^2 y}{dx^2}$ |
| Third | $f'''(x)$ | $y'''$ | $D_x^3 y$ | $\dfrac{d^3 y}{dx^3}$ |
| Fourth | $f''''(x)$ | $y''''$ | $D_x^4 y$ | $\dfrac{d^4 y}{dx^4}$ |
| Fifth | $f^{(5)}(x)$ | $y^{(5)}$ | $D_x^5 y$ | $\dfrac{d^5 y}{dx^5}$ |
| Sixth | $f^{(6)}(x)$ | $y^{(6)}$ | $D_x^6 y$ | $\dfrac{d^6 y}{dx^6}$ |
| $\vdots$ | $\vdots$ | $\vdots$ | $\vdots$ | $\vdots$ |
| $n$th | $f^{(n)}(x)$ | $y^{(n)}$ | $D_x^n y$ | $\dfrac{d^n y}{dx^n}$ |

**EXAMPLE 1**   If $y = \sin 2x$, find $d^3y/dx^3$, $d^4y/dx^4$, and $d^{12}y/dx^{12}$.

*Solution*

$$\frac{dy}{dx} = 2 \cos 2x$$

$$\frac{d^2 y}{dx^2} = -2^2 \sin 2x$$

$$\frac{d^3 y}{dx^3} = -2^3 \cos 2x$$

$$\frac{d^4 y}{dx^4} = 2^4 \sin 2x$$

$$\frac{d^5 y}{dx^5} = 2^5 \cos 2x$$

$$\vdots$$

$$\frac{d^{12} y}{dx^{12}} = 2^{12} \sin 2x$$

■

**Velocity and Acceleration**   In Section 3.1, we used the notion of instantaneous velocity to motivate the definition of the derivative. Let's review this notion by means of an example. Also, from now on, we will use the single word *velocity* in place of the more cumbersome phrase *instantaneous velocity*.

**EXAMPLE 2**   An object moves along a coordinate line so that its position $s$ satisfies $s = 2t^2 - 12t + 8$, where $s$ is measured in centimeters and $t$ in seconds with $t \geq 0$. Determine the velocity of the object when $t = 1$ and when $t = 6$. When is the velocity 0? When is it positive?

**Solution**   If we use the symbol $v(t)$ for the velocity at time $t$, then

$$v(t) = \frac{ds}{dt} = 4t - 12$$

Thus,

$$v(1) = 4(1) - 12 = -8 \text{ centimeters per second}$$
$$v(6) = 4(6) - 12 = 12 \text{ centimeters per second}$$

The velocity is 0 when $4t - 12 = 0$, that is, when $t = 3$. The velocity is positive when $4t - 12 > 0$, or when $t > 3$. All of this is shown schematically in Figure 1.

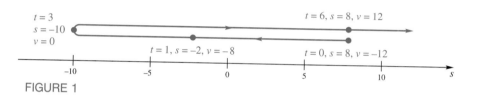

FIGURE 1

The object is, of course, moving along the $s$-axis, not on the colored path above it. But the colored path shows what is happening to the object. Between $t = 0$ and $t = 3$, the velocity is negative; the object is moving to the left (backing up). By the time $t = 3$, it has "slowed" to a zero velocity, and then starts moving to the right as its velocity becomes positive. Thus, negative velocity corresponds to moving in the direction of decreasing $s$; positive velocity corresponds to moving in the direction of increasing $s$. A rigorous discussion of these points will be given in Section 4.8.   ■

There is a technical distinction between the words *velocity* and *speed*. Velocity has a sign associated with it; it may be positive or negative. **Speed** is defined to be the absolute value of the velocity. Thus, in the example above, the speed at $t = 1$ is $|-8| = 8$ centimeters per second. The meter in most cars is a speedometer; it always gives nonnegative values.

Now we want to give a physical interpretation of the second derivative $d^2s/dt^2$. It is, of course, just the first derivative of the velocity. Thus, it measures the rate of change of velocity with respect to time, which has the name **acceleration**. If it is denoted by $a$, then

$$a = \frac{dv}{dt} = \frac{d^2s}{dt^2}$$

In Example 2, $s = 2t^2 - 12t + 8$. Thus,

$$v = \frac{ds}{dt} = 4t - 12$$

$$a = \frac{d^2s}{dt^2} = 4$$

This means that the velocity is increasing at a constant rate of 4 centimeters per second every second, which we write as 4 centimeters per second per second.

**EXAMPLE 3**    A point moves along a horizontal coordinate line in such a way that its position at time $t$ is specified by

$$s = t^3 - 12t^2 + 36t - 30$$

Here $s$ is measured in feet and $t$ in seconds.
(a) When is the velocity 0?
(b) When is the velocity positive?
(c) When is the point moving backward (that is, to the left)?
(d) When is the acceleration positive?

*Solution*
(a) $v = ds/dt = 3t^2 - 24t + 36 = 3(t - 2)(t - 6)$. Thus, $v = 0$ at $t = 2$ and $t = 6$.
(b) $v > 0$ when $(t - 2)(t - 6) > 0$. We learned how to solve quadratic inequalities in Section 1.3. The solution is $\{t : t < 2 \text{ or } t > 6\}$ or, in interval notation, $(-\infty, 2) \cup (6, \infty)$; see Figure 2.
(c) The point is moving left when $v < 0$—that is, when $(t - 2)(t - 6) < 0$. This inequality has as its solution the interval $(2, 6)$.
(d) $a = dv/dt = 6t - 24 = 6(t - 4)$. Thus, $a > 0$ when $t > 4$. The motion of the point is shown schematically in Figure 3.

FIGURE 2

Measuring Time

If $t = 0$ corresponds to the present moment, then $t < 0$ corresponds to the past and $t > 0$, to the future. In many problems, it will be obvious that we are concerned only with the future. However, since the statement of Example 3 does not specify, it seems reasonable to allow $t$ to have negative as well as positive values.

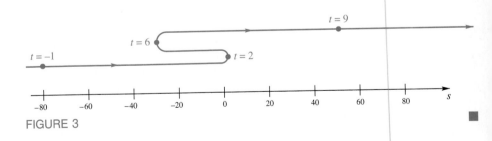

FIGURE 3

**Falling-Body Problems**    If an object is thrown straight up (or down) from an initial height of $s_0$ feet with an initial velocity of $v_0$ feet per second and if $s$ is its height above the ground in feet after $t$ seconds, then

$$s = -16t^2 + v_0t + s_0$$

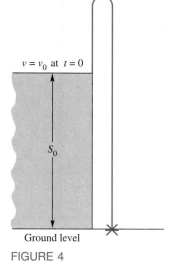

$v = v_0$ at $t = 0$

$s_0$

Ground level

FIGURE 4

This assumes that the experiment takes place near sea level and that air resistance can be neglected. The diagram in Figure 4 portrays the situation we have in mind.

**EXAMPLE 4**   From the top of a building 160 feet high, a ball was thrown upward with an initial velocity of 64 feet per second.
(a) When did it reach maximum height?
(b) What was its maximum height?
(c) When did it hit the ground?
(d) With what speed did it hit the ground?
(e) What was its acceleration at $t = 2$?

*Solution*   Let $t = 0$ correspond to the instant when the ball was thrown. Then $s_0 = 160$ and $v_0 = 64$, so

$$s = -16t^2 + 64t + 160$$

$$v = \frac{ds}{dt} = -32t + 64$$

$$a = \frac{dv}{dt} = -32$$

(a) The ball reached its maximum height at the time its velocity was 0—that is, when $-32t + 64 = 0$, or when $t = 2$ seconds.
(b) At $t = 2$, $s = -16(2)^2 + 64(2) + 160 = 224$ feet.
(c) The ball hit the ground when $s = 0$—that is, when

$$-16t^2 + 64t + 160 = 0$$

If we divide by $-16$ and use the quadratic formula, we obtain

$$t^2 - 4t - 10 = 0$$

$$t = \frac{4 \pm \sqrt{16 + 40}}{2} = \frac{4 \pm 2\sqrt{14}}{2} = 2 \pm \sqrt{14}$$

Only the positive answer makes sense. Thus, the ball hit the ground at $t = 2 + \sqrt{14} \approx 5.74$ seconds.
(d) At $t = 2 + \sqrt{14}$, $v = -32(2 + \sqrt{14}) + 64 \approx -119.73$. Thus, the ball hit the ground at a speed of 119.73 feet per second.
(e) The acceleration is always $-32$ feet per second per second. This is the acceleration of gravity near sea level.   ■

**Mathematical Modeling**   Galileo may have been right in claiming that the book of nature is written in mathematical language. Certainly, the scientific enterprise seems largely an effort to prove him correct. The task of taking a physical phenomenon and representing it in mathematical symbols is called **mathematical modeling**. One of its basic elements is translating word descriptions into mathematical language. Doing this, especially in connection with rates of change, will become increasingly important as we go on. Here are some simple illustrations.

---

The Book of Nature

"The great book of Nature lies ever open before our eyes and the true philosophy is written in it. . . . But we cannot read it unless we have first learned the language and the characters in which it is written. . . . It is written in mathematical language and the characters are triangles, circles, and other geometrical figures."

Galileo Galilei

| **Word Description** | **Mathematical Model** |
|---|---|
| Water is leaking from a cylindrical tank at a rate proportional to the depth of the water. | If $V$ denotes the volume of the water at time $t$, then $\dfrac{dV}{dt} = -kh$. |
| A wheel is spinning at a constant rate of 6 revolutions per minute—that is, at $6(2\pi)$ radians per minute. | $\dfrac{d\theta}{dt} = 6(2\pi)$ |
| The density (in grams per centimeter) of a wire at a point is twice its distance from the left end. | If $m$ denotes the mass of the left $x$ centimeters of the wire, then $\dfrac{dm}{dx} = 2x$. |
| The height of a tree continues to increase but at a slower and slower rate. | $\dfrac{dh}{dt} > 0, \ \dfrac{d^2h}{dt^2} < 0$ |

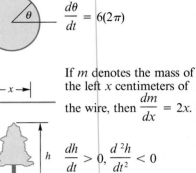

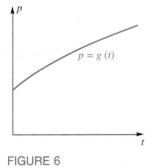

FIGURE 5

The use of mathematical language is not limited to the physical sciences; it is also appropriate in the social sciences, especially in economics.

**EXAMPLE 5** KTS News reported in May 1980, that unemployment was continuing to increase at an increasing rate. On the other hand, the price of food was increasing, but at a slower rate than before. Interpret these statements in mathematical language.

*Solution* Let $u = f(t)$ denote the number of people unemployed at time $t$. Though $u$ actually jumps by unit amounts, we follow standard practice in representing $u$ by a nice smooth curve, as in Figure 5. To say unemployment is increasing is to say $du/dt > 0$; to say that it is increasing at an increasing rate is to say $d^2u/dt^2 > 0$.

Similarly if $p = g(t)$ represents the price of food (for instance, the typical cost of one day's groceries for one person) at time $t$, then $dp/dt > 0$ but $d^2p/dt^2 < 0$; see Figure 6. ∎

FIGURE 6

## CONCEPTS REVIEW

**1.** If $y = f(x)$, then the third derivative of $y$ with respect to $x$ can be denoted by any one of the following three symbols: _____.

**2.** If $s = f(t)$ denotes the position of a particle on a coordinate line at time $t$, then its velocity is given by _____, its speed is given by _____, and its acceleration is given by _____.

**3.** Assume an object is thrown straight up (positive direction), so that its height $s$ at time $t$ is given by $s = f(t)$. The object reaches its maximum height when $ds/dt =$ _____, after which $ds/dt$ _____.

**4.** If the amount $W$ of water in a tank at time $t$ is increasing but at a slower and slower rate, then $dW/dt$ is _____ and $d^2W/dt^2$ is _____.

## PROBLEM SET 3.7

In Problems 1–8, find $d^3y/dx^3$.

**1.** $y = x^3 + 3x^2 + 6x$     **2.** $y = x^5 + x^4$

**3.** $y = (3x + 5)^3$     **4.** $y = (3 - 5x)^5$

**5.** $y = \sin(7x)$     **6.** $y = \sin(x^3)$

**7.** $y = \dfrac{1}{x - 1}$     **8.** $y = \dfrac{3x}{1 - x}$

In Problems 9–16, find $f''(2)$.

**9.** $f(x) = x^2 + 1$     **10.** $f(x) = 5x^3 + 2x^2 + x$

**11.** $f(t) = \dfrac{2}{t}$     **12.** $f(u) = \dfrac{2u^2}{5 - u}$

**13.** $f(\theta) = (\cos \theta\pi)^{-2}$     **14.** $f(t) = t \sin(\pi/t)$

**15.** $f(s) = s(1 - s^2)^3$     **16.** $f(x) = \dfrac{(x + 1)^2}{x - 1}$

**17.** Let $n! = n(n - 1)(n - 2) \cdots 3 \cdot 2 \cdot 1$. Thus, $4! = 4 \cdot 3 \cdot 2 \cdot 1 = 24$ and $5! = 5 \cdot 4 \cdot 3 \cdot 2 \cdot 1$. We give $n!$ the name **$n$ factorial**. Show that $D_x^n(x^n) = n!$.

**18.** Using the factorial symbol of Problem 17, find a formula for

$$D_x^n(a_n x^{n-1} + \cdots + a_1 x + a_0)$$

**19.** Without doing any calculating, find each derivative.

(a) $D_x^4(3x^3 + 2x - 19)$
(b) $D_x^{12}(100x^{11} - 79x^{10})$
(c) $D_x^{11}(x^2 - 3)^5$

**20.** Find a formula for $D_x^n(1/x)$.

**21.** If $f(x) = x^3 + 3x^2 - 45x - 6$, find the value of $f''$ at each zero of $f'$ —that is, at each point $c$ where $f'(c) = 0$.

**22.** Suppose $g(t) = at^2 + bt + c$ and $g(1) = 5$, $g'(1) = 3$, and $g''(1) = -4$. Find $a$, $b$, and $c$.

In Problems 23–28, an object is moving along a horizontal coordinate line according to the formula $s = f(t)$, where $s$, the directed distance from the origin, is in feet and $t$ is in seconds. In each case, answer the following questions (see Examples 2 and 3).

(a) What are $v(t)$ and $a(t)$, the velocity and acceleration at time $t$?
(b) When is the object moving right?
(c) When is it moving left?
(d) When is the acceleration negative?
(e) Draw a schematic diagram, showing the motion of the object.

**23.** $s = 12t - 2t^2$     **24.** $s = t^3 - 6t^2$

**25.** $s = t^3 - 9t^2 + 24t$     **26.** $s = 2t^3 - 6t + 5$

**27.** $s = t^2 + \dfrac{16}{t}, t > 0$     **28.** $s = t + \dfrac{4}{t}, t > 0$

**29.** If $s = \frac{1}{2}t^4 - 5t^3 + 12t^2$, find the velocity of the moving object when its acceleration is zero.

**30.** If $s = \frac{1}{10}(t^4 - 14t^3 + 60t^2)$, find the velocity of the moving object when its acceleration is zero.

**31.** Two particles move along a coordinate line. At the end of $t$ seconds their directed distances from the origin, in feet, are given by $s_1 = 4t - 3t^2$ and $s_2 = t^2 - 2t$, respectively.

(a) When do they have the same velocity?
(b) When do they have the same speed? (The speed of a particle is the absolute value of its velocity.)
(c) When do they have the same position?

**32.** The positions of two particles, $P_1$ and $P_2$, on a coordinate line at the end of $t$ seconds are given by $s_1 = 3t^3 - 12t^2 + 18t + 5$ and $s_2 = -t^3 + 9t^2 - 12t$, respectively. When do the two particles have the same velocity?

**33.** An object thrown directly upward is at a height $s = -16t^2 + 48t + 256$ feet after $t$ seconds (see Example 4).

(a) What was its initial velocity?
(b) When did it reach maximum height?
(c) What was its maximum height?
$\boxed{C}$ (d) When did it hit the ground?
$\boxed{C}$ (e) With what speed did it hit the ground?

**34.** An object thrown directly upward from ground level with an initial velocity of 48 feet per second is approximately $s = 48t - 16t^2$ feet high at the end of $t$ seconds.

(a) What is the maximum height attained?
(b) How fast is it moving, and in which direction, at the end of 1 second?
(c) How long does it take to return to its original position?

$\boxed{C}$ **35.** A projectile is fired directly upward from the ground with an initial velocity of $v_0$ feet per second. Its height in $t$ seconds is given by $s = v_0 t - 16t^2$ feet. What must its initial velocity be for the projectile to reach a maximum height of 1 mile?

**36.** An object thrown directly downward from the top of a cliff with an initial velocity of $v_0$ feet per second falls approximately $s = v_0 t + 16t^2$ feet in $t$ seconds. If it strikes the ocean below in 3 seconds with a velocity of 140 feet per second, how high is the cliff?

**37.** A point moves along a horizontal coordinate line in such a way that its position at time $t$ is specified by $s = t^3 - 3t^2 - 24t - 6$. Here $s$ is measured in centimeters and $t$ in seconds. When is the point slowing down, that is, when is its *speed* decreasing?

**38.** Convince yourself that a point moving along a line is slowing down when its velocity and acceleration have opposite signs (see Problem 37).

**39.** Translate each of the following into the language of first, second, and third derivatives of distance with respect to time.

(a) The speed of that car is proportional to the distance it has traveled.
(b) That car is speeding up.
(c) I didn't say that car was slowing down; I said its rate of increase in speed was slowing down.
(d) That car's speed is increasing 10 miles per hour every minute.
(e) That car is slowing very gently to a stop.
(f) That car always travels the same distance in equal time intervals.

**40.** Translate each of the following into the language of derivatives.

(a) Water is evaporating from that tank at a constant rate.
(b) Water is being poured into that tank at 3 gallons per minute but is also leaking out at $\frac{1}{2}$ gallon per minute.
(c) Since water is being poured into that conical tank at a constant rate, the water level is rising at a slower and slower rate.
(d) Inflation held steady this year but is expected to rise more and more rapidly in the years ahead.
(e) At present the price of oil is dropping but this trend is expected to slow and then reverse directions in 2 years.
(f) David's temperature is still rising but the penicillin seems to be taking effect.

**41.** Translate each of the following statements into mathematical language as in Example 5.

(a) The cost of a car continues to increase and at a faster and faster rate.
(b) During the last 2 years, the United States has continued to cut its consumption of oil, but at a slower and slower rate.
(c) World population continues to grow, but at a slower and slower rate.
(d) That car is going faster and faster at a constant rate.
(e) The angle the Leaning Tower of Pisa makes with the vertical is increasing more rapidly.
(f) Upper Midwest firm's profit growth slows.
(g) The XYZ Company has been losing money but will soon turn this situation around.

**42.** Translate each statement from the following newspaper column into a statement about derivatives.

(a) In the United States, the ratio $R$ of government debt to national income remained unchanged at around 28% up to 1981 but (b) then it began to increase more and more sharply, reaching 36% during 1983. (c) The IMF released a table showing that the speed of increase of $R$ was greater in the United States than in Japan.

**43.** Leibniz obtained a general formula for $D_x^n(uv)$, where $u$ and $v$ are both functions of $x$. See if you can find it. *Hint:* Begin by considering the cases $n = 1$, $n = 2$, and $n = 3$.

**44.** Use the formula of Problem 43 to find $D_x^4(x^4 \sin x)$.

[PC] **45.** Let $f(x) = x[\sin x - \cos(x/2)]$.

(a) Draw the graphs of $f(x), f'(x), f''(x)$ and $f'''(x)$ on $[0, 6]$ using the same axes.
(b) Evaluate $f'''(2.13)$.

[PC] **46.** Repeat Problem 45 for $f(x) = (x + 1)/(x^2 + 2)$.

**Answers to Concepts Review:** **1.** $f'''(x), D_x^3 y, d^3y/dx^3$ **2.** $ds/dt$; $|ds/dt|$; $d^2s/dt^2$ **3.** $0$; $< 0$ **4.** Positive; negative

---

**3.8**
**IMPLICIT**
**DIFFERENTIATION**

With a little work, you should be able to check that the graph of

$$y^3 + 7y = x^3$$

looks something like that shown in Figure 1. Certainly the point (2, 1) is on the graph, and there appears to be a well-defined tangent line at that point. How shall we find the slope of this tangent? Just calculate $dy/dx$ at this point, you may answer. But that is the rub; we do not know how to find $dy/dx$ in this situation.

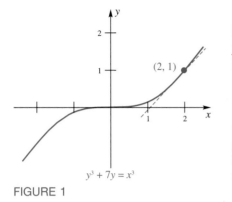

$y^3 + 7y = x^3$

FIGURE 1

The new element in this problem is that we are given an equation which is not explicitly solved for $y$. And try as we will, we cannot seem to solve it for $y$. Is it possible to find $dy/dx$ in circumstances like these? Yes. Differentiate both sides of the equation

$$y^3 + 7y = x^3$$

with respect to $x$ and equate the results. In doing this, we assume that the given equation does determine $y$ as a function of $x$ (it is just that we do not know how to find it explicitly). Thus, after using the Chain Rule on the first term, we get

$$3y^2 \cdot \frac{dy}{dx} + 7\frac{dy}{dx} = 3x^2$$

The latter can be solved for $dy/dx$ as follows.

$$\frac{dy}{dx}(3y^2 + 7) = 3x^2$$

$$\frac{dy}{dx} = \frac{3x^2}{3y^2 + 7}$$

Note that our expression for $dy/dx$ involves both $x$ and $y$, a fact that is often a nuisance. But if we wish only to find a slope at a point where we know both coordinates, no difficulty exists. At $(2, 1)$,

$$\frac{dy}{dx} = \frac{3(2)^2}{3(1)^2 + 7} = \frac{12}{10} = \frac{6}{5}$$

The slope is $\frac{6}{5}$.

The method just illustrated for finding $dy/dx$ without first solving the given equation for $y$ explicitly in terms of $x$ is called **implicit differentiation**. But is the method legitimate—does it give the right answer?

**An Example That Can Be Checked**    To give some evidence for the correctness of the method, consider the following example, which can be worked two ways.

**EXAMPLE 1**    Find $dy/dx$ if $4x^2y - 3y = x^3 - 1$.

*Solution*

METHOD 1    We can solve the given equation explicitly for $y$ as follows.

$$y(4x^2 - 3) = x^3 - 1$$

$$y = \frac{x^3 - 1}{4x^2 - 3}$$

Thus

$$\frac{dy}{dx} = \frac{(4x^2 - 3)(3x^2) - (x^3 - 1)(8x)}{(4x^2 - 3)^2} = \frac{4x^4 - 9x^2 + 8x}{(4x^2 - 3)^2}$$

METHOD 2 **(Implicit differentiation).** We equate the derivatives of the two sides of

$$4x^2 y - 3y = x^3 - 1$$

We obtain, after using the Product Rule on the first term,

$$4x^2 \cdot \frac{dy}{dx} + y \cdot 8x - 3\frac{dy}{dx} = 3x^2$$

$$\frac{dy}{dx}(4x^2 - 3) = 3x^2 - 8xy$$

$$\frac{dy}{dx} = \frac{3x^2 - 8xy}{4x^2 - 3}$$

Though this answer looks different from the result of Method 1, the two derivatives are equivalent. To see this, substitute $y = (x^3 - 1)/(4x^2 - 3)$ in the expression for $dy/dx$ just obtained.

$$\frac{dy}{dx} = \frac{3x^2 - 8xy}{4x^2 - 3} = \frac{3x^2 - 8x\dfrac{x^3 - 1}{4x^2 - 3}}{4x^2 - 3}$$

$$= \frac{12x^4 - 9x^2 - 8x^4 + 8x}{(4x^2 - 3)^2} = \frac{4x^4 - 9x^2 + 8x}{(4x^2 - 3)^2} \qquad \blacksquare$$

**Some Subtle Difficulties**    If an equation in $x$ and $y$ determines a function $y = f(x)$ and if this function is differentiable, then the method of implicit differentiation will yield a correct expression for $dy/dx$. But notice there are two big *ifs* in this statement.

Consider first the equation

$$x^2 + y^2 = -1$$

It has no solutions and therefore does not determine a function.

On the other hand,

$$x^2 + y^2 = 25$$

determines both the function $y = f(x) = \sqrt{25 - x^2}$ and the function $y = g(x) = -\sqrt{25 - x^2}$. Their graphs are shown in Figure 2.

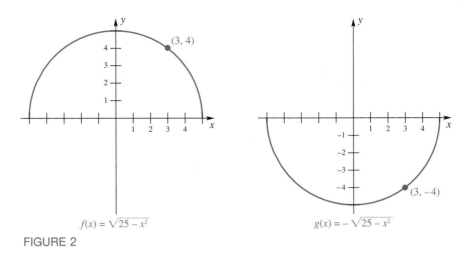

$f(x) = \sqrt{25 - x^2}$     $g(x) = -\sqrt{25 - x^2}$

FIGURE 2

Happily, both of these functions are differentiable on $(-5, 5)$. Consider $f$ first. It satisfies

$$x^2 + [f(x)]^2 = 25$$

When we differentiate implicitly and solve for $f'(x)$, we obtain

$$2x + 2f(x)f'(x) = 0$$
$$f'(x) = -\frac{x}{f(x)} = -\frac{x}{\sqrt{25 - x^2}}$$

A completely similar treatment of $g(x)$ yields

$$g'(x) = -\frac{x}{g(x)} = \frac{x}{\sqrt{25 - x^2}}$$

For practical purposes, we can obtain both of these results simultaneously by implicit differentiation of $x^2 + y^2 = 25$. This gives

$$2x + 2y\frac{dy}{dx} = 0$$

$$\frac{dy}{dx} = -\frac{x}{y} = \begin{cases} \dfrac{-x}{\sqrt{25 - x^2}} & \text{if } y = f(x) \\[2ex] \dfrac{-x}{-\sqrt{25 - x^2}} & \text{if } y = g(x) \end{cases}$$

Naturally, the results are identical with those obtained above.

Note that it is often enough to know $dy/dx = -x/y$ in order to apply our results. Suppose we want to know the slope of the tangent line to the circle $x^2 + y^2 = 25$ when $x = 3$. The corresponding $y$-values are 4 and $-4$. The slopes at $(3, 4)$ and $(3, -4)$, obtained by substituting in $-x/y$, are $-\frac{3}{4}$ and $\frac{3}{4}$, respectively (see Figure 2).

To complicate matters, we point out that

$$x^2 + y^2 = 25$$

determines many other functions. For example, consider the function $h$ defined by

$$h(x) = \begin{cases} \sqrt{25 - x^2} & \text{if } -5 \le x \le 3 \\ -\sqrt{25 - x^2} & \text{if } 3 < x \le 5 \end{cases}$$

It too satisfies $x^2 + y^2 = 25$, since $x^2 + [h(x)]^2 = 25$. But it is not even continuous at $x = 3$, so it certainly does not have a derivative there (see Figure 3).

While the subject of implicit functions leads to difficult technical questions (treated in advanced calculus), the problems we study have straight-forward solutions.

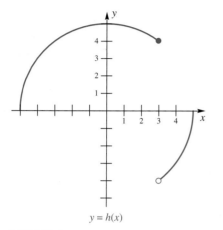

$y = h(x)$

FIGURE 3

**More Examples** In the examples that follow, we assume that the given equation determines one or more differentiable functions whose derivatives can be found by implicit differentiation. Note that in each case, we begin by taking the derivative of each side of the given equation with respect to the appropriate variable. Then we use the Chain Rule as needed.

**EXAMPLE 2** Find $dy/dx$ if $x^2 + 5y^3 = x + 9$.

*Solution*

$$\frac{d}{dx}(x^2 + 5y^3) = \frac{d}{dx}(x + 9)$$

$$2x + 15y^2\frac{dy}{dx} = 1$$

$$\frac{dy}{dx} = \frac{1 - 2x}{15y^2}$$

∎

**EXAMPLE 3** Find $D_t y$ if $t^3 + t^2 y - 10y^4 = 0$.

*Solution*

$$D_t(t^3 + t^2 y - 10y^4) = D_t(0)$$

$$3t^2 + t^2 D_t y + y(2t) - 40y^3 D_t y = 0$$

$$D_t y(t^2 - 40y^3) = -3t^2 - 2ty$$

$$D_t y = \frac{3t^2 + 2ty}{40y^3 - t^2}$$

∎

**EXAMPLE 4**    Find the equation of the tangent line to the curve

$$y^3 - xy^2 + \cos xy = 2$$

at the point $(0, 1)$.

*Solution*    For simplicity, let us use the notation $y'$ for $dy/dx$. When we differentiate both sides and equate the results, we obtain

$$3y^2y' - x(2yy') - y^2 - (\sin xy)(xy' + y) = 0$$
$$y'(3y^2 - 2xy - x \sin xy) = y^2 + y \sin xy$$
$$y' = \frac{y^2 + y \sin xy}{3y^2 - 2xy - x \sin xy}$$

At $(0, 1)$, $y' = \frac{1}{3}$. Thus, the equation of the tangent line at $(0, 1)$ is

$$y - 1 = \tfrac{1}{3}(x - 0)$$    ■

**The Power Rule Again**    We have learned that $D_x(x^n) = nx^{n-1}$, where $n$ is any integer. We now extend this to the case where $n$ is any rational number.

---

Theorem A

**(Power Rule).** Let $r$ be any rational number. Then

$$D_x(x^r) = rx^{r-1}$$

---

**Proof**    Since $r$ is rational, $r$ may be written as $p/q$, where $p$ and $q$ are integers with $q > 0$. Let

$$y = x^r = x^{p/q}$$

Then

$$y^q = x^p$$

and, by implicit differentiation,

$$qy^{q-1} D_x y = px^{p-1}$$

Thus,

$$D_x y = \frac{px^{p-1}}{qy^{q-1}} = \frac{p}{q} \frac{x^{p-1}}{(x^{p/q})^{q-1}} = \frac{p}{q} \frac{x^{p-1}}{x^{p-p/q}}$$

$$= \frac{p}{q} x^{p-1-p+p/q} = \frac{p}{q} x^{p/q-1} = rx^{r-1}$$

We have obtained the desired result, but—to be honest—we must point out a flaw in our argument. In the implicit differentiation step, we assumed that $D_x y$ exists—that is, that $y = x^{p/q}$ is differentiable. We can fill this gap but since it is hard work, we relegate the complete proof to the appendix (Section A.2, Theorem C). ■

**EXAMPLE 5**   Find $D_x y$ if $y = 2x^{11/3} + 4x^{3/4} - 6/x^{2/3}$.

*Solution*   First we write

$$D_x y = 2D_x(x^{11/3}) + 4D_x(x^{3/4}) - 6D_x(x^{-2/3})$$

Then, by the rule just proved,

$$D_x y = 2 \cdot \tfrac{11}{3} x^{8/3} + 4 \cdot \tfrac{3}{4} x^{-1/4} - 6(-\tfrac{2}{3})x^{-5/3}$$
$$= \tfrac{22}{3} x^{8/3} + 3x^{-1/4} + 4x^{-5/3}$$

■

**EXAMPLE 6**   If $y = \sqrt{t^4 - 3t + 17}$, find $dy/dt$.

*Solution*   Think of this as

$$y = u^{1/2} \quad \text{and} \quad u = t^4 - 3t + 17$$

and apply the Chain Rule.

$$\frac{dy}{dt} = \frac{dy}{du}\frac{du}{dt}$$
$$= \left(\frac{1}{2}u^{-1/2}\right)(4t^3 - 3)$$
$$= \left(\frac{1}{2\sqrt{u}}\right)(4t^3 - 3)$$
$$= \frac{4t^3 - 3}{2\sqrt{t^4 - 3t + 17}}$$

■

## CONCEPTS REVIEW

**1.** The implicit relation $yx^3 - 3y = 9$ can be solved explicitly for $y$ giving $y = $ _____.

**2.** Implicit differentiation of $y^3 + x^3 = 2x$ with respect to $x$ gives _____ $+ 3x^2 = 2$.

**3.** Implicit differentiation of $xy^2 + y^3 - y = x^3$ with respect to $x$ gives _____ $=$ _____.

**4.** The Power Rule with rational exponents says that $D_x(x^{p/q}) = $ _____. This rule together with the Chain Rule implies that $D_x[(x^2 - 5x)^{5/3}] = $ _____.

## PROBLEM SET 3.8

Assuming that each equation in Problems 1–12 defines a differentiable function of $x$, find $D_x y$ by implicit differentiation.

**1.** $y^2 - x^2 = 1$

**2.** $9x^2 + 4y^2 = 36$

**3.** $xy = 1$

**4.** $x^2 + \alpha^2 y^2 = 4\alpha^2$, where $\alpha$ is a constant.

5. $xy^2 = x - 8$

6. $x^2 + 2x^2y + 3xy = 0$

7. $4x^3 + 7xy^2 = 2y^3$

8. $x^2y = 1 + y^2x$

9. $\sqrt{5xy} + 2y = y^2 + xy^3$

10. $x\sqrt{y+1} = xy + 1$

11. $xy + \sin(xy) = 1$

12. $\cos(xy^2) = y^2 + x$

In Problems 13–18, find the equation of the tangent line at the indicated point (see Example 4).

13. $x^3y + y^3x = 30$; $(1, 3)$

14. $x^2y^2 + 4xy = 12y$; $(2, 1)$

15. $\sin(xy) = y$; $(\pi/2, 1)$

16. $y + \cos(xy^2) + 3x^2 = 4$; $(1, 0)$

17. $x^{2/3} - y^{2/3} - 2y = 2$; $(1, -1)$

18. $\sqrt{y} + xy^2 = 5$; $(4, 1)$

In Problems 19–32, find $dy/dx$ (see Examples 5 and 6).

19. $y = 3x^{5/3} + \sqrt{x}$     20. $y = \sqrt[3]{x} - 2x^{7/2}$

21. $y = \sqrt[3]{x} + \dfrac{1}{\sqrt[3]{x}}$     22. $y = \sqrt[4]{2x + 1}$

23. $y = \sqrt[4]{3x^2 - 4x}$     24. $y = (x^3 - 2x)^{1/3}$

25. $y = \dfrac{1}{(x^3 + 2x)^{2/3}}$     26. $y = (3x - 9)^{-5/3}$

27. $y = \sqrt{x^2 + \sin x}$     28. $y = \sqrt{x^2 \cos x}$

29. $y = \dfrac{1}{\sqrt[3]{x^2 \sin x}}$     30. $y = \sqrt[4]{1 + \sin 5x}$

31. $y = \sqrt[4]{1 + \cos(x^2 + 2x)}$

32. $y = \sqrt{\tan^2 x + \sin^2 x}$

33. If $s^2t + t^3 = 1$, find $ds/dt$ and $dt/ds$.

34. If $y = \sin(x^2) + 2x^3$, find $dx/dy$.

35. Sketch the graph of the circle $x^2 + 4x + y^2 + 3 = 0$ and then find equations of the two tangent lines that pass through the origin.

36. Find the equation of the **normal line** (line perpendicular to the tangent line) to the curve $8(x^2 + y^2)^2 = 100(x^2 - y^2)$ at $(3, 1)$.

37. Suppose $xy + y^3 = 2$. Then implicit differentiation twice with respect to $x$ yields in turn:

(a) $xy' + y + 3y^2y' = 0$;
(b) $xy'' + y' + y' + 3y^2y'' + 6y(y')^2 = 0$.

Solve (a) for $y'$ and substitute in (b), and then solve for $y''$.

38. Find $y''$ if $x^3 - 4y^2 + 3 = 0$ (see Problem 37).

39. Find $y''$ at $(2, 1)$ if $2x^2y - 4y^3 = 4$ (see Problem 37).

40. Use implicit differentiation twice to find $y''$ at $(3, 4)$ if $x^2 + y^2 = 25$.

41. Show that the normal line to $x^3 + y^3 = 3xy$ at $\left(\frac{3}{2}, \frac{3}{2}\right)$ passes through the origin.

42. Show that the hyperbolas $xy = 1$ and $x^2 - y^2 = 1$ intersect at right angles.

43. Show that the graphs of $2x^2 + y^2 = 6$ and $y^2 = 4x$ intersect at right angles.

44. Suppose curves $C_1$ and $C_2$ intersect at $(x_0, y_0)$ with slopes $m_1$ and $m_2$, respectively, as in Figure 4. Then (see Problem 40 of Section 2.3) the positive angle $\theta$ from $C_1$ (that is, from the tangent line to $C_1$ at $(x_0, y_0)$) to $C_2$ satisfies

$$\tan \theta = \frac{m_2 - m_1}{1 + m_1 m_2}$$

Find the angles from the circle $x^2 + y^2 = 1$ to the circle $(x - 1)^2 + y^2 = 1$ at the two points of intersection.

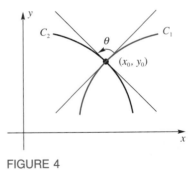

**FIGURE 4**

45. Find the angle from the line $y = 2x$ to the curve $x^2 - xy + 2y^2 = 28$ at their point of intersection in the first quadrant (see Problem 44).

46. A particle of mass $m$ moves along the $x$-axis so that its position $x$ and velocity $v = dx/dt$ satisfy

$$m(v^2 - v_0^2) = k(x_0^2 - x^2)$$

where $v_0$, $x_0$, and $k$ are constants. Show by implicit differentiation that

$$m\frac{dv}{dt} = -kx$$

whenever $v \neq 0$.

**47.** The curve $x^2 - xy + y^2 = 16$ is an ellipse centered at the origin and with the line $y = x$ as its major axis. Find the equations of the tangent lines at the two points where the ellipse intersects the $x$-axis.

**48.** Find any points on the curve $x^2y - xy^2 = 2$ where the tangent line is vertical, that is, where $dx/dy = 0$.

**49.** How high $h$ must the light bulb in Figure 5 be if the point $(1.25, 0)$ is right on the edge of the illuminated region?

---

**Answers to Concepts Review:** **1.** $y = 9/(x^3 - 3)$  **2.** $3y^2 \dfrac{dy}{dx}$

**3.** $x \cdot 2y \dfrac{dy}{dx} + y^2 + 3y^2 \dfrac{dy}{dx} - \dfrac{dy}{dx} = 3x^2$

**4.** $\dfrac{p}{q} x^{p/q - 1}$; $\dfrac{5}{3}(x^2 - 5x)^{2/3}(2x - 5)$

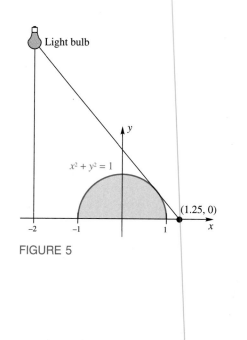

FIGURE 5

---

### 3.9
### RELATED RATES

If a variable $y$ depends on time $t$, then its derivative $dy/dt$ is called a **time rate of change**. Of course, if $y$ measures distance, then this time rate of change is also called velocity. We are interested in a wide variety of time rates: the rate at which water is flowing into a bucket, the rate at which the area of an oil spill is growing, the rate at which the value of a piece of real estate is increasing, and so on. If $y$ is given explicitly in terms of $t$, the problem is simple; we just differentiate and then evaluate the derivative at the required time.

It may be that, in place of knowing $y$ explicitly in terms of $t$, we know a relationship that connects $y$ and another variable $x$ and that we also know something about $dx/dt$. We may still be able to find $dy/dt$, since $dy/dt$ and $dx/dt$ are **related rates**. This will usually require implicit differentiation.

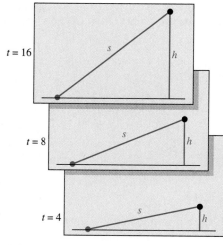

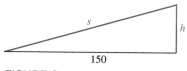

FIGURE 1

**Two Simple Examples** In preparation for outlining a systematic procedure for solving related rate problems, we discuss two examples.

**EXAMPLE 1** A small balloon is released at a point 150 feet away from an observer, who is on level ground. If the balloon goes straight up at a rate of 8 feet per second, how fast is the distance from the observer to the balloon increasing when the balloon is 50 feet high?

*Solution* Let $t$ denote the number of seconds after the balloon is released. Let $h$ denote the height of the balloon and $s$ its distance from the observer (see Figure 1). Both $h$ and $s$ are variables that depend on $t$; however, the base of the triangle (the distance from the observer to the point of release) remains unchanged as $t$ increases. Figure 2 shows the key quantities in one simple diagram.

⊡     Before going farther, we pick up a theme discussed earlier in the book, *estimating the answer*. Note that initially $s$ changes hardly at all

FIGURE 2

($ds/dt \approx 0$), but eventually $s$ changes about as fast as $h$ changes ($ds/dt \approx$ $dh/dt = 8$). An estimate for $ds/dt$ when $h = 50$ might be about one-third of $dh/dt$, or 2.7. If we get an answer far from this value, we will know we have made a mistake. For example, answers such as 17 and even 7 are clearly wrong.

We continue with the exact solution. For emphasis, we ask and answer two fundamental questions.

(a) What is given? *Answer*: $dh/dt = 8$.

(b) What do we want to know? *Answer*: We want to know $ds/dt$ at the instant when $h = 50$.

The variables $s$ and $h$ change with time (they are implicit functions of $t$), but they are always related by the Pythagorean equation

$$s^2 = h^2 + (150)^2$$

If we differentiate implicitly with respect to $t$ and use the Chain Rule, we obtain

$$2s \frac{ds}{dt} = 2h \frac{dh}{dt}$$

or

$$s \frac{ds}{dt} = h \frac{dh}{dt}$$

This relationship also holds for all $t > 0$.

Now, and *not before now*, we turn to the situation when $h = 50$. From the Pythagorean Theorem, we see that, when $h = 50$,

$$s = \sqrt{(50)^2 + (150)^2} = 50\sqrt{10}$$

Substituting in $s(ds/dt) = h(dh/dt)$ yields

$$50\sqrt{10} \frac{ds}{dt} = 50(8)$$

or

$$\frac{ds}{dt} = \frac{8}{\sqrt{10}} \approx 2.53$$

At the instant when $h = 50$, the distance between the balloon and the observer is increasing at 2.53 feet per second. ∎

**EXAMPLE 2** Water is pouring into a conical cistern at the rate of 8 cubic feet per minute. If the height of the cistern is 12 feet and the radius of its circular opening is 6 feet, how fast is the water level rising when the water is 4 feet deep?

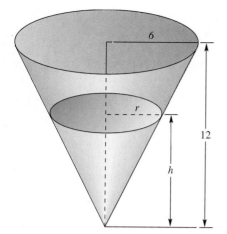

FIGURE 3

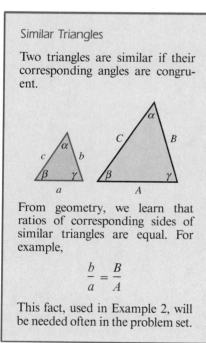

Similar Triangles

Two triangles are similar if their corresponding angles are congruent.

From geometry, we learn that ratios of corresponding sides of similar triangles are equal. For example,

$$\frac{b}{a} = \frac{B}{A}$$

This fact, used in Example 2, will be needed often in the problem set.

*Solution* · Denote the depth of the water in the cistern at any time $t$ by $h$ and let $r$ be the corresponding radius of the surface of the water (see Figure 3).

We are *given* that the volume, $V$, of water in the cistern is increasing at the rate of 8 cubic feet per minute; that is, $dV/dt = 8$. We *want to know* how fast the water is rising—that is, $dh/dt$—at the instant when $h = 4$.

We need to find an equation relating $V$ and $h$. The formula for the volume of water in the cistern, $V = \frac{1}{3}\pi r^2 h$, contains the unwanted variable $r$, unwanted because we do not know its rate $dr/dt$. However, by similar triangles (see the marginal box), we have $r/h = 6/12$, so $r = h/2$. Substituting this in $V = \frac{1}{3}\pi r^2 h$ gives

$$V = \frac{\pi h^3}{12}$$

a relation holding for all $t > 0$.

Now we differentiate implicitly, keeping in mind that $h$ depends on $t$. We obtain

$$\frac{dV}{dt} = \frac{3\pi h^2}{12}\frac{dh}{dt}$$

or

$$\frac{dV}{dt} = \frac{\pi h^2}{4}\frac{dh}{dt}$$

At this point, and not earlier, we consider the situation when $h = 4$. Substituting $h = 4$ and $dV/dt = 8$, we obtain

$$8 = \frac{\pi(4)^2}{4}\frac{dh}{dt}$$

from which

$$\frac{dh}{dt} = \frac{2}{\pi} \approx 0.637$$

When the depth of the water is 4 feet, the water level is rising at 0.637 foot per minute.

If you think about it for a moment, you realize that the water level will rise more and more slowly as time goes on. For example, when $h = 10$

$$8 = \frac{\pi(10)^2}{4}\frac{dh}{dt}$$

so $dh/dt = 32/100\pi \approx 0.102$ foot per minute.

What we are really saying is that the acceleration $d^2h/dt^2$ is negative. We can calculate an expression for it. At any time $t$,

$$8 = \frac{\pi h^2}{4}\frac{dh}{dt}$$

so

$$\frac{32}{\pi} = h^2 \frac{dh}{dt}$$

If we differentiate implicitly again, we get

$$0 = h^2 \frac{d^2h}{dt^2} + \frac{dh}{dt}\left(2h\frac{dh}{dt}\right)$$

from which

$$\frac{d^2h}{dt^2} = \frac{-2\left(\dfrac{dh}{dt}\right)^2}{h}$$

This is clearly negative. ■

**A Systematic Procedure**  Examples 1 and 2 suggest the following method for solving a related rates problem.

**Step 1**  Let $t$ denote the elapsed time. Draw a diagram that is valid for all $t > 0$. Label those quantities whose values do not change as $t$ increases with their given constant values. Assign letters to the quantities that vary with $t$, and label the appropriate parts of the figure with these variables.

**Step 2**  State what is given about the variables and what information is wanted about them. This information will be in the form of derivatives with respect to $t$.

**Step 3**  Write an equation relating the variables that is valid at all times $t > 0$, not just at some particular instant.

**Step 4**  Differentiate the equation found in Step 3 implicitly with respect to $t$. The resulting equation, containing derivatives with respect to $t$, is true for all $t > 0$.

**Step 5**  At this point, and no earlier, substitute in the equation found in Step 4 all data that are valid *at the particular instant* for which the answer to the problem is required. Solve for the desired derivative.

EXAMPLE 3  A plane flying north at 640 miles per hour passes over a certain town at noon, and a second plane going east at 600 miles per hour is directly over the same town 15 minutes later. If the planes are flying at the same altitude, how fast will they be separating at 1:15 P.M.?

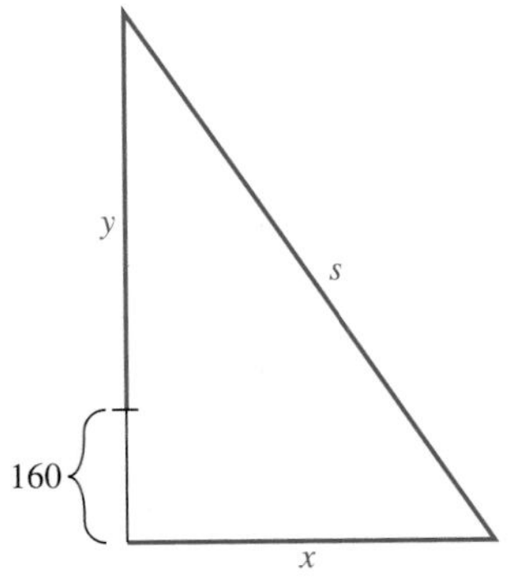

*Solution*

**Step 1**  Let $t$ denote the number of hours after 12:15 P.M. Figure 4 shows the situation for all $t > 0$. The distance in miles from the town to the northbound plane when $t = 0$ (12:15 P.M.) is labeled with the constant $\frac{640}{4} = 160$. For any $t > 0$, we let $y$ denote the distance in miles flown by the northbound plane (after 12:15 P.M.), $x$ the distance flown by the eastbound plane, and $s$ the distance between the planes.

**Step 2**  We are given that for all $t > 0$, $dy/dt = 640$ and $dx/dt = 600$. We want to know $ds/dt$ at $t = 1$, that is, at 1:15 P.M.

**Step 3**  By the Pythagorean Theorem,

$$s^2 = x^2 + (y + 160)^2$$

FIGURE 4

**Step 4** Differentiating implicitly with respect to $t$ and using the Chain Rule, we have

$$2s \frac{ds}{dt} = 2x \frac{dx}{dt} + 2(y + 160) \frac{dy}{dt}$$

or

$$s \frac{ds}{dt} = x \frac{dx}{dt} + (y + 160) \frac{dy}{dt}$$

**Step 5** For all $t > 0$, $dx/dt = 600$ and $dy/dt = 640$, while at the particular instant $t = 1$, $x = 600$, $y = 640$, and $s = \sqrt{(600)^2 + (640 + 160)^2} = 1000$. When we substitute these data in the equation of Step 4, we obtain

$$1000 \frac{ds}{dt} = (600)(600) + (640 + 160)(640)$$

from which

$$\frac{ds}{dt} = 872$$

At 1:15 P.M., the planes are separating at 872 miles per hour.

≈ Now let's see if our answer makes sense. Look at Figure 4 again. Clearly, $s$ is increasing faster than either $x$ or $y$ is increasing, so $ds/dt$ exceeds 640. On the other hand, $s$ is surely increasing more slowly than the sum of $x$ and $y$; that is, $ds/dt < 600 + 640$. Our answer, $ds/dt = 872$, is reasonable. ∎

**EXAMPLE 4** A woman standing on a cliff is watching a motor boat through a telescope as the boat approaches the shoreline directly below her. If the telescope is 250 feet above the water level and if the boat is approaching at 20 feet per second, at what rate is the angle of the telescope changing when the boat is 250 feet from shore?

**Solution**

**Step 1** We draw a figure (Figure 5) and introduce variables $x$ and $\theta$, as shown.
**Step 2** We are given that $dx/dt = -20$; the sign is negative because $x$ is decreasing with time. We want to know $d\theta/dt$ at the instant when $x = 250$.
**Step 3** From trigonometry,

$$\tan \theta = \frac{x}{250}$$

**Step 4** We differentiate implicitly using the fact that $D_\theta \tan \theta = \sec^2\theta$ (Example 2 of Section 3.4). We get

$$\sec^2\theta \frac{d\theta}{dt} = \frac{1}{250} \frac{dx}{dt}$$

**Step 5** At the instant when $x = 250$, $\theta$ is $\pi/4$ radians and $\sec^2\theta = \sec^2(\pi/4) = 2$. Thus,

$$2 \frac{d\theta}{dt} = \frac{1}{250}(-20)$$

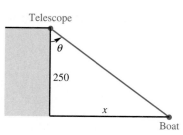

Telescope

$\theta$

250

$x$

Boat

FIGURE 5

or

$$\frac{d\theta}{dt} = \frac{-1}{25} = -0.04$$

The angle is changing at $-0.04$ radians per second. The sign is negative because $\theta$ is decreasing with time. ∎

**A Graphical Related Rate Problem**    Often in a real-life situation, we do not know a formula for a certain function, but rather have an empirically determined graph for it. We may still be able to answer questions about rates.

**EXAMPLE 5**    Webster City monitors the height of the water in its cylindrical water tank with an automatic recording device. Water is constantly pumped into the tank at a rate of 2400 cubic feet per hour, as shown in Figure 6. During a certain 12-hour period (beginning at midnight), the water level rose and fell according to the graph in Figure 7. If the radius of the tank is 20 feet, at what rate was water being used at 7:00 A.M.?

*Solution*    Let $t$ denote the number of hours past midnight, $h$ the height of the water in the tank at time $t$, and $V$ the volume of water in the tank at that time (see Figure 6). Then $dV/dt$ is the rate in minus the rate out, so $2400 - dV/dt$ is the rate at which water is being used at any time $t$. Since the slope of the tangent line at $t = 7$ is approximately $-3$ (Figure 7), we conclude that $dh/dt \approx -3$ at that time.

For a cylinder, $V = \pi r^2 h$, and so

$$V = \pi (20)^2 h$$

from which

$$\frac{dV}{dt} = 400\pi \frac{dh}{dt}$$

At $t = 7$,

$$\frac{dV}{dt} \approx 400\pi(-3) \approx -3770$$

Thus Webster City residents were using water at the rate of $2400 + 3770 = 6170$ cubic feet per hour at 7:00 A.M. ∎

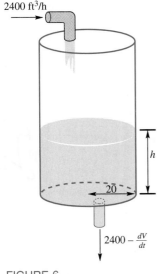

2400 ft³/h

$h$

20

$2400 - \dfrac{dV}{dt}$

FIGURE 6

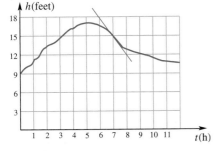

$h$(feet)

18
15
12
9
6
3

1  2  3  4  5  6  7  8  9  10  11    $t$(h)

FIGURE 7

## CONCEPTS REVIEW

**1.** To ask how fast $u$ is changing with respect to time $t$ after 2 hours is to ask the value of _____ at _____.

**2.** An airplane flew directly over an observer moving away at a constant airspeed of 400 miles per hour. The distance between the observer and plane grew at an increasing rate eventually approaching a rate of _____.

**3.** If $dh/dt$ is decreasing as time $t$ increases, then $d^2h/dt^2$ is _____.

**4.** If water is pouring into a spherical tank at a constant rate, then the height of the water grows at a variable and positive rate $dh/dt$, but $d^2h/dt^2$ is _____ until $h$ reaches half the height of the tank, after which $d^2h/dt^2$ becomes _____.

## PROBLEM SET 3.9

**1.** Each edge of a variable cube is increasing at a rate of 3 inches per second. How fast is the volume of the cube increasing when an edge is 12 inches long?

**2.** Assuming that a soap bubble retains its spherical shape as it expands, how fast is its radius increasing when its radius is 3 inches, if air is blown into it at a rate of 3 cubic inches a second?

≈ **3.** An airplane, flying horizontally at an altitude of 1 mile passes directly over an observer. If the constant speed of the plane is 400 miles per hour, how fast is its distance from the observer increasing 45 seconds later? *Hint*: Note that in 45 seconds $(\frac{3}{4} \frac{1}{60} = \frac{1}{80}$ hour), the plane goes 5 miles.

**4.** A student is using a straw to drink from a conical paper cup, whose axis is vertical, at a rate of 3 cubic centimeters a second. If the height of the cup is 10 centimeters and the diameter of its opening is 6 centimeters, how fast is the level of the liquid falling when the depth of the liquid is 5 centimeters?

≈ **5.** An airplane flying west at 300 miles per hour, goes over the control tower at noon, and a second plane at the same altitude, flying north at 400 miles per hour, goes over the tower an hour later. How fast is the distance between the planes changing at 2:00 P.M.? *Hint*: See Example 3.

≈ **6.** A woman on a dock is pulling in a rope fastened to the bow of a small boat. If the woman's hands are 10 feet higher than the point where the rope is attached to the boat and if she is retrieving the rope at a rate of 2 feet per second, how fast is the boat approaching the dock when 25 feet of rope are still out?

≈ **7.** A 20-foot ladder is leaning against a building. If the bottom of the ladder is sliding along the level pavement directly away from the building at 1 foot per second, how fast is the top of the ladder moving down when the foot of the ladder is 5 feet from the wall?

**8.** We assume that an oil spill is being cleaned up by deploying bacteria which eat up the oil at 4 cubic feet per hour. The oil spill itself is modeled in the form of a very thin cylinder whose height is the thickness of the oil slick. When the thickness of the slick is 0.001 feet the cylinder is 500 feet in diameter. If the height is decreasing at 0.0005 feet per hour, at what rate is the area of the slick changing?

**9.** Sand is pouring from a pipe at the rate of 16 cubic feet per second. If the falling sand forms a conical pile on the ground whose altitude is always $\frac{1}{4}$ the diameter of the base, how fast is the altitude increasing when the pile is 4 feet high? *Hint*: Refer to Figure 8 and use the fact that $V = \frac{1}{3}\pi r^2 h$.

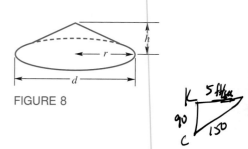

**FIGURE 8**

≈ **10.** A child is flying a kite. If the kite is 90 feet above the child's hand level and the wind is blowing it on a horizontal course at 5 feet per second, how fast is the child paying out cord when 150 feet of cord is out? (Assume the cord forms a line—actually an unrealistic assumption.)

**11.** A swimming pool is 40 feet long, 20 feet wide, 8 feet deep at the deep end, and 3 feet deep at the shallow end; the bottom is rectangular (see Figure 9). If the pool is filled by pumping water into it at the rate of 40 cubic feet per minute, how fast is the water level rising when it is 3 feet deep at the deep end?

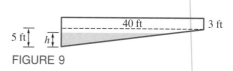

**FIGURE 9**

≈ **12.** A particle $P$ is moving along the graph of $y = \sqrt{x^2 - 4}$, $x \geq 2$, so that the $x$-coordinate of $P$ is increasing at the rate of 5 units per second. How fast is the $y$-coordinate of $P$ increasing when $x = 3$?

**13.** A metal disk expands during heating. If its radius increases at the rate of 0.02 inch per second, how fast is the area of one of its faces increasing when its radius is 8.1 inches?

≈ **14.** Two ships sail from the same island port, one going north at 24 knots (24 nautical miles per hour) and the other east at 30 knots. The northbound ship departed at 9:00 A.M. and the eastbound ship left at 11:00 A.M. How fast is the distance between them increasing at 2:00 P.M.? *Hint*: Let $t = 0$ at 11:00 A.M.

**15.** A light in a lighthouse 1 kilometer offshore from a straight shoreline is rotating at 2 revolutions per minute. How fast is the beam moving along the shoreline when it passes the point $\frac{1}{2}$ kilometer from the point opposite the lighthouse?

C **16.** An aircraft spotter observes a plane flying toward her at an altitude of 4000 feet. She notes that when the angle of elevation is $\frac{1}{2}$ radian, it is increasing at a rate of $\frac{1}{10}$ radian per second. What is the ground speed of the airplane?

**17.** Andy, who is 6 feet tall, is walking away from a street light pole 30 feet high at a rate of 2 feet per second.

(a) How fast is his shadow increasing in length when Andy is 24 feet from the pole? 30 feet?
(b) How fast is the tip of his shadow moving?
(c) To follow the tip of his shadow, at what angular rate must he lift his head when his shadow is 6 feet long?

**18.** The vertex angle opposite the base of an isosceles triangle with equal sides of length 100 centimeters is increasing at $\frac{1}{10}$ radian per minute. How fast is the area of the triangle increasing when the vertex angle measures $\pi/6$ radians? *Hint*: $A = \frac{1}{2}ab \sin$.

≈ **19.** A long, level highway bridge passes over a railroad track that is 100 feet below it and at right angles to it. If an automobile traveling 45 miles per hour (66 feet per second) is directly above a train going 60 miles per hour (88 feet per second), how fast will they be separating 10 seconds later?

**20.** Water is pumped at a uniform rate of 2 liters (1 liter = 1000 cubic centimeters) per minute into a tank shaped like a frustum of a right circular cone. The tank has altitude 80 centimeters and lower and upper radii of 20 and 40 centimeters, respectively (Figure 10). How fast is the water level rising when the depth of the water is 30 centimeters? *Note*: The volume, $V$, of a frustum of a right circular cone of altitude $h$ and lower and upper radii $a$ and $b$ is $V = \frac{1}{3}\pi h \cdot (a^2 + ab + b^2)$.

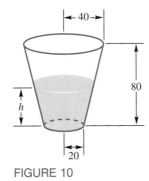

FIGURE 10

**21.** Water is leaking out the bottom of a hemispherical tank of radius 8 feet at a rate of 2 cubic feet per hour. The tank was full at a certain time. How fast is the water level

changing when its height $h$ is 3 feet? *Note*: The volume of a segment of height $h$ in a hemisphere of radius $r$ is $\pi h^2[r - (h/3)]$. (See Figure 11.)

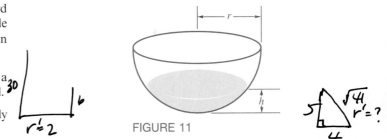

FIGURE 11

**22.** The hands on a clock are of length 5 inches (minute hand) and 4 inches (hour hand). How fast is the distance between the tips of the hands changing at 3:00?

**23.** A right circular cylinder with a piston at one end is filled with gas. Its volume is continually changing because of the movement of the piston. If the temperature of the gas is kept constant, then—by Boyle's Law—$PV = k$, where $P$ is the pressure (pounds per square inch), $V$ is the volume (cubic inches), and $k$ is a constant. The pressure was monitored by a recording device over one 10-minute period. The results are shown in Figure 12. Approximately how fast was the volume changing at $t = 6.5$ if its volume was 300 cubic inches at that instant? (See Example 5.)

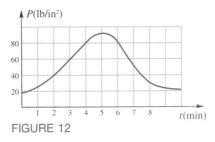

FIGURE 12

**24.** Rework Example 5 in the text assuming the water tank is a sphere of radius 20 feet. (See Problem 21 for the volume of a spherical segment.)

≈ **25.** An 18-foot ladder leans against a 12-foot vertical wall, its top extending over the wall. The bottom end of the ladder is pulled along the ground away from the wall at 2 feet per second.

(a) Find the vertical velocity of the top end when the ladder makes an angle of 60° with the ground.
(b) Find the vertical acceleration at the same instant.

**26.** A spherical steel ball rests at the bottom of the tank of Problem 21. Answer the question posed there if the ball has radius (a) 6 inches and (b) 2 feet.

**27.** A snowball melts at a rate proportional to its surface area.

(a) Show that its radius shrinks at a constant rate.
(b) If it melts to $\frac{8}{27}$ its original volume in one hour, how long will it take to melt completely?

**28.** A steel ball will drop $16t^2$ feet in $t$ seconds. Such a ball is dropped from a height of 64 feet at a horizontal distance 10 feet from a 48-foot street light. How fast is the ball's shadow moving when the ball hits the ground?

**29.** A girl 5 feet tall walks toward a street light 20 feet high at a rate of 4 feet per second. Her little brother, 3 feet tall, follows at a distance 4 feet behind her (Figure 13). Determine how fast the tip of the shadow is moving, that is, determine $dy/dt$. *Note*: When the girl is far from the light, she controls the tip of the shadow, whereas her brother controls it near the light.

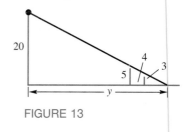

**FIGURE 13**

**Answers to Concepts Review: 1.** $du/dt$; $t = 2$
**2.** 400 mi/h **3.** Negative **4.** Negative; positive

---

**3.10**

**DIFFERENTIALS AND APPROXIMATIONS**

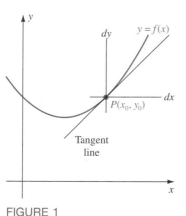

**FIGURE 1**

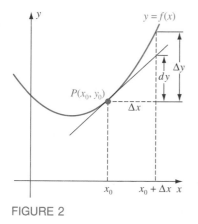

**FIGURE 2**

We have been using the Leibniz notation $dy/dx$ for the derivative of $y$ with respect to $x$. Up to now, we have treated $dy/dx$ as a single symbol and have not tried to give separate meanings to $dy$ and $dx$, as we will do next.

To motivate our definition, let $P(x_0, y_0)$ be a fixed point on the graph of $y = f(x)$, as shown in Figure 1. With $P$ as the origin, introduce new coordinate axes (the $dx$- and $dy$-axes) parallel to the old $x$- and $y$-axes. In this new coordinate system, the tangent line at $P$ has a particularly simple equation, namely, $dy = m\ dx$, where $m$ is its slope. But the slope $m$ in the new coordinate system is the same as it was in the old $xy$-system. Thus $m = f'(x_0)$, and so the equation of the tangent line may be written

$$dy = f'(x_0)dx$$

The usefulness of this notion rests on the fundamental fact that the tangent line is very close to the curve $y = f(x)$ in the vicinity of $P(x_0, y_0)$ (see Figure 2). Thus if $x$ is given a small increment $\Delta x = dx$, the corresponding increment in $y$ on the curve is $\Delta y = f(x_0 + \Delta x) - f(x_0)$, while on the tangent line it is $dy = f'(x_0)\Delta x$. But $dy$ is a good approximation to $\Delta y$ and, being just a constant times $\Delta x$, it is normally much easier to calculate. We explore the ramifications of this fact later in this section.

**Differential Defined** Here is the formal definition of *differential*.

---

**Definition**

**(Differential).** Let $y = f(x)$ be differentiable at $x$ and suppose that $dx$, the differential of the independent variable $x$, denotes an arbitrary increment of $x$. The corresponding differential $dy$ of the dependent variable $y$ is defined by

$$dy = f'(x)dx$$

**EXAMPLE 1**   Find $dy$ if (a) $y = x^3 - 3x + 1$, (b) $y = \sqrt{x^2 + 3x}$, (c) $y = \sin(x^4 - 3x^2 + 11)$.

*Solution*    If we know how to calculate derivatives, we know how to calculate differentials. We simply calculate the derivative and multiply it by $dx$.

(a)  $dy = (3x^2 - 3)dx$

(b)  $dy = \frac{1}{2}(x^2 + 3x)^{-1/2}(2x + 3)dx = \frac{2x + 3}{2\sqrt{x^2 + 3x}}dx$

(c)  $dy = \cos(x^4 - 3x^2 + 11) \cdot (4x^3 - 6x)dx$    ∎

Now we ask you to note several things. First, since $dy = f'(x)\,dx$, division of both sides by $dx$ yields

$$f'(x) = \frac{dy}{dx}$$

and we can, if we wish, interpret the derivative as a quotient of two differentials.

Second, corresponding to every derivative rule, there is a differential rule obtained from the former by multiplying through by $dx$. We illustrate the major rules in the table below.

| Derivative Rule | Differential Rule |
|---|---|
| 1. $\dfrac{dk}{dx} = 0$ | 1. $dk = 0$ |
| 2. $\dfrac{d(ku)}{dx} = k\dfrac{du}{dx}$ | 2. $d(ku) = k\,du$ |
| 3. $\dfrac{d(u + v)}{dx} = \dfrac{du}{dx} + \dfrac{dv}{dx}$ | 3. $d(u + v) = du + dv$ |
| 4. $\dfrac{d(uv)}{dx} = u\dfrac{dv}{dx} + v\dfrac{du}{dx}$ | 4. $d(uv) = u\,dv + v\,du$ |
| 5. $\dfrac{d(u/v)}{dx} = \dfrac{v(du/dx) - u(dv/dx)}{v^2}$ | 5. $d\left(\dfrac{u}{v}\right) = \dfrac{v\,du - u\,dv}{v^2}$ |
| 6. $\dfrac{d(u^n)}{dx} = nu^{n-1}\dfrac{du}{dx}$ | 6. $d(u^n) = nu^{n-1}\,du$ |

Third, while the definition of $dy$ assumes that $x$ is an independent variable, that assumption is not important. Suppose $y = f(x)$, where $x = g(t)$. Then $t$ is the independent variable and both $x$ and $y$ depend on it. Now

$$dx = g'(t)dt$$

and since

$$y = f(g(t))$$
$$dy = f'(g(t))g'(t)dt$$
$$= f'(x)dx$$

Note that $dy$ turns out to be $f'(x)\,dx$, as it would have if $x$ had been the independent variable.

Finally, we interject a warning. *Be careful to distinguish between derivatives and differentials.* They are not the same. When you write $D_x y$ or $dy/dx$, you are using a symbol for the derivative; when you write $dy$, you are denoting a differential. Do not be sloppy and write $dy$ when you mean to label a derivative. That would lead to boundless confusion.

**Approximations**  Differentials will play several roles in this book, but for now their chief use is in providing approximations. We hinted at this earlier.

Suppose $y = f(x)$, as shown in Figure 3. When $x$ is given an increment $\Delta x$, $y$ receives a corresponding increment $\Delta y$, which can be approximated by $dy$. Thus, $f(x + \Delta x)$ is approximated by

$$\boxed{f(x + \Delta x) \approx f(x) + dy = f(x) + f'(x)\Delta x}$$

This is the basis for the solutions to all the examples that follow.

**EXAMPLE 2**  Suppose you need good approximations to $\sqrt{4.6}$ and $\sqrt{8.2}$, but your calculator has died. What might you do?

***Solution***  Consider the graph of $y = \sqrt{x}$ sketched in Figure 4. When $x$ changes from 4 to 4.6, $\sqrt{x}$ changes from $\sqrt{4} = 2$ to (approximately) $\sqrt{4} + dy$. Now

$$dy = \frac{1}{2}x^{-1/2}\,dx = \frac{1}{2\sqrt{x}}dx$$

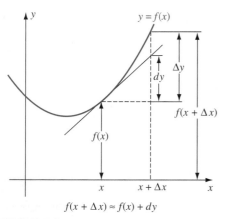

$f(x + \Delta x) \approx f(x) + dy$

**FIGURE 3**

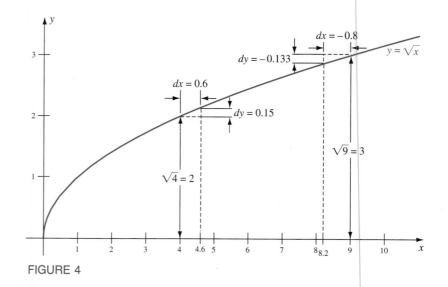

**FIGURE 4**

## Approximation

How good is the approximation
$$f(x + h) \approx f(x) + f'(x)h$$
If we let $\varepsilon(h)$ denote the error
$$\varepsilon(h) = f(x + h) - f(x) - f'(x)h$$
then
$$\lim_{h \to 0} \frac{\varepsilon(h)}{h} = 0$$
In other words, the error $\varepsilon(h)$ approaches 0 faster than $h$. A common terminology for this is to say that $\varepsilon(h)$ is of smaller order than $h$ and to write
$$\varepsilon(h) = o(h)$$
the so-called little-oh notation.

which, at $x = 4$ and $dx = 0.6$, has the value

$$dy = \frac{1}{2\sqrt{4}}(0.6) = \frac{0.6}{4} = 0.15$$

Thus,

$$\sqrt{4.6} \approx \sqrt{4} + dy = 2 + 0.15 = 2.15$$

Similarly, at $x = 9$ and $dx = -0.8$,

$$dy = \frac{1}{2\sqrt{9}}(-0.8) = \frac{-0.8}{6} = -0.133$$

Hence

$$\sqrt{8.2} \approx \sqrt{9} + dy = 3 - 0.133 = 2.867$$

Note that both $dx$ and $dy$ were negative in this case.

The approximate values 2.15 and 2.867 may be compared to the true values (to four decimal places) of 2.1448 and 2.8636. ∎

**EXAMPLE 3**   Use differentials to approximate the increase in the area of a soap bubble when its radius increases from 3 inches to 3.025 inches.

*Solution*   The area of a spherical soap bubble is given by $A = 4\pi r^2$. We may approximate the exact change, $\Delta A$, by the differential $dA$, where

$$dA = 8\pi r \, dr$$

At $r = 3$ and $dr = \Delta r = 0.025$,

$$dA = 8\pi(3)(0.025) \approx 1.885 \text{ square inches} \qquad ∎$$

**Estimating Errors**   Here is a typical problem in science. A researcher measures a certain variable $x$ to have a value $x_0$ with a possible error of size $\pm \Delta x$. The value $x_0$ is then used to calculate a value $y_0$ for $y$ that depends on $x$. The value $y_0$ is contaminated by the error in $x$, but how badly? The standard procedure is to estimate this error by means of differentials.

**EXAMPLE 4**   The side of a cube is measured as 11.4 centimeters with a possible error of $\pm 0.05$ centimeter. Evaluate the volume of the cube and give an estimate for the error in this value.

***Solution***   The volume $V$ of a cube of side $x$ is $V = x^3$. Thus $dV = 3x^2\, dx$. If $x = 11.4$ and $dx = 0.05$, then $V = (11.4)^3 \approx 1482$ and

$$dV = 3(11.4)^2(0.05) \approx 19.$$

Thus, we might report the volume of the cube as $1482 \pm 19$ cubic centimeters.    ∎

**EXAMPLE 5**   It is known that $y = 3 \sin 2t + 4 \cos^2 t$. If $t$ is measured as $1.13 \pm 0.005$, calculate $y$ and give an estimate for the error.

***Solution***

$$y = 3 \sin(2)(1.13) + 4 \cos^2(1.13) \approx 3.043$$
$$dy = [(3)(2)\cos 2t - (4)(2)\cos t \sin t]\, dt$$
$$= [6 \cos(2)(1.13) - 8 \cos(1.13)\sin(1.13)](0.005)$$
$$\approx -0.035$$

Thus, $y = 3.043 \pm 0.035$.    ∎

## CONCEPTS REVIEW

**1.** Let $y = f(x)$. The differential of $y$ in terms of $dx$ is defined by $dy =$ _____.

**2.** Consider the curve $y = f(x)$ and suppose $x$ is given an increment $\Delta x$. The corresponding change in $y$ on the curve is denoted by _____, whereas the corresponding change in $y$ on the tangent line is denoted by _____.

**3.** We can expect $dy \approx \Delta y$ to be a good approximation provided _____.

**4.** On the curve $y = \sqrt{x}$, we should expect $dy$ to be close to $\Delta y$ but always _____ than $\Delta y$. On the curve $y = x^2$, $x \geq 0$, we should expect $dy$ to be _____ than $\Delta y$.

## PROBLEM SET 3.10

In Problems 1–8, find $dy$.

**1.** $y = x^2 + x - 3$

**2.** $y = 7x^3 + 3x^2 + 1$

**3.** $y = (2x + 3)^{-4}$

**4.** $y = (3x^2 + x + 1)^{-2}$

**5.** $y = (\sin x + \cos x)^3$

**6.** $y = (\tan x + 1)^3$

**7.** $y = (7x^2 + 3x - 1)^{-\frac{3}{2}}$

**8.** $y = (x^{10} + \sqrt{\sin 2x})^2$

**9.** If $F(t) = \sqrt[5]{(t - 2)^2}$, find $dF$.

**10.** If $s = \sqrt{(t^2 - \cot t + 2)^3}$, find $ds$.

**11.** Let $y = f(x) = x^3$. Find the value of $dy$ in each case.

(a) $x = 0.5$, $dx = 1$         (b) $x = -1$, $dx = 0.75$

Make a careful drawing of the graph of $f$ for $-1.5 \leq x \leq 1.5$ and the tangents to the curve at $x = 0.5$ and $x = -1$; on this drawing label $dy$ and $dx$ for each of the given sets of data in (a) and (b).

**12.** Let $y = 1/x$. Find the value of $dy$ in each case.

(a) $x = 1$, $dx = 0.5$         (b) $x = -2$, $dx = 0.75$

Make a large scale drawing, as in Problem 11, for $-3 \le x < 0$ and $0 < x \le 3$.

[C] **13.** For the data of Problem 11, find the actual changes in $y$, namely, $\Delta y$.

[C] **14.** For the data of Problem 12, find the changes in $y$, namely, $\Delta y$.

**15.** If $y = x^2 - 3$, find the values of $\Delta y$ and $dy$ in each case.

    (a) $x = 2$ and $dx = \Delta x = 0.5$
[C] (b) $x = 3$ and $dx = \Delta x = -0.12$

**16.** If $y = x^4 + 2x$, find the values of $\Delta y$ and $dy$ in each case.

    (a) $x = 2$ and $dx = \Delta x = 1$
[C] (b) $x = 2$ and $dx = \Delta x = 0.005$

In Problems 17–20, use differentials to approximate the given number (see Example 2). Compare with calculator values.

**17.** $\sqrt{402}$            **19.** $\sqrt{35.9}$

**18.** $\sqrt[3]{26.91}$        **20.** $\sqrt[6]{64.05}$

[C] **21.** Approximate the volume of material in a spherical shell of inner radius 5 centimeters and outer radius 5.125 centimeters (see Example 3).

[C] **22.** All six sides of a cubical metal box are 0.25 inch thick, and the volume of the interior of the box is 40 cubic inches. Use differentials to find the approximate volume of metal used to make the box.

**23.** The outside diameter of a thin spherical shell is 12 feet. If the shell is 0.3 inch thick, use differentials to approximate the volume of the region interior to the shell.

**24.** The interior of an open cylindrical tank is 12 feet in diameter and 8 feet deep. The bottom is copper and the sides are steel. Use differentials to find approximately how many gallons of waterproofing paint are needed to apply a 0.05-inch coat to the steel part of the inside of the tank (1 gallon $\approx$ 231 cubic inches).

**25.** Assuming that the equator is a circle whose radius is approximately 4000 miles, how much longer than the equator would a concentric, coplanar circle be if each point on it were 2 feet above the equator? Use differentials.

**26.** The period of a simple pendulum of length $L$ feet is given by $T = 2\pi \sqrt{L/g}$ seconds. We assume that $g$, the acceleration due to gravity on (or very near) the surface of

the earth, is 32 feet per second per second. If the pendulum is that of a clock that keeps good time when $L = 4$ feet, how much time will the clock gain in 24 hours if the length of the pendulum is decreased to 3.97 feet?

**27.** The diameter of a sphere is measured as 20 ± 0.1 centimeters. Calculate the volume with an estimate for the error (see Examples 4 and 5).

**28.** A cylindrical roller is exactly 12 inches long and its diameter is measured as 6 ± 0.005 inches. Calculate its volume with an estimate for the error.

[C] **29.** The angle between the two equal sides of an isosceles triangle measures 0.53 ± 0.005 radians. The two equal sides are exactly 151 centimeters long. Calculate the length of the third side with an estimate for the error.

[C] **30.** Calculate the area of the triangle of Problem 29 with an estimate for the error. *Hint:* $A = \frac{1}{2} ab \sin$.

**31.** It can be shown that if $|d^2y/dx^2| \le M$ on a closed interval with $c$ and $c + \Delta x$ as endpoints, then

$$|\Delta y - dy| \le \tfrac{1}{2} M (\Delta x)^2$$

Find, using differentials, the change in $y = 3x^2 - 2x + 11$ when $x$ increases from 2 to 2.001 and then give a bound for the error you have made by using differentials.

**32.** Find $f(0.02)$ approximately if

$$f(x) = \sin[\sin(\sin 2x)]/\cos(\sin x).$$

**33.** A tank has the shape of a cylinder with hemispherical ends. If the cylindrical part is 100 centimeters long and has a radius of 10 centimeters, about how much paint is required to coat the outside of a tank to a thickness of 1 millimeter?

**34.** A conical cup, 10 centimeters high and 8 centimeters wide at the top, is filled with water to a depth of 9 centimeters. An ice cube 3 centimeters on a side is about to be dropped in. Use differentials to decide whether the cup will overflow.

[C] **35.** Einstein's Special Theory of Relativity says that mass $m$ is related to velocity $v$ by the formula

$$m = \frac{m_0}{\sqrt{1 - v^2/c^2}} = m_0 \left( 1 - \frac{v^2}{c^2} \right)^{-1/2}$$

Here $m_0$ is the rest mass and $c$ is the velocity of light. Use differentials to determine the percent increase in mass of an object when its velocity is increased from $0.9c$ to $0.92c$.

---

**Answers to Concepts Review:**    **1.** $f'(x)dx$    **2.** $\Delta y$; $dy$
**3.** $\Delta x$ is small.    **4.** Larger, smaller

## 3.11 CHAPTER REVIEW

### Concepts Test

Respond with true or false to each of the following assertions. Be prepared to justify your answer.

**1.** The tangent line to a curve at a point cannot cross the curve at that point.

**2.** The tangent line to a curve can touch the curve at only one point.

**3.** The slope of the tangent line to the curve $y = x^4$ is different at every point of the curve.

**4.** The slope of the tangent line to the curve $y = \cos x$ is different at every point on the curve.

**5.** It is possible for the velocity of an object to be increasing while its speed is decreasing.

**6.** It is possible for the speed of an object to be increasing while its velocity is decreasing.

**7.** If the tangent line to the graph of $y = f(x)$ is horizontal at $x = c$, then $f'(c) = 0$.

**8.** If $f'(x) = g'(x)$ for all $x$, then $f(x) = g(x)$ for all $x$.

**9.** If $g(x) = x$, then $f'(g(x)) = D_x f(g(x))$.

**10.** If $y = \pi^5$, then $D_x y = 5\pi^4$.

**11.** If $f'(c)$ exists, then $f$ is continuous at $c$.

**12.** The graph of $y = \sqrt[3]{x}$ has a tangent line at $x = 0$ and yet $D_x y$ does not exist there.

**13.** The derivative of a product is the product of the derivatives.

**14.** If the acceleration of an object is negative, then its velocity is decreasing.

**15.** If $x^3$ is a factor of the differentiable function $f(x)$, then $x^2$ is a factor of its derivative.

**16.** The equation of the line tangent to the graph of $y = x^3$ at $(1, 1)$ is $y - 1 = 3x^2(x - 1)$.

**17.** If $y = f(x)g(x)$, then $D_x^2 y = f(x)g''(x) + g(x)f''(x)$.

**18.** If $y = (x^3 + x)^8$, then $D_x^{25} y = 0$.

**19.** The derivative of a polynomial is a polynomial.

**20.** The derivative of a rational function is a rational function.

**21.** If $f'(c) = g'(c) = 0$ and $h(x) = f(x)g(x)$, then $h'(c) = 0$.

**22.** The expression

$$\lim_{x \to \pi/2} \frac{\sin x - 1}{x - \pi/2}$$

is the derivative of $f(x) = \sin x$ at $x = \pi/2$.

**23.** The operator $D^2$ is linear.

**24.** If $h(x) = f(g(x))$ where both $f$ and $g$ are differentiable, then $g'(c) = 0$ implies $h'(c) = 0$.

**25.** If $f'(2) = g'(2) = g(2) = 2$, then $(f \circ g)'(2) = 4$.

**26.** If $f$ is differentiable and increasing and if $dx = \Delta x > 0$, then $\Delta y > dy$.

**27.** If the radius of a sphere is increasing at 3 feet per second, then its volume is increasing at 27 cubic feet per second.

**28.** If the radius of a circle is increasing at 4 feet per second then its circumference is increasing at $8\pi$ feet per second.

**29.** $D_x^{n+4}(\sin x) = D_x^n(\sin x)$ for every positive integer $n$.

**30.** $D_x^{n+3}(\cos x) = -D_x^n(\sin x)$ for every positive integer $n$.

**31.** $\displaystyle\lim_{x \to 0} \frac{\tan x}{3x} = \frac{1}{3}$.

**32.** If $s = 5t^3 + 6t - 300$ gives the position of an object on a horizontal coordinate line at time $t$, then that object is always moving to the right (the direction of increasing $s$).

**33.** If air is being pumped into a spherical rubber ballon at a constant rate of 3 cubic inches per second, then the radius will increase but at a slower and slower rate.

**34.** If water is being pumped into a spherical tank of fixed radius at a rate of 3 gallons per second, the height of the water in the tank will increase more and more rapidly as the tank nears being full.

**35.** If an error $\Delta r$ is made in measuring the radius of a sphere, the corresponding error in the calculated volume will be approximately $S \cdot \Delta r$ where $S$ is the surface area of the sphere.

**36.** If $y = x^5$, then $dy \geq 0$.

## Sample Test Problems

1. Use $f'(x) = \lim\limits_{h \to 0} [f(x + h) - f(x)]/h$ to find the derivative of each of the following.

(a) $f(x) = 3x^3$

(b) $f(x) = 2x^5 + 3x$

(c) $f(x) = \dfrac{1}{3x}$

(d) $f(x) = \dfrac{1}{3x^2 + 2}$

(e) $f(x) = \sqrt{3x}$

(f) $f(x) = \sin 3x$

(g) $f(x) = \sqrt{x^2 + 5}$

(h) $f(x) = \cos \pi x$

2. Use $g'(x) = \lim\limits_{t \to x} \dfrac{g(t) - g(x)}{t - x}$ to find $g'(x)$ in each case.

(a) $g(x) = 2x^2$

(b) $g(x) = x^3 + x$

(c) $g(x) = \dfrac{1}{x}$

(d) $g(x) = \dfrac{1}{x^2 + 1}$

(e) $g(x) = \sqrt{x}$

(f) $g(x) = \sin \pi x$

(g) $g(x) = \sqrt{x^3 + C}$

(h) $g(x) = \cos 2x$

3. The given limit is a derivative, but of what function $f$ and at what point?

(a) $\lim\limits_{h \to 0} \dfrac{3(1 + h) - 3}{h}$

(b) $\lim\limits_{h \to 0} \dfrac{4(2 + h)^3 - 4(2)^3}{h}$

(c) $\lim\limits_{\Delta x \to 0} \dfrac{\sqrt{(1 + \Delta x)^3} - 1}{\Delta x}$

(d) $\lim\limits_{\Delta x \to 0} \dfrac{\sin(\pi + \Delta x)}{\Delta x}$

(e) $\lim\limits_{t \to x} \dfrac{4/t - 4/x}{t - x}$

(f) $\lim\limits_{t \to x} \dfrac{\sin 3x - \sin 3t}{t - x}$

(g) $\lim\limits_{h \to 0} \dfrac{\tan(\pi/4 + h) - 1}{h}$

(h) $\lim\limits_{h \to 0} \left( \dfrac{1}{\sqrt{5 + h}} - \dfrac{1}{\sqrt{5}} \right) \dfrac{1}{h}$

4.

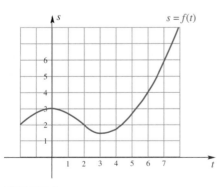

FIGURE 1

Use the sketch in the Figure 1 to approximate each of the following.

(a) $f'(2)$

(b) $f'(6)$

(c) $v_{\text{ave}}$ on $[3, 7]$

(d) $\dfrac{d}{dt} f(t^2)$ at $t = 2$

(e) $\dfrac{d}{dt} [f^2(t)]$ at $t = 2$

(f) $\dfrac{d}{dt} (f(f(t)))$ at $t = 2$

In Problems 5–14, find each derivative by using the rules we have developed.

5. $D_x(3x^5)$

6. $D_x(x^3 - 3x^2 + x^{-2})$

7. $D_z\,(z^3 + 4z^2 + 2z)$

8. $D_x \left( \dfrac{3x - 5}{x^2 + 1} \right)$

9. $D_t \left( \dfrac{4t - 5}{6t^2 + 2t} \right)$

10. $D_x^2(3x + 2)^{2/3}$

11. $\dfrac{d}{dx} \left( \dfrac{4x^2 - 2}{x^3 + x} \right)$

12. $D_t(t\sqrt{2t + 6})$

13. $\dfrac{d}{dx} \left( \dfrac{1}{\sqrt{x^2 + 4}} \right)$

14. $\dfrac{d}{dx} \sqrt{\dfrac{x^2 - 1}{x^3 - x}}$

15. $D_\theta^2(\sin \theta + \cos^3 \theta)$

16. $\dfrac{d}{dt} [\sin(t^2) - \sin^2(t)]$

17. $D_\theta(\sin(\theta^2))$

18. $\dfrac{d}{dx} (\cos^3 5x)$

19. $\dfrac{d}{d\theta} [\sin^2(\sin(\pi\theta))]$

20. $\dfrac{d}{dt} [\sin^2(\cos 4t)]$

21. $D_\theta \tan 3\theta$

22. $\dfrac{d}{dx} \left( \dfrac{\sin 3x}{\cos 5x^2} \right)$

23. $f'(2)$ if $f(x) = (x^2 - 1)^2(3x^3 - 4x)$

24. $g''(0)$ if $g(x) = \sin 3x + \sin^2 3x$

25. $\dfrac{d}{dx} \left( \dfrac{\cot x}{\sec x^2} \right)$

26. $D_t \left( \dfrac{4t \sin t}{\cos t - \sin t} \right)$

27. $f'(2)$ if $f(x) = (x - 1)^3(\sin \pi x - x)^2$

28. $h''(0)$ if $h(t) = (\sin 2t + \cos(3t))^5$

29. $g'''(1)$ if $g(r) = \cos^3 5r$

In Problems 30–34 assume that all the functions given are differentiable, and find the indicated derivative.

30. $f'(t)$ if $f(t) = h(g(t)) + g^2(t)$

31. $G''(x)$ if $G(x) = F(r(x) + s(x)) + s(x)$

**32.** If $F(x) = Q(R(x))$, $R(x) = \cos x$ and $Q(R) = R^3$ find $F'(x)$.

**33.** If $F(z) = r(s(z))$, $r(x) = \sin 3x$ and $s(t) = 3t^3$ find $F'(z)$.

**34.** If $F(\beta) = \Sigma(\Phi(\beta))$, $\Sigma(\Phi) = \sqrt{3\Phi}$ and $\Phi(\beta) = \beta^3 - \beta$ find $\dfrac{d}{d\beta} F(\beta)$.

**35.** Find the coordinates of the point on the curve $y = (x - 2)^2$ at which the tangent line is perpendicular to the line $2x - y + 2 = 0$.

**36.** A spherical balloon is expanding from the sun's heat. Find the rate of change of the volume of the balloon with respect to its radius when the radius is 5 meters.

**37.** Use differentials to approximate the change in volume of the balloon of Problem 36 when its radius increased from 5 to 5.1 meters.

**38.** If the volume of the balloon of Problem 36 is increasing at a constant rate of 10 cubic meters per hour, how fast is its radius increasing when the radius is 5 meters?

**39.** A trough 12 feet long has a cross section in the form of an isosceles triangle 4 feet deep and 6 feet across at the top. If water is filling the trough at the rate of 9 cubic feet per minute, how fast is the water level rising when the water is 3 feet deep?

**40.** An object is projected directly upward from the ground with an initial velocity of 128 feet per second. Its height $s$ at the end of $t$ seconds is approximately $s = 128t - 16t^2$ feet.

(a) When does it reach its maximum height and what is this height?
(b) When does it hit the ground and with what velocity?

**41.** An object moves on a horizontal coordinate line. Its directed distance $s$ from the origin at the end of $t$ seconds is $s = t^3 - 6t^2 + 9t$ feet.

(a) When is the object moving to the left?
(b) What is its acceleration when its velocity is zero?
(c) When is its acceleration positive?

**42.** Find $D_x^{20} y$ in each case.

(a) $y = x^{19} + x^{12} + x^5 + 100$
(b) $y = x^{20} + x^{19} + x^{18}$
(c) $y = 7x^{21} + 3x^{20}$
(d) $y = \sin x + \cos x$
(e) $y = \sin 2x$
(f) $y = \dfrac{1}{x}$

**43.** Find $dy/dx$ in each case.

(a) $(x - 1)^2 + y^2 = 5$
(b) $xy^2 + yx^2 = 1$
(c) $x^3 + y^3 = x^3 y^3$
(d) $x \sin(xy) = x^2 + 1$
(e) $x \tan(xy) = 2$

**44.** Show that the tangents to the curves $y^2 = 4x^3$ and $2x^2 + 3y^2 = 14$ at $(1, 2)$ are perpendicular to each other. *Hint*: Use implicit differentiation.

**45.** Let $y = \sin(\pi x) + x^2$. If $x$ changes from 2 to 2.01, approximately how much does $y$ change?

**46.** Let $xy^2 + 2y(x + 2)^2 + 2 = 0$

(a) If $x$ changes from $-2.00$ to $-2.01$ and $y > 0$, approximately how much does $y$ change?
(b) If $x$ changes from $-2.00$ to $-2.01$ and $y < 0$, approximately how much does $y$ change?

**47.** Suppose $f(2) = 3$, $f'(2) = 4$, $f''(2) = -1$, $g(2) = 2$, and $g'(2) = 5$. Find each value.

(a) $\dfrac{d}{dx} [f^2(x) + g^3(x)]$ at $x = 2$
(b) $\dfrac{d}{dx} [f(x)g(x)]$ at $x = 2$
(c) $\dfrac{d}{dx} [f(g(x))]$ at $x = 2$
(d) $D_x^2[f^2(x)]$ at $x = 2$

**48.** A 13-foot ladder is leaning against a vertical wall. If the bottom of the ladder is being pulled along the ground at a constant rate of 2 feet per second, how fast is the top end of the ladder moving down the wall when it is 5 feet above the ground?

**49.** An airplane is climbing at a $15°$ angle to the horizontal. How fast is it gaining altitude if its speed is 400 miles per hour?

**50.** Given that $D_x|x| = \dfrac{|x|}{x}$, $x \neq 0$, find a formula for

(a) $D_x(|x|^2)$
(b) $D_x^2|x|$
(c) $D_x^3 |x|$
(d) $D_x^2(|x|)^2$

**51.** Given that $D_t |t| = \dfrac{|t|}{t}$, $t \neq 0$, find a formula for

(a) $D_\theta|\sin \theta|$
(b) $D_\theta|\cos \theta|$

## 3.12 ADDITIONAL PROBLEMS

**1.** Explain what it means for a function $g(t)$ to be non-differentiable at the point $t = \alpha$ without using any form of the word "differentiable."

A function $y = f(x)$ is invertible if we can solve for $x$ in terms of $y$, for example $x = g(y)$ and $g$ is also a function. Problems 2–4 explore the notion of an inverse function and its relation to the derivative.

**2.** Show that all nonconstant linear functions are invertible. That is, show that if $y = f(x) = mx + b$, and $m \neq 0$, then there is a function $x = g(y)$ such that $f(g(y)) = y$ as well as $g(f(x)) = x$.

**3.** Show that the function $f(x) = (1 - x^n)^{1/n}$, for $0 \leq x \leq 1$ is its own inverse and verify that $f'(f(x))f'(x) = 1$.

**4.** The functions $y = F(x) = \sqrt{x^2 - 7}$, defined for $|x| \geq \sqrt{7}$ and $x = G(y) = \sqrt{y^2 + 7}$ defined for all $y$, have the property that $x = G(y) = G(F(x))$ for $x \geq \sqrt{7}$, and $y = F(x) = F(G(y))$ for $y \geq 0$.

(a) Check that these results are correct. This fact means that these functions are inverses of each other.
(b) Calculate $F'(x)$ and $G'(y)$.
(c) Verify that $F'(G(y))G'(y) = 1$, and that $G'(F(x))F'(x) = 1$

**5.** Let $m$ and $n$ be integers. Find the derivative of

$$\Psi(x) = x^{m/n}$$

by choosing $F = x^m$, $G(x) = x^{1/n}$ and computing the derivative of $\Psi = F \circ G$ using the Chain Rule.

**6.** For integer $n$ the expression $(r + 1)^n$ is a polynomial of degree $n$. For example $(r + 1)^3 = r^3 + 3r^2 + 3r + 1$. If the coefficients of $r^m$ are given by $C_n^m$ (in this case $C_3^3 = 1$, $C_3^2 = 3$, $C_3^1 = 3$, $C_3^0 = 1$) show that

$$C_n^m = \frac{n \cdot (n - 1) \ldots (n - m + 1)}{1 \cdot 2 \ldots m}$$

Do this by completing the following steps:

(a) Show that $D_r[(r + 1)^n] = n(r + 1)^{n-1}$.
(b) By equating like powers of $r$, show using (a) that $mC_n^m = nC_{n-1}^{m-1}$.
(c) Finally deduce the required expression for $C_n^m$ using the fact that $C_1^1 = 1$ and $C_n^0 = 1$.

The next set of problems expand on the idea of differentials as a form of best linear approximation. Given a function $f(x)$, $\Delta f = f(x + \Delta x) - f(x)$ is an exact measure of how $f(x)$ changes as $x$ changes by $\Delta x$. Generally $\Delta f$ may be difficult to calculate. On the other hand $df = f'(x)dx$ is usually much easier to compute and gives an accurate approximation to $\Delta f$ if $dx$ is small.

**7.** One important linearization frequently employed is

$$(1 + x)^\alpha \approx 1 + \alpha x, \text{ for } x \approx 0, \alpha \text{ any number.}$$

(a) Fill out a numerical table for $x$, $\sqrt{1 + x}$, $1 + \frac{x}{2}$ for the indicated values of $x$.

| $x$ | 0.01 | 0.02 | 0.03 | 0.04 | 0.05 |
|---|---|---|---|---|---|
| $\sqrt{1 + x}$ | | | | | |
| $1 + \dfrac{x}{2}$ | | | | | |

Comment on any remarkable behavior.

(b) Using the method of differentials, calculate $dy$ for the function $y = (1 + x)^{1/2}$ at the point $x = 0$ for $dx = 0.01$, 0.02, 0.03, 0.04, and 0.05.
(c) For $\alpha = \frac{1}{2}$, is the error in the linear approximation positive or negative. What happens if $\alpha = -\frac{1}{2}$?
(d) Construct a linear approximation for $y = (1 + x)^{1/2}$ about the point $x = 4$.

**8.** Find a linear approximation to the $\sin x$ and $\tan x$ for $x \approx 0$. Are the errors always of one sign?

**9.** Given the following data for a function $g(x) = x^2 + 3x - 2$

| $x$ | 5.5 | 6.0 | 6.5 | 7.0 | 7.5 | 8.0 |
|---|---|---|---|---|---|---|
| $g(x)$ | 45. | 52. | 60. | 68. | 77. | 86. |

(a) Estimate $g'(6.0)$, $g'(7.5)$, $g'(7.25)$.
(b) Estimate the rate of change of $g'$ at $x = 6.0$.
(c) Find as best as you can an approximation for the tangent line to the graph of $g(x)$ at $x = 7.0$.
(d) Estimate $g(7.1)$ using the result from (c).
(e) Knowing that $g(x)$ is actually $x^2 + 3x - 2$, check your answers in (a)–(d).

**10.** We can use tangent line approximations to get approximate solutions to a complicated equation. Find the solution to $f(x) = \sqrt{1 + x} - 2.1$ for $3 \le x \le 8$.

(a) Show that $f(3) < 0$ and $f(8) > 0$ which implies that at least one zero of $f$ must lie in the region.
(b) Linearize $f(x)$ about $x = 8$ and solve the resulting linear equation. Why does this solution have to be discarded?
(c) Linearize $f(x)$ about $x = 3$ and solve to find the zero of the resulting linear equation.
(d) Check your answer in (c) by using a calculator or computer to compute the zero of $f(x)$.

**11.** This problem provides a demonstration that tangent line approximation at a point $x = \alpha$ where $f(x)$ has a derivative given by

$$T(x) = f(\alpha) + f'(\alpha) \cdot (x - \alpha)$$

is the best approximation of the form $L(x) = c + m \cdot (x - \alpha)$. To put it another way, the best choice for $c$ is $f(\alpha)$, and the best choice for $m$ is $f'(\alpha)$. We assume that (i) $Error(x) = f(x) - L(x)$ between the function and the best linear approximation is zero at $x = \alpha$, namely $Error(\alpha) = 0$, and that (ii) $|Error(x)| \le |K(x) \cdot (x - \alpha)|$ where $\lim_{x \to \alpha} K(x) = 0$.

(a) Show that $Error(\alpha) = 0$ implies that $c = f(\alpha)$.
(b) Show that condition (ii) can be recast as $\lim_{x \to \alpha} \dfrac{Error(x)}{x - \alpha} = 0$ and combined with $c = f(\alpha)$ implies that $m = f'(\alpha)$.

The units of the derivative are inherited from the units of the difference quotient used to define the derivative. Thinking in terms of these units frequently gives insights into the meaning of the derivative. We explore this in Problems 12–14.

**12.** Suppose the cost in cents of a piece of paper is $P(A)$ where $A$ is the area of the paper in square inches.

(a) If $P(93.5) = 5.04$, what is the average cost of paper per square inch for a piece of paper 8.5 by 11 inches (= 93.5 square inches)?

(b) Assume that we know that $P'(93.5) = 10.5$. This is the "marginal" cost which has the units of cents per unit area. In this case it cost considerably less to add a little extra to an already large sheet of paper. Suppose that you want to increase the margins of the paper all around by 0.1 inches. Using differentials, compute the additional cost.

**13.** Given that $V(t)$ represents the volume of water at time $t$, attach a meaning to $\dfrac{dV}{dt}$ when

(a) $V(t)$ is the amount of water in a tank at time $t$.
(b) $V(t)$ is the amount of water flowing through a pipe at time $t$.

**14.** Explain why if $q(x)$ measures a distance $q$ as a function of time $x$, then it must be possible to interpret $q'(x)$ as a velocity, and $q''(x)$ must have an interpretation as an acceleration.

**15.** Newton's law of gravitational attraction states that the force of attraction between two masses $M_1$ and $M_2$ is directed along the line of their centers and is given by $F = GM_1M_2/r^2$ where $r$ is the distance between these centers. The $g$ of gravitational acceleration in Example 4 of Section 3.7, is actually $GM_1/r^2$ where $G$ is a constant, $M_1$ is the mass of the earth, and $r$ is the distance from the center of the earth to point where the object is located. As long as the object stays in a region close to the surface of the earth we can take $r$ as the constant $R$ which is the radius of the earth, and thus $g$ is approximately a constant. The force on an object (weight) with mass $M_2$ due to the earth's gravitational attraction is then given by $F = gM_2$. It is known that a much more accurate representation of $g$ is given by

$$g(r) = \begin{cases} \dfrac{GM_1}{r^2} & \text{if } r > R \\[2mm] g = 32 \dfrac{\text{feet}}{\text{second}^2} & \text{if } r = R \\[2mm] \dfrac{GM_1 r}{R^2} & \text{if } r < R \end{cases}$$

(a) Show that $g(r)$ is a continuous function of $r$.

(b) Plot $g(r)$ for $\dfrac{R}{2} < r < \dfrac{3R}{2}$.

(c) Plot $g'(r)$ for the same domain as in (b).
(d) Is $g(r)$ a differentiable function of $r$ in this region?

**16.** Changes can be measured as absolute, relative, or percent change:

| change: | absolute | relative | percent |
|---|---|---|---|
| Exact | $\Delta g$ | $\dfrac{\Delta g}{g}$ | $\dfrac{\Delta g}{g} \times 100$ |
| Approximate | $dg$ | $\dfrac{dg}{g}$ | $\dfrac{dg}{g} \times 100$ |

If we assume that the earth is a perfect sphere with a radius of 3960 miles or 20908800 feet and $g(r)$ is given by Problem 15, compute the absolute, relative, and percent change in $g(r)$ when you are

(a) 4 miles below the surface in a diamond mine.
(b) at 30,000 feet in a balloon.
(c) 2,000 feet above the surface in a building.

# TECHNOLOGY PROJECT

Your text shows that the derivative of $f(x) = x^2$ is $f'(x) = 2x$. This is a special case of the **Power Law**:

$$f(x) = x^n \Rightarrow f'(x) = nx^{n-1}, \text{ for } n = 1, 2, \ldots$$

You can verify this "law" graphically by plotting the function in question over a very small interval and measuring the resulting slope. For example, consider this figure:

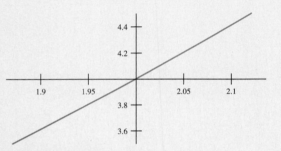

FIG. 1. Zooming in on $x^2$ near $x = 2$.

**Exercise 1**   Use Figure 1 to verify that the slope of $f(x) = x^2$ at $x = 2$ is what it should be. Make the best estimate you can of the slope and also state the error in your approximation to the exact slope.

**Exercise 2**   Use your computer to *analytically* verify the power law for $n = 3$ and $n = 4$. Method: Define a suitable $f(x)$ and simulate the standard simplification of the difference quotient by expanding out $(f(x + h) - f(x))/h$, fixing $x$ and $h$ if necessary. Then evaluate the resulting simple limit as $h \to 0$, any way you like.

**Exercise 3**   In this exercise, we find the tangent line to a curve using the idea that the tangent line is the *limit* of secant lines. As a nontrivial example, we use the function:

$$f(x) = \frac{(x^3 - 5)(x^2 - 1)}{x^2 + 1}$$

and find the tangent line at the specific point $(2, f(2)) = (2, 1.8)$.

To this end, consider two points on the graph, namely $(2, f(2))$ and $(b, f(b))$, where $b$ is close to, but not equal to, 2. The slope of the "secant" line (which connects the two points) is $(f(b) - f(2))/(b - 2)$. To calculate the secant line, you compute $slope = (f(b) - f(2))/(b - 2.0)$ and $secant = f(2.0) + slope(x - 2.0)$. Then simultaneously display the graph of $f$ and its secant line near the point of interest. You are to proceed, taking $b$ closer and closer to 2.0 until you have what looks like a tangent to the curve at $(2, f(2))$— that is, find a value of $b$ (close to, but not equal to, 2.0) such that you effectively have a tangent. Write down the values of $b$ and $f(b)$, and turn in a graph or sketch of the curve and its "tangent" you are most satisfied with.

**Exercise 4**   Use the approach of Exercise 1 to approximately determine the slope and tangent line to each of $f(x) = x^3$ and $f(x) = x^4$ at the point $x = 2$. (If your computer technology can select points off a graph, use this feature).

**Exercise 5**   Use the method of the previous exercise to again solve Exercise 3; that is, find the tangent line to $f(x) = (x^3 - 5)(x^2 - 1)/(x^2 + 1)$ at $x = 2$ by zooming in on $x = 2$.

**Exercise 6**   Use the method of the previous two exercises to explore the slope and tangent line to

$$f(x) = (x - 2)^{2/3} + 2x^3$$

at $x = 2$. You will need to "zoom" in much closer than previously to unearth the behavior near $x = 2$. Continue zooming until you are sure you've captured the true microscopic behavior near $x = 2$. Is there a unique tangent line at $x = 2$? Unique slope at $x = 2$? Why do you think so many zooms were needed?

**Exercise 7**   Find the (approximate) slopes of $f(x) = \sin x$ at the points $x = \pi/3$ and $x = \pi/6$.

**Exercise 8**   Use the power rule (as cited above) to check that the numerical values you found for the slopes of $x^3$ and $x^4$ were approximately correct.

**Exercise 9**   We haven't yet discovered the formula for the derivative of $\sin x$, but look it up in your textbook and check the results you got in Exercise 7.

# Applications of the Derivative

## 4.1 MAXIMA AND MINIMA

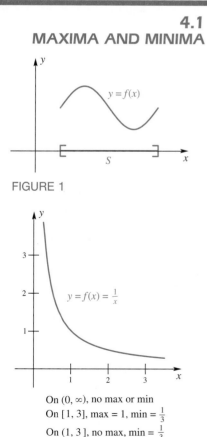

FIGURE 1

On $(0, \infty)$, no max or min
On $[1, 3]$, max = 1, min = $\frac{1}{3}$
On $(1, 3]$, no max, min = $\frac{1}{3}$

FIGURE 2

Often in life, we are faced with the problem of finding the *best* way to do something. For example, a farmer wants to choose the mix of crops that is likely to produce the largest profit. A doctor wishes to select the smallest dosage of a drug that will cure a certain disease. A manufacturer would like to minimize the cost of distributing its products. Sometimes a problem of this type can be formulated so it involves maximizing or minimizing a function over a specified set. If so, the methods of calculus provide a powerful tool for solving the problem.

Suppose then that we are given a function $f$ and a domain $S$ as in Figure 1. Our first job is to decide whether $f$ even has a maximum value or a minimum value on $S$. Assuming that such values exist, we want to know where on $S$ they are attained. Finally, we wish to determine the maximum and the minimum values. Analyzing these three tasks is the principal goal of this section.

We begin by introducing a precise vocabulary.

**Definition**

Let $S$, the domain of $f$, contain the point $c$. We say that:
**(i)** $f(c)$ is the **maximum value** of $f$ on $S$ if $f(c) \geq f(x)$ for all $x$ in $S$;
**(ii)** $f(c)$ is the **minimum value** of $f$ on $S$ if $f(c) \leq f(x)$ for all $x$ in $S$;
**(iii)** $f(c)$ is an **extreme value** of $f$ on $S$ if it is either the maximum value or the minimum value.

**The Existence Question** *Does $f$ have a maximum (or minimum) value on $S$?* The answer depends first of all on the set $S$. Consider $f(x) = 1/x$ on $S = (0, \infty)$; it has neither a maximum value nor a minimum value (Figure 2). On

175

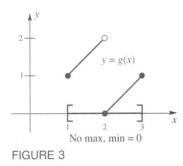

No max, min = 0

**FIGURE 3**

the other hand, the same function on $S = [1, 3]$ has the maximum value of $f(1) = 1$ and the minimum value of $f(3) = \frac{1}{3}$. On $S = (1, 3]$, $f$ has no maximum value and the minimum value of $f(3) = \frac{1}{3}$.

The answer also depends on the type of function. Consider the discontinuous function $g$ (Figure 3) defined by

$$g(x) = \begin{cases} x & \text{if } 1 \le x < 2 \\ x - 2 & \text{if } 2 \le x \le 3 \end{cases}$$

On $S = [1, 3]$, $g$ has no maximum value (it gets arbitrarily close to 2 but never attains it). However, $g$ has the minimum value $g(2) = 0$.

There is a nice theorem which answers the existence question for some of the problems that come up in practice. Though it is intuitively obvious, a rigorous proof is quite difficult; we leave that for more advanced textbooks.

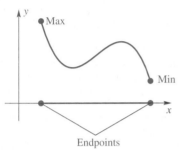

Endpoints

**FIGURE 4**

> ### Theorem A
>
> **(Max-Min Existence Theorem).** If $f$ is continuous on a closed interval $[a, b]$, then $f$ attains both a maximum value and a minimum value there.

Note the key words; $f$ is required to be *continuous* and the set $S$ is required to be a *closed interval*.

**Where Do Extreme Values Occur?**   Usually a function that we want to maximize or minimize will have an interval $I$ as its domain. But this interval may be any of the nine types discussed in Section 1.3. Some of them contain their **endpoints;** some do not. For instance, $I = [a, b]$ contains both its endpoints; $[a, b)$ contains only its left endpoint; $(a, b)$ contains neither endpoint. Extreme values of functions defined on closed intervals often occur at endpoints (see Figure 4).

If $c$ is a point at which $f'(c) = 0$, we call $c$ a **stationary point.** The name derives from the fact that at a stationary point, the graph of $f$ levels off, since the tangent line is horizontal. Extreme values often occur at stationary points (see Figure 5).

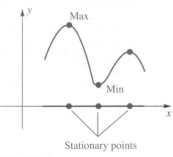

Stationary points

**FIGURE 5**

Finally, if $c$ is an interior point of $I$ where $f'$ fails to exist, we call $c$ a **singular point.** It is a point where the graph of $f$ has a sharp corner, a vertical tangent, or perhaps takes a jump (or near which it wiggles very badly). Extreme values can occur at singular points (Figure 6), though in practical problems this is quite rare.

These three kinds of points (endpoints, stationary points, and singular points) are the key points of max-min theory. Any point in the domain of a function $f$ of one of these three types is called a **critical point** of $f$.

**EXAMPLE 1**   Find the critical points of $f(x) = -2x^3 + 3x^2$ on $[-\frac{1}{2}, 2]$.

*Solution*   The endpoints are $-\frac{1}{2}$ and 2. To find the stationary points, we solve $f'(x) = -6x^2 + 6x = 0$ for $x$, obtaining 0 and 1. There are no singular points. Thus, the critical points are $-\frac{1}{2}$, 0, 1, 2. ■

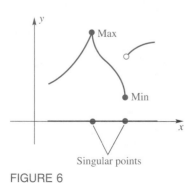

Singular points

**FIGURE 6**

Theorem B

**(Critical Point Theorem).** Let $f$ be defined on an interval $I$ containing the point $c$. If $f(c)$ is an extreme value, then $c$ must be a critical point; that is, either $c$ is:

**(i)**  an endpoint of $I$;
**(ii)**  a stationary point of $f$ ($f'(c) = 0$); or
**(iii)**  a singular point of $f$ ($f'(c)$ does not exist).

**Proof**  Consider first the case where $f(c)$ is the maximum value of $f$ on $I$ and suppose that $c$ is neither an endpoint nor a singular point. It will be enough to show that $c$ is then a stationary point.

Now since $f(c)$ is the maximum value, $f(x) \leq f(c)$ for all $x$ in $I$; that is,

$$f(x) - f(c) \leq 0$$

Thus if $x < c$, so that $x - c < 0$, then

(1)
$$\frac{f(x) - f(c)}{x - c} \geq 0$$

whereas if $x > c$, then

(2)
$$\frac{f(x) - f(c)}{x - c} \leq 0$$

But $f'(c)$ exists, since $c$ is not a singular point. Consequently, when we let $x \to c^-$ in (1) and $x \to c^+$ in (2), we obtain, respectively, $f'(c) \geq 0$ and $f'(c) \leq 0$. We conclude that $f'(c) = 0$, as desired.

The case where $f(c)$ is the minimum value is handled similarly.  ∎

In the proof just given, we used the fact that the inequality $\leq$ is preserved under the operation of taking limits (see Problem 26 of Section 2.5).

**What Are the Extreme Values?**  In view of Theorems A and B, we can now state a very simple procedure for finding the maximum value or minimum value of a continuous function $f$ on a *closed interval I*.

*Step 1*  Find the critical points of $f$ on $I$.
*Step 2*  Evaluate $f$ at each of these critical points. The largest of these values is the maximum value; the smallest is the minimum value.

**EXAMPLE 2**  Find the maximum and minimum values of

$$f(x) = -2x^3 + 3x^2$$

on $[-\frac{1}{2}, 2]$.

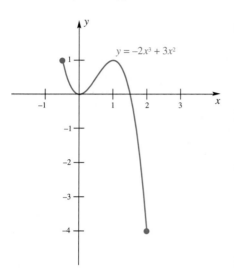

$$y = -2x^3 + 3x^2$$

FIGURE 7

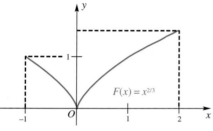

$$F(x) = x^{2/3}$$

FIGURE 8

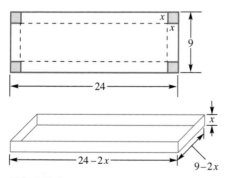

24

24 − 2x

9

9 − 2x

FIGURE 9

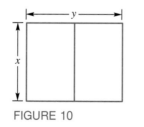

FIGURE 10

**Solution** In Example 1, we identified $-\frac{1}{2}$, 0, 1, 2 as the critical points. Now $f(-\frac{1}{2}) = 1, f(0) = 0, f(1) = 1$, and $f(2) = -4$. Thus the maximum value is 1 (attained at both $-\frac{1}{2}$ and 1) and the minimum value is $-4$ (attained at 2). The graph of $f$ is shown in Figure 7. ∎

**EXAMPLE 3** The function $F(x) = x^{2/3}$ is continuous everywhere. Find its maximum and minimum values on $[-1, 2]$.

**Solution** $F'(x) = \frac{2}{3}x^{-1/3}$, which is never 0. However, $F'(0)$ does not exist, so 0 is a critical point, as are the endpoints $-1$ and 2. Now $F(-1) = 1, F(0) = 0$, and $F(2) = \sqrt[3]{4} \approx 1.59$. Thus the maximum value is $\sqrt[3]{4}$; the minimum value is 0. The graph is shown in Figure 8. ∎

**Practical Problems** By a practical problem, we mean a problem that might arise in everyday life. Such problems rarely have singular points; in fact, for them, maximum and minimum values usually occur at stationary points, though endpoints should be checked. Here are two typical examples.

**EXAMPLE 4** A rectangular box is to be made from a piece of cardboard 24 inches long and 9 inches wide by cutting out identical squares from the four corners and turning up the sides, as in Figure 9. Find the dimensions of the box of maximum volume. What is this volume?

**Solution** Let $x$ be the width of the square to be cut out and $V$ the volume of the resulting box. Then

$$V = x(9 - 2x)(24 - 2x) = 216x - 66x^2 + 4x^3$$

Now $x$ cannot be less than 0 nor more than 4.5. Thus, our problem is to maximize $V$ on $[0, 4.5]$. The stationary points are found by setting $dV/dx$ equal to 0 and solving the resulting equation:

$$\frac{dV}{dx} = 216 - 132x + 12x^2 = 12(18 - 11x + x^2) = 12(9 - x)(2 - x) = 0$$

This gives $x = 2$ or $x = 9$, but 9 is not in the interval $[0, 4.5]$. We see that there are only three critical points, namely, 0, 2, and 4.5. At the endpoints 0 and 4.5, $V = 0$; at 2, $V = 200$. We conclude that the box has a maximum volume of 200 cubic inches if $x = 2$—that is, if the box is 20 inches long, 5 inches wide, and 2 inches deep. ∎

**EXAMPLE 5** Farmer Brown has 100 meters of wire fence with which he plans to build two identical adjacent pens, as shown in Figure 10. What are the dimensions of the total enclosure for which its area is a maximum?

**Solution** Let $x$ be the width and $y$ the length of the total enclosure, both in meters. Because there are 100 meters of fence, $3x + 2y = 100$—that is,

$$y = 50 - \tfrac{3}{2}x$$

The total area $A$ is given by

$$A = xy = 50x - \tfrac{3}{2}x^2$$

Since there must be three sides of length $x$, we see that $0 \le x \le \frac{100}{3}$. Thus, our problem is to maximize $A$ on $[0, \frac{100}{3}]$. Now

$$\frac{dA}{dx} = 50 - 3x$$

When we set $50 - 3x$ equal to 0 and solve, we get $x = \frac{50}{3}$ as a stationary point. Thus, there are three critical points: $0$, $\frac{50}{3}$, and $\frac{100}{3}$. The two endpoints $0$ and $\frac{100}{3}$ give $A = 0$, while $x = \frac{50}{3}$ yields $A = 416.67$. The desired dimensions are $x = \frac{50}{3} \approx 16.67$ meters and $y = 50 - \frac{3}{2}\left(\frac{50}{3}\right) = 25$ meters.

Is this answer sensible? Yes. We should expect to use more of the given fence in the $y$-direction than the $x$-direction because the former is fenced only twice, whereas the latter is fenced three times. ∎

Our next example illustrates a problem faced by a firm that delivers its products by truck. As the speed at which a truck is driven increases, the operating cost (gasoline, oil, and so on) increases, whereas the driver cost goes down. What is the most economical speed at which a truck should be driven?

**EXAMPLE 6**   The operating cost for a certain truck is estimated to be $(30 + v/2)$ cents per mile when it is driven at a speed of $v$ miles per hour. The driver is paid \$14 per hour. What speed will minimize the cost of making a delivery to a city $k$ miles away? Assume that the law restricts the speed to $40 \le v \le 60$.

*Solution*   Let $C$ be the total cost in cents of driving the truck $k$ miles. Then,

$$C = \text{driver cost} + \text{operating cost}$$

$$= \frac{k}{v}(1400) + k\left(30 + \frac{v}{2}\right) = 1400kv^{-1} + \left(\frac{k}{2}\right)v + 30k$$

Thus,

$$\frac{dC}{dv} = -1400kv^{-2} + \frac{k}{2} + 0$$

Setting $dC/dv$ equal to 0 gives

$$\frac{1400k}{v^2} = \frac{k}{2}$$

$$v^2 = 2800$$

$$v \approx 53$$

A speed of 53 miles per hour would appear to be optimum, but we must evaluate $C$ at the three critical points 40, 53, and 60 to be sure.

$$\text{At } v = 40, \ C = k\left(\frac{1400}{40}\right) + k(30 + 20) = 85k$$

$$\text{At } v = 53, \ C = k\left(\frac{1400}{53}\right) + k\left(30 + \frac{53}{2}\right) = 82.9k$$

$$\text{At } v = 60, \ C = k\left(\frac{1400}{60}\right) + k(30 + 30) = 83.3k$$

We conclude that a speed of 53 miles per hour is best. ■

We will have a good deal more to say about applied maxima and minima problems in Section 4.4.

## CONCEPTS REVIEW

**1.** A _____ function on a _____ interval will always have both a maximum value and a minimum value on that interval.

**2.** The term _____ value denotes either a maximum or a minimum value.

**3.** A function can attain an extreme value only at a critical point. Critical points are of three types: _____, _____, and _____.

**4.** A stationary point for $f$ is a number $c$ such that _____; a singular point for $f$ is a number $c$ such that _____.

## PROBLEM SET 4.1

In Problems 1–16, identify the critical points and find the maximum value and minimum value (see Examples 1, 2, and 3).

**1.** $f(x) = x^2 + 4x + 4; \ I = [-4, 0]$

**2.** $h(x) = x^2 + x; \ I = [-2, 2]$

**3.** $\Psi(x) = x^2 + 3x; \ I = [-2, 1]$

**4.** $G(x) = \frac{1}{5}(2x^3 + 3x^2 - 12x); \ I = [-3, 3]$

**5.** $f(x) = x^3 - 3x + 1; \ I = (-\frac{3}{2}, 3)$ *Hint:* Sketch the graph.

**6.** $f(x) = x^3 - 3x + 1; \ I = [-\frac{3}{2}, 3]$

**7.** $h(r) = \frac{1}{r}; \ I = [-1, 3]$

**8.** $g(x) = \frac{1}{1 + x^2}; \ I = [-3, 1]$

**9.** $g(x) = \frac{1}{1 + x^2}; \ I = (-\infty, \infty)$ *Hint:* Sketch the graph.

**10.** $f(x) = \frac{x}{1 + x^2}; \ I = [-1, 4]$

**11.** $r(\theta) = \sin \theta; \ I = \left[-\frac{\pi}{4}, \frac{\pi}{6}\right]$

**12.** $s(t) = \sin t - \cos t; \ I = [0, \pi]$

**13.** $a(x) = |x - 1|; \ I = [0, 3]$

**14.** $f(s) = |3s - 2|; \ I = [-1, 4]$

**15.** $g(x) = \sqrt[3]{x}; \ I = [-1, 27]$

**16.** $s(t) = t^{2/5}; \ I = [-1, 32]$

**17.** Find two nonnegative numbers whose sum is 10 and whose product is a maximum. *Hint:* If $x$ is one number, $10 - x$ is the other.

**18.** What number exceeds its square by the maximum amount? Begin by convincing yourself that this number is on the interval $[0, 1]$.

**19.** Joan has 200 feet of fence with which she plans to enclose a rectangular yard for her dog. If she wishes the area to be a maximum, what should the dimensions be?

**20.** Show that for a rectangle of given perimeter $K$, the one of maximum area is a square.

**21.** Find the volume of the largest open box that can be made from a piece of cardboard 24 inches square by cutting equal squares from the corners and turning up the sides (see Example 4).

**22.** A piece of wire 16 inches long is cut into two pieces; one piece is bent to form a square and the other is bent to form a circle. Where should the cut be made in order that the sum of the areas of the square and the circle be a minimum? A maximum? (Allow the possibility of no cut.)

≈  **23.** Farmer Brown has 80 feet of fence with which he plans to enclose a rectangular pen along one side of his 100-foot barn, as shown in Figure 11 (the side along the barn needs no fence). What are the dimensions of the pen that has maximum area?

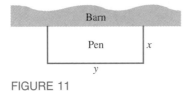

FIGURE 11

≈  **24.** Farmer Brown of Problem 23 decides to make three identical pens with his 80 feet of fence, as shown in Figure 12. What dimensions for the total enclosure make the area of the pens as large as possible?

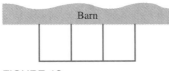

FIGURE 12

**25.** Suppose that Farmer Brown of Problem 23 has 180 feet of fence and wants the pen to adjoin to the whole side of the barn, as shown in Figure 13. What should the dimensions be for maximum area? Note that $0 \le x \le 40$ in this case.

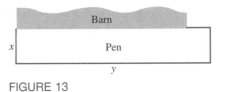

FIGURE 13

**26.** Suppose that Farmer Brown of Problem 23 decides to use his 80 feet of fence to make a rectangular pen to fit a 20-foot by 40-foot corner, as shown in Figure 14 (all the corner must be used and does not require fence). What dimensions give the pen a maximum area? *Hint*: Begin by deciding on the allowable values for $x$.

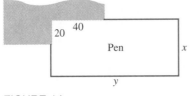

FIGURE 14

**27.** A rectangle has two corners on the $x$-axis and the other two on the parabola $y = 12 - x^2$, with $y \ge 0$. What are the dimensions of the rectangle of this type (Figure 15) with maximum area?

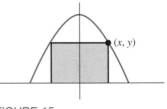

FIGURE 15

**28.** A rectangle is to be inscribed in a semicircle of radius $r$, as shown in Figure 16. What are the dimensions of the rectangle if its area is to be maximized?

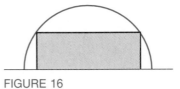

FIGURE 16

**29.** A metal rain gutter is to have 3-inch sides and a 3-inch horizontal bottom, the sides making an equal angle $\theta$ with the bottom (Figure 17). What should $\theta$ be in order to maximize the carrying capacity of the gutter? *Note*: $0 \le \theta \le \pi/2$.

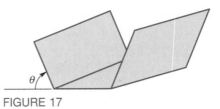

FIGURE 17

**30.** A huge conical tank is to be made from a circular piece of sheet metal of radius 10 meters by cutting out a sector with vertex angle $\theta$ and then welding the straight edges of the remaining piece together (Figure 18). Find $\theta$ so that the resulting cone has the largest possible volume.

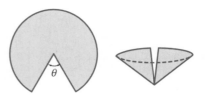

FIGURE 18

**31.** The operating cost of a certain truck is $25 + x/4$ cents per mile if the truck travels at $x$ miles per hour. In addition, the driver gets \$12 per hour. What is the most economical speed at which to operate the truck on a 400-mile run if the highway speed is required to be between 40 and 55 miles per hour?

**32.** Redo Problem 31 assuming the operating cost is $40 + 0.05x^{3/2}$ cents per mile.

**33.** Find the points $P$ and $Q$ on the curve $y = x^2/4$, $0 \le x \le 2\sqrt{3}$, which are closest to and farthest from the point $(0, 4)$. *Hint*: The algebra is simpler if you consider the square of the required distance rather than the distance itself.

**34.** A humidifier uses a rotating disk of radius $r$, which is partially submerged in water. The most evaporation occurs when the exposed wetted region (shown shaded in Figure 19) is maximized. Show that this happens when $h$

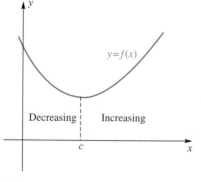

FIGURE 19

(the distance from the center to the water) is equal to $r/\sqrt{1 + \pi^2}$.

**35.** A covered box is to be made from a rectangular sheet of cardboard measuring 5 feet by 8 feet. This is done by cutting out the shaded regions of Figure 20 and then folding on the dotted lines. What are the dimensions $x$, $y$, and $z$ that maximize the volume?

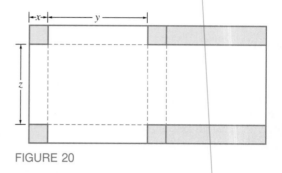

FIGURE 20

PC **36.** Identify the critical points and find the extreme values on $[-1, 5]$ for each function.

(a) $f(x) = x^3 - 6x^2 + x + 2$   (b) $g(x) = |f(x)|$

PC **37.** Follow the instructions of Problem 36 for $f(x) = \cos x + x \sin x + 2$.

**Answers to Concepts Review:** **1.** Continuous; closed **2.** Extreme **3.** Endpoints; stationary points; singular points **4.** $f'(c) = 0$; $f'(c)$ does not exist.

---

## 4.2 MONOTONICITY AND CONCAVITY

Consider the graph in Figure 1. No one will be surprised when we say that $f$ is decreasing to the left of $c$ and increasing to the right of $c$. But to make sure that we agree on terminology, we give precise definitions.

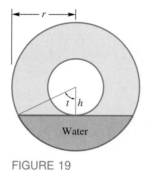

FIGURE 1

**Definition**

Let $f$ be defined on an interval $I$ (open, closed, or neither). We say that:

**(i)** $f$ is **increasing** on $I$ if for every pair of numbers $x_1$ and $x_2$ in $I$,

$$x_1 < x_2 \implies f(x_1) < f(x_2)$$

**(ii)** $f$ is **decreasing** on $I$ if for every pair of numbers $x_1$ and $x_2$ in $I$,

$$x_1 < x_2 \implies f(x_1) > f(x_2)$$

**(iii)** $f$ is **strictly monotonic** on $I$ if it is either increasing on $I$ or decreasing on $I$.

How shall we decide where a function is increasing? Someone might suggest that we draw its graph and look at it. But a graph is usually drawn by plotting a few points and connecting those points with a smooth curve. Who can be sure that the graph does not wiggle between the plotted points? We need a better procedure.

**The First Derivative and Monotonicity**    Recall that the first derivative $f'(x)$ gives us the slope of the tangent line to the graph of $f$ at the point $x$. Then, if $f'(x) > 0$, the tangent line is rising to the right (see Figure 2). Similarly, if $f'(x) < 0$, the tangent line is falling to the right. These facts make the following theorem intuitively clear. We postpone a rigorous proof until Section 4.8.

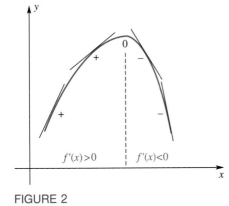

$f'(x)>0$    $f'(x)<0$

FIGURE 2

### Theorem A

**(Monotonicity Theorem).** Let $f$ be continuous on an interval $I$ and differentiable at every interior point of $I$.
**(i)** If $f'(x) > 0$ for all $x$ interior to $I$, then $f$ is increasing on $I$.
**(ii)** If $f'(x) < 0$ for all $x$ interior to $I$, then $f$ is decreasing on $I$.

This theorem usually allows us to determine precisely where a differentiable function increases and where it decreases. It is a matter of solving two inequalities.

**EXAMPLE 1**    If $f(x) = 2x^3 - 3x^2 - 12x + 7$, find where $f$ is increasing and where it is decreasing.

*Solution*    We begin by finding the derivative of $f$.

$$f'(x) = 6x^2 - 6x - 12 = 6(x + 1)(x - 2)$$

We need to determine where

$$(x + 1)(x - 2) > 0$$

and also where

$$(x + 1)(x - 2) < 0.$$

This problem was discussed in great detail in Section 1.3, a section worth reviewing now. The split points are $-1$ and 2; they split the $x$-axis into three intervals: $(-\infty, -1)$, $(-1, 2)$, and $(2, \infty)$. Using the test points $-2$, 0, and 3, we conclude that $f'(x) > 0$ on the first and last of these intervals and that $f'(x) < 0$ on the middle interval (Figure 3). Thus, by Theorem A, $f$ is increasing on $(-\infty, -1]$ and $[2, \infty)$; it is decreasing on $[-1, 2]$. Note that the theorem allows us to include the endpoints of these intervals, even though $f'(x) = 0$ at those points. The graph of $f$ is shown in Figure 4.  ∎

Values of $f'$

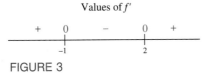

FIGURE 3

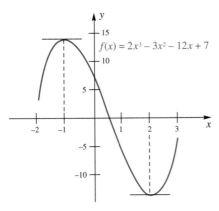

FIGURE 4

**EXAMPLE 2** Determine where $g(x) = x/(1 + x^2)$ is increasing and where it is decreasing.

**Solution**

$$g'(x) = \frac{(1 + x^2) - x(2x)}{(1 + x^2)^2} = \frac{1 - x^2}{(1 + x^2)^2} = \frac{(1 - x)(1 + x)}{(1 + x^2)^2}$$

Since the denominator is always positive, $g'(x)$ has the same sign as $(1 - x)(1 + x)$. The split points, $-1$ and $1$, determine the three intervals $(-\infty, -1), (-1, 1)$, and $(1, \infty)$. When we test them, we find that $g'(x) < 0$ on the first and last of these intervals and that $g'(x) > 0$ on the middle one (Figure 5). We conclude from Theorem A that $g$ is decreasing on $(-\infty, -1]$ and $[1, \infty)$, that it is increasing on $[-1, 1]$. We postpone graphing $g$ until later, but if you want to see the graph, turn to Figure 11 and Example 4. ∎

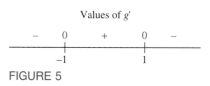

Values of $g'$

FIGURE 5

**The Second Derivative and Concavity** A function may be increasing and still have a very wiggly graph (Figure 6). To analyze wiggles, we need to study how the tangent line turns as we move from left to right along the graph. If the tangent line turns steadily in the counterclockwise direction, we say that the graph is *concave up*; if the tangent turns in the clockwise direction, the graph is *concave down*. Both definitions are better stated in terms of functions and their derivatives.

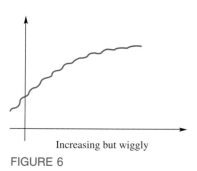

Increasing but wiggly

FIGURE 6

> **Definition**
>
> Let $f$ be differentiable on an open interval $I$. We say that $f$ (as well as its graph) is **concave up** on $I$ if $f'$ is increasing on $I$ and we say that $f$ is **concave down** on $I$ if $f'$ is decreasing on $I$.

The diagrams in Figure 7 will help to clarify these notions. Note that a curve that is concave *up* is shaped like a *cup*.

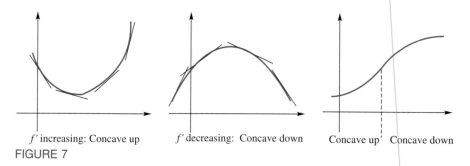

$f'$ increasing: Concave up        $f'$ decreasing: Concave down        Concave up   Concave down

FIGURE 7

In view of Theorem A, we have a simple criterion for deciding where a curve is concave up and where it is concave down. We simply keep in mind that the second derivative of $f$ is the first derivative of $f'$. Thus $f'$ is increasing if $f''$ is positive; it is decreasing if $f''$ is negative.

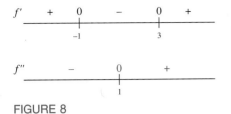

FIGURE 8

## Theorem B

**(Concavity Theorem).** Let $f$ be twice differentiable on the open interval $I$.
**(i)** If $f''(x) > 0$ for all $x$ in $I$, then $f$ is concave up on $I$.
**(ii)** If $f''(x) < 0$ for all $x$ in $I$, then $f$ is concave down on $I$.

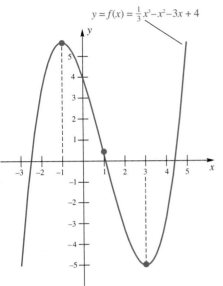

FIGURE 9

For most functions, this theorem reduces the problem of determining concavity to the problem of solving inequalities. We are experts at this.

**EXAMPLE 3**  Where is $f(x) = \frac{1}{3}x^3 - x^2 - 3x + 4$ increasing, decreasing, concave up, and concave down?

*Solution*

$$f'(x) = x^2 - 2x - 3 = (x + 1)(x - 3)$$
$$f''(x) = 2x - 2 = 2(x - 1)$$

By solving the inequality $(x + 1)(x - 3) > 0$ and its opposite we conclude that $f$ is increasing on $(-\infty, -1]$ and $[3, \infty)$ and decreasing on $[-1, 3]$ (Figure 8). Similarly, solving $2(x - 1) > 0$ and $2(x - 1) < 0$ shows that $f$ is concave up on $(1, \infty)$, concave down on $(-\infty, 1)$. The graph of $f$ is shown in Figure 9. ∎

**EXAMPLE 4**  Where is $g(x) = x/(1 + x^2)$ concave up and where is it concave down? Sketch the graph of $g$.

*Solution*  We began our study of this function in Example 2. There we learned that $g$ is decreasing on $(-\infty, -1]$ and $[1, \infty)$ and increasing on $[-1, 1]$. To analyze concavity, we calculate $g''$.

$$g'(x) = \frac{1 - x^2}{(1 + x^2)^2}$$

$$g''(x) = \frac{(1 + x^2)^2(-2x) - (1 - x^2)(2)(1 + x^2)(2x)}{(1 + x^2)^4}$$

$$= \frac{(1 + x^2)[(1 + x^2)(-2x) - (1 - x^2)(4x)]}{(1 + x^2)^4}$$

$$= \frac{2x^3 - 6x}{(1 + x^2)^3} = \frac{2x(x^2 - 3)}{(1 + x^2)^3}$$

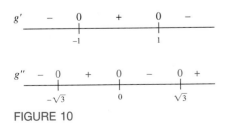

FIGURE 10

Since the denominator is always positive, we need only solve $x(x^2 - 3) > 0$ and its opposite. The split points are $-\sqrt{3}$, $0$, and $\sqrt{3}$. These three split points determine four intervals. After testing them (Figure 10), we conclude that $g$ is concave up on $(-\sqrt{3}, 0)$ and $(\sqrt{3}, \infty)$ and that it is concave down on $(-\infty, -\sqrt{3})$ and $(0, \sqrt{3})$.

To sketch the graph of $g$ we make use of all the information obtained so far, plus the fact that $g$ is an odd function whose graph is symmetric with respect to the origin (Figure 11). ■

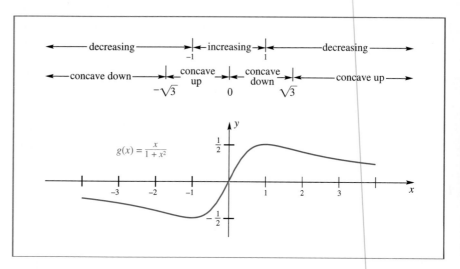

FIGURE 11

**Inflection Points**   Let $f$ be continuous at $c$. We call $(c, f(c))$ an **inflection point** of the graph of $f$ if $f$ is concave up on one side of $c$ and concave down on the other side. The graph in Figure 12 indicates a number of possibilities.

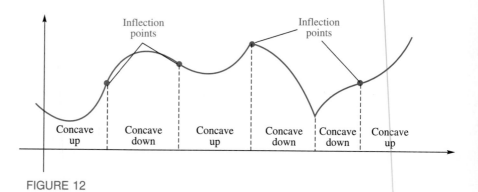

FIGURE 12

As you might guess, *points where $f''(x) = 0$ or where $f''(x)$ does not exist are the candidates for points of inflection.* We use the word *candidate* deliberately. Just as a candidate for political office may fail to be elected, so— for example—may a point where $f''(x) = 0$ fail to be a point of inflection. Consider $f(x) = x^4$, which has the graph shown in Figure 13. It is true that $f''(0) = 0$; yet the origin is not a point of inflection. However, in searching for inflection points, we begin by identifying those points where $f''(x) = 0$ (and where $f''(x)$ does not exist). Then we check to see if they really are inflection points.

Look back at the graph in Example 4. You will see that it has three inflection points. They are $(-\sqrt{3}, -\sqrt{3}/4)$, $(0, 0)$, and $(\sqrt{3}, \sqrt{3}/4)$.

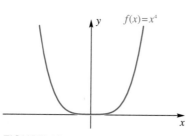

FIGURE 13

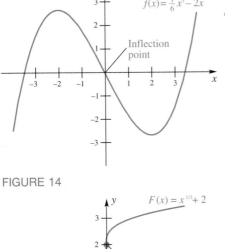

FIGURE 14

**EXAMPLE 5** Find all inflection points of the graph of $f(x) = \frac{1}{6}x^3 - 2x$.

**Solution**

$$f'(x) = \tfrac{1}{2}x^2 - 2$$
$$f''(x) = x$$

There is only one candidate for an inflection point, namely, the point where $f''(x) = 0$. This occurs at the origin, $(0, 0)$. That $(0, 0)$ is an inflection point follows from the fact that $f''(x) < 0$ for $x < 0$ and $f''(x) > 0$ for $x > 0$. Thus, the concavity changes direction at $(0, 0)$. The graph is shown in Figure 14. ∎

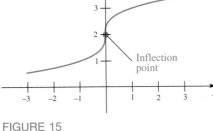

FIGURE 15

**EXAMPLE 6** Find all points of inflection for $F(x) = x^{1/3} + 2$.

**Solution**

$$F'(x) = \frac{1}{3x^{2/3}} \qquad F''(x) = \frac{-2}{9x^{5/3}}$$

The second derivative, $F''(x)$, is never 0; however, it fails to exist at $x = 0$. The point $(0, 2)$ is an inflection point since $F''(x) > 0$ for $x < 0$ and $F''(x) < 0$ for $x > 0$. The graph is sketched in Figure 15. ∎

## CONCEPTS REVIEW

**1.** If $f'(x) > 0$ everywhere, then $f$ is _____ everywhere; if $f''(x) > 0$ everywhere, then $f$ is _____ everywhere.

**2.** If _____ and _____ on an open interval $I$, then $f$ is both increasing and concave down on $I$.

**3.** A point on the graph of a continuous function where the concavity changes direction is called _____.

**4.** In trying to locate the inflection points for the graph of a function $f$, we should look at numbers $c$, where either _____ or _____.

## PROBLEM SET 4.2

In Problems 1–10, use the Monotonicity Theorem to find where the given function is increasing and where it is decreasing.

**1.** $f(x) = 3x + 3$

**2.** $g(x) = (x + 1)(x - 2)$

**3.** $h(t) = t^2 + 2t - 3$

**4.** $f(x) = x^3 - 1$

**5.** $G(x) = 2x^3 - 9x^2 + 12x$

**6.** $f(t) = t^3 + 3t^2 - 12$

**7.** $h(z) = \dfrac{z^4}{4} - \dfrac{4z^3}{6}$

**8.** $f(x) = \dfrac{x - 1}{x^2}$

**9.** $H(t) = \sin t,\ 0 \le t \le 2\pi$

**10.** $R(\theta) = \cos^2 \theta,\ 0 \le \theta \le 2\pi$

In Problems 11–18, use the Concavity Theorem to determine where the given function is concave up and where it is concave down. Also find all inflection points.

**11.** $f(x) = (x - 1)^2$

**12.** $G(w) = w^2 - 1$

**13.** $T(t) = 3t^3 - 18t$

**14.** $f(z) = z^2 - \dfrac{1}{z^2}$

15. $q(x) = x^4 - 6x^3 - 24x^2 + 3x + 1$

16. $f(x) = x^4 + 8x^3 - 2$

17. $F(x) = 2x^2 + \cos^2 x$

18. $G(x) = 24x^2 + 12\sin^2 x$

In Problems 19–28, determine where the graph of the given function is increasing, decreasing, concave up, and concave down. Then sketch the graph (see Example 4).

19. $f(x) = x^3 - 12x + 1$

20. $g(x) = 4x^3 - 3x^2 - 6x + 12$

21. $g(x) = 3x^4 - 4x^3 + 2$

22. $F(x) = x^6 - 3x^4$

23. $G(x) = 3x^5 - 5x^3 + 1$

24. $H(x) = \dfrac{x^2}{x^2 + 1}$

25. $f(x) = \sqrt{\sin x}$ on $[0, \pi]$

26. $g(x) = x\sqrt{x - 2}$

27. $f(x) = x^{2/3}(1 - x)$

28. $g(x) = 8x^{1/3} + x^{4/3}$

In Problems 29–32, on $[0, 6]$ sketch the graph of a continuous function $f$ that satisfies all of the stated conditions.

29. $f(0) = 3; f(3) = 0; f(6) = 4;$
    $f'(x) < 0$ on $(0, 3); f'(x) > 0$ on $(3, 6);$
    $f''(x) > 0$ on $(0, 5); f''(x) < 0$ on $(5, 6).$

30. $f(0) = 3; f(2) = 2; f(6) = 0;$
    $f'(x) < 0$ on $(0, 2) \cup (2, 6); f'(2) = 0;$
    $f''(x) < 0$ on $(0, 1) \cup (2, 6); f''(x) > 0$ on $(1, 2).$

31. $f(0) = f(4) = 1; f(2) = 2; f(6) = 0;$
    $f'(x) > 0$ on $(0, 2); f'(x) < 0$ on $(2, 4) \cup (4, 6);$
    $f'(2) = f'(4) = 0; f''(x) > 0$ on $(0, 1) \cup (3, 4);$
    $f''(x) < 0$ on $(1, 3) \cup (4, 6).$

32. $f(0) = f(3) = 3; f(2) = 4; f(4) = 2; f(6) = 0;$
    $f'(x) > 0$ on $(0, 2); f'(x) < 0$ on $(2, 4) \cup (4, 5);$
    $f'(2) = f'(4) = 0; f'(x) = -1$ on $(5, 6);$
    $f''(x) < 0$ on $(0, 3) \cup (4, 5); f''(x) > 0$ on $(3, 4).$

33. Prove that a quadratic function has no points of inflection.

34. Prove that a cubic function has exactly one point of inflection.

35. Prove that if $f'(x)$ exists and is continuous on an interval $I$ and if $f'(x) \neq 0$ at all interior points of $I$, then either $f$ is increasing throughout $I$ or decreasing throughout $I$. *Hint*: Use the Intermediate Value Theorem to show that

there cannot be two points $x_1$ and $x_2$ of $I$ where $f'$ has opposite signs.

36. Suppose $f$ is a function whose derivative is $f'(x) = (x^2 - x + 1)/(x^2 + 1)$. Use Problem 35 to prove that $f$ is increasing everywhere.

37. Use the Monotonicity Theorem to prove each statement if $0 < x < y$.

(a) $x^2 < y^2$      (b) $\sqrt{x} < \sqrt{y}$      (c) $\dfrac{1}{x} > \dfrac{1}{y}$

38. What conditions on $a$, $b$, and $c$ will make $f(x) = ax^3 + bx^2 + cx + d$ always increasing?

39. Determine $a$ and $b$ so that $f(x) = a\sqrt{x} + b/\sqrt{x}$ has the point $(4, 13)$ as an inflection point.

40. The general cubic function $f(x)$ has three zeros, $r_1$, $r_2$, and $r_3$. Show that its inflection point has $x$-coordinate $(r_1 + r_2 + r_3)/3$. *Hint*: $f(x) = a(x - r_1)(x - r_2)(x - r_3)$.

41. Suppose that $f'(x) > 0$ and $g'(x) > 0$ for all $x$. What simple additional conditions (if any) are needed to guarantee that:

(a) $f(x) + g(x)$ is increasing for all $x$;
(b) $f(x) \cdot g(x)$ is increasing for all $x$;
(c) $f(g(x))$ is increasing for all $x$?

42. Suppose that $f''(x) > 0$ and $g''(x) > 0$ for all $x$. What simple additional conditions (if any) are needed to guarantee that:

(a) $f(x) + g(x)$ is concave up for all $x$;
(b) $f(x) \cdot g(x)$ is concave up for all $x$;
(c) $f(g(x))$ is concave up for all $x$?

[PC] Use a computer to do Problems 43–46

43. Let $f(x) = \sin x + \cos(x/2)$ on the interval $I = (-2, 7)$.

(a) Draw the graph of $f$ on $I$.
(b) Use this graph to estimate where $f'(x) < 0$ on $I$.
(c) Use this graph to estimate where $f''(x) < 0$ on $I$.
(d) Draw the graph of $f'$ to confirm your answer to (b).
(e) Draw the graph of $f''$ to confirm your answer to (c).

44. Repeat Problem 43 for $f(x) = x\cos^2(x/3)$ on $(0, 10)$.

45. Let $f'(x) = x^3 - 5x^2 + 2$ on $I = [-2, 4]$. Where on $I$ is $f$ increasing?

46. Let $f''(x) = x^4 - 5x^3 + 4x^2 + 4$ on $I = [-2, 3]$. Where on $I$ is $f$ concave down?

**Answers to Concepts Review:** 1. Increasing; concave up
2. $f'(x) > 0; f''(x) < 0$   3. An inflection point
4. $f''(c) = 0; f''(c)$ does not exist.

**4.3**
**LOCAL MAXIMA**
**AND MINIMA**

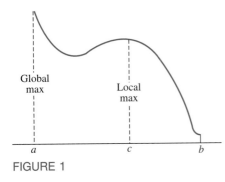

FIGURE 1

We recall from Section 4.1 that the maximum value (if it exists) of a function $f$ on a set $S$ is the largest value $f$ attains on the whole set $S$. It is sometimes referred to as the **global maximum value,** or the *absolute maximum value* of $f$. Thus for the function $f$ with domain $S = [a, b]$ whose graph is sketched in Figure 1, $f(a)$ is the global maximum value. But what about $f(c)$? It may not be king of the country, but at least it is chief of its own locality. We call it a **local maximum value,** or a *relative maximum value*. Of course a global maximum value is automatically a local maximum value. Figure 2 illustrates a number of possibilities. Note that the global maximum value (if it exists) is simply the largest of the local maximum values. Similarly, the global minimum value is the smallest of the local minimum values.

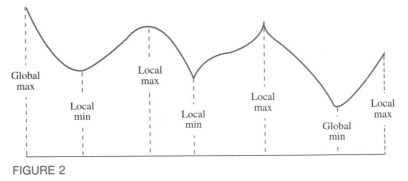

FIGURE 2

Here is the formal definition of local maxima and local minima. Recall that the symbol $\cap$ denotes the intersection (common part) of two sets.

---

**Definition**

Let $S$, the domain of $f$, contain the point $c$. We say that:

**(i)** $f(c)$ is a **local maximum value** of $f$ if there is an interval $(a, b)$ containing $c$ such that $f(c)$ is the maximum value of $f$ on $(a, b) \cap S$;

**(ii)** $f(c)$ is a **local minimum value** of $f$ if there is an interval $(a, b)$ containing $c$ such that $f(c)$ is the minimum value of $f$ on $(a, b) \cap S$;

**(iii)** $f(c)$ is a **local extreme value** of $f$ if it is either a local maximum or a local minimum value.

---

**Where Do Local Extreme Values Occur?**    The Critical Point Theorem (Theorem 4.1B) holds as stated, with the phrase *extreme value* replaced by *local extreme value*; the proof is essentially the same. Thus, the critical points (endpoints, stationary points, and singular points) are the candidates for points where local extrema may occur. We say *candidates* because we are not claiming that there must be a local extremum at every critical point. The left graph in Figure 3 makes this clear. However, if the derivative is positive on one side of the critical point and negative on the other, then we have a local extremum, as shown in the middle and right graphs of Figure 3.

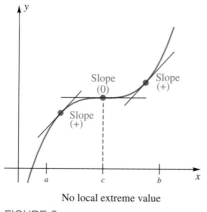

No local extreme value

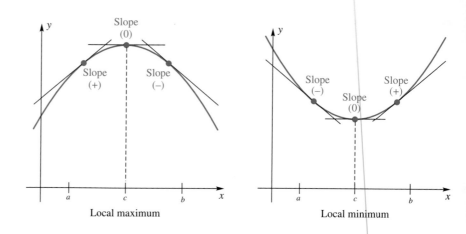

Local maximum

Local minimum

FIGURE 3

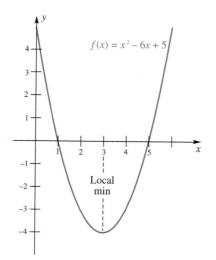

FIGURE 4

### Theorem A

**(First Derivative Test for Local Extrema).** Let $f$ be continuous on an open interval $(a, b)$ that contains a critical point $c$.

(i)   If $f'(x) > 0$ for all $x$ in $(a, c)$ and $f'(x) < 0$ for all $x$ in $(c, b)$, then $f(c)$ is a local maximum value of $f$.

(ii)  If $f'(x) < 0$ for all $x$ in $(a, c)$ and $f'(x) > 0$ for all $x$ in $(c, b)$, then $f(c)$ is a local minimum value of $f$.

(iii) If $f'(x)$ has the same sign on both sides of $c$, then $f(c)$ is not a local extreme value of $f$.

*Proof of (i)*   Since $f'(x) > 0$ for all $x$ in $(a, c)$, $f$ is increasing on $(a, c]$ by the Monotonicity Theorem. Again, since $f'(x) < 0$ for all $x$ in $(c, b)$, $f$ is decreasing on $[c, b)$. Thus, $f(x) < f(c)$ for all $x$ in $(a, b)$, except of course at $x = c$. We conclude that $f(c)$ is a local maximum.

The proofs of (ii) and (iii) are similar.   ∎

**EXAMPLE 1**   Find the local extreme values of the function $f(x) = x^2 - 6x + 5$ on $(-\infty, \infty)$.

*Solution*   The polynomial function $f$ is continuous everywhere, and its derivative, $f'(x) = 2x - 6$, exists for all $x$. Thus, the only critical point for $f$ is the single solution of $f'(x) = 0$, namely, $x = 3$.

Since $f'(x) = 2(x - 3) < 0$ for $x < 3$, $f$ is decreasing on $(-\infty, 3]$; and because $2(x - 3) > 0$ for $x > 3$, $f$ is increasing on $[3, \infty)$. Therefore, by the First Derivative Test, $f(3) = -4$ is a local minimum value of $f$. Since 3 is the only critical number, there are no other extreme values. The graph of $f$ is shown in Figure 4. Note that $f(3)$ is actually the (global) minimum value in this case.   ∎

**EXAMPLE 2**   Find the local extreme values of $f(x) = \frac{1}{3}x^3 - x^2 - 3x + 4$ on $(-\infty, \infty)$.

*Solution*   Since $f'(x) = x^2 - 2x - 3 = (x + 1)(x - 3)$, the only critical points of $f$ are $-1$ and 3. When we use the test points $-2$, 0, and 4, we learn that $(x + 1)(x - 3) > 0$ on $(-\infty, -1)$ and $(3, \infty)$ and $(x + 1)(x - 3) < 0$ on

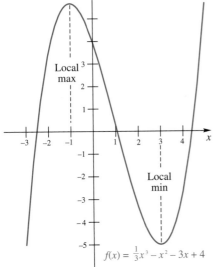

Local max

Local min

$$f(x) = \tfrac{1}{3}x^3 - x^2 - 3x + 4$$

**FIGURE 5**

$(-1, 3)$. By the First Derivative Test, we conclude that $f(-1) = \tfrac{17}{3}$ is a local maximum value and that $f(3) = -5$ is a local minimum value (Figure 5).

**EXAMPLE 3**  Find the local extreme values of $f(x) = (\sin x)^{2/3}$ on $(-\pi/6, 2\pi/3)$.

*Solution*

$$f'(x) = \frac{2 \cos x}{3(\sin x)^{1/3}}, \qquad x \neq 0$$

The points $0$ and $\pi/2$ are critical points, since $f'(0)$ does not exist and $f'(\pi/2) = 0$. Now $f'(x) < 0$ on $(-\pi/6, 0)$ and on $(\pi/2, 2\pi/3)$, while $f'(x) > 0$ on $(0, \pi/2)$. By the First Derivative Test, we conclude that $f(0) = 0$ is a local minimum value and that $f(\pi/2) = 1$ is a local maximum value. The graph of $f$ is shown in Figure 6.  ∎

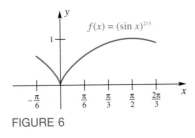

$$f(x) = (\sin x)^{2/3}$$

**FIGURE 6**

**The Second Derivative Test**    There is another test for local maxima and minima that is sometimes easier to apply than the First Derivative Test. It involves evaluating the second derivative at the stationary points. It does not apply to singular points.

> **Theorem B**
>
> **(Second Derivative Test for Local Extrema).** Let $f'$ and $f''$ exist at every point in an open interval $(a, b)$ containing $c$, and suppose $f'(c) = 0$.
> **(i)** If $f''(c) < 0$, $f(c)$ is a local maximum value of $f$.
> **(ii)** If $f''(c) > 0$, $f(c)$ is a local minimum value of $f$.

*Proof of (i)*    It is tempting to say that since $f''(c) < 0$, $f$ is concave downward near $c$ and to claim that this proves (i). However, to be sure that $f$ is concave downward in a neighborhood of $c$, we need $f''(x) < 0$ in that neighborhood (not just at $c$), and nothing in our hypothesis guarantees that. We take a different tack.

By definition and hypothesis,

$$f''(c) = \lim_{x \to c} \frac{f'(x) - f'(c)}{x - c} = \lim_{x \to c} \frac{f'(x) - 0}{x - c} < 0$$

so we can conclude that there is a (possibly small) interval $(\alpha, \beta)$ around $c$ where

$$\frac{f'(x)}{x - c} < 0, \qquad x \neq c$$

But this inequality implies that $f'(x) > 0$ for $\alpha < x < c$ and $f'(x) < 0$ for $c < x < \beta$. Thus, by the First Derivative Test, $f(c)$ is a local maximum value. The proof of (ii) is similar.  ∎

**EXAMPLE 4**   For $f(x) = x^2 - 6x + 5$, use the Second Derivative Test to identify local extrema.

*Solution*   This is the function of Example 1. Note that

$$f'(x) = 2x - 6 = 2(x - 3)$$
$$f''(x) = 2$$

Thus $f'(3) = 0$ and $f''(3) > 0$. Therefore, by the Second Derivative Test, $f(3)$ is a local minimum value.   ∎

**EXAMPLE 5**   For $f(x) = \frac{1}{3}x^3 - x^2 - 3x + 4$, use the Second Derivative Test to identify local extrema.

*Solution*   This is the function of Example 2.

$$f'(x) = x^2 - 2x - 3 = (x + 1)(x - 3)$$
$$f''(x) = 2x - 2$$

The critical points are $-1$ and $3$ $(f'(-1) = f'(3) = 0)$. Since $f''(-1) = -4$ and $f''(3) = 4$, we conclude—by the Second Derivative Test—that $f(-1)$ is a local maximum value and that $f(3)$ is a local minimum value.   ∎

Unfortunately, the Second Derivative Test sometimes fails, since $f''(x)$ may be 0 at a stationary point. For both $f(x) = x^3$ and $f(x) = x^4$, $f'(0) = 0$ and $f''(0) = 0$ (see Figure 7). The first does not have a local maximum or minimum value at 0; the second has a local minimum there. This shows that if $f''(x) = 0$ at a stationary point, we are unable to draw a conclusion about maxima or minima without more information.

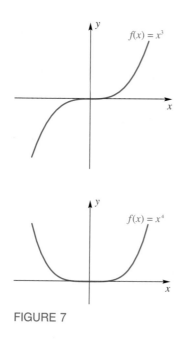

FIGURE 7

## CONCEPTS REVIEW

**1.** If $f$ is continuous at $c$, $f'(x) > 0$ near to $c$ on the left, and $f'(x) < 0$ near to $c$ on the right, then $f(c)$ is a local _____ value for $f$.

**2.** If $f'(x) = (x + 2)(x - 1)$, then $f(-2)$ is a local _____ value for $f$ and $f(1)$ is a local _____ value for $f$.

**3.** If $f'(x) = (x + 2)(x - 1)^2$, then $f(-2)$ is _____ for $f$ and $f(1)$ is _____ for $f$.

**4.** If $f'(c) = 0$ and $f''(c) < 0$, we expect to find a local _____ value for $f$ at $c$.

## PROBLEM SET 4.3

In Problems 1–6, identify the critical points. Then use (a) the First Derivative Test, and (if possible) (b) the Second Derivative Test to decide which of the critical points gives a local maximum and which gives a local minimum.

**1.** $f(x) = x^3 - 6x^2 + 4$

**2.** $f(x) = x^3 - 12x + \pi$

**3.** $f(\theta) = \sin 2\theta, 0 < \theta < \dfrac{\pi}{4}$

**4.** $f(x) = \frac{1}{2}x + \sin x, 0 < x < 2\pi$

**5.** $\Psi(\theta) = \sin^2 \theta, -\pi/2 < \theta < \pi/2$

**6.** $r(z) = z^4 + 4$

In Problems 7–16, find the critical points and use the test you prefer to decide which give a local maximum value and which give a local minimum value. What are these local maximum and minimum values?

7. $f(x) = x^3 - 3x$

8. $g(x) = x^4 + x^2 + 3$

9. $H(x) = x^4 - 2x^3$

10. $f(x) = (x - 2)^5$

11. $g(t) = \pi - (t - 2)^{2/3}$

12. $r(s) = 3s + s^{2/5}$

13. $f(t) = t - \dfrac{1}{t}, t \neq 0$

14. $f(x) = \dfrac{x^2}{\sqrt{x^2 + 4}}$

15. $\Lambda(\theta) = \dfrac{\cos \theta}{1 + \sin \theta}, 0 < \theta < 2\pi$

16. $g(\theta) = |\sin \theta|, 0 < \theta < 2\pi$

17. Find the (global) maximum and minimum values of $F(x) = 6\sqrt{x} - 4x$ on $[0, 4]$.

18. Do Problem 17 on the interval $[0, \infty)$.

19. Find (if possible) the maximum and minimum values of

$$f(x) = \frac{64}{\sin x} + \frac{27}{\cos x}$$

on $(0, \pi/2)$.

20. Find (if possible) the maximum and minimum values of

$$f(x) = x^2 + \frac{1}{x^2}$$

on $(0, \infty)$.

[C] 21. Find the minimum value of

$$g(x) = x^2 + \frac{16x^2}{(8 - x)^2}, x > 8$$

22. Consider $f(x) = Ax^2 + Bx + C$ with $A > 0$. Show that $f(x) \geq 0$ for all $x$ if and only if $B^2 - 4AC \leq 0$.

23. If $f'(x) = 2(x + 2)(x + 1)^2(x - 2)^4(x - 3)^3$, what values of $x$ make $f(x)$ a local maximum? A local minimum?

24. What conclusions can you draw about $f$ from the information that $f'(c) = f''(c) = 0$ and $f'''(c) > 0$?

25. Let $f$ be a continuous function and let $f'$ have the graph shown in Figure 8. Try to sketch a graph for $f$ and answer the following questions.

(a) Where is $f$ increasing? Decreasing?
(b) Where is $f$ concave up? Concave down?
(c) Where does $f$ attain a local maximum? A local minimum?
(d) Where are there inflection points for $f$?

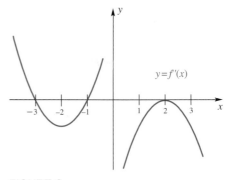

FIGURE 8

26. Repeat Problem 25 for Figure 9.

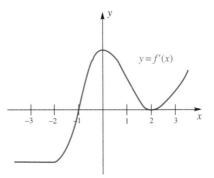

FIGURE 9

[PC] Problems 27–30 are intended for a computer. You will find the *Max/Min* section of *True BASIC Calculus* to be very useful in Problems 27 and 28. Experiment with this program using other intervals and other functions. Try reworking Problems 7–16.

27. Identify all local and global extrema for $f(x) = x^5 - 5x^3 + 4$ on $[-2, 2.5]$.

28. Identify all local and global extrema for $f(x) = \sin x - \cos(x/2)$ on $[-2\pi, 2\pi]$.

29. Suppose that $g'(x) = x^5 - 5x^3 + 4$. Approximately where on $(-2, 2.5)$ does $g$ attain local maximum values? Local minimum values?

30. Suppose that $g'(x) = \sin x - \cos(x/2)$. Approximately where on $(-2\pi, 2\pi)$ does $g$ attain local maximum values? Local minimum values?

**Answers to Concepts Review:**  **1.** Maximum
**2.** Maximum; minimum
**3.** Local minimum value; neither a maximum nor minimum value   **4.** Maximum

**4.4**
**MORE MAX-MIN**
**PROBLEMS**

The problems we studied in Section 4.1 usually assumed that the set on which we wanted to maximize or minimize a function was a *closed* interval. However, the intervals that arise in practice are not always closed; they are sometimes open or even half open, half closed. We can still handle these problems if we correctly apply the theory developed in Section 4.3. Keep in mind that maximum (minimum) with no qualifying adjective means global maximum (minimum).

**Extrema on Open Intervals** We give two examples to illustrate appropriate procedures for intervals that are open or half open.

**EXAMPLE 1** Find (if possible) the minimum and maximum values of $f(x) = x^4 - 4x$ on $(-\infty, \infty)$.

*Solution*

$$f'(x) = 4x^3 - 4 = 4(x^3 - 1) = 4(x - 1)(x^2 + x + 1)$$

Since $x^2 + x + 1 = 0$ has no real solutions (quadratic formula), there is only one critical point, namely, $x = 1$. For $x < 1, f'(x) < 0$, whereas for $x > 1, f'(x) > 0$. We conclude that $f(1) = -3$ is a local minimum value for $f$; and since $f$ is decreasing on the left of 1 and increasing on the right of 1, it must actually be the minimum value of $f$.

The facts stated above imply that $f$ cannot have a maximum value. The graph of $f$ is shown in Figure 1. ∎

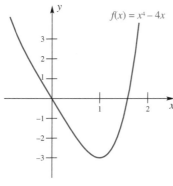

$f(x) = x^4 - 4x$

FIGURE 1

**EXAMPLE 2** Find (if possible) the maximum and minimum values of $g(x) = x/(x^3 + 2)$ on $[0, \infty)$.

*Solution*

$$g'(x) = \frac{x^3 + 2 - x(3x^2)}{(x^3 + 2)^2} = \frac{2 - 2x^3}{(x^3 + 2)^2} = \frac{2(1 - x)(1 + x + x^2)}{(x^3 + 2)^2}$$

On $[0, \infty)$, there are two critical points: the endpoint 0 and the stationary point 1. For $0 < x < 1, g'(x) > 0$, while for $x > 1, g'(x) < 0$. Thus, $g(1) = \frac{1}{3}$ is the maximum value of $g$ on $[0, \infty)$.

If $g$ has a minimum value, it must occur at the other critical point, namely, $x = 0$. Now $g(0) = 0$ and $g(x) > 0$ for $x > 0$, so $g(0) = 0$ is the minimum value of $g$ on $[0, \infty)$. The graph is shown in Figure 2. ∎

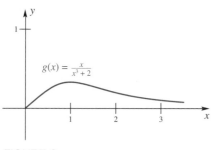

$g(x) = \frac{x}{x^3 + 2}$

FIGURE 2

**Practical Problems** Each of the examples below is different; yet there are common elements in the procedures that we use to solve them. At the end of the section, we will suggest a set of steps to use in solving any max-min problem.

**EXAMPLE 3** A handbill is to contain 50 square inches of printed matter, with 4-inch margins at top and bottom and 2-inch margins on each side. What dimensions for the handbill would use the least paper?

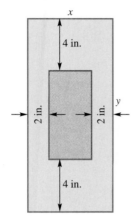

FIGURE 3

*Solution*  Let $x$ be the width and $y$ the height of the handbill (Figure 3). Its area is

$$A = xy$$

We wish to minimize $A$.

As it stands, $A$ is expressed in terms of two variables, a situation we do not know how to handle. However, we will find an equation connecting $x$ and $y$ so that one of these variables can be eliminated in the expression for $A$. The dimensions of the printed matter are $x - 4$ and $y - 8$ and its area is 50 square inches; so $(x - 4)(y - 8) = 50$. When we solve this equation for $y$, we obtain

$$y = \frac{50}{x - 4} + 8$$

Substituting this expression for $y$ in $A = xy$ gives $A$ in terms of $x$:

$$A = \frac{50x}{x - 4} + 8x$$

The allowable values for $x$ are $4 < x < \infty$; we want to minimize $A$ on the open interval $(4, \infty)$.

Now

$$\frac{dA}{dx} = \frac{(x - 4)50 - 50x}{(x - 4)^2} + 8 = \frac{8x^2 - 64x - 72}{(x - 4)^2} = \frac{8(x + 1)(x - 9)}{(x - 4)^2}$$

The only critical points are obtained by solving $\frac{dA}{dx} = 0$; this yields $x = 9$ and $x = -1$. We reject $x = -1$ because it is not in the interval $(4, \infty)$. Since $dA/dx < 0$ for $x$ in $(4, 9)$ and $dA/dx > 0$ for $x$ in $(9, \infty)$, we conclude that $A$ attains its minimum value at $x = 9$. This value of $x$ makes $y = 18$ (found by substituting in the equation connecting $x$ and $y$). So the dimensions for the handbill that will use the least amount of paper are 9 inches by 18 inches.  ■

## Common Sense

It would be hard to make any preliminary estimates in Example 3. However, common sense tells us that the handbill's height should be larger than its width. Why? Because we should capitalize on the narrower margins along the sides.

**EXAMPLE 4**  A rectangular beam is to be cut from a log with circular cross section. If the strength of the beam is proportional to the product of its width and the square of its depth, find the dimensions of the cross section that give the strongest beam.

*Solution*  Denote the diameter of the log by $a$ (constant) and the width and depth of the beam by $w$ and $d$, respectively (Figure 4). We wish to maximize $S$, the strength of the beam.

From the condition of the problem,

$$S = kwd^2$$

where $k$ is a constant of proportionality. The strength $S$ depends on the two variables $w$ and $d$, but there is a simple relationship between them,

$$d^2 + w^2 = a^2$$

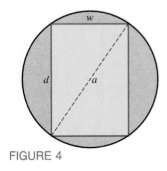

FIGURE 4

When we solve this equation for $d^2$ and substitute in the formula for $S$, we obtain $S$ in terms of the single variable $w$:

$$S = kw(a^2 - w^2) = ka^2w - kw^3$$

We consider the allowable values for $w$ to be $0 < w < a$, an open interval.

To find the critical points, we calculate $dS/dw$, set it equal to 0, and solve for $w$.

$$\frac{dS}{dw} = ka^2 - 3kw^2 = k(a^2 - 3w^2)$$

$$k(a^2 - 3w^2) = 0$$

$$w = \frac{a}{\sqrt{3}}$$

Since $a/\sqrt{3}$ is the only critical point in $(0, a)$, it is likely that it gives the maximum $S$. A check on the sign of $dS/dw$ to the left and right of $a/\sqrt{3}$ will confirm this.

When we substitute $w = a/\sqrt{3}$ in $d^2 + w^2 = a^2$, we learn that $d = \sqrt{2}a/\sqrt{3}$. The desired dimensions are $w = a/\sqrt{3}$ and $d = \sqrt{2}a/\sqrt{3}$. Note that $d = \sqrt{2}w$.  ∎

**EXAMPLE 5**　Henry, who is in a rowboat 2 miles from the nearest point $B$ on a straight shoreline, notices smoke billowing from his house, which is 6 miles down the shoreline from $B$. He figures he can row at 6 miles per hour and run at 10 miles per hour. How should he proceed in order to get to his house in the least time?

*Solution*　We interpret the problem to mean that we are to determine the $x$ in Figure 5 that will make Henry's traveling time a minimum. It is clear that we should restrict $x$ to the closed interval $[0, 6]$.

The distance $AD$ is $\sqrt{x^2 + 4}$ miles and the time to row it is $\sqrt{x^2 + 4}/6$ hours. The distance $DC$ is $6 - x$ miles and the time to run it is $(6 - x)/10$ hours. Thus the total time $T$ in hours is

$$T = \frac{\sqrt{x^2 + 4}}{6} + \frac{6 - x}{10}$$

We wish to minimize $T$ on $[0, 6]$.

This time there are three critical points, the endpoints 0 and 6 and a stationary point obtained by setting $dT/dx$ equal to 0.

$$\frac{dT}{dx} = \frac{1}{6} \cdot \frac{1}{2}(x^2 + 4)^{-1/2}(2x) - \frac{1}{10}$$

$$= \frac{x}{6\sqrt{x^2 + 4}} - \frac{1}{10}$$

$$= \frac{10x - 6\sqrt{x^2 + 4}}{60\sqrt{x^2 + 4}}$$

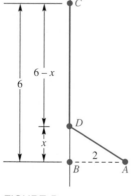

FIGURE 5

When we set $dT/dx$ equal to 0 and solve, we obtain, in successive steps,

$$10x - 6\sqrt{x^2 + 4} = 0$$
$$5x = 3\sqrt{x^2 + 4}$$
$$25x^2 = 9(x^2 + 4)$$
$$16x^2 = 36$$
$$x^2 = \tfrac{36}{16}$$
$$x = \tfrac{3}{2}$$

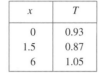

| $x$ | $T$ |
|-----|------|
| 0 | 0.93 |
| 1.5 | 0.87 |
| 6 | 1.05 |

FIGURE 6

Since the domain for $T$ is a closed interval, $T$ has a minimum (Theorem 4.1A), and this minimum must occur at a critical point (Theorem 4.1B). After consulting the table in Figure 6, we conclude that Henry should row to a point 1.5 miles down the shoreline and then run the rest of the way. It will take him about 0.87 hour, or 52 minutes. For a similar problem in which one of the endpoints produces the minimum time, see Problem 15. ∎

**EXAMPLE 6** Find the dimensions of the right circular cylinder of greatest volume that can be inscribed in a given right circular cone.

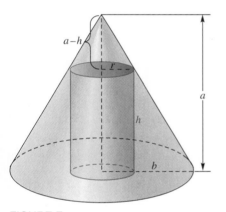

FIGURE 7

*Solution* Let $a$ be the altitude and $b$ the radius of the base of the given cone (both constants). Denote by $h$, $r$, and $V$ the altitude, radius, and volume, respectively, of an inscribed cylinder (see Figure 7).

The volume of the cylinder is

$$V = \pi r^2 h$$

From similar triangles,

$$\frac{a - h}{r} = \frac{a}{b}$$

which gives

$$h = a - \frac{a}{b} r$$

When we substitute this expression for $h$ in the formula for $V$, we obtain

$$V = \pi r^2 \left( a - \frac{a}{b} r \right) = \pi a r^2 - \pi \frac{a}{b} r^3$$

We wish to maximize $V$ for $r$ in the interval $[0, b]$. (Someone is sure to argue—and with good reason—that the appropriate interval is $(0, b)$. Actually, the answer is the same either way, though we have to use the First Derivative Test if we do the problem using $(0, b)$ as the domain.)

Now

$$\frac{dV}{dr} = 2\pi a r - 3\pi \frac{a}{b} r^2 = \pi a r \left( 2 - \frac{3}{b} r \right)$$

This yields the stationary point $r = 2b/3$, giving us three critical points on $[0, b]$ to consider: $0$, $2b/3$, and $b$. One look at the picture shows that $r = 0$ and $r = b$ both give a volume of $0$. Thus, $r = 2b/3$ has to give the maximum volume. When we substitute this value for $r$ in the equation connecting $r$ and $h$, we find that $h = a/3$. ■

**A Summary of the Method** Based on the examples above, we suggest a step-by-step method to use in applied max-min problems. Do not follow it slavishly; common sense may sometimes suggest an alternative approach or omission of some steps.

> **Step 1** Draw a picture for the problem and assign appropriate variables to key quantities.
> **Step 2** Write a formula for the quantity $Q$ to be maximized (minimized) in terms of those variables.
> **Step 3** Use the conditions of the problem to eliminate all but one of these variables and thereby express $Q$ as a function of a single variable, such as $x$.
> **Step 4** Determine the set of possible values for $x$, usually an interval.
> **Step 5** Find the critical points (endpoints, stationary points, singular points). Most often, the key critical points are the stationary points where $dQ/dx = 0$.
> **Step 6** Use the theory of this chapter to decide which critical point gives the maximum (minimum).

## CONCEPTS REVIEW

**1.** If a rectangle of area 100 has length $x$ and width $y$, then the allowable values for $x$ are _____.

**2.** The perimeter $P$ of the rectangle of Question 1 expressed in terms of $x$ is given by $P =$ _____.

**3.** If the rectangle of Question 1 is partitioned down the middle in both directions, the total length of fence $L$ required to enclose and partition it can be expressed in terms of $x$ by $L =$ _____.

**4.** If $f$ is continuous on $(0, \infty)$ and if $f'(x) < 0$ for $x < c$ and $f'(x) > 0$ for $x > c$, then $f$ has a _____ at $x = c$. It has no _____ on $(0, \infty)$.

## PROBLEM SET 4.4

**1.** Find two numbers whose product is $-16$ and the sum of whose squares is a minimum.

**2.** For what number does the principal square root exceed eight times the number by the largest amount?

**3.** For what number does the principal fourth root exceed twice the number by the largest amount?

**4.** Find two numbers whose product is $-12$ and the sum of whose squares is a minimum.

**5.** Find the points on the parabola $y = x^2$ that are closest to the point $(0, 5)$. *Hint*: Minimize the square of the distance between $(x, y)$ and $(0, 5)$.

**FIGURE 8**

**6.** Find the points on the parabola $x = 2y^2$ that are closest to the point $(10, 0)$. *Hint*: Minimize the square of the distance between $(x, y)$ and $(10, 0)$.

**7.** A farmer wishes to fence off two identical adjoining rectangular pens, each with 900 square feet of area, as shown in Figure 8. What are $x$ and $y$ so that the least amount of fence is required?

FIGURE 9

**8.** A farmer wishes to fence off three identical adjoining rectangular pens (see Figure 9), each with 300 square feet of area. What should the width and length of each pen be so that the least amount of fence is required?

**9.** Suppose that the outer boundary of the pens in Problem 8 requires heavy fence that costs $3 per foot, but that the two internal partitions require fence costing only $2 per foot. What dimensions $x$ and $y$ will produce the least expensive cost for the pens?

**10.** Solve Problem 8 assuming that the area of each pen is 900 square feet. Study the solution to this problem and to Problem 8 to see if you can make a conjecture about the ratio of $x/y$ in all problems of this type. You should try to prove your conjecture.

≈ **11.** If the strength of a rectangular beam is proportional to the product of its width and the square of its depth, find the dimensions of the strongest beam that can be cut from a log whose cross section has the form of the ellipse $9x^2 + 8y^2 = 72$. See Example 4.

**12.** The illumination at a point is inversely proportional to the square of the distance of the point from the light source and directly proportional to the intensity of the light source. If two light sources are $s$ feet apart and their intensities are $I_1$ and $I_2$, respectively, at what point between them will the sum of their illuminations be a minimum?

≈ **13.** A small island is 2 miles from the nearest point $P$ on the straight shoreline of a large lake. If a woman on the island can row a boat 3 miles per hour and can walk 4 miles per hour, where should the boat be landed in order to arrive at a town 10 miles down the shore from $P$ in the least time? See Example 5.

≈ **14.** In Problem 13, suppose the woman will be picked up by a car that will average 50 miles per hour when she gets to the shore. Then where should she land?

≈ **15.** In Problem 13, suppose the woman uses a motorboat that goes 20 miles per hour. Then where should she land?

**16.** A powerhouse is located on one bank of a straight river that is $w$ feet wide. A factory is situated on the opposite bank of the river, $L$ feet downstream from the point $A$ directly opposite the powerhouse. What is the most economical path for a cable connecting the powerhouse to the factory if it costs $a$ dollars per foot to lay the cable under water and $b$ dollars on land $(a > b)$?

**17.** At 7:00 A.M. one ship was 60 miles due east from a second ship. If the first ship sailed west at 20 miles per hour

and the second ship sailed southeast at 30 miles per hour, when were they closest together?

**18.** Find the equation of the line that is tangent to the ellipse $b^2x^2 + a^2y^2 = a^2b^2$ in the first quadrant and that forms with the coordinate axes the triangle with smallest possible area ($a$ and $b$ are positive constants).

**19.** Find the greatest volume that a right circular cylinder can have if it is inscribed in a sphere of radius $r$.

**20.** Show that the rectangle with maximum perimeter that can be inscribed in a circle is a square.

**21.** What are the relative dimensions of the right circular cylinder with greatest curved surface area that can be inscribed in a given sphere?

**22.** A right circular cone is to be inscribed in another right circular cone of given volume, with the same axis and with the vertex of the inner cone touching the base of the outer cone. What must be the ratio of their altitudes for the inscribed cone to have maximum volume?

**23.** A wire of length 100 centimeters is cut into two pieces; one is bent to form a square, the other an equilateral triangle. Where should the cut be made if (a) the sum of the two areas is to be a minimum; (b) a maximum? (Allow the possibility of no cut.)

**24.** A closed box in the form of a rectangular parallelepiped with a square base is to have a given volume. If the material used in the bottom costs 20% more per square inch than the material in the sides, and the material in the top costs 50% more per square inch than that of the sides, find the most economical proportions for the box.

**25.** An observatory is to be in the form of a right circular cylinder surmounted by a hemispherical dome. If the hemispherical dome costs twice as much per square foot as the cylindrical wall, what are the most economical proportions for a given volume?

**26.** A weight connected to a spring moves along the $x$-axis so that its $x$-coordinate at time $t$ is

$$x = \sin 2t + \sqrt{3}\cos 2t$$

What is the farthest the weight gets from the origin?

**27.** A flower bed will be in the shape of a sector of a circle (a pie-shaped region) of radius $r$ and vertex angle $\theta$. Find $r$ and $\theta$ if its area is a constant $A$ and the perimeter is a minimum.

**28.** A fence $h$ feet high runs parallel to a tall building and $w$ feet from it (Figure 10). Find the length of the shortest ladder that will reach from the ground across the top of the fence to the wall of the building.

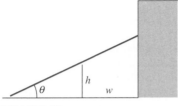

**FIGURE 10**

**29.** (Snell's Law). Fermat's Principle in optics says that light travels from point $A$ to point $B$ along the path that requires least time. Suppose that light travels in one medium at velocity $c_1$ and in a second medium at velocity $c_2$. If $A$ is in medium 1 and $B$ in medium 2 and the $x$-axis separates the two media, as shown in Figure 11, show that

$$\frac{\sin \theta_1}{c_1} = \frac{\sin \theta_2}{c_2}$$

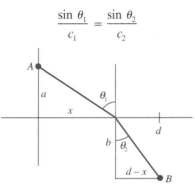

**FIGURE 11**

**30.** Light from $A$ is reflected to $B$ by a plane mirror. Use Fermat's Principle (Problem 29) to show that the angle of incidence is equal to the angle of reflection.

**31.** One end of a 27-foot ladder rests on the ground and the other end rests on the top of an 8-foot wall. As the bottom end is pushed along the ground, the top end extends beyond the wall. Find the maximum horizontal overhang of the top end.

**32.** I have enough pure silver to coat 1 square meter of surface area. I plan to coat a sphere and a cube. What dimensions should they be if the total volume of the silvered solids is to be a maximum? A minimum? (Allow the possibility of all the silver going onto one solid.)

**33.** One corner of a long narrow strip of paper is folded over so it just touches the opposite side, as shown in Figure

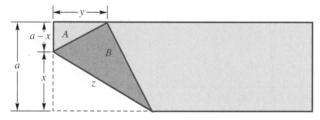

**FIGURE 12**

**12.** With parts labeled as indicated, determine $x$ in order to:
(a) maximize the area of triangle $A$;
(b) minimize the area of triangle $B$;
(c) minimize the length $z$.

**34.** Determine $\theta$ so that the area of the symmetric cross shown in Figure 13 is maximized. Then find this maximum area.

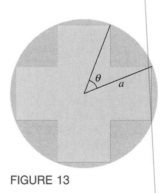

**FIGURE 13**

[PC] **35.** A clock has hour and minute hands of lengths $h$ and $m$, respectively, with $h \le m$. We wish to study this clock at times between 12:00 and 12:30. Let $\theta$, $\phi$, and $L$ be as in Figure 14 and note that $\theta$ increases at a constant rate. By the Law of Cosines, $L = L(\theta) = (h^2 + m^2 - 2hm \cos \theta)^{1/2}$ and so

$$L'(\theta) = hm(h^2 + m^2 - 2hm \cos \theta)^{-1/2} \sin \theta$$

(a) For $h = 3$ and $m = 5$, determine $L'$, $L$, and $\phi$ at the instant when $L'$ is largest.
(b) Rework part a when $h = 5$ and $m = 13$.
(c) Based on parts (a) and (b), make conjectures about the values of $L'$, $L$, and $\phi$ at the instant when the tips of the hands are separating most rapidly.
(d) Try to prove your conjectures.

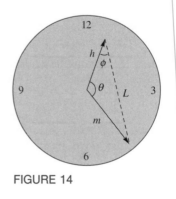

**FIGURE 14**

**Answers to Concepts Review:** **1.** $0 < x < \infty$
**2.** $2x + 200/x$ **3.** $3x + 300/x$
**4.** Global minimum; global maximum

## 4.5
## ECONOMIC
## APPLICATIONS

Each discipline has its own language. This is certainly true of economics, which has a highly developed special vocabulary. Once we learn this vocabulary, we will discover that many of the problems of economics are just ordinary calculus problems dressed in new clothes.

Consider a typical company, the ABC Company. For simplicity, assume that ABC produces and markets a single product; it might be television sets, car batteries, or bars of soap. If it sells $x$ units of the product in a fixed period of time (for example, a year), it will be able to charge a **price**, $p(x)$, for each unit. We indicate that $p$ depends on $x$ because if ABC increases its output, it will probably need to reduce the price per unit in order to sell its total output. The **total revenue** that ABC can expect is given by $R(x) = xp(x)$, the number of units times the price per unit.

To produce and market $x$ units, ABC will have a total cost, $C(x)$. This is normally the sum of a **fixed cost** (office utilities, real estate taxes, and so on) plus a **variable cost**, which depends directly on the number of units produced.

The key concept for a company is the **total profit**, $P(x)$. It is just the difference between revenue and cost—that is,

$$P(x) = R(x) - C(x) = xp(x) - C(x)$$

Generally, a company seeks to maximize its total profit.

Note a feature that tends to distinguish problems in economics from those in the physical sciences. In most cases, the product will be in discrete units (you can't make or sell 0.23 television sets or $\pi$ car batteries). Thus, the functions $R(x)$, $C(x)$, and $P(x)$ are usually defined only for $x = 0, 1, 2, \ldots$ and consequently, their graphs consist of discrete points (Figure 1). In order to make the tools of calculus available, we connect these points with a smooth curve (Figure 2), thereby pretending that $R$, $C$, and $P$ are nice differentiable functions. This illustrates an aspect of *mathematical modeling* that is almost always necessary, especially in economics. To model a real world problem, we must make simplifying assumptions. This means that the answers we get only approximate the answers we seek—one of the reasons economics is a less than perfect science.

A related problem for an economist is how to obtain formulas for the functions $C(x)$ and $p(x)$. In a simple case, $C(x)$ might have the form

$$C(x) = 10{,}000 + 50x$$

If so, $10,000 is the *fixed cost* and $50x$ is the *variable cost*, based on a $50 direct cost for each unit produced. Perhaps a more typical situation is

$$\hat{C}(x) = 10{,}000 + 45x + 100\sqrt{x}$$

Note that in this case the *average variable cost* per unit is

$$\frac{45x + 100\sqrt{x}}{x} = 45 + \frac{100}{\sqrt{x}}$$

### Discrete *versus* Continuous

Most problems in the social sciences are properly viewed as discrete in nature. Moreover, the digital computer is a fast, accurate tool for handling discrete quantities. A natural question arises: Why not study discrete problems using discrete tools rather than by first modeling them with continuous curves? For this reason, most colleges now offer courses in discrete mathematics. However, because of its beauty and power, calculus continues to enjoy popularity as a tool for analyzing social science as well as science problems.

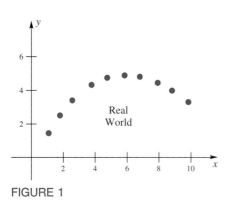

FIGURE 1

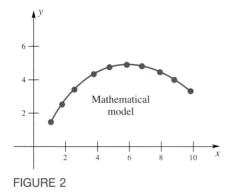

FIGURE 2

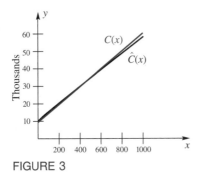

**FIGURE 3**

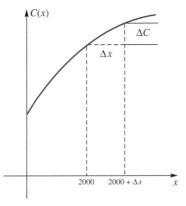

**FIGURE 4**

an amount that decreases as $x$ increases (efficiency of size). The cost functions $C(x)$ and $\hat{C}(x)$ are graphed together in Figure 3.

Selecting appropriate functions to model cost and price is a nontrivial task. Occasionally, they can be inferred from basic assumptions. In other cases, a careful study of the history of the firm will suggest reasonable choices. Sometimes, we must simply make intelligent guesses.

**Use of the Word *Marginal*** Suppose that ABC knows its cost function $C(x)$ and that it has tentatively planned to produce 2000 units this year. President Hornblower would like to determine the additional cost per unit if ABC increased production slightly. Would it, for example, be less than the additional revenue per unit? If so, it would make good economic sense to increase production.

If the cost function is the one shown in Figure 4, President Hornblower is asking for the value of $\Delta C/\Delta x$ when $\Delta x = 1$. But we expect that this will be very close to the value of

$$\lim_{\Delta x \to 0} \frac{\Delta C}{\Delta x}$$

when $x = 2000$. This is called the **marginal cost**. We mathematicians recognize it as $dC/dx$, the derivative of $C$ with respect to $x$.

In a similar vein, we define **marginal price** as $dp/dx$, **marginal revenue** as $dR/dx$, and **marginal profit** as $dP/dx$.

**Examples** We now illustrate how to solve a wide variety of economic problems.

**EXAMPLE 1** Suppose $C(x) = 8300 + 3.25x + 40\sqrt[3]{x}$ dollars. Find the average cost per unit and the marginal cost and then evaluate them when $x = 1000$.

***Solution***

$$\text{Average cost: } \frac{C(x)}{x} = \frac{8300 + 3.25x + 40x^{1/3}}{x}$$

$$\text{Marginal cost: } \frac{dC}{dx} = 3.25 + \frac{40}{3}x^{-2/3}$$

At $x = 1000$, these have the values 11.95 and 3.38, respectively. This means that it costs, on the average, \$11.95 per unit to produce the first 1000 units; to produce one additional unit beyond 1000 costs only about \$3.38. ∎

**EXAMPLE 2** A company estimates that it can sell 1000 units per week if it sets the unit price at \$3.00, but that its weekly sales will rise by 100 units for each 10¢ decrease in price. If $x$ is the number of units sold each week ($x \geq 1000$), find:
(a) the price function, $p(x)$;

---

**Economic Vocabulary**

Because economics tends to be a study of discrete phenomena, your economics professor may define marginal cost at $x$ as the cost of producing one additional unit— that is, as

$$C(x + 1) - C(x)$$

In the mathematical model, this number will be very close in value to $dC/dx$, and since the latter is a principal concept in calculus, we choose to take it as the definition of marginal cost. Similar statements hold for marginal revenue and marginal profit.

(b) the number of units and the corresponding price that will maximize weekly revenue;

(c) the maximum weekly revenue.

*Solution*

(a) We are given that

$$x = 1000 + \frac{3.00 - p(x)}{0.10} \, (100)$$

or, equivalently,

$$p(x) = 3.00 - (0.10) \frac{(x - 1000)}{100} = 4 - 0.001x$$

(b)

$$R(x) = xp(x) = 4x - 0.001x^2$$

$$\frac{dR}{dx} = 4 - 0.002x$$

The only critical points are the endpoint 1000 and the stationary point 2000, obtained by setting $dR/dx = 0$. The First Derivative Test ($R'(x) > 0$ for $1000 \le x < 2000$ and $R'(x) < 0$ for $x > 2000$) shows that $x = 2000$ gives the maximum revenue. This corresponds to a unit price $p(2000) = \$2.00$.

(c) The maximum weekly revenue is $R(2000) = \$4000$. ■

**EXAMPLE 3**    In manufacturing and selling $x$ units of a certain commodity, the price function $p$ and the cost function $C$ (in dollars) are given by

$$p(x) = 5.00 - 0.002x$$

$$C(x) = 3.00 + 1.10x$$

Find expressions for the marginal revenue, marginal cost, and marginal profit; determine the production level that will produce the maximum total profit.

*Solution*

$$R(x) = xp(x) = 5.00x - 0.002x^2$$

$$P(x) = R(x) - C(x) = -3.00 + 3.90x - 0.002x^2$$

Thus, we have the following derivatives.

Marginal revenue:  $\dfrac{dR}{dx} = 5 - 0.004x$

Marginal cost:  $\dfrac{dC}{dx} = 1.1$

Marginal profit:  $\dfrac{dP}{dx} = \dfrac{dR}{dx} - \dfrac{dC}{dx} = 3.9 - 0.004x$

To maximize profit, we set $dP/dx = 0$ and solve. This gives $x = 975$ as the only critical number to consider. It does provide a maximum, as may be checked by the First Derivative Test. The maximum profit is $P(975) = \$1898.25$. ∎

Note that at $x = 975$, both the marginal revenue and the marginal cost are $\$1.10$. In general, a company should expect to be at a maximum profit level when the cost of producing an additional unit equals the revenue from that unit.

The statement just made assumes that the cost function and the revenue function are nice, differentiable functions and that endpoints are not significant. In some situations, the cost function may take large jumps, as when a new employee or a new piece of equipment is added; also, a manufacturing plant may have a maximum capacity, thereby introducing an important endpoint. We illustrate these possibilities in the next two examples.

**EXAMPLE 4**   The XYZ Company manufactures wicker chairs. With its present machines, it has a maximum yearly output of 500 units. If it makes $x$ chairs, it can set a price of $p(x) = 200 - 0.15x$ dollars each and will have a total yearly cost of $C(x) = 4000 + 6x - (0.001)x^2$ dollars. What production level maximizes the total yearly profit?

*Solution*

$$R(x) = xp(x) = x(200 - 0.15x) = 200x - 0.15x^2$$

and so

$$P(x) = 200x - 0.15x^2 - (4000 + 6x - 0.001x^2)$$
$$= -4000 + 194x - 0.149x^2$$

Thus,

$$\frac{dP}{dx} = 194 - 0.298x$$

which yields the stationary point 651. However 651 is not in the interval $[0, 500]$, so the only critical points to check are the two endpoints, 0 and 500. If the maximum is at 0, the company better go out of the wicker-chair business fast. It is not. The maximum occurs at 500, and the maximum profit is $P(500) = \$55,750$. ∎

**EXAMPLE 5**   With the addition of a new machine, the XYZ Company of Example 4 could boost its yearly production of chairs to 750. However, its cost function $C(x)$ will then take the form

$$C(x) = \begin{cases} 4000 + 6x - (0.001)x^2 & \text{if } 0 \le x \le 500 \\ 6000 + 6x - (0.003)x^2 & \text{if } 500 < x \le 750 \end{cases}$$

What production level maximizes total yearly profit under these circumstances?

**Solution** The new cost function results in the new profit function

$$P(x) = \begin{cases} -4000 + 194x - 0.149x^2 & \text{if } 0 \le x \le 500 \\ -6000 + 194x - 0.147x^2 & \text{if } 500 < x \le 750 \end{cases}$$

On the interval $500 < x < 750$,

$$\frac{dP}{dx} = 194 - 0.294x$$

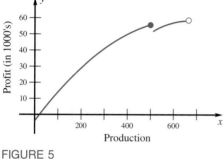

FIGURE 5

which gives the stationary point 660. There are four critical points, namely, 0, 500, 660, and 750. The corresponding values of $P$ are $-4000$, 55,750, 58,007, and 56,813. We conclude that a production level of 660 units gives the maximum profit. The graph in Figure 5 clarifies this example. ■

## CONCEPTS REVIEW

**1.** To use calculus to model economic problems that are usually discrete in nature, we must transform these discrete problems into _____ ones.

**2.** Total revenue is $R(x) = xp(x)$, where $x$ denotes _____ and $p(x)$ denotes _____.

**3.** Total cost $C(x)$ usually consists of two kinds of cost: _____ cost, consisting of the cost of utilities, real estate taxes, and so on, and _____ cost, which depends on the number of units produced.

**4.** In economics, $dR/dx$ is called _____ and $dC/dx$ is called _____.

## PROBLEM SET 4.5

**1.** It costs ACME $10 to make and sell one item. If the items sell at $y$ dollars a piece and the number sold is $n = 100/(y - 10) + 20(100 - y)$ what sale price will bring the maximum profit?

**2.** The fixed monthly cost of operating a plant that makes Zbars is $7000 and the direct costs of manufacturing each unit are $100. Write an expression for $C(x)$, the total cost of making $x$ Zbars in a month.

**3.** The manufacturer of Zbars estimates that 100 units per month can be sold if the unit price is $250, and the sales will increase by 10 units for each $5 decrease in price. Write an expression for the price $p(n)$ and the revenue $R(n)$ if $n$ units are sold in 1 month, $n \ge 100$ (see Example 2).

**4.** Use the information in Problems 2 and 3 to write an expression for the total monthly profit $P(n)$, $n \ge 100$.

**5.** Sketch the graph of $P(n)$ of Problem 4, and from it estimate the value of $n$ that maximizes $P$. Find this $n$ exactly by the methods of calculus.

[C] **6.** The total cost of producing and selling $x$ units of Xbars per month is $C(x) = 100 + 3.002x - 0.0001x^2$. If the production level is 1600 units per month, find the average cost, $C(x)/x$, of each unit and the marginal cost.

**7.** The total cost of producing and selling $n$ units of a certain commodity per week is $C(n) = 1000 + n^2/1200$. Find the average cost, $C(n)/n$, of each unit and the marginal cost at a production level of 800 units per week.

**8.** The total cost of producing and selling $100x$ units of a particular commodity per week is

$$C(x) = 1000 + 33x - 9x^2 + x^3$$

Find (a) the level of production at which the marginal cost is a minimum, and (b) the minimum marginal cost.

**9.** A price function, $p$, is defined by

$$p(x) = 20 + 4x - \frac{x^2}{3}$$

where $x \ge 0$ is the number of units.

(a) Find the total revenue function and the marginal revenue function.

(b) On what interval is the total revenue increasing?

(c) For what number $x$ is the marginal revenue a maximum?

[C]   **10.** For the price function defined by $p(x) = (182 - x/36)^{1/2}$, find the number of units $x_1$ that makes the total revenue a maximum and state the maximum possible revenue. What is the marginal revenue when the optimum number of units, $x_1$, are sold?

**11.** For the price function given by

$$p(x) = 800/(x + 3) - 3,$$

find the number of units $x_1$ that makes the total revenue a maximum and state the maximum possible revenue. What is the marginal revenue when the optimum number of units, $x_1$, is sold?

**12.** A river boat company offers a Fourth of July excursion to a fraternal organization with the understanding that there will be at least 400 passengers. The price of each ticket will be $12.00, and the company agrees to refund $0.20 to every passenger for each 10 passengers in excess of 400. Write an expression for the price function $p(x)$ and find the number $x_1$ of passengers that makes the total revenue a maximum.

**13.** A merchant finds that he can sell 4000 yards of a particular fabric each month if he prices it at $6.00 per yard, and that his monthly sales will increase by 250 yards for each $0.15 reduction in the price per yard. Write an expression for $p(x)$ and find the price per yard that would bring maximum revenue.

**14.** A manufacturer estimates that she can sell 500 articles per week if her unit price is $20.00, and that her weekly sales will rise by 50 with each $0.50 reduction in price. The cost of producing and selling $x$ articles a week is $C(x) = 4200 + 5.10x + 0.0001x^2$. Find each of the following.

(a) The price function.

(b) The level of weekly production for maximum profit.

(c) The price per article at the optimum level of production.

(d) The marginal price at that level of production.

**15.** The monthly overhead of a manufacturer of a certain commodity is $6000, and the cost of material is $1.00 per unit. If not more than 4500 units are manufactured per month, labor cost is $0.40 per unit; but for each unit over 4500, the manufacturer must pay time-and-a-half for labor. The manufacturer can sell 4000 units per month at $7.00 per unit and estimates that monthly sales will rise by 100 for each $0.10 reduction in price. Find (a) the total cost function, (b) the price function, and (c) the number of units that should be produced each month for maximum profit.

[C]   **16.** The ZEE Company makes zingos, which it markets at a price $p(x) = 10 - 0.001x$ dollars, where $x$ is the number produced each month. Its total monthly cost is $C(x) = 200 + 4x - 0.01x^2$. At peak production it can make 300 units. What is its maximum monthly profit and what level of production gives this profit?

[C]   **17.** If the company of Problem 16 expands its facilities so that it can produce up to 450 units each month, its monthly cost function takes the form $C(x) = 800 + 3x - 0.01x^2$ for $300 < x \le 450$. Find the production level that maximizes monthly profit and evaluate this profit. Sketch the graph of the monthly profit function $P(x)$ on $0 \le x \le 450$. (See Example 5.)

**18.** Let us suppose that a manufacturer has $m$ employees, who produce a total of $x$ units of product per week. These are sold at a price $p = p(x)$. Then the total weekly revenue $R(x) = x \cdot p$ can be thought of as depending on $m$. The derivative $dR/dm$ is called the *marginal revenue product*. It is (approximately) the change in revenue when the manufacturer adds one employee. Show that

$$\frac{dR}{dm} = \frac{dx}{dm}\left(p + x\frac{dp}{dx}\right)$$

*Hint*: Use the Product Rule and then the Chain Rule.

[C]   **19.** Refer to Problem 18. A manufacturer has determined that $m$ employees can produce $x = 5m^2/\sqrt{m^2 + 13}$ units in a week, which it can then sell at a price $p = 10x - (0.1)x^2$ dollars. Determine the marginal revenue product when $m = 6$.

**20.** To be successful, a retail store must control its inventory. Overstocking results in excessive interest costs, extra warehouse rental, and the danger of obsolescence. Too small an inventory involves more paperwork in reordering, extra delivery charges, and the greater likelihood of running out of stock. Carvers Appliance Outlet estimates that it costs $20 to hold a microwave oven in stock for a year. To reorder a lot of ovens costs $200 plus $3 for each oven. What lot size will result in the smallest inventory cost? Assume that Carvers sells 1000 ovens per year and that ordering lots of size $x$ means that, on the average, $x/2$ ovens will be in stock.

**21.** Suppose that a store expects to sell $N$ units of a particular item per year, that it costs $A$ dollars to keep one unit in stock for a year, and that to reorder a lot of size $x$ costs $(F + Bx)$ dollars. Show that $x = \sqrt{2FN/A}$ is the lot size that minimizes inventory cost. See Problem 20.

**22.** If expected sales of an item quadruple, what will happen to the ideal lot size? See Problem 21.

---

**Answers to Concepts Review:**   **1.** Continuous   **2.** Number of units sold; price per unit   **3.** Fixed; variable   **4.** Marginal revenue; marginal cost

## 4.6
### LIMITS AT INFINITY, INFINITE LIMITS

The concept of the "infinite" has inspired and bedeviled mathematicians from time immemorial. The deepest problems and most profound paradoxes of mathematics are often intertwined with use of this word. Yet mathematical progress can in part be measured in terms of understanding the concept of infinity. We have already used the symbols $\infty$ and $-\infty$ in our notation for certain intervals. Thus, $(3, \infty)$ is our way of denoting the set of all real numbers greater than 3. Please note that we have never referred to $\infty$ as a number. For example, we have never added it to a number or divided it by a number. We will use the symbols $\infty$ and $-\infty$ in a new way in this section, but they will still not represent numbers.

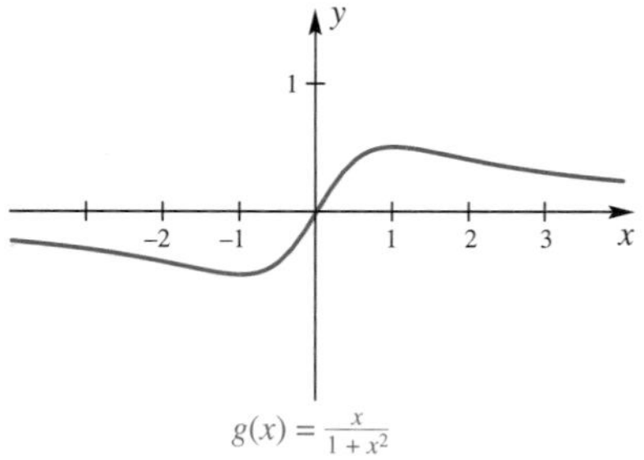

**An Example**   Consider the function $g(x) = x/(1 + x^2)$. It was graphed rather carefully in Section 4.2; a smaller version of that graph is shown in Figure 1. We ask this question: What happens to $g(x)$ as $x$ gets larger and larger? In symbols, we ask for the value of $\lim\limits_{x \to \infty} g(x)$.

When we write $x \to \infty$, we are *not* implying that somewhere far, far to the right on the $x$-axis, there is a number—bigger than all other numbers—which $x$ is approaching. Rather, we use $x \to \infty$ as a shorthand way of saying that $x$ gets larger and larger without bound.

In the table in Figure 2, we have listed values of $g(x) = x/(1 + x^2)$ for several values of $x$. It appears that $g(x)$ gets smaller and smaller as $x$ gets larger and larger. We write

$$\lim_{x \to \infty} \frac{x}{1 + x^2} = 0$$

Experimenting with large negative numbers would lead us to write

$$\lim_{x \to -\infty} \frac{x}{1 + x^2} = 0$$

**Rigorous Definitions of Limits as $x \to \pm \infty$**   In analogy with our $\varepsilon$, $\delta$ definition for ordinary limits, we make the following definition.

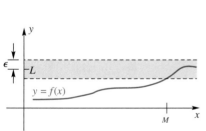

$$g(x) = \frac{x}{1 + x^2}$$

**FIGURE 1**

| $x$ | $\dfrac{x}{1 + x^2}$ |
|---|---|
| 10 | 0.099 |
| 100 | 0.010 |
| 1000 | 0.001 |
| 10000 | 0.0001 |
| ↓ | ↓ |
| $\infty$ | ? |

**FIGURE 2**

> **Definition**
>
> **(Limit as $x \to \infty$).** Let $f$ be defined on $[c, \infty)$ for some number $c$. We say that $\lim\limits_{x \to \infty} f(x) = L$ if for each $\varepsilon > 0$, there is a corresponding number $M$ such that
>
> $$x > M \Rightarrow |f(x) - L| < \varepsilon$$

You will note that $M$ can depend on $\varepsilon$. In general, the smaller $\varepsilon$ is, the larger $M$ will have to be. The graph in Figure 3 may help you understand what we are saying.

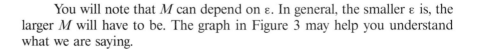

**FIGURE 3**

> **Definition**
>
> **(Limit as $x \rightarrow -\infty$).** Let $f$ be defined on $(-\infty, c]$ for some number $c$. We say that $\lim_{x \to -\infty} f(x) = L$ if for each $\varepsilon > 0$, there is a corresponding number $M$ such that
>
> $$x < M \Rightarrow |f(x) - L| < \varepsilon$$

**EXAMPLE 1**   Show that if $k$ is a positive integer, then

$$\lim_{x \to \infty} \frac{1}{x^k} = 0 \quad \text{and} \quad \lim_{x \to -\infty} \frac{1}{x^k} = 0$$

*Solution*   Let $\varepsilon > 0$ be given. After a preliminary analysis (as in Section 2.5), we choose $M = \sqrt[k]{1/\varepsilon}$. Then $x > M$ implies

$$\left| \frac{1}{x^k} - 0 \right| = \frac{1}{x^k} < \frac{1}{M^k} = \varepsilon$$

The proof of the second statement is similar.   ∎

Having given the definitions of these new limits, we must face the question of whether the Main Limit Theorem (Theorem 2.6A) holds for them. The answer is yes, and the proof is similar to the original one. Note how we use this theorem in the following examples.

**EXAMPLE 2**   Prove that $\lim_{x \to \infty} \dfrac{x}{1 + x^2} = 0$.

*Solution*   Here we use a standard trick: dividing numerator and denominator by the highest power of $x$ that appears in the denominator, namely, $x^2$.

$$\lim_{x \to \infty} \frac{x}{1 + x^2} = \lim_{x \to \infty} \frac{\dfrac{x}{x^2}}{\dfrac{1 + x^2}{x^2}} = \lim_{x \to \infty} \frac{\dfrac{1}{x}}{\dfrac{1}{x^2} + 1}$$

$$= \frac{\lim_{x \to \infty} \dfrac{1}{x}}{\lim_{x \to \infty} \dfrac{1}{x^2} + \lim_{x \to \infty} 1} = \frac{0}{0 + 1} = 0 \quad ∎$$

**EXAMPLE 3**   Find $\lim_{x \to \infty} \dfrac{2 - 3x + x^2}{7 + 4x - 5x^2}$.

*Solution*   Again, we divide numerator and denominator by $x^2$.

$$\lim_{x \to \infty} \frac{2 - 3x + x^2}{7 + 4x - 5x^2} = \lim_{x \to \infty} \frac{2/x^2 - 3/x + 1}{7/x^2 + 4/x - 5}$$

$$= \frac{\displaystyle\lim_{x \to \infty} (2/x^2 - 3/x + 1)}{\displaystyle\lim_{x \to \infty} (7/x^2 + 4/x - 5)} = \frac{0 - 0 + 1}{0 + 0 - 5} = -\frac{1}{5} \quad \blacksquare$$

**EXAMPLE 4**   Find $\displaystyle\lim_{x \to -\infty} \frac{2x^3}{1 + x^3}$.

*Solution*   Divide numerator and denominator by $x^3$.

$$\lim_{x \to -\infty} \frac{2x^3}{1 + x^3} = \lim_{x \to -\infty} \frac{2}{1/x^3 + 1} = \frac{2}{0 + 1} = 2 \quad \blacksquare$$

**Infinite Limits**   Consider the graph of $f(x) = 1/(x - 2)$, which is shown in Figure 4. It makes no sense to ask for $\displaystyle\lim_{x \to 2} 1/(x - 2)$, but we think it is reasonable to write

$$\lim_{x \to 2^-} \frac{1}{x - 2} = -\infty \qquad \lim_{x \to 2^+} \frac{1}{x - 2} = \infty$$

Here is the definition that relates to this situation.

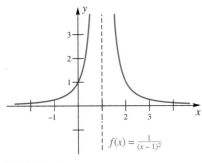

FIGURE 4

---

Definition

**(Infinite Limit).** We say that $\displaystyle\lim_{x \to c^+} f(x) = \infty$ if for each positive number $M$, there corresponds a $\delta > 0$ such that

$$0 < x - c < \delta \Rightarrow f(x) > M$$

---

There are corresponding definitions of

$$\lim_{x \to c^+} f(x) = -\infty, \ \lim_{x \to c^-} f(x) = \infty, \text{ and } \lim_{x \to c^-} f(x) = -\infty.$$

**EXAMPLE 5**   Find $\displaystyle\lim_{x \to 1^-} \frac{1}{(x - 1)^2}$ and $\displaystyle\lim_{x \to 1^+} \frac{1}{(x - 1)^2}$.

*Solution*   The graph of $f(x) = 1/(x - 1)^2$ is shown in Figure 5. We think it is quite clear that

$$\lim_{x \to 1^-} \frac{1}{(x - 1)^2} = \infty \qquad \lim_{x \to 1^+} \frac{1}{(x - 1)^2} = \infty$$

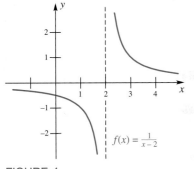

FIGURE 5

Since both limits are $\infty$, we could also write

$$\lim_{x \to 1} \frac{1}{(x-1)^2} = \infty$$

∎

**EXAMPLE 6**   Find $\displaystyle\lim_{x \to 2^+} \frac{x+1}{x^2 - 5x + 6}$.

*Solution*

$$\lim_{x \to 2^+} \frac{x+1}{x^2 - 5x + 6} = \lim_{x \to 2^+} \frac{x+1}{(x-3)(x-2)}$$

As $x \to 2^+$, we see that $x + 1 \to 3$, $x - 3 \to -1$, and $x - 2 \to 0^+$; thus, the numerator is approaching 3, but the denominator is negative and approaching 0. We conclude that

$$\lim_{x \to 2^+} \frac{x+1}{(x-3)(x-2)} = -\infty$$

∎

**Relation to Asymptotes**   Asymptotes were discussed briefly in Section 2.1, but now we can say more about them. The line $x = c$ is a **vertical asymptote** of the graph of $y = f(x)$ if any of the following four statements is true.

1. $\displaystyle\lim_{x \to c^+} f(x) = \infty$

2. $\displaystyle\lim_{x \to c^+} f(x) = -\infty$

3. $\displaystyle\lim_{x \to c^-} f(x) = \infty$

4. $\displaystyle\lim_{x \to c^-} f(x) = -\infty$

Thus, in Figure 4, the line $x = 2$ is a vertical asymptote and in Figure 5, the line $x = 1$ is a vertical asymptote. Likewise, the lines $x = 2$ and $x = 3$, although not shown graphically, are vertical asymptotes in Example 6.

In a similar vein, the line $y = b$ is a **horizontal asymptote** of the graph of $y = f(x)$ if either

$$\lim_{x \to \infty} f(x) = b \text{ or } \lim_{x \to -\infty} f(x) = b$$

The line $y = 0$ is a horizontal asymptote in both Figures 4 and 5.

**EXAMPLE 7**   Find the vertical and horizontal asymptotes of the graph of $y = f(x)$ if

$$f(x) = \frac{2x}{x-1}$$

*Solution*    We expect a vertical asymptote at a point where the denominator is zero, and we are right since

$$\lim_{x \to 1^+} \frac{2x}{x - 1} = \infty \quad \text{and} \quad \lim_{x \to 1^-} \frac{2x}{x - 1} = -\infty$$

On the other hand,

$$\lim_{x \to \infty} \frac{2x}{x - 1} = \lim_{x \to \infty} \frac{2}{1 - 1/x} = 2 \quad \text{and} \quad \lim_{x \to -\infty} \frac{2x}{x - 1} = 2$$

and so $y = 2$ is a horizontal asymptote. The graph of $y = 2x/(x - 1)$ is shown in Figure 6.

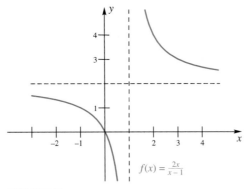

$$f(x) = \frac{2x}{x - 1}$$

FIGURE 6

## CONCEPTS REVIEW

**1.** To say that $x \to \infty$ means that _____; to say that $\lim_{x \to \infty} f(x) = L$ means that _____. Give your answers in informal language.

**2.** To say that $\lim_{x \to c^+} f(x) = \infty$ means that _____; to say that $\lim_{x \to c^-} f(x) = -\infty$ means that _____. Give your answers in informal language.

**3.** If $\lim_{x \to \infty} f(x) = 6$, then the line _____ is a _____ asymptote of the graph of $y = f(x)$.

**4.** If $\lim_{x \to 6^+} f(x) = \infty$, then the line _____ is a _____ asymptote of the graph of $y = f(x)$.

## PROBLEM SET 4.6

In Problems 1–36, find the limits.

**1.** $\lim_{x \to \infty} \dfrac{x}{x - 5}$

**2.** $\lim_{x \to \infty} \dfrac{x^2}{5 - x^3}$

**3.** $\lim_{t \to -\infty} \dfrac{t^2}{7 - t^2}$

**4.** $\lim_{t \to -\infty} \dfrac{t}{t - 5}$

**5.** $\lim_{x \to \infty} \dfrac{x^2}{(x - 5)(3 - x)}$

**6.** $\lim_{x \to \infty} \dfrac{x^2}{x^2 - 8x + 15}$

**7.** $\lim_{x \to \infty} \dfrac{x^3}{2x^3 - 100x^2}$

**8.** $\lim_{\theta \to -\infty} \dfrac{\pi \theta^5}{\theta^5 - 5\theta^4}$

**9.** $\lim_{x \to \infty} \dfrac{3x^3 - x^2}{\pi x^3 - 5x^2}$

**10.** $\lim_{\theta \to \infty} \dfrac{\sin^2 \theta}{\theta^2 - 5}$

**11.** $\lim_{x \to \infty} \dfrac{3\sqrt{x^3} + 3x}{\sqrt{2x^3}}$

**12.** $\lim_{x \to \infty} \sqrt[3]{\dfrac{\pi x^3 + 3x}{\sqrt{2x^3} + 7x}}$

13. $\lim\limits_{x \to \infty} \sqrt[3]{\dfrac{1 + 8x^2}{x^2 + 4}}$

14. $\lim\limits_{x \to \infty} \sqrt{\dfrac{x^2 + x + 3}{(x - 1)(x + 1)}}$

15. $\lim\limits_{x \to \infty} \dfrac{2x + 1}{\sqrt{x^2 + 3}}$. *Hint:* Divide numerator and denominator by $x$. Note that for $x > 0$, $\sqrt{x^2 + 3}/x = \sqrt{(x^2 + 3)/x^2}$.

16. $\lim\limits_{x \to \infty} \dfrac{\sqrt{2x + 1}}{x + 4}$

17. $\lim\limits_{x \to \infty} \left(\sqrt{2x^2 + 3} - \sqrt{2x^2 - 5}\right)$. *Hint:* Multiply and divide by $\sqrt{2x^2 + 3} + \sqrt{2x^2 - 5}$.

18. $\lim\limits_{x \to \infty} \left(\sqrt{x^2 + 2x} - x\right)$

19. $\lim\limits_{y \to -\infty} \dfrac{9y^3 + 1}{y^2 - 2y + 2}$. *Hint:* Divide numerator and denominator by $y^2$.

20. $\lim\limits_{x \to \infty} \dfrac{a_0 x^n + a_1 x^{n-1} + \cdots + a_{n-1} x + a_n}{b_0 x^n + b_1 x^{n-1} + \cdots + b_{n-1} x + b_n}$, where $a_0 \neq 0$, $b_0 \neq 0$, and $n$ is in $\mathbb{N}$.

21. $\lim\limits_{x \to 4^+} \dfrac{x}{x - 4}$

22. $\lim\limits_{t \to -3^+} \dfrac{t^2 - 9}{t + 3}$

23. $\lim\limits_{t \to 3^-} \dfrac{t^2}{9 - t^2}$

24. $\lim\limits_{x \to \sqrt[3]{5}^+} \dfrac{x^2}{5 - x^3}$

25. $\lim\limits_{x \to 5^-} \dfrac{x^2}{(x - 5)(3 - x)}$

26. $\lim\limits_{\theta \to \pi^+} \dfrac{\theta^2}{\sin \theta}$

27. $\lim\limits_{x \to 3^-} \dfrac{x^3}{x - 3}$

28. $\lim\limits_{\theta \to (\pi/2)^+} \dfrac{\pi \theta}{\cos \theta}$

29. $\lim\limits_{x \to 3^-} \dfrac{x^2 - x - 6}{x - 3}$

30. $\lim\limits_{x \to 2^+} \dfrac{x^2 + 2x - 8}{x^2 - 4}$

31. $\lim\limits_{x \to 0^+} \dfrac{[\![x]\!]}{x}$

32. $\lim\limits_{x \to 0^-} \dfrac{[\![x]\!]}{x}$

33. $\lim\limits_{x \to 0^-} \dfrac{|x|}{x}$

34. $\lim\limits_{x \to 0^+} \dfrac{|x|}{x}$

35. $\lim\limits_{x \to 0^-} \dfrac{1 + \cos x}{\sin x}$

36. $\lim\limits_{x \to \infty} \dfrac{\sin x}{x}$

In Problems 37–42, find the horizontal and vertical asymptotes for the graphs of the indicated functions. Then sketch their graphs.

37. $f(x) = \dfrac{3}{x + 1}$

38. $f(x) = \dfrac{3}{(x + 1)^2}$

39. $F(x) = \dfrac{2x}{x - 3}$

40. $F(x) = \dfrac{3}{9 - x^2}$

41. $g(x) = \dfrac{14}{2x^2 + 7}$

42. $g(x) = \dfrac{2x}{\sqrt{x^2 + 5}}$

43. The line $y = ax + b$ is called an **oblique asymptote** to the graph of $y = f(x)$ if either $\lim\limits_{x \to \infty} [f(x) - (ax + b)] = 0$ or $\lim\limits_{x \to -\infty} [f(x) - (ax + b)] = 0$. Find the oblique asymptote for

$$f(x) = \dfrac{2x^4 + 3x^3 - 2x - 4}{x^3 - 1}$$

*Hint:* Begin by dividing the denominator into the numerator.

44. Find the oblique asymptote for

$$f(x) = \dfrac{3x^3 + 4x^2 - x + 1}{x^2 + 1}$$

45. Using the symbols $M$ and $\delta$, give precise definitions of each expression.

(a) $\lim\limits_{x \to c^+} f(x) = -\infty$

(b) $\lim\limits_{x \to c^-} f(x) = \infty$

46. Using the symbols $M$ and $N$, give precise definitions of each expression.

(a) $\lim\limits_{x \to \infty} f(x) = \infty$

(b) $\lim\limits_{x \to -\infty} f(x) = \infty$

47. Give a rigorous proof that if $\lim\limits_{x \to \infty} f(x) = A$ and $\lim\limits_{x \to \infty} g(x) = B$, then

$$\lim\limits_{x \to \infty} [f(x) + g(x)] = A + B$$

48. We have given meaning to $\lim\limits_{x \to A} f(x)$ for $A = a, a^-$, $a^+$, $-\infty$, $\infty$. Moreover, in each case, this limit may be $L$ (finite), $-\infty$, $\infty$, or may fail to exist in any sense. Make a table illustrating each of the 20 possible cases.

49. Find each of the following limits or indicate that it does not exist even in the infinite sense.

(a) $\lim\limits_{x \to \infty} \sin x$

(b) $\lim\limits_{x \to \infty} \sin \dfrac{1}{x}$

(c) $\lim\limits_{x \to \infty} x \sin \dfrac{1}{x}$

(d) $\lim\limits_{x \to \infty} x^{3/2} \sin \dfrac{1}{x}$

(e) $\lim\limits_{x \to \infty} x^{-1/2} \sin x$

(f) $\lim\limits_{x \to \infty} \sin \left(\dfrac{\pi}{6} + \dfrac{1}{x}\right)$

(g) $\lim\limits_{x \to \infty} \sin \left(x + \dfrac{1}{x}\right)$

(h) $\lim\limits_{x \to \infty} \left[\sin \left(x + \dfrac{1}{x}\right) - \sin x\right]$

**50.** Let $f$ be differentiable for $x > a$. Prove or disprove.

(a) $\lim\limits_{x \to \infty} f(x) = 0 \Rightarrow \lim\limits_{x \to \infty} f'(x) = 0$

(b) $\lim\limits_{x \to a^+} f(x) = \infty \Rightarrow \lim\limits_{x \to a^+} f'(x) = -\infty$

[PC] Use a computer to find the limits in Problems 51–58. To do this using the *Limit* section of *True BASIC Calculus*, first transform a limit of the form $\lim\limits_{x \to \infty} f(x)$ to $\lim\limits_{x \to 0^+} f(1/x)$ and one of the form $\lim\limits_{x \to -\infty} f(x)$ to $\lim\limits_{x \to 0^-} f(1/x)$.

**51.** $\lim\limits_{x \to \infty} \dfrac{3x^2 + x + 1}{2x^2 - 1}$ **52.** $\lim\limits_{x \to -\infty} \sqrt{\dfrac{2x^2 - 3x}{5x^2 + 1}}$

**53.** $\lim\limits_{x \to -\infty} \left(\sqrt{2x^2 + 3x} - \sqrt{2x^2 - 5}\right)$

**54.** $\lim\limits_{x \to \infty} \dfrac{2x + 1}{\sqrt{3x^2 + 1}}$

**55.** $\lim\limits_{x \to \infty} \left(1 + \dfrac{1}{x}\right)^{10}$ **56.** $\lim\limits_{x \to \infty} \left(1 + \dfrac{1}{x}\right)^{x}$

**57.** $\lim\limits_{x \to \infty} \left(1 + \dfrac{1}{x}\right)^{x^2}$ **58.** $\lim\limits_{x \to \infty} \left(1 + \dfrac{1}{x}\right)^{\sin x}$

[PC] Find the one-sided limits in Problems 59–65. Your computer may indicate that some of these limits do not exist, but if so, you should be able to interpret the answer as either $+\infty$ or $-\infty$.

**59.** $\lim\limits_{x \to 3^-} \dfrac{\sin |x - 3|}{x - 3}$ **60.** $\lim\limits_{x \to 3^-} \dfrac{\sin |x - 3|}{\tan (x - 3)}$

**61.** $\lim\limits_{x \to 3^-} \dfrac{\cos(x - 3)}{x - 3}$ **62.** $\lim\limits_{x \to \frac{\pi}{2}^+} \dfrac{\cos x}{x - \pi/2}$

**63.** $\lim\limits_{x \to 0^+} (1 + \sqrt{x})^{1/\sqrt{x}}$ **64.** $\lim\limits_{x \to 0^+} (1 + \sqrt{x})^{1/x}$

**65.** $\lim\limits_{x \to 0^+} (1 + \sqrt{x})^{x}$

---

**Answers to Concepts Review:** **1.** $x$ increases without bound; $f(x)$ gets close to $L$ as $x$ increases without bound. **2.** $f(x)$ increases without bound as $x$ approaches $c$ from the right; $f(x)$ decreases without bound as $x$ approaches $c$ from the left. **3.** $y = 6$; horizontal. **4.** $x = 6$; vertical.

---

**4.7**
**SOPHISTICATED GRAPHING**

Our treatment of graphing in Section 1.7 was elementary. We proposed plotting enough points so the essential features of the graph were clear. We mentioned that symmetries of the graph could reduce the effort involved. We suggested that one should be alert to possible asymptotes. But if the equation to be graphed is complicated or if we want a very accurate graph, the techniques of Chapter 1 are inadequate.

Calculus provides a powerful tool for analyzing the fine structure of a graph, especially in identifying those points where the character of the graph changes. We can locate local maximum points, local minimum points, and inflection points; we can determine precisely where the graph is increasing or where it is concave up. Inclusion of all these ideas in our graphing procedure is the program for this section.

**Polynomial Functions** A polynomial of degree 1 or 2 is trivial to graph; one of degree 50 would be next to impossible. If the degree is of modest size, such as 3 to 6, we can use the tools of calculus to great advantage.

**EXAMPLE 1** Sketch the graph of $f(x) = \dfrac{3x^5 - 20x^3}{32}$.

*Solution* Since $f(-x) = -f(x)$, $f$ is an odd function, and therefore its graph is symmetric with respect to the origin. Setting $f(x) = 0$, we find the $x$-intercepts to be 0 and $\pm\sqrt{20/3} \approx \pm 2.6$. We can go this far without calculus.

When we differentiate $f$, we obtain

$$f'(x) = \frac{15x^4 - 60x^2}{32} = \frac{15x^2(x-2)(x+2)}{32}$$

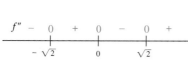

FIGURE 1

Thus, the critical points are $-2$, $0$, and $2$; we quickly discover (Figure 1) that $f'(x) > 0$ on $(-\infty, -2)$ and $(2, \infty)$ and that $f'(x) < 0$ on $(-2, 0)$ and $(0, 2)$. These facts tell us where $f$ is increasing and where it is decreasing; they also confirm that $f(-2) = 2$ is a local maximum value and that $f(2) = -2$ is a local minimum value.

Differentiating again, we get

$$f''(x) = \frac{60x^3 - 120x}{32} = \frac{15x(x-\sqrt{2})(x+\sqrt{2})}{8}$$

FIGURE 2

By studying the sign of $f''(x)$ (Figure 2), we deduce that $f$ is concave upward on $(-\sqrt{2}, 0)$ and $(\sqrt{2}, \infty)$ and concave downward on $(-\infty, -\sqrt{2})$ and $(0, \sqrt{2})$. Thus, there are three points of inflection, namely, $(-\sqrt{2}, 7\sqrt{2}/8) \approx (-1.4, 1.2)$, $(0, 0)$, and $(\sqrt{2}, -7\sqrt{2}/8) \approx (1.4, -1.2)$.

Much of this information is collected in the chart of Figure 3, which we use to sketch the graph directly below it.

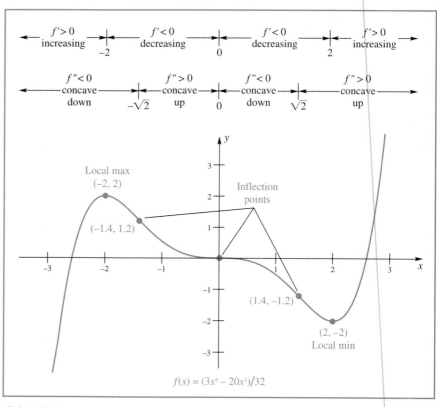

FIGURE 3

**Rational Functions**    A rational function, being the quotient of two polynomial functions, is considerably more complicated to graph than a polynomial. In particular, we can expect dramatic behavior wherever the denominator is zero.

**EXAMPLE 2**    Sketch the graph of $f(x) = \dfrac{x^2 - 2x + 4}{x - 2}$.

*Solution*    This function is neither even nor odd, so we do not expect any of the usual symmetries. There are no $x$-intercepts, since the solutions to $x^2 - 2x + 4 = 0$ are not real numbers. The $y$-intercept is $-2$. We anticipate a vertical asymptote at $x = 2$. In fact,

$$\lim_{x \to 2^-} \frac{x^2 - 2x + 4}{x - 2} = -\infty \qquad \lim_{x \to 2^+} \frac{x^2 - 2x + 4}{x - 2} = \infty$$

Differentiation twice gives

$$f'(x) = \frac{x(x - 4)}{(x - 2)^2} \qquad f''(x) = \frac{8}{(x - 2)^3}$$

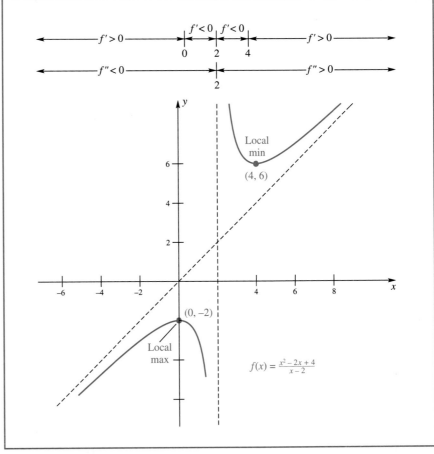

FIGURE 4

Thus, $f'(x) > 0$ on $(-\infty, 0) \cup (4, \infty)$ and $f'(x) < 0$ on $(0, 2) \cup (2, 4)$. Also, $f''(x) > 0$ on $(2, \infty)$ and $f''(x) < 0$ on $(-\infty, 2)$. Note that $f''(x)$ is never 0; there are no inflection points. On the other hand, $f(0) = -2$ and $f(4) = 6$ give local maximum and minimum values, respectively.

It is a good idea to check on the behavior of $f(x)$ for large $|x|$. Since

$$f(x) = \frac{x^2 - 2x + 4}{x - 2} = x + \frac{4}{x - 2}$$

the graph of $y = f(x)$ gets closer and closer to the line $y = x$ as $|x|$ gets larger and larger. We call the line $y = x$ an **oblique asymptote** for the graph of $f$ (see Problem 39 of Section 4.6).

With all this information, we are able to sketch a rather accurate graph (Figure 4 on the previous page). ∎

**Algebraic Functions**   There is an endless variety of algebraic functions. Here is one example.

**EXAMPLE 3**   Analyze the function

$$F(x) = \frac{\sqrt{x}(x - 5)^2}{4}$$

and sketch its graph.

*Solution*   The domain of $F$ is $[0, \infty)$ and the range is $[0, \infty)$, so the graph of $F$ is confined to the first quadrant. The $x$-intercepts are 0 and 5; the $y$-intercept is 0.

From

$$F'(x) = \frac{5(x - 1)(x - 5)}{8\sqrt{x}}, \qquad x > 0$$

we find the stationary points 1 and 5. Since $F'(x) > 0$ on $(0, 1)$ and $(5, \infty)$ while $F'(x) < 0$ on $(1, 5)$, we conclude that $F(1) = 4$ is a local maximum value and $F(5) = 0$ is a local minimum value.

So far, it has been clear sailing. But on calculating the second derivative, we obtain

$$F''(x) = \frac{5(3x^2 - 6x - 5)}{16x^{3/2}}, \qquad x > 0$$

which is quite complicated. However, $3x^2 - 6x - 5 = 0$ has one solution in $(0, \infty)$, namely $1 + 2\sqrt{6}/3 \approx 2.6$.

Using the test points 1 and 3, we conclude that $f''(x) < 0$ on $(0, 2.6)$ and $f''(x) > 0$ on $(2.6, \infty)$. It then follows that the point $(2.6, 2.3)$ is an inflection point.

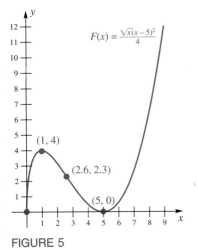

$$F(x) = \frac{\sqrt{x}(x-5)^2}{4}$$

(1, 4)

(2.6, 2.3)

(5, 0)

**FIGURE 5**

As $x$ grows large, $F(x)$ grows without bound and much faster than any linear function; there are no asymptotes. The graph is sketched in Figure 5. ∎

**Summary of the Method**    In graphing functions, there is no substitute for common sense. However, the following procedure will be helpful in most cases.

**Step 1**  Precalculus analysis.
(a) Check the *domain* and *range* of the function to see if any regions of the plane are excluded.
(b) Test for *symmetry* with respect to the y-axis and the origin. (Is the function even or odd?)
(c) Find the *intercepts*.

**Step 2**  Calculus analysis.
(a) Use the first derivative to find the critical points and to find out where the graph is *increasing* and *decreasing*.
(b) Test the critical points for *local maxima* and *minima*.
(c) Use the second derivative to find out where the graph is *concave upward* and *concave downward* and to locate *inflection points*.
(d) Find the *asymptotes*.

**Step 3**  Plot a few points (including all critical points and inflection points).

**Step 4**  Sketch the graph.

---

## CONCEPTS REVIEW

**1.** The graph of $f$ is symmetric with respect to the y-axis if $f(-x) =$ _____; the graph is symmetric with respect to the origin if $f(-x) =$ _____.

**2.** If $f'(x) < 0$ and $f''(x) > 0$ for all $x$ in an interval $I$, then the graph of $f$ is both _____ and _____ on $I$.

**3.** The graph of $f(x) = x^3/[(x + 1)(x - 2)(x - 3)]$ has

as vertical asymptotes the lines _____ and as a horizontal asymptote the line _____.

**4.** We call $f(x) = 3x^5 - 2x^2 + 6$ a(n) _____ function; we call $g(x) = (3x^5 - 2x^2 + 6)/(x^2 - 4)$ a(n) _____ function; and we call $h(x) = \sqrt{3x^5 - 2x^2 + 6}$ a(n) _____ function.

---

## PROBLEM SET 4.7

In Problems 1–27, make an analysis as suggested in the summary above and then sketch the graph.

**1.** $f(x) = x^3 - 3x + 5$

**2.** $f(x) = 2x^3 - 3x - 10$

**3.** $f(x) = 2x^3 - 3x^2 - 12x + 3$

**4.** $f(x) = (x - 1)^3$        **5.** $G(x) = (x - 1)^4$

**6.** $H(t) = t^2(t^2 - 1)$

**7.** $f(x) = x^3 - 3x^2 + 3x + 10$

**8.** $F(s) = \dfrac{4s^4 - 8s^2 - 12}{3}$

**9.** $g(x) = \dfrac{x}{x + 1}$

**10.** $g(s) = \dfrac{(s - \pi)^2}{s}$

**11.** $f(x) = \dfrac{x}{x^2 + 4}$

**12.** $\Lambda(\theta) = \dfrac{\theta^2}{\theta^2 + 1}$

**13.** $h(x) = \dfrac{x}{x - 1}$

**14.** $R(s) = \dfrac{s}{s^2 - 1}$

**15.** $f(x) = \dfrac{(x - 1)(x - 3)}{(x + 1)(x - 2)}$

**16.** $w(z) = z\sqrt{z + 1}$

**17.** $g(x) = \sqrt{1 - x} + 1$

**18.** $f(x) = |x|^3$ *Hint:* $\dfrac{d}{dx}|x| = \dfrac{x}{|x|}$

19. $R(z) = z\,|z|$

20. $H(q) = q^2\,|q|$

21. $g(x) = \dfrac{|x| + x}{2}\ 3x + 2$

22. $h(x) = \dfrac{|x| - x}{2}\ x^2 - x + 6$

23. $f(x) = |\sin x|$

24. $f(x) = \sqrt{\sin x}$

$\approx$ 25. $f(x) = \dfrac{5.235x^3 - 1.245x^2}{7.126x - 3.141}$

$\approx$ 26. $f(x) = \dfrac{2.4718x^2 - 10.965x + 12.508}{15.4035x - 3.4557x^2 - 14.344}$

$\approx$ 27. $g(x) = \dfrac{5.2347x^2 - 11.235x}{11.444x - 1.4223x^2 - 28.993}$

28. Sketch a possible graph of the function $f$ that has all the following properties:

(a) $f$ is everywhere continuous;
(b) $f(0) = 0, f(1) = 2$;
(c) $f$ is an even function;
(d) $f'(x) > 0$ for $x > 0$;
(e) $f''(x) > 0$ for $x > 0$.

29. Sketch a possible graph of the function $f$ that has all the following properties:

(a) $f$ is everywhere continuous;
(b) $f(2) = -3, f(6) = 1$;
(c) $f'(2) = 0, f'(x) > 0$ for $x \neq 2, f'(6) = 3$;
(d) $f''(6) = 0, f''(x) > 0$ for $2 < x < 6, f''(x) < 0$ for $x > 6$.

30. Sketch a possible graph of the function $g$ that has all the following properties:

(a) $g$ is everywhere smooth (continuous with a continuous first derivative);
(b) $g(0) = 0$;
(c) $g'(x) < 0$ for all $x$;
(d) $g''(x) < 0$ for $x < 0$ and $g''(x) > 0$ for $x > 0$.

31. Sketch a possible graph of a function $f$ that has all the following properties:

(a) $f$ is everywhere continuous;
(b) $f(-3) = 1$;
(c) $f'(x) < 0$ for $x < -3, f'(x) > 0$ for $x > -3, f''(x) < 0$ for $x \neq -3$.

32. Sketch a possible graph of a function $f$ that has all the following properties:

(a) $f$ is everywhere continuous;
(b) $f(-4) = -3, f(0) = 0, f(3) = 2$;
(c) $f'(-4) = 0, f'(3) = 0, f'(x) > 0$ for $x < -4, f'(x) > 0$ for $-4 < x < 3, f'(x) < 0$ for $x > 3$;
(d) $f''(-4) = 0, f''(0) = 0, f''(x) < 0$ for $x < -4, f''(x) > 0$ for $-4 < x < 0, f''(x) < 0$ for $x > 0$.

33. Sketch a possible curve $y = f(x)$ which is

(a) has a continuous first derivative;
(b) falling and concave up for $x < 3$;
(c) has an extremum at $(3, 1)$;
(d) rising and concave up for $3 < x < 5$;
(e) has an inflection point at $(5, 4)$;
(f) rising and concave down for $5 < x < 6$;
(g) extremum at $(6, 7)$;
(h) falling and concave down for $x > 6$.

Best linear approximations provide particularly good approximations at points of inflection. By using a graphing calculator one can easily investigate such behavior in Problems 34–36.

C 34. Graph $y = \sin x$ and its linear approximation $L(x) = x$ at the inflection point $x = 0$.

C 35. Graph $y = \cos x$ and its linear approximation $L(x) = -x + \pi/2$ at $x = \pi/2$.

C 36. Find the best linear approximation to the curve $y = (x - 1)^5 + 3$ at its point of inflection. Graph both the function and its linear approximation in the neighborhood of the inflection point.

37. Suppose $f'(x) = (x - 2)(x - 3)(x - 4)$ and $f(2) = 2, f(3) = 3$, and $f(4) = 1$. Sketch a possible graph of $y = f(x)$.

38. Suppose $f'(x) = (x - 3)(x - 2)^2(x - 1)$ and $f(2) = 0$. Sketch a possible graph of $f$.

39. Suppose $h'(x) = x^2(x - 1)^2(x - 2)$ and $h(0) = 0$, $h(1) = -1$, and $h(2) = -2$. Sketch a possible graph of $h(x)$.

40. Consider a general quadratic curve $y = ax^2 + bx + c$. Show using the concepts of this section that such a curve has no inflection points.

41. Consider the curve $y = ax^3 + bx^2 + cx + d$. Find conditions on $a$ which guarantee that this curve has one inflection point.

**42.** Consider a general quartic curve $y = ax^4 + bx^3 + cx^2 + dx + e$. What is the maximum number of inflection points that such a curve can have?

In Problems 43–46 the graph of $y = f(x)$ depends on a parameter $c$. Using a computer or graphing calculator, investigate how the extremum and inflection points depend on the value of $c$. Identify the values of $c$ at which the basic shape of the curves change.

**43.** $y = x^2 \sqrt{x^2 - c^2}$

**44.** $y = \dfrac{cx}{4 + (cx)^2}$

**45.** $y = \dfrac{1}{(cx^2 - 4)^2 + cx^2}$

**46.** $y = \dfrac{1}{x^2 + 4x + c}$

**47.** Let $g(x)$ be a function which has two derivatives and satisfies the following properties:

(a) $g(1) = 1$;
(b) $g'(x) > 0$ for all $x \neq 1$;
(c) $g$ is concave down for all $x < 1$ and concave up for all $x > 1$;
(d) Let $f(x) = g(x^4)$;

Sketch a possible graph of $f(x)$ and justify your answer.

**48.** Let $H(x)$ have three continuous derivatives, and be such that $H(1) = H'(1) = H''(1) = 0$ but $H'''(1) \neq 0$. Does $H(x)$ have a relative maximum, relative minimum, or a point of inflection at $x = 1$. Justify your answer.

**49.** In each case is it possible for a function with two continuous derivatives to satisfy all of the following properties? If so sketch such a function. If not, justify your answer.

(a) $F'(x) > 0$, $F''(x) > 0$ while $F(x) < 0$ for all $x$.
(b) $F''(x) < 0$ while $F(x) > 0$.
(c) $F''(x) < 0$ while $F'(x) > 0$.

PC **50.** Draw the graphs of each of the following functions on the indicated interval. Determine the coordinates of any of the global extrema and any inflection points. You should be able to give answers that are accurate to at least one decimal place. Restrict the $y$-axis window to $-5 \leq y \leq 5$.

(a) $f(x) = x^2 \tan x$; $\left(-\dfrac{\pi}{2}, \dfrac{\pi}{2}\right)$

(b) $f(x) = x^3 \tan x$; $\left(-\dfrac{\pi}{2}, \dfrac{\pi}{2}\right)$

(c) $f(x) = 2x + \sin x$; $[-\pi, \pi]$

(d) $f(x) = x - \dfrac{\sin x}{2}$; $[-\pi, \pi]$

PC **51.** In each case, the functions given are periodic on a fundamental interval. Draw the graphs of each of the following functions over one full period with the center of the interval located at the origin. Determine the coordinates of any of the global extrema and any inflection points. You should be able to give answers that are accurate to at least one decimal place.

(a) $f(x) = 2 \sin x + \cos^2 x$
(b) $f(x) = 2 \sin x + \sin^2 x$
(c) $f(x) = \cos 2x - 2 \cos x$
(d) $f(x) = \sin 3x - \sin x$
(e) $f(x) = \sin 2x - \cos 3x$

**52.** Let $f$ be a continuous function with $f(-3) = f(0) = 2$. If the graph of $y = f'(x)$ is as shown in Figure 6, sketch a possible graph for $y = f(x)$.

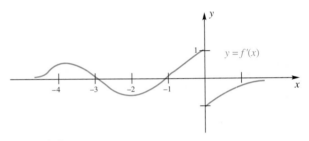

**FIGURE 6**

**53.** Let $f$ be a continuous function with $f(0) = f(2) = 0$. If the graph of $y = f'(x)$ is as shown in Figure 7, sketch a possible graph for $y = f(x)$.

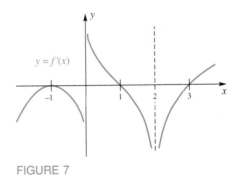

**FIGURE 7**

**54.** Suppose $f'(x) = (x - 3)(x - 1)^2(x + 2)$ and $f(1) = 2$. Sketch a possible graph of $f$.

PC **55.** Draw the graphs of each of the following functions on $[-1, 7]$. Determine the coordinates of any global extrema and any inflection points. You should be able to give answers that are accurate to at least one decimal place.

(a) $f(x) = x\sqrt{x^2 - 6x + 40}$
(b) $f(x) = \sqrt{|x|}(x^2 - 6x + 40)$
(c) $f(x) = \sqrt{x^2 - 6x + 40}/(x - 2)$
(d) $f(x) = \sin[(x^2 - 6x + 40)/6]$

PC **56.** Repeat Problem 55 for the following functions.

(a) $f(x) = x^3 - 8x^2 + 5x + 4$
(b) $f(x) = |x^3 - 8x^2 + 5x + 4|$
(c) $f(x) = (x^3 - 8x^2 + 5x + 4)/(x - 1)$
(d) $f(x) = (x^3 - 8x^2 + 5x + 4)/(x^3 + 1)$

**57.** Let $f$ be a continuous periodic function with an average value over a period of zero. If the graph of $y = f'(x)$ is as shown in Figure 8, sketch a possible graph of $y = f(x)$. Can you give an analytical representation of $f(x)$?

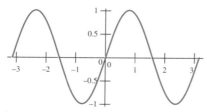

**FIGURE 8**

**58.** Let $f$ be a continuous periodic function with average value over a period of zero. If the graph of $y = f'(x)$ is as shown in Figure 9, sketch a possible graph of $y = f(x)$. Can you give an analytical representation of $f(x)$?

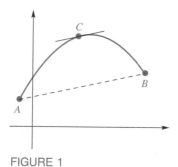

**FIGURE 9**

**59.** Let $f$ be a continuous periodic function with average value over a period of zero. If the graph of $y = f'(x)$ is as shown in Figure 10, sketch a possible graph of $y = f(x)$. Can you give an analytical representation of $f(x)$?

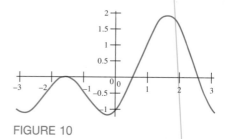

**FIGURE 10**

**60.** Let $f$ be a continuous periodic function with average value over a period of zero. If the graph of $y = f'(x)$ is as shown in Figure 11, sketch a possible graph of $y = f(x)$. Can you give an analytical representation of $f(x)$?

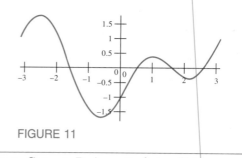

**FIGURE 11**

**Answers to Concepts Review:** **1.** $f(x)$, $-f(x)$
**2.** Decreasing; concave up **3.** $x = -1$, $x = 2$, $x = 3$; $y = 1$ **4.** Polynomial; rational; algebraic

**4.8**
**THE MEAN VALUE THEOREM**

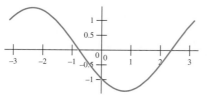

**FIGURE 1**

The Mean Value Theorem is the midwife of calculus—often helping to deliver other theorems that are of major significance. From now on, you will see the phrase "by the Mean Value Theorem" quite regularly, and later in this section we will use it to prove the Monotonicity Theorem, which we left unproved in Section 4.2.

In geometric language, the Mean Value Theorem is easy to state and understand. It says that if the graph of a continuous function has a nonvertical tangent line at every point between $A$ and $B$, then there is at least one point $C$ on the graph between $A$ and $B$ at which the tangent line is parallel to the secant line $AB$. In Figure 1, there is just one such point $C$; in Figure 2, there are several.

**The Theorem Proved** First we state the theorem in the language of functions; then we prove it.

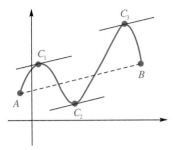

FIGURE 2

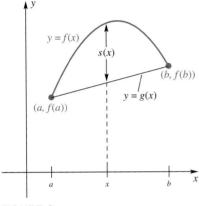

FIGURE 3

### Theorem A

**(Mean Value Theorem for Derivatives).** If $f$ is continuous on a closed interval $[a, b]$ and differentiable on its interior $(a, b)$, then there is at least one number $c$ in $(a, b)$ where

$$\frac{f(b) - f(a)}{b - a} = f'(c)$$

or, equivalently, where

$$f(b) - f(a) = f'(c)(b - a)$$

***Proof*** Our proof rests on a careful analysis of the function $s(x) = f(x) - g(x)$, introduced in Figure 3. Here $y = g(x)$ is the equation of the line through $(a, f(a))$ and $(b, f(b))$. Since this line has slope $[f(b) - f(a)]/(b - a)$ and goes through $(a, f(a))$, the point-slope form for its equation is

$$g(x) - f(a) = \frac{f(b) - f(a)}{b - a}(x - a)$$

This, in turn, yields a formula for $s(x)$, namely,

$$s(x) = f(x) - g(x) = f(x) - f(a) - \frac{f(b) - f(a)}{b - a}(x - a)$$

Note immediately that $s(b) = s(a) = 0$ and that for $x$ in $(a, b)$

$$s'(x) = f'(x) - \frac{f(b) - f(a)}{b - a}$$

Now we make a crucial observation. If we knew that there was a number $c$ in $(a, b)$ satisfying $s'(c) = 0$, we would be all done. For then the last equation would say

$$0 = f'(c) - \frac{f(b) - f(a)}{b - a}$$

which is equivalent to the conclusion of the theorem.

To see that $s'(c) = 0$ for some $c$ in $(a, b)$, reason as follows. Clearly $s$ is continuous on $[a, b]$, being the difference of two continuous functions. Thus, by the Max-Min Existence Theorem (Theorem 4.1A), $s$ must attain both a maximum and a minimum value on $[a, b]$. If both of these values happen to be 0, then $s(x)$ is identically 0 on $[a, b]$, and consequently $s'(x) = 0$ for all $x$ in $(a, b)$, much more than we need.

If either the maximum value or the minimum value is different from 0, then that value is attained at an interior point $c$, since $s(a) = s(b) = 0$. Now $s$ has a derivative at each point of $(a, b)$, and so by the Critical Point Theorem (Theorem 4.1B), $s'(c) = 0$. That is all we needed to know. ∎

### The Theorem Illustrated

**EXAMPLE 1**    Find the number $c$ guaranteed by the Mean Value Theorem for $f(x) = 2\sqrt{x}$ on $[1, 4]$.

*Solution*

$$f'(x) = 2 \cdot \frac{1}{2} x^{-1/2} = \frac{1}{\sqrt{x}}$$

and

$$\frac{f(4) - f(1)}{4 - 1} = \frac{4 - 2}{3} = \frac{2}{3}$$

Thus, we must solve

$$\frac{1}{\sqrt{c}} = \frac{2}{3}$$

The single solution is $c = \frac{9}{4}$ (Figure 4).    ∎

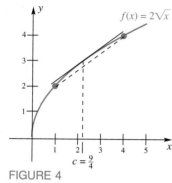

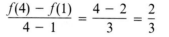

**FIGURE 4**

**EXAMPLE 2**    Let $f(x) = x^3 - x^2 - x + 1$ on $[-1, 2]$. Find all numbers $c$ satisfying the conclusion to the Mean Value Theorem.

*Solution*

$$f'(x) = 3x^2 - 2x - 1$$

and

$$\frac{f(2) - f(-1)}{2 - (-1)} = \frac{3 - 0}{3} = 1$$

Therefore, we must solve

$$3c^2 - 2c - 1 = 1$$

or, equivalently,

$$3c^2 - 2c - 2 = 0$$

By the Quadratic Formula, there are two solutions, $(2 \pm \sqrt{4 + 24})/6$, which correspond to $c_1 = -0.55$ and $c_2 = 1.22$. Both numbers are in the interval $(-1, 2)$. The appropriate graph is shown in Figure 5.    ∎

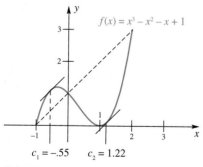

$c_1 = -.55 \quad c_2 = 1.22$

**FIGURE 5**

**EXAMPLE 3**    Let $f(x) = x^{2/3}$ on $[-8, 27]$. Show that the conclusion to the Mean Value Theorem fails and figure out why.

*Solution*

$$f'(x) = \frac{2}{3} x^{-1/3}, \ x \neq 0$$

and

$$\frac{f(27) - f(-8)}{27 - (-8)} = \frac{9 - 4}{35} = \frac{1}{7}$$

We must solve

$$\frac{2}{3} c^{-1/3} = \frac{1}{7}$$

which gives

$$c = \left(\frac{14}{3}\right)^3 \approx 102$$

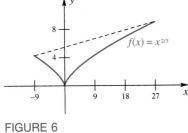

FIGURE 6

But $c = 102$ is not in the interval $(-8, 27)$ as required. The problem is, of course, that $f(x)$ is not differentiable everywhere on $(-8, 27)$; $f'(0)$ fails to exist (see Figure 6). ■

**The Theorem Used**    In Section 4.2, we promised a rigorous proof of the Monotonicity Theorem (Theorem 4.2A). This is the theorem that relates the sign of the derivative of a function to whether that function is increasing or decreasing.

***Proof of the Monotonicity Theorem***    We suppose that $f$ is continuous on $I$ and that $f'(x) > 0$ at each point $x$ in the interior of $I$. Consider any two points $x_1$ and $x_2$ of $I$ with $x_1 < x_2$. By the Mean Value Theorem applied on the interval $[x_1, x_2]$, there is a number $c$ in $(x_1, x_2)$ satisfying

$$f(x_2) - f(x_1) = f'(c)(x_2 - x_1)$$

Since $f'(c) > 0$, we see that $f(x_2) - f(x_1) > 0$—that is, $f(x_2) > f(x_1)$. This is what we mean when we say $f$ is increasing on $I$.

The case where $f'(x) < 0$ on $I$ is handled similarly. ■

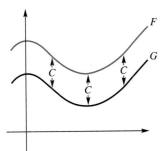

FIGURE 7

Our next theorem will be used repeatedly in the next chapter. In words, it says that *two functions with the same derivative differ by a constant*, possibly the zero constant (see Figure 7).

---

Theorem B

If $F'(x) = G'(x)$ for all $x$ in $(a, b)$, then there is a constant $C$ such that

$$F(x) = G(x) + C$$

for all $x$ in $(a, b)$.

---

***Proof***    Let $H(x) = F(x) - G(x)$. Then

$$H'(x) = F'(x) - G'(x) = 0$$

for all $x$ in $(a, b)$. Choose $x_1$ as some (fixed) point in $(a, b)$ and let $x$ be any other point there. The function $H$ satisfies the hypotheses of the Mean Value Theorem on the closed interval with endpoints $x_1$ and $x$. Thus there is a number $c$ between $x_1$ and $x$ such that

$$H(x) - H(x_1) = H'(c)(x - x_1)$$

But $H'(c) = 0$ by hypothesis. Therefore, $H(x) - H(x_1) = 0$ or, equivalently, $H(x) = H(x_1)$ for all $x$ in $(a, b)$. Since $H(x) = F(x) - G(x)$, we conclude that $F(x) - G(x) = H(x_1)$. Now let $C = H(x_1)$, and we have the conclusion $F(x) = G(x) + C$. ∎

## CONCEPTS REVIEW

**1.** The Mean Value Theorem says that if $f$ is _____ on $[a, b]$ and differentiable on _____, then there is a point $c$ in $(a, b)$ such that _____.

**2.** The function $f(x) = |\sin x|$ would satisfy the hypotheses of the Mean Value Theorem on the interval $[0, 1]$ but would not satisfy them on the interval $[-1, 1]$ because _____.

**3.** If two functions $F$ and $G$ have the same derivative on the interval $(a, b)$, then there is a constant $C$ such that _____.

**4.** Since $D_x(x^4) = 4x^3$, it follows that every function $F$ that satisfies $F'(x) = 4x^3$ has the form $F(x) =$ _____.

## PROBLEM SET 4.8

In each of the Problems 1–20, a function is defined and a closed interval is given. Decide whether the Mean Value Theorem applies to the given function on the given interval—if so, find all possible values of $c$; if not, state the reason. In each problem, sketch the graph of the given function on the given interval.

1. $f(x) = |x|$; $[1, 2]$

2. $g(x) = |x|$; $[-2, 2]$

3. $f(x) = x^2 + x$; $[-2, 2]$

4. $g(x) = (x + 1)^3$; $[-1, 1]$

5. $H(s) = s^2 + 3s - 1$; $[-3, 1]$

6. $F(x) = \dfrac{x^3}{3}$; $[-2, 2]$

7. $f(z) = \frac{1}{3}(z^3 + z - 4)$; $[-1, 2]$

8. $F(t) = \dfrac{1}{t - 1}$; $[0, 2]$

9. $h(x) = \dfrac{x}{x - 3}$; $[0, 2]$

10. $f(x) = \dfrac{x - 4}{x - 3}$; $[0, 4]$

11. $h(t) = t^{2/3}$; $[0, 2]$

12. $h(t) = t^{2/3}$; $[-2, 2]$

13. $g(x) = x^{5/3}$; $[0, 1]$

14. $g(x) = x^{5/3}$; $[-1, 1]$

15. $S(\theta) = \sin \theta$; $[-\pi, \pi]$

16. $C(\theta) = \csc \theta$; $[-\pi, \pi]$

17. $T(\theta) = \tan \theta$; $[0, \pi]$

18. $f(x) = x + \dfrac{1}{x}$; $[-1, \frac{1}{2}]$

19. $f(x) = x + \dfrac{1}{x}$; $[1, 2]$

20. $f(x) = [\![x]\!]$; $[1, 2]$

21. Johnny traveled 112 miles in 2 hours and claimed that he never exceeded 55 miles per hour. Use the Mean Value Theorem to prove that he lied. *Hint:* Let $f(t)$ be the distance traveled in time $t$.

22. **(Rolle's Theorem)** If $f$ *is continuous on* $[a, b]$ *and differentiable on* $(a, b)$ *and if* $f(a) = f(b)$, *then there is at least one number* $c$ *in* $(a, b)$ *such that* $f'(c) = 0$. Show that Rolle's Theorem is just a special case of the Mean Value Theorem. (Michel Rolle (1652–1719) was a French mathematician.)

**23.** For the function graphed on [0, 8] in Figure 8, find (approximately) all points $c$ that satisfy the conclusion to the Mean Value Theorem.

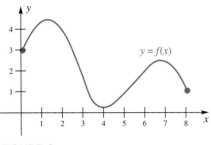

**FIGURE 8**

**24.** Show that if $f$ is the quadratic function defined by $f(x) = \alpha x^2 + \beta x + \gamma, \alpha \neq 0$, then the number $c$ of the Mean Value Theorem is always the midpoint of the given interval $[a, b]$.

**25.** Prove: If $f$ is continuous on $(a, b)$ and if $f'(x)$ exists and satisfies $f'(x) > 0$ except at one point $x_0$ in $(a, b)$, then $f$ is increasing on $(a, b)$. *Hint*: Consider $f$ on each of the intervals $(a, x_0]$ and $[x_0, b)$ separately.

**26.** Use Problem 25 to show that each of the following is increasing on $(-\infty, \infty)$.

(a) $f(x) = x^3$          (b) $f(x) = x^5$

(c) $f(x) = \begin{cases} x^3 & x \leq 0 \\ x & x > 0 \end{cases}$

**27.** Use the Mean Value Theorem to show that $s = 1/t$ decreases on any interval for which it is defined.

**28.** Use the Mean Value Theorem to show that $s = 1/t^2$ decreases for any interval to the right of the origin.

**29.** Prove that if $F'(x) = 0$ for all $x$ in $(a, b)$, then there is a constant $C$ such that $F(x) = C$ for all $x$ in $(a, b)$. *Hint*: Let $G(x) = 0$ and apply Theorem B.

**30.** Suppose that you know that $\cos(0) = 1$, $\sin(0) = 0$, $D_x \cos x = -\sin x$, and $D_x \sin x = \cos x$, but nothing else about the sine and cosine functions. Show that $\cos^2 x + \sin^2 x = 1$. *Hint*: Let $F(x) = \cos^2 x + \sin^2 x$ and use Problem 29.

**31.** Prove that if $F'(x) = D$ for all $x$ in $(a, b)$, then there is a constant $C$ such that $F(x) = Dx + C$ for all $x$ in $(a, b)$. *Hint*: Let $G(x) = Dx$ and apply Theorem B.

**32.** Suppose $F'(x) = 5$ and $F(0) = 4$. Find a formula for $F(x)$. *Hint*: See Problem 31.

**33.** Prove: Let $f$ have a derivative on $[a, b]$. If $f(a)$ and $f(b)$ have opposite signs and if $f'(x) \neq 0$ for all $x$ in $(a, b)$, then the equation $f(x) = 0$ has one and only one solution between $a$ and $b$. *Hint*: Use the Intermediate Value Theorem and Rolle's Theorem (Problem 22).

**34.** Show that $f(x) = 2x^3 - 9x^2 + 1 = 0$ has exactly one solution on each of the intervals $(-1, 0)$, $(0, 1)$, and $(4, 5)$. *Hint*: Apply Problem 33.

**35.** Prove: Let $f$ have a derivative on an interval $I$. Between successive distinct zeros of $f'$, there can be at most one zero of $f$. *Hint*: Try a proof by contradiction and use Rolle's Theorem (Problem 22).

**36.** Prove: Let $g$ be continuous on $[a, b]$ and suppose $g''(x)$ exists for all $x$ in $(a, b)$. If there are three values of $x$ in $[a, b]$ for which $g(x) = 0$, then there is at least one value of $x$ in $(a, b)$ such that $g''(x) = 0$.

**37.** Let $f(x) = (x - 1)(x - 2)(x - 3)$ prove by using Problem 36 that there is at least one value in the interval $[0, 4]$ where $f''(x) = 0$ and two values in the same interval where $f'(x) = 0$.

**38.** Prove that if $|f'(x)| \leq M$ for all $x$ in $(a, b)$ and if $x_1$ and $x_2$ are any two points in $(a, b)$, then

$$|f(x_2) - f(x_1)| \leq M |x_2 - x_1|$$

*Note*: A function satisfying the above inequality is said to satisfy a *Lipschitz condition* with constant $M$. (Rudolph Lipschitz (1832–1903) was a German mathematician.)

**39.** Show that $f(x) = \sin 2x$ satisfies a Lipschitz condition with constant 2 on the interval $(-\infty, \infty)$. See Problem 38.

**40.** A function $f$ is called **nondecreasing** on an interval $I$ if $x_1 < x_2 \Rightarrow f(x_1) \leq f(x_2)$ for $x_1$ and $x_2$ in $I$. Similarly, $f$ is **nonincreasing** on $I$ if $x_1 < x_2 \Rightarrow f(x_1) \geq f(x_2)$ for $x_1$ and $x_2$ in $I$.

(a) Sketch the graph of a function which is nondecreasing but not increasing.
(b) Sketch the graph of a function which is nonincreasing but not decreasing.

**41.** Prove that if $f$ is continuous on $I$ and if $f'(x)$ exists and satisfies $f'(x) \geq 0$ on the interior of $I$, then $f$ is nondecreasing on $I$. Similarly, if $f'(x) \leq 0$, then $f$ is nonincreasing on $I$.

**42.** Prove that if $f(x) \geq 0$ and $f'(x) \geq 0$ on $I$, then $f^2$ is nondecreasing on $I$.

**43.** Prove that if $g'(x) \leq h'(x)$ for all $x$ in $(a, b)$, then

$$x_1 < x_2 \quad \Rightarrow \quad g(x_2) - g(x_1) \leq h(x_2) - h(x_1)$$

for all $x_1$ and $x_2$ in $(a, b)$. *Hint*: Apply Problem 41 with $f(x) = h(x) - g(x)$.

**44.** Use the Mean Value Theorem to prove that

$$\lim_{x \to \infty} \left(\sqrt{x + 2} - \sqrt{x}\right) = 0$$

**45.** Use the Mean Value Theorem to show that

$$\left|\sin x - \sin y\right| \le \left|x - y\right|$$

**46.** Suppose that in a race, horse A and horse B finished in a dead heat. Prove that their speeds were identical at some instant of the race.

**47.** In Problem 46, suppose that the two horses crossed the finish line together at the same speed. Show that they had the same acceleration at some instant.

**48.** Use the Mean Value Theorem to show that the graph of a concave up function $f$ is always above its tangent line, that is, show that

$$f(x) > f(c) + f'(c)(x - c), \qquad x \ne c$$

**49.** Prove that if $\left|f(y) - f(x)\right| \le M(y - x)^2$ for all $x$ and $y$, then $f$ is a constant function.

**50.** Give an example of a function and an interval $(a, b)$ with $a < b$ such that $f$ is continuous on $[a, b]$, differentiable on $(a, b)$, *not* differentiable on $[a, b]$, and has a tangent line at every point of $[a, b]$.

---

**Answers to Concepts Review:** **1.** Continuous; $(a, b)$; $f(b) - f(a) = f'(c)(b - a)$ **2.** $f'(0)$ does not exist. **3.** $F(x) = G(x) + C$ **4.** $x^4 + C$

---

## 4.9 CHAPTER REVIEW

### Concepts Test

Respond with true or false to each of the following assertions. Be prepared to justify your answer.

**1.** A continuous function defined on a closed interval must attain a maximum value on that interval.

**2.** If a differentiable function $f$ attains a maximum value at an interior point $c$ of its domain, then $f'(c) = 0$.

**3.** It is possible for a function to have an infinite number of critical points.

**4.** A continuous function that increases for all $x$ must be differentiable everywhere.

**5.** If $f(x) = 3x^6 + 4x^4 + 2x^2$, then the graph of $f$ is concave up on the whole real line.

**6.** If $f$ is an increasing differentiable function on an interval $I$, then $f'(x) > 0$ for all $x$ in $I$.

**7.** If $f'(x) > 0$ for all $x$ in $I$, then $f$ is increasing on $I$.

**8.** If $f''(c) = 0$, then $f$ has an inflection point at $(c, f(c))$.

**9.** A quadratic function has no inflection points.

**10.** If $f'(x) > 0$ for all $x$ in $[a, b]$, then $f$ attains its maximum value on $[a, b]$ at $b$.

**11.** We use the symbol $\infty$ to denote a number that is bigger than any other number.

**12.** The function $y = 2x^3 + x$ has no maximum or minimum values.

**13.** The function $y = 2x^3 + x + \tan x$ has no maximum or minimum values.

**14.** The graph of $y = \dfrac{x^2 - x - 6}{x - 3} = \dfrac{(x + 2)(x - 3)}{x - 3}$ has a vertical asymptote at $x = 3$.

**15.** The graph of $y = \dfrac{x^2 + 1}{1 - x^2}$ has a horizontal asymptote of $y = -1$.

**16.** The graph of $y = \dfrac{3x^2 + 2x + \sin x}{x}$ has an oblique asymptote of $y = 3x + 2$.

**17.** The function $f(x) = \sqrt{x}$ satisfies the hypotheses of the Mean Value Theorem on $[0, 2]$.

**18.** The function $f(x) = \left|x\right|$ satisfies the hypotheses of the Mean Value Theorem on $[-1, 1]$.

**19.** On the interval $[-1, 1]$, there will be just one point where the tangent line to $y = x^3$ is parallel to the secant line.

**20.** If $f'(x) = 0$ for all $x$ in $(a, b)$, then $f$ is constant on this interval.

**21.** If $f'(c) = f''(c) = 0$, then $f(c)$ is neither a maximum nor minimum value.

**22.** The graph of $y = \sin x$ has infinitely many points of inflection.

**23.** Among rectangles of fixed area $K$, the one with maximum perimeter is a square.

**24.** If the graph of a differentiable function has three $x$-intercepts, then it must have at least two points where the tangent line is horizontal.

**25.** The sum of two increasing functions is an increasing function.

**26.** The product of two increasing functions is an increasing function.

**27.** If $f'(0) = 0$ and $f''(x) > 0$ for $x \ge 0$, then $f$ is increasing on $[0, \infty)$.

**28.** If $f'(x) \leq 2$ for all $x$ on the interval $[0, 3]$ and $f(0) = 1$, then $f(3) < 4$.

**29.** If $f$ is a differentiable function, then $f$ is nondecreasing on $(a, b)$ if and only if $f'(x) \geq 0$ on $(a, b)$.

**30.** Two differentiable functions have the same derivative on $(a, b)$ if and only if they differ by a constant on $(a, b)$.

**31.** If $f''(x) > 0$ for all $x$, then the graph of $y = f(x)$ cannot have a horizontal asymptote.

## Sample Test Problems

In Problems 1–12, a function $f$ and its domain are given. Determine the critical points, evaluate $f$ at these points, and find the (global) maximum and minimum values.

**1.** $f(x) = x^2 - 2x$; $[0, 4]$

**2.** $f(t) = \dfrac{1}{t}$; $[1, 4]$

**3.** $f(z) = \dfrac{1}{z^2}$; $[-2, -\frac{1}{2}]$

**4.** $f(x) = \dfrac{1}{x^2}$; $[-2, 0)$

**5.** $f(x) = |x|$; $[-\frac{1}{2}, 1]$

**6.** $f(s) = s + |s|$; $[-1, 1]$

**7.** $f(x) = 3x^4 - 4x^3$; $[-2, 3]$

**8.** $f(u) = u^2(u - 2)^{1/3}$; $[-1, 3]$

**9.** $f(x) = 2x^5 - 5x^4 + 7$; $[-1, 3]$

**10.** $f(x) = (x - 1)^3 (x + 2)^2$; $[-2, 2]$

**11.** $f(\theta) = \sin \theta$; $[\pi/4, 4\pi/3]$

**12.** $f(\theta) = \sin^2 \theta - \sin \theta$; $[0, \pi]$

In Problems 13–19, a function $f$ is given with domain $\mathbb{R}$. Indicate where $f$ is increasing and where it is concave down.

**13.** $f(x) = 3x - x^2$

**14.** $f(x) = x^9$

**15.** $f(x) = x^3 - 3x + 3$

**16.** $f(x) = -2x^3 - 3x^2 + 12x + 1$

**17.** $f(x) = x^4 - 4x^5$

**18.** $f(x) = x^3 - \frac{6}{5}x^5$      **19.** $f(x) = x^3 - x^4$

**20.** Find where the function $g$, defined by $g(t) = t^3 + 1/t$, is increasing and where it is decreasing. Find the local extreme values of $g$. Find the point of inflection. Sketch the graph.

**21.** Find where the function $f$, defined by $f(x) = x^2(x - 4)$, is increasing and where it is decreasing. Find the local extreme values of $f$. Find the point of inflection. Sketch the graph.

**32.** A global maximum value is always a local maximum value.

**33.** A cubic function $f(x) = ax^3 + bx^2 + cx + d, a \neq 0$ can have at most one local maximum value on any open interval.

**34.** The linear function $f(x) = ax + b, a \neq 0$ has no minimum value on any open interval.

**22.** Find the maximum and minimum values, if they exist, of the function defined by

$$f(x) = \frac{4}{x^2 + 1} + 2$$

In Problems 23–30, sketch the graph of the given function $f$, labeling all extrema (local and global) and the inflection points and showing any asymptotes. Be sure to make use of $f'$ and $f''$.

**23.** $f(x) = x^4 - 2x$      **24.** $f(x) = (x^2 - 1)^2$

**25.** $f(x) = x\sqrt{x - 3}$      **26.** $f(x) = \dfrac{x - 2}{x - 3}$

**27.** $f(x) = 3x^4 - 4x^3$      **28.** $f(x) = \dfrac{x^2 - 1}{x}$

**29.** $f(x) = \dfrac{3x^2 - 1}{x}$      **30.** $f(x) = \dfrac{2}{(x + 1)^2}$

In Problems 31–36, sketch the graph of the given function $f$, in the region $(-\pi, \pi)$ unless otherwise indicated, labeling all extrema (local and global) and the inflection points and showing any asymptotes. Be sure to make use of $f'$ and $f''$.

**31.** $f(x) = \cos x - \sin x$      **32.** $f(x) = \sin x - \tan x$

**33.** $f(x) = x \tan x$; $(-\pi/2, \pi/2)$

**34.** $f(x) = 2x - \cot x$; $(0, \pi)$

**35.** $f(x) = \sin x - \sin^2 x$      **36.** $f(x) = 2\cos x - 2\sin x$

**37.** Sketch a possible graph of a function $F$ that has all the following properties:

(a) $F$ is everywhere continuous;
(b) $F(-2) = 3, F(2) = -1$;
(c) $F'(x) = 0$ for $x > 2$;
(d) $F''(x) < 0$ for $x < 2$.

**38.** Sketch a possible graph of a function $F$ that has all the following properties:

(a) $F$ is everywhere continuous;
(b) $F(-1) = 6, F(3) = -2$;
(c) $F'(x) < 0$ for $x < -1, F'(-1) = F'(3) = -2, F'(7) = 0$;
(d) $F''(x) < 0$ for $x < -1, F''(x) = 0$ for $-1 < x < 3$, $F''(x) > 0$ for $x > 3$.

**39.** Sketch a possible graph of a function $F$ that has the following properties:

(a) $F$ is everywhere continuous;

(b) $F$ has period $\pi$;

(c) $0 \le F(x) \le 2$, $F(0) = 0$, $F\left(\dfrac{\pi}{2}\right) = 2$;

(d) $F'(x) > 0$ for $0 < x < \dfrac{\pi}{2}$, $F'(x) < 0$ for $\dfrac{\pi}{2} < x$;

(e) $F''(x) < 0$ for $0 < x < \pi$.

**40.** A long sheet of metal, 16 inches wide, is to be turned up at both sides to make a horizontal gutter with vertical sides. How many inches should be turned up at each side for maximum carrying capacity?

**41.** A fence, 8 feet high, is parallel to the wall of a building and 1 foot from the building. What is the shortest plank that can go over the fence, from the level ground, to prop the wall?

**42.** A page of a book is to contain 27 square inches of print. If the margins at the top, bottom, and one side are 2 inches and the margin at the other side is 1 inch, what size page would use the least paper?

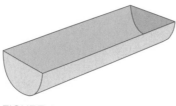

FIGURE 1

**43.** A metal water trough with equal semicircular ends and open top is to have a capacity of $128\pi$ cubic feet (Figure 1). Determine its radius $r$ and length $h$ if the trough is to require the least material for its construction.

**44.** Find the maximum and the minimum of the function defined on the closed interval $[-2, 2]$ by

$$f(x) = \begin{cases} \frac{1}{4}(x^2 + 6x + 8) & \text{if } -2 \le x \le 0 \\ -\frac{1}{6}(x^2 + 4x - 12) & \text{if } 0 \le x \le 2 \end{cases}$$

Find where the graph is concave up and where it is concave down. Sketch the graph.

**45.** For each of the following functions, decide whether the Mean Value Theorem applies on the indicated interval $I$. If so, find all possible values of $c$; if not, tell why. Make a sketch.

(a) $f(x) = \dfrac{x^3}{3}$; $I = [-3, 3]$

(b) $F(x) = x^{3/5} + 1$; $I = [-1, 1]$

(c) $g(x) = \dfrac{x + 1}{x - 1}$; $I = [2, 3]$

**46.** Find the equations of the tangent lines at the inflection points of the graph of

$$y = x^4 - 6x^3 + 12x^2 - 3x + 1$$

**47.** Determine each limit (possibly $\infty$ or $-\infty$) or state that it does not have a limit.

(a) $\displaystyle\lim_{x \to \infty} \dfrac{3x^2 - 2x + 7}{2x^2 + 5x + 9}$

(b) $\displaystyle\lim_{x \to \infty} \dfrac{3x + 9}{\sqrt{2x^2 + 1}}$

(c) $\displaystyle\lim_{x \to -\infty} \dfrac{\sin x}{x}$

(d) $\displaystyle\lim_{x \to -\infty} \cos x$

(e) $\displaystyle\lim_{x \to 3^+} \dfrac{x + 3}{x^2 - 9}$

(f) $\displaystyle\lim_{x \to 2} \dfrac{x + 2}{x^2 - 4}$

(g) $\displaystyle\lim_{x \to 1^-} \dfrac{|x - 1|}{x - 1}$

(h) $\displaystyle\lim_{x \to \infty} \cos\left(\dfrac{1}{x}\right)$

**48.** Sketch a possible graph of a function $G$ with all the following properties:

(a) $G(x)$ is continuous and $G''(x) > 0$ for all $x$ in $(-\infty, 0) \cup (0, \infty)$;

(b) $G(-2) = G(2) = 3$;

(c) $\displaystyle\lim_{x \to -\infty} G(x) = 2$, $\displaystyle\lim_{x \to \infty} [G(x) - x] = 0$;

(d) $\displaystyle\lim_{x \to 0^+} G(x) = \displaystyle\lim_{x \to 0^-} G(x) = \infty$.

**49.** Let $f$ be a continuous function with $f(1) = -1/4$, $f(2) = 0$, and $f(3) = -1/4$. If the graph of $y = f'(x)$ is as shown in Figure 12, sketch a possible graph for $y = f(x)$. Can you give an analytical representation of $f(x)$?

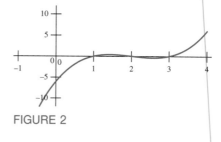

FIGURE 2

## 4.10 ADDITIONAL PROBLEMS

Inequalities are a very important topic in mathematics. In some cases one can use algebra to prove such inequalities, in other cases, calculus plays a very important role. We will explore the proof of a set of inequalities relating the *arithmetic mean* and *geometric mean*.

**1.** The *arithmetic mean* of two quantities is given by

$$\dfrac{a + b}{2}$$

and the *geometric mean* of two positive quantities is given by

$$\sqrt{ab}.$$

Show that the following inequality is true by squaring both sides of the expression

$$\sqrt{ab} \le \frac{a+b}{2}$$

**2.** Establish $\sqrt{ab} \le (a+b)/2$ for positive $a$ and $b$ by using the method of calculus. *Hint*: Squaring both sides and dividing by $b > 0$ gives the function $F(b) = \frac{(a+b)^2}{4b}$. If we can show that $F$ has its minimum at $a$ we are done.

(a) Show that for $b$ close to zero or very large $F(b)$ is not close to its hoped for minimum of $a$.
(b) By considering $F'(b)$ show that the minimum occurs in the interior when $b$ equals $a$, in which case $F(b)\big|_{b=a} = a$, thus proving the result.

For three positive numbers $a$, $b$, and $c$ the geometric mean is defined by $(abc)^{1/3}$ and the required inequality is given by

$$(abc)^{1/3} \le \frac{a+b+c}{3}.$$

We now construct a proof of this using calculus.

**3.** By considering the function

$$F(b) = \frac{1}{b}\left(\frac{a+b+c}{3}\right)^3$$

show using the calculus that the minimum occurs at $b = \frac{a+c}{2}$ and is given by $\left(\frac{a+b}{2}\right)^2$. Next using the result from Problem 2 conclude that

$$(abc)^{1/3} \le \frac{a+b+c}{3}.$$

**4.** Show that of all three dimensional boxes with a given surface area the cube has the greatest volume. *Hint*: The surface area $S = 2(lw + lh + hw)$, and the volume is $V = lwh$. Use Problem 3 to show that $(V^2)^{1/3} \le S/6$. When does equality hold?

**5.** Prove for any set of positive $x_j$ the inequality

$$(x_1 x_2 \ldots x_n)^{1/n} \le \frac{x_1 + x_2 + \ldots + x_n}{n}$$

holds with equality holding only in the case when all the $x_j$ are equal.

**6.** Show that for any $p > 1$ and $q > 1$ such that $\frac{1}{p} + \frac{1}{q} = 1$ that for positive $x$ and $y$

$$\frac{x}{p} + \frac{y}{q} \ge x^{1/p} y^{1/q}.$$

*Hint*: Consider $F(z) = r + (1-r)z - z^{1-r}$ where $z = y/x$ and $r = 1/p$. Show using calculus that $F$ has a minimum at $z = 1$.

**7.** Show using differential calculus that $f(x) = x - \sin x \ge 0$ for $x \ge 0$, with equality holding only at $x = 0$.

**8.** An important average in statistics is the root-mean-square. If $x$ and $y$ are non-negative their root-mean-square is

$$\sqrt{\frac{x^2+y^2}{2}}$$

Show that

$$\frac{x+y}{2} \le \sqrt{\frac{x^2+y^2}{2}}.$$

**9.** Show that for positive $x$ and $y$

$$\frac{2xy}{x+y} \le \frac{x+y}{2} \le \sqrt{\frac{x^2+y^2}{2}}.$$

**10.** The probability of an outcome is usually denoted by $p$ where $0 \le p \le 1$. The probability of the outcome not occurring is then $1 - p$. If $n$ attempts are made to get a given outcome, and a successful outcome does not depend on previous or future outcomes, then the probability of having $s$ successes in $n$ tries is given by

$$P_{ns} = \frac{n!}{s!(n-s)!} p^s (1-p)^{n-s},$$

where $n! = n(n-1)(n-2)\ldots 2 \cdot 1$. Using calculus find the value of $p$ which makes $P_{ns}$ a maximum.

**11.** Inscribe the rectangle of the greatest area in the ellipse

$$\frac{x^2}{a^2} + \frac{y^2}{b^2} = 1$$

**12.** Of all circular cylinders with a given surface area find the one with the maximum volume. *Note*: The ends of the cylinders are closed.

**13.** Of all rectangles with a given diagonal find the one with the maximum area.

# TECHNOLOGY PROJECT

**Law of Reflection.** The law of reflection states that the angle of incidence $\theta_1$ is equal to the angle of reflection $\theta_2$, that is, $\theta_1 = \theta_2$ or, equivalently, $\alpha = \beta$ (see Figure 1). Furthermore, if the speed in the medium is constant, then the reflection path from a point $P$ to a point $Q$ consists of two straight line segments meeting at the reflection point as in the Figure below.

**Exercise 1** Given two points $P$ and $Q$, we want to determine the location of the point of reflection on the interface (say $x$ measured horizontally from one of the points). As in Figure 1, assume that the two points are separated from each other by a total horizontal distance $L$ and that their vertical distances from the reflecting interface are respectively $a$ and $b$. Use the law of reflection to derive an expression for $x$ in terms of $a$, $b$, and $L$. **Check:** For $a = 50$, $b = 25$, and $L = 150$, the correct answer for $x$ is 100.

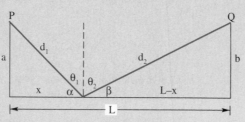

FIG. 1. Geometry of reflection.

**Snell's Law.** The law of refraction (or Snell's law) states that for an interface separating two media,

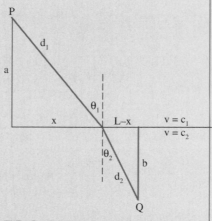

FIG. 2. Geometry of refraction.

the angle of incidence $\theta_1$ is related to the angle of refraction $\theta_2$ by the relation

$$\frac{\sin \theta_1}{c_1} = \frac{\sin \theta_2}{c_2}$$

(see Figure 2). Here $c_1$ and $c_2$ are the respective speeds above and below the interface and $\theta_1$ and $\theta_2$ are again the angles measured from the normal.

In contrast to the reflection case where the reflection point can be found explicitly, the location of the refracting point $x$ involves a fourth degree polynomial:

$$0 = (c_2^2 - c_1^2)\, x^4 - 2L\,(c_2^2 - c_1^2)\, x^3 + (L^2(c_2^2 - c_1^2) + c_2^2\, b^2 - c_1^2\, a^2)\, x^2 + 2Lc_1^2\, a^2 x - L^2 c_1^2\, a^2$$

**Exercise 2** Show that for the values $a = b = 1$, $L = 4$, $c_1 = 1$, and $c_2 = 1/2$, the above equation for the location of the refraction point reduces to

$$3x^4 - 24x^3 + 51x^2 - 32x + 64 = 0$$

This latter equation actually has *two* real roots:

(a) Hand in a graph that shows their approximate locations.

(b) Explain why only one of the roots is relevant to the refraction problem.

(c) Find the relevant zero to two significant figures, and explain how you know that your $x$ value is this accurate.

**Exercise 3** **A linear approximation to the *exact* zero of $f$.**

(a) Suppose you want $z$ such that $f(z) = 0$, and you have an $x_0$ that is close to $z$; but you want to get yet closer to $z$. Show that the tangent to the curve at $(z, f(z))$ intersects the $x$-axis at $x_1 = x_0 - f(x_0)/f'(x_0)$. This $x_1$ is called the linear (or Newton) approximation to the zero $z$. You will study this in more detail in Chap. 10. Hand in a sketch showing what's going on.

(b) Using your approximation in Exercise 2 as the $x_0$ value above, find the improved $x_1$ value as in part (a). Test the quality of these two values by evaluating $f(x_0)$ and $f(x_1)$.

# 5

# The Integral

**5.1
ANTIDERIVATIVES
(INDEFINITE INTEGRALS)**

I may put on my shoes and take them off again. The second operation undoes the first, restoring the shoes to their original position. We say the two operations are *inverse operations*. Mathematics has many pairs of inverse operations: addition and subtraction, multiplication and division, exponentiation and root taking, and taking logarithms and finding antilogarithms. We have been studying differentiation; its inverse is called *antidifferentiation*.

---

### Definition

We call $F$ an **antiderivative** of $f$ on the interval $I$ if $D_x F(x) = f(x)$ on $I$—that is, if $F'(x) = f(x)$ for all $x$ in $I$. (If $x$ is an endpoint of $I$, $F'(x)$ need be only a one-sided derivative.)

---

We used *an* antiderivative rather than *the* antiderivative in our definition. You will soon see why.

**EXAMPLE 1**   Find an antiderivative of the function $f(x) = 4x^3$ on $(-\infty, \infty)$.

***Solution***   We seek a function $F$ satisfying $F'(x) = 4x^3$ for all real $x$. From our experience with differentiation, we know that $F(x) = x^4$ is one such function.   ■

A moment's thought will suggest other solutions to Example 1. The function $F(x) = x^4 + 6$ also satisfies $F'(x) = 4x^3$; it too is an antiderivative of $f(x) = 4x^3$. In fact, $F(x) = x^4 + C$, where $C$ is any constant, is an antiderivative of $4x^3$ on $(-\infty, \infty)$ (see Figure 1).

Now we pose an important question. Is every antiderivative of $f(x) = 4x^3$ of the form $F(x) = x^4 + C$? The answer is yes. This follows from

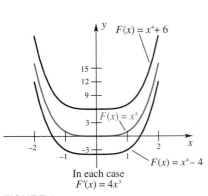

In each case
$F'(x) = 4x^3$

FIGURE 1

Theorem 4.8B, which says that if two functions have the same derivative they must differ by a constant.

Our conclusion is this. If a function $f$ has an antiderivative, it will have a whole family of them and each member of this family can be obtained from one of them by the addition of an appropriate constant. We call this family of functions the **general antiderivative** of $f$. After we get used to this notion, we will often omit the adjective *general*.

**EXAMPLE 2**   Find the general antiderivative of $f(x) = x^2$ on $(-\infty, \infty)$.

***Solution***   The function $F(x) = x^3$ will not do since its derivative is $3x^2$. But that suggests $F(x) = \frac{1}{3}x^3$, which satisfies $F'(x) = \frac{1}{3} \cdot 3x^2 = x^2$. However, the general antiderivative is $\frac{1}{3}x^3 + C$.   ∎

**Notation for Antiderivatives**   Since we used the symbol $D_x$ for the operation of taking a derivative, it would be natural to use $A_x$ for the operation of finding the antiderivative. Thus,

$$A_x(x^2) = \tfrac{1}{3}x^3 + C$$

This is the notation used by several authors and it was, in fact, used in earlier editions of this book. However, Leibniz's original notation continues to enjoy overwhelming popularity, and we therefore choose to follow him. Rather than $A_x$, Leibniz used the symbol $\int \ldots dx$. He wrote

$$\int x^2 \, dx = \tfrac{1}{3}x^3 + C$$

and

$$\int 4x^3 \, dx = x^4 + C$$

We will postpone explaining why Leibniz chose to use the elongated $s$, $\int$, and the $dx$ until later. For the moment, simply think of $\int \ldots dx$ as indicating the antiderivative with respect to $x$, just as $D_x$ indicates the derivative with respect to $x$. Note that $D_x \int f(x) \, dx = f(x)$.

---

### Theorem A

**(Power Rule).** If $r$ is any rational number except $-1$, then

$$\int x^r \, dx = \frac{x^{r+1}}{r+1} + C$$

---

***Proof***   To establish any result of the form

$$\int f(x) \, dx = F(x) + C$$

it is sufficient to show

$$D_x[F(x) + C] = f(x)$$

In our case,

$$D_x \left[ \frac{x^{r+1}}{r+1} + C \right] = \frac{1}{r+1}(r+1)x^r = x^r \qquad \blacksquare$$

We make two comments about Theorem A. First, it is meant to include the case $r = 0$, that is,

$$\int 1 \, dx = x + C$$

Second, since no interval $I$ is specified, the conclusion is understood to be valid only on intervals on which $x^r$ is defined. In particular, we must exclude any interval containing the origin if $r < 0$. A similar understanding holds in what follows.

**EXAMPLE 3**   Find the general antiderivative of $f(x) = x^{4/3}$.

*Solution*

$$\int x^{4/3} \, dx = \frac{x^{7/3}}{\frac{7}{3}} + C = \tfrac{3}{7}x^{7/3} + C$$

Note that *to antidifferentiate a power of $x$ we increase the exponent by 1 and divide by the new exponent.*   ∎

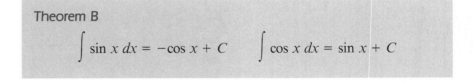

**Theorem B**

$$\int \sin x \, dx = -\cos x + C \qquad \int \cos x \, dx = \sin x + C$$

***Proof***   Simply note that $D_x(-\cos x) = \sin x$ and $D_x(\sin x) = \cos x$.   ∎

There is more to be said about notation. Following Leibniz, we shall often use the term **indefinite integral** in place of antiderivative. To antidifferentiate is also to **integrate**. In the symbol $\int f(x) \, dx$, $\int$ is called the **integral sign** and $f(x)$ is called the **integrand**. Thus, we integrate the integrand and thereby obtain the indefinite integral. Perhaps Leibniz used the adjective *indefinite* to suggest that the indefinite integral always involves an arbitrary constant.

**The Indefinite Integral is Linear**   Recall from Section 3.3 that $D_x$ is a linear operator. This means two things.

1. $D_x[kf(x)] = kD_xf(x)$
2. $D_x[f(x) + g(x)] = D_xf(x) + D_xg(x)$

From these two properties, a third follows automatically.

3. $D_x[f(x) - g(x)] = D_xf(x) - D_xg(x)$

What is true for derivatives is true for indefinite integrals (antiderivatives).

> ## Theorem C
>
> **(Linearity of $\int \ldots dx$).** Let $f$ and $g$ have antiderivatives (indefinite integrals) and let $k$ be a constant. Then:
> **(i)** $\int kf(x)\,dx = k \int f(x)\,dx$;
> **(ii)** $\int [f(x) + g(x)]\,dx = \int f(x)\,dx + \int g(x)\,dx$; and consequently
> **(iii)** $\int [f(x) - g(x)]\,dx = \int f(x)\,dx - \int g(x)\,dx$.

***Proof*** To show (i) and (ii), we simply differentiate the right side and observe that we get the integrand of the left side.

$$D_x \left[ k \int f(x)\,dx \right] = kD_x \int f(x)\,dx = kf(x)$$

$$D_x \left[ \int f(x)\,dx + \int g(x)\,dx \right] = D_x \int f(x)\,dx + D_x \int g(x)\,dx$$
$$= f(x) + g(x)$$

Property (iii) follows from (i) and (ii). ∎

**EXAMPLE 4** Find (a) $\int (3x^2 + 4x)\,dx$, (b) $\int (u^{3/2} - 3u + 14)\,du$, and c) $\int (1/t^2 + \sqrt{t})\,dt$ by using the linearity of $\int$.

***Solution***

(a)
$$\int (3x^2 + 4x)\,dx = \int 3x^2\,dx + \int 4x\,dx$$
$$= 3 \int x^2\,dx + 4 \int x\,dx$$
$$= 3 \left( \frac{x^3}{3} + C_1 \right) + 4 \left( \frac{x^2}{2} + C_2 \right)$$
$$= x^3 + 2x^2 + (3C_1 + 4C_2)$$
$$= x^3 + 2x^2 + C$$

Two arbitrary constants $C_1$ and $C_2$ appeared, but they were easily combined in one constant, $C$, a practice we consistently follow.

(b) Note the use of the variable $u$ rather than $x$. That is fine as long as the corresponding differential symbol is $du$, since we then have a complete change of notation.

$$\int (u^{3/2} - 3u + 14)\,du = \int u^{3/2}\,du - 3\int u\,du + 14\int 1\,du$$

$$= \tfrac{2}{5}u^{5/2} - \tfrac{3}{2}u^2 + 14u + C$$

(c) $\displaystyle \int \left(\frac{1}{t^2} + \sqrt{t}\right)dt = \int (t^{-2} + t^{1/2})\,dt = \int t^{-2}\,dt + \int t^{1/2}\,dt$

$$= \frac{t^{-1}}{-1} + \frac{t^{3/2}}{\tfrac{3}{2}} + C = -\frac{1}{t} + \frac{2}{3}t^{3/2} + C \qquad \blacksquare$$

**Generalized Power Rule**  Recall the Chain Rule as applied to a power of a function. If $u = g(x)$ is a differentiable function and $r$ is a rational number ($r \neq -1$), then

$$D_x\left[\frac{u^{r+1}}{r+1}\right] = u^r \cdot D_x u$$

or, in functional notation,

$$D_x\left(\frac{[g(x)]^{r+1}}{r+1}\right) = [g(x)]^r \cdot g'(x)$$

From this we obtain an important rule for indefinite integrals.

---

Theorem D

**(Generalized Power Rule).** Let $g$ be a differentiable function and $r$ a rational number different from $-1$. Then

$$\int [g(x)]^r g'(x)\,dx = \frac{[g(x)]^{r+1}}{r+1} + C$$

---

**EXAMPLE 5**  Find (a) $\int (x^4 + 3x)^{30}(4x^3 + 3)\,dx$ and (b) $\int \sin^{10}x \cos x\,dx$.

*Solution*
(a) Let $g(x) = x^4 + 3x$; then $g'(x) = 4x^3 + 3$. Thus, by Theorem D,

$$\int (x^4 + 3x)^{30}(4x^3 + 3)\,dx = \int [g(x)]^{30}g'(x)\,dx = \frac{[g(x)]^{31}}{31} + C$$

$$= \frac{(x^4 + 3x)^{31}}{31} + C$$

(b) Let $g(x) = \sin x$; then $g'(x) = \cos x$. Thus,

$$\int \sin^{10}x \cos x\,dx = \int [g(x)]^{10}g'(x)\,dx = \frac{[g(x)]^{11}}{11} + C$$

$$= \frac{\sin^{11}x}{11} + C \qquad \blacksquare$$

Now we can see why Leibniz used the differential $dx$ in his notation $\int \ldots dx$. If we let $u = g(x)$, then $du = g'(x)\,dx$. The conclusion of Theorem D is, therefore,

$$\int u^r\,du = \frac{u^{r+1}}{r+1} + C, \qquad r \neq -1$$

which is the ordinary power rule with $u$ as the variable. Thus, the generalized power rule is just the ordinary power rule applied to functions. But in applying it, we must always make sure that we have $du$ to go with $u^r$. The following examples illustrate what we mean.

**EXAMPLE 6**   Find (a) $\int (x^3 + 6x)^5(6x^2 + 12)\,dx$, (b) $\int (x^2 + 4)^{10}x\,dx$, and (c) $\int (x^2/2 + 3)^2 x^2\,dx$.

*Solution*

(a) Let $u = x^3 + 6x$; then $du = (3x^2 + 6)\,dx$. Thus, $(6x^2 + 12)\,dx = 2(3x^2 + 6)\,dx = 2\,du$, and so

$$\int (x^3 + 6x)^5(6x^2 + 12)\,dx = \int u^5 2\,du$$

$$= 2\int u^5\,du$$

$$= 2\left[\frac{u^6}{6} + C\right]$$

$$= \frac{u^6}{3} + 2C$$

$$= \frac{(x^3 + 6x)^6}{3} + K$$

Two things should be noted about our solution. First, the fact that $(6x^2 + 12)\,dx$ is $2\,du$ instead of $du$ caused no trouble; the factor 2 could be moved in front of the integral sign by linearity. Second, we wound up with an arbitrary constant of $2C$. This is still an arbitrary constant; we called it $K$.

(b) Let $u = x^2 + 4$; then $du = 2x\,dx$. Thus,

$$\int (x^2 + 4)^{10}x\,dx = \int (x^2 + 4)^{10} \cdot \frac{1}{2} \cdot 2x\,dx$$

$$= \frac{1}{2}\int u^{10}\,du$$

$$= \frac{1}{2}\left(\frac{u^{11}}{11} + C\right)$$

$$= \frac{(x^2 + 4)^{11}}{22} + K$$

(c) Let $u = x^2/2 + 3$; then $du = x\,dx$. The method illustrated in (a) and (b) fails because $x^2\,dx = x(x\,dx) = x\,du$, and the $x$ cannot be passed in front of the integral sign (that can be done only with a constant factor). However, we can expand the integrand by ordinary algebra and then use the Power Rule.

$$\int \left(\frac{x^2}{2} + 3\right)^2 x^2\,dx = \int \left(\frac{x^4}{4} + 3x^2 + 9\right) x^2\,dx$$

$$= \int \left(\frac{x^6}{4} + 3x^4 + 9x^2\right) dx$$

$$= \frac{x^7}{28} + \frac{3x^5}{5} + 3x^3 + C \qquad \blacksquare$$

## CONCEPTS REVIEW

**1.** The Power Rule for derivatives says that $d(x^r)/dx = $ _____. The Power Rule for integrals says that $\int x^r\,dx = $ _____.

**2.** The Generalized Power Rule for derivatives says that $d[f(x)]^r/dx = $ _____. The Generalized Power

Rule for integrals says that $\int$ _____ $dx = [f(x)]^{r+1}/(r+1) + C, r \neq -1$.

**3.** $\int (x^4 + 3x^2 + 1)^8(4x^3 + 6x)\,dx = $ _____

**4.** By linearity, $\int [c_1 f(x) + c_2 g(x)]\,dx = $ _____.

## PROBLEM SET 5.1

Find the general antiderivative $F(x) + C$ for each of the following.

**1.** $f(x) = 5$

**2.** $f(x) = x - 4$

**3.** $f(x) = x^2 + \pi$

**4.** $f(x) = 3x^2 + \sqrt{3}$

**5.** $f(x) = x^{5/4}$

**6.** $f(x) = 3x^{2/3}$

**7.** $f(x) = 1/\sqrt[3]{x^2}$

**8.** $f(x) = 7x^{-3/4}$

**9.** $f(x) = x^2 - x$

**10.** $f(x) = 3x^2 - \pi x$

**11.** $f(x) = 4x^5 - x^3$

**12.** $f(x) = x^{100} + x^{99}$

**13.** $f(x) = 27x^7 + 3x^5 - 45x^3 + \sqrt{2}x$

**14.** $f(x) = x^2(x^3 + 5x^2 - 3x + \sqrt{3})$

**15.** $f(x) = \dfrac{3}{x^2} - \dfrac{2}{x^3}$

**16.** $f(x) = \dfrac{\sqrt{2x}}{x} + \dfrac{3}{x^5}$

**17.** $f(x) = \dfrac{4x^6 + 3x^4}{x^3}$

**18.** $f(x) = \dfrac{x^6 - x}{x^3}$

In Problems 19–25, find the indicated integrals.

**19.** $\displaystyle\int (x^2 + x)\,dx$

**20.** $\displaystyle\int (x^3 + \sqrt{x})\,dx$

**21.** $\displaystyle\int (x + 1)^2\,dx$

**22.** $\displaystyle\int (z + \sqrt{2z})^2\,dz$

**23.** $\displaystyle\int \frac{(z^2 + 1)^2}{\sqrt{z}}\,dz$

**24.** $\displaystyle\int \frac{s(s + 1)^2}{\sqrt{s}}\,ds$

**25.** $\displaystyle\int (\sin\theta - \cos\theta)\,d\theta$

**26.** $\displaystyle\int (t^2 - 2\cos t)\,dt$

In Problems 27–32, use the methods of Examples 5 and 6 to find the indefinite integrals.

**27.** $\displaystyle\int (\sqrt{2}x + 1)^3\sqrt{2}\,dx$

**28.** $\displaystyle\int (\pi x^3 + 1)^4 3\pi x^2\,dx$

**29.** $\displaystyle\int (5x^2 + 1)(5x^3 + 3x - 8)^6\,dx$

**30.** $\displaystyle\int (5x^2 + 1)\sqrt{5x^3 + 3x - 2}\,dx$

**31.** $\displaystyle\int 3t\sqrt[3]{2t^2 - 11}\,dt$

**32.** $\displaystyle\int \frac{3y}{\sqrt{2y^2 + 5}}\,dy$

In Problems 33–38, $f''(x)$ is given. Find $f(x)$ by antidifferentiating twice. Note that in this case your answer should involve two arbitrary constants, one from each antidifferentiation. For example, if $f''(x) = x$, then $f'(x) = x^2/2 + C_1$ and $f(x) = x^3/6 + C_1 x + C_2$. The constants $C_1$ and $C_2$ cannot be combined.

**33.** $f''(x) = 3x + 1$

**34.** $f''(x) = -2x + 3$

**35.** $f''(x) = \sqrt{x}$

**36.** $f''(x) = x^{4/3}$

**37.** $f''(x) = \dfrac{x^4 + 1}{x^3}$

**38.** $f''(x) = 2\sqrt[3]{x + 1}$

**39.** Prove the formula

$$\int [f(x)g'(x) + g(x)f'(x)]\, dx = f(x)g(x) + C$$

**40.** Prove the formula

$$\int \frac{g(x)f'(x) - f(x)g'(x)}{g^2(x)}\, dx = \frac{f(x)}{g(x)} + C$$

**41.** Use the formula from Problem 39 to find

$$\int \left[ \frac{x^2}{2\sqrt{x-1}} + 2x\sqrt{x-1} \right] dx$$

**42.** Use the formula from Problem 39 to find

$$\int \left[ \frac{-x^3}{(2x+5)^{3/2}} + \frac{3x^2}{\sqrt{2x+5}} \right] dx$$

**43.** Prove the formula

$$\int \frac{x^4 + 1}{x^2\sqrt{x^4 - 1}}\, dx = \frac{\sqrt{x^4 - 1}}{x} + C$$

**44.** Prove the formula

$$\int \frac{1}{\cos^2 3x}\, dx = \frac{1}{3}\frac{\sin 3x}{\cos 3x} + C$$

**45.** Find $\int f''(x)\, dx$ if $f(x) = x\sqrt{x^3 + 1}$.

**46.** Prove the formula

$$\int \frac{2g(x)f'(x) - f(x)g'(x)}{2[g(x)]^{3/2}} = \frac{f(x)}{\sqrt{g(x)}} + C$$

**47.** Prove the formula

$$\int f^{m-1}(x)g^{n-1}(x)[nf(x)g'(x) + mg(x)f'(x)]\, dx$$

$$= f^m(x)g^n(x) + C$$

**48.** Find the indefinite integral

$$\int \sin^3[(x^2 + 1)^4]\cos[(x^2 + 1)^4](x^2 + 1)^3 x\, dx$$

*Hint:* Let $u = \sin[(x^2 + 1)^4]$.

**49.** Find $\displaystyle\int |x|\, dx$

**50.** Find $\displaystyle\int \sin^2 x\, dx$

PC **51.** Some software packages (including *True BASIC Calculus*, Version 4.0) can do a limited number of indefinite integrations. Experiment with your software on earlier problems in this set. Then find each of the following.

(a) $\displaystyle\int 6\sin(3(x-2))\, dx$    (b) $\displaystyle\int \sin^3(x/6)\, dx$

(c) $\displaystyle\int (x^2\cos 2x + x\sin 2x)\, dx$

PC **52.** Let $F_0(x) = x\sin x$ and $F_{n+1}(x) = \displaystyle\int F_n(x)\, dx$.

(a) Determine $F_1(x)$, $F_2(x)$, $F_3(x)$, and $F_4(x)$.
(b) On the basis of part (a), conjecture the form of $F_{16}(x)$.

---

**Answers to Concepts Review:**  **1.** $rx^{r-1}$; $x^{r+1}/(r+1) + C$, $r \neq -1$  **2.** $r[f(x)]^{r-1}f'(x)$; $[f(x)]^r f'(x)$
**3.** $(x^4 + 3x^2 + 1)^9/9 + C$  **4.** $c_1 \int f(x)\, dx + c_2 \int g(x)\, dx$

**5.2**
**INTRODUCTION TO**
**DIFFERENTIAL**
**EQUATIONS**

In the previous section, our task was to integrate (antidifferentiate) a function $f$ to obtain a new function $F$. We wrote

$$\int f(x)\, dx = F(x) + C$$

and this was correct, provided $F'(x) = f(x)$. Now $F'(x) = f(x)$ in the language of derivatives is equivalent to $dF(x) = f(x)\, dx$ in differential language (Section 3.10). Thus, we may look on the boxed formula as saying

$$\int dF(x) = F(x) + C$$

From this perspective, we integrate the differential of a function to obtain the function (plus a constant). This was Leibniz's viewpoint; adopting it will help us solve *differential equations.*

**What Is a Differential Equation?**   To motivate our answer, we begin with a simple example.

EXAMPLE 1   Find the *xy*-equation of the curve that passes through $(-1, 2)$ and whose slope at any point on the curve is equal to twice the abscissa (*x*-coordinate) of that point.

*Solution*   The condition that must hold at each point $(x, y)$ on the curve is

$$\frac{dy}{dx} = 2x$$

We are looking for a function $y = f(x)$ that satisfies this equation and the additional condition that $y = 2$ when $x = -1$. We suggest two ways of looking at this problem.

METHOD 1   When an equation has the form $dy/dx = g(x)$, we observe that $y$ must be an antiderivative of $g(x)$, that is,

$$y = \int g(x)\, dx$$

In our case,

$$y = \int 2x\, dx = x^2 + C$$

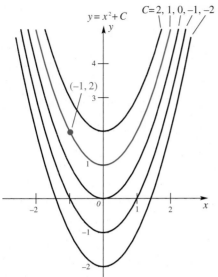

$y = x^2 + C$

$C = 2, 1, 0, -1, -2$

$(-1, 2)$

FIGURE 1

METHOD 2 Think of $dy/dx$ as a quotient of two differentials. When we multiply both sides of $dy/dx = 2x$ by $dx$, we get

$$dy = 2x\,dx$$

Next we integrate both sides, equate the results, and simplify.

$$\int dy = \int 2x\,dx$$

$$y + C_1 = x^2 + C_2$$

$$y = x^2 + C_2 - C_1$$

$$y = x^2 + C$$

The second method works in a wide variety of problems that are not of the simple form $dy/dx = g(x)$, as we shall see.

The solution $y = x^2 + C$ represents the family of curves illustrated in Figure 1. From this family, we must choose the one for which $y = 2$ when $x = -1$; thus, we want

$$2 = (-1)^2 + C$$

We conclude that $C = 1$ and therefore that $y = x^2 + 1$. ∎

The equation $dy/dx = 2x$ is called a *differential equation*. Other examples are

$$\frac{dy}{dx} = 2xy + \sin x$$

$$y\,dy = (x^3 + 1)\,dx$$

$$\frac{d^2y}{dx^2} + 3\frac{dy}{dx} - 2xy = 0$$

Any equation in which the unknown is a function and which involves derivatives (or differentials) of this unknown function is called a **differential equation**. To solve a differential equation is to find the unknown function. In general, this is a difficult job and one about which many thick books have been written. Here we consider only the simplest case, namely, **first-order separable** differential equations. These are equations involving just the first derivative of the unknown function and are such that the variables can be separated.

**Separation of Variables** Consider the differential equation

$$\frac{dy}{dx} = \frac{x + 3x^2}{y^2}$$

If we multiply both sides by $y^2\,dx$, we obtain

$$y^2\,dy = (x + 3x^2)\,dx$$

In this form, the differential equation has its variables separated—that is, the $y$ terms are on one side of the equation and the $x$ terms are on the other. In separated form, we can solve the differential equation using Method 2 (integrate both sides, equate the results, simplify), as we now illustrate.

**EXAMPLE 2**    Solve the differential equation

$$\frac{dy}{dx} = \frac{x + 3x^2}{y^2}$$

Then find that solution for which $y = 6$ when $x = 0$.

*Solution*    As noted earlier, the given equation is equivalent to

$$y^2\,dy = (x + 3x^2)\,dx$$

Thus,

$$\int y^2\,dy = \int (x + 3x^2)\,dx$$

$$\frac{y^3}{3} + C_1 = \frac{x^2}{2} + x^3 + C_2$$

$$y^3 = \frac{3x^2}{2} + 3x^3 + (3C_2 - 3C_1)$$

$$= \frac{3x^2}{2} + 3x^3 + C$$

$$y = \sqrt[3]{\frac{3x^2}{2} + 3x^3 + C}$$

To evaluate the constant $C$, we use the condition $y = 6$ when $x = 0$. This gives

$$6 = \sqrt[3]{C}$$

$$216 = C$$

Thus,

$$y = \sqrt[3]{\frac{3x^2}{2} + 3x^3 + 216}$$

The ultimate check on our work is to substitute this result in both sides of the original differential equation to see that this gives an equality.

Substituting in the left side, we get

$$\frac{dy}{dx} = \frac{1}{3}\left(\frac{3x^2}{2} + 3x^3 + 216\right)^{-2/3}(3x + 9x^2)$$

$$= \frac{x + 3x^2}{(\frac{3}{2}x^2 + 3x^3 + 216)^{2/3}}$$

On the right side, we get

$$\frac{x + 3x^2}{y^2} = \frac{x + 3x^2}{(\frac{3}{2}x^2 + 3x^3 + 216)^{2/3}}$$

As expected, the two expressions are equal. ∎

**Motion Problems**   Recall that if $s(t)$, $v(t)$, and $a(t)$ represent the position, velocity, and acceleration, respectively, at time $t$ of an object moving along a coordinate line, then

$$v(t) = s'(t) = \frac{ds}{dt}$$

$$a(t) = v'(t) = \frac{dv}{dt} = \frac{d^2s}{dt^2}$$

In earlier work (Section 3.7), we assumed that $s(t)$ was known, and from this we calculated $v(t)$ and $a(t)$. Now we want to consider the reverse process: given acceleration $a(t)$, find velocity $v(t)$ and position $s(t)$.

**EXAMPLE 3   (Falling-body problem).** Near the surface of the earth, the acceleration of a falling body due to gravity is 32 feet per second per second, provided we assume that air resistance can be neglected. If an object is thrown upward from an initial height of 1000 feet (Figure 2) with a velocity of 50 feet per second, find its velocity and height 4 seconds later.

*Solution*   Let us assume that the height $s$ is measured positively in the upward direction. Then $v = ds/dt$ is initially positive ($s$ is increasing) but $a = dv/dt$ is negative (the pull of gravity tends to decrease $v$). Thus, our starting point is the differential equation $dv/dt = -32$, with the additional conditions that $v = 50$ and $s = 1000$ when $t = 0$. Either Method 1 (direct antidifferentiation) or Method 2 (separation of variables) works well.

$$\frac{dv}{dt} = -32$$

$$v = \int -32\, dt = -32t + C$$

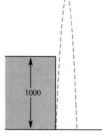

FIGURE 2

Since $v = 50$ at $t = 0$, we find that $C = 50$ and so

$$v = -32t + 50$$

Now $v = ds/dt$, and so we have another differential equation,

$$\frac{ds}{dt} = -32t + 50$$

When we integrate, we obtain

$$s = \int (-32t + 50)\, dt$$

$$s = -16t^2 + 50t + K$$

Since $s = 1000$ at $t = 0$, $K = 1000$ and

$$s = -16t^2 + 50t + 1000$$

Finally at $t = 4$,

$$v = -32(4) + 50 = -78 \text{ feet per second}$$
$$s = -16(4)^2 + 50(4) + 1000 = 944 \text{ feet} \qquad \blacksquare$$

We remark that if $v = v_0$ and $s = s_0$ at $t = 0$, the procedure of Example 3 leads to the well-known falling-body formulas:

$$a = -32$$
$$v = -32t + v_0$$
$$s = -16t^2 + v_0 t + s_0$$

**EXAMPLE 4**  The acceleration of an object moving along a coordinate line is given by $a(t) = (2t + 3)^{-3}$ in meters per second per second. If the velocity at $t = 0$ is 4 meters per second, find the velocity 2 seconds later.

*Solution*  To perform the integration in the second line of the solution, we multiply and divide by 2, thus preparing the integral for the Power Rule.

$$\frac{dv}{dt} = (2t + 3)^{-3}$$

$$v = \int (2t + 3)^{-3}\, dt = \frac{1}{2} \int \underbrace{(2t + 3)}_{u}{}^{-3}\underbrace{2\, dt}_{du}$$

$$= \frac{1}{2}\frac{(2t + 3)^{-2}}{-2} + C = -\frac{1}{4(2t + 3)^2} + C$$

Since $v = 4$ at $t = 0$,

$$4 = -\frac{1}{4(3)^2} + C$$

which gives $C = \frac{145}{36}$. Thus,

$$v = -\frac{1}{4(2t + 3)^2} + \frac{145}{36}$$

At $t = 2$,

$$v = -\frac{1}{4(49)} + \frac{145}{36} \approx 4.023 \text{ meters per second} \quad \blacksquare$$

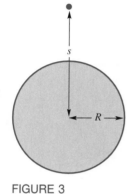

**FIGURE 3**

**EXAMPLE 5 (optional)   (Escape velocity).** The gravitational attraction $F$ exerted by the earth on an object of mass $m$ at a distance $s$ from the center of the earth is given by $F = -mgR^2/s^2$, where $-g$ ($g \approx 32$ feet per second per second) is the acceleration of gravity at the surface of the earth and $R$ ($R \approx 3960$ miles) is the radius of the earth (Figure 3). Show that an object projected outward from the earth with an initial velocity $v_0 \geq \sqrt{2gR}$ $\approx 6.93$ miles per second will not fall back to the earth. Neglect air resistance in making this calculation.

**Solution**   According to Newton's Second Law, $F = ma$; that is,

$$F = m\frac{dv}{dt} = m\frac{dv}{ds}\frac{ds}{dt} = m\frac{dv}{ds}v$$

Thus,

$$mv\frac{dv}{ds} = -mg\frac{R^2}{s^2}$$

Separating variables gives

$$v\,dv = -gR^2s^{-2}\,ds$$

$$\int v\,dv = -gR^2\int s^{-2}\,ds$$

$$\frac{v^2}{2} = \frac{gR^2}{s} + C$$

Now $v = v_0$ when $s = R$, and so $C = \frac{1}{2}v_0^2 - gR$. Consequently

$$v^2 = \frac{2gR^2}{s} + v_0^2 - 2gR$$

Finally, since $2gR^2/s$ gets small with increasing $s$, we see that $v$ remains positive if and only if $v_0 \geq \sqrt{2gR}$. $\quad \blacksquare$

# CONCEPTS REVIEW

**1.** $dy/dx = 3x^2 + 1$ and $dy/dx = x/y^2$ are examples of what is called a _____.

**2.** To solve the differential equation $dy/dx = g(x, y)$ is to find the _____ that, when substituted for $y$, makes the given equation true.

**3.** To solve the differential equation $dy/dx = x^2y^3$, the first step would be to _____.

**4.** To solve a falling-body problem near the surface of the earth, we start with the experimental fact that the acceleration $a$ of gravity is $-32$ feet per second per second—that is, $a = dv/dt = -32$. Solving this differential equation gives $v = ds/dt = $ _____, and solving the resulting differential equation gives $s = $ _____.

# PROBLEM SET 5.2

In Problems 1–4, show that the indicated function is a solution of the given differential equation; that is, substitute the indicated function for $y$ to see that it produces an equality.

**1.** $\dfrac{dy}{dx} + \dfrac{x}{y} = 0;\ y = \sqrt{1 - x^2}$

**2.** $-x\dfrac{dy}{dx} + y = 0;\ y = Cx$

**3.** $\dfrac{d^2y}{dx^2} + y = 0;\ y = C_1 \sin x + C_2 \cos x$

**4.** $\left(\dfrac{dy}{dx}\right)^2 + y^2 = 1;\ y = \sin(x + C)$ and $y = \pm 1$

In Problems 5–14, first find the general solution (involving a constant $C$) for the given differential equation. Then find the particular solution that satisfies the indicated condition. (See Example 2.)

**5.** $\dfrac{dy}{dx} = x^2 + 1;\ y = 1$ at $x = 1$

**6.** $\dfrac{dy}{dx} = x^{-3} + 2;\ y = 3$ at $x = 1$

**7.** $\dfrac{dy}{dx} = \dfrac{x}{y};\ y = 1$ at $x = 1$

**8.** $\dfrac{dy}{dx} = \sqrt{\dfrac{x}{y}};\ y = 4$ at $x = 1$

**9.** $\dfrac{dz}{dt} = t^2z^2;\ z = 1/3$ at $t = 1$

**10.** $\dfrac{dy}{dt} = y^4;\ y = 1$ at $t = 0$

**11.** $\dfrac{ds}{dt} = 16t^2 + 4t - 1;\ s = 100$ at $t = 0$

**12.** $\dfrac{du}{dt} = u^3(t^3 - t);\ u = 4$ at $t = 0$

**13.** $\dfrac{dy}{dx} = (2x + 1)^4;\ y = 6$ at $x = 0$

**14.** $\dfrac{dy}{dx} = -y^2x(x^2 + 2)^4;\ y = 1$ at $x = 0$

**15.** Find the $xy$-equation of the curve through $(1, 2)$ whose slope at any point is three times its $x$-coordinate (see Example 1).

**16.** Find the $xy$-equation of the curve through $(1, 2)$ whose slope at any point is three times the square of its $y$-coordinate.

In Problems 17–20, an object is moving along a coordinate line subject to the indicated acceleration $a$ (in centimeters per second per second) with the initial velocity $v_0$ (in centimeters per second) and directed distance $s_0$ (in centimeters). Find both the velocity $v$ and directed distance $s$ after 2 seconds (see Example 4).

**17.** $a = t;\ v_0 = 3,\ s_0 = 0$

**18.** $a = (1 + t)^{-4};\ v_0 = 0,\ s_0 = 10$

C **19.** $a = \sqrt[3]{2t + 1};\ v_0 = 0,\ s_0 = 10$

C **20.** $a = (3t + 1)^{-3};\ v_0 = 4,\ s_0 = 0$

**21.** A ball is thrown upward from the surface of the earth with an initial velocity of 96 feet per second. What is the maximum height it reaches? (See Example 3.)

**22.** A ball is thrown upward from the surface of a planet where the acceleration of gravity is $k$ (a negative constant) feet per second per second. If the initial velocity is $v_0$, show that the maximum height is $-v_0^2/2k$.

C **23.** On the surface of the moon, the acceleration of gravity is $-5.28$ feet per second per second. If an object is thrown upward from an initial height of 1000 feet with a velocity of 56 feet per second, find its velocity and height 4.5 seconds later.

C **24.** What is the maximum height that the object of Problem 23 reaches?

**25.** The rate of change of volume $V$ of a melting snowball is proportional to the surface area $S$ of the ball—that is, $dV/dt = -kS$, where $k$ is a positive constant. If the radius of the ball at $t = 0$ is $r = 2$ and at $t = 10$ is $r = 0.5$, show that $r = -\frac{3}{20}t + 2$.

**26.** From what height above the earth must a ball be dropped in order to strike the ground with a velocity of $-136$ feet per second?

C **27.** Determine the escape velocity for each of the following celestial bodies (see Example 5). Here $g \approx 32$ feet per second per second.

|        | Acceleration of Gravity | Radius (miles) |
|--------|-------------------------|----------------|
| Moon   | $-0.165g$               | 1,080          |
| Venus  | $-0.85g$                | 3,800          |
| Jupiter| $-2.6g$                 | 43,000         |
| Sun    | $-28g$                  | 432,000        |

**28.** If the brakes of a car, when fully applied, produce a constant deceleration of 11 feet per second per second, what is the shortest distance in which the car can be braked to a halt from a speed of 60 miles per hour?

**29.** What constant acceleration will cause a car to increase its velocity from 45 to 60 miles per hour in 10 seconds?

**30.** A block slides down an inclined plane with a constant acceleration of 8 feet per second per second. If the inclined plane is 75 feet long and the block reaches the bottom in 3.75 seconds, what was the initial velocity of the block?

**31.** A certain rocket shot straight up has an acceleration of $6t$ meters per second per second during the first 10 seconds after blast-off, after which the engine cuts out and the rocket is subject to gravitational acceleration of $-10$ meters per second per second. How high will the rocket go?

**32.** Starting at station A, a commuter train accelerates at 3 meters per second per second for 8 seconds, then travels at constant speed $v_m$ for 100 seconds, and finally brakes (decelerates) to a stop at station B at 4 meters per second per second. Find (a) $v_m$ and (b) the distance between A and B.

**33.** Starting from rest, a bus increases speed at constant acceleration $a_1$, then travels at constant speed $v_m$, and finally brakes to a stop at constant deceleration $a_2$. It took 4 minutes to travel the 2 miles between stop C and stop D and then 3 minutes to go the 1.4 miles between stop D and stop E.

(a) Sketch the graph of the speed $v$ as a function of time $t$, $0 \le t \le 7$.
(b) Find the maximum speed $v_m$.
(c) If $a_1 = a_2 = a$, evaluate $a$.

**34.** A hot-air balloon left the ground rising at 4 feet per second. Sixteen seconds later, Helena threw a ball straight up to her friend Janet in the balloon. At what speed did she throw the ball if it just made it to Janet?

**35.** According to Torricelli's Law, the time rate of change of the volume $V$ of water in a draining tank is proportional to the square root of the water's depth. A cylindrical tank of radius $10/\sqrt{\pi}$ centimeters and height 16 centimeters, which was full initially, took 40 seconds to drain.

(a) Write the differential equation for $V$ at time $t$ and the two corresponding conditions.
(b) Solve the differential equation.
(c) Find the volume of water after 10 seconds.

C **36.** The wolf population P in a certain state has been growing at a rate proportional to the cube root of the population size. The population was estimated at 1000 in 1970 and at 1700 in 1980.

(a) Write the differential equation for P at time $t$ with the two corresponding conditions.
(b) Solve the differential equation.
(c) When will the population reach 4000?

**37.** At $t = 0$, a ball was dropped from a height of 16 feet. It hit the floor and rebounded to a height of 9 feet (Figure 4).

(a) Find a two-part formula for the velocity $v(t)$, that is valid until the ball hits the floor a second time.
(b) At what two times was the ball at height 9 feet?

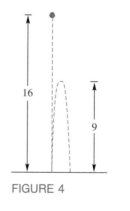

FIGURE 4

---

**Answers to Concepts Review:** **1.** Differential equation **2.** Function **3.** Separate variables **4.** $-32t + v_0$; $-16t^2 + v_0 t + s_0$

## 5.3
## SUMS AND SIGMA NOTATION

So far we have considered functions with domains that are intervals of real numbers. A typical example is $f(x) = x^2$, with—for example—domain the interval $[0, \infty)$. Its graph is shown in Figure 1. We used $x$ as the domain variable, but we could as well have used $s$, $t$, $u$, or $v$. By custom, mathematicians use letters near the end of the alphabet to name variables that take values on an interval of the real line.

When we want to name a variable that takes only integer values, we typically use letters near the middle of the alphabet, such as $i$, $j$, $k$, $m$, and (especially) $n$. Thus, in this section we wish to consider the function determined by $a(n) = n^2$, where $n$ takes positive integer values; its graph is shown in Figure 2. A function whose domain consists of just the positive integers (or some other subset of the integers) is called a **sequence**. In place of the standard functional notation $a(n)$, it is conventional to use $a_n$. Thus, we may say: Consider the sequence $\{a_n\}$ determined by $a_n = n^2$ and the sequence $\{b_n\}$ determined by $b_n = 1/n$. Sometimes we indicate a sequence by writing its first few values followed by dots as, for example,

$$a_1, a_2, a_3, a_4, \ldots$$

or even

$$1, 4, 9, 16, \ldots$$

Sequences will be studied in detail in Chapter 11. Here our main interest is in introducing notation for certain of their sums.

**Sigma Notation**   Consider the sums

$$1^2 + 2^2 + 3^2 + 4^2 + \cdots + 100^2$$

and

$$a_1 + a_2 + a_3 + a_4 + \cdots + a_n$$

To indicate these sums in a compact way, we write the first as

$$\sum_{i=1}^{100} i^2$$

and the second as

$$\sum_{i=1}^{n} a_i$$

Here $\sum$ (capital Greek sigma), which corresponds to our $S$, suggests that we are to sum (add) all numbers of the form indicated as the *index i* runs through the positive integers, starting with the integer shown below the $\sum$

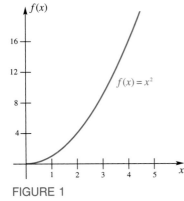

FIGURE 1

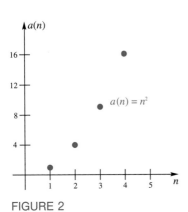

FIGURE 2

and ending with the integer at the top. Thus,

$$\sum_{i=2}^{5} b_i = b_2 + b_3 + b_4 + b_5$$

$$\sum_{j=1}^{n} \frac{1}{j} = \frac{1}{1} + \frac{1}{2} + \frac{1}{3} + \cdots + \frac{1}{n}$$

$$\sum_{k=1}^{4} \frac{k}{k^2 + 1} = \frac{1}{1^2 + 1} + \frac{2}{2^2 + 1} + \frac{3}{3^2 + 1} + \frac{4}{4^2 + 1}$$

and, for $n \geq m$,

$$\sum_{i=m}^{n} F(i) = F(m) + F(m + 1) + F(m + 2) + \cdots + F(n)$$

If all the $c_i$ in $\sum_{i=1}^{n} c_i$ have the same value, say $c$, then

$$\sum_{i=1}^{n} c_i = \underbrace{c + c + c + \cdots + c}_{n \text{ terms}} = nc$$

As a result, we adopt the convention

$$\boxed{\sum_{i=1}^{n} c = nc}$$

In particular,

$$\sum_{i=1}^{5} 2 = 5(2) = 10, \qquad \sum_{i=1}^{100} (-4) = 100(-4) = -400$$

**Properties of $\sum$**    Thought of as an operator, $\sum$ operates on sequences, and it does so in a linear way.

Theorem A

**(Linearity of $\sum$).** Let $\{a_i\}$ and $\{b_i\}$ denote two sequences and let $c$ be a constant. Then:

**(i)** $\displaystyle\sum_{i=1}^{n} ca_i = c \sum_{i=1}^{n} a_i;$

**(ii)** $\displaystyle\sum_{i=1}^{n} (a_i + b_i) = \sum_{i=1}^{n} a_i + \sum_{i=1}^{n} b_i;$ and consequently

**(iii)** $\displaystyle\sum_{i=1}^{n} (a_i - b_i) = \sum_{i=1}^{n} a_i - \sum_{i=1}^{n} b_i.$

**Proof**   The proofs are easy; we consider only (i).

$$\sum_{i=1}^{n} ca_i = ca_1 + ca_2 + \cdots + ca_n = c(a_1 + a_2 + \cdots + a_n)$$

$$= c\sum_{i=1}^{n} a_i \qquad \blacksquare$$

**EXAMPLE 1**   Suppose that $\sum_{i=1}^{100} a_i = 60$ and $\sum_{i=1}^{100} b_i = 11$. Calculate

$$\sum_{i=1}^{100} (2a_i - 3b_i + 4).$$

**Solution**

$$\sum_{i=1}^{100} (2a_i - 3b_i + 4) = \sum_{i=1}^{100} 2a_i - \sum_{i=1}^{100} 3b_i + \sum_{i=1}^{100} 4$$

$$= 2\sum_{i=1}^{100} a_i - 3\sum_{i=1}^{100} b_i + \sum_{i=1}^{100} 4$$

$$= 2(60) - 3(11) + 100(4) = 487 \qquad \blacksquare$$

**EXAMPLE 2**   **(Collapsing sums).**   Show that:

(a) $\displaystyle\sum_{i=1}^{n} (a_{i+1} - a_i) = a_{n+1} - a_1$;

(b) $\displaystyle\sum_{i=1}^{n} [(i+1)^2 - i^2] = (n+1)^2 - 1$.

**Solution**
  (a) Here we should resist our inclination to apply linearity and instead write out the sum, hoping for some nice cancellations.

$$\sum_{i=1}^{n} (a_{i+1} - a_i) = (a_2 - a_1) + (a_3 - a_2) + (a_4 - a_3) + \cdots + (a_{n+1} - a_n)$$

$$= -a_1 + a_2 - a_2 + a_3 - a_3 + a_4 - \cdots - a_n + a_{n+1}$$

$$= -a_1 + a_{n+1} = a_{n+1} - a_1$$

  (b) This follows immediately from part (a). $\qquad \blacksquare$

The symbol used for the index does not matter. Thus,

$$\sum_{i=1}^{n} a_i = \sum_{j=1}^{n} a_j = \sum_{k=1}^{n} a_k$$

and all of these are equal to $a_1 + a_2 + \cdots + a_n$. For this reason, the index is sometimes called a **dummy index**.

**Some Special Sum Formulas**   In the next section, we will need to consider the sum of the first $n$ positive integers, as well as the sums of their squares, cubes, and so on. There are nice formulas for these; proofs are indicated at the end of the section.

**1.** $\displaystyle\sum_{i=1}^{n} i = 1 + 2 + 3 + \cdots + n = \frac{n(n+1)}{2}$

**2.** $\displaystyle\sum_{i=1}^{n} i^2 = 1^2 + 2^2 + 3^2 + \cdots + n^2 = \frac{n(n+1)(2n+1)}{6}$

**3.** $\displaystyle\sum_{i=1}^{n} i^3 = 1^3 + 2^3 + 3^3 + \cdots + n^3 = \left[\frac{n(n+1)}{2}\right]^2$

**4.** $\displaystyle\sum_{i=1}^{n} i^4 = 1^4 + 2^4 + 3^4 + \cdots + n^4 = \frac{n(n+1)(6n^3 + 9n^2 + n - 1)}{30}$

**EXAMPLE 3**   Calculate (a) $\displaystyle\sum_{i=1}^{10} i$, (b) $\displaystyle\sum_{i=1}^{10} i^2$, and (c) $\displaystyle\sum_{i=2}^{10} i^4$.

**Solution**

(a) $\displaystyle\sum_{i=1}^{10} i = \frac{10(10+1)}{2} = 55$

(b) $\displaystyle\sum_{i=1}^{10} i^2 = \frac{10(10+1)(20+1)}{6} = 385$

(c) $\displaystyle\sum_{i=2}^{10} i^4 = \left(\sum_{i=1}^{10} i^4\right) - 1^4 = \frac{10(11)(6000 + 900 + 10 - 1)}{30} - 1$

$$= 25{,}332 \qquad\blacksquare$$

**EXAMPLE 4**   Calculate $\displaystyle\sum_{i=1}^{10} 2i(i-5)$.

**Solution**   We make use of Example 3 and linearity.

$$\sum_{i=1}^{10} 2i(i-5) = \sum_{i=1}^{10} (2i^2 - 10i) = 2\sum_{i=1}^{10} i^2 - 10\sum_{i=1}^{10} i$$

$$= 2(385) - 10(55) = 220 \qquad\blacksquare$$

**EXAMPLE 5**   Find a formula for $\displaystyle\sum_{j=1}^{n} (j+2)(j-5)$.

**Solution**

$$\sum_{j=1}^{n} (j+2)(j-5) = \sum_{j=1}^{n} (j^2 - 3j - 10) = \sum_{j=1}^{n} j^2 - 3\sum_{j=1}^{n} j - \sum_{j=1}^{n} 10$$

$$= \frac{n(n+1)(2n+1)}{6} - 3\frac{n(n+1)}{2} - 10n$$

$$= \frac{n}{6}\left[2n^2 + 3n + 1 - 9n - 9 - 60\right]$$

$$= \frac{n(n^2 - 3n - 34)}{3} \qquad\blacksquare$$

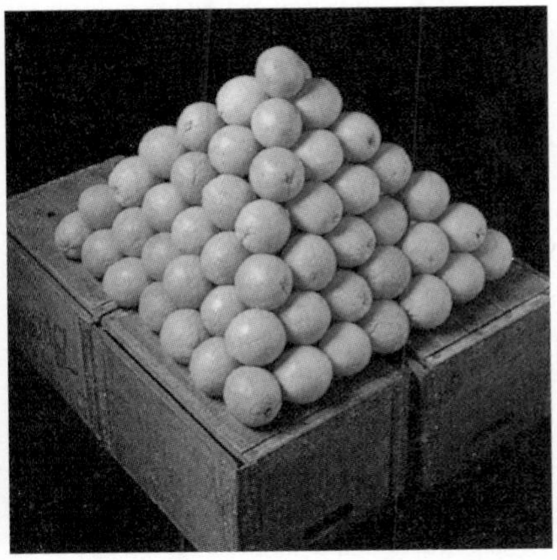

**EXAMPLE 6** How many oranges are in the pyramid shown in Figure 3?

*Solution*   $1^2 + 2^2 + \cdots + 7^2 = \dfrac{7(8)(15)}{6} = 140$   ∎

FIGURE 3

**Proofs of Special Sum Formulas**   Undoubtedly, many of our readers are expecting us to use *mathematical induction* to prove Formulas 1–4. Mathematical induction will certainly work (see Appendix A.1). However, we choose to offer alternative proofs that have the advantage of showing where the formulas come from.

To prove Formula 1, we start with the identity $(i + 1)^2 - i^2 = 2i + 1$, sum both sides, apply Example 2 on the left, and use linearity on the right.

$$(i + 1)^2 - i^2 = 2i + 1$$

$$\sum_{i=1}^{n} [(i + 1)^2 - i^2] = \sum_{i=1}^{n} (2i + 1)$$

$$(n + 1)^2 - 1^2 = 2 \sum_{i=1}^{n} i + \sum_{i=1}^{n} 1$$

$$n^2 + 2n = 2 \sum_{i=1}^{n} i + n$$

$$\frac{n^2 + n}{2} = \sum_{i=1}^{n} i$$

We use Formula 1 and a similar technique to obtain Formula 2.

$$(i + 1)^3 - i^3 = 3i^2 + 3i + 1$$

$$\sum_{i=1}^{n} [(i + 1)^3 - i^3] = \sum_{i=1}^{n} (3i^2 + 3i + 1)$$

$$(n + 1)^3 - 1^3 = 3 \sum_{i=1}^{n} i^2 + 3 \sum_{i=1}^{n} i + \sum_{i=1}^{n} 1$$

$$n^3 + 3n^2 + 3n = 3 \sum_{i=1}^{n} i^2 + 3 \frac{n(n + 1)}{2} + n$$

$$2n^3 + 6n^2 + 6n = 6 \sum_{i=1}^{n} i^2 + 3n^2 + 3n + 2n$$

$$\frac{2n^3 + 3n^2 + n}{6} = \sum_{i=1}^{n} i^2$$

$$\frac{n(n + 1)(2n + 1)}{6} = \sum_{i=1}^{n} i^2$$

Almost the same technique works to establish Formulas 3 and 4.

# CONCEPTS REVIEW

**1.** The value of $\sum_{i=1}^{5} 2i$ is _____ and the value of $\sum_{i=1}^{5} 2$ is _____.

**2.** If $\sum_{i=1}^{10} a_i = 9$ and $\sum_{i=1}^{10} b_i = 7$, then the value of $\sum_{i=1}^{10} (3a_i - 2b_i) =$ _____ and the value of $\sum_{i=1}^{10} (a_i + 4) =$ _____.

**3.** The value of the collapsing sum $\sum_{i=1}^{9} \left( \frac{1}{i} - \frac{1}{i+1} \right)$ is _____.

**4.** Since $\sum_{i=1}^{n} i = n(n+1)/2$ and $\sum_{i=1}^{n} i^2 = n(n+1)(2n+1)/6$, it follows that $\sum_{i=1}^{6} (2i - i^2) =$ _____.

# PROBLEM SET 5.3

In Problems 1–8, find the value of the indicated sum.

**1.** $\sum_{k=1}^{6} (k - 1)$

**2.** $\sum_{i=1}^{6} i^2$

**3.** $\sum_{k=1}^{7} \frac{1}{k+1}$

**4.** $\sum_{l=3}^{8} (l + 1)^2$

**5.** $\sum_{m=1}^{8} (-1)^m 2^{m-2}$

**6.** $\sum_{k=3}^{7} \frac{(-1)^k 2^k}{(k+1)}$

**7.** $\sum_{n=1}^{6} n \cos (n\pi)$

**8.** $\sum_{k=-1}^{6} k \sin (k\pi/2)$

In Problems 9–16, write the indicated sum in sigma notation.

**9.** $1 + 2 + 3 + \cdots + 41$

**10.** $2 + 4 + 6 + 8 + \cdots + 50$

**11.** $1 + \frac{1}{2} + \frac{1}{3} + \cdots + \frac{1}{100}$

**12.** $1 - \frac{1}{2} + \frac{1}{3} - \frac{1}{4} + \cdots - \frac{1}{100}$

**13.** $a_1 + a_3 + a_5 + a_7 + \cdots + a_{99}$

**14.** $b_{-1} + b_1 + b_3 + b_5 + \cdots + b_{1001}$

**15.** $f(c_1) + f(c_2) + \cdots + f(c_n)$

**16.** $f(w_1) \Delta x + f(w_2) \Delta x + \cdots + f(w_n) \Delta x$

In Problems 17–20, suppose that $\sum_{i=1}^{10} a_i = 40$ and $\sum_{i=1}^{10} b_i = 50$. Calculate each of the following (see Example 1).

**17.** $\sum_{i=1}^{10} (a_i + b_i)$

**18.** $\sum_{n=1}^{10} (3a_n + 2b_n)$

**19.** $\sum_{p=0}^{9} (a_{p+1} - b_{p+1})$

**20.** $\sum_{q=1}^{10} (a_q - b_q - q)$

In Problems 21–24, find the value of each of the (collapsing) sums (see Example 2).

**21.** $\sum_{k=1}^{40} \left( \frac{1}{k} - \frac{1}{k+1} \right)$

**22.** $\sum_{k=1}^{10} (2^k - 2^{k-1})$

**23.** $\sum_{k=3}^{20} \left( \frac{1}{(k+1)^2} - \frac{1}{k^2} \right)$

**24.** $\sum_{k=3}^{m+1} (a_k - a_{k-1})$

In Problems 25–30, use Formulas 1–4, page 250 to find each sum (see Examples 3–5).

**25.** $\sum_{i=1}^{100} (3i - 2)$

**26.** $\sum_{i=1}^{10} [(i - 1)(4i + 3)]$

**27.** $\sum_{k=1}^{10} (k^3 - k^2)$

**28.** $\sum_{k=1}^{10} 5k^2(k + 4)$

**29.** $\sum_{i=1}^{n} (2i^2 - 3i + 1)$

**30.** $\sum_{i=1}^{n} (2i - 3)^2$

Sometimes it is desirable to make a change of variable in the index for a sum. For example, the change of variable $k = i - 3$ gives

$$\sum_{i=4}^{13} (i - 3)^3 = \sum_{k=1}^{10} k^3$$

For Problems 31–34, make the indicated change of variable in the index.

**31.** $\sum_{i=3}^{19} i(i - 2); k = i - 2$

**32.** $\sum_{k=5}^{14} k2^{k-4}; i = k - 4$

**33.** $\sum_{k=0}^{10} \frac{k}{k+1}; i = k + 1$

**34.** $\sum_{k=4}^{13} (k - 3)\sin\left( \frac{\pi}{k-3} \right); i = k - 3$

**35.** Evaluate $\sum_{i=1}^{10} f(w_i) \Delta x$ if $f(x) = 3x$, $w_i = i/5$, and $\Delta x = \frac{1}{5}$.

**36.** Prove the following formula for a **geometric sum:**

$$\sum_{k=0}^{n} ar^k = a + ar + ar^2 + \cdots + ar^n = \frac{a - ar^{n+1}}{1 - r} \quad (r \neq 1)$$

*Hint*: Let $S = a + ar + \cdots + ar^n$. Simplify $S - rS$ and solve for $S$.

**37.** Use Problem 36 to calculate each sum.

(a) $\displaystyle\sum_{k=1}^{10} (\tfrac{1}{2})^k$  (b) $\displaystyle\sum_{k=1}^{10} 2^k$

**38.** Add the two equalities below, solve for $S$, and thereby give another proof of Formula 1.

$$S = 1 + 2 + 3 + \cdots + (n - 2) + (n - 1) + n$$
$$S = n + (n - 1) + (n - 2) + \cdots + 3 + 2 + 1$$

**39.** Use a derivation like that in Problem 38 to obtain a formula for the **arithmetic sum**:

$$\sum_{k=0}^{n} (a + kd) = a + (a + d) + (a + 2d) + \cdots + (a + nd)$$

**40.** Show that $\displaystyle\sum_{k=1}^{n} \frac{1}{\sqrt{k}} \geq \sqrt{n}$.

C **41.** In statistics we define the mean $\bar{x}$ and the variance $s^2$ of a sequence of numbers $x_1, x_2, \ldots, x_n$ by

$$\bar{x} = \frac{1}{n} \sum_{i=1}^{n} x_i, \qquad s^2 = \frac{1}{n} \sum_{i=1}^{n} (x_i - \bar{x})^2$$

Find $\bar{x}$ and $s^2$ for the sequence of numbers 2, 5, 7, 8, 9, 10, 14.

**42.** Use the definitions in Problem 41 to show that each is true.

(a) $\displaystyle\sum_{i=1}^{n} (x_i - \bar{x} \cdot) = 0$  (b) $s^2 = \left( \dfrac{1}{n} \displaystyle\sum_{i=1}^{n} x_i^2 \right) - \bar{x}^2$

**43.** Let

$$A = \sum_{i=1}^{n} a_i^2, \qquad B = \sum_{i=1}^{n} a_i b_i, \qquad C = \sum_{i=1}^{n} b_i^2$$

Prove that $At^2 + 2Bt + C \geq 0$ for all real numbers $t$. *Hint*: Show

$$At^2 + 2Bt + C = \sum_{i=1}^{n} (a_i t + b_i)^2.$$

**44.** Prove the **Cauchy Inequality**:

$$\left( \sum_{i=1}^{n} a_i b_i \right)^2 \leq \left( \sum_{i=1}^{n} a_i^2 \right) \left( \sum_{i=1}^{n} b_i^2 \right)$$

*Hint*: In the notation of Problem 43, we must show that $B^2 - AC \leq 0$. Note that this is one-fourth the discriminant of the quadratic equation $At^2 + 2Bt + C = 0$.

**45.** Establish the following identity, which is known as *summation by parts*:

$$\sum_{i=1}^{n} (a_i - a_{i-1})b_{i-1} = a_n b_n - a_0 b_0 - \sum_{i=1}^{n} a_i (b_i - b_{i-1})$$

**46.** Use the identity $(i + 1)^4 - i^4 = 4i^3 + 6i^2 + 4i + 1$ to prove Formula 3, namely,

$$1^3 + 2^3 + \cdots + n^3 = \left[ \frac{n(n + 1)}{2} \right]^2$$

**47.** Use the diagrams in Figure 4 to establish Formulas 1 and 3.

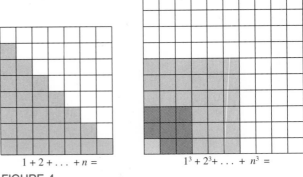

$1 + 2 + \ldots + n =$   $1^3 + 2^3 + \ldots + n^3 =$

**FIGURE 4**

**48.** Find a nice formula for

$$\frac{1}{1 \cdot 2} + \frac{1}{2 \cdot 3} + \frac{1}{3 \cdot 4} + \cdots + \frac{1}{n(n + 1)}$$

*Hint*: $\dfrac{1}{i(i + 1)} = \dfrac{1}{i} - \dfrac{1}{i + 1}$.

**49.** A grocer stacks oranges in a pyramid-like pile. If the bottom layer is rectangular with 10 rows of 16 oranges and the top layer has a single row of oranges, how many oranges are in the stack? Answer the same question if the bottom layer has 50 rows of 60 oranges. Generalize to the case of $m$ rows of $n$ oranges, $m \leq n$.

**50.** In the song *The Twelve Days of Christmas*, my true love gave me 1 gift on the first day, $1 + 2$ gifts on the second day, $1 + 2 + 3$ gifts on the third day, and so on for 12 days.

(a) Find the total number of gifts given in 12 days.
(b) Find a simple formula for $T_n$, the total number of gifts given during a Christmas of $n$ days.

**Answers to Concepts Review:** 1. 30; 10  2. 13; 49
3. 0.9  4. −49

---

**5.4**
**INTRODUCTION**
**TO AREA**

Two problems, both from geometry, motivate the two biggest ideas in calculus. The problem of the tangent line led us to the *derivative*. The problem of area will lead us to the *definite integral*.

For polygons (closed plane regions bounded by straight line segments), the area problem is hardly a problem at all. We start by defining the area of a rectangle to be its length times its width, and from this we successively derive the formulas for the area of a parallelogram, a triangle, and any polygon. The sequence of figures in Figure 1 suggests how this is done.

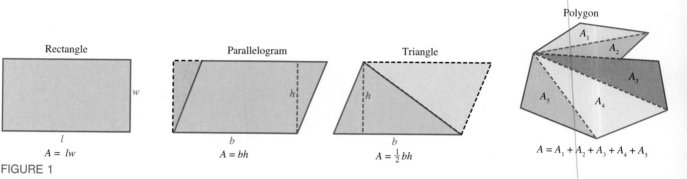

Rectangle
$A = lw$

Parallelogram
$A = bh$

Triangle
$A = \frac{1}{2}bh$

Polygon
$A = A_1 + A_2 + A_3 + A_4 + A_5$

FIGURE 1

Even in this simple setting, it is clear that we want area to satisfy five properties.

1. The area of a plane region is a nonnegative (real) number.
2. The area of a rectangle is the product of its length and width (both measured in the same units). The result is in square units—for example, square feet or square centimeters.
3. Congruent regions have equal areas.
4. The area of the union of two regions that overlap only in a line segment is the sum of the areas of the two regions.
5. If one region is contained in a second region, then the area of the first region is less than or equal to that of the second.

When we consider a region with a curved boundary, the problem of assigning area is significantly more difficult. However, over 2000 years ago, Archimedes provided the key to a solution. Consider, said he, a sequence of inscribed polygons that approximate the curved region with greater and greater accuracy. For example, for the circle of radius 1, consider regular inscribed polygons $P_1$, $P_2$, $P_3$, . . . with 4 sides, 8 sides, 16 sides, . . . , as shown in Figure 2. The area of the circle is the limit as $n \to \infty$ of the areas of $P_n$. Thus, if $A(F)$ denotes the area of a region $F$, then

$$A(\text{circle}) = \lim_{n \to \infty} A(P_n)$$

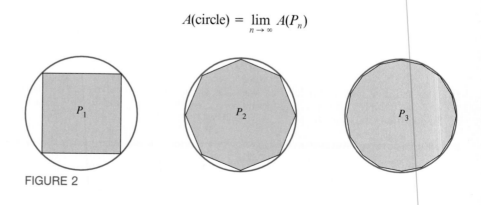

FIGURE 2

---

### Use of Language

Following common usage, we allow ourselves a certain abuse of language. The words triangle, rectangle, polygon, and circle will be used to denote both two-dimensional regions of the indicated shape and also their one-dimensional boundaries. Note that regions have areas, whereas curves have lengths. Thus, we say that a circle has area $\pi r^2$ and circumference $2\pi r$. The context will always make clear our intended usage.

Archimedes went further, considering also circumscribed polygons $T_1$, $T_2$, $T_3$, . . . (Figure 3). He showed that you get the same value for the area of the circle of radius 1 (namely, $\pi \approx 3.14159$) whether you use inscribed or circumscribed polygons. It is just a small step from what he did to our modern treatment of area.

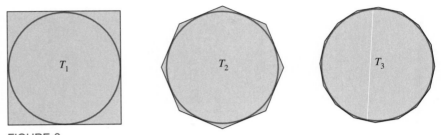

**FIGURE 3**

**Area by Inscribed Polygons**  Consider the region $R$ bounded by the parabola $y = f(x) = x^2$, the $x$-axis, and the vertical line $x = 2$ (Figure 4). We refer to $R$ as the region under the curve $y = x^2$ between $x = 0$ and $x = 2$. Our aim is to calculate its area $A(R)$.

Partition (as in Figure 5) the interval $[0, 2]$ into $n$ subintervals, each of length $\Delta x = 2/n$, by means of points.

$$0 = x_0 < x_1 < x_2 < \cdots < x_{n-1} < x_n = 2$$

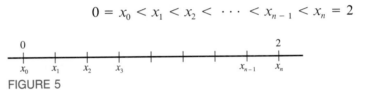

**FIGURE 5**

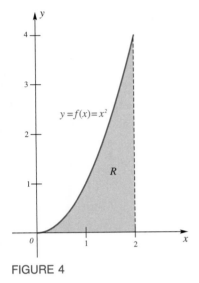

**FIGURE 4**

Thus,

$$x_0 = 0$$

$$x_1 = \Delta x = \frac{2}{n}$$

$$x_2 = 2 \cdot \Delta x = \frac{4}{n}$$

$$x_3 = 3 \cdot \Delta x = \frac{6}{n}$$

$$\cdots$$

$$x_i = i \cdot \Delta x = \frac{2i}{n}$$

$$\cdots$$

$$x_{n-1} = (n-1) \cdot \Delta x = \frac{(n-1)2}{n}$$

$$x_n = n \cdot \Delta x = n\left(\frac{2}{n}\right) = 2$$

Consider the typical rectangle with base $[x_{i-1}, x_i]$ and height $f(x_{i-1}) = x_{i-1}^2$. Its area is $f(x_{i-1}) \Delta x$ (see the upper left part of Figure 6). The union $R_n$ of all such rectangles forms the inscribed polygon shown in the lower right part of Figure 6.

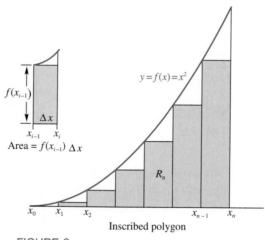

FIGURE 6

The area $A(R_n)$ can be calculated.

$$A(R_n) = f(x_0)\Delta x + f(x_1)\,\Delta x + f(x_2)\,\Delta x + \cdots + f(x_{n-1})\,\Delta x$$

Now

$$f(x_i)\,\Delta x = x_i^2\,\Delta x = \left(\frac{2i}{n}\right)^2 \cdot \frac{2}{n} = \left(\frac{8}{n^3}\right) i^2$$

Thus,

$$A(R_n) = \left[\frac{8}{n^3}(0^2) + \frac{8}{n^3}(1^2) + \frac{8}{n^3}(2^2) + \cdots + \frac{8}{n^3}(n-1)^2\right]$$

$$= \frac{8}{n^3}[1^2 + 2^2 + \cdots + (n-1)^2]$$

$$= \frac{8}{n^3}\left[\frac{(n-1)n(2n-1)}{6}\right] \qquad \text{(Formula 2, p. 250, with } n-1 \text{ replacing } n\text{)}$$

$$= \frac{8}{6}\left(\frac{2n^3 - 3n^2 + n}{n^3}\right)$$

$$= \frac{4}{3}\left(2 - \frac{3}{n} + \frac{1}{n^2}\right)$$

$$= \frac{8}{3} - \frac{4}{n} + \frac{4}{3n^2}$$

We conclude that

$$A(R) = \lim_{n \to \infty} A(R_n) = \lim_{n \to \infty} \left( \frac{8}{3} - \frac{4}{n} + \frac{4}{3n^2} \right) = \frac{8}{3}$$

The diagrams in Figure 7 should help you visualize what is happening as $n$ gets larger and larger.

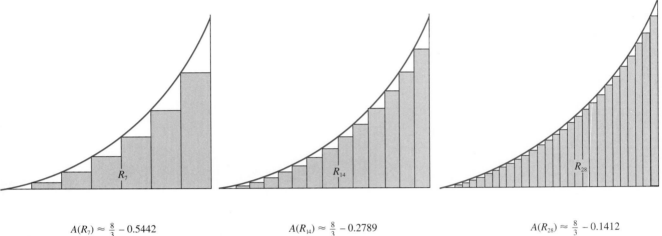

$$A(R_7) \approx \frac{8}{3} - 0.5442 \qquad A(R_{14}) \approx \frac{8}{3} - 0.2789 \qquad A(R_{28}) \approx \frac{8}{3} - 0.1412$$

**FIGURE 7**

**Area by Circumscribed Polygons**    Perhaps you are still not convinced that $A(R) = \frac{8}{3}$. We can give more evidence. Consider the rectangle with base $[x_{i-1}, x_i]$ and height $f(x_i) = x_i^2$ (shown at the upper left in Figure 8). Its area is $f(x_i) \Delta x$. The union $S_n$ of all such rectangles forms a circumscribed polygon for the region $R$, as shown at the lower right in Figure 8.

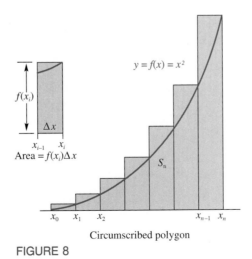

Circumscribed polygon

**FIGURE 8**

The area $A(S_n)$ is calculated in analogy with the calculation of $A(R_n)$.

$$A(S_n) = f(x_1)\Delta x + f(x_2)\,\Delta x + \cdots + f(x_n)\,\Delta x$$

As before, $f(x_i)\,\Delta x = x_i^2\,\Delta x = (8/n^3)i^2$, and so

$$A(S_n) = \left[\frac{8}{n^3}(1^2) + \frac{8}{n^3}(2^2) + \cdots + \frac{8}{n^3}(n^2)\right]$$

$$= \frac{8}{n^3}[1^2 + 2^2 + \cdots + n^2]$$

$$= \frac{8}{n^3}\left[\frac{n(n+1)(2n+1)}{6}\right] \quad \text{(Formula 2, p. 250)}$$

$$= \frac{8}{6}\left[\frac{2n^3 + 3n^2 + n}{n^3}\right]$$

$$= \frac{4}{3}\left[2 + \frac{3}{n} + \frac{1}{n^2}\right]$$

Thus,

$$\lim_{n \to \infty} A(S_n) = \lim_{n \to \infty} \frac{4}{3}\left[2 + \frac{3}{n} + \frac{1}{n^2}\right] = \frac{8}{3}$$

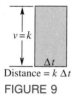

$v = k$

Distance $= k\,\Delta t$

FIGURE 9

**Another Problem—Same Theme**  Suppose that an object is traveling along the $t$-axis in such a way that its velocity at time $t$ is given by $v = f(t) = \frac{1}{4}t^3 + 1$ feet per second. How far did it travel between $t = 0$ and $t = 3$? This problem can be solved by the method of differential equations (Section 5.2), but we have something else in mind.

Our starting point is the familiar fact that if an object travels at constant velocity $k$ over a time interval of length $\Delta t$, then the distance traveled is $k\,\Delta t$. But this is just the area of a rectangle—the one shown in Figure 9.

Next consider the given problem, where $v = f(t) = \frac{1}{4}t^3 + 1$. The graph is shown in the left half of Figure 10. Partition the interval $[0, 3]$ into $n$ subintervals of length $\Delta t = 3/n$ by means of points $0 = t_0 < t_1 < t_2 < \cdots < t_n = 3$. Then consider the corresponding circumscribed polygon $S_n$, displayed in the right half of Figure 10 (we could as well have considered the inscribed polygon). Its area, $A(S_n)$, should be a good approximation of the distance traveled, especially if $\Delta t$ is small, since on each subinterval the actual velocity is almost equal to a constant (the value of $v$ at the end of the subinterval).

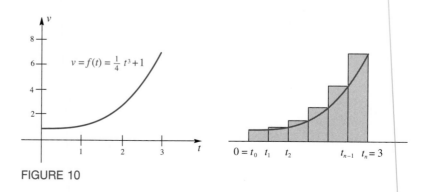

FIGURE 10

Moreover, this approximation should get better and better as $n$ gets larger. We are led to the conclusion that the exact distance traveled is $\lim_{n \to \infty} A(S_n)$; that is, it is the area of the region under the velocity curve between $t = 0$ and $t = 3$.

To calculate $A(S_n)$, note that $t_i = 3i/n$, and so the area of the $i$th rectangle is

$$f(t_i) \, \Delta t = \left[ \frac{1}{4} \left( \frac{3i}{n} \right)^3 + 1 \right] \frac{3}{n} = \frac{81}{4n^4} i^3 + \frac{3}{n}$$

Thus,

$$A(S_n) = f(t_1) \, \Delta t + f(t_2) \, \Delta t + \cdots + f(t_n) \, \Delta t$$

$$= \sum_{i=1}^{n} f(t_i) \, \Delta t$$

$$= \sum_{i=1}^{n} \left( \frac{81}{4n^4} i^3 + \frac{3}{n} \right)$$

$$= \frac{81}{4n^4} \sum_{i=1}^{n} i^3 + \sum_{i=1}^{n} \frac{3}{n}$$

$$= \frac{81}{4n^4} \left[ \frac{n(n+1)}{2} \right]^2 + \frac{3}{n} \cdot n \quad \text{(Formula 3, p. 250)}$$

$$= \frac{81}{16} \left[ n^2 \frac{(n^2 + 2n + 1)}{n^4} \right] + 3$$

$$= \frac{81}{16} \left( 1 + \frac{2}{n} + \frac{1}{n^2} \right) + 3$$

We conclude that

$$\lim_{n \to \infty} A(S_n) = \frac{81}{16} + 3 = \frac{129}{16} \approx 8.06$$

The object traveled about 8.06 feet between $t = 0$ and $t = 3$.

What was true in this example is true for any object moving with positive velocity. *The distance traveled is the area of the region under the velocity curve.*

## CONCEPTS REVIEW

**1.** If area($R$) = 7 and area($S$) = 9 and if $R$ and $S$ do not overlap (except possibly in a curve), then area($R \cup S$) = _____.

**2.** The area of a(n) _____ polygon underestimates the area of a region whereas the area of a(n) _____ polygon overestimates this area.

**3.** The exact area of the region under the curve $y = |x|$ between 0 and 4 is _____. Similarly the area under this curve between $-2$ and 4 is _____.

**4.** The exact area of the region under the curve $y = [[x]]$ between 0 and 4 is _____.

## PROBLEM SET 5.4

In Problems 1–6, find the area of the indicated inscribed or circumscribed polygon.

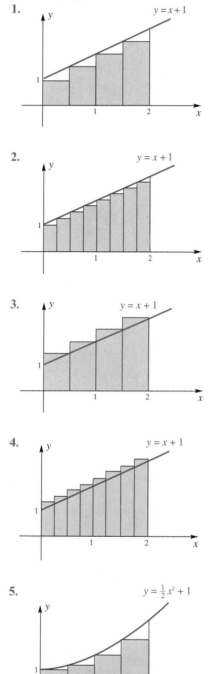

**1.** $y = x + 1$

**2.** $y = x + 1$

**3.** $y = x + 1$

**4.** $y = x + 1$

**5.** $y = \frac{1}{2}x^2 + 1$

**6.** $y = \frac{1}{2}x^2 + 1$

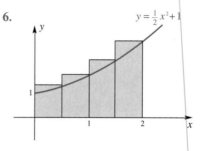

In Problems 7–10, sketch the graph of the given function over the interval $[a, b]$, then divide $[a, b]$ into $n$ equal subintervals, and finally calculate the area of the corresponding circumscribed polygon.

**7.** $f(x) = x + 1$; $a = -1$, $b = 2$, $n = 3$

**8.** $f(x) = 3x - 1$; $a = 1$, $b = 3$, $n = 4$

⊂ **9.** $f(x) = x^2 - 1$; $a = 2$, $b = 3$, $n = 6$

⊂ **10.** $f(x) = 3x^2 + x + 1$; $a = -1$, $b = 1$, $n = 10$

In Problems 11–16, find the area of the region under the curve $y = f(x)$ over the interval $[a, b]$. To do this, divide the interval $[a, b]$ into $n$ equal subintervals, calculate the area of the corresponding circumscribed polygon, and then let $n \to \infty$. (See the example for $y = x^2$ in the text.)

**11.** $y = x + 2$; $a = 0$, $b = 1$

**12.** $y = \frac{1}{2}x^2 + 1$; $a = 0$, $b = 1$

**13.** $y = 2x + 2$; $a = -1$, $b = 1$. *Hint:* $x_i = -1 + \dfrac{2i}{n}$

**14.** $y = x^2$; $a = -2$, $b = 2$

≈ **15.** $y = x^3$; $a = 0$, $b = 1$

≈ **16.** $y = x^3 + x$; $a = 0$, $b = 1$

**17.** Suppose that an object is traveling along the $t$-axis in such a way that its velocity at time $t$ seconds is $v = t + 2$ feet per second. How far did it travel between $t = 0$ and $t = 1$? *Hint:* See the discussion of the velocity problem at the end of this section and use the result of Problem 11.

**18.** Follow the directions of Problem 17 given that $v = \frac{1}{2}t^2 + 2$. You will need the result of Problem 12.

**19.** Let $A_a^b$ denote the area under the curve $y = x^2$ over the interval $[a, b]$.

(a) Prove that $A_0^b = b^3/3$. *Hint:* $\Delta x = b/n$ so $x_i = ib/n$; use circumscribed polygons.
(b) Show that $A_a^b = b^3/3 - a^3/3$. Assume $a \geq 0$.

**20.** Suppose that an object, moving along the $t$-axis, has velocity $v = t^2$ meters per second at time $t$ seconds. How far did it travel between $t = 3$ and $t = 5$? See Problem 19.

**21.** Use the results of Problem 19 to calculate the area under the curve $y = x^2$ over each of the following intervals.

(a) $[0, 5]$ (b) $[1, 4]$ (c) $[2, 5]$

**22.** From Formulas 1–4 of Section 5.3, you might guess that

$$1^m + 2^m + 3^m + \cdots + n^m = \frac{n^{m+1}}{m+1} + C_n$$

where $C_n$ is a polynomial in $n$ of degree $m$. Assume this is true (which it is) and, for $a \geq 0$, let $A_a^b(x^m)$ be the area under the curve $y = x^m$ over the interval $[a, b]$.

(a) Prove that $A_0^b(x^m) = \dfrac{b^{m+1}}{(m+1)}$.

(b) Show that $A_a^b(x^m) = \dfrac{b^{m+1}}{m+1} - \dfrac{a^{m+1}}{m+1}$

**23.** Use the results of Problem 22 to calculate each of the following areas.

(a) $A_0^2(x^3)$ (b) $A_1^2(x^3)$ (c) $A_1^2(x^5)$ (d) $A_0^2(x^9)$

**24.** Derive the formulas $A_n = \frac{1}{2}nr^2 \sin(2\pi/n)$ and $B_n = nr^2 \tan(\pi/n)$ for the areas of the inscribed and circumscribed regular $n$-sided polygons for a circle of radius $r$. Then show that $\lim_{n \to \infty} A_n$ and $\lim_{n \to \infty} B_n$ are both $\pi r^2$.

---

**Answers to Concepts Review: 1.** 16 **2.** Inscribed; circumscribed **3.** 8; 10 **4.** 6

---

| | |
|---|---|
| **5.5** | All the preparations have been made; we are ready to define the definite |
| **THE DEFINITE** | integral. Both Newton and Leibniz introduced early versions of this concept. |
| **INTEGRAL** | However, it was Riemann who gave us the modern definition. In formulating |

this definition, we are guided by the ideas we discussed in the previous section. The first notion is that of a Riemann sum.

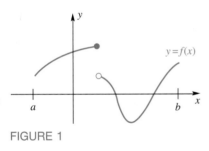

FIGURE 1

**Riemann Sums** Consider a function $f$ defined on a closed interval $[a, b]$. It may have both positive and negative values on the interval and it does not even need to be continuous. Its graph might look something like the one in Figure 1.

Consider a partition $P$ of the interval $[a, b]$ into $n$ subintervals (not necessarily of equal length) by means of points $a = x_0 < x_1 < x_2 < \cdots < x_{n-1} < x_n = b$ and let $\Delta x_i = x_i - x_{i-1}$. On each subinterval $[x_{i-1}, x_i]$, pick an arbitrary point $\bar{x}_i$ (which may be an endpoint); we call it a *sample point* for the $i$th subinterval. An example of these constructions is shown in Figure 2 for $n = 6$.

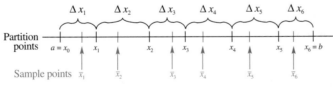

A Partition of $[a, b]$ with Sample Points $\bar{x}_i$

FIGURE 2

Form the sum

$$R_p = \sum_{i=1}^{n} f(\bar{x}_i) \, \Delta x_i$$

We call $R_p$ a **Riemann sum** for $f$ corresponding to the partition $P$. Its geometric interpretation is shown in Figure 3. Note that the contribution from a rectangle below the $x$-axis is the negative of its area, since in this case, $f(\overline{x}_i) < 0$.

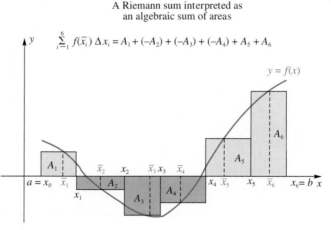

A Riemann sum interpreted as an algebraic sum of areas

$$\sum_{i=1}^{6} f(\overline{x}_i)\,\Delta x_i = A_1 + (-A_2) + (-A_3) + (-A_4) + A_5 + A_6$$

FIGURE 3

**EXAMPLE 1** Evaluate the Riemann sum $R_p$ for

$$f(x) = (x + 1)(x - 2)(x - 4) = x^3 - 5x^2 + 2x + 8$$

on the interval $[0, 5]$ using the partition $P$ with partition points $0 < 1.1 < 2 < 3.2 < 4 < 5$ and the corresponding sample points $\overline{x}_1 = 0.5$, $\overline{x}_2 = 1.5$, $\overline{x}_3 = 2.5$, $\overline{x}_4 = 3.6$, and $\overline{x}_5 = 5$.

*Solution*

$$\begin{aligned}
R_p &= \sum_{i=1}^{5} f(\overline{x}_i)\,\Delta x_i \\
&= f(\overline{x}_1)\,\Delta x_1 + f(\overline{x}_2)\,\Delta x_2 + f(\overline{x}_3)\,\Delta x_3 + f(\overline{x}_4)\,\Delta x_4 + f(\overline{x}_5)\,\Delta x_5 \\
&= f(0.5)(1.1 - 0) + f(1.5)(2 - 1.1) + f(2.5)(3.2 - 2) \\
&\quad + f(3.6)(4 - 3.2) + f(5)(5 - 4) \\
&= (7.875)(1.1) + (3.125)(0.9) + (-2.625)(1.2) \\
&\quad + (-2.944)(0.8) + 18(1) \\
&= 23.9698
\end{aligned}$$

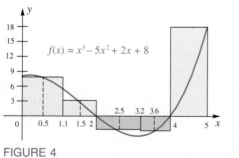

$f(x) = x^3 - 5x^2 + 2x + 8$

FIGURE 4

The corresponding geometric picture appears in Figure 4. ∎

**EXAMPLE 2** Evaluate the Riemann sum for $f(x) = x^2 + 1$ on the interval $[-1, 2]$ using the equally spaced partition points $-1 < -0.5 < 0 < 0.5 < 1 < 1.5 < 2$, with the sample point $\overline{x}_i$ being the midpoint of the $i$th subinterval.

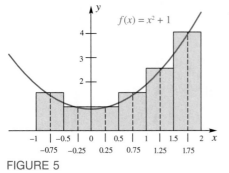

$f(x) = x^2 + 1$

FIGURE 5

*Solution* Note the picture in Figure 5.

$$R_p = \sum_{i=1}^{6} f(\overline{x}_i)\, \Delta x_i$$

$$= [f(-0.75) + f(-0.25) + f(0.25) + f(0.75) + f(1.25) + f(1.75)](0.5)$$

$$= [1.5625 + 1.0625 + 1.0625 + 1.5625 + 2.5625 + 4.0625](0.5)$$

$$= 5.9375 \qquad \blacksquare$$

**Definition of the Definite Integral**   Suppose now that $P$, $\Delta x_i$, and $\overline{x}_i$ have the meanings discussed above. Also let $|P|$, called the **norm** of $P$, denote the length of the longest of the subintervals of the partition $P$. For instance, in Example 1, $|P| = 3.2 - 2 = 1.2$; in Example 2, $|P| = 0.5$.

> **Definition**
>
> **(The definite integral).** Let $f$ be a function that is defined on the closed interval $[a, b]$. If
>
> $$\lim_{|P| \to 0} \sum_{i=1}^{n} f(\overline{x}_i)\, \Delta x_i$$
>
> exists, we say $f$ is **integrable** on $[a, b]$. Moreover, $\displaystyle\int_a^b f(x)\, dx$, called the **definite integral** (or Riemann integral) of $f$ from $a$ to $b$, is then given by
>
> $$\int_a^b f(x)\, dx = \lim_{|P| \to 0} \sum_{i=1}^{n} f(\overline{x}_i)\, \Delta x_i$$

The heart of the definition is the final line. The concept captured in that equation grows out of our discussion of area in the previous section. However, we have considerably modified the notion presented there. For example, we now allow $f$ to be negative on part or all of $[a, b]$, we use partitions with subintervals that may be of unequal length, and we allow $\overline{x}_i$ to be any point on the $i$th subinterval. Since we have made these changes, it is important to state precisely how the definite integral relates to area. In general, $\displaystyle\int_a^b f(x)\, dx$ gives the *signed area* of the region trapped between the curve $y = f(x)$ and the $x$-axis on the interval $[a, b]$, meaning that a plus sign is attached to areas of parts above the $x$-axis and a minus sign is attached to areas of parts below the $x$-axis. In symbols,

$$\boxed{\int_a^b f(x)\, dx = A_{\text{up}} - A_{\text{down}}}$$

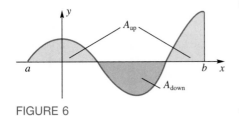

FIGURE 6

where $A_{\text{up}}$ and $A_{\text{down}}$ are as shown in Figure 6.

The meaning of the word *limit* in the definition of the definite integral is more general than in earlier usage and should be explained. The equality

$$\lim_{|P| \to 0} \sum_{i=1}^{n} f(\overline{x}_i) \Delta x_i = L$$

means that corresponding to each $\varepsilon > 0$, there is a $\delta > 0$ such that

$$\left| \sum_{i=1}^{n} f(\overline{x}_i) \Delta x_i - L \right| < \varepsilon$$

for all Riemann sums $\sum_{i=1}^{n} f(\overline{x}_i) \Delta x_i$ for $f$ on $[a, b]$ for which the norm $|P|$ of the associated partition is less than $\delta$. In this case, we say that the indicated limit exists and has the value $L$.

That was a mouthful, and we are not going to digest it just now. We simply assert that the usual limit theorems also hold for this kind of limit.

Returning to the symbol $\int_a^b f(x)\,dx$, we might call $a$ the lower endpoint and $b$ the upper endpoint for the integral. However, most authors use the terminology **lower limit** of integration and **upper limit** of integration, which is fine provided we realize that this usage of the word *limit* has nothing to do with its more technical meaning.

In our definition of $\int_a^b f(x)\,dx$, we implicitly assumed that $a < b$. We remove that restriction with the following definitions.

$$\int_a^a f(x)\,dx = 0$$

$$\int_a^b f(x)\,dx = -\int_b^a f(x)\,dx, \qquad a > b$$

Thus,

$$\int_2^2 x^3\,dx = 0, \qquad \int_6^2 x^3\,dx = -\int_2^6 x^3\,dx$$

Finally, we point out that $x$ is a **dummy variable** in the symbol $\int_a^b f(x)\,dx$. By this we mean that $x$ can be replaced by any other letter (provided, of course, that it is replaced in each place where it occurs). Thus,

$$\int_a^b f(x)\,dx = \int_a^b f(t)\,dt = \int_a^b f(u)\,du$$

**What Functions Are Integrable?**   Not every function is integrable. For example, the unbounded function

$$f(x) = \begin{cases} \dfrac{1}{x^2} & \text{if } x \neq 0 \\ \\ 1 & \text{if } x = 0 \end{cases}$$

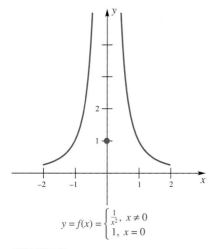

$$y = f(x) = \begin{cases} \frac{1}{x^2}, & x \neq 0 \\ 1, & x = 0 \end{cases}$$

**FIGURE 7**

which is graphed in Figure 7, is not integrable on $[-2, 2]$. This is because the contribution to any Riemann sum from the subinterval containing $x = 0$ can be made arbitrarily large by choosing the corresponding sample point $\overline{x}_i$ sufficiently close to zero. In fact, this reasoning shows that *any function that is integrable on $[a, b]$ must be bounded there*; that is, there must exist a constant $M$ such that $|f(x)| \leq M$ for all $x$ in $[a, b]$.

Even some bounded functions can fail to be integrable, but they have to be pretty complicated (see Problem 22 for one example). By all odds, Theorem A (below) is the most important theorem about integrability. Unfortunately, it is too difficult to prove here; we leave that for advanced calculus books.

---

### Theorem A

**(Integrability Theorem).** If $f$ is bounded on $[a, b]$ and if it is continuous there except at a finite number of points, then $f$ is integrable on $[a, b]$. In particular, if $f$ is continuous on the whole interval $[a, b]$, it is integrable on $[a, b]$.

---

As a consequence of this theorem, the following functions are integrable on every closed interval $[a, b]$.

1. Polynomial functions.
2. Sine and cosine functions.
3. Rational functions, provided the interval $[a, b]$ contains no points where a denominator is 0.

**Calculating Definite Integrals** Knowing that a function is integrable allows us to calculate its integral by using a **regular partition** (equal-length subintervals) and by picking the sample points $\overline{x}_i$ in any way convenient for us. Examples 3 and 4 involve polynomials, which we just learned are integrable.

**EXAMPLE 3** Evaluate $\displaystyle\int_{-2}^{3} (x + 3)\, dx$.

**Solution** Partition the interval $[-2, 3]$ into $n$ equal subintervals, each of length $\Delta x = 5/n$. In each subinterval $[x_{i-1}, x_i]$ use $\overline{x}_i = x_i$ as the sample point. Then

$$x_0 = -2$$
$$x_1 = -2 + \Delta x = -2 + \frac{5}{n}$$
$$x_2 = -2 + 2\,\Delta x = -2 + 2\left(\frac{5}{n}\right)$$
$$\vdots$$
$$x_i = -2 + i\,\Delta x = -2 + i\left(\frac{5}{n}\right)$$
$$\vdots$$
$$x_n = -2 + n\,\Delta x = -2 + n\left(\frac{5}{n}\right) = 3$$

Thus, $f(x_i) = x_i + 3 = 1 + i(5/n)$, and so

$$\sum_{i=1}^{n} f(\bar{x}_i)\,\Delta x_i = \sum_{i=1}^{n} f(x_i)\,\Delta x$$

$$= \sum_{i=1}^{n} \left[1 + i\left(\frac{5}{n}\right)\right]\frac{5}{n}$$

$$= \frac{5}{n} \sum_{i=1}^{n} 1 + \frac{25}{n^2} \sum_{i=1}^{n} i$$

$$= \frac{5}{n}(n) + \frac{25}{n^2}\left[\frac{n(n+1)}{2}\right] \qquad \text{(Formula 1, p. 250)}$$

$$= 5 + \frac{25}{2}\left(1 + \frac{1}{n}\right)$$

Since $P$ is a regular partition, $|P| \to 0$ is equivalent to $n \to \infty$. We conclude that

$$\int_{-2}^{3} (x + 3)\,dx = \lim_{|P| \to 0} \sum_{i=1}^{n} f(\bar{x}_i)\,\Delta x_i$$

$$= \lim_{n \to \infty} \left[5 + \frac{25}{2}\left(1 + \frac{1}{n}\right)\right]$$

$$= \frac{35}{2}$$

We can easily check our answer, since the required integral gives the area of the trapezoid in Figure 8. The familiar trapezoidal area formula $A = \frac{1}{2}(a + b)h$ gives $\frac{1}{2}(1 + 6)5 = 35/2$. ■

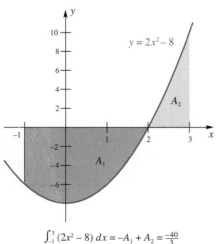

$\int_{-2}^{3} (x+3)\,dx = A = \frac{35}{2}$

FIGURE 8

**EXAMPLE 4** Evaluate $\int_{-1}^{3} (2x^2 - 8)\,dx$.

*Solution* No grade-school formulas will help here, though the integral does correspond to $-A_1 + A_2$, where $A_1$ and $A_2$ are the areas of the regions below and above the $x$-axis in Figure 9.

Let $P$ be a regular partition of $[-1, 3]$ into $n$ equal subintervals, each of length $\Delta x = 4/n$. In each subinterval $[x_{i-1}, x_i]$, choose $\bar{x}_i$ to be the right endpoint, so $\bar{x}_i = x_i$. Then

$$x_i = -1 + i\,\Delta x = -1 + i\left(\frac{4}{n}\right)$$

and

$$f(x_i) = 2x_i^2 - 8 = 2\left[-1 + i\left(\frac{4}{n}\right)\right]^2 - 8$$

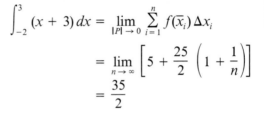

$$= -6 - \frac{16i}{n} + \frac{32i^2}{n^2}$$

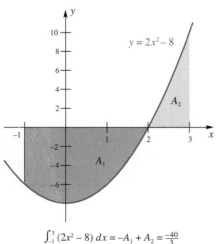

$\int_{-1}^{3} (2x^2 - 8)\,dx = -A_1 + A_2 = \frac{-40}{3}$

FIGURE 9

Consequently,

$$\sum_{i=1}^{n} f(\overline{x}_i)\,\Delta x_i = \sum_{i=1}^{n} f(x_i)\,\Delta x$$

$$= \sum_{i=1}^{n}\left[-6 - \frac{16}{n}\,i + \frac{32}{n^2}\,i^2\right]\frac{4}{n}$$

$$= -\frac{24}{n}\sum_{i=1}^{n}1 - \frac{64}{n^2}\sum_{i=1}^{n}i + \frac{128}{n^3}\sum_{i=1}^{n}i^2$$

$$= \frac{-24}{n}\,(n) - \frac{64}{n^2}\,\frac{n(n+1)}{2} + \frac{128}{n^3}\,\frac{n(n+1)(2n+1)}{6}$$

$$= -24 - 32\left(1 + \frac{1}{n}\right) + \frac{128}{6}\left(2 + \frac{3}{n} + \frac{1}{n^2}\right)$$

We conclude that

$$\int_{-1}^{3}(2x^2 - 8)\,dx = \lim_{|P|\to 0}\sum_{i=1}^{n} f(\overline{x}_i)\,\Delta x_i$$

$$= \lim_{n\to\infty}\left[-24 - 32\left(1 + \frac{1}{n}\right) + \frac{128}{6}\left(2 + \frac{3}{n} + \frac{1}{n^2}\right)\right]$$

$$= -24 - 32 + \frac{128}{3} = \frac{-40}{3}$$

That the answer is negative is not surprising, since the region below the x-axis is larger than the one above the x-axis (Figure 9). Our answer is close to the estimate given in COMMON SENSE; this reassures us that our answer is correct. ∎

> ≈ **Common Sense**
>
> Given the graph of a function, we can always make a rough estimate for the value of a definite integral by using the fact that it is the signed area
>
> $$A_{\text{up}} - A_{\text{down}}$$
>
> Thus, in Example 4, we might estimate the value of the integral by pretending the part above the x-axis is a triangle and the part below is a rectangle. Our estimate is
>
> $$\tfrac{1}{2}(1)(10) - (3)(6) = -13$$

## CONCEPTS REVIEW

**1.** A sum of the form $\sum_{i=1}^{n} f(\overline{x}_i)\,\Delta x_i$ is called a _____ .

**2.** The limit of the sum above is called a _____ and is symbolized by _____ .

**3.** Geometrically, the definite integral corresponds to a signed area. In terms of $A_{\text{up}}$ and $A_{\text{down}}$, $\int_{a}^{b} f(x)\,dx =$ _____ .

**4.** Thus, the value of $\int_{-1}^{4} x\,dx$ is _____ .

## PROBLEM SET 5.5

In Problems 1 and 2, calculate the indicated Riemann sum (see Example 1).

**1.**

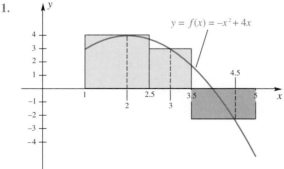

**2.**

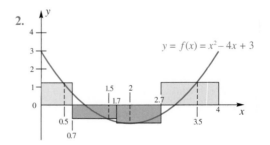

In Problems 3–6, calculate the Riemann sum $\sum_{i=1}^{n} f(\bar{x}_i)\,\Delta x_i$ for the given data.

**3.** $f(x) = x - 1$; $P$: $3 < 3.75 < 4.25 < 5.5 < 6 < 7$; $\bar{x}_1 = 3, \bar{x}_2 = 4, \bar{x}_3 = 4.75, \bar{x}_4 = 6, \bar{x}_5 = 6.5$

**4.** $f(x) = -x/2 + 3$; $P$: $-3 < -1.3 < 0 < 0.9 < 2$; $\bar{x}_1 = -2, \bar{x}_2 = -0.5, \bar{x}_3 = 0, \bar{x}_4 = 2$.

C **5.** $f(x) = x^2/2 + x$; $[-2, 2]$ is divided into eight equal subintervals, $\bar{x}_i$ is the midpoint (see Example 2).

C **6.** $f(x) = 4x^3 + 1$; $[0, 3]$ is divided into six equal subintervals, $\bar{x}_i$ is the right endpoint.

In Problems 7–10, use the given values of $a$ and $b$ and express the given limit as a definite integral.

**7.** $\lim\limits_{|P| \to 0} \sum\limits_{i=1}^{n} (\bar{x}_i)^3 \Delta x_i$; $a = 1, b = 3$

**8.** $\lim\limits_{|P| \to 0} \sum\limits_{i=1}^{n} (\bar{x}_i + 1)^3 \Delta x_i$; $a = 0, b = 2$

**9.** $\lim\limits_{|P| \to 0} \sum\limits_{i=1}^{n} \dfrac{\bar{x}_i^2}{1 + \bar{x}_i} \Delta x_i$; $a = -1, b = 1$

**10.** $\lim\limits_{|P| \to 0} \sum\limits_{i=1}^{n} (\sin \bar{x}_i)^2 \Delta x_i$; $a = 0, b = \pi$

≈ In Problems 11–16, evaluate the definite integrals using the definition, as in Examples 3 and 4.

**11.** $\displaystyle\int_0^2 (x + 1)\,dx$     **12.** $\displaystyle\int_0^2 (x^2 + 1)\,dx$

*Hint:* Use $\bar{x}_i = \dfrac{2i}{n}$.

**13.** $\displaystyle\int_{-2}^1 (2x + \pi)\,dx$     **14.** $\displaystyle\int_{-2}^1 (3x^2 + 2)\,dx$

*Hint:* Use $\bar{x}_i = \dfrac{3i}{n}$.

**15.** $\displaystyle\int_0^5 (x + 1)\,dx$     **16.** $\displaystyle\int_{-10}^{10} (x^2 + x)\,dx$

In Problems 17–20, calculate $\displaystyle\int_a^b f(x)\,dx$, where $a$ and $b$ are the left- and right-endpoints for which $f$ is defined, by using appropriate area formulas from plane geometry. Begin by graphing the given function.

**17.** $f(x) = \begin{cases} 2x & \text{if } 0 \le x \le 1 \\ 2 & \text{if } 1 < x \le 2 \\ x & \text{if } 2 < x \le 5 \end{cases}$

**18.** $f(x) = \begin{cases} 2x & \text{if } 0 \le x \le 1 \\ 2(x - 1) + 2 & \text{if } 1 < x \le 2 \end{cases}$

**19.** $f(x) = \begin{cases} \sqrt{1 - x^2} & \text{if } 0 \le x \le 1 \\ x - 1 & \text{if } 1 < x \le 2 \end{cases}$

**20.** $f(x) = \begin{cases} -\sqrt{4 - x^2} & \text{if } -2 \le x \le 0 \\ -2x - 2 & \text{if } 0 < x \le 2 \end{cases}$

**21.** Which of the following functions are integrable on $[-2, 2]$ and which are not? Give reasons.

(a) $f(x) = x^3 + \sin x$

(b) $f(x) = \dfrac{1}{x + 3}$

(c) $f(x) = \dfrac{1}{x - 1}, f(1) = 3$

(d) $f(x) = \tan x$

(e) $f(x) = \begin{cases} \sin \dfrac{1}{x} & \text{if } x \ne 0 \\ 0 & \text{if } x = 0 \end{cases}$

(f) $f(x) = \begin{cases} x^2 & \text{if } -2 \le x \le 0 \\ 1 & \text{if } 0 < x \le 2 \end{cases}$

**22.** Let $f(x) = \begin{cases} 1 & \text{if } x \text{ is rational} \\ 0 & \text{if } x \text{ is irrational} \end{cases}$

Prove that $f$ is not integrable on $[0, 1]$. *Hint:* Show that no matter how small the norm of the partition, $|P|$, the Riemann sum $\sum_{i=1}^{n} f(\bar{x}_i)\,\Delta x_i$ can be made to have value either 0 or 1.

**23.** Recall that $[\![x]\!]$ denotes the greatest integer less than or equal to $x$. Calculate each of the following integrals. You may use geometric reasoning and the fact that $\displaystyle\int_0^b x^2\,dx = b^3/3$. (The latter is shown in Problem 26.)

(a) $\displaystyle\int_{-3}^3 [\![x]\!]\,dx$     (b) $\displaystyle\int_{-3}^3 [\![x]\!]^2\,dx$

(c) $\displaystyle\int_{-3}^3 (x - [\![x]\!])\,dx$     (d) $\displaystyle\int_{-3}^3 (x - [\![x]\!])^2\,dx$

(e) $\displaystyle\int_{-3}^3 |x|\,dx$     (f) $\displaystyle\int_{-3}^3 x|x|\,dx$

(g) $\displaystyle\int_{-1}^2 |x|[\![x]\!]\,dx$     (h) $\displaystyle\int_{-1}^2 x^2 [\![x]\!]\,dx$

**24.** Let $f$ be an odd function and $g$ be an even function, and suppose $\displaystyle\int_0^1 |f(x)|\,dx = \int_0^1 g(x)\,dx = 3$. Use geometric reasoning to calculate each of the following.

(a) $\displaystyle\int_{-1}^1 f(x)\,dx$     (b) $\displaystyle\int_{-1}^1 g(x)\,dx$

(c) $\int_{-1}^{1} |f(x)|\, dx$

(d) $\int_{-1}^{1} [-g(x)]\, dx$

(e) $\int_{-1}^{1} x g(x)\, dx$

(f) $\int_{-1}^{1} f^{3}(x) g(x)\, dx$

**25.** Show that $\int_{a}^{b} x\, dx = \tfrac{1}{2}(b^{2} - a^{2})$ by completing the following argument. For the partition $a = x_{0} < x_{1} < \cdots < x_{n} = b$, choose $\bar{x}_{i} = \tfrac{1}{2}(x_{i-1} + x_{i})$. Then, $R_{p} = \sum_{i=1}^{n} \bar{x}_{i}\, \Delta x_{i} = \tfrac{1}{2}\sum_{i=1}^{n} (x_{i} + x_{i-1})(x_{i} - x_{i-1})$. Now simplify $R_{p}$ (collapsing sum) and take a limit.

**26.** Show that $\int_{a}^{b} x^{2}\, dx = \tfrac{1}{3}(b^{3} - a^{3})$ by an argument like that in Problem 25 but using $\bar{x}_{i} = [\tfrac{1}{3}(x_{i-1}^{2} + x_{i-1}x_{i} + x_{i}^{2})]^{1/2}$

PC Many computer algebra systems (for example the *student* package in *Maple* or the *Area* section in *True BASIC Calculus*) permit the evaluation of Riemann sums for left endpoint, right endpoint, or midpoint evaluations of the function. Using such a system in Problems 27–30, evaluate the 10 subinterval Riemann sums using left endpoint, right endpoint, and midpoint evaluations.

**27.** $\int_{0}^{2} (x^{3} + 1)\, dx$

**28.** $\int_{0}^{1} \tan x\, dx$

**29.** $\int_{0}^{1} \cos x\, dx$

**30.** $\int_{1}^{3} (1/x)\, dx$

PC Many computer algebra systems and graphing calculators can be used for the numerical evaluation of definite integrals. Use such an aid to evaluate the definite integrals in Problems 31–36 (or state that the function is not integrable on the given interval).

**31.** $\int_{-2}^{4} (-1 + |x|)\, dx$

**32.** $\int_{0}^{6} \sin x\, dx$

**33.** $\int_{-1}^{2} (x^{4} - 3x^{2} + 1)\, dx$

**34.** $\int_{-2}^{2} [(x - 1)/(x^{2} + 1)]\, dx$

**35.** $\int_{0}^{1} (1/x)\, dx$

**36.** $\int_{0}^{2} \tan x\, dx$

**Answers to Concepts Review:** **1.** Riemann sum
**2.** Definite integral, $\int_{a}^{b} f(x)\, dx$ **3.** $A_{\text{up}} - A_{\text{down}}$ **4.** $\frac{15}{2}$

## 5.6
## THE FUNDAMENTAL THEOREM OF CALCULUS

We have been able to evaluate a few definite integrals directly from the definition, but only because we have nice formulas for $1 + 2 + \cdots + n, 1^{2} + 2^{2} + \cdots + n^{2}$, and so on. Calculating definite integrals this way is always tedious, usually difficult, and sometimes practically impossible. Fortunately, there is a better way; that is the subject of this section.

We have credited Isaac Newton and Gottfried Leibniz with the simultaneous, but independent, discovery of calculus. Yet the concepts of the slope of the tangent line (derivative) and the area of a curved region (definite integral) were known earlier. Why then do Newton and Leibniz figure so prominently in the history of calculus? They do so because they understood and exploited the intimate relationship that exists between antiderivatives and definite integrals, a relationship that enables us to compute easily the exact values of many definite integrals without ever using Riemann sums. This connection is so important that it is called the *Fundamental Theorem of Calculus*.

**The Fundamental Theorem** You have met several fundamental theorems before in your mathematical career. The Fundamental Theorem of Arithmetic says that a composite whole number factors uniquely into a product of primes. The Fundamental Theorem of Algebra says that an *n*th-degree polynomial equation has exactly *n* solutions, counting multiplicities. Any theorem with the title Fundamental Theorem should be studied carefully and then permanently committed to memory.

### Is It Fundamental?

The Fundamental Theorem of Calculus is important in providing a powerful tool for evaluating definite integrals. But its deepest significance lies in the link it makes between differentiation and integration, between derivatives and integrals. This link appears in sparkling clarity when we rewrite the conclusion to the theorem with $f(x)$ replaced by $g'(x)$.

$$\int_a^b g'(x)\, dx = g(b) - g(a)$$

### Theorem A

**(Fundamental Theorem of Calculus).** Let $f$ be continuous (hence integrable) on $[a, b]$ and let $F$ be *any* antiderivative of $f$ there. Then,

$$\int_a^b f(x)\, dx = F(b) - F(a)$$

**Proof** Let $P: a = x_0 < x_1 < x_2 < \cdots < x_{n-1} < x_n = b$ be an arbitrary partition of $[a, b]$. Then the standard "subtract-and-add" trick gives

$$F(b) - F(a) = F(x_n) - F(x_{n-1}) + F(x_{n-1}) - F(x_{n-2}) + \cdots + F(x_1) - F(x_0)$$
$$= \sum_{i=1}^n [F(x_i) - F(x_{i-1})]$$

By the Mean Value Theorem for Derivatives (Theorem 4.8A) applied to $F$ on the interval $[x_{i-1}, x_i]$,

$$F(x_i) - F(x_{i-1}) = F'(\overline{x}_i)(x_i - x_{i-1}) = f(\overline{x}_i)\,\Delta x_i$$

for some choice of $\overline{x}_i$ in the open interval $(x_{i-1}, x_i)$. Thus,

$$F(b) - F(a) = \sum_{i=1}^n f(\overline{x}_i)\,\Delta x_i$$

On the left we have a constant; on the right we have a Riemann sum for $f$ on $[a, b]$. When we take limits of both sides as $|P| \to 0$, we obtain

$$F(b) - F(a) = \lim_{|P| \to 0} \sum_{i=1}^n f(\overline{x}_i)\,\Delta x_i = \int_a^b f(x)\, dx \qquad \blacksquare$$

Before going on to examples, ask yourself why we can use the word *any* in the statement of the theorem.

**EXAMPLE 1** Show that $\int_a^b k\, dx = k(b - a)$, $k$ a constant.

**Solution** $F(x) = kx$ is an antiderivative of $f(x) = k$. Thus, by the Fundamental Theorem of Calculus,

$$\int_a^b k\, dx = F(b) - F(a) = kb - ka = k(b - a) \qquad \blacksquare$$

**EXAMPLE 2** Show that $\int_a^b x\, dx = \dfrac{b^2}{2} - \dfrac{a^2}{2}$.

**Solution** $F(x) = x^2/2$ is an antiderivative of $f(x) = x$. Therefore,

$$\int_a^b x\, dx = F(b) - F(a) = \frac{b^2}{2} - \frac{a^2}{2} \qquad \blacksquare$$

**EXAMPLE 3**    Show that if $r$ is a rational number different from $-1$, then

$$\int_a^b x^r\, dx = \frac{b^{r+1}}{r+1} - \frac{a^{r+1}}{r+1}$$

*Solution*    $F(x) = x^{r+1}/(r+1)$ is an antiderivative of $f(x) = x^r$. Thus, by the Fundamental Theorem of Calculus,

$$\int_a^b x^r\, dx = F(b) - F(a) = \frac{b^{r+1}}{r+1} - \frac{a^{r+1}}{r+1}$$

*Technical point*: If $r < 0$, we require that $0$ is not in $[a, b]$. Why?    ∎

It is convenient to introduce a special symbol for $F(b) - F(a)$. We write

$$F(b) - F(a) = [F(x)]_a^b$$

Thus, for example,

$$\int_2^5 x^2\, dx = \left[\frac{x^3}{3}\right]_2^5 = \frac{125}{3} - \frac{8}{3} = \frac{117}{3} = 39$$

**EXAMPLE 4**    Evaluate $\displaystyle\int_{-1}^2 (4x - 6x^2)\, dx$.

*Solution*

$$\begin{aligned}
\int_{-1}^2 (4x - 6x^2)\, dx &= [2x^2 - 2x^3]_{-1}^2 \\
&= (8 - 16) - (2 + 2) = -12 \quad\quad ∎
\end{aligned}$$

**EXAMPLE 5**    Evaluate $\displaystyle\int_1^8 (x^{1/3} + x^{4/3})\, dx$.

*Solution*

$$\begin{aligned}
\int_1^8 (x^{1/3} + x^{4/3})\, dx &= [\tfrac{3}{4}x^{4/3} + \tfrac{3}{7}x^{7/3}]_1^8 \\
&= (\tfrac{3}{4}\cdot 16 + \tfrac{3}{7}\cdot 128) - (\tfrac{3}{4}\cdot 1 + \tfrac{3}{7}\cdot 1) \\
&= \tfrac{45}{4} + \tfrac{381}{7} \approx 65.68 \quad\quad ∎
\end{aligned}$$

**EXAMPLE 6**    Evaluate $\displaystyle\int_0^\pi 3 \sin x\, dx$.

*Solution*

$$\int_0^\pi 3 \sin x\, dx = [-3 \cos x]_0^\pi = 3 + 3 = 6 \quad\quad ∎$$

In terms of the symbol for indefinite integrals, we may write the conclusion of the Fundamental Theorem of Calculus as

$$\int_a^b f(x)\,dx = \left[\int f(x)\,dx\right]_a^b$$

The nontrivial part of applying the theorem is always to find the indefinite integral $\int f(x)\,dx$. To do that we may want to use the substitution technique (Generalized Power Rule) we learned in Section 5.1.

**EXAMPLE 7**  Evaluate $\displaystyle\int_0^4 \sqrt{x^2 + x}\,(2x + 1)\,dx.$

*Solution*  Let $u = x^2 + x$; then $du = (2x + 1)\,dx.$

Thus,

$$\int \sqrt{x^2 + x}\,(2x + 1)\,dx = \int u^{1/2}\,du = \tfrac{2}{3}u^{3/2} + C$$

$$= \tfrac{2}{3}(x^2 + x)^{3/2} + C$$

Therefore, by the Fundamental Theorem of Calculus,

$$\int_0^4 \sqrt{x^2 + x}\,(2x + 1)\,dx = \left[\tfrac{2}{3}(x^2 + x)^{3/2} + C\right]_0^4$$

$$= \left[\tfrac{2}{3}\,(20)^{3/2} + C\right] - \left[0 + C\right]$$

$$= \tfrac{2}{3}\,(20)^{3/2} \approx 59.63 \qquad \blacksquare$$

Note that the $C$ of the indefinite integration cancels out, as it always will, in the definite integration. That is why in the statement of the Fundamental Theorem we could use the phrase *any antiderivative*. In particular, we may always choose $C = 0$ in applying the Fundamental Theorem.

**EXAMPLE 8**  Evaluate $\displaystyle\int_0^{\pi/4} \sin^3 2x \cos 2x\,dx.$

*Solution*  Let $u = \sin 2x$; then $du = 2 \cos 2x\,dx$. Thus,

$$\int \sin^3 2x \cos 2x\,dx = \frac{1}{2}\int (\sin^3 2x)(2 \cos 2x)\,dx = \frac{1}{2}\int u^3\,du$$

$$= \frac{1}{2}\frac{u^4}{4} + C = \frac{\sin^4 2x}{8} + C$$

Therefore, by the Fundamental Theorem of Calculus,

$$\int_0^{\pi/4} \sin^3 2x \cos 2x\,dx = \left[\frac{\sin^4 2x}{8}\right]_0^{\pi/4} = \frac{1}{8} - 0 = \frac{1}{8} \qquad \blacksquare$$

**The Definite Integral is a Linear Operator**   Earlier we learned that $D_x$, $\int \cdots dx$, and $\Sigma$ are linear operators. You can add $\int_a^b \cdots dx$ to the list.

---

Theorem B

**(Linearity of the Definite Integral).** Suppose that $f$ and $g$ are integrable on $[a, b]$ and that $k$ is a constant. Then $kf$ and $f + g$ are integrable and:

**(i)** $\displaystyle\int_a^b kf(x)\,dx = k\int_a^b f(x)\,dx;$

**(ii)** $\displaystyle\int_a^b [f(x) + g(x)]\,dx = \int_a^b f(x)\,dx + \int_a^b g(x)\,dx;$ and consequently

**(iii)** $\displaystyle\int_a^b [f(x) - g(x)]\,dx = \int_a^b f(x)\,dx - \int_a^b g(x)\,dx.$

---

**Proof**  The proofs of (i) and (ii) depend on the linearity of $\Sigma$ and the properties of limits. We show (ii).

$$
\begin{aligned}
\int_a^b [f(x) + g(x)]\,dx &= \lim_{|P| \to 0} \sum_{i=1}^{n} [f(\overline{x}_i) + g(\overline{x}_i)]\,\Delta x_i \\
&= \lim_{|P| \to 0} \left[ \sum_{i=1}^{n} f(\overline{x}_i)\,\Delta x_i + \sum_{i=1}^{n} g(\overline{x}_i)\,\Delta x_i \right] \\
&= \lim_{|P| \to 0} \sum_{i=1}^{n} f(\overline{x}_i)\,\Delta x_i + \lim_{|P| \to 0} \sum_{i=1}^{n} g(\overline{x}_i)\,\Delta x_i \\
&= \int_a^b f(x)\,dx + \int_a^b g(x)\,dx
\end{aligned}
$$

Part (iii) follows from (i) and (ii) on writing $f(x) - g(x)$ as $f(x) + (-1)g(x)$. ∎

**EXAMPLE 9**   Evaluate $\displaystyle\int_{-1}^{2} (4x - 6x^2)\,dx.$

**Solution**   This is the same as Example 4, but now we will do it another way making use of the linearity of the definite integral.

$$
\begin{aligned}
\int_{-1}^{2} (4x - 6x^2)\,dx &= 4\int_{-1}^{2} x\,dx - 6\int_{-1}^{2} x^2\,dx \\
&= 4\left[\frac{x^2}{2}\right]_{-1}^{2} - 6\left[\frac{x^3}{3}\right]_{-1}^{2} \\
&= 4\left(\frac{4}{2} - \frac{1}{2}\right) - 6\left(\frac{8}{3} + \frac{1}{3}\right) = -12
\end{aligned}
$$
■

**EXAMPLE 10**   Evaluate $\displaystyle\int_0^1 [x^2 + (x^2 + 1)^4 x]\,dx$.

**Solution**

$$\int_0^1 [x^2 + (x^2 + 1)^4 x]\,dx = \int_0^1 x^2\,dx + \int_0^1 (x^2 + 1)^4 x\,dx$$

The first integral is easy to do directly. To handle the second, we let $u = x^2 + 1$, so $du = 2x\,dx$ and write

$$\int (x^2 + 1)^4 x\,dx = \frac{1}{2}\int (x^2 + 1)^4 2x\,dx = \frac{1}{2}\int u^4\,du$$

$$= \frac{1}{2}\frac{u^5}{5} + C = \frac{(x^2 + 1)^5}{10} + C$$

Therefore,

$$\int_0^1 x^2\,dx + \int_0^1 (x^2 + 1)^4 x\,dx = \left[\frac{x^3}{3}\right]_0^1 + \left[\frac{(x^2 + 1)^5}{10}\right]_0^1$$

$$= \left(\frac{1}{3} - 0\right) + \left(\frac{32}{10} - \frac{1}{10}\right)$$

$$= \frac{1}{3} + \frac{31}{10} = \frac{103}{30} \qquad \blacksquare$$

## CONCEPTS REVIEW

**1.** If $f$ is continuous on $[a, b]$ and if $F$ is any _____ of $f$ there, then $\displaystyle\int_a^b f(x)\,dx =$ _____.

**2.** The symbol $[F(x)]_a^b$ stands for the expression _____.

**3.** $\displaystyle\int_1^3 x^2\,dx = [x^3/3]_1^3 =$ _____.

**4.** $\displaystyle\int_{-2}^{-1} x^{-2}\,dx = [$_____$]_{-2}^{-1} =$ _____.

## PROBLEM SET 5.6

In Problems 1–14, use the Fundamental Theorem of Calculus to evaluate each definite integral (see Examples 1–6).

**1.** $\displaystyle\int_0^2 x^3\,dx$

**2.** $\displaystyle\int_{-1}^2 x^4\,dx$

**3.** $\displaystyle\int_{-1}^2 (3x^2 - 2x + 3)\,dx$

**4.** $\displaystyle\int_1^2 (4x^3 + 7)\,dx$

**5.** $\displaystyle\int_1^4 \frac{1}{w^2}\,dw$

**6.** $\displaystyle\int_1^3 \frac{2}{t^3}\,dt$

**7.** $\displaystyle\int_0^4 \sqrt{t}\,dt$

**8.** $\displaystyle\int_1^8 \sqrt[3]{w}\,dw$

**9.** $\displaystyle\int_{-4}^{-2} \left(y^2 + \frac{1}{y^3}\right)dy$

**10.** $\displaystyle\int_1^4 \frac{s^4 - 8}{s^2}\,ds$

**11.** $\displaystyle\int_0^{\pi/2} \cos x\,dx$

**12.** $\displaystyle\int_{\pi/6}^{\pi/2} 2\sin t\,dt$

**13.** $\displaystyle\int_0^1 (2x^4 - 3x^2 + 5)\,dx$

**14.** $\displaystyle\int_0^1 (x^{4/3} - 2x^{1/3})\,dx$

In Problems 15–30, use the Fundamental Theorem of Calculus combined with the Generalized Power Rule to evaluate the given definite integral (see Examples 7–10).

**15.** $\int_0^1 (x^2 + 1)^{10}(2x)\, dx$

**16.** $\int_{-1}^0 \sqrt{x^3 + 1}(3x^2)\, dx$

**17.** $\int_{-1}^3 \dfrac{1}{(t + 2)^2}\, dt$

**18.** $\int_2^{10} \sqrt{y - 1}\, dy$

**19.** $\int_5^8 \sqrt{3x + 1}\, dx$

**20.** $\int_1^7 \dfrac{1}{\sqrt{2x + 2}}\, dx$

**21.** $\int_{-3}^3 \sqrt{7 + 2t^2}\,(8t)\, dt$

**22.** $\int_1^3 \dfrac{x^2 + 1}{\sqrt{x^3 + 3x}}\, dx$

**23.** $\int_0^{\pi/2} \cos^2 x \sin x\, dx$

**24.** $\int_0^{\pi/2} \sin^2 3x \cos 3x\, dx$

**25.** $\int_0^{\pi/2} (2x + \sin x)\, dx$

**26.** $\int_0^{\pi/2} [4x + 3 + \cos x]\, dx$

**27.** $\int_0^4 \left[\sqrt{x} + \sqrt{2x + 1}\right] dx$

**28.** $\int_{-4}^{-1} \dfrac{1 - s^4}{2s^2}\, ds$

**29.** $\int_0^1 (x^2 + 2x)^2\, dx$

**30.** $\int_a^{8a} (a^{1/3} - x^{1/3})^3\, dx$

In Problems 31–34, first recognize the given limit as a definite integral and then evaluate that integral by the Fundamental Theorem of Calculus.

**31.** $\lim\limits_{n \to \infty} \sum\limits_{i=1}^n \left(\dfrac{3i}{n}\right)^2 \dfrac{3}{n}$

**32.** $\lim\limits_{n \to \infty} \sum\limits_{i=1}^n \left(\dfrac{2i}{n}\right)^3 \dfrac{2}{n}$

**33.** $\lim\limits_{n \to \infty} \sum\limits_{i=1}^n \left[\sin\left(\dfrac{\pi i}{n}\right)\right] \dfrac{\pi}{n}$

**34.** $\lim\limits_{n \to \infty} \sum\limits_{i=1}^n \left[1 + \dfrac{2i}{n} + \left(\dfrac{2i}{n}\right)^2\right] \dfrac{2}{n}$

**35.** Explain why $(1/n^3) \sum\limits_{i=1}^n i^2$ should be a good approximation to $\int_0^1 x^2\, dx$ for large $n$. Now calculate the summa-

tion expression for $n = 10$ and the integral by the Fundamental Theorem of Calculus and compare their values.

Suppose that $\int_0^1 f(x)\, dx = 4$ and $\int_0^1 g(x)\, dx = -2$. Use the linearity of the integral to calculate each of the integrals in Problems 36–40.

**36.** $\int_0^1 3f(x)\, dx$

**37.** $\int_0^1 [f(x) - g(x)]\, dx$

**38.** $\int_0^1 [3f(x) + 2g(x)]\, dx$

**39.** $\int_0^1 [2g(x) - 3f(x)]\, dx$

**40.** $\int_1^0 [2f(x) - 3x^2]\, dx$

**41.** Evaluate $\int_{-2}^4 (2[\![x]\!] - 3|x|)\, dx$

**42.** Show that $\frac{1}{2}x|x|$ is an antiderivative of $|x|$ and use this fact to get a simple formula for $\int_a^b |x|\, dx$.

**43.** Find a nice formula for $\int_0^b [\![x]\!]\, dx$, $b > 0$.

**44.** We claim that

$$\int_a^b x^n\, dx + \int_{a^n}^{b^n} \sqrt[n]{y}\, dy = b^{n+1} - a^{n+1}$$

(a) Use Figure 1 to justify this by a geometric argument.
(b) Prove the result using the Fundamental Theorem of Calculus.
(c) Show that $A_n = nB_n$.

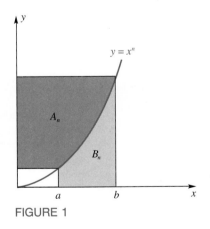

FIGURE 1

**5.7**
**MORE PROPERTIES OF**
**THE DEFINITE INTEGRAL**

Our definition of the definite integral was motivated by the problem of area for curved regions. Consider the two curved regions $R_1$ and $R_2$ in Figure 1 and let $R = R_1 \cup R_2$. It is clear that

$$A(R) = A(R_1 \cup R_2) = A(R_1) + A(R_2)$$

which suggests that

$$\int_a^c f(x)\,dx = \int_a^b f(x)\,dx + \int_b^c f(x)\,dx$$

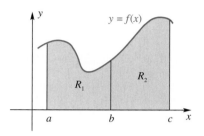

FIGURE 1

We quickly point out that this does not constitute a proof of the fact about integrals, since—first of all—our discussion of area in Section 5.4 was rather informal, and—second—our diagram supposes that $f$ is positive, which it need not be. Nevertheless, definite integrals do satisfy this interval additive property and they do it no matter how the three points $a$, $b$, and $c$ are arranged. We leave the rigorous proof to more advanced works (however, see Problem 45).

---

**Theorem A**

**(Interval Additive Property).** If $f$ is integrable on an interval containing the three points $a$, $b$, and $c$, then

$$\int_a^c f(x)\,dx = \int_a^b f(x)\,dx + \int_b^c f(x)\,dx$$

no matter what the order of $a$, $b$, and $c$.

---

For example,

$$\int_0^2 x^2\,dx = \int_0^1 x^2\,dx + \int_1^2 x^2\,dx$$

which most people readily believe. But it is also true that

$$\int_0^2 x^2\,dx = \int_0^3 x^2\,dx + \int_3^2 x^2\,dx$$

which may seem surprising. If you mistrust the theorem, you might actually evaluate each of the above integrals to see that the equality holds.

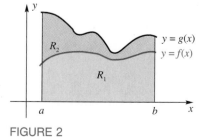

FIGURE 2

**Comparison Properties**　Consideration of the areas of the regions $R_1$ and $R_2$ in Figure 2 suggests another property of definite integrals.

### Theorem B

**(Comparison Property).** If $f$ and $g$ are integrable on $[a, b]$ and if $f(x) \le g(x)$ for all $x$ in $[a, b]$, then

$$\int_a^b f(x)\, dx \le \int_a^b g(x)\, dx$$

In informal but descriptive language, we say that the definite integral preserves inequalities.

***Proof*** Let $P: a = x_0 < x_1 < x_2 < \cdots < x_n = b$ be an arbitrary partition of $[a, b]$, and for each $i$ let $\bar{x}_i$ be any sample point on the $i$th subinterval $[x_{i-1}, x_i]$. We may conclude successively that

$$f(\bar{x}_i) \le g(\bar{x}_i)$$

$$f(\bar{x}_i)\, \Delta x_i \le g(\bar{x}_i)\, \Delta x_i$$

$$\sum_{i=1}^n f(\bar{x}_i)\, \Delta x_i \le \sum_{i=1}^n g(\bar{x}_i)\, \Delta x_i$$

$$\lim_{|P| \to 0} \sum_{i=1}^n f(\bar{x}_i)\, \Delta x_i \le \lim_{|P| \to 0} \sum_{i=1}^n g(\bar{x}_i)\, \Delta x_i$$

$$\int_a^b f(x)\, dx \le \int_a^b g(x)\, dx \qquad \blacksquare$$

### Theorem C

**(Boundedness Property).** If $f$ is integrable on $[a, b]$ and if $m \le f(x) \le M$ for all $x$ in $[a, b]$, then

$$m(b - a) \le \int_a^b f(x)\, dx \le M(b - a)$$

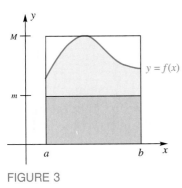

FIGURE 3

***Proof*** The picture in Figure 3 helps us understand the theorem. Note that $m(b - a)$ is the area of the lower, small rectangle, $M(b - a)$ is the area of the large rectangle, and $\int_a^b f(x)\, dx$ is the area under the curve.

To prove the right-hand inequality, let $g(x) = M$ for all $x$ in $[a, b]$. Then, by Theorem B,

$$\int_a^b f(x)\, dx \le \int_a^b g(x)\, dx$$

However,

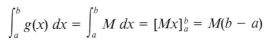

$$\int_a^b g(x)\, dx = \int_a^b M\, dx = [Mx]_a^b = M(b - a)$$

The left-hand inequality is handled similarly. $\qquad \blacksquare$

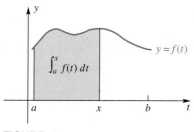

**FIGURE 4**

**Fundamental Too**

You will use Theorem D much less frequently than what we earlier called the Fundamental Theorem of Calculus. Yet it is a fundamental theorem too because it links differentiation and integration in another way. Differentiating a definite integral of a function (with respect to the upper limit) returns the original function. Differentiation and integration are inverse processes; they undo each other.

**Differentiating a Definite Integral with Respect to a Variable Upper Limit** The next theorem is important—so important that some authors call it the Second Fundamental Theorem of Calculus.

Let $f$ be integrable on $[a, b]$ and let $x$ be any point in $[a, b]$ (Figure 4). Then for each $x$, $\int_a^x f(t)\,dt$ is a unique number, and so

$$G(x) = \int_a^x f(t)\,dt$$

determines a function $G$ with domain $[a, b]$. We call $\int_a^x f(t)\,dt$ an integral with variable upper limit. Note our use of $t$ rather than $x$ as the dummy variable to avoid confusion with the upper limit.

**Theorem D**

**(Differentiating a Definite Integral).** Let $f$ be continuous on the closed interval $[a, b]$ and let $x$ be a (variable) point in $(a, b)$. Then

$$D_x \left[ \int_a^x f(t)\,dt \right] = f(x)$$

This should be learned in words. *The derivative of a definite integral with respect to its upper limit is the integrand at the upper limit.*

**Proof** If $G(x) = \int_a^x f(t)\,dt$, we must show that

$$G'(x) = \lim_{h \to 0} \frac{G(x + h) - G(x)}{h} = f(x)$$

Now

$$G(x + h) - G(x) = \int_a^{x+h} f(t)\,dt - \int_a^x f(t)\,dt = \int_x^{x+h} f(t)\,dt$$

by Theorem A.

Assume for the moment that $h > 0$ and let $m$ and $M$ be the minimum value and maximum value, respectively, of $f$ on the interval $[x, x + h]$ (Figure 5). By Theorem C,

$$mh \le \int_x^{x+h} f(t)\,dt \le Mh$$

or

$$mh \le G(x + h) - G(x) \le Mh$$

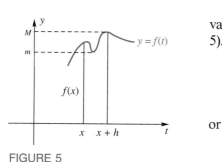

**FIGURE 5**

Dividing by $h$, we obtain

$$m \le \frac{G(x + h) - G(x)}{h} \le M$$

Now $m$ and $M$ really depend on $h$. Moreover, since $f$ is continuous, both $m$ and $M$ must approach $f(x)$ as $h \to 0$. Thus, by the Squeeze Theorem,

$$\lim_{h \to 0} \frac{G(x + h) - G(x)}{h} = f(x)$$

The case where $h < 0$ is handled similarly. ■

One theoretical consequence of this theorem is that every continuous function $f$ has an antiderivative $F$ given by

$$F(x) = \int_a^x f(t)\, dt$$

However, this fact is not helpful in getting a simple formula for any particular antiderivative.

**EXAMPLE 1**  Find $D_x \left[ \int_1^x t^2\, dt \right]$ two ways.

*Solution*  First, we will do it the hard way—that is, by evaluating the integral and then taking the derivative.

$$\int_1^x t^2\, dt = \left[ \frac{t^3}{3} \right]_1^x = \frac{x^3}{3} - \frac{1}{3}$$

Therefore,

$$D_x \left[ \int_1^x t^2\, dt \right] = D_x \left[ \frac{x^3}{3} - \frac{1}{3} \right] = x^2$$

Now we do it the easy way. By Theorem D,

$$D_x \left[ \int_1^x t^2\, dt \right] = x^2$$

■

**EXAMPLE 2**  Find $D_x \left[ \int_2^x \frac{t^{3/2}}{\sqrt{t^2 + 17}}\, dt \right]$.

*Solution*  We challenge anyone to do this example by first evaluating the integral. However, by Theorem D, it is a trivial problem:

$$D_x \left[ \int_2^x \frac{t^{3/2}}{\sqrt{t^2 + 17}}\, dt \right] = \frac{x^{3/2}}{\sqrt{x^2 + 17}}$$

■

**EXAMPLE 3** Find $D_x \left[ \int_x^4 \tan^2 u \cos u \, du \right]$, $\frac{\pi}{2} < x < \frac{3\pi}{2}$.

*Solution* Use of the dummy variable $u$ rather than $t$ should not bother anyone. However, the fact that $x$ is the lower limit, rather than the upper limit, is troublesome. Here is how we handle this difficulty.

$$D_x \left[ \int_x^4 \tan^2 u \cos u \, du \right] = D_x \left[ -\int_4^x \tan^2 u \cos u \, du \right]$$

$$= -D_x \left[ \int_4^x \tan^2 u \cos u \, du \right] = -\tan^2 x \cos x$$

The interchange of the upper and lower limits is allowed if we prefix a minus sign (see the definition on p. 264). ■

**EXAMPLE 4** Find $D_x \left[ \int_0^{x^2} (3t - 1) \, dt \right]$

*Solution* Now we have a new complication; the upper limit is $x^2$ rather than $x$ (we need $x$ there in order to apply Theorem D). This problem is handled by the Chain Rule. We may think of the expression in brackets as

$$\int_0^u (3t - 1) \, dt \qquad \text{where } u = x^2$$

By the Chain Rule, the derivative with respect to $x$ of this composite function is

$$D_u \left[ \int_0^u (3t - 1) \, dt \right] \cdot D_x u = (3u - 1)(2x) = (3x^2 - 1)(2x) = 6x^3 - 2x$$

If you are skeptical of what we have just done, try doing the problem by first evaluating the integral. ■

**EXAMPLE 5** Find $D_x \left[ \int_{2x}^5 \sqrt{t^2 + 2} \, dt \right]$.

*Solution* Here we first interchange limits and then use Theorem D in conjunction with the Chain Rule.

$$D_x \left[ \int_{2x}^5 \sqrt{t^2 + 2} \, dt \right] = D_x \left[ -\int_5^{2x} \sqrt{t^2 + 2} \, dt \right]$$

$$= -\sqrt{(2x)^2 + 2} \, (2) = -2\sqrt{4x^2 + 2}$$ ■

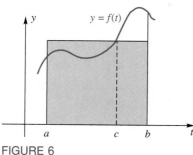

FIGURE 6

**The Mean Value Theorem for Integrals** By this time you realize that the Mean Value Theorem for Derivatives plays an important role in calculus. There is a theorem by the same name for integrals, which is of less importance but still worth knowing. Geometrically, it says that in Figure 6, the area under the curve is equal to the area of the shaded rectangle.

≈ Estimating Integrals

Theorem E with its accompanying Figure 6 suggests a good way to estimate the value of a definite integral. The area of the region under a curve is equal to the area of a rectangle. One can make a good guess at this rectangle by simply "eyeballing" the region. In Figure 6, the area of the shaded part above the curve should match the area of the white part below the curve.

### Theorem E

**(Mean Value Theorem for Integrals).** If $f$ is continuous on $[a, b]$, there is a number $c$ between $a$ and $b$ such that

$$\int_a^b f(t)\, dt = f(c)(b - a)$$

**Proof**  Let

$$G(x) = \int_a^x f(t)\, dt \qquad a \leq x \leq b$$

By the Mean Value Theorem for Derivatives applied to $G$, there is a point $c$ in $(a, b)$ such that

$$G(b) - G(a) = G'(c)(b - a)$$

that is,

$$\int_a^b f(t)\, dt - 0 = G'(c)(b - a)$$

But by Theorem D, $G'(c) = f(c)$. The conclusion follows.  ■

Note that if we solve for $f(c)$ in the conclusion of Theorem E, we get

$$f(c) = \frac{\int_a^b f(t)\, dt}{b - a}$$

The number $\int_a^b f(x)\, dx/(b - a)$ is called the mean value, or **average value**, of $f$ on $[a, b]$. To see why it has this name, consider a regular partition $P: a = x_0 < x_1 < x_2 < \cdots < x_n = b$ with $\Delta x = (b - a)/n$. The average of the $n$ values $f(x_1), f(x_2), \ldots, f(x_n)$ is

$$\frac{f(x_1) + f(x_2) + \cdots + f(x_n)}{n} = \sum_{i=1}^n f(x_i) \frac{1}{n}$$

$$= \frac{1}{b - a} \sum_{i=1}^n f(x_i) \frac{b - a}{n}$$

$$= \frac{1}{b - a} \sum_{i=1}^n f(x_i) \Delta x$$

The sum in the last expression is a Riemann sum for $f$ on $[a, b]$ and therefore approaches $\int_a^b f(x)\, dx$ as $n \to \infty$. Thus $\left(\int_a^b f(x)\, dx\right)/(b - a)$ appears as the natural extension of the familiar notion of average value.

## CONCEPTS REVIEW

1. $\int_1^6 f(x)\,dx = \int_1^8 f(x)\,dx + \int_a^6 f(x)\,dx$, where $a = $ _____.

2. Since $4 \le x^2 \le 16$ for all $x$ in $[2, 4]$, the Boundedness Property of the definite integral allows us to say _____ $\le \int_2^4 x^2\,dx \le$ _____.

3. $D_x\left[\int_1^x \sin^3 t\,dt\right] = $ _____, but

$D_x\left[\int_1^{x^2} \sin^3 t\,dt\right] = $ _____.

4. According to the Mean Value Theorem for Integrals, there is a number $c$ between $-1$ and $3$ such that $\int_{-1}^3 x^2\,dx = $ _____.

## PROBLEM SET 5.7

Suppose $\int_0^1 f(x)\,dx = 2$, $\int_1^2 f(x)\,dx = 3$, $\int_0^1 g(x)\,dx = -1$, and $\int_0^2 g(x)\,dx = 4$. Use properties of definite integrals (linearity, interval additivity, and so on) to calculate each of the integrals in Problems 1–10.

1. $\int_1^2 2f(x)\,dx$

2. $\int_0^2 2f(x)\,dx$

3. $\int_0^2 [2f(x) + g(x)]\,dx$

4. $\int_0^2 [2f(s) + g(s)]\,ds$

5. $\int_2^1 [2f(s) + 5g(s)]\,ds$

6. $\int_1^1 [3f(x) + 2g(x)]\,dx$

7. $\int_0^2 [3f(t) + 2g(t)]\,dt$

8. $\int_0^2 [\sqrt{3}f(t) + \sqrt{2}g(t) + \pi]\,dt$

9. $\int_0^1 [f(t) - g(t) + \pi t]\,dt$

10. $\int_0^1 tf(t)\,dt + \int_1^0 tf(t)\,dt + \int_1^0 \pi g(t)\,dt$

In Problems 11–20, find $G'(x)$ (see Examples 1–5).

11. $G(x) = \int_1^x 2t\,dt$

12. $G(x) = \int_x^1 2t\,dt$

13. $G(x) = \int_0^x (2t^2 + \sqrt{t})\,dt$

14. $G(x) = \int_1^x \cos^3(2t)\tan(t)\,dt$; $-\pi/2 < x < \pi/2$

15. $G(x) = \int_x^{\pi/2} (s - 2)\cot(2s)\,ds$; $0 < x < \pi$

16. $G(x) = \int_1^x xt\,dt$. (Be careful.)

17. $G(x) = \int_1^{x^2} \sin t\,dt$

18. $G(x) = \int_1^{x^2 + x} \sqrt{2z + \sin z}\,dz$

19. $G(x) = \int_{-x^2}^x \exp t^2\,dt$; Hint: $\int_{-x^2}^x = \int_0^x + \int_{-x^2}^0$

20. $G(x) = \int_{\cos x}^{\sin x} t^5\,dt$

21. Show that the graph of $y = f(x)$ is concave up everywhere if

$$f(x) = \int_0^x \frac{s}{\sqrt{a^2 + s^2}}\,ds, \qquad a \ne 0$$

Hint: Show $f''(x) > 0$ for all real $x$.

22. Find the interval on which the graph of $y = f(x)$ is concave up if

$$f(x) = \int_0^x \frac{1 + t}{1 + t^2}\,dt$$

In Problems 23–26, use the interval additive property (Theorem A) to evaluate $\int_0^4 f(x)\,dx$. Begin by drawing a graph of $f$.

23. $f(x) = \begin{cases} x^2 & \text{if } 0 \leq x < 2 \\ x & \text{if } 2 \leq x \leq 4 \end{cases}$

24. $f(x) = \begin{cases} 1 & \text{if } 0 \leq x < 1 \\ x & \text{if } 1 \leq x < 2 \\ 4 - x & \text{if } 2 \leq x \leq 4 \end{cases}$

25. $f(x) = |x - 2|$

26. $f(x) = |\cos x|$

27. Show that $1 \leq \int_0^1 \sqrt{1 + x^4}\, dx \leq \frac{6}{5}$. *Hint:* $1 \leq \sqrt{1 + x^4} \leq 1 + x^4$; use the comparison property (Theorem B).

28. Let $f$ be continuous on $[a, b]$ and thus integrable there. Show that

$$\left| \int_a^b f(x)\, dx \right| \leq \int_a^b |f(x)|\, dx$$

*Hint:* $-|f(x)| \leq f(x) \leq |f(x)|$; use Theorem B.

In Problems 29–32, find the average value of the given function on the given interval.

29. $f(x) = 4x^3$; $[1, 3]$

30. $f(x) = \dfrac{x}{\sqrt{x^2 + 16}}$; $[0, 3]$

31. $f(x) = 2 + |x|$; $[-1, 3]$

32. $f(x) = \sin^2 x \cos x$; $[0, \pi/2]$

≈ Sketch the graph of the integrand in Problems 33–36. Then estimate the integral as suggested in the marginal box accompanying Theorem E.

33. $\displaystyle\int_0^2 2^x\, dx$

34. $\displaystyle\int_0^2 [1 + \sin(x^2)]\, dx$

35. $\displaystyle\int_{-1}^1 \dfrac{2}{1 + x^2}\, dx$

36. $\displaystyle\int_{10}^{20} \left(1 + \dfrac{1}{x}\right)^5 dx$

In Problems 37–42, decide whether the given statement is true or false. Then justify your answer.

37. If $f$ is continuous and $f(x) \geq 0$ for all $x$ in $[a, b]$, then $\displaystyle\int_a^b f(x)\, dx \geq 0$.

38. If $\displaystyle\int_a^b f(x)\, dx \geq 0$, then $f(x) \geq 0$ for all $x$ in $[a, b]$.

39. If $\displaystyle\int_a^b f(x)\, dx = 0$, then $f(x) = 0$ for all $x$ in $[a, b]$.

40. If $f(x) \geq 0$ and $\displaystyle\int_a^b f(x)\, dx = 0$, then $f(x) = 0$ for all $x$ in $[a, b]$.

41. If $\displaystyle\int_a^b f(x)\, dx > \int_a^b g(x)\, dx$, then

$$\int_a^b [f(x) - g(x)]\, dx > 0.$$

42. If $f$ and $g$ are continuous and $f(x) > g(x)$ for all $x$ in $[a, b]$, then $\left| \displaystyle\int_a^b f(x)\, dx \right| > \left| \displaystyle\int_a^b g(x)\, dx \right|$.

43. Let $s(t)$ and $v(t)$ denote the position and velocity at time $t$ of an object moving along a coordinate line. In Section 3.1, we defined the average velocity on the interval $[a, b]$ as $[s(b) - s(a)]/(b - a)$; in this section, we defined it as $\displaystyle\int_a^b v(t)\, dt/(b - a)$. Show the two definitions are equivalent.

44. Suppose that $f'$ is integrable and $|f'(x)| \leq M$ for all $x$. Prove that $|f(x)| \leq |f(a)| + M|x - a|$.

45. Prove Theorem A in the case where $f$ is continuous on the interval $[a, c]$ and $a < b < c$ as follows. Let $P: a = x_0 < x_1 < \cdots < x_n = c$ be any partition of $[a, c]$ with $b$ as one of the partition points and let $x_m = b$. Then

$$\sum_{i=1}^n f(\overline{x}_i)\, \Delta x_i = \sum_{i=1}^m f(\overline{x}_i)\, \Delta x_i + \sum_{i=m+1}^n f(\overline{x}_i)\, \Delta x_i$$

Finish.

46. Give an alternate proof of the Fundamental Theorem of Calculus (Theorem 5.6A) as follows: Let $G(x) = \displaystyle\int_a^x f(t)\, dt$ and let $F$ be any antiderivative of $f$. Then $F'(x) = G'(x)$. What can we conclude about $G(x) - F(x)$? Now evaluate $G(a) - F(a)$ and then $G(b) - F(b)$.

47. Give a proof of Theorem E that does not use Theorem D. *Hint:* Use the Intermediate Value Theorem.

48. Use Theorem E to give an alternate proof of Theorem D.

PC Use a computer to redo Problems 29–32. Then find the average value of each of the following functions on the given interval.

49. $f(x) = \sin(\cos(x))$; $[0, 10]$

50. $f(x) = \tan(x^4)$; $[-1, 1]$

**Answers to Concepts Review:** **1.** 8  **2.** 8; 32  **3.** $\sin^3 x$; $2x \sin^3(x^2)$  **4.** $4c^2$

**5.8**
**AIDS IN EVALUATING**
**DEFINITE INTEGRALS**

Evaluating definite integrals is generally a two-step process. First, we find an indefinite integral; then we apply the Fundamental Theorem of Calculus. If the indefinite integration is easy, we can combine the two steps, as in

$$\int_1^2 x^2 \, dx = \left[\frac{x^3}{3}\right]_1^2 = \frac{8}{3} - \frac{1}{3} = \frac{7}{3}$$

However, if the indefinite integration is complicated enough to require a substitution, we typically separate the two steps. Thus, to calculate

$$\int_0^4 x\sqrt{x^2 + 9} \, dx$$

we first write (using $u = x^2 + 9$ and $du = 2x \, dx$)

$$\int x \sqrt{x^2 + 9} \, dx = \tfrac{1}{2} \int \sqrt{x^2 + 9}(2x \, dx)$$

$$= \tfrac{1}{2} \int u^{1/2} \, du$$

$$= \tfrac{1}{2} \cdot \tfrac{2}{3} u^{3/2} + C$$

$$= \tfrac{1}{3}(x^2 + 9)^{3/2} + C$$

Then, we apply the Fundamental Theorem of Calculus.

$$\int_0^4 x\sqrt{x^2 + 9} \, dx = [\tfrac{1}{3}(x^2 + 9)^{3/2}]_0^4 = \tfrac{125}{3} - \tfrac{27}{3} = \tfrac{98}{3}$$

The substitution method just illustrated generalizes in two ways. First, though it was introduced in Section 5.1 only for power functions, its application extends far beyond that use. Second, there is a way of using substitution directly in a definite integral. We take up these matters now.

**The Method of Substitution**    Consider the problem of finding

$$\int (2x + 3)\cos(x^2 + 3x) \, dx$$

If we let $u = x^2 + 3x$ so that $du = (2x + 3) \, dx$, the integral above transforms to $\int \cos u \, du$, which you will note is not the integral of a power function. However, formally,

$$\int (2x + 3)\cos(x^2 + 3x) \, dx = \int \cos u \, du$$

$$= \sin u + C$$

$$= \sin(x^2 + 3x) + C$$

In this example, we can easily check our answer by differentiating the result. But will the method of substitution always work? Yes, provided we can prove the Substitution Rule for Indefinite Integrals. This rule plays the same role for integrals that the Chain Rule plays for derivatives. Both deal with composite functions.

---

**How A Theorem Grew**

The Power Rule

$$\int x^r \, dx = \frac{x^{r+1}}{r+1} + C$$

led via the substitution $u = g(x)$ to the Generalized Power Rule

$$\int [g(x)]^r g'(x) \, dx = \int u^r \, du$$

$$= \frac{u^{r+1}}{r+1} + C$$

$$= \frac{[g(x)]^{r+1}}{r+1} + C$$

But this, in turn, is just the very special case of the Substitution Rule in which $f(u) = u^r$. Note how an acorn has grown to a sapling and then to a full-size tree.

---

**Theorem A**

**(Substitution Rule for Indefinite Integrals).** Let $g$ be a differentiable function and suppose that $F$ is an antiderivative of $f$. Then if $u = g(x)$,

$$\int f(g(x))g'(x) \, dx = \int f(u) \, du = F(u) + C = F(g(x)) + C$$

**Proof**  It is enough to show that the derivative of the right member is the integrand of the left member. But that is a simple application of the Chain Rule combined with the fact that $F' = f$.

$$D_x[F(g(x)) + C] = F'(g(x))g'(x) = f(g(x))g'(x) \qquad \blacksquare$$

**EXAMPLE 1**  Find $\int \dfrac{\sin \sqrt{x}}{\sqrt{x}} \, dx$.

**Solution**  Let $u = \sqrt{x} = x^{1/2}$, so $du = \frac{1}{2}x^{-1/2} \, dx$. Then,

$$\int \frac{\sin \sqrt{x}}{\sqrt{x}} \, dx = 2 \int \sin \sqrt{x} \, (\tfrac{1}{2} x^{-1/2} \, dx)$$

$$= 2 \int \sin u \, du$$

$$= -2 \cos u + C$$

$$= -2 \cos \sqrt{x} + C \qquad \blacksquare$$

**EXAMPLE 2**  Evaluate $\displaystyle\int_0^{\sqrt{\pi/2}} x \sin^3(x^2)\cos(x^2) \, dx$.

**Solution**  Let $u = \sin(x^2)$, so $du = 2x \cos(x^2) \, dx$. Then

$$\int x \sin^3(x^2)\cos(x^2) \, dx = \frac{1}{2} \int \sin^3(x^2) \cdot 2x \cos(x^2) \, dx$$

$$= \frac{1}{2} \int u^3 \, du$$

$$= \frac{1}{2} \frac{u^4}{4} + C$$

$$= \frac{1}{8} \sin^4(x^2) + C$$

Then, by the Fundamental Theorem of Calculus,

$$\int_0^{\sqrt{\pi/2}} x \sin^3(x^2)\cos(x^2)\,dx = \left[\frac{1}{8}\sin^4(x^2)\right]_0^{\sqrt{\pi/2}}$$

$$= \frac{1}{8}\sin^4\left(\frac{\pi}{4}\right) - \frac{1}{8}\cdot 0 = \frac{1}{32} \qquad \blacksquare$$

Note that in the two-step procedure illustrated in Example 2, we must be sure to express the indefinite integral in terms of $x$ before we apply the Fundamental Theorem. This is because the limits 0 and $\sqrt{\pi}/2$ apply to $x$, not $u$. But what if, in making the substitution $u = \sin(x^2)$, we also made the corresponding changes in the limits of integration to $u = \sin(0^2) = 0$ and $u = \sin[(\sqrt{\pi/2})^2] = \sqrt{2}/2$? Could we then finish the integration with $u$ as variable? The answer is yes.

$$\int_0^{\sqrt{\pi/2}} x \sin^3(x^2)\cos(x^2)\,dx = \frac{1}{2}\int_0^{\sqrt{2}/2} u^3\,du = \left[\frac{1}{2}\frac{u^4}{4}\right]_0^{\sqrt{2}/2} = \frac{1}{32}$$

Here is the general result, which lets us substitute in the limits of integration as well as in the integrand, thereby producing a one-step procedure.

---

### Theorem B

**(Substitution Rule for Definite Integrals).** Let $g$ have a continuous derivative on $[a, b]$ and let $f$ be continuous on the range of $g$. Then

$$\int_a^b f(g(x))g'(x)\,dx = \int_{g(a)}^{g(b)} f(u)\,du$$

---

**Proof** Let $F$ be an antiderivative of $f$ (the existence of $F$ is guaranteed by Theorem 5.7D). Then by the Fundamental Theorem of Calculus,

$$\int_{g(a)}^{g(b)} f(u)\,du = \left[F(u)\right]_{g(a)}^{g(b)} = F(g(b)) - F(g(a))$$

On the other hand, by the Substitution Theorem for Indefinite Integrals (Theorem A),

$$\int f(g(x))g'(x)\,dx = F(g(x)) + C$$

and so, again by the Fundamental Theorem of Calculus,

$$\int_a^b f(g(x))g'(x)\,dx = [F(g(x))]_a^b = F(g(b)) - F(g(a)) \qquad \blacksquare$$

**EXAMPLE 3**   Evaluate $\displaystyle\int_0^1 \frac{x+1}{(x^2+2x+6)^2}\,dx.$

***Solution*** Let $u = x^2 + 2x + 6$, so $du = (2x + 2)\, dx = 2(x + 1)\, dx$, and note that $u = 6$ when $x = 0$ and $u = 9$ when $x = 1$. Thus,

$$
\int_0^1 \frac{x + 1}{(x^2 + 2x + 6)^2}\, dx = \frac{1}{2} \int_0^1 \frac{2x + 2}{(x^2 + 2x + 6)^2}\, dx
$$

$$
= \frac{1}{2} \int_6^9 u^{-2}\, du = \left[ -\frac{1}{2u} \right]_6^9
$$

$$
= -\frac{1}{18} - \left( -\frac{1}{12} \right) = \frac{1}{36} \qquad \blacksquare
$$

**EXAMPLE 4** Evaluate $\displaystyle\int_{\pi^2/9}^{\pi^2/4} \frac{\cos\sqrt{x}}{\sqrt{x}}\, dx$.

***Solution*** Let $u = \sqrt{x}$, so $du = dx/(2\sqrt{x})$. Thus,

$$
\int_{\pi^2/9}^{\pi^2/4} \frac{\cos\sqrt{x}}{\sqrt{x}}\, dx = 2 \int_{\pi^2/9}^{\pi^2/4} \cos\sqrt{x} \cdot \frac{1}{2\sqrt{x}}\, dx
$$

$$
= 2 \int_{\pi/3}^{\pi/2} \cos u\, du
$$

$$
= [2 \sin u]_{\pi/3}^{\pi/2} = 2 - \sqrt{3}
$$

Note the change in the limits of integration at the second equality. When $x = \pi^2/9$, $u = \sqrt{\pi^2/9} = \pi/3$; when $x = \pi^2/4$, $u = \pi/2$. $\qquad \blacksquare$

**Use of Symmetry** Recall that an even function is one satisfying $f(-x) = f(x)$, whereas an odd function satisfies $f(-x) = -f(x)$. The graph of the former is symmetric with respect to the $y$-axis; the graph of the latter is symmetric with respect to the origin. Here is a useful integration theorem for such functions.

Theorem C

**(Symmetry Theorem).** If $f$ is an even function, then

$$
\int_{-a}^a f(x)\, dx = 2 \int_0^a f(x)\, dx
$$

If $f$ is an odd function, then

$$
\int_{-a}^a f(x)\, dx = 0
$$

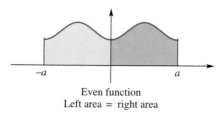

Even function
Left area = right area

FIGURE 1

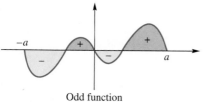

Odd function
Left area neutralizes right area

FIGURE 2

---

**Strong Conditions**

Be sure to note the hypotheses of the Symmetry Theorem. The integrand must be even or odd and the interval of integration must be symmetric about the origin. These are restrictive conditions, but it is surprising how often they hold in the applications we will meet in the next chapter. When they do hold, they can greatly simplify integrations.

---

***Proof***  The geometric interpretation of this theorem is shown in Figures 1 and 2. To justify the results analytically, we first write

$$\int_{-a}^{a} f(x)\,dx = \int_{-a}^{0} f(x)\,dx + \int_{0}^{a} f(x)\,dx$$

In the first of the integrals on the right, we make the substitution $u = -x$, $du = -dx$. If $f$ is even, $f(-x) = f(x)$ and

$$\int_{-a}^{0} f(x)\,dx = -\int_{-a}^{0} f(-x)(-dx) = -\int_{a}^{0} f(u)\,du$$

$$= \int_{0}^{a} f(u)\,du = \int_{0}^{a} f(x)\,dx$$

On the other hand, if $f$ is odd, then $f(-x) = -f(x)$, and we can write

$$\int_{-a}^{0} f(x)\,dx = \int_{-a}^{0} -f(-x)\,dx = \int_{a}^{0} f(u)\,du$$

$$= -\int_{0}^{a} f(u)\,du = -\int_{0}^{a} f(x)\,dx \qquad \blacksquare$$

**EXAMPLE 5**  Evaluate $\displaystyle\int_{-\pi}^{\pi} \cos\left(\frac{x}{4}\right) dx$.

***Solution***  Since $\cos(-x/4) = \cos(x/4)$, $f(x) = \cos(x/4)$ is an even function. Thus,

$$\int_{-\pi}^{\pi} \cos\left(\frac{x}{4}\right) dx = 2\int_{0}^{\pi} \cos\left(\frac{x}{4}\right) dx = 8\int_{0}^{\pi} \cos\left(\frac{x}{4}\right) \cdot \frac{1}{4}\,dx$$

$$= 8\int_{0}^{\pi/4} \cos u\,du = [8 \sin u]_{0}^{\pi/4} = 4\sqrt{2} \qquad \blacksquare$$

**EXAMPLE 6**  Evaluate $\displaystyle\int_{-5}^{5} \frac{x^5}{x^2 + 4}\,dx$.

***Solution***  $f(x) = x^5/(x^2 + 4)$ is an odd function. Thus the above integral has the value 0. $\qquad \blacksquare$

**EXAMPLE 7**  Evaluate $\displaystyle\int_{-2}^{2} (x \sin^4 x + x^3 - x^4)\,dx$.

***Solution***  The first two terms in the integrand are odd, the last is even. Thus, we may write the integral as

$$\int_{-2}^{2} (x \sin^4 x + x^3)\,dx - \int_{-2}^{2} x^4\,dx = 0 - 2\int_{0}^{2} x^4\,dx$$

$$= \left[-2\frac{x^5}{5}\right]_{0}^{2} = -\frac{64}{5} \qquad \blacksquare$$

**Use of Periodicity**    A function $f$ is **periodic** if there is a number $p$ such that

$$f(x + p) = f(x)$$

for all real numbers $x$ in the domain of $f$. The smallest such positive number $p$ is the **period** of a periodic function. The trigonometric functions are the best known examples of periodic functions.

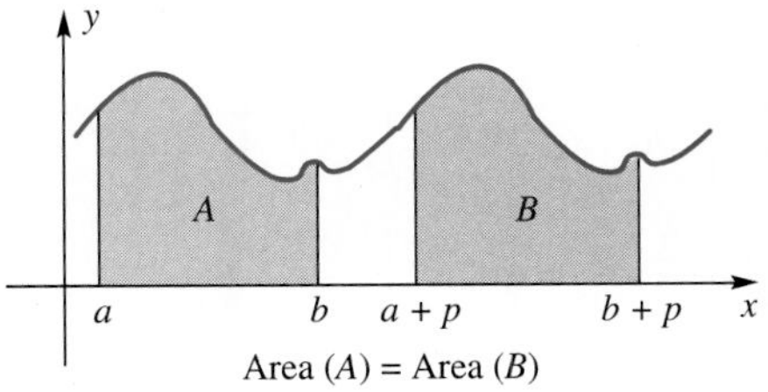

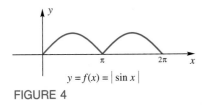

$y$

$A$        $B$

$a$        $b$    $a + p$        $b + p$    $x$

Area ($A$) = Area ($B$)

**FIGURE 3**

> **Theorem D**
>
> If $f$ is periodic with period $p$, then
>
> $$\int_{a+p}^{b+p} f(x)\,dx = \int_{a}^{b} f(x)\,dx$$

***Proof***    The geometric interpretation can be seen in Figure 3. To prove the result, let $u = x - p$ so $x = u + p$ and $du = dx$. Then

$$\int_{a+p}^{b+p} f(x)\,dx = \int_{a}^{b} f(u + p)\,du = \int_{a}^{b} f(u)\,du = \int_{a}^{b} f(x)\,dx$$

We could replace $f(u + p)$ by $f(u)$ at the second stage because $f$ is periodic.    ∎

**EXAMPLE 8**    Evaluate $\int_{0}^{2\pi} |\sin x|\,dx$.

***Solution***    Note that $f(x) = |\sin x|$ is periodic with period $\pi$ (Figure 4).

$$\int_{0}^{2\pi} |\sin x|\,dx = \int_{0}^{\pi} |\sin x|\,dx + \int_{\pi}^{2\pi} |\sin x|\,dx$$

$$= \int_{0}^{\pi} |\sin x|\,dx + \int_{0}^{\pi} |\sin x|\,dx$$

$$= 2 \int_{0}^{\pi} \sin x\,dx = [-2 \cos x]_{0}^{\pi} = 2 - (-2) = 4 \quad ∎$$

$y$

$\pi$    $2\pi$    $x$

$y = f(x) = |\sin x|$

**FIGURE 4**

## CONCEPTS REVIEW

**1.** Under the substitution $u = x^3 + 1$, the definite integral $\int_{0}^{1} x^2(x^3 + 1)^4\,dx$ transforms to the new definite integral _____.

**2.** If $f$ is an odd function, $\int_{-2}^{2} f(x)\,dx =$ _____; if $f$ is an even function, $\int_{-2}^{2} f(x)\,dx =$ _____.

**3.** Using Question 2, we may write

$$\int_{-2}^{2} (2 + x + x^2 + 3x^5)\,dx = \int_{-2}^{2} (2 + x^2)\,dx$$
$$+ \int_{-2}^{2} (x + 3x^5)\,dx$$
$$= 2(\underline{\quad\quad})$$
$$+ \underline{\quad\quad}.$$

**4.** The function $f$ is periodic if there is a number $p$ such that _____ for all $x$ in the domain of $f$. The smallest such positive number $p$ is called the _____ of the function.

## PROBLEM SET 5.8

Use the method of substitution to find each of the following indefinite integrals.

**1.** $\int \sqrt{3x + 2} \, dx$

**2.** $\int \sqrt[3]{2x - 4} \, dx$

**3.** $\int (6x - 7)^{1/8} \, dx$

**4.** $\int (5u - \pi)^{21/8} \, du$

**5.** $\int \cos(3x + 2) \, dx$

**6.** $\int \sin(2x - 4) \, dx$

**7.** $\int \sin(6x - 7) \, dx$

**8.** $\int \cos(\pi v - \sqrt{7}) \, dv$

**9.** $\int x\sqrt{x^2 + 4} \, dx$

**10.** $\int x^2(x^3 + 5)^9 \, dx$

**11.** $\int x(x^2 + 3)^{-12/7} \, dx$

**12.** $\int v(\sqrt{3}v^2 + \pi)^{7/8} \, dv$

**13.** $\int x \sin(x^2 + 4) \, dx$

**14.** $\int x^2 \cos(x^3 + 5) \, dx$

**15.** $\int x^2\sin(6x^3 - 7) \, dx$

**16.** $\int v^4 \cos(\pi v^5 - \sqrt{7}) \, dv$

**17.** $\int \dfrac{x \sin\sqrt{x^2 + 4}}{\sqrt{x^2 + 4}} \, dx$

**18.** $\int \dfrac{z \cos(\sqrt[3]{z^2 + 3})}{(\sqrt[3]{z^2 + 3})^2} \, dz$

**19.** $\int x^2(x^3 + 5)^8 \cos[(x^3 + 5)^9] \, dx$

**20.** $\int x^6(7x^7 + \pi)^8 \sin[(7x^7 + \pi)^9] \, dx$

**21.** $\int x \cos(x^2 + 4)\sqrt{\sin(x^2 + 4)} \, dx$

**22.** $\int x^6 \sin(3x^7 + 9)\sqrt[3]{\cos(3x^7 + 9)} \, dx$

**23.** $\int x^2 \sin(x^3 + 5)\cos^9(x^3 + 5) \, dx$

**24.** $\int x^{-4}\sec^2(x^{-3} + 1)\sqrt[5]{\tan(x^{-3} + 1)} \, dx$

**25.** $\int (x + 1) \sec^{1/2}(x^2 + 2x) \tan(x^2 + 2x) \, dx$

**26.** $\int \dfrac{(\sqrt{x} + 4)^2}{\sqrt{x}} \, dx$

**27.** $\int \dfrac{(\sqrt{t} + 4)^3}{\sqrt{t}} \, dt$

**28.** $\int \left(1 + \dfrac{1}{t}\right)^{-2}\left(\dfrac{1}{t^2}\right) dt$

**29.** $\int \dfrac{(\sqrt[5]{32z} + \pi)^7}{(\sqrt[5]{z})^4} \, dz$

**30.** $\int \dfrac{(1/s + s)^5 (s^2 - 1)}{s^2} \, ds$

Use the method of substitution in definite integrals to evaluate each of the following (see Examples 3 and 4).

**31.** $\int_0^1 (3x + 1)^3 \, dx$

**32.** $\int_0^4 \sqrt{2t + 1} \, dt$

**33.** $\int_0^2 \dfrac{t}{(t^2 + 9)^2} \, dt$

**34.** $\int_0^{\sqrt{5}} \sqrt{9 - x^2}x \, dx$

**35.** $\int_0^1 \dfrac{x + 2}{(x^2 + 4x + 1)^2} \, dx$

**36.** $\int_0^2 \dfrac{x^2}{(9 - x^3)^{3/2}} \, dx$

**37.** $\int_0^{\pi/6} \sin^3\theta \cos\theta \, d\theta$

**38.** $\int_0^{\pi/6} \dfrac{\sin\theta}{\cos^3\theta} \, d\theta$

**39.** $\int_0^1 \cos(3x - 3) \, dx$

**40.** $\int_0^{1/2} \sin(2\pi x) \, dx$

**41.** $\int_0^1 x \sin(\pi x^2) \, dx$

**42.** $\int_0^\pi x^4 \cos(2x^5) \, dx$

**43.** $\int_0^{\pi/4} (\cos 2x + \sin 2x) \, dx$

**44.** $\int_{-\pi/2}^{\pi/2} (\cos 3x + \sin 5x)$

**45.** $\int_0^{\pi/2} \sin x \sin(\cos x) \, dx$

**46.** $\int_{-\pi/2}^{\pi/2} \cos\theta\cos(\pi\sin\theta) \, d\theta$

**47.** $\int_0^1 x \cos^3(x^2)\sin(x^2) \, dx$

**48.** $\int_{-\pi/2}^{\pi/2} x^2\sin^2(x^3)\cos(x^3) \, dx$

**49.** $\int_1^4 \dfrac{1}{\sqrt{t}(\sqrt{t} + 1)^3} \, dt$

**50.** $\int_1^2 \left(1 + \dfrac{1}{t}\right)^2\left(\dfrac{1}{t^2}\right) dt$

In Problems 51–62, use symmetry to help you evaluate the given integral.

**51.** $\int_{-\pi}^\pi (\sin x + \cos x) \, dx$

**52.** $\int_{-1}^1 \dfrac{x^3}{(1 + x^2)^4} \, dx$

**53.** $\int_{-\pi/2}^{\pi/2} \dfrac{\sin x}{1 + \cos x} \, dx$

**54.** $\int_{-\sqrt[3]{\pi}}^{\sqrt[3]{\pi}} x^2 \cos(x^3) \, dx$

**55.** $\int_{-\pi}^\pi (\sin x + \cos x)^2 \, dx$

**56.** $\int_{-\pi/2}^{\pi/2} z \sin^2(z^3) \cos(z^3) \, dz$

**57.** $\int_{-1}^1 (1 + x + x^2 + x^3) \, dx$

**58.** $\int_{-100}^{100} (v + \sin v + v \cos v + \sin^3 v)^5 \, dv$

**59.** $\displaystyle\int_{-1}^{1} (|x^3| + x^3)\, dx$

**60.** $\displaystyle\int_{-\pi/4}^{\pi/4} (|x|\sin^5 x + |x|^2 \tan x)\, dx$

**61.** $\displaystyle\int_{-\pi}^{\pi} (x^5 + |\sin x|)\, dx$    **62.** $\displaystyle\int_{-\pi}^{\pi} |\sin^5 \theta|\cos\theta\, d\theta$

**63.** How does $\displaystyle\int_{-b}^{-a} f(x)\, dx$ compare with $\displaystyle\int_{a}^{b} f(x)\, dx$ in case $f$ is an even function? An odd function?

**64.** Prove (by a substitution) that

$$\int_{a}^{b} f(-x)\, dx = \int_{-b}^{-a} f(x)\, dx$$

**65.** Use periodicity (see Example 8) to calculate $\displaystyle\int_{0}^{4\pi} |\cos x|\, dx.$

**66.** Calculate $\displaystyle\int_{0}^{4\pi} |\sin 2x|\, dx.$

**67.** If $f$ is periodic with period $p$, then

$$\boxed{\int_{a}^{a+p} f(x)\, dx = \int_{0}^{p} f(x)\, dx}$$

Convince yourself that this is true by drawing a picture and then use the result to calculate $\displaystyle\int_{1}^{1+\pi} |\sin x|\, dx.$

**68.** Use the result in Problem 67 to calculate $\displaystyle\int_{2}^{2+\pi/2} |\sin 2x|\, dx.$

**69.** The temperature $T$ on a certain day satisfied

$$T(t) = 70 + 8\sin\left[\frac{\pi}{12}(t - 9)\right]$$

where $t$ was the number of hours after midnight. Find the average temperature from 6 A.M. to 6 P.M.

**70.** Complete the generalization of the Pythagorean Theorem begun in Problem 37 of Section 1.5 by showing that $A + B = C$ in Figure 5, these being the areas of similar regions built on the two legs and the hypotenuse of a right triangle.

(a) Convince yourself that similarity means

$$g(x) = \frac{a}{c}f\left(\frac{c}{a}x\right) \text{ and } h(x) = \frac{b}{c}f\left(\frac{c}{b}x\right)$$

(b) Show that $\displaystyle\int_{0}^{a} g(x)\, dx + \int_{0}^{b} h(x)\, dx = \int_{0}^{c} f(x)\, dx$

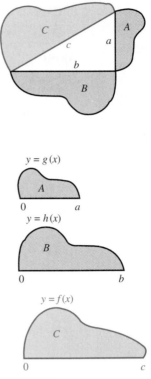

$y = g(x)$

$y = h(x)$

$y = f(x)$

**FIGURE 5**

**71.** Find the area of the shaded region shown in Figure 6.

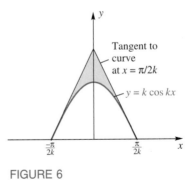

**FIGURE 6**

**72.** Integrals which occur frequently in applications are
$\displaystyle\int_{0}^{2\pi} \cos^2 x\, dx$ and $\displaystyle\int_{0}^{2\pi} \sin^2 x\, dx.$

(a) Using a trigonometric identity show that
$$\int_{0}^{2\pi} (\sin^2 x + \cos^2 x)\, dx = 2\pi$$

(b) Show from graphical considerations that $\displaystyle\int_{0}^{2\pi} \cos^2 x\, dx = \int_{0}^{2\pi} \sin^2 x\, dx$

(c) Conclude that $\displaystyle\int_{0}^{2\pi} \cos^2 x\, dx = \int_{0}^{2\pi} \sin^2 x\, dx = \pi$

PC 73. To illustrate some of the results of this section, let $f(x) = |\sin x| \sin(\cos x)$.

(a) Is $f$ even, odd, or neither?
(b) Note that $f$ is periodic. What is its period?
(c) Evaluate the definite integral of $f$ for each of the following intervals: $[0, \pi/2]$, $[-\pi/2, \pi/2]$, $[0, 3\pi/2]$, $[-3\pi/2, 3\pi/2]$, $[0, 2\pi]$, $[\pi/6, 13\pi/6]$, $[\pi/6, 4\pi/3]$, $[13\pi/6, 10\pi/3]$.

74. Repeat Problem 73 for $f(x) = \sin x|\sin(\sin x)|$.

---

**Answers to Concepts Review:** **1.** $\frac{1}{3} \int_{1}^{2} u^4 \, du$

**2.** $0; 2 \int_{0}^{2} f(x) \, dx$  **3.** $\int_{0}^{2} (2 + x^2) \, dx; 0$

**4.** $f(x + p) = f(x)$; period

## 5.9 CHAPTER REVIEW

### Concepts Test

Respond with true or false to each of the following assertions. Be prepared to justify your answer.

1. The indefinite integral is a linear operator.

2. $\int [f(x)g'(x) + g(x)f'(x)] \, dx = f(x)g(x) + C$.

3. $y = \cos x$ is a solution to the differential equation $(dy/dx)^2 = 1 - y^2$.

4. All continuous functions have antiderivatives.

5. All functions which are antiderivatives must have derivatives.

6. If the second derivatives of two functions are equal then the functions differ at most by a constant.

7. $\int f'(x) \, dx = f(x)$ for every differentiable function $f$.

8. If $s = -16t^2 + v_0 t$ gives the height at time $t$ of a ball thrown straight up from the surface of the earth, then the ball will hit the ground with velocity $-v_0$.

9. $\sum_{i=1}^{n} (a_i + a_{i-1}) = a_0 + a_n + 2 \sum_{i=1}^{n-1} a_i$.

10. $\sum_{i=1}^{100} (2i - 1) = 10{,}000$.

11. If $\sum_{i=1}^{10} a_i^2 = 100$ and $\sum_{i=1}^{10} a_i = 20$, then $\sum_{i=1}^{10} (a_i + 1)^2 = 150$.

12. If $f$ is bounded on $[a, b]$, then $f$ is integrable there.

13. If $f$ is integrable on $[a, b]$, then $f$ must be bounded on $[a, b]$.

14. If $\int_{a}^{b} f(x) \, dx = 0$, then $f(x) = 0$ for all $x$ in $[a, b]$.

15. If $\int_{a}^{b} [f(x)]^2 \, dx = 0$ then $f(x) = 0$ for all $x$ in $[a, b]$.

16. If $a > x$ and $G(x) = \int_{a}^{x} f(z) \, dz$ then $G'(x) = -f(x)$.

17. The value of $\int_{x}^{x + 2\pi} (\sin x + \cos x) \, dx$ is independent of $x$.

18. The operator lim is linear.

19. $\int_{-\pi}^{\pi} \sin^{13} x \, dx = 0$.

20. $\int_{1}^{5} \sin^2 x \, dx = \int_{1}^{7} \sin^2 x \, dx + \int_{7}^{5} \sin^2 x \, dx$.

21. If $f$ is continuous and positive everywhere, then $\int_{c}^{d} f(x) \, dx$ is positive.

22. $D_x \left[ \int_{0}^{x^2} \frac{1}{1 + t^2} \, dt \right] = \frac{1}{1 + x^4}$

23. $\int_{0}^{2\pi} |\sin x| \, dx = \int_{0}^{2\pi} |\cos x| \, dx$

24. $\int_{0}^{2\pi} |\sin x| \, dx = 4 \int_{0}^{\pi/2} \sin x \, dx$

25. The antiderivatives of odd functions are even functions.

26. If $F(x)$ is the antiderivative of $f(x)$ then $F(5x)$ is an antiderivative of $f(5x)$.

27. If $F(x)$ is the antiderivative of $f(x)$ then $F(2x + 1)$ is an antiderivative of $f(2x + 1)$.

28. If $F(x)$ is the antiderivative of $f(x)$ then $F(x) + 1$ is an antiderivative of $f(x) + 1$.

29. If $F(x)$ is the antiderivative of $f(x)$ then $\int f(v(x)) \, dx = F(v(x)) + C$.

30. If $F(x)$ is the antiderivative of $f(x)$ then $\int f^2(x) \, dx = \frac{1}{3} F^3(x) + C$.

31. If $F(x)$ is the antiderivative of $f(x)$ then $\int f(x) \frac{df}{dx} \, dx = \frac{1}{2} F^2(x) + C$.

32. If $f(x) = 4$ on $[0, 3]$, then every Riemann sum for $f$ on the given interval has the value 12.

33. If $F'(x) = G'(x)$ for all $x$ in $[a, b]$, then $F(b) - F(a) = G(b) - G(a)$.

34. If $f(x) = f(-x)$ for all $x$ in $[-a, a]$, then $\int_{-a}^{a} f(x)\,dx = 0$.

35. If $\bar{z} = \frac{1}{2}\int_{-1}^{1} z(t)\,dt$ then $z(t) - \bar{z}$ is an odd function for $-1 \le t \le 1$.

36. If $F'(x) = f(x)$ for all $x$ in $[0, b]$, then $\int_{0}^{b} f(x)\,dx = F(b)$.

37. $\int_{-99}^{99} (ax^3 + bx^2 + cx)\,dx = 2\int_{0}^{99} bx^2\,dx$.

38. If $f(x) \le g(x)$ on $[a, b]$, then $\int_{a}^{b} |f(x)|\,dx \le \int_{a}^{b} |g(x)|\,dx$.

39. If $f(x) \le g(x)$ on $[a, b]$, then $\left|\int_{a}^{b} f(x)\,dx\right| \le \left|\int_{a}^{b} g(x)\,dx\right|$.

40. $\left|\sum_{i=1}^{n} a_i\right| \le \sum_{i=1}^{n} |a_i|$.

41. If $f$ is continuous on $[a, b]$, then $\left|\int_{a}^{b} f(x)\,dx\right| \le \int_{a}^{b} |f(x)|\,dx$.

42. $\lim_{n \to \infty} \sum_{i=1}^{n} \sin\left(\frac{2i}{n}\right) \cdot \frac{2}{n} = \int_{0}^{2} \sin x\,dx$.

43. To say that $|P| \to 0$ is the same as to say the number of subintervals in the partition $P$ tends to $\infty$.

## Sample Test Problems

In Problems 1–11, find the indicated integrals.

1. $\int_{0}^{1} (x^3 - 3x^2 + 3\sqrt{x})\,dx$

2. $\int \dfrac{2x^4 - 3x^2 + 1}{x^2}\,dx$

3. $\int \dfrac{y^3 - 9y \sin y + 26y^{-1}}{y}\,dy$

4. $\int y\sqrt{y^2 - 4}\,dy$

5. $\int z(2z^2 - 3)^{1/3}\,dz$

6. $\int_{0}^{\pi/2} \cos^4 x \sin x\,dx$

7. $\int_{0}^{\pi} (x + 1)\tan^2(3x^2 + 6x)\sec^2(3x^2 + 6x)\,dx$

8. $\int_{0}^{2} \dfrac{t^3}{\sqrt{t^4 + 9}}\,dt$

9. $\int_{1}^{2} t^4(t^5 + 5)^{2/3}\,dt$

10. $\int \dfrac{y^2 - 1}{(y^3 - 3y)^2}\,dy$

11. $\int \dfrac{(y^2 + y + 1)}{\sqrt[5]{2y^3 + 3y^2 + 6y}}\,dy$

In Problems 12–18, solve the differential equation subject to the indicated condition.

12. $\dfrac{dy}{dx} = \sin x$; $y = 2$ at $x = 0$

13. $\dfrac{dy}{dx} = \dfrac{1}{\sqrt{x + 1}}$; $y = 18$ at $x = 3$

14. $\dfrac{dy}{dx} = \csc y$; $y = \pi$ at $x = 0$

15. $\dfrac{dy}{dt} = \sqrt{2t - 1}$; $y = -1$ at $t = \frac{1}{2}$

16. $\dfrac{dy}{dt} = t^2 y^4$; $y = 1$ at $t = 1$

17. $\dfrac{dy}{dx} = \dfrac{6x - x^3}{2y}$; $y = 3$ at $x = 0$

18. $\dfrac{dy}{dx} = x \sec y$; $y = \pi$ at $x = 0$

19. Find the equation of the curve through $(-2, -\frac{1}{3})$ if its slope at each $x$ is the negative reciprocal of the slope of the curve with equation $xy = 2$.

20. If a particle moving on the $x$-axis has acceleration $a = 15\sqrt{t} + 8$ at time $t$ and if $v_0 = -6$, $x_0 = -44$, find its position $x$ at $t = 4$. Assume $x$ is measured in feet and $t$ in seconds.

21. A ball is thrown directly upward from a tower 448 feet high with an initial velocity of 48 feet per second. In how many seconds will it strike the ground and with what velocity? Assume $g = 32$ feet per second per second and neglect air resistance.

22. What constant acceleration will cause a car to increase its velocity from 45 to 60 miles per hour in 10 seconds?

23. Let $P$ be a regular partition of the interval $[0, 2]$ into four equal subintervals, and let $f(x) = x^2 - 1$. Write out the Riemann sum for $f$ on $P$, in which $\bar{x}_i$ is the right endpoint of each subinterval of $P$, $i = 1, 2, 3, 4$. Find the value of this Riemann sum and make a sketch.

24. If $f(x) = \int_{-2}^{x} \dfrac{1}{t + 3}\,dt$, $-2 \le x$, find $f'(7)$.

25. Find $\int_{0}^{3} (2 - \sqrt{x + 1})^2\,dx$

26. If $f(x) = 3x^2\sqrt{x^3 - 4}$, find the average value of $f$ on $[2, 5]$.

**27.** Find $\int_2^4 \dfrac{5x^2 - 1}{x^2}\, dx$.

**28.** Evaluate $\displaystyle\sum_{i=1}^{n} (3^i - 3^{i-1})$.

**29.** Evaluate $\displaystyle\sum_{i=1}^{10} (6i^2 - 8i)$.

**30.** Evaluate each sum.

(a) $\displaystyle\sum_{m=2}^{4} \left(\dfrac{1}{m}\right)$

(b) $\displaystyle\sum_{i=1}^{6} (2 - i)$

(c) $\displaystyle\sum_{k=0}^{4} \cos\left(\dfrac{k\pi}{4}\right)$

**31.** Write in sigma notation.

(a) $\frac{1}{2} + \frac{1}{3} + \frac{1}{4} + \cdots + \frac{1}{78}$
(b) $x^2 + 2x^4 + 3x^6 + 4x^8 + \cdots + 50x^{100}$

**32.** Sketch the region under the curve $y = 16 - x^2$ between $x = 0$ and $x = 3$, showing the inscribed polygon corresponding to a regular partition of $[0, 3]$ into $n$ subintervals. Find a formula for the area of this polygon and then find the area under the curve by taking a limit.

**33.** If $\int_0^1 f(x)\, dx = 4$, $\int_0^2 f(x)\, dx = 2$, and $\int_0^2 g(x)\, dx = -3$, evaluate each integral.

(a) $\int_1^2 f(x)\, dx$

(b) $\int_1^0 f(x)\, dx$

(c) $\int_0^2 3f(u)\, du$

(d) $\int_0^2 [2g(x) - 3f(x)]\, dx$

(e) $\int_0^{-2} f(-x)\, dx$

**34.** Evaluate each integral.

(a) $\int_0^4 |x - 1|\, dx$

(b) $\int_0^4 [\![x]\!]\, dx$

(c) $\int_0^4 (x - [\![x]\!])\, dx$

*Hint:* In (a) and (b), first sketch a graph.

**35.** Suppose that $f(x) = f(-x)$, $f(x) \le 0$, $g(-x) = -g(x)$, $\int_0^2 f(x)\, dx = -4$, and $\int_0^2 g(x)\, dx = 5$. Evaluate each integral.

(a) $\int_{-2}^2 f(x)\, dx$

(b) $\int_{-2}^2 |f(x)|\, dx$

(c) $\int_{-2}^2 g(x)\, dx$

(d) $\int_{-2}^2 [f(x) + f(-x)]\, dx$

(e) $\int_0^2 [2g(x) + 3f(x)]\, dx$

(f) $\int_{-2}^0 g(x)\, dx$

**36.** Evaluate $\int_{-100}^{100} (x^3 + \sin^5 x)\, dx$

**37.** Find $c$ of the Mean Value Theorem for Integrals for $f(x) = 3x^2$ on $[-4, -1]$.

**38.** Find $G'(x)$ for each function $G$.

(a) $G(x) = \int_1^x \dfrac{1}{t^2 + 1}\, dt$

(b) $G(x) = \int_1^{x^2} \dfrac{1}{t^2 + 1}\, dt$

(c) $G(x) = \int_x^{x^3} \dfrac{1}{t^2 + 1}\, dt$

**39.** Find $G'(x)$ for each function $G$.

(a) $G(x) = \int_1^x \sin^2 z\, dz$

(b) $G(x) = \int_x^{x+1} f(z)\, dz$

(c) $G(x) = \dfrac{1}{x}\int_0^x f(z)\, dz$

(d) $G(x) = \int_0^x \left(\int_0^u f(t)\, dt\right) du$

(e) $G(x) = \int_0^{g(x)} \dfrac{dg(u)}{du}\, du$

(f) $G(x) = \int_0^{-x} f(-t)\, dt$

**40.** Evaluate each of the following limits by recognizing it as a definite integral.

(a) $\displaystyle\lim_{n \to \infty} \sum_{i=1}^n \sqrt{\dfrac{4i}{n}} \cdot \dfrac{4}{n}$

(b) $\displaystyle\lim_{n \to \infty} \sum_{i=1}^n \left(1 + \dfrac{2i}{n}\right)^2 \dfrac{2}{n}$

**41.** Show that if $f(x) = \int_{2x}^{5x} \dfrac{1}{t}\, dt$, then $f$ is a constant function on $(0, \infty)$.

---

## 5.10  ADDITIONAL PROBLEMS

**1.** Which of the Figures 1, 2, or 3 shows the solution of the initial value problem $\dfrac{dy}{dx} = -\dfrac{x}{y}$; $y(0) = 1$. Give reasons for your answer.

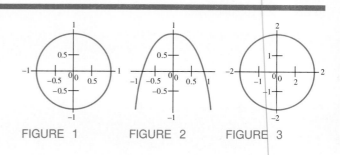

FIGURE 1          FIGURE 2          FIGURE 3

**2.** In Figure 4 below, the velocity $v(t)$ for $0 \le t \le 10$ is given. Use the graph to sketch a rough graph of the position $s(t)$ and the acceleration $a(t)$ which has the indicated velocity for $0 \le t \le 10$ in the case when

(a) $s(0) = 0$
(b) $s(10) = 0$
(c) $s(5) = 0$

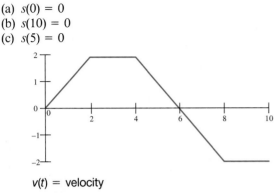

$v(t) = $ velocity

FIGURE 4

**3.** A very long freight train is attempting to come to as quick a stop as it can. Such long trains can take several miles or more to come to a stop. At the time the engineer put on the brakes the train is going 60 miles per hour (88 feet per second). Here is a table of the speedometer readings taken in the 4 minutes it takes the train to stop.

| 0 sec | | 30 | 60 | 90 | 120 | 150 | 180 | 210 | 240 |
|---|---|---|---|---|---|---|---|---|---|
| 88ft/sec | | 80 | 66 | 51 | 37 | 26 | 14 | 5 | 0 |

By using Riemann sums

(a) estimate the maximum distance in feet that train needed to come to rest.
(b) estimate the minimum distance in feet that the train needed to come to rest.
(c) estimate of the distance the train needs to stop based upon the average speed between each speedometer reading.
(d) explain why the value in (c) is the average of the values in (a) and (b).

**4.** For each of the functions below with its indicated interval

(a) graph the function
(b) find its average value on the indicated interval
(c) indicate all the points at which the function takes on its average value.
(i) $f(x) = x^4 - 3x^2 + 1$; $[-\sqrt{3}, \sqrt{3}]$
(ii) $f(x) = -\dfrac{x^2 + x + 1}{2}$; $[0, 4]$

**5.** Prove or disprove that the integral of the average value equals the integral of the function on the interval:
$$\int_b^a \bar{f}\,dx = \int_a^b f(x)\,dx,$$
where $\bar{f}$ is the average value of a function $f$ over the interval $[a, b]$.

**6.** Assuming that $u$ and $v$ can be integrated over the interval $[a, b]$ and that the average value over the interval is denoted by $\bar{u}$ and $\bar{v}$, prove or disprove that

(a) $\overline{u + v} = \bar{u} + \bar{v}$,
(b) $\overline{ku} = k\bar{u}$, where $k$ is any constant,
(c) if $u \le v$ then $\bar{u} \le \bar{v}$ no matter if $a < b$ or $b < a$.

**7.** Household electric current can be modelled by a voltage $V = \hat{V} \sin(120\pi t + \phi)$, where $t$ is measured in seconds, $\hat{V}$ is the maximum value that $V$ can attain, and $\phi$ is the phase angle. Such a voltage is usually said to be 60 cycle, since in 1 second the voltage goes through 60 oscillations. The root-mean-square voltage, usually denoted by $V_{rms}$ is defined to be the square root of the average of $V^2$. Hence

$$V_{rms} = \sqrt{\int_\phi^{1+\phi} \left(\hat{V} \sin(120\pi t + \phi)\right)^2 dt}$$

A good measure of how much heat a given voltage can produce is given by $V_{rms}$.

(a) compute the average voltage over 1 second.
(b) compute the average voltage over $1/120$ of a second.
(c) show that $V_{rms} = \dfrac{\hat{V}\sqrt{2}}{2}$ by computing the integral for $V_{rms}$. Hint: $\int \sin^2 t\,dt = -\dfrac{1}{2}\cos t \sin t + \dfrac{1}{2} t$
(d) If the $V_{rms}$ for household current is usually 120 Volts, what is the value $\hat{V}$ in this case?

**8.** Consider the function $G(x) = \int_0^x f(t)\,dt$ where $f(t)$ oscillates about the line $y = 2$ over the $x$-region $[0, 10]$ and is given by Figure 5.

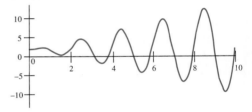

FIGURE 5

(a) At what values of $x$ over this region do the local maximum and miniums occur?
(b) Where does $G(x)$ attain its absolute maximum and absolute minimum?
(c) On what intervals is $G(x)$ concave downwards?
(d) Sketch a graph of $G(x)$.

**9.** Perform this same analysis as you did in Problem 8 for the function $G(x) = \int_0^x f(t)\,dt$ given by Figure 6, where $f(t)$ oscillates about the line $y = 2$ for the interval $[0, 10]$.

FIGURE 6

# TECHNOLOGY PROJECT

If the function $f$ has an explicit antiderivative $F$ in terms of elementary functions, then the Fundamental Theorem of Calculus tells us that

$$\int_a^b f(x)\, dx = F(b) - F(a),$$

so that evaluating the definite integral of $f$ can be achieved by merely evaluating $F$ at $a$ and $b$. For a continuous $f$, the Existence Theorem and Fundamental Theorem together guarantee that the definite integral and antiderivative exist (for example, $F(x) = \int_a^x f(t)\, dt$). However, for many elementary functions $f$, the antiderivative $F$ is not itself an elementary function that we can evaluate directly and hence the Fundamental Theorem does not provide a means for evaluation of the definite integral. In such cases, we can still get accurate numerical approximations to the definite integral. In this project, we examine the most elementary of these approximations, the direct use of Riemann sums. In this project, you will see that some Riemann sums provide better approximations than others, even when using the same number of subintervals.

**Exercise 1** Use the Fundamental Theorem to check that

$$F(x) = \int_a^x f(t)\, dt$$

is an antiderivative of $f(x)$.

**Exercise 2** Show that the exact area between the $x$-axis and the function $f(x) = x^4 + 1$ on the interval $[0, 5]$ is 630.

**Exercise 3** Explain the geometrical meaning of the curve $y = \sqrt{100 - x^2}$ on the interval $[0, 10]$ and thus evaluate exactly the definite integral,

$$I = \int_0^{10} \sqrt{100 - x^2}\, dx.$$

As a "warm-up," we compute some Riemann sums for $f(x) = x^4 + 1$ on $[0, 5]$. We start with $n = 4$ and will increase $n$ and observe what happens. Define the function $f$ in your technology and set limits $a = 0.0$ and $b = 5.0$. All the common technologies have a "Sum" command—here we will use *Mathematica*'s as an example to set up the Riemann sum with $n = 5$. Since $\mathbf{\Delta}x = $ delx is a constant, we take it outside the Sum :

```
n = 5; delx = (b - a)/n;
Rsum = delx Sum[f[a + i delx],
{i, 0, n-1}]359.
```

Notice that the answer is awful!

**Exercise 4** After repeating the above run as a check, compute the Riemann sums for $n = 10$ and $n = 100$.

We now improve the above Riemann sum code, which is so far restricted to using the left endpoint of each subinterval. Introduce a parameter $p$ (standing for "proportion"), where $p$ is a value between 0 and 1. Notice that when $p = 0$ the new Sum command below acts just like the above Sum command. But when $p = 1$, we get the Riemann sum using the *right* endpoint of each subinterval. The intermediate choice $p = 0.5$ gives the "midpoint rule." So consider the following *Mathematica* code in which we first set $p = 1$:

```
exact = 630;
n = 10;
p = 1.0;
Rsum = delx Sum[f[a +
(i + p)delx], {i, 0, n - 1}]
```

**Exercise 5** Repeat Exercise 4, this time using *right*-endpoint sums. Contrast the results.

**Exercise 6** Add the line `exact - Rsum / N` to the improved Riemann sum code and make your study systematic by filling in an *error* table with column headings for left-endpoint ($p = 0$), midpoint ($p = 0.5$), and right-endpoint ($p = 1$), and rows labeled by $n = 10$, 100, and 1000. The nine entries in the table should be the errors in the corresponding Riemann sum. (It is not necessary to do anything fancy unless you want to, just get the nine values and write them in a table by hand.)

**Exercise 7** Do the analog of the previous problem for the function and interval in Exercise 3.

**Exercise 8** The errors in numerical approximations tend to be proportional to $1/n$, or to $1/n^2$, or to $1/n^3$ and so on, where $n$ is the number of subintervals. Try to detect this pattern on the basis of the tables you made in Exercises 6 and 7.

# 6

# Applications of the Integral

**6.1**

**THE AREA OF A PLANE REGION**

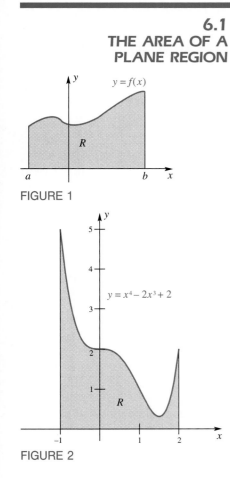

FIGURE 1

FIGURE 2

The brief discussion of area in Section 5.4 served to motivate the definition of the definite integral. With the latter notion now firmly established, we reverse directions and use the definite integral to calculate areas of regions of more and more complicated shapes. As is our practice, we begin with simple cases.

**A Region Above the X-Axis**   Let $y = f(x)$ determine a curve in the $xy$-plane and suppose $f$ is continuous and nonnegative on the interval $a \le x \le b$ (as in Figure 1). Consider the region $R$ bounded by the graphs of $y = f(x)$, $x = a$, $x = b$, and $y = 0$. We refer to $R$ as the region under $y = f(x)$ between $x = a$ and $x = b$. Its area $A(R)$ is given by

$$A(R) = \int_a^b f(x)\, dx$$

**EXAMPLE 1**   Find the area of the region $R$ under $y = x^4 - 2x^3 + 2$ between $x = -1$ and $x = 2$.

**Solution**   The graph of $R$ is shown in Figure 2. A reasonable estimate for the area of $R$ is its base times an average height, say $(3)(2) = 6$. The exact value is

$$A(R) = \int_{-1}^{2} (x^4 - 2x^3 + 2)\, dx = \left[ \frac{x^5}{5} - \frac{x^4}{2} + 2x \right]_{-1}^{2}$$

$$= \left( \frac{32}{5} - \frac{16}{2} + 4 \right) - \left( -\frac{1}{5} - \frac{1}{2} - 2 \right) = \frac{51}{10} = 5.1$$

297

The calculated value 5.1 is close enough to our estimate, 6, to give us confidence in its correctness.    ■

**A Region Below the X-Axis**    Area is a nonnegative number. If the graph of $y = f(x)$ is below the $x$-axis, then $\int_a^b f(x)\,dx$ is a negative number and therefore cannot be an area. However, it is just the negative of the area of the region bounded by $y = f(x)$, $x = a$, $x = b$, and $y = 0$.

**EXAMPLE 2**    Find the area of the region $R$ bounded by $y = x^3/3 - 4$, the $x$-axis, $x = -2$, and $x = 3$.

≈ **Solution**    The region $R$ is shown in Figure 3. Our estimate for its area is $(5)(3) = 15$. The exact value is

$$A(R) = -\int_{-2}^{3}\left(\frac{x^2}{3} - 4\right)dx = \int_{-2}^{3}\left(-\frac{x^2}{3} + 4\right)dx$$

$$= \left[-\frac{x^3}{9} + 4x\right]_{-2}^{3} = \left(-\frac{27}{9} + 12\right) - \left(\frac{8}{9} - 8\right) = \frac{145}{9} \approx 16.11$$

We are reassured by the nearness of 16.11 to our estimate.    ■

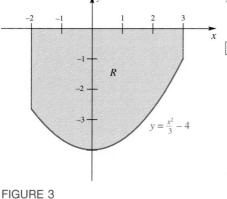

FIGURE 3

**EXAMPLE 3**    Find the area of the region $R$ bounded by $y = x^3 - 3x^2 - x + 3$, the segment of the $x$-axis between $x = -1$ and $x = 2$, and the line $x = 2$.

**Solution**    The region $R$ is shaded in Figure 4. Note that part of it is above the $x$-axis and part is below. The areas of these two parts, $R_1$ and $R_2$, must be calculated separately. You can check that the curve crosses the $x$-axis at $-1$, 1, and 3. Thus,

$$A(R) = A(R_1) + A(R_2)$$

$$= \int_{-1}^{1}(x^3 - 3x^2 - x + 3)\,dx - \int_{1}^{2}(x^3 - 3x^2 - x + 3)\,dx$$

$$= \left[\frac{x^4}{4} - x^3 - \frac{x^2}{2} + 3x\right]_{-1}^{1} - \left[\frac{x^4}{4} - x^3 - \frac{x^2}{2} + 3x\right]_{1}^{2}$$

$$= 4 - \left(-\frac{7}{4}\right) = \frac{23}{4}$$

Notice that we could have written this area as one integral using the absolute value symbol.

$$A(R) = \int_{-1}^{2}|x^3 - 3x^2 - x + 3|\,dx$$

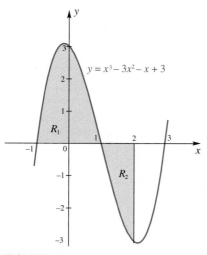

FIGURE 4

But this is no real simplification since, in order to evaluate this integral, we have to split it into two parts just as we did above.    ■

**A Helpful Way of Thinking**    So far so good. For simple regions of the type considered above, it is quite easy to write down the correct integral. When we consider more complicated regions (for example, regions between two curves), the task of selecting the right integral is more difficult. However, there is a way of thinking that can be very helpful. It goes back to the definition of area and of the definite integral. Here it is in five steps.

*Step 1*    Sketch the region.
*Step 2*    Slice it into thin pieces (strips); label a typical piece.
*Step 3*    Approximate the area of this typical piece, pretending it is a rectangle.
*Step 4*    Add up the approximations to the areas of the pieces.
*Step 5*    Take the limit as the width of the pieces approaches zero, thus getting a definite integral.

To illustrate we consider yet another simple example.

**EXAMPLE 4**    Set up the integral for the area of the region under $y = 1 + \sqrt{x}$ between $x = 0$ and $x = 4$ (Figure 5).

*Solution*

1. Sketch

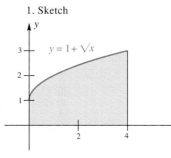

3. Approximate area of typical piece:

$$\Delta A_i \approx (1 + \sqrt{x_i})\, \Delta x_i$$

4. Add up: $A \approx \sum (1 + \sqrt{x_i})\, \Delta x_i$

5. Take limit: $A = \int_0^4 (1 + \sqrt{x})\, dx$

2. Slice

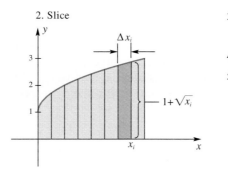

FIGURE 5

Once we understand this five-step procedure, we can abbreviate it to three: *slice, approximate, integrate*. Think of the word *integrate* as meaning add up and take the limit as the piece width tends to zero. In this process $\sum \ldots \Delta x$ transforms into $\int \ldots dx$. Figure 6 gives the abbreviated form for the same problem.

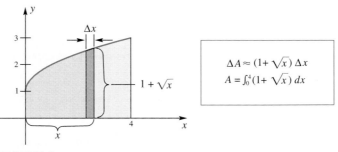

$$\Delta A \approx (1 + \sqrt{x})\, \Delta x$$
$$A = \int_0^4 (1 + \sqrt{x})\, dx$$

FIGURE 6    ■

**A Region Between Two Curves**   Consider curves $y = f(x)$ and $y = g(x)$ with $g(x) \le f(x)$ on $a \le x \le b$. They determine the region shown in Figure 7. We use the *slice, approximate, integrate* method to find its area. Be sure to note that $f(x) - g(x)$ gives the correct height for the thin slice, even when the graph of $g$ goes below the $x$-axis. For in this case, $g(x)$ is negative; so subtracting $g(x)$ is the same as adding a positive number. You can check that $f(x) - g(x)$ also gives the correct height, even when both $f(x)$ and $g(x)$ are negative.

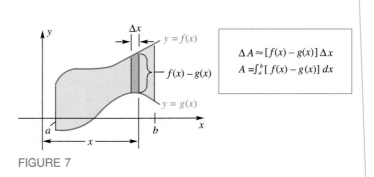

FIGURE 7

**EXAMPLE 5**   Find the area of the region between the curves $y = x^4$ and $y = 2x - x^2$.

*Solution*   We start by finding where the two curves intersect. To do this, we need to solve $2x - x^2 = x^4$, a fourth-degree equation, which would usually be difficult to solve. However, in this case, $x = 0$ and $x = 1$ are rather obvious solutions. Our sketch of the region, together with the appropriate approximation and the corresponding integral, is shown in Figure 8.

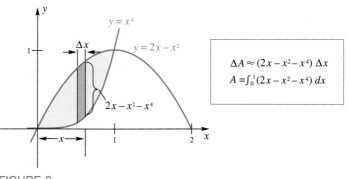

FIGURE 8

One job remains—to evaluate the integral.

$$\int_0^1 (2x - x^2 - x^4)\, dx = \left[ x^2 - \frac{x^3}{3} - \frac{x^5}{5} \right]_0^1 = 1 - \frac{1}{3} - \frac{1}{5} = \frac{7}{15} \qquad \blacksquare$$

**EXAMPLE 6**   **(Horizontal slicing).** Find the area of the region between the parabola $y^2 = 4x$ and the line $4x - 3y = 4$.

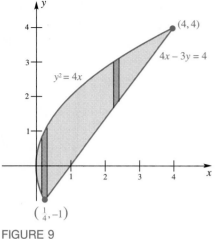

$(\frac{1}{4}, -1)$

FIGURE 9

*Solution*    We will need the points of intersection of these two curves. The $y$-coordinates of these points can be found by writing the second equation as $4x = 3y + 4$ and then equating the two expressions for $4x$.

$$y^2 = 3y + 4$$
$$y^2 - 3y - 4 = 0$$
$$(y - 4)(y + 1) = 0$$
$$y = 4, -1$$

From this, we conclude that the points of intersection are $(4, 4)$ and $(\frac{1}{4}, -1)$. The required region is sketched in Figure 9.

Now imagine slicing this region vertically. We face a problem, because the lower boundary consists of two different curves. Slices at the extreme left extend from the lower branch of the parabola to its upper branch. For the rest of the region, slices extend from the line to the parabola. To do the problem with vertical slices requires that we first split our region into two parts, set up an integral for each part, and then evaluate both integrals.

A far better approach is to slice the region horizontally as shown in Figure 10, thus using $y$ rather than $x$ as the integration variable. Note that horizontal slices always go from the parabola (at the left) to the line (at the right). The width of such a slice is the largest $x$-value $(x = \frac{1}{4}(3y + 4))$ minus the smallest $x$-value $(x = \frac{1}{4}y^2)$.

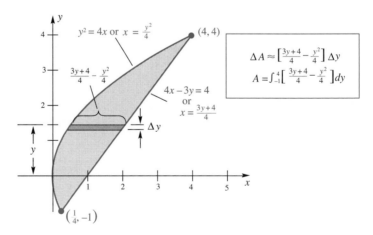

FIGURE 10

$$A = \int_{-1}^{4} \left[ \frac{3y + 4 - y^2}{4} \right] dy = \frac{1}{4} \int_{-1}^{4} (3y + 4 - y^2) \, dy$$

$$= \frac{1}{4} \left[ \frac{3y^2}{2} + 4y - \frac{y^3}{3} \right]_{-1}^{4}$$

$$= \frac{1}{4} \left[ \left( 24 + 16 - \frac{64}{3} \right) - \left( \frac{3}{2} - 4 + \frac{1}{3} \right) \right]$$

$$= \frac{125}{24} \approx 5.21$$

There are two items to note: (1) The integrand resulting from a horizontal slicing involves $y$, not $x$; and (2) to get the integrand, solve both equations for $x$ and subtract the smaller $x$-value from the larger. ∎

**Distance and Displacement** Consider an object moving along a straight line with velocity $v(t)$ at time $t$. If $v(t) \geq 0$, then $\int_a^b v(t)\, dt$ gives the distance traveled during the time interval $a \leq t \leq b$ (see Section 5.4). However, if $v(t)$ is sometimes negative (which corresponds to the object moving in reverse), then

$$\int_a^b v(t)\, dt = s(b) - s(a)$$

measures the **displacement** of the object, that is, the directed distance from its starting position $s(a)$ to its ending position $s(b)$. To get the **total distance** that the object traveled during $a \leq t \leq b$, we must calculate $\int_a^b |v(t)|\, dt$, the area between the velocity curve and the $t$-axis. Problems 31–33 illustrate these ideas.

## CONCEPTS REVIEW

**1.** Let $R$ be the region between the curve $y = f(x)$ and the $x$-axis on the interval $[a, b]$. If $f(x) \geq 0$ for all $x$ in $[a, b]$, then $A(R) =$ _____, but if $f(x) \leq 0$ for all $x$ in $[a, b]$, then $A(R) =$ _____.

**2.** To find the area of the region between two curves, it is wise to think of the following three-word motto: _____.

**3.** Suppose the curves $y = f(x)$ and $y = g(x)$ bound a region $R$ on which $f(x) \leq g(x)$. Then the area of $R$ is given by $A(R) = \int_a^b$ _____ $dx$, where $a$ and $b$ are determined by solving the equation _____.

**4.** If $p(y) \leq q(y)$ for all $y$ in $[c, d]$, then the area $A(R)$ of the region $R$ bounded by the curves $x = p(y)$ and $x = q(y)$ between $y = c$ and $y = d$ is given by $A(R) =$ _____.

## PROBLEM SET 6.1

In Problems 1–10, use the three-step procedure (*slice, approximate, integrate*) to set up an integral (or integrals) for the area of the indicated region.

**1.**

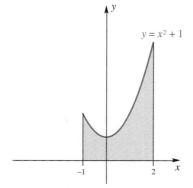

$y = x^2 + 1$

**2.**

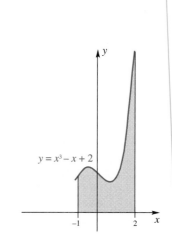

$y = x^3 - x + 2$

**3.**

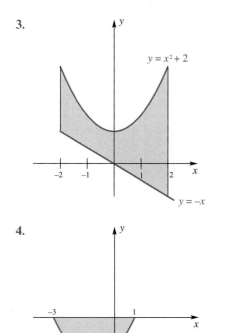

$y = x^2 + 2$

$y = -x$

**4.**

$y = x^2 + 2x - 3$

**5.**

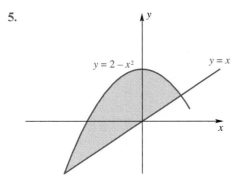

$y = 2 - x^2$

$y = x$

*Hint*: To find the intersection points, solve $x = 2 - x^2$.

**6.**

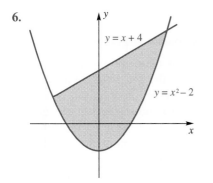

$y = x + 4$

$y = x^2 - 2$

**7.**

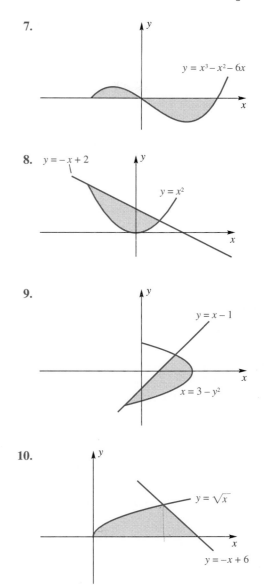

$y = x^3 - x^2 - 6x$

**8.**  $y = -x + 2$

$y = x^2$

**9.**

$y = x - 1$

$x = 3 - y^2$

**10.**

$y = \sqrt{x}$

$y = -x + 6$

≈ In Problems 11–28, sketch the region bounded by the graphs of the given equations, show a typical rectangle, approximate its area, set up an integral, and calculate the area of the region. Make an estimate of the area to confirm your answer.

11.  $y = 3 - \frac{1}{3}\,x^2$, $y = 0$, between $x = 0$ and $x = 3$

12.  $y = 5x - x^2$, $y = 0$, between $x = 1$ and $x = 3$

13.  $y = (x - 4)(x + 2)$, $y = 0$, between $x = 0$ and $x = 3$

14.  $y = x^2 - 4x - 5$, $y = 0$, between $x = -1$ and $x = 4$

15.  $y = \frac{1}{4}\,(x^2 - 7)$, $y = 0$, between $x = 0$ and $x = 2$

16.  $y = x^3$, $y = 0$, between $x = -3$ and $x = 3$

17.  $y = \sqrt[3]{x}$, $y = 0$, between $x = -2$ and $x = 2$

**18.** $y = \sqrt{x} - 10$, $y = 0$, between $x = 0$ and $x = 9$

**19.** $y = (x - 3)(x - 1)$, $y = x$

**20.** $y = 3\sqrt{x}$, $y = x - 4$, $x = 0$

**21.** $y = x^2 - 2x$, $y = -x^2$

**22.** $y = x^2 - 9$, $y = (2x - 1)(x + 3)$

**23.** $x = 8y - y^2$, $x = 0$

**24.** $x = (3 - y)(y + 1)$, $x = 0$

**25.** $x = -6y^2 + 4y$, $x + 3y - 2 = 0$

**26.** $x = y^2 - 2y$, $x - y - 4 = 0$

**27.** $4y^2 - 2x = 0$, $4y^2 + 4x - 12 = 0$

**28.** $x = 4y^4$, $x = 8 - 4y^4$

**29.** Sketch the region $R$ bounded by $y = x + 6$, $y = x^3$, and $2y + x = 0$. Then find its area. *Hint*: You will have to divide $R$ into two pieces.

**30.** Find the area of the triangle with vertices at $(-1, 4)$, $(2, -2)$, and $(5, 1)$ by integration.

**31.** An object moves along a line so that its velocity at time $t$ is $v(t) = 3t^2 - 24t + 36$ feet per second (see Example 3 of Section 3.7). Find the displacement and total distance traveled by the object for $-1 \le t \le 9$.

**32.** Follow the directions of Problem 31 if $v(t) = \frac{1}{2} + \sin 2t$ and the interval is $0 \le t \le 3\pi/2$.

**33.** Starting at $s = 0$ when $t = 0$, an object moves along a line so that its velocity at time $t$ is $v(t) = 2t - 4$ centimeters per second. How long will it take to get to $s = 12$? To travel a total distance of 12 centimeters?

**34.** Consider the curve $y = 1/x^2$ for $1 \le x \le 6$.

(a) Calculate the area under this curve.
(b) Determine $c$ so that the line $x = c$ bisects the area of part (a).
(c) Determine $d$ so that the line $y = d$ bisects the area of part (a).

**35.** Calculate areas $A$, $B$, $C$, and $D$ in Figure 11. Check by calculating $A + B + C + D$ in one integration.

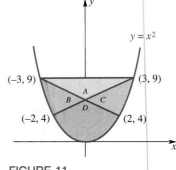

FIGURE 11

**36.** Prove Cavalieri's Principle. If two regions have the same width at every $x$ in $[a, b]$, then they have the same area (see Figure 12).

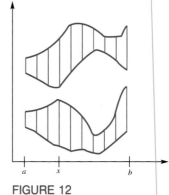

FIGURE 12

**37.** Find the area of the region trapped between $y = \sin x$ and $y = \frac{1}{2}$, $0 \le x \le 17\pi/6$.

---

**Answers to Concepts Review:** **1.** $\displaystyle\int_a^b f(x)\, dx; \ -\int_a^b f(x)\, dx$

**2.** Slice, approximate, integrate **3.** $g(x) - f(x); f(x) = g(x)$

**4.** $\displaystyle\int_c^d [q(y) - p(y)]\, dy$

---

**6.2**
**VOLUMES OF SOLIDS:**
**SLABS, DISKS, WASHERS**

That the definite integral can be used to calculate areas is not surprising; it was invented for that purpose. But uses of the integral go far beyond that application. Almost any quantity that can be thought of as a result of chopping something into small pieces, approximating each piece, adding up, and taking the limit as the pieces shrink in size can be interpreted as a definite integral. In particular, this is true for the volumes of solids that can be cut into thin slices, where the volume of each slice is easy to approximate.

## The Volume of a Coin

Consider an ordinary coin, say a quarter.

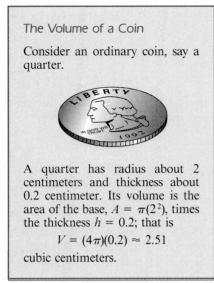

A quarter has radius about 2 centimeters and thickness about 0.2 centimeter. Its volume is the area of the base, $A = \pi(2^2)$, times the thickness $h = 0.2$; that is

$$V = (4\pi)(0.2) \approx 2.51$$

cubic centimeters.

What is volume? We start with simple solids called *right cylinders*, four of which are shown in Figure 1. In each case the solid is generated by moving a plane region (the base) through a distance $h$ in a direction perpendicular to that region. And in each case, the volume of the solid is defined to be the area $A$ of the base times the height $h$—that is,

$$V = A \cdot h$$

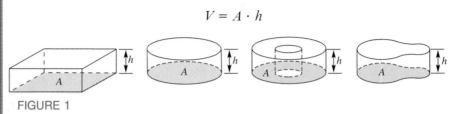

FIGURE 1

Next consider a solid which has the property that cross sections perpendicular to a given line have known area. In particular, suppose that line is the $x$-axis and that the area of the cross section at $x$ is $A(x)$, $a \le x \le b$ (Figure 2). Partition the interval $[a, b]$ by inserting points $a = x_0 < x_1 < x_2 < \ldots < x_n = b$ and pass planes through these points perpendicular to the $x$-axis, thus slicing the solid into thin **slabs** (Figure 3). The "volume" $\Delta V_i$ of a slab should be approximately that of a cylinder, namely,

$$\Delta V_i \approx A(\bar{x}_i)\Delta x_i, \qquad x_{i-1} \le \bar{x}_i \le x_i$$

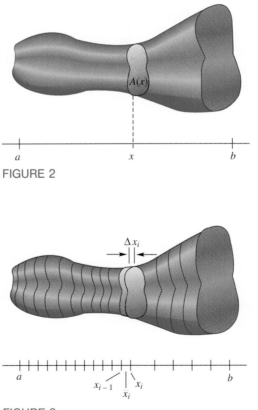

FIGURE 2

FIGURE 3

and the "volume" $V$ of the solid should be given approximately by the Riemann sum

$$V \approx \sum_{i=1}^{n} A(\bar{x}_i)\Delta x_i$$

When we let the norm of the partition approach zero, we obtain a definite integral; this integral is defined to be the **volume** of the solid.

$$V = \int_{a}^{b} A(x)dx$$

Rather than routinely applying the boxed formula to obtain volumes, we suggest that in each problem you go through the process that led to it, at least in summary form. Just as for areas, we call this process *slice, approximate, integrate*. It is illustrated in the examples that follow.

**Solids of Revolution: Method of Disks** When a plane region, lying entirely on one side of a fixed line in its plane, is revolved about that line, it generates a **solid of revolution.** The fixed line is called the **axis** of the solid of revolution.

As an illustration, if the region bounded by a semicircle and its diameter is revolved about that diameter, it sweeps out a spherical solid. If the region inside a right triangle is revolved about one of its legs, it generates a conical solid (Figure 4). When a circular region is revolved about a line in its plane that does not intersect the circle (Figure 5), it sweeps out a torus (doughnut). In each case, it is possible to represent the volume as a definite integral.

**EXAMPLE 1** Find the volume of the solid of revolution obtained by revolving the plane region $R$ bounded by $y = \sqrt{x}$, the $x$-axis, and the line $x = 4$ about the $x$-axis.

*Solution* The region $R$, with a typical slice, is displayed below as the left part of Figure 6. When revolved about the $x$-axis, this region generates a solid of revolution and the slice generates a disk, a thin coin-shaped object.

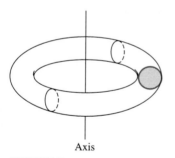

FIGURE 4

FIGURE 5

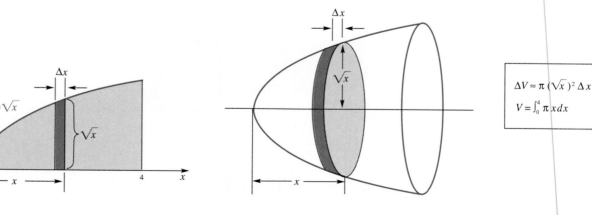

$$\Delta V \approx \pi \left( \sqrt{x} \right)^2 \Delta x$$
$$V = \int_{0}^{4} \pi\, x\, dx$$

FIGURE 6

Recalling that the volume of a circular cylinder is $\pi r^2 h$, we approximate the volume $\Delta V$ of this disk, $\Delta V \approx \pi(\sqrt{x})^2\,\Delta x$, and then integrate.

$$V = \pi \int_0^4 x\,dx = \pi\left[\frac{x^2}{2}\right]_0^4 = \pi\frac{16}{2} = 8\pi \approx 25.13 \quad \blacksquare$$

**EXAMPLE 2** Find the volume of the solid generated by revolving the region bounded by the curve $y = x^3$, the $y$-axis, and the line $y = 3$ about the $y$-axis (Figure 7).

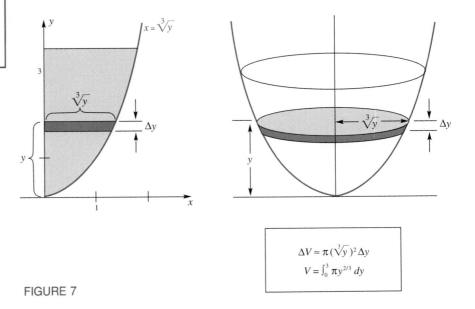

$$\Delta V \approx \pi(\sqrt[3]{y})^2\Delta y$$
$$V = \int_0^3 \pi y^{2/3}\,dy$$

**FIGURE 7**

**Solution** Here we slice horizontally, which makes $y$ the appropriate choice for the integration variable. Note that $y = x^3$ is equivalent to $x = \sqrt[3]{y}$ and $\Delta V \approx \pi(\sqrt[3]{y})^2\Delta y$.

$$V = \pi \int_0^3 y^{2/3}\,dy = \pi\left[\frac{3}{5}y^{5/3}\right]_0^3 = \pi\frac{9\sqrt[3]{9}}{5} \approx 11.76 \quad \blacksquare$$

**Method of Washers** Sometimes, slicing a solid of revolution results in disks with holes in the middle. We call them **washers**. See the diagram and accompanying volume formula shown in Figure 8.

**EXAMPLE 3** Find the volume of the solid generated by revolving the region bounded by the parabolas $y = x^2$ and $y^2 = 8x$ about the $x$-axis.

**Solution** The key words are still *slice, approximate, integrate* (see Figure 9).

$$V = \pi \int_0^2 (8x - x^4)\,dx = \pi\left[\frac{8x^2}{2} - \frac{x^5}{5}\right]_0^2 = \frac{48\pi}{5} \approx 30.16 \quad \blacksquare$$

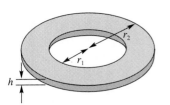

$$V = A \cdot h$$
$$= \pi(r_2^2 - r_1^2)h$$

**FIGURE 8**

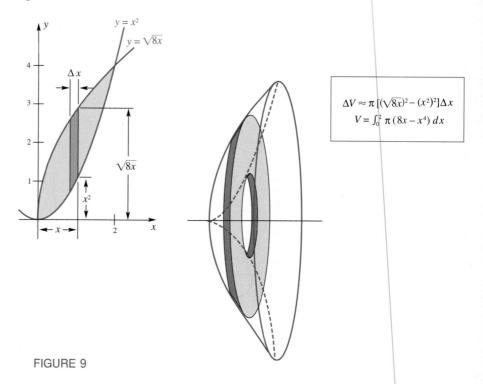

$$\Delta V \approx \pi \left[(\sqrt{8x})^2 - (x^2)^2\right] \Delta x$$
$$V = \int_0^2 \pi (8x - x^4)\, dx$$

FIGURE 9

**EXAMPLE 4**  The semicircular region bounded by $x = \sqrt{4 - y^2}$ and the $y$-axis is revolved about the line $x = -1$. Set up the integral that represents its volume.

*Solution*  Here the outer radius of the washer is $\sqrt{4 - y^2} + 1$ and the inner radius is 1. Figure 10 exhibits the solution. The integral can be simplified.

$$\Delta V \approx \pi \left[(1 + \sqrt{4 - y^2})^2 - 1^2\right] \Delta y$$
$$V = \int_{-2}^{2} \pi \left[(1 + \sqrt{4 - y^2})^2 - 1^2\right] dy$$

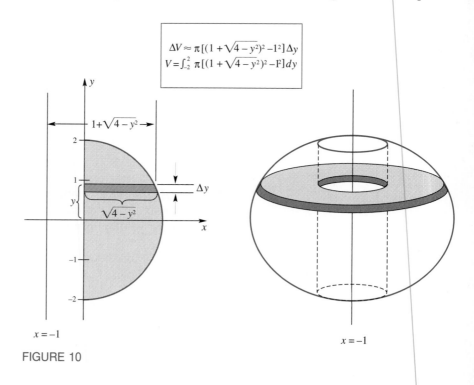

FIGURE 10

The part above the $x$-axis has the same volume as the part below it (which manifests itself in an even integrand). Thus, we may integrate from 0 to 2 and double the result. Also, the integrand simplifies.

$$V = \pi \int_{-2}^{2} \left[ (1 + \sqrt{4 - y^2})^2 - 1^2 \right] dy$$

$$= 2\pi \int_{0}^{2} \left[ 2\sqrt{4 - y^2} + 4 - y^2 \right] dy$$

Now see Problem 31 for a way of evaluating this integral.    ■

**Other Solids with Known Cross Sections**    So far, our solids have had circular cross sections. However, our method works just as well for solids whose cross sections are squares or triangles. In fact, all that is really needed is that the areas of the cross sections can be calculated, since, in this case, we can also calculate the volume of the slice—a slab—with this cross section.

**EXAMPLE 5**    Let the base of a solid be the first quadrant plane region bounded by $y = 1 - x^2/4$, the $x$-axis, and the $y$-axis. Suppose that cross sections perpendicular to the $x$-axis are squares. Find the volume of the solid.

*Solution*    When we slice this solid perpendicularly to the $x$-axis, we get thin square boxes (Figure 11), like slices of cheese.

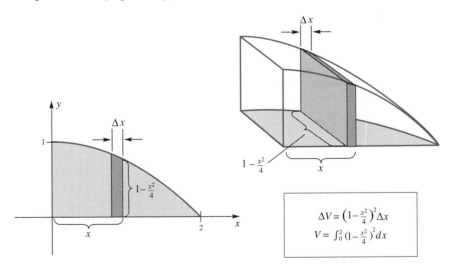

FIGURE 11

$$V = \int_{0}^{2} \left( 1 - \frac{x^2}{2} + \frac{x^4}{16} \right) dx = \left[ x - \frac{x^3}{6} + \frac{x^5}{80} \right]_{0}^{2}$$

$$= 2 - \frac{8}{6} + \frac{32}{80} = \frac{16}{15} \approx 1.07 \quad ■$$

**EXAMPLE 6**    The base of a solid is the region between one arch of $y = \sin x$ and the $x$-axis. Each cross section perpendicular to the $x$-axis is an equilateral triangle sitting on this base. Find the volume of the solid.

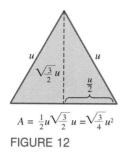

$$A = \frac{1}{2}u\frac{\sqrt{3}}{2}u = \frac{\sqrt{3}}{4}u^2$$

**FIGURE 12**

***Solution*** We need the fact that the area of an equilateral triangle of side $u$ is $\sqrt{3}u^2/4$ (see Figure 12). We proceed as shown in Figure 13.

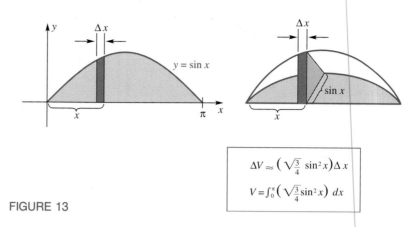

$$\Delta V \approx \left(\frac{\sqrt{3}}{4}\sin^2 x\right)\Delta x$$

$$V = \int_0^\pi \left(\frac{\sqrt{3}}{4}\sin^2 x\right)\,dx$$

**FIGURE 13**

To perform the indicated integration, we use the half-angle formula $\sin^2 x = (1 - \cos 2x)/2$.

$$V = \frac{\sqrt{3}}{4}\int_0^\pi \frac{1 - \cos 2x}{2}\,dx = \frac{\sqrt{3}}{8}\int_0^\pi (1 - \cos 2x)\,dx$$

$$= \frac{\sqrt{3}}{8}\left[\int_0^\pi 1\,dx - \frac{1}{2}\int_0^\pi \cos 2x \cdot 2\,dx\right]$$

$$= \frac{\sqrt{3}}{8}\left[x - \frac{1}{2}\sin 2x\right]_0^\pi = \frac{\sqrt{3}}{8}\pi \approx 0.68 \quad \blacksquare$$

## CONCEPTS REVIEW

**1.** The volume of a disk of radius $r$ and thickness $h$ is _____.

**2.** The volume of a washer of inner radius $r$, outer radius $R$, and thickness $h$ is _____.

**3.** If the region $R$ bounded by $y = x^2$, $y = 0$, and $x =$

3 is revolved about the $x$-axis, the disk at $x$ will have volume $\Delta V \approx$ _____.

**4.** If the region $R$ of Question 3 is revolved about the line $y = -2$, the washer at $x$ will have volume $\Delta V \approx$ _____.

## PROBLEM SET 6.2

In Problems 1–4, find the volume of the solid generated when the indicated region is revolved about the specified axis; *slice, approximate, integrate.*

**1.** $x$-axis

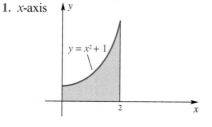

**2.** $x$-axis

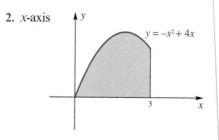

3. (a) x-axis
   (b) y-axis

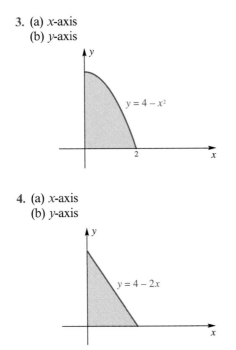

$y = 4 - x^2$

4. (a) x-axis
   (b) y-axis

$y = 4 - 2x$

≈ In Problems 5–10, sketch the region $R$ bounded by the graphs of the given equations, showing a typical vertical rectangle. Then find the volume of the solid generated by revolving $R$ about the x-axis.

5. $y = \dfrac{x^2}{\pi}$, $x = 4$, $y = 0$

6. $y = x^3$, $x = 3$, $y = 0$

7. $y = \dfrac{1}{x}$, $x = 2$, $x = 4$, $y = 0$

8. $y = x^{3/2}$, $y = 0$, between $x = 2$ and $x = 3$.

9. $y = \sqrt{9 - x^2}$, $y = 0$, between $x = -2$ and $x = 3$.

10. $y = x^{2/3}$, $y = 0$, between $x = 1$ and $x = 27$.

≈ In Problems 11–16, sketch the region $R$ bounded by the graphs of the given equations and show a typical horizontal rectangle. Find the volume of the solid generated by revolving $R$ about the y-axis.

11. $x = y^2$, $x = 0$, $y = 3$

12. $x = \dfrac{2}{y}$, $y = 2$, $y = 6$, $x = 0$

13. $x = 2\sqrt{y}$, $y = 4$, $x = 0$

14. $x = y^{2/3}$, $y = 27$, $x = 0$

15. $x = y^{3/2}$, $y = 9$, $x = 0$

16. $x = \sqrt{4 - y^2}$, $x = 0$

17. Find the volume of the solid generated by revolving about the x-axis the region bounded by the upper half of the ellipse

$$\frac{x^2}{a^2} + \frac{y^2}{b^2} = 1$$

and the x-axis and thus find the volume of a *prolate spheroid*. Here $a$ and $b$ are positive constants, with $a > b$.

18. Find the volume of the solid generated by revolving about the x-axis the region bounded by the line $y = 6x$ and the parabola $y = 6x^2$.

19. Find the volume of the solid generated by revolving about the x-axis the region bounded by the line $x - 2y = 0$ and the parabola $y^2 = 4x$.

20. Find the volume of the solid generated by revolving about the x-axis the region in the first quadrant bounded by the circle $x^2 + y^2 = r^2$, the x-axis, and the line $x = r - h$, $0 < h < r$, and thus find the volume of a *spherical segment* of height $h$, radius of sphere $r$.

21. Find the volume of the solid generated by revolving about the y-axis the region bounded by the line $y = 4x$ and the parabola $y = 4x^2$.

22. Find the volume of the solid generated by revolving about the line $y = 2$ the region in the first quadrant bounded by the parabolas $3x^2 - 16y + 48 = 0$ and $x^2 - 16y + 80 = 0$, and the y-axis.

23. The base of a solid is the region inside the circle $x^2 + y^2 = 4$. Find the volume of the solid if every cross section by a plane perpendicular to the x-axis is a square. *Hint*: See Examples 5 and 6.

24. Do Problem 23 assuming every cross section by a plane perpendicular to the x-axis is an isosceles triangle with base on the xy-plane and altitude 4. *Hint*: To complete the

evaluation, interpret $\displaystyle\int_{-2}^{2} \sqrt{4 - x^2}\, dx$ as the area of a

semicircle.

25. The base of a solid is bounded by one arch of $y = \sqrt{\cos x}$, $-\pi/2 \le x \le \pi/2$, and the x-axis. Each cross section perpendicular to the x-axis is a square sitting on this base. Find the volume of the solid.

26. The base of a solid is the region bounded by $y = 1 - x^2$ and $y = 1 - x^4$. Cross sections of the solid that are perpendicular to the x-axis are squares. Find the volume of the solid.

27. Find the volume of one octant (one-eighth) of the solid region common to two right circular cylinders of radius 1 whose axes intersect at right angles. *Hint*: Horizontal cross sections are squares. See Figure 14.

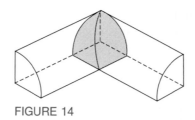

FIGURE 14

**28.** The base of a solid is the region $R$ bounded by $y = \sqrt{x}$ and $y = x^2$. Each cross section perpendicular to the $x$-axis is a semicircle with diameter extending across $R$. Find the volume of the solid.

**29.** Find the volume of the solid generated by revolving the region in the first quadrant bounded by the curve $y^2 = x^3$, the line $x = 4$, and the $x$-axis

(a) about the line $x = 4$;     (b) about the line $y = 8$.

**30.** Find the volume of the solid generated by revolving the region bounded by the curve $y^2 = x^3$, the line $y = 8$, and the $y$-axis

(a) about the line $x = 4$;     (b) about the line $y = 8$.

**31.** Complete the evaluation of the integral in Example 4 by noting that

$$\int_0^2 [2\sqrt{4 - y^2} + 4 - y^2]\, dy$$

$$= 2\int_0^2 \sqrt{4 - y^2}\, dy + \int_0^2 (4 - y^2)\, dy$$

Now interpret the first integral as the area of a quarter circle.

**32.** An open barrel of radius $r$ and height $h$ is initially full of water. It is tilted until the water level coincides with a diameter of the base and just touches the rim of the top. Find the volume of water left in the barrel. See Figure 15.

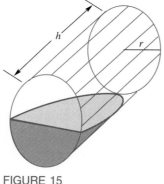

FIGURE 15

**33.** A wedge is cut from a right circular cylinder of radius $r$ (Figure 16). The upper surface of the wedge is in a plane through a diameter of the circular base and makes an angle $\theta$ with the base. Find the volume of the wedge.

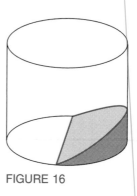

FIGURE 16

**34.** (The Water Clock) A water tank is obtained by revolving the curve $y = kx^4$, $k > 0$, about the $y$-axis.

(a) Find $V(y)$, the volume of water in the tank as a function of its depth $y$.

(b) Water drains through a small hole according to Torricelli's Law $(dV/dt = -m\sqrt{y})$. Show that the water level falls at a constant rate.

**35.** Show that the volume of a general cone (Figure 17) is $\frac{1}{3}Ah$, where $A$ is the area of the base and $h$ is the height. Use this result to give the formula for the volume of:

(a) a right circular cone of radius $r$ and height $h$;

(b) a regular tetrahedron with edge length $r$.

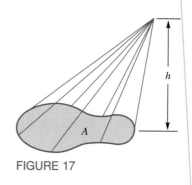

FIGURE 17

**36.** State the version of Cavalieri's Principle for volume (see Problem 36 of Section 6.1).

**Answers to Concepts Review:**  **1.** $\pi r^2 h$   **2.** $\pi(R^2 - r^2)h$
**3.** $\pi x^4\, \Delta x$   **4.** $\pi[(x^2 + 2)^2 - 4]\, \Delta x$

## 6.3 VOLUMES OF SOLIDS OF REVOLUTION: SHELLS

There is another method for finding the volume of a solid of revolution—the method of cylindrical shells. For many problems, it is easier to apply than the methods of disks or washers.

A cylindrical shell is a solid bounded by two concentric right circular cylinders (Figure 1). If the inner radius is $r_1$, the outer radius is $r_2$, and the height is $h$, then its volume is given by

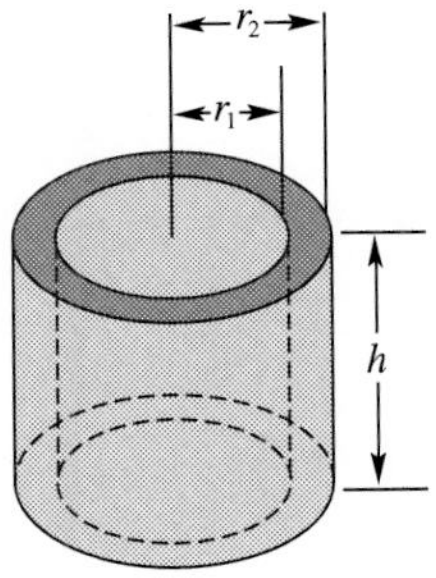

FIGURE 1

$$
\begin{aligned}
V &= (\text{area of base}) \cdot (\text{height}) \\
&= (\pi r_2^2 - \pi r_1^2)h \\
&= \pi(r_2 + r_1)(r_2 - r_1)h \\
&= 2\pi\left(\frac{r_2 + r_1}{2}\right)h(r_2 - r_1)
\end{aligned}
$$

Thus,

$$
\begin{aligned}
V &= 2\pi \cdot (\text{average radius}) \cdot (\text{height}) \cdot (\text{thickness}) \\
&= 2\pi rh\,\Delta r
\end{aligned}
$$

Here is a good way to remember this formula: If the shell were very thin and flexible (like paper), we could slit it down the side, open it up to form a rectangular sheet, and then calculate its volume by pretending that this sheet forms a thin box of length $2\pi r$, height $h$, and thickness $\Delta r$ (Figure 2).

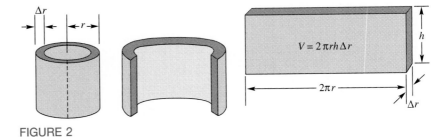

FIGURE 2

**The Method of Shells** Consider now a region of the type shown in Figure 3. Slice it vertically and revolve it about the $y$-axis. It will generate a solid of revolution and each slice will generate a piece that is approximately a cylindrical shell. To get the volume of this solid, we calculate the volume $\Delta V$ of a typical shell, add, and take the limit as the thickness of the shells tends to zero. The latter is, of course, an integral.

**EXAMPLE 1** The region bounded by $y = 1/\sqrt{x}$, the $x$-axis, $x = 1$, and $x = 4$ is revolved about the $y$-axis. Find the volume of the resulting solid.

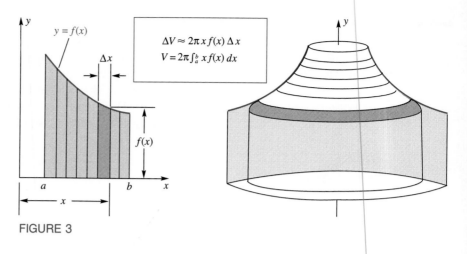

FIGURE 3

*Solution* The diagrams in Figure 3 are approximately correct. Thus,

$$V = 2\pi \int_1^4 x \frac{1}{\sqrt{x}} \, dx = 2\pi \int_1^4 x^{1/2} \, dx$$

$$= 2\pi \left[ \frac{2}{3} x^{3/2} \right]_1^4 = 2\pi \left( \frac{2}{3} \cdot 8 - \frac{2}{3} \cdot 1 \right) = \frac{28\pi}{3} \approx 29.32 \quad \blacksquare$$

**EXAMPLE 2** The region bounded by the line $y = (r/h)x$, the $x$-axis, and $x = h$ is revolved about the $x$-axis, thereby generating a cone (assume $r > 0$, $h > 0$). Find its volume by the disk method and by the shell method.

*Solution* **(Disk method).** Follow the steps suggested by Figure 4; that is, *slice, approximate, integrate.*

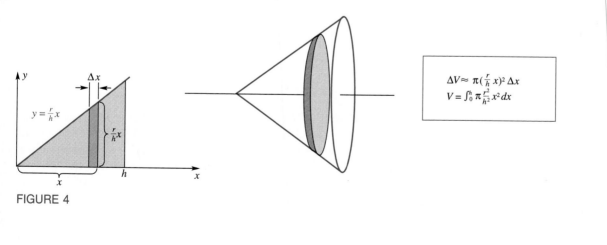

FIGURE 4

$$V = \pi \frac{r^2}{h^2} \int_0^h x^2 \, dx = \pi \frac{r^2}{h^2} \left[ \frac{x^3}{3} \right]_0^h = \frac{\pi r^2 h^3}{3h^2} = \frac{1}{3} \pi r^2 h$$

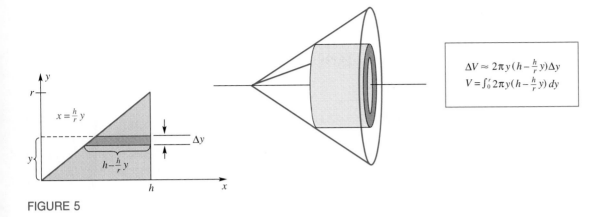

$$\Delta V \approx 2\pi y \left(h - \frac{h}{r}y\right)\Delta y$$
$$V = \int_0^r 2\pi y \left(h - \frac{h}{r}y\right) dy$$

FIGURE 5

**(Shell method).** See Figure 5.

$$V = \int_0^r 2\pi y \left(h - \frac{h}{r}y\right) dy = 2\pi h \int_0^r \left(y - \frac{1}{r}y^2\right) dy$$

$$= 2\pi h \left[\frac{y^2}{2} - \frac{y^3}{3r}\right]_0^r = 2\pi h \left[\frac{r^2}{2} - \frac{r^2}{3}\right] = \frac{1}{3}\pi r^2 h$$

Naturally, both methods yield the well-known formula for the volume of a cone. ∎

**EXAMPLE 3**    Find the volume of the solid generated by revolving the region in the first quadrant that is above the parabola $y = x^2$ and below the parabola $y = 2 - x^2$ about the $y$-axis.

*Solution*    One look at the region (left part of Figure 6) should convince you that horizontal slices leading to the disk method is not the best choice (because the right boundary consists of parts of two curves, making it necessary to use two integrals). However, vertical slices, resulting in cylindrical shells, will work fine.

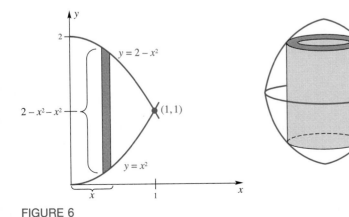

$$\Delta V \approx 2\pi x(2 - x^2 - x^2)\,\Delta x$$
$$V = \int_0^1 2\pi x\,(2 - 2x^2)\,dx$$

FIGURE 6

$$V = \int_0^1 2\pi x(2 - 2x^2)\, dx = 4\pi \int_0^1 (x - x^3)\, dx$$

$$= 4\pi \left[ \frac{x^2}{2} - \frac{x^4}{4} \right]_0^1 = 4\pi \left[ \frac{1}{2} - \frac{1}{4} \right] = \pi \approx 3.14 \qquad \blacksquare$$

**Putting It All Together**    Although most of us can draw a reasonably good plane figure, some of us do less well at drawing three-dimensional solids. But no law says we have to draw a solid in order to calculate its volume. Usually a plane figure will do, provided we can visualize the corresponding solid in our minds. In the example below, we are going to imagine revolving the region $R$ of Figure 7 about various axes. Our job is to set up an integral for the volume of the resulting solid, and we are going to do it by looking at a plane figure. Be sure to study the example carefully.

**EXAMPLE 4**    Set up an integral for the volume of the solid that results when the region $R$ shown in Figure 7 is revolved about (a) the $x$-axis, (b) the $y$-axis, (c) the line $y = -1$, and (d) the line $x = 4$.

*Solution*

(a)

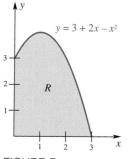

FIGURE 7

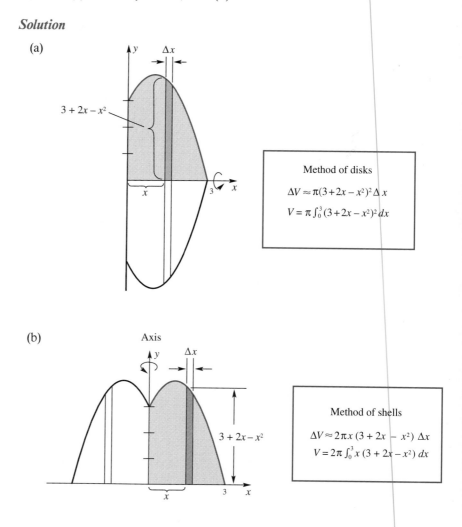

(b)

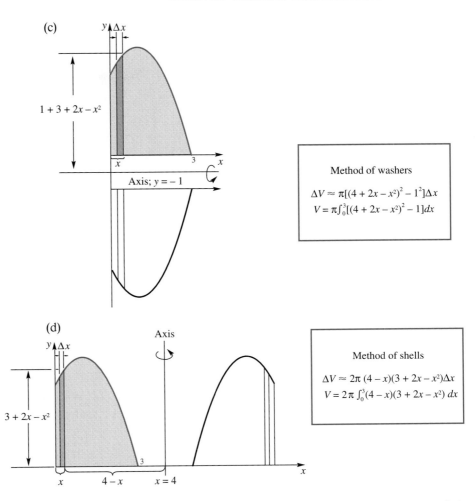

(c)

$1 + 3 + 2x - x^2$

Axis; $y = -1$

Method of washers

$\Delta V \approx \pi[(4 + 2x - x^2)^2 - 1^2]\Delta x$
$V = \pi\int_0^3[(4 + 2x - x^2)^2 - 1]dx$

(d)

Axis

$3 + 2x - x^2$

$4 - x$    $x = 4$

Method of shells

$\Delta V \approx 2\pi (4 - x)(3 + 2x - x^2)\Delta x$
$V = 2\pi \int_0^3(4 - x)(3 + 2x - x^2) \, dx$

Note that in all four cases the limits of integration are the same; it is the original plane region that determines these limits. ■

## CONCEPTS REVIEW

**1.** The volume $\Delta V$ of a thin cylindrical shell of radius $x$, height $f(x)$, and thickness $\Delta x$ is given by $\Delta V \approx$ _____.

**2.** The triangular region $R$ bounded by $y = x$, $y = 0$, and $x = 2$ is revolved about the $y$-axis, generating a solid. The method of shells gives the integral _____ as its volume; the method of washers gives the integral _____ as its volume.

**3.** The region $R$ of Question 2 is revolved about the line $x = -1$, generating a solid. The method of shells gives the integral _____ as its volume.

**4.** The region $R$ of Question 2 is revolved about the line $y = -1$, generating a solid. The method of shells gives the integral _____ as its volume.

## PROBLEM SET 6.3

In Problems 1–12, you are to find the volume of the solid generated when the region $R$ bounded by the given curves is revolved about the indicated axis. Do this by performing the following steps.

(a) Sketch the region $R$.
(b) Show a typical rectangular slice properly labeled.
(c) Write a formula for the approximate volume of the shell generated by this slice.

(d) Set up the corresponding integral.
(e) Evaluate this integral.

**1.** $y = \dfrac{1}{x}$, $x = 1$, $x = 4$, $y = 0$; about the $y$-axis.

**2.** $y = x^2$, $x = 1$, $y = 0$; about the $y$-axis.

**3.** $y = \sqrt{x}$, $x = 3$, $y = 0$; about the $y$-axis.

**4.** $y = 9 - x^2$ $(x \geq 0)$ $x = 0$, $y = 0$; about the $y$-axis.

**5.** $y = \sqrt{x}$, $x = 5$, $y = 0$; about the line $x = 5$.

**6.** $y = 9 - x^2$ $(x \geq 0)$ $x = 0$, $y = 0$; about the line $x = 3$.

**7.** $y = \frac{1}{4}x^3 + 1$, $y = 1 - x$, $x = 1$; about the $y$-axis.

**8.** $y = x^2$, $y = 3x$; about the $y$-axis.

**9.** $x = y^2$, $y = 1$, $x = 0$; about the $x$-axis.

**10.** $x = \sqrt{y} + 1$, $y = 4$, $x = 0$, $y = 0$; about the $x$-axis.

**11.** $x = y^2$, $y = 2$, $x = 0$; about the line $y = 2$.

**12.** $x = \sqrt{2y} + 1$, $y = 2$, $x = 0$, $y = 0$; about the line $y = 3$.

**13.** Consider the region $R$ (Figure 8). Set up an integral for the volume of the solid obtained when $R$ is revolved about the given line using the indicated method.

(a) The $x$-axis (washers).     (b) The $y$-axis (shells).
(c) The line $x = a$ (shells).     (d) The line $x = b$ (shells).

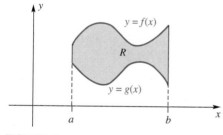

FIGURE 8

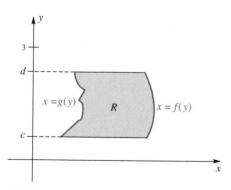

FIGURE 9

**14.** A region $R$ is shown in Figure 9. Set up an integral for the volume of the solid obtained when $R$ is revolved about each line. Use the indicated method.

(a) The $y$-axis (washers).     (b) The $x$-axis (shells).
(c) The line $y = 3$ (shells).

**15.** Sketch the region $R$ bounded by $y = 1/x^3$, $x = 1$, $x = 3$, and $y = 0$. Set up (but do not evaluate) integrals for each of the following.

(a) Area of $R$.
(b) Volume of the solid obtained when $R$ is revolved about the $y$-axis.
(c) Volume of the solid obtained when $R$ is revolved about $y = -1$.
(d) Volume of the solid obtained when $R$ is revolved about $x = 4$.

**16.** Follow the directions of Problem 15 for the region $R$ bounded by $y = x^3 + 1$ and $y = 0$ and between $x = 0$ and $x = 2$.

**17.** Find the volume of the solid generated by revolving the region $R$ bounded by the curves $x = \sqrt{y}$ and $x = y^3/32$ about the $x$-axis.

**18.** Follow the directions of Problem 17, but revolve $R$ about the line $y = 4$.

**19.** A round hole of radius $a$ is drilled through the center of a solid sphere of radius $b$ (assume $b > a$). Find the volume of the solid that remains.

**20.** Set up the integral (using shells) for the volume of the torus obtained by revolving the region inside the circle $x^2 + y^2 = a^2$ about the line $x = b$, where $b > a$. Then evaluate this integral. *Hint:* As you simplify, it may help to think of part of this integral as an area.

**21.** The region in the first quadrant bounded by $x = 0$, $y = \sin(x^2)$, and $y = \cos(x^2)$ is revolved about the $y$-axis. Find the volume of the resulting solid.

**22.** The region bounded by $y = 2 + \sin x$, $y = 0$, $x = 0$, and $x = 2\pi$ is revolved about the $y$-axis. Find the volume that results. *Hint:* $\int x \sin x \, dx = \sin x - x \cos x + C$.

**23.** Let $R$ be the region bounded by $y = x^2$ and $y = x$. Find the volume of the solid that results when $R$ is revolved around: (a) the $x$-axis; (b) the $y$-axis; (c) the line $y = x$.

**24.** Suppose we know the formula $S = 4\pi r^2$ for the surface area of a sphere but do not know the corresponding formula for its volume $V$. Obtain this formula by slicing the solid sphere into thin concentric *spherical shells* (Figure 10). *Hint:* The volume $\Delta V$ of a thin spherical shell of outer radius $x$ is $\Delta V \approx 4\pi x^2 \, \Delta x$.

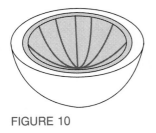

FIGURE 10

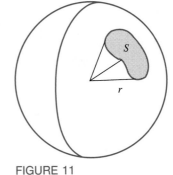

FIGURE 11

**25.** Consider an irregular region of area $S$ on the surface of a sphere of radius $r$. Find the volume of the solid that results when each point of this region is connected to the center of the sphere by a line segment (Figure 11). *Hint*: Use the method of spherical shells mentioned in Problem 24.

**Answers to Concepts Review:  1.** $2\pi x\, f(x)\, \Delta x$

**2.** $2\pi \displaystyle\int_0^2 x^2\, dx; \ \pi \displaystyle\int_0^2 (4 - y^2)\, dy$   **3.** $2\pi \displaystyle\int_0^2 (1 + x)x\, dx$

**4.** $2\pi \displaystyle\int_0^2 (1 + y)(2 - y)\, dy$

---

**6.4**
**LENGTH OF A PLANE CURVE**

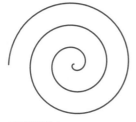

FIGURE 1

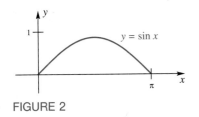

FIGURE 2

How long is the spiral curve shown in Figure 1? If it were a piece of string, most of us would stretch it taut and measure it with a ruler. But if it is the graph of an equation, that is a little hard to do.

A little reflection suggests a prior question. What is a plane curve? We have used the term *curve* informally until now; it is time to be more precise. We begin with several examples.

The graph of $y = \sin x$, $0 \le x \le \pi$, is a plane curve (Figure 2). So is the graph of $x = y^2$, $-2 \le y \le 2$ (Figure 3). In both cases, the curve is the graph of a function, the first of the form $y = f(x)$, the second of the form $x = g(y)$. However, the spiral curve does not fit this pattern. Neither does the circle $x^2 + y^2 = a^2$, though in this case we could think of it as the combined graph of the two functions $y = f(x) = \sqrt{a^2 - x^2}$ and $y = g(x) = -\sqrt{a^2 - x^2}$.

The circle suggests another way of thinking about curves. Recall from trigonometry that $x = a \cos t$, $y = a \sin t$, $0 \le t \le 2\pi$, describe the circle $x^2 + y^2 = a^2$ (Figure 4). Here $t$ is an auxiliary variable, from now on called a **parameter**. Both $x$ and $y$ are expressed in terms of this parameter. We say that $x = a \cos t$, $y = a \sin t$, $0 \le t \le 2\pi$, are **parametric equations** describing the circle.

If we were to graph the parametric equations $x = t \cos t$, $y = t \sin t$, $0 \le t \le 5\pi$, we would get a curve something like the spiral with which we started. And we can even think of the sine curve (Figure 2) and the parabola (Figure 3) in parametric form. We write

$$y = \sin x, \qquad x = x, \qquad\qquad 0 \le x \le \pi$$
$$y = y, \qquad\quad x = y^2, \qquad\quad -2 \le y \le 2$$

In the first case, $x$ is the parameter; in the second, $y$ is the parameter.

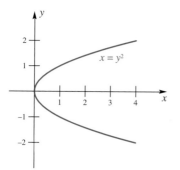

FIGURE 3

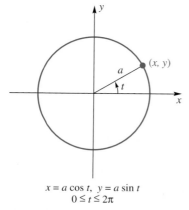

$x = a \cos t, \ y = a \sin t$
$0 \le t \le 2\pi$

FIGURE 4

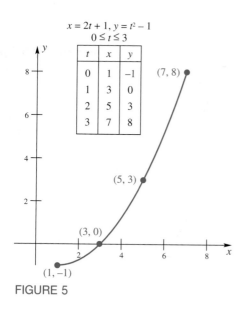

FIGURE 5

Thus, for us, a **plane curve** is determined by a pair of parametric equations $x = f(t)$, $y = g(t)$, $a \le t \le b$, where we assume that $f$ and $g$ are continuous on the given interval. Think of $t$ as a measuring time. As $t$ increases from $a$ to $b$, the point $(x, y)$ traces out a curve in the plane. Here is another example.

**EXAMPLE 1** Sketch the curve determined by the parametric equations $x = 2t + 1$, $y = t^2 - 1$, $0 \le t \le 3$.

*Solution* We make a three-column table of values, then plot the ordered pairs $(x, y)$, and—finally—connect these points in the order of increasing $t$, as shown in Figure 5. ■

Actually, the definition we have given is too broad for the purposes we have in mind, so we immediately restrict it to what is called a *smooth curve*.

> **Definition**
>
> A plane curve is **smooth** if it is determined by a pair of parametric equations $x = f(t)$, $y = g(t)$, $a \le t \le b$, where $f'$ and $g'$ exist and are continuous on $[a, b]$, and $f'(t)$ and $g'(t)$ are not simultaneously zero on $(a, b)$.

The adjective *smooth* is chosen to indicate that an object moving along the curve so that its position at time $t$ is $(x, y)$ would suffer no sudden changes of direction (continuity of $f'$ and $g'$ insures this) and would not stop or double back ($f'(t)$ and $g'(t)$ not simultaneously zero insures this).

**Length** Finally, we are ready for the main question. What is meant by the length of the smooth curve given parametrically by $x = f(t)$, $y = g(t)$, $a \le t \le b$?

Partition the interval $[a, b]$ into $n$ subintervals by means of points

$$a = t_0 < t_1 < t_2 < \cdots < t_n = b$$

This cuts the curve into $n$ pieces with corresponding endpoints $Q_0$, $Q_1$, $Q_2$, ..., $Q_{n-1}$, $Q_n$, as shown in Figure 6.

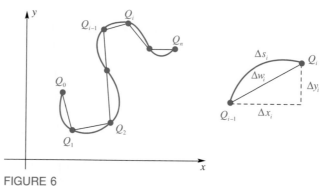

FIGURE 6

Our idea is to approximate the curve by the indicated polygonal line, calculate its length, and then take the limit as the norm of the partition approaches zero. In particular, we approximate the "length" $\Delta s_i$ of the typical piece (see Figure 6) by

$$\Delta w_i = \sqrt{(\Delta x_i)^2 + (\Delta y_i)^2}$$
$$= \sqrt{[f(t_i) - f(t_{i-1})]^2 + [g(t_i) - g(t_{i-1})]^2}$$

From the Mean Value Theorem for Derivatives, we know that there are points $\bar{t}_i$ and $\hat{t}_i$ in $(t_{i-1}, t_i)$ such that

$$f(t_i) - f(t_{i-1}) = f'(\bar{t}_i)\Delta t_i$$
$$g(t_i) - g(t_{i-1}) = g'(\hat{t}_i)\Delta t_i$$

where $\Delta t_i = t_i - t_{i-1}$. Thus,

$$\Delta w_i = \sqrt{[f'(\bar{t}_i)\Delta t_i]^2 + [g'(\hat{t}_i)\Delta t_i]^2}$$
$$= \sqrt{[f'(\bar{t}_i)]^2 + [g'(\hat{t}_i)]^2}\,\Delta t_i$$

and the length of the polygonal line is

$$\sum_{i=1}^{n} \Delta w_i = \sum_{i=1}^{n} \sqrt{[f'(\bar{t}_i)]^2 + [g'(\hat{t}_i)]^2}\,\Delta t_i$$

The latter expression is almost a Riemann sum, the only difficulty being that $\bar{t}_i$ and $\hat{t}_i$ are not likely to be the same point. However, it is shown in advanced calculus books that in the limit, this makes no difference. Thus, we may define the length $L$ of the curve to be the limit of the expression above as the norm of the partition approaches zero; that is,

$$L = \int_a^b \sqrt{[f'(t)]^2 + [g'(t)]^2}\, dt = \int_a^b \sqrt{\left(\frac{dx}{dt}\right)^2 + \left(\frac{dy}{dt}\right)^2}\, dt$$

Two special cases are of great interest. If this curve is given by $y = f(x)$, $a \le x \le b$, we treat $x$ as the parameter and the boxed result takes the form

$$L = \int_a^b \sqrt{1 + \left(\frac{dy}{dx}\right)^2}\, dx$$

Similarly, if the curve is given by $x = g(y)$, $c \le y \le d$, we treat $y$ as the parameter, obtaining

$$L = \int_c^d \sqrt{1 + \left(\frac{dx}{dy}\right)^2}\, dy$$

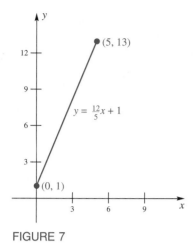

FIGURE 7

Something would be amiss if these formulas failed to yield the familiar results for circles and line segments. They do, as the following two examples illustrate.

**EXAMPLE 2** Find the circumference of the circle $x^2 + y^2 = a^2$.

*Solution* We write the equation of the circle in parametric form: $x = a \cos t$, $y = a \sin t$, $0 \leq t \leq 2\pi$. Then $dx/dt = -a \sin t$, $dy/dt = a \cos t$, and by the first of our formulas,

$$L = \int_0^{2\pi} \sqrt{a^2 \sin^2 t + a^2 \cos^2 t}\, dt = \int_0^{2\pi} a\, dt = [at]_0^{2\pi} = 2\pi a \qquad \blacksquare$$

**EXAMPLE 3** Find the length of the line segment from $A(0, 1)$ to $B(5, 13)$.

*Solution* The given line segment is shown in Figure 7. Note that the equation of the corresponding line is $y = \frac{12}{5}x + 1$, so $dy/dx = \frac{12}{5}$ and so by the second of the three length formulas,

$$L = \int_0^5 \sqrt{1 + \left(\frac{12}{5}\right)^2}\, dx = \int_0^5 \sqrt{\frac{5^2 + 12^2}{5^2}}\, dx = \frac{13}{5} \int_0^5 1\, dx$$

$$= \left[\frac{13}{5}x\right]_0^5 = 13$$

This does agree with the result obtained by use of the distance formula.

$\blacksquare$

$x = 2 \cos t,\ y = 4 \sin t$
$0 \leq t \leq \pi$

| $t$ | $x$ | $y$ |
|-----|-----|-----|
| 0 | 2 | 0 |
| $\pi/6$ | $\sqrt{3}$ | 2 |
| $\pi/3$ | 1 | $2\sqrt{3}$ |
| $\pi/2$ | 0 | 4 |
| $2\pi/3$ | $-1$ | $2\sqrt{3}$ |
| $5\pi/6$ | $-\sqrt{3}$ | 2 |
| $\pi$ | $-2$ | 0 |

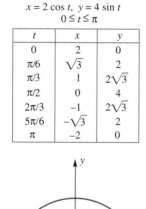

FIGURE 8

**EXAMPLE 4** Sketch the graph of the curve given parametrically by $x = 2 \cos t$, $y = 4 \sin t$, $0 \leq t \leq \pi$, and find its length.

*Solution* The graph (Figure 8) is drawn, as in Example 1, by first making a three-column table of values.

The length $L$ of this curve is given by

$$L = \int_0^\pi \sqrt{(-2 \sin t)^2 + (4 \cos t)^2}\, dt = 2 \int_0^\pi \sqrt{\sin^2 t + 4 \cos^2 t}\, dt$$

$$= 2 \int_0^\pi \sqrt{1 + 3 \cos^2 t}\, dt$$

We are unable to carry the evaluation further because we cannot find an antiderivative for $\sqrt{1 + 3 \cos^2 t}$. In fact, it has been shown that this function does not have an antiderivative that can be written in terms of the elementary functions of calculus. If we want a value for the integral, we will have to find it by an approximation method (see Chapter 10). $\blacksquare$

The situation in Example 4 is quite typical. To get an integration that we can complete, we must pick our curves very carefully.

Arc-Length Problems

How do authors choose arc-length problems? Very, very carefully. Pick a curve $y = f(x)$ at random and it is almost certain that you will be unable to integrate $\sqrt{1 + [f'(x)]^2}$ by hand. For example, we can find the length of the curve $y = x^r$ by the Fundamental Theorem of Calculus only if $r = 1$ or $r = 1 + 1/n$ for some integer $n$. We have chosen the curve of Example 5 and those in the problem set very, very carefully.

**EXAMPLE 5**  Find the length of the arc of the curve $y = x^{3/2}$ from the point $(1, 1)$ to the point $(4, 8)$ (see Figure 9).

☲ **Solution**  We begin by estimating this length. Pretend the curve is a straight line. Then its length would be $\sqrt{(4 - 1)^2 + (8 - 1)^2} = \sqrt{58} \approx 7.6$.

For the exact calculation, we note that $dy/dx = \frac{3}{2}x^{1/2}$, so

$$L = \int_1^4 \sqrt{1 + \left(\frac{3}{2}x^{1/2}\right)^2} \, dx = \int_1^4 \sqrt{1 + \frac{9}{4}x} \, dx$$

Let $u = 1 + \frac{9}{4}x$; then $du = \frac{9}{4}dx$. Hence,

$$\int \sqrt{1 + \frac{9}{4}x} \, dx = \frac{4}{9} \int \sqrt{u} \, du = \frac{4}{9}\frac{2}{3}u^{3/2} + C$$

$$= \frac{8}{27}\left(1 + \frac{9}{4}x\right)^{3/2} + C$$

Therefore,

$$\int_1^4 \sqrt{1 + \frac{9}{4}x} \, dx = \left[\frac{8}{27}\left(1 + \frac{9}{4}x\right)^{3/2}\right]_1^4 = \frac{8}{27}\left(10^{3/2} - \frac{13^{3/2}}{8}\right) \approx 7.63 \quad ■$$

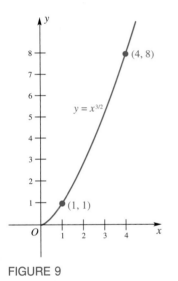

FIGURE 9

**Differential of Arc Length**  Let $f$ be continuously differentiable on $[a, b]$. For each $x$ in $(a, b)$, define $s(x)$ by

$$s(x) = \int_a^x \sqrt{1 + [f'(u)]^2} \, du$$

Then $s(x)$ gives the length of the arc of the curve $y = f(u)$ from the point $(a, f(a))$ to $(x, f(x))$ (see Figure 10). By the theorem on differentiating an integral with respect to its upper limit (Theorem 5.7D),

$$s'(x) = \frac{ds}{dx} = \sqrt{1 + [f'(x)]^2} = \sqrt{1 + \left(\frac{dy}{dx}\right)^2}$$

Thus, $ds$—the differential of arc length—can be written as

$$ds = \sqrt{1 + \left(\frac{dy}{dx}\right)^2} \, dx$$

In fact, depending on how a graph is parametrized, we are led to three formulas for $ds$, namely,

$$ds = \sqrt{1 + \left(\frac{dy}{dx}\right)^2} \, dx = \sqrt{1 + \left(\frac{dx}{dy}\right)^2} \, dy = \sqrt{\left(\frac{dx}{dt}\right)^2 + \left(\frac{dy}{dt}\right)^2} \, dt$$

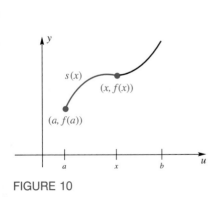

FIGURE 10

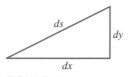

**FIGURE 11**

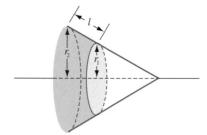

**FIGURE 12**

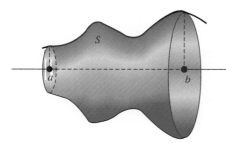

**FIGURE 13**

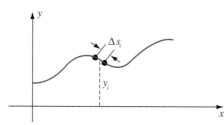

**FIGURE 14**

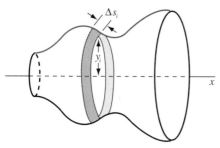

**FIGURE 15**

Some people prefer to remember these formulas by writing (see Figure 11)

$$(ds)^2 = (dx)^2 + (dy)^2$$

The three forms arise by dividing and multiplying the right-hand side by $(dx)^2$, $(dy)^2$, and $(dt)^2$, respectively. For example,

$$(ds)^2 = \left[\frac{(dx)^2}{(dx)^2} + \frac{(dy)^2}{(dx)^2}\right](dx)^2 = \left[1 + \left(\frac{dy}{dx}\right)^2\right](dx)^2$$

which gives the first of the three formulas.

**Area of a Surface of Revolution**   If a smooth plane curve is revolved about an axis in its plane, it generates a surface of revolution as illustrated in Figure 12. Our aim is to determine the area of such a surface.

To get started, we introduce the formula for the area of the frustum of a cone. A **frustum** of a cone is the surface between two planes perpendicular to the axis of the cone (shaded in Figure 13). If a frustum has base radii $r_1$ and $r_2$ and slant height $\ell$, then its area $A$ is given by

$$A = 2\pi \left(\frac{r_1 + r_2}{2}\right) \ell = 2\pi \text{ (average radius)} \cdot \text{(slant height)}$$

The derivation of this result depends only on the formula for the area of a circle (see Problem 25).

Let $x = f(t)$, $y = g(t)$, $a \le t \le b$, determine a smooth curve in the upper half of the $xy$-plane, as shown in Figure 14. Partition the interval $[a, b]$ into $n$ pieces by means of points $a = t_0 < t_1 < \cdots < t_n = b$, thereby also dividing the curve into $n$ pieces. Let $\Delta s_i$ denote the length of the typical piece and let $y_i$ be the $y$-coordinate of a point on this piece. When the curve is revolved about the $x$-axis, it generates a surface, and the typical piece generates a narrow band. The "area" of this band ought to be approximately that of a frustum—that is, approximately $2\pi y_i \Delta s_i$. When we add the contributions of all the pieces and take the limit as the norm of the partition approaches zero, we get what we define to be the area of the surface of revolution. All this is indicated in Figure 15 and the boxed formula below.

$$A = \lim_{|P| \to 0} \sum_{i=1}^{n} 2\pi y_i \Delta s_i = \int_{*}^{**} 2\pi y \, ds$$

In using the formula for $A$, we must make an appropriate interpretation of $y$, $ds$, and the limits * and **. Our discussion of the differential of arc length is helpful here. Thus, if the surface is obtained by revolving the curve $y = f(x)$, $a \le x \le b$, about the $x$-axis, the formula takes the form

$$A = 2\pi \int_{*}^{**} y \, ds = 2\pi \int_{a}^{b} f(x) \sqrt{1 + [f'(x)]^2} \, dx$$

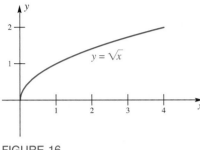

**FIGURE 16**

**EXAMPLE 6** Find the area of the surface of revolution generated by revolving the curve $y = \sqrt{x}$, $0 \le x \le 4$, about the $x$-axis (Figure 16).

**Solution** Here, $f(x) = \sqrt{x}$ and $f'(x) = 1/(2\sqrt{x})$. Thus,

$$A = 2\pi \int_0^4 \sqrt{x} \sqrt{1 + \frac{1}{4x}}\, dx = 2\pi \int_0^4 \sqrt{x} \sqrt{\frac{4x + 1}{4x}}\, dx$$

$$= \pi \int_0^4 \sqrt{4x + 1}\, dx = \left[ \pi \cdot \frac{1}{4} \cdot \frac{2}{3}\, (4x + 1)^{3/2} \right]_0^4$$

$$= \frac{\pi}{6}\, (17^{3/2} - 1^{3/2}) \approx 36.18 \qquad \blacksquare$$

If the curve is given parametrically by $x = f(t)$, $y = g(t)$, $a \le t \le b$, then the surface area formula becomes

$$A = 2\pi \int_*^{**} y\, ds = 2\pi \int_a^b g(t)\, \sqrt{[f'(t)]^2 + [g'(t)]^2}\, dt$$

## CONCEPTS REVIEW

**1.** The graph of the parametric equations $x = 4 \cos t$, $y = 4 \sin t$, $0 \le t \le 2\pi$, is a curve called a _____.

**2.** The curve determined by $y = x^2 + 1$, $0 \le x \le 4$, can be put in parametric form using $x$ as the parameter by writing $y = $ _____, $x = $ _____.

**3.** The formula for the length $L$ of the curve $x = f(t)$, $y = g(t)$, $a \le t \le b$, is $L = $ _____.

**4.** The proof of the formula for the length of a curve depends strongly on an earlier theorem with the name _____.

## PROBLEM SET 6.4

**1.** Use an $x$-integration to find the length of the segment of the line $y = 2x + 3$ between $x = 1$ and $x = 3$. Check by the distance formula.

**2.** Use a $y$-integration to find the length of the segment of the line $2y - 2x + 3 = 0$ between $y = 1$ and $y = 3$. Check by the distance formula.

≈ In Problems 3–8, find the length of the indicated curve.
**3.** $y = 4x^{3/2}$ between $x = 1/3$ and $x = 5$.

**4.** $y = \frac{2}{3}\, (x^2 + 1)^{3/2}$ between $x = 1$ and $x = 2$.

**5.** $y = (4 - x^{2/3})^{3/2}$ between $x = 1$ and $x = 27$.

**6.** $y = (x^4 + 3)/(6x)$ between $x = 1$ and $x = 3$.

**7.** $x = y^4/16 + 1/(2y^2)$ between $y = -3$ and $y = -2$.

*Hint:* Watch signs; $\sqrt{u^2} = -u$ when $u < 0$.

**8.** $30xy^3 - y^8 = 15$ between $y = 1$ and $y = 3$.

In Problems 9–12, sketch the graph of the given parametric equations and find its length.
**9.** $x = t^3/3$, $y = t^2/2$; $0 \le t \le 1$

**10.** $x = 3t^2 + 2$, $y = 2t^3 - 1/2$; $1 \le t \le 4$

**11.** $x = 4 \sin t$, $y = 4 \cos t - 5$; $0 \le t \le \pi$

**12.** $x = \sqrt{5} \sin 2t - 2$, $y = \sqrt{5} \cos 2t - \sqrt{3}$; $0 \le t \le \pi/4$

**13.** Sketch the graph of the four-cusped *hypocycloid* $x = a \sin^3 t$, $y = a \cos^3 t$, $0 \le t \le 2\pi$, and find its length. *Hint:* By symmetry, you can quadruple the length of the first quadrant portion.

**14.** A point $P$ on the rim of a wheel of radius $a$ is initially at the origin. As the wheel rolls to the right along the $x$-axis,

P traces out a curve called a *cycloid* (see Figure 17). Derive parametric equations for the cycloid as follows.

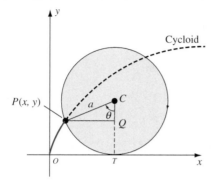

**FIGURE 17**

(a) Show that $\overline{OT} = a\theta$.
(b) Convince yourself that $\overline{PQ} = a \sin \theta$, $\overline{QC} = a \cos \theta$, $0 \le \theta \le \pi/2$.
(c) Show that $x = a(\theta - \sin \theta)$, $y = a(1 - \cos \theta)$.

**15.** Find the length of one arch of the cycloid of Problem 14. *Hint*: First show that

$$\left(\frac{dx}{d\theta}\right)^2 + \left(\frac{dy}{d\theta}\right)^2 = 4a^2 \sin^2 \left(\frac{\theta}{2}\right)$$

**16.** Suppose the wheel of Problem 14 turns at a constant rate $\omega = d\theta/dt$, where $t$ is time. Then $\theta = \omega t$.

(a) Show that the speed $ds/dt$ of $P$ along the cycloid is

$$\frac{ds}{dt} = 2a\omega\left|\sin \frac{\omega t}{2}\right|$$

(b) When is the speed a maximum and when is it a minimum?
(c) Explain why a bug on a wheel of a car going 60 miles per hour sometimes is traveling at 120 miles per hour.

**17.** Find the length of each curve.

(a) $y = \int_1^x \sqrt{u^3 - 1} \, du$, $1 \le x \le 2$
(b) $x = t - \sin t$, $y = 1 - \cos t$, $0 \le t \le 4\pi$

**18.** Find the length of each curve.

(a) $y = \int_{\pi/6}^x \sqrt{64 \sin^2 u \cos^4 u - 1} \, du$, $\frac{\pi}{6} \le x \le \frac{\pi}{3}$
(b) $x = a \cos t + at \sin t$, $y = a \sin t - at \cos t$, $-1 \le t \le 1$

In Problems 19–24, find the area of the surface generated by revolving the given curve about the *x*-axis.

**19.** $y = 6x$, $0 \le x \le 1$

**20.** $y = \sqrt{25 - x^2}$, $-2 \le x \le 3$

**21.** $y = x^3/3$, $1 \le x \le \sqrt{7}$

**22.** $y = (x^6 + 2)/(8x^2)$, $1 \le x \le 3$

**23.** $x = t$, $y = t^3$, $0 \le t \le 1$

**24.** $x = 1 - t^2$, $y = 2t$, $0 \le t \le 1$

**25.** If the surface of a cone of slant height $\ell$ and base radius $r$ is cut along a lateral edge and laid flat, it becomes the sector of a circle of radius $\ell$ and central angle $\theta$ (see Figure 18).

(a) Show that $\theta = 2\pi r/\ell$ radians.
(b) Use the formula $\frac{1}{2}\ell^2\theta$ for the area of a sector of radius $\ell$ and central angle $\theta$ to show that the lateral surface area of a cone is $\pi r\ell$.
(c) Use the result of (b) to obtain the formula $A = 2\pi[(r_1 + r_2)/2]\ell$ for the lateral area of a frustum of a cone with base radii $r_1$ and $r_2$ and slant height $\ell$ (see the discussion near the end of this section).

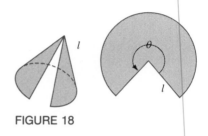

**FIGURE 18**

**26.** Show that the area of the part of the surface of a sphere of radius $a$ between two parallel planes $h$ units apart ($h < 2a$) is $2\pi ah$. Thus, show that if a right circular cylinder is circumscribed about a sphere, then two planes parallel to the base of the cylinder bound regions of the same area on the sphere and the cylinder.

**27.** Figure 19 shows one arch of a cycloid. Its parametric equations (see Problem 14) are given by

$$x = a(t - \sin t) \qquad y = a(1 - \cos t)$$

$$0 \le t \le 2\pi$$

(a) Show that the area of the surface generated when this curve is revolved about the *x*-axis is

$$A = 2\sqrt{2}\pi a^2 \int_0^{2\pi} (1 - \cos t)^{3/2} \, dt$$

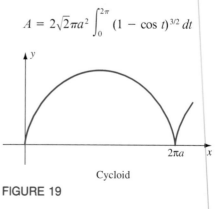

Cycloid

**FIGURE 19**

(b) With the help of the half-angle formula $1 - \cos t = 2 \sin^2(t/2)$, evaluate $A$.

**28.** The circle $x = a \cos t, y = a \sin t, 0 \le t \le 2\pi$, is revolved about the line $x = b, 0 < a < b$, thus generating a torus (doughnut). Find its surface area.

PC **29.** Sketch the graphs of each of the following parametric equations.

(a) $x = 3 \cos t, y = 3 \sin t, 0 \le t \le 2\pi$
(b) $x = 3 \cos t, y = \sin t, 0 \le t \le 2\pi$
(c) $x = t \cos t, y = t \sin t, 0 \le t \le 6\pi$
(d) $x = \cos t, y = \sin 2t, 0 \le t \le 2\pi$

(e) $x = \cos 3t, y = \sin 2t, 0 \le t \le 2\pi$
(f) $x = \cos t, y = \sin \pi t, 0 \le t \le 40$

PC **30.** Find the lengths of each of the curves in Problem 29. You will first have to set up the appropriate integral and then use a computer to evaluate it.

PC **31.** Using the same axes, draw the graphs of $y = x^n$ on $[0, 1]$ for $n = 1, 2, 4, 10$, and $100$. Find the length of each of these curves. Guess at the length when $n = 10,000$.

---

**Answers to Concepts Review:** **1.** Circle **2.** $x^2 + 1; x$
**3.** $\int_a^b \{[f'(t)]^2 + [g'(t)]^2\}^{1/2} \, dt$ **4.** Mean Value Theorem (for derivatives)

---

## 6.5 WORK

We turn now to an application of the definite integral that does not concern length, area, or volume. In physics, we learn that if an object moves a distance $d$ along a line while subjected to a *constant* force $F$ in the direction of the motion, then the work $W$ done by that force is given by

$$\text{Work} = (\text{force}) \cdot (\text{distance})$$

that is,

$$W = F \cdot d$$

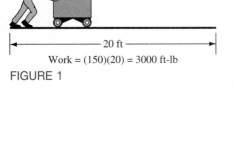

Force = 150 lb

20 ft

Work = (150)(20) = 3000 ft-lb

FIGURE 1

If force is measured in pounds and distance in feet, then work is in foot-pounds. If force is in dynes (the force required to give a mass of 1 gram an acceleration of 1 centimeter per second per second) and distance is in centimeters, then work is in dyne-centimeters, also called ergs. If force is in newtons and distance is in meters, then work is in newton-meters, also called joules. For example, a worker pushing a cart with a constant force of 150 pounds through a distance of 20 feet does $(150)(20) = 3000$ foot-pounds of work (Figure 1). A person lifting a weight (force) of 2 newtons a distance of 3 meters does $(2)(3) = 6$ joules of work.

In most practical situations, force is not constant, but rather varies as the object moves along the line. Suppose, in fact, that the object is being moved along the $x$-axis from $a$ to $b$ subject to a variable force of magnitude $F(x)$ at the point $x$, where $F$ is a continuous function. Then how much work is done? Once again, the words *slice, approximate, integrate* lead us to an answer. Here *slice* means to partition the interval $[a, b]$ into small pieces; *approximate* means to suppose that on a typical piece from $x$ to $x + \Delta x$, the force is constant with value $F(x)$, so that the bit of work done is $\Delta W \approx F(x)\Delta x$; *integrate* means to add up all the bits of work corresponding to the pieces $\Delta x$ and then take the limit as the length of the pieces approaches zero (Figure 2). We conclude that the work done in moving the object from $a$ to

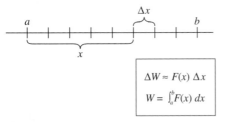

$\Delta W \approx F(x) \Delta x$

$W = \int_a^b F(x) \, dx$

FIGURE 2

*b* is given by

$$W = \int_a^b F(x)\,dx$$

**Application to Springs**   According to Hooke's Law in physics, the force $F(x)$ necessary to keep a spring stretched (or compressed) $x$ units beyond (or short of) its natural length (Figure 3) is given by

$$F(x) = kx$$

Here, the constant $k$, the so-called spring constant, is positive and depends on the particular spring under consideration. The stiffer the spring, the greater the value of $k$.

**EXAMPLE 1**   If the natural length of a spring is 10 inches and if it takes a force of 3 pounds to keep it extended 2 inches, find the work done in stretching the spring from its natural length to a length of 15 inches.

*Solution*   By Hooke's Law, the force $F(x)$ required to keep the spring stretched $x$ inches is given by $F(x) = kx$. To evaluate $k$ for this particular spring, we note that $F(2) = 3$. Thus, $k \cdot 2 = 3$, or $k = \frac{3}{2}$, and so

$$F(x) = \tfrac{3}{2}x$$

When the spring is at its natural length of 10 inches, $x = 0$; when it is 15 inches long, $x = 5$. Therefore, the work done in stretching the spring is

$$W = \int_0^5 \frac{3}{2}\,x\,dx = \left[\frac{3}{2} \cdot \frac{x^2}{2}\right]_0^5 = \frac{75}{4} = 18.75 \text{ inch-pounds} \qquad \blacksquare$$

**Application to Pumping a Liquid**   To pump water out of a tank requires work, as anyone who has ever tried a hand pump will know (Figure 4). But how much work? That is the question.

    While the problem does not quite fit within the previous discussion, its solution rests on the same basic principles. We illustrate.

**EXAMPLE 2**   A tank in the shape of a right circular cone (Figure 5) is full of water. If the height of the tank is 10 feet and the radius of its top is 4 feet, find the work done in pumping the water (a) over the top edge of the tank, and (b) to a height 10 feet above the top of the tank.

*Solution*
(a) Position the tank in a coordinate system, as shown in Figure 5. Both a three-dimensional view and a two-dimensional cross section are shown. Imagine slicing the water into thin horizontal disks, each of which must be lifted over the edge of the tank. A disk of thickness

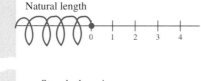
Natural length

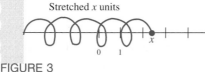
Stretched *x* units

**FIGURE 3**

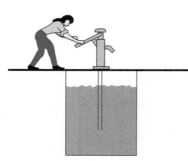
**FIGURE 4**

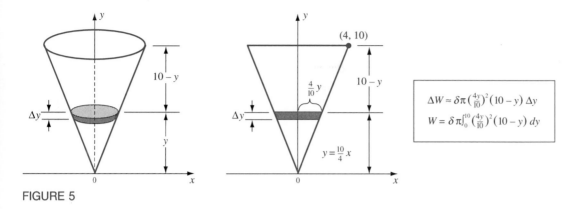

FIGURE 5

$$\Delta W \approx \delta\pi\left(\tfrac{4y}{10}\right)^2(10-y)\,\Delta y$$

$$W = \delta\pi\int_0^{10}\left(\tfrac{4y}{10}\right)^2(10-y)\,dy$$

$\Delta y$ at height $y$ has radius $4y/10$. Thus, its volume is approximately $\pi(4y/10)^2\,\Delta y$ and its weight is about $\delta\pi(4y/10)^2\,\Delta y$, where $\delta = 62.4$ is the (weight) density of water in pounds per cubic foot. The force necessary to lift this disk of water is its weight, and it must be lifted a distance $10 - y$. Thus, the work $\Delta W$ done on this disk is approximately

$$\Delta W = (\text{force}) \cdot (\text{distance}) \approx \delta\pi\left(\frac{4y}{10}\right)^2\Delta y \cdot (10 - y)$$

$$W = \int_0^{10}\delta\pi\left(\frac{4y}{10}\right)^2(10 - y)\,dy = \delta\pi\frac{4}{25}\int_0^{10}(10y^2 - y^3)\,dy$$

$$= \frac{(4\pi)(62.4)}{25}\left[\frac{10y^3}{3} - \frac{y^4}{4}\right]_0^{10} \approx 26{,}138 \text{ foot-pounds}$$

(b) Part b is just like part a, except that each disk of water must now be lifted a distance $20 - y$, rather than $10 - y$. Thus,

$$W = \delta\pi\int_0^{10}\left(\frac{4y}{10}\right)^2(20 - y)\,dy = \frac{4\delta\pi}{25}\int_0^{10}(20y^2 - y^3)\,dy$$

$$= \frac{4(62.4)(\pi)}{25}\left[\frac{20y^3}{3} - \frac{y^4}{4}\right]_0^{10} \approx 130{,}690 \text{ foot-pounds}$$

Note that the limits are still 0 and 10 (not 0 and 20). Why?    ■

**EXAMPLE 3**  Find the work done in pumping the water over the rim of a tank, which is 50 feet long and has a semicircular end of radius 10 feet, if the tank is filled to a depth of 7 feet (Figure 6).

*Solution*  We position the end of the tank in a coordinate system, as shown in Figure 7. A typical horizontal slice is shown both on this two-dimensional picture and on the three-dimensional one in Figure 6. This slice is approximately a thin box, so we calculate its volume by multiplying length, width, and thickness. Its weight is its density $\delta = 62.4$ times its volume. Finally we note that this slice must be lifted through a distance $-y$ (the minus sign results from the fact that $y$ is negative in our diagram).

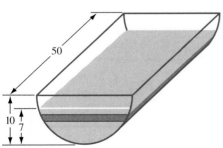

FIGURE 6

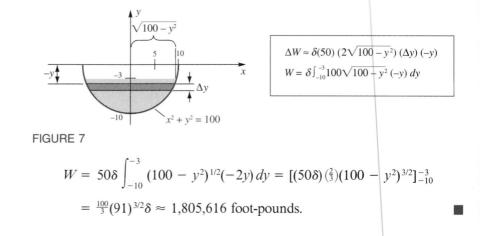

FIGURE 7

$$W = 50\delta \int_{-10}^{-3} (100 - y^2)^{1/2}(-2y)\,dy = [(50\delta)(\tfrac{2}{3})(100 - y^2)^{3/2}]_{-10}^{-3}$$

$$= \tfrac{100}{3}(91)^{3/2}\delta \approx 1{,}805{,}616 \text{ foot-pounds.} \qquad \blacksquare$$

## CONCEPTS REVIEW

**1.** The work done by a force $F$ in moving an object along a straight line from $a$ to $b$ is _____ if $F$ is constant but is _____ if $F = F(x)$ is variable.

**2.** Hooke's Law says that the force $F$ required to keep a spring stretched $x$ units beyond its natural length is _____.

**3.** The work done in lifting an object weighing 30 pounds from ground level to a height of 10 feet is _____ foot-pounds.

**4.** The work done in lifting a thin horizontal disk of water of radius 5 feet and thickness $\Delta y$ feet a distance of $12 - y$ feet is _____ foot-pounds, assuming that water weighs 62.4 pounds per cubic foot. Thus, if a cylindrical tank of radius 5 feet and height 12 feet is full of water, the total work done in pumping all the water over the top rim is given by the integral _____.

## PROBLEM SET 6.5

**1.** A force of 6 pounds is required to keep a spring stretched $\frac{1}{2}$ foot beyond its normal length. Find the value of the spring constant and the work done in stretching the spring $\frac{1}{2}$ foot beyond its natural length.

**2.** For the spring of Problem 1, how much work is done in stretching the spring 2 feet?

**3.** A force of 100 dynes is required to keep a spring of natural length of 7 centimeters compressed to a length of 8 centimeters. Find the work done in compressing the spring from its natural length to a length of 6 centimeters. (Hooke's Law applies to compressing as well as stretching.)

**4.** It requires 50 ergs (dyne-centimeters) of work to stretch a spring from a length of 8 centimeters to 9 centimeters and another 100 ergs to stretch it from 9 centimeters to 10 centimeters. Evaluate the spring constant and find the natural length of the spring.

**5.** For any spring obeying Hooke's Law, show that the work done in stretching a spring a distance $d$ is given by $W = \frac{1}{2}kd^2$.

**6.** For a certain type of nonlinear spring, the force required to keep the spring stretched a distance $s$ is given by the formula $F = ks^{4/3}$. If the force required to keep it stretched 8 inches is 2 pounds how much work is done in stretching this spring 27 inches?

**7.** A spring is such that the force required to keep it stretched $s$ feet is given by $F = 9s$ pounds. How much work is done in stretching it 2 feet?

**8.** Two similar springs $S_1$ and $S_2$, each 3 feet long, are such that the force required to keep either of them stretched a distance of $s$ feet is $F = 6s$ pounds. One end of one spring is fastened to an end of the other, and the combination is stretched between the walls of a room 10 feet wide (Figure 8). What work is done in moving the midpoint, $P$, 1 foot to the right?

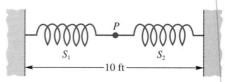

FIGURE 8

In each of Problems 9–12, the vertical end of a tank is shown. Assume the tank is 10 feet long and that it is full of water and that the water is to be pumped to a height 5 feet above the top of the tank. Find the work done in emptying the tank.

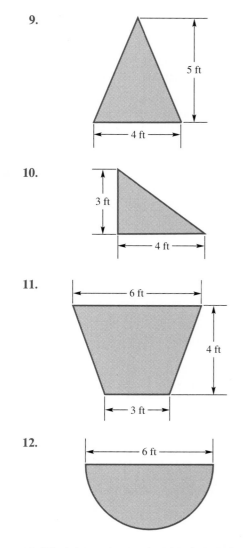

**9.**

5 ft

4 ft

**10.**

3 ft

4 ft

**11.**

6 ft

4 ft

3 ft

**12.**

6 ft

**13.** Find the work done in pumping all the oil (density $\delta = 50$ pounds per cubic foot) over the edge of a cylindrical tank which stands on end. Assume that the radius of the base is 4 feet, that the height is 10 feet, and that the tank is full of oil.

**14.** Do Problem 13 assuming that the tank has circular cross sections of radius $4 + x$ feet at height $x$ feet above the base.

**15.** A volume $v$ of gas is confined in a cylinder, one end of which is closed by a movable piston. If $A$ is the area in square inches of the face of the piston and $x$ is the distance in inches from the cylinder head to the piston, then $v = Ax$.

The pressure of the confined gas is a continuous function $p$ of the volume, and $p(v) = p(Ax)$ will be denoted by $f(x)$. Show that the work done by the piston in compressing the gas from a volume $v_1 = Ax_1$ to a volume $v_2 = Ax_2$ is

$$W = A \int_{x_2}^{x_1} f(x)\, dx$$

*Hint*: The total force on the face of the piston is $p(v) \cdot A = p(Ax) \cdot A = A \cdot f(x)$.

⌐C⌐  **16.** A cylinder and piston, whose cross-sectional area is 1 square inch, contain 16 cubic inches of gas under a pressure of 40 pounds per square inch. If the pressure and the volume of the gas are related adiabatically (that is, without loss of heat) by the law $pv^{1.4} = c$ (a constant), how much work is done by the piston in compressing the gas to 2 cubic inches?

⌐C⌐  **17.** If the area of the face of the piston in Problem 16 is 2 square inches, find the work done by the piston.

**18.** One cubic foot of air under a pressure of 80 pounds per square inch expands adiabatically to 4 cubic feet according to the law $pv^{1.4} = c$. Find the work done by the gas.

**19.** A cable weighing 2 pounds per foot is used to haul a 200-pound load to the top of a shaft that is 500 feet deep. How much work is done?

**20.** A 10-pound monkey hangs at the end of a 20-foot chain that weighs $\frac{1}{2}$ pound per foot. How much work does it do in climbing the chain to the top? Assume that the end of the chain is attached to the monkey.

**21.** A space capsule weighing 5000 pounds is propelled to an altitude of 200 miles above the surface of the earth. How much work is done against the force of gravity? Assume the earth is a sphere of radius 4000 miles and that the force of gravity is $f(x) = -k/x^2$, where $x$ is the distance from the center of the earth to the capsule (the inverse-square law). Thus the lifting force required is $k/x^2$, and this equals 5000 when $x = 4000$.

**22.** According to Coulomb's Law, two like electrical charges repel each other with a force that is inversely proportional to the square of the distance between them. If the force of repulsion is 10 dynes when they are 2 centimeters apart, find the work done in bringing the charges from 5 centimeters apart to 1 centimeter apart.

**23.** A bucket weighing 100 pounds is filled with sand weighing 500 pounds. A crane lifts the bucket from the ground to a point 80 feet in the air at a rate of 2 feet per second, but sand simultaneously leaks out through a hole at 3 pounds per second. Neglecting friction and the weight of the cable, determine how much work is done. *Hint*: Begin by estimating $\Delta W$, the work required to lift the bucket from $y$ to $y + \Delta y$.

C 24. Center City has just built a new water tower (Figure 9). Its main elements consist of a spherical tank having an inner radius of 10 feet and weighing 10,000 pounds; four 40-foot pillars of linear density $100 \cos(x^2/10{,}000)$ pounds per foot, with $x$ measuring the number of feet from the base; and a 30-foot filler pipe weighing 40 pounds per foot. Initially, all elements were on the ground with the pillars and pipe lying flat. The tank was lifted to position and then the pillars and pipe were pivoted about their bases to the vertical. Find the work done.

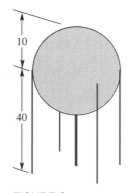

FIGURE 9

25. The filler pipe for the water tower of Problem 24 has inner diameter 1 foot. Assume that the water was pumped from ground level up through the pipe into the tank. How much work was done in filling the pipe and the tank with water?

26. A conical buoy weighs $m$ pounds and floats with its vertex $V$ down and $h$ feet below the surface of the water (Figure 10). A boat crane lifts the buoy to the deck so that $V$ is 15 feet above the water surface. How much work is done? *Hint:* Use Archimedes's Principle, which says that the force required to hold the buoy $y$ feet above its original position ($0 \le y \le h$) is equal to its weight minus the weight of the water displaced by the buoy.

27. Rather than lifting the buoy of Problem 26 and Figure 10 out of the water, suppose we attempt to push it

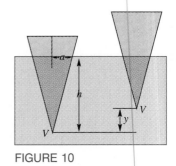

FIGURE 10

down until its top is even with the water level. Assume $h = 8$, that the top is originally 2 feet above water level, and that the buoy weighs 300 pounds. How much work is required? *Hint:* You do not need to know $a$ (the radius at water level), but it is helpful to know that $\delta \left(\frac{1}{3}\pi a^2\right)(8) = 300$. Archimedes's Principle implies that the force needed to hold the buoy $z$ feet ($0 \le z \le 2$) below floating position is equal to the weight of the additional water displaced.

28. Initially the bottom tank in Figure 11 was full of water and the top tank was empty. Find the work done in pumping all the water into the top tank. The dimensions are in feet.

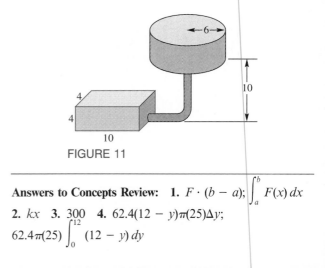

FIGURE 11

---

**Answers to Concepts Review: 1.** $F \cdot (b - a)$; $\displaystyle\int_a^b F(x)\, dx$

**2.** $kx$ **3.** 300 **4.** $62.4(12 - y)\pi(25)\Delta y$;

$62.4\pi(25)\displaystyle\int_0^{12}(12 - y)\, dy$

---

**6.6**

**MOMENTS, CENTER OF MASS**

Suppose that two masses of sizes $m_1$ and $m_2$ are placed on a seesaw at distances $d_1$ and $d_2$ from the fulcrum and on opposite sides of it (Figure 1). The seesaw will balance if and only if $d_1 m_1 = d_2 m_2$.

A good mathematical model for this situation is obtained by replacing the seesaw with a horizontal coordinate line having its origin at the fulcrum (Figure 2). Then the coordinate $x_1$ of $m_1$ is $x_1 = -d_1$, that of $m_2$ is $x_2 = d_2$, and the condition for balance is

$$x_1 m_1 + x_2 m_2 = 0$$

The product of the mass $m$ of a particle by its *directed* distance from a point (its lever arm) is called the **moment** of the particle with respect to that

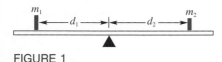

FIGURE 1

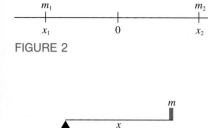

FIGURE 2

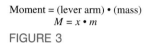

Moment = (lever arm) • (mass)
$$M = x \cdot m$$

FIGURE 3

point (Figure 3). It measures the tendency of the mass to produce a rotation about that point. The condition for two masses along a line to balance at a point on that line is that the sum of their moments with respect to the point be zero.

The situation just described can be generalized. The total moment $M$ (with respect to the origin) of a system of $n$ masses of sizes $m_1, m_2, \ldots, m_n$ located at points $x_1, x_2, \ldots, x_n$ along the $x$-axis is the sum of individual moments—that is,

$$M = x_1 m_1 + x_2 m_2 + \cdots + x_n m_n = \sum_{i=1}^{n} x_i m_i$$

The condition for balance at the origin is that $M = 0$. Of course, we should not expect balance at the origin except in special circumstances. But surely any system of masses will balance somewhere. The question is where. What is the $x$-coordinate of the point where the fulcrum should be placed to make the system in Figure 4 balance?

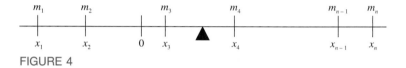

FIGURE 4

Call the desired coordinate $\bar{x}$. The total moment with respect to it should be zero—that is,

$$(x_1 - \bar{x})m_1 + (x_2 - \bar{x})m_2 + \cdots + (x_n - \bar{x})m_n = 0$$

or

$$x_1 m_1 + x_2 m_2 + \cdots + x_n m_n = \bar{x}m_1 + \bar{x}m_2 + \cdots + \bar{x}m_n$$

When we solve for $\bar{x}$, we obtain

$$\bar{x} = \frac{M}{m} = \frac{\sum_{i=1}^{n} x_i m_i}{\sum_{i=1}^{n} m_i}$$

The point $\bar{x}$, called the **center of mass,** is the balance point. Notice that it is just the total moment with respect to the origin divided by the total mass.

**EXAMPLE 1**   Masses of 4, 2, 6, and 7 pounds are located at points 0, 1, 2, and 4, respectively, along the $x$-axis (Figure 5). Find the center of mass.

*Solution*

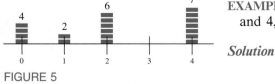

FIGURE 5

$$x = \frac{(0)(4) + (1)(2) + (2)(6) + (4)(7)}{4 + 2 + 6 + 7} = \frac{42}{19} \approx 2.21$$

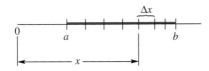

$$\Delta m \approx \delta(x)\,\Delta x \qquad \Delta M \approx x\,\delta(x)\,\Delta x$$

$$m = \int_a^b \delta(x)\,dx \qquad M = \int_a^b x\,\delta(x)\,dx$$

FIGURE 6

Your intuition should confirm that $x = 2.21$ is about right for the balance point. ∎

**Continuous Mass Distribution Along a Line**  Consider now a straight segment of thin wire of varying density (mass per unit length) for which we desire to find the balance point. We impose a coordinate line along the wire and suppose the density at $x$ is $\delta(x)$. (For the meaning of density at a point, see Problem 19 of Section 3.1.) Following our standard procedure (*slice, approximate, integrate*), we first obtain the total mass $m$ and then the total moment $M$ with respect to the origin (Figure 6). This leads to the formula

$$\overline{x} = \frac{M}{m} = \frac{\displaystyle\int_a^b x\,\delta(x)\,dx}{\displaystyle\int_a^b \delta(x)\,dx}$$

Two comments are in order. First, remember this formula by analogy with the point-mass formula.

$$\frac{\sum x_i m_i}{\sum m_i} \sim \frac{\sum x\,\Delta m}{\sum \Delta m} \sim \frac{\int x\,\delta(x)\,dx}{\int \delta(x)\,dx}$$

Second, note that we have assumed that moments of small pieces of wire add together to give the total moment, just as was the case for point masses. This should seem reasonable to you if you imagine the mass of the typical piece of length $\Delta x$ to be concentrated at the point $x$.

**EXAMPLE 2**  The density $\delta(x)$ of a wire at the point $x$ centimeters from one end is given by $\delta(x) = 3x^2$ grams per centimeter. Find the center of mass of the piece between $x = 0$ and $x = 10$.

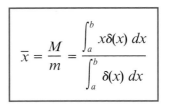

0                                                                  10
FIGURE 7

*Solution*  We expect $\overline{x}$ to be nearer 10 than 0, since the wire is much heavier (denser) toward the right end (Figure 7).

$$\overline{x} = \frac{\displaystyle\int_0^{10} x \cdot 3x^2\,dx}{\displaystyle\int_0^{10} 3x^2\,dx} = \frac{[3x^4/4]_0^{10}}{[x^3]_0^{10}} = \frac{7500}{1000} = 7.5 \text{ cm} \qquad ∎$$

**Mass Distributions in the Plane**  Consider $n$ point masses of sizes $m_1$, $m_2, \ldots, m_n$ situated at points $(x_1,\ y_1)$, $(x_2,\ y_2), \ldots, (x_n,\ y_n)$ in the coordinate plane (Figure 8). Then the total moments $M_y$ and $M_x$ with respect to the $y$-axis and $x$-axis, respectively, are given by

$$M_y = \sum_{i=1}^{n} x_i m_i \qquad M_x = \sum_{i=1}^{n} y_i m_i$$

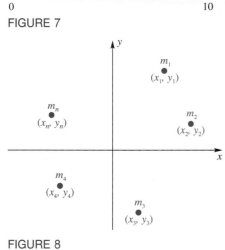

FIGURE 8

The coordinates $(\bar{x}, \bar{y})$ of the center of mass (balance point) are

$$\bar{x} = \frac{M_y}{m} = \frac{\displaystyle\sum_{i=1}^{n} x_i m_i}{\displaystyle\sum_{i=1}^{n} m_i} \qquad \bar{y} = \frac{M_x}{m} = \frac{\displaystyle\sum_{i=1}^{n} y_i m_i}{\displaystyle\sum_{i=1}^{n} m_i}$$

**EXAMPLE 3** Five particles, having masses 1, 4, 2, 3, and 2 units are located at points $(6, -1), (2, 3), (-4, 2), (-7, 4)$, and $(2, -2)$, respectively. Find the center of mass.

*Solution*

$$\bar{x} = \frac{(6)(1) + (2)(4) + (-4)(2) + (-7)(3) + (2)(2)}{1 + 4 + 2 + 3 + 2} = \frac{-11}{12}$$

$$\bar{y} = \frac{(-1)(1) + (3)(4) + (2)(2) + (4)(3) + (-2)(2)}{1 + 4 + 2 + 3 + 2} = \frac{23}{12} \qquad\blacksquare$$

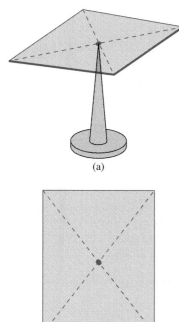

(a)

(b)

FIGURE 9

We next consider the problem of finding the center of mass of a lamina (thin planar sheet). For simplicity, we suppose it is homogeneous—that is, that it has constant mass density $\delta$. For a homogeneous rectangular sheet, the center of mass is at the geometric center, as diagrams (a) and (b) in Figure 9 suggest.

Consider the homogeneous lamina bounded by $x = a$, $x = b$, $y = f(x)$, and $y = g(x)$, with $g(x) \le f(x)$. *Slice* this lamina into narrow strips parallel to the $y$-axis, which are therefore nearly rectangular in shape, and imagine the mass of each strip to be concentrated at its geometric center. Then *approximate* and *integrate* (Figure 10). From this we can calculate the coordi-

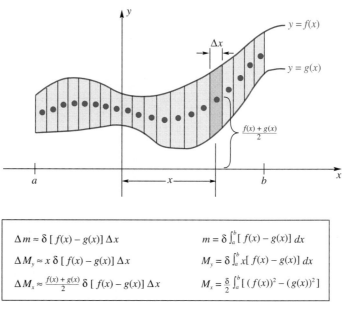

| | |
|---|---|
| $\Delta m \approx \delta\,[\,f(x) - g(x)]\,\Delta x$ | $m = \delta \int_a^b [\,f(x) - g(x)]\,dx$ |
| $\Delta M_y \approx x\,\delta\,[\,f(x) - g(x)]\,\Delta x$ | $M_y = \delta \int_a^b x[\,f(x) - g(x)]\,dx$ |
| $\Delta M_x \approx \frac{f(x) + g(x)}{2}\,\delta\,[\,f(x) - g(x)]\,\Delta x$ | $M_x = \frac{\delta}{2} \int_a^b [\,(f(x))^2 - (g(x))^2\,]$ |

FIGURE 10

nates $(\bar{x}, \bar{y})$ of the center of mass using the formulas

$$\bar{x} = \frac{M_y}{m} \qquad \bar{y} = \frac{M_x}{m}$$

When we do, the factor $\delta$ cancels between numerator and denominator, and we obtain

$$\bar{x} = \frac{\int_a^b x[f(x) - g(x)] \, dx}{\int_a^b [f(x) - g(x)] \, dx} \qquad \bar{y} = \frac{\frac{1}{2} \int_a^b [(f(x))^2 - (g(x))^2] \, dx}{\int_a^b [f(x) - g(x)] \, dx}$$

Sometimes, slicing parallel to the $x$-axis works better than slicing parallel to the $y$-axis. This leads to formulas for $\bar{x}$ and $\bar{y}$ in which $y$ is the variable of integration. Do not try to memorize all these formulas. It is much better to remember how they were derived.

The center of mass of a homogeneous lamina does not depend on its density or its mass, but only on the shape of the corresponding region in the plane. Thus, our problem becomes a geometric problem rather than a physical one. Accordingly, we often speak of the **centroid** of a planar region rather than the center of mass of a homogeneous lamina.

**EXAMPLE 4**  Find the centroid of the region bounded by the curves $y = x^3$ and $y = \sqrt{x}$.

*Solution*  Note the diagram in Figure 11.

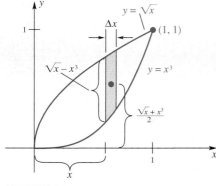

FIGURE 11

$$\bar{x} = \frac{\int_0^1 x[\sqrt{x} - x^3] \, dx}{\int_0^1 [\sqrt{x} - x^3] \, dx} = \frac{\left[\frac{2}{5}x^{5/2} - \frac{x^5}{5}\right]_0^1}{\left[\frac{2}{3}x^{3/2} - \frac{x^4}{4}\right]_0^1} = \frac{\frac{1}{5}}{\frac{5}{12}} = \frac{12}{25}$$

$$\bar{y} = \frac{\int_0^1 \frac{1}{2}(\sqrt{x} + x^3)(\sqrt{x} - x^3) \, dx}{\int_0^1 (\sqrt{x} - x^3) \, dx} = \frac{\frac{1}{2}\int_0^1 [(\sqrt{x})^2 - (x^3)^2] \, dx}{\int_0^1 (\sqrt{x} - x^3) \, dx}$$

$$= \frac{\frac{1}{2}\left[\frac{x^2}{2} - \frac{x^7}{7}\right]_0^1}{\frac{5}{12}} = \frac{\frac{5}{28}}{\frac{5}{12}} = \frac{3}{7}$$

These answers should seem reasonable to you.  ∎

**EXAMPLE 5**  Find the centroid of the region under the curve $y = \sin x$, $0 \le x \le \pi$ (Figure 12).

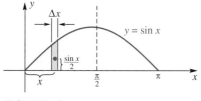

FIGURE 12

*Solution*  This region is symmetric about the line $x = \pi/2$, from which we conclude (without an integration) that $\bar{x} = \pi/2$. In fact, it is both intuitively obvious and true that if a region has a vertical or horizontal line of symmetry, then the centroid will lie on that line.

Your intuition should also tell you that $\bar{y}$ will be less than $\frac{1}{2}$, since more of the area is near the $x$-axis. But to find this number exactly, we must calculate

$$\bar{y} = \frac{\int_0^\pi \frac{1}{2} \sin x \cdot \sin x \, dx}{\int_0^\pi \sin x \, dx} = \frac{\frac{1}{2} \int_0^\pi \sin^2 x \, dx}{\int_0^\pi \sin x \, dx}$$

The denominator is easy to calculate; it has value 2. To calculate the numerator, we use the half-angle formula $\sin^2 x = (1 - \cos 2x)/2$.

$$\int_0^\pi \sin^2 x \, dx = \frac{1}{2} \left( \int_0^\pi 1 \, dx - \int_0^\pi \cos 2x \, dx \right)$$

$$= \frac{1}{2} \left[ x - \frac{1}{2} \sin 2x \right]_0^\pi = \frac{\pi}{2}$$

Thus,

$$\bar{y} = \frac{\frac{1}{2} \cdot \frac{\pi}{2}}{2} = \frac{\pi}{8} \approx 0.39 \qquad \blacksquare$$

**The Theorem of Pappus**  About A.D. 300, the Greek geometer Pappus stated a novel result, which connects centroids with volumes of solids of revolution (Figure 13).

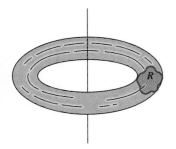

Theorem A

**(Pappus's Theorem).** If a region $R$, lying on one side of a line in its plane, is revolved about that line, then the volume of the resulting solid is equal to the area of $R$ multiplied by the distance traveled by its centroid.

FIGURE 13

Rather than prove this theorem, which is really quite easy (see Problem 22), we illustrate it.

**EXAMPLE 6**  Illustrate the correctness of Pappus's Theorem for the region under $y = \sin x$, $0 \le x \le \pi$, when it is revolved about the $x$-axis (Figure 14).

**Solution**  This is the region of Example 5, for which $\bar{y} = \pi/8$. The area $A$ of this region is

$$A = \int_0^\pi \sin x \, dx = [-\cos x]_0^\pi = 2$$

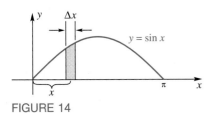

FIGURE 14

The volume $V$ of the corresponding solid of revolution is

$$V = \pi \int_0^\pi \sin^2 x\, dx = \frac{\pi}{2} \int_0^\pi [1 - \cos 2x]\, dx = \frac{\pi}{2} \left[ x - \frac{1}{2}\sin 2x \right]_0^\pi = \frac{\pi^2}{2}$$

We must show that

$$A(2\pi\bar{y}) = V$$

But this amounts to showing

$$2\left(2\pi\frac{\pi}{8}\right) = \frac{\pi^2}{2}$$

which is clearly true.                                   ∎

## CONCEPTS REVIEW

**1.** An object of mass 4 is at $x = 1$ and a second object of mass 6 is at $x = 3$. Simple geometric intuition tells us that the center of mass will be to the _____ of $x = 2$. In fact, it is at $\bar{x} =$ _____.

**2.** A homogeneous wire lying along the $x$-axis between $x = 0$ and $x = 5$ will balance at $\bar{x} =$ _____. However, if the wire has density $\delta(x) = 1 + x$, it will balance to the _____ of 2.5. In fact it will balance at $\bar{x}$, where
$$\bar{x} = \int_0^5 \underline{\quad\quad} dx \Big/ \int_0^5 \underline{\quad\quad} dx.$$

**3.** The homogeneous rectangular lamina with corner points $(0, 0)$, $(2, 0)$, $(2, 6)$, and $(0, 6)$ will balance at $\bar{x} =$ _____, $\bar{y} =$ _____.

**4.** A rectangular lamina with corners at $(2, 0)$, $(4, 0)$, $(4, 2)$, and $(2, 2)$ is attached to the lamina of Question 3. Assuming both laminas have the same constant density, the resulting $L$-shaped lamina will balance at $\bar{x} =$ _____, $\bar{y} =$ _____.

## PROBLEM SET 6.6

**1.** Particles of mass $m_1 = 5$, $m_2 = 7$, and $m_3 = 9$ are located at $x_1 = 2$, $x_2 = -2$, and $x_3 = 1$ along a line. Where is the center of mass?

**2.** John and Mary—weighing 180 and 110 pounds, respectively—sit at opposite ends of a 12-foot teeter board with the fulcrum in the middle. Where should their 80-pound son Tom sit in order for the board to balance?

**3.** A straight wire 7 units long has density $\delta(x) = \sqrt{x}$ at a point $x$ units from one end. Find the distance from this end to the center of mass.

**4.** Do Problem 3 if $\delta(x) = 1 + x^3$.

**5.** The masses and coordinates of a system of particles in the coordinate plane are given by the following: 2, $(1, 1)$; 3, $(7, 1)$; 4, $(-2, -5)$; 6, $(-1, 0)$; 2, $(4, 6)$. Find the moments of this system with respect to the coordinate axes, and find the coordinates of the center of mass.

**6.** The masses and coordinates of a system of particles are given by the following: 5, $(-3, 2)$; 6, $(-2, -2)$; 2, $(3, 5)$; 7, $(4, 3)$; 1, $(7, -1)$. Find the moments of this system with respect to the coordinate axes, and find the coordinates of the center of mass.

In Problems 7–10, divide the indicated region into rectangular pieces and assume that the moments $M_x$ and $M_y$ of the whole region can be found by adding the corresponding moments of the pieces. Use this to find the centroid of each region.

**7.**

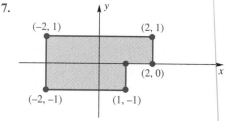

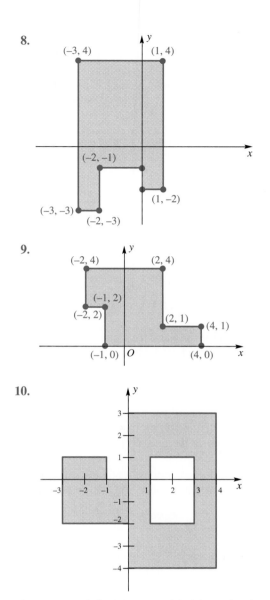

**8.**

**9.**

**10.**

In Problems 11–18, find the centroid of the region bounded by the given curves. Make a sketch and use symmetry where possible.

**11.** $y = 2 - x^2$, $y = 0$

**12.** $y = \frac{1}{3}x^2$, $y = 0$, $x = 4$

**13.** $y = x^3$, $y = 0$, $x = 1$

**14.** $y = \frac{1}{2}(x^2 - 10)$, $y = 0$, and between $x = -2$ and $x = 2$

**15.** $y = 2x - 4$, $y = 2\sqrt{x}$, $x = 1$

**16.** $y = x^2$, $y = x + 3$

**17.** $x = y^2$, $x = 2$

**18.** $x = y^2 - 3y - 4$, $x = -y$

**19.** Use Pappus's Theorem to find the volume of the

solid obtained when the region bounded by $y = x^3$, $y = 0$, and $x = 1$ is revolved about the $y$-axis (see Problem 13 for the centroid). Do the same problem by the method of cylindrical shells to check your answer.

**20.** Use Pappus's Theorem to find the volume of the torus obtained when the region inside the circle $x^2 + y^2 = a^2$ is revolved about the line $x = 2a$.

**21.** Use Pappus's Theorem together with the known volume of a sphere to find the centroid of a semicircular region of radius $a$.

**22.** Prove Pappus's Theorem by assuming that the region of area $A$ in Figure 15 is to be revolved about the $y$-axis. *Hint:* $V = 2\pi \int_a^b xh(x)\,dx$ and $\bar{x} = \int_a^b (xh(x)\,dx)/A$.

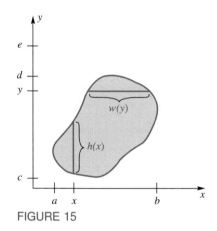

FIGURE 15

**23.** The region of Figure 15 is revolved about the line $y = e$, generating a solid.

(a) Use cylindrical shells to write a formula for the volume in terms of $w(y)$.
(b) Show that Pappus's formula, when simplified, gives the same result.

**24.** Consider the triangle $T$ of Figure 16.

(a) Show that $\bar{y} = h/3$ (and thus that the centroid of a triangle is at the intersection of the medians).
(b) Find the volume of the solid obtained when $T$ is revolved around $y = k$ (Pappus's Theorem).

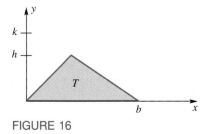

FIGURE 16

**25.** A regular polygon $P$ of $2n$ sides is inscribed in a circle of radius $r$.
(a) Find the volume of the solid obtained when $P$ is revolved about one of its sides.
(b) Check your answer by letting $n \to \infty$.

**26.** Let $f$ be a nonnegative continuous function on $[0, 1]$.

(a) Show that $\displaystyle\int_0^\pi xf(\sin x)\,dx = (\pi/2)\int_0^\pi f(\sin x)\,dx$.

(b) Use part (a) to evaluate $\displaystyle\int_0^\pi x \sin x \cos^4 x\,dx$.

**27.** Let $0 \le f(x) \le g(x)$ for all $x$ in $[0, 1]$ and let $R$ and $S$ be the regions under the graphs of $f$ and $g$, respectively. Prove or disprove that $\bar{y}_R \le \bar{y}_S$.

---

**Answers to Concepts Review:** **1.** Right; $(4 \cdot 1 + 6 \cdot 3)/10 = 2.2$ **2.** 2.5; right; $x(1 + x)$; $1 + x$ **3.** 1; 3 **4.** $\frac{24}{16}$; $\frac{40}{16}$

---

## 6.7 CHAPTER REVIEW

### Concepts Test

Respond with true or false to each of the following assertions. Be prepared to justify your answer.

**1.** The area of the region bounded by $y = \cos x$, $y = 0$, $x = 0$, and $x = \pi$ is $\displaystyle\int_0^\pi \cos x\,dx$.

**2.** The area of a circle of radius $a$ is $4\displaystyle\int_0^a \sqrt{a^2 - x^2}\,dx$.

**3.** The area of the region bounded by $y = f(x)$, $y = g(x)$, $x = a$, and $x = b$ is either $\displaystyle\int_a^b [f(x) - g(x)]dx$ or its negative.

**4.** All right cylinders whose bases have the same area and whose heights are the same have identical volumes.

**5.** If two solids with bases in the same plane have cross sections of the same area in all planes parallel to their bases, then they have the same volume.

**6.** If the radius of the base of a cone is doubled while the height is halved, the volume will remain the same.

**7.** To calculate the volume of the solid obtained by revolving the region bounded by $y = -x^2 + x$ and $y = 0$ about the $y$-axis, one should use the method of washers in preference to the method of shells.

**8.** The solids obtained by revolving the region of Problem 7 about $x = 0$ and $x = 1$ have the same volume.

**9.** Any smooth curve in the plane that lies entirely within the unit circle will have finite length.

**10.** The work required to stretch a spring 2 inches beyond its natural length is twice that required to stretch it 1 inch (assume Hooke's Law).

**11.** It will require the same amount of work to empty a cone-shaped tank and a cylindrical tank of water by pumping it over the rim if both tanks have the same height and volume.

**12.** Two weights of 100 pounds at distances 10 and 15 feet from the fulcrum will just balance a 200-pound weight on the other side of the fulcrum and 12.5 feet from it.

**13.** If $\bar{x}$ is the center of mass of a system of masses $m_1$, $m_2, \ldots, m_n$ distributed along a line at points with coordinates $x_1, x_2, \ldots, x_n$, respectively, then $\displaystyle\sum_{i=1}^n (x_i - \bar{x})m_i = 0$.

**14.** The centroid of the region bounded by $y = \cos x$, $y = 0$, $x = 0$, and $x = 2\pi$ is at $(\pi, 0)$.

**15.** According to the theorem of Pappus, the volume of the solid obtained by revolving the region (of area 2) bounded by $y = \sin x$, $y = 0$, $x = 0$, and $x = \pi$ about the $y$-axis is $2(2\pi)\left(\dfrac{\pi}{2}\right) = 2\pi^2$.

**16.** The area of the region bounded by $y = \sqrt{x}$, $y = 0$, and $x = 9$ is $\displaystyle\int_0^3 (9 - y^2)\,dy$.

**17.** If the density of a wire is proportional to the square of the distance from its midpoint, then its center of mass is at the midpoint.

**18.** The centroid of a triangle with base on the $x$-axis has $y$-coordinate equal to one-third the altitude of the triangle.

## Sample Test Problems

Problems 1–7 refer to the plane region $R$ bounded by the curve $y = x - x^2$ and the $x$-axis (Figure 1).

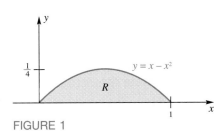

FIGURE 1

**1.** Find the area of $R$.

**2.** Find the volume of the solid $S_1$ generated by revolving the region $R$ about the $x$-axis.

**3.** Use the cylindrical shell method to find the volume of the solid $S_2$ generated by revolving $R$ about the $y$-axis.

**4.** Find the volume of the solid $S_3$ generated by revolving $R$ about the line $y = -2$.

**5.** Find the volume of the solid $S_4$ generated by revolving $R$ about the line $x = 3$.

**6.** Find the coordinates of the centroid of $R$.

**7.** Use Pappus's Theorem and Problems 1 and 6 to find the volume of the solids $S_1$, $S_2$, $S_3$, and $S_4$ above.

**8.** The natural length of a certain spring is 16 inches, and a force of 8 pounds is required to keep it stretched 8 inches. Find the work done in each case.

(a) Stretching it from a length of 18 inches to a length of 24 inches.
(b) Compressing it from its natural length to a length of 12 inches.

**9.** An upright cylindrical tank is 10 feet in diameter and 10 feet high. If water in the tank is 6 feet deep, how much work is done in pumping all the water over the edge of the top of the tank?

**10.** An object weighing 200 pounds is suspended from the top of a building by a uniform cable. If the cable is 100 feet long and weighs 120 pounds, how much work is done in pulling the object to the top?

**11.** A region $R$ is bounded by the line $y = 4x$ and the parabola $y = x^2$. Find the area of $R$ by (a) taking $x$ as the integration variable, and (b) taking $y$ as the integration variable.

**12.** Find the centroid of $R$ in Problem 11.

**13.** Find the volume of the solid of revolution generated by revolving the region $R$ of Problem 11 about the $x$-axis. Check by using Pappus's Theorem.

**14.** Find the volume of the solid generated by revolving the region $R$ of Problem 11 about the $y$-axis. Check by using Pappus's Theorem.

**15.** Find the length of the arc of the curve $y = x^3/3 + 1/(4x)$ from $x = 1$ to $x = 3$.

**16.** Sketch the graph of the parametric equations

$$x = t^2, \qquad y = \tfrac{1}{3}(t^3 - 3t)$$

Then find the length of the loop of the resulting curve.

**17.** A solid with the semicircular base bounded by $y = \sqrt{9 - x^2}$ and $y = 0$ has cross sections perpendicular to the $x$-axis that are squares. Find the volume of this solid.

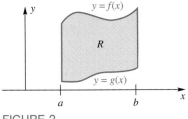

FIGURE 2

In Problems 18–22, write an expression involving integrals that represents the required concept. Refer to Figure 2.

**18.** The area of $R$.

**19.** The volume of the solid obtained when $R$ is revolved about the $x$-axis.

**20.** The volume of the solid obtained when $R$ is revolved about $x = a$.

**21.** The moments $M_x$ and $M_y$ of a homogeneous lamina with shape $R$, assuming its density is $\delta$.

**22.** The *total* length of the boundary of $R$.

**23.** The *total* surface area of the solid of Problem 19.

## 6.8 ADDITIONAL PROBLEMS

**1.** When functions whose units are distance are differentiated with respect to time, the resulting derivatives have units of distance/time or velocity. Similarly, integrating a function which specifies velocity with respect to time will give a result which is distance. In each case below give the units of the function $F$ and a physical description of the quantity $F$.

(a) $F(T) = \displaystyle\int_6^T a(t)\, dt$, where $a$ is measured in feet/(second)$^2$ and $T$ is measured in seconds.

(b) $F(s) = \displaystyle\int_3^s f(z)\, dz$, where $f$ is measured in pounds and $s$ is measured in feet.

(c) $F(r) = \dfrac{\displaystyle\int_0^r xf(x)\,dx}{\displaystyle\int_0^r f(x)\,dx}$, where $f$ is measured in slugs and $x$ is measured in inches.

2. Show that if two solids have cross-sectional areas $A_1(x)$ and $A_2(x)$ such that $A_1(x) = A_2(x)$ for $a < x < b$, then the volume of the solids are equal. Give an example using a deck of cards (where $A_i(x) = $ constant) as an application of this result. Sketch the cross sections of two objects which satisfy the condition that $A_1(x) = A_2(x)$ for $a < x < b$, but whose overall shape is different. This result is known as Cavalieri's Theorem.

Integration plays a central role in the theory of probability. In Problems 3–10 we explore that role. A *probability distribution function* $P(t)$ satisfies the properties $P(0) = 0$, $P(t) \geq 0$, $P(t)$ is an increasing function of $t$, and $\lim\limits_{t \to \infty} P(t) = 1$. (In what follows we will just assume that $P(t) = 1$ when $t$ is sufficiently large.) The derivative $P'(t)$ of the probability distribution is called the *probability density function* $p(t) = P'(t)$.

3. Given the following functions, choose the correct value for the constant $A$ for them to be probability density functions.

(a) $p(t) = \begin{cases} A & \text{if } 0 \leq t \leq 3 \\ 0 & \text{otherwise} \end{cases}$

(b) $p(t) = \begin{cases} A(t^2 - 7t) & \text{if } 0 \leq t \leq 7 \\ 0 & \text{otherwise} \end{cases}$

(c) $p(t) = \begin{cases} At^2 & \text{if } 0 \leq t < 3 \\ A(9 - t) & \text{if } 3 \leq t < 9 \\ 0 & \text{otherwise} \end{cases}$

4. In reliability theory the probability distribution function $P(t)$ is known as the *failure distribution function* and gives the percentage of the items that have failed at time $t$. Interpreting

$$p(t) = \begin{cases} 15At^2(4 - t)^2 & \text{if } 0 \leq t \leq 4 \\ 0 & \text{if } t > 4 \end{cases}$$

as the *failure density function* for laptop computer batteries ($t$ measured in years)

(a) what value should $A$ be given to make $p(t)$ a failure density function?
(b) out of 100,000 batteries how many will have failed after 3 years?
(c) sketch a graph of failure distribution function ($P(t) = \int_0^t p(u)\,du$) as well as the failure density function in this case.

5. The average time of failure (also called the *expected time of failure*) of the batteries in Problem 4 is of concern to

users and producers alike. This can be calculated approximately as a sum, which can then be represented exactly as an integral.

(a) Justify why the average time to failure $\bar{t}$ is given approximately by $t_{approx} = \sum\limits_{i=1}^{n} t_i^* \cdot \{\text{percentage of the product failing in } [t_{i-1}, t_i)\}$ where $t_i^*$ is some time in the interval $[t_{i-1}, t_i)$.
(b) Show, using the Mean Value Theorem for $P(t)$, that

$$t_{approx} = \sum_{i=1}^{n} t_i^* \cdot (P(t_i) - P(t_{i-1})) = \sum_{i=1}^{n} t_i^* \cdot P'(\hat{t}_i)\,\Delta t_i,$$

where $t_{i-1} < \hat{t}_i < t_i$.
(c) Letting $\Delta t_i \to 0$ and $n \to \infty$ conclude that $t_{approx} \to \bar{t} = \int_0^T tp(t)\,dt$, where $T$ is the time at which all of the product fails, and $P'(t) = p(t)$.
(d) Explain the connection between the expected value and the center of mass. Discuss why $\bar{t}$ plays the same role as $\bar{x}$ when $\delta = p$ and $m = \int p(t)\,dt = 1$.

6. Using concepts developed in Problems 4 and 5, with the failure density given by

$$p(t) = \begin{cases} 6At^2(5 - t)^2 & \text{otherwise} \\ 0 & \text{if } t > 5 \end{cases}$$

(a) Determine the constant $A$ so that $p(t)$ is a failure density function.
(b) Find the failure distribution function $P(t)$.
(c) Find the expected time of failure.
(d) How many of the units will have failed by the expected time of failure?

7. Using concepts developed in Problems 4 and 5, with the failure density given by

$$p(t) = \begin{cases} (6/125)At^2(6 - t)^2 & \text{if } 0 \leq t \leq 5 \\ 2A(8 - t) & \text{if } 5 < t < 8 \\ 0 & \text{otherwise} \end{cases}$$

(a) determine the constant $A$ so that $p(t)$ is a failure density function
(b) find the failure distribution function $P(t)$
(c) find the expected time of failure
(d) How many of the units will have failed by the expected time of failure?

Have you ever noticed that pressure at the bottom of a pool does not depend on which way your ear is pointing but just upon your depth below the surface? The physical principle here is that the pressure $p$ at a depth $h$ is the fluid's weight per unit volume, $w$, (62.4 lb/ft$^3$ for water) times $h$ or $p = wh$. Pressure obviously has units of weight/(length)$^2$. If this pressure is applied over an area, then the result will be a force (or weight) applied to that area. We can use the calculus to compute such forces.

**8.** If the area of your eardrum is 0.0004 feet$^2$, find the total force in pounds on your eardrum and the pressure on your eardrum at 10 feet below the surface of the pool.

**9.** Suppose we want to find the total force that the pool water exerts against the side of the pool. The force at each point along the wall depends only on the height of the water above it.

(a) Explain why the total force on a wall 75 feet long in a pool which has a depth of 10 feet is given by

$$F \cong 62.4 \cdot 75$$
$$\cdot \sum \Delta h \cdot (\text{depth of the water strip of width } \Delta h)$$

where for example $\Delta h = 1/100$, the sum is from 1 to 100.

(b) Show that in the limit

$$F = 62.4 \cdot 75 \int_0^{10} (10 - h)\, dh$$

for the origin of the coordinate system taken at the bottom of the pool. The water at the surface contributes the least force while the water at the bottom contributes the most.

(c) If the side of the pool is an equilateral triangle (one side along the top) with an altitude of 10 feet (Figure 1) show that the total force on that pool face is $F =$ $F = 62.4 \int_0^{10} (10 - h)W(h)\, dh$, where $W(h)$ is the width of the pool at a height $h$ above the bottom and is given by $\frac{2}{\sqrt{3}}h$.

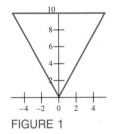

FIGURE 1

**10.** If we insert a flat plate vertically into a pool then the total force on one face of the plate can become large. Of course, there is an equal total force on the other face of the plate, and so the plate will not deform when inserted into the water. Find the total force exerted on one face of a flat equilateral triangular plate whose sides are 6 feet long when one side is kept parallel to the bottom of the pool and the top of the plate is just submerged.

**11.** If the plate in Problem 10 is lowered another 3 feet into the pool, what is the total force on one face?

**12.** A cubical tank has a patch located at the bottom made up of the curves $y = x^2$ and $y = 2$ (Figure 2) which is designed to withstand a total force of 200 pounds without coming off. What is the largest size cube that can be filled to the top with water without the patch coming off?

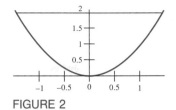

FIGURE 2

**13.** An aquarium tank has a window located at the bottom in the shape of an isosceles triangle (Figure 3) which is designed to withstand a total force of 400 pounds without breaking. What is the maximum height of the water that the window will withstand without breaking.

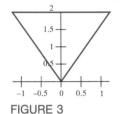

FIGURE 3

*Power*, $P(t)$, is defined as the rate of doing work, or as force $F(t)$ times velocity $v(t)$. If we integrate a force over a distance we get the work that the force does. If we integrate power over a time interval we also get the work performed. Perhaps the most familiar units of power are given by horsepower. One horsepower = 550 foot-pounds/second, taken literally this means that one horse can lift 550 pounds a distance of 1 foot in 1 second. If $F = F(s)$ is the force as a function of position $s$ with $s$ parameterized by time $t$:

$$F(s) = F(s(t)) = \tilde{F}(t), \qquad P(t) = \tilde{F}(t) \cdot v(t)$$
$$\text{Work} = W(s) = \int_{s_1}^{s} F(x)\, dx =$$
$$W(s(t)) = \tilde{W}(t) = \int_{t_1}^{t} P(\tau)\, d\tau = \int_{t_1}^{t} F(\tau)v(\tau)\, d\tau$$
$$\frac{d\tilde{W}(t)}{dt} = P(t)$$

We investigate the relationship between these quantities in Problems 14–15.

**14.** A 3,200 pound car, starting from a standstill, uniformly accelerates to a final speed of 45 miles per hour (66 feet per second) in 6 seconds.

(a) How much horsepower is needed to do this?
(b) Can you do this problem if you assume the car does not accelerate uniformly? If so what is the answer?

**15.** Suppose this 3,200-pound car is cruising along at 45 miles per hour down a level road. The car comes to an incline in the road (hill) which has a slope of 0.05

(a) How much additional horsepower is required to maintain its speed up the incline?
(b) How much additional horsepower is required to increase its speed to 60 miles per hour in 4 seconds when starting to climb the hill?
(c) How much work is done by the engine if it takes the car going 45 miles per hour 3 minutes to get to the top of the hill?

# TECHNOLOGY PROJECT

**Exercise 1** We know that the equation $x^2 + y^2 = r^2$ represents a circle of radius $r$. Use this fact to justify the formula:

$$A_{circle} = 4r \int_0^r \sqrt{1 - (x/r)^2}\, dx = \pi r^2$$

for the area of this circle.

**Exercise 2** Similarly, the equation

$$\left(\frac{x}{a}\right)^2 + \left(\frac{y}{b}\right)^2 = 1$$

represents an ellipse with axes $a$ and $b$. Justify both of the formulas:

$$A_{ellipse} = 4b \int_0^a \sqrt{1 - (x/a)^2}\, dx$$

$$= 4a \int_0^b \sqrt{1 - (y/b)^2}\, dy$$

for the area of this ellipse.

**Exercise 3** Make the substitution $u = x/a$ in Equation 2 and use Equation 1 to compute $A_{ellipse}$ explicitly. Check your result in each of the following two ways:

(a) derive the result again using Equation 3,

(b) show that your result specializes correctly in the case when the ellipse becomes a circle of radius $r$.

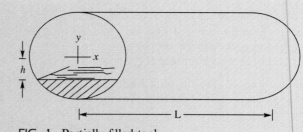

FIG. 1. Partially filled tank.

Figure 1 shows an underground fuel tank (e.g., as at a service station) in the shape of an elliptical cylinder. The tank is lying on its side (i.e., the cylinder is horizontal) and has length $L$ ft. The elliptical cross-section has axes $a$ ft and $b$ ft. The owner has only a crude way of measuring the contents: he lowers a stick down into the tank and measures how high the fuel level is on the stick. You are to help him convert this measurement, call it $h$ (in ft), into cubic ft of fuel in the tank.

**Exercise 4** Derive the following formula for the volume $V = V(h)$, where $-b \leq h \leq b$:

$$V = 2La \int_{-b}^h \sqrt{1 - (y/b)^2}\, dy$$

*Hint*: The volume is an area times the length of the tank.

**Exercise 5** Use your computing technology to check your evaluation of the integrals in Equations 1, 2, and 3. For some technologies, only specific numerical checks are possible.

**Exercise 6** In reference to the tank volume in Equation 4 with the specific parameter values $L = 20$, $a = 10$, and $b = 5$, make a table with two columns, containing respectively $h$ and $V(h)$ from $h = -5$ to $h = 5$ with unit spacing.

**Exercise 7 Challenge Problem:** Your friend would like to put a mark on the stick as a warning to indicate when only 500 ft$^3$ are left. Using the same numerical parameter values as in Exercise 6, figure out the height $h$ at which the stick should be marked as accurately as you can. **Suggestions**: Your technology may have root finding capability, or you could use Newton's method (cf. the Chapter 4 project or the index), or you could use the "zooming" techniques of the first projects.

**Exercise 8** Explain why Exercise 5 implies that the total volume of the cylindrical tank is $\pi abL$ and use this result to check your output in Exercise 6 for the values $h = -5$, $h = 0$, and $h = 5$.

# 7

# Transcendental Functions

**7.1**
**THE NATURAL LOGARITHM FUNCTION**

The power of calculus, both that of derivatives and integrals, has already been amply demonstrated. Yet we have only scratched the surface of potential applications. To dig deeper, we need to expand the class of functions with which we can work. That is the object of this chapter.

We begin by asking you to notice a peculiar gap in our knowledge of derivatives.

$$
\begin{aligned}
D_x(x^3/3) &= x^2 \\
D_x(x^2/2) &= x^1 \\
D_x(x) = 1 &= x^0 \\
D_x(????) &= x^{-1} \\
D_x(-x^{-1}) &= x^{-2} \\
D_x(-x^{-2}/2) &= x^{-3}
\end{aligned}
$$

Is there a function whose derivative is $1/x$? Alternatively, is there an antiderivative $\int (1/x)\, dx$? We will have to answer no if we restrict our attention to the functions studied so far. However, we are about to launch the business of creating new functions.

The first function to be introduced is chosen to fill the gap noticed above. We call it the natural logarithm function and it does have something to do with the logarithm studied in elementary algebra, but the relationship will only appear later. For the time being, just accept the fact that we are going to define a new function and study its properties.

> **Definition**
>
> The **natural logarithm function**, denoted by ln, is defined by
>
> $$\ln x = \int_1^x \frac{1}{t}\, dt, \qquad x > 0$$
>
> Its domain is the set of positive real numbers.

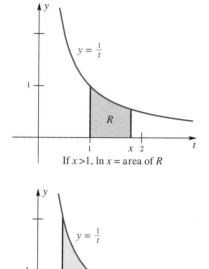

If $x > 1$, $\ln x =$ area of $R$

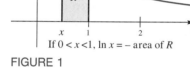

If $0 < x < 1$, $\ln x = -$ area of $R$

**FIGURE 1**

The diagrams in Figure 1 indicate the geometric meaning of ln $x$. It measures the area under the curve $y = 1/t$ between 1 and $x$ if $x > 1$ and the negative of this area if $0 < x < 1$. Also ln $1 = 0$. Clearly, ln $x$ is well defined for $x > 0$. And what is the derivative of this new function? Just exactly what we want.

**The Derivative of the Natural Logarithm** Recall that the derivative of an integral with respect to its upper limit is the integrand evaluated at the upper limit (Theorem 5.7D). Thus

$$\boxed{D_x \ln x = \frac{1}{x}, \qquad x > 0}$$

This can be combined with the Chain Rule. If $u = f(x) > 0$ and if $f$ is differentiable, then

$$D_x \ln u = \frac{1}{u} D_x u$$

**EXAMPLE 1** Find $D_x \ln \sqrt{x}$.

**Solution** Let $u = \sqrt{x} = x^{1/2}$. Then

$$D_x \ln \sqrt{x} = \frac{1}{x^{1/2}} \cdot D_x(x^{1/2}) = \frac{1}{x^{1/2}} \cdot \frac{1}{2} x^{-1/2} = \frac{1}{2x} \qquad \blacksquare$$

**EXAMPLE 2** Find $D_x \ln(x^2 - x - 2)$.

**Solution** This problem makes sense, provided $x^2 - x - 2 > 0$. Now $x^2 - x - 2 = (x - 2)(x + 1)$, which is positive provided $x < -1$ or $x > 2$. Thus, the domain of $\ln(x^2 - x - 2)$ is $(-\infty, -1) \cup (2, \infty)$. On this domain,

$$D_x \ln(x^2 - x - 2) = \frac{1}{x^2 - x - 2} D_x(x^2 - x - 2) = \frac{2x - 1}{x^2 - x - 2} \qquad \blacksquare$$

**EXAMPLE 3** Show that

$$\boxed{D_x \ln|x| = \frac{1}{x}, \qquad x \neq 0}$$

*Solution*   Two cases are to be considered. If $x > 0$, $|x| = x$ and

$$D_x \ln|x| = D_x \ln x = \frac{1}{x}$$

If $x < 0$, $|x| = -x$, and so

$$D_x \ln|x| = D_x \ln(-x) = \frac{1}{-x} D_x(-x) = \left(\frac{1}{-x}\right)(-1) = \frac{1}{x} \qquad \blacksquare$$

We know that for every differentiation formula, there is a corresponding integration formula. Thus, Example 3 implies that

$$\int \frac{1}{x} dx = \ln|x| + C, \qquad x \neq 0$$

or, with $u$ replacing $x$,

$$\boxed{\int \frac{1}{u} du = \ln|u| + C, \qquad u \neq 0}$$

This fills the long-standing gap in the Power Rule: $\int u^r \, du = u^{r+1}/(r+1)$, from which we had to exclude the exponent $r = -1$.

**EXAMPLE 4**   Find $\int \frac{5}{2x + 7} dx$.

*Solution*   Let $u = 2x + 7$, so $du = 2 \, dx$. Then

$$\int \frac{5}{2x + 7} dx = \frac{5}{2} \int \frac{1}{2x + 7} 2 \, dx = \frac{5}{2} \int \frac{1}{u} du$$

$$= \frac{5}{2} \ln|u| + C = \frac{5}{2} \ln|2x + 7| + C \qquad \blacksquare$$

**EXAMPLE 5**   Evaluate $\int_{-1}^{3} \frac{x}{10 - x^2} dx$.

*Solution*   Let $u = 10 - x^2$, so $du = -2x \, dx$. Then

$$\int \frac{x}{10 - x^2} dx = -\frac{1}{2} \int \frac{-2x}{10 - x^2} dx = -\frac{1}{2} \int \frac{1}{u} du$$

$$= -\frac{1}{2} \ln|u| + C = -\frac{1}{2} \ln|10 - x^2| + C$$

Thus, by the Fundamental Theorem of Calculus,

$$\int_{-1}^{3} \frac{x}{10 - x^2}\, dx = \left[ -\frac{1}{2} \ln|10 - x^2| \right]_{-1}^{3}$$

$$= -\frac{1}{2} \ln 1 + \frac{1}{2} \ln 9 = \frac{1}{2} \ln 9$$

For the above calculation to be valid, $10 - x^2$ must never be 0 on the interval $[-1, 3]$. It is easy to see that this is true.  ∎

**Properties of the Natural Logarithm** You should be reminded of common logarithms by the results of our next theorem.

---

### Common Logarithms

Properties (ii) and (iii) for common logarithms (base 10 logarithms) were what motivated the invention of logarithms. John Napier (1550–1617) wanted to simplify the complicated calculations that arise in astronomy and navigation. To replace multiplication by addition and division by subtraction was his goal—exactly what (ii) and (iii) accomplish. For over 350 years, common logarithms were an essential aid in computation, but today we use calculators and computers for this purpose. However, natural logarithms retain their importance for other reasons, as you will see.

---

### Theorem A

If $a$ and $b$ are positive numbers and $r$ is any rational number, then
(i)   $\ln 1 = 0$;
(ii)  $\ln ab = \ln a + \ln b$;
(iii) $\ln \dfrac{a}{b} = \ln a - \ln b$;
(iv)  $\ln a^r = r \ln a$.

**Proof**

(i)   $\ln 1 = \displaystyle\int_{1}^{1} \frac{1}{t}\, dt = 0$

(ii)  Since, for $x > 0$,

$$D_x \ln ax = \frac{1}{ax} \cdot a = \frac{1}{x}$$

and

$$D_x \ln x = \frac{1}{x}$$

it follows from the theorem about two functions with the same derivative (Theorem 4.8B) that

$$\ln ax = \ln x + C$$

To evaluate $C$, let $x = 1$, obtaining $\ln a = C$. Thus,

$$\ln ax = \ln x + \ln a$$

Finally, let $x = b$.

**(iii)** Replace $a$ by $1/b$ in (ii) to obtain

$$\ln \frac{1}{b} + \ln b = \ln \left( \frac{1}{b} \cdot b \right) = \ln 1 = 0$$

Thus,

$$\ln \frac{1}{b} = -\ln b$$

Applying (ii) again, we get

$$\ln \frac{a}{b} = \ln \left( a \cdot \frac{1}{b} \right) = \ln a + \ln \frac{1}{b} = \ln a - \ln b$$

**(iv)** Since, for $x > 0$,

$$D_x(\ln x^r) = \frac{1}{x^r} \cdot rx^{r-1} = \frac{r}{x}$$

and

$$D_x(r \ln x) = r \cdot \frac{1}{x} = \frac{r}{x}$$

it follows by Theorem 4.8B, which we used in (ii), that

$$\ln x^r = r \ln x + C$$

Let $x = 1$, which gives $C = 0$. Thus,

$$\ln x^r = r \ln x$$

a result equivalent to (iv). ■

**EXAMPLE 6** Find $dy/dx$ if $y = \ln \sqrt[3]{(x-1)/x^2}$, $x > 1$.

*Solution* Our problem is easier if we first use the properties of natural logarithms to simplify $y$.

$$y = \ln \left( \frac{x-1}{x^2} \right)^{1/3} = \frac{1}{3} \ln \left( \frac{x-1}{x^2} \right)$$
$$= \frac{1}{3} [\ln(x-1) - \ln x^2] = \frac{1}{3} [\ln(x-1) - 2 \ln x]$$

Thus,

$$\frac{dy}{dx} = \frac{1}{3} \left[ \frac{1}{x-1} - \frac{2}{x} \right] = \frac{2-x}{3x^2 - 3x}$$

■

**Logarithmic Differentiation**    The labor of differentiating expressions involving quotients, products, or powers can often be substantially reduced by first applying the natural logarithm function and using its properties. This method, called **logarithmic differentiation**, is illustrated in Example 7.

**EXAMPLE 7**    Differentiate $y = \dfrac{\sqrt{1 - x^2}}{(x + 1)^{2/3}}$.

*Solution*    First we take natural logarithms; then we differentiate implicitly with respect to $x$ (recall Section 3.8).

$$\ln y = \frac{1}{2} \ln(1 - x^2) - \frac{2}{3} \ln(x + 1)$$

$$\frac{1}{y} \frac{dy}{dx} = \frac{-2x}{2(1 - x^2)} - \frac{2}{3(x + 1)} = \frac{-(x + 2)}{3(1 - x^2)}$$

Thus,

$$\frac{dy}{dx} = \frac{-y(x + 2)}{3(1 - x^2)} = \frac{-\sqrt{1 - x^2}(x + 2)}{3(x + 1)^{2/3}(1 - x^2)}$$

$$= \frac{-(x + 2)}{3(x + 1)^{2/3}(1 - x^2)^{1/2}}$$    ■

Example 7 could have been done directly without first taking logarithms, and we suggest you try it. You should be able to make the two answers agree.

**The Graph of the Natural Logarithm**    The domain of $\ln x$ consists of the positive real numbers, so the graph of $y = \ln x$ is in the right half-plane. Also, for $x > 0$,

$$D_x \ln x = \frac{1}{x} > 0$$

and

$$D_x^2 \ln x = -\frac{1}{x^2} < 0$$

The first formula tells us that the graph is continuous (why?) and rises as $x$ increases; the second tells us that the graph is everywhere concave downward. In Problems 39 and 40, it will be shown that

$$\lim_{x \to \infty} \ln x = \infty$$

and

$$\lim_{x \to 0^+} \ln x = -\infty$$

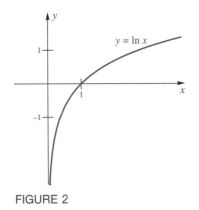

FIGURE 2

Finally, $\ln 1 = 0$. These facts imply that the graph of $y = \ln x$ is similar in shape to that shown in Figure 2.

Values of $\ln x$ have been tabulated by methods that will be explained later. For example,

$$\ln 2 \approx 0.6931$$

$$\ln 3 \approx 1.0986$$

$$\ln 10 \approx 2.3026$$

Table III of the appendix gives a fairly complete listing. If your calculator has an $\boxed{\ln}$ button, values for the natural logarithm are at your fingertips.

## CONCEPTS REVIEW

**1.** The function ln is defined by $\ln x = $ _____. The domain of this function is _____ and its range is _____.

**2.** From the preceding definition, it follows that $D_x \ln x = $ _____ for $x > 0$.

**3.** More generally, for $x \neq 0$, $D_x \ln|x| = $ _____ and so $\int (1/x)\, dx = $ _____.

**4.** Some common properties of ln are: $\ln(xy) = $ _____, $\ln(x/y) = $ _____ and $\ln(x^r) = $ _____.

## PROBLEM SET 7.1

**1.** Use the approximations $\ln 2 = 0.693$ and $\ln 3 = 1.099$ together with the properties stated in Theorem A to calculate approximations to each of the following. For example, $\ln 6 = \ln(2 \cdot 3) = \ln 2 + \ln 3 = 0.693 + 1.099 = 1.792$.

(a) $\ln 6$      (b) $\ln 1.5$      (c) $\ln 81$
(d) $\ln \sqrt{2}$      (e) $\ln \left(\frac{1}{36}\right)$      (f) $\ln 48$

$\boxed{C}$ **2.** Use your calculator to make the computations in Problem 1 directly.

In Problems 3–14, find the indicated derivative (see Examples 1 and 2). Assume in each case that $x$ is restricted so that ln is defined.

**3.** $D_x \ln(x^2 + 3x + \pi)$      **4.** $D_x \ln(3x^3 + 2x)$

**5.** $D_x \ln(x - 4)^3$      **6.** $D_x \ln \sqrt{3x - 2}$

**7.** $\dfrac{dy}{dx}$ if $y = 3 \ln x$      **8.** $\dfrac{dy}{dx}$ if $y = x^2 \ln x$

**9.** $\dfrac{dz}{dx}$ if $z = x^2 \ln x^2 + (\ln x)^3$

**10.** $\dfrac{dr}{dx}$ if $r = \dfrac{\ln x}{x^2 \ln x^2} + \left(\ln \dfrac{1}{x}\right)^3$

**11.** $g'(x)$ if $g(x) = \ln\left(x + \sqrt{x^2 + 1}\right)$

**12.** $h'(x)$ if $h(x) = \ln\left(x + \sqrt{x^2 - 1}\right)$

**13.** $f'(81)$ if $f(x) = \ln \sqrt[3]{x}$

**14.** $f'\left(\frac{\pi}{2}\right)$ if $f(x) = \ln(\cos x)$

In Problems 15–22, find the integrals (see Examples 4 and 5).

**15.** $\displaystyle\int \dfrac{1}{2x + 1}\, dx$      **16.** $\displaystyle\int \dfrac{1}{1 - 2x}\, dx$

**17.** $\displaystyle\int \dfrac{6v + 9}{3v^2 + 9v}\, dv$      **18.** $\displaystyle\int \dfrac{z}{2z^2 + 8}\, dz$

**19.** $\displaystyle\int \dfrac{2 \ln x}{x}\, dx$      **20.** $\displaystyle\int \dfrac{-1}{x(\ln x)^2}\, dx$

**21.** $\displaystyle\int_0^3 \dfrac{x^4}{2x^5 + \pi}\, dx$      **22.** $\displaystyle\int_0^1 \dfrac{t + 1}{2t^2 + 4t + 3}\, dt$

In Problems 23–26, use Theorem A to write the expressions as the logarithm of a single quantity.

**23.** $2 \ln(x + 1) - \ln x$

**24.** $\frac{1}{2} \ln(x - 9) + \frac{1}{2} \ln x$

**25.** $\ln(x - 2) - \ln(x + 2) + 2 \ln x$

**26.** $\ln(x^2 - 9) - 2 \ln(x - 3) - \ln(x + 3)$

In Problems 27–30, find $dy/dx$ by logarithmic differentiation (see Example 7).

**27.** $y = \dfrac{x + 11}{\sqrt{x^3 - 4}}$

**28.** $y = (x^2 + 3x)(x - 2)(x^2 + 1)$

**29.** $y = \dfrac{\sqrt{x + 13}}{(x - 4)\sqrt[3]{2x + 1}}$

**30.** $y = \dfrac{(x^2 + 3)^{2/3}(3x + 2)^2}{\sqrt{x + 1}}$

In Problems 31–34, make use of the known graph of $y = \ln x$ to sketch the graphs of the equations.

**31.** $y = \ln|x|$

**32.** $y = \ln\sqrt{x}$

**33.** $y = \ln\left(\dfrac{1}{x}\right)$

**34.** $y = \ln(x - 2)$

**35.** Sketch the graph of $y = \ln \cos x + \ln \sec x$ on $(-\pi/2, \pi/2)$, but think before you begin.

**36.** Explain why $\displaystyle\lim_{x \to 0} \ln \dfrac{\sin x}{x} = 0$.

**37.** Find all local extreme values of $f(x) = 2x^2 \ln x - x^2$ on its domain.

**38.** The rate of transmission in a telegraph cable is observed to be proportional to $x^2 \ln(1/x)$, where $x$ is the ratio of the radius of the core to the thickness of the insulation ($0 < x < 1$). What value of $x$ gives the maximum rate of transmission?

**39.** Use the fact that $\ln 4 > 1$ to show that $\ln 4^m > m$ for $m > 0$. Conclude that $\ln x$ can be made as large as desired by choosing $x$ sufficiently large. What does this imply about $\displaystyle\lim_{x \to \infty} \ln x$?

**40.** Use the fact that $\ln x = -\ln(1/x)$ and Problem 39 to show that $\displaystyle\lim_{x \to 0^+} \ln x = -\infty$.

**41.** Solve for $x$: $\displaystyle\int_{1/3}^{x} \frac{1}{t}\, dt = 2\int_{1}^{x} \frac{1}{t}\, dt$

**42.** Prove.
(a) Since $1/t < 1/\sqrt{t}$ for $t > 1$, $\ln x < 2(\sqrt{x} - 1)$ for $x > 1$.
(b) $\displaystyle\lim_{x \to \infty} (\ln x)/x = 0$.

**43.** Calculate

$$\lim_{n \to \infty} \left[ \frac{1}{n + 1} + \frac{1}{n + 2} + \cdots + \frac{1}{2n} \right]$$

by writing the expression in brackets as

$$\left[ \frac{1}{1 + 1/n} + \frac{1}{1 + 2/n} + \cdots + \frac{1}{1 + n/n} \right] \frac{1}{n}$$

and recognizing the latter as a Riemann sum.

C **44.** A famous theorem (the Prime Number Theorem) says that the number of primes less than $n$ for large $n$ is approximately $n/(\ln n)$. About how many primes are there less than 1,000,000?

**45.** Find and simplify $f'(1)$.

(a) $f(x) = \ln\left(\dfrac{ax - b}{ax + b}\right)^c$, where $c = \dfrac{a^2 - b^2}{2ab}$.

(b) $f(x) = \displaystyle\int_{1}^{u} \cos^2 t\, dt$, where $u = \ln(x^2 + x - 1)$.

**46.** Evaluate

(a) $\displaystyle\int_{0}^{\pi/3} \tan x\, dx$
(b) $\displaystyle\int_{\pi/4}^{\pi/3} \sec x \csc x\, dx$

**47.** The region bounded by $y = (x^2 + 4)^{-1}$, $y = 0$, $x = 1$, and $x = 4$ is revolved about the $y$-axis, generating a solid. Find its volume.

**48.** Find the length of the curve $y = x^2/4 - \ln\sqrt{x}$, $1 \le x \le 2$.

**49.** By appealing to the graph of $y = 1/x$, show that

$$\frac{1}{2} + \frac{1}{3} + \cdots + \frac{1}{n} < \ln n < 1 + \frac{1}{2} + \frac{1}{3} + \cdots + \frac{1}{n-1}$$

**50.** Prove Napier's Inequality, which says that for $0 < x < y$,

$$\frac{1}{y} < \frac{\ln y - \ln x}{y - x} < \frac{1}{x}$$

PC **51.** Let $f(x) = \ln(1.5 + \sin x)$.

(a) Find the absolute extreme points on $[0, 3\pi]$.
(b) Find any inflection points on $[0, 3\pi]$.
(c) Evaluate $\displaystyle\int_{0}^{3\pi} \ln(1.5 + \sin x)\, dx$.

PC **52.** Let $f(x) = \cos(\ln x)$.

(a) Find the absolute extreme points on $[0.1, 20]$.
(b) Find the absolute extreme points on $[0.01, 20]$.
(c) Evaluate $\displaystyle\int_{0.1}^{20} \cos(\ln x)\, dx$.

PC **53.** Draw the graphs of $f(x) = x \ln(1/x)$ and $g(x) = x^2 \ln(1/x)$ on $(0, 1]$.

(a) Find the area of the region between these curves on $(0, 1]$.
(b) Find the absolute maximum value of $|f(x) - g(x)|$ on $(0, 1]$.

PC **54.** Follow the directions of Problem 53 for $f(x) = x \ln x$ and $g(x) = \sqrt{x} \ln x$.

---

**Answers to Concepts Review:** **1.** $\displaystyle\int_{1}^{x} (1/t)\, dt$; $(0, \infty)$; $(-\infty, \infty)$ **2.** $1/x$ **3.** $1/x$; $\ln|x| + C$ **4.** $\ln x + \ln y$; $\ln x - \ln y$; $r \ln x$

**7.2**
**INVERSE FUNCTIONS AND**
**THEIR DERIVATIVES**

Our stated aim for this chapter is to expand the number of functions in our repertoire. One way to manufacture new functions is to take old ones and "reverse" them. When we do this for the natural logarithm function, we will be led to the natural exponential functions, the subject of Section 7.3. In this section, we study the general problem of reversing (or inverting) a function. Here is the idea.

A function $f$ takes a number $x$ from its domain $D$ and assigns to it a single value $y$ from its range $R$. If we are lucky, as in the case of the two functions graphed in Figures 1 and 2, we can reverse $f$; that is, for any given $y$ in $R$, we can unambiguously go back and find the $x$ from which it came. This new function that takes $y$ and assigns $x$ to it is denoted by $f^{-1}$. Note that its domain is $R$ and its range is $D$. It is called the **inverse** of $f$, or simply $f$-inverse. Here we are using the superscript $-1$ in a new way. The symbol $f^{-1}$ does not denote $1/f$, as you might have expected. We, and all mathematicians, use it to name the inverse function.

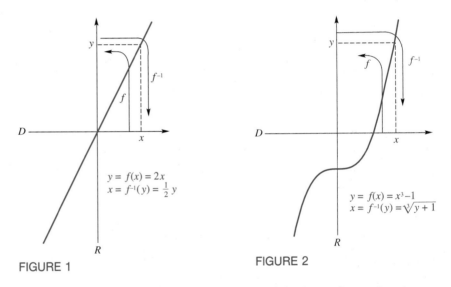

$$y = f(x) = 2x$$
$$x = f^{-1}(y) = \tfrac{1}{2}y$$

FIGURE 1

$$y = f(x) = x^3 - 1$$
$$x = f^{-1}(y) = \sqrt[3]{y + 1}$$

FIGURE 2

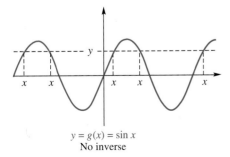

$$y = f(x) = x^2$$
No inverse

FIGURE 3

$$y = g(x) = \sin x$$
No inverse

FIGURE 4

Sometimes, we can give a formula for $f^{-1}$. If $y = f(x) = 2x$, then $x = f^{-1}(y) = \tfrac{1}{2}y$ (see Figure 1). Similarly, if $y = f(x) = x^3 - 1$, then $x = f^{-1}(y) = \sqrt[3]{y + 1}$ (Figure 2). In each case, we simply solve the equation that determines $f$ for $x$ in terms of $y$. The result is $x = f^{-1}(y)$.

But life is more complicated than the two examples above indicate. Not every function can be reversed in an unambiguous way. Consider $y = f(x) = x^2$, for example. For a given $y$ there are *two* $x$'s that correspond to it (Figure 3). The function $y = g(x) = \sin x$ is even worse. For each $y$, there are infinitely many $x$'s that correspond to it (Figure 4). Such functions do not have inverses; at least, they do not unless we somehow restrict the set of $x$-values, a subject we will take up later.

**Existence of Inverse Functions**   It would be nice to have a simple criterion for deciding whether a function $f$ has an inverse. One such criterion is that the function be **one-to-one**; that is, $x_1 \neq x_2$ implies $f(x_1) \neq f(x_2)$. This is equivalent to the geometric condition that every horizontal line meet the graph of $y = f(x)$ in at most one point. But in a given situation, this criterion may be very hard to apply, since it demands that we have complete knowledge of the graph. A more practical criterion that covers most examples that arise in this book is that a function be **strictly monotonic**. By this we mean that it is either increasing or decreasing on its domain.

### Theorem A

If $f$ is strictly monotonic on its domain, then $f$ has an inverse.

This is a practical result, because we have an easy way of deciding whether a function $f$ is strictly monotonic. We simply examine the sign of $f'$.

**EXAMPLE 1** Show that $f(x) = x^5 + 2x + 1$ has an inverse.

***Solution*** $f'(x) = 5x^4 + 2 > 0$ for all $x$. Thus $f$ is increasing on the whole real line and so has an inverse there. ∎

We do not claim that we can always give a formula for $f^{-1}$. In the example just considered, this would require that we be able to solve $y = x^5 + 2x + 1$ for $x$, a task beyond our capabilities.

There is a way of salvaging the notion of inverse for functions that do not have inverses on their natural domain. We simply *restrict the domain* to a set on which the graph is either increasing or decreasing. Thus for $y = f(x) = x^2$, we may restrict the domain to $x \geq 0$ ($x \leq 0$ would also work). For $y = g(x) = \sin x$, we restrict the domain to the interval $[-\pi/2, \pi/2]$. Then both functions have inverses (see Figure 5), and we can even give a formula for the first one, namely, $f^{-1}(y) = \sqrt{y}$.

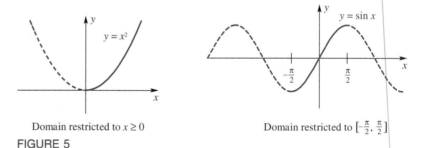

Domain restricted to $x \geq 0$          Domain restricted to $\left[-\frac{\pi}{2}, \frac{\pi}{2}\right]$

**FIGURE 5**

If $f$ has an inverse $f^{-1}$, then $f^{-1}$ also has an inverse, namely, $f$. Thus, we may call $f$ and $f^{-1}$ a pair of inverse functions. One function undoes (or reverses) what the other did—that is,

$$f^{-1}(f(x)) = x \quad \text{and} \quad f(f^{-1}(y)) = y$$

**EXAMPLE 2** Show that $f(x) = 2x + 6$ has an inverse, find a formula for $f^{-1}(y)$, and verify the results in the box above.

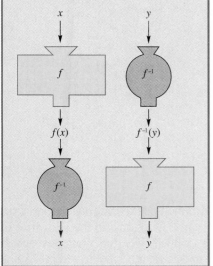

### Undoing Machines

We may view a function as a machine that accepts an input and produces an output. If the $f$ machine and the $f^{-1}$ machine are hooked together in tandem, they undo each other.

***Solution*** Since $f$ is an increasing function, it has an inverse. To find $f^{-1}(y)$, we solve $y = 2x + 6$ for $x$, which gives $x = (y - 6)/2 = f^{-1}(y)$. Finally, note that

$$f^{-1}(f(x)) = f^{-1}(2x + 6) = \frac{(2x + 6) - 6}{2} = x$$

and

$$f(f^{-1}(y)) = f\left(\frac{y - 6}{2}\right) = 2\left(\frac{y - 6}{2}\right) + 6 = y$$ ■

**The Graph of $y = f^{-1}(x)$**   Suppose $f$ has an inverse. Then

$$x = f^{-1}(y) \quad \Leftrightarrow \quad y = f(x)$$

Consequently, $y = f(x)$ and $x = f^{-1}(y)$ determine the same $(x, y)$ pairs, and so have identical graphs. However, it is conventional to use $x$ as the domain variable for functions, so we now inquire about the graph of $y = f^{-1}(x)$ (note that we have interchanged the roles of $x$ and $y$). A little thought convinces us that to interchange the roles of $x$ and $y$ on a graph is to reflect the graph across the line $y = x$. Thus the graph of $y = f^{-1}(x)$ is just the graph of $y = f(x)$ reflected across the line $y = x$ (Figure 6).

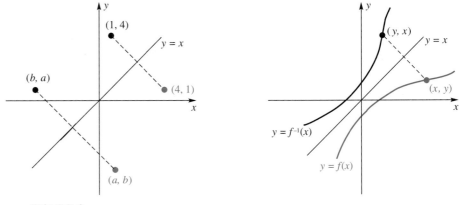

FIGURE 6

A related matter is that of finding a formula for $f^{-1}(x)$. To do it, we first find $f^{-1}(y)$ and then replace $y$ by $x$ in the resulting formula. Thus, we propose the following three-step process for finding $f^{-1}(x)$.

***Step 1***   Solve the equation $y = f(x)$ for $x$ in terms of $y$.
***Step 2***   Use $f^{-1}(y)$ to name the resulting expression in $y$.
***Step 3***   Replace $y$ by $x$ to get the formula for $f^{-1}(x)$.

Before trying the three-step process on a particular function $f$, you might think we should first verify that $f$ has an inverse. However, if we can actually carry out the first step and get a single $x$ for each $y$, then $f^{-1}$ does exist. (Note that when we try this for $y = f(x) = x^2$, we get $x = \pm\sqrt{y}$, which immediately shows that $f^{-1}$ does not exist, unless, of course, we have restricted the domain to eliminate one of the two signs, $+$ or $-$.)

**EXAMPLE 3**   Find a formula for $f^{-1}(x)$ if $y = f(x) = x/(1 - x)$.

*Solution*   Here are the three steps for this example.

**Step 1**
$$y = \frac{x}{1 - x}$$
$$(1 - x)y = x$$
$$y - xy = x$$
$$x + xy = y$$
$$x(1 + y) = y$$
$$x = \frac{y}{1 + y}$$

**Step 2**   $f^{-1}(y) = \dfrac{y}{1 + y}$

**Step 3**   $f^{-1}(x) = \dfrac{x}{1 + x}$   ∎

**Derivatives of Inverse Functions**   We conclude this section by investigating the relationship between the derivative of a function and the derivative of its inverse. Consider first what happens to a line $l_1$ when it is reflected across the line $y = x$. As the left half of Figure 7 makes clear, $l_1$ is reflected into a line $l_2$; moreover, their respective slopes $m_1$ and $m_2$ are related by $m_2 = 1/m_1$, provided $m_1 \neq 0$. If $l_1$ happens to be the tangent line to the graph of $f$ at the point $(c, d)$, then $l_2$ is the tangent line to the graph of $f^{-1}$ at the point $(d, c)$ (see the right half of Figure 7). We are led to the conclusion that

$$(f^{-1})'(d) = m_2 = \frac{1}{m_1} = \frac{1}{f'(c)}$$

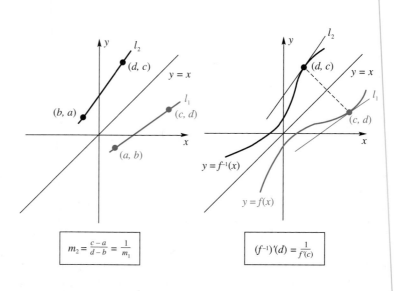

FIGURE 7

Pictures are sometimes deceptive, so we claim only to have made the following result plausible. For a formal proof, see any advanced calculus book.

> ### Theorem B
>
> **(Inverse Function Theorem).** Let $f$ be differentiable and strictly monotonic on an interval $I$. If $f'(x) \neq 0$ at a certain $x$ in $I$, then $f^{-1}$ is differentiable at the corresponding point $y = f(x)$ in the range of $f$ and
>
> $$(f^{-1})'(y) = \frac{1}{f'(x)}$$

The conclusion to Theorem B is often written symbolically as

$$\frac{dx}{dy} = \frac{1}{dy/dx}$$

**EXAMPLE 4**  Let $y = f(x) = x^5 + 2x + 1$, as in Example 1. Find $(f^{-1})'(4)$.

*Solution*  Even though we cannot find a formula for $f^{-1}$ in this case, we note that $y = 4$ corresponds to $x = 1$ and, since $f'(x) = 5x^4 + 2$,

$$(f^{-1})'(4) = \frac{1}{f'(1)} = \frac{1}{5 + 2} = \frac{1}{7}$$ ∎

## CONCEPTS REVIEW

**1.** A function is one-to-one if $x_1 \neq x_2$ implies _____ .

**2.** A one-to-one function $f$ has an inverse $f^{-1}$ satisfying $f^{-1}(f(x)) =$ _____ and $f($_____$) = y$.

**3.** A useful criterion for $f$ to be one-to-one (and so have an inverse) on a domain is that $f$ be strictly _____ there. This means that $f$ is either _____ or _____ .

**4.** Let $y = f(x)$, where $f$ has the inverse $f^{-1}$. The relation connecting the derivatives of $f$ and $f^{-1}$ is _____ .

## PROBLEM SET 7.2

In Problems 1–6, the graph of $y = f(x)$ is shown. In each case, decide whether $f$ has an inverse, and if so, estimate $f^{-1}(2)$.

**1.**

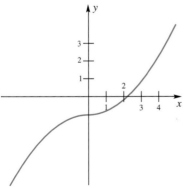

**2.**

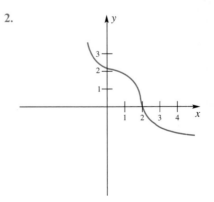

**3.**

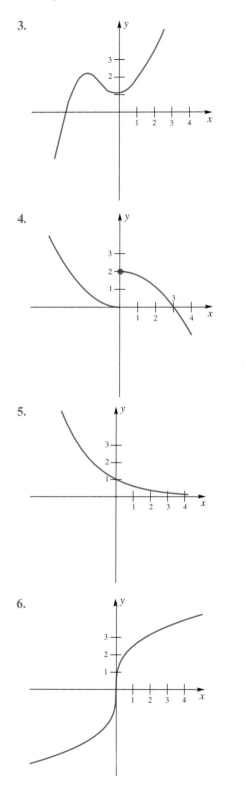

**4.**

**5.**

**6.**

In Problems 7–14, show that $f$ has an inverse by showing that it is strictly monotonic (see Example 1).

7. $f(x) = -x^5 - x^3$

8. $f(x) = x^7 + x^5$

9. $f(\theta) = \cos \theta,\ 0 \le \theta \le \pi$

10. $f(x) = \cot x = \dfrac{\cos x}{\sin x},\ 0 < x < \dfrac{\pi}{2}$

11. $f(z) = (z - 1)^2,\ z \ge 1$

12. $f(x) = x^2 + x - 6,\ x \ge 2$

13. $f(x) = \displaystyle\int_0^x \sqrt{t^4 + t^2 + 10}\, dt$     *Hint:* Theorem 5.7D

14. $f(r) = \displaystyle\int_r^1 \cos^4 t\, dt$

In Problems 15–28, find a formula for $f^{-1}(x)$ and then verify that $f^{-1}(f(x)) = x$ and $f(f^{-1}(x)) = x$ (see Examples 2 and 3).

15. $f(x) = x + 1$  16. $f(x) = -\dfrac{x}{3} + 1$

17. $f(x) = \sqrt{x + 1}$  18. $f(x) = -\sqrt{1 - x}$

19. $f(x) = -\dfrac{1}{x - 3}$  20. $f(x) = \sqrt{\dfrac{1}{x - 2}}$

21. $f(x) = 4x^2,\ x \le 0$

22. $f(x) = (x - 3)^2,\ x \ge 3$

23. $f(x) = (x - 1)^3$  24. $f(x) = x^{5/2},\ x \ge 0$

25. $f(x) = \dfrac{x - 1}{x + 1}$  26. $f(x) = \left(\dfrac{x - 1}{x + 1}\right)^3$

27. $f(x) = \dfrac{x^3 + 2}{x^3 + 1}$  28. $f(x) = \left(\dfrac{x^3 + 2}{x^3 + 1}\right)^5$

In Problems 29 and 30, restrict the domain of $f$ so that $f$ has an inverse, yet keeping its range as large as possible. Then find $f^{-1}(x)$. *Suggestion:* First graph $f$.

29. $f(x) = 2x^2 + x - 4$  30. $f(x) = x^2 - 3x + 1$

In each of Problems 31–34, the graph of $y = f(x)$ is shown. Sketch the graph of $y = f^{-1}(x)$ and estimate $(f^{-1})'(3)$.

**31.**

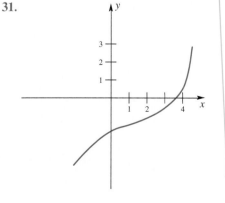

32.

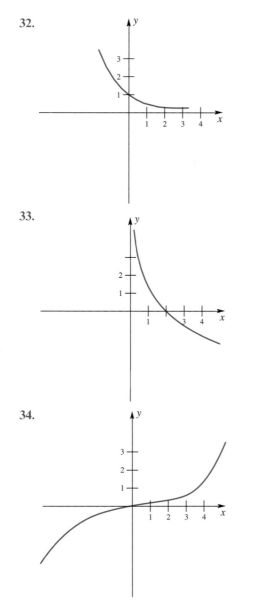

33.

34.

**37.** $f(x) = 2 \tan x, -\dfrac{\pi}{2} < x < \dfrac{\pi}{2}$

**38.** $f(x) = \sqrt{x + 1}$

**39.** Suppose that both $f$ and $g$ have inverses and that $h(x) = (f \circ g)(x) = f(g(x))$. Show that $h$ has an inverse given by $h^{-1} = g^{-1} \circ f^{-1}$.

**40.** Verify the result of Problem 39 for $f(x) = 1/x$, $g(x) = 3x + 2$.

**41.** If $f(x) = \displaystyle\int_0^x \sqrt{1 + \cos^2 t}\, dt$, then $f$ has an inverse. (Why?) Let $A = f(\pi/2)$ and $B = f(5\pi/6)$. Find (a) $(f^{-1})'(A)$, (b) $(f^{-1})'(B)$, and (c) $(f^{-1})'(0)$.

**42.** Let $f(x) = \dfrac{ax + b}{cx + d}$ and assume $bc - ad \neq 0$.

(a) Find the formula for $f^{-1}(x)$.
(b) Why is the condition $bc - ad \neq 0$ needed?
(c) What condition on $a$, $b$, $c$, and $d$ will make $f = f^{-1}$?

**43.** Suppose $f$ is continuous and strictly increasing on $[0, 1]$ with $f(0) = 0$ and $f(1) = 1$. If $\displaystyle\int_0^1 f(x)\, dx = \frac{2}{5}$, calculate $\displaystyle\int_0^1 f^{-1}(y)\, dy$. *Hint:* Draw a picture.

**44.** Let $f$ be continuous and strictly increasing on $[0, \infty)$ with $f(0) = 0$ and $f(x) \to \infty$ as $x \to \infty$. Use geometric reasoning to establish **Young's Inequality**. For $a > 0, b > 0$,

$$ab \leq \int_0^a f(x)\, dx + \int_0^b f^{-1}(y)\, dy$$

What is the condition for equality?

**45.** Let $p > 1, q > 1$, and $1/p + 1/q = 1$. Show that the inverse of $f(x) = x^{p-1}$ is $f^{-1}(y) = y^{q-1}$ and use this together with Problem 44 to prove **Minkowski's Inequality**:

$$ab \leq \frac{a^p}{p} + \frac{b^q}{q}, \qquad a > 0, b > 0$$

In Problems 35–38, find $(f^{-1})'(2)$ by using Theorem B (see Example 4). Note that you can find the $x$ corresponding to $y = 2$ by inspection.

**35.** $f(x) = 3x^5 + x - 2$  **36.** $f(x) = x^5 + 5x - 4$

**Answers to Concepts Review:** **1.** $f(x_1) \neq f(x_2)$
**2.** $x$; $f^{-1}(y)$  **3.** Monotonic; increasing; decreasing
**4.** $(f^{-1})'(y) = 1/f'(x)$

**7.3**
**THE NATURAL EXPONENTIAL FUNCTION**

The graph of $y = f(x) = \ln x$ was obtained at the end of Section 7.1 and is reproduced in Figure 1. The natural logarithm function is differentiable (hence continuous) and increasing on its domain $D = (0, \infty)$; its range is $R = (-\infty, \infty)$. It is, in fact, precisely the kind of function studied in Section 7.2, and therefore has an inverse $\ln^{-1}$ with domain $(-\infty, \infty)$ and range $(0, \infty)$. This function is so important that it is given a special name and a special symbol.

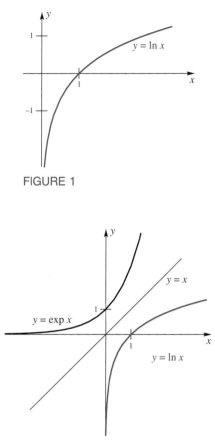

FIGURE 1

FIGURE 2

**Definition**

The inverse of ln is called the **natural exponential function** and is denoted by exp. Thus

$$x = \exp y \quad \Leftrightarrow \quad y = \ln x$$

It follows immediately from this definition that:

**(i)**  $\exp(\ln x) = x, \quad x > 0$

**(ii)**  $\ln(\exp y) = y, \quad \text{all } y$

Since exp and ln are inverse functions, the graph of $y = \exp x$ is just the graph of $y = \ln x$ reflected across the line $y = x$ (Figure 2).

But why the name *exponential function*? You will see.

**Properties of the Exponential Function**  We begin by introducing a new number, which—like $\pi$—is so important in mathematics that it gets a special symbol, namely, $e$. The letter $e$ is appropriate since Leonhard Euler first recognized the significance of this number.

**Definition**

The letter $e$ denotes the unique positive real number such that $\ln e = 1$.

Since $\ln e = 1$, it is also true that $\exp 1 = e$. The number $e$, like $\pi$, is irrational. Its decimal expansion is known to thousands of places; the first few digits are

$$e \approx 2.718281828459045$$

Now we make a crucial observation, one that depends only on facts already demonstrated: (i) above and Theorem 7.1A. If $r$ is any rational number

$$e^r = \exp(\ln e^r) = \exp(r \ln e) = \exp r$$

Let us emphasize the result. For rational $r$, exp $r$ *is identical with $e^r$*. What was introduced in the most abstract way as the inverse of the natural logarithm, which itself was defined by an integral, has turned out to be a simple power.

But what if $r$ is irrational? Here we remind you of a gap in all elementary algebra books. Never are irrational powers defined in anything approaching a rigorous manner. What is meant by $e^{\sqrt{2}}$? You will have a hard time pinning that number down, based on elementary algebra. But we must pin it down if we are to talk of such things as $D_x e^x$. Guided by what we

**Definition of $e$**

Authors choose different ways to define $e$.

**1.** $e = \ln^{-1} 1$ (our definition)

**2.** $e = \lim_{h \to 0} (1 + h)^{1/h}$

**3.** $e = \lim_{n \to \infty} \left( 1 + \dfrac{1}{1!} + \dfrac{1}{2!} + \right.$
$$\left. \cdots + \dfrac{1}{n!} \right)$$

In our book, definitions 2 and 3 become theorems (See Section 7.5, Theorem A, and Section 11.7, Example 3).

learned above, we simply define $e^x$ for all $x$ (rational or irrational) by

$$e^x = \exp x$$

Note that (i) and (ii) at the beginning of this section now take the form:

| | |
|---|---|
| **(i)′** | $e^{\ln x} = x,$     $x > 0$ |
| **(ii)′** | $\ln(e^y) = y,$     all $y$ |

Also, we can easily prove two of the familiar laws of exponents.

### Theorem A

Let $a$ and $b$ be any real numbers. Then $e^a e^b = e^{a+b}$ and $e^a/e^b = e^{a-b}$.

***Proof***   To prove the first, we write

$$
\begin{aligned}
e^a e^b &= \exp(\ln e^a e^b) && \text{(by (i))} \\
&= \exp(\ln e^a + \ln e^b) && \text{(Theorem 7.1A)} \\
&= \exp(a + b) && \text{(by (ii)′)} \\
&= e^{a+b} && \text{(since } \exp x = e^x)
\end{aligned}
$$

The second fact is proved similarly.    ■

**A Phoenix**

The number $e$ appears throughout mathematics but its importance rests most securely on its use as the base for the natural exponential function. And what makes this function so significant?

"Who has not been amazed to learn that the function $y = e^x$, like a phoenix rising again from its own ashes, is its own derivative?"

François Le Lionnais

**The Derivative of $e^x$**   Since exp and ln are inverses, we know from Theorem 7.2B that $\exp x = e^x$ is differentiable. To find a formula for $D_x e^x$, we could use that theorem. Alternatively, let $y = e^x$, so that

$$x = \ln y$$

Now differentiate both sides with respect to $x$, obtaining

$$1 = \frac{1}{y} D_x y \qquad \text{(Chain Rule)}$$

Thus,

$$D_x y = y = e^x$$

We have proved the remarkable fact that $e^x$ is its own derivative; that is,

$$\boxed{D_x e^x = e^x}$$

If $u = f(x)$ is differentiable, then the Chain Rule yields

$$D_x e^u = e^u D_x u$$

**EXAMPLE 1**    Find $D_x e^{\sqrt{x}}$.

***Solution***    Using $u = \sqrt{x}$, we obtain

$$D_x e^{\sqrt{x}} = e^{\sqrt{x}} D_x \sqrt{x} = e^{\sqrt{x}} \cdot \frac{1}{2} x^{-1/2} = \frac{e^{\sqrt{x}}}{2\sqrt{x}}$$    ∎

**EXAMPLE 2**    Find $D_x e^{x^2 \ln x}$.

***Solution***

$$D_x e^{x^2 \ln x} = e^{x^2 \ln x} D_x(x^2 \ln x)$$

$$= e^{x^2 \ln x}\left(x^2 \cdot \frac{1}{x} + 2x \ln x\right)$$

$$= x e^{x^2 \ln x}(1 + \ln x^2)$$    ∎

**EXAMPLE 3**    Let $f(x) = xe^{x/2}$. Find where $f$ is increasing and decreasing and where it is concave upward and downward. Also, identify all extreme values and points of inflection. Then, sketch the graph of $f$.

***Solution***

$$f'(x) = \frac{xe^{x/2}}{2} + e^{x/2} = e^{x/2}\left(\frac{x+2}{2}\right)$$

and

$$f''(x) = \frac{e^{x/2}}{2} + \left(\frac{x+2}{2}\right)\frac{e^{x/2}}{2} = e^{x/2}\left(\frac{x+4}{4}\right)$$

Keeping in mind that $e^{x/2} > 0$ for all $x$, we see that $f'(x) < 0$ for $x < -2$, $f'(-2) = 0$, and $f'(x) > 0$ for all $x > -2$. Thus, $f$ is decreasing on $(-\infty, -2]$, increasing on $[-2, \infty)$, and has its minimum value at $x = -2$ of $f(-2) = -2/e \approx -0.7$.

Also, $f''(x) < 0$ for $x < -4$, $f''(-4) = 0$, and $f''(x) > 0$ for $x > -4$; so the graph of $f$ is concave downward on $(-\infty, -4)$, concave upward on $(-4, \infty)$, and has a point of inflection at $(-4, -4e^{-2}) \approx (-4, -0.54)$. This information, plus a few plotted points, yields the graph in Figure 3.    ∎

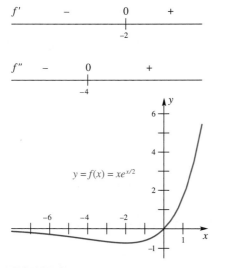

FIGURE 3

The derivative formula $D_x e^x = e^x$ automatically yields the integral formula $\int e^x\, dx = e^x + C$, or, with $u$ replacing $x$,

$$\int e^u\, du = e^u + C$$

**EXAMPLE 4**  Find $\int e^{-4x}\, dx$

*Solution*  Let $u = -4x$, so $du = -4\, dx$. Then,

$$\int e^{-4x}\, dx = -\tfrac{1}{4} \int e^{-4x}(-4\, dx) = -\tfrac{1}{4} \int e^u\, du = -\tfrac{1}{4} e^u + C$$

$$= -\tfrac{1}{4} e^{-4x} + C \qquad\blacksquare$$

**EXAMPLE 5**  Find $\int x^2 e^{-x^3}\, dx$.

*Solution*  Let $u = -x^3$, so $du = -3x^2\, dx$. Then,

$$\int x^2 e^{-x^3}\, dx = -\tfrac{1}{3} \int e^{-x^3}(-3x^2\, dx)$$

$$= -\tfrac{1}{3} \int e^u\, du = -\tfrac{1}{3} e^u + C$$

$$= -\tfrac{1}{3} e^{-x^3} + C \qquad\blacksquare$$

**EXAMPLE 6**  Evaluate $\int_1^3 xe^{-3x^2}\, dx$.

*Solution*  Let $u = -3x^2$, so $du = -6x\, dx$. Then,

$$\int xe^{-3x^2}\, dx = -\tfrac{1}{6} \int e^{-3x^2}(-6x\, dx) = -\tfrac{1}{6} \int e^u\, du$$

$$= -\tfrac{1}{6} e^u + C = -\tfrac{1}{6} e^{-3x^2} + C$$

Thus, by the Fundamental Theorem of Calculus,

$$\int_1^3 xe^{-3x^2}\, dx = \left[ -\frac{1}{6} e^{-3x^2} \right]_1^3 = -\frac{1}{6}(e^{-27} - e^{-3})$$

$$= \frac{e^{-3} - e^{-27}}{6} \approx 0.0082978$$

The last result can be obtained by using Table IV or a calculator. $\qquad\blacksquare$

Although the symbol $e^y$ will largely supplant exp $y$ throughout the rest of this book, exp occurs frequently in scientific writing, especially when the exponent $y$ is complicated. For example, in the study of statistics, one often encounters the normal curve, which is the graph of

$$f(x) = \frac{1}{\sigma \sqrt{2\pi}} \exp\left[ -\frac{(x - \mu)^2}{2\sigma^2} \right]$$

## CONCEPTS REVIEW

1. The function ln is strictly _____ on $(0, \infty)$ and so has an inverse denoted by $\ln^{-1}$ or by _____.

2. The number $e$ is defined in terms of ln by _____; its value to two decimal places is _____.

3. Since $e^x = \exp x = \ln^{-1}x$, it follows that $e^{\ln x} =$ _____ and $\ln(e^x) =$ _____.

4. Two remarkable facts about $e^x$ are that $D_x(e^x) =$ _____ and $\int e^x \, dx =$ _____.

## PROBLEM SET 7.3

[C] 1. Use your calculator or Tables II–IV of the appendix to calculate each of the following. *Note*: On some calculators there is an $\boxed{e^x}$ button. On others you must press the $\boxed{\text{INV}}$ and $\boxed{\ln x}$ buttons.

(a) $e^3$      (b) $e^{-4}$
(c) $e^{2.1}$      (d) $e^{6.3}$
(e) $e^{\sqrt{2}}$      (f) $e^{\sin 4}$
(g) $e^{\cos(\ln 4)}$

[C] 2. Calculate

(a) $e^{3 \ln 2}$      (b) $e^{(\ln 64)/2}$

Explain why your answers are not surprising.

In Problems 3–10, simplify the given expression.

3. $e^{3 \ln x}$      4. $e^{-2 \ln x}$

5. $\ln e^{\cos x}$      6. $\ln e^{-2x - 3}$

7. $\ln (x^3 e^{-3x})$      8. $e^{x - \ln x}$

9. $e^{\ln 3 + 2 \ln x}$      10. $e^{\ln x^2 - y \ln x}$

In Problems 11–22, find $D_x y$ (see Examples 1 and 2).

11. $y = e^{x + 2}$      12. $y = e^{2x^2 - x}$

13. $y = e^{\sqrt{x + 2}}$      14. $y = e^{-1/x^2}$

15. $y = e^{2 \ln x}$      16. $y = e^{x/\ln x}$

17. $y = x^3 e^x$      18. $y = e^{x^3 \ln x}$

19. $y = \sqrt{e^{x^2}} + e^{\sqrt{x^2}}$      20. $y = e^{1/x^2} + 1/e^{x^2}$

21. $e^{xy} + xy = 2$ *Hint*: Use implicit differentiation.

22. $e^{x + y} = x + y$

23. Use your knowledge of the graph of $y = e^x$ to sketch the graphs of (a) $y = -e^x$ and (b) $y = e^{-x}$.

24. Explain why $a < b \Rightarrow e^{-a} > e^{-b}$.

In Problems 25–28, discuss and sketch the graph of the given function, as in Example 3.

25. $f(x) = xe^{-x}$      26. $f(x) = e^x + x$

27. $f(x) = e^{-(x - 2)^2}$      28. $f(x) = e^x - e^{-x}$

In Problems 29–36, find the integrals (see Examples 4–6).

29. $\displaystyle\int e^{3x + 1} \, dx$      30. $\displaystyle\int xe^{x^2 - 3} \, dx$

31. $\displaystyle\int (x + 3)e^{x^2 + 6x} \, dx$      32. $\displaystyle\int \frac{e^x}{e^x - 1} \, dx$

33. $\displaystyle\int \frac{e^{-1/x}}{x^2} \, dx$      34. $\displaystyle\int e^{x + e^x} \, dx$

35. $\displaystyle\int_0^1 e^{2x + 3} \, dx$      36. $\displaystyle\int_1^2 \frac{e^{3/x}}{x^2} \, dx$

37. Find the volume of the solid generated by revolving the region bounded by $y = e^x$, $y = 0$, $x = 0$, and $x = \ln 3$ about the $x$-axis.

38. The region bounded by $y = e^{-x^2}$, $y = 0$, $x = 0$, and $x = 1$ is revolved about the $y$-axis. Find the volume of the resulting solid.

39. Find the area of the region bounded by the curve $y = e^{-x}$ and the line through the points $(0, 1)$ and $(1, 1/e)$.

40. Show that $f(x) = \dfrac{x}{e^x - 1} - \ln(1 - e^{-x})$ is decreasing for $x > 0$.

[C] 41. Stirling's Formula says that for large $n$, we can approximate $n! = 1 \cdot 2 \cdot 3 \cdots n$ by

$$n! \approx \sqrt{2\pi n} \left(\frac{n}{e}\right)^n$$

(a) Calculate 10! exactly and then approximately using the above formula.
(b) Approximate 60!.

[C] 42. It will be shown later (Section 10.1) that for small $x$,

$$e^x \approx 1 + x + \frac{x^2}{2!} + \frac{x^3}{3!} + \frac{x^4}{4!}$$

$$= \left\{ \left[ \left(\frac{x}{4} + 1\right)\frac{x}{3} + 1\right]\frac{x}{2} + 1 \right\} x + 1$$

Use this result to calculate $e^{0.3}$ and compare your result with what you get by calculating it directly.

**43.** Find the length of the curve given parametrically by $x = e^t \sin t$, $y = e^t \cos t$, $0 \le t \le \pi$.

⟦C⟧ **44.** If customers arrive at a check-out counter at the average rate of $k$ per minute, then (see books on probability theory) the probability that exactly $n$ customers will arrive in a period of $x$ minutes is given by the formula

$$P_n(x) = \frac{(kx)^n e^{-kx}}{n!}$$

Find the probability that exactly 8 customers will arrive during a 30-minute period if the average rate for this checkout counter is 1 customer every 4 minutes.

**45.** Let $f(x) = \dfrac{\ln x}{1 + (\ln x)^2}$ for $x$ in $(0, \infty)$. Find:

(a) $\lim\limits_{x \to 0^+} f(x)$ and $\lim\limits_{x \to \infty} f(x)$;

(b) the maximum and minimum values of $f(x)$;

(c) $F'(\sqrt{e})$ if $F(x) = \displaystyle\int_1^{x^2} f(t)\, dt$.

**46.** Let $R$ be the region bounded by $x = 0$, $y = e^x$, and the tangent line to $y = e^x$ that goes through the origin. Find:

(a) the area of $R$;

(b) the volume of the solid obtained when $R$ is revolved around the $x$-axis.

**47.** Evaluate $\lim\limits_{n \to \infty} \dfrac{e^{1/n} + e^{2/n} + \cdots + e^{n/n}}{n}$.

**48.** The **normal curve** with mean $\mu$ and standard deviation $\sigma$ is defined by

$$y = f(x) = \frac{1}{\sigma\sqrt{2\pi}} \exp\left[ -\frac{1}{2} \left( \frac{x - \mu}{\sigma} \right)^2 \right]$$

Show that:

(a) its graph is symmetric about the line $x = \mu$;

(b) it has a maximum at $x = \mu$ and inflection points at $x = \mu \pm \sigma$.

⟦PC⟧ Use a computer to do Problems 49–53.

**49.** Evaluate.

(a) $\displaystyle\int_{-3}^3 \exp(-1/x^2)\, dx$

(b) $\displaystyle\int_0^{8\pi} e^{-0.1x} \sin x\, dx$

**50.** Evaluate.

(a) $\lim\limits_{x \to 0} (1 + x)^{1/x}$

(b) $\lim\limits_{x \to 0} (1 + x)^{-1/x}$

**51.** Find the area of the region between the graphs of $y = f(x) = \exp(-x^2)$ and $y = f''(x)$ on $[-3, 3]$.

**52.** Draw the graphs of $y = x^p e^{-x}$ for various positive values of $p$ using the same axes. Make conjectures about:

(a) $\lim\limits_{x \to \infty} x^p e^{-x}$;

(b) the $x$-coordinate of the maximum point for $f(x) = x^p e^{-x}$.

**53.** Describe the behavior of $\ln(x^2 + e^{-x})$ for large negative $x$. For large positive $x$.

**54.** Draw the graphs of $f$ and $f'$, where $f(x) = 1/(1 + e^{1/x})$. Then determine each of the following.

(a) $\lim\limits_{x \to 0^+} f(x)$

(b) $\lim\limits_{x \to 0^-} f(x)$

(c) $\lim\limits_{x \to \pm\infty} f(x)$

(d) $\lim\limits_{x \to 0} f'(x)$

(e) The maximum and minimum values of $f$ (if they exist).

---

**Answers to Concepts Review:   1.** Increasing; exp
**2.** $\ln e = 1$; 2.72   **3.** $x$; $x$   **4.** $e^x$; $e^x + C$

---

| **7.4** | We defined $e^{\sqrt{2}}$, $e^\pi$, and all other irrational powers of $e$ in the previous section. |
|---|---|
| **GENERAL EXPONENTIAL AND LOGARITHMIC FUNCTIONS** | But what about $2^{\sqrt{2}}$, $\pi^\pi$, $\pi^e$, $\sqrt{2}^\pi$, and similar irrational powers of other numbers? In fact, we want to give meaning to $a^x$ for $a > 0$ and $x$ any real number. Now if $r = p/q$ is a rational number, $a^r = (\sqrt[q]{a})^p$. But we also know that |

$$a^r = \exp(\ln a^r) = \exp(r \ln a) = e^{r \ln a}$$

This suggests a definition.

**What is $2^\pi$?**

When asked this equation, one student wrote

$$2^\pi = 2 \cdot 2 \cdot 2 \cdot 2$$

which is nonsense. Another responded, $2^\pi$ is the limit of the sequence

$$2^3, \; 2^{3.1}, \; 2^{3.14}, \; 2^{3.141}, \ldots$$

which is a correct but complicated notion. For us,

$$2^\pi = e^{\pi \ln 2}$$

Although calculators vary, it is likely that your calculator uses this formula in finding $2^\pi$. Ours gives 8.8249778 as the rounded 8-digit value.

> **Definition**
>
> For $a > 0$ and any real number $x$,
>
> $$a^x = e^{x \ln a}$$

Of course, this definition will be appropriate only if the usual properties of exponents are valid for it, a matter we take up shortly. To shore up our confidence in the definition, we use it to calculate $3^2$ (with a little help from our calculator).

$$3^2 = e^{2 \ln 3} \approx e^{2(1.0986123)} \approx 9.0000002$$

The slight discrepancy is due to the round-off characteristics of our calculator.

Now we can fill a small gap in the properties of the natural logarithm left over from Section 7.1.

$$\ln(a^x) = \ln(e^{x \ln a}) = x \ln a$$

Thus, Property (iv) of Theorem 7.1A holds for all real $x$, not just rational $x$ as claimed there. We will need this fact in the proof of Theorem A below.

**Properties of $a^x$**   Theorem A summarizes the familiar properties of exponents, which can all be proved now in a completely rigorous manner. Theorem B shows us how to differentiate and integrate $a^x$.

> **Theorem A**
>
> If $a > 0$, $b > 0$, and $x$ and $y$ are real numbers,
> (i)   $a^x a^y = a^{x+y}$;
> (ii)   $\dfrac{a^x}{a^y} = a^{x-y}$;
> (iii)   $(a^x)^y = a^{xy}$;
> (iv)   $(ab)^x = a^x b^x$;
> (v)   $\left(\dfrac{a}{b}\right)^x = \dfrac{a^x}{b^x}$.

***Proof***   We content ourselves with proving (ii) and (iii), leaving the others for you.

(ii)   $\dfrac{a^x}{a^y} = e^{\ln(a^x/a^y)} = e^{\ln a^x - \ln a^y}$

$$= e^{x \ln a - y \ln a} = e^{(x-y)\ln a} = a^{x-y}$$

(iii)   $(a^x)^y = e^{y \ln a^x} = e^{yx \ln a} = a^{yx} = a^{xy}$  ∎

> **Theorem B**
>
> $$D_x a^x = a^x \ln a$$
>
> $$\int a^x \, dx = \left(\frac{1}{\ln a}\right) a^x + C, \qquad a \neq 1$$

***Proof***

$$D_x a^x = D_x(e^{x \ln a}) = e^{x \ln a} D_x(x \ln a)$$
$$= a^x \ln a$$

The integral formula follows immediately from the derivative formula. ∎

**EXAMPLE 1** Find $D_x(3^{\sqrt{x}})$.

***Solution*** We use the Chain Rule with $u = \sqrt{x}$.

$$D_x(3^{\sqrt{x}}) = 3^{\sqrt{x}} \ln 3 \cdot D_x \sqrt{x} = \frac{3^{\sqrt{x}} \ln 3}{2\sqrt{x}}$$  ∎

**EXAMPLE 2** Find $dy/dx$ if $y = (x^4 + 2)^5 + 5^{x^4 + 2}$.

***Solution***

$$\frac{dy}{dx} = 5(x^4 + 2)^4 \cdot 4x^3 + 5^{x^4 + 2} \ln 5 \cdot 4x^3$$
$$= 4x^3[5(x^4 + 2)^4 + 5^{x^4 + 2} \ln 5]$$
$$= 20x^3[(x^4 + 2)^4 + 5^{x^4 + 1} \ln 5]$$  ∎

**EXAMPLE 3** Find $\displaystyle\int 2^{x^3} x^2 \, dx$.

***Solution*** Let $u = x^3$, so $du = 3x^2 \, dx$. Then

$$\int 2^{x^3} x^2 \, dx = \frac{1}{3} \int 2^{x^3} (3x^2 \, dx) = \frac{1}{3} \int 2^u \, du$$
$$= \frac{1}{3} \frac{2^u}{\ln 2} + C = \frac{2^{x^3}}{3 \ln 2} + C$$  ∎

---

**Why Other Bases?**

Are bases other than $e$ really needed? No. The formulas

$$a^x = e^{x \ln a}$$

and

$$\log_a x = \frac{\ln x}{\ln a}$$

allow us to turn any problem involving exponential functions or logarithmic functions with base $a$ to corresponding functions with base $e$. This supports our terminology—natural exponential and natural logarithmic functions. It also explains the universal use of the latter functions in advanced work.

---

**The Function $\log_a$** Finally, we are ready to make a connection with the logarithms you studied in algebra. We note that if $0 < a < 1$, $f(x) = a^x$ is a decreasing function; if $a > 1$, it is an increasing function, as you may check by considering the derivative. In either case, $f$ has an inverse. We call this inverse the **logarithmic function to the base $a$**. This is equivalent to the following definition.

> **Definition**
>
> Let $a$ be a positive number different from 1. Then
>
> $$y = \log_a x \iff x = a^y$$

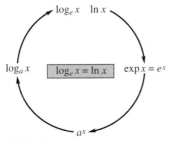

FIGURE 1

Historically, the most commonly used base was base 10 and the resulting logarithms were called **common logarithms**. But in calculus and all of advanced mathematics, the significant base is $e$. Notice that $\log_e$, being the inverse of $f(x) = e^x$, is just another symbol for ln; that is,

$$\log_e x = \ln x$$

We have come full circle (Figure 1). The function ln, which we introduced in Section 7.1, has turned out to be an ordinary logarithm but to a rather surprising base, $e$.

Now observe that if $y = \log_a x$ so $x = a^y$, then

$$\ln x = y \ln a$$

from which we conclude that

$$\boxed{\log_a x = \frac{\ln x}{\ln a}}$$

From this, it follows that $\log_a$ satisfies the usual properties associated with logarithms (see Theorem 7.1A). Also,

$$\boxed{D_x \log_a x = \frac{1}{x \ln a}}$$

**EXAMPLE 4** If $y = \log_{10}(x^4 + 13)$, find $\dfrac{dy}{dx}$.

***Solution*** Let $u = x^4 + 13$ and apply the Chain Rule.

$$\frac{dy}{dx} = \frac{1}{(x^4 + 13)\ln 10} \cdot 4x^3 = \frac{4x^3}{(x^4 + 13)\ln 10}$$ ∎

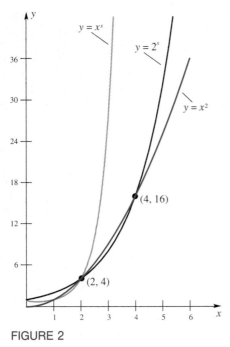

FIGURE 2

**The Functions $a^x$, $x^a$, and $x^x$** Begin by comparing the three graphs in Figure 2. More generally, let $a$ be a constant. Do not confuse $f(x) = a^x$, an **exponential function**, with $g(x) = x^a$, a **power function**. And do not confuse their derivatives. We have just learned that

$$D_x(a^x) = a^x \ln a$$

What about $D_x(x^a)$? For $a$ rational, we proved in Chapter 3 that

$$D_x(x^a) = ax^{a-1}$$

Now we assert that this is true even if $a$ is irrational. To see this, write

$$D_x(x^a) = D_x(e^{a \ln x}) = e^{a \ln x} \cdot \frac{a}{x}$$

$$= x^a \cdot \frac{a}{x} = ax^{a-1}$$

Finally, we consider $f(x) = x^x$, a variable to a variable power. There is a formula for $D_x(x^x)$, but we do not recommend that you memorize it. Rather, we suggest that you learn two methods for finding it, as illustrated below.

**EXAMPLE 5**   If $y = x^x$, $x > 0$, find $D_x y$ by two different methods.

*Solution*

METHOD 1   We may write

$$y = x^x = e^{x \ln x}$$

Thus, by the Chain Rule,

$$D_x y = e^{x \ln x} D_x(x \ln x) = x^x \left( x \cdot \frac{1}{x} + \ln x \right) = x^x(1 + \ln x)$$

METHOD 2   Recall the *logarithmic differentiation* technique from Section 7.1.

$$y = x^x$$

$$\ln y = x \ln x$$

$$\frac{1}{y} D_x y = x \cdot \frac{1}{x} + \ln x$$

$$D_x y = y(1 + \ln x) = x^x(1 + \ln x) \quad \blacksquare$$

**EXAMPLE 6**   If $y = (x^2 + 1)^\pi + \pi^{\sin x}$, find $dy/dx$.

*Solution*

$$\frac{dy}{dx} = \pi(x^2 + 1)^{\pi - 1}(2x) + \pi^{\sin x} \ln \pi \cdot \cos x \quad \blacksquare$$

**From $a^x$ to $[f(x)]^{g(x)}$**

Note the increasing complexity of the functions we have considered. The progression $a^x$ to $x^a$ to $x^x$ is one chain. A more complex chain is $a^{f(x)}$ to $[f(x)]^a$ to $[f(x)]^{g(x)}$. We now know how to find the derivatives of all these functions. Finding the derivative of the last of these is best accomplished by logarithmic differentiation, a technique introduced in Section 7.1 and illustrated in Examples 5 and 7.

**EXAMPLE 7** If $y = (x^2 + 1)^{\sin x}$, find $\dfrac{dy}{dx}$.

**Solution** We use logarithmic differentiation.

$$\ln y = \sin x \, \ln(x^2 + 1)$$

$$\frac{1}{y}\frac{dy}{dx} = (\sin x)\frac{2x}{x^2 + 1} + (\cos x)\ln(x^2 + 1)$$

$$\frac{dy}{dx} = (x^2 + 1)^{\sin x}\left[\frac{2x \sin x}{x^2 + 1} + (\cos x)\ln(x^2 + 1)\right]$$

∎

**EXAMPLE 8** Evaluate $\displaystyle\int_{1/2}^{1} \frac{5^{1/x}}{x^2} \, dx$.

**Solution** Let $u = 1/x$, so $du = (-1/x^2)\,dx$. Then

$$\int \frac{5^{1/x}}{x^2}\,dx = -\int 5^{1/x}\left(-\frac{1}{x^2}\,dx\right) = -\int 5^u\,du$$

$$= -\frac{5^u}{\ln 5} + C = -\frac{5^{1/x}}{\ln 5} + C$$

Thus, by the Fundamental Theorem of Calculus,

$$\int_{1/2}^{1} \frac{5^{1/x}}{x^2}\,dx = \left[-\frac{5^{1/x}}{\ln 5}\right]_{1/2}^{1} = \frac{1}{\ln 5}(5^2 - 5)$$

$$= \frac{20}{\ln 5} \approx 12.426699$$

∎

## CONCEPTS REVIEW

**1.** In terms of $e$ and $\ln$, $\pi^{\sqrt{3}} = $ _____. More generally $a^x = $ _____.

**2.** $\ln x = \log_a x$, where $a = $ _____.

**3.** $\log_a x$ can be expressed in terms of $\ln$ by $\log_a x = $ _____.

**4.** The derivative of the power function $f(x) = x^a$ is $f'(x) = $ _____; the derivative of the exponential function $g(x) = a^x$ is $g'(x) = $ _____.

## PROBLEM SET 7.4

In Problems 1–8, solve for $x$. *Hint*: $\log_a b = c \iff a^c = b$.

**1.** $\log_2 8 = x$      **2.** $\log_5 x = 2$

**3.** $\log_4 x = \frac{3}{2}$      **4.** $\log_x 64 = 4$

**5.** $2\log_9\left(\dfrac{x}{3}\right) = 1$      **6.** $\log_4\left(\dfrac{1}{2x}\right) = 3$

**7.** $\log_2(x + 3) - \log_2 x = 2$

**8.** $\log_5(x + 3) - \log_5 x = 1$

[C] Use your calculator (or Table III) and $\log_a x = (\ln x)/(\ln a)$ to calculate each of the logarithms in Problems 9–12.

**9.** $\log_5 12$      **10.** $\log_7(0.11)$

**11.** $\log_{11}(8.12)^{1/5}$      **12.** $\log_{10}(8.57)^7$

C In Problems 13–16, use natural logarithms to solve each of the exponential equations. *Hint*: To solve $3^x = 11$, take ln of both sides, obtaining $x \ln 3 = \ln 11$; then $x = (\ln 11)/(\ln 3) \approx 2.1827$.

**13.** $2^x = 17$      **14.** $5^z = 13$

**15.** $5^{2s-3} = 4$      **16.** $12^{1/(\theta-1)} = 4$

Find the indicated derivative or integral (see Examples 1–4).

**17.** $D_x(6^{2x})$      **18.** $D_x(3^{2x^2-3x})$

**19.** $D_x \log_3 e^x$      **20.** $D_x \log_{10}(x^3+9)$

**21.** $D_z[3^z \ln(z+5)]$      **22.** $D_\theta \sqrt{\log_{10}(3^{\theta^2-\theta})}$

**23.** $\int x 2^{x^2}\, dx$      **24.** $\int 10^{5x-1}\, dx$

**25.** $\int_1^4 \frac{5^{\sqrt{x}}}{\sqrt{x}}\, dx$      **26.** $\int_0^1 (10^{3x}+10^{-3x})\, dx$

In Problems 27–32, find $dy/dx$. Be sure to distinguish between problems of the type $a^x$, $x^a$, and $x^x$ as in Examples 5–7.

**27.** $y = 10^{(x^2)} + (x^2)^{10}$      **28.** $y = \sin^2 x + 2^{\sin x}$

**29.** $y = x^{\pi+1} + (\pi+1)^x$      **30.** $y = 2^{(e^x)} + (2^e)^x$

**31.** $y = (x^2+1)^{\ln x}$      **32.** $y = (\ln x^2)^{2x+3}$

**33.** If $f(x) = x^{\sin x}$, find $f'(1)$.

C **34.** Let $f(x) = \pi^x$ and $g(x) = x^\pi$. Which is larger, $f(e)$ or $g(e)$? $f'(e)$ or $g'(e)$?

**35.** How are $\log_{1/2} x$ and $\log_2 x$ related?

**36.** Sketch the graphs of $\log_{1/3} x$ and $\log_3 x$ using the same coordinate axes.

C **37.** The magnitude $M$ of an earthquake on the Richter scale is

$$M = 0.67 \log_{10}(0.37E) + 1.46$$

where $E$ is the energy of the earthquake in kilowatt-hours. Find the energy of an earthquake of magnitude 7. Of magnitude 8.

C **38.** The loudness of sound is measured in decibels in honor of Alexander Graham Bell (1847–1922), inventor of the telephone. If the variation in pressure is $P$ pounds per square inch, then the loudness $L$ in decibels is

$$L = 20 \log_{10}(121.3P)$$

Find the variation in pressure caused by a rock band at 115 decibels.

C **39.** In the equally tempered scale to which keyed instruments have been tuned since the days of J. S. Bach (1685–

1750), the frequencies of successive notes C, C#, D, D#, E, F, F#, G, G#, A, A#, B, $\overline{C}$ form a geometric sequence (progression), with $\overline{C}$ having twice the frequency of C. What is the ratio $r$ between the frequencies of successive notes? If the frequency of $A$ is 440, find the frequency of $\overline{C}$.

**40.** Prove that $\log_2 3$ is irrational. *Hint*: Use proof by contradiction.

**41.** You are suspicious that the $xy$-data you have collected lie on either an exponential curve, $y = A \cdot b^x$, or a power curve, $y = A \cdot x^b$. To check, you plot $\ln y$ against $x$ (using semilog graph paper) and also $\ln y$ against $\ln x$ (using log log graph paper). Explain how this will help you come to a conclusion.

**42.** (An Amusement) Given the problem of finding $y'$ if $y = x^x$, student A did the following:

$$
\begin{array}{ll}
& y = x^x \\
\text{WRONG 1} \quad & y' = x \cdot x^{x-1} \cdot 1 \qquad \left(\begin{array}{l}\text{misapplying the}\\ \text{Power Rule}\end{array}\right)\\
& = x^x
\end{array}
$$

Student B did this:

$$
\begin{array}{ll}
& y = x^x \\
\text{WRONG 2} \quad & y' = x^x \cdot \ln x \cdot 1 \qquad \left(\begin{array}{l}\text{misapplying the}\\ \text{Exponential}\\ \text{Function Rule}\end{array}\right)\\
& = x^x \ln x
\end{array}
$$

The sum $x^x + x^x \ln x$ is correct (Example 5), so

$$\text{WRONG 1} + \text{WRONG 2} = \text{RIGHT}$$

Show that the same procedure yields a correct answer for $y = f(x)^{g(x)}$.

**43.** Convince yourself that $f(x) = (x^x)^x$ and $g(x) = x^{(x^x)}$ are not the same function. Then find $f'(x)$ and $g'(x)$. *Note*: When mathematicians write $x^{x^x}$, they mean $x^{(x^x)}$.

**44.** Consider $f(x) = \dfrac{a^x - 1}{a^x + 1}$ for fixed $a$, $a > 0$, $a \neq 1$. Show that $f$ has an inverse and find a formula for $f^{-1}(x)$.

**45.** For fixed $a > 1$, let $f(x) = x^a/a^x$ on $[0, \infty)$. Show:
(a) $\lim_{x \to \infty} f(x) = 0$ (study $\ln f(x)$);
(b) $f(x)$ is maximized at $x_0 = a/\ln a$;
(c) $x^a = a^x$ has two positive solutions if $a \neq e$, only one such solution if $a = e$;
(d) $\pi^e < e^\pi$.

**46.** Let $f_u(x) = x^u e^{-x}$ for $x \geq 0$. Show that for any fixed $u > 0$:

(a) $f_u(x)$ attains its maximum at $x_0 = u$;
(b) $f_u(u) > f_u(u+1)$ and $f_{u+1}(u+1) > f_{u+1}(u)$ imply

$$\left(\frac{u+1}{u}\right)^u < e < \left(\frac{u+1}{u}\right)^{u+1}$$

(c) $\dfrac{u}{u+1}\, e < \left(\dfrac{u+1}{u}\right)^{u} < e.$

Conclude from (c) that $\lim\limits_{u \to \infty} \left(1 + \dfrac{1}{u}\right)^{u} = e.$

PC **47.** Find $\lim\limits_{x \to 0^{+}} x^{x}$. Also find the coordinates of the minimum point for $f(x) = x^{x}$ on $[0, 4]$.

PC **48.** Draw the graphs of $y = x^{3}$ and $y = 3^{x}$ using the same axes and find all their intersection points.

PC **49.** Evaluate $\displaystyle\int_{0}^{4\pi} x^{\sin x}\, dx.$

PC **50.** Refer to Problem 43. Draw the graphs of $f$ and $g$ using the same axes. Then draw the graphs of $f'$ and $g'$ using the same axes.

---

**Answers to Concepts Review:** **1.** $e^{\sqrt{3}\ln \pi}$; $e^{x \ln a}$ **2.** $e$ **3.** $(\ln x)/(\ln a)$ **4.** $ax^{a-1}$; $a^{x} \ln a$

---

**7.5
EXPONENTIAL
GROWTH AND DECAY**

At the beginning of 1987, the world's population was about 5 billion. It is said that by the year 2010, it will reach 7.7 billion. How are such predictions made?

To treat the problem mathematically, let $y = f(t)$ denote the size of the population at time $t$, where $t$ is the number of years after 1987. Actually $f(t)$ is an integer, and its graph "jumps" when someone is born or dies. However, for a large population, these jumps are so small relative to the total population that we will not go far wrong if we pretend that $f$ is a nice differentiable function.

It seems reasonable to suppose that the increase $y$ in population (births minus deaths) during a short time period $t$ is proportional to the size of the population at the beginning of the period and to the length of that period. Thus, $y = ky\ t$, or

$$\frac{y}{t} = ky$$

In its limiting form, this gives the differential equation

$$\boxed{\dfrac{dy}{dt} = ky}$$

If $k > 0$ the population is growing, if $k < 0$, it is shrinking. For world population, history indicates that $k$ is about 0.019 (assuming that $t$ is measured in years), though some statisticians report a considerably lower figure.

**Solving the Differential Equation** We began our study of differential equations in Section 5.2, and you might refer to that section now. We want to solve $dy/dt = ky$ subject to the condition that $y = y_0$ when $t = 0$. Separating variables and integrating, we obtain

$$\frac{dy}{y} = k\,dt$$

$$\int \frac{dy}{y} = \int k\,dt$$

$$\ln y = kt + C$$

The condition $y = y_0$ at $t = 0$ gives $C = \ln y_0$. Thus,

$$\ln y - \ln y_0 = kt$$

or

$$\ln \frac{y}{y_0} = kt$$

Changing to exponential form yields

$$\frac{y}{y_0} = e^{kt}$$

or, finally,

$$\boxed{y = y_0 e^{kt}}$$

Returning to the problem of world population, we choose to measure time $t$ in years after January 1, 1987, and $y$ in billions of people. Thus, $y_0 = 5$ and, since $k = 0.019$,

$$y = 5e^{0.019t}$$

By the year 2010, when $t = 23$, we can predict that $y$ will be about

$$y = 5e^{0.019(23)} \approx 7.7 \text{ billion}$$

Naturally, we used our calculator to make the last calculation.

EXAMPLE 1    How long will it take world population to double under the assumptions above?

*Solution*    The question is equivalent to asking how many years after 1987 it will take for the population to reach 10 billion. We need to solve

$$10 = 5e^{0.019t}$$

for $t$. After dividing both sides by 5, we take logarithms.

$$\ln 2 = 0.019t$$
$$t = \frac{\ln 2}{0.019} \approx 36 \text{ years} \qquad \blacksquare$$

If world population will double in the first 36 years after 1987, it will double in any 36-year period; so, for example, it will quadruple in 72 years. More generally, if an exponentially growing quantity doubles in an initial interval of length $T$, it will double in *any* interval of length $T$, since

$$\frac{y(t + T)}{y(t)} = \frac{y_0 e^{k(t + T)}}{y_0 e^{kt}} = \frac{y_0 e^{kT}}{y_0} = 2$$

We call the number $T$ the **doubling time**.

**EXAMPLE 2**   The number of bacteria in a rapidly growing culture was estimated to be 10,000 at noon and 40,000 after 2 hours. How many bacteria will there be at 5 P.M.?

*Solution*   We assume that the differential equation $dy/dt = ky$ is applicable, so $y = y_0 e^{kt}$. Now we have two conditions ($y_0 = 10,000$ and $y = 40,000$ at $t = 2$), from which we conclude that

$$40,000 = 10,000 e^{k(2)}$$

or

$$4 = e^{2k}$$

Taking logarithms yields

$$\ln 4 = 2k$$

or

$$k = \tfrac{1}{2} \ln 4 = \ln \sqrt{4} = \ln 2 \approx 0.693$$

Thus,

$$y = 10,000 e^{0.693t}$$

and, at $t = 5$, this gives

$$y = 10,000 e^{0.693(5)} \approx 320,000 \qquad \blacksquare$$

The exponential model for population growth is flawed since it projects faster and faster growth indefinitely far into the future (Figure 1). In most cases (including that of world population), the limited amount of space and resources will eventually force a slowing of the growth rate. This suggests another model for population growth, called the **logistic model**, in which we assume the rate of growth is proportional both to the population size $y$ and to the difference $L - y$, where $L$ is the maximum population that can be supported. This leads to the differential equation

$$\frac{dy}{dt} = ky(L - y)$$

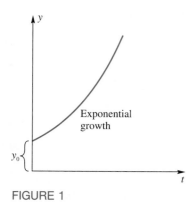

FIGURE 1

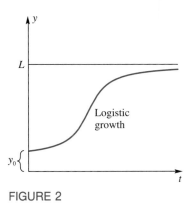

FIGURE 2

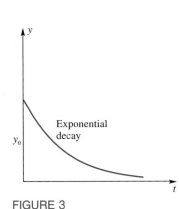

FIGURE 3

Shroud of Turin

Carbon dating was used in 1988 to show that the Shroud of Turin was made between A.D. 1260 and 1390. This seems to refute the belief that it is the burial cloth of Jesus Christ.

Note that for small $y$, $dy/dt \approx kLy$ which gives exponential type growth. But as $y$ nears $L$, $dy/dt$ gets smaller and smaller, curtailing the growth and producing a growth curve like Figure 2. This model is explored in Problems 24, 25, and 33 of this section and again in Problems 26 and 27 of Section 8.5.

**Radioactive Decay**   Not everything grows; some things decrease over time. In particular, the radioactive elements decay, and they do it at a rate proportional to the amount present. Thus, their change rates also satisfy the differential equation

$$\frac{dy}{dt} = ky$$

but now with $k$ negative. It is still true that $y = y_0 e^{kt}$ is the solution to this equation. A typical graph appears in Figure 3.

**EXAMPLE 3**   Carbon-14, one of the three isotopes of carbon, is radioactive and decays at a rate proportional to the amount present. Its **half-life** is 5730 years; that is, it takes 5730 years for a given amount of carbon-14 to decay to one-half its original size. If 10 grams were present originally, how much will be left after 2000 years?

*Solution*   The half-life of 5730 allows us to determine $k$, since it implies that

$$\frac{1}{2} = 1e^{k(5730)}$$

or after taking logarithms,

$$-\ln 2 = 5730k$$

$$k = \frac{-\ln 2}{5730} \approx -0.000121$$

Thus,

$$y = 10e^{-0.000121t}$$

At $t = 2000$, this gives

$$y = 10e^{-0.000121(2000)} \approx 7.85 \; grams \qquad \blacksquare$$

In Problem 13, we show how Example 3 may be used to determine the age of fossils and other once-living things.

**Compound Interest**   If we put $100 in the bank at 12% interest compounded monthly, it will be worth $100(1.01) at the end of 1 month, $100(1.01)^2$ at the end of 2 months, and $100(1.01)^{12}$ at the end of 12 months, or 1 year. More generally if we put $A_0$ dollars in the bank at 100$r$ percent compounded $n$ times per year, it will be worth $A(t)$ dollars at the end of $t$ years, where

$$A(t) = A_0\left(1 + \frac{r}{n}\right)^{nt}$$

**EXAMPLE 4** Suppose that John put $500 in the bank at 13% interest, compounded daily. How much will it be worth at the end of 2 years?

*Solution* Here $r = 0.13$ and $n = 365$, so

$$A = 500\left(1 + \frac{0.13}{365}\right)^{365(2)} \approx \$648.43$$

Of course, we did this calculation on a calculator using the $\boxed{y^x}$ key. ∎

Now let us consider what happens when interest is **compounded continuously**—that is, when $n$, the number of compounding periods in a year, tends to infinity. Then we claim

$$A(t) = \lim_{n \to \infty} A_0\left(1 + \frac{r}{n}\right)^{nt} = A_0 \lim_{n \to \infty}\left[\left(1 + \frac{r}{n}\right)^{n/r}\right]^{rt}$$
$$= A_0 \left[\lim_{h \to 0}(1 + h)^{1/h}\right]^{rt} = A_0 e^{rt}$$

Here we replaced $r/n$ by $h$ and noted that $n \to \infty$ corresponds to $h \to 0$. But the big step is knowing that the expression in brackets is the number $e$. This result is important enough to be called a theorem.

**Theorem A**

$$\lim_{h \to 0}(1 + h)^{1/h} = e$$

**Another View of Continuity**

Recall that to say a function is continuous at $x_0$ means that
$$\lim_{x \to x_0} f(x) = f(x_0)$$
That is,
$$\lim_{x \to x_0} f(x) = f\left(\lim_{x \to x_0} x\right)$$
Thus, continuity for a function means that we can pass a limit inside the function. This is what we did for the function $f(x) = \exp(x)$ near the end of the proof of Theorem A.

*Proof* First recall that if $f(x) = \ln x$, then $f'(x) = 1/x$ and, in particular, $f'(1) = 1$. Then from the definition of derivative and properties of ln, we get

$$1 = f'(1) = \lim_{h \to 0} \frac{f(1 + h) - f(1)}{h} = \lim_{h \to 0} \frac{\ln(1 + h) - \ln 1}{h}$$
$$= \lim_{h \to 0} \frac{1}{h} \ln(1 + h) = \lim_{h \to 0} \ln(1 + h)^{1/h}$$

Thus, $\lim_{h \to 0} \ln(1 + h)^{1/h} = 1$, a result we will use in a moment. Now, $g(x) = e^x = \exp x$ is a continuous function, and it therefore follows that we can pass the limit inside the exponential function in the following argument.

$$\lim_{h \to 0}(1 + h)^{1/h} = \lim_{h \to 0} \exp[\ln(1 + h)^{1/h}] = \exp\left[\lim_{h \to 0} \ln(1 + h)^{1/h}\right]$$
$$= \exp 1 = e \quad ∎$$

For another proof of Theorem A, see Problem 46 of Section 7.4.

**EXAMPLE 5**   Suppose the bank of Example 4 compounded interest continuously. How much would John then have at the end of 2 years?

*Solution*

$$A(t) = A_0 e^{rt} = 500 e^{(0.13)(2)} \approx \$648.47$$

Note that, though some banks try to get advertising mileage out of offering continuous compounding of interest, the difference in yields between continuous and daily compounding (which many banks offer) is miniscule.   ■

Here is another approach to the problem of continuous compounding of interest. Let $A$ be the value at time $t$ of $A_0$ dollars invested at the interest rate $r$. To say that interest is compounded continuously is to say that the instantaneous rate of change of $A$ with respect to time is $rA$—that is,

$$\frac{dA}{dt} = rA$$

This differential equation was solved at the beginning of the section; its solution is $A = A_0 e^{rt}$.

## CONCEPTS REVIEW

**1.** The rate of change $dy/dt$ of a quantity $y$ growing exponentially satisfies the differential equation $dy/dt = $ _____. In contrast, if $y$ is growing logistically toward an upper bound $L$, $dy/dt = $ _____.

**2.** If a quantity growing exponentially doubles after $T$ years, it will be _____ times as large after $3T$ years.

**3.** The time for an exponentially decaying quantity $y$ to go from size $y_0$ to size $y_0/2$ is called the _____.

**4.** The number $e$ can be expressed as a limit by $e = \lim\limits_{h \to 0}$ _____.

## PROBLEM SET 7.5

In most of these problems, you will need to use a calculator (or Tables III and IV) to complete the calculation. In Problems 1–4, solve the given differential equation subject to the given condition. Note that $y(a)$ denotes the value of $y$ at $t = a$.

**1.** $\dfrac{dy}{dt} = -6y$, $y(0) = 4$   **2.** $\dfrac{dy}{dt} = 6y$, $y(0) = 1$

**3.** $\dfrac{dy}{dt} = 0.005y$, $y(10) = 2$

**4.** $\dfrac{dy}{dt} = -0.003y$, $y(-2) = 3$

**5.** A bacterial population grows at a rate proportional to its size. Initially, it is 10,000 and after 10 days it is 20,000. What is the population after 25 days? See Example 2.

**6.** How long will it take the population of Problem 5 to double? See Example 1.

**7.** How long will it take the population of Problem 5 to triple? See Example 1.

**8.** The population of the United States was 4 million in 1790 and 180 million in 1960. If the rate of growth is assumed proportional to the number present, what estimate would you give for the population in 2020?

**9.** The population of a certain country is growing at 3.2% per year; that is, if it is $A$ at the beginning of a year, it is $1.032A$ at the end of that year. Assuming it is 4.5 million now, what will it be at the end of 1 year? 2 years? 10 years? 100 years?

**10.** Determine the proportionality constant $k$ in $dy/dt = ky$ for Problem 9. Then use $y = 5e^{kt}$ to find the population after 100 years.

**11.** A radioactive substance has a half-life of 700 years. If there were 10 grams initially, how much would be left after 300 years?

**12.** If a radioactive substance loses 15% of its radioactivity in 2 days, what is its half-life?

**13.** (Carbon Dating) All living things contain carbon-12, which is stable, and carbon-14, which is radioactive. While a plant or animal is alive, the ratio of these two isotopes of carbon remains unchanged since the carbon-14 is constantly renewed; after death, no more carbon-14 is absorbed. The half-life of carbon-14 is 5730 years. If charred logs of an old fort showed only 70% of the carbon-14 expected in living matter, when did the fort burn down? Assume the fort burned soon after it was built of freshly sawn logs.

**14.** Human hair from a grave in Africa proved to have only 51% of the carbon-14 of living tissue. When was the body buried? See Problem 13.

**15.** Newton's law of cooling states that the rate at which an object cools is proportional to the difference in temperature between the object and the surrounding medium. Thus, if an object is taken from an oven at 300°F and left to cool in a room at 75°F, its temperature $T$ after $t$ hours will satisfy the differential equation

$$\frac{dT}{dt} = k(T - 75)$$

If the temperature fell to 200°F in $\frac{1}{2}$ hour, what will it be after 3 hours?

**16.** A thermometer registered $-20°C$ outside and then was brought in the house where the temperature was 24°C. After 5 minutes, it registered 0°C. When will it register 20°C? See Problem 15.

**17.** If $375 is put in the bank today, what will it be worth at the end of 2 years if interest is at 9.5% and is compounded as specified?

(a) Annually         (b) Monthly
(c) Daily            (d) Continuously

*Hint:* See Examples 4 and 5.

**18.** Do Problem 17 assuming the interest rate is 14.4%.

**19.** How long does it take money to double in value for the specified interest?

(a) 12% compounded monthly
(b) 12% compounded continuously

**20.** Inflation between 1977 and 1981 ran at about 11.5% per year. On this basis, what would you expect a car that cost $4000 in 1977 to cost in 1981?

**21.** Manhattan Island is said to have been bought from the Indians by Peter Minuit in 1626 for $24. Suppose Minuit had instead put the $24 in the bank at 6% interest compounded continuously. What was that $24 worth in 1992? It would be interesting to compare this result with the actual value of Manhattan Island in 1992.

**22.** If Methuselah's parents had put $100 in the bank for him at birth and left it there, what would Methuselah have had at his death (969 years later) if interest was 8% compounded annually?

**23.** It will be shown later that for small $x$, $\ln(1 + x) \approx x$. Use this fact to show that the doubling time for money invested at $p$ percent compounded annually is about $70/p$ years.

**24.** The equation for logistic growth is

$$\frac{dy}{dt} = ky(L - y)$$

Show that this differential equation has the solution

$$y = \frac{Ly_0}{y_0 + (L - y_0)e^{-Lkt}}$$

*Hint:* $\dfrac{1}{y(L - y)} = \dfrac{1}{Ly} + \dfrac{1}{L(L - y)}.$

**25.** Sketch the graph of the solution in Problem 24 when $y_0 = 5$, $L = 16$, $k = 0.00186$ (a *logistic model* for world population—see the discussion at the beginning of this section). Note that $\lim_{t \to \infty} y = 16$.

**26.** Find each of the following limits.

(a) $\lim_{x \to 0} (1 + x)^{1000}$          (b) $\lim_{x \to 0} (1)^{1/x}$

(c) $\lim_{x \to 0^+} (1 + \varepsilon)^{1/x}, \varepsilon > 0$     (d) $\lim_{x \to 0^-} (1 + \varepsilon)^{1/x}, \varepsilon > 0$

(e) $\lim_{x \to 0} (1 + x)^{1/x}$

**27.** Use the fact that $e = \lim_{h \to 0} (1 + h)^{1/h}$ to find each limit.

(a) $\lim_{x \to 0} (1 - x)^{1/x}$   *Hint:* $(1 - x)^{1/x} = \left[(1 - x)^{1/-x}\right]^{-1}$

(b) $\lim_{x \to 0} (1 + 3x)^{1/x}$          (c) $\lim_{n \to \infty} \left(\frac{n + 2}{n}\right)^n$

(d) $\lim_{n \to \infty} \left(\frac{n - 1}{n}\right)^{2n}$

**28.** Show that the differential equation

$$\frac{dy}{dt} = ay + b$$

has solution

$$y = \left(y_0 + \frac{b}{a}\right)e^{at} - \frac{b}{a}$$

Assume $a \neq 0$.

**29.** Consider a state which had a population of 10 million in 1985, a natural exponential growth rate of 1.2% per year, and immigration from other states and countries of 60,000 per year. Use the differential equation of Problem 28 to model this situation and predict the population in 2010. Take $a = 0.012$.

**30.** Important news is said to diffuse through an adult population of fixed size $L$ at a time rate proportional to the number of people who have not heard the news. Five days after a scandal in city hall was reported, a poll showed that half the people had heard it. How long will it take for 99% of the people to hear it?

**31.** Assume (1) world population will continue to grow exponentially with growth constant $k = 0.019$ indefinitely into the future, (2) it takes $\frac{1}{2}$ acre of land to supply food for a person, and (3) there are 13,500,000 square miles of arable land in the world. How long will it be before the

world reaches the maximum population? *Note*: There were 5 billion people in 1987 and 1 square mile is 640 acres.

**32.** Let $E$ be a differentiable function satisfying $E(u + v) = E(u)E(v)$ for all $u$ and $v$. Find a formula for $E(x)$. *Hint*: First find $E'(x)$.

PC **33.** Using the same axes, draw the graphs for $0 \leq t \leq 100$ of the following two models for the growth of world population (both described in this section).

(a) Exponential growth: $y = 5e^{0.019t}$
(b) Logistic growth: $y = 80/(5 + 11e^{-0.030t})$

Compare what the two models predict for world population in 2000, 2040, and 2090. *Note*: Both models assume world population was 5 billion in 1987 ($t = 0$).

PC **34.** Evaluate:

(a) $\lim\limits_{x \to 0}(1 + x)^{1/x}$        (b) $\lim\limits_{x \to 0}(1 - x)^{1/x}$

The limit in (a) determines $e$. What special number does the limit in (b) determine?

---

**Answers to Concepts Review:**   **1.** $ky; ky(L - y)$   **2.** 8   **3.** Half-life   **4.** $(1 + h)^{1/h}$

---

| **7.6** | |
|---|---|
| **THE INVERSE TRIGONOMETRIC FUNCTIONS** | The six basic trigonometric functions (sine, cosine, tangent, cotangent, secant, and cosecant) were defined in Section 2.3 and we have used them occasionally in examples and problems. With respect to the notion of inverse, they are miserable functions, since for each $y$ in their range, there are infinitely many $x$'s that correspond to it (Figure 1). Nonetheless, we are going to introduce a notion of inverse for them. That this is possible rests on a procedure called **restricting the domain**, which was discussed briefly in Section 7.2. |

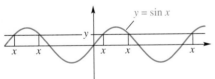

FIGURE 1

**Inverse Sine and Inverse Cosine**    In the case of sine and cosine, we restrict the domain, keeping the range as large as possible while insisting that the resulting function have an inverse. This can be done in many ways, but the agreed procedure is shown in Figures 2 and 3. We show also the graph

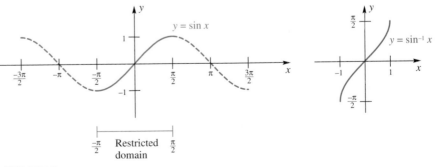

FIGURE 2

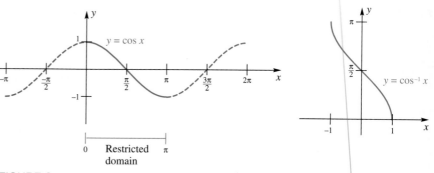

FIGURE 3

of the corresponding inverse function, obtained—as usual—by reflecting across the line $y = x$.

We formalize what we have shown in a definition.

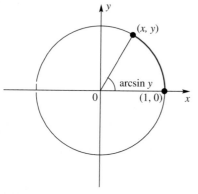

FIGURE 4

> **Definition**
>
> To obtain inverses for sine and cosine, we restrict their domains to $[-\pi/2, \pi/2]$ and $[0, \pi]$, respectively. Thus
>
> $$x = \sin^{-1} y \iff y = \sin x \quad \text{and} \quad \frac{-\pi}{2} \le x \le \frac{\pi}{2}$$
>
> $$x = \cos^{-1} y \iff y = \cos x \quad \text{and} \quad 0 \le x \le \pi$$

The symbol arcsin is often used for $\sin^{-1}$, and arccos is similarly used for $\cos^{-1}$. Think of arcsin as meaning "the arc whose sine is," or "the angle whose sine is" (Figure 4).

**EXAMPLE 1** Calculate (a) $\sin^{-1}(\sqrt{2}/2)$, (b) $\sin^{-1}(-\tfrac{1}{2})$, (c) $\cos^{-1}(\sqrt{3}/2)$, (d) $\cos^{-1}(-\tfrac{1}{2})$, (e) $\cos(\cos^{-1} 0.6)$, and (f) $\sin^{-1}(\sin 3\pi/2)$.

*Solution*

(a) $\sin^{-1}\left(\dfrac{\sqrt{2}}{2}\right) = \dfrac{\pi}{4}$        (b) $\sin^{-1}\left(-\dfrac{1}{2}\right) = -\dfrac{\pi}{6}$

(c) $\cos^{-1}\left(\dfrac{\sqrt{3}}{2}\right) = \dfrac{\pi}{6}$        (d) $\cos^{-1}\left(-\dfrac{1}{2}\right) = \dfrac{2\pi}{3}$

(e) $\cos(\cos^{-1} 0.6) = 0.6$        (f) $\sin^{-1}\left(\sin\dfrac{3\pi}{2}\right) = -\dfrac{\pi}{2}$

The only one of these that is tricky is (f). Note that it would be wrong to give $3\pi/2$ as the answer, since $\sin^{-1} y$ is always in the interval $[-\pi/2, \pi/2]$. Work the problem in steps, as follows.

$$\sin^{-1}\left(\sin\frac{3\pi}{2}\right) = \sin^{-1}(-1) = -\pi/2 \qquad \blacksquare$$

**EXAMPLE 2**   Calculate (a) $\cos^{-1}(-0.61)$, (b) $\sin^{-1}(-0.87)$, (c) $\sin^{-1}(1.21)$, and (d) $\sin^{-1}(\sin 4.13)$.

*Solution*   Use a calculator in radian mode. It has been programmed to give answers that are consistent with the definitions we have given.
(a) $\cos^{-1}(-0.61) = 2.2268569$
(b) $\sin^{-1}(-0.87) = -1.0552023$
(c) Your machine should indicate an error, since $\sin^{-1}(1.21)$ does not exist.
(d) $\sin^{-1}(\sin 4.13) = -0.9884073$   ■

**Inverse Tangent**   In Figure 5, we have shown the graph of the tangent function, its restricted domain, and the graph of $y = \tan^{-1}x$.

---

*Another Way To Say It*

$$\sin^{-1}y$$

is the number in the interval $[-\pi/2, \pi/2]$ whose sine is $y$.

$$\cos^{-1}y$$

is the number in the interval $[0, \pi]$ whose cosine is $y$.

$$\tan^{-1}y$$

is the number in the interval $(-\pi/2, \pi/2)$ whose tangent is $y$.

---

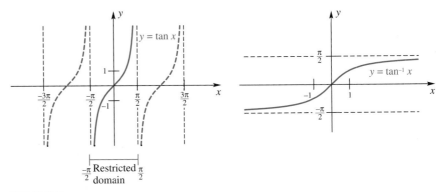

FIGURE 5

---

**Definition**

To obtain an inverse for tangent, we restrict the domain to $(-\pi/2, \pi/2)$. Thus,

$$x = \tan^{-1}y \iff y = \tan x \quad \text{and} \quad -\frac{\pi}{2} < x < \frac{\pi}{2}$$

---

There is a standard way to restrict the domain of the cotangent function, namely to $(0, \pi)$, so that it has an inverse. However, this function does not play a significant role in calculus, so we will say no more about it.

**EXAMPLE 3**   Calculate (a) $\tan^{-1}(1)$, (b) $\tan^{-1}(-\sqrt{3})$, (c) $\tan^{-1}(-0.145)$, and (d) $\tan^{-1}(\tan 5.236)$.

*Solution*

(a) $\tan^{-1}(1) = \dfrac{\pi}{4}$

**(b)** $\tan^{-1}(-\sqrt{3}) = -\dfrac{\pi}{3}$

**(c)** $\tan^{-1}(-0.145) = -0.1439964$

**(d)** $\tan^{-1}(\tan 5.236) = -1.0471853$

The results in (c) and (d) were obtained on a calculator. ■

**Inverse Secant**  To obtain an inverse for secant, we graph $y = \sec x$, restrict its domain appropriately and then graph $y = \sec^{-1}x$ (Figure 6).

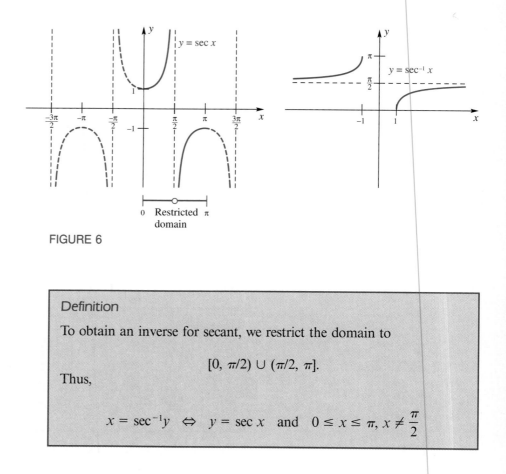

FIGURE 6

---

**Definition**

To obtain an inverse for secant, we restrict the domain to

$$[0, \pi/2) \cup (\pi/2, \pi].$$

Thus,

$$x = \sec^{-1}y \iff y = \sec x \quad \text{and} \quad 0 \le x \le \pi, x \ne \frac{\pi}{2}$$

---

Some authors restrict the domain of the secant in a different way. Thus, if you refer to another book, you must check that author's definition. We will have no need to define $\csc^{-1}$, though this can also be done.

**EXAMPLE 4**  Calculate (a) $\sec^{-1}(-1)$, (b) $\sec^{-1}(2)$, and (c) $\sec^{-1}(-1.32)$.

*Solution*  Most of us have trouble remembering our secants; moreover, most calculators fail to have a secant button. Therefore, we suggest that you remember that $\sec x = 1/\cos x$. From this, it follows that

$$\sec^{-1}y = \cos^{-1}\left(\frac{1}{y}\right)$$

and this allows us to use known facts about the cosine.

(a)  $\sec^{-1}(-1) = \cos^{-1}(-1) = \pi$

(b)  $\sec^{-1}(2) = \cos^{-1}\left(\frac{1}{2}\right) = \frac{\pi}{3}$

(c)  $\sec^{-1}(-1.32) = \cos^{-1}\left(-\frac{1}{1.32}\right) = \cos^{-1}(-0.7575758)$
$$= 2.4303875 \qquad \blacksquare$$

**Four Useful Identities**    In the next section, we will need each of the following identities. You can recall them by reference to the triangles in Figure 7.

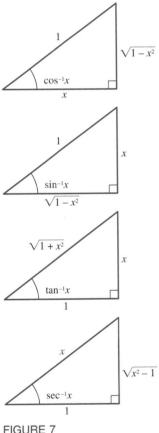

(i)    $\sin(\cos^{-1}x) = \sqrt{1 - x^2}$

(ii)   $\cos(\sin^{-1}x) = \sqrt{1 - x^2}$

(iii)  $\sec(\tan^{-1}x) = \sqrt{1 + x^2}$

(iv)   $\tan(\sec^{-1}x) = \pm\sqrt{x^2 - 1}$

To prove (i), recall that $\sin^2\theta + \cos^2\theta = 1$; in particular, if $0 \le \theta \le \pi$, then

$$\sin\theta = \sqrt{1 - \cos^2\theta}$$

Now apply this with $\theta = \cos^{-1}x$ and use the fact that $\cos(\cos^{-1}x) = x$ to get

$$\sin(\cos^{-1}x) = \sqrt{1 - \cos^2(\cos^{-1}x)} = \sqrt{1 - x^2}$$

Identity (ii) is proved in a completely similar manner. To prove (iii) and (iv), use the identity $\sec^2\theta = 1 + \tan^2\theta$ in place of $\sin^2\theta + \cos^2\theta = 1$. We note that we cannot eliminate the $\pm$ in (iv). If $x \ge 1$, we must use the plus sign; if $x \le -1$, we must use the minus sign.

**EXAMPLE 5**    Calculate $\sin[2\cos^{-1}\left(\frac{2}{3}\right)]$.

*Solution*    Recall the double-angle identity $\sin 2\theta = 2\sin\theta\cos\theta$. Thus,

$$\sin\left[2\cos^{-1}\left(\frac{2}{3}\right)\right] = 2\sin\left[\cos^{-1}\left(\frac{2}{3}\right)\right]\cos\left[\cos^{-1}\left(\frac{2}{3}\right)\right]$$

$$= 2 \cdot \sqrt{1 - \left(\frac{2}{3}\right)^2} \cdot \frac{2}{3} = \frac{4\sqrt{5}}{9} \qquad \blacksquare$$

**FIGURE 7**

**EXAMPLE 6** Show that $\cos(2 \tan^{-1}x) = \dfrac{1 - x^2}{1 + x^2}$.

***Solution*** We apply the double-angle identity $\cos 2\theta = 2 \cos^2\theta - 1$, with $\theta = \tan^{-1}x$. Thus,

$$
\begin{aligned}
\cos(2 \tan^{-1}x) &= \cos 2\theta \\
&= 2 \cos^2\theta - 1 \\
&= \frac{2}{\sec^2\theta} - 1 \\
&= \frac{2}{1 + \tan^2\theta} - 1 \\
&= \frac{2}{1 + x^2} - 1 \\
&= \frac{1 - x^2}{1 + x^2}
\end{aligned}
$$
∎

## CONCEPTS REVIEW

**1.** To obtain an inverse for the sine function, we restrict the domain to _____. The resulting inverse function is denoted by $\sin^{-1}$ or by _____.

**2.** To obtain an inverse for the cosine function, we restrict the domain to _____. The resulting inverse function is denoted by $\cos^{-1}$ or by _____.

**3.** To obtain an inverse for the tangent function, we restrict the domain to _____. The resulting inverse function is denoted by $\tan^{-1}$ or by _____.

**4.** $\arcsin(-1) =$ _____; $\arccos(-1) =$ _____; $\arctan(-1) =$ _____.

## PROBLEM SET 7.6

In Problems 1–12, find exact values without use of a calculator.

**1.** $\arccos\left(\dfrac{\sqrt{2}}{2}\right)$   **2.** $\arcsin\left(-\dfrac{\sqrt{3}}{2}\right)$

**3.** $\sin^{-1}\left(-\dfrac{\sqrt{3}}{2}\right)$   **4.** $\sin^{-1}\left(-\dfrac{\sqrt{2}}{2}\right)$

**5.** $\arctan(\sqrt{3})$   **6.** $\operatorname{arcsec}(2)$

**7.** $\arcsin(-\tfrac{1}{2})$   **8.** $\tan^{-1}\left(-\dfrac{\sqrt{3}}{3}\right)$

**9.** $\operatorname{arccsc}(-2)$   **10.** $\operatorname{arccsc}\left(\dfrac{2\sqrt{3}}{3}\right)$

**11.** $\sin(\sin^{-1}0.4567)$   **12.** $\cos(\sin^{-1}0.56)$

C In Problems 13–24, use a calculator (or tables) to approximate each value.

**13.** $\sin^{-1}(0.1113)$   **14.** $\arccos(0.6341)$

**15.** $\arctan(-6.235)$   **16.** $\arcsin(-0.5422)$

**17.** $\cos(\operatorname{arccot} 3.212)$   **18.** $\sec(\arccos 0.5111)$

**19.** $\sec^{-1}(-2.222)$   **20.** $\tan^{-1}(-60.11)$

**21.** $\cos(\sin(\tan^{-1}2.001))$   **22.** $\sin^2(\ln(\cos 0.5555))$

**23.** $\dfrac{\sin^{-1}(-0.9812)}{1 + \cos^3(5.321)}$

**24.** $\dfrac{2 \arccos(-0.1567)}{\tan 5.193 + \tan^{-1}5.193}$

In Problems 25–30, express $\theta$ in terms of $x$ using the inverse trigonometric functions $\sin^{-1}$, $\cos^{-1}$, $\tan^{-1}$, and $\sec^{-1}$.

**25.**

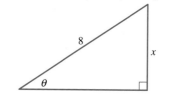

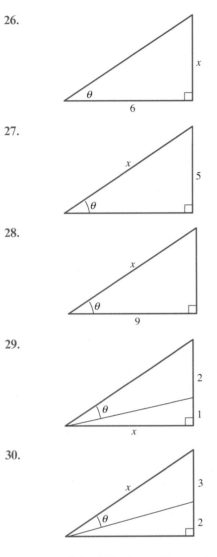

26.

27.

28.

29.

30.

Calculate each of the following without using a calculator (see Example 5).

31. $\cos[2 \sin^{-1}(-\frac{2}{3})]$

32. $\tan[2 \tan^{-1}(\frac{1}{3})]$

33. $\sin[\cos^{-1}(\frac{3}{5}) + \cos^{-1}(\frac{5}{13})]$

34. $\cos[\cos^{-1}(\frac{4}{5}) + \sin^{-1}(\frac{12}{13})]$

In Problems 35–40, show that the equations are identities (see Example 6).

35. $\tan(\sin^{-1}x) = \dfrac{x}{\sqrt{1 - x^2}}$

36. $\sin(\tan^{-1}x) = \dfrac{x}{\sqrt{1 + x^2}}$

37. $\cos(2 \sin^{-1}x) = 1 - 2x^2$

38. $\tan(2 \tan^{-1}x) = \dfrac{2x}{1 - x^2}$

39. $\cos(\tan^{-1}x) = \dfrac{1}{\sqrt{1 + x^2}}$

40. $\sec(2 \cos^{-1}x) = \dfrac{1}{2x^2 - 1}$

41. Find each limit.

(a) $\lim\limits_{x \to \infty} \tan^{-1}x$  (b) $\lim\limits_{x \to -\infty} \tan^{-1}x$

42. Find each limit.

(a) $\lim\limits_{x \to \infty} \sec^{-1}x$  (b) $\lim\limits_{x \to -\infty} \sec^{-1}x$

43. Sketch the graph of $y = \cot^{-1}x$, assuming it has been obtained by restricting the domain of the cotangent to $(0, \pi)$.

C  44. A picture 5 feet high is hung on a wall so that its bottom is 8 feet from the floor, as shown in Figure 8. A viewer with eye level at 5.4 feet stands $b$ feet from the wall. Express $\theta$, the vertical angle subtended by the picture at her eye, in terms of $b$ and then find $\theta$ if $b = 12.9$ feet.

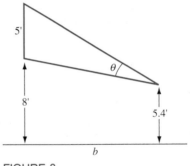

FIGURE 8

45. Find formulas for $f^{-1}(x)$ for each of the following functions $f$, first indicating how you would restrict the domain so that $f$ has an inverse. For example, if $f(x) = 3 \sin 2x$ and we restrict the domain to $-\pi/4 \le x \le \pi/4$, then $f^{-1}(x) = \frac{1}{2} \sin^{-1}(x/3)$.

(a) $f(x) = 3 \cos 2x$  (b) $f(x) = 2 \sin 3x$

(c) $f(x) = \frac{1}{2} \tan x$  (d) $f(x) = \sin \dfrac{1}{x}$

46. Show by repeated use of the addition formula $\tan(x + y) = (\tan x + \tan y)/(1 - \tan x \tan y)$ that

$$\frac{\pi}{4} = 3 \tan^{-1}\left(\frac{1}{4}\right) + \tan^{-1}\left(\frac{5}{99}\right)$$

47. Verify that

$$\frac{\pi}{4} = 4 \tan^{-1}\left(\frac{1}{5}\right) - \tan^{-1}\left(\frac{1}{239}\right)$$

a result discovered by John Machin in 1706 and used by him to calculate the first 100 decimal places of $\pi$.

**48.** Sketch the graph of $y = \sin^{-1}(\sin x)$ for $-\pi/2 \le x \le 6\pi$ and from it deduce that

$$D_x \sin^{-1}(\sin x) = \frac{\cos x}{|\cos x|}$$

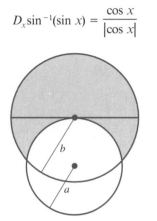

FIGURE 9

**49.** Without using calculus, find a formula for the area of the shaded region in Figure 9 in terms of $a$ and $b$. Note that the center of the larger circle is on the rim of the smaller.

PC **50.** Draw the graphs of $y = \arcsin x$ and $y = \arctan\left(x/\sqrt{1 - x^2}\right)$ using the same axes. Make a conjecture. Prove it.

PC **51.** Draw the graph of $y = \pi/2 - \arcsin x$. Make a conjecture. Prove it.

PC **52.** Draw the graph of $y = \sin(\arcsin x)$ on $[-1, 1]$. Then draw the graph of $y = \arcsin(\sin x)$ on $[-2\pi, 2\pi]$. Explain the differences you observe.

---

**Answers to Concepts Review:** **1.** $[-\pi/2, \pi/2]$; arcsin **2.** $[0, \pi]$; arccos **3.** $(-\pi/2, \pi/2)$; arctan **4.** $-\pi/2; \pi; -\pi/4$

---

## 7.7
## DERIVATIVES OF TRIGONOMETRIC FUNCTIONS

We learned in Section 3.4 that $D_x \sin x = \cos x$ and $D_x \cos x = -\sin x$. The other four trigonometric functions are defined in terms of sine and cosine by

$$\tan x = \frac{\sin x}{\cos x} \qquad \cot x = \frac{\cos x}{\sin x}$$

$$\sec x = \frac{1}{\cos x} \qquad \csc x = \frac{1}{\sin x}$$

These facts, together with the Quotient Rule for derivatives, allow us to find the derivatives of each of these functions. For example,

$$D_x \cot x = D_x\left(\frac{\cos x}{\sin x}\right) = \frac{\sin x(-\sin x) - \cos x \cos x}{\sin^2 x}$$

$$= \frac{-1}{\sin^2 x} = -\csc^2 x$$

Here is a summary of the six derivative formulas. They should be memorized.

$$D_x \sin x = \cos x \qquad\qquad D_x \cos x = -\sin x$$
$$D_x \tan x = \sec^2 x \qquad\qquad D_x \cot x = -\csc^2 x$$
$$D_x \sec x = \sec x \tan x \qquad\qquad D_x \csc x = -\csc x \cot x$$

**Composite Functions** We can combine the rules above with the Chain Rule and with the other differentiation rules in more and more complicated ways. For example, if $u = f(x)$ is differentiable, then

$$D_x \sin u = \cos u \cdot D_x u$$

Here are several examples.

**EXAMPLE 1** Find $D_x \sin(3x^2 + 4)$.

*Solution* Let $u = 3x^2 + 4$.

$$D_x \sin(3x^2 + 4) = D_x \sin u = \cos u D_x u$$
$$= [\cos(3x^2 + 4)]6x = 6x \cos(3x^2 + 4) \qquad \blacksquare$$

**EXAMPLE 2** Find $D_x \tan^2(9x)$.

*Solution* This requires a double use of the Chain Rule.

$$D_x \tan^2(9x) = 2 \tan(9x)D_x \tan(9x)$$
$$= 2 \tan(9x)\sec^2(9x)D_x(9x)$$
$$= 2 \tan(9x)\sec^2(9x) \cdot 9$$
$$= 18 \tan(9x)\sec^2(9x) \qquad \blacksquare$$

**EXAMPLE 3** If $y = (\sin^2 x)/(1 - \cot x)$, find $dy/dx$.

*Solution*

$$\frac{dy}{dx} = \frac{(1 - \cot x)D_x(\sin^2 x) - \sin^2 x D_x(1 - \cot x)}{(1 - \cot x)^2}$$
$$= \frac{(1 - \cot x)2 \sin x \cos x - \sin^2 x \csc^2 x}{(1 - \cot x)^2}$$
$$= \frac{2 \sin x \cos x - 2 \cos^2 x - 1}{(1 - \cot x)^2} \qquad \blacksquare$$

**Inverse Trigonometric Functions** From the Inverse Function Theorem (Theorem 7.2B), we conclude that $\sin^{-1}$, $\cos^{-1}$, $\tan^{-1}$, and $\sec^{-1}$ are differentiable. Our aim is to find formulas for their derivatives. We state the results and then show from where they come.

**DERIVATIVES OF FOUR INVERSE TRIGONOMETRIC FUNCTIONS**

**(i)** $D_x \sin^{-1}x = \dfrac{1}{\sqrt{1 - x^2}}, \qquad -1 < x < 1$

**(ii)** $D_x \cos^{-1}x = \dfrac{-1}{\sqrt{1 - x^2}}, \qquad -1 < x < 1$

**(iii)** $D_x \tan^{-1}x = \dfrac{1}{1 + x^2}$

**(iv)** $D_x \sec^{-1}x = \dfrac{1}{|x| \sqrt{x^2 - 1}}, \qquad |x| > 1$

Our derivations follow the same pattern in each case. To see (i), let $y = \sin^{-1}x$, so that

$$x = \sin y$$

Now differentiate both sides with respect to $x$, using the Chain Rule on the right-hand side. Then

$$1 = \cos y \, D_x y = \cos(\sin^{-1}x)D_x(\sin^{-1}x)$$
$$= \sqrt{1 - x^2}D_x(\sin^{-1}x)$$

At the last step, we used Identity (ii) from the end of Section 7.6. We conclude that $D_x(\sin^{-1}x) = 1/\sqrt{1 - x^2}$.

Results (ii), (iii), and (iv) are proved similarly, but (iv) has a little twist—so we will go through its demonstration. Let $y = \sec^{-1}x$, so

$$x = \sec y$$

Differentiating both sides with respect to $x$, we obtain

$$1 = \sec y \tan y D_x y$$
$$= \sec(\sec^{-1}x)\tan(\sec^{-1}x)D_x(\sec^{-1}x)$$
$$= x(\pm \sqrt{x^2 - 1})D_x(\sec^{-1}x)$$

At the last step, we used Identity (iv) from the end of Section 7.6. But as explained in that section, we use the plus sign if $x > 1$ and the minus sign if $x < -1$. Thus,

$$1 = \pm x \sqrt{x^2 - 1}D_x(\sec^{-1}x)$$
$$= |x| \sqrt{x^2 - 1}D_x(\sec^{-1}x)$$

The desired result follows immediately.

---

$D_x \sec^{-1}x$

Here is another way to derive the formula for the derivative of $\sec^{-1}x$.

$$D_x \sec^{-1}x = D_x \cos^{-1}\left(\frac{1}{x}\right)$$
$$= \frac{-1}{\sqrt{1 - 1/x^2}} \cdot \frac{-1}{x^2}$$
$$= \frac{1}{\sqrt{x^2 - 1}} \cdot \frac{\sqrt{x^2}}{x^2}$$
$$= \frac{1}{\sqrt{x^2 - 1}} \cdot \frac{|x|}{x^2}$$
$$= \frac{1}{|x| \sqrt{x^2 - 1}}$$

---

**EXAMPLE 4**   Find $D_x \sin^{-1}(3x - 1)$.

*Solution*   We use (i) and the Chain Rule.

$$D_x \sin^{-1}(3x - 1) = \frac{1}{\sqrt{1 - (3x - 1)^2}} D_x(3x - 1)$$
$$= \frac{3}{\sqrt{-9x^2 + 6x}}$$    ■

**EXAMPLE 5**   Find $D_x \tan^{-1} \sqrt{x + 1}$.

*Solution*   We use (iii) and the Chain Rule.

$$D_x \tan^{-1} \sqrt{x + 1} = \frac{1}{1 + (\sqrt{x + 1})^2} D_x \sqrt{x + 1}$$

$$= \frac{1}{x + 2} \cdot \frac{1}{2}(x + 1)^{-1/2}$$

$$= \frac{1}{2(x + 2) \sqrt{x + 1}} \qquad \blacksquare$$

Of course, every differentiation formula leads to an integration formula, a matter we will say much more about in the next chapter. In particular:

(i) $\displaystyle\int \frac{1}{\sqrt{1 - x^2}} \, dx = \sin^{-1}x + C;$

(ii) $\displaystyle\int \frac{1}{1 + x^2} \, dx = \tan^{-1}x + C;$

(iii) $\displaystyle\int \frac{1}{x \sqrt{x^2 - 1}} \, dx = \sec^{-1}|x| + C.$

**EXAMPLE 6**   Evaluate $\displaystyle\int_0^{1/2} \frac{dx}{\sqrt{1 - x^2}}.$

*Solution*

$$\int_0^{1/2} \frac{1}{\sqrt{1 - x^2}} \, dx = [\sin^{-1}x]_0^{1/2} = \sin^{-1}\frac{1}{2} - \sin^{-1}0$$

$$= \frac{\pi}{6} - 0 = \frac{\pi}{6} \qquad \blacksquare$$

**EXAMPLE 7**   A man standing on top of a vertical cliff is 200 feet above a lake. As he watches, a motorboat moves directly away from the foot of the cliff at a rate of 25 feet per second. How fast is the angle of depression of his line of sight changing when the boat is 150 feet from the foot of the cliff?

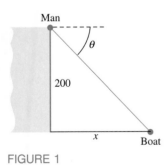

Man

$\theta$

200

$x$

Boat

FIGURE 1

*Solution*   The essential details are shown in Figure 1. Note that $\theta$, the angle of depression, is

$$\theta = \tan^{-1}\left(\frac{200}{x}\right)$$

Thus,

$$\frac{d\theta}{dt} = \frac{1}{1 + (200/x)^2} \cdot \frac{-200}{x^2} \cdot \frac{dx}{dt} = \frac{-200}{x^2 + 40{,}000} \cdot \frac{dx}{dt}$$

When we substitute $x = 150$ and $dx/dt = 25$, we obtain $d\theta/dt = -0.08$ radians per second. $\qquad \blacksquare$

## CONCEPTS REVIEW

1. $D_x \sin x = $ _____; $D_x \cos x = $ _____; $D_x \sin(\cos x) = $ _____.

2. $D_x \sin(\arcsin x) = $ _____.

3. Since $D_x \arctan x = 1/(1 + x^2)$, it follows that $4\int_0^1 1/(1 + x^2)\,dx = $ _____.

4. Since $D_x \arcsin x = 1/\sqrt{1 - x^2}$, it follows that
$$\int_0^{1/2} \left(1/\sqrt{1 - x^2}\right) dx = \text{_____}.$$

## PROBLEM SET 7.7

In Problems 1–22, find $dy/dx$.

1. $y = \sin^2(x - 1)$
2. $y = \cos \sqrt{2 - 7x}$
3. $y = \cot x \sec x$
4. $y = \dfrac{x^2}{1 - \tan x}$
5. $y = e^{\tan x}$
6. $y = e^x \sin x$
7. $y = (\cos x)e^{\cos x}$
8. $y = \ln(\sin x + \cos x)$
9. $y = \ln(\sec x + \tan x)$
10. $y = -\ln(\csc x + \cot x)$
11. $y = \sin^{-1}(2x^2)$
12. $y = \arccos(e^x)$
13. $y = x^3 \tan^{-1}(e^x)$
14. $y = e^x \arcsin x^2$
15. $y = x \operatorname{arccot} x$
16. $y = (\tan^{-1} x)^3$
17. $y = \cos(\tan^{-1} x)$
18. $y = \tan(\cos^{-1} x)$
19. $y = \sec^{-1}(x^3)$
20. $y = (\sec^{-1} x)^3$
21. $y = \tan^{-1}\left(\dfrac{1 + x}{1 - x}\right)$
22. $y = (1 + \sin^{-1} x)^3$

In Problems 23–36, find each of the integrals.

23. $\displaystyle\int x \sin(x^2)\,dx$

24. $\displaystyle\int \sin 2x \cos 2x\,dx$

25. $\displaystyle\int \tan x\,dx = \int \dfrac{\sin x}{\cos x}\,dx$

26. $\displaystyle\int \cot x\,dx$

27. $\displaystyle\int \dfrac{\sec^2 x}{\tan x}\,dx$

28. $\displaystyle\int \dfrac{\sin x}{\cos^2 x}\,dx$

29. $\displaystyle\int_0^1 e^{2x} \cos(e^{2x})\,dx$

30. $\displaystyle\int_0^{\pi/2} \sin^2 x \cos x\,dx$

31. $\displaystyle\int_0^{\sqrt{2}/2} \dfrac{1}{\sqrt{1 - x^2}}\,dx$

32. $\displaystyle\int_{\sqrt{2}}^2 \dfrac{dx}{x\sqrt{x^2 - 1}}$

33. $\displaystyle\int_{-1}^1 \dfrac{1}{1 + x^2}\,dx$

34. $\displaystyle\int_0^{\pi/2} \dfrac{\sin\theta}{1 + \cos^2\theta}\,d\theta$

35. $\displaystyle\int \dfrac{1}{1 + 4x^2}\,dx$

36. $\displaystyle\int \dfrac{e^x}{1 + e^{2x}}\,dx$

37. Show that
$$\int \dfrac{dx}{\sqrt{a^2 - x^2}} = \sin^{-1}\dfrac{x}{a} + C, \qquad a > 0$$
by writing $a^2 - x^2 = a^2[1 - (x/a)^2]$ and making the substitution $u = x/a$.

38. Show the result in Problem 37 by differentiating the right side to get the integrand.

39. Show
$$\int \dfrac{dx}{a^2 + x^2} = \dfrac{1}{a}\tan^{-1}\dfrac{x}{a} + C, \qquad a \neq 0$$

40. Show
$$\int \dfrac{dx}{x\sqrt{x^2 - a^2}} = \dfrac{1}{a}\sec^{-1}\dfrac{|x|}{a}, \qquad a > 0$$

41. Show, by differentiating the right side, that
$$\int \sqrt{a^2 - x^2}\,dx = \dfrac{x}{2}\sqrt{a^2 - x^2} + \dfrac{a^2}{2}\sin^{-1}\dfrac{x}{a} + C, \qquad a > 0$$

42. Use the result of Problem 41 to show
$$\int_{-a}^a \sqrt{a^2 - x^2}\,dx = \dfrac{\pi a^2}{2}$$
Why is this result expected?

43. Find the area of the region bounded by $x^2 y + y - 4 = 0$, the coordinate axes, and the line $x = 1$.

44. Find the volume of the solid generated by revolving about the $x$-axis the region bounded by $y = 5(x^2 + 1)^{-1/2}$, $y = 0$, $x = 0$, and $x = 4$.

45. The lower edge of a wall hanging, 10 feet in height, is 2 feet above the observer's eye level. Find the ideal

distance $b$ to stand from the wall for viewing the hanging; that is, find $b$ that maximizes the angle subtended at the viewer's eye. (Compare with Problem 44 of Section 7.6.)

**46.** Express $d\theta/dt$ in terms of $x$, $dx/dt$, and the constants $a$ and $b$.

(a)

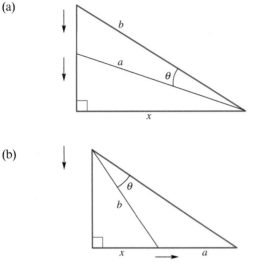

(b)

**47.** The structural steel work of a new office building is finished. Across the street, 60 feet from the ground floor of the freight elevator shaft in the building, a spectator is standing and watching the freight elevator ascend at a constant rate of 15 feet per second. How fast is the angle of elevation of the spectator's line of sight to the elevator increasing 6 seconds after his line of sight passes the horizontal?

**48.** An airplane is flying at a constant altitude of 2 miles and a constant speed of 600 miles per hour on a straight course that will take it directly over an observer on the ground. How fast is the angle of elevation of the observer's line of sight increasing when the distance from her to the plane is 3 miles? Give your result in radians per minute.

**49.** A revolving beacon light is located on an island and is 2 miles away from the nearest point $P$ of the straight shoreline of the mainland. The beacon throws a spot of light that moves along the shoreline as the beacon revolves. If the speed of the spot of light on the shoreline is $5\pi$ miles per minute when the spot is 1 mile from $P$, how fast is the beacon revolving?

**50.** A man on a dock is pulling in a rope attached to a rowboat at a rate of 5 feet per second. If the man's hands are 8 feet higher than the point where the rope is attached to the boat, how fast is the angle of depression of the rope changing when there are still 17 feet of rope out?

$\boxed{C}$ **51.** A visitor from outer space is approaching the earth (radius = 6376 kilometers) at 2 kilometers per second. How fast is the angle $\theta$ subtended by the earth at her eye increasing when she is 3000 kilometers from the surface?

**52.** An object starts at the origin and moves along the $y$-axis so that its velocity in centimeters per second is $dy/dt = (0.2)\cos^2(0.1y)$. How long will it take to get to $y = 2.5\pi$ centimeters? To get to $y = 5\pi$ centimeters?

$\boxed{PC}$ **53.** Where (if at all) on the interval $[0, \infty)$ is $f(x) = 1.25e^x + 0.135 \sin(12x)$ decreasing?

---

**Answers to Concepts Review:** **1.** $\cos x$; $-\sin x$; $-\sin x \cos(\cos x)$ **2.** 1 **3.** $\pi$ **4.** $\pi/6$

---

**7.8**
**THE HYPERBOLIC FUNCTIONS AND THEIR INVERSES**

In both mathematics and science, certain combinations of $e^x$ and $e^{-x}$ occur so often that they are given special names.

> **Definition**
>
> **(The hyperbolic functions).** The hyperbolic sine, hyperbolic cosine, and four related functions are defined by
>
> $$\sinh x = \frac{1}{2}(e^x - e^{-x}) \qquad \cosh x = \frac{1}{2}(e^x + e^{-x})$$
>
> $$\tanh x = \frac{\sinh x}{\cosh x} \qquad \coth x = \frac{\cosh x}{\sinh x}$$
>
> $$\operatorname{sech} x = \frac{1}{\cosh x} \qquad \operatorname{csch} x = \frac{1}{\sinh x}$$

The terminology suggests that there must be some connection with the trigonometric functions; there is. First, the fundamental identity for the hyperbolic functions (reminiscent of $\cos^2 x + \sin^2 x = 1$ in trigonometry) is

$$\cosh^2 x - \sinh^2 x = 1$$

To verify it, we write

$$\cosh^2 x - \sinh^2 x = \frac{e^{2x} + 2 + e^{-2x}}{4} - \frac{e^{2x} - 2 + e^{-2x}}{4} = 1$$

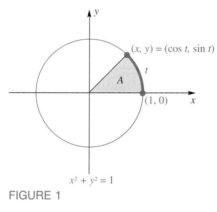

$x^2 + y^2 = 1$

FIGURE 1

Second, recall that the trigonometric functions are intimately related to the unit circle (Figure 1), so much so that they are sometimes called the circular functions. In fact, the parametric equations $x = \cos t$, $y = \sin t$ describe the unit circle. In parallel fashion, the parametric equations $x = \cosh t$, $y = \sinh t$ describe the right branch of the unit hyperbola $x^2 - y^2 = 1$ (Figure 2). Moreover, in both cases the parameter $t$ is related to the shaded area $A$ by $t = 2A$, though this is not obvious in the second case (Problem 60).

Since $\sinh(-x) = -\sinh x$, sinh is an odd function; $\cosh(-x) = \cosh x$, so cosh is an even function. Correspondingly, the graph of $y = \sinh x$ is symmetric with respect to the origin and the graph of $y = \cosh x$ is symmetric with respect to the $y$-axis. The graphs are shown in Figure 3.

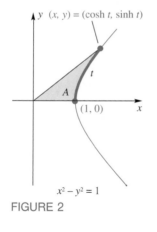

$x^2 - y^2 = 1$

FIGURE 2

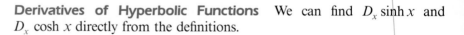

FIGURE 3

**Derivatives of Hyperbolic Functions** We can find $D_x \sinh x$ and $D_x \cosh x$ directly from the definitions.

$$D_x \sinh x = D_x \left( \frac{e^x - e^{-x}}{2} \right) = \frac{e^x + e^{-x}}{2} = \cosh x$$

and

$$D_x \cosh x = D_x \left( \frac{e^x + e^{-x}}{2} \right) = \frac{e^x - e^{-x}}{2} = \sinh x$$

Note that these facts confirm the character of the graphs we drew. For example, since $D_x(\sinh x) = \cosh x > 0$, the graph of hyperbolic sine is always increasing. Similarly $D_x^2(\cosh x) = \cosh x > 0$, which means that the graph of hyperbolic cosine is concave upward

The derivatives of the other four hyperbolic functions follow from those for the first two, combined with the Quotient Rule. Here is a list of all six differentiation formulas.

$$D_x \sinh x = \cosh x \qquad D_x \cosh x = \sinh x$$
$$D_x \tanh x = \operatorname{sech}^2 x \qquad D_x \coth x = -\operatorname{csch}^2 x$$
$$D_x \operatorname{sech} x = -\operatorname{sech} x \tanh x$$
$$D_x \operatorname{csch} x = -\operatorname{csch} x \coth x$$

**EXAMPLE 1**  Find $D_x \tanh(\sin x)$.

*Solution*

$$D_x \tanh(\sin x) = \operatorname{sech}^2(\sin x)D_x(\sin x)$$
$$= \cos x \cdot \operatorname{sech}^2(\sin x) \qquad \blacksquare$$

**EXAMPLE 2**  Find $D_x \cosh^2(3x - 1)$.

*Solution*  We apply the Chain Rule twice.

$$D_x \cosh^2(3x - 1) = 2 \cosh(3x - 1)D_x \cosh(3x - 1)$$
$$= 2 \cosh(3x - 1)\sinh(3x - 1)D_x(3x - 1)$$
$$= 6 \cosh(3x - 1)\sinh(3x - 1) \qquad \blacksquare$$

**EXAMPLE 3**  Find $\int \tanh x \, dx$.

*Solution*  Let $u = \cosh x$, so $du = \sinh x \, dx$.

$$\int \tanh x \, dx = \int \frac{\sinh x}{\cosh x} \, dx = \int \frac{1}{u} \, du$$
$$= \ln|u| + C = \ln|\cosh x| + C = \ln(\cosh x) + C$$

We could drop the absolute value signs because $\cosh x > 0$. $\qquad \blacksquare$

**Inverse Hyperbolic Functions**  Since hyperbolic sine and hyperbolic tangent have positive derivatives, they are increasing functions and automatically have inverses. To obtain inverses for hyperbolic cosine and hyperbolic secant, we restrict their domains to $x \geq 0$. Thus,

$$x = \sinh^{-1}y \quad \Leftrightarrow \quad y = \sinh x$$
$$x = \cosh^{-1}y \quad \Leftrightarrow \quad y = \cosh x \quad \text{and } x \geq 0$$
$$x = \tanh^{-1}y \quad \Leftrightarrow \quad y = \tanh x$$
$$x = \text{sech}^{-1}y \quad \Leftrightarrow \quad y = \text{sech } x \quad \text{and } x \geq 0$$

Since the hyperbolic functions are defined in terms of $e^x$ and $e^{-x}$, it is not too surprising that the inverse hyperbolic functions can be expressed in terms of the natural logarithm. For example, consider $y = \cosh x$ for $x \geq 0$; that is, consider

$$y = \frac{e^x + e^{-x}}{2}, \qquad x \geq 0$$

Our goal is to solve this equation for $x$, which will give $\cosh^{-1}y$. Multiplying both members by $2e^x$, we get $2ye^x = e^{2x} + 1$, or

$$(e^x)^2 - 2ye^x + 1 = 0, \qquad x \geq 0$$

If we solve this quadratic equation in $e^x$, we obtain

$$e^x = \frac{2y \pm \sqrt{(2y)^2 - 4}}{2} = y \pm \sqrt{y^2 - 1}$$

or, after taking natural logarithms,

$$x = \ln\left(y \pm \sqrt{y^2 - 1}\right)$$

The condition $x \geq 0$ forces us to choose the plus sign, so

$$x = \cosh^{-1}y = \ln\left(y + \sqrt{y^2 - 1}\right)$$

Similar arguments apply to each of the inverse hyperbolic functions. We obtain the following results (note that the roles of $x$ and $y$ have been interchanged).

$$\sinh^{-1}x = \ln\left(x + \sqrt{x^2 + 1}\right)$$
$$\cosh^{-1}x = \ln\left(x + \sqrt{x^2 - 1}\right), \qquad x \geq 1$$
$$\tanh^{-1}x = \frac{1}{2}\ln\frac{1 + x}{1 - x}, \qquad -1 < x < 1$$
$$\text{sech}^{-1}x = \ln\left(\frac{1 + \sqrt{1 - x^2}}{x}\right), \qquad 0 < x \leq 1$$

Each of these functions is differentiable. In fact,

$$D_x \sinh^{-1}x = \frac{1}{\sqrt{x^2 + 1}}$$

$$D_x \cosh^{-1}x = \frac{1}{\sqrt{x^2 - 1}}, \qquad x > 1$$

$$D_x \tanh^{-1}x = \frac{1}{1 - x^2}, \qquad -1 < x < 1$$

$$D_x \operatorname{sech}^{-1}x = \frac{-1}{x\sqrt{1 - x^2}}, \qquad 0 < x < 1$$

**EXAMPLE 4**   Show that $D_x \sinh^{-1}x = 1/\sqrt{x^2 + 1}$ by two different methods.

*Solution*

METHOD 1   Let $y = \sinh^{-1}x$, so

$$x = \sinh y$$

Now differentiate both sides with respect to $x$.

$$1 = (\cosh y)D_x y$$

Thus,

$$D_x y = D_x(\sinh^{-1}x) = \frac{1}{\cosh y} = \frac{1}{\sqrt{1 + \sinh^2 y}} = \frac{1}{\sqrt{1 + x^2}}$$

METHOD 2   Use the logarithmic expression for $\sinh^{-1}x$.

$$D_x(\sinh^{-1}x) = D_x \ln(x + \sqrt{x^2 + 1})$$

$$= \frac{1}{x + \sqrt{x^2 + 1}} D_x(x + \sqrt{x^2 + 1})$$

$$= \frac{1}{x + \sqrt{x^2 + 1}}\left(1 + \frac{x}{\sqrt{x^2 + 1}}\right)$$

$$= \frac{1}{\sqrt{x^2 + 1}} \qquad \blacksquare$$

**Applications: The Catenary**   If a homogeneous flexible cable or chain is suspended between two fixed points at the same height, it forms a curve called a **catenary** (Figure 4). Furthermore (see Problem 57), a catenary can be placed in a coordinate system so that its equation takes the form

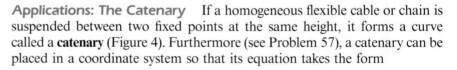

$$y = a \cosh \frac{x}{a}$$

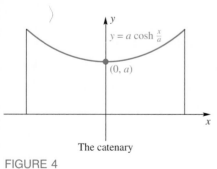

$y = a \cosh \frac{x}{a}$

$(0, a)$

The catenary

FIGURE 4

An Inverted Catenary

**EXAMPLE 5** Find the length of the catenary $y = a \cosh(x/a)$ between $x = -a$ and $x = a$.

*Solution* The desired length (see Section 6.4) is given by

$$
\int_{-a}^{a} \sqrt{1 + \left(\frac{dy}{dx}\right)^2}\, dx = \int_{-a}^{a} \sqrt{1 + \sinh^2\left(\frac{x}{a}\right)}\, dx
$$

$$
= \int_{-a}^{a} \sqrt{\cosh^2\left(\frac{x}{a}\right)}\, dx
$$

$$
= 2 \int_{0}^{a} \cosh\left(\frac{x}{a}\right) dx
$$

$$
= 2a \int_{0}^{a} \cosh\left(\frac{x}{a}\right)\left(\frac{1}{a}\, dx\right)
$$

$$
= \left[ 2a \sinh \frac{x}{a} \right]_{0}^{a}
$$

$$
= 2a \sinh 1 \approx 2.35a \qquad \blacksquare
$$

## CONCEPTS REVIEW

**1.** sinh and cosh are defined by sinh $x =$ _____ and cosh $x =$ _____.

**2.** In *hyperbolic* trigonometry, the identity corresponding to $\sin^2 x + \cos^2 x = 1$ is _____.

**3.** Because of the identity in Question 2, the graph of

the parametric equations $x = \cosh t$, $y = \sinh t$ is _____.

**4.** The graph of $y = a \cosh(x/a)$ is a curve called a _____; this curve is important as a model for _____.

## PROBLEM SET 7.8

In Problems 1–12, verify that the given equations are identities.

1. $e^x = \cosh x + \sinh x$

2. $e^{2x} = \cosh 2x + \sinh 2x$

3. $e^{-x} = \cosh x - \sinh x$

4. $e^{-2x} = \cosh 2x - \sinh 2x$

5. $\sinh(x + y) = \sinh x \cosh y + \cosh x \sinh y$

6. $\sinh(x - y) = \sinh x \cosh y - \cosh x \sinh y$

7. $\cosh(x + y) = \cosh x \cosh y + \sinh x \sinh y$

8. $\cosh(x - y) = \cosh x \cosh y - \sinh x \sinh y$

9. $\tanh(x + y) = \dfrac{\tanh x + \tanh y}{1 + \tanh x \tanh y}$

10. $\tanh(x - y) = \dfrac{\tanh x - \tanh y}{1 - \tanh x \tanh y}$

11. $\sinh 2x = 2 \sinh x \cosh x$

12. $\cosh 2x = \cosh^2 x + \sinh^2 x$

In Problems 13–40, find $D_x y$.

13. $y = \sinh^2 x$

14. $y = \cosh^2 x$

15. $y = 5 \sinh^2 x$

16. $y = \cosh^3 x$

17. $y = \cosh(3x + 1)$

18. $y = \sinh(x^2 + x)$

19. $y = \ln(\sinh x)$

20. $y = \ln(\coth x)$

21. $y = x^2 \cosh x$

22. $y = x^{-2} \sinh x$

23. $y = \cosh 3x \sinh x$

24. $y = \sinh x \cosh 4x$

25. $y = \tanh x \sinh 2x$

26. $y = \coth 4x \sinh x$

27. $y = \sinh^{-1}(x^2)$

28. $y = \cosh^{-1}(x^3)$

29. $y = \tanh^{-1}(2x - 3)$

30. $y = \coth^{-1}(x^5)$

31. $y = x \cosh^{-1}(3x)$

32. $y = x^2 \sinh^{-1}(x^5)$

33. $y = \ln(\cosh^{-1} x)$

34. $y = \ln(\sinh^{-1} x)$

35. $y = \sinh(\sin x)$

36. $y = \sinh(\cos x)$

37. $y = \sinh^{-1}(\cos x)$    38. $y = \cosh^{-1}(\cos x)$

39. $y = \tanh(\cot x)$    40. $y = \coth^{-1}(\tanh x)$

41. Find the area of the region bounded by $y = \cosh 2x$, $y = 0$, $x = 0$, and $x = \ln 3$.

In Problems 42–49, find the integrals.

42. $\int \sinh(3x + 2)\, dx$    43. $\int x \cosh(\pi x^2 + 5)\, dx$

44. $\int \dfrac{\cosh \sqrt{z}}{\sqrt{z}}\, dz$    45. $\int \dfrac{\sinh(2z^{1/4})}{\sqrt[4]{z^3}}\, dz$

46. $\int e^x \sinh e^x\, dx$    47. $\int \cos x \sinh(\sin x)\, dx$

48. $\int \tanh x \ln(\cosh x)\, dx$

49. $\int x \coth x^2 \ln(\sinh x^2)\, dx$

50. Find the area of the region bounded by $y = \cosh 2x$, $y = 0$, $x = -\ln 5$, and $x = \ln 5$.

51. Find the area of the region bounded by $y = \sinh x$, $y = 0$, and $x = \ln 2$.

52. Find the area of the region bounded by $y = \tanh x$, $y = 0$, $x = -8$, and $x = 8$.

53. The region bounded by $y = \cosh x$, $y = 0$, $x = 0$, and $x = 1$ is revolved about the x-axis. Find the volume of the resulting solid. *Hint*: $\cosh^2 x = (1 + \cosh 2x)/2$.

54. The region bounded by $y = \sinh x$, $y = 0$, $x = 0$, and $x = \ln 10$ is revolved about the x-axis. Find the volume of the resulting solid.

55. The curve $y = \cosh x$, $0 \le x \le 1$, is revolved about the x-axis. Find the area of the resulting surface.

56. The curve $y = \sinh x$, $0 \le x \le 1$, is revolved about the x-axis. Find the area of the resulting surface.

57. To derive the equation of a hanging cable (catenary), we consider the section $AP$ from the lowest point $A$ to a general point $P(x, y)$ (see Figure 5) and imagine the rest of the cable to have been removed.

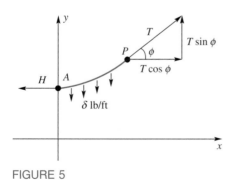

FIGURE 5

The forces acting on the cable are:

1. $H$ = horizontal tension pulling at $A$;

2. $T$ = tangential tension pulling at $P$;

3. $W = \delta s$ = weight of $s$ feet of cable of density $\delta$ pounds per foot.

To be in equilibrium, the horizontal and vertical components of $T$ must just balance $H$ and $W$, respectively. Thus, $T \cos \phi = H$ and $T \sin \phi = W = \delta s$, and so

$$\frac{T \sin \phi}{T \cos \phi} = \tan \phi = \frac{\delta s}{H}$$

But since $\tan \phi = dy/dx$, we get

$$\frac{dy}{dx} = \frac{\delta s}{H}$$

and therefore

$$\frac{d^2y}{dx^2} = \frac{\delta}{H}\frac{ds}{dx} = \frac{\delta}{H}\sqrt{1 + \left(\frac{dy}{dx}\right)^2}$$

Now show that $y = a \cosh(x/a) + C$ satisfies this differential equation with $a = H/\delta$.

[c] 58. Call the graph of $y = b - a \cosh(x/a)$ an inverted catenary and imagine it to be an arch sitting on the x-axis. Show that if the width of this arch along the x-axis is $2a$, then each of the following is true.

(a) $b = a \cosh 1 \approx 1.54308a$.
(b) The height of the arch is approximately $0.54308a$.
(c) The height of an arch of width 48 is approximately 13.

[c] 59. A farmer built a large hayshed of length 100 feet and width 48 feet. A cross section has the shape of an inverted catenary (see Problem 58) with equation $y = 37 - 24 \cosh(x/24)$.

(a) Draw a picture of this shed.
(b) Find the volume of the shed.
(c) Find the surface area of the roof of the shed.

60. Show that $A = t/2$, where $A$ denotes the area in Figure 2 of this section. *Hint*: At some point you will need to use Formula 44 from the back of the book.

61. Demonstrate for $r$ any real number:

(a) $(\sinh x + \cosh x)^r = \sinh rx + \cosh rx$
(b) $(\cosh x - \sinh x)^r = \cosh rx - \sinh rx$
(c) $(\cos x + i \sin x)^r = \cos rx + i \sin rx$
(d) $(\cos x - i \sin x)^r = \cos rx - i \sin rx$

62. The **gudermannian** of $t$ is defined by

$$\text{gd}(t) = \tan^{-1}(\sinh t)$$

Show:

(a) gd is odd and increasing with an inflection point at the origin;

(b) $\text{gd}(t) = \sin^{-1}(\tanh t) = \int_0^t \text{sech } u \, du$

63. Show that the area under the curve $y = \cosh t$, $0 \le t \le x$ is equal to its arc length.

64. Find the equation of the Gateway Arch in St. Louis, Missouri, given that it is an inverted catenary. Assume that it stands on the $x$-axis, that it is symmetric with respect to the $y$-axis, and that it is 630 feet wide at the base and 630 feet high at the center.

[PC] 65. Draw the graphs of $y = \sinh x$, $y = \ln\left(x + \sqrt{x^2 + 1}\right)$, and $y = x$ using the same axes and scaled so that $-3 \le x \le 3$ and $-3 \le y \le 3$. What does this demonstrate?

[PC] 66. Refer to Problem 62. Derive a formula for $gd^{-1}(x)$. Draw its graph and also that of $gd(x)$ using the same axes and thereby confirm your formula.

---

**Answers to Concepts Review:** **1.** $(e^x - e^{-x})/2$; $(e^x + e^{-x})/2$ **2.** $\cosh^2 x - \sinh^2 x = 1$ **3.** A hyperbola **4.** Catenary; a hanging chain

---

## 7.9 CHAPTER REVIEW

### Concepts Test

Respond with true or false to each of the following assertions. Be prepared to justify your answer.

1. $\ln|x|$ is defined for all real $x$.

2. The graph of $y = \ln x$ has no inflection points.

3. $\int_1^{e^3} \frac{1}{t} dt = 3$

4. The graph of an invertible function $y = f(x)$ is intersected exactly once by every horizontal line.

5. The domain of $\ln^{-1}$ is the set of all real numbers.

6. $\ln x/\ln y = \ln x - \ln y$

7. $(\ln x)^4 = 4 \ln x$

8. $\ln(2e^{x+1}) - \ln(2e^x) = 1$ for all $x$ in $\mathbb{R}$.

9. The functions $f(x) = 4 + e^x$ and $g(x) = \ln(x - 4)$ are a pair of inverse functions.

10. $\exp x + \exp y = \exp(x + y)$.

11. If $x > y$ then $\ln x > \ln y$.

12. If $a \ln x < b \ln x$ then $a < b$.

13. If $a < b$ then $ae^z < be^z$.

14. If $a < b$ then $e^a < e^b$.

15. $\lim_{x \to 0^+} (\ln \sin x - \ln x) = 0$.

16. $\pi^{\sqrt{2}} = e^{\sqrt{2} \ln \pi}$    17. $\frac{d}{dx} (\ln \pi) = \frac{1}{\pi}$

18. $\int \frac{1}{x} dx = \ln 3|x| + C$    19. $D_x(x^e) = ex^{e-1}$

20. If $f(x) \cdot \exp[g(x)] = 0$ for $x = x_0$, then $f(x_0) = 0$.

21. $D_x(x^x) = x^x \ln x$.

22. The graphs of $y = \sin x$ and $y = \cos x$ intersect at right angles.

23. $\text{Arcsin}(\sin x) = x$ for all real numbers $x$.

24. If $a < b$ then $\sinh a < \sinh b$.

25. If $a < b$ then $\cosh a < \cosh b$.

26. $\cosh x \le e^{|x|}$.

27. $|\sinh x| \le e^{|x|}/2$.

28. $\tan^{-1} x = \sin^{-1} x / \cos^{-1} x$.

29. $\cosh(\ln 3) = \frac{5}{6}$

30. $\lim_{x \to 0} \ln\left(\frac{\sin x}{x}\right) = 1$.

31. $\lim_{x \to -\infty} \tan^{-1} x = -\frac{\pi}{2}$

32. $\sin^{-1}(\cosh x)$ is defined for all real $x$.

33. $f(x) = \tanh x$ is an odd function.

34. Both $y = \sinh x$ and $y = \cosh x$ satisfy the differential equation $y'' + y = 0$.

35. $\ln(3^{100}) > 100$.

36. $\ln(2x^2 - 18) - \ln(x - 3) - \ln(x + 3) = \ln 2$ for all $x$ in $\mathbb{R}$.

37. If $y$ is growing exponentially and if $y$ triples between $t = 0$ and $t = t_1$, then $y$ will also triple between $t = 2t_1$ and $t = 3t_1$.

38. The time necessary for $x(t) = Ce^{-kt}$ to drop to half its value is $\frac{\ln 2}{\ln k}$.

39. If $y'(t) = ky(t)$ and $z'(t) = kz(t)$, then $(y(t) + z(t))' = k(y(t) + z(t))$.

**40.** If $y_1(t)$ and $y_2(t)$ both satisfy $y'(t) = ky(t) + C$, then so does $(y_1(t) + y_2(t))$.

**41.** $\lim_{h \to 0} (1 - h)^{-1/h} = e^{-1}$.

## Sample Test Problems

In Problems 1–24, differentiate the indicated functions.

**1.** $\ln \dfrac{x^4}{2}$

**2.** $\sin^2(x^3)$

**3.** $e^{x^2 - 4x}$

**4.** $\log_{10}(x^5 - 1)$

**5.** $\tan(\ln e^x)$

**6.** $e^{\ln \cot x}$

**7.** $2 \tanh \sqrt{x}$

**8.** $\tanh^{-1}(\sin x)$

**9.** $\sinh^{-1}(\tan x)$

**10.** $2 \sin^{-1} \sqrt{3x}$

**11.** $\sec^{-1} e^x$

**12.** $\ln \sin^2\left(\dfrac{x}{2}\right)$

**13.** $3 \ln(e^{5x} + 1)$

**14.** $\ln(2x^3 - 4x + 5)$

**15.** $\cos e^{\sqrt{x}}$

**16.** $\ln(\tanh x)$

**17.** $2 \cos^{-1} \sqrt{x}$

**18.** $4^{3x} + (3x)^4$

**19.** $2 \csc e^{\ln \sqrt{x}}$

**20.** $(\log_{10} 2x)^{2/3}$

**21.** $4 \tan 5x \sec 5x$

**22.** $x \tan^{-1} \dfrac{x^2}{2}$

**23.** $x^{1+x}$

**24.** $(1 + x^2)^e$

In Problems 25–34, find the antiderivatives of the indicated functions and verify your results by differentiation.

**25.** $e^{3x - 1}$

**26.** $6 \cot 3x$

**27.** $e^x \sin e^x$

**28.** $\dfrac{6x + 3}{x^2 + x - 5}$

**29.** $\dfrac{e^{x+2}}{e^{x+3} + 1}$

**30.** $4x \cos x^2$

**31.** $\dfrac{4}{\sqrt{1 - 4x^2}}$

**32.** $\dfrac{\cos x}{1 + \sin^2 x}$

**33.** $\dfrac{-1}{x + x(\ln x)^2}$

**34.** $\operatorname{sech}^2(x - 3)$

**42.** It is to a saver's advantage to have money invested at 11% compounded continuously rather than 12% compounded monthly.

**43.** If $D_x(a^x) = a^x$ with $a > 0$, then $a = e$.

In Problems 35 and 36, find the intervals on which $f$ is increasing and the intervals on which $f$ is decreasing. Find where the graph of $f$ is concave upward and where it is concave downward. Find any extreme values and points of inflection. Then sketch the graph of $f$.

**35.** $f(x) = \sin x + \cos x, \ \dfrac{-\pi}{2} \le x \le \dfrac{\pi}{2}$

**36.** $f(x) = \dfrac{x^2}{e^x}, \ x \in \mathbb{R}$

**37.** Let $f(x) = x^5 + 2x^3 + 4x, \ -\infty < x < \infty$

(a) Prove that $f$ has an inverse $g = f^{-1}$.
(b) Evaluate $g(7) = f^{-1}(7)$.
(c) Evaluate $g'(7)$.

**38.** A certain radioactive substance has a half-life of 10 years. How long will it take for 100 grams to decay to 1 gram?

**39.** If $100 is put in the bank today at 12% interest, how much will it be worth at the end of 1 year if interest is compounded as indicated?

(a) Annually        (b) Monthly
(c) Daily           (d) Continuously

**40.** An airplane is flying horizontally at an altitude of 500 feet at the speed of 300 feet per second directly away from a searchlight on the ground. The searchlight is kept directed at the plane. At what rate is the angle between the light beam and the ground changing when this angle is 30°?

**41.** Find the equation of the tangent line to $y = (\cos x)^{\sin x}$ at $(0, 1)$.

**42.** A town grew exponentially from 10,000 in 1970 to 14,000 in 1980. Assuming the same type of growth continues, what will the population be in 2000?

## 7.10 ADDITIONAL PROBLEMS

Here we investigate the miracle of compound interest. We saw in section 7.5 (Example 5) that an amount of money $A_0$ in the bank at $r$% interest per year compounded continuously grows to $A(t) = A_0 e^{rt}$, when $t$ is measured in years. We investigate other financial scenarios in Problems 1–5.

**1.** Find the value of $1000 at the end of 1 year when the interest is compounded continuously at 5%. This is called the future value.

**2.** Suppose that after 1 year you will have $1000 in the bank if the interest was compounded continuously at 5%. At the end of the year, how much money did you have to start with one year ago? This is called the present value.

**3.** In many cases, as you earn money you deposit it to a bank account at regular time intervals. The money in the bank generally earns interest which is compounded continuously. Suppose you deposit $100 every 30 days and it is earning interest of 5% compounded continuously. After 60 days you would have $100e^{0.05 \cdot (60/365)} + 100e^{0.05 \cdot (30/365)}$.

(a) How much money would you have after 360 days?
(b) Can you create a compact formula which gives this amount?

4. An annuity is money which you can take out of the bank at regular intervals. Often you want to calculate the amount of money you need at the start to be able to remove a given amount of money for a set number of months. Let us suppose that you want to remove $144 every 30 days for the next 3,600 days ($\approx$ 10 years) from your account which is earning 5% interest. Assume that at the end of that time you will have no money left in the bank.

(a) Show that if the interest is compounded every 30 days then the deposit $A_0$ needed would be

$$A_0 = 144 \left( \frac{1 - (1 + 0.05 \cdot (30/365))^{-10 \cdot (365/30)}}{0.05/(30/365)} \right)$$

(b) Find a formula for $A_0$ if the payments are made every 30 days, but the interest is compounded at 5% continuously.

5. The *rule of 70* gives a quick method of computing how long it takes your money to double when it is earning interest compounded continuously. The rule states that it takes about $T = 70/$(interest rate in percent) years for money in the bank to double.

(a) Show that $T$ satisfies the equation $A_0 e^{rT} = 2A_0$, where $r$ is the continuous rate of interest
(b) Show that $T \approx 70/(100r)$
(c) Find the number of years that it would take money earning interest at 7% to double.

Besides providing an easy way to differentiate products (see Section 7.1 Example 7) the logarithmic derivative also provides a measure of the *relative or fractional rate of change*. We explore this concept in Problems 6–10.

6. Show that the relative rate of change of $e^{kt}$ as a function of $t$ is $k$.

7. Show that the relative rate of change of any polynomial approaches zero as the independent variable approaches infinity.

8. Prove that if the relative rate of change is a positive constant, then the function must represent exponential growth.

9. Prove that if the relative rate of change is a negative constant then the function must represent exponential decay.

10. The condition number for a function provides a measure of how errors in the input are reflected as errors in the output. Suppose that the input to $f(x)$ at $x = 1.00$ is only known to $\pm 0.01$. How is this margin of error reflected in the computation of $f(x)$?

(a) Show that the relative error in $f(x)$ due to a relative error in the input of size $\Delta x$ is

$$\frac{f(x + \Delta x) - f(x)}{f(x)} = \frac{f(x + \Delta x) - f(x)}{\Delta x \cdot f(x)} \cdot \frac{(x + \Delta x) - x}{x} \cdot x$$

(b) By taking limits of the above expression, justify why one could say that if $\rho_f$ = *relative error in $f(x)$*, and $\rho_x$ = *relative error in x* then

$$\rho_f \approx x \cdot \frac{d\ln(f(x))}{dx} \cdot \rho_x$$

The quantity $x \cdot \dfrac{d\ln(f(x))}{dx}$ is called the condition number of the computation.

(c) Find the condition number for computing $e^x$.

Our graphing experience so far has been restricted to using standard (linear) coordinate spacings. When working with exponential and logarithmic functions is may be more natural to use logarithmic and log-log scales. We explore these techniques in Problems 11–12.

11. On a single set of axis use your calculator to draw the graphs of $y = 2^x$, $y = 3^x$, and $y = 4^x$ over the region $0 < x < 4$. Do the same for the inverse functions $y = \log_2 x$, $y = \log_3 x$ and $y = \log_4 x$. If we use a computer graphing program which permits the graphing using semilog axis (a logarithmic scale on the $y$-axis, and a normal scale on the $x$-axis) to graph the functions $y = 2^x$, $y = 3^x$, and $y = 4^x$ over the region $-5 < x < 5$ (Figure 1) we get 3 straight lines.

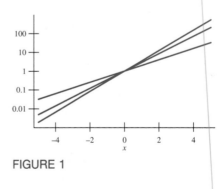

**FIGURE 1**

(a) Identify each of the curves in Figure 1.
(b) Noting that if $y = Cb^x$ then $\ln y = \ln C + x \ln b$, explain why all the curves in Figure 1 are straight lines going through the point $(0, 1)$.
(c) Based upon the semilog plot given by Figure 2, determine the $C$ and $b$ in the equation.

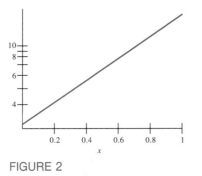

FIGURE 2

**12.** If we use a log scaling for the $x$-axis as well as the $y$-axis (called a log-log plot) and graph several power functions, we will also get straight lines. Using the result that if $y = Cx^r$ becomes upon taking logs $\log y = \log C + r \log x$ identify each of the equations which give rise to the graphs in Figure 3.

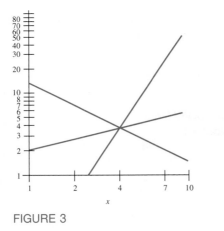

FIGURE 3

C **13.** In many situations we want to see the entire graph of a function over the domain $(-\infty, \infty)$. Clearly this is not possible. One trick frequently used is to map the domain $(-\infty, \infty)$ into a finite region $(-1, 1)$ using the map $\tilde{x} = \tan(x\pi/2)$. To get a feeling for such graphs we consider the graph of $y = \tilde{x}^2$ over the domain $(-\infty, \infty)$, but instead plot the function $y = (\tan(x\pi/2))^2$ over the region $(-1, 1)$ in Figure 4.

Notice that an $x$-value of 0.5 actually corresponds to tan $(0.5 \cdot \pi/2) = 1.0$ where $y = 1$. If we wanted to compute what $x$-value on the graph corresponded to the point $\tilde{x} = 1$, then we have $x = (2/\pi) \arctan(1) = 1/2$. Using your graphing calculator or computer, plot the following functions on the domain $(-\infty, \infty)$ after mapping the domain into $(-1, 1)$ using $\tilde{x} = \tan(x\pi/2)$. Make sure that you restrict the range in $y$ to get appropriate windows.

(a) $y = 2x^3 + 2x^2 - x + 3$
(b) $y = x \sin x$
(c) $y = \dfrac{x^2}{(x - 2)(x - 1)}$

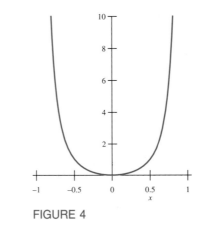

FIGURE 4

**14.** In 1895 an equation was derived by Korteweg and de Vries for a finite amplitude water wave which moves unchanged in form down a channel. Such waves are called solitons and also play a very important role in fiber optical signal propagation. If you move with the wave, the shape satisfies the nonlinear ordinary differential equation $\eta'(x) = \eta(x)\eta'(x) + \eta'''(x)$. This equation can be integrated once and then multiplied by $\eta'(x)$ and integrated again to give $3\eta^2 = \eta^3 + 3(\eta')^2$.

(a) Verify that the shape of the wave is given by the sech$^2$ function by showing by substitution $\eta = 3 \operatorname{sech}^2(x/2)$ satisfies the equation.
C (b) Plot the function $\eta = 3 \operatorname{sech}^2(x/2)$ for the domain $(-5, 5)$ and verify that it does look like a water wave.

# TECHNOLOGY PROJECT

We've seen several functions that are defined by integrals. When a continuous function $f$ that does not possess an elementary function antiderivative arises in applications and yet its antiderivative is needed, the Fundamental Theorem tells us that the "special function" antiderivative, $F(z)$, is well-defined by

$$F(z) = c \int_a^z f(x)\, dx \qquad (1)$$

Here, the lower limit $a$ and constant $c$ are chosen as some fixed values, compatible with the particular historical tradition for the specific $F$.

**Log.** We've seen that the "cleanest" way to define the exponential and logarithmic functions is to begin with the definition

$$\ln z = \int_1^z \frac{dx}{x} \qquad (2)$$

**Exercise 1** Identify $f$, $F$, $a$, and $c$ in Equation 1 definition for the log function as defined in Equation 2.

Most calculus texts show how all the familiar properties of the log follow from the integral definition. The next problem gives a start on this program.

**Exercise 2** From Equation 2 evaluate $\dfrac{d}{dz}\ln(z)$ and $\ln 1$.

**Erf.** The special function, erf, arises in connection with the probability

$$P(z) = \frac{1}{\sqrt{2\pi}} \int_{-z}^{z} e^{-x^2/2} dx$$

$$= \sqrt{\frac{2}{\pi}} \int_0^z e^{-x^2/2} dx \qquad (3)$$

of being no more than $z$ standard deviations from the mean in a normal distribution. For example, in *Maple*, the erf function arises when we try to evaluate this integral:

```
P := sqrt(1/Pi) * int( exp(-x^2/2),
    x = 0 .. z);
             1/2              1/2
P := 1/2 2      erf (1/2 2      z)
```

The function erf is defined as a multiple of the antiderivative of $e^{-x^2}$. Erf is not an elementary function and is defined by:

$$\mathrm{erf}(z) = \frac{2}{\sqrt{\pi}} \int_0^z e^{-t^2}\, dt \qquad (4)$$

**Exercise 3** What are $f$, $F$, $a$, and $c$ in Equation 1 for the erf function as defined in Equation 4?

**Exercise 4** Use Equation 4 and a change of variables to prove that *Maple's* output above for P (as defined by Equation 3) is correct.

**Exercise 5** Evaluate $\mathrm{erf}(0)$ and $\dfrac{d}{dz}\,\mathrm{erf}(z)$.

**The Bible for Special Functions.** Milton Abramowitz and Irene A. Stegun edited the *Handbook of Mathematical Functions with Formulas, Graphs and Mathematical Tables* under the auspices of the National Science Foundation and Massachusetts Institute of Tech-

nology. This work is cited a *vast* number of times in the scientific literature and you may want to consider buying your own copy of the Dover Edition.

The Handbook has over 20 chapters discussing the "special functions" common in scientific applications, many of which are defined by integrals, for example,

**Gamma Function** $\Gamma(z) = \int_0^\infty t^{z-1} e^{-t} dt$

**Cosine Integral** $C(z) = \int_0^z \cos\left(\frac{\pi}{2} t^2\right) dt$

**Exercise 6** Verify the results in Exercise 2 using your technologies' built-in natural log function.

**Exercise 7** Make a table with three columns. In the first column, show the numbers between 1 and 2 at spacing 0.1, in the second column show the corresponding values of ln using your computing technology on Equation 2, and in the third column show the values of your built-in ln.

**Exercise 8** Estimate $\mathrm{erf}(\infty)$ with your integral definition (Equation 4). **Hint:** The exact answer is a simple integer.

**Exercise 9** Hand in a plot of erf on the interval $[-5, 5]$ and discuss how your plot bears out the other results you've obtained for the erf function, in particular, those of Exercises 5 and 8.

# Techniques of Integration

**8.1**

**INTEGRATION BY SUBSTITUTION**

Our repertoire of functions now includes all the so-called elementary functions. These are the constant functions, the power functions, the logarithmic and exponential functions, the trigonometric and inverse trigonometric functions, and all functions obtained from them by addition, subtraction, multiplication, division, and composition. Thus,

$$f(x) = \frac{e^x + e^{-x}}{2} = \cosh x$$

$$g(x) = (1 + \cos^4 x)^{1/2}$$

$$h(x) = \frac{3^{x^2 - 2x}}{\ln(x^2 + 1)} - \sin[\cos(\cosh x)]$$

are elementary functions.

Differentiation of an elementary function is straightforward, requiring only a systematic use of the rules we have learned. And the result is always an elementary function. Integration (antidifferentiation) is a far different matter. It involves a few techniques and a large bag of tricks; what is worse, it does not always yield an elementary function. For example, it is known that the antiderivatives of $e^{-x^2}$ and $(\sin x)/x$ are not elementary functions.

The two principal techniques for integration are *substitution* (Sections 8.1–8.3) and *integration by parts* (Section 8.4). The method of substitution was introduced in Sections 5.1 and 5.8; we have used it occasionally in the intervening chapters. Here, we review the method and apply it in a wide variety of situations.

Finally in Section 8.5, we discuss the problem of integrating a rational function, that is, a quotient of two polynomial functions. We learn that, in theory, we can always perform such an integration and the result will be an elementary function.

**Standard Forms**   Effective use of the method of substitution depends on the ready availability of a list of known integrals. One such list (but too long to memorize) appears inside the back cover of this book. The short list shown below is so useful that we think every calculus student should memorize it.

---

**STANDARD INTEGRAL FORMS**

*Constants, Powers*

1. $\int k\, du = ku + C$

2. $\int u^r\, du = \begin{cases} \dfrac{u^{r+1}}{r+1} + C & r \neq -1 \\ \ln|u| + C & r = -1 \end{cases}$

*Exponentials*

3. $\int e^u\, du = e^u + C$

4. $\int a^u\, du = \dfrac{a^u}{\ln a} + C,\ a \neq 1,\ a > 0$

*Trigonometric Functions*

5. $\int \sin u\, du = -\cos u + C$

6. $\int \cos u\, du = \sin u + C$

7. $\int \sec^2 u\, du = \tan u + C$

8. $\int \csc^2 u\, du = -\cot u + C$

9. $\int \sec u \tan u\, du = \sec u + C$

10. $\int \csc u \cot u\, du = -\csc u + C$

11. $\int \tan u\, du = -\ln|\cos u| + C$

12. $\int \cot u\, du = \ln|\sin u| + C$

*Algebraic Functions*

13. $\int \dfrac{du}{\sqrt{a^2 - u^2}} = \sin^{-1}\left(\dfrac{u}{a}\right) + C$

14. $\int \dfrac{du}{a^2 + u^2} = \dfrac{1}{a}\tan^{-1}\left(\dfrac{u}{a}\right) + C$

15. $\int \dfrac{du}{u\sqrt{u^2 - a^2}} = \dfrac{1}{a}\sec^{-1}\left(\dfrac{|u|}{a}\right) + C = \dfrac{1}{a}\cos^{-1}\left(\dfrac{a}{|u|}\right) + C$

---

**Substitution in Indefinite Integrals**   Suppose that you face an indefinite integration. If it is a standard form, simply write the answer. If not, look for a substitution that will change it to a standard form. If the first substitution you try does not work, try another. Skill at this, like most worthwhile activities, depends on practice.

The method of substitution was given in Theorem 5.8A and is restated here for easy reference.

---

Theorem A

**(Substitution in Indefinite Integrals).** Let $g$ be a differentiable function and suppose that $F$ is an antiderivative of $f$. Then if $u = g(x)$,

$$\int f(g(x))g'(x)\, dx = \int f(u)\, du = F(u) + C = F(g(x)) + C$$

**EXAMPLE 1**　Find $\displaystyle\int \frac{x}{\cos^2(x^2)}\,dx$.

*Solution*　Stare at this integral for a few moments. Since $1/\cos^2 x = \sec^2 x$, you may be reminded of the standard form $\displaystyle\int \sec^2 u\,du$. Let $u = x^2$, $du = 2x\,dx$. Then

$$\int \frac{x}{\cos^2(x^2)}\,dx = \frac{1}{2}\int \frac{1}{\cos^2(x^2)} \cdot 2x\,dx = \frac{1}{2}\int \sec^2 u\,du$$

$$= \frac{1}{2}\tan u + C = \frac{1}{2}\tan(x^2) + C \qquad \blacksquare$$

**EXAMPLE 2**　Find $\displaystyle\int \frac{3}{\sqrt{5 - 9x^2}}\,dx$.

*Solution*　Think of $\displaystyle\int \frac{du}{\sqrt{a^2 - u^2}}$. Let $u = 3x$, so $du = 3\,dx$. Then,

$$\int \frac{3}{\sqrt{5 - 9x^2}}\,dx = \int \frac{1}{\sqrt{5 - u^2}}\,du = \sin^{-1}\left(\frac{u}{\sqrt{5}}\right) + C$$

$$= \sin^{-1}\left(\frac{3x}{\sqrt{5}}\right) + C \qquad \blacksquare$$

**EXAMPLE 3**　Find $\displaystyle\int \frac{6e^{1/x}}{x^2}\,dx$.

*Solution*　Think of $\displaystyle\int e^u\,du$. Let $u = 1/x$, so $du = (-1/x^2)\,dx$. Then,

$$\int \frac{6e^{1/x}}{x^2}\,dx = -6\int e^{1/x}\left(\frac{-1}{x^2}\,dx\right) = -6\int e^u\,du$$

$$= -6e^u + C = -6e^{1/x} + C \qquad \blacksquare$$

**EXAMPLE 4**　Find $\displaystyle\int \frac{e^x}{4 + 9e^{2x}}\,dx$.

*Solution*　Think of $\displaystyle\int \frac{1}{a^2 + u^2}\,du$. Let $u = 3e^x$, so $du = 3e^x\,dx$. Then,

$$\int \frac{e^x}{4 + 9e^{2x}}\,dx = \frac{1}{3}\int \frac{1}{4 + 9e^{2x}}\,(3e^x\,dx) = \frac{1}{3}\int \frac{1}{4 + u^2}\,du$$

$$= \frac{1}{3} \cdot \frac{1}{2}\tan^{-1}\left(\frac{u}{2}\right) + C = \frac{1}{6}\tan^{-1}\left(\frac{3e^x}{2}\right) + C \qquad \blacksquare$$

　　No law says you have to write out the *u*-substitution. If you can do the substitution mentally, that is fine. Here are two illustrations.

**EXAMPLE 5** Find $\int x^3 \sqrt{x^4 + 11}\, dx$.

*Solution* Mentally, substitute $u = x^4 + 11$.

$$\int x^3 \sqrt{x^4 + 11}\, dx = \frac{1}{4} \int (x^4 + 11)^{1/2} (4x^3\, dx)$$

$$= \frac{1}{4} \int (x^4 + 11)^{1/2}\, d(x^4 + 11)$$

$$= \frac{1}{6}(x^4 + 11)^{3/2} + C \qquad\blacksquare$$

**EXAMPLE 6** Find $\int \dfrac{a^{\tan t}}{\cos^2 t}\, dt$.

*Solution* Mentally, substitute $u = \tan t$.

$$\int \frac{a^{\tan t}}{\cos^2 t}\, dt = \int a^{\tan t} \sec^2 t\, dt$$

$$= \int a^{\tan t} d(\tan t) = \frac{a^{\tan t}}{\ln a} + C \qquad\blacksquare$$

**Manipulating the Integrand** Before you make a substitution, you may need to rewrite the integrand in a more convenient form.

**EXAMPLE 7** Find $\int \dfrac{7}{x^2 - 6x + 25}\, dx$.

*Solution* Integrals with quadratic expressions in the denominator can often be reduced to standard forms by *completing the square*. Recall that $x^2 + bx$ becomes a perfect square by the addition of $(b/2)^2$.

$$\int \frac{7}{x^2 - 6x + 25}\, dx = \int \frac{7}{x^2 - 6x + 9 + 16}\, dx$$

$$= 7 \int \frac{1}{(x - 3)^2 + 4^2}\, d(x - 3)$$

$$= \frac{7}{4} \tan^{-1}\left(\frac{x - 3}{4}\right) + C$$

We made the mental substitution $u = x - 3$ at the final stage. $\qquad\blacksquare$

**EXAMPLE 8** Find $\int \dfrac{x^2 - x}{x + 1}\, dx$.

*Solution* When the integrand is the quotient of two polynomials (that is, a rational function) and the numerator is of equal or greater degree than the denominator, always *divide the denominator into the numerator first*. By

$$\begin{array}{r} x - 2 \\ x + 1 \overline{)\begin{array}{l} x^2 - x \\ \underline{x^2 + x} \\ -2x \\ \underline{-2x - 2} \\ 2 \end{array}} \end{array}$$

FIGURE 1

long division (Figure 1),

$$\frac{x^2 - x}{x + 1} = x - 2 + \frac{2}{x + 1}$$

Hence

$$\int \frac{x^2 - x}{x + 1}\, dx = \int (x - 2)\, dx + 2 \int \frac{1}{x + 1}\, dx$$

$$= \frac{x^2}{2} - 2x + 2 \int \frac{1}{x + 1}\, d(x + 1)$$

$$= \frac{x^2}{2} - 2x + 2 \ln|x + 1| + C \qquad \blacksquare$$

**EXAMPLE 9** Find $\int \sec x\, dx$.

*Solution* The manipulations in Examples 7 and 8 seemed quite natural and reasonable. But the one we now use is more like a magician pulling a rabbit out of a hat.

$$\int \sec x\, dx = \int \sec x \frac{\sec x + \tan x}{\sec x + \tan x}\, dx$$

$$= \int \frac{\sec^2 x + \sec x \tan x}{\sec x + \tan x}\, dx$$

$$= \int \frac{1}{\sec x + \tan x}\, d(\sec x + \tan x)$$

$$= \ln|\sec x + \tan x| + C \qquad \blacksquare$$

See Problem 72 for another way to do Example 9.

**Substitution in Definite Integrals** This topic was covered in Section 5.8. It is just like substitution in indefinite integrals, but we must remember to make the appropriate change in the limits of integration.

**EXAMPLE 10** Evaluate $\int_2^5 t\sqrt{t^2 - 4}\, dt$.

*Solution* Let $u = t^2 - 4$, so $du = 2t\, dt$; note that $u = 0$ when $t = 2$ and $u = 21$ when $t = 5$. Thus,

$$\int_2^5 t\sqrt{t^2 - 4}\, dt = \frac{1}{2} \int_2^5 (t^2 - 4)^{1/2}(2t\, dt)$$

$$= \frac{1}{2} \int_0^{21} u^{1/2}\, du$$

$$= \left[\frac{1}{3}u^{3/2}\right]_0^{21} = \frac{1}{3}(21)^{3/2} \approx 32.08 \qquad \blacksquare$$

**Use of Tables of Integrals** Our list of standard forms is very short (15 formulas); the list inside the back cover of this book is longer (113 formulas) and potentially more useful. Notice that the integrals listed there are grouped according to type. We illustrate the use of this list.

**EXAMPLE 11** Find $\int \sqrt{6x - x^2}\, dx$ and $\int_0^{\pi/2} (\cos x) \sqrt{6 \sin x - \sin^2 x}\, dx$.

*Solution* We use Formula 102 with $a = 3$.

$$\int \sqrt{6x - x^2}\, dx = \frac{x - 3}{2} \sqrt{6x - x^2} + \frac{9}{2} \sin^{-1}\left(\frac{x - 3}{3}\right) + C$$

In the second integral, let $u = \sin x$, so $du = \cos x\, dx$. Then apply Formula 102 again.

$$\int_0^{\pi/2} \cos x \sqrt{6 \sin x - \sin^2 x}\, dx = \int_0^1 \sqrt{6u - u^2}\, du$$

$$= \left[\frac{u - 3}{2} \sqrt{6u - u^2} + \frac{9}{2} \sin^{-1}\left(\frac{u - 3}{3}\right)\right]_0^1$$

$$= -\sqrt{5} + \frac{9}{2} \sin^{-1}\left(\frac{-2}{3}\right) - \frac{9}{2} \sin^{-1}(-1)$$

$$\approx 1.55 \qquad \blacksquare$$

Much more extensive tables of integrals may be found in most libraries. One of the better known is *Standard Mathematical Tables*, published by the Chemical Rubber Company.

## CONCEPTS REVIEW

**1.** The substitution $u = 1 + x^3$ transforms $\int 3x^2(1 + x^3)^5\, dx$ to _____.

**2.** The substitution $u =$ _____ transforms $\int e^x/(4 + e^{2x})\, dx$ to $\int 1/(4 + u^2)\, du$.

**3.** The substitution $u = 1 + \sin x$ transforms $\int_0^{\pi/2} (1 + \sin x)^3 \cos x\, dx$ to _____.

**4.** The initial trick in handling $\int 1/(x^2 + 2x + 2)\, dx$ is to _____.

## PROBLEM SET 8.1

In Problems 1–58, perform the indicated integrations.

**1.** $\int (x - 2)^5\, dx$

**2.** $\int \sqrt{3x}\, dx$

**3.** $\int x(x^2 + 1)^5\, dx$

**4.** $\int x \sqrt{1 - x^2}\, dx$

**5.** $\int \frac{dx}{x^2 + 4}$

**6.** $\int \frac{e^x}{2 + e^x}\, dx$

**7.** $\int \frac{x}{x^2 + 4}\, dx$

**8.** $\int \frac{2t^2}{2t^2 + 1}\, dt$

9. $\int 6z\sqrt{4+z^2}\,dz$

10. $\int \dfrac{5}{\sqrt{2t+1}}\,dt$

11. $\int \dfrac{\tan z}{\cos^2 z}\,dz$

12. $\int \dfrac{e^{\cos z}}{\csc z}\,dz$

13. $\int \dfrac{\sin\sqrt{t}}{\sqrt{t}}\,dt$

14. $\int \dfrac{2x\,dx}{\sqrt{1-x^4}}$

15. $\int_0^{\pi/4} \dfrac{\cos x}{1+\sin^2 x}\,dx$

16. $\int_0^{3/4} \dfrac{\sin\sqrt{1-x}}{\sqrt{1-x}}\,dx$

17. $\int \dfrac{3x^2+2x}{x+1}\,dx$

18. $\int \dfrac{x^3+7x}{x-1}\,dx$

19. $\int \dfrac{2\sqrt{\tan x}}{1-\sin^2 x}\,dx$

20. $\int e^{\sin^2\theta}\sin 2\theta\,d\theta$

21. $\int \dfrac{\sin(\ln 4x^2)}{x}\,dx$

22. $\int \dfrac{\sec^2(\ln x)}{2x}\,dx$

23. $\int \dfrac{6e^x}{\sqrt{1-e^{2x}}}\,dx$

24. $\int \dfrac{x}{x^4+4}\,dx$

25. $\int \dfrac{3e^{2x}}{\sqrt{1-e^{2x}}}\,dx$

26. $\int \dfrac{x^3}{x^4+4}\,dx$

27. $\int_0^1 t3^{t^2}\,dt$

28. $\int_0^{\pi/6} 2^{\cos x}\sin x\,dx$

29. $\int \dfrac{\sin x-\cos x}{\sin x}\,dx$

30. $\int \dfrac{\sin(4t-1)}{1-\sin^2(4t-1)}\,dt$

31. $\int \dfrac{z+2}{\cot(z^2+4z-3)}\,dz$

32. $\int \csc 2t\,dt$

33. $\int e^x\sec e^x\,dx$

34. $\int e^x\sec^2(e^x)\,dx$

35. $\int \dfrac{\sec^3 x+e^{\sin x}}{\sec x}\,dx$

36. $\int \dfrac{(6t-1)\sin\sqrt{3t^2-t-1}}{\sqrt{3t^2-t-1}}\,dt$

37. $\int \dfrac{t^2\cos(t^3-2)}{\sin^2(t^3-2)}\,dt$

38. $\int \dfrac{1+\cos 2x}{\sin^2 2x}\,dx$

39. $\int \dfrac{t^2\cos^2(t^3-2)}{\sin^2(t^3-2)}\,dt$

40. $\int \dfrac{\csc^2 2t}{\sqrt{1+\cot 2t}}\,dt$

41. $\int \dfrac{e^{\tan^{-1}2t}}{1+4t^2}\,dt$

42. $\int (t+1)e^{-t^2-2t-5}\,dt$

43. $\int \dfrac{y}{\sqrt{16-9y^4}}\,dy$

44. $\int \dfrac{\sec^2 2y}{9+\tan^2 2y}\,dy$

45. $\int \dfrac{\sec x\tan x}{1+\sec^2 x}\,dx$

46. $\int \dfrac{5}{\sqrt{9-4x^2}}\,dx$

47. $\int \dfrac{e^{3t}}{\sqrt{4-e^{6t}}}\,dt$

48. $\int \dfrac{dt}{2t\sqrt{4t^2-1}}$

49. $\int_0^{\pi/2} \dfrac{\sin x}{16+\cos^2 x}\,dx$

50. $\int_0^1 \dfrac{e^{2x}-e^{-2x}}{e^{2x}+e^{-2x}}\,dx$

51. $\int \dfrac{1}{x^2+2x+5}\,dx$

52. $\int \dfrac{1}{x^2-4x+9}\,dx$

53. $\int \dfrac{dx}{9x^2+18x+10}$

54. $\int \dfrac{dx}{\sqrt{16+6x-x^2}}$

55. $\int \dfrac{x+1}{9x^2+18x+10}\,dx$

56. $\int \dfrac{3-x}{\sqrt{16+6x-x^2}}\,dx$

57. $\int \dfrac{dt}{t\sqrt{2t^2-9}}$

58. $\int \dfrac{\tan x}{\sqrt{\sec^2 x-4}}\,dx$

In Problems 59–70, use the table of integrals on the back end pages, perhaps combined with a substitution, to perform the indicated integrations.

59. $\int x\sqrt{3x+2}\,dx$

60. $\int 2t\sqrt{3-4t}\,dt$

61. $\int \dfrac{dx}{9-16x^2}$

62. $\int \dfrac{dx}{5x^2-11}$

63. $\int x^2\sqrt{9-2x^2}\,dx$

64. $\int \dfrac{\sqrt{16-3t^2}}{t}\,dt$

65. $\int \dfrac{dx}{\sqrt{5+3x^2}}$

66. $\int t^2\sqrt{3+5t^2}\,dt$

67. $\int \dfrac{dt}{\sqrt{t^2+2t-3}}$

68. $\int \dfrac{\sqrt{x^2+2x-3}}{x+1}\,dx$

69. $\int \dfrac{\sin t\cos t}{\sqrt{3\sin t+5}}\,dt$

70. $\int \dfrac{\sin x}{\cos x\sqrt{5-4\cos x}}\,dx$

71. Find the length of the curve $y=\ln(\cos x)$ between $x=0$ and $x=\pi/4$.

72. Establish the identity

$$\sec x=\dfrac{\sin x}{\cos x}+\dfrac{\cos x}{1+\sin x}$$

and then use it to give a new derivation of the formula for $\int \sec x\,dx$.

**73.** Evaluate $\int_0^{2\pi} \dfrac{x|\sin x|}{1 + \cos^2 x}\, dx$. *Hint*: Make the substitution $u = x - \pi$ in the definite integral and then use symmetry properties.

**74.** Let $R$ be the region bounded by $y = \sin x$ and $y = \cos x$ between $x = -\pi/4$ and $x = 3\pi/4$. Find the volume of the solid gotten when $R$ is revolved about $x = -\pi/4$. *Hint*:

Use cylindrical shells to write a single integral, make the substitution $u = x - \pi/4$, and apply symmetry properties.

---

**Answers to Concepts Review: 1.** $\int u^5\, du$ **2.** $e^x$

**3.** $\int_1^2 u^3\, du$ **4.** Complete the square

---

## 8.2
## SOME TRIGONOMETRIC INTEGRALS

When we combine the method of substitution with a clever use of trigonometric identities, we can integrate a wide variety of trigonometric forms. We consider five commonly encountered types.

1. $\displaystyle\int \sin^n x\, dx$ and $\displaystyle\int \cos^n x\, dx$

2. $\displaystyle\int \sin^m x \cos^n x\, dx$

3. $\displaystyle\int \tan^n x\, dx$ and $\displaystyle\int \cot^n x\, dx$

4. $\displaystyle\int \tan^m x \sec^n x\, dx$ and $\displaystyle\int \cot^m x \csc^n x\, dx$

5. $\displaystyle\int \sin mx \cos nx\, dx,\ \int \sin mx \sin nx\, dx,\ \int \cos mx \cos nx\, dx$

**Type 1 ($\int \sin^n x\, dx$, $\int \cos^n x\, dx$)**  Consider first the case where $n$ is an odd positive integer. After taking out either the factor $\sin x$ or $\cos x$, use the identity $\sin^2 x + \cos^2 x = 1$.

**EXAMPLE 1  ($n$ odd).** Find $\displaystyle\int \sin^5 x\, dx$.

*Solution*

$$\int \sin^5 x\, dx = \int \sin^4 x \sin x\, dx$$

$$= \int (1 - \cos^2 x)^2 \sin x\, dx$$

$$= \int (1 - 2\cos^2 x + \cos^4 x)\sin x\, dx$$

$$= -\int (1 - 2\cos^2 x + \cos^4 x)\, d(\cos x)$$

$$= -\cos x + \tfrac{2}{3}\cos^3 x - \tfrac{1}{5}\cos^5 x + C$$

■

---

**Useful Identities**

Some trigonometric identities needed in this section are the following.

*Pythagorean Identities*
$$\sin^2 x + \cos^2 x = 1$$
$$1 + \tan^2 x = \sec^2 x$$
$$1 + \cot^2 x = \csc^2 x$$

*Half-Angle Identities*
$$\sin^2 x = \frac{1 - \cos 2x}{2}$$
$$\cos^2 x = \frac{1 + \cos 2x}{2}$$

**EXAMPLE 2** (**n even**). Find $\displaystyle\int \sin^2 x\, dx$ and $\displaystyle\int \cos^4 x\, dx$.

*Solution*   Here we make use of half-angle identities.

$$\int \sin^2 x\, dx = \int \frac{1 - \cos 2x}{2}\, dx$$

$$= \frac{1}{2}\int dx - \frac{1}{4}\int (\cos 2x)(2)\, dx$$

$$= \frac{1}{2}\int dx - \frac{1}{4}\int \cos 2x\, d(2x)$$

$$= \frac{1}{2}x - \frac{1}{4}\sin 2x + C$$

$$\int \cos^4 x\, dx = \int \left(\frac{1 + \cos 2x}{2}\right)^2 dx$$

$$= \frac{1}{4}\int (1 + 2\cos 2x + \cos^2 2x)\, dx$$

$$= \frac{1}{4}\int dx + \frac{1}{4}\int (\cos 2x)(2)\, dx + \frac{1}{8}\int (1 + \cos 4x)\, dx$$

$$= \frac{3}{8}\int dx + \frac{1}{4}\int \cos 2x\, d(2x) + \frac{1}{32}\int \cos 4x\, d(4x)$$

$$= \frac{3}{8}x + \frac{1}{4}\sin 2x + \frac{1}{32}\sin 4x + C \qquad ■$$

**Type 2 ($\int \sin^m x \cos^n x\, dx$)**   If either $m$ or $n$ is an odd positive integer and the other exponent is any number, we factor out $\sin x$ or $\cos x$ and use the identity $\sin^2 x + \cos^2 x = 1$.

**EXAMPLE 3** (**m or n odd**). Find $\displaystyle\int \sin^3 x \cos^{-4} x\, dx$.

*Solution*

$$\int \sin^3 x \cos^{-4} x\, dx = \int (1 - \cos^2 x)(\cos^{-4} x)(\sin x)\, dx$$

$$= -\int (\cos^{-4} x - \cos^{-2} x)\, d(\cos x)$$

$$= -\left[\frac{(\cos x)^{-3}}{-3} - \frac{(\cos x)^{-1}}{-1}\right] + C$$

$$= \frac{1}{3}\sec^3 x - \sec x + C \qquad ■$$

---

**Are They Different?**

Indefinite integrations may lead to different looking answers. By one method,

$$\int \sin x \cos x\, dx$$

$$= -\int \cos x\, d(\cos x)$$

$$= -\tfrac{1}{2}\cos^2 x + C$$

By a second method,

$$\int \sin x \cos x\, dx = \int \sin x\, d(\sin x)$$

$$= \tfrac{1}{2}\sin^2 x + C$$

But two such answers should differ by at most an additive constant. Note, however, that

$$\tfrac{1}{2}\sin^2 x + C = \tfrac{1}{2}(1 - \cos^2 x) + C$$

$$= -\tfrac{1}{2}\cos^2 x + (\tfrac{1}{2} + C)$$

Now reconcile these answers with a third answer.

$$\int \sin x \cos x\, dx = \tfrac{1}{2}\int \sin 2x\, dx$$

$$= -\tfrac{1}{4}\cos 2x + C$$

If both $m$ and $n$ are even positive integers, we use half-angle identities to reduce the degree of the integrand. Example 4 gives an illustration.

**EXAMPLE 4** **(Both $m$ and $n$ even).** Find $\int \sin^2x \cos^4x\, dx$.

*Solution*

$$\int \sin^2x \cos^4x\, dx$$

$$= \int \left(\frac{1 - \cos 2x}{2}\right)\left(\frac{1 + \cos 2x}{2}\right)^2 dx$$

$$= \frac{1}{8} \int (1 + \cos 2x - \cos^2 2x - \cos^3 2x)\, dx$$

$$= \frac{1}{8} \int \left[1 + \cos 2x - \frac{1}{2}(1 + \cos 4x) - (1 - \sin^2 2x)\cos 2x\right] dx$$

$$= \frac{1}{8} \int \left[\frac{1}{2} - \frac{1}{2}\cos 4x + \sin^2 2x \cos 2x\right] dx$$

$$= \frac{1}{8} \left[\int \frac{1}{2}\, dx - \frac{1}{8} \int \cos 4x\, d(4x) + \frac{1}{2}\int \sin^2 2x\, d(\sin 2x)\right]$$

$$= \frac{1}{8} \left[\frac{1}{2}x - \frac{1}{8}\sin 4x + \frac{1}{6}\sin^3 2x\right] + C \qquad \blacksquare$$

**Type 3 ($\int \tan^n x\, dx$, $\int \cot^n x\, dx$)** In the tangent case, factor out $\tan^2x = \sec^2x - 1$; in the cotangent case, factor out $\cot^2x = \csc^2x - 1$.

**EXAMPLE 5** Find $\int \cot^4x\, dx$.

*Solution*

$$\int \cot^4x\, dx = \int \cot^2x(\csc^2x - 1)\, dx$$

$$= \int \cot^2x \csc^2x\, dx - \int \cot^2x\, dx$$

$$= -\int \cot^2x\, d(\cot x) - \int (\csc^2x - 1)\, dx$$

$$= -\tfrac{1}{3}\cot^3x + \cot x + x + C \qquad \blacksquare$$

**EXAMPLE 6**   Find $\int \tan^5 x\, dx$.

*Solution*

$$\int \tan^5 x\, dx = \int \tan^3 x (\sec^2 x - 1)\, dx$$

$$= \int \tan^3 x \sec^2 x\, dx - \int \tan^3 x\, dx$$

$$= \int \tan^3 x\, d(\tan x) - \int \tan x (\sec^2 x - 1)\, dx$$

$$= \int \tan^3 x\, d(\tan x) - \int \tan x\, d(\tan x) + \int \tan x\, dx$$

$$= \tfrac{1}{4} \tan^4 x - \tfrac{1}{2} \tan^2 x - \ln|\cos x| + C \qquad \blacksquare$$

**Type 4 ($\int \tan^m x \sec^n x\, dx$, $\int \cot^m x \csc^n x\, dx$)**

**EXAMPLE 7**   (***n* even, *m* any number**). Find $\int \tan^{-3/2} x \sec^4 x\, dx$.

*Solution*

$$\int \tan^{-3/2} x \sec^4 x\, dx = \int (\tan^{-3/2} x)(1 + \tan^2 x)\sec^2 x\, dx$$

$$= \int (\tan^{-3/2} x)\sec^2 x\, dx + \int (\tan^{1/2} x)\sec^2 x\, dx$$

$$= -2 \tan^{-1/2} x + \tfrac{2}{3} \tan^{3/2} x + C \qquad \blacksquare$$

**EXAMPLE 8**   (***m* odd, *n* any number**). Find $\int \tan^3 x \sec^{-1/2} x\, dx$.

*Solution*

$$\int \tan^3 x \sec^{-1/2} x\, dx = \int (\tan^2 x)(\sec^{-3/2} x)(\sec x \tan x)\, dx$$

$$= \int (\sec^2 x - 1)\sec^{-3/2} x\, d(\sec x)$$

$$= \int \sec^{1/2} x\, d(\sec x) - \int \sec^{-3/2} x\, d(\sec x)$$

$$= \tfrac{2}{3} \sec^{3/2} x + 2 \sec^{-1/2} x + C \qquad \blacksquare$$

**Type 5 ($\int \sin mx \cos nx\, dx$, $\int \sin mx \sin nx\, dx$, $\int \cos mx \cos nx\, dx$)**
Integrals of this type occur in alternating current theory, heat-transfer problems, and any place where Fourier series are used. To handle these integrals, we use the product identities.

$$\sin mx \cos nx = \tfrac{1}{2}[\sin(m + n)x + \sin(m - n)x]$$
$$\sin mx \sin nx = -\tfrac{1}{2}[\cos(m + n)x - \cos(m - n)x]$$
$$\cos mx \cos nx = \tfrac{1}{2}[\cos(m + n)x + \cos(m - n)x]$$

**EXAMPLE 9**  Find $\displaystyle\int \sin 2x \cos 3x\, dx$.

*Solution*

$$\int \sin 2x \cos 3x\, dx = \frac{1}{2}\int [\sin 5x + \sin(-x)]\, dx$$

$$= \frac{1}{10}\int \sin 5x\, d(5x) - \frac{1}{2}\int \sin x\, dx$$

$$= -\frac{1}{10}\cos 5x + \frac{1}{2}\cos x + C \qquad \blacksquare$$

**EXAMPLE 10**  If $m$ and $n$ are positive integers, show that

$$\int_{-\pi}^{\pi} \sin mx \sin nx\, dx = \begin{cases} 0 & \text{if } n \neq m \\ \pi & \text{if } n = m \end{cases}$$

*Solution*   If $m \neq n$,

$$\int_{-\pi}^{\pi} \sin mx \sin nx\, dx = -\frac{1}{2}\int_{-\pi}^{\pi} [\cos(m + n)x - \cos(m - n)x]\, dx$$

$$= -\frac{1}{2}\left[\frac{1}{m + n}\sin(m + n)x - \frac{1}{m - n}\sin(m - n)x\right]_{-\pi}^{\pi}$$

$$= 0$$

If $m = n$,

$$\int_{-\pi}^{\pi} \sin mx \sin nx\, dx = -\frac{1}{2}\int_{-\pi}^{\pi} [\cos 2mx - 1]\, dx$$

$$= -\frac{1}{2}\left[\frac{1}{2m}\sin 2mx - x\right]_{-\pi}^{\pi}$$

$$= -\frac{1}{2}[-2\pi] = \pi \qquad \blacksquare$$

## CONCEPTS REVIEW

**1.** To handle $\int \cos^2 x\, dx$, we first rewrite it as _____.

**2.** To handle $\int \cos^3 x\, dx$, we first rewrite it as _____.

**3.** To handle $\int \sin^2 x \cos^3 x\, dx$, we first rewrite it as _____.

**4.** To handle $\int \tan^3 x\, dx$, we first rewrite it as _____.

## PROBLEM SET 8.2

In Problems 1–30, perform the indicated integrations.

1. $\displaystyle\int \sin^2 x \, dx$

2. $\displaystyle\int \sin^4 6x \, dx$

3. $\displaystyle\int \sin^3 x \, dx$

4. $\displaystyle\int \cos^3 x \, dx$

5. $\displaystyle\int_0^{\pi/2} \cos^5 \theta \, d\theta$

6. $\displaystyle\int_0^{\pi/2} \sin^6 \theta \, d\theta$

7. $\displaystyle\int \tan^3 2z \, dz$

8. $\displaystyle\int \tan^6 \theta \, d\theta$

9. $\displaystyle\int \sin^5 4x \cos^2 4x \, dx$

10. $\displaystyle\int (\sin^3 2t) \sqrt{\cos 2t} \, dt$

11. $\displaystyle\int \cos^3 3\theta \sin^{-2} 3\theta \, d\theta$

12. $\displaystyle\int \sin^{1/2} 2z \cos^3 2z \, dz$

13. $\displaystyle\int \sin^4 3t \cos^4 3t \, dt$

14. $\displaystyle\int \cos^6 \theta \sin^2 \theta \, d\theta$

15. $\displaystyle\int \tan^3 4y \sec^3 4y \, dy$

16. $\displaystyle\int \sin^4\!\left(\frac{w}{2}\right) \cos^2\!\left(\frac{w}{2}\right) dw$

17. $\displaystyle\int \cot x \csc^4 x \, dx$

18. $\displaystyle\int \cot 2x \sec^2 2x \, dx$

19. $\displaystyle\int \tan^{-4} q \sec^2 q \, dq$

20. $\displaystyle\int \tan^5 2\theta \sec^{-3/2} 2\theta \, d\theta$

21. $\displaystyle\int \sin 4y \cos 5y \, dy$

22. $\displaystyle\int \cos y \cos 4y \, dy$

23. $\displaystyle\int \cot^4 2x \, dx$

24. $\displaystyle\int \csc^4 3y \, dy$

25. $\displaystyle\int \sec^4 7x \, dx$

26. $\displaystyle\int \cot^6 4w \, dw$

27. $\displaystyle\int \cot^3 x \, dx$

28. $\displaystyle\int \sin 3t \sin t \, dt$

29. $\displaystyle\int (\tan x + \cot x)^2 \, dx$

30. $\displaystyle\int_{-\pi}^{\pi} \cos mx \cos nx \, dx, \; m \neq n; \, m, \, n \text{ integers.}$

≈ **31.** The region bounded by $y = x + \sin x$, $y = 0$, $x = \pi$, is revolved about the $x$-axis. Find the volume of the resulting solid.

≈ **32.** The region bounded by $y = \sin^2(x^2)$, $y = 0$, and $x = \sqrt{\pi/2}$ is revolved about the $y$-axis. Find the volume of the resulting solid.

**33.** Let $f(x) = \displaystyle\sum_{n=1}^{N} a_n \sin(nx)$. Use Example 10 to show each of the following.

(a) $\dfrac{1}{\pi} \displaystyle\int_{-\pi}^{\pi} f(x)\sin(mx) \, dx = \begin{cases} a_m & \text{if } m \leq N \\ 0 & \text{if } m > N \end{cases}$

(b) $\dfrac{1}{\pi} \displaystyle\int_{-\pi}^{\pi} f^2(x) \, dx = \displaystyle\sum_{n=1}^{N} a_n^2$

*Note*: Integrals of this type occur in a subject called *Fourier series*, which has applications to heat, vibrating strings, and other physical phenomena.

**34.** Show that

$$\lim_{n \to \infty} \cos \frac{x}{2} \cos \frac{x}{4} \cos \frac{x}{8} \cdots \cos \frac{x}{2^n} = \frac{\sin x}{x}$$

by completing the following steps.

(a) $\cos \dfrac{x}{2} \cos \dfrac{x}{4} \cdots \cos \dfrac{x}{2^n}$

$$= \left[\cos \frac{1}{2^n} x + \cos \frac{3}{2^n} x + \cdots + \cos \frac{2^n - 1}{2^n} x\right]\frac{1}{2^{n-1}}$$

(See Problem 36 of Section 2.3.)
(b) Recognize a Riemann sum leading to a definite integral.
(c) Evaluate the definite integral.

**35.** Use the result of Problem 34 to obtain a famous formula of François Vieta (1540–1603), namely

$$\frac{2}{\pi} = \frac{\sqrt{2}}{2} \cdot \frac{\sqrt{2 + \sqrt{2}}}{2} \cdot \frac{\sqrt{2 + \sqrt{2 + \sqrt{2}}}}{2} \cdots$$

**36.** The shaded region (Figure 1) between one arch of $y = \sin x$, $0 \leq x \leq \pi$, and the line $y = k$, $0 \leq k \leq 1$, is revolved about the line $y = k$, generating a solid $S$. Determine $k$ so that $S$ has (a) minimum volume and (b) maximum volume.

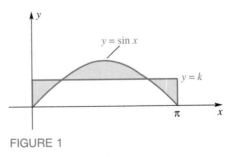

FIGURE 1

**Answers to Concepts Review:** 1. $\int [(1 + \cos 2x)/2] \, dx$
2. $\int (1 - \sin^2 x)\cos x \, dx$  3. $\int \sin^2 x(1 - \sin^2 x)\cos x \, dx$
4. $\int \tan x(\sec^2 x - 1) \, dx$

**8.3**
**RATIONALIZING**
**SUBSTITUTIONS**

Radicals in an integrand are always troublesome and we usually try to get rid of them. Often an appropriate substitution will rationalize the integrand.

**Integrands Involving $\sqrt[n]{ax + b}$**   If $\sqrt[n]{ax + b}$ appears in an integral, the substitution $u = \sqrt[n]{ax + b}$ will eliminate the radical.

**EXAMPLE 1**   Find $\displaystyle\int \frac{dx}{x - \sqrt{x}}$.

*Solution*   Let $u = \sqrt{x}$ so $u^2 = x$ and $2u\,du = dx$. Then

$$\int \frac{dx}{x - \sqrt{x}} = \int \frac{2u}{u^2 - u}\,du = 2\int \frac{1}{u - 1}\,du = 2\int \frac{1}{u - 1}\,d(u - 1)$$

$$= 2\,\ln|u - 1| + C = 2\,\ln|\sqrt{x} - 1| + C \qquad\blacksquare$$

**EXAMPLE 2**   Find $\displaystyle\int x\,\sqrt[3]{x - 4}\,dx$.

*Solution*   Let $u = \sqrt[3]{x - 4}$, so $u^3 = x - 4$ and $3u^2\,du = dx$. Then

$$\int x\,\sqrt[3]{x - 4}\,dx = \int (u^3 + 4)u \cdot 3u^2\,du = 3\int (u^6 + 4u^3)\,du$$

$$= 3\left[\frac{u^7}{7} + u^4\right] + C = \frac{3}{7}(x - 4)^{7/3} + 3(x - 4)^{4/3} + C \qquad\blacksquare$$

**EXAMPLE 3**   Find $\displaystyle\int x\,\sqrt[5]{(x + 1)^2}\,dx$.

*Solution*   Let $u = (x + 1)^{1/5}$, so $u^5 = x + 1$ and $5u^4\,du = dx$. Then

$$\int x(x + 1)^{2/5}\,dx = \int (u^5 - 1)u^2 \cdot 5u^4\,du$$

$$= 5\int (u^{11} - u^6)\,du = \tfrac{5}{12}u^{12} - \tfrac{5}{7}u^7 + C$$

$$= \tfrac{5}{12}(x + 1)^{12/5} - \tfrac{5}{7}(x + 1)^{7/5} + C \qquad\blacksquare$$

**Integrands Involving $\sqrt{a^2 - x^2}$, $\sqrt{a^2 + x^2}$, and $\sqrt{x^2 - a^2}$**   To rationalize these three expressions, we make the following trigonometric substitutions.

| Radical | Substitution | Restriction on $t$ |
|---|---|---|
| 1. $\sqrt{a^2 - x^2}$ | $x = a\sin t$ | $-\pi/2 \le t \le \pi/2$ |
| 2. $\sqrt{a^2 + x^2}$ | $x = a\tan t$ | $-\pi/2 < t < \pi/2$ |
| 3. $\sqrt{x^2 - a^2}$ | $x = a\sec t$ | $0 \le t \le \pi, t \ne \pi/2$ |

Now note the simplifications that these substitutions achieve.

1. $\sqrt{a^2 - x^2} = \sqrt{a^2 - a^2 \sin^2 t} = \sqrt{a^2 \cos^2 t} = |a \cos t| = a \cos t$
2. $\sqrt{a^2 + x^2} = \sqrt{a^2 + a^2 \tan^2 t} = \sqrt{a^2 \sec^2 t} = |a \sec t| = a \sec t$
3. $\sqrt{x^2 - a^2} = \sqrt{a^2 \sec^2 t - a^2} = \sqrt{a^2 \tan^2 t} = |a \tan t| = \pm a \tan t$

The restrictions on $t$ allowed us to remove the absolute value signs in the first two cases but they also achieved something else. These restrictions are exactly the ones we introduced in Section 7.6 in order to make sine, tangent, and secant invertible. This means that we can solve the substitution equations for $t$ in each case, and this will allow us to write our final answers in the following examples in terms of $x$.

**EXAMPLE 4**    Find $\displaystyle\int \sqrt{a^2 - x^2}\, dx$.

*Solution*    We make the substitution

$$x = a \sin t, \qquad -\frac{\pi}{2} \le t \le \frac{\pi}{2}$$

Then $dx = a \cos t\, dt$ and $\sqrt{a^2 - x^2} = a \cos t$. Thus,

$$\int \sqrt{a^2 - x^2}\, dx = \int a \cos t \cdot a \cos t\, dt = a^2 \int \cos^2 t\, dt$$

$$= \frac{a^2}{2} \int (1 + \cos 2t)\, dt$$

$$= \frac{a^2}{2} \left( t + \frac{1}{2} \sin 2t \right) + C$$

$$= \frac{a^2}{2} (t + \sin t \cos t) + C$$

Now $x = a \sin t$ is equivalent to $x/a = \sin t$ and, since $t$ was restricted so sine is invertible,

$$t = \sin^{-1} \left( \frac{x}{a} \right)$$

Also, from an identity at the end of Section 7.6,

$$\cos t = \cos \left[ \sin^{-1} \left( \frac{x}{a} \right) \right] = \sqrt{1 - \frac{x^2}{a^2}} = \frac{1}{a} \sqrt{a^2 - x^2}$$

a fact that you can also see from the triangle in Figure 1. Thus,

$$\int \sqrt{a^2 - x^2}\, dx = \frac{a^2}{2} \sin^{-1} \left( \frac{x}{a} \right) + \frac{x}{2} \sqrt{a^2 - x^2} + C \qquad \blacksquare$$

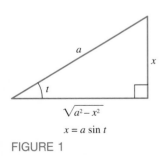

$\sqrt{a^2 - x^2}$

$x = a \sin t$

FIGURE 1

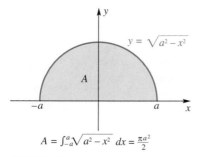

$$A = \int_{-a}^{a}\sqrt{a^2 - x^2}\, dx = \frac{\pi a^2}{2}$$

FIGURE 2

The result in Example 4 is interesting. In particular, it allows us to calculate the following definite integral, which represents the area of a semicircle (Figure 2). Thus, calculus confirms a result we already know.

$$\int_{-a}^{a}\sqrt{a^2 - x^2}\, dx = \left[\frac{a^2}{2}\sin^{-1}\left(\frac{x}{a}\right) + \frac{x}{2}\sqrt{a^2 - x^2}\right]_{-a}^{a} = \frac{a^2}{2}\left[\frac{\pi}{2} + \frac{\pi}{2}\right] = \frac{\pi a^2}{2}$$

**EXAMPLE 5** Find $\displaystyle\int \frac{\sqrt{4 - x^2}}{x^2}\, dx$.

**Solution** Let $x = 2\sin t$, $-\pi/2 \le t \le \pi/2$. Then $dx = 2\cos t\, dt$ and $\sqrt{4 - x^2} = 2\cos t$. Thus,

$$\int \frac{\sqrt{4 - x^2}}{x^2}\, dx = \int \frac{2\cos t}{4\sin^2 t}(2\cos t)\, dt$$

$$= \int \cot^2 t\, dt = \int (\csc^2 t - 1)\, dt$$

$$= -\cot t - t + C$$

Now $\sin t = x/2$, from which $t = \sin^{-1}(x/2)$. Also (see Figure 3), $\cot t = \sqrt{4 - x^2}/x$. Thus,

$$\int \frac{\sqrt{4 - x^2}}{x^2}\, dx = -\frac{\sqrt{4 - x^2}}{x} - \sin^{-1}\left(\frac{x}{2}\right) + C \qquad \blacksquare$$

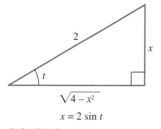

$x = 2\sin t$

FIGURE 3

**EXAMPLE 6** Find $\displaystyle\int \frac{dx}{\sqrt{9 + x^2}}$.

**Solution** Let $x = 3\tan t$, $-\pi/2 < t < \pi/2$. Then $dx = 3\sec^2 t\, dt$ and $\sqrt{9 + x^2} = 3\sec t$.

$$\int \frac{dx}{\sqrt{9 + x^2}} = \int \frac{3\sec^2 t}{3\sec t}\, dt = \int \sec t\, dt$$

$$= \ln|\sec t + \tan t| + C$$

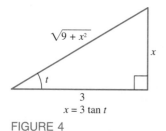

$x = 3\tan t$

FIGURE 4

The last step—namely, the integration of $\sec t$—was handled in Example 9 of Section 8.1. Now $\tan t = x/3$, which suggests the triangle in Figure 4, from which we conclude that $\sec t = \sqrt{9 + x^2}/3$. Finally,

$$\int \frac{dx}{\sqrt{9 + x^2}} = \ln\left|\frac{\sqrt{9 + x^2} + x}{3}\right| + C$$

$$= \ln|\sqrt{9 + x^2} + x| - \ln 3 + C$$

$$= \ln|\sqrt{9 + x^2} + x| + K \qquad \blacksquare$$

**EXAMPLE 7** Calculate $\displaystyle\int_2^4 \frac{\sqrt{x^2 - 4}}{x}\,dx$.

***Solution*** Let $x = 2 \sec t$, where $0 \le t < \pi/2$. Note that the restriction of $t$ to the interval above is acceptable, since $x$ is in the interval $2 \le x \le 4$. This is important because it allows us to remove the absolute value sign that normally appears when we simplify $\sqrt{x^2 - a^2}$. In our case,

$$\sqrt{x^2 - 4} = \sqrt{4 \sec^2 t - 4} = \sqrt{4 \tan^2 t} = 2|\tan t| = 2 \tan t$$

We now use the theorem on substituting in a definite integral (which requires changing the limits of integration) to write

$$\int_2^4 \frac{\sqrt{x^2 - 4}}{x}\,dx = \int_0^{\pi/3} \frac{2 \tan t}{2 \sec t} \, 2 \sec t \tan t\,dt$$

$$= \int_0^{\pi/3} 2 \tan^2 t\,dt = 2 \int_0^{\pi/3} (\sec^2 t - 1)\,dt$$

$$= 2[\tan t - t]_0^{\pi/3} = 2\sqrt{3} - \frac{2\pi}{3} \approx 1.37 \qquad \blacksquare$$

**Completing the Square** When a quadratic expression of the type $x^2 + Bx + C$ appears under a radical, completing the square will prepare it for a trigonometric substitution.

**EXAMPLE 8** Find $\displaystyle\int \frac{dx}{\sqrt{x^2 + 2x + 26}}$ and $\displaystyle\int \frac{2x}{\sqrt{x^2 + 2x + 26}}\,dx$.

***Solution*** $x^2 + 2x + 26 = x^2 + 2x + 1 + 25 = (x + 1)^2 + 25$. Let $u = x + 1$ and $du = dx$. Then,

$$\int \frac{dx}{\sqrt{x^2 + 2x + 26}} = \int \frac{du}{\sqrt{u^2 + 25}}$$

Next let $u = 5 \tan t$, $-\pi/2 < t < \pi/2$. Then $du = 5 \sec^2 t\,dt$ and $\sqrt{u^2 + 25} = \sqrt{25(\tan^2 t + 1)} = 5 \sec t$, so

$$\int \frac{du}{\sqrt{u^2 + 25}} = \int \frac{5 \sec^2 t\,dt}{5 \sec t} = \int \sec t\,dt$$

$$= \ln|\sec t + \tan t| + C$$

$$= \ln\left|\frac{\sqrt{u^2 + 25}}{5} + \frac{u}{5}\right| + C \qquad \text{(by Figure 5)}$$

$$= \ln|\sqrt{u^2 + 25} + u| - \ln 5 + C$$

$$= \ln|\sqrt{x^2 + 2x + 26} + x + 1| + K$$

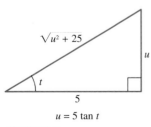

$u = 5 \tan t$

**FIGURE 5**

To handle the second integral, we write

$$\int \frac{2x}{\sqrt{x^2 + 2x + 26}}\,dx = \int \frac{2x + 2}{\sqrt{x^2 + 2x + 26}}\,dx - 2\int \frac{1}{\sqrt{x^2 + 2x + 26}}\,dx$$

The first of the integrals on the right is handled by the substitution $u = x^2 + 2x + 26$; the second was just done. We obtain

$$\int \frac{2x}{\sqrt{x^2 + 2x + 26}}\,dx$$
$$= 2\sqrt{x^2 + 2x + 26} - 2\ln\left|\sqrt{x^2 + 2x + 26} + x + 1\right| + K \quad\blacksquare$$

## CONCEPTS REVIEW

**1.** To handle $\int x\sqrt{x - 3}\,dx$, make the substitution $u = $ _____.

**2.** To handle an integral involving $\sqrt{4 - x^2}$, make the substitution $x = $ _____.

**3.** To handle an integral involving $\sqrt{4 + x^2}$, make the substitution $x = $ _____.

**4.** To handle an integral involving $\sqrt{x^2 - 4}$, make the substitution $x = $ _____.

## PROBLEM SET 8.3

In Problems 1–20, perform the indicated integrations.

**1.** $\int x\sqrt{x + 1}\,dx$

**2.** $\int x\sqrt[3]{x + \pi}\,dx$

**3.** $\int \frac{t\,dt}{\sqrt{3t + 4}}$

**4.** $\int \frac{x^2 + 3x}{\sqrt{x + 4}}\,dx$

**5.** $\int_1^2 \frac{dt}{\sqrt{t + e}}$

**6.** $\int_0^1 \frac{\sqrt{t}}{t + 1}\,dt$

**7.** $\int t(3t + 2)^{3/2}\,dt$

**8.** $\int x(1 - x)^{2/3}\,dx$

**9.** $\int \frac{\sqrt{4 - x^2}}{x}\,dx$

**10.** $\int \frac{x^2\,dx}{\sqrt{16 - x^2}}$

**11.** $\int \frac{dx}{x\sqrt{x^2 - 9}}$

**12.** $\int \frac{dx}{(x^2 + 4)^{3/2}}$

**13.** $\int_2^3 \frac{dt}{t^2\sqrt{t^2 - 1}}$

**14.** $\int_{-2}^{-3} \frac{\sqrt{t^2 - 1}}{t^3}\,dt$

**15.** $\int \frac{t}{\sqrt{1 - t^2}}\,dt$

**16.** $\int_5^1 \frac{t}{\sqrt{t^2 + 16}}\,dt$

**17.** $\int \frac{2z - 3}{\sqrt{1 - z^2}}\,dz$

**18.** $\int_0^\pi \frac{\pi x - 1}{\sqrt{x^2 + \pi^2}}\,dx$

**19.** $\int \frac{y^3}{(y^2 + 9)^{3/2}}\,dy$

**20.** $\int \frac{y^2}{(16 - y^2)^{5/2}}\,dy$

In Problems 21–30, use the method of completing the square, along with a trigonometric substitution if needed, to do the indicated integration.

**21.** $\int \frac{dx}{\sqrt{x^2 + 2x + 5}}$

**22.** $\int \frac{dx}{\sqrt{x^2 + 4x + 5}}$

**23.** $\int \frac{3x}{\sqrt{x^2 + 2x + 5}}\,dx$

**24.** $\int \frac{2x - 1}{\sqrt{x^2 + 4x + 5}}\,dx$

**25.** $\int \sqrt{5 - 4x - x^2}\,dx$

**26.** $\int \frac{dx}{\sqrt{16 + 6x - x^2}}$

**27.** $\int \frac{dx}{\sqrt{4x - x^2}}$

**28.** $\int \frac{x}{\sqrt{4x - x^2}}\,dx$

**29.** $\int \frac{2x + 1}{x^2 + 2x + 2}\,dx$

**30.** $\int \frac{2x - 1}{x^2 - 6x + 18}\,dx$

**31.** The region bounded by $y = 1/(x^2 + 2x + 5)$, $y = 0$, $x = 0$, and $x = 1$ is revolved about the $x$-axis. Find the volume of the resulting solid.

**32.** The region of Problem 31 is revolved about the $y$-axis. Find the volume of the resulting solid.

**33.** Find $\int \frac{x\,dx}{x^2 + 9}$ by (a) an algebraic substitution and (b) a trigonometric substitution. Then reconcile your answers.

**34.** Find $\int_0^3 \dfrac{x^3\,dx}{\sqrt{9+x^2}}$ by making the substitutions

$$u = \sqrt{9+x^2},\ u^2 = 9+x^2,\ 2u\,du = 2x\,dx$$

**35.** Find $\int \dfrac{\sqrt{4-x^2}}{x}\,dx$ by (a) the substitution $u = \sqrt{4-x^2}$ and (b) a trigonometric substitution. Then reconcile your answers.

**36.** Two circles of radius $b$ intersect as shown in Figure 6 with their centers $2a$ apart ($0 \le a \le b$). Find the area of the region in which they overlap.

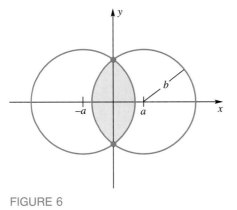

FIGURE 6

**37.** Hippocrates of Chios (ca. 430 B.C.) showed that the two shaded regions in Figure 7 have the same area (he squared the lune). Note that $C$ is the center of the lower arc. Show Hippocrates's result (a) using calculus and (b) without calculus.

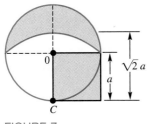

FIGURE 7

**38.** Generalize Problem 37 by finding a formula for the area of the shaded lune shown in Figure 8.

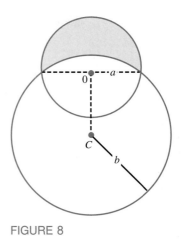

FIGURE 8

**39.** Starting at $(a, 0)$, an object is pulled along by a string of length $a$ with the pulling end moving along the positive $y$-axis. The path of the object is a curve called a **tractrix** and has the property that the string is always tangent to the curve (see Figure 9). Set up a differential equation for the curve and solve it.

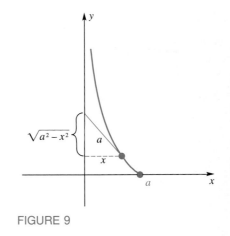

FIGURE 9

**Answers to Concepts Review:** **1.** $\sqrt{x-3}$ **2.** $2\sin t$
**3.** $2\tan t$ **4.** $2\sec t$

---

**8.4**
**INTEGRATION BY PARTS**

If integration by substitution fails, it may be possible to use a double substitution, better known as *integration by parts*. This method is based on the integration of the formula for the derivative of a product of two functions. Let $u = u(x)$ and $v = v(x)$. Then

$$D_x[u(x)v(x)] = u(x)v'(x) + v(x)u'(x)$$

By integrating both members of this equation, we obtain

$$u(x)v(x) = \int u(x)v'(x)\,dx + \int v(x)u'(x)\,dx$$

or, after rearrangement,

$$\int u(x)v'(x)\,dx = u(x)v(x) - \int v(x)u'(x)\,dx$$

Since $dv = v'(x)\,dx$ and $du = u'(x)\,dx$, the preceding equation is usually written symbolically as follows.

A Geometric Interpretation
of
Integration by Parts

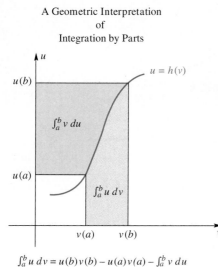

$$\int_a^b u\,dv = u(b)v(b) - u(a)v(a) - \int_a^b v\,du$$

FIGURE 1

### INTEGRATION BY PARTS—INDEFINITE INTEGRALS

$$\int u\,dv = uv - \int v\,du$$

The corresponding formula for definite integrals is

$$\int_a^b u(x)v'(x)\,dx = [u(x)v(x)]_a^b - \int_a^b v(x)u'(x)\,dx$$

We abbreviate this as follows (see also Figure 1).

### INTEGRATION BY PARTS—DEFINITE INTEGRALS

$$\int_a^b u\,dv = [uv]_a^b - \int_a^b v\,du$$

These formulas allow us to shift the problem of integrating $u\,dv$ to that of integrating $v\,du$. Success depends on the proper choice of $u$ and $dv$, which comes with practice.

### Simple Examples

**EXAMPLE 1**    Find $\displaystyle\int x\cos x\,dx$.

*Solution*    We wish to write $x\cos x\,dx$ as $u\,dv$. One possibility is to let $u = x$ and $dv = \cos x\,dx$. Then $du = dx$ and $v = \int \cos x\,dx = \sin x$ (we can omit the arbitrary constant at this stage). Here is a summary of this double substitution in a convenient format.

$$u = x \qquad dv = \cos x\,dx$$
$$du = dx \qquad v = \sin x$$

The integration-by-parts formula gives

$$\int \underbrace{x}_{u} \underbrace{\cos x \, dx}_{dv} = \underbrace{x}_{u} \underbrace{\sin x}_{v} - \int \underbrace{\sin x}_{v} \underbrace{dx}_{du}$$

$$= x \sin x + \cos x + C$$

We were successful on our first try. Another substitution would be

$$u = \cos x \qquad dv = x \, dx$$

$$du = -\sin x \, dx \qquad v = \frac{x^2}{2}$$

The integration-by-parts formula gives

$$\int \underbrace{\cos x}_{u} \underbrace{x \, dx}_{dv} = \underbrace{(\cos x)}_{u} \underbrace{\frac{x^2}{2}}_{v} - \int \underbrace{\frac{x^2}{2}}_{v} \underbrace{(-\sin x \, dx)}_{du}$$

which is correct but not helpful. The new integral on the right-hand side is more complicated than the original one. Thus, we see the importance of a wise choice for $u$ and $dv$. ∎

**EXAMPLE 2**    Find $\int_{1}^{2} \ln x \, dx$.

*Solution*    We make the following substitutions.

$$u = \ln x \qquad dv = dx$$

$$du = \left(\frac{1}{x}\right) dx \qquad v = x$$

Then, according to the integration-by-parts formula,

$$\int_{1}^{2} \ln x \, dx = [x \ln x]_{1}^{2} - \int_{1}^{2} x \frac{1}{x} \, dx$$

$$= 2 \ln 2 - \int_{1}^{2} dx$$

$$= 2 \ln 2 - 1$$ ∎

**EXAMPLE 3**    Find $\int \arcsin x \, dx$.

*Solution*    We make the substitutions

$$u = \arcsin x \qquad dv = dx$$

$$du = \frac{1}{\sqrt{1 - x^2}} \, dx \qquad v = x$$

Then

$$\int \arcsin x \, dx = x \arcsin x - \int \frac{x}{\sqrt{1 - x^2}} \, dx$$

$$= x \arcsin x + \frac{1}{2} \int (1 - x^2)^{-1/2}(-2x \, dx)$$

$$= x \arcsin x + \frac{1}{2} \cdot 2(1 - x^2)^{1/2} + C$$

$$= x \arcsin x + \sqrt{1 - x^2} + C$$ ∎

**Repeated Integration by Parts**  Sometimes it is necessary to apply integration by parts several times.

**EXAMPLE 4**  Find $\int x^2 \sin x \, dx$.

*Solution*  Let

$$u = x^2 \qquad dv = \sin x \, dx$$
$$du = 2x \, dx \qquad v = -\cos x$$

Then

$$\int x^2 \sin x \, dx = -x^2 \cos x + 2 \int x \cos x \, dx$$

We have improved our situation (the exponent on $x$ has gone from 2 to 1), which suggests reapplying integration by parts to the integral on the right. Actually, we did this integration in Example 1, so we will make use of the result obtained there.

$$\int x^2 \sin x \, dx = -x^2 \cos x + 2(x \sin x + \cos x + C)$$

$$= -x^2 \cos x + 2x \sin x + 2 \cos x + K$$ ∎

**EXAMPLE 5**  Find $\int e^x \sin x \, dx$.

*Solution*  Take $u = e^x$ and $dv = \sin x \, dx$. Then $du = e^x \, dx$ and $v = -\cos x$. Thus,

$$\int e^x \sin x \, dx = -e^x \cos x + \int e^x \cos x \, dx$$

which does not seem to have improved things. But do not give up; let us integrate by parts again. In the right integral, let $u = e^x$ and $dv = \cos x \, dx$, so $du = e^x\,dx$ and $v = \sin x$. Then

$$\int e^x \cos x\,dx = e^x \sin x - \int e^x \sin x\,dx$$

When we substitute this in our first result, we get

$$\int e^x \sin x\,dx = -e^x \cos x + e^x \sin x - \int e^x \sin x\,dx$$

By transposing the last term to the left side and combining terms, we obtain

$$2\int e^x \sin x\,dx = e^x(\sin x - \cos x) + C$$

from which

$$\int e^x \sin x\,dx = \tfrac{1}{2}e^x(\sin x - \cos x) + K \qquad\blacksquare$$

The fact that the integral we wanted to find reappeared on the right side is what made Example 5 work. Here is another illustration of this technique.

**EXAMPLE 6**  Find $\displaystyle\int \sec^3\theta\,d\theta$.

***Solution***  Let $u = \sec\theta$ and $dv = \sec^2\theta\,d\theta$. Then $du = \sec\theta\tan\theta\,d\theta$, $v = \tan\theta$, and

$$\int \sec^3\theta\,d\theta = \sec\theta\tan\theta - \int \sec\theta\tan^2\theta\,d\theta$$

$$= \sec\theta\tan\theta - \int \sec\theta\,(\sec^2\theta - 1)\,d\theta$$

$$= \sec\theta\tan\theta - \int \sec^3\theta\,d\theta + \int \sec\theta\,d\theta$$

Consequently,

$$2\int \sec^3\theta\,d\theta = \sec\theta\tan\theta + \int \sec\theta\,d\theta$$

$$= \sec\theta\tan\theta + \ln|\sec\theta + \tan\theta| + C$$

the latter by Example 9 of Section 8.1. Finally,

$$\int \sec^3\theta\,d\theta = \tfrac{1}{2}\sec\theta\tan\theta + \tfrac{1}{2}\ln|\sec\theta + \tan\theta| + K \qquad\blacksquare$$

See Problem 82 for another way to do Example 6.

**Reduction Formulas**    A formula of the form

$$\int f^n(x)\, dx = g(x) + \int f^k(x)\, dx,$$

where $k < n$ is called a **reduction formula** (the exponent on $f$ is reduced). Such formulas can often be obtained via integration by parts.

**EXAMPLE 7**   Derive a reduction formula for $\int \sin^n x\, dx$.

*Solution*   Let $u = \sin^{n-1} x$ and $dv = \sin x\, dx$. Then

$$du = (n-1)\sin^{n-2} x \cos x\, dx \quad \text{and} \quad v = -\cos x$$

from which

$$\int \sin^n x\, dx = -\sin^{n-1} x \cos x + (n-1)\int \sin^{n-2} x \cos^2 x\, dx$$

If we replace $\cos^2 x$ by $1 - \sin^2 x$ in the last integral, we obtain

$$\int \sin^n x\, dx = -\sin^{n-1} x \cos x + (n-1)\int \sin^{n-2} x\, dx - (n-1)\int \sin^n x\, dx$$

After combining the first and last integrals above, and solving for $\int \sin^n x\, dx$, we get the reduction formula (valid for $n \geq 2$)

$$\int \sin^n x\, dx = \frac{-\sin^{n-1} x \cos x}{n} + \frac{n-1}{n}\int \sin^{n-2} x\, dx \qquad \blacksquare$$

**EXAMPLE 8**   Use the reduction formula above to evaluate

$$\int_0^{\pi/2} \sin^8 x\, dx$$

*Solution*   Note first that

$$\int_0^{\pi/2} \sin^n x\, dx = \left[\frac{-\sin^{n-1} x \cos x}{n}\right]_0^{\pi/2} + \frac{n-1}{n}\int_0^{\pi/2} \sin^{n-2} x\, dx$$

$$= 0 + \frac{n-1}{n}\int_0^{\pi/2} \sin^{n-2} x\, dx$$

Thus,

$$\int_0^{\pi/2} \sin^8 x \, dx = \frac{7}{8} \int_0^{\pi/2} \sin^6 x \, dx$$

$$= \frac{7}{8} \cdot \frac{5}{6} \int_0^{\pi/2} \sin^4 x \, dx$$

$$= \frac{7}{8} \cdot \frac{5}{6} \cdot \frac{3}{4} \int_0^{\pi/2} \sin^2 x \, dx$$

$$= \frac{7}{8} \cdot \frac{5}{6} \cdot \frac{3}{4} \cdot \frac{1}{2} \int_0^{\pi/2} 1 \, dx$$

$$= \frac{7}{8} \cdot \frac{5}{6} \cdot \frac{3}{4} \cdot \frac{1}{2} \cdot \frac{\pi}{2} = \frac{35}{256} \pi \qquad \blacksquare$$

The general formula for $\int_0^{\pi/2} \sin^n x \, dx$ can be found in a similar way (Formula 113 at the back of the book).

## CONCEPTS REVIEW

**1.** The integration-by-parts formula says $\int u \, dv = $ _____.

**2.** To apply this formula to $\int x \sin x \, dx$, let $u = $ _____ and $dv = $ _____.

**3.** Applying the integration-by-parts formula yields the value _____ for $\int_0^{\pi/2} x \sin x \, dx$.

**4.** A formula that expresses $\int f^n(x) \, dx$ in terms of $\int f^k(x) \, dx$, where $k < n$, is called a _____ formula.

## PROBLEM SET 8.4

In Problems 1–42, use integration by parts to perform the indicated integrations.

**1.** $\int xe^x \, dx$

**2.** $\int xe^{3x} \, dx$

**3.** $\int te^{5t + \pi} \, dt$

**4.** $\int (t + 7)e^{2t + 3} \, dt$

**5.** $\int x \cos x \, dx$

**6.** $\int x \sin 2x \, dx$

**7.** $\int (t - 3) \cos(t - 3) \, dt$

**8.** $\int (x - \pi) \sin(x) \, dx$

**9.** $\int t\sqrt{t + 1} \, dt$

**10.** $\int t\sqrt[3]{2t + 7} \, dt$

**11.** $\int \ln 3x \, dx$

**12.** $\int \ln(7x^5) \, dx$

**13.** $\int \arctan x \, dx$

**14.** $\int \arctan 5x \, dx$

**15.** $\int \frac{\ln x}{x^2} \, dx$

**16.** $\int_2^3 \frac{\ln 2x^5}{x^2} \, dx$

**17.** $\int \sqrt{t} \ln t \, dt$

**18.** $\int_5^1 \sqrt{2x} \ln x^3 \, dx$

**19.** $\int z^3 \ln z \, dz$

**20.** $\int t \arctan t \, dt$

**21.** $\int \arctan(1/t) \, dt$

**22.** $\int t^5 \ln(t^7) \, dt$

**23.** $\int x \cos^2 x \sin x \, dx$

**24.** $\int x \sin^3 \pi x \cos \pi x \, dx$

**25.** $\int_{\pi/6}^{\pi/2} x \csc^2 x \, dx$

**26.** $\int_{\pi/4}^{\pi/2} \csc^3 x \, dx$

**27.** $\int_{\pi/6}^{\pi/4} x \sec^2 x \, dx$

**28.** $\int \sec^{-1} \sqrt{x} \, dx$

**29.** $\int x^5 \sqrt{x^3 + 4} \, dx$

**30.** $\int x^{13} \sqrt{x^7 + 1} \, dx$

**31.** $\int \frac{t^7}{(7 - 3t^4)^{3/2}} \, dt$

**32.** $\int x^3 \sqrt{4 - x^2} \, dx$

**33.** $\int \frac{z^7}{(4 - z^4)^2} \, dz$

**34.** $\int x \cosh x \, dx$

35. $\int x \sinh x \, dx$

36. $\int \tan^{-1} \sqrt{z} \, dz$

37. $\int \sec^3 x \, dx$

38. $\int e^{\sqrt{z}} \, dz$

39. $\int \dfrac{\ln x}{\sqrt{x}} \, dx$

40. $\int x(3x + 10)^{49} \, dx$

41. $\int x2^x \, dx$

42. $\int za^z \, dz$

In Problems 43–54, use the formula for integration by parts twice to perform the indicated integrations (see Examples 4 and 5).

43. $\int x^2 e^x \, dx$

44. $\int x^5 e^{x^2} \, dx$

45. $\int \ln^2 z \, dz$

46. $\int \ln^2 x^{20} \, dx$

47. $\int e^t \cos t \, dt$

48. $\int e^{at} \sin t \, dt$

49. $\int x^2 \cos x \, dx$

50. $\int r^2 \sin r \, dr$

51. $\int \sin(\ln x) \, dx$

52. $\int \cos(\ln x) \, dx$

53. $\int (\ln x)^3 \, dx$. *Hint*: Use Problem 45.

54. $\int (\ln x)^4 \, dx$. *Hint*: Use Problems 45 and 53.

In Problems 55–60 derive the given integration by using integration by parts.

55. $\int \sin(x) \sin(3x) \, dx$
$= -\tfrac{3}{8} \sin x \cos 3x + \tfrac{1}{8} \cos x \sin 3x + C$

56. $\int \cos(5x) \sin(7x) \, dx$
$= -\tfrac{7}{24} \cos 5x \cos 7x - \tfrac{5}{24} \sin 5x \sin 7x + C$

57. $\int e^{az} \sin \beta z \, dz = \dfrac{e^{az}(\alpha \sin \beta z - \beta \cos \beta z)}{\alpha^2 + \beta^2} + C$

58. $\int e^{az} \cos \beta z \, dz = \dfrac{e^{az}(\alpha \cos \beta z + \beta \sin \beta z)}{\alpha^2 + \beta^2} + C$

59. $\int x^\alpha \ln x \, dx = \dfrac{x^{\alpha+1}}{\alpha+1} \ln x - \dfrac{x^{\alpha+1}}{(\alpha+1)^2} + C, \alpha \neq -1$

60. $\int x^\alpha (\ln x)^2 \, dx = \dfrac{x^{\alpha+1}}{\alpha+1} (\ln x)^2$
$- 2\dfrac{x^{\alpha+1}}{(\alpha+1)^2} \ln x + 2 \dfrac{x^{\alpha+1}}{(\alpha+1)^3} + C, \alpha \neq -1$

Integration by parts can frequently be used to derive reduction formulas (Example 7). Computer algebra systems in their present state are not usually able to do this. In Problems 61–69 we ask you to derive the given reduction formula using integration by parts.

61. $\int x^\alpha e^{\beta x} \, dx = \dfrac{x^\alpha e^{\beta x}}{\beta} - \dfrac{\alpha}{\beta} \int x^{\alpha-1} e^{\beta x} \, dx$

62. $\int x^\alpha \sin \beta x \, dx$
$= -\dfrac{x^\alpha \cos \beta x}{\beta} + \dfrac{\alpha}{\beta} \int x^{\alpha-1} \cos \beta x \, dx$

63. $\int x^\alpha \cos \beta x \, dx = \dfrac{x^\alpha \sin \beta x}{\beta} - \dfrac{\alpha}{\beta} \int x^{\alpha-1} \sin \beta x \, dx$

64. $\int (\ln x)^\alpha \, dx = x(\ln x)^\alpha - \alpha \int (\ln x)^{\alpha-1} \, dx$

65. $\int (a^2 - x^2)^\alpha \, dx$
$= x(a^2 - x^2)^\alpha + 2\alpha \int x^2 (a^2 - x^2)^{\alpha-1} \, dx$

66. $\int \cos^\alpha x \, dx$
$= \dfrac{\cos^{\alpha-1} x \sin x}{\alpha} + \dfrac{\alpha-1}{\alpha} \int \cos^{\alpha-2} x \, dx$

67. $\int \cos^\alpha \beta x \, dx =$
$\dfrac{\cos^{\alpha-1} \beta x \sin \beta x}{\alpha \beta} + \dfrac{\alpha-1}{\alpha} \int \cos^{\alpha-2} \beta x \, dx$

68. $\int \sin^\alpha x \, dx$
$= -\dfrac{\sin^{\alpha-1} x \cos x}{\alpha} + \dfrac{\alpha-1}{\alpha} \int \sin^{\alpha-2} x \, dx$

69. $\int \sin^\alpha \beta x \, dx$
$= -\dfrac{\sin^{\alpha-1} \beta x \cos \beta x}{\alpha \beta} + \dfrac{\alpha-1}{\alpha} \int \sin^{\alpha-2} \beta x \, dx$

70. Use the result of Problem 61 to fully evaluate
$$\int x^4 e^{3x} \, dx = \tfrac{1}{3} x^4 e^{3x} - \tfrac{4}{9} x^3 e^{3x} + \tfrac{4}{9} e^{3x} x^2 - \tfrac{8}{27} e^{3x} x + \tfrac{8}{81} e^{3x} + C$$

71. Use the result of Problems 62 and 63 to fully evaluate $\int x^4 \cos 3x \, dx$
$= \tfrac{1}{3} x^4 \sin 3x + \tfrac{4}{9} x^3 \cos 3x - \tfrac{4}{9} x^2 \sin 3x + \tfrac{8}{81} \sin 3x -$
$\tfrac{8}{27} x \cos 3x + C$

72. Use the result of Problem 67 to fully evaluate
$\int \cos^6 3x \, dx = \tfrac{1}{18} \sin 3x \cos^5 3x + \tfrac{5}{72} \sin 3x \cos^3 3x +$
$\tfrac{5}{48} \sin 3x \cos 3x + \tfrac{5}{16} x + C$

≈ **73.** Find the area of the region bounded by the curve $y = \ln x$, the $x$-axis, and the line $x = e$.

**74.** Find the volume of the solid generated by revolving the region of Problem 73 about the $x$-axis.

**75.** Find the area of the region bounded by the curves $y = 3e^{-x/3}$, $y = 0$, and $x = 9$. Make a sketch.

**76.** Find the volume of the solid generated by revolving the region described in Problem 75 about the $x$-axis.

**77.** Find the area of the region bounded by the graphs of $y = x \sin x$ and $y = x \cos x$ from $x = 0$ to $x = \pi/4$.

**78.** Find the volume of the solid obtained by revolving the region under the graph of $y = \sin(x/2)$ from $x = 0$ to $x = 2\pi$ about the $y$-axis.

**79.** Find the centroid of the region bounded by $y = \ln x^2$ and the $x$-axis from $x = 1$ to $x = e$.

**80.** Evaluate the integral $\int \cot x \csc^2 x \, dx$ by parts in two different ways:

(a) By differentiating $\cot x$
(b) By differentiating $\csc x$
(c) Show that the two results are equivalent up to a constant.

**81.** If $p(x)$ is a polynomial of degree $n$ and $G_1, G_2, \dots,$ $G_{n+1}$ are successive antiderivatives of $g$, then by repeated integration by parts

$$\int p(x)g(x) \, dx = p(x)G_1(x) - p'(x)G_2(x) + p''(x)G_3(x) - \cdots$$
$$+ (-1)^n p^{(n)}(x)G_{n+1}(x) + C$$

Use this result to find each of the following.

(a) $\int (x^3 - 2x)e^x \, dx$

(b) $\int (x^2 - 3x + 1)\sin x \, dx$

**82.** Establish the identity

$$2 \sec^3 x \, dx = \sec x \, dx + d(\sec x \tan x)$$

and use it to give a new derivation of the formula for $\int \sec^3 x \, dx$ (Example 6).

**83.** Evaluate $\int_0^{2\pi} \sin^n x \, dx$. Hint: First rewrite this expression in terms of $\int_0^{\pi/2} \sin^n x \, dx$ and then use Formula 113 from the back of the book.

≈ **84.** The graph of $y = x \sin x$ for $x \geq 0$ is sketched in Figure 2.

(a) Find a formula for the area of the $n$th hump.

(b) The second hump is revolved about the $y$-axis. Find the volume of the resulting solid.

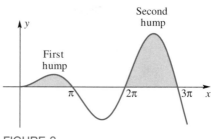

**FIGURE 2**

**85.** The Fourier coefficient, $a_n = \dfrac{1}{\pi}\displaystyle\int_{-\pi}^{\pi} f(x)\sin nx \, dx$, plays an important role in applied mathematics. Show that if $f'(x)$ is continuous on $[-\pi, \pi]$, then $\lim\limits_{n \to \infty} a_n = 0$. Hint: Integration by parts.

**86.** Let $G_n = \sqrt[n]{(n + 1)(n + 2) \cdots (n + n)}$. Show that $\lim\limits_{n \to \infty} (G_n/n) = 4/e$. Hint: Consider $\ln(G_n/n)$, recognize it as a Riemann sum, and use Example 2.

**87.** Find the error in the following "proof" that $0 = 1$. In $\int (1/t) \, dt$, set $u = 1/t$ and $dv = dt$. Then $du = -t^{-2} \, dt$, and $uv = 1$. Integration by parts gives

$$\int (1/t) \, dt = 1 - \int (-1/t) \, dt$$

or $0 = 1$.

**88.** Suppose that you want to compute the integral of $\int e^{5x}(4 \cos 7x + 6 \sin 7x) \, dx$, and you know from experience that the result will be of the form $e^{5x}(C_1 \cos 7x + C_2 \sin 7x) + C_3$, compute $C_1$ and $C_2$ by differentiating the result and setting it equal to the integrand.

Many surprising theoretical results can be derived through the use of integration by parts. In all cases one starts with an integral. We explore some of those results here.

**89.** Show that $\displaystyle\int_a^b f(x) \, dx = xf(x) \Big|_a^b - \int_a^b xf'(x) \, dx =$

$$(x - a) f(x) \Big|_a^b - \int_a^b (x - a) f'(x) \, dx$$

**90.** Using Problem 89 and replacing $f$ by $f'$ show that $f(b) - f(a) = \displaystyle\int_a^b f'(x) \, dx =$

$$f'(b)(b - a) - \int_a^b (x - a) f''(x) \, dx = f'(a)(b - a) - \int_a^b (x - b) f''(x) \, dx$$

**91.** Show that $f(t) = f(a) + \sum_{i=1}^{n} \frac{f^{(i)}(a)}{i!}(t - a)^i +$

$\int_a^t \frac{(t - a)^n}{n!} f^{(n+1)}(x) dx$ provided that $f$ can be differentiated $n + 1$ times.

**92.** The *Beta function*, which is important in many branches of mathematics, is defined as

$$B(\alpha, \beta) = \int_0^1 x^\alpha(1 - x)^\beta dx = \int_0^1 x^\beta(1 - x)^\alpha dx = B(\beta, \alpha)$$

with the condition that $\alpha > 0$, and $\beta > 0$.

(a) Show by a change of variables that $B(\alpha, \beta) = \int_0^1 x^\beta(1 - x)^\alpha dx\ B(\beta, \alpha)$

(b) Integrate by parts to show that

$B(\alpha, \beta) = \frac{\alpha}{\beta + 1} B(\alpha - 1, \beta + 1) = \frac{\beta}{\alpha + 1} B(\alpha + 1, \beta - 1)$

(c) Assume now that $\alpha = n$ and $\beta = m$ and that $n$ and $m$ are positive integers. By using the result in (b) repeatedly show that

$$B(n, m) = \frac{n!m!}{(n + m + 1)!}.$$

This result is even valid in the case where $n$ and $m$ are not integers, provided we can give a proper meaning to $n!$, $m!$ and $(n + m + 1)!$.

**93.** Suppose that $f(t)$ has the property that $f'(a) = f'(b) = 0$ and that $f(t)$ has two continuous derivatives. Use integration by parts to prove that $\int_a^b f''(t)f(t) dt \le 0$. *Hint:* Use integration by parts by differentiating $f(t)$ and integrating $f''(t)$. This result has many applications in the field of applied mathematics and partial differential equations.

**94.** Derive the formula

$$\int_0^x \left( \int_0^t f(z) dz \right) dt = \int_0^x f(t)(x - t) dt$$

using integration by parts.

**95.** Generalize the formula given in Problem 94 to one for an $n$-fold iterated integral

$$\int_0^x \int_0^{t_1} \cdots \int_0^{t_{n-1}} f(t_n) dt_n \ldots dt_1 = \frac{1}{(n - 1)!} \int_0^x f(t_1)(x - t_1)^{n-1} dt_1$$

**96.** If $P_n(x)$ is a polynomial of degree $n$, show that

$$\int e^x P_n(x) dx = e^x \sum_{j=0}^{n} (-1)^j \frac{d^j P_n(x)}{dx^j}.$$

**97.** Use the result from Problem 96 to evaluate

$$\int (3x^4 + 2x^2)e^x dx$$

---

**Answers to Concepts Review:** **1.** $uv - \int v\, du$ **2.** $x$; $\sin x\, dx$ **3.** 1 **4.** Reduction

---

**8.5**
**INTEGRATION OF**
**RATIONAL FUNCTIONS**

A **rational function** is by definition the quotient of two polynomial functions. Examples are

$$f(x) = \frac{2}{(x + 1)^3}, \quad g(x) = \frac{2x + 2}{x^2 - 4x + 8}, \quad h(x) = \frac{x^5 + 2x^3 - x + 1}{x^3 + 5x}$$

Of these, $f$ and $g$ are **proper rational functions**, meaning that the degree of the numerator is less than that of the denominator. An improper (not proper) rational function can always be written as a sum of a polynomial function and a proper rational function. Thus, for example,

$$h(x) = \frac{x^5 + 2x^3 - x + 1}{x^3 + 5x} = x^2 - 3 + \frac{14x + 1}{x^3 + 5x}$$

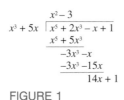

$$\begin{array}{r} x^2 - 3 \\ x^3 + 5x\ \overline{\smash{\big)}\ x^5 + 2x^3 - x + 1} \\ \underline{x^5 + 5x^3} \\ -3x^3 - x \\ \underline{-3x^3 - 15x} \\ 14x + 1 \end{array}$$

**FIGURE 1**

a result obtained by long division (Figure 1). Since polynomials are easy to integrate, the problem of integrating rational functions is really that of integrating proper rational functions. But can we always integrate proper rational functions? In theory, the answer is yes, though the practical details may overwhelm us. Consider first the cases of $f$ and $g$ above.

**EXAMPLE 1** Find $\int \dfrac{2}{(x + 1)^3} \, dx$.

*Solution* Think of the substitution $u = x + 1$.

$$\int \frac{2}{(x + 1)^3} \, dx = 2 \int (x + 1)^{-3} d(x + 1) = \frac{2(x + 1)^{-2}}{-2} + C$$

$$= \frac{-1}{(x + 1)^2} + C$$ ■

**EXAMPLE 2** Find $\int \dfrac{2x + 2}{x^2 - 4x + 8} \, dx$.

*Solution* Think first of the substitution $u = x^2 - 4x + 8$ for which $du = (2x - 4) \, dx$. Then write the given integral as a sum of two integrals.

$$\int \frac{2x + 2}{x^2 - 4x + 8} \, dx = \int \frac{2x - 4}{x^2 - 4x + 8} \, dx + \int \frac{6}{x^2 - 4x + 8} \, dx$$

$$= \ln|x^2 - 4x + 8| + 6 \int \frac{1}{x^2 - 4x + 8} \, dx$$

In the second integral, complete the square.

$$\int \frac{1}{x^2 - 4x + 8} \, dx = \int \frac{1}{x^2 - 4x + 4 + 4} \, dx = \int \frac{1}{(x - 2)^2 + 4} \, dx$$

$$= \int \frac{1}{(x - 2)^2 + 4} \, d(x - 2) = \frac{1}{2} \tan^{-1}\left(\frac{x - 2}{2}\right) + C$$

We conclude that

$$\int \frac{2x + 2}{x^2 - 4x + 8} \, dx = \ln|x^2 - 4x + 8| + 3 \tan^{-1}\left(\frac{x - 2}{2}\right) + K$$ ■

It is a remarkable fact that any proper rational function can be written as a sum of *simple* proper rational functions something like those illustrated in Examples 1 and 2. We must be more precise.

**Partial Fraction Decomposition (Linear Factors)** To add fractions is a standard algebraic exercise. For example,

$$\frac{2}{x - 1} + \frac{3}{x + 1} = \frac{5x - 1}{(x - 1)(x + 1)} = \frac{5x - 1}{x^2 - 1}$$

It is the reverse process of decomposing a fraction into a sum of simpler fractions that interests us now. We focus on the denominator and consider cases.

EXAMPLE 3 **(Distinct linear factors).** Decompose $(3x - 1)/(x^2 - x - 6)$ and then find its indefinite integral.

*Solution* Since the denominator factors as $(x + 2)(x - 3)$, it seems reasonable to hope for a decomposition of the following form.

(1) $$\frac{3x - 1}{(x + 2)(x - 3)} = \frac{A}{x + 2} + \frac{B}{x - 3}$$

Our job is, of course, to determine $A$ and $B$ so (1) is an identity, a task we find easier after we have multiplied both sides by $(x + 2)(x - 3)$. We obtain

(2) $$3x - 1 = A(x - 3) + B(x + 2)$$

or, equivalently,

(3) $$3x - 1 = (A + B)x + (-3A + 2B)$$

However (3) is an identity if and only if coefficients of like powers of $x$ on both sides are equal, that is,

$$A + B = 3$$
$$-3A + 2B = -1$$

By solving this pair of equations for $A$ and $B$, we obtain $A = \frac{7}{5}$, $B = \frac{8}{5}$. Consequently,

$$\frac{3x - 1}{x^2 - x - 6} = \frac{3x - 1}{(x + 2)(x - 3)} = \frac{\frac{7}{5}}{x + 2} + \frac{\frac{8}{5}}{x - 3}$$

and

$$\int \frac{3x - 1}{x^2 - x - 6}\, dx = \frac{7}{5} \int \frac{1}{x + 2}\, dx + \frac{8}{5} \int \frac{1}{x - 3}\, dx$$

$$= \frac{7}{5} \ln|x + 2| + \frac{8}{5} \ln|x - 3| + C \qquad \blacksquare$$

If there was anything difficult about this process, it was the determination of $A$ and $B$. We found their values by "brute force"; there is an easier way. In (2), which we wish to be an identity, substitute the convenient values $x = 3$ and $x = -2$, obtaining

$$8 = A \cdot 0 + B \cdot 5$$

$$-7 = A \cdot (-5) + B \cdot 0$$

This immediately gives $B = \frac{8}{5}$ and $A = \frac{7}{5}$.

You have just witnessed an odd, but correct, mathematical maneuver. Equation (1) turns out to be an identity (true for all $x$ except $-2$ and 3) if and only if the essentially equivalent equation (2) is true precisely at $-2$ and 3.

---

Solve This D.E.

". . . often, there is little resemblance between a differential equation and its solution. Who would suppose that an expression as simple as

$$\frac{dy}{dx} = \frac{1}{a^2 - x^2}$$

could be transformed into

$$y = \frac{1}{2a} \log_e \left(\frac{a + x}{a - x}\right) + C$$

This resembles the transformation of a chrysalis into a butterfly."

*Silvanus P. Thompson*

The method of partial fractions makes this an easy transformation. Do you see how it is done?

Ask yourself why this is so. Ultimately it depends on the fact that the two sides of equation (2), both linear polynomials, are identical if they have the same values at any two points.

**EXAMPLE 4** **(Distinct linear factors).** Find $\displaystyle\int \frac{5x + 3}{x^3 - 2x^2 - 3x}\, dx.$

*Solution* Since the denominator factors as $x(x + 1)(x - 3)$, we write

$$\frac{5x + 3}{x(x + 1)(x - 3)} = \frac{A}{x} + \frac{B}{x + 1} + \frac{C}{x - 3}$$

and seek to determine $A$, $B$, and $C$. Clearing of fractions gives

$$5x + 3 = A(x + 1)(x - 3) + Bx(x - 3) + Cx(x + 1)$$

Substitution of the values $x = 0$, $x = -1$, and $x = 3$ results in

$$3 = A(-3)$$
$$-2 = B(4)$$
$$18 = C(12)$$

or $A = -1$, $B = -\frac{1}{2}$, $C = \frac{3}{2}$. Thus,

$$\int \frac{5x + 3}{x^3 - 2x^2 - 3x}\, dx = -\int \frac{1}{x}\, dx - \frac{1}{2}\int \frac{1}{x + 1}\, dx + \frac{3}{2}\int \frac{1}{x - 3}\, dx$$

$$= -\ln|x| - \frac{1}{2}\ln|x + 1| + \frac{3}{2}\ln|x - 3| + C \quad \blacksquare$$

**EXAMPLE 5** **(Repeated linear factors).** Find $\displaystyle\int \frac{x}{(x - 3)^2}\, dx.$

*Solution* Now the decomposition takes the form

$$\frac{x}{(x - 3)^2} = \frac{A}{x - 3} + \frac{B}{(x - 3)^2}$$

with $A$ and $B$ to be determined. After clearing of fractions, we get

$$x = A(x - 3) + B$$

If we now substitute the convenient value $x = 3$ and any other value, such as $x = 0$, we obtain $B = 3$ and $A = 1$. Thus,

$$\int \frac{x}{(x - 3)^2}\, dx = \int \frac{1}{x - 3}\, dx + 3\int \frac{1}{(x - 3)^2}\, dx$$

$$= \ln|x - 3| - \frac{3}{x - 3} + C \quad \blacksquare$$

EXAMPLE 6    **(Some distinct, some repeated linear factors).** Find

$$\int \frac{3x^2 - 8x + 13}{(x + 3)(x - 1)^2} \, dx.$$

*Solution*   We decompose the integrand in the following way.

$$\frac{3x^2 - 8x + 13}{(x + 3)(x - 1)^2} = \frac{A}{x + 3} + \frac{B}{x - 1} + \frac{C}{(x - 1)^2}$$

Clearing of fractions changes this to

$$3x^2 - 8x + 13 = A(x - 1)^2 + B(x + 3)(x - 1) + C(x + 3)$$

Substitution of $x = 1$, $x = -3$, and $x = 0$ yields $C = 2$, $A = 4$, and $B = -1$. Thus,

$$\int \frac{3x^2 - 8x + 13}{(x + 3)(x - 1)^2} \, dx = 4 \int \frac{dx}{x + 3} - \int \frac{dx}{x - 1} + 2 \int \frac{dx}{(x - 1)^2}$$

$$= 4 \ln|x + 3| - \ln|x - 1| - \frac{2}{x - 1} + C \quad \blacksquare$$

Be sure to note the inclusion of the two fractions $B/(x - 1)$ and $C/(x - 1)^2$ in the decomposition above. The general rule for decomposing fractions with repeated linear factors in the denominator is this: For each factor $(ax + b)^k$ of the denominator, there are $k$ terms in the partial fraction decomposition, namely,

$$\frac{A_1}{ax + b} + \frac{A_2}{(ax + b)^2} + \frac{A_3}{(ax + b)^3} + \cdots + \frac{A_k}{(ax + b)^k}$$

**Partial Fraction Decomposition (Quadratic Factors)**   In factoring the denominator of a fraction, we may well get some quadratic factors (such as $x^2 + 1$), which cannot be factored into linear factors without introducing complex numbers.

EXAMPLE 7    **(A single quadratic factor).** Decompose $\dfrac{6x^2 - 3x + 1}{(4x + 1)(x^2 + 1)}$ and then find its indefinite integral.

*Solution*   The best we can hope for is a decomposition of the form

$$\frac{6x^2 - 3x + 1}{(4x + 1)(x^2 + 1)} = \frac{A}{4x + 1} + \frac{Bx + C}{x^2 + 1}$$

To determine the constants $A$, $B$, and $C$, we multiply both sides by $(4x + 1)(x^2 + 1)$ and obtain

$$6x^2 - 3x + 1 = A(x^2 + 1) + (Bx + C)(4x + 1)$$

Substitution of $x = -\frac{1}{4}$, $x = 0$, and $x = 1$ yields

$$\frac{6}{16} + \frac{3}{4} + 1 = A\left(\frac{17}{16}\right) \qquad\qquad \Rightarrow \quad A = 2$$

$$1 = 2 + C \qquad\qquad \Rightarrow \quad C = -1$$

$$4 = 4 + (B - 1)5 \quad \Rightarrow \quad B = 1$$

Thus,

$$\int \frac{6x^2 - 3x + 1}{(4x + 1)(x^2 + 1)}\, dx = \int \frac{2}{4x + 1}\, dx + \int \frac{x - 1}{x^2 + 1}\, dx$$

$$= \frac{1}{2}\int \frac{4\, dx}{4x + 1} + \frac{1}{2}\int \frac{2x\, dx}{x^2 + 1} - \int \frac{dx}{x^2 + 1}$$

$$= \frac{1}{2}\ln|4x + 1| + \frac{1}{2}\ln(x^2 + 1) - \tan^{-1}x + C \quad\blacksquare$$

**EXAMPLE 8** **(A repeated quadratic factor).** Find $\displaystyle\int \frac{6x^2 - 15x + 22}{(x + 3)(x^2 + 2)^2}\, dx.$

*Solution*   Here the appropriate decomposition is

$$\frac{6x^2 - 15x + 22}{(x + 3)(x^2 + 2)^2} = \frac{A}{x + 3} + \frac{Bx + C}{x^2 + 2} + \frac{Dx + E}{(x^2 + 2)^2}$$

After considerable work, we discover that $A = 1$, $B = -1$, $C = 3$, $D = -5$, and $E = 0$. Thus,

$$\int \frac{6x^2 - 15x + 22}{(x + 3)(x^2 + 2)^2}\, dx$$

$$= \int \frac{dx}{x + 3} - \int \frac{x - 3}{x^2 + 2}\, dx - 5\int \frac{x}{(x^2 + 2)^2}\, dx$$

$$= \int \frac{dx}{x + 3} - \frac{1}{2}\int \frac{2x}{x^2 + 2}\, dx + 3\int \frac{dx}{x^2 + 2} - \frac{5}{2}\int \frac{2x\, dx}{(x^2 + 2)^2}$$

$$= \ln|x + 3| - \frac{1}{2}\ln(x^2 + 2) + \frac{3}{\sqrt{2}}\tan^{-1}\left(\frac{x}{\sqrt{2}}\right) + \frac{5}{2(x^2 + 2)} + C$$

$\blacksquare$

**Summary**   To decompose a rational function $f(x) = p(x)/q(x)$ into partial fractions, proceed as follows.

   ***Step 1***   If $f(x)$ is improper—that is, if $p(x)$ is of degree at least that of $q(x)$—divide $p(x)$ by $q(x)$, obtaining

$$f(x) = \text{a polynomial} + \frac{N(x)}{D(x)}$$

   ***Step 2***   Factor $D(x)$ into a product of linear and irreducible quadratic factors with real coefficients. By a theorem of algebra, this is always (theoretically) possible.

***Step 3*** For each factor of the form $(ax + b)^k$, expect the decomposition to have the terms

$$\frac{A_1}{(ax + b)} + \frac{A_2}{(ax + b)^2} + \cdots + \frac{A_k}{(ax + b)^k}$$

***Step 4*** For each factor of the form $(ax^2 + bx + c)^m$, expect the decomposition to have the terms

$$\frac{B_1 x + C_1}{ax^2 + bx + c} + \frac{B_2 x + C_2}{(ax^2 + bx + c)^2} + \cdots + \frac{B_m x + C_m}{(ax^2 + bx + c)^m}$$

***Step 5*** Set $N(x)/D(x)$ equal to the sum of all the terms found in Steps 3 and 4. The number of constants to be determined should equal the degree of the denominator, $D(x)$.

***Step 6*** Multiply both sides of the equation found in Step 5 by $D(x)$ and solve for the unknown constants. This can be done by either of two methods: (1) Equate coefficients of like-degree terms; or (2) assign convenient values to the variable $x$.

## CONCEPTS REVIEW

**1.** If the degree of the polynomial $p(x)$ is less than the degree of $q(x)$, then $f(x) = p(x)/q(x)$ is called a _____ rational function.

**2.** To integrate the improper rational function $f(x) = (x^2 + 4)/(x + 1)$, we first rewrite it as $f(x) =$ _____.

**3.** If $(x - 1)(x + 1) + 3x + x^2 = ax^2 + bx + c$, then $a =$ _____, $b =$ _____, and $c =$ _____.

**4.** $(3x + 1)/[(x - 1)^2(x^2 + 1)]$ can be decomposed in the form _____.

## PROBLEM SET 8.5

In Problems 1–40, use the method of partial fraction decomposition to perform the required integration.

**1.** $\displaystyle\int \frac{1}{x(x + 1)}\, dx$

**2.** $\displaystyle\int \frac{2}{x^2 + 3x}\, dx$

**3.** $\displaystyle\int \frac{3}{x^2 - 1}\, dx$

**4.** $\displaystyle\int \frac{5x}{2x^3 + 6x^2}\, dx$

**5.** $\displaystyle\int \frac{x - 11}{x^2 + 3x - 4}\, dx$

**6.** $\displaystyle\int \frac{x - 7}{x^2 - x - 12}\, dx$

**7.** $\displaystyle\int \frac{3x - 13}{x^2 + 3x - 10}\, dx$

**8.** $\displaystyle\int \frac{x + \pi}{x^2 - 3\pi x + 2\pi^2}\, dx$

**9.** $\displaystyle\int \frac{2x + 21}{2x^2 + 9x - 5}\, dx$

**10.** $\displaystyle\int \frac{2x^2 - x - 20}{x^2 + x - 6}\, dx$

**11.** $\displaystyle\int \frac{17x - 3}{3x^2 + x - 2}\, dx$

**12.** $\displaystyle\int \frac{5 - x}{x^2 - x(\pi + 4) + 4\pi}\, dx$

**13.** $\displaystyle\int \frac{2x^2 + x - 4}{x^3 - x^2 - 2x}\, dx$

**14.** $\displaystyle\int \frac{7x^2 + 2x - 3}{(2x - 1)(3x + 2)(x - 3)}\, dx$

**15.** $\displaystyle\int \frac{6x^2 + 22x - 23}{(2x - 1)(x^2 + x - 6)}\, dx$

**16.** $\displaystyle\int \frac{x^3 - 6x^2 + 11x - 6}{4x^3 - 28x^2 + 56x - 32}\, dx$

**17.** $\displaystyle\int \frac{x^3}{x^2 + x - 2}\, dx$

**18.** $\displaystyle\int \frac{x^3 + x^2}{x^2 + 5x + 6}\, dx$

19. $\displaystyle\int \frac{x^4 + 8x^2 + 8}{x^3 - 4x}\,dx$

20. $\displaystyle\int \frac{x^6 + 4x^3 + 4}{x^3 - 4x^2}\,dx$

21. $\displaystyle\int \frac{x + 1}{(x - 3)^2}\,dx$

22. $\displaystyle\int \frac{5x + 7}{x^2 + 4x + 4}\,dx$

23. $\displaystyle\int \frac{3x + 2}{x^3 + 3x^2 + 3x + 1}\,dx$

24. $\displaystyle\int \frac{x^6}{(x - 2)^2(1 - x)^5}\,dx$

25. $\displaystyle\int \frac{3x^2 - 21x + 32}{x^3 - 8x^2 + 16x}\,dx$

26. $\displaystyle\int \frac{x^2 + 19x + 10}{2x^4 + 5x^3}\,dx$

27. $\displaystyle\int \frac{2x^2 + x - 8}{x^3 + 4x}\,dx$

28. $\displaystyle\int \frac{3x + 2}{x(x + 2)^2 + 16x}\,dx$

29. $\displaystyle\int \frac{2x^2 - 3x - 36}{(2x - 1)(x^2 + 9)}\,dx$

30. $\displaystyle\int \frac{1}{x^4 - 16}\,dx$

31. $\displaystyle\int \frac{1}{(x - 1)^2(x + 4)^2}\,dx$

32. $\displaystyle\int \frac{x^3 - 8x^2 - 1}{(x + 3)(x^2 - 4x + 5)}\,dx$

33. $\displaystyle\int \frac{(\sin^3 t - 8\sin^2 t - 1)\cos t}{(\sin t + 3)(\sin^2 t - 4\sin t + 5)}\,dt$

34. $\displaystyle\int \frac{\cos t}{\sin^4 t - 16}\,dt$

35. $\displaystyle\int \frac{x^3 - 4x}{(x^2 + 1)^2}\,dx$

36. $\displaystyle\int \frac{(\sin t)(4\cos^2 t - 1)}{(\cos t)(1 + 2\cos^2 t + \cos^4 t)}\,dt$

37. $\displaystyle\int \frac{2x^3 + 5x^2 + 16x}{x^5 + 8x^3 + 16x}\,dx$ *Hint:* To integrate

$\displaystyle\int (x^2 + 4)^{-2}\,dx$, let $x = 2\tan\theta$.

38. Evaluate $\displaystyle\int_4^6 \frac{x - 17}{x^2 + x - 12}\,dx$

39. Evaluate $\displaystyle\int_0^{\pi/4} \frac{\cos\theta}{(1 - \sin^2\theta)(\sin^2\theta + 1)^2}\,d\theta$

40. Find the value of $\displaystyle\int_1^5 \frac{3x + 13}{x^2 + 4x + 3}\,dx$

41. The ordinary differential equation

$$\frac{dy}{dt} = \varepsilon y(t) - \sigma(y(t))^2, \quad \varepsilon > 0 \text{ and } \sigma > 0, \; y(0) = y_0 > 0$$

is known in ecology as the logistic or *Verhulst* equation.

(a) By separating variables and integrating, show that the solution is

$$y(t) = \frac{\varepsilon}{\sigma + \left(\dfrac{\varepsilon - \sigma y_0}{y_0}\right) e^{-\varepsilon t}}$$

(b) Verify that a special solution is given by $y(t) = y_0 =$ constant $= \dfrac{\varepsilon}{\sigma}$.

42. The Law of Mass Action in chemistry results in the differential equation

$$\frac{dx}{dt} = k(a - x)(b - x) \qquad k > 0, \quad a > 0, \quad b > 0$$

where $x$ is the amount of a substance at time $t$ resulting from the reaction of two others. Assume that $x = 0$ when $t = 0$.

(a) Solve this differential equation in the case $b > a$.
(b) Show that $x \to a$ as $t \to \infty$ (if $b > a$).
(c) Suppose $a = 2$ and $b = 4$ and that 1 gram of the substance is formed in 20 minutes. How much will be present in 1 hour?
(d) Solve the differential equation if $a = b$.

[C]  43. In many population growth problems, there is an upper limit beyond which the population cannot grow. Let us suppose that the earth will not support a population of more than 16 billion and that there were 2 billion people in 1925 and 4 billion people in 1975. Then if $y$ is the population $t$ years after 1925, an appropriate model is the differential equation

$$\frac{dy}{dt} = ky(16 - y)$$

(a) Solve this differential equation.
(b) Find the population in 2015.
(c) When will the population be 9 billion?

This model, called the logistic model, was discussed in Section 7.5.

44. Do Problem 43 assuming the upper limit for the population is 10 billion.

**45.** The differential equation

$$\frac{dy}{dt} = k(y - m)(M - y)$$

with $k > 0$ and $0 \le m < y_0 < M$ has been used to model some growth problems. Solve the equation and find $\lim_{t \to \infty} y$.

**46.** As a model for the production of trypsin from trypsinogen in digestion, biochemists have proposed the model

$$\frac{dy}{dt} = k(A - y)(B + y)$$

where $k > 0$, $A$ is the initial amount of trypsinogen, and $B$ is the original amount of trypsin. Solve this differential equation.

**47.** Evaluate

$$\int_{\pi/6}^{\pi/2} \frac{\cos x}{\sin x \, (\sin^2 x + 1)^2} \, dx$$

---

**Answers to Concepts Review:** **1.** Proper **2.** $x - 1 + \dfrac{5}{x + 1}$

**3.** $2; 3; -1$ **4.** $\dfrac{A}{x - 2} + \dfrac{B}{(x - 2)^2} + \dfrac{Cx + D}{x^2 + 1}$

---

## 8.6 CHAPTER REVIEW

### Concepts Test

Respond with true or false to each of the following assertions. Be prepared to justify your answer.

**1.** To find $\displaystyle\int x \sin(x^2) \, dx$, try the substitution $u = x^2$.

**2.** To find $\displaystyle\int \frac{x}{9 + x^4} \, dx$, try the substitution $u = x^2$.

**3.** To find $\displaystyle\int \frac{x^3}{9 + x^4} \, dx$, try the substitution $u = x^2$.

**4.** To find $\displaystyle\int \frac{2x - 3}{x^2 - 3x + 5} \, dx$, begin by completing the square of the denominator.

**5.** To find $\displaystyle\int \frac{3}{x^2 - 3x + 5} \, dx$, begin by completing the square of the denominator.

**6.** To find $\displaystyle\int \frac{1}{\sqrt{4 - 5x^2}} \, dx$, try the substitution $u = \sqrt{5} \, x$.

**7.** To find $\displaystyle\int \frac{t + 2}{t^3 - 9t} \, dt$, try a partial fraction decomposition.

**8.** To find $\displaystyle\int \frac{t^4}{t^2 - 1} \, dt$, use integration by parts.

**9.** To find $\displaystyle\int \sin^6 x \cos^2 x \, dx$, use half-angle formulas.

**10.** To find $\displaystyle\int \frac{e^x}{1 + e^x} \, dx$, use integration by parts.

**11.** To find $\displaystyle\int \frac{x + 2}{\sqrt{-x^2 - 4x}} \, dx$, use a trigonometric substitution.

**12.** To find $\displaystyle\int x^2 \sqrt[3]{3 - 2x} \, dx$, let $u = \sqrt[3]{3 - 2x}$.

**13.** To find $\displaystyle\int \sin^2 x \cos^5 x \, dx$, rewrite the integrand as $\sin^2 x (1 - \sin^2 x)^2 \cos x$.

**14.** To find $\displaystyle\int \frac{1}{x^2 \sqrt{9 - x^2}} \, dx$, try a trigonometric substitution.

**15.** To find $\displaystyle\int x^2 \ln x \, dx$, try integration by parts.

**16.** To find $\displaystyle\int \sin 2x \cos 4x \, dx$, use half-angle formulas.

**17.** $\dfrac{x^2}{x^2 - 1}$ can be expressed in the form $\dfrac{A}{x - 1} + \dfrac{B}{x + 1}$.

**18.** $\dfrac{x^2 + 2}{x(x^2 - 1)}$ can be expressed in the form $\dfrac{A}{x} + \dfrac{B}{x - 1} + \dfrac{C}{x + 1}$.

**19.** $\dfrac{x^2 + 2}{x(x^2 + 1)}$ can be expressed in the form $\dfrac{A}{x} + \dfrac{Bx + C}{x^2 + 1}$.

**20.** $\dfrac{x + 2}{x^2(x^2 - 1)}$ can be expressed in the form $\dfrac{A}{x^2} + \dfrac{B}{x - 1} + \dfrac{C}{x + 1}$.

**21.** The integration by parts formula is the integral version of the Product Rule for derivatives.

**22.** To complete the square of $ax^2 + bx$, add $(b/2)^2$.

**23.** Any polynomial with real coefficients can be factored into a product of linear polynomials with real coefficients.

**24.** Two polynomials in $x$ have the same values for all $x$ only if the coefficients of like degree terms are identical.

**25.** If $a_n x^n + a_{n-1} x^{n-1} + \cdots + a_1 x + a_0 = 0$ has $n + 1$ solutions, then all the coefficients are 0.

## Sample Test Problems

In Problems 1–42, perform the indicated integrations by any correct method.

1. $\displaystyle\int_0^4 \frac{t}{\sqrt{9+t^2}}\, dt$

2. $\displaystyle\int \cot^2(2\theta)\, d\theta$

3. $\displaystyle\int_0^{\pi/2} e^{\cos x} \sin x\, dx$

4. $\displaystyle\int_0^{\pi/4} x \sin 2x\, dx$

5. $\displaystyle\int \frac{y^3 + y}{y+1}\, dy$

6. $\displaystyle\int \sin^3(2t)\, dt$

7. $\displaystyle\int \frac{(y-2)}{y^2 - 4y + 2}\, dy$

8. $\displaystyle\int_0^{3/2} \frac{dy}{\sqrt{2y+1}}$

9. $\displaystyle\int \frac{e^{2t}}{e^t - 2}\, dt$

10. $\displaystyle\int \frac{\sin x + \cos x}{\tan x}\, dx$

11. $\displaystyle\int \frac{dx}{\sqrt{16 + 4x - 2x^2}}$

12. $\displaystyle\int x^2 e^x\, dx$

13. $\displaystyle\int \frac{dy}{\sqrt{2 + 3y^2}}$

14. $\displaystyle\int \frac{w^3}{1 - w^2}\, dw$

15. $\displaystyle\int \frac{\tan x}{\ln|\cos x|}\, dx$

16. $\displaystyle\int \frac{3\, dt}{t^3 - 1}$

17. $\displaystyle\int \sinh x\, dx$

18. $\displaystyle\int \frac{(\ln y)^5}{y}\, dy$

19. $\displaystyle\int x \cot^2 x\, dx$

20. $\displaystyle\int \frac{\sin \sqrt{x}}{\sqrt{x}}\, dx$

21. $\displaystyle\int \frac{\ln t^2}{t}\, dt$

22. $\displaystyle\int \ln(y^2 + 9)\, dy$

23. $\displaystyle\int e^{t/3} \sin 3t\, dt$

24. $\displaystyle\int \frac{t+9}{t^3 + 9t}\, dt$

25. $\displaystyle\int \sin \frac{3x}{2} \cos \frac{x}{2}\, dx$

26. $\displaystyle\int \cos^4\!\left(\frac{x}{2}\right) dx$

27. $\displaystyle\int \tan^3 2x \sec 2x\, dx$

28. $\displaystyle\int \frac{\sqrt{x}}{1 + \sqrt{x}}\, dx$

29. $\displaystyle\int \tan^{3/2} x \sec^4 x\, dx$

30. $\displaystyle\int \frac{dt}{t(t^{1/6} + 1)}$

31. $\displaystyle\int \frac{e^{2y}\, dy}{\sqrt{9 - e^{2y}}}$

32. $\displaystyle\int \cos^5 x \sqrt{\sin x}\, dx$

33. $\displaystyle\int e^{\ln(3\cos x)}\, dx$

34. $\displaystyle\int \frac{\sqrt{9 - y^2}}{y}\, dy$

35. $\displaystyle\int \frac{e^{4x}}{1 + e^{8x}}\, dx$

36. $\displaystyle\int \frac{\sqrt{x^2 + a^2}}{x^4}\, dx$

37. $\displaystyle\int \frac{w}{\sqrt{w+5}}\, dw$

38. $\displaystyle\int \frac{\sin t\, dt}{\sqrt{1 + \cos t}}$

39. $\displaystyle\int \frac{\sin y \cos y}{9 + \cos^4 y}\, dy$

40. $\displaystyle\int \frac{dx}{\sqrt{1 - 6x - x^2}}$

41. $\displaystyle\int \frac{4x^2 + 3x + 6}{x^2(x^2 + 3)}\, dx$

42. $\displaystyle\int \frac{dx}{(16 + x^2)^{3/2}}$

43. For the following problems express each of the rational functions as a partial fraction without computing the exact coefficients. For example $\dfrac{3x + 1}{(x-1)^2} =$

$$\frac{A}{(x-1)} + \frac{B}{(x-1)^2}$$

(a) $\dfrac{3 - 4x^2}{(2x+1)^3}$

(b) $\dfrac{7x - 41}{(x-1)^2(2-x)^3}$

(c) $\dfrac{3x + 1}{(x^2 + x + 10)^2}$

(d) $\dfrac{(x+1)^2}{(x^2 - x + 10)^2 (1 - x^2)^2}$

(e) $\dfrac{x^5}{(x+3)^4 (x^2 + 2x + 10)^2}$

(f) $\dfrac{(3x^2 + 2x - 1)^2}{(2x^2 + x + 10)^3}$

44. Find the volume of the solid generated by revolving the region under the graph of

$$y = \frac{1}{\sqrt{3x - x^2}}$$

from $x = 1$ to $x = 2$ about

(a) the $x$-axis
(b) the $y$-axis

45. Find the length of the curve $y = x^2/16$ from $x = 0$ to $x = 4$.

46. The region under the curve

$$y = \frac{1}{(x^2 + 5x + 6)}$$

from $x = 0$ to $x = 3$ is rotated about the $x$-axis. Compute the volume of solid.

47. If the curve given in Problem 46 is rotated about the $y$-axis, find the volume of the solid.

48. Find the volume of the solid created by revolving the area created by the $x$-axis and the curve $y = 4x \sqrt{2 - x}$ about the $y$-axis.

**49.** Find the volume when the area created by the $x$-axis, $y$-axis, the curve $y = 2(e^x - 1)$ and the curve $x = \ln 3$ is revolved about the line $x = \ln 3$.

**50.** Find the area of the region bounded by the $x$-axis, the curve $y = 18/(x^2 \sqrt{x^2 + 9})$, and the lines $x = \sqrt{3}$ and $x = 3\sqrt{3}$.

**51.** Find the area of the region bounded by the curve $s = t/(t - 1)^2$, $s = 0$, $t = -6$, and $t = 0$.

≈ **52.** Find the volume of the solid generated by revolving the region

$$\left\{ (x, y): -3 \le x \le -1, \frac{6}{x\sqrt{x + 4}} \le y \le 0 \right\}$$

about the $x$-axis. Make a sketch.

≈ **53.** Find the length of the segment of the curve $y = \ln(\sin x)$ from $x = \pi/6$ to $x = \pi/3$.

## 8.7   INTEGRATION USING COMPUTER ALGEBRA SYSTEMS (CAS)

We assume that you have available one of the Computer Algebra Systems (CAS) Maple or Mathematica which you can use to help compute integrals. We will investigate how these systems can be used to the best advantage as well as some of their pitfalls.

**1.** All these systems have the ability to perform partial fraction decompositions. Let the integrand *expr* be a function of $x$ to which you want to apply partial fraction decomposition. In Maple the command would be *convert(expr, parfrac, x)*. In Mathematica the command would be *Apart[expr]*. For example, to compute the partial fraction decomposition of $\dfrac{x}{x^2 - 4}$, you would enter

(a) In Maple: *convert (x/(x^2 − 4), parfrac, x)*;
(b) In Mathematica: *Apart[x/(x^2 − 4)]*

Using the CAS you have available, compute the partial fraction decomposition of

$$\frac{x^4 + 3x^3 - 2x^2 + x - 1}{4x^5 - 60x^4 + 309x^3 - 688x^2 + 687x - 252}$$

and then finish the integration by integrating the resulting expression by hand.

**2.** Use the CAS of your choice to compute the partial fraction decomposition of the following expressions. Use this result to compute the indefinite integral of each expression.

(a) $\dfrac{x^3 + 3x^2 - 2x + 6}{706x^3 - 435x^2 - 273x^4 - 118x + 120}$

(b) $\dfrac{3x^3 + 5x^2 + 7x + 16}{x^4 + 8x^3 + 48x^2 + 119x + 230}$

(c) $\dfrac{3x^3 + 5x^2 + 7x + 16}{x^6 + 29x^4 - 31x^3 + 180x^2 - 389x + 210}$

(d) $\dfrac{x^5 - 5x^4 + 10x^3 - 10x^2 + 5x - 1}{x^5 - 57x^4 + 1299x^3 - 14795x^2 + 84216x - 191664}$

(e) $\dfrac{1}{217x^4 - 60x^5 + 36x^6 - 556x^3 + 586x^2 - 1056x + 1089}$

Most CAS can be asked to give the integral directly. In Maple the command is *int(expr, x)*; for indefinite integrals and *int(expr, x = a..b)*; for the definite integral from a to b. In Mathematica the command is *Integrate[expr, x]* for indefinite integrals and *Integrate[expr, {x, a, b}]* for definite integrals. In many cases, the results are not exactly what you would have expected if you did the computation by hand. In particular, for integrals of the form $\int x(x^2 + a)^4 \, dx$ both systems will give $\frac{1}{10}x^{10} + \frac{1}{2}ax^8 + a^2x^6 + a^3x^4 + \frac{1}{2}a^4x^2$. Clearly they have not computed the integral by using the substitution $u = (x^2 + a)$, but have expanded the integrand and integrated term by term. In such cases you will have to perform algebra on their answer to get it into a form where you can compare it to your answer.

**3.** Using the CAS of your choice compute $\int x^2(2x^3 + 4) \, dx$ and check that the answer you computed using substitution techniques agrees with the CAS answer. Mathematica and Maple will give two seemingly different answers for $\int \sin^3 x \cos^5 x \, dx$. Maple computes $-\frac{1}{8} \sin^2 x \cos^6 x - \frac{1}{24} \cos^6 x$, while Mathematica produces $\frac{3}{384} \cos 6x - \frac{1}{256} \cos 4x - \frac{3}{128} \cos 2x + \frac{1}{1024} \cos 8x$. By using the proper commands in each CAS you can transform one answer into the other.

**4.** After researching the commands *?combine* in Maple or *?TrigReduce* which is in the *Algebra* 'Trigonometry' package of Mathematica show that the answer to $\int \sin^3 x \cos^5 x \, dx$ is the same in each CAS (up to a constant of integration.)

**5.** Most CAS are notorious for leaving out absolute value signs. This can cause incorrect results. Consider the integral $\displaystyle\int \sqrt{9x^2 + x^4} \, dx$

(a) Explain why $\sqrt{9x^2 + x^4} = |x| \sqrt{9 + x^2}$

(b) Explain why $\displaystyle\int_{-1}^{1} \sqrt{9x^2 + x^4} \, dx = 2 \int_{0}^{1} \sqrt{9x^2 + x^4} \, dx \ne 0$

(c) Compute the value of this integral by hand.

(d) Use a CAS to evaluate both $\displaystyle\int \sqrt{9x^2 + x^4} \, dx$ and $\displaystyle\int_{-1}^{1} \sqrt{9x^2 + x^4} \, dx$.

**6.** Use the CAS of your choice to evaluate the following integrals. Compare your answers to those found inside the back cover. If they are not the same, try to show that they are equivalent up to a constant.

(a) $\displaystyle\int \sqrt{u^2 + 9}\, du$

(b) $\displaystyle\int \sinh^2(5u)\, du$

(c) $\displaystyle\int (u^2 + a^2)^{3/2}\, du$

(d) $\displaystyle\int \frac{5}{u\sqrt{3u+2}}\, du$

(e) $\displaystyle\int \operatorname{csch}(\pi u)\, du$

(f) $\displaystyle\int_0^{\pi/2} \sin^{12} u\, du$

**7.** Maple has a command which permits you to perform integration by parts and change variables of integration (substitution). To integrate $\int x^2 e^{2x}\, dx$ by parts you would use *student*[*intparts*](*Int*(*expr*, *x*), *expr_diff*); where *expr_diff* is the portion of expr that you want to differentiate. e.g., *student*[*intparts*](*Int*(*x*^2 * exp(2 * *x*), *x*), *x*^2); If you have access to Maple use this command to evaluate $\int x^3 \sin 7x\, dx$ repeatedly by parts, always differentiating the $x$ term in any resultant integrals.

# TECHNOLOGY PROJECT

As you know, if a constant force $F$ is applied over a distance $d$, the work done is simply the product $Fd$. More frequently the force is changing due to any number of things (e.g., the terrain, the wind, a magnetic field, etc.). Moreover, often the force is not applied in a single direction, and this complicates things a bit. You will be exploring this important notion in a simple setting in which we assume:

(a) The force is a function only of the horizontal variable $x$ (i.e., $F = F(x)$).

(b) The force is exerted directly along the curve of motion.

(c) The curve of motion is a planar curve explicitly defined by $y = h(x)$ on some interval $[a, b]$.

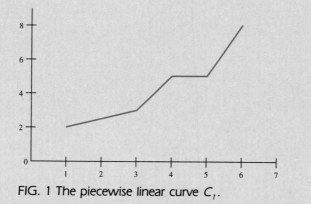

FIG. 1 The piecewise linear curve $C_1$.

**Exercise 1**  Without using calculus,

(a) Calculate the length of the curve $C_1$, the piecewise linear curve connecting the points $(1, 2)$, $(3, 3)$, $(4, 5)$, $(5, 5)$, and $(6, 8)$, shown in Figure 1.

(b) Calculate the length of the curve $C_2$, defined by:

$$y = \sqrt{4 - x^2}, \text{ for } -2 \le x \le 2.$$

*Hint:* Just use some geometry.

(c) By approximating with a few inscribed chords, *estimate* the length of the curve $C_3$, defined by:

$$y = x^2 - 4, \text{ for } 0 \le x \le 2.$$

**Exercise 2**  In this problem, $k$ is a constant, $m(x)$ is the slope of the curves $C_1$, $C_2$, and $C_3$ defined in Exercise 1. Find $m(x)$ for each of these curves and then without doing any integrals,

(a) Calculate the work done if the force $F = k \cdot m(x)$ is applied along the curve $C_1$.

(b) Calculate the work done if the force $F = k$ is applied along the curve $C_2$.

(c) *Estimate* the work done if the force $F = k \cdot m(x)$ is applied along the curve $C_3$.

**Exercise 3**  Now let $C$ represent a general planar curve defined by $y = h(x)$ on the interval $[a, b]$ and let $F(x)$ be a force always exerted in the direction of $C$.

(a) By considering the element of work done on a small interval, $[x, x + dx]$, carefully justify the following formula:

$$\text{Work} = \int_a^b F(x)\sqrt{1 + (h'(x))^2}\, dx$$

(b) Use this formula to compute the work done by $F(x) = 3x$ along the curve $C_3$ defined above.

Recall that the length of the curve $C$ discussed above is given by:

$$\text{Length} = \int_a^b \sqrt{1 + (h'(x))^2}\, dx$$

On your computing technology, set up the statements to compute the length of a curve $C$ and the work for the situation described in Exercise 3. (You may have to do a numerical integraion in some cases to follow.)

**Exercise 4**  Using your computing technology, find the lengths of $C_2$ and $C_3$. Compare with your earlier results.

**Exercise 5**  Using your computing technology, find the work associated with each of the following curves and forces:

(a) Check your result for the second part of Exercise 3.

(b) Let $C$ be defined by

$$h(x) = \frac{\sin 3x}{1 + x^2}$$

on the interval $[0, \pi]$ and let $F(x) = 2x^2$. Compute both the length and work.

(c) Compute the work for the last curve, but now with a force equal to 3 times the slope of the curve.

# Indeterminate Forms and Improper Integrals

Here are three familiar limit problems:

$$\lim_{x \to 0} \frac{\sin x}{x}, \qquad \lim_{x \to 3} \frac{x^2 - 9}{x^2 - x - 6}, \qquad \lim_{x \to a} \frac{f(x) - f(a)}{x - a}$$

The first was treated at length in Section 3.4 and the third actually defines the derivative $f'(a)$. The three limits have a common feature. In each case, a quotient is involved and, in each case, both numerator and denominator have 0 as their limits. An attempt to apply the quotient rule for limits leads to the nonsensical answer 0/0. In fact, the quotient rule does not apply, since that rule requires that the limit of the denominator be different from 0. We are not saying that these limits do not exist, only that the quotient rule will not determine them.

You may recall that an intricate geometric argument led us to the conclusion $\lim_{x \to 0} (\sin x)/x = 1$. On the other hand, the algebraic technique of factoring yields

$$\lim_{x \to 3} \frac{x^2 - 9}{x^2 - x - 6} = \lim_{x \to 3} \frac{(x - 3)(x + 3)}{(x - 3)(x + 2)} = \lim_{x \to 3} \frac{x + 3}{x + 2} = \frac{6}{5}$$

Would it not be nice to have a standard procedure for handling all problems of this type? That is too much to hope for. However, there is a simple rule that works beautifully on a wide variety of such problems. It is known as *l'Hôpital's Rule* (pronounced Lō′pē-täl).

**L'Hôpital's Rule**   In 1696, Guillaume François Antoine de l'Hôpital published the first textbook on differential calculus; it included the following rule, which he had learned from his teacher Johann Bernoulli.

Theorem A

**(L'Hôpital's Rule for forms of type 0/0).** Suppose that $\lim\limits_{x \to u} f(x) = \lim\limits_{x \to u} g(x) = 0$. If $\lim\limits_{x \to u} [f'(x)/g'(x)]$ exists in either the finite or infinite sense (that is, if this limit is a finite number or $-\infty$ or $+\infty$), then

$$\lim_{x \to u} \frac{f(x)}{g(x)} = \lim_{x \to u} \frac{f'(x)}{g'(x)}$$

Here $u$ may stand for any of the symbols $a$, $a^-$, $a^+$, $-\infty$, or $+\infty$.

Common Sense

Study the diagrams below. They should make l'Hôpital's Rule seem quite reasonable.

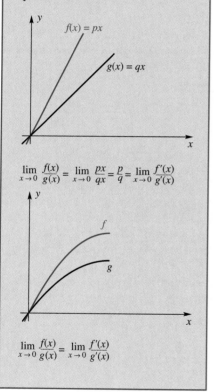

$$\lim_{x \to 0} \frac{f(x)}{g(x)} = \lim_{x \to 0} \frac{px}{qx} = \frac{p}{q} = \lim_{x \to 0} \frac{f'(x)}{g'(x)}$$

$$\lim_{x \to 0} \frac{f(x)}{g(x)} = \lim_{x \to 0} \frac{f'(x)}{g'(x)}$$

Before attempting to prove this theorem, we illustrate it. Note that l'Hôpital's Rule allows us to replace one limit by another, which may be simpler and—in particular—may not have the 0/0 form.

**EXAMPLE 1** Use l'Hôpital's Rule to show that

$$\lim_{x \to 0} \frac{\sin x}{x} = 1 \quad \text{and} \quad \lim_{x \to 0} \frac{1 - \cos x}{x} = 0$$

*Solution* We worked pretty hard to demonstrate these two facts in Section 3.4. After noting that both limits have the 0/0 form, we can now establish the desired results in two lines (but see Problem 25). By l'Hôpital's Rule,

$$\lim_{x \to 0} \frac{\sin x}{x} = \lim_{x \to 0} \frac{D_x \sin x}{D_x x} = \lim_{x \to 0} \frac{\cos x}{1} = 1$$

$$\lim_{x \to 0} \frac{1 - \cos x}{x} = \lim_{x \to 0} \frac{D_x(1 - \cos x)}{D_x x} = \lim_{x \to 0} \frac{\sin x}{1} = 0 \quad \blacksquare$$

**EXAMPLE 2** Find $\lim\limits_{x \to 3} \dfrac{x^2 - 9}{x^2 - x - 6}$ and $\lim\limits_{x \to 2^+} \dfrac{x^2 + 3x - 10}{x^2 - 4x + 4}$.

*Solution* Both limits have the 0/0 form, so by l'Hôpital's Rule,

$$\lim_{x \to 3} \frac{x^2 - 9}{x^2 - x - 6} = \lim_{x \to 3} \frac{2x}{2x - 1} = \frac{6}{5}$$

$$\lim_{x \to 2^+} \frac{x^2 + 3x - 10}{x^2 - 4x + 4} = \lim_{x \to 2^+} \frac{2x + 3}{2x - 4} = \infty$$

The first of these limits was handled at the beginning of this section by the method of factoring. Of course, we get the same answer either way. $\blacksquare$

**EXAMPLE 3** Find $\lim\limits_{x \to 0} \dfrac{\tan 2x}{\ln(1 + x)}$.

*Solution* Both numerator and denominator have limit 0. Hence,

$$\lim_{x \to 0} \frac{\tan 2x}{\ln(1 + x)} = \lim_{x \to 0} \frac{2 \sec^2 2x}{1/(1 + x)} = \frac{2}{1} = 2$$
$\blacksquare$

Sometimes $\lim f'(x)/g'(x)$ also has the indeterminate form 0/0. Then we may apply l'Hôpital's Rule again, as we now illustrate. Each application of l'Hôpital's Rule is flagged with the symbol $\bigcirc\!\!\!L$.

**EXAMPLE 4**   Find $\lim\limits_{x \to 0} \dfrac{\sin x - x}{x^3}$.

*Solution*   By l'Hôpital's Rule applied three times in succession,

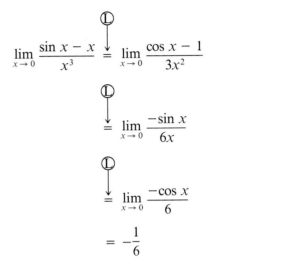

$$\lim_{x \to 0} \frac{\sin x - x}{x^3} \overset{\bigcirc\!\!\!L}{=} \lim_{x \to 0} \frac{\cos x - 1}{3x^2}$$

$$\overset{\bigcirc\!\!\!L}{=} \lim_{x \to 0} \frac{-\sin x}{6x}$$

$$\overset{\bigcirc\!\!\!L}{=} \lim_{x \to 0} \frac{-\cos x}{6}$$

$$= -\frac{1}{6} \qquad\blacksquare$$

Just because we have an elegant rule does not mean we should use it indiscriminately. In particular, we must always make sure it applies. Otherwise we will be led into all kinds of errors, as we now illustrate.

**EXAMPLE 5**   Find $\lim\limits_{x \to 0} \dfrac{1 - \cos x}{x^2 + 3x}$.

*Solution*

$$\lim_{x \to 0} \frac{1 - \cos x}{x^2 + 3x} \overset{\bigcirc\!\!\!L}{=} \lim_{x \to 0} \frac{\sin x}{2x + 3} \overset{\bigcirc\!\!\!L}{=} \lim_{x \to 0} \frac{\cos x}{2} = \frac{1}{2} \qquad \text{WRONG}$$

The first application of l'Hôpital's Rule was correct; the second was not since, at that stage, the limit did not have the 0/0 form. Here is what we should have done.

$$\lim_{x \to 0} \frac{1 - \cos x}{x^2 + 3x} \overset{\bigcirc\!\!\!L}{=} \lim_{x \to 0} \frac{\sin x}{2x + 3} = 0 \qquad \text{RIGHT}$$

We stop differentiating as soon as either the numerator or denominator has a nonzero limit.   $\blacksquare$

Even if l'Hôpital's Rule applies, it may not help us; witness the following example.

**EXAMPLE 6** Find $\lim\limits_{x \to \infty} \dfrac{e^{-x}}{x^{-1}}$.

*Solution* We may apply l'Hôpital's Rule indefinitely.

$$\lim_{x \to \infty} \frac{e^{-x}}{x^{-1}} \overset{\text{\textcircled{L}}}{=} \lim_{x \to \infty} \frac{e^{-x}}{x^{-2}} \overset{\text{\textcircled{L}}}{=} \lim_{x \to \infty} \frac{e^{-x}}{2x^{-3}} = \cdots$$

Clearly, we are only complicating the problem. A better approach is to do a bit of algebra first.

$$\lim_{x \to \infty} \frac{e^{-x}}{x^{-1}} = \lim_{x \to \infty} \frac{x}{e^x}$$

Written this way, the limit is indeterminate of the form $\infty/\infty$, the subject of the next section. However, you should be able to guess that the limit is 0 by considering how much faster $e^x$ grows than $x$ (see Figure 1). A rigorous demonstration will come later (Example 1 of Section 9.2). ∎

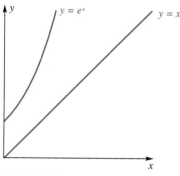

FIGURE 1

**Cauchy's Mean Value Theorem** The modern proof of l'Hôpital's Rule depends on an extension of the Mean Value Theorem due to Augustin-Louis Cauchy (1789–1857).

> **Theorem B**
>
> **(Cauchy's Mean Value Theorem).** Let the functions $f$ and $g$ be differentiable on $(a, b)$ and continuous on $[a, b]$. If $g'(x) \neq 0$ for all $x$ in $(a, b)$, then there exists a number $c$ in $(a, b)$ such that
>
> $$\frac{f(b) - f(a)}{g(b) - g(a)} = \frac{f'(c)}{g'(c)}$$

Note that this theorem reduces to the ordinary Mean Value Theorem (Theorem 4.8A) when $g(x) = x$.

*Proof* It is tempting to apply the ordinary Mean Value Theorem to both numerator and denominator of the left side of the conclusion. If we do this, we obtain

(1)    $f(b) - f(a) = f'(c_1)(b - a)$

and

(2)    $g(b) - g(a) = g'(c_2)(b - a)$

for appropriate choices of $c_1$ and $c_2$. If only $c_1$ and $c_2$ were equal, we could divide the first equality by the second and be done; but there is no reason to hope for such a coincidence. However, this attempt is not a complete failure since (2) yields the valuable information that $g(b) - g(a) \neq 0$, a fact we will need later (this follows from the hypothesis that $g'(x) \neq 0$ for all $x$ in $(a, b)$).

Recall that the proof of the Mean Value Theorem (Theorem 4.8A) rested on the introduction of an auxiliary function $s$. If we try to mimic that proof, we are led to the following choice for $s(x)$. Let

$$s(x) = f(x) - f(a) - \frac{f(b) - f(a)}{g(b) - g(a)} [g(x) - g(a)]$$

No division by zero is involved since we earlier established that $g(b) - g(a) \neq 0$. Note further that $s(a) = 0 = s(b)$. Also, $s$ is continuous on $[a, b]$ and differentiable on $(a, b)$, this following from the corresponding facts for $f$ and $g$. Thus, by the Mean Value Theorem, there is a number $c$ in $(a, b)$ such that

$$s'(c) = \frac{s(b) - s(a)}{b - a} = \frac{0 - 0}{b - a} = 0$$

but

$$s'(c) = f'(c) - \frac{f(b) - f(a)}{g(b) - g(a)} g'(c) = 0$$

or

$$\frac{f'(c)}{g'(c)} = \frac{f(b) - f(a)}{g(b) - g(a)}$$

which is what we wished to prove.  ∎

### Proof of L'Hôpital's Rule

**Proof**  Refer back to Theorem A, which actually states several theorems at once. We will prove only the case where $L$ is finite and the limit is the one-sided limit $\lim\limits_{x \to a^+}$.

The hypotheses for Theorem A imply more than they say explicitly. In particular, the existence of $\lim\limits_{x \to a^+} [f'(x)/g'(x)]$ implies that both $f'(x)$ and $g'(x)$ exist in at least a small interval $(a, b]$ and that $g'(x) \neq 0$ there. At $a$, we do not even know that $f$ and $g$ are defined, but we do know that $\lim\limits_{x \to a^+} f(x) = 0$ and $\lim\limits_{x \to a^+} g(x) = 0$. Thus, we may define (or redefine if necessary) both $f(a)$ and $g(a)$ to be zero, thereby making both $f$ and $g$ (right) continuous at $a$. All of this is to say that $f$ and $g$ satisfy the hypotheses of Cauchy's Mean Value Theorem on $[a, b]$. Consequently, there is a number $c$ in $(a, b)$ such that

$$\frac{f(b) - f(a)}{g(b) - g(a)} = \frac{f'(c)}{g'(c)}$$

or, since $f(a) = 0 = g(a)$,

$$\frac{f(b)}{g(b)} = \frac{f'(c)}{g'(c)}$$

When we let $b \to a^+$, thereby forcing $c \to a^+$, we obtain

$$\lim_{b \to a^+} \frac{f(b)}{g(b)} = \lim_{c \to a^+} \frac{f'(c)}{g'(c)}$$

which is equivalent to what we wanted to prove.

A very similar proof works for the case of left-hand limits, and thus for two-sided limits. The proofs for the cases where $a$ or $L$ is infinite are harder, and we omit them. ∎

## CONCEPTS REVIEW

**1.** L'Hôpital's Rule is useful in finding $\lim_{x \to a} [f(x)/g(x)]$, where both _____ and _____ are zero.

**2.** L'Hôpital's Rule says that under appropriate conditions $\lim_{x \to a} f(x)/g(x) = \lim_{x \to a}$ _____.

**3.** From l'Hôpital's Rule, we can conclude that

$\lim_{x \to 0} (\tan x)/x = \lim_{x \to 0}$ _____ $=$ _____, but l'Hôpital's Rule gives us no information about $\lim_{x \to 0} (\cos x)/x$

because _____.

**4.** The proof of l'Hôpital's Rule depends on _____ Theorem.

## PROBLEM SET 9.1

In Problems 1–24, find the indicated limit. Make sure l'Hôpital's Rule applies before you use it.

**1.** $\lim_{x \to 0} \dfrac{2x - \sin x}{x}$

**2.** $\lim_{x \to (1/2)\pi} \dfrac{\cos x}{\frac{1}{2}\pi - x}$

**3.** $\lim_{x \to 0} \dfrac{x - \sin 2x}{\tan x}$

**4.** $\lim_{x \to 0} \dfrac{\tan^{-1} 3x}{\sin^{-1} x}$

**5.** $\lim_{x \to -2} \dfrac{x^2 + 6x + 8}{x^2 - 3x - 10}$

**6.** $\lim_{x \to 0} \dfrac{x^3 - 3x^2 + x}{x^3 - 2x}$

**7.** $\lim_{x \to 1^-} \dfrac{x^2 - 2x + 2}{x^2 - 1}$

**8.** $\lim_{x \to 1} \dfrac{\ln x^2}{x^2 - 1}$

**9.** $\lim_{x \to (1/2)\pi} \dfrac{\ln (\sin x)^3}{\frac{1}{2}\pi - x}$

**10.** $\lim_{x \to 0} \dfrac{e^x - e^{-x}}{2 \sin x}$

**11.** $\lim_{t \to 1} \dfrac{\sqrt{t} - t^2}{\ln t}$

**12.** $\lim_{x \to 0^+} \dfrac{7^{\sqrt{x}} - 1}{2^{\sqrt{x}} - 1}$

**13.** $\lim_{x \to 0} \dfrac{\ln \cos 2x}{7x^2}$

**14.** $\lim_{x \to 0^-} \dfrac{3 \sin x}{\sqrt{-x}}$

**15.** $\lim_{x \to 0} \dfrac{\tan x - x}{\sin 2x - 2x}$

**16.** $\lim_{x \to 0} \dfrac{\sin x - \tan x}{x^2 \sin x}$

**17.** $\lim_{x \to 0^+} \dfrac{x^2}{\sin x - x}$

**18.** $\lim_{x \to 0} \dfrac{e^x - \ln(1 + x) - 1}{x^2}$

**19.** $\lim_{x \to 0} \dfrac{\tan^{-1}x - x}{8x^3}$

**20.** $\lim_{x \to 0} \dfrac{\cosh x - 1}{x^2}$

**21.** $\lim_{x \to 0^+} \dfrac{1 - \cos x - x \sin x}{2 - 2 \cos x - \sin^2 x}$

**22.** $\lim_{x \to 0^-} \dfrac{\sin x + \tan x}{e^x + e^{-x} - 2}$

**23.** $\lim_{x \to 0} \dfrac{\displaystyle\int_0^x \sqrt{1 + \sin t}\, dt}{x}$

**24.** $\lim_{x \to 0^+} \dfrac{\displaystyle\int_0^x \sqrt{t}\, \cos t\, dt}{x^2}$

**25.** In Section 3.4, we worked very hard to prove that $\lim_{x \to 0} (\sin x)/x = 1$; l'Hôpital's Rule allows us to show this fact in one line. However, even if we had l'Hôpital's Rule available at that stage, it would not have helped us. Explain why.

**26.** Find $\lim_{x \to 0} \dfrac{x^2 \sin(1/x)}{\tan x}$

*Hint*: Begin by deciding why l'Hôpital's Rule is not applicable. Then find the limit by other means.

**27.** For Figure 2, compute the following limits.

(a) $\lim_{t \to 0^+} \dfrac{\text{area of triangle } ABC}{\text{area of curved region } ABC}$

(b) $\lim_{t \to 0^+} \dfrac{\text{area of curved region } BCD}{\text{area of curved region } ABC}$

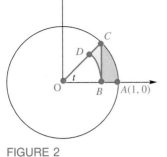

FIGURE 2

28. In Figure 3, $\overline{CD} = DE = \overline{DF} = t$. Find each limit.

(a) $\lim_{t \to 0^+} y$

(b) $\lim_{t \to 0^+} x$

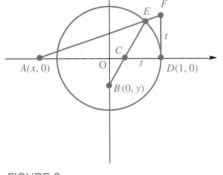

FIGURE 3

29. Using the concepts of Section 6.4, you can show that the surface area of the prolate spheroid gotten by rotating the ellipse $x^2/a^2 + y^2/b^2 = 1$ $(a > b)$ about the

*x*-axis is

$$A = 2\pi b^2 + 2\pi ab \left[ \frac{a}{\sqrt{a^2 - b^2}} \arcsin \frac{\sqrt{a^2 - b^2}}{a} \right]$$

What should $A$ approach as $a \to b^+$? Use l'Hôpital's Rule to show that this does happen.

30. Determine constants $a$, $b$, and $c$ so that

$$\lim_{x \to 1} \frac{ax^4 + bx^3 + 1}{(x - 1)\sin \pi x} = c$$

31. L'Hôpital's Rule in its 1696 form said this: If $\lim_{x \to a} f(x) = \lim_{x \to a} g(x) = 0$, then $\lim_{x \to a} f(x)/g(x) = f'(a)/g'(a)$, provided $f'(a)$ and $g'(a)$ both exist and $g'(a) \neq 0$. Prove this result without recourse to Cauchy's Mean Value Theorem. [This result can be extended to cover the case where $f'(a) = g'(a) = 0$. Then $\lim_{x \to a} f(x)/g(x) = f''(a)/g''(a)$.]

PC Use a Computer Algebra System of your choice to evaluate the limits in Problems 1–22; then use it to do Problems 32 and 33.

32. Find each limit.

(a) $\lim_{x \to 0} \frac{\cos x - 1 + x^2/2}{x^4}$

(b) $\lim_{x \to 0} \frac{e^x - 1 - x - x^2/2 - x^3/6}{x^4}$

33. Find each limit.

(a) $\lim_{x \to 0} \frac{1 - \cos(x^2)}{x^3 \sin x}$

(b) $\lim_{x \to 0} \frac{\tan x - x}{\arcsin x - x}$

---

**Answers to Concepts Review:** **1.** $\lim_{x \to a} f(x); \lim_{x \to a} g(x)$
**2.** $f'(x)/g'(x)$ **3.** $\sec^2 x$; 1; $\lim_{x \to 0} \cos x \neq 0$ **4.** Cauchy's Mean Value

---

## 9.2
## OTHER INDETERMINATE FORMS

In the solution to Example 6 of the previous section, we faced the following limit problem.

$$\lim_{x \to \infty} \frac{x}{e^x}$$

This is typical of a class of problems of the form $\lim_{x \to \infty} f(x)/g(x)$, where both numerator and denominator are growing indefinitely large; we call it an indeterminate form of type $\infty/\infty$. It turns out that l'Hôpital's Rule also applies in this situation, that is,

$$\lim_{x \to \infty} \frac{f(x)}{g(x)} = \lim_{x \to \infty} \frac{f'(x)}{g'(x)}$$

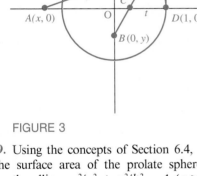

FIGURE 1

A rigorous proof is quite difficult, but there is an intuitive way of seeing that the result has to be true. Imagine that $f(t)$ and $g(t)$ represent the positions of two cars on the $t$-axis at time $t$ (Figure 1). These two cars, the $f$-car and the $g$-car, are on endless journeys with respective velocities $f'(t)$ and $g'(t)$. Now, if

$$\lim_{t \to \infty} \frac{f'(t)}{g'(t)} = L$$

then ultimately the $f$-car travels about $L$ times as fast as the $g$-car. It is therefore reasonable to say that in the long run, it will travel about $L$ times as far; that is,

$$\lim_{t \to \infty} \frac{f(t)}{g(t)} = L$$

We do not call this a proof, but it does lend plausibility to a result that we now state formally.

---

**Theorem A**

**(L'Hôpital's Rule for forms of type $\infty/\infty$).** Suppose that $\lim_{x \to u} |f(x)| = \lim_{x \to u} |g(x)| = \infty$. If $\lim_{x \to u} [f'(x)/g'(x)]$ exists in either the finite or infinite sense, then

$$\lim_{x \to u} \frac{f(x)}{g(x)} = \lim_{x \to u} \frac{f'(x)}{g'(x)}$$

Here $u$ may stand for any of the symbols $a$, $a^-$, $a^+$, $-\infty$, or $+\infty$.

---

**The Indeterminate Form $\infty/\infty$** We use Theorem A to finish Example 6 of the previous section.

**EXAMPLE 1** Find $\lim\limits_{x \to \infty} \dfrac{x}{e^x}$.

*Solution* Both $x$ and $e^x$ tend to $\infty$ as $x \to \infty$. Hence, by l'Hôpital's Rule,

$$\lim_{x \to \infty} \frac{x}{e^x} = \lim_{x \to \infty} \frac{D_x x}{D_x e^x} = \lim_{x \to \infty} \frac{1}{e^x} = 0 \qquad \blacksquare$$

Here is a general result of the same type.

**EXAMPLE 2** Show that if $a$ is any positive real number, $\lim\limits_{x \to \infty} \dfrac{x^a}{e^x} = 0$.

*Solution* Suppose as a special case that $a = 2.5$. Then three applications of l'Hôpital's Rule give

$$\lim_{x \to \infty} \frac{x^{2.5}}{e^x} \overset{\textcircled{L}}{=} \lim_{x \to \infty} \frac{2.5x^{1.5}}{e^x} \overset{\textcircled{L}}{=} \lim_{x \to \infty} \frac{(2.5)(1.5)x^{0.5}}{e^x} \overset{\textcircled{L}}{=} \lim_{x \to \infty} \frac{(2.5)(1.5)(0.5)}{x^{0.5}e^x} = 0$$

A similar argument works for any $a > 0$. $\qquad \blacksquare$

## See How They Grow

In computer science, one pays careful attention to the amount of time needed to perform a task. For example, to sort $x$ items using the "bubble sort" algorithm takes time proportional to $x^2$, whereas the "quick sort" algorithm does the same task in time proportional to $x \ln x$, a major improvement. Here is a chart illustrating how some common functions grow as $x$ increases from 10 to 100 to 1000.

| $\ln x$ | 2.3 | 4.6 | 6.9 |
|---|---|---|---|
| $\sqrt{x}$ | 3.2 | 10 | 31.6 |
| $x$ | 10 | 100 | 1000 |
| $x \ln x$ | 23 | 461 | 6908 |
| $x^2$ | 100 | 10000 | $10^6$ |
| $e^x$ | $2.2 \times 10^4$ | $2.7 \times 10^{43}$ | $10^{434}$ |

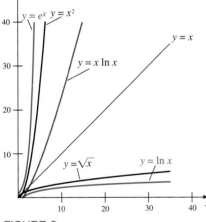

FIGURE 2

**EXAMPLE 3** Show that if $a$ is any positive real number, $\displaystyle\lim_{x \to \infty} \frac{\ln x}{x^a} = 0$.

**Solution** Both $\ln x$ and $x^a$ tend to $\infty$ as $x \to \infty$. Hence by one application of l'Hôpital's Rule,

$$\lim_{x \to \infty} \frac{\ln x}{x^a} \overset{L}{=} \lim_{x \to \infty} \frac{1/x}{ax^{a-1}} = \lim_{x \to \infty} \frac{1}{ax^a} = 0 \quad\blacksquare$$

Examples 2 and 3 say something that is worth remembering, namely, *for large $x$, $e^x$ grows faster as $x$ increases than any constant power of $x$, whereas $\ln x$ grows more slowly than any constant power of $x$.* For example, $e^x$ grows faster than $x^{100}$ and $\ln x$ grows more slowly than $\sqrt[100]{x}$. The chart in the margin and Figure 2 offer additional illustration.

**EXAMPLE 4** Find $\displaystyle\lim_{x \to 0^+} \frac{\ln x}{\cot x}$.

**Solution** As $x \to 0^+$, $\ln x \to -\infty$ and $\cot x \to \infty$, so l'Hôpital's Rule applies.

$$\lim_{x \to 0^+} \frac{\ln x}{\cot x} \overset{L}{=} \lim_{x \to 0^+} \left[ \frac{1/x}{-\csc^2 x} \right]$$

This is still indeterminate as it stands, but rather than apply l'Hôpital's Rule again (which only makes things worse), we rewrite the bracketed expression as

$$\frac{1/x}{-\csc^2 x} = -\frac{\sin^2 x}{x} = -\sin x \frac{\sin x}{x}$$

Thus,

$$\lim_{x \to 0^+} \frac{\ln x}{\cot x} = \lim_{x \to 0^+} \left[ -\sin x \frac{\sin x}{x} \right] = 0 \cdot 1 = 0 \quad\blacksquare$$

**The Indeterminate Forms $0 \cdot \infty$ And $\infty - \infty$** Suppose that $A(x) \to 0$, but $B(x) \to \infty$. What is going to happen to the product $A(x)B(x)$? Two competing forces are at work, tending to pull the product in opposite directions. Which will win this battle, $A$ or $B$ or neither? It depends on whether one is stronger (that is, doing its job at a faster rate) or if they are evenly matched. L'Hôpital's Rule will help us decide, but only after we have transformed the problem to a 0/0 or $\infty/\infty$ form.

**EXAMPLE 5** Find $\displaystyle\lim_{x \to \pi/2} (\tan x \cdot \ln \sin x)$.

**Solution** Since $\displaystyle\lim_{x \to \pi/2} \ln \sin x = 0$ and $\displaystyle\lim_{x \to \pi/2} |\tan x| = \infty$, this is a $0 \cdot \infty$ indeterminate form. We can rewrite it as a 0/0 form by the simple device of changing $\tan x$ to $1/\cot x$. Thus,

$$\lim_{x \to \pi/2} (\tan x \cdot \ln \sin x) = \lim_{x \to \pi/2} \frac{\ln \sin x}{\cot x}$$

$$\overset{\text{\textcircled{L}}}{=} \lim_{x \to \pi/2} \frac{\frac{1}{\sin x} \cdot \cos x}{-\csc^2 x} = \lim_{x \to \pi/2} (-\cos x \cdot \sin x) = 0 \qquad \blacksquare$$

**EXAMPLE 6** Find $\displaystyle\lim_{x \to 1^+} \left( \frac{x}{x - 1} - \frac{1}{\ln x} \right)$.

*Solution*   The first term is growing without bound; so is the second. We say that the limit is an $\infty - \infty$ indeterminate form. Again a battle rages. L'Hôpital's Rule will determine the result, but only after we rewrite the problem in a form to which the rule applies. In this case, the two fractions must be combined, a procedure that changes the problem to a 0/0 form. A double application of l'Hôpital's Rule yields

$$\lim_{x \to 1^+} \left( \frac{x}{x - 1} - \frac{1}{\ln x} \right) = \lim_{x \to 1^+} \frac{x \ln x - x + 1}{(x - 1)\ln x} \overset{\text{\textcircled{L}}}{=} \lim_{x \to 1^+} \frac{x \cdot 1/x + \ln x - 1}{(x - 1)(1/x) + \ln x}$$

$$= \lim_{x \to 1^+} \frac{x \ln x}{x - 1 + x \ln x} \overset{\text{\textcircled{L}}}{=} \lim_{x \to 1^+} \frac{1 + \ln x}{2 + \ln x} = \frac{1}{2} \qquad \blacksquare$$

**The Indeterminate Forms $0^0$, $\infty^0$, $1^\infty$**   We turn now to three indeterminate forms of exponential type. Here the trick is to consider not the original expression, but rather its logarithm. Usually l'Hôpital's Rule will apply to the logarithm.

**EXAMPLE 7** Find $\displaystyle\lim_{x \to 0^+} (x + 1)^{\cot x}$.

*Solution*   This takes the indeterminate form $1^\infty$. Let $y = (x + 1)^{\cot x}$, so

$$\ln y = \cot x \ln(x + 1) = \frac{\ln(x + 1)}{\tan x}$$

By l'Hôpital's Rule for 0/0 forms,

$$\lim_{x \to 0^+} \ln y = \lim_{x \to 0^+} \frac{\ln(x + 1)}{\tan x} \overset{\text{\textcircled{L}}}{=} \lim_{x \to 0^+} \frac{\frac{1}{x + 1}}{\sec^2 x} = 1$$

Now $y = e^{\ln y}$, and since the exponential function $f(x) = e^x$ is continuous,

$$\lim_{x \to 0^+} y = \lim_{x \to 0^+} \exp(\ln y) = \exp\left( \lim_{x \to 0^+} \ln y \right) = \exp 1 = e \qquad \blacksquare$$

**EXAMPLE 8** Find $\lim\limits_{x \to \pi/2^-} (\tan x)^{\cos x}$.

*Solution* This has the indeterminate form $\infty^0$. Put $y = (\tan x)^{\cos x}$, so

$$\ln y = \cos x \cdot \ln \tan x = \frac{\ln \tan x}{\sec x}$$

Then,

$$\lim_{x \to \pi/2^-} \ln y = \lim_{x \to \pi/2^-} \frac{\ln \tan x}{\sec x} \overset{\text{\textcircled{L}}}{=} \lim_{x \to \pi/2^-} \frac{\frac{1}{\tan x} \cdot \sec^2 x}{\sec x \tan x}$$

$$= \lim_{x \to \pi/2^-} \frac{\sec x}{\tan^2 x} = \lim_{x \to \pi/2^-} \frac{\cos x}{\sin^2 x} = 0$$

Therefore,

$$\lim_{x \to \pi/2^-} y = e^0 = 1 \qquad \blacksquare$$

**Summary** We have classified certain limit problems as indeterminate forms, using the seven symbols $0/0$, $\infty/\infty$, $0 \cdot \infty$, $\infty - \infty$, $0^0$, $\infty^0$, and $1^\infty$. Each involves a competition of opposing forces, which means that the result is not obvious. However, with the help of l'Hôpital's Rule, which applies directly only to the $0/0$ and $\infty/\infty$ forms, we can usually determine the correct limit.

There are many other possibilities symbolized by, for example, $0/\infty$, $\infty/0$, $\infty + \infty$, $\infty \cdot \infty$, $0^\infty$, and $\infty^\infty$. Why don't we call these indeterminate forms? Because, in each of these cases, the forces are in collusion, not competition.

**EXAMPLE 9** Find $\lim\limits_{x \to 0^+} (\sin x)^{\cot x}$.

*Solution* We might call this a $0^\infty$ form, but it is not indeterminate. Note that $\sin x$ is approaching zero, and raising it to the exponent $\cot x$, an increasingly large number, serves only to make it approach zero faster. Thus,

$$\lim_{x \to 0^+} (\sin x)^{\cot x} = 0 \qquad \blacksquare$$

## CONCEPTS REVIEW

**1.** If $\lim\limits_{x \to a} f(x) = \lim\limits_{x \to a} g(x) = \infty$, then $\lim\limits_{x \to a} f(x)/g(x) = \lim\limits_{x \to a}$ _____ .

**2.** If $\lim\limits_{x \to a} f(x) = 0$ and $\lim\limits_{x \to a} g(x) = \infty$, then $\lim\limits_{x \to a} f(x)g(x)$ is an indeterminate form. To apply l'Hôpital's

Rule, we may rewrite this latter limit as _____ .

**3.** Seven indeterminate forms are discussed in this book. They are symbolized by $0/0$, $\infty/\infty$, $0 \cdot \infty$ and _____ .

**4.** $e^x$ grows faster than any power of $x$ but _____ grows more slowly than any power of $x$.

## PROBLEM SET 9.2

Find the limits in Problems 1–40. Be sure l'Hôpital's Rule applies before you use it.

1. $\lim\limits_{x \to \infty} \dfrac{\ln x^{10000}}{x}$

2. $\lim\limits_{x \to \infty} \dfrac{(\ln x)^2}{2^x}$

3. $\lim\limits_{x \to \infty} \dfrac{x^{10000}}{e^x}$

4. $\lim\limits_{x \to \infty} \dfrac{3x}{\ln(100x + e^x)}$

5. $\lim\limits_{x \to \pi/2} \dfrac{3 \sec x + 5}{\tan x}$

6. $\lim\limits_{x \to 0^+} \dfrac{\ln \sin^2 x}{3 \ln \tan x}$

7. $\lim\limits_{x \to \infty} \dfrac{\ln(\ln x^{1000})}{\ln x}$

8. $\lim\limits_{x \to \left(\frac{1}{2}\right)^-} \dfrac{\ln(4 - 8x)^2}{\tan \pi x}$

9. $\lim\limits_{x \to 0^+} \dfrac{\cot x}{\sqrt{-\ln x}}$

10. $\lim\limits_{x \to 0} \dfrac{2 \csc^2 x}{\cot^2 x}$

11. $\lim\limits_{x \to 0} (x \ln x^{1000})$

12. $\lim\limits_{x \to 0} 3x^2 \csc^2 x$

13. $\lim\limits_{x \to 0} (\csc^2 x - \cot^2 x)$

14. $\lim\limits_{x \to \pi/2} (\tan x - \sec x)$

15. $\lim\limits_{x \to 0^+} (3x)^{x^2}$

16. $\lim\limits_{x \to 0} (\cos x)^{\csc x}$

17. $\lim\limits_{x \to (\pi/2)^-} (5 \cos x)^{\tan x}$

18. $\lim\limits_{x \to 0} \left( \csc^2 x - \dfrac{1}{x^2} \right)^2$

19. $\lim\limits_{x \to 0} (x + e^{x/3})^{3/x}$

20. $\lim\limits_{x \to (\pi/2)^-} (\cos 2x)^{x - \pi/2}$

21. $\lim\limits_{x \to \pi/2} (\sin x)^{\cos x}$

22. $\lim\limits_{x \to \infty} x^x$

23. $\lim\limits_{x \to \infty} x^{1/x}$

24. $\lim\limits_{x \to 0} (\cos x)^{1/x^2}$

25. $\lim\limits_{x \to 0^+} (\tan x)^{2/x}$

26. $\lim\limits_{x \to -\infty} (e^{-x} - x)$

27. $\lim\limits_{x \to 0^+} (\sin x)^x$

28. $\lim\limits_{x \to 0} (\cos x - \sin x)^{1/x}$

29. $\lim\limits_{x \to 0} \left( \csc x - \dfrac{1}{x} \right)$

30. $\lim\limits_{x \to \infty} \left( 1 + \dfrac{1}{x} \right)^x$

31. $\lim\limits_{x \to 0^+} (1 + 2e^x)^{1/x}$

32. $\lim\limits_{x \to 1} \left( \dfrac{1}{x - 1} - \dfrac{x}{\ln x} \right)$

33. $\lim\limits_{x \to 0} (\cos x)^{1/x}$

34. $\lim\limits_{x \to 0^+} (x^{1/2} \ln x)$

35. $\lim\limits_{x \to \infty} e^{\cos x}$

36. $\lim\limits_{x \to \infty} [\ln(x + 1) - \ln(x - 1)]$

37. $\lim\limits_{x \to 0^+} \dfrac{x}{\ln x}$

38. $\lim\limits_{x \to 0^+} (\ln x \cot x)$

39. $\lim\limits_{x \to \infty} \dfrac{\displaystyle\int_1^x \sqrt{1 + e^{-t}}\, dt}{x}$

40. $\lim\limits_{x \to 1^+} \dfrac{\displaystyle\int_1^x \sin t\, dt}{x - 1}$

41. Find each limit.

(a) $\lim\limits_{n \to \infty} \sqrt[n]{a}$

(b) $\lim\limits_{n \to \infty} \sqrt[n]{n}$

(c) $\lim\limits_{n \to \infty} n(\sqrt[n]{a} - 1)$

(d) $\lim\limits_{n \to \infty} n(\sqrt[n]{n} - 1)$

*Hint*: Transform to problems involving a continuous variable $x$.

42. Find each limit.

(a) $\lim\limits_{x \to 0^+} x^x$

(b) $\lim\limits_{x \to 0^+} (x^x)^x$

(c) $\lim\limits_{x \to 0^+} x^{(x^x)}$

(d) $\lim\limits_{x \to 0^+} ((x^x)^x)^x$

(e) $\lim\limits_{x \to 0^+} x^{(x^{(x^x)})}$

43. Graph $y = x^{1/x}$ for $x > 0$. Show what happens for very small $x$ and very large $x$. Indicate the maximum value.

44. Find each limit.

(a) $\lim\limits_{x \to 0^+} (1^x + 2^x)^{1/x}$

(b) $\lim\limits_{x \to 0^-} (1^x + 2^x)^{1/x}$

(c) $\lim\limits_{x \to \infty} (1^x + 2^x)^{1/x}$

(d) $\lim\limits_{x \to -\infty} (1^x + 2^x)^{1/x}$

45. For $k \geq 0$, find

$$\lim\limits_{n \to \infty} \dfrac{1^k + 2^k + \cdots + n^k}{n^{k + 1}}$$

*Hint*: Though this has the $\infty/\infty$ form, l'Hôpital's Rule is not helpful. Think of another often-used technique.

46. Let $c_1, c_2, \ldots, c_n$ be positive constants with $\displaystyle\sum_{i=1}^n c_i = 1$ and let $x_1, x_2, \ldots, x_n$ be positive numbers. Take natural logarithms and then use l'Hôpital's Rule to show that

$$\lim\limits_{t \to 0^+} \left( \sum_{i=1}^n c_i x_i^t \right)^{1/t} = x_1^{c_1} x_2^{c_2} \cdots x_n^{c_n} = \prod_{i=1}^n x_i^{c_i}$$

In particular, if $a, b, x, y$ are positive and $a + b = 1$,

$$\lim\limits_{t \to 0^+} (ax^t + by^t)^{1/t} = x^a y^b$$

[PC] 47. Verify the last statement in Problem 46 by calculating each of the following.

(a) $\lim\limits_{t \to 0^+} \left( \tfrac{1}{2} 2^t + \tfrac{1}{2} 5^t \right)^{1/t}$

(b) $\lim\limits_{t \to 0^+} \left( \tfrac{1}{5} 2^t + \tfrac{4}{5} 5^t \right)^{1/t}$

(c) $\lim\limits_{t \to 0^+} \left( \tfrac{1}{10} 2^t + \tfrac{9}{10} 5^t \right)^{1/t}$

[PC] 48. Consider $f(x) = n^2 x e^{-nx}$.

(a) Graph $f(x)$ for $n = 1, 2, 3, 4, 5, 6$ on $[0, 1]$ using the same axes.

(b) For $x > 0$, find $\lim\limits_{n \to \infty} f(x)$.

(c) Evaluate $\int_0^1 f(x)\,dx$ for $n = 1, 2, 3, 4, 5, 6$.

(d) Guess at $\lim\limits_{n \to \infty} \int_0^1 f(x)\,dx$. Then justify your answer rigorously.

---

### 9.3
### IMPROPER INTEGRALS: INFINITE LIMITS

In the definition of $\int_a^b f(x)\,dx$, it was assumed that the interval $[a, b]$ was finite. However, there are many applications in physics, economics, and probability in which we wish to allow $a$ or $b$ (or both) to be infinite. We must therefore find a way to give meaning to symbols like

$$\int_0^\infty \frac{1}{1 + x^2}\,dx, \qquad \int_{-\infty}^{-1} xe^{-x^2}\,dx, \qquad \int_{-\infty}^\infty x^2 e^{-x^2}\,dx$$

These integrals are called **improper integrals** with infinite limits.

**One Infinite Limit** The graph of $f(x) = e^{-x}$ on $[0, \infty)$ is shown in Figure 1. The integral $\int_0^b e^{-x}\,dx$ makes perfectly good sense no matter how large we make $b$; in fact, we can evaluate this integral explicitly.

$$\int_0^b e^{-x}\,dx = -\int_0^b e^{-x}(-dx) = \left[-e^{-x}\right]_0^b = 1 - e^{-b}$$

Now $\lim\limits_{b \to \infty}(1 - e^{-b}) = 1$, so it seems natural to define

$$\int_0^\infty e^{-x}\,dx = 1$$

Here is the general definition.

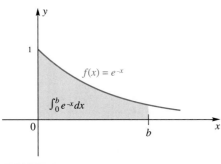

FIGURE 1

> **Definition**
>
> $$\int_{-\infty}^b f(x)\,dx = \lim_{a \to -\infty} \int_a^b f(x)\,dx$$
>
> $$\int_a^\infty f(x)\,dx = \lim_{b \to \infty} \int_a^b f(x)\,dx$$
>
> If the limits on the right exist and have finite values, then we say the corresponding improper integrals **converge** and have those values. Otherwise, the integrals are said to **diverge.**

**EXAMPLE 1** Find, if possible, $\int_{-\infty}^{-1} xe^{-x^2} \, dx$.

*Solution*

$$\int_{a}^{-1} xe^{-x^2} \, dx = -\frac{1}{2} \int_{a}^{-1} e^{-x^2}(-2x \, dx) = \left[ -\frac{1}{2} e^{-x^2} \right]_{a}^{-1}$$

$$= -\frac{1}{2} e^{-1} + \frac{1}{2} e^{-a^2}$$

Thus,

$$\int_{-\infty}^{-1} xe^{-x^2} \, dx = \lim_{a \to -\infty} \left[ -\frac{1}{2} e^{-1} + \frac{1}{2} e^{-a^2} \right] = -\frac{1}{2e}$$

We say the integral converges and has value $-1/2e$. ∎

**EXAMPLE 2** Find, if possible, $\int_{0}^{\infty} \sin x \, dx$.

*Solution*

$$\int_{0}^{\infty} \sin x \, dx = \lim_{b \to \infty} \int_{0}^{b} \sin x \, dx = \lim_{b \to \infty} [-\cos x]_{0}^{b}$$

$$= \lim_{b \to \infty} [1 - \cos b]$$

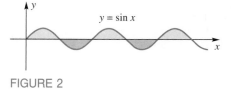

FIGURE 2

The latter limit does not exist; we conclude that the given integral diverges. Think about the geometric meaning of $\int_{0}^{\infty} \sin x \, dx$ to confirm this result (Figure 2). ∎

**EXAMPLE 3** According to Newton's Inverse-Square Law, the force exerted by the earth on a space capsule is $-k/x^2$, where $x$ is the distance (in miles, for instance) from the capsule to the center of the earth (Figure 3). The force $F(x)$ required to lift the capsule is therefore $F(x) = k/x^2$. How much work is done in propelling a 1000-pound capsule out of the earth's gravitational field?

*Solution* We can evaluate $k$ by noting that at $x = 3960$ miles (the radius of the earth), $F = 1000$ pounds. This yields $k = 1000(3960)^2 \approx 1.568 \times 10^{10}$. The work done in mile-pounds is therefore

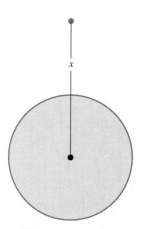

FIGURE 3

$$1.568 \times 10^{10} \int_{3960}^{\infty} \frac{1}{x^2} \, dx = \lim_{b \to \infty} 1.568 \times 10^{10} \left[ -\frac{1}{x} \right]_{3960}^{b}$$

$$= \lim_{b \to \infty} 1.568 \times 10^{10} \left[ -\frac{1}{b} + \frac{1}{3960} \right]$$

$$= \frac{1.568 \times 10^{10}}{3960} \approx 3.96 \times 10^{6} \quad ∎$$

**Both Limits Infinite**   First we need a definition.

Definition

If both $\int_{-\infty}^{0} f(x)\,dx$ and $\int_{0}^{\infty} f(x)\,dx$ converge, then $\int_{-\infty}^{\infty} f(x)\,dx$ is said to converge and have value

$$\int_{-\infty}^{\infty} f(x)\,dx = \int_{-\infty}^{0} f(x)\,dx + \int_{0}^{\infty} f(x)\,dx$$

Otherwise, $\int_{-\infty}^{\infty} f(x)\,dx$ diverges.

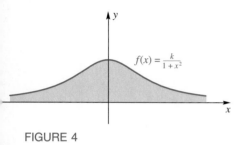

FIGURE 4

**EXAMPLE 4**   The function $f(x) = k/(1 + x^2)$ occurs in probability theory, where it is called the Cauchy density function. The constant $k$ is to be chosen so that the total area under the curve on $(-\infty, \infty)$ is 1 (see Figure 4). Determine $k$.

*Solution*   We want

$$\int_{-\infty}^{\infty} \frac{k}{1 + x^2}\,dx = 1$$

Now,

$$\int_{0}^{\infty} \frac{k}{1 + x^2}\,dx = \lim_{b \to \infty} \int_{0}^{b} \frac{k}{1 + x^2}\,dx$$

$$= \lim_{b \to \infty} [k \tan^{-1}x]_{0}^{b}$$

$$= \lim_{b \to \infty} k \tan^{-1}b = k\frac{\pi}{2}$$

Similarly, from symmetry,

$$\int_{-\infty}^{0} \frac{k}{1 + x^2}\,dx = k\frac{\pi}{2}$$

Thus,

$$\int_{-\infty}^{\infty} \frac{k}{1 + x^2}\,dx = \int_{-\infty}^{0} \frac{k}{1 + x^2}\,dx + \int_{0}^{\infty} \frac{k}{1 + x^2}\,dx = k\pi$$

We conclude that $k = 1/\pi$.  ■

**EXAMPLE 5**   The most important density function in probability theory is the standard **normal** density function, defined by

$$f(x) = \frac{1}{\sqrt{2\pi}}\, e^{-x^2/2}$$

It is surprisingly difficult to show that

$$\frac{1}{\sqrt{2\pi}} \int_{-\infty}^{\infty} e^{-x^2/2} \, dx = 1$$

though we will do so later (Section 16.4). Use this fact to show that this density function has mean 0 and variance 1; that is, show each of the following.

(a) $\dfrac{1}{\sqrt{2\pi}} \displaystyle\int_{-\infty}^{\infty} xe^{-x^2/2} \, dx = 0$  (b) $\dfrac{1}{\sqrt{2\pi}} \displaystyle\int_{-\infty}^{\infty} x^2 e^{-x^2/2} \, dx = 1$

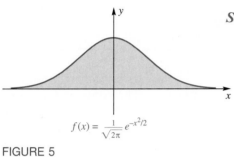

$$f(x) = \frac{1}{\sqrt{2\pi}} e^{-x^2/2}$$

FIGURE 5

*Solution*
(a) Since $e^{-x^2/2}$ is an even function, the graph of $y = f(x)$ is symmetric with respect to the $y$-axis (Figure 5). Thus,

$$\frac{1}{\sqrt{2\pi}} \int_{0}^{\infty} e^{-x^2/2} \, dx = \frac{1}{2}$$

a fact we will need shortly. Now

$$\frac{1}{\sqrt{2\pi}} \int_{0}^{\infty} xe^{-x^2/2} \, dx = \lim_{b \to \infty} \left[ -\frac{1}{\sqrt{2\pi}} \int_{0}^{b} e^{-x^2/2} (-x) \, dx \right]$$

$$= \lim_{b \to \infty} \left[ -\frac{1}{\sqrt{2\pi}} e^{-x^2/2} \right]_{0}^{b}$$

$$= \frac{1}{\sqrt{2\pi}}$$

Since $xe^{-x^2/2}$ is an odd function,

$$\frac{1}{\sqrt{2\pi}} \int_{-\infty}^{0} xe^{-x^2/2} \, dx = -\frac{1}{\sqrt{2\pi}} \int_{0}^{\infty} xe^{-x^2/2} \, dx = -\frac{1}{\sqrt{2\pi}}$$

which establishes (a).
(b) The demonstration of (b) uses integration by parts and l'Hôpital's Rule.

$$\frac{1}{\sqrt{2\pi}} \int_{0}^{\infty} x^2 e^{-x^2/2} \, dx = \lim_{b \to \infty} \frac{1}{\sqrt{2\pi}} \int_{0}^{b} (-x)(-e^{-x^2/2}x) \, dx$$

$$= \lim_{b \to \infty} \frac{1}{\sqrt{2\pi}} \left( [-xe^{-x^2/2}]_{0}^{b} + \int_{0}^{b} e^{-x^2/2} \, dx \right)$$

$$= \frac{1}{\sqrt{2\pi}} \left( 0 + \int_{0}^{\infty} e^{-x^2/2} \, dx \right) = \frac{1}{2}$$

Since $x^2 e^{-x^2/2}$ is an even function, we get a similar contribution to the left of zero, and so

$$\frac{1}{\sqrt{2\pi}} \int_{-\infty}^{\infty} x^2 e^{-x^2/2} \, dx = 1 \qquad \blacksquare$$

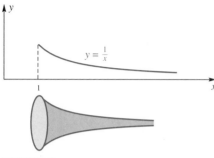

**The Paradox of Gabriel's Horn** Let the curve $y = 1/x$ on $[1, \infty)$ be revolved about the $x$-axis, thereby generating a surface called Gabriel's horn (Figure 6). We claim:

1. the volume $V$ of this horn is finite;
2. the surface area $A$ of the horn is infinite.

To put the results in practical terms, they seem to say that the horn can be filled with a finite amount of paint, and yet there is not enough to paint its inside surface. Before we try to unravel this paradox, let us establish (1) and (2). We use results for volume from Section 6.2 and for surface area from Section 6.4.

$$V = \int_1^\infty \pi \left(\frac{1}{x}\right)^2 dx = \lim_{b \to \infty} \pi \int_1^b x^{-2}\, dx$$

$$= \lim_{b \to \infty} \left[-\frac{\pi}{x}\right]_1^b = \pi$$

$$A = \int_1^\infty 2\pi y\, ds = \int_1^\infty 2\pi y \sqrt{1 + \left(\frac{dy}{dx}\right)^2}\, dx$$

$$= 2\pi \int_1^\infty \frac{1}{x} \sqrt{1 + \left(\frac{-1}{x^2}\right)^2}\, dx$$

$$= \lim_{b \to \infty} 2\pi \int_1^b \frac{\sqrt{x^4 + 1}}{x^3}\, dx$$

Now,

$$\frac{\sqrt{x^4 + 1}}{x^3} > \frac{\sqrt{x^4}}{x^3} = \frac{1}{x}$$

Thus,

$$\int_1^b \frac{\sqrt{x^4 + 1}}{x^3}\, dx > \int_1^b \frac{1}{x}\, dx = \ln b$$

FIGURE 6

and since $\ln b \to \infty$ as $b \to \infty$, we conclude that $A$ is infinite.

Is something wrong with our mathematics? No. Imagine the horn to be slit down the side, opened up, and laid flat. Given a finite amount of paint, we could not possibly paint this surface with a paint coat of *uniform* thickness. However, we could do it if we allowed the paint coat to get thinner and thinner as we moved farther and farther from the horn's fat end. And, of course, that is exactly what happens when we fill the unslit horn with $\pi$ cubic units of paint. (Imaginary paint can be spread to arbitrary thinness.)

This problem involved study of two integrals of the form $\int_1^\infty 1/x^p\, dx$.

For later reference, we now analyze this integral for all values of $p$.

---

**Gabriel Paves a Street**

When told to pave the infinite street $0 \le x < \infty$, $0 \le y \le 1$ with pure gold, Gabriel obeyed but made the thickness $h$ of the gold at $x$ satisfy

$$h = e^{-x}$$

How much gold did it take?

$$V = \int_0^\infty e^{-x} dx = \lim_{b \to \infty} \int_0^b e^{-x} dx$$

$$= \lim_{b \to \infty} [-e^{-x}]_0^b = 1$$

Just 1 cubic unit.

---

**EXAMPLE 6** Show that $\int_1^\infty 1/x^p \, dx$ diverges for $p \le 1$ and converges for $p > 1$.

*Solution* We showed in our solution to Gabriel's horn that the integral diverges for $p = 1$. If $p \ne 1$,

$$\int_1^\infty \frac{1}{x^p} dx = \lim_{b \to \infty} \int_1^b x^{-p} dx = \lim_{b \to \infty} \left[ \frac{x^{-p+1}}{-p+1} \right]_1^b$$

$$= \lim_{b \to \infty} \left[ \frac{1}{1-p} \right] \left[ \frac{1}{b^{p-1}} - 1 \right] = \begin{cases} \infty & \text{if } p < 1 \\ \dfrac{1}{p-1} & \text{if } p > 1 \end{cases}$$

The conclusion follows. ∎

## CONCEPTS REVIEW

1. $\int_a^\infty f(x)\,dx$ is said to _____ if $\lim_{b \to \infty} \int_a^b f(x)\,dx$ exists and is finite.

2. $\int_0^\infty \cos x \, dx$ does not converge because _____ does not exist.

3. $\int_{-\infty}^\infty f(x)\,dx$ is said to diverge if either _____ or _____ diverges.

4. $\int_1^\infty (1/x^p)\,dx$ converges if and only if _____ .

## PROBLEM SET 9.3

In Problems 1–24, evaluate the given improper integral or show that it is divergent.

1. $\int_{100}^\infty e^x \, dx$

2. $\int_{-\infty}^{-5} \frac{dx}{x^4}$

3. $\int_1^\infty 2xe^{-x^2} \, dx$

4. $\int_{-\infty}^1 e^{4x} \, dx$

5. $\int_9^\infty \frac{x\,dx}{\sqrt{1+x^2}}$

6. $\int_1^\infty \frac{dx}{\sqrt{\pi x}}$

7. $\int_1^\infty \frac{dx}{x^{1.00001}}$

8. $\int_{10}^\infty \frac{x}{1+x^2} \, dx$

9. $\int_1^\infty \frac{dx}{x^{0.99999}}$

10. $\int_1^\infty \frac{x}{(1+x^2)^2} \, dx$

11. $\int_e^\infty \frac{1}{x \ln x} \, dx$

12. $\int_e^\infty \frac{\ln x}{x} \, dx$

13. $\int_2^\infty \frac{\ln x}{x^2} \, dx$

14. $\int_1^\infty xe^{-x} \, dx$

15. $\int_{-\infty}^1 \frac{dx}{(2x-3)^3}$

16. $\int_4^\infty \frac{dx}{(\pi - x)^{2/3}}$

17. $\int_{-\infty}^\infty \frac{x}{\sqrt{x^2+9}} \, dx$

18. $\int_{-\infty}^\infty \frac{dx}{(x^2+16)^2}$

19. $\int_{-\infty}^\infty \frac{1}{x^2+2x+10} \, dx$

20. $\int_{-\infty}^\infty \frac{x}{e^{2|x|}} \, dx$

21. $\int_{-\infty}^\infty \text{sech } x \, dx$.

*Hint:* $\int \text{sech } x \, dx = \tan^{-1}(\sinh x) + C$.

22. $\int_1^\infty \text{csch } x \, dx$

23. $\int_0^\infty e^{-x} \cos x \, dx$. *Hint:* Use table of integrals.

24. $\int_0^\infty e^{-x} \sin x \, dx$

25. Find the area of the region under the curve $y = 2/(4x^2 - 1)$ to the right of $x = 1$. *Hint:* Use partial fractions.

26. Find the area of the region under the curve $y = 1/(x^2 + x)$ to the right of $x = 1$.

**27.** Suppose that Newton's law for the force of gravity had the form $-k/x$ rather than $-k/x^2$ (see Example 3). Show that it would then be impossible to send anything out of the earth's gravitational field.

**28.** If a 1000-pound capsule weighs only 165 pounds on the moon (radius 1080 miles), how much work is done in propelling this capsule out of the moon's gravitational field (see Example 3)?

**29.** Suppose a company expects its annual profits $t$ years from now to be $f(t)$ dollars and that interest is considered to be compounded continuously at an annual rate $r$. Then the present value of all future profits can be shown to be

$$FP = \int_0^\infty e^{-rt}f(t)\,dt$$

Find $FP$ if $r = 0.08$ and $f(t) = 100{,}000$.

**30.** Do Problem 29 assuming $f(t) = 100{,}000 + 1000t$.

**31.** In probability theory, *waiting times* tend to have the exponential density given for $\alpha > 0$ by $f(x) = \alpha e^{-\alpha x}$ on $[0, \infty)$. Show each of the following.

(a) $\displaystyle\int_0^\infty f(x)\,dx = 1$

(b) $\displaystyle\int_0^\infty xf(x)\,dx = 1/\alpha$ (the mean)

**32.** In electromagnetic theory, the magnetic potential $u$ at a point on the axis of a circular coil is given by

$$u = Ar\int_a^\infty \frac{dx}{(r^2 + x^2)^{3/2}}$$

where $A$, $r$, and $a$ are constants. Evaluate $u$.

**33.** There is a subtlety in the definition of $\displaystyle\int_{-\infty}^\infty f(x)\,dx$ illustrated by the following. Show that (a) $\displaystyle\int_{-\infty}^\infty \sin x\,dx$ diverges and (b) $\displaystyle\lim_{a\to\infty}\int_{-a}^a \sin x\,dx = 0$.

**34.** Consider an infinitely long wire coinciding with the positive $x$-axis and having mass density $\delta(x) = (1 + x^2)^{-1}$ at $x$, $0 \le x < \infty$.

(a) Calculate the total mass of the wire (Example 4).
(b) Show that this wire does not have a center of mass. (In probability language, the Cauchy distribution does not have a mean.)

**35.** Give an example of a region in the first quadrant that gives a solid of finite volume when revolved about the $x$-axis but gives a solid of infinite volume when revolved about the $y$-axis.

**36.** Let $f$ be a nonnegative continuous function defined on $0 \le x < \infty$ with $\displaystyle\int_0^\infty f(x)\,dx < \infty$. Show:

(a) if $\displaystyle\lim_{x\to\infty} f(x)$ exists, it must be 0;

(b) it is possible that $\displaystyle\lim_{x\to\infty} f(x)$ does not exist.

PC **37.** We can use a computer to calculate $\displaystyle\int_1^\infty f(x)\,dx$ by taking $b$ very large in $\displaystyle\int_1^b f(x)\,dx$ *provided* we know the first integral converges. Calculate $\displaystyle\int_1^{100} (1/x^p)\,dx$ for $p = 2$, 1.1, 1.01, 1, and 0.99. Note that this gives no hint that the integral $\displaystyle\int_1^\infty (1/x^p)\,dx$ converges for $p > 1$ and diverges for $p \le 1$.

PC **38.** Calculate $\displaystyle\int_{-a}^a \frac{1}{\pi}(1 + x^2)^{-1}\,dx$ for $a = 10$, 50, and 100 and thereby add support to the result of Example 4.

PC **39.** Calculate $\displaystyle\int_{-a}^a \frac{1}{\sqrt{2\pi}}\exp(-0.5x^2)\,dx$ for $a = 1, 2, 3,$ and 4 (Example 5).

---

**Answers to Concepts Review:** **1.** Converge **2.** $\displaystyle\lim_{b\to\infty}\int_0^b \cos x\,dx$ **3.** $\displaystyle\int_{-\infty}^0 f(x)\,dx;\ \int_0^\infty f(x)\,dx$ **4.** $p > 1$

---

**9.4**
**IMPROPER INTEGRALS:**
**INFINITE INTEGRANDS**

Considering the many complicated integrations we have done, here is one that looks simple enough.

$$\int_{-2}^1 \frac{1}{x^2}\,dx = \left[-\frac{1}{x}\right]_{-2}^1 = -1 - \frac{1}{2} = -\frac{3}{2} \qquad ???$$

One glance at the graph in Figure 1 tells us that something is terribly wrong. The answer (if there is one) has to be a positive number. (Why?)

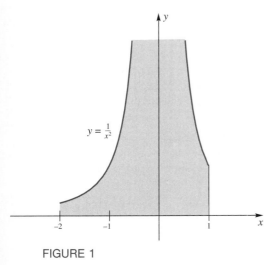

FIGURE 1

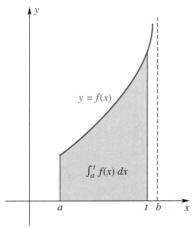

FIGURE 2

Where is our mistake? To answer, we refer back to Section 5.5. Recall that for a function to be integrable in the standard (or proper) sense, it must be bounded. Our function, $f(x) = 1/x^2$, is not bounded, so it is not integrable in the proper sense. We say that $\int_{-2}^{1} x^{-2}\, dx$ is an improper integral with an infinite integrand (*unbounded integrand* is a more accurate but less colorful term).

Until now, we have carefully avoided infinite integrands in all our examples and problems. We could continue to do this, but that would be to avoid a kind of integral which has important applications. Our task for this section is to define and analyze this new kind of integral.

**Integrands That Are Infinite at an Endpoint**    We give the definition for the case where $f$ tends to infinity at the right endpoint of the interval of integration. There is a completely analogous definition for the case where the left endpoint is the troublesome one.

---

**Definition**

Let $f$ be continuous on the half-open interval $[a, b)$ and suppose $\lim_{x \to b^-} |f(x)| = \infty$. Then,

$$\int_{a}^{b} f(x)\, dx = \lim_{t \to b^-} \int_{a}^{t} f(x)\, dx$$

provided this limit exists and is finite, in which case we say the integral converges. Otherwise, we say the integral diverges.

---

Note the geometric interpretation shown in Figure 2.

**EXAMPLE 1**    Evaluate, if possible, the improper integral $\int_{0}^{2} \dfrac{dx}{\sqrt{4 - x^2}}$.

**Solution**    Note that the integrand tends to infinity at 2.

$$\int_{0}^{2} \frac{dx}{\sqrt{4 - x^2}} = \lim_{t \to 2^-} \int_{0}^{t} \frac{dx}{\sqrt{4 - x^2}} = \lim_{t \to 2^-} \left[ \sin^{-1}\left(\frac{x}{2}\right) \right]_{0}^{t}$$

$$= \lim_{t \to 2^-} \left[ \sin^{-1}\left(\frac{t}{2}\right) - \sin^{-1}\left(\frac{0}{2}\right) \right] = \frac{\pi}{2} \qquad \blacksquare$$

**EXAMPLE 2**    Evaluate, if possible, $\int_{0}^{16} \dfrac{1}{\sqrt[4]{x}}\, dx$.

**Solution**

$$\int_{0}^{16} x^{-1/4}\, dx = \lim_{t \to 0^+} \int_{t}^{16} x^{-1/4}\, dx = \lim_{t \to 0^+} \left[ \frac{4}{3} x^{3/4} \right]_{t}^{16}$$

$$= \lim_{t \to 0^+} \left[ \frac{32}{3} - \frac{4}{3} t^{3/4} \right] = \frac{32}{3} \qquad \blacksquare$$

Two Key Examples

From Example 6 of Section 9.3, we learned that

$$\int_1^\infty \frac{1}{x^p}\, dx$$

converges if and only if $p > 1$. From Example 4 of the present section, we learn that

$$\int_0^1 \frac{1}{x^p}\, dx$$

converges if and only if $p < 1$. The first has an infinite limit; the second has an infinite integrand. If you feel at home with these two integrals, you should also be at ease with any other improper integrals you may meet.

**EXAMPLE 3**   Evaluate, if possible, $\displaystyle\int_0^1 \frac{1}{x}\, dx$.

**Solution**

$$\int_0^1 \frac{1}{x}\, dx = \lim_{t \to 0^+} \int_t^1 \frac{1}{x}\, dx = \lim_{t \to 0^+} [\ln x]_t^1$$

$$= \lim_{t \to 0^+} [-\ln t] = \infty$$

We conclude that the integral diverges.   ∎

**EXAMPLE 4**   Show that $\displaystyle\int_0^1 \frac{1}{x^p}\, dx$ converges if $p < 1$, but diverges if $p \geq 1$.

**Solution**   Example 3 took care of the case $p = 1$. If $p \neq 1$,

$$\int_0^1 \frac{1}{x^p}\, dx = \lim_{t \to 0^+} \int_t^1 x^{-p}\, dx = \lim_{t \to 0^+} \left[\frac{x^{-p+1}}{-p+1}\right]_t^1$$

$$= \lim_{t \to 0^+} \left[\frac{1}{1-p} - \frac{1}{1-p} \cdot \frac{1}{t^{p-1}}\right] = \begin{cases} \dfrac{1}{1-p} & \text{if } p < 1 \\ \infty & \text{if } p > 1 \end{cases} \quad ∎$$

**EXAMPLE 5**   Sketch the graph of the hypocycloid of four cusps, $x^{2/3} + y^{2/3} = 1$, and find its perimeter.

≈ **Solution**   The graph is shown in Figure 3. To find the perimeter, it is enough to find the length $L$ of the first quadrant portion and quadruple it. We estimate $L$ to be a bit more than $\sqrt{2} \approx 1.4$. Its exact value is (see Section 6.4).

$$L = \int_0^1 \sqrt{1 + (y')^2}\, dx$$

By implicit differentiation of $x^{2/3} + y^{2/3} = 1$, we obtain

$$\frac{2}{3} x^{-1/3} + \frac{2}{3} y^{-1/3} y' = 0$$

or

$$y' = -\frac{y^{1/3}}{x^{1/3}}$$

Thus,

$$1 + (y')^2 = 1 + \frac{y^{2/3}}{x^{2/3}} = 1 + \frac{1 - x^{2/3}}{x^{2/3}} = \frac{1}{x^{2/3}}$$

and so

$$L = \int_0^1 \sqrt{1 + (y')^2}\, dx = \int_0^1 \frac{1}{x^{1/3}}\, dx$$

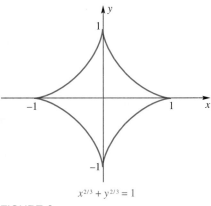

$x^{2/3} + y^{2/3} = 1$

FIGURE 3

The value of this improper integral can be read from the solution to Example 4; it is $L = 1/(1 - \frac{1}{3}) = \frac{3}{2}$. We conclude that the hypocycloid has perimeter $4L = 6$. ∎

**Integrands That Are Infinite at an Interior Point** The integral $\int_{-2}^{1} 1/x^2\, dx$ of our introduction has an integrand that tends to infinity at $x = 0$, an interior point of the interval $[-2, 1]$. Here is the appropriate definition to give meaning to such an integral.

---

**Definition**

Let $f$ be continuous on $[a, b]$ except at a number $c$, where $a < c < b$, and suppose $\lim_{x \to c} |f(x)| = \infty$. Then we define

$$\int_a^b f(x)\, dx = \int_a^c f(x)\, dx + \int_c^b f(x)\, dx$$

provided both integrals on the right converge. Otherwise, we say $\int_a^b f(x)\, dx$ diverges.

---

**EXAMPLE 6** Show that $\int_{-2}^{1} 1/x^2\, dx$ diverges.

*Solution*

$$\int_{-2}^{1} \frac{1}{x^2}\, dx = \int_{-2}^{0} \frac{1}{x^2}\, dx + \int_{0}^{1} \frac{1}{x^2}\, dx$$

The second of the integrals on the right diverges by Example 4. This is enough to give the conclusion. ∎

**EXAMPLE 7** Evaluate, if possible, the improper integral $\int_0^3 \dfrac{dx}{(x - 1)^{2/3}}$.

*Solution* The integrand tends to infinity at $x = 1$ (see Figure 4). Thus,

$$\int_0^3 \frac{dx}{(x - 1)^{2/3}} = \int_0^1 \frac{dx}{(x - 1)^{2/3}} + \int_1^3 \frac{dx}{(x - 1)^{2/3}}$$

$$= \lim_{t \to 1^-} \int_0^t \frac{dx}{(x - 1)^{2/3}} + \lim_{s \to 1^+} \int_s^3 \frac{dx}{(x - 1)^{2/3}}$$

$$= \lim_{t \to 1^-} [3(x - 1)^{1/3}]_0^t + \lim_{s \to 1^+} [3(x - 1)^{1/3}]_s^3$$

$$= 3 \lim_{t \to 1^-} [(t - 1)^{1/3} + 1] + 3 \lim_{s \to 1^+} [2^{1/3} - (s - 1)^{1/3}]$$

$$= 3 + 3(2^{1/3}) \approx 6.78 \qquad ∎$$

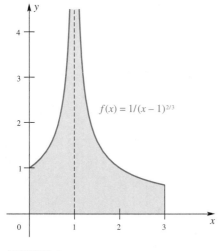

$f(x) = 1/(x - 1)^{2/3}$

**FIGURE 4**

## CONCEPTS REVIEW

**1.** The integral $\int_0^1 (1/\sqrt{x})\,dx$ does not exist in the proper sense because the function $f(x) = 1/\sqrt{x}$ is _____ on the interval $(0, 1]$.

**2.** Considered as an improper integral, $\int_0^1 (1/\sqrt{x})\,dx =$

$\lim\limits_{a \to 0^+} \int_a^1 x^{-1/2}\,dx =$ _____.

**3.** The improper integral $\int_0^4 (1/\sqrt{4 - x})\,dx$ is defined by _____.

**4.** The improper integral $\int_0^1 (1/x^p)\,dx$ converges if and only if _____.

## PROBLEM SET 9.4

In Problems 1–32, evaluate the given improper integral or show that it diverges.

**1.** $\int_1^3 \dfrac{dx}{(x - 1)^{1/3}}$

**2.** $\int_1^3 \dfrac{dx}{(x - 1)^{4/3}}$

**3.** $\int_3^{10} \dfrac{dx}{\sqrt{x - 3}}$

**4.** $\int_0^9 \dfrac{dx}{\sqrt{9 - x}}$

**5.** $\int_0^1 \dfrac{dx}{\sqrt{1 - x^2}}$

**6.** $\int_{100}^{\infty} \dfrac{x}{\sqrt{1 + x^2}}\,dx$

**7.** $\int_{-1}^3 \dfrac{1}{x^3}\,dx$

**8.** $\int_5^{-5} \dfrac{1}{x^{2/3}}\,dx$

**9.** $\int_{-1}^{128} x^{-5/7}\,dx$

**10.** $\int_0^1 \dfrac{x}{\sqrt[3]{1 - x^2}}\,dx$

**11.** $\int_0^4 \dfrac{dx}{(2 - 3x)^{1/3}}$

**12.** $\int_{\sqrt{5}}^{\sqrt{8}} \dfrac{x}{(16 - 2x^2)^{2/3}}\,dx$

**13.** $\int_0^{-4} \dfrac{x}{16 - 2x^2}\,dx$

**14.** $\int_0^3 \dfrac{x}{\sqrt{9 - x^2}}\,dx$

**15.** $\int_{-2}^{-1} \dfrac{dx}{(x + 1)^{4/3}}$

**16.** $\int_0^3 \dfrac{dx}{x^2 + x - 2}$

**17.** $\int_0^3 \dfrac{dx}{x^3 - x^2 - x + 1}$

**18.** $\int_0^{27} \dfrac{x^{1/3}}{x^{2/3} - 9}\,dx$

**19.** $\int_0^{\pi/4} \tan 2x\,dx$

**20.** $\int_0^{\pi/2} \csc x\,dx$

**21.** $\int_0^{\pi/2} \dfrac{\sin x}{1 - \cos x}\,dx$

**22.** $\int_0^{\pi/2} \dfrac{\cos x}{\sqrt[3]{\sin x}}\,dx$

**23.** $\int_0^{\pi/2} \tan^2 x \sec^2 x\,dx$

**24.** $\int_0^{\pi/4} \dfrac{\sec^2 x}{(\tan x - 1)^2}\,dx$

**25.** $\int_0^{\pi} \dfrac{dx}{\cos x - 1}$

**26.** $\int_{-3}^{-1} \dfrac{dx}{x\sqrt{\ln(-x)}}$

**27.** $\int_0^{\ln 3} \dfrac{e^x\,dx}{\sqrt{e^x - 1}}$

**28.** $\int_2^4 \dfrac{dx}{\sqrt{4x - x^2}}$

**29.** $\int_1^e \dfrac{dx}{x \ln x}$

**30.** $\int_1^{10} \dfrac{dx}{x \ln^{100} x}$

**31.** $\int_{2c}^{4c} \dfrac{dx}{\sqrt{x^2 - 4c^2}}$

**32.** $\int_c^{2c} \dfrac{x\,dx}{\sqrt{x^2 + xc - 2c^2}},\ c > 0$

**33.** It is possible to change an improper integral into a proper one by using integration by parts. Consider $\lim\limits_{c \to 0} \int_c^1 \dfrac{dx}{\sqrt{x}(1 + x)}$. Use integration by parts on the interval $[c, 1]$ where $c > 0$ to show that

$$\int_c^1 \dfrac{dx}{\sqrt{x}(1 + x)} = 1 - \dfrac{2\sqrt{c}}{c + 1} + 2\int_c^1 \dfrac{\sqrt{x}\,dx}{(1 + x)^2}$$

and thus conclude that upon taking the limit as $c \to 0$ an improper integral can be turned into a proper integral.

**34.** Use integration by parts and the technique of Problem 33 to transform the improper integral $\int_0^1 \dfrac{dx}{\sqrt{x}(1 + x)}$ into a proper integral.

**35.** If $f(x)$ tends to infinity at both $a$ and $b$, then we define

$$\int_a^b f(x)\,dx = \int_a^c f(x)\,dx + \int_c^b f(x)\,dx,$$

where $c$ is any point between $a$ and $b$, provided of course that both the latter integrals converge. Otherwise, we say the given integral diverges. Use this to evaluate $\int_{-3}^3 \dfrac{x}{\sqrt{9 - x^2}}\,dx$ or show that it diverges.

**36.** Evaluate $\int_{-3}^3 \dfrac{x}{9 - x^2}\,dx$ or show that it diverges. See Problem 35.

**37.** Evaluate $\int_{-4}^4 \dfrac{1}{16 - x^2}\,dx$ or show that it diverges. See Problem 35.

**38.** Evaluate $\int_{-1}^1 \dfrac{1}{x\sqrt{-\ln|x|}}\,dx$ or show that it diverges.

**39.** Show that $\int_0^\infty \frac{1}{x^p} \, dx$ diverges for all $p$. *Hint*: Write

$$\int_0^\infty = \int_0^1 + \int_1^\infty.$$

**40.** Suppose $f$ is continuous on $[0, \infty)$ except at $x = 1$, where $\lim_{x \to 1} |f(x)| = \infty$. How would you define $\int_0^\infty f(x) \, dx$?

**41.** Find the area of the region between the curves $y = (x - 8)^{-2/3}$ and $y = 0$ for $0 \le x < 8$.

**42.** Find the area of the region between the curves $y = 1/x$ and $y = 1/(x^3 + x)$ for $0 < x \le 1$.

**43.** Let $R$ be the region in the first quadrant below the curve $y = x^{-2/3}$ and to the left of $x = 1$.

(a) Show that the area of $R$ is finite by finding its value.
(b) Show that the volume of the solid generated by revolving $R$ about the $x$-axis is infinite.

**44.** Find $b$ so that $\int_0^b \ln x \, dx = 0$.

**45.** Is $\int_0^1 \frac{\sin x}{x} \, dx$ an improper integral? Explain.

**46.** (Comparison Test) If $0 \le f(x) \le g(x)$ on $[a, \infty)$, it can be shown that the convergence of $\int_a^\infty g(x) \, dx$ implies the convergence of $\int_a^\infty f(x) \, dx$ and the divergence of $\int_a^\infty f(x) \, dx$ implies the divergence of $\int_a^\infty g(x) \, dx$. Use this to show that $\int_1^\infty \frac{1}{x^4(1 + x^4)} \, dx$ converges. *Hint*: On $[1, \infty)$, $1/x^4(1 + x^4) \le 1/x^4$.

**47.** Use the comparison test of Problem 46 to show that $\int_1^\infty e^{-x^2} \, dx$ converges. *Hint*: $e^{-x^2} \le e^{-x}$ on $[1, \infty)$.

**48.** Formulate a comparison test for improper integrals with infinite integrands.

**49.** (a) Use Example 2 of Section 9.2 to show that for any positive number $n$, there is a number $M$ such that

$$0 < \frac{x^{n-1}}{e^x} \le \frac{1}{x^2} \quad \text{for } x \ge M$$

(b) Use (a) and Problem 46 to show that $\int_1^\infty x^{n-1}e^{-x} \, dx$ converges.

**50.** Using Problem 48, prove that $\int_0^1 x^{n-1}e^{-x} \, dx$ converges for $n > 0$.

**51.** (Gamma Function) Let $\Gamma(n) = \int_0^\infty x^{n-1}e^{-x} \, dx$, $n > 0$. This integral converges by Problems 49 and 50. Show each of the following.

(a) $\Gamma(1) = 1$
(b) $\Gamma(n + 1) = n\Gamma(n)$
(c) $\Gamma(n + 1) = n!$, $n$ a positive integer.

**52.** Suppose $0 < p < q$ and $\int_0^\infty \frac{1}{x^p + x^q} \, dx$ converges. What can you say about $p$ and $q$?

**53.** By interpreting each of the following integrals as an area and then calculating this area by a $y$-integration, evaluate:

(a) $\int_0^1 \sqrt{\frac{1 - x}{x}} \, dx$;

(b) $\int_{-1}^1 \sqrt{\frac{1 + x}{1 - x}} \, dx$.

PC **54.** Evaluate $\int_0^\infty x^{n-1}e^{-x} \, dx$ for $n = 1, 2, 3, 4,$ and $5$, thereby confirming Problem 51(c).

---

**Answers to Concepts Review:** 1. Unbounded 2. 2
3. $\lim_{b \to 4^-} \int_0^b (1/\sqrt{4 - x}) \, dx$ 4. $p < 1$

---

## 9.5 CHAPTER REVIEW

### Concepts Test

Respond with true or false to each of the following assertions. Be prepared to justify your answer.

**1.** $\lim_{x \to \infty} \frac{x^{100}}{e^x} = 0$.

**2.** $\lim_{x \to \infty} \frac{x^{1/10}}{\ln x} = \infty$.

**3.** $\lim_{x \to \infty} \frac{1000x^4 + 1000}{0.001x^4 + 1} = \infty$.

**4.** $\lim_{x \to \infty} xe^{-1/x} = 0$.

**5.** If $\lim_{x \to a} f(x) = \lim_{x \to a} g(x) = \infty$, then $\lim_{x \to a} \frac{f(x)}{g(x)} = 1$.

**6.** If $\lim_{x \to a} f(x) = 1$ and $\lim_{x \to a} g(x) = \infty$, then $\lim_{x \to a} [f(x)]^{g(x)} = 1$.

**7.** If $\lim_{x \to a} f(x) = 1$, then $\lim_{n \to \infty} \left\{ \lim_{x \to a} [f(x)]^n \right\} = 1$.

**8.** If $\lim\limits_{x \to a} f(x) = 0$ and $\lim\limits_{x \to a} g(x) = \infty$, then $\lim\limits_{x \to a} [f(x)]^{g(x)} = 0$. (Assume $f(x) \geq 0$ for $x \neq a$.)

**9.** If $\lim\limits_{x \to a} f(x) = -1$ and $\lim\limits_{x \to a} g(x) = \infty$, then $\lim\limits_{x \to a} [f(x)g(x)] = -\infty$.

**10.** If $\lim\limits_{x \to a} f(x) = 0$ and $\lim\limits_{x \to a} g(x) = \infty$, then $\lim\limits_{x \to a} [f(x)g(x)] = 0$.

**11.** If $\lim\limits_{x \to \infty} \dfrac{f(x)}{g(x)} = 3$, then $\lim\limits_{x \to \infty} [f(x) - 3g(x)] = 0$.

**12.** If $\lim\limits_{x \to a} f(x) = 2$ and $\lim\limits_{x \to a} g(x) = 0$, then $\lim\limits_{x \to a} \dfrac{f(x)}{|g(x)|} = \infty$. (Assume $g(x) \neq 0$ for $x \neq a$.)

**13.** If $\lim\limits_{x \to \infty} \ln f(x) = 2$, then $\lim\limits_{x \to \infty} f(x) = e^2$.

**14.** If $f(x) \neq 0$ for $x \neq a$ and $\lim\limits_{x \to a} f(x) = 0$, then $\lim\limits_{x \to a} [1 + f(x)]^{1/f(x)} = e$.

**15.** If $p(x)$ is a polynomial, then $\lim\limits_{x \to \infty} \dfrac{p(x)}{e^x} = 0$.

**16.** If $p(x)$ is a polynomial, then $\lim\limits_{x \to 0} \dfrac{p(x)}{e^x} = p(0)$.

**17.** If $f(x)$ and $g(x)$ are both differentiable and $\lim\limits_{x \to 0} \dfrac{f'(x)}{g'(x)} = L$, then $\lim\limits_{x \to 0} \dfrac{f(x)}{g(x)} = L$.

**18.** $\displaystyle\int_0^1 \dfrac{1}{x^{1.001}}\, dx$ converges.

**19.** $\displaystyle\int_0^\infty \dfrac{1}{x^p}\, dx$ diverges for all $p > 0$.

**20.** If $f$ is continuous on $[0, \infty]$ and $\lim\limits_{x \to \infty} f(x) = 0$, then $\displaystyle\int_0^\infty f(x)\, dx$ converges.

**21.** If $f$ is an even function and $\displaystyle\int_0^\infty f(x)\, dx$ converges, then $\displaystyle\int_{-\infty}^\infty f(x)\, dx$ converges.

**22.** If $\lim\limits_{b \to \infty} \displaystyle\int_{-b}^b f(x)\, dx$ exists and is finite, then $\displaystyle\int_{-\infty}^\infty f(x)\, dx$ converges.

**23.** If $f'$ is continuous on $[0, \infty)$ and $\lim\limits_{x \to \infty} f(x) = 0$, then $\displaystyle\int_0^\infty f'(x)\, dx$ converges.

**24.** If $0 \leq f(x) \leq e^{-x}$ on $[0, \infty)$, then $\displaystyle\int_0^\infty f(x)\, dx$ converges.

**25.** $\displaystyle\int_0^{\pi/4} \dfrac{\tan x}{x}\, dx$ is an improper integral.

## Sample Test Problems

Find the limits in Problems 1–18.

**1.** $\lim\limits_{x \to 0} \dfrac{4x}{\tan x}$

**2.** $\lim\limits_{x \to 0} \dfrac{\tan 2x}{\sin 3x}$

**3.** $\lim\limits_{x \to 0} \dfrac{\sin x - \tan x}{\frac{1}{3}x^2}$

**4.** $\lim\limits_{x \to 0} \dfrac{\cos x}{x^2}$

**5.** $\lim\limits_{x \to 0} 2x \cot x$

**6.** $\lim\limits_{x \to 1^-} \dfrac{\ln(1 - x)}{\cot \pi x}$

**7.** $\lim\limits_{t \to \infty} \dfrac{\ln t}{t^2}$

**8.** $\lim\limits_{x \to \infty} \dfrac{2x^3}{\ln x}$

**9.** $\lim\limits_{x \to 0^+} (\sin x)^{1/x}$

**10.** $\lim\limits_{x \to 0^+} x \ln x$

**11.** $\lim\limits_{x \to 0^+} x^x$

**12.** $\lim\limits_{x \to 0} (1 + \sin x)^{2/x}$

**13.** $\lim\limits_{x \to 0^+} \sqrt{x} \ln x$

**14.** $\lim\limits_{t \to \infty} t^{1/t}$

**15.** $\lim\limits_{x \to 0^+} \left(\dfrac{1}{\sin x} - \dfrac{1}{x}\right)$

**16.** $\lim\limits_{x \to \pi/2} \dfrac{\tan 3x}{\tan x}$

**17.** $\lim\limits_{x \to \pi/2} (\sin x)^{\tan x}$

**18.** $\lim\limits_{x \to \pi/2} \left(x \tan x - \dfrac{\pi}{2} \sec x\right)$

In Problems 19–38, evaluate the given improper integral or show that it diverges.

**19.** $\displaystyle\int_0^\infty \dfrac{dx}{(x + 1)^2}$

**20.** $\displaystyle\int_0^\infty \dfrac{dx}{1 + x^2}$

**21.** $\displaystyle\int_{-\infty}^1 e^{2x}\, dx$

**22.** $\displaystyle\int_{-1}^1 \dfrac{dx}{1 - x}$

**23.** $\displaystyle\int_0^\infty \dfrac{dx}{x + 1}$

**24.** $\displaystyle\int_{1/2}^2 \dfrac{dx}{x(\ln x)^{1/5}}$

**25.** $\displaystyle\int_1^\infty \dfrac{dx}{x^2 + x^4}$

**26.** $\displaystyle\int_{-\infty}^1 \dfrac{dx}{(2 - x)^2}$

**27.** $\int_{-2}^{0} \frac{dx}{2x + 3}$

**28.** $\int_{1}^{4} \frac{dx}{\sqrt{x - 1}}$

**29.** $\int_{2}^{\infty} \frac{dx}{x(\ln x)^2}$

**30.** $\int_{0}^{\infty} \frac{dx}{e^{x/2}}$

**31.** $\int_{3}^{5} \frac{dx}{(4 - x)^{2/3}}$

**32.** $\int_{2}^{\infty} xe^{-x^2} dx$

**33.** $\int_{-\infty}^{\infty} \frac{x}{x^2 + 1} dx$

**34.** $\int_{-\infty}^{\infty} \frac{x}{1 + x^4} dx$

**35.** $\int_{0}^{\infty} \frac{e^x}{e^{2x} + 1} dx$

**36.** $\int_{-\infty}^{\infty} x^2 e^{-x^3} dx$

**37.** $\int_{-3}^{3} \frac{x}{\sqrt{9 - x^2}} dx$

**38.** $\int_{\pi/3}^{\pi/2} \frac{\tan x}{(\ln \cos x)^2} dx$

**39.** For what values of $p$ does the integral $\int_{1}^{\infty} \frac{1}{x^p} dx$ converge and for what values does it diverge?

**40.** For what values of $p$ does the integral $\int_{0}^{1} \frac{1}{x^p} dx$ converge and for what values does it diverge?

In Problems 41–44 use a comparison test (see Problem 46 of Section 9.4) to decide whether each of the following converges or diverges.

**41.** $\int_{1}^{\infty} \frac{dx}{\sqrt{x^6 + x}}$

**42.** $\int_{1}^{\infty} \frac{\ln x}{e^{2x}} dx$

**43.** $\int_{3}^{\infty} \frac{\ln x}{x} dx$

**44.** $\int_{1}^{\infty} \frac{\ln x}{x^3} dx$

## 9.6  ADDITIONAL PROBLEMS

**1.** One can change an improper integral into a proper integral by changing the variable of integration (Sections 5.8 and 8.1). If you use a change of variable given by $u = g(x)$ in the integral $\int_{a}^{b} f(x) dx$ then $g(x)$ must be a differentiable function for all $x$ such that $a < x < b$.

(a) Show that the integral $\int_{1}^{c} \frac{1}{1 + x^2} dx$ using the substitution $u = 1/x$ (which is a monotonic differentiable function whose derivative does not vanish anywhere on the interval of integration) is transformed into $\int_{1/c}^{1} \frac{1}{1 + u^2} du$.

(b) Show that under the transformation given in part (a) the improper integral $\int_{1}^{\infty} \frac{1}{1 + x^2} dx$ is equal to $\int_{0}^{1} \frac{1}{1 + u^2} du$ and evaluate the integral.

**2.** Use an appropriate change of variables to convert the improper integral $\int_{1}^{\infty} \frac{x}{x^3 + 1} dx$ into the proper integral $\int_{0}^{1} \frac{1}{u^3 + 1} du$.

There are many different probability density functions besides the Cauchy density function and the normal density function given in Section 9.3, Examples 4 and 5. In fact, any function which is positive and has the property that $\int_{-\infty}^{\infty} f(x) dx = 1$ can be interpreted as a *probability density function*. Such functions also have an associated *mean* $\mu = \int_{-\infty}^{\infty} xf(x) dx$, and *variance* $\sigma^2 = \int_{-\infty}^{\infty} (x - \mu)^2 f(x) dx$, ($\sigma$ is

called the *standard deviation*) which are often used to fit a given type of distribution to a specific situation. In addition, the expected value of $g(x)$ is defined to be $E(g(x)) = \int_{-\infty}^{\infty} g(x)f(x) dx$. The expected value of $x$, $E(x)$ is the mean. The probability that a random variable $n$ falls between $a$ and $b$ is found by integrating the probability density function $f(x)$:

$$P(a \le x \le b) = \int_{a}^{b} f(x) dx.$$

We explore various probability density functions in Problems 3–8.

**3.** Define a density function by $f(x) = C |x| e^{-kx^2}$ for positive $k$.

(a) Find the constant $C$ which makes $f(x)$ a probability density function.
(b) Using the value of $C$ found in part (a), find the mean $\mu = \int_{-\infty}^{\infty} xf(x) dx$.

**4.** Sketch the graph of the normal probability density function

$$f(x) = \sigma \frac{1}{\sqrt{2\pi}} e^{-(x - \mu)^2/2\sigma^2}$$

and show using calculus that $\sigma$ is the distance from the mean $\mu$ to the inflection points on the curve which lie above and below the mean. Confirm that $\mu$ and $\sigma$ are the mean and standard deviation.

**5.** A random variable $x$ ($x \geq 0$) has an exponential distribution if its probability density function has an exponential behavior of the form $Ce^{-xK}$. Such distributions are often used in queueing theory to model the time elapsed between independent successive events such as customers arriving to be serviced.

(a) Find $C$ so that

$$f(x) = \begin{cases} Ce^{-xK} & \text{if } x \geq 0 \\ 0 & \text{if } x < 0 \end{cases}$$

is an exponential probability density function.
(b) Using the $C$ found in part (a), find the mean and standard deviation.
(c) Using the $C$ found in part (a), find $P(x > 1)$ which gives the probability that no customer will arrive in the next minute.

**6.** The *Weibull probability density function* is often used to represent the lifetime of plants and animals. It has the following form

$$f(x) = \begin{cases} Cmx^{m-1}e^{-x^m K} & \text{if } x > 0 \\ 0 & \text{if } x \leq 0 \end{cases}$$

(a) Find $C$ so that $f(x)$ is a probability density function.
(b) If $K = 1/6$ and $m = 3$ find the mean $\mu$ and standard deviation $\sigma$ for the value of $C$ found in part (a).
(c) Find $P(x > 5)$ for the values of $C$, $K$ and $m$ given in (b).

**7.** The *Pareto probability density function* has the form

$$f(x) = \begin{cases} \dfrac{CM^k}{x^{k+1}} & \text{if } x \geq M \\ 0 & \text{if } x < M \end{cases}$$

where $k$ and $M$ are positive constants.

(a) Find the value of $C$ which makes $f(x)$ a probability density function.
(b) For the value of $C$ found in part (a) find the value of the mean $\mu$. Is the mean finite for all positive $k$? If not, how does the mean depend on $k$?
(c) For the value of $C$ found in part (a) find the variance $\sigma^2$. How does the variance depend on $k$?

**8.** The *gamma probability density function* has the form

$$f(x) = \begin{cases} Cx^{\alpha-1}e^{-\beta x} & \text{if } x > 0 \\ 0 & \text{if } x \leq 0 \end{cases}$$

where $\alpha$ and $\beta$ are positive constants.

(a) Remembering that the gamma function is defined by $\Gamma(\alpha) = \int_0^\infty x^{\alpha-1}e^{-x}\,dx$ for $\alpha > 0$, find the value of $C$, depending on both $\alpha$ and $\beta$, that makes $f(x)$ a probability density function.
(b) For the value of $C$ found in part (a) find the value of the mean $\mu$.
(c) For the value of $C$ found in part (a) find the variance $\sigma^2$.

**9.** The function

$$f(x) = \begin{cases} e^{-1/x} & \text{if } x > 0 \\ 0 & \text{if } x \leq 0 \end{cases}$$

has an infinite number of derivatives. Denote $\dfrac{d^n}{dx^n}f(x)$ by $f^{(n)}(x)$.

(a) Sketch the graphs of $f(x)$, $f^{(1)}(x)$, $f^{(2)}(x)$, and $f^{(3)}(x)$ for $-2 < x < 5$.
(b) Compute $\lim_{x \to 0^+} f^{(i)}(x)$, for $i = 0, 1, 2, 3$ and show that $f^{(i)}(x)$ is a continuous function of $x$ for these values of $i$.
(c) Give a plausible argument for the conclusion that $f^{(n)}(x)$ is a continuous function for all positive integral values of $n$.

**10.** The Laplace transform of a function $f(x)$ is given by $L\{f(t)\}(s) = \int_0^\infty f(t)e^{-st}\,dt$. Laplace transforms are useful for solving linear ordinary differential equations.

(a) Show that the Laplace transform of $t^\alpha$ is given by $\Gamma(\alpha+1)/s^{\alpha+1}$ and is defined as long as $s > 0$.
(b) Show that the Laplace transform of $e^{\alpha t}$ is given by $1/(s-\alpha)$ and is defined as long as $s > \alpha$.
(c) Show that the Laplace transform of $\sin(\alpha t)$ is given by $\alpha/(s^2 + \alpha^2)$ and is defined as long as $s > 0$.
(d) Show that the Laplace transform of $\cos(\alpha t)$ is given by $s/(s^2 + \alpha^2)$ and is defined as long as $s > 0$.

# TECHNOLOGY PROJECT

**Exercise 1** Suppose that $f$ is an *even* function, that is: $f(-x) = f(x)$. Prove that $\int_{-a}^{a} f(x)\,dx = 2\int_{0}^{a} f(x)\,dx$. This has practical use because evaluating at $x = 0$ is less error prone, in hand calculations, than evaluating at a negative number.

The function

$$n(x) = \frac{1}{\sqrt{2\pi\sigma^2}} e^{-(x-\mu)^2/2\sigma^2}$$

is the famous and important probability density for the *normal* distribution with mean value $\mu$ and standard deviation $\sigma$, see Figure 1.

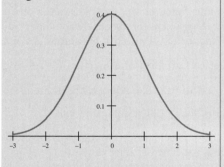

FIG. 1. The normal probability density.

If the "random variable" $x$ follows the normal distribution, then the probability that $x$ lies in the interval $[a, b]$ is defined as

$$P = \int_{a}^{b} n(x)\,dx$$

It is often of interest to know the probability of lying within 1, 2, or in general, $k$, standard deviations of the mean. This is given by the integral

$$P(k) = \int_{\mu-k\sigma}^{\mu+k\sigma} n(x)\,dx \quad (1)$$

Everyone taking a statistics course knows the:

**Rule of Thumb for the Normal Distribution:**
2/3 of the observed values lie within one standard deviation from the mean, 95% lie within two standard deviations, and 99% lie within three standard deviations.

Among other things, we'll be checking out this Rule of Thumb in this project. One famous (or infamous?)application is "grading on the curve": those students scoring within one standard deviation of the mean earn C's, while those scoring between one and two standard deviations above the mean earn B's, and those scoring over two standard deviations above the mean earn A's and similarly at the low end for the D's and F's.

**Exercise 2** Perform the substitution $z = (x - \mu)/\sigma$ to show that

$$P(k) = \frac{1}{\sqrt{2\pi}} \int_{-k}^{k} e^{-z^2/2}\,dz$$
$$= \sqrt{\frac{2}{\pi}} \int_{0}^{k} e^{-z^2/2}\,dz \quad (2)$$

Also explain the "folding" of the integral to the interval $[0, k]$. Thus, $P(k)$ does not depend on $\mu$ or $\sigma$. This is important if you want to make tables for $P$—you need only one table instead of many.

**Exercise 3** As $k \to \infty$, we include all the probability, so you'd expect $\lim_{k \to \infty} P(k) = 1$. Check this out on your technology by computing $P(k)$ for $k = \infty$, or just "large" if you cannot accommodate $\infty$. (Expect some round-off error in your answer.)

**Exercise 4** The *standard* normal distribution is the normal distribution with mean $\mu = 0$ and standard deviation $\sigma = 1$. Show that the probability density function for the standard normal is $p(x) = \frac{1}{\sqrt{2\pi}} e^{-x^2/2}$. Then show that the simplified form in Equation 2 of $P(k)$ is just the original form of Equation 1 with $n$ replaced by $p$.

**Exercise 5** Using the definition in Equation 2, show that $P(k) = \sqrt{2/\pi} \int_{0}^{k} e^{-z^2/2}\,dz$. Use your computing technology to numerically evaluate $P(1)$, $P(2)$, and $P(3)$. Explain what these results mean in terms of probabilities and compare them to the rule of thumb given above. Explain how one can use Figure 1 to get a rough check that all is well.

**Exercise 6** In statistical studies it is important to know the value of $k$ such that $P(k)$ equals a given value $v$; e.g., $v = .95$ ("significant") and $v = .99$ ("highly significant") are especially popular. To at least three significant figures, compute the $k$ values corresponding to the three $v$ values: .90, .95, and .99. Compare with the above "Rule of Thumb."

*Hint:* For more accurate values of $k$, consider Newton's method (cf. the Chapter 4 project or the index of your text).

# 10

# Numerical Methods, Approximations

## 10.1
## TAYLOR'S APPROXIMATION TO FUNCTIONS

So far in this book we have emphasized what might be called exact methods. There have been some exceptions, however. The simplest example is illustrated by rounding decimals, as when we write $\frac{1}{3} \approx 0.333$ or $\pi \approx 3.1416$. More notable was our discussion of differentials in Section 3.10. There we used the differential $dy$ to approximate the actual change $\Delta y$ in $y = f(x)$ when $x$ changed by an amount $\Delta x$. That example, in fact, illustrates the kind of approximation methods we want to highlight in this chapter.

Two factors contribute to the importance of approximation methods. First is the fact that many of the mathematical entities that occur in applications cannot be calculated by exact methods. We mention, for example, the integrals $\int_0^b \sin(x^2)\, dx$, which is used extensively in optics, and $\int_a^b e^{-x^2}\, dx$, which plays a key role in statistics. Second, the invention of high-speed electronic computers and calculators has made approximate numerical methods practical. In fact, it is often easier to calculate something approximately using a calculator (and get the answer to desired accuracy) than to use exact methods, even when the latter are available.

**Linear Approximation**   The idea behind the differential approximation introduced in Section 3.10 was to approximate a curve near a point by its tangent line at that point. Our idea here is exactly the same, only we do not use differential notation. Examine the diagram in Figure 1. The equation of the tangent line to the curve $y = f(x)$ at $(a, f(a))$ is

$$y = f(a) + f'(a)(x - a)$$

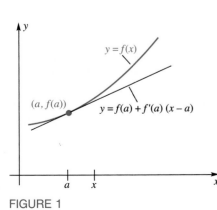

FIGURE 1

471

This leads directly to the linear approximation

$$\boxed{f(x) \approx f(a) + f'(a)(x - a)}$$

The linear polynomial $P_1(x) = f(a) + f'(a)(x - a)$ is called the **Taylor polynomial of order 1 based at** *a* for $f(x)$, after the English mathematician Brook Taylor (1685–1731). Clearly, we can expect $P_1(x)$ to be a good approximation to $f(x)$ only near $x = a$.

**EXAMPLE 1**    Find $P_1(x)$ based at $a = 1$ for $f(x) = \ln x$ and use it to approximate $\ln(0.9)$ and $\ln(1.5)$.

*Solution*    Since $f(x) = \ln x$, $f'(x) = 1/x$; thus, $f(1) = 0$ and $f'(1) = 1$. Therefore,

$$P_1(x) = 0 + 1(x - 1) = x - 1$$

Consequently (see Figure 2),

$$\ln x \approx x - 1$$

and

$$\ln(0.9) \approx 0.9 - 1 = -0.1$$
$$\ln(1.5) \approx 1.5 - 1 = 0.5$$

These approximations should be compared with the correct four-place values of $-0.1054$ and $0.4055$. As expected, the approximation is much better in the first case, since 0.9 is closer to 1 than is 1.5.    ■

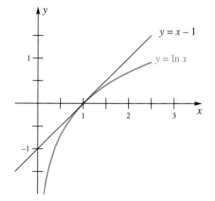

FIGURE 2

**Taylor Polynomials of Order _n_**    Polynomials are the easiest of all functions to evaluate, since they involve only the three arithmetic operations of addition, subtraction, and multiplication. This is why we use polynomials to approximate other functions. If a linear polynomial gives a certain approximation to $f(x)$, we should expect a quadratic polynomial (with its curved graph) to do better, a cubic polynomial to do still better, and so on. Our aim is to find the polynomial of degree $n$ that "best" approximates $f(x)$. We consider the quadratic case first.

A significant observation to make about the linear case is that $f$ and its approximation $P_1$, as well as their derivatives $f'$ and $P_1'$, agree at $x = a$. In generalizing to the quadratic polynomial $P_2$, we impose three conditions, namely,

$$f(a) = P_2(a), \qquad f'(a) = P_2'(a), \qquad f''(a) = P_2''(a)$$

The unique quadratic polynomial satisfying these conditions (the **Taylor polynomial of order 2 based at** $a$) is

$$P_2(x) = f(a) + f'(a)(x - a) + \frac{f''(a)}{2}(x - a)^2$$

as you may check. The corresponding quadratic approximation is

$$f(x) \approx f(a) + f'(a)(x - a) + \frac{f''(a)}{2}(x - a)^2$$

**EXAMPLE 2**   Find $P_2(x)$ based at $a = 1$ for $f(x) = \ln x$ and use it to approximate $\ln(0.9)$ and $\ln(1.5)$.

*Solution*   Here $f(x) = \ln x$, $f'(x) = 1/x$, and $f''(x) = -1/x^2$, and so $f(1) = 0$, $f'(1) = 1$, and $f''(1) = -1$. Thus,

$$P_2(x) = 0 + 1(x - 1) - \tfrac{1}{2}(x - 1)^2$$

Consequently,

$$\ln x \approx (x - 1) - \tfrac{1}{2}(x - 1)^2$$

and

$$\ln(0.9) \approx (0.9 - 1) - \tfrac{1}{2}(0.9 - 1)^2 = -0.1050$$
$$\ln(1.5) \approx (1.5 - 1) - \tfrac{1}{2}(1.5 - 1)^2 = 0.3750$$

As expected, these are better approximations than we got in the linear case (Example 1). Note the diagram in Figure 3.    ■

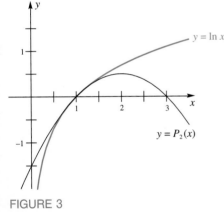

FIGURE 3

With a little effort, we could now derive $P_3(x)$. And from this, it is a simple matter to guess the form of the **Taylor polynomial of order** $n$ **based at** $a$; that is, the $n$th order polynomial $P_n$, which—together with its first $n$ derivatives—agrees with $f$ and its derivatives at $x = a$. It is

$$P_n(x) = f(a) + f'(a)(x - a) + \frac{f''(a)}{2!}(x - a)^2 + \cdots + \frac{f^{(n)}(a)}{n!}(x - a)^n$$

Here $n!$ (read $n$ **factorial**) is the standard symbol for the product of the first $n$ positive integers—that is,

$$n! = n(n - 1)(n - 2) \cdots 3 \cdot 2 \cdot 1$$

The corresponding approximation is

$$f(x) \approx f(a) + f'(a)(x - a) + \frac{f''(a)}{2!} (x - a)^2 + \cdots + \frac{f^{(n)}(a)}{n!} (x - a)^n$$

**Maclaurin Polynomials**   In case $a = 0$, the Taylor polynomial of order $n$ simplifies to the **Maclaurin polynomial of order $n$**, named in honor of the Scottish mathematician Colin Maclaurin (1698–1746). It gives a particularly useful approximation valid near $x = 0$, namely,

$$f(x) \approx f(0) + f'(0)x + \frac{f''(0)}{2!} x^2 + \cdots + \frac{f^{(n)}(0)}{n!} x^n$$

**EXAMPLE 3**   Find the Maclaurin polynomials of order $n$ for $e^x$ and $\cos x$. Then approximate $e^{0.2}$ and $\cos(0.2)$ using $n = 4$.

*Solution*   The calculation of the required derivatives is shown in the table.

|          |       | At $x = 0$ |          | At $x = 0$ |
|----------|-------|------------|----------|------------|
| $f(x)$      | $e^x$ | 1 | $\cos x$  | 1  |
| $f'(x)$     | $e^x$ | 1 | $-\sin x$ | 0  |
| $f''(x)$    | $e^x$ | 1 | $-\cos x$ | -1 |
| $f^{(3)}(x)$ | $e^x$ | 1 | $\sin x$  | 0  |
| $f^{(4)}(x)$ | $e^x$ | 1 | $\cos x$  | 1  |
| $f^{(5)}(x)$ | $e^x$ | 1 | $-\sin x$ | 0  |
| $\vdots$ | $\vdots$ | $\vdots$ | $\vdots$ | $\vdots$ |

---

**Order versus Degree**

We have chosen the terminology Taylor (and Maclaurin) polynomial of *order n* because the highest-order derivative involved in its construction is of order $n$. Note that this polynomial can have degree less than $n$ if $f^{(n)}(a) = 0$. If $n$ is odd in Example 3, then the Maclaurin polynomial of order $n$ for $\cos x$ will be of degree $n - 1$. For example, the Maclaurin polynomial of order 5 for $\cos x$ is

$$1 - \tfrac{1}{2}x^2 + \tfrac{1}{24}x^4$$

a polynomial of degree 4.

---

It follows that

$$e^x \approx 1 + x + \frac{1}{2!} x^2 + \frac{1}{3!} x^3 + \frac{1}{4!} x^4 + \cdots + \frac{1}{n!} x^n$$

$$\cos x \approx 1 - \frac{1}{2!} x^2 + \frac{1}{4!} x^4 - \cdots + (-1)^{n/2} \frac{1}{n!} x^n \qquad (n \text{ even})$$

Thus, using $n = 4$ and $x = 0.2$, we obtain

$$e^{0.2} \approx 1 + 0.2 + \frac{(0.2)^2}{2} + \frac{(0.2)^3}{6} + \frac{(0.2)^4}{24} = 1.2214$$

$$\cos(0.2) \approx 1 - \frac{(0.2)^2}{2} + \frac{(0.2)^4}{24} \approx 0.9800667$$

These results should be compared with the correct seven-place values of 1.2214028 and 0.9800666. ∎

To give you a visual idea of how the Maclaurin polynomials provide approximation to cos $x$, we have sketched the graphs of $P_1(x)$ through $P_5(x)$ and $P_8(x)$ in Figure 4.

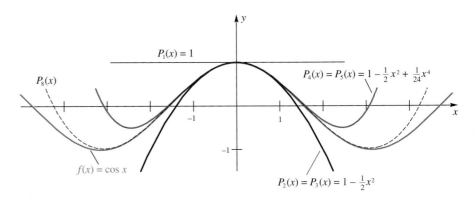

Maclaurin approximations to $f(x) = \cos x$

FIGURE 4

**Horner's Method for Evaluating Polynomials**    To evaluate the polynomial

$$p(x) = a_n x^n + a_{n-1} x^{n-1} + \cdots + a_1 x + a_0$$

at $x = c$ is, of course, to evaluate

$$p(c) = a_n c^n + a_{n-1} c^{n-1} + \cdots + a_1 c + a_0$$

One way to do this on a calculator is to proceed directly from left to right using the $\boxed{y^x}$ key. However, on most calculators this key is both slow and notably inaccurate.

A better procedure goes by the name **Horner's Method**. Begin by writing $p(x)$ in nested form.

$$p(x) = \{ \ldots [(a_n x + a_{n-1})x + a_{n-2}]x + \cdots + a_1\}x + a_0$$

For example,

$$2x^4 + 3x^3 - 10x^2 + 2x - 5 = \{[(2x+3)x - 10]x + 2\}x - 5$$

which avoids exponents completely. The whole calculation of $p(c)$ is easily done on a calculator without even using parentheses, as we now illustrate.

**EXAMPLE 4**    Evaluate $3x^3 + 2x^2 - 4x + 8$ at $x = 2.123$.

**Solution**    Note that

$$3x^3 + 2x^2 - 4x + 8 = [(3x + 2)x - 4]x + 8$$

---

**Synthetic Division**

Those who remember the synthetic division method of evaluating a polynomial will recognize it as Horner's Method in disguise. Check that the steps in the evaluation at $x = 2$ of

$$3x^3 + 2x^2 - 4x + 8$$

by synthetic division are exactly those in Horner's Method.

$$\begin{array}{r|rrrr} & 3 & 2 & -4 & 8 \\ & & 6 & 16 & 24 \\ \hline 2 & 3 & 8 & 12 & \textcircled{32} \end{array}$$

Let $\boxed{\text{S}}$ denote the key (or keys) required to store a number and $\boxed{\text{R}}$ the corresponding key (or keys) to recall it. Then, on an algebraic logic calculator, we may perform the required evaluation by pressing in succession

$$2.123 \;\boxed{\text{S}}\; 3$$
$$\boxed{\times}\;\boxed{\text{R}}\;\boxed{+}\; 2 \;\boxed{=}$$
$$\boxed{\times}\;\boxed{\text{R}}\;\boxed{-}\; 4 \;\boxed{=}$$
$$\boxed{\times}\;\boxed{\text{R}}\;\boxed{+}\; 8 \;\boxed{=}$$

Our calculator responds with 37.228163. ∎

**EXAMPLE 5**   Evaluate $2x^5 - 4x^4 + 5x^2 - 12x - 9$ at $x = 1.95$.

*Solution*   Press the following keys in succession.

$$1.95 \;\boxed{\text{S}}\; 2$$
$$\boxed{\times}\;\boxed{\text{R}}\;\boxed{-}\; 4 \;\boxed{=}$$
$$\boxed{\times}\;\boxed{\text{R}}\;\boxed{+}\; 0 \;\boxed{=}$$
$$\boxed{\times}\;\boxed{\text{R}}\;\boxed{+}\; 5 \;\boxed{=}$$
$$\boxed{\times}\;\boxed{\text{R}}\;\boxed{-}\; 12 \;\boxed{=}$$
$$\boxed{\times}\;\boxed{\text{R}}\;\boxed{-}\; 9 \;\boxed{=}$$

This yields the answer $-14.833401$. Note the insertion of 0 corresponding to the missing term in $x^3$. ∎

The calculations in Examples 4 and 5 are even more easily done on a reverse Polish logic calculator. If one has to do many evaluations for the same polynomial, then the advantage of having a programmable calculator becomes obvious. In any case, practice making calculations of the above type on your calculator.

## CONCEPTS REVIEW

**1.** If $P_2(x)$ is the Taylor polynomial of order 2 based at 1 for $f(x)$, then $P_2(1) =$ _____, $P_2'(1) =$ _____, and $P_2''(1) =$ _____.

**2.** The Maclaurin polynomial of order $n$ is just a special name for the Taylor polynomial of order $n$ based at _____.

**3.** The Maclaurin polynomial of order 4 for $e^x$ is _____.

**4.** The coefficient of $x^6$ in the Maclaurin polynomial of order 9 for $f(x)$ is _____.

## PROBLEM SET 10.1

In Problems 1–4, calculate $f(3.21)$ first by using the $\boxed{y^x}$ key, and then by using Horner's Method.

**1.** $f(x) = 2x^3 - 3x^2 - 2x + 5$

**2.** $f(x) = 4x^3 + 1.2x^2 - 3x - 6$

**3.** $f(x) = 2x^4 - 3x^3 - 2x^2 + 5x - 2$

4. $f(x) = 3x^4 - 11x^2 + 2x - 3$

[C] 5. Calculate $p(4.567)$ if $p(x) = x^5 - 3x^4 + 2x^3 + 4x^2 + 5x + 1$ using Horner's Method.

[C] 6. Calculate $p(6.543)$ if $p(x) = x^5 - 3.12x^4 + 2.53x^3 - 6.32$.

[C] In Problems 7–14, find the Maclaurin polynomial of order 4 for $f(x)$ and use it to approximate $f(0.12)$.

7. $f(x) = e^{2x}$      8. $f(x) = e^{-3x}$

9. $f(x) = \sin 2x$      10. $f(x) = \tan x$

11. $f(x) = \ln(1 + x)$      12. $f(x) = \sqrt{1 + x}$

13. $f(x) = \tan^{-1} x$      14. $f(x) = \sinh x$

[C] In Problems 15–20, find the Taylor polynomial of order 3 based at $a$ for the given function.

15. $e^x;\ a = 1$      16. $\sin x;\ a = \dfrac{\pi}{4}$

17. $\tan x;\ a = \dfrac{\pi}{6}$      18. $\sec x;\ a = \dfrac{\pi}{4}$

19. $\cot^{-1} x;\ a = 1$      20. $\sqrt{x};\ a = 2$

21. Find the Taylor polynomial of order 3 based at 1 for $f(x) = x^3 - 2x^2 + 3x + 5$ and show that it is an exact representation of $f(x)$.

22. Find the Taylor polynomial of order 4 based at 2 for $f(x) = x^4$ and show that it represents $f(x)$ exactly.

23. Find the Maclaurin polynomial of order $n$ for $f(x) = 1/(1 - x)$. Then use it with $n = 4$ to approximate each of the following.

(a) $f(0.1)$    (b) $f(0.5)$    (c) $f(0.9)$    (d) $f(2)$

This example should convince you that the Maclaurin approximation can be exceedingly poor if $x$ is far from zero.

[C] 24. Find the Maclaurin polynomial of order $n$ ($n$ odd) for $\sin x$. Then use it with $n = 5$ to approximate each of the following.

(a) $\sin(0.1)$      (b) $\sin(0.5)$
(c) $\sin(1)$      (d) $\sin(10)$

Compare your answers with those given by your calculator. What conclusion do you draw?

25. Use a Maclaurin polynomial to obtain the approximation $A \approx r^2t^3/12$ for the area of the shaded region in Figure 5. First express $A$ exactly; then approximate.

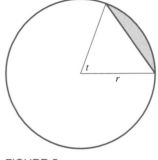

**FIGURE 5**

26. If an object of rest mass $m_0$ has velocity $v$, then (according to the theory of relativity) its mass $m$ is given by $m = m_0/\sqrt{1 - v^2/c^2}$, where $c$ is the velocity of light. Show how physicists get the approximation

$$m \approx m_0 + \frac{m_0}{2}\left(\frac{v}{c}\right)^2$$

27. If money is invested at interest rate $r$ compounded monthly, it will double in $n$ years, where $n$ satisfies

$$\left(1 + \frac{r}{12}\right)^{12n} = 2$$

(a) Show that

$$n = \ln 2 \left[\frac{1}{12 \ln(1 + r/12)}\right]$$

(b) Use the Maclaurin polynomial of order 2 for $\ln(1 + x)$ and a partial fraction decomposition to obtain the approximation

$$n \approx \frac{0.693}{r} + 0.029$$

[C] (c) Some people use the *rule of 72*, $n \approx 72/100r$, to approximate $n$. Fill in the table to compare the values obtained from these three formulas.

| $r$ | $n$ (*exact*) | $n$ (*approx.*) | $n$ (*rule 72*) |
|---|---|---|---|
| 0.05 | | | |
| 0.10 | | | |
| 0.15 | | | |
| 0.20 | | | |

28. The author of a biology text claimed that the smallest positive solution to $x = 1 - e^{-(1 + k)x}$ is approximately $x = 2k$, provided $k$ is very small. Show how she reached this conclusion and check on it for $k = 0.01$.

PC Use a computer alegebra system such as *Maple*, *Mathematica* or *True BASIC* Calculus to work Problems 29–31.

**29.** Draw using the same axes the graphs of the Maclaurin polynomials of orders 1, 2, 3, and 4 for each of the following

(a) $\sin(e^x)$                    (b) $(\sin x)/(2 + \sin x)$

**30.** Follow the directions of Problem 29.

(a) $\exp(-x^2)$                    (b) $\sin(\ln(1 + x))$

**31.** Find the Maclaurin polynomial of order 4 for each of the following.

(a) $2x^3 - 3x^2 + x$              (b) $-x^4 + 2x^3 - 3x^2 + x$
(c) $x^5 - x^4 + 2x^3 - 3x^2 + x$
(d) $x^6 + x^5 - x^4 + 2x^3 - 3x^2 + x$
(e) $1/(1 - x)$                    (f) $\tan^{-1} x$
(g) $1/(1 - x^2)$                  (h) $\tan^{-1} x + 1/(1 - x)$

On the basis of the preceding (and perhaps other examples), make several conjectures about Maclaurin polynomials. Then see how many of your conjectures you can prove.

---

**Answers to Concepts Review:** **1.** $f(1)$; $f'(1)$; $f''(1)$  **2.** 0
**3.** $1 + x + \frac{1}{2}x^2 + \frac{1}{6}x^3 + \frac{1}{24}x^4$  **4.** $f^{(6)}(0)/6!$

---

## 10.2
## BOUNDING THE ERRORS

In Example 3 of the previous section, we used the Maclaurin polynomial of order 4 to approximate $\cos(0.2)$ as follows:

$$\cos(0.2) \;\overset{\text{first error}}{\approx}\; 1 - \frac{1}{2!}(0.2)^2 + \frac{1}{4!}(0.2)^4 \;\overset{\text{second error}}{\approx}\; 0.9800667$$

This example illustrates the two kinds of errors that occur in approximation processes. First, there is the **error of the method**. In this case, we approximated $\cos x$ by a fourth-degree polynomial. Second, there is the **error of calculation**. This includes errors due to round-off, as when we replaced the unending decimal 0.0000666 . . . by 0.0000667 in the last term above. It may also include errors due to the characteristics of the calculating device we are using. For example, the $\boxed{y^x}$ key tends to be inaccurate on many hand-held calculators.

Now notice a sad fact of the numerical analyst's life. We can reduce the error of the method by taking more terms in the Maclaurin expansion. But taking more terms means more calculations, which potentially increases the error of calculation. To be a good numerical analyst is to know how to compromise between these two types of error. That unfortunately is more of an art than a science. However, we can say something definite about the first type of error, the subject to which we now turn.

**The Error in the Method**    For the problem of approximating a function by its Taylor polynomial, we can actually give a formula for the error. This formula is due to the French-Italian mathematician, Joseph-Louis Lagrange (1736–1813).

Theorem A

**(Taylor's Formula with Remainder).** Let $f$ be a function whose $(n + 1)$th derivative, $f^{(n + 1)}(x)$, exists for each $x$ in an open interval $I$ containing $a$. Then for each $x$ in $I$,

$$f(x) = f(a) + f'(a)(x - a) + \frac{f''(a)}{2!} (x - a)^2 + \cdots$$

$$+ \frac{f^{(n)}(a)}{n!} (x - a)^n + R_n(x)$$

where the remainder (or error) $R_n(x)$ is given by the formula

$$R_n(x) = \frac{f^{(n + 1)}(c)}{(n + 1)!} (x - a)^{n + 1}$$

and $c$ is some point between $x$ and $a$.

We postpone the proof until later in the section, choosing to begin with several illustrations. The case $a = 0$ occurs most often in practice, and Taylor's Formula is then called Maclaurin's Formula.

**EXAMPLE 1** Approximate $e^{0.8}$ with an error of less than 0.001.

*Solution* For $f(x) = e^x$, Maclaurin's Formula gives the remainder

$$R_n(x) = \frac{f^{(n + 1)}(c)}{(n + 1)!} x^{n + 1} = \frac{e^c}{(n + 1)!} x^{n + 1}$$

and so

$$R_n(0.8) = \frac{e^c}{(n + 1)!} (0.8)^{n + 1}$$

where $0 < c < 0.8$. Our goal is to choose $n$ large enough so $|R_n(0.8)| < 0.001$. Now, $e^c < e^{0.8} < 3$ and $(0.8)^{n + 1} < (1)^{n + 1}$, and so

$$|R_n(0.8)| < \frac{3(1)^{n + 1}}{(n + 1)!} = \frac{3}{(n + 1)!}$$

It is easy to check that $3/(n + 1)! < 0.001$ when $n \geq 6$, and so we can obtain the desired accuracy by using the Maclaurin polynomial of order 6.

$$e^{0.8} \approx 1 + (0.8) + \frac{(0.8)^2}{2!} + \frac{(0.8)^3}{3!} + \frac{(0.8)^4}{4!} + \frac{(0.8)^5}{5!} + \frac{(0.8)^6}{6!}$$

Our calculator gives 2.2254948.

Can we be sure that this answer is within 0.001 of the true result? Certainly the error of the method is less than 0.001. But could the error of calculation have distorted our answer? Possibly so; however, so few calculations are involved that we feel confident in reporting an answer of 2.2255 accurate within 0.001. ∎

**EXAMPLE 2**  Use the Taylor polynomial of order 4 based at $a = 1$ to approximate $\ln(0.9)$ and give a bound for error that is made.

*Solution*  We will need the first five derivatives of $f(x) = \ln x$.

$$f(x) = \ln x \qquad f(1) = 0$$
$$f'(x) = x^{-1} \qquad f'(1) = 1$$
$$f''(x) = -x^{-2} \qquad f''(1) = -1$$
$$f'''(x) = 2x^{-3} \qquad f'''(1) = 2$$
$$f^{(4)}(x) = -6x^{-4} \qquad f^{(4)}(1) = -6$$
$$f^{(5)}(x) = 24x^{-5} \qquad f^{(5)}(c) = \frac{24}{c^5}$$

Thus, by Taylor's Formula,

$$\ln(x) = (x - 1) - \frac{1}{2}(x - 1)^2 + \frac{1}{3}(x - 1)^3 - \frac{1}{4}(x - 1)^4 + R_4(x)$$

and

$$\ln(0.9) = -0.1 - \frac{1}{2}(0.1)^2 - \frac{1}{3}(0.1)^3 - \frac{1}{4}(0.1)^4 + R_4(0.9)$$
$$\approx -0.1053583 + R_4(0.9)$$

Also,

$$R_4(0.9) = \frac{24}{c^5}\frac{(-0.1)^5}{5!} = \frac{(-0.1)^5}{5c^5}$$

where $0.9 < c < 1$, and so (using the fact that a fraction gets bigger when its denominator is made smaller)

$$|R_4(0.9)| < \frac{(0.1)^5}{5(0.9)^5} \approx 0.0000034$$

We conclude that $\ln(0.9) = -0.1053583$ with an error of less than 0.0000034. In saying this, we assume that the error of calculation is insignificant. ∎

**EXAMPLE 3** Use a Taylor polynomial of order 2 to approximate $\cos 62°$ and then give a bound for the error of the approximation.

*Solution* Since $62°$ is near $60°$ (whose cosine and sine are known), we use the Taylor polynomial based at $a = \pi/3$.

$$f(x) = \cos x \qquad f\left(\frac{\pi}{3}\right) = \frac{1}{2}$$

$$f'(x) = -\sin x \qquad f'\left(\frac{\pi}{3}\right) = -\frac{\sqrt{3}}{2}$$

$$f''(x) = -\cos x \qquad f''\left(\frac{\pi}{3}\right) = -\frac{1}{2}$$

$$f'''(x) = \sin x \qquad f'''(c) = \sin c$$

Now

$$62° = \frac{\pi}{3} + \frac{\pi}{90} \text{ radians}$$

Thus,

$$\cos x = \frac{1}{2} - \frac{\sqrt{3}}{2}\left(x - \frac{\pi}{3}\right) - \frac{1}{4}\left(x - \frac{\pi}{3}\right)^2 + R_2(x)$$

and

$$\cos\left(\frac{\pi}{3} + \frac{\pi}{90}\right) = \frac{1}{2} - \frac{\sqrt{3}}{2}\left(\frac{\pi}{90}\right) - \frac{1}{4}\left(\frac{\pi}{90}\right)^2 + R_2\left(\frac{\pi}{3} + \frac{\pi}{90}\right)$$

$$\approx 0.4694654 + R_2$$

Now

$$|R_2| = \left|\frac{\sin c}{3!}\left(\frac{\pi}{90}\right)^3\right| < \frac{1}{6}\left(\frac{\pi}{90}\right)^3 \approx 0.0000071$$

Again the number of calculations is small, so we feel safe in reporting $\cos 62° = 0.4694654$ with an error of less than $0.0000071$. ■

**Useful Tools for Bounding $|R_n|$** The precise value of $R_n$ is almost never obtainable, since we do not know $c$, only that $c$ lies on a certain interval. Our task is therefore to find the maximum possible value of $|R_n|$ for $c$ in the given interval. To do this exactly is often difficult, so we usually content ourselves with getting a good upper bound for $|R_n|$. This involves a sensible use of inequalities. Our chief tools are the triangle inequality $|a \pm b| \le |a| + |b|$ and the fact that a fraction gets larger when we make its numerator larger or its denominator smaller.

**EXAMPLE 4** If $c$ is known to be in $[2, 4]$, give a good bound for the maximum value of

$$\left| \frac{c^2 - \sin c}{c} \right|$$

**Solution**

$$\left| \frac{c^2 - \sin c}{c} \right| = \frac{|c^2 - \sin c|}{|c|} \leq \frac{|c^2| + |\sin c|}{|c|} \leq \frac{4^2 + 1}{2} = 8.5$$

A different and better bound is obtained as follows.

$$\left| \frac{c^2 - \sin c}{c} \right| = \left| c - \frac{\sin c}{c} \right| \leq |c| + \left| \frac{\sin c}{c} \right| \leq 4 + \frac{1}{2} = 4.5 \qquad \blacksquare$$

**Proof of Taylor's Formula**   Recall that $R_n(x)$ is defined on $I$ by

$$R_n(x) = f(x) - f(a) - f'(a)(x - a) - \frac{f''(a)(x - a)^2}{2!} - \cdots - \frac{f^{(n)}(a)}{n!}(x - a)^n$$

Now think of $x$ as a constant and define a new function $g$ on $I$ by

$$g(t) = f(x) - f(t) - f'(t)(x - t) - \frac{f''(t)(x - t)^2}{2!} - \cdots$$

$$- \frac{f^{(n)}(t)}{n!}(x - t)^n - R_n(x) \frac{(x - t)^{n+1}}{(x - a)^{n+1}}$$

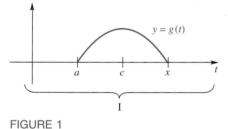

$y = g(t)$

FIGURE 1

Clearly $g(x) = 0$, and using the definition of $R_n(x)$, it is almost as easy to see that $g(a) = 0$ (Figure 1). By the Mean Value Theorem for Derivatives applied to $g$ (on the interval with endpoints $a$ and $x$), there is a point $c$ between $a$ and $x$ such that $g'(c) = 0$. Now if we differentiate the expression for $g(t)$ with respect to $t$ (keeping $x$ fixed), we find that most of the terms cancel. The result is

$$g'(t) = -\frac{f^{(n+1)}(t)}{n!}(x - t)^n + R_n(x)(n + 1)\frac{(x - t)^n}{(x - a)^{n+1}}$$

When we set $g'(c) = 0$, we obtain Taylor's Formula for $R_n(x)$.

**The Error of Calculation**   In all of our examples so far, we have assumed that the error of calculation is small enough so that it can be ignored. We will ordinarily make that assumption in this book, since our problems will always involve a small number of calculations. We feel obligated, however, to warn you that when high-speed computers are used to do thousands or millions of operations, these errors of calculation may well accumulate and distort an answer.

There are two sources of calculation errors that may be significant even in using a hand-held calculator. Consider calculating

$$a + b_1 + b_2 + b_3 + \cdots + b_m$$

where $a$ is very much larger than any of the $b$'s; for example $a = 10{,}000{,}000$ and $b_i = 0.4$, $i = 1, 2, \ldots, m$. If we use an eight-digit calculator and proceed from left to right, first adding $b_1$ to $a$, then adding $b_2$ to the result, and so on, we will simply get 10,000,000 at each stage. Yet a sum of just 25 of the $b$'s ought to affect the seventh digit of our answer. The moral here is that in adding a large number of small terms to one or more large ones, it is wise to find the sum of the small terms first.

A more likely source of calculation error is due to the loss of significant digits in a subtraction of nearly equal numbers. For example, subtracting 0.823421 from 0.823445, each with six significant digits, results in 0.000024, which has only two significant digits. That this can cause trouble is easily illustrated by calculating a numerical approximation to a derivative.

Consider calculating $f'(2)$ for $f(x) = x^4$ by using the difference quotient

$$\frac{f(2 + h) - f(2)}{h} = \frac{(2 + 10^{-n})^4 - 2^4}{10^{-n}}$$

Theoretically, as $n$ increases (and $h = 10^{-n}$ correspondingly decreases) the result should get closer and closer to the correct value, namely, 32. But note what happens on one eight-digit calculator when $n$ gets too large.

| $n$ | $(2 + 10^{-n})^4 - 2^4$ | $[(2 + 10^{-n})^4 - 2^4]10^n$ |
|---|---|---|
| 2 | 0.32240801 | 32.240801 |
| 3 | 0.03202401 | 32.024010 |
| 4 | 0.00320024 | 32.002400 |
| 5 | 0.00032000 | 32.000000 |
| 6 | 0.00003200 | 32.000000 |
| 7 | 0.00000320 | 32.000000 |
| 8 | 0.00000032 | 32.000000 |
| 9 | 0.00000003 | 30.000000 |
| 10 | 0.00000000 | 0.000000 |
| $\vdots$ | $\vdots$ | $\vdots$ |

## CONCEPTS REVIEW

**1.** The two types of errors that arise in approximation theory are called _____ and _____.

**2.** The remainder $R_n(x)$ in Taylor's Formula has the form _____.

**3.** The number _____ is a good upper bound for $1/(1 + x^2)$ on $[-3, -1]$.

**4.** Calculation errors in using Taylor's Formula tend to _____ as $n$ increases, whereas errors of the method tend to _____ as $n$ increases.

## PROBLEM SET 10.2

As illustrated in Example 4, find a good bound for the maximum value of a given expression, $c$ being in the given interval. Answers may vary depending on the technique used.

1. $\left| e^{2c} + e^{-2c} \right|$; $[0, 3]$

2. $\left| \tan c + \sec c \right|$; $\left[ 0, \dfrac{\pi}{4} \right]$

3. $\left| \dfrac{4c}{\sin c} \right|$; $\left[ \dfrac{\pi}{4}, \dfrac{\pi}{2} \right]$.

4. $\left| \dfrac{4c}{c + 4} \right|$; $[0, 1]$

5. $\left| \dfrac{e^c}{c + 5} \right|$; $[-2, 4]$

6. $\left| \dfrac{\cos c}{c + 2} \right|$; $\left[ 0, \dfrac{\pi}{4} \right]$

7. $\left| \dfrac{c^2 + \sin c}{10 \ln c} \right|$; $[2, 4]$

8. $\left| \dfrac{c^2 - c}{\cos c} \right|$; $\left[ 0, \dfrac{\pi}{4} \right]$

In Problems 9–12, find a formula for $R_6(x)$, the remainder for the Taylor polynomial of order 6 based at $a$. Then obtain a good bound for $\left| R_6(0.5) \right|$. See Examples 1 and 2.

9. $\ln(2 + x)$; $a = 0$

10. $e^{-x}$; $a = 1$

11. $\sin x$; $a = \pi/4$

12. $\dfrac{1}{x - 3}$; $a = 1$

In the formula for the remainder $R_n(x) = \dfrac{f^{(n + 1)}(c)}{(n + 1)!} (x - a)^{n + 1}$ there is a value of $c$ for which $R_n(x)$ is the exact value of the remainder. Sometimes it is useful to know the minimum as well as the maximum estimate for $R_n(x)$. In Problems 13–14 we explore this situation.

[C] 13. Consider the Maclaurin polynomial of order 3 for $e^x$. Estimate the minimum and maximum value of the error made in calculating $e^{-0.1}$ using that polynomial. Compare the estimated maximum and minimum errors with the actual error at that point.

[C] 14. Consider the Taylor polynomial of order 3 for $\sin x$ about the point $a = \dfrac{\pi}{4}$. Estimate the minimum and maximum value of the error made in calculating $\sin \dfrac{\pi}{8}$ using that polynomial. Compare the estimated maximum and minimum errors with the actual error at that point.

15. Determine the order $n$ of the Maclaurin polynomial for $e^x$ that is required to approximate $e$ to five decimal places—that is, so $\left| R_n(1) \right| \le 0.000005$ (see Example 1).

16. Find the third-order Maclaurin polynomial for $(1 + x)^{3/2}$ and bound the error $R_3(x)$ if $-0.1 \le x \le 0$.

17. Find the third-order Maclaurin polynomial for $(1 + x)^{-1/2}$ and bound the error $R_3(x)$ if $-0.05 \le x \le 0.05$.

18. Find the fourth-order Maclaurin polynomial for $\ln[(1 + x)/(1 - x)]$ and bound the error $R_4(x)$ for $-0.5 \le x \le 0.5$.

19. Note that the fourth-order Maclaurin polynomial for $\sin x$ is really of third degree since the coefficient of $x^4$ is 0. Thus,

$$\sin x = x - \dfrac{x^3}{6} + R_4(x)$$

Show that if $0 \le x \le 0.5$, $\left| R_4(x) \right| \le 0.0002605$. Use this result to approximate $\displaystyle\int_0^{0.5} \sin x \, dx$ and give a bound for the error.

20. In analogy with Problem 19,

$$\cos x = 1 - \dfrac{x^2}{2} + \dfrac{x^4}{24} + R_5(x)$$

If $0 \le x \le 1$, give a good bound for $\left| R_5(x) \right|$. Then use your result to calculate $\displaystyle\int_0^1 \cos x \, dx$ with a bound for the error made.

21. Expand $x^4 - 3x^3 + 2x^2 + x - 2$ in a Taylor polynomial of order 4 based at 1 and show that $R_4(x) = 0$ for all $x$.

22. Show that an $n$th degree polynomial can be expressed exactly as an $n$th degree polynomial in $x - a$ (see Problem 21).

[C] 23. Calculate $\sin 43° = \sin 43\pi/180$ by using the Taylor polynomial of order 3 based at $\pi/4$ for $\sin x$. Then obtain a good bound for the error made. See Example 3.

[C] 24. Calculate $\cos 63°$ by the method illustrated in Example 3. Choose $n$ large enough so $\left| R_n \right| \le 0.0005$.

[C] 25. Show that if $x$ is in $[0, \pi/2]$, the error in using

$$\sin x \approx x - \dfrac{x^3}{3!} + \dfrac{x^5}{5!} - \dfrac{x^7}{7!} + \dfrac{x^9}{9!}$$

is less than $5 \times 10^{-6}$ and, therefore, that this formula is good enough to build a five-place sine table.

26. Use Maclaurin's Formula rather than l'Hôpital's Rule to find:

(a) $\displaystyle\lim_{x \to 0} \dfrac{\sin x - x + x^3/6}{x^5}$

(b) $\displaystyle\lim_{x \to 0} \dfrac{\cos x - 1 + x^2/2 - x^4/24}{x^6}$

**27.** Let $g(x) = p(x) + x^{n+1} f(x)$, where $p(x)$ is a polynomial of degree at most $n$ and $f$ has derivatives through order $n$. Show that $p(x)$ is the Maclaurin polynomial of order $n$ for $g$.

**28.** Recall that the Second Derivative Test (Section 4.3) for maxima and minima fails if $f''(c) = 0$. Prove the following generalization, which may help in this case. Suppose that

$$f'(c) = f''(c) = f'''(c) = \cdots = f^{(n)}(c) = 0$$

where $n$ is odd and $f^{(n+1)}(x)$ is continuous near $c$.

(i) If $f^{(n+1)}(c) < 0$, $f(c)$ is a local maximum value.
(ii) If $f^{(n+1)}(c) > 0$, $f(c)$ is a local minimum value.

Test this result on $f(x) = x^4$.

**29.** Many other polynomial approximations to functions exist besides Taylor and Maclaurin polynomials. Here we consider the *Lagrange Interpolating Polynomials* as a specific example.

(a) Show that the polynomial

$$L_{51}(x) = \frac{(x - x_2)(x - x_3)(x - x_4)(x - x_5)}{(x_1 - x_2)(x_1 - x_3)(x_1 - x_4)(x_1 - x_5)}$$

is of degree 4 and has the property that $L_{51}(x_1) = 1$ while $L_{51}(x_j) = 0$ for $j = 2, 3, 4, 5$.

(b) Using $L_{51}(x)$ as a model, construct fourth degree polynomials $L_{5i}(x)$ which are 1 at $x_i$ and 0 at $x_j$ for $j \neq i$, where $i = 2, 3, 4, 5$.
(c) Consider the polynomial

$$L_5(x) \equiv L_{51}(x) \cdot y_1 + L_{52}(x) \cdot y_2 + L_{53}(x) \cdot y_3$$
$$+ L_{54}(x) \cdot y_4 + L_{55}(x) \cdot y_5$$

and show that $L_5$ is a polynomial of degree less than or equal to 4 which takes the value $y_i$ at $x = x_i$, $i = 1, \ldots, 5$. Such a polynomial is called the *Lagrange interpolating polynomial* which interpolates the points $(x_i, y_i)$, $i = 1 \ldots,$ 5 where the $x_i$ are all distinct.
(d) Construct the second degree Lagrange interpolating polynomial which goes through the points $(1, 2), (2, 2.5)$ and $(0, 0)$.

**30.** Plot the points $(1, 2), (2, 3), (3, 4), (4, 5), (5, 6)$. Construct the Lagrange interpolating polynomial which goes through these points and show that after some algebra the answer reduces to $x + 1$.

**31.** The Lagrange interpolating polynomial through a set of points is unique. This follows from the *Fundamental Theorem of Algebra* which states that a polynomial of degree $\leq n$ which is zero at $n + 1$ points must be identically zero.

(a) Using this result show that if two polynomials $L_5(x)$ and $\hat{L}_5(x)$ which must be of degree 4 or less interpolate the same 5 points, then they must be identical. *Hint*: Consider $L(x) \equiv L_5(x) - \hat{L}_5(x)$ and show that it is of degree 4 or less, and has 5 zeros.
(b) Consider the fourth degree Lagrange interpolating polynomial which goes through the five points $(0, 2), (1, 3), (2, 4),$ $(3, 5), (4, 6)$. Notice that the polynomial $x + 2$ interpolates these points. Explain why it is the only Lagrange polynomial of degree less than or equal to four which interpolates these points. Conclude that it must be true that $L_5(x) = x + 2$.
(c) Consider the interpolating polynomial at the points $(x_1, x_2, \ldots, x_n)$ given by

$$L_n^*(x) = 1 \cdot L_{n1}(x) + 1 \cdot L_{n2}(x) + \ldots + 1 \cdot L_{nn}(x)$$

show that $L_n^*(x) \equiv 1$. *Hint*: The interpolating polynomial $y = 1$ does the job, and the answer is unique.
(d) Consider the interpolating polynomial at the points $(x_1, x_2, \ldots, x_n)$ given by

$$L_n^{**}(x) = x_1 \cdot L_{n1}(x) + x_2 \cdot L_{n2}(x) + \ldots + x_n \cdot L_{nn}(x)$$

show that $L_n^{**}(x) \equiv x$.

**32.** Suppose we write a fourth degree *Newton Interpolating Polynomial* in the form

$$N_5(x) = \xi_1 + \xi_2 \cdot (x - x_1) + \xi_3 \cdot (x - x_1)(x - x_2)$$
$$+ \xi_4 \cdot (x - x_1)(x - x_2)(x - x_3)$$
$$+ \xi_5 \cdot (x - x_1)(x - x_2)(x - x_3)(x - x_4)$$

and require that it go through the points $(x_i, y_i)$ for $i = 1,$ $\ldots$ 5 where the $x_i$ are all distinct. Compute the coefficients $\xi_1, \xi_2$ in terms of $(x_1, y_1)$ and $(x_2, y_2)$. Indicate how the rest of the $\xi_i$ can be computed.

**33.** Using the result of Problem 31 on the uniqueness of the interpolating polynomial, and the definition of the $L_5(x)$ in Problem 29 and $N_5(x)$ in Problem 32, show that

(a) $L_5(x) = N_5(x)$ if they both go through the points $(x_i, y_i)$ for $i = 1, \ldots, 5$, where the $x_i$ are all distinct.
(b) Observe that if we require that the interpolating polynomial $L_5(x)$ be changed into $L_6(x)$ which goes through a sixth point $(x_6, y_6)$ then each of the $L_{6i}(x)$ must be computed. However, in computing $N_6(x)$ we need only compute $\xi_6$, and can use the previously computed values for $\xi_1, \ldots, \xi_5$. That is $N_6(x) = N_5(x) + \xi_6 \cdot (x - x_1)(x - x_2)(x - x_3)(x - x_4)(x - x_5)$. This is a distinct advantage to the Newton form of the interpolating polynomial.

**34.** The divided difference of the first order for the function $y = f(x)$ is defined by

$$f[x_1, x_2] \equiv \frac{f(x_1) - f(x_2)}{x_1 - x_2}$$

where $x_1 - x_2 \neq 0$. In a similar fashion we can define higher order divided differences by

$$f[x_1, x_2, x_3] \equiv \frac{f[x_1, x_2] - f[x_2, x_3]}{x_1 - x_3}, \ldots,$$

$$f[x_1, x_2, \ldots, x_n] \equiv \frac{f[x_1, x_2, \ldots, x_{n-1}] - f[x_2, x_3, \ldots, x_n]}{x_1 - x_n}$$

(a) Show that $f[x_1, x_2] = f[x_2, x_1]$, and $f[x_1, x_2, x_3] = f[x_2, x_1, x_3] = f[x_3, x_1, x_2] = \ldots = f[x_3, x_2, x_1]$ and consequently the divided difference is independent of the order in which the points are taken.

(b) Show that if $f(x) = \alpha \cdot g(x) + \beta \cdot h(x)$ then $f[x_1, x_2, \ldots, x_n] = \alpha \cdot g[x_1, x_2, \ldots, x_n] + \beta \cdot h[x_1, x_2, \ldots, x_n]$, thus showing that the Newton divided difference is linear.

(c) If $f(x)$ has a third derivative at $x_1$ show that

$$f'(x_1) = \lim_{x_2 \to x_1} f[x_1, x_2]$$

$$\frac{f''(x_1)}{2!} = \lim_{x_2, x_3 \to x_1} f[x_1, x_2, x_3]$$

and

$$\frac{f'''(x_1)}{3!} = \lim_{x_2, x_3, x_4 \to x_1} f[x_1, x_2, x_3, x_4].$$

(d) Show by multiplying out terms that the Newton polynomial $N_3(x) = \xi_1 + \xi_2 \cdot (x - x_1) + \xi_3 \cdot (x - x_1)(x - x_2)$ going through the points $(x_1, f(x_1))$, $(x_2, f(x_2))$, and $(x_3, f(x_3))$ is given by

$$N_3(x) = f(x_1) + f[x_1, x_2] \cdot (x - x_1) + f[x_1, x_2, x_3] \cdot (x - x_1)(x - x_2)$$

thus establishing the relationship between the Newton interpolating polynomial and the Newton divided difference.

**35.** If $y = F(x) \equiv ax^2 + bx + c$ is a second degree polynomial:

(a) Compute its Taylor series about the point $x = 1$.
(b) Compute $N_3(x)$ as given in Problem 34.
(c) Show that both results give the function $F(x)$ when fully expanded.

**36.** This problem establishes the formula for the error in the Lagrange interpolating polynomial $P_n(x)$ of degree $n$ and uses many of the same techniques which were used to derive the error term in the Taylor expansion.

(a) Since $P_n(x)$ interpolates the function $f(x)$ at $n + 1$ points, the error at these points must be zero. Use this fact to justify the formula for the error $R_n(x)$ as $R_n(x) = f(x) - P_n(x) = (x - x_1)(x - x_2) \ldots (x - x_{n+1}) M(x)$ where $M(x)$ models the behavior of $R_n(x)$ at the points where it is not zero. Show that $f(x) - P_n(x) - R_n(x) = 0$. If we can determine $M(x)$ we will have a formula for the error.

(b) Using the auxiliary function

$$g(t) \equiv f(t) - P_n(t) - (t - x_1)(t - x_2) \ldots (t - x_{n+1}) M(x)$$

(noting that $x$ has not been replaced by $t$ in $M(x)$) we now investigate the zeros of $g(t)$ (which is really a function of both $x$ and $t$.) Show that $g(t)$ has $n + 2$ zeros which occur at $t = x_1, x_2, \ldots, x_{n+1}$ as well as at $x$.

(c) Under what conditions can we apply the Mean Value Theorem to $g(t)$ and its first $n + 1$ derivatives? Show that if $f(t)$ has $n + 1$ continuous derivatives then so must $g(t)$. In what follows, we assume that $g(t)$ meets those conditions.

(d) Show using the Mean Value Theorem that if $g(t)$ has $n + 2$ distinct zeros then $g'(t)$ must have a zero between each of these zeros. Continuing in this fashion, show that $\dfrac{d^{n+1} g(t)}{dt^{n+1}}$ must have at least one zero, say at $t = \eta$, on the interval which contains all of the points $x_1, x_2, \ldots, x_{n+1}, x$.

(e) Show that

$$\left. \frac{d^{n+1} g(t)}{dt^{n+1}} \right|_{t=\eta} = \left. \frac{d^{n+1} f(t)}{dt^{n+1}} \right|_{t=\eta} - (n+1)! M(x)$$

(f) Conclude that $M(x) = \dfrac{1}{(n+1)!} \left. \dfrac{d^{n+1} f(t)}{dt^{n+1}} \right|_{t=\eta}$

and consequently that

$$R_n(x) = \frac{(x - x_1)(x - x_2) \ldots (x - x_{n+1})}{(n+1)!} \left. \frac{d^{n+1} f(t)}{dt^{n+1}} \right|_{t=\eta}$$

where $\eta$ is some point contained in the interval which includes all of the points $x_1, x_2, \ldots, x_{n+1}, x$.

© **37.** Given that $\ln(1) = 0$, $\ln(3) = 1.098$, and $\ln(5) = 1.609$ write a second degree polynomial which interpolates these values. Use this interpolation to compute an approximation for $\ln(2)$. Use the error formula given in Problem 36 to give a maximum estimate for the error. Compare your estimate for the error to the actual error in your computed result.

© **38.** Given that $e^0 = 1$, $e^{0.2} = 1.221$ and $e^{0.3} = 1.350$ write a second degree polynomial which interpolates these values. Use this interpolation to compute an approximation for $e^{0.25}$. Use the error formula given in Problem 36 to give a maximum and a minimum estimate for the error. Compare your estimate for the error to the actual error in your computed result.

© **39.** Construct a Maclaurin polynomial of $e^x$ which is of second degree. Also construct a second order polynomial interpolation using the values $e^0 = 1$, $e^{0.2} = 1.221$ and $e^{0.3} = 1.349$. Using the expressions for $R_2(x)$ for the Maclaurin expansion and $R_2(x)$ for the polynomial approximation (see Problem 36) compute an estimate for the maximum error for $0 \leq x \leq 0.3$. Compare your answers with the actual error at $x = 0.1$.

---

**Answers to Concepts Review:** **1.** Error of the method: error of calculation   **2.** $f^{(n+1)}(c)(x - a)^{n+1}/(n+1)!$   **3.** $\frac{1}{2}$
**4.** Increase; decrease

## 10.3
## NUMERICAL INTEGRATION

We know that if $f$ is continuous on a closed interval $[a, b]$, then the definite integral $\int_a^b f(x)\,dx$ must exist. Existence is one thing; evaluation is a very different matter. There are many definite integrals that cannot be evaluated by the methods we have learned—that is, by use of the Fundamental Theorem of Calculus. For example, the indefinite integrals of such integrands as

$$e^{-x^2}, \qquad \sin(x^2), \qquad \sqrt{1 - x^4}, \qquad \frac{\sin x}{x}$$

cannot be expressed algebraically in terms of elementary functions, that is, in terms of functions studied in a first calculus course (see Section 8.1). Even when elementary indefinite integrals can be found, it is often advantageous to use the approximation methods soon to be discussed, since they are well suited to use on a calculator or computer. We present two such methods: the Trapezoidal Rule and the Parabolic Rule.

**The Trapezoidal Rule**   Consider the graph of $y = f(x)$ on $[a, b]$; it might look something like the curve in Figure 1. Partition the interval $[a, b]$ into $n$ subintervals, each of length $h = (b - a)/n$, by means of points $a = x_0 < x_1 < x_2 < \cdots < x_n = b$. Join the pairs of points $(x_{i-1}, f(x_{i-1}))$ and $(x_i, f(x_i))$ by line segments, as shown in the figure, thus forming $n$ trapezoids.

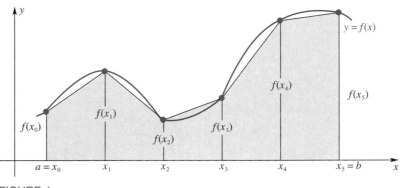

FIGURE 1

Recalling the area formula shown in Figure 2, we can write the area of the $i$th trapezoid as

$$A_i = \frac{h}{2}[f(x_{i-1}) + f(x_i)]$$

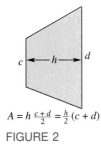

$A = h\,\frac{c+d}{2} = \frac{h}{2}(c + d)$

FIGURE 2

More accurately, we should say *signed* area, since $A_i$ will be negative for a subinterval where $f$ is negative. The definite integral $\int_a^b f(x)\,dx$ is approximately equal to $A_1 + A_2 + \cdots + A_n$, that is, to

$$\frac{h}{2}[f(x_0) + f(x_1)] + \frac{h}{2}[f(x_1) + f(x_2)] + \cdots + \frac{h}{2}[f(x_{n-1}) + f(x_n)]$$

This simplifies to the Trapezoidal Rule.

Trapezoidal Rule

$$\int_a^b f(x)\, dx \approx \frac{h}{2}\, [f(x_0) + 2f(x_1) + 2f(x_2) + \cdots + 2f(x_{n-1}) + f(x_n)]$$

We illustrate this rule first for a definite integral where we know the exact answer.

**EXAMPLE 1** Use the Trapezoidal Rule to find an approximate value for

$$\int_1^3 x^4\, dx,$$

taking $n = 8$.

*Solution* Since $n = 8$, $h = (3 - 1)/8 = 0.25$, and

$$
\begin{aligned}
x_0 &= 1.00 & f(x_0) &= (1.00)^4 = & 1.0000 \\
x_1 &= 1.25 & f(x_1) &= (1.25)^4 \approx & 2.4414 \\
x_2 &= 1.50 & f(x_2) &= (1.50)^4 = & 5.0625 \\
x_3 &= 1.75 & f(x_3) &= (1.75)^4 \approx & 9.3789 \\
x_4 &= 2.00 & f(x_4) &= (2.00)^4 = & 16.0000 \\
x_5 &= 2.25 & f(x_5) &= (2.25)^4 \approx & 25.6289 \\
x_6 &= 2.50 & f(x_6) &= (2.50)^4 = & 39.0625 \\
x_7 &= 2.75 & f(x_7) &= (2.75)^4 \approx & 57.1914 \\
x_8 &= 3.00 & f(x_8) &= (3.00)^4 = & 81.0000
\end{aligned}
$$

Thus,

$$\int_1^3 x^4\, dx \approx \frac{0.25}{2}\, [1.0000 + 2(2.4414) + \cdots + 2(57.1914) + 81.0000]$$

$$= 48.9414$$

This may be compared with the exact value

$$\int_1^3 x^4\, dx = \left[\frac{x^5}{5}\right]_1^3 = \frac{242}{5} = 48.4000$$

Presumably we could get a better approximation by taking $n$ larger; this would be easy to do using a computer. However, while taking $n$ larger reduces the error of the method, it at least potentially increases the error of calculation. It would be unwise, for example, to take $n = 1,000,000$, since the potential round-off errors would more than compensate for the fact that the error of the method would be minuscule. We will have more to say about errors shortly. ■

**EXAMPLE 2** Use the Trapezoidal Rule with $n = 6$ to approximate

$$\int_0^1 e^{-x^2}\, dx$$

*Solution* Since $h = \frac{1}{6}$ and $f(x) = e^{-x^2}$,

$$\begin{aligned}
x_0 &= 0.0000 & f(x_0) &= 1.0000 \\
x_1 &\approx 0.1667 & f(x_1) &\approx 0.9726 \\
x_2 &\approx 0.3333 & f(x_2) &\approx 0.8948 \\
x_3 &= 0.5000 & f(x_3) &\approx 0.7788 \\
x_4 &\approx 0.6667 & f(x_4) &\approx 0.6412 \\
x_5 &\approx 0.8333 & f(x_5) &\approx 0.4994 \\
x_6 &= 1.0000 & f(x_6) &\approx 0.3679
\end{aligned}$$

$$\int_0^1 e^{-x^2}\, dx \approx \tfrac{1}{12}\,[1.000 + 2(0.9726) + \cdots + 2(0.4994) + 0.3679]$$

$$= 0.7451$$

As we mentioned earlier, this integral, important in statistical theory, cannot be evaluated by a direct application of the Fundamental Theorem of Calculus. ■

**The Error in the Trapezoidal Rule** In any practical use of the Trapezoidal Rule, we need to have some idea of the size of the error involved. Fortunately, we can give a formula for the error of the method.

---

**Theorem A**

Suppose that $f''$ exists on $[a, b]$. Then,

$$\int_a^b f(x)\, dx = \frac{h}{2}\,[f(x_0) + 2f(x_1) + \cdots + 2f(x_{n-1}) + f(x_n)] + E_n$$

where the error $E_n$ is given by

$$E_n = -\frac{(b-a)^3}{12n^2}\, f''(c)$$

and $c$ is some point between $a$ and $b$.

We omit the proof of this theorem, which may be found in more advanced books (such as J. M. H. Olmsted, *Advanced Calculus* (New York: Prentice Hall, 1961) pp. 118–19), choosing rather to illustrate its use.

**EXAMPLE 3**   Give a bound for the possible error in Example 2.

*Solution*   Since $f(x) = e^{-x^2}, f'(x) = -2xe^{-x^2}$, and $f''(x) = e^{-x^2}(4x^2 - 2)$. We could actually find the maximum value of $f''(x)$ on [0, 1], but it is sufficient to bound it there using properties of absolute value.

$$|f''(x)| = |e^{-x^2}(4x^2 - 2)| = e^{-x^2}|4x^2 - 2|$$
$$\leq e^{-x^2}(4x^2 + 2) \leq 1(4 + 2) = 6$$

Thus,

$$|E_n| = \frac{(b - a)^3}{12n^2} |f''(c)| \leq \frac{1^3}{12(6^2)} (6) = \frac{1}{72} \approx 0.0139$$

This is, of course, a bound for the error of the method. However, $n$ is so small that the error of calculation can safely be ignored. We feel confident in reporting

$$\int_0^1 e^{-x^2} dx = 0.7451 \pm 0.0139$$

A slightly more sophisticated analysis based on the fact that $4x^2 - 2$ is increasing on [0, 1] shows that $|4x^2 - 2| \leq 2$ on [0, 1], and hence $|E_n| \leq 0.0047$.   ∎

**EXAMPLE 4**   How large must $n$ be to insure that the error of the method in Example 2 is less than 0.0001?

*Solution*   From Example 3,

$$|E_n| = \frac{(b - a)^3}{12n^2} |f''(c)| \leq \frac{1^3 \cdot 6}{12n^2} = \frac{1}{2n^2}$$

We want $1/2n^2 < 0.0001$, which holds if $n \geq 71$.   ∎

**The Parabolic Rule (Simpson's Rule)**   In the Trapezoidal Rule, we approximated the curve $y = f(x)$ by line segments. It seems likely that we could do better using parabolic segments. Just as before, partition the interval $[a, b]$ into $n$ subintervals of length $h = (b - a)/n$, but this time with $n$ an *even* number. Then fit parabolic segments to neighboring triples of points, as shown in Figure 3.

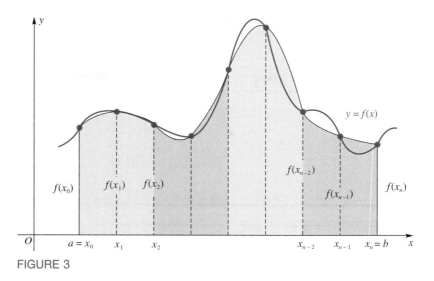

FIGURE 3

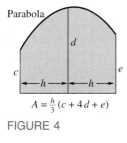

$$A = \frac{h}{3}(c + 4d + e)$$

FIGURE 4

Using the area formula in Figure 4 (see Problem 15 for the derivation) leads to an approximation called the **Parabolic Rule**. It is also called **Simpson's Rule**, after the English mathematician Thomas Simpson (1710–1761).

> **Parabolic Rule (*n* even)**
>
> $$\int_a^b f(x)\,dx \approx \frac{h}{3}\,[f(x_0) + 4f(x_1) + 2f(x_2) + \cdots + 4f(x_{n-1}) + f(x_n)]$$

The pattern of coefficients is 1, 4, 2, 4, 2, 4, 2, . . . , 2, 4, 1.

**EXAMPLE 5**    Use the Parabolic Rule with $n = 8$ to approximate

$$\int_1^3 x^4\,dx$$

*Solution*    We can make use of the calculations in Example 1. We obtain

$$\int_1^3 x^4\,dx \approx \frac{0.25}{3}[1.0000 + 4(2.4414) + 2(5.0625) + 4(9.3789) + 2(16.0000)$$

$$+ 4(25.6289) + 2(39.0625) + 4(57.1914) + 81.0000]$$

$$\approx 48.4010 \qquad \blacksquare$$

As expected, the Parabolic Rule gives an answer closer to the true value 48.4000 than did the Trapezoidal Rule, which gave 48.9414. Because it requires so little extra work, we prefer the Parabolic Rule over the Trapezoidal Rule in most problems. That its error term is generally smaller is borne out by the following theorem (note the factor of $n^4$ in the denominator). For a proof, see J. M. H. Olmsted, *Advanced Calculus* (New York: Prentice Hall, 1961), pp. 118–19.

### Theorem B

Suppose that the fourth derivative $f^{(4)}(x)$ exists on $[a, b]$. Then the error $E_n$ in the Parabolic Rule is given by

$$E_n = -\frac{(b - a)^5}{180n^4} f^{(4)}(c)$$

for some $c$ between $a$ and $b$.

**EXAMPLE 6**  Use the Parabolic Rule with $n = 8$ to approximate

$$\int_1^2 (1 + x)^{-1} dx$$

and find a bound for the error made.

*Solution*  Since $h = \frac{1}{8}$ and $f(x) = 1/(1 + x)$, we calculate the results shown in the table.

| $i$ | $x_i$ | $f(x_i)$ | $c_i$ | $c_i f(x_i)$ |
|---|---|---|---|---|
| 0 | 1 | 0.50000 | 1 | 0.50000 |
| 1 | 1.125 | 0.47059 | 4 | 1.88236 |
| 2 | 1.25 | 0.44444 | 2 | 0.88888 |
| 3 | 1.375 | 0.42105 | 4 | 1.68420 |
| 4 | 1.5 | 0.40000 | 2 | 0.80000 |
| 5 | 1.625 | 0.38095 | 4 | 1.52380 |
| 6 | 1.75 | 0.36364 | 2 | 0.72728 |
| 7 | 1.875 | 0.34783 | 4 | 1.39132 |
| 8 | 2 | 0.33333 | 1 | 0.33333 |
| | | | | Sum = 9.73117 |

### Computers and Integration

Computer packages that claim to evaluate definite integrals do so by some form of numerical integration. They may use one of the rules discussed in this section or they may use something more sophisticated. In any case, they use approximation methods, so you should not expect the answers they give to be exact.

Thus,

$$\int_1^2 \frac{dx}{1 + x} \approx \frac{1}{24} (9.73117) \approx 0.4055$$

To find a bound for $E_8$, we first calculate $f^{(4)}(x)$ to be $24/(1 + x)^5$, and then

$$E_8 = -\frac{(2 - 1)^5}{180(8^4)} \cdot \frac{24}{(1 + c)^5}$$

Consequently

$$|E_8| \le \frac{24}{180(8^4)(1 + 1)^5} \approx 0.00000102$$

Assuming the calculation errors to be negligible, we feel very confident in reporting that the answer 0.4055 is accurate to four decimal places.  ∎

## CONCEPTS REVIEW

**1.** The pattern of coefficients in the Trapezoidal Rule is _____.

**2.** The pattern of coefficients in the Parabolic Rule is _____.

**3.** The error in the Trapezoidal Rule has $n^2$ in the denominator, whereas the error in the Parabolic Rule has

_____ in the denominator, so we expect the latter to give a better approximation to a definite integral.

**4.** If $f$ is concave up, the Trapezoidal Rule will always give a value for $\int_a^b f(x)\,dx$ that is too _____.

## PROBLEM SET 10.3

[C] In Problems 1–4, use the Trapezoidal Rule and the Parabolic Rule, both with $n = 8$, to approximate each integral. Then calculate the integral using the Fundamental Theorem of Calculus.

**1.** $\int_1^3 \dfrac{1}{x^2}\,dx$

**2.** $\int_1^3 \dfrac{1}{x}\,dx$

**3.** $\int_0^2 \sqrt{x}\,dx$

**4.** $\int_1^3 x\sqrt{x^2 + 1}\,dx$

[C] **5.** Use the Trapezoidal Rule with $n = 2, 6, 12$ to approximate $\int_0^\pi \sin x\,dx$. Note how these approximations do get closer to the actual value, which is 2.

[C] **6.** Follow the instructions of Problem 5 using the Parabolic Rule.

[C] **7.** Approximate $\pi$ by calculating

$$\int_0^1 \frac{4}{1 + x^2}\,dx$$

using the Parabolic Rule with $n = 10$.

[C] **8.** Use the Trapezoidal Rule with $n = 10$ to approximate $\int_0^1 \cos(\sin x)\,dx$.

[C] In Problems 9–12, determine an $n$ so that the Trapezoidal Rule will approximate the integral with an error $E_n$ satisfying $|E_n| \le 0.01$ (see Example 4). Then, using that $n$, approximate the integral.

**9.** $\int_0^1 e^{-x^2}\,dx$

**10.** $\int_0^{0.6} e^{x^2}\,dx$

**11.** $\int_1^{1.5} \sqrt{\cos x}\,dx$

**12.** $\int_1^2 \cos\sqrt{x}\,dx$

[C] In Problems 13 and 14, determine an $n$ so that the Parabolic Rule will approximate the integral with an error $E_n$ satisfying $|E_n| \le 0.005$. Then, using that $n$, approximate the integral.

**13.** $\int_2^6 \dfrac{1 + x}{1 - x}\,dx$

**14.** $\int_1^3 \ln x\,dx$

**15.** Let $f(x) = ax^2 + bx + c$. Show that

$$\int_{m-h}^{m+h} f(x)\,dx \text{ and } (h/3)[f(m - h) + 4f(m) + f(m + h)]$$

both have the value $(h/3)[a(6m^2 + 2h^2) + b(6m) + 6c]$. This establishes the area formula on which the Parabolic Rule is based.

**16.** Show that the Parabolic Rule is exact for any cubic polynomial in two different ways.

(a) By direct calculation.
(b) By showing that $E_n = 0$.

**17.** We know that $\ln 2 = \int_1^2 (1/x)\,dx$. If we wish to estimate $\ln 2$ using the Trapezoidal Rule with an error less than $10^{-10}$, how large an $n$ would we need?

**18.** Answer the question of Problem 17 for the Parabolic Rule.

**19.** Show that the Parabolic Rule gives the exact value of $\int_{-a}^a x^k\,dx$ provided $k$ is odd.

**20.** It is interesting that a modified version of the Trapezoidal Rule turns out to be in general more accurate than the Parabolic Rule. This version says that

$$\int_a^b f(x)\,dx \approx T - \frac{[f'(b) - f'(a)]h^2}{12}$$

where $T$ is the standard trapezoidal estimate.

(a) Use this formula with $n = 8$ to estimate $\int_1^3 x^4\,dx$ and note its remarkable accuracy (see Example 1 for $T$ and also the exact value of this integral).

(b) Use this formula with $n = 12$ to estimate $\int_0^\pi \sin x \, dx$ (the true value is 2 and $T$ was calculated in Problem 9).

That this rule is so accurate does not seem so surprising after one shows that it is exact for all cubic polynomials, a result we will not ask you to prove since it is quite tedious.

**21.** Use the Trapezoidal Rule to approximate the area of the lakeside lot shown in Figure 5. Dimensions are in feet.

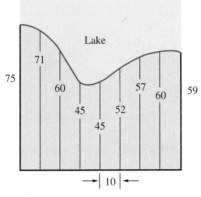

FIGURE 5

**22.** Use the Parabolic Rule to approximate the amount of water required to fill a pool shaped like Figure 6 to a depth of 6 feet. All dimensions are in feet.

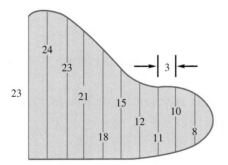

FIGURE 6

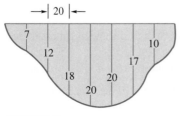

FIGURE 7

C **23.** Figure 7 shows the depth in feet of the water in a river measured at 20-foot intervals across the width of the river. If the river flows at 4 miles per hour, how much water (in cubic feet) flows past the place where these measurements were taken in one day? Use the Parabolic Rule.

C **24.** Another commonly used numerical integration rule is the **Midpoint Rule**. Let $n$ and $h$ have their usual meanings, but assume $n$ is even. Then $\int_a^b f(x) \, dx \approx M_n$, where

$$M_n = 2h[f(x_1) + f(x_3) + f(x_5) + \cdots + f(x_{n-1})]$$

(a) Draw a picture to interpret this rule.

(b) Use this rule with $n = 16$ to approximate $\int_1^3 x^4 \, dx$ (see Example 1).

(c) Let $T_n$ and $P_n$ denote the corresponding trapezoidal and parabolic approximations. Show that $P_n = \frac{1}{3}M_n + \frac{2}{3}T_n$.

Special numerical methods need to be used to evaluate improper integrals. In Problems 25–30 we develop techniques to handle such integrals which may have infinite limits of integration and infinite integrands.

**25.** By numerically computing the integral over the finite part of the interval of integration where most of the contribution to the integral is made and estimating the error of the portion that is neglected we can accurately compute integrals over infinite intervals. Since $\int_0^\infty e^{-x^2} \, dx = \int_0^X e^{-x^2} \, dx + \int_X^\infty e^{-x^2} \, dx$, we have to choose $X$ so that the second integral is small.

(a) Show that for $x \geq X$ we have that $\int_X^\infty e^{-x^2} \, dx < \int_X^\infty e^{-Xx} \, dx = \frac{1}{X} e^{-X^2}.$

C (b) Show that for $X = 5$ that $\int_X^\infty e^{-x^2} \, dx < 10^{-11}$.

C (c) Use the Parabolic rule with $n = 10$ to compute $\int_0^5 e^{-x^2} \, dx$ and use this result to compute $\frac{2}{\sqrt{\pi}} \int_0^\infty e^{-x^2} \, dx$ with an estimate for the error in your answer.

**26.** An integral over an infinite or half-infinite interval can be transformed into one over a finite interval by the technique of substitution. This technique frequently leads to problems where the integrand is infinite at some point in the interval of integration.

(a) Show that the substitution $t = e^{-x}$ transforms the interval $[0, \infty)$ into $(0, 1]$, and $t = \frac{x}{1+x}$ transforms the interval $[0, \infty)$ into $[0, 1)$.

(b) Show that the substitution $t = \dfrac{e^x - 1}{e^x + 1}$ transforms the interval $(-\infty, \infty)$ into $(-1, 1)$.

(c) Convert the integral $\displaystyle\int_0^\infty \dfrac{e^{-x}}{\sqrt{1 + e^{-2x}}}\, dx$ to an integral over a finite region using the substitution $t = e^{-x}$. Use the Parabolic Rule with $n = 10$ on the resulting integral to get an approximate answer with an estimate for the error.

C (d) Use the substitution $t = 1/x$ and the Parabolic Rule with $n = 10$ to evaluate the improper integral $\displaystyle\int_1^\infty \dfrac{x}{1 + x^3}\, dx$.

**27.** Singular integrands can sometimes be transformed using integration by parts into non-singular integrals.

(a) Integrate by parts to show that the improper integral $\displaystyle\int_0^2 \dfrac{dx}{\sqrt{4x + x^2}}$ can be transformed into a proper one.

(b) Integrate by parts to show that the improper integral $\displaystyle\int_1^\infty \dfrac{\sin x}{x}\, dx$ exists.

(c) Integrate by parts to show that $\displaystyle\int_0^1 \dfrac{1}{\sqrt{x}}\dfrac{1}{4 + x}\, dx$ can be turned into a non-singular integral. *Hint*: Differentiate the factor $1/(4 + x)$.

**28.** Integration by parts can be used to transform an integral into a form which permits more accurate results using numerical integration routines. As a specific example we consider $\displaystyle\int_0^2 x \cdot x\,(4 - x^2)^{1/4}\, dx$.

(a) Integrate by parts by differentiating $x$, to show that the integral is equivalent to $\displaystyle\int_0^2 (2/5)\,(4 - x^2)^{5/4}\, dx$.

(b) Plot the functions $x^2\,(4 - x^2)^{1/4}$ and $(2/5)\,(4 - x^2)^{5/4}$ over the domain $[0, 2]$ and pay particular attention to the slope of the curves. Notice that it is difficult to perform accurate numerical integration when a function has a large and rapidly varying slope. This is reflected in the error estimate for numerical integration which depends on the higher derivatives of the integrand.

C (c) Break up the original integral into two parts $\displaystyle\int_0^1 (2/5)\,(4 - x^2)^{5/4}\, dx + (2/5)\int_1^2 (1/x) \cdot x(4 - x^2)^{5/4}\, dx$ and integrate the second integral by parts again. Use the Trapezoidal Rule (with $n = 4$) with an estimate for the error to integrate both integrals. Explain why the error estimate would fail to give any reasonable result for the original integral.

C **29.** Integrate the singular integral $\displaystyle\int_0^{\pi/2} \ln(\sin x)\, dx$ numerically using the Parabolic Rule with $n = 4$. Hint: Use $\ln(\sin x) = \ln x + \ln\!\left(\dfrac{\sin x}{x}\right)$, and then use the Parabolic Rule to integrate $\ln\!\left(\dfrac{\sin x}{x}\right)$.

C **30.** Use the method of truncating the interval of integration to find an $X$ such that $\left|\displaystyle\int_X^\infty \dfrac{\cos x}{1 + x^4}\, dx\right| < 10^{-5}$, by choosing an $X$ which is $(2n + 1)\pi/2$ and using the fact that the cosine function alternates in sign. Explain how to use this result to evaluate the integral $\displaystyle\int_0^\infty \dfrac{\cos x}{1 + x^4}\, dx$ numerically. Be sure to indicate how you would estimate the error in your result.

**31.** The Parabolic Rule has an error term which depends on the fourth derivative of the function being integrated. That means that the rule is exact for cubics. This leads to the following amazing result. Let a polynomial function $P_3(x)$ go through the 3 equally spaced points $(-h, y_1)$, $(0, y_2)$, $(h, y_3)$ and any other point $(x^*, y^*)$. Such a polynomial can be constructed by using the Newton form (see Problem 32 in Section 10.2)

$$P_3(x) = \gamma_1 + \gamma_2(x + h) + \gamma_3\,(x + h)x + \gamma_4(x + h)x(x - h)$$

where the $\gamma$'s can be determined from the condition that $P_3(x)$ go through the required points.

(a) Show that $\gamma_1 = y_1$, $\gamma_2 = \dfrac{y_2 - y_1}{h}$, and $\gamma_3 = \dfrac{y_3 - 2y_2 + y_1}{2h^2}$.

(b) Show that $\displaystyle\int_{-h}^h P_3(x)\, dx$ does not depend on the value of $\gamma_4$ and consequently is independent of the value $(x^*, y^*)$.

(c) Compute the value of $\displaystyle\int_{-h}^h P_3(x)\, dx$ using the Parabolic Rule and explain why the answer is exact.

(d) Explain why the above analysis demonstrates that two fourth degree polynomials going through the same three equally spaced points centered about the origin must have the same integral over the region covered by the equally spaced points. (See Figure 8 where the polynomials $26 + 5(x - 2) + (x - 2)x + 3(x - 2)x(x + 2)$ and $26 + 5(x - 2) + (x - 2)x - 3(x - 2)x(x + 2)$ are plotted.)

(e) Is this result valid if the points are not centered about the origin? Justify your answer

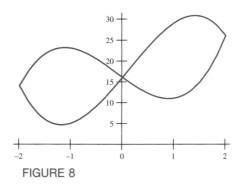

FIGURE 8

Another whole class of numerical integration techniques were developed by Gauss. These rules are more efficient than either the Trapezoidal or Parabolic Rules in the sense that they give more accurate results (integrate a given degree polynomial exactly) with fewer function evaluations. We explore the derivation and use of these rules in Problems 32–38.

**32.** Assume that we are trying to construct a numerical integration formula which integrates all linear functions exactly. We strive for a formula which gives exactly correct results for $f_0(x) = 1$ and $f_1(x) = x$ and any linear combination of these functions such as $\alpha f_0(x) + \beta f_1(x)$. We perform all our integration on the interval $[-1, 1]$ with the knowledge that we can map all other finite intervals into that interval by a simple linear substitution in the integrand.

(a) Assuming a formula of the form $\int_{-1}^{1} f(x)\,dx = c_0 f(x_0)$ we seek $c_0$ and $x_0$ so that the formula is exact for $f_0(x)$ and $f_1(x)$ as well as any linear combination of them of the form $\alpha f_0(x) + \beta f_1(x)$. Hence $\int_{-1}^{1} 1\,dx = c_0 f_0(x_0)$ and $\int_{-1}^{1} x\,dx = c_0 f_1(x_0) = c_0 x_0$. Show, by evaluating the integrals, that $\int_{-1}^{1} f(x)\,dx = 2f(0)$ integrates any expression of the form $f(x) = \alpha + \beta x$ exactly. This is also known as the Midpoint Rule (see Problem 24).

(b) Assuming a formula of the form $\int_{-1}^{1} f(x)\,dx = c_1 f(x_1) + c_2 f(x_2)$ show that if $c_1 = 1$, $c_2 = 1$, $x_1 = -\frac{\sqrt{3}}{3}$, and $x_2 = \frac{\sqrt{3}}{3}$ then the functions $1$, $x$, $x^2$, and $x^3$ are integrated exactly. Notice that this means that the error must depend on $f^{(4)}(x)$. This is usually called the Two Point Gauss Rule.

(c) Why is the formula
$$\int_{-1}^{1} f(x)\,dx \approx f\left(-\frac{\sqrt{3}}{3}\right) + f\left(\frac{\sqrt{3}}{3}\right)$$
more efficient than the Parabolic Rule? Justify your answer.

(d) Use a substitution which maps the interval $[-1, 1]$ into $[a, b]$ given by $t = \frac{a + b}{2} + x\frac{b - a}{2}$ to establish the Gauss integration formula
$$\int_{a}^{b} f(t)\,dt \approx \frac{b - a}{2} \times$$
$$\left[ f\left(\frac{a + b}{2} - \frac{\sqrt{3}}{3}\frac{b - a}{2}\right) + f\left(\frac{a + b}{2} + \frac{\sqrt{3}}{3}\frac{b - a}{2}\right)\right]$$
which is exact for all polynomials up to and including degree three.

**33.** We investigate the evaluation of the integral $\int_0^2 \sin t\,dt$ using various methods of integration.

(a) Evaluate this integral using the Fundamental Theorem of Calculus. Evaluate the answer to 5 decimal places.
(b) Use the Parabolic Rule with $n = 2$ to compute this integral to 5 decimal places.
(c) Use the Two Point Gauss Rule (see Problem 32 (d)) to compute this integral to 5 decimal places.

C **34.** We investigate the numerical evaluation of the integral $\int_0^2 \frac{\sin t}{t}\,dt = \int_{-1}^{1} \frac{\sin(x + 1)}{x + 1}\,dx$.

(a) Use the Parabolic Rule with $n = 2$ to compute this integral to 5 decimal places.
(b) Use the Parabolic Rule with $n = 4$ to compute this integral to 5 decimal places.
(c) Use the Two Point Gauss Rule (see Problem 32 (d)) to compute this integral to 5 decimal places.
(d) Use the Two Point Gauss Rule, breaking up the region of integration into 4 segments, to compute this integral to 5 decimal places.

PC Most computer algebra systems permit you to compute numerical integration using the trapezoidal and parabolic (Simpson's) approximations. Experiment with such a program; in particular, find these approximations with $n = 10$ and $n = 20$ for the following integrals and note the accuracy in each case.

PC **35.** We have developed formulas for the error terms in the Trapezoidal and Parabolic Rules, but they are not easy to apply in hand calculations because of the difficulty in finding and bounding higher-order derivatives. Computer algebra systems can help us out by first finding complicated derivatives and then graphing them, thus allowing a visual determination of bounds. Consider $f(x) = (1 + 16x^6)^{1/2}$ on $[0, 1]$, a function that must be integrated to determine the length of the curve $y = x^4$ on $[0, 1]$.

(a) Determine a good bound for $|f''(x)|$ on this interval.
(b) How large must $n$ be to guarantee that the error term in the Trapezoidal Rule is less than 0.001?
(c) Determine a good bound for $|f''''(x)|$ on this interval.
(d) How large must $n$ be to guarantee that the error term in the Parabolic Rule is less than 0.001?
(e) Evaluate $\int_0^1 f(x)\,dx$ using the Parabolic Rule with $n = 10$ and give a good bound for the error.

PC **36.** Follow the instructions of Problem 35 for $f(x) = (\sin x)/(2 + \sin x)$, with the interval $[0, 2\pi]$ replacing $[0, 1]$.

---

**Answers to Concepts Review:** **1.** 1, 2, 2, 2, . . . , 2, 1 **2.** 1, 4, 2, 4, 2, . . . , 4, 1 **3.** $n^4$ **4.** Large

## 10.4
## SOLVING EQUATIONS NUMERICALLY

In mathematics and science, we often need to find the roots (solutions) of an equation $f(x) = 0$. To be sure, if $f(x)$ is a linear or quadratic polynomial, formulas for writing exact solutions exist and are well known. But for other algebraic equations, and certainly for transcendental equations, formulas for exact solutions are rarely available. What can be done in such cases?

There is a general method of solving problems known to all resourceful people. Given a cup of tea, we add sugar a bit at a time until it tastes just right. Given a stopper too large for a hole, we whittle it down until it fits. We change the solution a bit at a time, improving the accuracy, until we are satisfied. Mathematicians call it the *method of successive approximations*, or the *method of iterations*.

In this section, we present two such methods for solving equations: the Bisection Method and Newton's Method. Both are designed to find the real roots of $f(x) = 0$. Both require many computations. You will want to keep your calculator handy.

**The Bisection Method** This method has two great virtues—simplicity and reliability. It also has a major vice—the large number of steps needed to achieve desired accuracy (otherwise known as slowness of convergence).

Begin the process by sketching the graph of $f$, which is assumed to be a continuous function. A real root $r$ of $f(x) = 0$ is a point (technically, the $x$-coordinate of a point) where the graph crosses the $x$-axis. As a first step in pinning down this point, locate two points, $a_1 < b_1$, at which you are sure that $f$ has opposite signs. (Try choosing $a_1$ and $b_1$ on opposite sides of your best guess at $r$.) The Intermediate Value Theorem guarantees the existence of a root between $a_1$ and $b_1$. Now evaluate $f$ at the midpoint $m_1 = (a_1 + b_1)/2$ of $[a_1, b_1]$. The number $m_1$ is our first approximation to $r$ (see Figure 1).

Either $f(m_1) = 0$, in which case we are done, or $f(m_1)$ differs in sign from $f(a_1)$ or $f(b_1)$. Denote the one of the subintervals $[a_1, m_1]$ or $[m_1, b_1]$ on which the sign change occurs by the symbol $[a_2, b_2]$ and evaluate $f$ at its midpoint $m_2 = (a_2 + b_2)/2$. The number $m_2$ is our second approximation to $r$ (Figure 2).

Repeat the process, thus determining a sequence of approximations $m_1$, $m_2$, $m_3$, . . . and subintervals $[a_1, b_1]$, $[a_2, b_2]$, $[a_3, b_3]$, . . . , each subinterval containing the root $r$ and each half the length of its predecessor. Stop when $r$ is determined to desired accuracy—that is, when $(b_n - a_n)/2$ is less than the allowable error.

To streamline the process for a calculator, we first suppose that the graph of $y = f(x)$ is rising as it crosses the $x$-axis; if not, consider instead the graph of $y = -f(x)$, which crosses at the same point. Then let

$$m_n = \frac{a_n + b_n}{2}, \qquad h_n = \frac{b_n - a_n}{2}$$

and note that $m_n$ is the midpoint of $[a_n, b_n]$ and $h_n$ is its half-length. Assuming that $m_n$ and $h_n$ are known, use the following **algorithm** to determine $m_{n+1}$ and $h_{n+1}$.

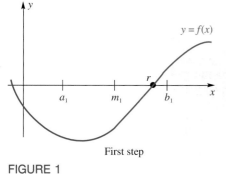

First step

**FIGURE 1**

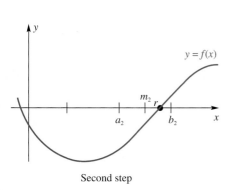

Second step

**FIGURE 2**

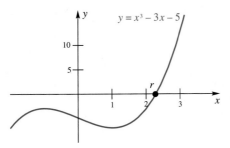

FIGURE 3

1. Calculate $f(m_n)$ and if $f(m_n) = 0$, stop.
2. Put $h_{n+1} = \dfrac{h_n}{2}$.
3. If $f(m_n) < 0$, put $m_{n+1} = m_n + h_{n+1}$.
4. If $f(m_n) > 0$, put $m_{n+1} = m_n - h_{n+1}$.

**EXAMPLE 1** Determine the real root $r$ of $f(x) = x^3 - 3x - 5 = 0$ to accuracy within 0.0000001.

**Solution** We first sketch the graph of $y = x^3 - 3x - 5$ (Figure 3) and, noting that it crosses the $x$-axis between 2 and 3, we use $a_1 = 2$, $b_1 = 3$, $m_1 = 2.5$, and $h_1 = 0.5$. Then we construct the following table.

| $n$ | $h_n$ | $m_n$ | $f(m_n)$ |
|---|---|---|---|
| 1 | 0.5 | 2.5 | 3.125 |
| 2 | 0.25 | 2.25 | −0.359 |
| 3 | 0.125 | 2.375 | 1.271 |
| 4 | 0.0625 | 2.3125 | 0.429 |
| 5 | 0.03125 | 2.28125 | 0.02811 |
| 6 | 0.015625 | 2.265625 | −0.16729 |
| 7 | 0.0078125 | 2.2734375 | −0.07001 |
| 8 | 0.0039063 | 2.2773438 | −0.02106 |
| 9 | 0.0019532 | 2.2792969 | 0.00350 |
| 10 | 0.0009766 | 2.2783203 | −0.00878 |
| 11 | 0.0004883 | 2.2788086 | −0.00264 |
| 12 | 0.0002442 | 2.2790528 | 0.00043 |
| 13 | 0.0001221 | 2.2789307 | −0.00111 |
| 14 | 0.0000611 | 2.2789918 | −0.00034 |
| 15 | 0.0000306 | 2.2790224 | 0.00005 |
| 16 | 0.0000153 | 2.2790071 | −0.00015 |
| 17 | 0.0000077 | 2.2790148 | −0.00005 |
| 18 | 0.0000039 | 2.2790187 | −0.000001 |
| 19 | 0.0000020 | 2.2790207 | 0.000024 |
| 20 | 0.0000010 | 2.2790197 | 0.000011 |
| 21 | 0.0000005 | 2.2790192 | 0.000005 |
| 22 | 0.0000003 | 2.2790189 | 0.0000014 |
| 23 | 0.0000002 | 2.2790187 | −0.0000011 |
| 24 | 0.0000001 | 2.2790188 | 0.0000001 |

We conclude that $r = 2.2790188$ with an error of at most 0.0000001. ∎

Example 1 illustrates the shortcoming of the Bisection Method. The approximations $m_1$, $m_2$, $m_3$, . . . converge very slowly to the root $r$. But they do converge; that is, $\lim\limits_{n \to \infty} m_n = r$. The method works and we have at step $n$ a good bound for the error $E_n = r - m_n$, namely, $|E_n| \le h_n$.

**Newton's Method** We are still considering the problem of solving the equation $f(x) = 0$ for a root $r$. Suppose that $f$ is differentiable, so that the graph of $y = f(x)$ has a tangent line at each point. If we can find a first approximation $x_1$ to $r$ by graphing or any other means, then a better approximation $x_2$ ought to lie at the intersection of the tangent at $(x_1, f(x_1))$ with

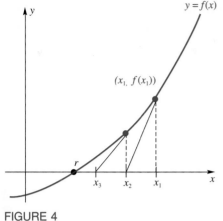

**FIGURE 4**

the $x$-axis (see Figure 4). Using $x_2$ as an approximation, we can then find a still better approximation $x_3$, and so on.

The process can be mechanized so that it is easy to do on a calculator. The equation of the tangent at $(x_1, f(x_1))$ is

$$y - f(x_1) = f'(x_1)(x - x_1)$$

and its $x$-intercept $x_2$ is found by setting $y = 0$ and solving for $x$. The result is

$$x_2 = x_1 - \frac{f(x_1)}{f'(x_1)}$$

More generally we have the following algorithm, also called a *recursion formula* or an *iteration scheme*,

$$x_{n+1} = x_n - \frac{f(x_n)}{f'(x_n)}$$

**EXAMPLE 2**   Use Newton's Method to find the real root $r$ of $f(x) = x^3 - 3x - 5 = 0$ to seven decimal places.

**Solution**   This is the same equation considered in Example 1. Let's use $x_1 = 2.5$ as our first approximation to $r$, as we did there. Since $f(x) = x^3 - 3x - 5$ and $f'(x) = 3x^2 - 3$, the algorithm is

$$x_{n+1} = x_n - \frac{x_n^3 - 3x_n - 5}{3x_n^2 - 3} = \frac{2x_n^3 + 5}{3x_n^2 - 3}$$

We obtain the data in the following table.

| $n$ | $x_n$ |
|---|---|
| 1 | 2.5 |
| 2 | 2.30 |
| 3 | 2.2793 |
| 4 | 2.2790188 |
| 5 | 2.2790188 |

After just four steps, we get a repetition of the first eight digits. We feel confident in reporting that $r \approx 2.2790188$, with perhaps some question about the last digit.   ∎

**EXAMPLE 3**   Use Newton's Method to find the real root $r$ of $f(x) = x - e^{-x} = 0$ to seven decimal places.

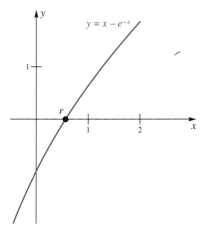

**FIGURE 5**

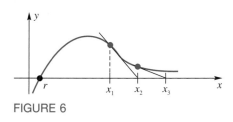

**FIGURE 6**

***Solution*** The graph of $y = x - e^{-x}$ is sketched in Figure 5. We use $x_1 = 0.5$ and $x_{n+1} = x_n - (x_n - e^{-x_n})/(1 + e^{-x_n}) = (x_n + 1)/(e^{x_n} + 1)$ to obtain the table below.

| $n$ | $x_n$ |
|-----|-----------|
| 1 | 0.5 |
| 2 | 0.566 |
| 3 | 0.56714 |
| 4 | 0.5671433 |
| 5 | 0.5671433 |

After just four steps, we get a repetition of the seven digits after the decimal point. We conclude that $r \approx 0.5671433$. ∎

**Convergence of Newton's Method** It is not always obvious that Newton's Method yields approximations that converge to the root $r$, though our two examples give evidence for that assertion. Could it be that we were just lucky in our choice of these two examples? As a matter of fact, the method does not always lead to convergence, as the diagram in Figure 6 shows (see also Problem 21). In this case, the difficulty is that $x_1$ is not close enough to $r$ to get a convergent process started. Another obvious difficulty arises if $f'(x)$ is zero at or near $r$ since $f'(x_n)$ occurs in the denominator of the algorithm. However, we have the following theorem.

---

**Theorem A**

Let $f$ be twice differentiable on an interval $I$, having as its midpoint a root $r$ of $f(x) = 0$. Suppose there are positive numbers $m$ and $M$ such that $|f'(x)| \geq m$ and $|f''(x)| \leq M$ on $I$. If $x_1$ is in $I$ and sufficiently close to $r$ ($|x_1 - r| < 2m/M$ will do), then

**(i)** $|x_{n+1} - r| \leq \dfrac{M}{2m} (x_n - r)^2$;

**(ii)** $x_n$ converges to $r$ as $n \to \infty$.

---

Another way to state conclusion (ii) is to say $\lim\limits_{n \to \infty} x_n = r$. The concept of limit for a sequence is studied in detail in the next chapter.

***Proof*** From Taylor's Formula with Remainder (Theorem 10.2A), there is a number $c$ between $x_n$ and $r$ such that

$$f(r) = f(x_n) + f'(x_n)(r - x_n) + \frac{f''(c)}{2} (r - x_n)^2$$

After dividing both sides by $f'(x_n)$ and using the fact that $f(r) = 0$, we obtain

$$0 = \frac{f(x_n)}{f'(x_n)} + r - x_n + \frac{f''(c)}{2f'(x_n)} (r - x_n)^2$$

and then, successively,

$$x_n - \frac{f(x_n)}{f'(x_n)} - r = \frac{f''(c)}{2f'(x_n)}(r - x_n)^2$$

$$|x_{n+1} - r| = \left|\frac{f''(c)}{2f'(x_n)}\right|(x_n - r)^2$$

$$|x_{n+1} - r| \leq \frac{M}{2m}(x_n - r)^2$$

which is (i).

From (i), one may show by induction (Problem 18) that

$$|x_n - r| \leq \frac{2m}{M}\left(\frac{M}{2m}|x_1 - r|\right)^{2^{n-1}}$$

Since $(M/2m)|x_1 - r| < 1$, the right side of the last inequality approaches 0 as $n \to \infty$. This implies that $|x_n - r|$ also tends to 0 as $n \to \infty$, which is equivalent to (ii). ∎

The speed of convergence of Newton's Method is truly remarkable, tending in fact to double the number of decimal places of accuracy at each step. To see why this is so, suppose $M/2m \leq 2$. Then if the error $|x_n - r|$ at the $n$th step is less than 0.005, the error $|x_{n+1} - r|$ at the next step satisfies (by (i))

$$|x_{n+1} - r| \leq \frac{M}{2m}|x_n - r|^2 \leq 2(0.005)^2 = 0.00005$$

Thus the accuracy of $x_n$ to two decimal places is doubled to an accuracy of $x_{n+1}$ to four decimal places. Of course, we should not expect quite such spectacular results if $M/2m$ is substantially greater than 2.

## CONCEPTS REVIEW

**1.** The virtues of the Bisection Method are its simplicity and reliability; its vice is its _____.

**2.** If $f$ is continuous on $[a, b]$ and $f(a)$ and $f(b)$ have opposite signs, then there is a _____ of $f(x) = 0$ between $a$ and $b$. This follows from the _____ Theorem.

**3.** Both the Bisection Method and Newton's Method are examples of _____; that is, they provide a finite sequence of steps that—if followed—will produce an answer (a root of an equation to desired accuracy).

**4.** Newton's Method can fail to yield a root of $f(x) = 0$. This can happen if _____ is too far from the root $r$ or if _____.

## PROBLEM SET 10.4

C In Problems 1–4, use the Bisection Method to find the real root of the given equation on the given interval. Your answers should be accurate to two decimal places.

1. $x^3 + 2x - 6 = 0$; $[1, 2]$

2. $x^4 + 5x^3 + 1 = 0$; $[-1, 0]$

3. $2\cos x - e^{-x} = 0$; $[1, 2]$

4. $x - 2 + 2\ln x = 0$; $[1, 2]$

[C] In Problems 5–14, use Newton's Method to approximate the indicated root of the given equation accurate to five decimal places. Begin by sketching a graph.

5. The largest root of $x^3 + 6x^2 + 9x + 1 = 0$.

6. The real root of $7x^3 + x - 5 = 0$.

7. The root of $x - 2 + 2 \ln x = 0$ (see Problem 4).

8. The smallest positive root of $2 \cos x - e^{-x} = 1$ (see Problem 3).

9. The root of $\cos x = 2x$.

10. The root of $x \ln x = 2$.

11. All real roots of $x^4 - 8x^3 + 22x^2 - 24x + 8 = 0$.

12. All real roots of $x^4 + 6x^3 + 2x^2 + 24x - 8 = 0$.

13. The positive root of $2x^2 - \sin^{-1}x = 0$.

14. The positive root of $2 \tan^{-1} x = x$.

[C] 15. Use Newton's Method to calculate $\sqrt[3]{6}$ to five decimal places. *Hint*: Solve $x^3 - 6 = 0$.

[C] 16. Use Newton's Method to calculate $\sqrt[4]{47}$ to five decimal places.

[C] 17. Where on $(\pi, 2\pi)$ does $(\sin x)/x$ attain a minimum and what is its minimum value?

18. Show by induction that if

$$|x_{n+1} - r| \le \frac{M}{2m}(x_n - r)^2, \qquad n = 1, 2, \dots$$

then

$$|x_n - r| \le \frac{2m}{M}\left(\frac{M}{2m}|x_1 - r|\right)^{2^{n-1}}, \qquad n = 1, 2, \dots$$

[C] 19. Suppose we use Newton's Method to find the positive root of $x^2 - 2 = 0$—that is, to approximate $\sqrt{2}$. Suppose further that we know this root is on the interval $[1, 2]$. Calculate $m$ and $M$ of Theorem A. Use the second inequality of Problem 18 to estimate $|x_6 - \sqrt{2}|$, given that $x_1 = 1.5$.

[C] 20. How large should we take $n$ in Problem 19 to make sure $|x_n - \sqrt{2}| \le 5 \times 10^{-41}$?

[C] 21. Consider finding the real root of $(1 + \ln x)/x = 0$ by Newton's Method. Show that this leads to the algorithm

$$x_{n+1} = 2x_n + \frac{x_n}{\ln x_n}$$

Apply this algorithm with $x_1 = 1.2$. Next try it with $x_1 = 0.5$. Finally, graph $y = (1 + \ln x)/x$ to understand your results.

22. Sketch the graph of $y = x^{1/3}$. Obviously, its only $x$-intercept is zero. Convince yourself that Newton's Method fails to converge. Explain this failure.

23. In installment buying, one would like to figure out the real interest rate (effective rate), but unfortunately this involves solving a complicated equation. If one buys an item worth $\$P$ today and agrees to pay for it with payments of $\$R$ at the end of each month for $k$ months, then

$$P = \frac{R}{i}\left[1 - \frac{1}{(1 + i)^k}\right]$$

where $i$ is the interest rate per month. Tom bought a used car for $2000 and agreed to pay for it with $100 payments at the end of each of the next 24 months.

(a) Show that $i$ satisfies the equation

$$20i(1 + i)^{24} - (1 + i)^{24} + 1 = 0$$

(b) Derive Newton's Algorithm for this equation, namely,

$$i_{n+1} = i_n - \left[\frac{20i_n^2 + 19i_n - 1 + (1 + i_n)^{-23}}{500i_n - 4}\right]$$

[C] (c) Find $i$ accurate to five decimal places starting with $i_1 = 0.012$ and then give the annual rate $r$ as a percent ($r = 1200i$).

24. In applying Newton's Algorithm to solve $f(x) = 0$, one can usually tell by simply looking at the numbers $x_1, x_2, x_3, \dots$ whether the sequence is converging. But even if it converges, say to $\bar{x}$, can we be sure that $\bar{x}$ is a solution? Show that the answer is yes provided $f$ and $f'$ are continuous at $\bar{x}$ and $f'(\bar{x}) \ne 0$.

25. Experiment with the algorithm

$$x_{n+1} = 2x_n - ax_n^2$$

using several different values of $a$.

(a) Make a conjecture about what this algorithm computes.
(b) Prove your conjecture.

[PC] Some computer packages (including *True BASIC Calculus*, Version 4.0) implement Newton's Method. Experiment with your software on some of the earlier problems in this list. Then find all real roots of the following equations.

26. $x^6 - 4 = 0$        27. $x^3 - 3x + 1 = 0$

28. $x^2 - 2x = \cos 3x$        29. $x^3 - 3x + 1 = 2 \sin 4x$

30. $\sqrt{|x - 1|} = 0$

---

**Answers to Concepts Review:** **1.** Slowness of convergence **2.** Root; Intermediate Value **3.** Algorithms **4.** $x_1$; $f'(r) = 0$

**10.5**
**FIXED-POINT METHODS**

We offer next a method of solving equations that is so simple it has no right to work. Yet it does work in a large number of cases. Moreover, this method has a host of applications in advanced mathematics.

Suppose an equation that interests us can be written in the form $x = g(x)$. To solve this equation is to find a number $r$ that is unchanged by the function $g$. We call such a number a **fixed point** of $g$. To find this number, we propose the following algorithm. Make a first guess $x_1$. Then let $x_2 = g(x_1)$, $x_3 = g(x_2)$, and, in general,

$$x_{n+1} = g(x_n)$$

If we are lucky, $x_n$ will converge to the root $r$ as $n \to \infty$.

**The Method Illustrated**  We begin with an example treated in the previous section (Example 3).

**EXAMPLE 1**  Solve $x - e^{-x} = 0$ using the preceding algorithm.

**Solution**  We write the equation as $x = e^{-x}$ and apply the algorithm $x_{n+1} = e^{-x_n}$ with $x_1 = 0.5$. The results (obtained on a hand-held calculator) are shown in the accompanying table.

| $n$ | $x_n$ | $n$ | $x_n$ | $n$ | $x_n$ |
|---|---|---|---|---|---|
| 1 | 0.5 | 10 | 0.5675596 | 19 | 0.5671408 |
| 2 | 0.6065307 | 11 | 0.5669072 | 20 | 0.5671447 |
| 3 | 0.5452392 | 12 | 0.5672772 | 21 | 0.5671425 |
| 4 | 0.5797031 | 13 | 0.5670674 | 22 | 0.5671438 |
| 5 | 0.5600646 | 14 | 0.5671864 | 23 | 0.5671430 |
| 6 | 0.5711722 | 15 | 0.5671189 | 24 | 0.5671435 |
| 7 | 0.5648630 | 16 | 0.5671572 | 25 | 0.5671432 |
| 8 | 0.5684381 | 17 | 0.5671354 | 26 | 0.5671433 |
| 9 | 0.5664095 | 18 | 0.5671478 | 27 | 0.5671433 |

Although it took 27 steps to get a repetition of the first seven digits, the process did produce a sequence that converges and to the right value. Moreover, the process was very easy to carry out on a calculator.  ■

**EXAMPLE 2**  Solve $x = 2 \cos x$.

**Solution**  Note first that solving this equation is equivalent to solving the pair of equations $y = x$ and $y = 2 \cos x$. Thus, to get our initial seed value, we graph these two equations (Figure 1) and observe that the two curves cross at approximately $x = 1$. Taking $x_1 = 1$ and applying the algorithm $x_{n+1} = 2 \cos x_n$, we obtain the results in the following table.

| $n$ | $x_n$ | $n$ | $x_n$ |
|---|---|---|---|
| 1 | 1 | 6 | 1.4394614 |
| 2 | 1.0806046 | 7 | 0.2619155 |
| 3 | 0.9415902 | 8 | 1.9317916 |
| 4 | 1.1770062 | 9 | −0.7064109 |
| 5 | 0.7673820 | 10 | 1.5213931 |

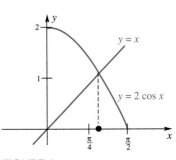

FIGURE 1

Quite clearly the process is unstable, even though our initial guess is very close to the actual root.

Let's take a different tack. Rewrite the equation $x = 2 \cos x$ as $x = (x + 2 \cos x)/2$ and use the algorithm

$$x_{n+1} = \frac{x_n + 2 \cos x_n}{2}$$

This yields the data in the table below.

| $n$ | $x_n$ | $n$ | $x_n$ | $n$ | $x_n$ |
|---|---|---|---|---|---|
| 1 | 1 | 7 | 1.0298054 | 13 | 1.0298665 |
| 2 | 1.0403023 | 8 | 1.0298883 | 14 | 1.0298666 |
| 3 | 1.0261107 | 9 | 1.0298588 | 15 | 1.0298665 |
| 4 | 1.0312046 | 10 | 1.0298693 | 16 | 1.0298666 |
| 5 | 1.0293881 | 11 | 1.0298655 | | |
| 6 | 1.0300374 | 12 | 1.0298668 | | |

The process has produced a convergent sequence (the oscillation in the last digit is probably due to round-off errors). ∎

Now we raise an obvious question. Why did the second algorithm yield a convergent sequence, whereas the first one failed to do so? And can we be sure that we have obtained the correct answer in the second case? To this latter question we can offer an affirmative answer. The nature of the calculation suggests that when we get a repetition of the first seven digits, we have a solution to at least six-digit accuracy. We will answer the first question after we have considered another example.

**EXAMPLE 3** Solve $x^3 + 6x - 3 = 0$ by the Fixed-Point Method.

*Solution* The given equation is equivalent to $x = (-x^3 + 3)/6$, so we use the algorithm

$$x_{n+1} = \frac{-x_n^3 + 3}{6}$$

The graph in Figure 2 suggests an initial value of $x_1 = 0.5$, but we also consider what happens with $x_1 = 1.5$, $x_1 = 2.2$, and $x_1 = 2.7$. The results are shown in the table.

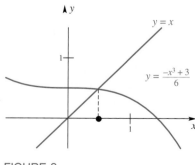

FIGURE 2

| $n$ | $x_n$ | $n$ | $x_n$ | $n$ | $x_n$ | $n$ | $x_n$ |
|---|---|---|---|---|---|---|---|
| 1 | 0.5 | 1 | 1.5 | 1 | 2.2 | 1 | 2.7 |
| 2 | 0.4791667 | 2 | -0.0625 | 2 | -1.2744667 | 2 | -2.7805 |
| 3 | 0.4816638 | 3 | 0.5000407 | 3 | 0.8451745 | 3 | 4.0827578 |
| 4 | 0.4813757 | 4 | 0.4791616 | 4 | 0.3993792 | 4 | -10.842521 |
| 5 | 0.4814091 | 5 | 0.4816644 | 5 | 0.4893829 | 5 | 212.9416 |
| 6 | 0.4814052 | 6 | 0.4813756 | 6 | 0.4804658 | 6 | -16909274.5 |
| 7 | 0.4814057 | 7 | 0.4814091 | 7 | 0.4815143 | | |
| 8 | 0.4814056 | 8 | 0.4814052 | 8 | 0.4813930 | | |
| 9 | 0.4814056 | 9 | 0.4814057 | 9 | 0.4814071 | | |
| | | 10 | 0.4814056 | 10 | 0.4814054 | | |
| | | 11 | 0.4814056 | 11 | 0.4814056 | | |
| | | | | 12 | 0.4814056 | | |

It appears that if our initial guess $x_1$ is close enough to the fixed point $r$, the sequence will converge, but if we begin too far away from $r$, it will diverge.    ■

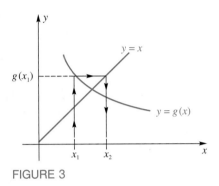

FIGURE 3

**Convergence of the Method**    Sometimes the method works; sometimes it fails. We can get a pretty good idea about which is which by executing an appropriate number of steps of the algorithm. But wouldn't it be nice to have a way of telling in advance whether there will be convergence? And to be sure of our conclusion?

To get a feeling for the problem, let's look at it geometrically. Note that we can get $x_2$ from $x_1$ by locating $g(x_1)$, sending it horizontally to the line $y = x$, and projecting down to the $x$-axis (see Figure 3). When we use this process repeatedly, we are faced with one of the situations in Figure 4.

What determines success or failure, convergence or divergence? It appears to depend on the slope of the curve $y = g(x)$, namely, $g'(x)$, near the root $r$. If $|g'(x)|$ is too large, the method fails; if $|g'(x)|$ is small enough, the method works. Here is a general result.

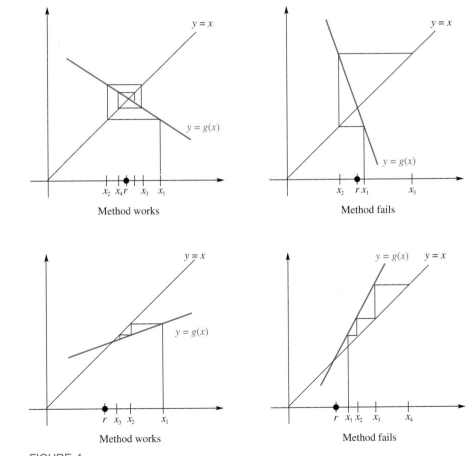

FIGURE 4

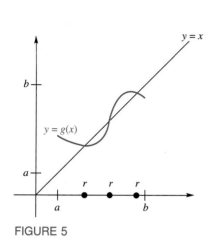

FIGURE 5

Making It Work

Can we make the fixed-point algorithm work on every equation? No. First, it may be impossible to write the equation in the form $x = g(x)$. Second, even if the equation can be written in the right form, the fixed-point algorithm may produce a divergent sequence. However, equations can be written in many equivalent ways, as Example 4 illustrates. Our goal then is to find one for which $|g'(r)| < 1$. Moreover, we would prefer to have this derivative as small as possible, since the smaller it is, the faster is the convergence. In summary, making the fixed-point algorithm work requires more ingenuity than a first impression suggests.

### Theorem A

**(Fixed-Point Theorem).** Let $g$ be a continuous function taking $[a, b]$ into itself—that is, which satisfies $a \le g(x) \le b$ whenever $a \le x \le b$. Then $g$ has at least one fixed point $r$ on $[a, b]$. If, in addition, $g$ is differentiable and satisfies $|g'(x)| \le M < 1$ for all $x$ in $[a, b]$, $M$ a constant, then the fixed point is unique and the algorithm

$$x_{n+1} = g(x_n), \qquad x_1 \text{ in } [a, b]$$

yields a sequence which converges to $r$ as $n \to \infty$.

**Proof** A typical graph of a continuous function mapping $[a, b]$ into $[a, b]$ is shown in Figure 5. If either $g(a) = a$ or $g(b) = b$, we have our fixed point, so suppose neither is true. Let $h(x) = g(x) - x$ and note that $h(a) > 0$ and $h(b) < 0$. By the Intermediate Value Theorem, there is a point $r$ (possibly several points) such that $h(r) = 0$, that is, $r = g(r)$. The first assertion of our theorem is proved.

Next suppose $|g'(x)| \le M < 1$ for all $x$ in $[a, b]$ and let $r$ be a fixed point of $g$. By the Mean Value Theorem for Derivatives, we may write

$$g(x) - g(r) = g'(c)(x - r)$$

with $c$ some point between $x$ and $r$. Thus,

$$|g(x) - g(r)| = |g'(c)|\,|x - r| \le M|x - r|$$

Applying this inequality successively to $x_1, x_2, \ldots$ yields

$$|x_2 - r| = |g(x_1) - g(r)| \le M|x_1 - r|$$
$$|x_3 - r| = |g(x_2) - g(r)| \le M|x_2 - r| \le M^2|x_1 - r|$$
$$|x_4 - r| = |g(x_3) - g(r)| \le M|x_3 - r| \le M^3|x_1 - r|$$
$$\vdots$$
$$|x_n - r| = |g(x_{n-1}) - g(r)| \le M|x_{n-1} - r| \le M^{n-1}|x_1 - r|$$

Since $M^{n-1} \to 0$ as $n \to \infty$, we conclude that $x_n \to r$ as $n \to \infty$.

Finally, if $r$ and $s$ are two fixed points of $g$, we have just shown that $x_n \to r$ and $x_n \to s$ as $n \to \infty$. This is impossible unless $r = s$. Thus there is only one fixed point. ∎

Now we can understand the behavior in Example 2. If $g(x) = 2 \cos x$, then $|g'(x)| = |-2 \sin x|$, which is greater than 1 in a neighborhood of the fixed point $x \approx 1.03$. On the other hand, if $g(x) = (x + 2 \cos x)/2$, then $|g'(x)| = |\frac{1}{2} - \sin x| < 1$ near $x = 1.03$. We should not expect convergence in the first case; we can guarantee it in the second.

In Example 3, $g(x) = (-x^3 + 3)/6$ and $g'(x) = -x^2/2$. Clearly $|g'(x)| \le 1$ near the fixed point $x \approx 0.48$. We can be confident of convergence so long as we choose $x_1$ on the interval where $|g'(x)| \le 1$. Actually, our experiments in Example 3 showed that we can start as far away as $x_1 = 2.2$ (which is more than we have a right to expect), but $x_1 = 2.7$ is too far away.

One final remark: The closer $|g'(x)|$ is to zero near the root, the faster will be the convergence of the fixed-point algorithm.

**EXAMPLE 4**  The equation $x^3 - 3x + 1 = 0$ has three real roots (see Figure 6). Use the fixed-point method to find the root between 1 and 2.

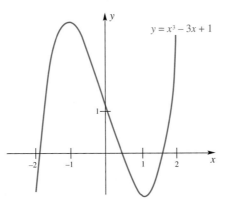

$y = x^3 - 3x + 1$

FIGURE 6

*Solution*  Proceeding as in Example 3, we write

$$x = \frac{x^3 + 1}{3} = g(x)$$

but unfortunately $g'(x) = x^2 \geq 1$ on $[1, 2]$. Another way to write the equation is

$$x = -\frac{1}{x^2 - 3} = g(x)$$

Now $g'(x) = 2x/(x^2 - 3)^2$. This too can be greater than 1 on the interval $[1, 2]$ (for example, $g'(1.5) \approx 5.33$). But there are still other possibilities. Consider

$$x = \frac{3}{x} - \frac{1}{x^2} = g(x)$$

for which $g'(x) = (-3x + 2)/x^3$. After a bit of work (we study $g''$), we see that $g'$ is increasing on $[1, 2]$, varying from $g'(1) = -1$ to $g'(2) = -\frac{1}{2}$. Thus, $|g'(x)|$ is strictly less than 1 as long as we stay strictly away from the left endpoint $x = 1$.

With the algorithm

$$x_{n+1} = \frac{3}{x_n} - \frac{1}{x_n^2}$$

we obtain the data in the table in the margin. ■

| $n$ | $x_n$ |
|-----|-------|
| 1 | 1.5 |
| 2 | 1.5555556 |
| 3 | 1.5153061 |
| ⋮ | ⋮ |
| 21 | 1.5320411 |
| 22 | 1.5321234 |
| ⋮ | ⋮ |
| 33 | 1.5320871 |
| 34 | 1.5320902 |
| ⋮ | ⋮ |
| 43 | 1.5320888 |
| 44 | 1.5320889 |
| 45 | 1.5320889 |

## CONCEPTS REVIEW

1. A point $x$ satisfying $g(x) = x$ is called a _____ of $g$.

2. The fixed-point algorithm for $g$ is _____ $= g(x_n)$.

3. The critical condition on $g$ needed to make the fixed-point algorithm converge is that _____ in a neighborhood of the fixed point.

4. The equation $x = g(x) = x^2 - 2$ has 2 as a root. Yet the algorithm $x_{n+1} = x_n^2 - 2$ will not converge to this root because _____.

## PROBLEM SET 10.5

C In Problems 1–4, use the fixed-point algorithm with $x_1$ as indicated to solve the equations to five decimal places.

1. $x = \frac{1}{9}e^{-2x}$; $x_1 = 1$

2. $x = 2 \tan^{-1} x$; $x_1 = 2$

3. $x = \sqrt{2.7 + x}$; $x_1 = 1$

4. $x = \sqrt{3.2 + x}$; $x_1 = 47$

C 5. Consider the equation $x = 2(x - x^2) = g(x)$.

(a) Sketch the graph of $y = x$ and $y = g(x)$ using the same coordinate system, and thereby approximately locate the positive root of $x = g(x)$.
(b) Try solving the equation by the fixed point algorithm starting with $x_1 = 0.7$.
(c) Solve the equation algebraically.
(d) Find $g'(x)$ and evaluate it at the root.

C 6. Follow the directions of Problem 5 for $x = 5(x - x^2) = g(x)$. Explain your results.

C 7. Consider the equation $x = (3/2) \sin \pi x = g(x)$.

(a) Sketch the graphs of $y = x$ and $y = g(x)$.
(b) Try solving the equation by the fixed point algorithm.
(c) Find $g'(x)$ and use it to explain your results.

8. Follow the directions for Problem 7 for $x = \frac{1}{2} \sin \pi x = g(x)$

9. Consider $x = (3/2) \sin \pi x$ of Problem 7 again.

(a) Show that it can be written as $x = \frac{5}{6}x + \frac{1}{4} \sin \pi x = g(x)$.
(b) Now solve the latter equation by the fixed-point algorithm.
(c) Why is the convergence so rapid? *Hint*: Evaluate $g'(x)$ at the root.

10. Consider $x = 5(x - x^2) = g(x)$ of Problem 6

(a) Rewrite this equation so that the fixed-point algorithm will converge (see Problem 9).
(b) Use this algorithm to solve the equation.

C 11. Find the positive root of $x^3 - x^2 - x - 1 = 0$. *Hint*: See Example 4.

12. Evaluate

$$\sqrt{5 + \sqrt{5 + \sqrt{5 + \cdots}}}$$

*Hint*: Consider solving $x = \sqrt{5 + x}$ by the fixed-point algorithm, starting with $x_1 = 0$. What are $x_2$, $x_3$, $x_4$, and $x_5$? Are you sure that this fixed-point algorithm converges? To what does it converge? (You can find this number algebraically if you wish.)

C 13. Note that $\sqrt{a}$ is a solution of $x = \frac{1}{2}(x + a/x) = g(x)$. Use the fixed-point algorithm to find $\sqrt{\pi}$. Calculate $g'(a)$ to see why the convergence is so rapid.

C 14. Kepler's equation $x = m + E \sin x$ is important in astronomy. Use the fixed-point algorithm to solve this equation when $m = 0.8$ and $E = 0.2$.

C 15. If an item selling today for $P$ dollars is purchased on the installment plan with monthly payments of $R$ dollars at the end of each of the next $k$ months with interest at the rate of $i$ per month, then

$$P = \frac{R}{i}[1 - (1 + i)^{-k}]$$

(a) A new car costing $10,000 is purchased on the installment plan with payments of $R$ dollars at the end of each of the next 48 months and stated interest of 18% (which means $i = 0.18/12 = 0.015$). Use algebra to find $R$.
(b) Suppose the monthly payment in (a) is $300. Then what is $i$? *Hint*: Use the fixed-point algorithm with $i_1 = 0.015$.

C 16. A television set costing $500 is purchased on the installment plan with payments of $30 at the end of each of the next 24 months. What is $i$, the monthly interest rate? See Problem 15.

C 17. Consider the equation $x = x - f(x)/f'(x)$ and suppose $f'(x) \neq 0$ in an interval $[a, b]$.

(a) Show that if $r$ is in $[a, b]$, then $r$ is a root if and only if $f(r) = 0$.
(b) Show that Newton's Method is a special case of the fixed-point method, in which $g'(r) = 0$.

C 18. Sketch graphs to convince yourself that each of the following equations has a unique solution. Decide whether the Fixed-Point Method will work and, if so, use it. Otherwise, solve by Newton's Method.

(a) $\sin^{-1} x = \dfrac{1}{\sin x}$   (b) $\cos^{-1} x = \dfrac{1}{\cos x}$

(c) $\tan^{-1} x = \dfrac{1}{\tan x}$

PC The iterative method that we have called the Fixed-Point Method leads to an area of contemporary research and serves as a possible model for turbulence, one of the least-understood phenomena in science. The remaining problems will introduce you to this exciting area, but you will need a programmable calculator or a computer to carry out the required calculations. Each problem deals with the equation

(*) $\qquad\qquad x = f(x) = \lambda(x - x^2)$

which was introduced in Problems 5 and 6 for $\lambda = 2$ and $\lambda = 5$. To find its positive root, consider the corresponding fixed-point iteration

$$x_{n+1} = \lambda(x_n - x_n^2)$$

as we gradually increase $\lambda$ from 2 to 5.

19. ($\lambda = 2.5$) Sketch $y = x$ and $y = 2.5(x - x^2)$ using the same axes and solve equation (*) by iteration.

20. Solve equation (*) for $\lambda = 2.5$ by simple algebra, thus confirming your answer to Problem 19.

21. ($\lambda = 3.1$) Sketch $y = x$ and $y = f(x) = 3.1(x - x^2)$ using the same axes and attempt to solve (*) by iteration (note that $|f'(x)| > 1$ at the root). You will find that $x_n$ bounces back and forth but gets closer and closer to two values $r_1$ and $r_2$, called *attractors*. Find $r_1$ and $r_2$ to five decimal places. Superimpose the graph of $y = g(x) = f(f(x)) = 9.61x - 39.40x^2 + 59.58x^3 - 29.79x^4$ on your earlier graph and observe that $r_1$ and $r_2$ appear to be the two roots of $x = g(x)$, where $|g'(x)| < 1$.

22. Note that $f(r_1) = r_2$ and $f(r_2) = r_1$. Use this to show that $g'(r_1) = g'(r_2)$.

23. ($\lambda = 3.5$) In this case, use the iteration to obtain four attractors, $s_1, s_2, s_3$, and $s_4$. Guess to what equation they are the solutions.

24. ($\lambda = 3.56$) Use the iteration to get eight attractors.

25. ($\lambda = 3.57$) As you keep increasing $\lambda$ by smaller and smaller amounts, you will double the number of attractors at each stage until at approximately 3.57 you should get chaos. Beyond $\lambda = 3.57$, other strange things happen, which you may want to investigate.

26. Try similar experiments on $x = \lambda \sin \pi x$ (see Problems 7 and 8).

For very readable accounts of this strange phenomenon, see:

Robert M. May, *Simple mathematical models with very complicated dynamics*, Nature, vol. 261 (1976), 459–467.

Douglas R. Hofstadter, *Strange attractors: mathematical patterns delicately poised between order and chaos*, Scientific American, vol. 245 (November 1981), 22–43.

**Answers to Concepts Review:** 1. Fixed point   2. $x_{n+1}$
3. $|g'(x)| \le M < 1$   4. $|2x| > 1$ near $x = 2$

## 10.6 CHAPTER REVIEW

### Concepts Test

Respond with true or false to each of the following assertions. Be prepared to justify your answer.

1. If $P(x)$ is the Maclaurin polynomial of order 2 for $f(x)$, then $P(0) = f(0)$, $P'(0) = f'(0)$, and $P''(0) = f''(0)$.

2. The Taylor polynomial of order $n$ based at $a$ for $f(x)$ is unique; that is, $f(x)$ has only one such polynomial.

3. $f(x) = x^{5/2}$ has a second-order Maclaurin polynomial.

4. The Maclaurin polynomial of order 3 for $f(x) = 2x^3 - x^2 + 7x - 11$ is an exact representation of $f(x)$.

5. The Maclaurin polynomial of order 16 for $\cos x$ involves only even powers of $x$.

**6.** If $f'(0)$ exists for an even function, then $f'(0) = 0$.

**7.** Taylor's Formula with Remainder contains the Mean Value Theorem for Derivatives as a special case.

**8.** With a calculator and the formula $[f(a + h) - f(a)]/h$ one can approximate $f'(a)$ to any desired degree of accuracy by taking $h$ small enough.

**9.** We can always express the indefinite integral of an elementary function in terms of elementary functions.

**10.** The Trapezoidal Rule with $n = 10$ will give a value for $\int_0^5 x^3\, dx$ that is smaller than the true value.

**11.** The Parabolic Rule with $n = 10$ will give the exact value of $\int_0^5 x^3\, dx$.

**12.** With a computer and the Parabolic Rule one can always approximate $\int_a^b f(x)\, dx$ to any desired degree of accuracy by taking $h$ small enough.

**13.** The function $f(x) = e^{-x^2} + x^2 + \sin(x + 1)$ satisfies $|f(x)| \le 6$ on $[-1, 2]$.

**14.** If $f(x) = ax^2 + bx + c$, then $\int_{-2}^2 f(x)\, dx = \frac{2}{3}[f(-2) + 4f(0) + f(2)]$.

**15.** If $f$ is continuous on $[a, b]$ and $f(a)f(b) < 0$, then $f(x) = 0$ has a root between $a$ and $b$.

**16.** One of the virtues of the Bisection Method is its rapid convergence.

**17.** Newton's Method will produce a convergent sequence for the function $f(x) = x^{1/3}$.

**18.** If $f'(x) > 1$ on an open interval containing a root $r$ of $x = f(x)$, then the Fixed-Point Method will fail to produce a sequence converging to $r$ (unless the first guess happens to be $r$).

**19.** The Fixed-Point Method will work to find the largest root of $x = 5(x - x^2) + 0.01$.

**20.** The Fixed-Point Method will produce a convergent sequence for $x = \frac{1}{2}\left(x + \frac{a}{x}\right)$ if $a > 0$ and the first guess is greater than $\sqrt{a/3}$.

## Sample Test Problems

C **1.** Use Horner's Method to calculate $p(2.31)$ if $p(x) = 3x^4 - 2x^3 - 5x^2 + 7x - 3$.

**2.** Find the Maclaurin polynomial of order 4 for $f(x)$ and use it to approximate $f(0.1)$.

(a) $f(x) = xe^x$          (b) $f(x) = \cosh x$

**3.** Find the Taylor polynomial of order 3 based at 2 for $g(x) = x^3 - 2x^2 + 5x - 7$ and show that it is an exact representation of $g(x)$.

**4.** Use the result of Problem 3 to calculate $g(2.1)$.

**5.** Find the Taylor polynomial of order 4 based at 1 for $f(x) = 1/(x + 1)$.

**6.** Obtain an expression for the error term $R_4(x)$ in Problem 5 and find a bound for it if $x = 1.2$.

**7.** Find the Maclaurin polynomial of order 4 for $f(x) = \sin^2 x = \frac{1}{2}(1 - \cos 2x)$ and find a bound for the error $R_4(x)$ if $|x| \le 0.2$. *Note:* A sharper bound is obtained if you observe that $R_4(x) = R_5(x)$ and then bound $R_5(x)$.

C **8.** If $f(x) = \ln x$, then $f^{(n)}(x) = (-1)^{n-1}(n - 1)!/x^n$. Thus, the Taylor polynomial of order $n$ based at 1 for $\ln x$ is

$$\ln x = (x - 1) - \frac{1}{2}(x - 1)^2 + \frac{1}{3}(x - 1)^3 + \cdots$$

$$+ \frac{(-1)^{n-1}}{n}(x - 1)^n + R_n(x)$$

How large would $n$ have to be for us to know $|R_n(x)| \le 0.00005$ if $0.8 \le x \le 1.2$?

C **9.** Refer to Problem 8. Use the Taylor polynomial of order 5 based at 1 to find

$$\int_{0.8}^{1.2} \ln x\, dx$$

and give a good bound for the error that is made.

C **10.** Use the Trapezoidal Rule with $n = 8$ to calculate

$$\int_{0.8}^{1.2} \ln x\, dx$$

and give an error bound.

C **11.** Use the Parabolic Rule with $n = 8$ to calculate

$$\int_{0.8}^{1.2} \ln x\, dx$$

and give a good error bound.

C **12.** Compute

$$\int_{0.8}^{1.2} \ln x\, dx$$

using the Fundamental Theorem of Calculus. *Hint:* $D_x[x \ln x - x] = \ln x$.

C  **13.** Use Newton's Method to solve $3x - \cos 2x = 0$ accurate to six decimal places. Use $x_1 = 0.5$.

C  **14.** Use the Fixed-Point Method to solve $3x - \cos 2x = 0$, starting with $x_1 = 0.5$.

C  **15.** Use Newton's Method to find the solution of $x - \tan x = 0$ in the interval $(\pi, 2\pi)$ accurate to four decimal

places. *Hint:* Sketch graphs of $y = x$ and $y = \tan x$ using the same axes to get a good initial guess for $x_1$.

C  **16.** Try the Fixed-Point Method on the equation of Problem 15. Why doesn't it work?

C  **17.** Use Newton's Method to find the largest solution of $e^x - \sin x = 0$. *Hint:* Begin by sketching $y = e^x$ and $y = \sin x$ to get an initial guess, $x_1$.

---

## 10.7  ADDITIONAL PROBLEM SET

Most first order ordinary differential equations of the form

$$\frac{dy}{dx} = f(x, y(x)), \; y(x_0) = y_0$$

can not be solved exactly. Even if they can be solved, the solutions may be so complicated that their form provides little insight into the nature of the solution. In such cases, a numerical solution may provide the only means of obtaining a solution or of gaining insight. We explore such numerical techniques here. For a more complete development of this material see a book on ordinary differential equations (such as C. H. Edwards, Jr. and D. E. Penny, *Differential Equations and Modeling* (New York: Prentice Hall, 1996) pp. 96–128).

C  **1.** If the function $f(x, y(x)) \equiv f(x) = \sin x$, then the problem reduces to one of numerical integration using the formula $y(x) = y_0 + \int_{x_0}^{x_1} \sin t \, dt$. Solve the ordinary differential equation $\frac{dy}{dx} = \sin x$, $y(\pi) = 1$ using the Trapezoidal rule to construct a table for $(x_i, y_i)$ where $y_i = y(x_i)$ for $\pi \le x \le 2\pi$ at 11 evenly spaced points. Compute all values to 4 decimal digits. *Hint:* $x_i = \pi + i \cdot \frac{2\pi - \pi}{10}$ for $i = 0, 1, \ldots, 10$. Compute $y_1 = y_0 + \int_{x_0}^{x_1} \sin t \, dt$, then $y_2 = y_1 + \int_{x_0}^{x_1} \sin t \, dt$ and so on.

C  **2.** *Euler's Method* provides a means of computing a table of numerical approximations to the solution of such first order ordinary differential equations.

$$\frac{dy}{dx} = f(x, y(x)), \; y(x_0) = y_0$$

In this case we write $y_1 = y_0 + \int_{x_0}^{x_1} f(t, y(t)) \, dt$, $y_2 = y_1 + \int_{x_1}^{x_2} f(t, y(t)) \, dt, \ldots, y_n = y_{n-1} + \int_{x_{n-1}}^{x_n} f(t, y(t)) \, dt$ and use the Riemann sums with the left end point approximation to compute each of the integrals:

$$y_1 = y_0 + \int_{x_0}^{x_1} f(t, y(t)) \, dt \approx y_0 + (x_1 - x_0) \cdot f(x_0, y_0)$$

,

$$y_2 = y_1 + \int_{x_1}^{x_2} f(t, y(t)) \, dt \approx y_1 + (x_2 - x_1) \cdot f(x_1, y_1)$$

and so on. This method can be succinctly summarized by the formula

$$y_{n+1} = y_n + h \cdot f(x_n, y_n); \; h = x_{n+1} - x_n.$$

where $h$ is called the stepsize.

(a)  Compute the solution to $y'(x) = x + y(x)$, $y(0) = 1$ with stepsize. $h = 0.1$ at the points $(0.1, y_1)$, $(0.2, y_2)$, $(0.3, y_3)$, $(0.4, y_4)$.

(b)  Compute the solution to $y'(x) = y(x)$, $y(0) = 1$ using a stepsize of $h = 0.25$. Compute the solution up to $x = 1$ and compare your answer to the exact answer $e \approx 2.71828183$.

(c)  Compute the solution to $y'(x) = y(x)$, $y(0) = 1$ using a stepsize of $h = 0.1$. Compute the solution up to $x = 1$ and compare your answer to the exact answer.

C  **3.** In some problems the Euler Method does not provide sufficiently accurate results even if the stepsize is taken smaller. One possibility is to use a better numerical method to compute the integral of the right-hand side of the ordinary differential equation. For example, if we use the trapezoidal rule we would have

$$y_{n+1} = y_n + h/2 \cdot [f(x_n, y_n) + f(x_{n+1}, y_{n+1})];$$
$$h = x_{n+1} - x_n$$

which clearly requires the knowledge of $f(x_{n+1}, y_{n+1})$ which we do not yet know since we do not know $y_{n+1}$. One way to get around this problem is to use the Euler Method to compute $y_{n+1}^{\text{predicted}} = y_n + h \cdot f(x_n, y_n)$ and then compute $y_{n+1} = y_n + h/2 \cdot [f(x_n, y_n) + f(x_{n+1}, y_{n+1}^{\text{predicted}})]$. This is the *Improved Euler Method.*

(a)  Use the Improved Euler Method to compute the solution to $y'(x) = y(x)$, $y(0) = 1$. Use a stepsize of $h = 0.25$ and $h = 0.1$. Compute the solution up to $x = 1$ and compare your answer to the exact answer $e \approx 2.71828183$.

(b)  Use the Euler Method and the Improved Euler Method to compute the solution to $y'(x) = x + y(x)$, $y(0) = 1$ at the point $x = 1$. Use a step size of $h = 0.05$ for the Euler Method and $h = 0.1$ for the Improved Euler Method. Compare the results with those of the exact solution $y(x) = 2e^x - 1 - x$.

## TECHNOLOGY PROJECT

**Prerequisites** Read the material in your text about Newton's method for solving the equation $f(x) = 0$, paying special attention to the derivation of the Newton iteration,

$$x_{n+1} = x_n - \frac{f(x_n)}{f'(x_n)}$$

**Exercise 1** For the specific equation $x^2 = c$, where $c$ is a given constant, use Newton's method to derive the Babylonian square root iteration scheme of averaging the current guess (say $a$) and $c/a$ to obtain

$$\frac{1}{2}\left(a + \frac{c}{a}\right)$$

as the new approximation. Using the starting approximation $x = 2$, carry out two iteration steps by hand for the case $c = 5$. Explain your calculations clearly and check the accuracy of your result.

We now discuss the implementation of Newton's method in the context of determining a numerical approximation for $\sqrt{5}$. Thus, consider the function $f = x^2 - 5$, where the positive solution of $f = 0$ gives us the desired value. Let's illustrate geometrically how Newton's method generates the first iterate. Take the (poor) starting value of $x = 6$. Figure 1 shows that the tangent at $(6, f(6))$ cuts the $x$-axis between 6 and the true root, thus producing an improved $x$-estimate of $\sqrt{5}$ that is about $x = 3.5$.

Implementation of Newton's method is quite simple on any technology. For example, in *Mathematica* after defining $f$ and

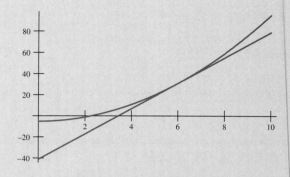

FIG. 1. The first Newton iteration for $x^2 - 5 = 0$, with starting value $x = 6$.

setting the starting value $xn = 6.0$ one iteration is done by:

$$xn = xn - f[xn]/f'[xn].$$

**Exercise 2** Program the above idea on your computing technology and execute the first iteration. You should obtain a value consistent with Figure 1. To get the *next* iterate, just execute the above command again. And so on. Try it for a total of 4 or 5 iterations with the starting value 6.0. Report on the accuracy attained.

**Exercise 3** Now try an experiment with taking an even poorer starting value, $x = 15.0$. How many iterations does it take to converge to $x = 2.23607$?

**Exercise 4** One problem with the above method is that we erase each approximation in the process of producing the next one. A simple way to avoid doing this is to use a "loop"; for example in *Mathematica* there is a "Do" loop to get 8 iterations. Note the print command asks for 16 digits, a good idea when checking for high precision.

```
Do[
  xn = xn - f[xn]/f'[xn];
  Print [ N[nx, 16] ],
{8}]
```

Create a similar looping structure on your technology; and then alter it to answer the question: how many iterations does it take to get 10 correct significant figures with the (absolutely insane) starting value 200.0?

**Exercise 5** This Lab concludes with the study of the more challenging function

$$f = (x^3 - 2.1x^2 + x - 2)/(x^6 + 1)$$

(a) *Prove* that $f$ has a zero in the interval $[-10, 10]$.

(b) Graphically show that $f$ has only one zero in this interval.

(c) Pick a reasonable starting $x_n$-value, e.g., the closest integer, and do enough Newton iterations to get at least six significant figure precision. In your writeup show all the iterates.

(d) Now pick $xn$ bigger; e.g., $xn = 3.0$ or larger. Do several iterations and explain the results.

(e) Try $xn$ around 0.5 and tell what happens, and why it happens.

(f) Try $xn$ around $-1.0$ and tell what happens, and why.

# 11

# Infinite Series

**11.1**
**INFINITE SEQUENCES**

The word *sequence* was introduced in Chapter 5 and reappeared in Chapter 10. In simple language, a sequence

$$a_1, a_2, a_3, a_4, \ldots$$

is an ordered arrangement of real numbers, one for each positive integer. More formally, an **infinite sequence** is a function whose domain is the set of positive integers and whose range is a set of real numbers. We may denote a sequence by $a_1, a_2, a_3, \ldots$, by $\{a_n\}_{n=1}^{\infty}$, or simply by $\{a_n\}$. Occasionally, we will extend the notion slightly by allowing the domain to consist of all integers greater than or equal to a specified integer, as in $b_0, b_1, b_2, \ldots$ and $c_8, c_9, c_{10}, \ldots$, which are also denoted by $\{b_n\}_{n=0}^{\infty}$ and $\{c_n\}_{n=8}^{\infty}$.

A sequence may be specified by giving enough initial terms to establish a pattern, as in

$$1, 4, 7, 10, 13, \ldots$$

by an **explicit formula** for the $n$th term, as in

$$a_n = 3n - 2, \qquad n \geq 1$$

or by a **recursion formula**

$$a_n = a_{n-1} + 3, \qquad n \geq 2, a_1 = 1$$

Note that each of our three illustrations describes the same sequence. Here are four more explicit formulas and the first few terms of the sequences they generate.

---

### PATTERNS

Someone is sure to argue that there are many different sequences that begin

$$1, 4, 7, 10, 13$$

and we agree. For example, the formula

$$3n - 2 + (n - 1) \cdot (n - 2) \cdots (n - 5)$$

generates those five numbers. Who but an expert would think of this formula? When we ask you to look for a pattern, we mean a simple and obvious pattern.

$$(1) \quad a_n = 1 - \frac{1}{n}, \qquad n \geq 1: \qquad 0, \frac{1}{2}, \frac{2}{3}, \frac{3}{4}, \frac{4}{5}, \ldots$$

$$(2) \quad b_n = 1 + (-1)^n \frac{1}{n}, \qquad n \geq 1: \qquad 0, \frac{3}{2}, \frac{2}{3}, \frac{5}{4}, \frac{4}{5}, \frac{7}{6}, \frac{6}{7}, \cdots$$

$$(3) \quad c_n = (-1)^n + \frac{1}{n}, \qquad n \geq 1: \qquad 0, \frac{3}{2}, \frac{-2}{3}, \frac{5}{4}, \frac{-4}{5}, \frac{7}{6}, \frac{-6}{7}, \ldots$$

$$(4) \quad d_n = 0.999, \qquad n \geq 1: \qquad 0.999, 0.999, 0.999, 0.999, \ldots$$

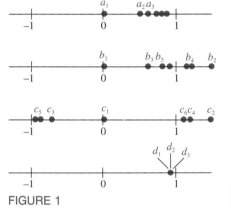

FIGURE 1

**Convergence**   Consider the four sequences just defined. Each of them has values that pile up near 1 (see the diagrams in Figure 1). But do they all *converge* to 1? The correct response is that sequences $\{a_n\}$ and $\{b_n\}$ converge to 1, but $\{c_n\}$ and $\{d_n\}$ do not.

For a sequence to converge to 1 means first that values of the sequence should get close to 1. But they must do more than get close; they must remain close, which rules out sequence $\{c_n\}$. And close means arbitrarily close, that is, within *any* specified degree of accuracy, which rules out sequence $\{d_n\}$. While sequence $\{d_n\}$ does not converge to 1, it is correct to say that it converges to 0.999. Sequence $\{c_n\}$ does not converge at all; we say it diverges.

Here is a formal definition; it should sound vaguely familiar.

---

**Definition**

The sequence $\{a_n\}$ is said to **converge** to $L$ and we write

$$\lim_{n \to \infty} a_n = L$$

if for each positive number $\varepsilon$, there is a corresponding positive number $N$ such that

$$n \geq N \implies |a_n - L| < \varepsilon$$

A sequence that fails to converge to any finite number $L$ is said to **diverge,** or to be divergent.

---

To see a relationship with something studied earlier (Section 4.6), consider graphing $a_n = 1 - 1/n$ and $a(x) = 1 - 1/x$. The only difference is that in the sequence case, the domain is restricted to the positive integers. In the first case, we write $\lim_{n \to \infty} a_n = 1$; in the second, $\lim_{x \to \infty} a(x) = 1$. Note the interpretation of $\varepsilon$ and $N$ in the diagrams in Figure 2.

**EXAMPLE 1**   Show that if $p$ is a positive integer, then

$$\lim_{n \to \infty} \frac{1}{n^p} = 0$$

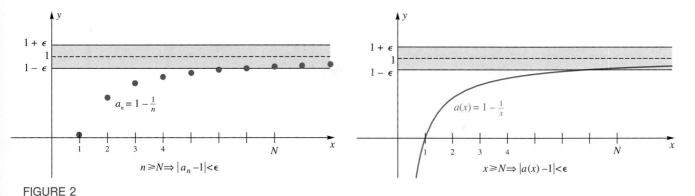

FIGURE 2

*Solution* This is almost obvious from earlier work, but we can give a formal demonstration. Let an arbitrary $\varepsilon > 0$ be given. Choose $N$ to be any number greater than $\sqrt[p]{1/\varepsilon}$. Then $n \geq N$ implies

$$|a_n - L| = \left|\frac{1}{n^p} - 0\right| = \frac{1}{n^p} \leq \frac{1}{N^p} < \frac{1}{(\sqrt[p]{1/\varepsilon})^p} = \varepsilon \quad \blacksquare$$

All the familiar limit theorems hold for convergent sequences. We state them without proof.

---

### Theorem A

Let $\{a_n\}$ and $\{b_n\}$ be convergent sequences and $k$ a constant. Then:

1. $\displaystyle\lim_{n \to \infty} k = k$;

2. $\displaystyle\lim_{n \to \infty} k a_n = k \lim_{n \to \infty} a_n$;

3. $\displaystyle\lim_{n \to \infty} (a_n \pm b_n) = \lim_{n \to \infty} a_n \pm \lim_{n \to \infty} b_n$;

4. $\displaystyle\lim_{n \to \infty} (a_n \cdot b_n) = \lim_{n \to \infty} a_n \cdot \lim_{n \to \infty} b_n$;

5. $\displaystyle\lim_{n \to \infty} \frac{a_n}{b_n} = \frac{\displaystyle\lim_{n \to \infty} a_n}{\displaystyle\lim_{n \to \infty} b_n}$ provided $\displaystyle\lim_{n \to \infty} b_n \neq 0$.

---

**EXAMPLE 2** Find $\displaystyle\lim_{n \to \infty} \frac{3n^2}{7n^2 + 1}$.

*Solution* To decide what is happening to a quotient of two polynomials in $n$ as $n$ gets large, it is wise to divide numerator and denominator by the largest power of $n$ that occurs in the denominator. This justifies our first step below; the others are justified by appealing to statements from Theorem A as indicated by the circled numbers.

$$\lim_{n \to \infty} \frac{3n^2}{7n^2 + 1} = \lim_{n \to \infty} \frac{3}{7 + (1/n^2)}$$

$$\overset{\text{⑤}}{=} \frac{\lim_{n \to \infty} 3}{\lim_{n \to \infty} [7 + (1/n^2)]}$$

$$\overset{\text{③}}{=} \frac{\lim_{n \to \infty} 3}{\lim_{n \to \infty} 7 + \lim_{n \to \infty} 1/n^2}$$

$$\overset{\text{①}}{=} \frac{3}{7 + \lim_{n \to \infty} 1/n^2} = \frac{3}{7 + 0} = \frac{3}{7}$$

By this time, the limit theorems are so familiar that we will normally jump directly from the first step to the final result. ∎

**EXAMPLE 3** Does the sequence $\{(\ln n)/e^n\}$ converge, and if so, to what number?

*Solution* Here and in many sequence problems, it is convenient to use the following almost obvious fact (see Figure 2).

$$\boxed{\text{If } \lim_{x \to \infty} f(x) = L, \text{ then } \lim_{n \to \infty} f(n) = L.}$$

This is convenient because we can apply l'Hôpital's Rule to the continuous variable problem. In particular, by l'Hôpital's Rule,

$$\lim_{x \to \infty} \frac{\ln x}{e^x} = \lim_{x \to \infty} \frac{1/x}{e^x} = 0$$

Thus,

$$\lim_{n \to \infty} \frac{\ln n}{e^n} = 0$$

that is, $\{(\ln n)/e^n\}$ converges to 0. ∎

Here's another theorem that we've seen before in a slightly different guise (Theorem 2.6C).

**Theorem B**

**(Squeeze Theorem).** Suppose $\{a_n\}$ and $\{c_n\}$ both converge to $L$ and that $a_n \le b_n \le c_n$ for $n \ge K$ ($K$ a fixed integer). Then $\{b_n\}$ also converges to $L$.

**EXAMPLE 4**   Show that $\lim_{n \to \infty} \dfrac{\sin^3 n}{n} = 0$.

**Solution**   For $n \geq 1$, $-1/n \leq (\sin^3 n)/n \leq 1/n$. Since $\lim_{n \to \infty} (-1/n) = 0$ and $\lim_{n \to \infty}$ $(1/n) = 0$, the result follows by the Squeeze Theorem.   ∎

For sequences of variable sign, it is helpful to have the following result.

> ### Theorem C
>
> If $\lim_{n \to \infty} |a_n| = 0$, then $\lim_{n \to \infty} a_n = 0$.

**Proof**   Since $-|a_n| \leq a_n \leq |a_n|$, the result follows from the Squeeze Theorem.   ∎

What happens to the numbers in the sequence $\{0.999^n\}$ as $n \to \infty$? We suggest that you calculate $0.999^n$ for $n = 10, 100, 1000$, and $10,000$ on your calculator to make a good guess. Then note the following example.

**EXAMPLE 5**   Show that if $-1 < r < 1$, then $\lim_{n \to \infty} r^n = 0$.

**Solution**   If $r = 0$, the result is trivial, so suppose otherwise. Then $1/|r| > 1$, and so $1/|r| = 1 + p$ for some number $p > 0$. By the Binomial Theorem,

$$\frac{1}{|r|^n} = (1 + p)^n = 1 + pn + (\text{positive terms}) \geq pn$$

Thus,

$$0 \leq |r|^n \leq \frac{1}{pn}.$$

Since $\lim_{n \to \infty} (1/pn) = (1/p) \lim_{n \to \infty} (1/n) = 0$, it follows from the Squeeze Theorem that $\lim_{n \to \infty} |r|^n = 0$, or—equivalently—$\lim_{n \to \infty} |r^n| = 0$. By Theorem C, $\lim_{n \to \infty} r^n = 0$.   ∎

What if $r > 1$; for example, $r = 1.5$? Then $r^n$ will march off toward $\infty$. In this case, we write

$$\lim_{n \to \infty} r^n = \infty, \qquad r > 1$$

However, we say the sequence $\{r^n\}$ diverges. To converge, it would have to approach a *finite* limit. The sequence $\{r^n\}$ also diverges when $r \leq -1$.

**Monotonic Sequences**   Consider now an arbitrary **nondecreasing sequence** $\{a_n\}$, by which we mean $a_n \leq a_{n+1}, n \geq 1$. One example is the sequence

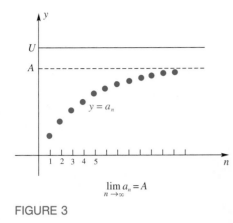

$$\lim_{n \to \infty} a_n = A$$

**FIGURE 3**

$a_n = n^2$; another is $a_n = 1 - 1/n$. If you think about it a little, you may convince yourself that such a sequence can only do one of two things. Either it marches off to infinity or, if it cannot do that because it is bounded above, then it must bump against a lid (see Figure 3). Here is the formal statement of this very important result.

---

### Theorem D

**(Monotonic Sequence Theorem).** If $U$ is an upper bound for a nondecreasing sequence $\{a_n\}$, then the sequence converges to a limit $A$ that is less than or equal to $U$. Similarly, if $L$ is a lower bound for a nonincreasing sequence $\{b_n\}$, then the sequence $\{b_n\}$ converges to a limit $B$ that is greater than or equal to $L$.

---

The phrase **monotonic sequence** is used to describe either a nondecreasing or nonincreasing sequence; hence the name for this theorem.

Theorem D describes a very deep property of the real number system. It is equivalent to the *completeness property* of the real numbers, which in simple language says that the real line has no "holes" in it (see Problems 47–48). It is this property that distinguishes the real number line from the rational number line (which is full of holes). A great deal more could be said about this topic; we hope Theorem D appeals to your intuition and that you will accept it on faith until you take a more advanced course.

We make one more comment about Theorem D. It is not necessary that the sequences $\{a_n\}$ and $\{b_n\}$ be monotonic initially, only that they be monotonic from some point on—that is, for $n \geq K$. In fact, *the convergence or divergence of a sequence does not depend on the character of the initial terms but rather on what is true for large n.*

**EXAMPLE 6** Show that the sequence $b_n = n^2/2^n$ converges by using Theorem D.

*Solution* The first few terms of this sequence are

$$\frac{1}{2}, 1, \frac{9}{8}, 1, \frac{25}{32}, \frac{36}{64}, \frac{49}{128}, \ldots$$

For $n \geq 3$ the sequence appears to be decreasing ($b_n > b_{n+1}$), a fact that we now establish. Each of the following inequalities is equivalent to the others.

$$\frac{n^2}{2^n} > \frac{(n+1)^2}{2^{n+1}}$$
$$n^2 > \frac{(n+1)^2}{2}$$
$$2n^2 > n^2 + 2n + 1$$
$$n^2 - 2n > 1$$
$$n(n-2) > 1$$

The last inequality is clearly true for $n \geq 3$. Since the sequence is decreasing (a stronger condition than nonincreasing) and is bounded below by zero, the Monotonic Sequence Theorem guarantees that it has a limit.

It would be easy using l'Hôpital's Rule to show that the limit is zero. ∎

## CONCEPTS REVIEW

**1.** An arrangement of numbers $a_1, a_2, a_3, \ldots$ is called _____ .

**2.** We say the sequence $\{a_n\}$ converges if _____ .

**3.** An increasing sequence that is also _____ must converge.

**4.** The sequence $\{r^n\}$ converges if and only if _____ $< r \leq$ _____ .

## PROBLEM SET 11.1

In Problems 1–20, an explicit formula for $a_n$ is given. Write out the first five terms, determine if the sequence converges or diverges, and if it converges, find $\lim a_n$.

**1.** $a_n = \dfrac{n}{3n - 1}$

**2.** $a_n = \dfrac{3n + 2}{n + 1}$

**3.** $a_n = \dfrac{4n^2 + 2}{n^2 + 3n - 1}$

**4.** $a_n = \dfrac{3n^2 + 2}{2n - 1}$

**5.** $a_n = \dfrac{n^3 + 3n^2 + 3n}{(n + 1)^3}$

**6.** $a_n = \dfrac{\sqrt{3n^2 + 2}}{2n + 1}$

**7.** $a_n = (-1)^n \dfrac{n}{n + 2}$

**8.** $a_n = \dfrac{n \cos(n\pi)}{2n - 1}$

**9.** $a_n = \dfrac{\cos(n\pi)}{n}$

**10.** $a_n = e^{-n} \sin n$

**11.** $a_n = \dfrac{e^{2n}}{n^2 + 3n - 1}$

**12.** $a_n = \dfrac{e^{2n}}{4^n}$

**13.** $a_n = \dfrac{(-\pi)^n}{5^n}$

**14.** $a_n = \left(\tfrac{1}{4}\right)^n + 3^{n/2}$

**15.** $a_n = 2 + (0.99)^n$

**16.** $a_n = \dfrac{n^{100}}{e^n}$

**17.** $a_n = \dfrac{\ln n}{\sqrt{n}}$

**18.** $a_n = \dfrac{\ln(1/n)}{\sqrt{2n}}$

**19.** $a_n = \left(1 + \dfrac{2}{n}\right)^{n/2}$

**20.** $a_n = (2n)^{1/2n}$

*Hint*: Theorem 7.5A

In Problems 21–30, find an explicit formula $a_n =$ _____ for each sequence, determine if the sequence converges or diverges, and if it converges, find $\displaystyle\lim_{n \to \infty} a_n$.

**21.** $\frac{1}{2}, \frac{2}{3}, \frac{3}{4}, \frac{4}{5}, \ldots$

**22.** $\dfrac{1}{2^2}, \dfrac{2}{2^3}, \dfrac{3}{2^4}, \dfrac{4}{2^5}, \ldots$

**23.** $-1, \frac{2}{3}, -\frac{3}{5}, \frac{4}{7}, -\frac{5}{9}, \ldots$

**24.** $1, \dfrac{1}{1 - \frac{1}{2}}, \dfrac{1}{1 - \frac{2}{3}}, \dfrac{1}{1 - \frac{3}{4}}, \ldots$

**25.** $1, \dfrac{2}{2^2 - 1^2}, \dfrac{3}{3^2 - 2^2}, \dfrac{4}{4^2 - 3^2}, \ldots$

**26.** $\dfrac{1}{2 - \frac{1}{2}}, \dfrac{2}{3 - \frac{1}{3}}, \dfrac{3}{4 - \frac{1}{4}}, \dfrac{4}{5 - \frac{1}{5}}, \ldots$

**27.** $\sin 1, 2 \sin \frac{1}{2}, 3 \sin \frac{1}{3}, 4 \sin \frac{1}{4}, \ldots$

**28.** $-\frac{1}{3}, \frac{4}{9}, -\frac{9}{27}, \frac{16}{81}, \ldots$

**29.** $2, 1, \dfrac{2^3}{3^2}, \dfrac{2^4}{4^2}, \dfrac{2^5}{5^2}, \ldots$

**30.** $1 - \frac{1}{2}, \frac{1}{2} - \frac{1}{3}, \frac{1}{3} - \frac{1}{4}, \frac{1}{4} - \frac{1}{5}, \ldots$

In Problems 31–36, write the first four terms of the sequence $\{a_n\}$. Then use Theorem D to show that the sequence converges.

**31.** $a_n = \dfrac{4n - 3}{2^n}$

**32.** $a_n = \dfrac{n}{n + 1} \left(2 - \dfrac{1}{n^2}\right)$

**33.** $a_n = \left(1 - \dfrac{1}{4}\right)\left(1 - \dfrac{1}{9}\right) \cdots \left(1 - \dfrac{1}{n^2}\right)$, $n \geq 2$

**34.** $a_n = 1 + \dfrac{1}{2!} + \dfrac{1}{3!} + \cdots + \dfrac{1}{n!}$

35. $a_{n+1} = 1 + \frac{1}{2}a_n$, $a_1 = 1$

36. $a_{n+1} = \frac{1}{2}\left(a_n + \frac{2}{a_n}\right)$, $a_1 = 2$

C 37. Assuming that $u_{n+1} = \sqrt{3 + u_n}$ and $u_1 = \sqrt{3}$, determine a convergent sequence, find $\lim_{n \to \infty} u_n$ to four decimal places, using a calculator.

38. Show that $\{u_n\}$ of Problem 37 is bounded above and increasing. Conclude from Theorem D that $\{u_n\}$ converges. *Hint*: Use mathematical induction.

39. Find $\lim_{n \to \infty} u_n$ of Problem 37 algebraically. *Hint*: Let $u = \lim_{n \to \infty} u_n$. Then since $u_{n+1} = \sqrt{3 + u_n}$, $u = \sqrt{3 + u}$. Now square both sides and solve for $u$.

40. Use the technique of Problem 39 to find $\lim_{n \to \infty} a_n$ of Problem 36.

C 41. Assuming that $u_{n+1} = 1.1^{u_n}$ and $u_1 = 0$, determine a convergent sequence, find $\lim_{n \to \infty} u_n$ to four decimal places using a calculator.

42. Show that $\{u_n\}$ of Problem 41 is increasing and bounded above by 2.

43. Find

$$\lim_{n \to \infty} \sum_{k=1}^{n} \left(\sin \frac{k}{n}\right) \frac{1}{n}$$

*Hint*: Write an equivalent definite integral.

44. Show that

$$\lim_{n \to \infty} \sum_{k=1}^{n} \left[\frac{1}{1 + \left(\frac{k}{n}\right)^2}\right] \frac{1}{n} = \frac{\pi}{4}$$

45. Using the definition of limit, prove that $\lim_{n \to \infty} n/(n+1) = 1$; that is, for a given $\varepsilon > 0$ find $N$ such that $n \geq N \Rightarrow |n/(n+1) - 1| < \varepsilon$.

46. As in Problem 45, prove that $\lim_{n \to \infty} n/(n^2 + 1) = 0$.

47. Let $S = \{x : x \text{ rational}, x^2 < 2\}$. Convince yourself that $S$ does not have a *least* upper bound in the rational numbers, but does have such a bound in the real numbers. Alternatively, the sequence of rational numbers 1, 1.4, 1.41, 1.414, . . . has no limit within the rational numbers.

48. The *completeness property* of the real numbers says that every set of real numbers that is bounded above has a least upper bound, which is a real number. This property is usually taken as an axiom for the real numbers. Prove Theorem D using this property.

49. Prove that if $\lim_{n \to \infty} a_n = 0$ and $\{b_n\}$ is bounded, then $\lim_{n \to \infty} a_n b_n = 0$.

50. Prove that if $\{a_n\}$ converges and $\{b_n\}$ diverges, then $\{a_n + b_n\}$ diverges.

51. If $\{a_n\}$ and $\{b_n\}$ both diverge, does it follow that $\{a_n + b_n\}$ diverges?

52. A famous sequence $\{f_n\}$, called the *Fibonacci Sequence* after Leonardo Fibonacci, who introduced it around A.D. 1200, is defined by the recursion formula

$$f_{n+2} = f_{n+1} + f_n, \qquad f_1 = f_2 = 1$$

(a) Find $f_3$ through $f_{10}$.
(b) Let $\phi = \frac{1}{2}(1 + \sqrt{5}) \approx 1.618034$. The Greeks called this number the *golden ratio*, claiming that a rectangle whose dimensions were in this ratio was perfect. It can be shown that

$$f_n = \frac{1}{\sqrt{5}}\left[\left(\frac{1 + \sqrt{5}}{2}\right)^n - \left(\frac{1 - \sqrt{5}}{2}\right)^n\right]$$

$$= \frac{1}{\sqrt{5}}\left[\phi^n - (-1)^n \phi^{-n}\right]$$

and you should check that this gives the right result for $n = 1$ and $n = 2$. The general result can be proved by induction (it is a nice challenge). More in line with this section, use this explicit formula to prove that $\lim_{n \to \infty} f_{n+1}/f_n = \phi$.

(c) Show, using the limit just proved, that $\phi$ satisfies the equation $x^2 - x - 1 = 0$. Then, in another interesting twist, use the Quadratic Formula to show that the two roots of this equation are $\phi$ and $-1/\phi$, the two numbers that occur in the explicit formula for $f_n$.

53. Consider an equilateral triangle containing $1 + 2 + 3 + \cdots + n = n(n+1)/2$ circles each of diameter 1 and stacked as indicated in Figure 4 for the case $n = 4$. Find $\lim_{n \to \infty} A_n/B_n$, where $A_n$ is the total area of the circles and $B_n$ is the area of the triangle.

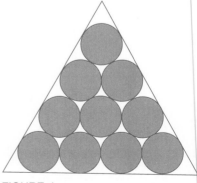

FIGURE 4

PC Recall that $\lim_{x \to \infty} f(x)$ can be transformed to $\lim_{x \to 0^{++}} f(1/x)$. Use this fact and a computer to find the following limits.

54. $\lim_{n \to \infty} \left(1 + \dfrac{1}{n}\right)^n$

55. $\lim_{n \to \infty} \left(1 + \dfrac{1}{2n}\right)^n$

56. $\lim_{n \to \infty} \left(1 + \dfrac{1}{n^2}\right)^n$

57. $\lim_{n \to \infty} \left(\dfrac{n-1}{n+1}\right)^n$

58. $\lim_{n \to \infty} \left(\dfrac{2+n^2}{3+n^2}\right)^n$

59. $\lim_{n \to \infty} \left(\dfrac{2+n^2}{3+n^2}\right)^{n^2}$

---

**Answers to Concepts Review:** **1.** A sequence **2.** $\lim_{n \to \infty} a_n$ exists (finite sense) **3.** Bounded above **4.** $-1$; $1$

---

## 11.2
## INFINITE SERIES

In a famous paradox announced some 2400 years ago, Zeno of Elea said that a runner cannot finish a race because he must first cover half the distance, then half the remaining distance, then half the still remaining distance, and so on, forever. Since the runner's time is finite, he cannot traverse the infinite number of segments of the course. Yet we all know that runners do finish races.

Imagine a race course to be 1 mile long. The segments of Zeno's argument would then have length $\frac{1}{2}$ mile, $\frac{1}{4}$ mile, $\frac{1}{8}$ mile, and so on (Figure 1). In mathematical language, finishing the race would amount to evaluating the sum

$$\tfrac{1}{2} + \tfrac{1}{4} + \tfrac{1}{8} + \tfrac{1}{16} + \tfrac{1}{32} + \cdots$$

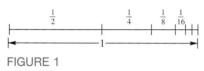

**FIGURE 1**

which might seem impossible. But wait. Up to now, the word *sum* has been defined only for the addition of a finite number of terms. The indicated "infinite sum" has, as yet, no meaning for us.

Consider the partial sums

$$S_1 = \frac{1}{2}$$

$$S_2 = \frac{1}{2} + \frac{1}{4} = \frac{3}{4}$$

$$S_3 = \frac{1}{2} + \frac{1}{4} + \frac{1}{8} = \frac{7}{8}$$

$$S_n = \frac{1}{2} + \frac{1}{4} + \frac{1}{8} + \cdots + \frac{1}{2^n} = 1 - \frac{1}{2^n}$$

Clearly these partial sums get increasingly close to 1. In fact

$$\lim_{n \to \infty} S_n = \lim_{n \to \infty} \left(1 - \frac{1}{2^n}\right) = 1$$

This we define to be the value of the infinite sum.

More generally, consider

$$a_1 + a_2 + a_3 + a_4 + \cdots$$

which is also denoted by $\sum\limits_{k=1}^{\infty} a_k$ or $\sum a_k$ and is called an **infinite series** (or series for short). Then $S_n$, the **nth partial sum**, is given by

$$S_n = a_1 + a_2 + a_3 + \cdots + a_n = \sum_{k=1}^{n} a_k$$

We make the following formal definition.

---

**Definition**

The infinite series $\sum\limits_{k=1}^{\infty} a_k$ **converges** and has sum $S$ if the sequence of partial sums $\{S_n\}$ converges to $S$. If $\{S_n\}$ diverges, then the series **diverges.** A divergent series has no sum.

---

**Geometric Series**    A series of the form

$$\sum_{k=1}^{\infty} ar^{k-1} = a + ar + ar^2 + ar^3 + \cdots$$

where $a \neq 0$ is called a **geometric series**.

**EXAMPLE 1**    Show that a geometric series converges with sum $S = a/(1 - r)$ if $|r| < 1$, but diverges if $|r| \geq 1$.

**Solution**    Let $S_n = a + ar + ar^2 + \cdots + ar^{n-1}$. If $r = 1$, $S_n = na$, which grows without bound, and so $\{S_n\}$ diverges. If $r \neq 1$, we may write

$$S_n - rS_n = (a + ar + \cdots + ar^{n-1}) - (ar + ar^2 + \cdots + ar^n) = a - ar^n$$

and so

$$S_n = \frac{a - ar^n}{1 - r} = \frac{a}{1 - r} - \frac{a}{1 - r} r^n$$

If $|r| < 1$, then $\lim\limits_{n \to \infty} r^n = 0$ (Section 11.1, Example 5) and thus

$$S = \lim_{n \to \infty} S_n = \frac{a}{1 - r}$$

If $|r| > 1$ or $r = -1$, the sequence $\{r^n\}$ diverges, and consequently so does $\{S_n\}$. ∎

**EXAMPLE 2**    Use the result of Example 1 to sum the following two geometric series.

(a) $\dfrac{4}{3} + \dfrac{4}{9} + \dfrac{4}{27} + \dfrac{4}{81} + \cdots$

(b) $0.515151 \ldots = \dfrac{51}{100} + \dfrac{51}{10,000} + \dfrac{51}{1,000,000} + \cdots$

*Solution*

(a) $S = \dfrac{a}{1-r} = \dfrac{\frac{4}{3}}{1-\frac{1}{3}} = \dfrac{\frac{4}{3}}{\frac{2}{3}} = 2$

(b) $S = \dfrac{\frac{51}{100}}{1-\frac{1}{100}} = \dfrac{\frac{51}{100}}{\frac{99}{100}} = \dfrac{51}{99} = \dfrac{17}{33}$

Incidently, the procedure in (b) shows that any repeating decimal represents a rational number. ∎

**EXAMPLE 3**  The diagram in Figure 2 represents an equilateral triangle containing infinitely many circles, tangent to the triangle and each other and reaching into the corners. What fraction of the area of the triangle is occupied by the circles?

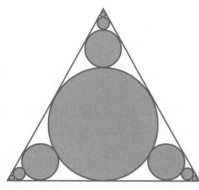

FIGURE 2

*Solution*  Suppose for convenience that the triangle has sides of length $2\sqrt{3}$, which gives it an altitude of 3. Concentrate attention on the vertical stack of circles. With a bit of geometric reasoning (the center of the large circle is two-thirds of the way from the upper vertex to the base), we see that the radii of these circles are $1, \frac{1}{3}, \frac{1}{9}, \ldots$ and conclude that the vertical stack has area

$$\pi\left[1^2 + \left(\frac{1}{3}\right)^2 + \left(\frac{1}{9}\right)^2 + \left(\frac{1}{27}\right)^2 + \cdots\right]$$

$$= \pi\left[1 + \frac{1}{9} + \frac{1}{81} + \frac{1}{729} + \cdots\right] = \pi\left[\frac{1}{1-\frac{1}{9}}\right] = \frac{9\pi}{8}$$

The total area of all the circles is three times this number minus twice the area of the big circle, that is, $27\pi/8 - 2\pi$, or $11\pi/8$. Since the triangle has area $3\sqrt{3}$, the fraction of this area occupied by the circles is

$$\frac{11\pi}{24\sqrt{3}} \approx 0.83$$

∎

**A General Test for Divergence**  Consider the geometric series $a + ar + ar^2 + \cdots + ar^{n-1} + \cdots$ once more. Its *n*th term $a_n$ is given by $a_n = ar^{n-1}$. Example 1 shows that a geometric series converges *if and only if* $\lim_{n\to\infty} a_n = 0$.

Could this possibly be true of all series? The answer is no, although half of the statement (the "only-if" half) is correct. This leads to an important divergence test for series.

---

Theorem A

(***n*th-Term Test for Divergence**). If the series $\displaystyle\sum_{n=1}^{\infty} a_n$ converges, then $\displaystyle\lim_{n\to\infty} a_n = 0$. Equivalently, if $\displaystyle\lim_{n\to\infty} a_n \neq 0$ (or $\displaystyle\lim_{n\to\infty} a_n$ does not exist), the series diverges.

***Proof*** Let $S_n$ be the $n$th partial sum and $S = \lim\limits_{n \to \infty} S_n$. Since $a_n = S_n - S_{n-1}$, it follows that

$$\lim_{n \to \infty} a_n = \lim_{n \to \infty} S_n - \lim_{n \to \infty} S_{n-1} = S - S = 0 \qquad \blacksquare$$

**EXAMPLE 4** Show that $\displaystyle\sum_{n=1}^{\infty} \frac{n^3}{3n^3 + 2n^2}$ diverges.

***Solution***

$$\lim_{n \to \infty} a_n = \lim_{n \to \infty} \frac{n^3}{3n^3 + 2n^2} = \lim_{n \to \infty} \frac{1}{3 + 2/n} = \frac{1}{3}$$

Thus by the $n$th-Term Test, the series diverges. $\qquad \blacksquare$

**The Harmonic Series** Students invariably want to turn Theorem A around and make it say that $a_n \to 0$ implies convergence of $\sum a_n$. The **harmonic series**

$$\sum_{n=1}^{\infty} \frac{1}{n} = 1 + \frac{1}{2} + \frac{1}{3} + \cdots + \frac{1}{n} + \cdots$$

shows that this is false. Clearly $\lim\limits_{n \to \infty} a_n = \lim\limits_{n \to \infty}(1/n) = 0$. However, the series diverges, as we now show.

**EXAMPLE 5** Show that the harmonic series diverges.

***Solution*** We show that $S_n$ grows without bound. Imagine $n$ to be large and write

$$
\begin{aligned}
S_n &= 1 + \frac{1}{2} + \frac{1}{3} + \frac{1}{4} + \frac{1}{5} + \cdots + \frac{1}{n} \\
&= 1 + \frac{1}{2} + \left(\frac{1}{3} + \frac{1}{4}\right) + \left(\frac{1}{5} + \frac{1}{6} + \frac{1}{7} + \frac{1}{8}\right) + \left(\frac{1}{9} + \cdots + \frac{1}{16}\right) + \cdots + \frac{1}{n} \\
&> 1 + \frac{1}{2} + \frac{2}{4} + \frac{4}{8} + \frac{8}{16} + \cdots + \frac{1}{n} \\
&= 1 + \frac{1}{2} + \frac{1}{2} + \frac{1}{2} + \frac{1}{2} + \cdots + \frac{1}{n}
\end{aligned}
$$

It is clear that by taking $n$ sufficiently large, we can introduce as many $\frac{1}{2}$'s into the last expression as we wish. Thus, $\{S_n\}$ diverges; hence, so does the harmonic series. $\qquad \blacksquare$

**Collapsing Series** A geometric series is one of the few series where we can actually give an explicit formula for $S_n$; a **collapsing series** is another (see Example 2 of Section 5.3).

**EXAMPLE 6** Show that the following series converges and find its sum.

$$\sum_{k=1}^{\infty} \frac{1}{(k+2)(k+3)}$$

*Solution* Use a partial fraction decomposition to write

$$\frac{1}{(k+2)(k+3)} = \frac{1}{k+2} - \frac{1}{k+3}$$

Then,

$$S_n = \sum_{k=1}^{n} \left( \frac{1}{k+2} - \frac{1}{k+3} \right) = \left( \frac{1}{3} - \frac{1}{4} \right) + \left( \frac{1}{4} - \frac{1}{5} \right) + \cdots + \left( \frac{1}{n+2} - \frac{1}{n+3} \right)$$

$$= \frac{1}{3} - \frac{1}{n+3}$$

Therefore,

$$\lim_{n \to \infty} S_n = \tfrac{1}{3}$$

The series converges and has sum $\tfrac{1}{3}$. ∎

**Properties of Convergent Series** Convergent series behave much like finite sums; what you expect to be true usually is true.

---

**Theorem B**

**(Linearity of Convergent Series).** If $\sum\limits_{k=1}^{\infty} a_k$ and $\sum\limits_{k=1}^{\infty} b_k$ both converge and $c$ is a constant, then $\sum\limits_{k=1}^{\infty} ca_k$ and $\sum\limits_{k=1}^{\infty} (a_k + b_k)$ also converge and, moreover,

(i) $\sum\limits_{k=1}^{\infty} ca_k = c \sum\limits_{k=1}^{\infty} a_k$;

(ii) $\sum\limits_{k=1}^{\infty} (a_k + b_k) = \sum\limits_{k=1}^{\infty} a_k + \sum\limits_{k=1}^{\infty} b_k$.

---

*Proof* This theorem introduces a subtle shift in language. The symbol $\sum\limits_{k=1}^{\infty} a_k$ is now being used both for the infinite series $a_1 + a_2 + \cdots$ and for the sum of this series, which is a number.

By hypothesis, $\lim\limits_{n \to \infty} \sum\limits_{k=1}^{n} a_k$ and $\lim\limits_{n \to \infty} \sum\limits_{k=1}^{n} b_k$ both exist. Thus, use the properties of sums with finitely many terms and the properties of limits.

(i) $\displaystyle \sum_{k=1}^{\infty} ca_k = \lim_{n \to \infty} \sum_{k=1}^{n} ca_k = \lim_{n \to \infty} c \sum_{k=1}^{n} a_k$

$$= c \lim_{n \to \infty} \sum_{k=1}^{n} a_k = c \sum_{k=1}^{\infty} a_k$$

**(ii)**
$$\sum_{k=1}^{\infty} (a_k + b_k) = \lim_{n \to \infty} \sum_{k=1}^{n} (a_k + b_k) = \lim_{n \to \infty} \left[ \sum_{k=1}^{n} a_k + \sum_{k=1}^{n} b_k \right]$$

$$= \lim_{n \to \infty} \sum_{k=1}^{n} a_k + \lim_{n \to \infty} \sum_{k=1}^{n} b_k = \sum_{k=1}^{\infty} a_k + \sum_{k=1}^{\infty} b_k \quad \blacksquare$$

**EXAMPLE 7** Calculate $\sum_{k=1}^{\infty} [3 \left(\frac{1}{8}\right)^k - 5 \left(\frac{1}{3}\right)^k]$.

*Solution*   By Theorem B and Example 1,

$$\sum_{k=1}^{\infty} \left[ 3 \left(\frac{1}{8}\right)^k - 5 \left(\frac{1}{3}\right)^k \right] = 3 \sum_{k=1}^{\infty} \left(\frac{1}{8}\right)^k - 5 \sum_{k=1}^{\infty} \left(\frac{1}{3}\right)^k$$

$$= 3 \frac{\frac{1}{8}}{1 - \frac{1}{8}} - 5 \frac{\frac{1}{3}}{1 - \frac{1}{3}} = \frac{3}{7} - \frac{5}{2} = -\frac{29}{14} \quad \blacksquare$$

> **Theorem C**
>
> If $\sum_{k=1}^{\infty} a_k$ diverges and $c \neq 0$, then $\sum_{k=1}^{\infty} c a_k$ diverges.

We leave the proof of this theorem to you (Problem 35). It implies, for example, that

$$\sum_{k=1}^{\infty} \frac{1}{3k} = \sum_{k=1}^{\infty} \frac{1}{3} \cdot \frac{1}{k}$$

diverges, since we know that the harmonic series diverges.

The Associative Property of Addition allows us to group terms in a finite sum in any way we please. For example,

$$2 + 7 + 3 + 4 + 5 = (2 + 7) + (3 + 4) + 5 = 2 + (7 + 3) + (4 + 5)$$

But notice what happens to the infinite series

$$1 - 1 + 1 - 1 + 1 - 1 + \cdots + (-1)^{n+1} + \cdots$$

when we group terms in two different ways.

$$(1 - 1) + (1 - 1) + (1 - 1) + \cdots = 0 + 0 + 0 + \cdots = 0$$
$$1 - (1 - 1) - (1 - 1) - (1 - 1) - \cdots = 1 - 0 - 0 - 0 - \cdots = 1$$

The original series was divergent (since $\lim_{n \to \infty} a_n \neq 0$), whereas the first series of grouped terms converges with sum 0 and the second series of grouped terms converges with sum 1. The contradictions that can arise from grouping are pretty obvious; fortunately, if the original series converges, there are no problems at all.

> **Theorem D**
>
> **(Grouping).** The terms of a convergent series can be grouped in any way (provided the order of the terms is maintained) and the new series will converge with the same sum as the original series.

***Proof*** Let $\sum a_n$ be the original convergent series and let $\{S_n\}$ be its sequence of partial sums. If $\sum b_m$ is a series formed by grouping the terms of $\sum a_n$ and if $\{T_m\}$ is its sequence of partial sums, then each $T_m$ is one of the $S_n$'s. For example, $T_4$ might be

$$T_4 = a_1 + (a_2 + a_3) + (a_4 + a_5 + a_6) + (a_7 + a_8)$$

in which case $T_4 = S_8$. Thus, $\{T_m\}$ is a "subsequence" of $\{S_n\}$. A moment's thought should convince you that if $S_n \to S$, then $T_m \to S$. ∎

## CONCEPTS REVIEW

**1.** An expression of the form $a_1 + a_2 + a_3 + \cdots$ is called _____.

**2.** A series $a_1 + a_2 + \cdots$ is said to converge if the sequence $\{S_n\}$ converges, where $S_n =$ _____.

**3.** The geometric series $a + ar + ar^2 + \cdots$ converges if _____; in this case the sum of the series is _____.

**4.** If $\lim_{n \to \infty} a_n \neq 0$, we can be sure that the series $\sum_{n=1}^{\infty} a_n$ _____.

## PROBLEM SET 11.2

In Problems 1–14, indicate whether the given series converges or diverges. If it converges, find its sum. *Hint*: It may help you to write out the first few terms of the series.

**1.** $\sum_{k=1}^{\infty} \left(\frac{1}{7}\right)^k$

**2.** $\sum_{k=1}^{\infty} \left(-\frac{1}{4}\right)^{-k-2}$

**3.** $\sum_{k=0}^{\infty} \left[2\left(\frac{1}{4}\right)^k + 3\left(-\frac{1}{5}\right)^k\right]$

**4.** $\sum_{k=1}^{\infty} \left[5\left(\frac{1}{2}\right)^k - 3\left(\frac{1}{7}\right)^{k+1}\right]$

**5.** $\sum_{k=1}^{\infty} \frac{k-5}{k+2}$

**6.** $\sum_{k=1}^{\infty} \left(\frac{9}{8}\right)^k$

**7.** $\sum_{k=2}^{\infty} \left(\frac{1}{k} - \frac{1}{k-1}\right)$ *Hint*: Example 6.

**8.** $\sum_{k=1}^{\infty} \frac{3}{k}$

**9.** $\sum_{k=1}^{\infty} \frac{k!}{100^k}$

**10.** $\sum_{k=1}^{\infty} \frac{2}{(k+2)k}$

**11.** $\sum_{k=1}^{\infty} \left(\frac{e}{\pi}\right)^{k+1}$

**12.** $\sum_{k=1}^{\infty} \frac{4^{k+1}}{7^{k-1}}$

**13.** $\sum_{k=2}^{\infty} \left(\frac{3}{(k-1)^2} - \frac{3}{k^2}\right)$

**14.** $\sum_{k=6}^{\infty} \frac{2}{k-5}$

In Problems 15–20, write the given decimal as an infinite series, then find the sum of the series, and finally use the result to write the decimal as a ratio of two integers (see Example 2).

**15.** 0.22222 . . .

**16.** 0.21212121 . . .

**17.** 0.013013013 . . .

**18.** 0.125125125 . . .

**19.** 0.49999 . . .

**20.** 0.36717171 . . .

**21.** Evaluate $\sum_{k=0}^{\infty} r(1-r)^k$, $0 < r < 2$.

**22.** Evaluate $\sum_{k=0}^{\infty} (-1)^k x^k$, $-1 < x < 1$.

**23.** Show that $\sum_{k=1}^{\infty} \ln \frac{k}{k+1}$ diverges. *Hint*: Obtain a formula for $S_n$.

**24.** Show that $\sum_{k=2}^{\infty} \ln\left(1 - \frac{1}{k^2}\right) = -\ln 2$.

**25.** A ball is dropped from a height of 100 feet. Each time it hits the floor, it rebounds to $\frac{2}{3}$ its previous height. Find the total distance it travels.

**26.** Three people, A, B, and C, divide an apple as follows. First they divide it in fourths, each taking a quarter. Then they divide the left over quarter in fourths, each taking a quarter, and so on. Show that each gets a third of the apple.

**27.** Suppose the government pumps an extra $1 billion into the economy. Assume that each business and individual saves 25% of its income and spends the rest, so that of the initial $1 billion, 75% is respent by individuals and businesses. Of that amount, 75% is spent, and so forth. What is the total increase in spending due to the government action? (This is called the *multiplier effect* in economics.)

**28.** Do Problem 27 assuming that only 10% of the income is saved at each stage.

≈ **29.** Assume square $ABCD$ (Figure 3) has sides of length 1 and that $E$, $F$, $G$, and $H$ are midpoints of their respective sides. If the indicated pattern is continued indefinitely, what will be the area of the painted region?

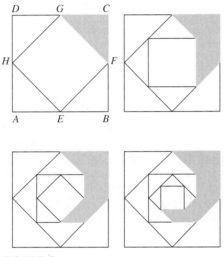

**FIGURE 3**

≈ **30.** If the pattern shown in Figure 4 is continued indefinitely, what fraction of the original square will be painted?

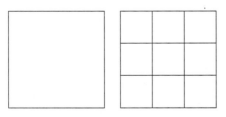

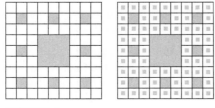

**FIGURE 4**

≈ **31.** Each triangle in the descending chain (Figure 5) has its vertices at the midpoints of the sides of the next larger one. If the indicated pattern of painting is continued indefinitely, what fraction of the area of the original triangle will be painted? Does the original triangle need to be equilateral for this to be true?

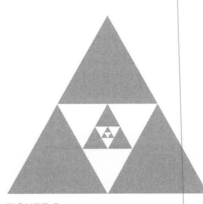

**FIGURE 5**

≈ **32.** Circles are inscribed in the triangles of Problem 31 as indicated in Figure 6. If the original triangle is equilateral, what fraction of the area is painted?

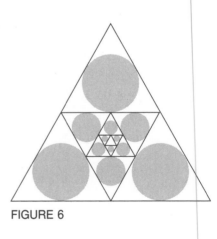

**FIGURE 6**

**33.** In another version of Zeno's paradox, Achilles can run ten times as fast as the tortoise, but the tortoise has a 100-yard headstart. Achilles cannot catch the tortoise, says Zeno, because when he runs 100 yards, the tortoise will have moved 10 yards ahead, when Achilles runs another 10 yards, the tortoise will have moved 1 yard ahead, and so on. Convince Zeno that Achilles will catch the tortoise and tell him exactly how many yards Achilles will have to run to do it.

**34.** Tom and Joel are good runners, both able to run at a constant speed of 10 miles per hour. Their amazing dog Trot can do even better; he runs at 20 miles per hour. Starting from towns 60 miles apart, Tom and Joel run toward each other while Trot runs back and forth between them. How far does Trot run by the time the boys meet? Assume Trot started with Tom running toward Joel and that he is able to make instant turnarounds. Solve the problem two ways.

(a) Use a geometric series.
(b) Find a trivial way to do the problem.

**35.** Prove: If $\sum_{k=1}^{\infty} a_k$ diverges, so does $\sum_{k=1}^{\infty} ca_k$ for $c \neq 0$.

**36.** Use Problem 35 to conclude that $\frac{1}{2} + \frac{1}{4} + \frac{1}{6} + \frac{1}{8} + \cdots$ diverges.

**37.** Suppose one has an unlimited supply of identical blocks each 1 unit long.

(a) Convince yourself that they may be stacked as in Figure 7 without toppling. *Hint*: Consider centers of mass.
(b) How far can one make the top block protrude to the right of the bottom block using this method of stacking?

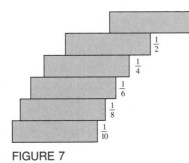

**FIGURE 7**

[C] **38.** How large must $N$ be in order for $S_N = \sum_{k=1}^{N} (1/k)$

just to exceed 4? *Note*: Computer calculations show that for $S_N$ to exceed 20, $N = 272{,}400{,}600$, and for $S_N$ to exceed 100, $N \approx 1.5 \times 10^{43}$.

**39.** Prove that if $\sum a_n$ diverges and $\sum b_n$ converges, then $\sum (a_n + b_n)$ diverges.

**40.** Show that it is possible for $\sum a_n$ and $\sum b_n$ both to diverge, and yet for $\sum (a_n + b_n)$ to converge.

**41.** By looking at the region in Figure 8 first vertically and then horizontally, conclude that

$$1 + \tfrac{1}{2} + \tfrac{1}{4} + \tfrac{1}{8} + \cdots = \tfrac{1}{2} + \tfrac{2}{4} + \tfrac{3}{8} + \tfrac{4}{16} + \cdots$$

and use this fact to calculate:

(a) $\displaystyle\sum_{k=1}^{\infty} \frac{k}{2^k}$

(b) $\bar{x}$, the horizontal coordinate of the centroid of the region

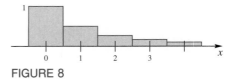

**FIGURE 8**

**42.** Let $r$ be a fixed number with $|r| < 1$. Then it can be shown that $\displaystyle\sum_{k=1}^{\infty} kr^k$ converges, say with sum $S$. Use the properties of $\sum$ to show that

$$(1 - r)S = \sum_{k=1}^{\infty} r^k$$

and then obtain a formula for $S$ thus generalizing Problem 41(a).

**43.** Many drugs are eliminated from the body in an exponential manner. Thus, if a drug is given in dosages of size $C$ at time intervals of length $t$, the amount $A_n$ of the drug in the body just after the $(n + 1)$st dose is

$$A_n = C + Ce^{-kt} + Ce^{-2kt} + \cdots + Ce^{-nkt}$$

where $k$ is a positive constant that depends on the type of drug.

(a) Derive a formula for $A$, the amount of drug in the body just after a dose if a person has been on the drug for a very long time (assume an infinitely long time).
(b) Evaluate $A$ if it is known that one-half of a dose is eliminated from the body in 6 hours and doses of size 2 milligrams are given every 12 hours.

**44.** Find the sum of the series

$$\sum_{k=1}^{\infty} \frac{2^k}{(2^{k+1} - 1)(2^k - 1)}$$

**45.** Evaluate $\displaystyle\sum_{k=1}^{\infty} \frac{1}{f_k f_{k+2}}$, where $\{f_k\}$ is the Fibonacci sequence introduced in Problem 52 of Section 11.1. *Hint*: First show

$$\frac{1}{f_k f_{k+2}} = \frac{1}{f_k f_{k+1}} - \frac{1}{f_{k+1} f_{k+2}}$$

**Answers to Concepts Review:** **1.** An infinite series **2.** $a_1 + a_2 + a_3 + \cdots + a_n$ **3.** $|r| < 1$; $a/(1 - r)$ **4.** Diverges

## 11.3
## POSITIVE SERIES:
## THE INTEGRAL TEST

---

**Important Reminders**

$$a_1, a_2, a_3, \ldots$$

is a *sequence*.

$$a_1 + a_2 + a_3 + \cdots$$

is a *series*.

$$S_n = a_1 + a_2 + a_3 + \cdots + a_n$$

is the $n$th *partial sum* of the series.

$$S_1, S_2, S_3, \ldots$$

is the *sequence of partial sums* of the series. The series *converges* if and only if

$$S = \lim_{n \to \infty} S_n$$

exists and is finite, in which case $S$ is called the *sum* of series.

---

We introduced some very big ideas in Section 11.2, but we illustrated them mainly for two very special types of series, namely, geometric series and collapsing series. These are series for which we can give exact formulas for the partial sums $S_n$, something that we can rarely do for most other types of series. Our task now is to begin a study of very general infinite series.

There are always two important questions to ask about a series.

**1.** Does the series converge?
**2.** If it converges, what is its sum?

How shall we answer these questions? Someone is sure to suggest that we use a computer. To answer the first question, simply add up more and more terms of the series watching the numbers you get as partial sums. If these numbers seem to settle down on a fixed number $S$, the series converges. And in that case, $S$ is the sum of the series, answering the second question. This response is plain wrong for question 1 and only partially adequate for question 2. Let us see why.

Consider the harmonic series

$$1 + \tfrac{1}{2} + \tfrac{1}{3} + \tfrac{1}{4} + \cdots$$

introduced in Section 11.2 and discussed in Example 5 and Problem 38 of that section. We know that this series diverges, but a computer would not help us to discover this fact. The partial sums $S_n$ of this series grow without bound, but they grow so slowly that it takes over 272 million terms for $S_n$ to reach 20 and over $10^{43}$ terms for $S_n$ to reach 100. Because of the inherent limitation in the number of digits it can handle, a computer would rather soon give repeated values for $S_n$, suggesting wrongly that the $S_n$'s were converging. What is true for the harmonic series is true for any slowly diverging series. We state it emphatically: A computer is no substitute for good mathematical tests of convergence and divergence, a subject to which we now turn.

In this and the next section, we restrict our attention to series with positive (or at least nonnegative) terms. With this restriction, we will be able to give some remarkably simple convergence tests. Tests for series with terms of arbitrary sign will follow in Section 11.5.

**Bounded Partial Sums**    Our first result flows directly from the Monotonic Sequence Theorem.

---

### Theorem A

**(Bounded-Sum Test).** A series $\sum a_k$ of nonnegative terms converges if and only if its partial sums are bounded above.

---

***Proof***    As usual, let $S_n = a_1 + a_2 + \cdots + a_n$. Since $a_k \geq 0$, $S_{n+1} \geq S_n$; that is, $\{S_n\}$ is a nondecreasing sequence. Thus, by Theorem 11.1D, the sequence $\{S_n\}$ will converge provided there is a number $U$ such that $S_n \leq U$ for all $n$. Otherwise, the $S_n$'s will grow without bound, in which case $\{S_n\}$ diverges. ∎

**EXAMPLE 1**  Show that the series $\dfrac{1}{1!} + \dfrac{1}{2!} + \dfrac{1}{3!} + \cdots$ converges.

**Solution**  We aim to show that the partial sums $S_n$ are bounded above. Note first that

$$n! = 1 \cdot 2 \cdot 3 \cdots n \geq 1 \cdot 2 \cdot 2 \cdots 2 = 2^{n-1}$$

and so $1/n! \leq 1/2^{n-1}$. Thus,

$$S_n = \frac{1}{1!} + \frac{1}{2!} + \frac{1}{3!} + \cdots + \frac{1}{n!}$$

$$\leq 1 + \frac{1}{2} + \frac{1}{4} + \cdots + \frac{1}{2^{n-1}}$$

These latter terms come from a geometric series with $r = \frac{1}{2}$. They can be added by a formula in Example 1 of Section 11.2. We obtain

$$S_n \leq \frac{1 - \left(\frac{1}{2}\right)^n}{1 - \frac{1}{2}} = 2\left[1 - \left(\frac{1}{2}\right)^n\right] < 2$$

Thus, by the Bounded-Sum Test, the given series converges. The argument also shows that its sum $S$ is at most 2. Later we will show that $S = e - 1 \approx 1.71828$.  ∎

**Series and Improper Integrals**  The behavior of $\displaystyle\sum_{k=1}^{\infty} f(k)$ and $\displaystyle\int_{1}^{\infty} f(x)\, dx$ with respect to convergence is similar and gives a very powerful test.

---

Theorem B

**(Integral Test).** Let $f$ be a continuous, positive, nonincreasing function on the interval $[1, \infty)$ and suppose $a_k = f(k)$ for all positive integers $k$. Then the infinite series

$$\sum_{k=1}^{\infty} a_k$$

converges if and only if the improper integral

$$\int_{1}^{\infty} f(x)\, dx$$

converges.

---

We remark that the integer 1 may be replaced by any positive integer $M$ throughout this theorem (see Example 4).

**Proof**   The diagrams in Figure 1 indicate how we may interpret the partial sums of the series $\Sigma a_k$ as areas and thereby relate the series to a corresponding integral. Note that the area of each rectangle is equal to its height, since the width is 1 in each case. From these diagrams, we easily see that

$$\sum_{k=2}^{n} a_k \le \int_{1}^{n} f(x)\,dx \le \sum_{k=1}^{n-1} a_k$$

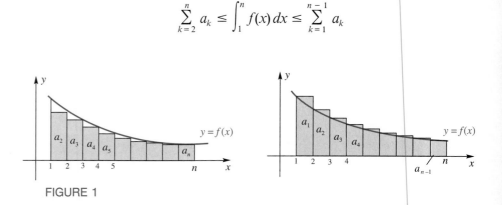

FIGURE 1

Now suppose $\int_{1}^{\infty} f(x)\,dx$ converges. Then, by the left inequality above,

$$S_n = a_1 + \sum_{k=2}^{n} a_k \le a_1 + \int_{1}^{n} f(x)\,dx \le a_1 + \int_{1}^{\infty} f(x)\,dx$$

Therefore, by the Bounded-Sum Test, $\displaystyle\sum_{k=1}^{\infty} a_k$ converges.

On the other hand, suppose $\displaystyle\sum_{k=1}^{\infty} a_k$ converges. Then, by the right inequality above, if $t \le n$,

$$\int_{1}^{t} f(x)\,dx \le \int_{1}^{n} f(x)\,dx \le \sum_{k=1}^{n-1} a_k \le \sum_{k=1}^{\infty} a_k$$

Since $\int_{1}^{t} f(x)\,dx$ increases with $t$ and is bounded above, $\displaystyle\lim_{t \to \infty} \int_{1}^{t} f(x)\,dx$ must exist; that is, $\int_{1}^{\infty} f(x)\,dx$ converges.   ∎

The conclusion to Theorem B is often stated this way. *The series $\displaystyle\sum_{k=1}^{\infty} f(k)$ and the improper integral $\int_{1}^{\infty} f(x)\,dx$ converge or diverge together.* You should see that this is equivalent to our statement.

**EXAMPLE 2   (p-Series Test).** The series

$$\sum_{k=1}^{\infty} \frac{1}{k^p} = 1 + \frac{1}{2^p} + \frac{1}{3^p} + \frac{1}{4^p} + \cdots$$

where $p$ is a constant, is called a **$p$-series.** Show each of the following.
(a) The $p$-series converges if $p > 1$.
(b) The $p$-series diverges if $p \leq 1$.

*Solution*　If $p \geq 0$, the function $f(x) = 1/x^p$ is continuous, positive, and nonincreasing on $[1, \infty)$ and $f(k) = 1/k^p$. Thus, by the Integral Test, $\sum (1/k^p)$ converges if and only if $\lim\limits_{t \to \infty} \int_1^t x^{-p} \, dx$ exists (as a finite number).

If $p \neq 1$,

$$\int_1^t x^{-p} \, dx = \left[ \frac{x^{1-p}}{1-p} \right]_1^t = \frac{t^{1-p} - 1}{1 - p}$$

If $p = 1$,

$$\int_1^t x^{-1} \, dx = [\ln x]_1^t = \ln t$$

Since $\lim\limits_{t \to \infty} t^{1-p} = 0$ if $p > 1$ and $\lim\limits_{t \to \infty} t^{1-p} = \infty$ if $p < 1$ and since $\lim\limits_{t \to \infty} \ln t = \infty$, we conclude that the $p$-series converges if $p > 1$ and diverges if $0 \leq p \leq 1$.

We still have the case $p < 0$ to consider. In that case, the $n$th term of $\sum (1/k^p)$, namely, $1/n^p$ does not even tend toward 0. Thus by the $n$th-Term Test, the series diverges.

Note that the case $p = 1$ gives the harmonic series, which was treated in Section 11.2. Our results here and there are consistent. The harmonic series diverges. ∎

**EXAMPLE 3**　Does $\sum\limits_{k=4}^{\infty} \dfrac{1}{k^{1.001}}$ converge or diverge?

*Solution*　By the $p$-Series Test, $\sum\limits_{k=1}^{\infty} (1/k^{1.001})$ converges. *The insertion or deletion of a finite number of terms in a series cannot affect its convergence or divergence* (*though it may affect the sum*). Thus, the given series converges. ∎

---

### The Tail of a Series

The beginning of a series plays no role in its convergence or divergence. Only the tail is important (the tail really does wag the dog). By the *tail* of a series, we mean

$$a_N + a_{N+1} + a_{N+2} + \cdots$$

where $N$ denotes an arbitrarily large number. Hence, in testing for convergence or divergence of a series, we can ignore the beginning terms or even change them. Clearly, however, the sum of a series does depend on all its terms, including the initial ones.

---

**EXAMPLE 4**　Determine whether $\sum\limits_{k=2}^{\infty} \dfrac{1}{k \ln k}$ converges or diverges.

*Solution*　The hypotheses of the Integral Test are satisfied for $f(x) = 1/(x \ln x)$ on $[2, \infty)$. That the interval is $[2, \infty)$ rather than $[1, \infty)$ is inconsequential, as we noted right after Theorem B. Now,

$$\int_2^{\infty} \frac{1}{x \ln x} \, dx = \lim_{t \to \infty} \int_2^t \frac{1}{\ln x} \, d(\ln x) = \lim_{t \to \infty} [\ln \ln x]_2^t = \infty$$

Thus, $\sum 1/(k \ln k)$ diverges. ∎

**EXAMPLE 5** By means of an improper integral, find a good upper bound for the error in using the sum of the first five terms of the convergent series

$$\sum_{n=1}^{\infty} \frac{n}{e^{n^2}}$$

to approximate the sum of the series.

**Solution** The error $E$ is

$$E = \sum_{n=6}^{\infty} \frac{n}{e^{n^2}}$$

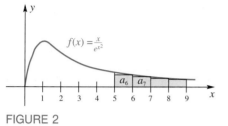

FIGURE 2

The function $f(x) = x/e^{x^2}$ is continuous, positive, and nonincreasing on $[5, \infty)$ (see Figure 2). Thus,

$$E = \sum_{n=6}^{\infty} \frac{n}{e^{n^2}} < \int_5^{\infty} xe^{-x^2}\, dx$$

$$= \lim_{t \to \infty} -\frac{1}{2} \int_5^t e^{-x^2}(-2x\, dx)$$

$$= \lim_{t \to \infty} -\frac{1}{2} \left[ e^{-x^2} \right]_5^t = \frac{1}{2} e^{-25} \approx 6.94 \times 10^{-12}$$ ∎

## CONCEPTS REVIEW

**1.** A series of nonnegative terms converges if and only if its partial sums are _____ .

**2.** The Integral Test relates the convergence of $\displaystyle\sum_{k=1}^{\infty} a_k$ and $\displaystyle\int_1^{\infty} f(x)\, dx$, assuming $a_k =$ _____ and $f$ is _____ , _____ , and _____ on $[1, \infty)$.

**3.** The insertion or deletion of a finite number of terms in a series does not affect its _____ , although it may affect its sum.

**4.** The $p$-series $\displaystyle\sum_{k=1}^{\infty} (1/k^p)$ converges if and only if _____ .

## PROBLEM SET 11.3

Use the Integral Test to decide the convergence or divergence of each of the following series.

**1.** $\displaystyle\sum_{k=0}^{\infty} \frac{1}{k + 3}$

**2.** $\displaystyle\sum_{k=1}^{\infty} \frac{3}{2k - 3}$

**3.** $\displaystyle\sum_{k=0}^{\infty} \frac{k}{k^2 + 3}$

**4.** $\displaystyle\sum_{k=1}^{\infty} \frac{3}{2k^2 + 1}$

**5.** $\displaystyle\sum_{k=1}^{\infty} \frac{-2}{\sqrt{k + 2}}$

**6.** $\displaystyle\sum_{k=100}^{\infty} \frac{3}{(k + 2)^2}$

**7.** $\displaystyle\sum_{k=2}^{\infty} \frac{7}{4k + 2}$

**8.** $\displaystyle\sum_{k=1}^{\infty} \frac{k^2}{e^k}$

**9.** $\displaystyle\sum_{k=1}^{\infty} \frac{3}{(4 + 3k)^{7/6}}$

**10.** $\displaystyle\sum_{k=1}^{\infty} \frac{1000k^2}{1 + k^3}$

**11.** $\displaystyle\sum_{k=1}^{\infty} ke^{-3k^2}$

**12.** $\displaystyle\sum_{k=5}^{\infty} \frac{1000}{k(\ln k)^2}$

In Problems 13–22, use any test developed so far including those in Section 11.2 to decide about the convergence or divergence of the series. Give a reason for your conclusion.

**13.** $\displaystyle\sum_{k=1}^{\infty} \frac{k^2 + 1}{k^2 + 5}$

**14.** $\displaystyle\sum_{k=1}^{\infty} \left(\frac{3}{\pi}\right)^k$

**15.** $\displaystyle\sum_{k=1}^{\infty} \left[\left(\frac{1}{2}\right)^k + \frac{k - 1}{2k + 1}\right]$

**16.** $\displaystyle\sum_{k=1}^{\infty} \left(\frac{1}{k^2} + \frac{1}{2^k}\right)$

17. $\displaystyle\sum_{k=1}^{\infty} \sin\left(\frac{k\pi}{2}\right)$

18. $\displaystyle\sum_{k=1}^{\infty} k \sin\frac{1}{k}$

19. $\displaystyle\sum_{k=1}^{\infty} k^2 e^{-k^3}$

20. $\displaystyle\sum_{k=1}^{\infty} \left(\frac{1}{k} - \frac{1}{k+1}\right)$

21. $\displaystyle\sum_{k=1}^{\infty} \frac{\tan^{-1} k}{1 + k^2}$

22. $\displaystyle\sum_{k=1}^{\infty} \frac{1}{1 + 4k^2}$

In Problems 23–26, estimate the error that is made by approximating the sum of the given series by the sum of the first five terms (see Example 5).

23. $\displaystyle\sum_{k=1}^{\infty} \frac{k}{e^k}$

24. $\displaystyle\sum_{k=1}^{\infty} \frac{1}{k\sqrt{k}}$

25. $\displaystyle\sum_{k=1}^{\infty} \frac{1}{1 + k^2}$

26. $\displaystyle\sum_{k=1}^{\infty} \frac{1}{k(k+1)} = \sum_{k=1}^{\infty} \left(\frac{1}{k} - \frac{1}{k+1}\right)$

27. For what values of $p$ does $\displaystyle\sum_{n=2}^{\infty} 1/[n(\ln n)^p]$ converge?

28. Does $\displaystyle\sum_{n=3}^{\infty} 1/[n \cdot \ln n \cdot \ln(\ln n)]$ converge or diverge?

29. Use diagrams, as in Figure 1, to show

$$\ln(n + 1) < 1 + \frac{1}{2} + \frac{1}{3} + \cdots + \frac{1}{n} < 1 + \ln n$$

*Hint:* $\displaystyle\int_1^n (1/x)\,dx = \ln n.$

30. Show, using Problem 29, that the sequence

$$B_n = 1 + \frac{1}{2} + \frac{1}{3} + \cdots + \frac{1}{n} - \ln(n + 1)$$

is increasing and bounded above by 1.

31. From Problem 30, conclude that $\lim_{n \to \infty} B_n$ exists; call it $\gamma$ ($\gamma$ is known as *Euler's constant*). It is easy to show that $\gamma \approx 0.5772$. Decide whether $\gamma$ is rational or irrational (but before you work too hard on this, consult the answer key).

32. Use Problem 29 to get good upper and lower bounds for the sum of the first 10 million terms of the harmonic series.

33. From Problem 31, we infer that

$$1 + \frac{1}{2} + \frac{1}{3} + \cdots + \frac{1}{n} \approx \gamma + \ln(n + 1)$$

Use this to estimate the number of terms of the harmonic series that are needed to get a sum greater than 20 and compare with the result reported in Problem 38 of Section 11.2.

34. Now that we have shown the existence of Euler's constant the hard way (Problems 29–31), we will solve a much more general problem the easy way and watch $\gamma$ appear out of thin air, so to speak. Let $f$ be continuous and decreasing on $[1, \infty)$ and let

$$B_n = f(1) + f(2) + \cdots + f(n) - \int_1^{n+1} f(x)\,dx$$

Note that $B_n$ is the area of the shaded region in Figure 3.

(a) Why is it obvious that $B_n$ increases with $n$?
(b) Show that $B_n \le f(1)$. *Hint:* Simply shift all the little shaded pieces leftward into the outlined rectangle.
(c) Conclude that $\lim_{n \to \infty} B_n$ exists.
(d) How do we get $\gamma$ out of this?

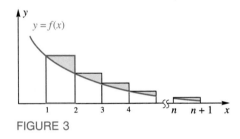

FIGURE 3

35. Let $f$ be continuous, increasing, and concave down on $[1, \infty)$ as in Figure 4. Further, let $A_n$ be the area of the shaded region. Show that $A_n$ is increasing with $n$, that $A_n \le T$ where $T$ is the area of the heavily outlined triangle, and thus that $\lim_{n \to \infty} A_n$ exists.

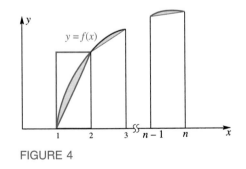

FIGURE 4

36. Specialize $f$ of Problem 35 to $f(x) = \ln x$.

(a) Show that

$$A_n = \int_1^n \ln x\,dx - \left[\frac{\ln 1 + \ln 2}{2} + \cdots + \frac{\ln(n-1) + \ln n}{2}\right]$$

$$= n \ln n - n + 1 - \ln n! + \ln \sqrt{n}$$

$$= 1 + \ln \frac{(n/e)^n \sqrt{n}}{n!}$$

(b) Conclude from (a) and Problem 35 that

$$k = \lim_{n \to \infty} \frac{n!}{(n/e)^n \sqrt{n}}$$

exists. It can be shown that $k = \sqrt{2\pi}$.

(c) This means that $n! \approx \sqrt{2\pi n}(n/e)^n$, which is called

**Stirling's Formula.** Use it to approximate $15!$ and compare it with the value your calculator gives for $15!$

---

**Answers to Concepts Review:** **1.** Bounded above **2.** $f(k)$; continuous; positive; nonincreasing **3.** Convergence or divergence **4.** $p > 1$

---

## 11.4
## POSITIVE SERIES: OTHER TESTS

We have completely analyzed the convergence and divergence of two series—the geometric series and the $p$-series.

$$\sum_{n=1}^{\infty} r^n \text{ converges if } -1 < r < 1, \text{ diverges otherwise}$$

$$\sum_{n=1}^{\infty} \frac{1}{n^p} \text{ converges if } p > 1, \text{ diverges otherwise}$$

These series provide standards, or models, against which we can measure other series. Keep in mind that we are still considering series whose terms are positive (or at least nonnegative).

**Comparing One Series with Another** A series with terms less than the corresponding terms of a convergent series ought to converge; a series with terms greater than the corresponding terms of a divergent series ought to diverge. What ought to be true is true.

### Theorem A

**(Ordinary Comparison Test).** Suppose $0 \le a_n \le b_n$ for $n \ge N$.

**(i)** If $\sum b_n$ converges, so does $\sum a_n$.
**(ii)** If $\sum a_n$ diverges, so does $\sum b_n$.

**Proof** We suppose $N = 1$; the case $N > 1$ is only slightly harder. To prove (i), let $S_n = a_1 + a_2 + \cdots + a_n$ and note that $\{S_n\}$ is a nondecreasing sequence. If $\sum b_n$ converges—for instance, with sum $B$—then

$$S_n \le b_1 + b_2 + \cdots + b_n \le \sum_{n=1}^{\infty} b_n = B$$

By the Bounded-Sum Test (Theorem 11.3A), $\sum a_n$ converges.

Property (ii) follows from (i); for if $\sum b_n$ converged, then $\sum a_n$ would have to converge. ∎

**EXAMPLE 1**   Does $\sum \dfrac{n}{5n^2 - 4}$ converge or diverge?

*Solution*   A good guess would be that it diverges, since the $n$th term behaves like $1/5n$ for large $n$. In fact,

$$\frac{n}{5n^2 - 4} > \frac{n}{5n^2} = \frac{1}{5} \cdot \frac{1}{n}$$

We know $\sum \dfrac{1}{5} \cdot \dfrac{1}{n}$ diverges since it is one-fifth of the harmonic series (Theorem 11.2C). Thus, by the Ordinary Comparison Test, the given series also diverges. ∎

**EXAMPLE 2**   Does $\sum \dfrac{n}{2^n(n + 1)}$ converge or diverge?

*Solution*   A good guess would be that it converges, since the $n$th term behaves like $(1/2)^n$ for large $n$. To substantiate our guess, we note that

$$\frac{n}{2^n(n + 1)} = \left(\frac{1}{2}\right)^n \frac{n}{n + 1} < \left(\frac{1}{2}\right)^n$$

Since $\sum \left(\frac{1}{2}\right)^n$ converges—it is a geometric series with $r = \frac{1}{2}$—we conclude that the given series converges. ∎

If there is a problem in applying the Ordinary Comparison Test, it is in finding exactly the right known series with which to compare the series to be tested. Suppose we wish to determine the convergence or divergence of

$$\sum_{n=3}^{\infty} \frac{1}{(n - 2)^2} = \sum_{n=3}^{\infty} \frac{1}{n^2 - 4n + 4}$$

Our inclination is to compare $1/(n - 2)^2$ with $1/n^2$, but unfortunately

$$\frac{1}{(n - 2)^2} > \frac{1}{n^2}$$

which gives no test at all (the inequality goes the wrong way for what we want). After some experimenting, we discover that

$$\frac{1}{(n - 2)^2} \leq \frac{9}{n^2}$$

for $n \geq 3$; since $\sum 9/n^2$ converges, so does $\sum 1/(n - 2)^2$.

Can we avoid these contortions with inequalities? Our intuition tells us that $\sum a_n$ and $\sum b_n$ converge or diverge together, provided $a_n$ and $b_n$ are approximately the same size for large $n$ (give or take a multiplicative constant). That is the essential content of our next theorem.

Theorem B

**(Limit Comparison Test).** Suppose $a_n \geq 0$, $b_n > 0$, and

$$\lim_{n \to \infty} \frac{a_n}{b_n} = L$$

If $0 < L < \infty$, then $\sum a_n$ and $\sum b_n$ converge or diverge together. If $L = 0$ and $\sum b_n$ converges, then $\sum a_n$ converges.

**Proof** Begin by taking $\varepsilon = L/2$ in the definition of limit of a sequence (Section 11.1). There is a number $N$ such that $n \geq N \Rightarrow |(a_n/b_n) - L| < L/2$; that is,

$$-\frac{L}{2} < \frac{a_n}{b_n} - L < \frac{L}{2}$$

This inequality is equivalent (by adding $L$ throughout) to

$$\frac{L}{2} < \frac{a_n}{b_n} < \frac{3L}{2}$$

Hence for $n \geq N$,

$$b_n < \frac{2}{L} a_n \quad \text{and} \quad a_n < \frac{3L}{2} b_n$$

These two inequalities, together with the Ordinary Comparison Test, show that $\sum a_n$ and $\sum b_n$ converge or diverge together. We leave the proof of the final statement of the theorem to the reader (Problem 37). ∎

**EXAMPLE 3** Determine the convergence or divergence of each series.

(a) $\displaystyle\sum_{n=1}^{\infty} \frac{3n - 2}{n^3 - 2n^2 + 11}$    (b) $\displaystyle\sum_{n=1}^{\infty} \frac{1}{\sqrt{n^2 + 19n}}$

*Solution* We apply the Limit Comparison Test, but we still must decide to what we should compare the $n$th term. We see what the $n$th term is like for large $n$ by looking at the largest-degree terms in numerator and denominator. In the first case, the $n$th term is like $3/n^2$; in the second, it is like $1/n$.

(a) $\displaystyle\lim_{n \to \infty} \frac{a_n}{b_n} = \lim_{n \to \infty} \frac{(3n - 2)/(n^3 - 2n^2 + 11)}{3/n^2} = \lim_{n \to \infty} \frac{3n^3 - 2n^2}{3n^3 - 6n^2 + 33} = 1$

(b) $\displaystyle\lim_{n \to \infty} \frac{a_n}{b_n} = \lim_{n \to \infty} \frac{1/\sqrt{n^2 + 19n}}{1/n} = \lim_{n \to \infty} \sqrt{\frac{n^2}{n^2 + 19n}} = 1$

We conclude that the series in (a) converges, and that the series in (b) diverges. ∎

**EXAMPLE 4**   Does $\displaystyle\sum_{n=1}^{\infty} \frac{\ln n}{n^2}$ converge or diverge?

*Solution*   To what shall we compare $(\ln n)/n^2$? If we try $1/n^2$, we get

$$\lim_{n \to \infty} \frac{a_n}{b_n} = \lim_{n \to \infty} \frac{\ln n}{n^2} \div \frac{1}{n^2} = \lim_{n \to \infty} \ln n = \infty$$

The test fails because its conditions are not satisfied. On the other hand, if we use $1/n$, we get

$$\lim_{n \to \infty} \frac{a_n}{b_n} = \lim_{n \to \infty} \frac{\ln n}{n^2} \div \frac{1}{n} = \lim_{n \to \infty} \frac{\ln n}{n} = 0$$

Again, the test fails. Possibly something between $1/n^2$ and $1/n$ will work, such as $1/n^{3/2}$.

$$\lim_{n \to \infty} \frac{a_n}{b_n} = \lim_{n \to \infty} \frac{\ln n}{n^2} \div \frac{1}{n^{3/2}} = \lim_{n \to \infty} \frac{\ln n}{\sqrt{n}} = 0$$

(The last equality follows from l'Hôpital's Rule.) We conclude from the second part of the Limit Comparison Test that $\sum (\ln n)/n^2$ converges (since $\sum 1/n^{3/2}$ converges). ∎

**Comparing a Series with Itself**   To apply the comparison tests requires real insight. We must choose wisely among known series to find one that is just right for comparison with the series we wish to test. Wouldn't it be nice if we could somehow compare a series with itself and thereby determine convergence or divergence? Roughly speaking, this is what we do in the Ratio Test.

---

Theorem C

**(Ratio Test).** Let $\sum a_n$ be a series of positive terms and suppose

$$\lim_{n \to \infty} \frac{a_{n+1}}{a_n} = \rho$$

**(i)**   If $\rho < 1$, the series converges.
**(ii)**  If $\rho > 1$, the series diverges.
**(iii)** If $\rho = 1$, the test is inconclusive.

---

*Proof*   Here is what is behind the Ratio Test. Since $\lim_{n \to \infty} a_{n+1}/a_n = \rho, a_{n+1} \approx \rho a_n$; that is, the series behaves like a geometric series with ratio $\rho$. A geometric series converges when its ratio is less than 1 and diverges when its ratio is greater than 1. Tying down this argument is the task before us.

**(i)**   Since $\rho < 1$, we may choose a number $r$ such that $\rho < r < 1$—for example, $r = (\rho + 1)/2$. Next choose $N$ so large that $n \geq N$ implies $a_{n+1}/a_n < r$.

Then,

$$\begin{aligned}
a_{N+1} &< ra_N \\
a_{N+2} &< ra_{N+1} < r^2 a_N \\
a_{N+3} &< ra_{N+2} < r^3 a_N \\
&\vdots
\end{aligned}$$

Since $ra_N + r^2 a_N + r^3 a_N + \cdots$ is a geometric series with $0 < r < 1$, it converges. By the Ordinary Comparison Test, $\sum_{n=N+1}^{\infty} a_n$ converges, and hence so does $\sum_{n=1}^{\infty} a_n$.

(ii) Since $\rho > 1$, there is a number $N$ such that $a_{n+1}/a_n > 1$ for all $n \geq N$. Thus

$$\begin{aligned}
a_{N+1} &> a_N \\
a_{N+2} &> a_{N+1} > a_N \\
&\vdots
\end{aligned}$$

Hence, $a_n > a_N > 0$ for all $n > N$, which means that $\lim_{n \to \infty} a_n$ cannot be zero. By the $n$th-Term Test for Divergence, $\sum a_n$ diverges.

(iii) We know that $\sum 1/n$ diverges, whereas $\sum 1/n^2$ converges. For the first series,

$$\lim_{n \to \infty} \frac{a_{n+1}}{a_n} = \lim_{n \to \infty} \frac{1}{n+1} \div \frac{1}{n} = \lim_{n \to \infty} \frac{n}{n+1} = 1$$

For the second series,

$$\lim_{n \to \infty} \frac{a_{n+1}}{a_n} = \lim_{n \to \infty} \frac{1}{(n+1)^2} \div \frac{1}{n^2} = \lim_{n \to \infty} \frac{n^2}{(n+1)^2} = 1$$

Thus, the Ratio Test does not distinguish between convergence and divergence when $\rho = 1$. ∎

The Ratio Test will always fail for a series whose $n$th term is a rational expression in $n$, since in this case $\rho = 1$ (the cases $a_n = 1/n$ and $a_n = 1/n^2$ were treated above). However for a series whose $n$th term involves $n!$ or $r^n$, the Ratio Test usually works beautifully.

**EXAMPLE 5**  Test for convergence or divergence: $\sum_{n=1}^{\infty} \dfrac{2^n}{n!}$.

*Solution*

$$\rho = \lim_{n \to \infty} \frac{a_{n+1}}{a_n} = \lim_{n \to \infty} \frac{2^{n+1}}{(n+1)!} \frac{n!}{2^n} = \lim_{n \to \infty} \frac{2}{n+1} = 0$$

We conclude by the Ratio Test that the series converges. ∎

**EXAMPLE 6**  Test for convergence or divergence: $\sum_{n=1}^{\infty} \dfrac{2^n}{n^{20}}$.

*Solution*

$$\rho = \lim_{n \to \infty} \frac{a_{n+1}}{a_n} = \lim_{n \to \infty} \frac{2^{n+1}}{(n+1)^{20}} \frac{n^{20}}{2^n}$$

$$= \lim_{n \to \infty} \left( \frac{n}{n+1} \right)^{20} \cdot 2 = 2$$

We conclude that the given series diverges. ■

**EXAMPLE 7** Test for convergence or divergence: $\sum_{n=1}^{\infty} \frac{n!}{n^n}$.

*Solution* We will need the fact that

$$\lim_{n \to \infty} \left( 1 + \frac{1}{n} \right)^n = e$$

which follows from Theorem 7.5A. Taking this as known, we may write

$$\rho = \lim_{n \to \infty} \frac{a_{n+1}}{a_n} = \lim_{n \to \infty} \frac{(n+1)!}{(n+1)^{n+1}} \frac{n^n}{n!} = \lim_{n \to \infty} \left( \frac{n}{n+1} \right)^n$$

$$= \lim_{n \to \infty} \frac{1}{((n+1)/n)^n} = \lim_{n \to \infty} \frac{1}{(1+1/n)^n} = \frac{1}{e} < 1$$

Therefore, the given series converges. ■

**Summary** To test a series $\sum a_n$ of positive terms for convergence or divergence, look carefully at $a_n$.

---

1. If $\lim_{n \to \infty} a_n \neq 0$, conclude from the $n$th-Term Test that the series diverges.
2. If $a_n$ involves $n!$, $r^n$, or $n^n$, try the Ratio Test.
3. If $a_n$ involves only constant powers of $n$, try the Limit Comparison Test. In particular, if $a_n$ is a rational expression in $n$, use this test with $b_n$ as the quotient of the leading terms from numerator and denominator.
4. If the tests above do not work, try the Ordinary Comparison Test, the Integral Test, or the Bounded-Sum Test.
5. Some series require a clever manipulation or a neat trick to determine convergence or divergence.

---

## CONCEPTS REVIEW

**1.** The Ordinary Comparison Test says that if _____ and if $\sum b_k$ converges, than $\sum a_k$ also converges.

**2.** Assume $a_k \geq 0$ and $b_k > 0$. The Limit Comparison Test says that if $0 < $ _____ $< \infty$, then $\sum a_k$ and $\sum b_k$ converge or diverge together.

**3.** The Ratio Test says that a series $\sum a_k$ of positive terms converges if _____, diverges if _____, and may do either if _____. Here $\rho =$ _____.

**4.** $\sum (3^k/k!)$ is an obvious candidate for the _____ Test, whereas $\sum k/(k^3 - k - 1)$ is an obvious candidate for the _____ Test.

## PROBLEM SET 11.4

In Problems 1–4, use the Limit Comparison Test to determine convergence or divergence.

**1.** $\displaystyle\sum_{n=1}^{\infty} \frac{n}{n^2 + 2n + 3}$    **2.** $\displaystyle\sum_{n=1}^{\infty} \frac{3n + 1}{n^3 - 4}$

**3.** $\displaystyle\sum_{n=1}^{\infty} \frac{1}{n\sqrt{n + 1}}$    **4.** $\displaystyle\sum_{n=1}^{\infty} \frac{\sqrt{2n + 1}}{n^2}$

In Problems 5–10, use the Ratio Test to determine convergence or divergence.

**5.** $\displaystyle\sum_{n=1}^{\infty} \frac{8^n}{n!}$    **6.** $\displaystyle\sum_{n=1}^{\infty} \frac{5^n}{n^5}$

**7.** $\displaystyle\sum_{n=1}^{\infty} \frac{n!}{n^{100}}$    **8.** $\displaystyle\sum_{n=1}^{\infty} n \left(\tfrac{1}{3}\right)^n$

**9.** $\displaystyle\sum_{n=1}^{\infty} \frac{n^3}{(2n)!}$    **10.** $\displaystyle\sum_{k=1}^{\infty} \frac{3^k + k}{k!}$

In Problems 11–34, determine convergence or divergence for each of the series. Indicate the test you used.

**11.** $\displaystyle\sum_{n=1}^{\infty} \frac{n}{n + 200}$    **12.** $\displaystyle\sum_{n=1}^{\infty} \frac{5 + n}{n!}$

**13.** $\displaystyle\sum_{n=1}^{\infty} \frac{n + 3}{n^2\sqrt{n}}$    **14.** $\displaystyle\sum_{n=1}^{\infty} \frac{\sqrt{n + 1}}{n^2 + 1}$

**15.** $\displaystyle\sum_{n=1}^{\infty} \frac{n^2}{n!}$    **16.** $\displaystyle\sum_{n=1}^{\infty} \frac{\ln n}{2^n}$

**17.** $\displaystyle\sum_{n=1}^{\infty} \frac{4n^3 + 3n}{n^5 - 4n^2 + 1}$    **18.** $\displaystyle\sum_{n=1}^{\infty} \frac{n^2 + 1}{3^n}$

**19.** $\dfrac{1}{1 \cdot 2} + \dfrac{1}{2 \cdot 3} + \dfrac{1}{3 \cdot 4} + \dfrac{1}{4 \cdot 5} + \cdots.$

*Hint:* $a_n = \dfrac{1}{n(n + 1)}.$

**20.** $\dfrac{1}{2^2} + \dfrac{2}{3^2} + \dfrac{3}{4^2} + \dfrac{4}{5^2} + \cdots$

**21.** $\dfrac{2}{1 \cdot 3 \cdot 4} + \dfrac{3}{2 \cdot 4 \cdot 5} + \dfrac{4}{3 \cdot 5 \cdot 6} + \dfrac{5}{4 \cdot 6 \cdot 7} + \cdots$

**22.** $\dfrac{1}{1^2 + 1} + \dfrac{2}{2^2 + 1} + \dfrac{3}{3^2 + 1} + \dfrac{4}{4^2 + 1} + \cdots$

**23.** $\dfrac{1}{3} + \dfrac{2}{3^2} + \dfrac{3}{3^3} + \dfrac{4}{3^4} + \cdots$

**24.** $3 + \dfrac{3^2}{2!} + \dfrac{3^3}{3!} + \dfrac{3^4}{4!} + \cdots$

**25.** $1 + \dfrac{1}{2\sqrt{2}} + \dfrac{1}{3\sqrt{3}} + \dfrac{1}{4\sqrt{4}} + \cdots$

**26.** $\dfrac{\ln 2}{2^2} + \dfrac{\ln 3}{3^2} + \dfrac{\ln 4}{4^2} + \dfrac{\ln 5}{5^2} + \cdots$

**27.** $\displaystyle\sum_{n=1}^{\infty} \frac{1}{2 + \sin^2 n}$    **28.** $\displaystyle\sum_{n=1}^{\infty} \frac{5}{3^n + 1}$

**29.** $\displaystyle\sum_{n=1}^{\infty} \frac{4 + \cos n}{n^3}$    **30.** $\displaystyle\sum_{n=1}^{\infty} \frac{5^{2n}}{n!}$

**31.** $\displaystyle\sum_{n=1}^{\infty} \frac{n^n}{(2n)!}$    **32.** $\displaystyle\sum_{n=2}^{\infty} \left(1 - \frac{1}{n}\right)^n$

**33.** $\displaystyle\sum_{n=1}^{\infty} \frac{4^n + n}{n!}$    **34.** $\displaystyle\sum_{n=1}^{\infty} \frac{n}{2 + n5^n}$

**35.** Let $a_n > 0$ and suppose $\sum a_n$ converges. Prove that $\sum a_n^2$ converges.

**36.** Prove that $\lim\limits_{n \to \infty} (n!/n^n) = 0$ by considering the series $\sum n!/n^n$. *Hint:* Example 7, followed by $n$th-Term Test.

**37.** Prove that if $a_n \geq 0$, $b_n > 0$, $\lim\limits_{n \to \infty} a_n/b_n = 0$, and $\sum b_n$ converges, then $\sum a_n$ converges.

**38.** Prove that if $a_n \geq 0$, $b_n > 0$, $\lim\limits_{n \to \infty} a_n/b_n = \infty$, and $\sum b_n$ diverges, then $\sum a_n$ diverges.

**39.** Suppose $\lim\limits_{n \to \infty} na_n = 1$. Prove $\sum a_n$ diverges.

**40.** Prove that if $\sum a_n$ is a convergent series of positive terms, then $\sum \ln(1 + a_n)$ converges.

**41. (Root Test)** Prove that if $a_n > 0$ and $\lim\limits_{n \to \infty} (a_n)^{1/n} = R$, then $\sum a_n$ converges if $R < 1$ and diverges if $R > 1$.

**42.** Test for convergence or divergence using the Root Test.

(a) $\displaystyle\sum_{n=2}^{\infty} \left(\frac{1}{\ln n}\right)^n$    (b) $\displaystyle\sum_{n=1}^{\infty} \left(\frac{n}{3n + 2}\right)^n$

(c) $\displaystyle\sum_{n=1}^{\infty} \left(\frac{1}{2} + \frac{1}{n}\right)^n$

**43.** Test for convergence or divergence. In some cases, a clever manipulation using the properties of logarithms will simplify the problem.

(a) $\displaystyle\sum_{n=1}^{\infty} \ln\left(1 + \frac{1}{n}\right)$    (b) $\displaystyle\sum_{n=1}^{\infty} \ln\left[\frac{(n + 1)^2}{n(n + 2)}\right]$

(c) $\displaystyle\sum_{n=2}^{\infty} \frac{1}{(\ln n)^{\ln n}}$

(d) $\displaystyle\sum_{n=3}^{\infty} \frac{1}{[\ln(\ln n)]^{\ln n}}$

(e) $\displaystyle\sum_{n=2}^{\infty} \frac{1}{(\ln n)^4}$

(f) $\displaystyle\sum_{n=1}^{\infty} \left[\frac{\ln n}{n}\right]^2$

**44.** Let $p(n)$ and $q(n)$ be polynomials in $n$ with nonnegative coefficients. Give simple conditions that determine the convergence or divergence of $\displaystyle\sum_{n=1}^{\infty} \frac{p(n)}{q(n)}$.

**45.** Give conditions on $p$ which determine the convergence or divergence of $\displaystyle\sum_{n=1}^{\infty} \frac{1}{n^p}\left(1 + \frac{1}{2^p} + \frac{1}{3^p} + \cdots + \frac{1}{n^p}\right)$.

**46.** Test for convergence or divergence.

(a) $\displaystyle\sum_{n=1}^{\infty} \sin^2\!\left(\frac{1}{n}\right)$

(b) $\displaystyle\sum_{n=1}^{\infty} \tan\!\left(\frac{1}{n}\right)$

(c) $\displaystyle\sum_{n=1}^{\infty} \sqrt{n}\left[1 - \cos\!\left(\frac{1}{n}\right)\right]$

---

**Answers to Concepts Review: 1.** $0 \le a_k \le b_k$
**2.** $\displaystyle\lim_{k \to \infty} (a_k/b_k)$ **3.** $\rho < 1; \rho > 1; \rho = 1; \displaystyle\lim_{n \to \infty} (a_{n+1}/a_n)$
**4.** Ratio; Limit Comparison

---

## 11.5
## ALTERNATING SERIES, ABSOLUTE CONVERGENCE

In the last two sections, we have considered series of nonnegative terms. Now we remove that restriction, allowing some terms to be negative. In particular, we study **alternating series**—that is, series of the form

$$a_1 - a_2 + a_3 - a_4 + \cdots$$

where $a_n > 0$ for all $n$. An important example is the **alternating harmonic series**

$$1 - \tfrac{1}{2} + \tfrac{1}{3} - \tfrac{1}{4} + \cdots$$

We have seen that the harmonic series diverges; we shall soon see that the alternating harmonic series converges.

**A Convergence Test** Let us suppose that the sequence $\{a_n\}$ is decreasing; that is, $a_{n+1} < a_n$ for all $n$. Also let $S_n$ have its usual meaning,

$$S_1 = a_1$$
$$S_2 = a_1 - a_2 = S_1 - a_2$$
$$S_3 = a_1 - a_2 + a_3 = S_2 + a_3$$
$$S_4 = a_1 - a_2 + a_3 - a_4 = S_3 - a_4$$

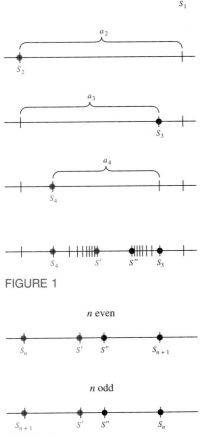

FIGURE 1

*n even*

*n odd*

FIGURE 2

and so on. A geometric interpretation of these partial sums is shown in Figure 1. Note that the even numbered terms $S_2, S_4, S_6, \ldots$ are increasing and bounded above and hence must converge to a limit, call it $S'$. Similarly, the odd numbered terms $S_1, S_3, S_5, \ldots$ are decreasing and bounded below. They also converge, say to $S''$.

Both $S'$ and $S''$ are between $S_n$ and $S_{n+1}$ for all $n$ (see Figure 2) and so

$$|S'' - S'| \le |S_{n+1} - S_n| = a_{n+1}$$

Thus, the condition $a_{n+1} \to 0$ as $n \to \infty$ will guarantee that $S' = S''$ and, consequently, the convergence of the series to their common value, which we call $S$. Finally, we note that since $S$ is between $S_n$ and $S_{n+1}$,

$$|S - S_n| \leq |S_{n+1} - S_n| = a_{n+1}$$

that is, the error made by using $S_n$ as an approximation to the sum $S$ of the whole series is not more than the magnitude of the first neglected term. We have proved the following theorem.

---

**Theorem A**

**(Alternating-Series Test).** Let

$$a_1 - a_2 + a_3 - a_4 + \cdots$$

be an alternating series with $a_n > a_{n+1} > 0$. If $\lim_{n \to \infty} a_n = 0$, then the series converges. Moreover, the error made by using the sum $S_n$ of the first $n$ terms to approximate the sum $S$ of the series is not more than $a_{n+1}$.

---

**EXAMPLE 1** Show that the alternating harmonic series

$$1 - \tfrac{1}{2} + \tfrac{1}{3} - \tfrac{1}{4} + \cdots$$

converges. How many terms of this series would we need to take in order to get a partial sum $S_n$ within 0.01 of the sum $S$ of the whole series?

*Solution* The alternating harmonic series satisfies the hypotheses of Theorem A and so converges. We want $|S - S_n| \leq 0.01$, and this will hold if $a_{n+1} \leq 0.01$. Since $a_{n+1} = 1/(n + 1)$, we require $1/(n + 1) \leq 0.01$, which is satisfied if $n \geq 99$. Thus we need to take 99 terms to make sure we have the desired accuracy. This gives you an idea of how slowly the alternating harmonic series converges. (See Problem 45 for a clever way to find the exact sum of this series.) ∎

**EXAMPLE 2** Show that

$$\frac{1}{1!} - \frac{1}{2!} + \frac{1}{3!} - \frac{1}{4!} + \cdots$$

converges. Calculate $S_5$ and estimate the error made by using this as a value for the sum of the whole series.

*Solution* The Alternating-Series Test (Theorem A) applies and guarantees convergence.

$$S_5 = 1 - \frac{1}{2} + \frac{1}{6} - \frac{1}{24} + \frac{1}{120} \approx 0.6333$$

$$|S - S_5| \leq a_6 = \frac{1}{6!} \approx 0.0014$$

∎

**EXAMPLE 3** Show that $\sum_{n=1}^{\infty} (-1)^{n+1} \dfrac{n^2}{2^n}$ converges.

**Solution**   To get a feeling for this series, we write the first few terms:

$$\tfrac{1}{2} - 1 + \tfrac{9}{8} - 1 + \tfrac{25}{32} - \tfrac{36}{64} + \cdots$$

The series is alternating and $\lim_{n \to \infty} n^2/2^n = 0$ (l'Hôpital's Rule), but unfortunately the terms are not decreasing initially. However, they do appear to be decreasing after the first two terms; that is good enough, since what happens at the beginning of a series never affects convergence or divergence. To show that the sequence $\{n^2/2^n\}$ is decreasing from the third term on, consider the function

$$f(x) = \frac{x^2}{2^x}$$

Note that if $x \geq 3$, the derivative

$$f'(x) = \frac{2x \cdot 2^x - x^2 2^x \ln 2}{2^{2x}} = \frac{x 2^x (2 - x \ln 2)}{2^{2x}}$$

$$\approx \frac{x(2 - 0.69x)}{2^x} < 0$$

Thus, $f$ is decreasing on $[3, \infty)$, and so $\{n^2/2^n\}$ is decreasing for $n \geq 3$. For a different demonstration of this last fact, see Example 6 of Section 11.1. ∎

**Absolute Convergence**   Does a series such as

$$1 + \tfrac{1}{4} - \tfrac{1}{9} + \tfrac{1}{16} + \tfrac{1}{25} - \tfrac{1}{36} + \cdots$$

in which there is a pattern of two positive terms followed by one negative term, converge or diverge? The Alternating-Series Test does not apply. However, since the corresponding series of all positive terms

$$1 + \tfrac{1}{4} + \tfrac{1}{9} + \tfrac{1}{16} + \tfrac{1}{25} + \tfrac{1}{36} + \cdots$$

converges ($p$-series with $p = 2$), it seems plausible to think that the same series with some terms negative should converge (even better). That is the content of our next theorem.

---

**Theorem B**

**(Absolute Convergence Test).** If $\sum |u_n|$ converges, then $\sum u_n$ converges.

---

**Proof**   We use a trick. Let $v_n = u_n + |u_n|$, so

$$u_n = v_n - |u_n|$$

Now $0 \leq v_n \leq 2|u_n|$, and so $\sum v_n$ converges by the Ordinary Comparison Test. It follows from the Linearity Theorem (Theorem 11.2B) that $\sum u_n = \sum (v_n - |u_n|)$ converges. ∎

A series $\sum u_n$ is said to **converge absolutely** if $\sum |u_n|$ converges. Theorem B asserts that absolute convergence implies convergence. All our tests for convergence of series of positive terms are automatically tests for the absolute convergence of a series in which some terms are negative. In particular, this is true of the Ratio Test, which we now restate.

---

Theorem C

**(Absolute Ratio Test).** Let $\sum u_n$ be a series of nonzero terms and suppose

$$\lim_{n \to \infty} \frac{|u_{n+1}|}{|u_n|} = \rho$$

**(i)** If $\rho < 1$, the series converges absolutely (hence converges).
**(ii)** If $\rho > 1$, the series diverges.
**(iii)** If $\rho = 1$, the test is inconclusive.

---

**Proof** Only (ii) requires proof. From the original Ratio Test, we could conclude that $\sum |u_n|$ diverges, but here we are claiming more—namely, that $\sum u_n$ diverges. Since

$$\lim_{n \to \infty} \frac{|u_{n+1}|}{|u_n|} > 1$$

it follows that for $n$ sufficiently large, say $n \geq N$, $|u_{n+1}| > |u_n|$. This, in turn, implies that $|u_n| > |u_N| > 0$ for all $n \geq N$ and so $\lim_{n \to \infty} u_n$ cannot be 0. We conclude by the $n$th-Term Test that $\sum u_n$ diverges. ∎

**EXAMPLE 4**   Show that $\sum_{n=1}^{\infty} (-1)^{n+1} \frac{3^n}{n!}$ converges absolutely.

**Solution**

$$\rho = \lim_{n \to \infty} \frac{|u_{n+1}|}{|u_n|} = \lim_{n \to \infty} \frac{3^{n+1}}{(n+1)!} \div \frac{3^n}{n!}$$

$$= \lim_{n \to \infty} \frac{3}{n+1} = 0$$

We conclude from the Absolute Ratio Test that the series converges absolutely (and therefore converges). ∎

**EXAMPLE 5**   Test for the convergence or divergence of $\sum_{n=1}^{\infty} \frac{\cos(n!)}{n^2}$.

**Solution**   If you write out the first hundred terms of this series, you will discover that the signs of the terms vary in a rather random way. The series is in fact a difficult one to analyze directly. However,

$$\left| \frac{\cos(n!)}{n^2} \right| \leq \frac{1}{n^2}$$

and so the series converges absolutely by the Ordinary Comparison Test. We conclude from the Absolute Convergence Test (Theorem B) that the series converges. ■

**Conditional Convergence**   A common error is to try to turn Theorem B around. It does *not* say that convergence implies absolute convergence. That is clearly false; witness the alternating harmonic series. We know that

$$1 - \tfrac{1}{2} + \tfrac{1}{3} - \tfrac{1}{4} + \cdots$$

converges but that

$$1 + \tfrac{1}{2} + \tfrac{1}{3} + \tfrac{1}{4} + \cdots$$

diverges. A series $\sum u_n$ is called **conditionally convergent** if $\sum u_n$ converges but $\sum |u_n|$ diverges. The alternating harmonic series is the premier example of a conditionally convergent series, but there are many others.

**EXAMPLE 6**   Show that $\displaystyle\sum_{n=1}^{\infty} (-1)^{n+1} \frac{1}{\sqrt{n}}$ is conditionally convergent.

*Solution*   $\displaystyle\sum_{n=1}^{\infty} (-1)^{n+1} [1/\sqrt{n}]$ converges by the Alternating-Series Test. However, $\displaystyle\sum_{n=1}^{\infty} 1/\sqrt{n}$ diverges, since it is a $p$-series with $p = \tfrac{1}{2}$. ■

Absolutely convergent series behave much better than do conditionally convergent ones. Here is a nice (but hard-to-prove) theorem about absolutely convergent series. It is spectacularly false for conditionally convergent series (see Problems 35–38).

> **Theorem D**
>
> **(Rearrangement Theorem).** The terms of an absolutely convergent series can be rearranged without affecting either the convergence or the sum of the series.

For example, the series

$$1 + \tfrac{1}{4} - \tfrac{1}{9} + \tfrac{1}{16} + \tfrac{1}{25} - \tfrac{1}{36} + \tfrac{1}{49} + \tfrac{1}{64} - \tfrac{1}{81} + \cdots$$

converges absolutely. The rearrangement

$$1 + \tfrac{1}{4} + \tfrac{1}{16} - \tfrac{1}{9} + \tfrac{1}{25} + \tfrac{1}{49} + \tfrac{1}{64} - \tfrac{1}{36} + \cdots$$

converges and has the same sum as the original series.

## CONCEPTS REVIEW

1. The alternating series $a_1 - a_2 + a_3 - \cdots$ will converge provided the terms are decreasing in size and _____.

2. If $\sum |u_k|$ converges, we say the series $\sum u_k$ converges _____; if $\sum u_k$ converges but $\sum |u_k|$ diverges, we say $\sum u_k$ converges _____.

3. The premier example of a conditionally convergent series is _____.

4. The terms of an absolutely convergent series may be _____ at will without affecting its convergence or its sum.

## PROBLEM SET 11.5

In Problems 1–6, show that each of the alternating series converges and then estimate the error made by using the partial sum $S_9$ as an approximation to the sum $S$ of the series (see Examples 1–3).

1. $\displaystyle\sum_{n=1}^{\infty} (-1)^{n+1} \frac{2}{3n+1}$    2. $\displaystyle\sum_{n=1}^{\infty} (-1)^{n+1} \frac{1}{\sqrt{n}}$

3. $\displaystyle\sum_{n=1}^{\infty} (-1)^{n+1} \frac{1}{\ln(n+1)}$    4. $\displaystyle\sum_{n=1}^{\infty} (-1)^{n+1} \frac{n}{n^2+1}$

5. $\displaystyle\sum_{n=1}^{\infty} (-1)^{n+1} \frac{\ln n}{n}$    6. $\displaystyle\sum_{n=1}^{\infty} (-1)^{n+1} \frac{\ln n}{\sqrt{n}}$

In Problems 7–12, show that each of the series converges absolutely.

7. $\displaystyle\sum_{n=1}^{\infty} \left(-\frac{3}{4}\right)^n$    8. $\displaystyle\sum_{n=1}^{\infty} (-1)^n \frac{1}{n\sqrt{n}}$

9. $\displaystyle\sum_{n=1}^{\infty} (-1)^{n+1} \frac{n}{2^n}$    10. $\displaystyle\sum_{n=1}^{\infty} (-1)^{n+1} \frac{n^2}{e^n}$

11. $\displaystyle\sum_{n=1}^{\infty} (-1)^{n+1} \frac{1}{n(n+1)}$    12. $\displaystyle\sum_{n=1}^{\infty} (-1)^{n+1} \frac{2^n}{n!}$

In Problems 13–30, classify the series as absolutely convergent, conditionally convergent, or divergent.

13. $\displaystyle\sum_{n=1}^{\infty} (-1)^{n+1} \frac{1}{5n}$    14. $\displaystyle\sum_{n=1}^{\infty} (-1)^{n+1} \frac{1}{5n^{1.1}}$

15. $\displaystyle\sum_{n=1}^{\infty} (-1)^{n+1} \frac{n}{10n+1}$

16. $\displaystyle\sum_{n=1}^{\infty} (-1)^{n+1} \frac{n}{10n^{1.1}+1}$

17. $\displaystyle\sum_{n=2}^{\infty} (-1)^n \frac{1}{n \ln n}$    18. $\displaystyle\sum_{n=1}^{\infty} (-1)^{n+1} \frac{1}{n(1+\sqrt{n})}$

19. $\displaystyle\sum_{n=1}^{\infty} (-1)^{n+1} \frac{n^4}{2^n}$    20. $\displaystyle\sum_{n=2}^{\infty} (-1)^n \frac{1}{\sqrt{n^2-1}}$

21. $\displaystyle\sum_{n=1}^{\infty} (-1)^{n+1} \frac{n}{n^2+1}$    22. $\displaystyle\sum_{n=1}^{\infty} (-1)^{n+1} \frac{n-1}{n}$

23. $\displaystyle\sum_{n=1}^{\infty} \frac{\cos n\pi}{n}$    24. $\displaystyle\sum_{n=1}^{\infty} \frac{\sin(n\pi/2)}{n^2}$

25. $\displaystyle\sum_{n=1}^{\infty} (-1)^n \frac{\sin n}{n\sqrt{n}}$    26. $\displaystyle\sum_{n=1}^{\infty} n \sin\left(\frac{1}{n}\right)$

27. $\displaystyle\sum_{n=1}^{\infty} (-1)^{n+1} \frac{1}{\sqrt{n(n+1)}}$    28. $\displaystyle\sum_{n=1}^{\infty} \frac{(-1)^{n+1}}{\sqrt{n+1}+\sqrt{n}}$

29. $\displaystyle\sum_{n=1}^{\infty} \frac{(-3)^{n+1}}{n^2}$    30. $\displaystyle\sum_{n=1}^{\infty} (-1)^{n+1} \sin \frac{\pi}{n}$

31. Prove that if $\sum a_n$ diverges, so does $\sum |a_n|$.

32. Give an example of two series $\sum a_n$ and $\sum b_n$, both convergent, such that $\sum a_n b_n$ diverges.

33. Show that the positive terms of the alternating harmonic series form a divergent series. Show the same for the negative terms.

34. Show that the results in Problem 33 hold for any conditionally convergent series.

35. Show that the alternating harmonic series

$$1 - \tfrac{1}{2} + \tfrac{1}{3} - \tfrac{1}{4} + \tfrac{1}{5} - \tfrac{1}{6} + \cdots$$

(whose sum is actually $\ln 2 \approx 0.69$) can be rearranged to converge to 1.3 by using the following steps.

(a) Take enough of the positive terms $1 + \frac{1}{3} + \frac{1}{5} + \cdots$ to just exceed 1.3.
(b) Now add enough of the negative terms $-\frac{1}{2} - \frac{1}{4} - \frac{1}{6} - \cdots$ so that the partial sum $S_n$ falls just below 1.3.
(c) Add just enough more positive terms to again exceed 1.3, and so on.

[C] 36. Use your calculator to help you find the first 20 terms of the series described in Problem 35. Calculate $S_{20}$.

**37.** Convince yourself that a conditionally convergent series can be rearranged to converge to any given number.

**38.** Show that a conditionally convergent series can be rearranged so as to diverge.

**39.** Show that $\lim_{n \to \infty} a_n = 0$ is not sufficient to guarantee the convergence of the alternating series $\sum (-1)^{n+1}a_n$. *Hint:* Alternate the terms of $\sum 1/n$ and $\sum -1/n^2$.

**40.** Discuss the convergence or divergence of

$$\frac{1}{\sqrt{2}-1} - \frac{1}{\sqrt{2}+1} + \frac{1}{\sqrt{3}-1} - \frac{1}{\sqrt{3}+1} + \frac{1}{\sqrt{4}+1}$$
$$- \frac{1}{\sqrt{4}-1} + \cdots$$

**41.** Prove that if $\sum\limits_{k=1}^{\infty} a_k^2$ and $\sum\limits_{k=1}^{\infty} b_k^2$ both converge, then $\sum\limits_{k=1}^{\infty} a_k b_k$ converges absolutely. *Hint:* First show that $2|a_k b_k| \le a_k^2 + b_k^2$.

**42.** Sketch the graph of $y = (\sin x)/x$ and then show that $\int_0^\infty (\sin x)/x\, dx$ converges.

**43.** Show that $\int_0^\infty |\sin x|/x\, dx$ diverges.

**44.** Show that the graph of $y = x \sin \dfrac{\pi}{x}$ on $(0, 1]$ has infinite length.

**45.** Note that

$$1 - \frac{1}{2} + \frac{1}{3} - \frac{1}{4} + \cdots - \frac{1}{2n} = 1 + \frac{1}{2} + \frac{1}{3} + \cdots + \frac{1}{2n}$$
$$- \left(1 + \frac{1}{2} + \frac{1}{3} + \cdots + \frac{1}{n}\right)$$
$$= \frac{1}{n+1} + \frac{1}{n+2} + \cdots +$$

Recognize the latter expression as a Riemann sum and use it to find the sum of the alternating harmonic series.

---

**Answers to Concepts Review:** **1.** $\lim_{n \to \infty} a_n = 0$ **2.** Absolutely; conditionally **3.** The alternating harmonic series **4.** Rearranged.

---

## 11.6
## POWER SERIES

So far, we have been studying what might be called *series of constants*—that is, series of the form $\sum u_n$, where each $u_n$ is a number. Now we consider *series of functions*, series of the form $\sum u_n(x)$. A typical example of such a series is

$$\sum_{n=1}^{\infty} \frac{\sin nx}{n^2} = \frac{\sin x}{1} + \frac{\sin 2x}{4} + \frac{\sin 3x}{9} + \cdots$$

Of course, as soon as we substitute a value for $x$ (such as $x = 2.1$), we are back to familiar territory; we have a series of constants.

There are two important questions to ask about a series of functions.

**1.** For what $x$'s does the series converge?
**2.** To what function does it converge; that is, what is the sum $S(x)$ of the series?

The general situation is a proper subject for an advanced calculus course. However, even in elementary calculus, we can learn a good deal about the special case of a power series. A **power series in *x*** has the form

$$\sum_{n=0}^{\infty} a_n x^n = a_0 + a_1 x + a_2 x^2 + \cdots$$

(Here we interpret $a_0 x^0$ to be $a_0$ even if $x = 0$.) We can immediately answer our two questions for one such power series.

### Fourier Series

The series of sine functions mentioned in the introduction is an example of a *Fourier series*, named after Jean Baptiste Joseph Fourier (1768–1830). Fourier series are of immense importance in the study of wave phenomena, since they allow us to represent a complicated wave as a sum of its fundamental components (called the pure tones in the case of sound waves). It is a large field, which we leave to other authors and other books.

**EXAMPLE 1**  For what $x$'s does the power series

$$\sum_{n=0}^{\infty} ax^n = a + ax + ax^2 + ax^3 + \cdots$$

converge and what is its sum? Assume $a \neq 0$.

*Solution*  We actually studied this series in Section 11.2 (with $r$ in place of $x$) and called it a geometric series. It converges for $-1 < x < 1$ and has sum $S(x)$ given by

$$S(x) = \frac{a}{1 - x}, \quad -1 < x < 1 \qquad \blacksquare$$

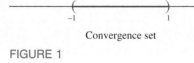

Convergence set

FIGURE 1

**The Convergence Set**  We call the set on which a power series converges its **convergence set**. What kind of set can be a convergence set? Example 1 shows that it can be an open interval (see Figure 1). Are there other possibilities?

**EXAMPLE 2**  What is the convergence set for

$$\sum_{n=0}^{\infty} \frac{x^n}{(n+1)2^n} = 1 + \frac{1}{2}\frac{x}{2} + \frac{1}{3}\frac{x^2}{2^2} + \frac{1}{4}\frac{x^3}{2^3} + \cdots$$

*Solution*  Note that some of the terms may be negative (if $x$ is negative). Let's test for absolute convergence using the Absolute Ratio Test (Theorem 11.5C).

$$\rho = \lim_{n \to \infty} \left| \frac{x^{n+1}}{(n+2)2^{n+1}} \div \frac{x^n}{(n+1)2^n} \right| = \lim_{n \to \infty} \frac{|x|}{2} \cdot \frac{n+1}{n+2} = \frac{|x|}{2}$$

The series converges absolutely (hence converges) when $\rho = |x|/2 < 1$ and diverges when $|x|/2 > 1$. Consequently, it converges when $|x| < 2$ and diverges when $|x| > 2$.

If $x = 2$ or $x = -2$, the Ratio Test fails. However, when $x = 2$, the series is the harmonic series, which diverges; and when $x = -2$, it is the alternating harmonic series, which converges. We conclude that the convergence set for the given series is the interval $-2 \le x < 2$ (Figure 2). $\blacksquare$

Convergence set

FIGURE 2

**EXAMPLE 3**  Find the convergence set for $\displaystyle\sum_{n=0}^{\infty} \frac{x^n}{n!}$.

*Solution*

$$\rho = \lim_{n \to \infty} \left| \frac{x^{n+1}}{(n+1)!} \div \frac{x^n}{n!} \right| = \lim_{n \to \infty} \frac{|x|}{n+1} = 0$$

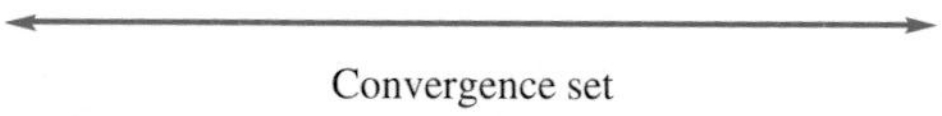

Convergence set

FIGURE 3

We conclude from the Absolute Ratio Test that the series converges for all $x$ (Figure 3). $\blacksquare$

**EXAMPLE 4**  Find the convergence set for $\displaystyle\sum_{n=0}^{\infty} n!x^n$.

**Solution**

$$\rho = \lim_{n \to \infty} \left| \frac{(n + 1)!x^{n + 1}}{n!x^n} \right| = \lim_{n \to \infty} (n + 1)|x| = \begin{cases} 0 & \text{if } x = 0 \\ \infty & \text{if } x \neq 0 \end{cases}$$

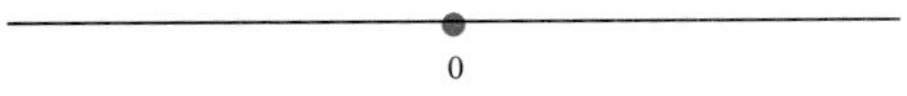

0

Convergence set

FIGURE 4

We conclude that the series converges only at $x = 0$ (Figure 4).  ∎

In each of our examples, the convergence set was an interval (a degenerate interval in the last example). This will always be the case. For example, it is impossible for a power series to have a convergence set consisting of two disconnected parts (like $[0, 1] \cup [2, 3]$). Our next theorem tells the whole story.

---

Theorem A

The convergence set for a power series $\sum a_n x^n$ is always an interval of one of the following three types.
**(i)**  The single point $x = 0$.
**(ii)**  An interval $(-R, R)$, plus possibly one or both endpoints.
**(iii)**  The whole real line.

---

In (i), (ii), and (iii), the series is said to have **radius of convergence** $0$, $R$, and $\infty$, respectively.

**Proof**  Suppose that the series converges at $x = x_1 \neq 0$. Then $\lim_{n \to \infty} a_n x_1^n = 0$, and so there is certainly a number $N$ such that $|a_n x_1^n| < 1$ for $n \geq N$. Then, for any $x$ for which $|x| < |x_1|$,

$$|a_n x^n| = |a_n x_1^n| \left| \frac{x}{x_1} \right|^n < \left| \frac{x}{x_1} \right|^n$$

this holding for $n \geq N$. Now $\sum |x/x_1|^n$ converges, since it is a geometric series with ratio less than 1. Thus, by the Ordinary Comparison Test, $\sum |a_n x^n|$ converges. We have shown that if a power series converges at $x_1$, it converges (absolutely) for all $x$ such that $|x| < |x_1|$.

On the other hand, suppose a power series diverges at $x_2$. Then it must diverge for all $x$ for which $|x| > |x_2|$. For if it converged at $x_1$ such that $|x_1| > |x_2|$, then, by what we have already shown, it would converge at $x_2$, contrary to hypothesis.

These two paragraphs together eliminate all convergence sets except the three types mentioned in the theorem.  ∎

Actually we have proved slightly more than we have claimed in Theorem A, and it is worth stating this as another theorem.

> **Theorem B**
>
> A power series $\sum a_n x^n$ converges absolutely on the interior of its interval of convergence.

Of course, it might even converge absolutely at the endpoints of the interval of convergence, but of that we cannot be sure; witness Example 2.

**Power Series in $x - a$**   A series of the form

$$\sum a_n(x - a)^n = a_0 + a_1(x - a) + a_2(x - a)^2 + \cdots$$

is called a **power series in $x - a$**. All that we have said about power series in $x$ applies equally well for series in $x - a$. In particular, its convergence set is always one of the following kinds of intervals.

**FIGURE 5**

1. The single point $x = a$.
2. An interval $(a - R, a + R)$, plus possibly one or both endpoints (Figure 5).
3. The whole real line.

**EXAMPLE 5**   Find the convergence set for $\displaystyle\sum_{n=0}^{\infty} \frac{(x - 1)^n}{(n + 1)^2}$.

*Solution*   We apply the Absolute Ratio Test.

$$\rho = \lim_{n \to \infty} \left| \frac{(x - 1)^{n + 1}}{(n + 2)^2} \div \frac{(x - 1)^n}{(n + 1)^2} \right| = \lim_{n \to \infty} |x - 1| \frac{(n + 1)^2}{(n + 2)^2}$$

$$= |x - 1|$$

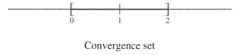

Convergence set

**FIGURE 6**

Thus, the series converges if $|x - 1| < 1$—that is, if $0 < x < 2$; it diverges if $|x - 1| > 1$. It also converges (even absolutely) at both of the endpoints 0 and 2, as we see by substitution of these values. The convergence set is the closed interval $[0, 2]$ (Figure 6).   ■

**EXAMPLE 6**   Determine the convergence set for

$$\frac{x + 2}{3} + \frac{(x + 2)^2 \ln 2}{2 \cdot 9} + \frac{(x + 2)^3 \ln 3}{3 \cdot 27} + \frac{(x + 2)^4 \ln 4}{4 \cdot 81} + \cdots$$

*Solution*   The $n$th term is $u_n = \dfrac{(x + 2)^n \ln n}{n \cdot 3^n}$. Thus,

$$\rho = \lim_{n \to \infty} \left| \frac{(x + 2)^{n + 1} \ln(n + 1)}{(n + 1)3^{n + 1}} \cdot \frac{n3^n}{(x + 2)^n \ln n} \right|$$

$$= \frac{|x + 2|}{3} \lim_{n \to \infty} \frac{n}{n + 1} \frac{\ln(n + 1)}{\ln n} = \frac{|x + 2|}{3}$$

We know the series converges when $\rho < 1$, that is, when $|x + 2| < 3$ or, equivalently, $-5 < x < 1$, but we must check the endpoints $-5$ and $1$.
At $x = -5$,

$$u_n = \frac{(-3)^n \ln n}{n3^n} = (-1)^n \frac{\ln n}{n}$$

and $\sum (-1)^n (\ln n)/n$ converges by the Alternating Series Test.

At $x = 1$, $u_n = (\ln n)/n$ and $\sum (\ln n)/n$ diverges by comparison with the harmonic series.

We conclude that the given series converges on the interval $-5 \leq x < 1$. ∎

## CONCEPTS REVIEW

**1.** A series of the form $a_0 + a_1 x + a_2 x^2 + \cdots$ is called a _____.

**2.** Rather than asking whether a power series converges, we should ask _____.

**3.** A power series always converges on an _____, which may or may not include its _____.

**4.** The series $5 + x + x^2 + x^3 + \cdots$ converges on the interval _____.

## PROBLEM SET 11.6

In Problems 1–20, find the convergence set of the given power series. *Hint*: First find a formula for the $n$th term; then use the Absolute Ratio Test.

**1.** $\dfrac{x}{1 \cdot 2} - \dfrac{x^2}{2 \cdot 3} + \dfrac{x^3}{3 \cdot 4} - \dfrac{x^4}{4 \cdot 5} + \dfrac{x^5}{5 \cdot 6} - \cdots$

**2.** $1 + x + \dfrac{x^2}{2!} + \dfrac{x^3}{3!} + \dfrac{x^4}{4!} + \cdots$

**3.** $x - \dfrac{x^3}{3!} + \dfrac{x^5}{5!} - \dfrac{x^7}{7!} + \dfrac{x^9}{9!} - \cdots$

**4.** $1 - \dfrac{x^2}{2!} + \dfrac{x^4}{4!} - \dfrac{x^6}{6!} + \dfrac{x^8}{8!} - \dfrac{x^{10}}{10!} + \cdots$

**5.** $x + 2x^2 + 3x^3 + 4x^4 + \cdots$

**6.** $x + 2^2 x^2 + 3^2 x^3 + 4^2 x^4 + \cdots$

**7.** $1 - x + \dfrac{x^2}{2} - \dfrac{x^3}{3} + \dfrac{x^4}{4} - \cdots$

**8.** $1 + x + \dfrac{x^2}{\sqrt{2}} + \dfrac{x^3}{\sqrt{3}} + \dfrac{x^4}{\sqrt{4}} + \dfrac{x^5}{\sqrt{5}} + \cdots$

**9.** $1 - \dfrac{x}{1 \cdot 3} + \dfrac{x^2}{2 \cdot 4} - \dfrac{x^3}{3 \cdot 5} + \dfrac{x^4}{4 \cdot 6} - \cdots$

**10.** $\dfrac{x}{2^2 - 1} + \dfrac{x^2}{3^2 - 1} + \dfrac{x^3}{4^2 - 1} + \dfrac{x^4}{5^2 - 1} + \cdots$

**11.** $1 - \dfrac{x}{2} + \dfrac{x^2}{2^2} - \dfrac{x^3}{2^3} + \dfrac{x^4}{2^4} - \cdots$

**12.** $1 + 2x + 2^2 x^2 + 2^3 x^3 + 2^4 x^4 + \cdots$

**13.** $1 + 2x + \dfrac{2^2 x^2}{2!} + \dfrac{2^3 x^3}{3!} + \dfrac{2^4 x^4}{4!} + \cdots$

**14.** $\dfrac{x}{2} + \dfrac{2x^2}{3} + \dfrac{3x^3}{4} + \dfrac{4x^4}{5} + \dfrac{5x^5}{6} + \cdots$

**15.** $\dfrac{(x-1)}{1} + \dfrac{(x-1)^2}{2} + \dfrac{(x-1)^3}{3} + \dfrac{(x-1)^4}{4} + \cdots$

**16.** $1 + (x+2) + \dfrac{(x+2)^2}{2!} + \dfrac{(x+2)^3}{3!} + \cdots$

**17.** $1 + \dfrac{(x+1)}{2} + \dfrac{(x+1)^2}{2^2} + \dfrac{(x+1)^3}{2^3} + \cdots$

**18.** $\dfrac{(x-2)}{1^2} + \dfrac{(x-2)^2}{2^2} + \dfrac{(x-2)^3}{3^2} + \dfrac{(x-2)^4}{4^2} + \cdots$

**19.** $\dfrac{(x+5)}{1 \cdot 2} + \dfrac{(x+5)^2}{2 \cdot 3} + \dfrac{(x+5)^3}{3 \cdot 4} + \dfrac{(x+5)^4}{4 \cdot 5} + \cdots$

**20.** $(x + 3) - 2(x + 3)^2 + 3(x + 3)^3 - 4(x + 3)^4 + \cdots$

**21.** From Example 3, we know that $\sum x^n/n!$ converges for all $x$. Why can we conclude that $\lim_{n \to \infty} x^n/n! = 0$ for all $x$?

**22.** Let $k$ be an arbitrary number and $-1 < x < 1$. Prove that

$$\lim_{n \to \infty} \frac{k(k - 1)(k - 2) \cdots (k - n)}{n!} \, x^n = 0$$

*Hint*: See Problem 21.

**23.** Find the radius of convergence of

$$\sum_{n=1}^{\infty} \frac{1 \cdot 2 \cdot 3 \cdots n}{1 \cdot 3 \cdot 5 \cdots (2n - 1)} \, x^{2n+1}$$

**24.** Find the radius of convergence of

$$\sum_{n=0}^{\infty} \frac{(pn)!}{(n!)^p} \, x^n$$

where $p$ is a positive integer.

**25.** Find the sum $S(x)$ of $\sum_{n=0}^{\infty} (x - 3)^n$. Where is it valid?

**26.** Suppose $\sum_{n=0}^{\infty} a_n(x - 3)^n$ converges at $x = -1$. Why can you conclude that it converges at $x = 6$? Can you be sure it converges at $x = 7$? Explain.

**27.** Find the convergence set for each series.

(a) $\displaystyle\sum_{n=1}^{\infty} \frac{(3x + 1)^n}{n \cdot 2^n}$        (b) $\displaystyle\sum_{n=1}^{\infty} (-1)^n \frac{(2x - 3)^n}{4^n \sqrt{n}}$

**28.** Refer to Problem 52 of Section 11.1, where the Fibonacci sequence $f_1, f_2, f_3, \ldots$ was defined. Find the radius of convergence of $\sum_{n=1}^{\infty} f_n x^n$.

**29.** Suppose that $a_{n+3} = a_n$ and let $S(x) = \sum_{n=0}^{\infty} a_n x^n$. Show that the series converges for $|x| < 1$ and give a formula for $S(x)$.

**30.** Follow the directions of Problem 29 for the case where $a_{n+p} = a_n$ for some fixed positive integer $p$.

---

**Answers to Concepts Review:** **1.** Power series **2.** Where it converges **3.** Interval; endpoints **4.** $(-1, 1)$

---

## 11.7
### OPERATIONS ON POWER SERIES

We know from the previous section that the convergence set of a power series $\sum a_n x^n$ is an interval $I$. This interval is the domain for a new function $S(x)$, the sum of the series. The most obvious question to ask about $S(x)$ is whether we can give a simple formula for it. We have done this for one series, a geometric series.

$$\sum_{n=0}^{\infty} ax^n = \frac{a}{1 - x}, \qquad -1 < x < 1$$

Actually, there is little reason to hope that the sum of an arbitrarily given power series will be one of the elementary functions studied earlier in this book, though we will make a little progress in that direction in this section and more in Section 11.8.

A better question to ask now is whether we can say anything about the properties of $S(x)$. For example, is it differentiable? Is it integrable? The answer to both questions is yes.

**Term-By-Term Differentiation and Integration** Think of a power series as a polynomial with infinitely many terms. It behaves just like a polynomial under both integration and differentiation; these operations can be performed term by term, as follows.

Theorem A

Suppose that $S(x)$ is the sum of a power series on an interval $I$; that is,

$$S(x) = \sum_{n=0}^{\infty} a_n x^n = a_0 + a_1 x + a_2 x^2 + a_3 x^3 + \cdots$$

Then, if $x$ is interior to $I$,

(i) $\quad S'(x) = \sum_{n=0}^{\infty} D_x(a_n x^n) = \sum_{n=1}^{\infty} n a_n x^{n-1}$

$$= a_1 + 2a_2 x + 3a_3 x^2 + \cdots$$

(ii) $\quad \int_0^x S(t)\,dt = \sum_{n=0}^{\infty} \int_0^x a_n t^n\,dt = \sum_{n=0}^{\infty} \frac{a_n}{n+1} x^{n+1}$

$$= a_0 x + \frac{1}{2} a_1 x^2 + \frac{1}{3} a_2 x^3 + \frac{1}{4} a_3 x^4 + \cdots$$

The theorem entails several things. It asserts that $S$ is both differentiable and integrable, it shows how the derivative and integral may be calculated, and it implies that the radius of convergence of both the differentiated and integrated series is the same as for the original series (though it says nothing about the endpoints of the interval of convergence). The theorem is hard to prove. We leave the proof to more advanced books.

A nice consequence of Theorem A is that we can apply it to a power series with a known sum formula to obtain sum formulas for other series.

**EXAMPLE 1** Apply Theorem A to the geometric series

$$\frac{1}{1-x} = 1 + x + x^2 + x^3 + \cdots, \qquad -1 < x < 1$$

to obtain formulas for two new series.

*Solution* Differentiating term by term yields

$$\frac{1}{(1-x)^2} = 1 + 2x + 3x^2 + 4x^3 + \cdots, \qquad -1 < x < 1$$

Integrating term by term gives

$$\int_0^x \frac{1}{1-t}\,dt = \int_0^x 1\,dt + \int_0^x t\,dt + \int_0^x t^2\,dt + \cdots$$

that is,

$$-\ln(1-x) = x + \frac{x^2}{2} + \frac{x^3}{3} + \cdots, \qquad -1 < x < 1$$

*An Endpoint Result*

The question of what is true at an endpoint of the interval of convergence of a power series is tricky. One result is due to Norway's greatest mathematician, Niels Henrik Abel (1802–1829). Suppose

$$f(x) = \sum_{n=0}^{\infty} a_n x^n$$

for $|x| < R$. If $f$ is continuous at an endpoint ($R$ or $-R$) and if the series converges there, then the formula also holds at that endpoint.

If we replace $x$ by $-x$ in the latter and multiply both sides by $-1$, we obtain

$$\ln(1 + x) = x - \frac{x^2}{2} + \frac{x^3}{3} - \frac{x^4}{4} + \cdots, \qquad -1 < x < 1$$

From Problem 45 of Section 11.5, we learn that this result is valid at the endpoint $x = 1$ (also see the note in the margin). ∎

**EXAMPLE 2**   Find the power series representation for $\tan^{-1}x$.

*Solution*   Recall that

$$\tan^{-1}x = \int_0^x \frac{1}{1 + t^2}\, dt$$

From the geometric series for $1/(1 - x)$ with $x$ replaced by $-t^2$, we get

$$\frac{1}{1 + t^2} = 1 - t^2 + t^4 - t^6 + \cdots, \qquad -1 < t < 1$$

Thus,

$$\tan^{-1}x = \int_0^x (1 - t^2 + t^4 - t^6 + \cdots)\, dt$$

that is,

$$\tan^{-1}x = x - \frac{x^3}{3} + \frac{x^5}{5} - \frac{x^7}{7} + \cdots, \qquad -1 < x < 1$$

(By the note in the margin, this also holds at $x = \pm 1$.) ∎

**EXAMPLE 3**   Find a formula for the sum of the series

$$S(x) = 1 + x + \frac{x^2}{2!} + \frac{x^3}{3!} + \cdots$$

*Solution*   We saw earlier (Section 11.6, Example 3) that this series converges for all $x$. Differentiating term by term, we obtain

$$S'(x) = 1 + x + \frac{x^2}{2!} + \frac{x^3}{3!} + \cdots$$

that is, $S'(x) = S(x)$ for all $x$. Furthermore, $S(0) = 1$. This differential equation has the unique solution $S(x) = e^x$ (see Section 7.5). Thus,

$$e^x = 1 + x + \frac{x^2}{2!} + \frac{x^3}{3!} + \cdots$$

∎

**EXAMPLE 4**   Obtain the power series representation for $e^{-x^2}$.

*Solution*   Simply substitute $-x^2$ for $x$ in the series for $e^x$.

$$e^{-x^2} = 1 - x^2 + \frac{x^4}{2!} - \frac{x^6}{3!} + \cdots$$

$\blacksquare$

**Algebraic Operations**   Power series behave like polynomials under the operations of addition and subtraction, as we know from Theorem 11.2B. The same is true for multiplication and division, as we now illustrate.

**EXAMPLE 5**   Multiply and divide the power series for $\ln(1 + x)$ by that for $e^x$.

*Solution*   We refer to Examples 1 and 3 for the required series. The key to multiplication is to first find the constant term, then the $x$-term, then the $x^2$-term, and so on. We arrange our work as follows.

$$0 + x - \frac{x^2}{2} + \frac{x^3}{3} - \frac{x^4}{4} + \cdots$$

$$1 + x + \frac{x^2}{2!} + \frac{x^3}{3!} + \frac{x^4}{4!} + \cdots$$

$$0 + (0 + 1)x + \left(0 + 1 - \frac{1}{2}\right)x^2 + \left(0 + \frac{1}{2!} - \frac{1}{2} + \frac{1}{3}\right)x^3$$

$$+ \left(0 + \frac{1}{3!} - \frac{1}{2!2} + \frac{1}{3} - \frac{1}{4}\right)x^4 + \cdots$$

$$= 0 + x + \frac{1}{2}x^2 + \frac{1}{3}x^3 + 0 \cdot x^4 + \cdots$$

Here is how division is done.

$$
\begin{array}{r}
x - \frac{3}{2}x^2 + \frac{4}{3}x^3 - \quad x^4 + \cdots \\
1 + x + \frac{1}{2}x^2 + \frac{1}{6}x^3 + \cdots \enclose{longdiv}{x - \frac{1}{2}x^2 + \frac{1}{3}x^3 - \frac{1}{4}x^4 + \cdots} \\
\underline{x + x^2 + \frac{1}{2}x^3 + \frac{1}{6}x^4 + \cdots} \\
-\frac{3}{2}x^2 - \frac{1}{6}x^3 - \frac{5}{12}x^4 + \cdots \\
\underline{-\frac{3}{2}x^2 - \frac{3}{2}x^3 - \frac{3}{4}x^4 + \cdots} \\
\frac{4}{3}x^3 + \frac{1}{3}x^4 + \cdots \\
\underline{\frac{4}{3}x^3 + \frac{4}{3}x^4 + \cdots} \\
-x^4 + \cdots
\end{array}
$$

$\blacksquare$

The real question relative to Example 5 is whether the two series we have obtained converge to $[\ln(1 + x)]e^x$ and $[\ln(1 + x)]/e^x$, respectively. Our next theorem, stated without proof, answers this question.

Theorem B

Let $f(x) = \sum a_n x^n$ and $g(x) = \sum b_n x^n$, with both of these series converging at least for $|x| < r$. If the operations of addition, subtraction, and multiplication are performed on these series as if they were polynomials, the resulting series will converge for $|x| < r$ and represent $f(x) + g(x)$, $f(x) - g(x)$, and $f(x) \cdot g(x)$, respectively. If $b_0 \neq 0$, the corresponding result holds for division, but we can guarantee its validity only for $|x|$ sufficiently small.

We mention that the operation of substituting one power series in another is also legitimate for $|x|$ sufficiently small, provided the constant term of the substituted series is zero. Here is an illustration.

**EXAMPLE 6** Find the power series for $e^{\tan^{-1} x}$ through terms of degree 4.

**Solution** Since

$$e^u = 1 + u + \frac{u^2}{2!} + \frac{u^3}{3!} + \frac{u^4}{4!} + \cdots$$

$$e^{\tan^{-1} x} = 1 + \tan^{-1} x + \frac{(\tan^{-1} x)^2}{2!} + \frac{(\tan^{-1} x)^3}{3!} + \frac{(\tan^{-1} x)^4}{4!} + \cdots$$

Now substitute the series for $\tan^{-1} x$ from Example 2 and combine like terms.

$$e^{\tan^{-1} x} = 1 + \left(x - \frac{x^3}{3} + \cdots\right) + \frac{\left(x - \dfrac{x^3}{3} + \cdots\right)^2}{2!} + \frac{\left(x - \dfrac{x^3}{3} + \cdots\right)^3}{3!}$$

$$+ \frac{\left(x - \dfrac{x^3}{3} + \cdots\right)^4}{4!} + \cdots$$

$$= 1 + \left(x - \frac{x^3}{3} + \cdots\right) + \frac{(x^2 - \frac{2}{3}x^4 + \cdots)}{2} + \frac{(x^3 + \cdots)}{6}$$

$$+ \frac{(x^4 + \cdots)}{24} + \cdots$$

$$= 1 + x + \frac{x^2}{2} - \frac{x^3}{6} - \frac{7x^4}{24} + \cdots$$ ∎

**Power Series in $x - a$** We have stated the theorems of this section for power series in $x$, but with obvious modifications they are equally valid for power series in $x - a$.

## CONCEPTS REVIEW

**1.** A power series may be differentiated or _____ term by term on the _____ of its interval of convergence.

**2.** The first five terms in the power series expansion for $\ln(1 - x)$ are _____.

**3.** The first four terms in the power series expansion for $\exp(x^2)$ are _____.

**4.** The first five terms in the power series expansion for $\exp(x^2) - \ln(1 - x)$ are _____.

## PROBLEM SET 11.7

In Problems 1–10, find the power series representation for $f(x)$ and specify the radius of convergence. Each is somehow related to a geometric series (see Examples 1 and 2).

**1.** $f(x) = \dfrac{1}{1 + x}$

**2.** $f(x) = \dfrac{1}{(1 + x)^2}$  *Hint*: Differentiate Problem 1.

**3.** $f(x) = \dfrac{1}{(1 - x)^3}$

**4.** $f(x) = \dfrac{x}{(1 + x)^2}$

**5.** $f(x) = \dfrac{1}{2 - 3x} = \dfrac{\frac{1}{2}}{1 - \frac{3}{2}x}$

**6.** $f(x) = \dfrac{1}{3 + 2x}$

**7.** $f(x) = \dfrac{x^2}{1 - x^4}$   **8.** $f(x) = \dfrac{x^3}{2 - x^3}$

**9.** $f(x) = \displaystyle\int_0^x \ln(1 + t)\, dt$   **10.** $f(x) = \displaystyle\int_0^x \tan^{-1} t\, dt$

**11.** Obtain the power series in $x$ for $\ln[(1 + x)/(1 - x)]$ and specify its radius of convergence. *Hint*: $\ln[(1 + x)/(1 - x)] = \ln(1 + x) - \ln(1 - x)$.

**12.** Show that any positive number $M$ can be represented by $(1 + x)/(1 - x)$, where $x$ lies within the interval of convergence of the series of Problem 11. Hence conclude that the natural logarithm of any positive number can be found by means of this series. Find $\ln 8$ this way to three decimal places.

In Problems 13–16, use the result of Example 3 to find the power series in $x$ for the given functions.

**13.** $f(x) = e^{-x}$   **14.** $f(x) = xe^{x^2}$

**15.** $f(x) = e^x + e^{-x}$   **16.** $f(x) = e^{2x} - 1 - 2x$

In Problems 17–24, use the methods of Example 5 to find power series in $x$ for each function $f$.

**17.** $f(x) = e^{-x} \cdot \dfrac{1}{1 - x}$   **18.** $f(x) = e^x \tan^{-1} x$

**19.** $f(x) = \dfrac{\tan^{-1} x}{e^x}$   **20.** $f(x) = \dfrac{e^x}{1 + \ln(1 + x)}$

**21.** $f(x) = (\tan^{-1} x)(1 + x^2 + x^4)$

**22.** $f(x) = \dfrac{\tan^{-1} x}{1 + x^2 + x^4}$

**23.** $f(x) = \displaystyle\int_0^x \dfrac{e^t}{1 + t}\, dt$   **24.** $f(x) = \displaystyle\int_0^x \dfrac{\tan^{-1} t}{t}\, dt$

**25.** Find the sum of each of the following series by recognizing how it is related to something familiar.

(a) $x - x^2 + x^3 - x^4 + x^5 - \cdots$

(b) $\dfrac{1}{2!} + \dfrac{x}{3!} + \dfrac{x^2}{4!} + \dfrac{x^3}{5!} + \cdots$

(c) $2x + \dfrac{4x^2}{2} + \dfrac{8x^3}{3} + \dfrac{16x^4}{4} + \cdots$

**26.** Follow the directions of Problem 25.

(a) $1 + x^2 + x^4 + x^6 + x^8 + \cdots$

(b) $\cos x + \cos^2 x + \cos^3 x + \cos^4 x + \cdots$

(c) $\dfrac{x^2}{2} + \dfrac{x^4}{4} + \dfrac{x^6}{6} + \dfrac{x^8}{8} + \cdots$

**27.** Find the sum of $\displaystyle\sum_{n=1}^{\infty} nx^n$.

**28.** Find the sum of $\displaystyle\sum_{n=1}^{\infty} n(n + 1)x^n$.

**29.** Use the method of substitution (Example 6) to find power series through terms of degree 3.

(a) $\tan^{-1}(e^x - 1)$       (b) $e^{e^x - 1}$

**30.** Suppose that $f(x) = \sum_{n=0}^{\infty} a_n x^n = \sum_{n=0}^{\infty} b_n x^n$ for $|x| < R$. Show that $a_n = b_n$ for all $n$. *Hint:* Let $x = 0$; then differentiate and let $x = 0$ again. Continue.

**31.** Find the power series representation of $x/(x^2 - 3x + 2)$. *Hint:* Use partial fractions.

**32.** Let $y = y(x) = x - \frac{x^3}{3!} + \frac{x^5}{5!} - \frac{x^7}{7!} + \cdots$. Show that $y$ satisfies the differential equation $y'' + y = 0$ with the conditions $y(0) = 0$ and $y'(0) = 1$. From this, guess at a simple formula for $y$.

**33.** Let $\{f_n\}$ be the Fibonacci sequence defined by

$$f_{n+2} = f_{n+1} + f_n, \qquad f_0 = 0, f_1 = 1$$

(See Problem 52 of Section 11.1 and Problem 28 of Section 11.6.) If $F(x) = \sum_{n=0}^{\infty} f_n x^n$, show that

$$F(x) - xF(x) - x^2 F(x) = x$$

and then use this fact to obtain a simple formula for $F(x)$.

**34.** Let $y = y(x) = \sum_{n=0}^{\infty} \frac{f_n}{n!} x^n$, where $f_n$ is as in Problem 33. Show that $y$ satisfies the differential equation $y'' - y' - y = 0$.

$\boxed{C}$ **35.** Did you ever wonder how people find the decimal expansion of $\pi$ to a large number of places? One method depends on the following identity (see Problem 47 of Section 7.6).

$$\pi = 16 \tan^{-1}\left(\tfrac{1}{5}\right) - 4 \tan^{-1}\left(\tfrac{1}{239}\right)$$

Find the first 6 digits of $\pi$ using this identity and the series for $\tan^{-1}x$. (You will need terms through $x^9/9$ for $\tan^{-1}\left(\tfrac{1}{5}\right)$

but only the first term for $\tan^{-1}(1/239)$.) In 1706, John Machin used this method to calculate the first 100 digits of $\pi$, while in 1973, Jean Guilloud and Martine Bouyer found the first million digits using the related identity

$$\pi = 48 \tan^{-1}\left(\tfrac{1}{18}\right) + 32 \tan^{-1}\left(\tfrac{1}{57}\right) - 20 \tan^{-1}\left(\tfrac{1}{239}\right)$$

In 1983, $\pi$ was calculated to over 16 million digits by a somewhat different method. Of course, computers were used in these recent calculations.

**36.** The number $e$ is readily calculated to as many digits as desired using the rapidly converging series

$$e = 1 + 1 + \frac{1}{2!} + \frac{1}{3!} + \frac{1}{4!} + \cdots$$

This series can also be used to show that $e$ is irrational. Do so by completing the following argument. Suppose $e = p/q$, where $p$ and $q$ are positive integers. Choose $n > q$ and let

$$M = n!\left(e - 1 - 1 - \frac{1}{2!} - \frac{1}{3!} - \cdots - \frac{1}{n!}\right)$$

Now $M$ is a positive integer. (Why?) Also

$$M = n!\left[\frac{1}{(n+1)!} + \frac{1}{(n+2)!} + \frac{1}{(n+3)!} + \cdots\right]$$

$$= \frac{1}{n+1} + \frac{1}{(n+1)(n+2)} + \frac{1}{(n+1)(n+2)(n+3)} + \cdots$$

$$< \frac{1}{n+1} + \frac{1}{(n+1)^2} + \frac{1}{(n+1)^3} + \cdots$$

$$= \frac{1}{n}$$

which gives a contradiction (to what?).

---

**Answers to Concepts Review:** **1.** Integrated; interior
**2.** $-x - \frac{1}{2}x^2 - \frac{1}{3}x^3 - \frac{1}{4}x^4 - \frac{1}{5}x^5$ **3.** $1 + x^2 + \frac{1}{2}x^4 + \frac{1}{6}x^6$
**4.** $1 + x + \frac{3}{2}x^2 + \frac{1}{3}x^3 + \frac{3}{4}x^4$

---

**11.8**
**TAYLOR AND MACLAURIN SERIES**

The big question still dangling is this: Given a function $f$ (for example, $\sin x$ or $\ln(\cos^2 x)$), can we represent it as a power series in $x$ or, more generally, in $x - a$? More precisely, can we find numbers $c_0, c_1, c_2, c_3, \ldots$ such that

$$f(x) = c_0 + c_1(x - a) + c_2(x - a)^2 + c_3(x - a)^3 + \cdots$$

on some interval around $a$?

Suppose such a representation exists. Then, by the theorem on differentiating series (Theorem 11.7A),

$$f'(x) = c_1 + 2c_2(x - a) + 3c_3(x - a)^2 + 4c_4(x - a)^3 + \cdots$$
$$f''(x) = 2!c_2 + 3!c_3(x - a) + 4 \cdot 3c_4(x - a)^2 + \cdots$$
$$f'''(x) = 3!c_3 + 4!c_4(x - a) + 5 \cdot 4 \cdot 3c_5(x - a)^2 + \cdots$$
$$\vdots$$

When we substitute $x = a$ and solve for $c_n$, we get

$$c_0 = f(a)$$
$$c_1 = f'(a)$$
$$c_2 = \frac{f''(a)}{2!}$$
$$c_3 = \frac{f'''(a)}{3!}$$

and, more generally,

$$c_n = \frac{f^{(n)}(a)}{n!}$$

(To make this valid for $n = 0$, we assume $f^{(0)}(a)$ means $f(a)$ and $0! = 1$.) Thus, the coefficients $c_n$ are determined by the function $f$. This also shows that a function $f$ cannot be represented by two different power series in $x - a$, an important point that we have glossed over until now. We summarize in the following theorem.

### Theorem A

**(Uniqueness Theorem).** Suppose $f$ satisfies

$$f(x) = c_0 + c_1(x - a) + c_2(x - a)^2 + c_3(x - a)^3 + \cdots$$

for all $x$ in some interval around $a$. Then,

$$c_n = \frac{f^{(n)}(a)}{n!}$$

Thus, a function cannot have more than one power series in $x - a$ that represents it.

The form of the coefficients $c_n$ should remind us of Taylor's Formula. Because of this, we call the power series representation of a function in $x - a$ its **Taylor series**. If $a = 0$, the corresponding series is called the **Maclaurin series**.

**Convergence of the Taylor Series**  But still the big question remains. Given a function $f$, can we represent it in a power series in $x - a$ (which must necessarily be the Taylor series)? Here is the answer.

### Theorem B

**(Taylor's Theorem).** Let $f$ be a function with derivatives of all orders in some interval $(a - r, a + r)$. A necessary and sufficient condition that the Taylor series

$$f(a) + f'(a)(x - a) + \frac{f''(a)}{2!}(x - a)^2 + \frac{f'''(a)}{3!}(x - a)^3 + \cdots$$

represents the function $f$ on that interval is that

$$\lim_{n \to \infty} R_n(x) = 0$$

where $R_n(x)$ is the remainder in Taylor's Formula; that is,

$$R_n(x) = \frac{f^{(n+1)}(c)}{(n+1)!}(x - a)^{n+1}$$

$c$ being some point in $(a - r, a + r)$.

**Proof**  We need only recall Taylor's Formula (Theorem 10.2A),

$$f(x) = f(a) + f'(a)(x - a) + \cdots + \frac{f^{(n)}(a)}{n!}(x - a)^n + R_n(x)$$

and the result is obvious.  ∎

Note that if $a = 0$, we get the Maclaurin series

$$f(0) + f'(0)x + \frac{f''(0)}{2!}x^2 + \frac{f'''(0)}{3!}x^3 + \cdots$$

> **Warning**
>
> Here is a fact that surprises many students. It is possible that the Taylor series for $f(x)$ converges on an interval but does not represent $f(x)$ there. This is shown by example in Problem 38. Of course,
>
> $$\lim_{n \to \infty} R_n(x) \neq 0$$
>
> in this example.

**EXAMPLE 1**  Find the Maclaurin series for $\sin x$ and prove that it represents $\sin x$ for all $x$.

**Solution**

$$
\begin{aligned}
f(x) &= \sin x & f(0) &= 0 \\
f'(x) &= \cos x & f'(0) &= 1 \\
f''(x) &= -\sin x & f''(0) &= 0 \\
f'''(x) &= -\cos x & f'''(0) &= -1 \\
f^{(4)}(x) &= \sin x & f^{(4)}(0) &= 0 \\
&\vdots & &\vdots
\end{aligned}
$$

Thus,

$$\sin x = x - \frac{x^3}{3!} + \frac{x^5}{5!} - \frac{x^7}{7!} + \cdots$$

and this is valid for all $x$, provided we can show

$$\lim_{n \to \infty} R_n(x) = \lim_{n \to \infty} \frac{f^{(n+1)}(c)}{(n+1)!} x^{n+1} = 0$$

Now, $|f^{(n+1)}(x)| = |\cos x|$ or $|f^{(n+1)}(x)| = |\sin x|$ and so

$$|R_n(x)| \le \frac{|x|^{n+1}}{(n+1)!}$$

But $\lim_{n \to \infty} x^n/n! = 0$, since $x^n/n!$ is the $n$th term of a convergent series (see Example 3 and Problem 21 of Section 11.6). As a consequence, we see that $\lim_{n \to \infty} R_n(x) = 0$. ■

**EXAMPLE 2**    Find the Maclaurin series for $\cos x$ and show that it represents $\cos x$ for all $x$.

*Solution*    We could proceed as in Example 1. However, it is easier to get the result by differentiating the series of that example (a valid procedure according to Theorem 11.7A). We obtain

$$\cos x = 1 - \frac{x^2}{2!} + \frac{x^4}{4!} - \frac{x^6}{6!} + \cdots$$

■

**EXAMPLE 3**    Find the Maclaurin series for $f(x) = \cosh x$ in two different ways, and show that it represents $\cosh x$ for all $x$.

*Solution*

*Method 1* is the direct method.

$$\begin{aligned}
f(x) &= \cosh x & f(0) &= 1 \\
f'(x) &= \sinh x & f'(0) &= 0 \\
f''(x) &= \cosh x & f''(0) &= 1 \\
f'''(x) &= \sinh x & f'''(0) &= 0 \\
&\;\;\vdots & &\;\;\vdots
\end{aligned}$$

Thus,

$$\cosh x = 1 + \frac{x^2}{2!} + \frac{x^4}{4!} + \frac{x^6}{6!} + \cdots$$

provided we can show $\lim_{n \to \infty} R_n(x) = 0$ for all $x$.

Now let $B$ be an arbitrary number and suppose $|x| \le B$. Then,

$$|\cosh x| = \left| \frac{e^x + e^{-x}}{2} \right| \le \frac{e^x}{2} + \frac{e^{-x}}{2} \le \frac{e^B}{2} + \frac{e^B}{2} = e^B$$

By similar reasoning, $|\sinh x| \le e^B$. Since $f^{(n+1)}(x)$ is either $\cosh x$ or $\sinh x$, we conclude that

$$|R_n(x)| = \left| \frac{f^{(n+1)}(c)x^{n+1}}{(n+1)!} \right| \le \frac{e^B |x|^{n+1}}{(n+1)!}$$

The latter expression tends to zero as $n \to \infty$, just as in Example 1.

*Method 2.* We use the fact that $\cosh x = (e^x + e^{-x})/2$. From Example 3 of Section 11.7,

$$e^x = 1 + x + \frac{x^2}{2!} + \frac{x^3}{3!} + \frac{x^4}{4!} + \cdots$$

$$e^{-x} = 1 - x + \frac{x^2}{2!} - \frac{x^3}{3!} + \frac{x^4}{4!} + \cdots$$

The previously obtained result follows by adding these two series and dividing by 2. ■

**EXAMPLE 4** Find the Maclaurin series for $\sinh x$ and show that it represents $\sinh x$ for all $x$.

*Solution* We do both jobs at once when we differentiate the series for $\cosh x$ (Example 3) term by term and use Theorem 11.7A.

$$\sinh x = x + \frac{x^3}{3!} + \frac{x^5}{5!} + \frac{x^7}{7!} + \cdots$$

■

**The Binomial Series** We are all familiar with the Binomial Formula. For a positive integer $p$,

$$(1 + x)^p = 1 + \binom{p}{1}x + \binom{p}{2}x^2 + \cdots + \binom{p}{p}x^p$$

where

$$\binom{p}{k} = \frac{p(p-1)(p-2)\cdots(p-k+1)}{k!}$$

Note that the symbol $\binom{p}{k}$ makes sense for any real number $p$, provided $k$ is a positive integer. This suggests a theorem.

---

### Theorem C

**(Binomial Series).** For any real number $p$ and for $|x| < 1$,

$$(1+x)^p = 1 + \binom{p}{1}x + \binom{p}{2}x^2 + \binom{p}{3}x^3 + \cdots$$

with $\binom{p}{k}$ defined as above.

---

***Partial proof***  Let $f(x) = (1 + x)^p$. Then,

$$
\begin{aligned}
f(x) &= (1+x)^p & f(0) &= 1 \\
f'(x) &= p(1+x)^{p-1} & f'(0) &= p \\
f''(x) &= p(p-1)(1+x)^{p-2} & f''(0) &= p(p-1) \\
f'''(x) &= p(p-1)(p-2)(1+x)^{p-3} & f'''(0) &= p(p-1)(p-2) \\
&\;\;\vdots & &\;\;\vdots
\end{aligned}
$$

Thus, the Maclaurin series for $(1 + x)^p$ is as indicated in the theorem. To show that it represents $(1 + x)^p$, we need to show that $\lim_{n \to \infty} R_n(x) = 0$. That, unfortunately, is difficult, and we leave it for more advanced courses. (See Problem 36 for a completely different way to prove Theorem C.)  ∎

If $p$ is a positive integer, $\binom{p}{k} = 0$ for $k > p$, and so the infinite series collapses to a series with finitely many terms; the series reduces to the usual Binomial Formula in this case.

**EXAMPLE 5**  Represent $(1 - x)^{-2}$ in a Maclaurin series for $-1 < x < 1$.

***Solution***  By Theorem C,

$$(1+x)^{-2} = 1 + (-2)x + \frac{(-2)(-3)}{2!}x^2 + \frac{(-2)(-3)(-4)}{3!}x^3 + \cdots$$

$$= 1 - 2x + 3x^2 - 4x^3 + \cdots$$

Thus,

$$(1 - x)^{-2} = 1 + 2x + 3x^2 + 4x^3 + \cdots$$

Naturally, this agrees with a result we obtained by a different method in Example 1 of Section 11.7.  ∎

**EXAMPLE 6**  Represent $\sqrt{1 + x}$ in a Maclaurin series and use it to approximate $\sqrt{1.1}$ to five decimal places.

*Solution*  By Theorem C,

$$(1 + x)^{1/2} = 1 + \frac{1}{2}x + \frac{\left(\frac{1}{2}\right)\left(-\frac{1}{2}\right)}{2!}x^2 + \frac{\left(\frac{1}{2}\right)\left(-\frac{1}{2}\right)\left(-\frac{3}{2}\right)}{3!}x^3$$

$$+ \frac{\left(\frac{1}{2}\right)\left(-\frac{1}{2}\right)\left(-\frac{3}{2}\right)\left(-\frac{5}{2}\right)}{4!}x^4 + \cdots$$

$$= 1 + \frac{1}{2}x - \frac{1}{8}x^2 + \frac{1}{16}x^3 - \frac{5}{128}x^4 + \cdots$$

Thus,

$$\sqrt{1.1} = 1 + \frac{0.1}{2} - \frac{0.01}{8} + \frac{0.001}{16} - \frac{5(0.0001)}{128} + \cdots$$

$$\approx 1.04881$$  ∎

**EXAMPLE 7**  Compute $\displaystyle\int_0^{0.4} \sqrt{1 + x^4}\, dx$ to five decimal places.

*Solution*  From Example 6,

$$\sqrt{1 + x^4} = 1 + \frac{1}{2}x^4 - \frac{1}{8}x^8 + \frac{1}{16}x^{12} - \frac{5}{128}x^{16} + \cdots$$

Thus,

$$\int_0^{0.4} \sqrt{1 + x^4}\, dx = \left[ x + \frac{x^5}{10} - \frac{x^9}{72} + \frac{x^{13}}{208} + \cdots \right]_0^{0.4} \approx 0.40102$$  ∎

**Summary**   We conclude our discussion of series with a list of the important Maclaurin series we have found. These series will be useful in doing the problem set but, what is more significant, they find application throughout mathematics and science.

**IMPORTANT MACLAURIN SERIES**

**1.** $\dfrac{1}{1-x} = 1 + x + x^2 + x^3 + x^4 + \cdots$    $-1 < x < 1$

**2.** $\ln(1+x) = x - \dfrac{x^2}{2} + \dfrac{x^3}{3} - \dfrac{x^4}{4} + \dfrac{x^5}{5} - \cdots$    $-1 < x \leq 1$

**3.** $\tan^{-1} x = x - \dfrac{x^3}{3} + \dfrac{x^5}{5} - \dfrac{x^7}{7} + \dfrac{x^9}{9} + \cdots$    $-1 \leq x \leq 1$

**4.** $e^x = 1 + x + \dfrac{x^2}{2!} + \dfrac{x^3}{3!} + \dfrac{x^4}{4!} + \cdots$

**5.** $\sin x = x - \dfrac{x^3}{3!} + \dfrac{x^5}{5!} - \dfrac{x^7}{7!} + \dfrac{x^9}{9!} - \cdots$

**6.** $\cos x = 1 - \dfrac{x^2}{2!} + \dfrac{x^4}{4!} - \dfrac{x^6}{6!} + \dfrac{x^8}{8!} - \cdots$

**7.** $\sinh x = x + \dfrac{x^3}{3!} + \dfrac{x^5}{5!} + \dfrac{x^7}{7!} + \dfrac{x^9}{9!} + \cdots$

**8.** $\cosh x = 1 + \dfrac{x^2}{2!} + \dfrac{x^4}{4!} + \dfrac{x^6}{6!} + \dfrac{x^8}{8!} + \cdots$

**9.** $(1+x)^p = 1 + \dbinom{p}{1}x + \dbinom{p}{2}x^2 + \dbinom{p}{3}x^3 + \dbinom{p}{4}x^4 + \cdots$    $-1 < x < 1$

## CONCEPTS REVIEW

**1.** If a function $f(x)$ is represented by the power series $\sum c_k x^k$, then $c_k = $ _____.

**2.** The Taylor series for a function will represent the function for those $x$ for which the remainder $R_n(x)$ in Taylor's Formula satisfies _____.

**3.** The Maclaurin series for $\sin x$ represents $\sin x$ for _____ $< x <$ _____.

**4.** The first four terms in the Maclaurin series for $(1+x)^{1/3}$ are _____.

## PROBLEM SET 11.8

In Problems 1–16, find the terms through $x^5$ in the Maclaurin series for $f(x)$. *Hint*: It may be easiest to use known Maclaurin series and then perform multiplications, divisions, and so on. For example, $\tan x = (\sin x)/(\cos x)$.

**1.** $f(x) = \tan x$

**2.** $f(x) = \tanh x$

**3.** $f(x) = e^x \sin x$

**4.** $f(x) = e^{-x} \cos x$

**5.** $f(x) = \cos x \ln(1+x)$

**6.** $f(x) = (\sin x)\sqrt{1+x}$

**7.** $f(x) = e^x + x + \sin x$

**8.** $f(x) = \dfrac{\cos x - 1 + x^2/2}{x^4}$

**9.** $f(x) = \dfrac{1}{1-x} \cosh x$

**10.** $f(x) = \dfrac{1}{1+x} \ln\left(\dfrac{1}{1+x}\right) = \dfrac{-\ln(1+x)}{1+x}$

**11.** $f(x) = \dfrac{1}{1+x+x^2}$

**12.** $f(x) = \dfrac{1}{1 - \sin x}$

**13.** $f(x) = \sin^3 x$

**14.** $f(x) = x(\sin 2x + \sin 3x)$

**15.** $f(x) = x \sec(x^2) + \sin x$

**16.** $f(x) = \dfrac{\cos x}{\sqrt{1 + x}}$

**17.** $f(x) = (1 + x)^{3/2}$　　　**18.** $f(x) = (1 - x^2)^{2/3}$

In Problems 19–24, find the Taylor series in $x - a$ through the term in $(x - a)^3$.

**19.** $e^x,\ a = 1$　　　　　　**20.** $\sin x,\ a = \dfrac{\pi}{6}$

**21.** $\cos x,\ a = \dfrac{\pi}{3}$　　　**22.** $\tan x,\ a = \dfrac{\pi}{4}$

**23.** $1 + x^2 + x^3,\ a = 1$

**24.** $2 - x + 3x^2 - x^3,\ a = -1$

**25.** Let $f(x) = \sum a_n x^n$ be an even function ($f(-x) = f(x)$) for $x$ in $(-R, R)$. Prove that $a_n = 0$ if $n$ is odd. *Hint:* Use the Uniqueness Theorem.

**26.** State and prove a theorem analogous to that in Problem 25 for odd functions.

**27.** Recall that

$$\sin^{-1} x = \int_0^x \frac{1}{\sqrt{1 - t^2}}\, dt$$

Find the first four nonzero terms in the Maclaurin series for $\sin^{-1} x$.

**28.** Given that

$$\sinh^{-1} x = \int_0^x \frac{1}{\sqrt{1 + t^2}}\, dt$$

find the first four nonzero terms in the Maclaurin series for $\sinh^{-1} x$.

☐ **29.** Calculate, accurate to four decimal places,

$$\int_0^1 \cos(x^2)\, dx.$$

☐ **30.** Calculate, accurate to five decimal places,

$$\int_0^{0.5} \sin \sqrt{x}\, dx.$$

**31.** By writing $1/x = 1/[1 - (1 - x)]$ and using the known expansion of $1/(1 - x)$, find the Taylor series for $1/x$ in powers of $x - 1$.

**32.** Let $f(x) = (1 + x)^{1/2} + (1 - x)^{1/2}$. Find the Maclaurin series for $f$ and use it to find $f^{(4)}(0)$ and $f^{(51)}(0)$.

**33.** In each case, find the Maclaurin series for $f(x)$ by use of known series and then use it to calculate $f^{(4)}(0)$.

(a) $f(x) = e^{x + x^2}$　　　　　(b) $f(x) = e^{\sin x}$

(c) $f(x) = \displaystyle\int_0^x \frac{e^{t^2} - 1}{t^2}\, dt$　　(d) $f(x) = e^{\cos x} = e \cdot e^{\cos x - 1}$

(e) $f(x) = \ln(\cos^2 x)$

**34.** One can sometimes find a Maclaurin series by the *method of equating coefficients*. For example, let

$$\tan x = \frac{\sin x}{\cos x} = a_0 + a_1 + a_2 x^2 + \cdots$$

Then multiply by $\cos x$ and replace $\sin x$ and $\cos x$ by their series to obtain

$$x - \frac{x^3}{6} + \cdots = (a_0 + a_1 x + a_2 x^2 + \cdots)\left(1 - \frac{x^2}{2} + \cdots\right)$$

$$= a_0 + a_1 + \left(a_2 - \frac{a_0}{2}\right)x^2 + \left(a_3 - \frac{a_1}{2}\right)x^3 + \cdots$$

Thus,

$$a_0 = 0, \quad a_1 = 1, \quad a_2 - \frac{a_0}{2} = 0, \quad a_3 - \frac{a_1}{2} = -\frac{1}{6}, \cdots$$

so

$$a_0 = 0, \quad a_1 = 1, \quad a_2 = 0, \quad a_3 = \tfrac{1}{3}, \cdots$$

and therefore

$$\tan x = 0 + x + 0 + \tfrac{1}{3}x^3 + \cdots$$

which agrees with Problem 1. Use this method to find the terms through $x^4$ in the series for $\sec x$.

**35.** Use the method of Problem 34 to find the terms through $x^5$ in the Maclaurin series for $\tanh x$.

**36.** Prove Theorem C as follows. Let

$$f(x) = 1 + \sum_{n=1}^{\infty} \binom{p}{n} x^n.$$

(a) Show the series converges for $|x| < 1$.
(b) Show $(1 + x)f'(x) = pf(x)$ and $f(0) = 1$.
(c) Solve this differential equation to get $f(x) = (1 + x)^p$.

**37.** Let

$$f(t) = \begin{cases} 0 & t < 0 \\ t^4 & t \geq 0 \end{cases}$$

Explain why $f(t)$ cannot be represented by a Maclaurin series. Also show that if $g(t)$ gives the distance traveled by a car that is stationary for $t < 0$ and moving ahead for $t \geq 0$, then $g(t)$ cannot be represented by a Maclaurin series.

**38.** Let

$$f(x) = \begin{cases} e^{-1/x^2} & x \neq 0 \\ 0 & x = 0 \end{cases}$$

(a) Show that $f'(0) = 0$ by using the definition of the derivative.
(b) Show that $f''(0) = 0$.
(c) Assuming the known fact that $f^{(n)}(0) = 0$ for all $n$, find the Maclaurin series for $f(x)$.
(d) Does the Maclaurin series represent $f(x)$?
(e) What is the remainder in Maclaurin's Formula in this case?

This shows that a Maclaurin series may exist and yet not represent the given function (the remainder does not tend to 0 as $n \to \infty$).

PC Use a computer (Section *Taylor* of *True BASIC Calculus*) to find the first four nonzero terms in the Maclaurin series for each of the following. Check Problems 41–46 to see that you get the same answers using the results of Section 11.7.

**39.** $\sin x$      **40.** $\exp x$

**41.** $3 \sin x - 2 \exp x$      **42.** $\exp(x^2)$

**43.** $\sin(\exp x - 1)$      **44.** $\exp(\sin x)$

**45.** $(\sin x)(\exp x)$      **46.** $(\sin x)/(\exp x)$

---

**Answers to Concepts Review:** **1.** $f^{(k)}(0)/k!$
**2.** $\lim\limits_{n \to \infty} R_n(x) = 0$   **3.** $-\infty; \infty$   **4.** $1 + \frac{1}{3}x - \frac{1}{9}x^2 + \frac{5}{81}x^3$

---

## 11.9 CHAPTER REVIEW

### Concepts Test

Respond with true or false to each of the following assertions. Be prepared to justify your answer.

**1.** If $0 \leq a_n \leq b_n$ for all $n$ in $\mathbb{N}$ and $\lim\limits_{n \to \infty} b_n$ exists, then $\lim\limits_{n \to \infty} a_n$ exists.

**2.** For every positive integer $n$, it is true that $n! \leq n^n \leq (2n - 1)!$.

**3.** If $\lim\limits_{n \to \infty} a_n = L$, then $\lim\limits_{n \to \infty} a_{3n+4} = L$.

**4.** If $\lim\limits_{n \to \infty} a_{2n} = L$ and $\lim\limits_{n \to \infty} a_{3n} = L$, then $\lim\limits_{n \to \infty} a_n = L$.

**5.** If $\lim\limits_{n \to \infty} a_{mn} = L$ for every positive integer $m \geq 2$, then $\lim\limits_{n \to \infty} a_n = L$.

**6.** If $\lim\limits_{n \to \infty} a_{2n} = L$ and $\lim\limits_{n \to \infty} a_{2n+1} = L$, then $\lim\limits_{n \to \infty} a_n = L$.

**7.** If $\lim\limits_{n \to \infty} (a_n - a_{n+1}) = 0$, then $\lim\limits_{n \to \infty} a_n$ exists and is finite.

**8.** If $\{a_n\}$ and $\{b_n\}$ both diverge, then $\{a_n + b_n\}$ diverges.

**9.** If $\{a_n\}$ converges, then $\{a_n/n\}$ converges to 0.

**10.** If $\sum\limits_{n=1}^{\infty} a_n$ converges, so does $\sum\limits_{n=1}^{\infty} a_n^2$.

**11.** If $0 < a_{n+1} < a_n$ for all $n$ in $\mathbb{N}$ and $\lim\limits_{n \to \infty} a_n = 0$, then $\sum\limits_{n=1}^{\infty} (-1)^{n+1} a_n$ converges and has sum $S$ satisfying $0 < S < a_1$.

**12.** $\sum\limits_{n=1}^{\infty} \left(\frac{1}{n}\right)^n$ converges and has sum $S$ satisfying $1 < S < 2$.

**13.** If a series $\sum a_n$ diverges, then its sequence of partial sums is unbounded.

**14.** If $0 \leq a_n \leq b_n$ for all $n$ in $\mathbb{N}$ and if $\sum\limits_{n=1}^{\infty} b_n$ diverges, then $\sum\limits_{n=1}^{\infty} a_n$ diverges.

**15.** The Ratio Test will not help in determining the convergence or divergence of $\sum\limits_{n=1}^{\infty} \dfrac{2n + 3}{3n^4 + 2n^3 + 3n + 1}$.

**16.** If $a_n > 0$ for all $n$ in $\mathbb{N}$ and $\sum\limits_{n=1}^{\infty} a_n$ converges, then $\lim\limits_{n \to \infty} (a_{n+1}/a_n) < 1$.

**17.** $\sum\limits_{n=1}^{\infty} \left(1 - \frac{1}{n}\right)^n$ converges.

**18.** $\sum\limits_{n=1}^{\infty} \dfrac{1}{\ln(n^4 + 1)}$ converges.

**19.** $\sum\limits_{n=2}^{\infty} \dfrac{n + 1}{(n \ln n)^2}$ converges.

**20.** $\sum\limits_{n=1}^{\infty} \dfrac{\sin^2(n\pi/2)}{n}$ converges.

**21.** If $0 \leq a_{n+100} \leq b_n$ for all $n$ in $\mathbb{N}$ and $\sum\limits_{n=1}^{\infty} b_n$ converges, then $\sum\limits_{n=1}^{\infty} a_n$ converges.

22. If for some $c > 0$, $ca_n \geq 1/n$ for all $n$ in $\mathbb{N}$, then $\sum_{n=1}^{\infty} a_n$ diverges.

23. $\frac{1}{3} + \left(\frac{1}{3}\right)^2 + \left(\frac{1}{3}\right)^3 + \cdots + \left(\frac{1}{3}\right)^{1000} < \frac{1}{2}$.

24. If $\sum_{n=1}^{\infty} a_n$ converges, then $\sum_{n=1}^{\infty} (-1)^n a_n$ converges.

25. If $b_n \leq a_n \leq 0$ for all $n$ in $\mathbb{N}$ and $\sum_{n=1}^{\infty} b_n$ converges, then $\sum_{n=1}^{\infty} a_n$ converges.

26. If $0 \leq a_n$ for all $n$ in $\mathbb{N}$ and $\sum_{n=1}^{\infty} a_n$ converges, then $\sum_{n=1}^{\infty} (-1)^n a_n$ converges.

27. $\left| \sum_{n=1}^{\infty} (-1)^{n+1} \frac{1}{n} - \sum_{n=1}^{\infty} (-1)^{n+1} \frac{1}{n} \right| < 0.01$.

28. If $\sum_{n=1}^{\infty} a_n$ diverges, then $\sum_{n=1}^{\infty} |a_n|$ diverges.

29. If the power series $\sum_{n=0}^{\infty} a_n(x - 3)^n$ converges at $x = -1.1$, it also converges at $x = 7$.

30. If $\sum_{n=0}^{\infty} a_n x^n$ converges at $x = -2$, it also converges at $x = 2$.

31. If $f(x) = \sum_{n=0}^{\infty} a_n x^n$ and the series converges at $x = 1.5$, then $\int_0^1 f(x)\, dx = \sum_{n=0}^{\infty} a_n/(n + 1)$.

32. Every power series converges for at least two values of the variable.

33. If $f(0), f'(0), f''(0), \ldots$ all exist, then the Maclaurin series for $f(x)$ converges to $f(x)$ in a neighborhood of $x = 0$.

34. The function $f(x) = 1 + x + x^2 + x^3 + \cdots$ satisfies the differential equation $y' = y^2$ on the interval $(-1, 1)$.

35. The function $f(x) = \sum_{n=0}^{\infty} (-1)^n x^n/n!$ satisfies the differential equation $y' + y = 0$ on the whole real line.

## Sample Test Problems

In Problems 1–8, determine whether the given sequence converges or diverges and, if it converges, find $\lim_{n \to \infty} a_n$.

1. $a_n = \dfrac{9n}{\sqrt{9n^2 + 1}}$

2. $a_n = \dfrac{\ln n}{\sqrt{n}}$

3. $a_n = \left(1 + \dfrac{4}{n}\right)^n$

4. $a_n = \dfrac{n!}{3^n}$

5. $a_n = \sqrt[n]{n}$

6. $a_n = \dfrac{1}{\sqrt[3]{n}} + \dfrac{1}{\sqrt[3]{3}}$

7. $a_n = \dfrac{\sin^2 n}{\sqrt{n}}$

8. $a_n = \cos\left(\dfrac{n\pi}{6}\right)$

In Problems 9–18, determine whether the given series converges or diverges and, if it converges, find its sum.

9. $\sum_{k=1}^{\infty} \left(\dfrac{1}{\sqrt{k}} - \dfrac{1}{\sqrt{k + 1}}\right)$

10. $\sum_{k=1}^{\infty} \left(\dfrac{1}{k} - \dfrac{1}{k + 2}\right)$

11. $\ln \frac{1}{2} + \ln \frac{2}{3} + \ln \frac{3}{4} + \cdots$

12. $\sum_{k=1}^{\infty} \cos k\pi$

13. $\sum_{k=0}^{\infty} e^{-2k}$

14. $\sum_{k=0}^{\infty} \left(\dfrac{3}{2^k} + \dfrac{4}{3^k}\right)$

15. $0.91919191\ldots = \sum_{k=1}^{\infty} 91 \left(\dfrac{1}{100}\right)^k$

16. $\sum_{k=1}^{\infty} \left(\dfrac{1}{\ln 2}\right)^k$

17. $1 - \dfrac{2^2}{2!} + \dfrac{2^4}{4!} - \dfrac{2^6}{6!} + \cdots$

18. $1 - \dfrac{1}{1!} + \dfrac{1}{2!} - \dfrac{1}{3!} + \dfrac{1}{4!} - \cdots$

In Problems 19–32, indicate whether the given series converges or diverges and give a reason for your conclusion.

19. $\sum_{n=1}^{\infty} \dfrac{n}{1 + n^2}$

20. $\sum_{n=1}^{\infty} \dfrac{n + 5}{1 + n^3}$

21. $\sum_{n=1}^{\infty} (-1)^{n+1} \dfrac{1}{\sqrt[3]{n}}$

22. $\sum_{n=1}^{\infty} (-1)^{n+1} \dfrac{1}{\sqrt[n]{3}}$

23. $\sum_{n=1}^{\infty} \dfrac{2^n + 3^n}{4^n}$

24. $\sum_{n=1}^{\infty} \dfrac{n}{e^{n^2}}$

25. $\sum_{n=1}^{\infty} (-1)^{n+1} \dfrac{n + 1}{10n + 12}$

26. $\sum_{n=1}^{\infty} \dfrac{\sqrt{n}}{n^2 + 7}$

27. $\sum_{n=1}^{\infty} \dfrac{n^2}{n!}$

28. $\sum_{n=1}^{\infty} \dfrac{n^3 3^n}{(n + 1)!}$

29. $\sum_{n=1}^{\infty} \dfrac{2^n n!}{(n + 2)!}$

30. $\sum_{n=2}^{\infty} \left(1 - \dfrac{1}{n}\right)^n$

31. $\sum_{n=1}^{\infty} n^2 \left(\dfrac{2}{3}\right)^n$

32. $\sum_{n=1}^{\infty} \dfrac{(-1)^n}{1 + \ln n}$

In Problems 33–36, state whether the given series is absolutely convergent, conditionally convergent, or divergent.

**33.** $\displaystyle\sum_{n=1}^{\infty} (-1)^n \frac{1}{3n-1}$

**34.** $\displaystyle\sum_{n=1}^{\infty} \frac{(-1)^n n^3}{2^n}$

**35.** $\displaystyle\sum_{n=1}^{\infty} (-1)^n \frac{3^n}{2^{n+8}}$

**36.** $\displaystyle\sum_{n=2}^{\infty} \frac{(-1)^n \sqrt[n]{n}}{\ln n}$

In Problems 37–42, find the convergence set for the power series.

**37.** $\displaystyle\sum_{n=0}^{\infty} \frac{x^n}{n^3+1}$

**38.** $\displaystyle\sum_{n=0}^{\infty} \frac{(-2)^{n+1} x^n}{2n+3}$

**39.** $\displaystyle\sum_{n=0}^{\infty} \frac{(-1)^n (x-4)^n}{n+1}$

**40.** $\displaystyle\sum_{n=0}^{\infty} \frac{3^n x^{3n}}{(3n)!}$

**41.** $\displaystyle\sum_{n=0}^{\infty} \frac{(x-3)^n}{2^n+1}$

**42.** $\displaystyle\sum_{n=0}^{\infty} \frac{n!(x+1)^n}{3^n}$

**43.** By differentiating the geometric series

$$\frac{1}{1+x} = 1 - x + x^2 - x^3 + x^4 - \cdots, \qquad |x| < 1,$$

find a power series that represents $1/(1+x)^2$. What is its interval of convergence?

**44.** Find a power series that represents $1/(1+x)^3$ on the interval $(-1, 1)$.

**45.** Find the Maclaurin series for $\sin^2 x$. For what values of $x$ does the series represent the function?

**46.** Find the first five terms of the Taylor series for $e^x$ based at the point $x = 2$.

**47.** Write the Maclaurin series for $f(x) = \sin x + \cos x$. For what values of $x$ does it represent $f$?

C **48.** Write the Maclaurin series for $f(x) = \cos x^2$ and use it to approximate

$$\int_0^1 \cos x^2 \, dx$$

How many terms of the series are needed to compute the value of this integral correct to four decimal places?

C **49.** Calculate the following integral correct to five decimal places.

$$\int_0^{0.2} \frac{e^x - 1}{x} \, dx$$

**50.** How many terms do we have to take in the convergent series

$$1 - \frac{1}{\sqrt{2}} + \frac{1}{\sqrt{3}} - \frac{1}{\sqrt{4}} + \frac{1}{\sqrt{5}} - \frac{1}{\sqrt{6}} + \cdots$$

to be sure that we have approximated its sum to within 0.001?

**51.** Give a good bound for the maximum error made in approximating $\cos x$ by $1 - x^2/2$ for $-0.1 \le x \le 0.1$.

**52.** Use the simplest method you can think of to find the first three nonzero terms of the Maclaurin series for each of the following.

(a) $\dfrac{1}{1-x^3}$

(b) $\sqrt{1+x^2}$

(c) $e^{-x} - 1 + x$

(d) $x \sec x$

(e) $e^{-x} \sin x$

(f) $\dfrac{1}{1+\sin x}$

# TECHNOLOGY PROJECT

**Exercise 1**  Do the following "by hand."

(a) Find the Taylor polynomial of degree three, $P_3(x)$, by expanding the function $f(x) = e^x \sin x$ about $x = 0$. **A check:** You should get $f^{(iv)}(x) = -4e^x \sin x$ for the fourth derivative.

(b) Using the remainder term in Taylor's theorem, determine an upper bound for the absolute value of the maximum error in $P_3(x)$ on the interval $[-1, 1]$.

**Exercise 2**  The following idea of building on known expansions is often used to get an approximation for a more complicated function. Consider, for example, the function $g(x) = \sin(\sin x)$.

(a) Explain why $\sin x \approx x - \frac{1}{6} x^3$ is a decent approximation for "small" $x$.

(b) Suppose that the above small $x$ approximation is good enough for your purposes. You can get a similar approximation for $g(x)$ by substituting $u = x - \frac{1}{6} x^3$ (your approximation to $\sin x$), and similarly approximate $\sin u$. Do this.

Most modern computing technologies have the ability to compute the Taylor polynomial $P_n$ of degree $n$ about $x = a$ for a given function $f$. For example, here is how *Maple* defines $f = x \cos(2x)$ and finds a polynomial expansion about $a = 0$:

```
f := x -> x * cos(2*x):
a := 0:  n := 6:
ourSeries := taylor(f(x), x = a, n);
            3        5        7
     x - 2 x  + 2/3 x  + O(x )
Pn := convert (ourSeries, polynom);
            3        5
Pn := x - 2 x  + 2/3 x
```

Note that the O-term at the end of the output of the Taylor command designates that the error term is of "order" 7. In order to get a usable polynomial, Pn, we used the **convert** command. Let's see how good our approximation is on the interval $[-1, 1]$. Figure 1 shows that the approximation is fair, although it is breaking down at the ends of the interval.

**Exercise 3**  For the function $f(x) = x \cos 2x$ discussed above, set up for a Taylor expansion on your computing technology.

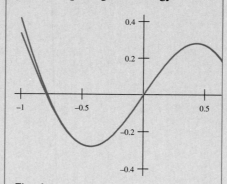

Fig. 1.
$x \cos 2x$ and its 5th degree Taylor polynomial near $x = 0$.

(a) Gradually increase $n$ until you have a good graphical approximation. Show your final graph.

(b) Suppose you wish to approximate $f(x)$ on the interval $[-1, 1]$ by an $n$th degree polynomial $P_n(x)$ to within an error of 0.005. By plotting the difference $f(x) - P_n(x)$, experimentally find the smallest $n$ that is adequate. Show your final graph of $f(x) - P_n(x)$. (Note: if your technology cannot produce this much accuracy, do the best you can.)

**Exercise 4**  Following up on Exercises 1 and 2:

(a) Use your computing technology to find the $P_3(x)$ in Exercise 1. Graph the error $f(x) - P_3(x)$ and comment on how the actual error compares with your upper bound in Exercise 1.

(b) Use your technology to find a Taylor polynomial approximating $g(x) = \sin(\sin x)$, thus checking your earlier work in Exercise 2.

**Exercise 5**  Consider the two functions $\sin(\tan x)$ and $\tan(\sin x)$.

(a) Obtain series approximations to these functions about $x = 0$ and determine the first power of $x$ for which the series differ.

(b) Since these two series look very much alike, you may think they represent essentially the same function. Hand in plots of $\sin(\tan x)$ and $\tan(\sin x)$ on intervals $[0, \pi/4]$ and $[0, \pi]$ and comment, especially on their behavior near $\pi/2$.

# 12

# Conics and Polar Coordinates

## 12.1
### THE PARABOLA

Take a right circular cone with two nappes and pass planes through it at various angles, as shown in Figure 1. As sections, you will obtain curves called, respectively, an ellipse, a parabola, and a hyperbola. (You may also obtain various limiting forms: a circle, a point, intersecting lines, and one line.) These curves are called *conic sections*, or simply *conics*. This definition, which is due to the Greeks, is cumbersome and we shall immediately adopt a different one. It can be shown that the two notions are consistent.

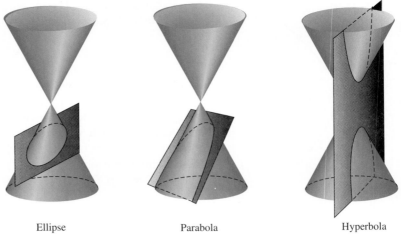

Ellipse      Parabola      Hyperbola

FIGURE 1

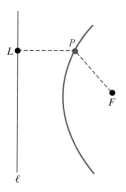

FIGURE 2

In the plane let $\ell$ be a fixed line (the **directrix**) and $F$ be a fixed point (the **focus**) not on the line, as in Figure 2. The set of points $P$ for which the ratio of the distance $|PF|$ from the focus to the distance $|PL|$ from the line is a positive constant $e$ (the **eccentricity**)—that is, which satisfies

$$\boxed{|PF| = e|PL|}$$

is called a **conic**. If $0 < e < 1$, the conic is an **ellipse**; if $e = 1$, it is a **parabola**; if $e > 1$, it is a **hyperbola**.

When we draw the curves corresponding to $e = \frac{1}{2}$, $e = 1$, and $e = 2$, we get the three curves shown in Figure 3.

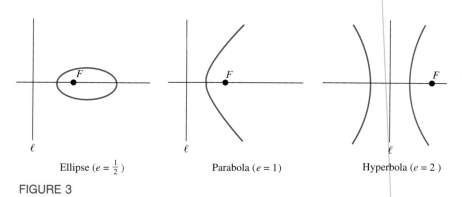

Ellipse $(e = \frac{1}{2})$     Parabola $(e = 1)$     Hyperbola $(e = 2)$

FIGURE 3

In each case, the curves are symmetric with respect to the line through the focus perpendicular to the directrix. We call this line the **major axis** (or simply the *axis*) of the conic. A point where the conic crosses the axis is called a **vertex**. The parabola has one vertex, while the ellipse and hyperbola have two vertices.

**The Parabola ($e = 1$)**  A **parabola** is the set of points $P$ that are equidistant from the directrix $\ell$ and the focus $F$—that is, which satisfy

$$|PF| = |PL|$$

From this definition, we wish to derive the $xy$-equation, and we want it to be as simple as possible. The position of the coordinate axes has no effect on the curve, but it does affect the simplicity of the curve's equation. Since a parabola is symmetric with respect to its axis, it is natural to place one of the coordinate axes—for instance, the $x$-axis—along the axis. Let the focus $F$ be to the right of the origin, say at $(p, 0)$, and the directrix to the left with equation $x = -p$. Then the vertex is at the origin. All this is shown in Figure 4.

From the condition $|PF| = |PL|$ and the distance formula, we get

$$\sqrt{(x - p)^2 + (y - 0)^2} = \sqrt{(x + p)^2 + (y - y)^2}$$

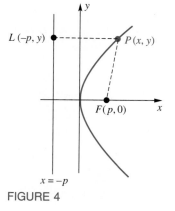

FIGURE 4

After squaring both sides and simplifying, we obtain

$$y^2 = 4px$$

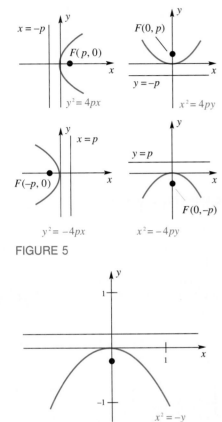

FIGURE 5

This is called the **standard equation** of a horizontal parabola (horizontal axis) opening to the right. Note that $p > 0$ and that $p$ is the distance from the focus to the vertex.

**EXAMPLE 1**  Find the focus and directrix of the parabola with equation $y^2 = 12x$.

*Solution*  Since $y^2 = 4(3)x$, we see that $p = 3$. The focus is at $(3, 0)$; the directrix is the line $x = -3$.  ∎

There are three variants of the standard equation. If we interchange the roles of $\boldsymbol{x}$ and $y$, we obtain the equation $x^2 = 4py$. It is the equation of a vertical parabola with focus at $(0, p)$ and directrix $y = -p$. Finally, introducing a minus sign on one side of the equation causes the parabola to open in the opposite direction. All four cases are shown in Figure 5.

**EXAMPLE 2**  Determine the focus and directrix of the parabola $x^2 = -y$ and sketch the graph.

*Solution*  We write $x^2 = -4\left(\frac{1}{4}\right)y$, from which we conclude that $p = \frac{1}{4}$. The form of the equation tells us that the parabola is vertical and opens down. The focus is at $\left(0, -\frac{1}{4}\right)$; the directrix is the line $y = \frac{1}{4}$. The graph is shown in Figure 6.  ∎

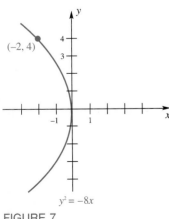

FIGURE 6

**EXAMPLE 3**  Find the equation of the parabola with vertex at the origin and focus at $(0, 5)$.

*Solution*  The parabola opens up and $p = 5$. The equation is $x^2 = 4(5)y$— that is, $x^2 = 20y$.  ∎

**EXAMPLE 4**  Find the equation of the parabola with vertex at the origin that goes through $(-2, 4)$ and opens left. Sketch the graph.

*Solution*  The equation has the form $y^2 = -4px$. Because $(-2, 4)$ is on the graph, $(4)^2 = -4p(-2)$, from which $p = 2$. The desired equation is $y^2 = -8x$ and its graph is sketched in Figure 7.  ∎

**The Optical Property**  A simple geometric property of a parabola is the basis of many important applications. If $F$ is the focus and $P$ is any point on the parabola, the tangent line at $P$ makes equal angles with $FP$ and the line $GP$, which is parallel to the axis of the parabola (see Figure 8). A principle from physics says that when a light ray strikes a reflecting surface, the angle of incidence is equal to the angle of reflection. It follows that if a parabola is revolved about its axis to form a hollow reflecting shell, all light rays from

FIGURE 7

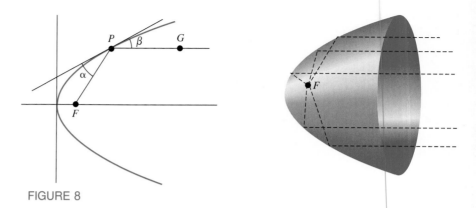

FIGURE 8

the focus after hitting the shell are reflected outward parallel to the axis. This property of the parabola is used in designing search lights, with the light source placed at the focus. Conversely, it is used in certain telescopes where incoming parallel rays from a distant star are focused at a single point.

**EXAMPLE 5**  Prove the optical property of the parabola.

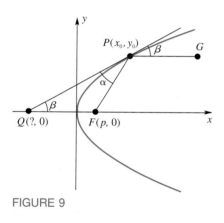

FIGURE 9

**Solution**  In Figure 9, let $QP$ be the tangent line at $P$ and let $GP$ be the line parallel to the $x$-axis. We must show that $\alpha = \beta$. After noting that $\angle FQP = \beta$, we reduce the problem to showing that triangle $FQP$ is isosceles.

First, we obtain the $x$-coordinate of $Q$. Differentiating $y^2 = 4px$ implicitly gives $2y'y = 4p$, from which we conclude that the slope of the tangent line at $P(x_0, y_0)$ is $2p/y_0$. The equation of this line is

$$y - y_0 = \frac{2p}{y_0}(x - x_0)$$

Setting $y = 0$ and solving for $x$ gives $-y_0 = (2p/y_0)(x - x_0)$, or $x - x_0 = -y_0^2/2p$. Now $y_0^2 = 4px_0$, which gives $x - x_0 = -2x_0$—that is, $x = -x_0$; $Q$ has coordinates $(-x_0, 0)$.

To show that the segments $FP$ and $FQ$ have equal length, we use the distance formula

$$|FP| = \sqrt{(x_0 - p)^2 + y_0^2} = \sqrt{x_0^2 - 2x_0p + p^2 + 4px_0}$$
$$= \sqrt{x_0^2 + 2x_0p + p^2} = x_0 + p = |FQ| \qquad \blacksquare$$

Sound obeys the same laws of reflection as light, and parabolic microphones are used to pick up and concentrate sounds from, for example, a distant part of a football stadium. Radar and radio telescopes are also based on these same principles.

There are many other applications of parabolas. For example, the path of a projectile is a parabola if air resistance and other minor factors are neglected. The cable of an evenly loaded suspension bridge takes the form of a parabola. Arches are often parabolic. The paths of a few comets are parabolic.

# CONCEPTS REVIEW

**1.** The set of points $P$ satisfying $|PF| = e|PL|$ (that is, distance to the focus equals $e$ times distance to the directrix) is an ellipse if _____, a parabola if _____, and a hyperbola if _____.

**2.** The standard equation of a parabola, vertex at the origin and opening right, is _____.

**3.** The parabola $y = \frac{1}{4}x^2$ has focus _____ and directrix _____.

**4.** The rays from a light source at the focus of a parabolic mirror will be reflected in a direction _____.

# PROBLEM SET 12.1

In Problems 1–8, find the coordinates of the focus and the equation of the directrix for each parabola. Make a sketch showing the parabola, its focus, and its directrix.

**1.** $y^2 = 4x$

**2.** $y^2 = -12x$

**3.** $x^2 = -12y$

**4.** $x^2 = -16y$

**5.** $y^2 = x$

**6.** $y^2 + 3x = 0$

**7.** $6y - 2x^2 = 0$

**8.** $3x^2 - 9y = 0$

In Problems 9–14, find the standard equation of each parabola from the given information. Assume the vertex is at the origin.

**9.** Focus is at $(2, 0)$.

**10.** Directrix is $x = 3$.

**11.** Directrix is $y - 2 = 0$.

**12.** Focus is $(0, -\frac{1}{9})$.

**13.** Focus is $(-4, 0)$.

**14.** Directrix is $y = \frac{7}{2}$.

**15.** Find the equation of the parabola with vertex at the origin and axis along the $x$-axis if the parabola passes through the point $(3, -1)$. Make a sketch.

**16.** Find the equation of the parabola through the point $(-2, 4)$ if its vertex is at the origin and its axis is along the $x$-axis. Make a sketch.

**17.** Find the equation of the parabola through the point $(6, -5)$ if its vertex is at the origin and its axis is along the $y$-axis. Make a sketch.

**18.** Find the equation of the parabola whose vertex is the origin and whose axis is the $y$-axis if the parabola passes through the point $(-3, 5)$. Make a sketch.

In Problems 19–26, find the equations of the tangent and the normal to the given parabola at the given point. Sketch the parabola, the tangent, and the normal in 19–22.

**19.** $y^2 = 16x$, $(1, -4)$

**20.** $x^2 = -10y$, $(2\sqrt{5}, -2)$

**21.** $x^2 = 2y$, $(4, 8)$

**22.** $y^2 = -9x$, $(-1, -3)$

**23.** $y^2 = -15x$, $(-3, -3\sqrt{5})$

**24.** $x^2 = 4y$, $(4, 4)$

**25.** $x^2 = -6y$, $(3\sqrt{2}, -3)$

**26.** $y^2 = 20x$, $(2, -2\sqrt{10})$

**27.** The slope of the tangent to the parabola $y^2 = 5x$ at a certain point on the parabola is $\sqrt{5}/4$. Find the coordinates of that point. Make a sketch.

**28.** The slope of the tangent to the parabola $x^2 = -14y$ at a certain point on the parabola is $-2\sqrt{7}/7$. Find the coordinates of that point.

**29.** Find the equation of the tangent to the parabola $y^2 = -18x$ that is parallel to the line $3x - 2y + 4 = 0$.

**30.** Any line segment through the focus of a parabola, with endpoints on the parabola, is a **focal chord**. Prove that the tangents to a parabola at the extremities of any focal chord intersect on the directrix.

**31.** Prove that the tangents to a parabola at the extremities of any focal chord are perpendicular to each other (see Problem 30).

**32.** A chord of a parabola that is perpendicular to the axis and 1 unit from the vertex has length 1 unit. How far is it from the vertex to the focus?

**33.** Prove that the vertex is the point on a parabola closest to the focus.

**34.** A space ship from outer space is sighted from earth moving on a parabolic path with the earth at the focus. When the line from the earth to the space ship first makes an angle of 90° with the axis of the parabola, it is measured to be 40 million miles away. How close will the space ship come to the earth (see Problem 33)? Treat the earth as a point.

C **35.** Work Problem 34 assuming the angle is 75° rather than 90°.

**36.** The cables for the central span of a suspension bridge take the shape of a parabola (see Problem 41). If the towers are 800 meters apart and the cables are attached to them at points 400 meters above the floor of the bridge, how long must the vertical strut be that is 100 meters from the tower? Assume that the cable touches the floor at the midpoint of the bridge (Figure 10).

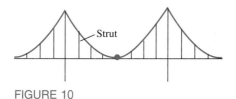

FIGURE 10

**37.** The focal chord that is perpendicular to the axis of a parabola is called the **latus rectum**. For the parabola $y^2 = 4px$ in Figure 11, let $F$ be the focus, $R$ be any point on the parabola to the left of the latus rectum, and $G$ be the intersection of the latus rectum with the line through $R$ parallel to the axis. Find $|FR| + |RG|$ and note that it is a constant.

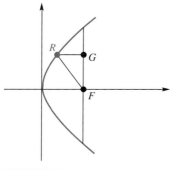

FIGURE 11

**38.** Show that the set of points equidistant from a circle and a line outside the circle is a parabola.

**39.** Show that the focal chord of the parabola $y^2 = 4px$ with endpoints $(x_1, y_1)$ and $(x_2, y_2)$ has length $x_1 + x_2 + 2p$. Specialize to find the length $L$ of the latus rectum.

**40.** For the parabola $y^2 = 4px$ in Figure 12, $P$ is any of its points except the vertex, $PB$ is the normal line at $P$,

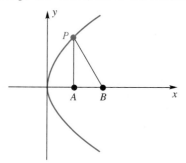

FIGURE 12

$PA$ is perpendicular to the axis of the parabola, $A$ and $B$ are on the axis. Find $|AB|$ and note that it is a constant.

**41.** Consider a bridge deck weighing $\delta$ pounds per lineal foot and supported by a cable, which is assumed to be of negligible weight compared to the bridge deck. The section $OP$ from the lowest point (the origin) to a general point $P(x, y)$ is shown in Figure 13.

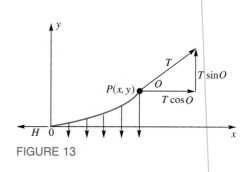

FIGURE 13

The forces acting on this section of cable are:

$$H = \text{horizontal tension pulling at } O$$
$$T = \text{tangential tension pulling at } P$$
$$W = \delta x = \text{weight of } x \text{ feet of bridge deck}$$

For equilibrium, the horizontal and vertical components of $T$ must balance $H$ and $W$, respectively. Thus,

$$\frac{T \sin \phi}{T \cos \phi} = \tan \phi = \frac{\delta x}{H}$$

that is,

$$\frac{dy}{dx} = \frac{\delta x}{H}, \qquad y(0) = 0$$

Solve this differential equation to show that the cable hangs in the shape of a parabola. (Compare this result with that for the unloaded hanging cable of Problem 29 of Section 7.8.)

**42.** Consider the parabola $y = x^2$ over the interval $[a, b]$ and let $c = (a + b)/2$ be the midpoint of $[a, b]$, $d$ be the midpoint of $[a, c]$, and $e$ be the midpoint of $[c, b]$. Let $T_1$ be the triangle with vertices on the parabola at $a$, $c$, and $b$, and let $T_2$ be the union of the two triangles with vertices on the parabola at $a$, $d$, $c$ and $c$, $e$, $b$, respectively (Figure 14). Continue to build triangles on triangles in this manner, thus obtaining sets $T_3$, $T_4$, . . . .

(a) Show that $A(T_1) = (b - a)^3/8$.
(b) Show that $A(T_2) = A(T_1)/4$.
(c) Let $S$ be the parabolic segment cut off by the chord $PQ$.

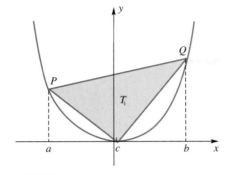

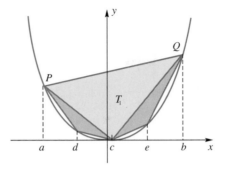

FIGURE 14

Show that the area of $S$ satisfies

$$A(S) = A(T_1) + A(T_2) + A(T_3) + \cdots = \left(\tfrac{4}{3}\right)A(T_1)$$

This is a famous result of Archimedes, which he obtained without coordinates.

(d) Use these results to show that the area under $y = x^2$ between $a$ and $b$ is $b^3/3 - a^3/3$.

$\boxed{\text{PC}}$ **43.** Illustrate Problems 30 and 31 for the parabola $y = \tfrac{1}{4}x^2 + 2$ by drawing (using the same axes) the parabola, its directrix, its focal chord parallel to the $x$-axis and the tangent lines at the ends of the focal chord.

---

**Answers to Concepts Review:** **1.** $e < 1$; $e = 1$; $e > 1$
**2.** $y^2 = 4px$ **3.** $(0, 1)$; $y = -1$ **4.** Parallel to the axis.

---

## 12.2
## ELLIPSES AND
## HYPERBOLAS

Recall that the conic determined by the condition $|PF| = e|PL|$ is an **ellipse** if $0 < e < 1$ and a **hyperbola** if $e > 1$ (see the introduction to Section 12.1). In either case, the conic has two vertices, which we label $A'$ and $A$. Call the point on the major axis midway between $A'$ and $A$ the **center** of the conic. Ellipses and hyperbolas are symmetric with respect to their centers (as we shall demonstrate soon) and are, therefore, called *central conics*.

To derive the equation of a central conic, place the $x$-axis along the major axis with the origin at the center. We may suppose the focus to be $F(c, 0)$, the directrix $x = k$, and the vertices $A'(-a, 0)$ and $A(a, 0)$, with $c$, $k$, and $a$ all positive. The two possible arrangements are shown in Figures 1 and 2.

The defining condition $|PF| = e|PL|$ applied first with $P = A$ and then with $P = A'$ yields

$$a - c = e(k - a) = ek - ea$$
$$a + c = e(k + a) = ek + ea$$

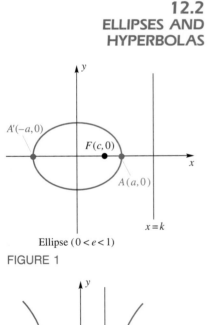

Ellipse $(0 < e < 1)$

FIGURE 1

When these two equations are solved for $c$ and $k$, we get

$$c = ea \qquad \text{and} \qquad k = \frac{a}{e}$$

Now let $P(x, y)$ be any point on the ellipse (or hyperbola). Then $L(a/e, y)$ is its projection on the directrix (see Figure 3 for the case of the ellipse). The condition $|PF| = e|PL|$ becomes

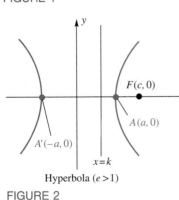

Hyperbola $(e > 1)$

FIGURE 2

$$\sqrt{(x - ae)^2 + y^2} = e\sqrt{\left(x - \frac{a}{e}\right)^2}$$

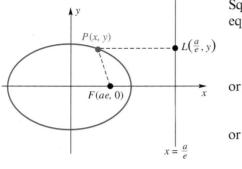

FIGURE 3

Squaring both members and collecting terms, we obtain the equivalent equation (why is it equivalent?)

$$x^2 - 2aex + a^2e^2 + y^2 = e^2\left(x^2 - \frac{2a}{e}x + \frac{a^2}{e^2}\right)$$

or

$$(1 - e^2)x^2 + y^2 = a^2(1 - e^2)$$

or

$$\frac{x^2}{a^2} + \frac{y^2}{a^2(1 - e^2)} = 1$$

Because this last equation contains $x$ and $y$ only to even powers, it corresponds to a curve that is symmetric with respect to both the $x$- and $y$-axes and the origin. Also, because of this symmetry, there must be a second focus at $(-ae, 0)$ and a second directrix at $x = -a/e$. The axis containing the two vertices (and the two foci) is the **major axis** and the axis perpendicular to it (through the center) is the **minor axis**.

**The Standard Equation of the Ellipse** For the ellipse, $0 < e < 1$, and so $(1 - e^2)$ is positive. To simplify notation, let $b = a\sqrt{1 - e^2}$. Then the equation derived above takes the form

$$\boxed{\frac{x^2}{a^2} + \frac{y^2}{b^2} = 1}$$

which is called the **standard equation of the ellipse**. The number $2a$ is the **major diameter**, whereas $2b$ is the **minor diameter**. Moreover since $c = ae$, the numbers $a$, $b$, and $c$ satisfy the Pythagorean relationship $a^2 = b^2 + c^2$, as is easily verified. All this is summarized in Figure 4. Note the role of the shaded triangle, which captures the condition $a^2 = b^2 + c^2$.

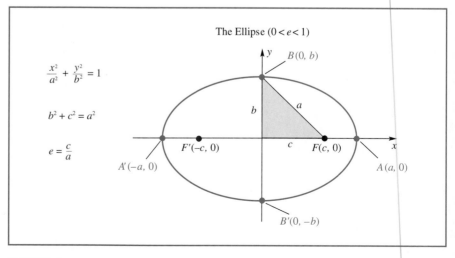

FIGURE 4

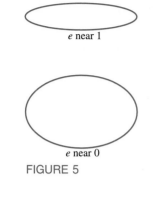

*e* near 1

*e* near 0

FIGURE 5

Consider the effect of changing the value of *e*. If *e* is near 1, then $b = a\sqrt{1 - e^2}$ is small relative to *a*; the ellipse is thin and very eccentric. On the other hand, if *e* is near 0 (near zero eccentricity), *b* is almost as large as *a*; the ellipse is fat and well rounded (Figure 5). In the limiting case where $b = a$, the equation takes the form

$$\frac{x^2}{a^2} + \frac{y^2}{a^2} = 1$$

which is equivalent to $x^2 + y^2 = a^2$. This is the equation of a circle of radius *a* centered at the origin.

**EXAMPLE 1** Sketch the graph of

$$\frac{x^2}{36} + \frac{y^2}{4} = 1$$

and determine its foci and eccentricity.

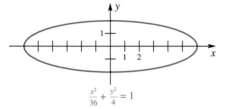

$\frac{x^2}{36} + \frac{y^2}{4} = 1$

FIGURE 6

*Solution* Since $a = 6$ and $b = 2$, we calculate

$$c = \sqrt{a^2 - b^2} = \sqrt{36 - 4} = 4\sqrt{2} \approx 5.66$$

The foci are at $(\pm c, 0) = (\pm 4\sqrt{2}, 0)$, and $e = c/a \approx 0.94$. The graph is sketched in Figure 6. ∎

We call the ellipses sketched so far *horizontal ellipses* because the major axis is the *x*-axis. If we interchange the roles of *x* and *y*, we obtain a vertical ellipse with equation

$$\frac{x^2}{b^2} + \frac{y^2}{a^2} = 1$$

**EXAMPLE 2** Sketch the graph of

$$\frac{x^2}{16} + \frac{y^2}{25} = 1$$

and determine its foci and eccentricity.

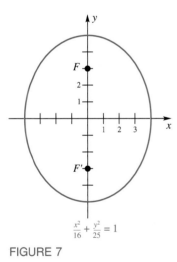

$\frac{x^2}{16} + \frac{y^2}{25} = 1$

FIGURE 7

*Solution* The larger square is now under $y^2$, which tells us that the major axis is vertical. Noting that $a = 5$ and $b = 4$, we conclude that $c = \sqrt{25 - 16} = 3$. Thus, the foci are $(0, \pm 3)$ and $e = c/a = \frac{3}{5} = 0.6$ (Figure 7). ∎

**The Standard Equation of the Hyperbola** For the hyperbola, $e > 1$ and so $e^2 - 1$ is positive. If we let $b = a\sqrt{e^2 - 1}$, then the equation $x^2/a^2 + y^2/(1 - e^2)a^2 = 1$, which was derived earlier, takes the form

$$\boxed{\frac{x^2}{a^2} - \frac{y^2}{b^2} = 1}$$

This is called the **standard equation of the hyperbola**. Since $c = ae$, we now obtain $c^2 = a^2 + b^2$. (Note how this differs from the corresponding relationship for an ellipse.)

To interpret $b$, observe that if we solve for $y$ in terms of $x$, we get

$$y = \pm \frac{b}{a} \sqrt{x^2 - a^2}$$

For large $x$, $\sqrt{x^2 - a^2}$ behaves like $x$ (that is, $(\sqrt{x^2 - a^2} - x) \to 0$ as $x \to \infty$; see Problem 38) and hence $y$ behaves like

$$y = \pm \frac{b}{a} x$$

More precisely, the graph of the given hyperbola has these two lines as asymptotes.

The important facts for the hyperbola are summarized in Figure 8. Once again, there is an important triangle (shaded in our diagram); it determines the asymptotes mentioned above.

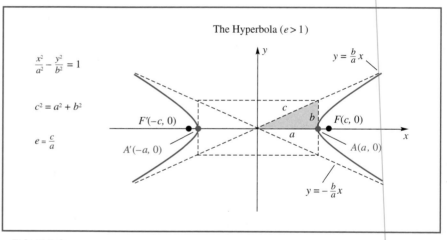

The Hyperbola $(e > 1)$

$\dfrac{x^2}{a^2} - \dfrac{y^2}{b^2} = 1$

$c^2 = a^2 + b^2$

$e = \dfrac{c}{a}$

$y = \dfrac{b}{a} x$

$y = -\dfrac{b}{a} x$

$F'(-c, 0)$   $A'(-a, 0)$   $F(c, 0)$   $A(a, 0)$

FIGURE 8

**EXAMPLE 3**   Sketch the graph of

$$\frac{x^2}{9} - \frac{y^2}{16} = 1$$

showing the asymptotes. What are the equations of the asymptotes? What are the foci?

***Solution***   We begin by determining the fundamental triangle; it has horizontal leg 3 and vertical leg 4. After drawing it, we can indicate the asymptotes and sketch the graph (Figure 9). The asymptotes are $y = \pm \frac{4}{3} x$. Since $c = \sqrt{a^2 + b^2} = \sqrt{9 + 16} = 5$, the foci are at $(\pm 5, 0)$. ∎

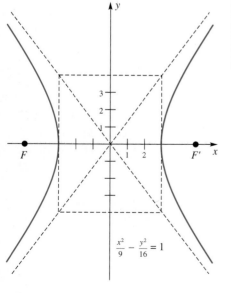

$\dfrac{x^2}{9} - \dfrac{y^2}{16} = 1$

FIGURE 9

Again, we should consider the effect of interchanging the roles of $x$ and $y$. The equation takes the form

$$\frac{y^2}{a^2} - \frac{x^2}{b^2} = 1$$

This is the equation of a vertical hyperbola (vertical major axis). Its vertices are at $(0, \pm a)$; its foci are at $(0, \pm c)$.

For both the ellipse and the hyperbola, $a$ is always the distance from the center to a vertex. For the ellipse, $a > b$; for the hyperbola, there is no such requirement.

**EXAMPLE 4**   Determine the foci of

$$-\frac{x^2}{4} + \frac{y^2}{9} = 1$$

and sketch its graph.

*Solution*   We note immediately that this is a vertical hyperbola, which is determined by the fact that the plus sign is associated with the $y^2$ term. Thus, $a = 3$, $b = 2$, and $c = \sqrt{9 + 4} = \sqrt{13} \approx 3.61$. The foci are at $(0, \pm\sqrt{13})$ (Figure 10).  ∎

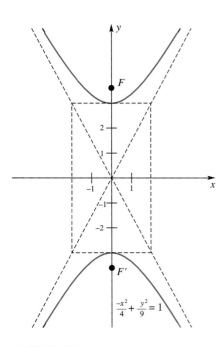

$$\frac{-x^2}{4} + \frac{y^2}{9} = 1$$

**FIGURE 10**

**Applications**   According to Johannes Kepler, the planets revolve around the sun in elliptical orbits, with the sun at one focus. Other examples of elliptical orbits are satellites orbiting the earth and electrons orbiting the nucleus of an atom.

**EXAMPLE 5**   The earth's maximum distance from the sun is 94.56 million miles, and its minimum distance is 91.45 million miles. What is the eccentricity of the orbit and what are the major and minor diameters?

*Solution*   Using the notation in Figure 11, we see that

$$a + c = 94.56 \qquad a - c = 91.45$$

When we solve these equations for $a$ and $c$, we obtain $a = 93.01$ and $c = 1.56$. Thus,

$$e = \frac{c}{a} = \frac{1.56}{93.01} \approx 0.017$$

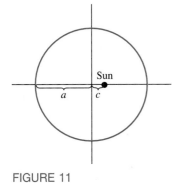

**FIGURE 11**

and the major diameter and minor diameter (in millions of miles) are, respectively,

$$2a \approx 186.02 \qquad 2b = 2\sqrt{a^2 - c^2} \approx 185.99 \qquad ∎$$

Other applications of ellipses and hyperbolas result from the optical properties of these curves. They are discussed in the next section.

## CONCEPTS REVIEW

**1.** The standard equation of the horizontal ellipse centered at (0, 0) is _____ .

**2.** The $xy$-equation of the vertical ellipse centered at (0, 0) that has major diameter 8 and minor diameter 6 is _____ .

**3.** The standard equation of the horizontal hyperbola centered at (0, 0) is _____ .

**4.** The hyperbola $x^2/9 - y^2/4 = 1$ has asymptotes _____ .

## PROBLEM SET 12.2

In Problems 1–8, name the conic (horizontal ellipse, vertical hyperbola, and so on) corresponding to the given equation.

**1.** $\dfrac{x^2}{9} + \dfrac{y^2}{4} = 1$

**2.** $\dfrac{x^2}{9} - \dfrac{y^2}{4} = 1$

**3.** $\dfrac{-x^2}{9} + \dfrac{y^2}{4} = 1$

**4.** $\dfrac{-x^2}{9} + \dfrac{y^2}{4} = -1$

**5.** $\dfrac{-x^2}{9} + \dfrac{y}{4} = 0$

**6.** $\dfrac{-x^2}{9} = \dfrac{y}{4}$

**7.** $9x^2 + 4y^2 = 9$

**8.** $x^2 - 4y^2 = 4$

In Problems 9–16, sketch the graph of the given equation indicating vertices, foci, and asymptotes (if it is a hyperbola).

**9.** $\dfrac{x^2}{16} + \dfrac{y^2}{4} = 1$

**10.** $\dfrac{x^2}{16} - \dfrac{y^2}{4} = 1$

**11.** $\dfrac{-x^2}{9} + \dfrac{y^2}{4} = 1$

**12.** $\dfrac{x^2}{7} + \dfrac{y^2}{4} = 1$

**13.** $16x^2 + 4y^2 = 32$

**14.** $4x^2 + 25y^2 = 100$

**15.** $10x^2 - 25y^2 = 100$

**16.** $x^2 - 4y^2 = 8$

In Problems 17–30, find the equation of the given conic. Assume the center is at the origin.

**17.** The ellipse with a focus at $(-3, 0)$ and a vertex at $(6, 0)$.

**18.** The ellipse with a focus at $(6, 0)$ and eccentricity $\frac{2}{3}$.

**19.** The ellipse with a focus at $(0, -5)$ and eccentricity $\frac{1}{3}$.

**20.** The ellipse with a focus at $(0, 3)$ and minor diameter 8.

**21.** The ellipse with a vertex at $(5, 0)$ and passing through $(2, 3)$.

**22.** The hyperbola with a focus at $(5, 0)$ and a vertex at $(4, 0)$.

**23.** The hyperbola with a vertex at $(0, -4)$ and a focus at $(0, -5)$.

**24.** The hyperbola with a vertex at $(0, -3)$ and eccentricity $\frac{3}{2}$.

**25.** The hyperbola with asymptotes $2x \pm 4y = 0$ and a vertex at $(8, 0)$.

**26.** The vertical hyperbola with eccentricity $\sqrt{6}/2$ which passes through $(2, 4)$.

**27.** The ellipse with foci $(\pm 2, 0)$ and directrices $x = \pm 8$.

**28.** The hyperbola with foci $(\pm 4, 0)$ and directrices $x = \pm 1$.

**29.** The hyperbola whose asymptotes are $x \pm 2y = 0$ and that goes through the point $(4, 3)$.

**30.** The horizontal ellipse which goes through $(-5, 1)$ and $(-4, -2)$.

**31.** A door has the shape of an elliptical arch (a half-ellipse) that is 10 feet wide and 4 feet high at the center. A box 2 feet high is to be pushed through the door. How wide can the box be?

**32.** How high is the arch of Problem 31 at a distance 2 feet from the center?

**33.** How long is the *latus rectum* (chord through the focus perpendicular to the major axis) for the ellipse $x^2/a^2 + y^2/b^2 = 1$?

**34.** Determine the length of the latus rectum (see Problem 33) of the hyperbola $x^2/a^2 - y^2/b^2 = 1$.

C **35.** Halley's comet has an elliptical orbit with major and minor diameters of 36.18 AU and 9.12 AU, respectively (1 AU is 1 astronomical unit, the earth's mean distance from the sun). What is its closest approach to the sun (assuming the sun is at a focus)?

C **36.** The orbit of the comet Kahoutek is an ellipse with eccentricity $e = 0.999925$ with the sun at a focus. If its minimum distance to the sun is 0.13 AU, what is its maximum distance from the sun? See Problem 35.

C **37.** In October, 1957, Russia launched Sputnik I. Its elliptical orbit around the earth reached maximum and minimum distances from the earth of 583 miles and 132 miles, respectively. Assuming the earth is a sphere of radius 4000 miles, find the eccentricity of the orbit.

**38.** Show that $\left(\sqrt{x^2 - a^2} - x\right) \to 0$ as $x \to \infty$. *Hint:* Rationalize the numerator.

**39.** For an ellipse, let $p$ and $q$ be the distances from a focus to the two vertices. Show that $b = \sqrt{pq}$, $2b$ being the minor diameter.

**40.** The wheel in Figure 12 is turning at $t$ radians per second so that $Q$ has coordinates $(a \cos t, a \sin t)$. Find the

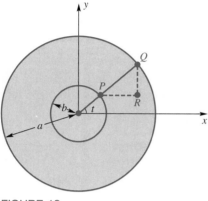

FIGURE 12

coordinates $(x, y)$ of $R$ at time $t$ and show that it is traveling an elliptical path. *Note:* $PQR$ is a right triangle.

**41.** Let $P$ be a point on a ladder of length $a + b$, $P$ being $a$ units from the top end. As the ladder slides with its top end on the $y$-axis and its bottom end on the $x$-axis, $P$ traces out a curve. Find the equation of this curve.

**42.** Show that a line through a focus of a hyperbola and perpendicular to an asymptote intersects that asymptote on the directrix nearest the focus.

**43.** If a horizontal hyperbola and a vertical hyperbola have the same asymptotes, show that their eccentricities $e$ and $E$ satisfy $e^{-2} + E^{-2} = 1$.

**44.** Let $C$ be the curve of intersection of a right circular cylinder and a plane making an angle $\phi$ ($0 < \phi < \pi/2$) with the axis of the cylinder. Show that $C$ is an ellipse.

[PC] **45.** Using the same axes, draw the conics $y = \pm(ax^2 + 1)^{1/2}$ for $-2 \le x \le 2$ and $-2 \le y \le 2$ using $a = -2, -1, -0.5, -0.1, 0, 0.1, 0.6, 1$. Make a conjecture.

**Answers to Concepts Review:**  **1.** $x^2/a^2 + y^2/b^2 = 1$
**2.** $x^2/9 + y^2/16 = 1$   **3.** $x^2/a^2 - y^2/b^2 = 1$   **4.** $y = \pm\frac{2}{3}x$

---

### 12.3
### MORE ON ELLIPSES AND HYPERBOLAS

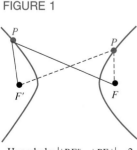

Ellipse: $|PF'| + |PF| = 2a$

FIGURE 1

Hyperbola: $||PF'| - |PF|| = 2a$

FIGURE 2

In Section 12.1, we gave the so-called eccentricity definition of the ellipse and hyperbola. The condition $|PF| = e|PL|$ determines an ellipse if $0 < e < 1$ and a hyperbola if $e > 1$. This definition allowed us to treat these two curves in a unified way. Many authors prefer to introduce these curves via the following alternative definitions.

An **ellipse** is the set of all points $P$ in the plane, the sum of whose distances from two fixed points (the foci) is a given positive constant $2a$.

A **hyperbola** is the set of all points $P$ in the plane, the difference of whose distances from two fixed points (the foci) is a given positive constant $2a$. Here, the word *difference* is taken to mean the larger distance minus the smaller distance.

To interpret these definitions geometrically, study Figures 1 and 2. For the ellipse, imagine a string of length $2a$ tacked down at its two endpoints. If a pencil is held tight against the string with its tip at $P$, it can be used to trace the ellipse (also see Problem 29). We refer to the properties described in the new definitions as the *string properties* of the ellipse and hyperbola. For us, these properties should be consequences of the eccentricity definition. We derive them now.

**Derivation of String Properties**   We pointed out in Section 12.2 that the ellipse and hyperbola have two foci and two directrices. When these curves are placed in the coordinate system with major axis along the $x$-axis and center at the origin, the foci have coordinates $(\pm ae, 0)$ and the directrices have equations $x = \pm a/e$. These facts are indicated in Figure 3.

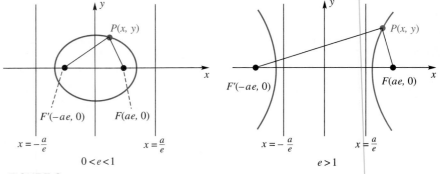

FIGURE 3

If we take an arbitrary point $P(x, y)$ on the ellipse, then—from the condition $|PF| = e|PL|$, applied first to the left focus and directrix and then to the right ones—we get

$$|PF'| = e\left(x + \frac{a}{e}\right) = ex + a \qquad |PF| = e\left(\frac{a}{e} - x\right) = a - ex$$

and so

$$\boxed{|PF'| + |PF| = 2a}$$

Next consider the hyperbola with $P(x, y)$ on its right branch, as shown in the right part of Figure 3. Then,

$$|PF'| = e\left(x + \frac{a}{e}\right) = ex + a \qquad |PF| = e\left(x - \frac{a}{e}\right) = ex - a$$

and so $|PF'| - |PF| = 2a$. If $P(x, y)$ had been on the left branch, we would have gotten $-2a$ in place of $2a$. In either case,

$$\boxed{\big||PF'| - |PF|\big| = 2a}$$

**EXAMPLE 1**   Find the equation of the set of points, the sum of whose distances from $(\pm 3, 0)$ is equal to 10.

***Solution***   This is a horizontal ellipse with $a = 5$ and $c = 3$. Thus $b = \sqrt{a^2 - c^2} = 4$, and the equation is

$$\frac{x^2}{25} + \frac{y^2}{16} = 1$$

∎

**EXAMPLE 2**   Find the equation of the set of points, the difference of whose distances from $(0, \pm 6)$ is equal to 4.

***Solution*** This is a vertical hyperbola with $a = 2$ and $c = 6$. Thus, $b = \sqrt{c^2 - a^2} = \sqrt{32} = 4\sqrt{2}$, and the equation is

$$-\frac{x^2}{32} + \frac{y^2}{4} = 1$$

∎

**Optical Properties** Consider mirrors with the shapes of an ellipse and hyperbola, respectively. If a light ray emanating from one focus strikes the mirror, it will be reflected back to the other focus in the case of the ellipse and directly away from the other focus in the case of the hyperbola. These facts are shown in Figure 4.

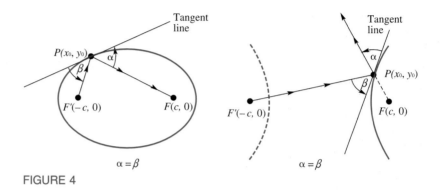

**FIGURE 4**

To demonstrate these optical properties (that is, to show $\alpha = \beta$ in Figure 4), we suppose the curves to be in standard position, so that their equations are $x^2/a^2 + y^2/b^2 = 1$ and $x^2/a^2 - y^2/b^2 = 1$, respectively. For the ellipse, we differentiate implicitly and then substitute $(x_0, y_0)$, thereby obtaining the slope $m$ of the tangent line.

$$\frac{2x}{a^2} + \frac{2yy'}{b^2} = 0$$

$$y' = -\frac{b^2}{a^2}\frac{x}{y}$$

$$m = -\frac{b^2}{a^2}\frac{x_0}{y_0}$$

The equation of the tangent line may be written successively as

$$y - y_0 = -\frac{b^2 x_0}{a^2 y_0}(x - x_0)$$

$$\frac{x_0}{a^2}(x - x_0) + \frac{y_0}{b^2}(y - y_0) = 0$$

$$\frac{x_0 x}{a^2} + \frac{y_0 y}{b^2} = \frac{x_0^2}{a^2} + \frac{y_0^2}{b^2} = 1$$

A similar derivation for the hyperbola leads to similar results. We summarize in the following table.

|  | **Ellipse** | **Hyperbola** |
|---|---|---|
| Equation | $\dfrac{x^2}{a^2} + \dfrac{y^2}{b^2} = 1$ | $\dfrac{x^2}{a^2} - \dfrac{y^2}{b^2} = 1$ |
| Slope of tangent at $(x_0, y_0)$ | $m = \dfrac{-b^2 x_0}{a^2 y_0}$ | $m = \dfrac{b^2 x_0}{a^2 y_0}$ |
| Equation of tangent at $(x_0, y_0)$ | $\dfrac{x_0 x}{a^2} + \dfrac{y_0 y}{b^2} = 1$ | $\dfrac{x_0 x}{a^2} - \dfrac{y_0 y}{b^2} = 1$ |

To calculate $\tan \alpha$ for the ellipse, we recall a formula for the tangent of the counterclockwise angle from one line $\ell_1$ to another $\ell$ in terms of their respective slopes $m_1$ and $m$, namely,

$$\tan \alpha = \frac{m - m_1}{1 + m m_1}$$

Now refer to Figure 4 and let $\ell_1$ be the line $FP$ and $\ell$ be the tangent line at $P$. Then,

$$\tan \alpha = \frac{\dfrac{-b^2 x_0}{a^2 y_0} - \dfrac{y_0 - 0}{x_0 - c}}{1 + \left(\dfrac{-b^2 x_0}{a^2 y_0}\right)\left(\dfrac{y_0 - 0}{x_0 - c}\right)} = \frac{-b^2 x_0 (x_0 - c) - a^2 y_0^2}{a^2 y_0 (x_0 - c) - b^2 x_0 y_0}$$

$$= \frac{b^2 c x_0 - (b^2 x_0^2 + a^2 y_0^2)}{(a^2 - b^2) x_0 y_0 - a^2 c y_0} = \frac{b^2 c x_0 - a^2 b^2}{c^2 x_0 y_0 - a^2 c y_0}$$

$$= \frac{b^2 (c x_0 - a^2)}{c y_0 (c x_0 - a^2)} = \frac{b^2}{c y_0}$$

The same calculation with $c$ replaced by $-c$ gives

$$\tan(-\beta) = \frac{b^2}{-c y_0}$$

and so $\tan \beta = b^2/c y_0$. We conclude that $\tan \alpha = \tan \beta$, and consequently $\alpha = \beta$.

A similar derivation establishes the corresponding result for the hyperbola.

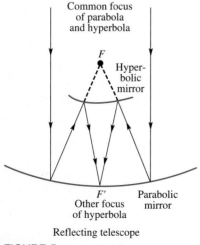

Common focus of parabola and hyperbola

$F$

Hyper-bolic mirror

$F'$    Parabolic mirror
Other focus of hyperbola

Reflecting telescope

FIGURE 5

**Applications** The reflecting property of the ellipse is the basis of the "whispering gallery" effect that can be observed, for example, in the U.S.

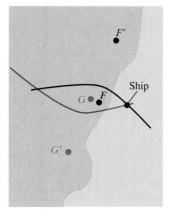

FIGURE 6

Capitol and the Mormon Tabernacle. A speaker standing at one focus can be heard whispering by a listener at the other focus, even though his or her voice is inaudible in other parts of the room.

The optical properties of the parabola and hyperbola are combined in one design for a reflecting telescope (Figure 5). The parallel rays from a star are finally focused at the eyepiece at $F''$.

The string property of the hyperbola is used in navigation. A ship at sea can determine the difference $2a$ in its distance from two fixed transmitters by measuring the difference in reception times of synchronized radio signals. This puts its path on a hyperbola, with the two transmitters $F$ and $F'$ as foci. If another pair of transmitters $G$ and $G'$ are used, the ship must lie at the intersection of the two corresponding hyperbolas (see Figure 6). LORAN, a system of long-range navigation, is based on this principle.

## CONCEPTS REVIEW

**1.** An ellipse is the set of points $P$ satisfying $|PF| + |PF'| = 2a$, where $F$ and $F'$ are fixed points called the _____ of the ellipse.

**2.** Similarly, a hyperbola is the set of points $P$ satisfying _____.

**3.** A ray from a light source at one focus of an elliptical mirror will be reflected _____.

**4.** A ray from a light source at one focus of a hyperbolic mirror will be reflected _____.

## PROBLEM SET 12.3

In Problems 1–4, find the equation of the set of points $P$ satisfying the given conditions.

**1.** The sum of the distances of $P$ from $(0, \pm9)$ is 26.

**2.** The sum of the distances of $P$ from $(\pm4, 0)$ is 14.

**3.** The difference of the distances of $P$ from $(\pm7, 0)$ is 12.

**4.** The difference of the distances of $P$ from $(0, \pm6)$ is 10.

In Problems 5–12, find the equation of the tangent line to the given curve at the given point (see table on page 588).

**5.** $\dfrac{x^2}{27} + \dfrac{y^2}{9} = 1$ at $(3, \sqrt{6})$

**6.** $\dfrac{x^2}{24} + \dfrac{y^2}{16} = 1$ at $(3\sqrt{2}, -2)$

**7.** $\dfrac{x^2}{27} + \dfrac{y^2}{9} = 1$ at $(3, -\sqrt{6})$

**8.** $\dfrac{x^2}{2} - \dfrac{y^2}{4} = 1$ at $(\sqrt{3}, \sqrt{2})$

**9.** $x^2 + y^2 = 169$ at $(5, 12)$

**10.** $x^2 - y^2 = -1$ at $(\sqrt{2}, \sqrt{3})$

**11.** The curve of Problem 1 at $(0, 13)$.

**12.** The curve of Problem 2 at $(7, 0)$.

**13.** If two tangent lines to the ellipse $9x^2 + 4y^2 = 36$ intersect the $y$-axis at $(0, 6)$, find the points of tangency.

**14.** If the tangent lines to the hyperbola $9x^2 - y^2 = 36$ intersect the $y$-axis at $(0, 6)$, find the points of tangency.

**15.** The slope of the tangent to the hyperbola $2x^2 - 7y^2 - 35 = 0$ at a certain point on the hyperbola is $-\frac{2}{3}$. What are the coordinates of the point of tangency (two solutions)?

**16.** Find the equations of the tangents to the ellipse $x^2 + 2y^2 - 2 = 0$ that are parallel to the line $3x - 3\sqrt{2}y - 7 = 0$.

**17.** Find the area of the ellipse $b^2x^2 + a^2y^2 = a^2b^2$.

**18.** Find the volume of the solid obtained by revolving the ellipse $b^2x^2 + a^2y^2 = a^2b^2$ about the $y$-axis.

**19.** The region bounded by the hyperbola $b^2x^2 - a^2y^2 = a^2b^2$ and a vertical line through a focus is revolved about the $x$-axis. Find the volume of the resulting solid.

**20.** If the ellipse of Problem 18 is revolved about the x-axis, find the volume of the resulting solid.

**21.** Find the dimensions of the rectangle having the greatest possible area that can be inscribed in the ellipse $b^2x^2 + a^2y^2 = a^2b^2$. Assume that the sides of the rectangle are parallel to the axes of the ellipse.

**22.** Show that the point of contact of any tangent to a hyperbola is midway between the points in which the tangent intersects the asymptotes.

**23.** Find the point in the first quadrant where the two hyperbolas $25x^2 - 9y^2 = 225$ and $-25x^2 + 18y^2 = 450$ intersect.

**24.** Find the points of intersection of $x^2 + 4y^2 = 20$ and $x + 2y = 6$.

**25.** Sketch the design for a reflecting telescope that uses a parabola and an ellipse rather than the parabola and hyperbola described in the text.

**26.** A ball placed at a focus of an elliptical billiard table is shot with tremendous force so that it continues to bounce off the cushions indefinitely. What is its ultimate path? *Hint*: Draw a picture.

**27.** If the ball of Problem 26 is initially on the major axis between a focus and the neighboring vertex, what can you say about its path?

**28.** Show that an ellipse and a hyperbola with the same foci intersect at right angles. *Hint*: Draw a picture and use the optical properties.

**29.** Describe a string apparatus for constructing a hyperbola. (There are several possibilities).

**30.** Sound travels at $u$ feet per second and a rifle bullet at $v > u$ feet per second. The sound of the firing of a rifle and the impact of the bullet hitting the target were heard simultaneously. If the rifle was at $A(-c, 0)$, the target was at $B(c, 0)$, and the listener was at $P(x, y)$, find the equation of the curve on which $P$ lies (in terms of $u$, $v$, and $c$).

**31.** Listeners $A(-8, 0)$, $B(8, 0)$, and $C(8, 10)$ recorded the exact times at which they heard an explosion. If $B$ and $C$ heard the explosion at the same time and $A$ heard it 12 seconds later, where was the explosion? Assume that distances are in kilometers and that sound travels $\frac{1}{3}$ kilometer per second.

---

**Answers to Concepts Review:** **1.** Foci
**2.** $\left\| PF \right| - \left| PF' \right\| = 2a$ **3.** To the other focus
**4.** Directly away from the other focus

---

<table>
<tr><td>**12.4**<br>**TRANSLATION OF AXES**</td></tr>
</table>

So far we have placed the conics in the coordinate system in very special ways—always with the major axis along one of the coordinate axes and either the vertex (in the case of a parabola) or the center (in the case of an ellipse or hyperbola) at the origin. Now we place our conics in a more general position, though we still require that the major axis be parallel to one of the coordinate axes. Even this restriction will be removed in Section 12.5.

The case of a circle is instructive. The circle of radius 5 centered at (2, 3) has equation

$$(x - 2)^2 + (y - 3)^2 = 25$$

or, in equivalent expanded form,

$$x^2 + y^2 - 4x - 6y = 12$$

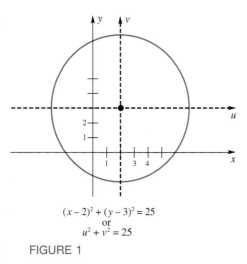

$(x - 2)^2 + (y - 3)^2 = 25$
or
$u^2 + v^2 = 25$

FIGURE 1

The same circle with its center at the origin of the $uv$-coordinate system (Figure 1) has the simple equation

$$u^2 + v^2 = 25$$

The introduction of new axes does not change the shape or size of a curve, but it may greatly simplify its equation. It is this so-called translation of axes and the corresponding change of variables in an equation that we wish to investigate.

**Translations**    If new axes are chosen in the plane, every point will have two sets of coordinates, the old ones, $(x, y)$, relative to the old axes and the new ones, $(u, v)$, relative to the new axes. The original coordinates are said to undergo a **transformation**. If the new axes are parallel, respectively, to the original axes and have the same directions, the transformation is called a **translation of axes**.

From Figure 2, it is easy to see how the new coordinates $(u, v)$ relate to the old ones $(x, y)$. Let $(h, k)$ be the old coordinates of the new origin. Then,

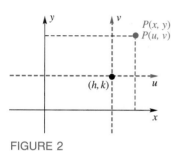

FIGURE 2

$$u = x - h, \qquad v = y - k$$

or, equivalently,

$$x = u + h, \qquad y = v + k$$

**EXAMPLE 1**    Find the new coordinates of $P(-6, 5)$ after a translation of axes to a new origin at $(2, -4)$.

*Solution*    Since $h = 2$ and $k = -4$, it follows that

$$u = x - h = -6 - 2 = -8 \qquad v = y - k = 5 - (-4) = 9$$

The new coordinates are $(-8, 9)$.    ■

**EXAMPLE 2**    Given the equation $4x^2 + y^2 + 40x - 2y + 97 = 0$, find the equation of its graph after a translation with new origin $(-5, 1)$.

*Solution*    In the equation, we replace $x$ by $u + h = u - 5$ and $y$ by $v + k = v + 1$. We obtain

$$4(u - 5)^2 + (v + 1)^2 + 40(u - 5) - 2(v + 1) + 97 = 0$$

or

$$4u^2 - 40u + 100 + v^2 + 2v + 1 + 40u - 200 - 2v - 2 + 97 = 0$$

This simplifies to

$$4u^2 + v^2 = 4$$

or

$$u^2 + \frac{v^2}{4} = 1$$

which we recognize as the equation of an ellipse.    ■

**Completing the Square** Given a complicated second-degree equation, how do we know what translation will simplify the equation and bring it to a recognizable form? Here a familiar algebraic process called **completing the square** provides the answer. In particular, we can use this process to eliminate the first-degree terms of any expression of the form

$$Ax^2 + Cy^2 + Dx + Ey + F = 0, \qquad A \neq 0, C \neq 0$$

**EXAMPLE 3** Make a translation which will eliminate the first-degree terms of

$$4x^2 + 9y^2 + 8x - 90y + 193 = 0$$

and use this information to sketch the graph of the given equation.

*Solution* Recall that to complete the square of $x^2 + ax$, we must add $a^2/4$ (the square of half the coefficient of $x$). Using this, we rewrite the given equation by adding the same numbers to both sides.

$$4(x^2 + 2x \quad ) + 9(y^2 - 10y \quad ) = -193$$
$$4(x^2 + 2x + 1) + 9(y^2 - 10y + 25) = -193 + 4 + 225$$
$$4(x + 1)^2 + 9(y - 5)^2 = 36$$
$$\frac{(x + 1)^2}{9} + \frac{(y - 5)^2}{4} = 1$$

The translation $u = x + 1$ and $v = y - 5$ transforms this to

$$\frac{u^2}{9} + \frac{v^2}{4} = 1$$

which is the standard form of a horizontal ellipse. The graph is shown in Figure 3. ∎

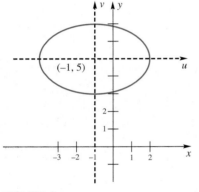

FIGURE 3

(−1, 5)

**EXAMPLE 4** Use a translation to simplify

$$y^2 - 4x - 12y + 28 = 0$$

Then determine which conic it represents, list the important characteristics of this conic, and sketch its graph.

*Solution* We complete the square.

$$y^2 - 12y = 4x - 28$$
$$y^2 - 12y + 36 = 4x - 28 + 36$$
$$(y - 6)^2 = 4(x + 2)$$

The translation $u = x + 2$, $v = y - 6$ transforms this to $v^2 = 4u$, which we recognize as a horizontal parabola opening right with $p = 1$ (Figure 4). ∎

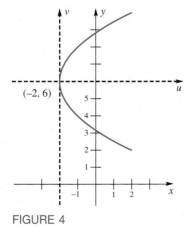

FIGURE 4

(−2, 6)

**General Second-Degree Equations**    Now we ask an important question. Is the graph of an equation of the form

$$Ax^2 + Cy^2 + Dx + Ey + F = 0$$

always a conic? The answer is no, unless we admit certain limiting forms. The table below indicates the possibilities with a sample equation for each.

| **Conics** | | **Limiting Forms** | |
|---|---|---|---|
| 1. $(AC = 0)$ Parabola: $y^2 = 4x$ | | Parallel lines: $y^2 = 4$ | |
| | | Single line: $y^2 = 0$ | |
| | | Empty set: $y^2 = -1$ | |
| 2. $(AC > 0)$ Ellipse: $\dfrac{x^2}{9} + \dfrac{y^2}{4} = 1$ | | Circle: $x^2 + y^2 = 4$ | |
| | | Point: $2x^2 + y^2 = 0$ | |
| | | Empty set: $2x^2 + y^2 = -1$ | |
| 3. $(AC < 0)$ Hyperbola: $\dfrac{x^2}{9} - \dfrac{y^2}{4} = 1$ | | Intersecting lines: $x^2 - y^2 = 0$ | |

Thus, the graphs of the general quadratic equation above fall into three general categories but yield nine different possibilities, including limiting forms.

**EXAMPLE 5**    Use a translation to simplify

$$4x^2 - y^2 - 8x - 6y - 5 = 0$$

and sketch its graph.

*Solution*    We rewrite the equation as follows.

$$4(x^2 - 2x \qquad) - (y^2 + 6y \qquad) = 5$$
$$4(x^2 - 2x + 1) - (y^2 + 6y + 9) = 5 + 4 - 9$$
$$4(x - 1)^2 - (y + 3)^2 = 0$$

Let $u = x - 1$ and $v = y + 3$, which results in

$$4u^2 - v^2 = 0$$

or

$$(2u - v)(2u + v) = 0$$

This is the equation of two intersecting lines (Figure 5).

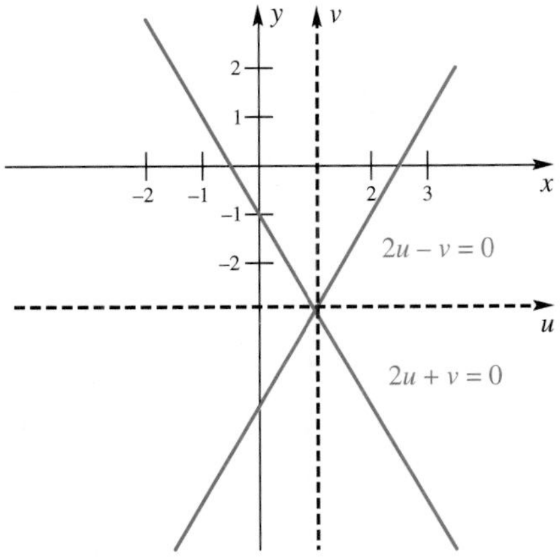

FIGURE 5

(graph labels: $2u - v = 0$, $2u + v = 0$)

**EXAMPLE 6** Write the equation of a hyperbola with foci at $(1, 1)$ and $(1, 11)$ and vertices at $(1, 3)$ and $(1, 9)$.

*Solution* The center is $(1, 6)$, midway between the vertices. Thus, $a = 3$ and $c = 5$, and so $b = \sqrt{c^2 - a^2} = 4$. The equation is

$$\frac{(y - 6)^2}{9} - \frac{(x - 1)^2}{16} = 1 \qquad \blacksquare$$

**Summary** Consider the general equation

$$Ax^2 + Cy^2 + Dx + Ey + F = 0$$

If both $A$ and $C$ are zero, we have the equation of a line (provided, of course, that $D$ and $E$ are not both zero). If at least one of $A$ and $C$ is different from zero, we may apply the process of completing the square. We obtain one of several forms, the most typical being:

(1) $$(y - k)^2 = \pm 4p(x - h)$$

(2) $$\frac{(x - h)^2}{a^2} + \frac{(y - k)^2}{b^2} = 1$$

(3) $$\frac{(x - h)^2}{a^2} - \frac{(y - k)^2}{b^2} = 1$$

These can be recognized even in this form as the equations of a horizontal parabola with vertex at $(h, k)$, a horizontal ellipse (if $a^2 > b^2$) with center at $(h, k)$, and a horizontal hyperbola with center at $(h, k)$. But to remove any doubt, we may translate the axes by the substitutions $u = x - h$, $v = y - k$, thereby obtaining:

(1) $$v^2 = \pm 4pu$$

(2) $$\frac{u^2}{a^2} + \frac{v^2}{b^2} = 1$$

(3) $$\frac{u^2}{a^2} - \frac{v^2}{b^2} = 1$$

Our work may also yield these equations with $u$ and $v$ interchanged, or we may get one of the six limiting forms illustrated in the table on page 593. There are no other possibilities.

## CONCEPTS REVIEW

**1.** The quadratic form $x^2 + ax$ is made a square by adding _____ .

**2.** $x^2 + 6x + 2(y^2 - 2y) = 3$ is (after completing the squares) equivalent to $(x + 3)^2 + 2(y - 1)^2 =$ _____ , which is the equation of an _____ .

**3.** Besides circle, ellipse, parabola, and hyperbola, other possible graphs for a second-degree equation in $x$ and $y$ are _____ .

**4.** The graph of $4x^2 - 9y^2 = 0$ is _____ .

---

## PROBLEM SET 12.4

In Problems 1–16, name the conic or limiting form represented by the given equation. Usually you will need to use the process of completing the square (see Examples 3–5).

1. $x^2 + y^2 - 2x + 2y + 1 = 0$

2. $x^2 + y^2 + 6x - 2y + 6 = 0$

3. $9x^2 + 4y^2 + 72x - 16y + 124 = 0$

4. $16x^2 - 9y^2 + 192x + 90y - 495 = 0$

5. $9x^2 + 4y^2 + 72x - 16y + 160 = 0$

6. $16x^2 + 9y^2 + 192x + 90y + 1000 = 0$

7. $y^2 - 5x - 4y - 6 = 0$

8. $4x^2 + 4y^2 + 8x - 28y - 11 = 0$

9. $3x^2 + 3y^2 - 6x + 12y + 60 = 0$

10. $4x^2 - 4y^2 - 2x + 2y + 1 = 0$

11. $4x^2 - 4y^2 + 8x + 12y - 5 = 0$

12. $4x^2 - 4y^2 + 8x + 12y - 6 = 0$

13. $4x^2 - 24x + 36 = 0$

14. $4x^2 - 24x + 35 = 0$

15. $25x^2 + 4y^2 + 150x - 8y + 129 = 0$

16. $4x^2 - 25y^2 - 8x + 150y + 129 = 0$

In Problems 17–30, sketch the graph of the given equation.

17. $\dfrac{(x + 3)^2}{4} + \dfrac{(y + 2)^2}{16} = 1$

18. $(x + 3)^2 + (y - 4)^2 = 25$

19. $\dfrac{(x + 3)^2}{4} - \dfrac{(y + 2)^2}{16} = 1$

20. $4(x + 3) = (y + 2)^2$

21. $(x + 2)^2 = 8(y - 1)$

22. $(x + 2)^2 = 4$

23. $(y - 1)^2 = 16$

24. $\dfrac{(x + 3)^2}{4} + \dfrac{(y - 2)^2}{8} = 0$

25. $x^2 + 4y^2 - 2x + 16y + 1 = 0$

26. $25x^2 + 9y^2 + 150x - 18y + 9 = 0$

27. $9x^2 - 16y^2 + 54x + 64y - 127 = 0$

28. $x^2 - 4y^2 - 14x - 32y - 11 = 0$

29. $4x^2 + 16x - 16y + 32 = 0$

30. $x^2 - 4x + 8y = 0$

31. Find the focus and directrix of the parabola
$$2y^2 - 4y - 10x = 0.$$

32. Determine the distance between the vertices of
$$-9x^2 + 18x + 4y^2 + 24y = 9.$$

33. Find the foci of the ellipse
$$16(x - 1)^2 + 25(y + 2)^2 = 400.$$

34. Find the focus and directrix of the parabola
$$x^2 - 6x + 4y + 3 = 0.$$

In Problems 35–44, find the equation of the given conic.

35. The horizontal ellipse with center $(5, 1)$, major diameter 10, minor diameter 8.

36. The hyperbola with center $(2, -1)$, vertex at $(4, -1)$, and focus at $(5, -1)$.

37. The parabola with vertex $(2, 3)$ and focus $(2, 5)$.

38. The ellipse with center $(2, 3)$ passing through $(6, 3)$ and $(2, 5)$.

39. The hyperbola with vertices at $(0, 0)$ and $(0, 6)$ and a focus at $(0, 8)$.

40. The ellipse with foci at $(2, 0)$ and $(2, 12)$ and a vertex at $(2, 14)$.

41. The parabola with focus $(2, 5)$ and directrix $x = 10$.

42. The parabola with focus $(2, 5)$ and vertex $(2, 6)$.

43. The ellipse with foci $(\pm 2, 2)$ that passes through the origin.

44. The hyperbola with foci $(0, 0)$ and $(0, 4)$ that passes through $(12, 9)$.

45. A curve $C$ goes through the three points, $(-1, 2)$, $(0, 0)$, and $(3, 6)$. Find an equation for $C$ if $C$ is:

(a) a vertical parabola;  (b) a horizontal parabola;
(c) a circle.

46. The ends of an elastic string with a knot at $K(x, y)$ are attached to a fixed point $A(a, b)$ and a point $P$ on the rim of a wheel of radius $r$ centered at $(0, 0)$. As the wheel turns, $K$ traces a curve $C$. Find the equation of $C$. Assume the string stays taut and stretches uniformly (that is, $\alpha = |KP|/|AP|$ is constant).

**47.** Name the conic $y^2 = Lx + Kx^2$ according to the value of $K$ and then show that in every case, $|L|$ is the length of the latus rectum of the conic. Assume $L \neq 0$.

**48.** Show that the equations of the parabola and hyperbola with vertex $(a, 0)$ and focus $(c, 0)$, $c > a > 0$, can be written as $y^2 = 4(c - a)(x - a)$ and $y^2 = (b^2/a^2)(x^2 - a^2)$,

respectively. Then use these expressions for $y^2$ to show that the parabola is always "inside" the right branch of the hyperbola.

---

**Answers to Concepts Review: 1.** $a^2/4$ **2.** 14; ellipse **3.** A line, parallel lines, intersecting lines, a point, the empty set **4.** Intersecting lines

---

## 12.5
## ROTATION OF AXES

Consider the most general second-degree equation in $x$ and $y$

$$Ax^2 + Bxy + Cy^2 + Dx + Ey + F = 0$$

The new feature is the appearance of the cross-product term $Bxy$. Is it still true that the graph is a conic or one of the limiting forms? The answer is yes, but the axis (or axes) of the conic are rotated with respect to the coordinate axes.

**Rotations**   Introduce a new pair of coordinate axes, the $u$- and $v$-axes, with the same origin as the $x$- and $y$-axes but rotated through an angle $\theta$, as shown in Figure 1. A point $P$ then has two sets of coordinates: $(x, y)$ and $(u, v)$. How are they related?

Let $r$ denote the length of $OP$ and let $\phi$ denote the angle from the positive $u$-axis to $OP$. Then, $x$, $y$, $u$, and $v$ have the geometric interpretations shown in the diagram.

Looking at the right triangle $OPM$, we see that

$$\cos(\phi + \theta) = \frac{x}{r}$$

so

$$x = r\cos(\phi + \theta) = r(\cos\phi\cos\theta - \sin\phi\sin\theta)$$
$$= (r\cos\phi)\cos\theta - (r\sin\phi)\sin\theta$$

Consideration of triangle $OPN$ shows that $u = r\cos\phi$ and $v = r\sin\phi$. Thus,

$$\boxed{x = u\cos\theta - v\sin\theta}$$

Similar reasoning leads to

$$\boxed{y = u\sin\theta + v\cos\theta}$$

These formulas determine a transformation called a **rotation of axes**.

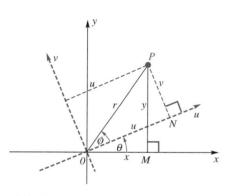

FIGURE 1

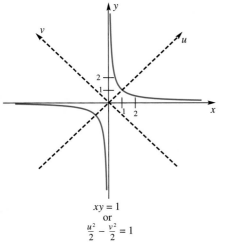

$xy = 1$
or
$\dfrac{u^2}{2} - \dfrac{v^2}{2} = 1$

FIGURE 2

**EXAMPLE 1**  Find the new equation that results from $xy = 1$ after a rotation of axes through $\theta = \pi/4$. Sketch the graph.

**Solution**  The required substitutions are

$$x = u \cos \frac{\pi}{4} - v \sin \frac{\pi}{4} = \frac{\sqrt{2}}{2}(u - v)$$

$$y = u \sin \frac{\pi}{4} + v \cos \frac{\pi}{4} = \frac{\sqrt{2}}{2}(u + v)$$

The equation $xy = 1$ takes the form

$$\frac{\sqrt{2}}{2}(u - v)\frac{\sqrt{2}}{2}(u + v) = 1$$

which simplifies to

$$\frac{u^2}{2} - \frac{v^2}{2} = 1$$

This we recognize as the equation of a hyperbola with $a = b = \sqrt{2}$. Note that the cross-product term has disappeared as a result of the rotation. The choice of the angle $\theta = \pi/4$ was just right to make this happen. The graph is shown in Figure 2.  ∎

**Determining the Angle $\theta$**  How do we know what rotation to make in order to eliminate the cross-product term? Consider the equation

$$Ax^2 + Bxy + Cy^2 + Dx + Ey + F = 0$$

If we make the substitutions

$$x = u \cos \theta - v \sin \theta$$
$$y = u \sin \theta + v \cos \theta$$

this equation takes the form

$$au^2 + buv + cv^2 + du + ev + f = 0$$

where $a$, $b$, $c$, $d$, $e$, and $f$ are numbers that depend upon $\theta$. We could find expressions for all of them, but we really care only about $b$. When we do the necessary algebra, we find

$$b = B(\cos^2\theta - \sin^2\theta) - 2(A - C)\sin \theta \cos \theta$$
$$= B \cos 2\theta - (A - C)\sin 2\theta$$

To make $b = 0$, we require

$$B \cos 2\theta = (A - C)\sin 2\theta$$

or

$$\boxed{\cot 2\theta = \frac{A - C}{B}}$$

This formula answers our question. To eliminate the cross-product term, choose $\theta$ so it satisfies this formula. In the equation $xy = 1$ of Example 1, $A = 0$, $B = 1$, and $C = 0$, so we choose $\theta$ satisfying $\cot 2\theta = 0$. One angle that works is $\theta = \pi/4$. We could also use $\theta = 3\pi/4$ or $\theta = -5\pi/4$, but it is customary to choose a first-quadrant angle; that is, we choose $2\theta$ satisfying $0 \le 2\theta < \pi$, so that $0 \le \theta < \pi/2$.

**EXAMPLE 2** Make a rotation of axes to eliminate the cross-product term in

$$4x^2 + 2\sqrt{3}xy + 2y^2 + 10\sqrt{3}x + 10y = 5$$

Then sketch the graph.

*Solution*

$$\cot 2\theta = \frac{A - C}{B} = \frac{4 - 2}{2\sqrt{3}} = \frac{1}{\sqrt{3}}$$

which means that $2\theta = \pi/3$ and $\theta = \pi/6$. The appropriate substitutions are

$$x = u\frac{\sqrt{3}}{2} - v\frac{1}{2} = \frac{\sqrt{3}u - v}{2}$$

$$y = u\frac{1}{2} + v\frac{\sqrt{3}}{2} = \frac{u + \sqrt{3}v}{2}$$

Our equation transforms first to

$$4\frac{(\sqrt{3}u - v)^2}{4} + 2\sqrt{3}\frac{(\sqrt{3}u - v)(u + \sqrt{3}v)}{4} + 2\frac{(u + \sqrt{3}v)^2}{4}$$

$$+ 10\sqrt{3}\frac{\sqrt{3}u - v}{2} + 10\frac{u + \sqrt{3}v}{2} = 5$$

and, after simplifying, to

$$5u^2 + v^2 + 20u = 5$$

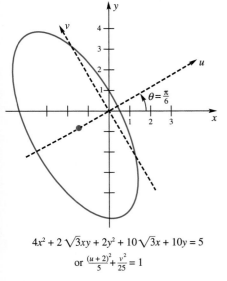

$$4x^2 + 2\sqrt{3}xy + 2y^2 + 10\sqrt{3}x + 10y = 5$$

$$\text{or } \frac{(u+2)^2}{5} + \frac{v^2}{25} = 1$$

**FIGURE 3**

To put this equation in recognizable form, we complete the square.

$$5(u^2 + 4u + 4) + v^2 = 5 + 20$$

$$\frac{(u+2)^2}{5} + \frac{v^2}{25} = 1$$

We identify the last equation as that of a vertical ellipse with center at $u = -2$ and $v = 0$ and with $a = 5$ and $b = \sqrt{5}$. This allows us to draw the graph shown in Figure 3. If we wanted to carry the simplifying process further, we would make the translation $r = u + 2$, $s = v$, which results in the standard equation $r^2/5 + s^2/25 = 1$. ∎

**Rotation Through a Nonspecial Angle** Our two examples involved rotations through the special angles of $\pi/4$ and $\pi/6$, respectively. Nonspecial angles are handled through use of the half-angle formulas.

$$\sin\theta = \pm\sqrt{\frac{1 - \cos 2\theta}{2}} \qquad \cos\theta = \pm\sqrt{\frac{1 + \cos 2\theta}{2}}$$

**EXAMPLE 3** Use a rotation to eliminate the cross-product term in

$$x^2 + 24xy + 8y^2 = 136$$

Then sketch the graph.

*Solution* We choose $\theta$ satisfying

$$\cot 2\theta = \frac{A - C}{B} = \frac{1 - 8}{24} = -\frac{7}{24}$$

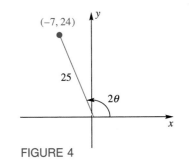

**FIGURE 4**

We take $2\theta$ to be a second-quadrant angle with $(-7, 24)$ on its terminal side, a point with distance $r = \sqrt{(-7)^2 + (24)^2} = 25$ from the origin (Figure 4). It follows that $\cos 2\theta = -\frac{7}{25}$. Applying the half-angle formulas gives

$$\sin\theta = \sqrt{\frac{1 + \frac{7}{25}}{2}} = \frac{4}{5} \qquad \cos\theta = \sqrt{\frac{1 - \frac{7}{25}}{2}} = \frac{3}{5}$$

Next we use the rotation formulas

$$x = \frac{3}{5}u - \frac{4}{5}v \qquad y = \frac{4}{5}u + \frac{3}{5}v$$

When these expressions are substituted in our equation, we obtain

$$\left(\frac{3u - 4v}{5}\right)^2 + 24\left(\frac{3u - 4v}{5}\right)\left(\frac{4u + 3v}{5}\right) + 8\left(\frac{4u + 3v}{5}\right)^2 = 136$$

This simplifies to

$$425u^2 - 200v^2 = (136)(25)$$

or

$$\frac{u^2}{8} - \frac{v^2}{17} = 1$$

The graph of the last equation is a horizontal hyperbola in the $uv$-plane with center at the origin (Figure 5).

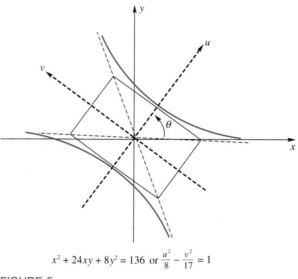

$$x^2 + 24xy + 8y^2 = 136 \quad \text{or} \quad \frac{u^2}{8} - \frac{v^2}{17} = 1$$

FIGURE 5

## CONCEPTS REVIEW

**1.** The most general second degree equation in $x$ and $y$ has the form _____.

**2.** The cross-product term (the $xy$-term) can be eliminated by a rotation of axes through an angle $\theta$ satisfying $\cot 2\theta =$ _____.

**3.** The graph of the equation $xy = 1$ is a _____.

**4.** To put a general second-degree equation in standard form, we first make a _____ of axes and then a _____ of axes.

## PROBLEM SET 12.5

In Problems 1–12, eliminate the cross-product term by a suitable rotation of axes and then, if necessary, translate axes (complete the squares) to put the equation in standard form. Finally graph the equation showing the rotated axes.

**1.** $x^2 + xy + y^2 = 6$

**2.** $3x^2 + 10xy + 3y^2 + 10 = 0$

**3.** $4x^2 + xy + 4y^2 = 56$

**4.** $4xy - 3y^2 = 64$

**5.** $4x^2 - 3xy = 18$

**6.** $11x^2 + 96xy + 39y^2 + 240x + 570y + 875 = 0$

**7.** $-\frac{1}{2}x^2 + 7xy - \frac{1}{2}y^2 - 6\sqrt{2}x - 6\sqrt{2}y = 0$

8. $\frac{3}{2}x^2 + xy + \frac{3}{2}y^2 + \sqrt{2}x + \sqrt{2}y = 13$

9. $34x^2 + 24xy + 41y^2 + 250y = -325$

10. $16x^2 + 24xy + 9y^2 - 20x - 15y - 150 = 0$

11. $5x^2 - 3xy + y^2 + 65x - 25y + 203 = 0$

12. $6x^2 - 5xy - 6y^2 + 78x + 52y + 26 = 0$

13. The graph of $x \cos \alpha + y \sin \alpha = d$ is a line. Show that the perpendicular distance from the origin to this line is $|d|$ by making a rotation of axes through the angle $\alpha$.

14. Transform the equation $x^{1/2} + y^{1/2} = a^{1/2}$ by a rotation of axes through $45°$ and then square twice to eliminate radicals on variables. Identify the corresponding curve.

15. Solve the rotation formulas for $u$ and $v$ in terms of $x$ and $y$.

16. Use the results of Problem 15 to find the $uv$-coordinates corresponding to $(x, y) = (5, -3)$ after a rotation of axes through $60°$.

17. Find the points of $x^2 + 14xy + 49y^2 = 100$ that is closest to the origin.

18. Recall that $Ax^2 + Bxy + Cy^2 + Dx + Ey + F = 0$ transforms to $au^2 + buv + cv^2 + du + ev + f = 0$ under a rotation of axes. Find formulas for $a$ and $c$ and show that $a + c = A + C$.

19. Show that $b^2 - 4ac = B^2 - 4AC$ (see Problem 18).

20. Use the result of Problem 19 to convince yourself that the graph of the general second-degree equation will be:

(a) a parabola if $B^2 - 4AC = 0$,
(b) an ellipse if $B^2 - 4AC < 0$,
(c) a hyperbola if $B^2 - 4AC > 0$,

or limiting forms of the above conics.

21. Let $Ax^2 + Bxy + Cy^2 = 1$ be transformed to $au^2 + cv^2 = 1$ by a rotation of axes and suppose $\Delta = 4AC - B^2 \neq 0$. Use Problems 18 and 19 to show that:

(a) $1/ac = 4/\Delta$,
(b) $1/a + 1/c = (4/\Delta)(A + C)$,
(c) $1/a$ and $1/c$ are the two values of

$$(2/\Delta)(A + C \pm \sqrt{(A - C)^2 + B^2})$$

22. Show that if $A + C$ and $\Delta = 4AC - B^2$ are both positive, then the graph of $Ax^2 + Bxy + Cy^2 = 1$ is an ellipse (or circle) with area $2\pi/\sqrt{\Delta}$. (Recall from Problem 17 of Section 12.3 that the area of the ellipse $x^2/p^2 + y^2/q^2 = 1$ is $\pi pq$.)

23. For what values of $B$ is the graph of $x^2 + Bxy + y^2 = 1$:

(a) an ellipse,
(b) a circle,
(c) a hyperbola,
(d) two parallel lines?

24. Use the results of Problems 21 and 22 to find the distance between the foci and the area of the ellipse

$$25x^2 + 8xy + y^2 = 1$$

---

**Answers to Concepts Review:  1.** $Ax^2 + Bxy + Cy^2 + Dx + Ey + F = 0$  **2.** $(A - C)/B$  **3.** Hyperbola  **4.** Rotation, translation

---

## 12.6
## THE POLAR COORDINATE SYSTEM

Two Frenchmen, Pierre Fermat and René Descartes, introduced what we now call the *Cartesian*, or *rectangular*, coordinate system. Their idea was to specify each point $P$ in the plane by giving two numbers $(x, y)$ the directed distances from a pair of perpendicular axes (Figure 1). This notion is by now so familiar that we use it almost without thinking. Yet it is the fundamental idea in analytic geometry and makes possible the development of calculus as we have given it so far.

Giving the directed distances from a pair of perpendicular axes is not the only way to specify a point. Another way to do this is by giving the so-called polar coordinates.

**Polar Coordinates**  We start with a fixed half-line, called the **polar axis**, emanating from a fixed point $O$, called the **pole** or **origin**. By custom, the polar axis is chosen to be horizontal and pointing to the right and may therefore be identified with the positive $x$-axis in the rectangular coordinate system. Any point $P$ (other than the pole) is the intersection of a unique circle with center at $O$ and a unique ray emanating from $O$. If $r$ is the radius

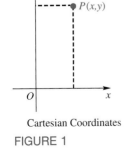

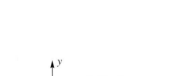
$P(x,y)$

Cartesian Coordinates

FIGURE 1

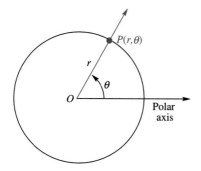

Polar Coordinates

FIGURE 2

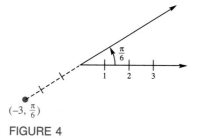

FIGURE 4

of the circle and $\theta$ is one of the angles the ray makes with the polar axis, then $(r, \theta)$ is a pair of **polar coordinates** for $P$ (Figure 2).

Points specified by polar coordinates are easiest to plot if we use polar graph paper. The grid on such paper consists of concentric circles and rays emanating from their common center. A simple version of a polar grid is shown in Figure 3, where we have also plotted several points.

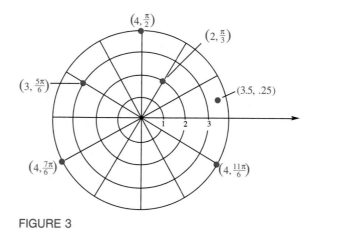

FIGURE 3

Notice a phenomenon that did not occur with Cartesian coordinates. Each point has many sets of polar coordinates, due to the fact that the angles $\theta + 2\pi n$, $n = 0, \pm 1, \pm 2, \ldots$, have the same terminal sides. For example, the point with polar coordinates $(4, \pi/2)$ also has coordinates $(4, 5\pi/2)$, $(4, 9\pi/2)$, $(4, -3\pi/2)$, and so on. This is especially true because we allow $r$ to be negative. In this case, $(r, \theta)$ is on the ray oppositely directed from the terminal side of $\theta$ and $|r|$ units from the origin. Thus, the point with polar coordinates $(-3, \pi/6)$ is as shown in Figure 4, and $(-4, 3\pi/2)$ is another set of coordinates for $(4, \pi/2)$. The origin has coordinates $(0, \theta)$, where $\theta$ is any angle.

**Polar Equations**   Examples of polar equations are

$$r = 8 \sin \theta \qquad \text{and} \qquad r = \frac{2}{1 - \cos \theta}$$

Polar equations, like rectangular ones, are best visualized from their graphs. The **graph of a polar equation** is the set of points, each of which has at least one pair of polar coordinates that satisfy the equation. The most basic way to sketch a graph is to construct a table of values, plot the corresponding points, and then connect these points with a smooth curve.

**EXAMPLE 1**   Sketch the graph of the polar equation $r = 8 \sin \theta$.

*Solution*   We substitute multiples of $\pi/6$ for $\theta$ and calculate the corresponding $r$-values. Note that as $\theta$ increases from 0 to $2\pi$, the graph is traced twice (Figure 5).   ∎

| $\theta$ | $r$ |
|------|-------|
| 0 | 0 |
| $\pi/6$ | 4 |
| $\pi/3$ | 6.93 |
| $\pi/2$ | 8 |
| $2\pi/3$ | 6.93 |
| $5\pi/6$ | 4 |
| $\pi$ | 0 |
| $7\pi/6$ | −4 |
| $4\pi/3$ | −6.93 |
| $3\pi/2$ | −8 |
| $5\pi/3$ | −6.93 |
| $11\pi/6$ | −4 |

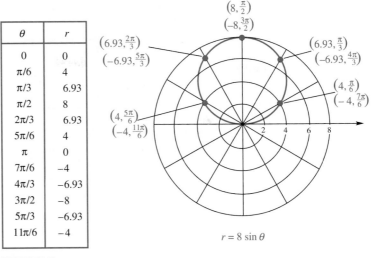

$r = 8 \sin \theta$

**FIGURE 5**

**EXAMPLE 2** Sketch the graph of $r = \dfrac{2}{1 - \cos \theta}$.

*Solution* See Figure 6.

| $\theta$ | $r$ |
|------|-------|
| 0 | − |
| $\pi/4$ | 6.8 |
| $\pi/2$ | 2 |
| $3\pi/4$ | 1.2 |
| $\pi$ | 1 |
| $5\pi/4$ | 1.2 |
| $3\pi/2$ | 2 |
| $7\pi/4$ | 6.8 |
| $2\pi$ | − |

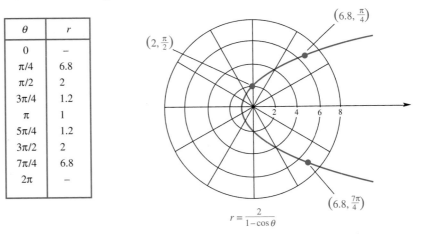

$r = \dfrac{2}{1-\cos\theta}$

**FIGURE 6**

Note a phenomenon that does not occur with rectangular coordinates. The coordinates $(-2, 3\pi/2)$ do not satisfy the equation. Yet the point $P(-2, 3\pi/2)$ is on the graph, due to the fact that $(2, \pi/2)$ specifies the same point and does satisfy the equation. We conclude that *in polar coordinates, failure of a particular set of coordinates to satisfy a given equation is no guarantee that the corresponding point is not on the graph of that equation.* This fact causes many difficulties; we must learn to live with them. ■

**Relation to Cartesian Coordinates**    We suppose that the polar axis coincides with the positive *x*-axis of the Cartesian system. Then the polar coordinates $(r, \theta)$ of a point *P* and the Cartesian coordinates $(x, y)$ of the same

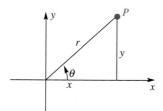

**FIGURE 7**

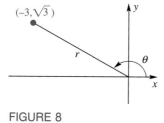

**FIGURE 8**

point are related by the equations

$$x = r \cos \theta \qquad r^2 = x^2 + y^2$$
$$y = r \sin \theta \qquad \tan \theta = \frac{y}{x}$$

That this is true for a point $P$ in the first quadrant is clear from Figure 7 and is easy to show for points in the other quadrants.

**EXAMPLE 3**   Find the Cartesian coordinates corresponding to $(4, \pi/6)$ and polar coordinates corresponding to $(-3, \sqrt{3})$.

*Solution*   If $(r, \theta) = (4, \pi/6)$, then

$$x = 4 \cos \frac{\pi}{6} = 4 \cdot \frac{\sqrt{3}}{2} = 2\sqrt{3}$$

$$y = 4 \sin \frac{\pi}{6} = 4 \cdot \frac{1}{2} = 2$$

If $(x, y) = (-3, \sqrt{3})$, then (see Figure 8)

$$r^2 = (-3)^2 + (\sqrt{3})^2 = 12$$

$$\tan \theta = \frac{\sqrt{3}}{-3}$$

One value of $(r, \theta)$ is $(2\sqrt{3}, 5\pi/6)$. Another is $(-2\sqrt{3}, -\pi/6)$. ∎

Sometimes we can identify the graph of a polar equation by finding its equivalent Cartesian form. Here is an illustration.

**EXAMPLE 4**   Show that the graph of $r = 8 \sin \theta$ (Example 1) is a circle and that the graph of $r = 2/(1 - \cos \theta)$ (Example 2) is a parabola by changing to Cartesian coordinates.

*Solution*   If we multiply $r = 8 \sin \theta$ by $r$, we get

$$r^2 = 8r \sin \theta$$

which, in Cartesian coordinates, is

$$x^2 + y^2 = 8y$$

and may be written successively as

$$x^2 + y^2 - 8y = 0$$
$$x^2 + y^2 - 8y + 16 = 16$$
$$x^2 + (y - 4)^2 = 16$$

The latter is the equation of a circle of radius 4 centered at $(0, 4)$.

**Caution**

Since $r$ can be 0, there is a potential danger in multiplying both sides of a polar equation by $r$ or in dividing both sides by $r$. In the first case, we might add the pole to the graph; in the second, we might delete the pole from the graph. In Example 4, we multiplied both sides of $r = 8 \sin \theta$ by $r$ but no harm was done since the pole was already on the graph as the point with $\theta$-coordinate 0.

The second equation is handled by the following steps.

$$r = \frac{2}{1 - \cos \theta}$$

$$r - r \cos \theta = 2$$

$$r - x = 2$$

$$r = x + 2$$

$$r^2 = x^2 + 4x + 4$$

$$x^2 + y^2 = x^2 + 4x + 4$$

$$y^2 = 4(x + 1)$$

We recognize the last equation as that of a parabola with vertex at $(-1, 0)$ and focus at the origin. ∎

**Polar Equations for Lines, Circles, and Conics**    If a line passes through the pole, it has the simple equation $\theta = \theta_0$. If the line does not go through the pole, it is some distance $d > 0$ from it. Let $\theta_0$ be the angle from the polar axis to the perpendicular from the pole to the given line (Figure 9). Then, if $P(r, \theta)$ is any point on the line, $\cos(\theta - \theta_0) = d/r$, or

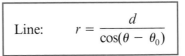

$$\text{Line:} \quad r = \frac{d}{\cos(\theta - \theta_0)}$$

If a circle of radius $a$ is centered at the pole, its equation is simply $r = a$. If it is centered at $(r_0, \theta_0)$, its equation is quite complicated unless we choose $r_0 = a$, as in Figure 10. Then, by the Law of Cosines, $a^2 = r^2 + a^2 - 2ra \cos(\theta - \theta_0)$, which simplifies to

$$\text{Circle:} \quad r = 2a \cos(\theta - \theta_0)$$

The cases $\theta_0 = 0$ and $\theta_0 = \pi/2$ are particularly nice. The first gives $r = 2a \cos \theta$; the second gives $r = 2a \cos(\theta - \pi/2)$; that is, $r = 2a \sin \theta$. The latter should be compared with Example 1.

Finally, if a conic (ellipse, parabola, or hyperbola) is placed so that its focus is at the pole, and its directrix is $d$ units away, as in Figure 11, then the familiar defining equation $|PF| = e|PL|$ takes the form

$$r = e[d - r \cos(\theta - \theta_0)]$$

or, equivalently,

$$\text{Conic:} \quad r = \frac{ed}{1 + e \cos(\theta - \theta_0)}$$

Again, there is special interest in the cases $\theta_0 = 0$ and $\theta_0 = \pi/2$. Note in particular that if $e = 1$, $d = 2$, and $\theta_0 = 0$, we have the equation of Example 2.

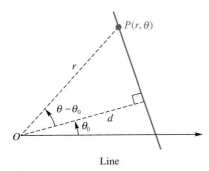

Line

FIGURE 9

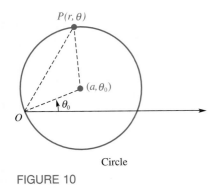

Circle

FIGURE 10

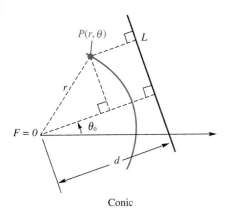

Conic

FIGURE 11

Our results are summarized in the following chart.

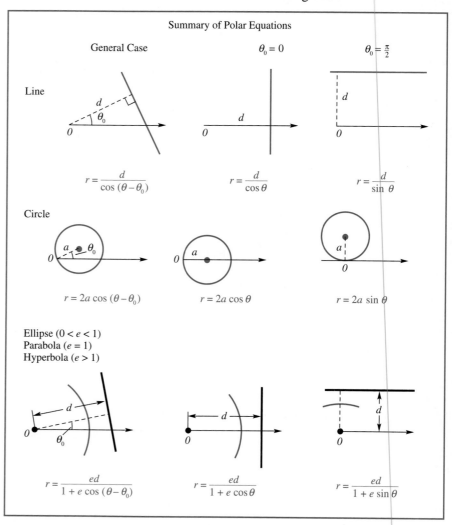

**Summary of Polar Equations**

**EXAMPLE 5** Find the equation of the horizontal ellipse with eccentricity $\frac{1}{2}$, focus at the pole, and vertical directrix 10 units to the right of the pole.

***Solution***

$$r = \frac{\frac{1}{2} \cdot 10}{1 + \frac{1}{2} \cos \theta} = \frac{10}{2 + \cos \theta}$$

$\blacksquare$

**EXAMPLE 6** Identify and sketch the graph of $r = \dfrac{7}{2 + 4 \sin \theta}$.

***Solution*** We put this equation in standard form.

$$r = \frac{7}{2 + 4 \sin \theta} = \frac{\frac{7}{2}}{1 + 2 \sin \theta} = \frac{2 \left( \frac{7}{4} \right)}{1 + 2 \sin \theta}$$

which we recognize as the polar equation of a hyperbola with $e = 2$, focus at the pole, and horizontal directrix $\frac{7}{4}$ units above the polar axis (Figure 12).

$\blacksquare$

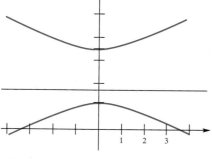

FIGURE 12

## CONCEPTS REVIEW

**1.** Every point in the plane has a unique pair $(x, y)$ of Cartesian coordinates but _____ pairs $(r, \theta)$ of polar coordinates.

**2.** The relations $x =$ _____ and $y =$ _____ connect Cartesian and polar coordinates; also _____ $=$ $x^2 + y^2$.

**3.** The graph of the polar equation $r = 5$ is a _____; the graph of $\theta = 5$ is a _____.

**4.** The graph of the polar equation $r = ed/(1 + e \cos \theta)$ is a _____.

## PROBLEM SET 12.6

**1.** Plot the points whose polar coordinates are $(3, \frac{1}{3}\pi)$, $(1, \frac{1}{2}\pi)$, $(4, \frac{1}{3}\pi)$, $(0, \pi)$, $(1, 4\pi)$, $(3, \frac{11}{4}\pi)$, $(\frac{5}{3}, \frac{1}{2}\pi)$, and $(4, 0)$.

**2.** Plot the points whose polar coordinates are $(3, 2\pi)$, $(2, \frac{1}{2}\pi)$, $(4, -\frac{1}{3}\pi)$, $(0, 0)$, $(1, 54\pi)$, $(3, -\frac{1}{6}\pi)$, $(1, \frac{1}{2}\pi)$, and $(3, -\frac{3}{2}\pi)$.

**3.** Plot the points whose polar coordinates are $(3, 2\pi)$, $(-2, \frac{1}{3}\pi)$, $(-2, -\frac{1}{4}\pi)$, $(-1, 1)$, $(1, -4\pi)$, $(\sqrt{3}, -\frac{7}{6}\pi)$, $(-2, \frac{1}{4}\pi)$, and $(-1, -\frac{1}{2}\pi)$.

**4.** Plot the points whose polar coordinates are $(3, \frac{9}{4}\pi)$, $(-2, \frac{1}{2}\pi)$, $(-2, -\frac{1}{3}\pi)$, $(-1, -1)$, $(1, -7\pi)$, $(-3, -\frac{1}{6}\pi)$, $(-2, -\frac{1}{2}\pi)$, and $(3, -\frac{33}{2}\pi)$.

**5.** Plot the points whose polar coordinates follow. For each, give four other pairs of polar coordinates, two with positive $r$ and two with negative $r$.

(a) $(1, \frac{1}{2}\pi)$        (b) $(-1, \frac{1}{4}\pi)$

(c) $(\sqrt{2}, -\frac{1}{3}\pi)$      (d) $(-\sqrt{2}, \frac{5}{2}\pi)$

**6.** Plot the points whose polar coordinates follow. For each, give four other pairs of polar coordinates, two with positive $r$ and two with negative $r$.

(a) $(3\sqrt{2}, \frac{7}{2}\pi)$      (b) $(-1, \frac{15}{4}\pi)$

(c) $(-\sqrt{2}, -\frac{2}{3}\pi)$     (d) $(-2\sqrt{2}, \frac{29}{2}\pi)$

**7.** Find the Cartesian coordinates of the points in Problem 5.

**8.** Find the Cartesian coordinates of the points in Problem 6.

**9.** Find the polar coordinates of the points whose Cartesian coordinates are given.

(a) $(3\sqrt{3}, 3)$       (b) $(-2\sqrt{3}, 2)$

(c) $(-\sqrt{2}, -\sqrt{2})$    (d) $(0, 0)$

**10.** Find the polar coordinates of the points whose Cartesian coordinates are given.

(a) $(-3/\sqrt{3}, 1/\sqrt{3})$    (b) $(-\sqrt{3}/2, \sqrt{3}/2)$

(c) $(0, -2)$           (d) $(3, -4)$

In each of the Problems 11–16, sketch the graph of the given Cartesian equation, and then find the polar equation for it.

**11.** $x - 3y + 2 = 0$       **12.** $x = 0$

**13.** $y = -2$            **14.** $x - y = 0$

**15.** $x^2 + y^2 = 4$       **16.** $x^2 = 4py$

In Problems 17–22, find the Cartesian equations of the graphs of the given polar equations.

**17.** $\theta = \frac{1}{2}\pi$          **18.** $r = 3$

**19.** $r \cos \theta + 3 = 0$    **20.** $r - 5 \cos \theta = 0$

**21.** $r \sin \theta - 1 = 0$

**22.** $r^2 - 6r \cos \theta - 4r \sin \theta + 9 = 0$

In Problems 23–36, name the curve with the given polar equation. If it is a conic, give its eccentricity. Sketch the graph.

**23.** $r = 6$             **24.** $\theta = \dfrac{2\pi}{3}$

**25.** $r = \dfrac{3}{\sin \theta}$       **26.** $r = \dfrac{-4}{\cos \theta}$

**27.** $r = 4 \sin \theta$       **28.** $r = -4 \cos \theta$

**29.** $r = \dfrac{4}{1 + \cos \theta}$    **30.** $r = \dfrac{4}{1 + 2 \sin \theta}$

**31.** $r = \dfrac{6}{2 + \sin \theta}$    **32.** $r = \dfrac{6}{4 - \cos \theta}$

**33.** $r = \dfrac{4}{2 + 2 \cos \theta}$   **34.** $r = \dfrac{4}{2 + 2 \cos(\theta - \pi/3)}$

**35.** $r = \dfrac{4}{\frac{1}{2} + \cos(\theta - \pi)}$   **36.** $r = \dfrac{4}{3 \cos(\theta - \pi/3)}$

**37.** Show that the polar equation of the circle with center $(c, \alpha)$ and radius $a$ is $r^2 + c^2 - 2rc \cos(\theta - \alpha) = a^2$.

**38.** Prove that $r = a \sin \theta + b \cos \theta$ represents a circle and find its center and radius.

**39.** Find the length of the latus rectum for the general conic $r = ed/[1 + e\cos(\theta - \theta_0)]$ in terms of $e$ and $d$.

**40.** Let $r_1$ and $r_2$ be the minimum and maximum distances (**perihelion** and **aphelion**) of the ellipse $r = ed/[1 + e\cos(\theta - \theta_0)]$ from a focus. Show that:

(a) $r_1 = ed/(1 + e)$, $r_2 = ed/(1 - e)$,
(b) major diameter = $2ed/(1 - e^2)$ and minor diameter = $2ed/\sqrt{1 - e^2}$.

**41.** The perihelion and aphelion for the astroid Icarus are 17 and 183 million miles, respectively. What is the eccentricity of its elliptical orbit?

**42.** The earth's orbit around the sun is an ellipse of eccentricity 0.0167 and major diameter 185.8 million miles. Find its perihelion.

**43.** The orbit of a certain comet is a parabola with the sun at a focus. The angle between the axis of the parabola (assumed pointing into the concave side) and a ray from the sun to the comet is 120° when the comet is 100 million miles from the sun. How close does the comet get to the sun?

**44.** The position of a comet with a highly eccentric elliptical orbit ($e$ very near 1) is measured with respect to a fixed polar axis (sun is at focus but polar axis is not axis of ellipse) at two times, giving the two points $(4, \pi/2)$ and $(3, \pi/4)$ of the orbit. Here distances are measured in astronomical units (1 AU ≈ 93 million miles). For the part of the orbit near the sun, assume $e = 1$, so the orbit is given by $r = d/[1 + \cos(\theta - \theta_0)]$.

(a) The two points give two conditions for $d$ and $\theta_0$. Use them to show that $4.24\cos\theta_0 - 3.76\sin\theta_0 - 2 = 0$.
(b) Solve for $\theta_0$ using Newton's method.
(c) How close does the comet get to the sun?

PC **45.** In order to graph a polar equation such as $r = f(t)$ using a parametric equation grapher you must replace this equation by $x = f(t)\cos t$ and $y = f(t)\sin t$. These equations can be obtained by multiplying $r = f(t)$ by $\cos t$ and $\sin t$ respectively. Confirm the discussions of conics in the text by graphing $r = 4e/(1 + e\cos t)$ for $e = 0.1, 0.5, 0.9, 1, 1.1$ and 1.3 on $[-\pi, \pi]$.

---

**Answers to Concepts Review:** **1.** Infinitely many **2.** $r\cos\theta$; $r\sin\theta$; $r^2$ **3.** Circle; line **4.** Conic

---

## 12.7
## GRAPHS OF POLAR EQUATIONS

The polar equations considered in the previous section led to familiar graphs—mainly lines, circles, and conics. Now we turn our attention to more exotic graphs—cardioids, limaçons, roses, and spirals. The polar equations for these curves are still rather simple; the corresponding Cartesian equations are quite complicated. Thus, we see one of the advantages of having more than one coordinate system available. Some curves have simple equations in one system; other curves have simple equations in a second system. We will exploit this later in the book when we often begin the solution of a problem by choosing a convenient coordinate system.

Symmetry properties can help us draw a graph. Here are some *sufficient* tests for symmetry in polar coordinates. The diagrams in the margin will help you establish their validity.

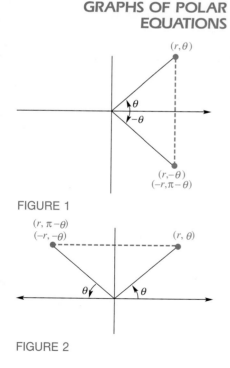

FIGURE 1

FIGURE 2

**1.** The graph of a polar equation is symmetric about the x-axis (the polar axis) if replacing $(r, \theta)$ by $(r, -\theta)$ (or by $(-r, \pi - \theta)$) produces an equivalent equation (Figure 1).

**2.** The graph of a polar equation is symmetric about the y-axis (the line $\theta = \pi/2$) if replacing $(r, \theta)$ by $(-r, -\theta)$ (or by $(r, \pi - \theta)$) produces an equivalent equation (Figure 2).

**3.** The graph of a polar equation is symmetric about the origin (pole) if replacing $(r, \theta)$ by $(-r, \theta)$ (or by $(r, \pi + \theta)$) produces an equivalent equation (Figure 3).

Because of the multiple representation of points in polar coordinates, symmetries may exist that are not identified by these three tests (see Problem 39).

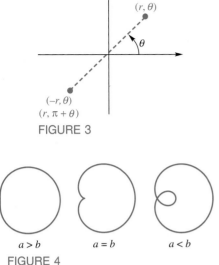

FIGURE 3

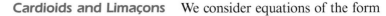

**Cardioids and Limaçons**   We consider equations of the form

$$r = a \pm b \cos \theta \qquad r = a \pm b \sin \theta$$

with $a$ and $b$ positive. Their graphs are called **limaçons**, with the special cases in which $a = b$ referred to as **cardioids**. Typical graphs are shown in Figure 4.

**EXAMPLE 1**   Analyze the equation $r = 2 + 4 \cos \theta$ for symmetry and sketch its graph.

*Solution*   Since cosine is an even function ($\cos(-\theta) = \cos \theta$), the graph is symmetric with respect to the $x$-axis. The other symmetry tests fail. A table of values and the graph appear in Figure 5.

$a > b$       $a = b$       $a < b$

FIGURE 4

| $\theta$ | $r$ |
|----------|-----|
| 0 | 6 |
| $\pi/6$ | 5.5 |
| $\pi/3$ | 4 |
| $\pi/2$ | 2 |
| $7\pi/12$ | 1.0 |
| $2\pi/3$ | 0 |
| $3\pi/4$ | $-0.8$ |
| $5\pi/6$ | $-1.5$ |
| $\pi$ | $-2$ |

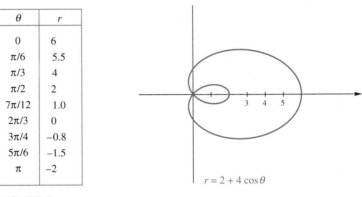

$r = 2 + 4 \cos \theta$

FIGURE 5

**Lemniscates**   The graphs of

$$r^2 = \pm a \cos 2\theta \qquad r^2 = \pm a \sin 2\theta$$

are figure-eight-shaped curves called **lemniscates.**

**EXAMPLE 2**   Analyze the equation $r^2 = 8 \cos 2\theta$ for symmetry and sketch its graph.

*Solution*   Since $\cos(-2\theta) = \cos 2\theta$ and

$$\cos[2(\pi - \theta)] = \cos(2\pi - 2\theta) = \cos(-2\theta) = \cos 2\theta$$

the graph is symmetric with respect to both axes. Clearly, it is also symmetric with respect to the origin. A table of values and the graph are shown in Figure 6.

| $\theta$ | $r$ |
|----------|----------|
| 0 | $\pm2.8$ |
| $\pi/12$ | $\pm2.6$ |
| $\pi/6$ | $\pm2$ |
| $\pi/4$ | 0 |

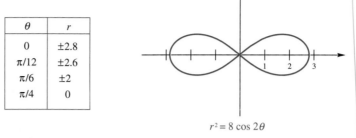

$r^2 = 8 \cos 2\theta$

FIGURE 6 ■

**Roses** Polar equations of the form

$$r = a \cos n\theta \qquad r = a \sin n\theta$$

represent flower-shaped curves called **roses.** The rose has $n$ leaves if $n$ is odd and $2n$ leaves if $n$ is even.

**EXAMPLE 3** Analyze $r = 4 \sin 2\theta$ for symmetry and sketch its graph.

**Solution** You can check that $r = 4 \sin 2\theta$ satisfies all three symmetry tests. For example, it meets Test 1 since

$$\sin 2(\pi - \theta) = \sin(2\pi - 2\theta) = -\sin 2\theta$$

and so replacing $(r, \theta)$ by $(-r, \pi - \theta)$ produces an equivalent equation.

To construct the correct graph, we begin by making a rather extensive table of values for $0 \le \theta \le \pi/2$ and a somewhat briefer one for $\pi/2 \le \theta \le 2\pi$. This table and the corresponding graph are shown in Figure 7. The arrows on the curve indicate the direction $P(r, \theta)$ moves as $\theta$ increases from 0 to $2\pi$.

| $\theta$ | $r$ | $\theta$ | $r$ |
|----------|----------|----------|----------|
| 0 | 0 | $2\pi/3$ | $-3.5$ |
| $\pi/12$ | 2 | $5\pi/6$ | $-3.5$ |
| $\pi/8$ | 2.8 | $\pi$ | 0 |
| $\pi/6$ | 3.5 | $7\pi/6$ | 3.5 |
| $\pi/4$ | 4 | $4\pi/3$ | 3.5 |
| $\pi/3$ | 3.5 | $3\pi/2$ | 0 |
| $3\pi/8$ | 2.8 | $5\pi/3$ | $-3.5$ |
| $5\pi/12$ | 2 | $11\pi/6$ | $-3.5$ |
| $\pi/2$ | 0 | $2\pi$ | 0 |

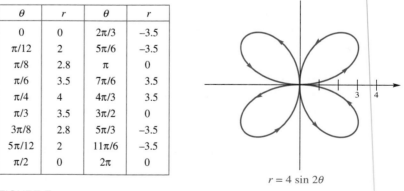

$r = 4 \sin 2\theta$

FIGURE 7 ■

**Spirals** The graph of $r = a\theta$ is called a **spiral of Archimedes**; the graph of $r = ae^{b\theta}$ is called a **logarithmic spiral.**

**EXAMPLE 4** Sketch the graph of $r = \theta$ for $\theta \ge 0$.

**Solution** We omit a table of values, but note that the graph crosses the polar axis at $(0, 0)$, $(2\pi, 2\pi)$, $(4\pi, 4\pi)$, . . . and crosses its extension to the left at $(\pi, \pi)$, $(3\pi, 3\pi)$, $(5\pi, 5\pi)$, . . . , as in Figure 8.

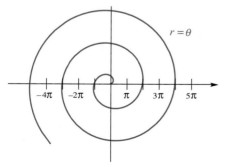

$r = \theta$

FIGURE 8

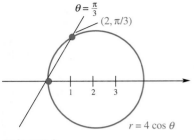

**FIGURE 9**

**Intersection of Curves in Polar Coordinates** In Cartesian coordinates, all points of intersection of two curves can be found by solving the equations of the curves simultaneously. But in polar coordinates, this is not always the case. This is because a point $P$ has many pairs of polar coordinates, and one pair may satisfy the polar equation of one curve and a different pair may satisfy the polar equation of the other curve. For instance (see Figure 9), the circle $r = 4 \cos \theta$ intersects the line $\theta = \pi/3$ in two points, the pole and $(2, \pi/3)$, and yet only the latter is a common solution of the two equations. This happens because the coordinates of the pole that satisfy the equation of the line are $(0, \pi/3)$ and those that satisfy the equation of the circle are $(0, \pi/2 + n\pi)$.

Our conclusion is this. In order to find all intersections of two curves whose polar equations are given, solve the equations simultaneously; then graph the two equations carefully to discover other possible points of intersection.

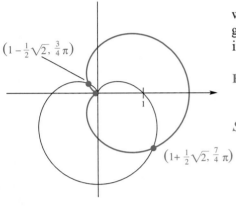

**FIGURE 10**

**EXAMPLE 5** Find the points of intersection of the two cardioids $r = 1 + \cos \theta$ and $r = 1 - \sin \theta$.

*Solution* If we eliminate $r$ between the two equations, we get $1 + \cos \theta = 1 - \sin \theta$. Thus $\cos \theta = -\sin \theta$, or $\tan \theta = -1$. We conclude that $\theta = \frac{3}{4}\pi$ or $\theta = \frac{7}{4}\pi$, which yields the two intersection points $(1 - \frac{1}{2}\sqrt{2}, \frac{3}{4}\pi)$ and $(1 + \frac{1}{2}\sqrt{2}, \frac{7}{4}\pi)$. The graphs in Figure 10 show, however, that we have missed a third intersection point, namely, the pole. The reason we missed it is that $r = 0$ in $r = 1 + \cos \theta$ when $\theta = \pi$, but $r = 0$ in $r = 1 - \sin \theta$ when $\theta = \pi/2$. ∎

---

## CONCEPTS REVIEW

1. The graph of $r = 3 + 2 \cos \theta$ is a _____.
2. The graph of $r = 2 + 2 \cos \theta$ is a _____.
3. The graph of $r = 4 \sin n\theta$ is a _____ with $n$

leaves if $n$ is _____ and $2n$ leaves if $n$ is _____.
4. The graph of $r = \theta/3$ is a _____.

---

## PROBLEM SET 12.7

In Problems 1–32, sketch the graph of the given polar equation.

1. $\theta^2 - 4 = 0$

2. $(r - 3)(\theta - \frac{\pi}{4}) = 0$

3. $r \sin \theta + 4 = 0$

4. $r = -4 \sec \theta$

5. $r = 2 \cos \theta$

6. $r = 4 \sin \theta$

7. $r = \dfrac{2}{1 - \cos \theta}$

8. $r = \dfrac{4}{1 + \sin \theta}$

9. $r = 3 - 3 \cos \theta$ (cardioid)

10. $r = 5 - 5 \sin \theta$ (cardioid)

11. $r = 1 - \sin \theta$ (cardioid)

12. $r = \sqrt{2} - \sqrt{2} \sin \theta$ (cardioid)

13. $r = 1 - 2 \sin \theta$ (limaçon)

14. $r = 4 - 3 \cos \theta$ (limaçon)

15. $r = 2 - 3 \sin \theta$ (limaçon)

16. $r = 5 - 3 \cos \theta$ (limaçon)

17. $r^2 = 4 \cos 2\theta$ (lemniscate)

18. $r^2 = 9 \sin 2\theta$ (lemniscate)

19. $r^2 = -9 \cos 2\theta$ (lemniscate)

20. $r^2 = -16 \cos 2\theta$ (lemniscate)

21. $r = 5 \cos 3\theta$ (three-leaved rose)

22. $r = 3 \sin 3\theta$ (three-leaved rose)

23. $r = 6 \sin 2\theta$ (four-leaved rose)

24. $r = 4 \cos 2\theta$ (four-leaved rose)

25. $r = 7 \cos 5\theta$ (five-leaved rose)

26. $r = 3 \sin 5\theta$ (five-leaved rose)

27. $r = \frac{1}{2}\theta$, $\theta \geq 0$ (spiral of Archimedes)

28. $r = 2\theta$, $\theta \geq 0$ (spiral of Archimedes)

29. $r = e^{\theta}$, $\theta \geq 0$ (logarithmic spiral)

30. $r = e^{\theta/2}$, $\theta \geq 0$ (logarithmic spiral)

31. $r = \dfrac{2}{\theta}$, $\theta > 0$ (reciprocal spiral)

32. $r = -\dfrac{1}{\theta}$, $\theta > 0$ (reciprocal spiral)

In Problems 33–38, sketch the given curves and find their points of intersection.

33. $r = 6$, $r = 4 + 4 \cos \theta$

34. $r = 1 - \cos \theta$, $r = 1 + \cos \theta$

35. $r = 3\sqrt{3} \cos \theta$, $r = 3 \sin \theta$

36. $r = 5$, $r = \dfrac{5}{1 - 2 \cos \theta}$

37. $r = 6 \sin \theta$, $r = \dfrac{6}{1 + 2 \sin \theta}$

38. $r^2 = 4 \cos 2\theta$, $r = 2\sqrt{2} \sin \theta$

39. The conditions for symmetry given in the text are sufficient conditions, not necessary conditions. Give an example of a polar equation $r = f(\theta)$ whose graph is symmetric with respect to the $y$-axis, even though replacing $(r, \theta)$ by either $(-r, -\theta)$ or $(r, \pi - \theta)$ fails to yield an equivalent equation.

40. Let $a$ and $b$ be fixed positive numbers and suppose $OAP$ is a line segment with $A$ on the line $x = a$ and $|AP| = b$. Find both the polar equation and the rectangular equation for the set of points $P$ (called a *conchoid*) and sketch its graph.

41. Let $F$ and $F'$ be fixed points with polar coordinates $(a, 0)$ and $(-a, 0)$, respectively. Show that the set of points $P$ satisfying $|PF|\,|PF'| = a^2$ is a lemniscate by finding its polar equation.

42. A line segment $L$ of length $2a$ has its two endpoints on the $x$- and $y$-axes, respectively. The point $P$ is on $L$ and is such that $OP$ is perpendicular to $L$. Show that the set of points $P$ satisfying this condition is a four-leaved rose by finding its polar equation.

43. Find the polar equation for the curve described by the following Cartesian equations.

(a) $y = 45$

(b) $x^2 + y^2 = 36$

(c) $x^2 + 2x + y^2 - 4y - 25 = 0$

(d) $4xy = 1$

(e) $y = 3x + 2$

(f) $3x^2 + 4y = 2$

(g) $x^2 - y^2 = 1$

(h) $\dfrac{x^2}{4} + \dfrac{y^2}{9} = 1$

Computers and graphing calculators offer a wonderful opportunity to experiment with the graphing of polar equations of the form $r = f(\theta)$. In some cases these aids require that the equations be recast in a parametric form. Since $x = r \cos \theta = f(\theta) \cos \theta$ and $y = r \sin \theta = f(\theta) \sin \theta$ you can use the parametric graphing capabilities to graph $x = f(t) \cos t$ and $y = f(t) \sin t$ as a set of parametric equations. In Problems 44–58 we explore such graphs.

44. Graph the curve $r = \cos(8\theta/5)$ using the parametric graphing facility of a graphing calculator or computer. Notice that it is necessary to determine the proper domain for $\theta$. Assuming you start at $\theta = 0$ you have to determine the value of $\theta$ which makes the curve start to repeat itself. Since the cos is periodic with period $2\pi$ show that the correct domain is $0 \leq \theta \leq 10\pi$.

45. Polar curves given by the equation $r = a - b \sin(\theta + \phi)$ are called limaçons.

(a) Taking $\phi = 0$ and writing the equation in the form $r = a(1 - (b/a)) \sin \theta$ give reasons why it will be sufficient to investigate curves of the form $r = 1 - c \sin \theta$.

(b) Determine the domain of $\theta$ necessary for the equation to give a complete closed curve.

(c) Show by graphing several examples that the effect of $\phi$ is to cause a rotation of the figure by an angle of $-\phi$.

(d) Considering the limaçons given by $r = 1 + c \sin \theta$, graph the curve for the values of $c = 2.4, 1.5, 1, 0.75, 0.5, 0.25, 0, -0.25, -0.5, -1, 2$.

**46.** Match the polar equations to the graphs labeled I–VIII in each case giving reasons for your choices.

(a) $r = \cos(\theta/2)$
(b) $r = \sec(3\theta)$
(c) $r = 2 - 3\sin(5\theta)$
(d) $r = 1 - 2\sin(5\theta)$
(e) $r = \cos(\theta/4)$
(f) $r = \theta\cos\theta$
(g) $r = 1/(\theta)^{3/2}$
(h) $r = 2\cos 3\theta$

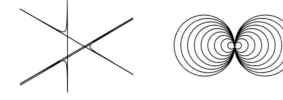

I        II

III        IV

V        VI

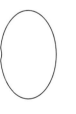

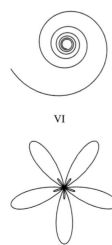

VII        VIII

Use a computer or calculator graphing device to draw the following polar curves. Make sure that you choose a sufficiently large interval for the parameter so that the entire curve is drawn.

**47.** $r = 1 + 2\cos(\theta/2)$

**48.** $r = \sqrt{1 - 0.5\sin^2\theta}$

**49.** $r = \cos(13\theta/5)$

**50.** $r = \sin(5\theta/7)$

**51.** $r = 1 + 3\cos(\theta/3)$

**52.** In many cases polar graphs are related to each other by rotation. We explore that concept here.

(a) How are the graphs of $r = 1 + \sin(\theta - \pi/3)$ and $r = 1 + \sin(\theta + \pi/3)$ related to the graph of $r = 1 + \sin\theta$?
(b) How is the graph of $r = 1 + \sin\theta$ related to the graph of $r = 1 - \sin\theta$?
(c) How is the graph of $r = 1 + \sin\theta$ related to the graph of $r = 1 + \cos\theta$?
(d) How is the graph of $r = f(\theta)$ related to the graph of $r = f(\theta - \alpha)$?

**53.** Use the graph to estimate the $y$-coordinate of the highest point on the curve $r = \sin 2\theta$. Confirm, by using calculus, that this value is correct.

**54.** Observe that if a curve has a continuous turning tangent the points on a polar curve which are closest and furthest from the origin will have the property that $\frac{df(\theta)}{d\theta} = 0$. Identify such points by graphing on the curve $r = 1 - 0.5\sin\theta$ and confirm your findings using calculus.

**55.** Investigate the family of curves defined by the polar equations $r = \cos n\theta$ where $n$ is some positive integer. How do the number of loops depend on $n$?

**56.** Investigate the family of curves defined by the polar equations $r = |\cos n\theta|$ where $n$ is some positive integer. How do the number of loops depend on $n$?

**57.** Investigate the family of curves given by $r = a + b\cos(n(\theta + \phi))$ where $a$, $b$ and $\phi$ are real numbers and $n$ is a positive integer.

(a) How are the graphs for $\phi = 0$ related to those for which $\phi \neq 0$?
(b) How does the graph change as $n$ increases?
(c) How do the relative magnitude and sign of $a$ and $b$ change the nature of the graph?

Be sure that you graph a sufficient number of examples to justify your conclusions.

**58.** Consider the family of polar curves given by

$$r = \frac{1 - c\sin\theta}{1 + c\sin\theta}$$

and investigate how the graph changes for values of $c$. Pay particular attention to what happens for values of $|c| < 1$, $|c| = 1$, and $|c| > 1$.

614 **59.** Polar graphs can be used to easily represent different spirals. The spirals can unwind clockwise or counterclockwise. Find the condition on $c$ to make the *spiral of Archimedes*, $r = c\theta$, unwind clockwise and counterclockwise for $0 \le \theta \le 6\pi$.

**60.** Sketch the *hyperbolic spiral* given by $r = c/\theta$ for $\pi \le \theta \le 6\pi$. For $c > 0$, does it unwind in the clockwise direction?

**61.** Two types of spirals are given by the *parabolic spiral* $r^2 = c\theta$ and the *logarithmic spiral* $r = ce^\theta$. Compare these spirals with the hyperbolic and Archimedian spirals given in Problems 59 and 61 for the range $\pi \le \theta \le 3\pi$, by sketching each for $c = 1$.

**62.** The following polar equations are represented by four graphs. Match the proper graph with the correct equation.

(a) $r = \sin\theta + \sin^2\theta$      (b) $r = \cos\theta + \cos^2\theta$
(c) $r = \sin\theta + \cos^2\theta$      (d) $r = \cos\theta + \sin^2\theta$

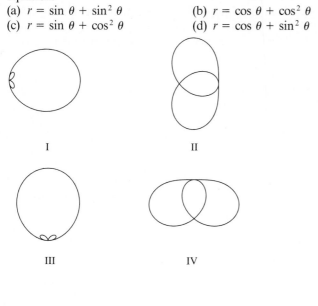

**63.** The following polar equations are represented by four graphs. Match the proper graph with the correct equation.

(a) $r = \sin 3\theta + \sin^2 2\theta$      (b) $r = \cos 2\theta + \cos^2 4\theta$
(c) $r = \sin 4\theta + \sin^2 5\theta$      (d) $r = \cos 2\theta + \cos^2 3\theta$
(e) $r = \cos 4\theta + \cos^2 4\theta$      (f) $r = \sin 4\theta + \sin^2 4\theta$

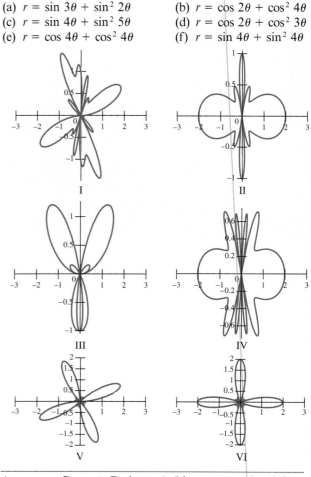

**Answers to Concepts Review:**    **1.** Limaçon    **2.** Cardioid    **3.** Rose; odd, even    **4.** Spiral

---

**12.8**
**CALCULUS IN POLAR COORDINATES**

The two most basic problems in calculus are the determinations of the slope of a tangent line and the area of a curved region. Here we consider both problems, but in the context of polar coordinates. The area problem plays a larger role in the rest of the book, so we consider it first.

In Cartesian coordinates, the fundamental building block in area problems was the rectangle. In polar coordinates, it is the sector of a circle (a pie-shaped region like that in Figure 1). From the fact that the area of a circle is $\pi r^2$, we infer that the area of a sector with central angle $\theta$ radians is $(\theta/2\pi)\pi r^2$; that is,

$$\text{Area of a sector:} \qquad A = \frac{1}{2}\theta r^2$$

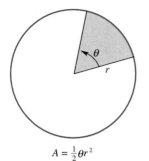

$$A = \tfrac{1}{2}\theta r^2$$

FIGURE 1

**Area in Polar Coordinates**   To begin, let $r = f(\theta)$ determine a curve in the plane, where $f$ is a continuous, nonnegative function for $\alpha \le \theta \le \beta$ and $\beta - \alpha \le 2\pi$. The curves $r = f(\theta)$, $\theta = \alpha$, and $\theta = \beta$ bound a region $R$ (the one shown at the left in Figure 2), whose area $A(R)$ we wish to determine.

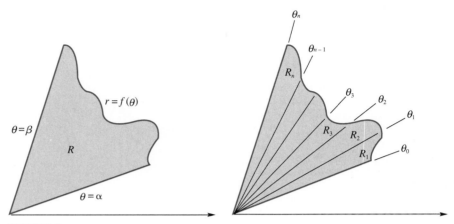

FIGURE 2

Partition the interval $[\alpha, \beta]$ into $n$ subintervals by means of numbers $\alpha = \theta_0 < \theta_1 < \theta_2 < \cdots < \theta_n = \beta$, thereby slicing $R$ into $n$ smaller pie-shaped regions $R_1, R_2, \ldots, R_n$, as shown in the right half of Figure 2. Clearly $A(R) = A(R_1) + A(R_2) + \cdots + A(R_n)$.

We approximate the area $A(R_i)$ of the $i$th slice; in fact, we do it in two ways. On the $i$th interval $[\theta_{i-1}, \theta_i]$, $f$ achieves its minimum value and maximum value—for instance, at $u_i$ and $v_i$, respectively (Figure 3). Thus, if $\Delta\theta_i = \theta_i - \theta_{i-1}$,

$$\tfrac{1}{2}[f(u_i)]^2 \, \Delta\theta_i \le A(R_i) \le \tfrac{1}{2}[f(v_i)]^2 \, \Delta\theta_i$$

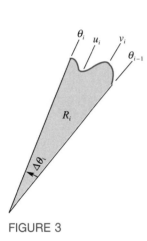

FIGURE 3

and so

$$\sum_{i=1}^{n} \tfrac{1}{2}[f(u_i)]^2 \, \Delta\theta_i \le \sum_{i=1}^{n} A(R_i) \le \sum_{i=1}^{n} \tfrac{1}{2}[f(v_i)]^2 \, \Delta\theta_i$$

The first and third members of this inequality are Riemann sums for the same integral, namely, $\int_{\alpha}^{\beta} \tfrac{1}{2}[f(\theta)]^2 \, d\theta$. When we let the norm of the partition tend toward zero, we obtain (Squeeze Principle) the area formula

$$\boxed{A = \tfrac{1}{2} \int_{\alpha}^{\beta} [f(\theta)]^2 \, d\theta}$$

This formula can, of course, be memorized. We prefer that you remember how it was derived. In fact, you will note that the three familiar words, *slice, approximate, integrate,* are also the key to area problems in polar coordinates. We illustrate what we mean.

**EXAMPLE 1** Find the area of the region inside the limaçon $r = 2 + \cos \theta$.

**Solution** The graph is sketched in Figure 4; note that $\theta$ varies from 0 to $2\pi$. We slice, approximate, integrate.

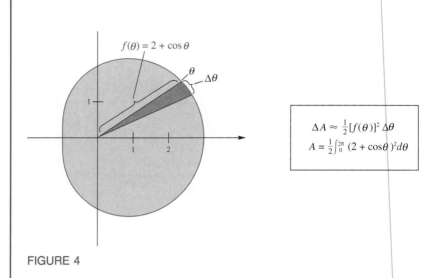

$$\Delta A \approx \tfrac{1}{2}[f(\theta)]^2 \, \Delta\theta$$
$$A = \tfrac{1}{2}\int_0^{2\pi} (2 + \cos\theta)^2 d\theta$$

FIGURE 4

By symmetry, we can double the integral from 0 to $\pi$. Thus,

$$A = \int_0^\pi (2 + \cos\theta)^2 \, d\theta = \int_0^\pi (4 + 4\cos\theta + \cos^2\theta) \, d\theta$$

$$= \int_0^\pi 4 \, d\theta + 4\int_0^\pi \cos\theta \, d\theta + \frac{1}{2}\int_0^\pi (1 + \cos 2\theta) \, d\theta$$

$$= \int_0^\pi \frac{9}{2} \, d\theta + 4\int_0^\pi \cos\theta \, d\theta + \frac{1}{4}\int_0^\pi \cos 2\theta \cdot 2 \, d\theta$$

$$= \left[\frac{9}{2}\theta\right]_0^\pi + \left[4\sin\theta\right]_0^\pi + \left[\frac{1}{4}\sin 2\theta\right]_0^\pi$$

$$= \frac{9\pi}{2}$$

$\blacksquare$

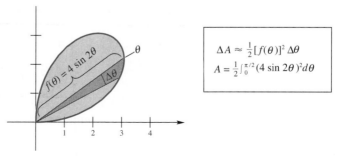

$$\Delta A \approx \tfrac{1}{2}[f(\theta)]^2 \, \Delta\theta$$
$$A = \tfrac{1}{2}\int_0^{\pi/2}(4\sin 2\theta)^2 d\theta$$

FIGURE 5

**EXAMPLE 2**   Find the area of one leaf of the four-leaved rose $r = 4 \sin 2\theta$.

⊟ *Solution*   The complete rose was sketched in Example 3 of the previous section. Here we show only the first-quadrant leaf (Figure 5). This leaf is 4 units long and averages about 1.5 units in width, giving 6 as an estimate for its area. The exact area $A$ is given by

$$A = \frac{1}{2}\int_0^{\pi/2} 16 \sin^2 2\theta \, d\theta = 8\int_0^{\pi/2} \frac{1 - \cos 4\theta}{2} \, d\theta$$

$$= 4\int_0^{\pi/2} d\theta - \int_0^{\pi/2} \cos 4\theta \cdot 4 \, d\theta$$

$$= \left[ 4\theta \right]_0^{\pi/2} - \left[ \sin 4\theta \right]_0^{\pi/2} = 2\pi \qquad \blacksquare$$

**EXAMPLE 3**   Find the area of the region outside the cardioid $r = 1 + \cos \theta$ and inside the circle $r = \sqrt{3} \sin \theta$.

*Solution*   The graphs of the two curves are sketched in Figure 6. We will need the $\theta$-coordinates of the points of intersection. Let's try solving the two equations simultaneously.

$$1 + \cos \theta = \sqrt{3} \sin \theta$$

$$1 + 2 \cos \theta + \cos^2\theta = 3 \sin^2\theta$$

$$1 + 2 \cos \theta + \cos^2\theta = 3(1 - \cos^2\theta)$$

$$4 \cos^2\theta + 2 \cos \theta - 2 = 0$$

$$2 \cos^2\theta + \cos \theta - 1 = 0$$

$$(2 \cos \theta - 1)(\cos \theta + 1) = 0$$

$$\cos \theta = \frac{1}{2}, -1 \qquad \theta = \frac{\pi}{3}, \pi$$

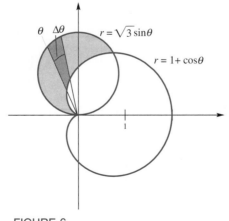

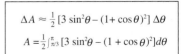

$$\Delta A \approx \frac{1}{2}[3 \sin^2\theta - (1 + \cos\theta)^2] \, \Delta\theta$$

$$A = \frac{1}{2}\int_{\pi/3}^{\pi} [3 \sin^2\theta - (1 + \cos\theta)^2] \, d\theta$$

FIGURE 6

$$A = \frac{1}{2} \int_{\pi/3}^{\pi} [3 \sin^2\theta - (1 + \cos\theta)^2] \, d\theta$$

$$= \frac{1}{2} \int_{\pi/3}^{\pi} [3 \sin^2\theta - 1 - 2\cos\theta - \cos^2\theta] \, d\theta$$

$$= \frac{1}{2} \int_{\pi/3}^{\pi} \left[\frac{3}{2}(1 - \cos 2\theta) - 1 - 2\cos\theta - \frac{1}{2}(1 + \cos 2\theta)\right] d\theta$$

$$= \frac{1}{2} \int_{\pi/3}^{\pi} [-2\cos\theta - 2\cos 2\theta] \, d\theta$$

$$= \frac{1}{2} \left[-2\sin\theta - \sin 2\theta\right]_{\pi/3}^{\pi}$$

$$= \frac{1}{2} \left[2\frac{\sqrt{3}}{2} + \frac{\sqrt{3}}{2}\right] = \frac{3\sqrt{3}}{4} \approx 1.299$$ ∎

**Tangents in Polar Coordinates**  In Cartesian coordinates, the slope $m$ of the tangent line to a curve is given by $m = dy/dx$. We quickly reject $dr/d\theta$ as the corresponding slope formula in polar coordinates. Rather, if $r = f(\theta)$ determines the curve, we write

$$y = r \sin\theta = f(\theta)\sin\theta$$

$$x = r \cos\theta = f(\theta)\cos\theta$$

Thus,

$$\frac{dy}{dx} = \lim_{\Delta x \to 0} \frac{\Delta y}{\Delta x} = \lim_{\Delta\theta \to 0} \frac{\Delta y/\Delta\theta}{\Delta x/\Delta\theta} = \frac{dy/d\theta}{dx/d\theta}$$

that is,

$$m = \frac{f(\theta)\cos\theta + f'(\theta)\sin\theta}{-f(\theta)\sin\theta + f'(\theta)\cos\theta}$$

The formula just derived simplifies when the graph of $r = f(\theta)$ passes through the pole. Suppose, for example, that for some angle $\alpha$, $r = f(\alpha) = 0$ and $f'(\alpha) \neq 0$. Then (at the pole) our formula for $m$ is

$$m = \frac{f'(\alpha)\sin\alpha}{f'(\alpha)\cos\alpha} = \tan\alpha$$

Since the line $\theta = \alpha$ also has slope $\tan\alpha$, we conclude that this line is tangent to the curve at the pole. We infer the useful fact that *tangent lines at the pole can be found by solving the equation* $f(\theta) = 0$. We illustrate below.

**EXAMPLE 4**  Consider the polar equation $r = 4\sin 3\theta$.
  (a) Find the slope of the tangent line at $\theta = \pi/6$ and $\theta = \pi/4$.
  (b) Find the tangent lines at the pole.
  (c) Sketch the graph.
  (d) Find the area of one leaf.

*Solution*

(a) $m = \dfrac{f(\theta)\cos\theta + f'(\theta)\sin\theta}{-f(\theta)\sin\theta + f'(\theta)\cos\theta} = \dfrac{4\sin 3\theta\cos\theta + 12\cos 3\theta\sin\theta}{-4\sin 3\theta\sin\theta + 12\cos 3\theta\cos\theta}$

At $\theta = \pi/6$,

$$m = \frac{4 \cdot 1 \cdot \dfrac{\sqrt{3}}{2} + 12 \cdot 0 \cdot \dfrac{1}{2}}{-4 \cdot 1 \cdot \dfrac{1}{2} + 12 \cdot 0 \cdot \dfrac{\sqrt{3}}{2}} = -\sqrt{3}$$

At $\theta = \pi/4$,

$$m = \frac{4 \cdot \dfrac{\sqrt{2}}{2} \cdot \dfrac{\sqrt{2}}{2} - 12 \cdot \dfrac{\sqrt{2}}{2} \cdot \dfrac{\sqrt{2}}{2}}{-4 \cdot \dfrac{\sqrt{2}}{2} \cdot \dfrac{\sqrt{2}}{2} - 12 \cdot \dfrac{\sqrt{2}}{2} \cdot \dfrac{\sqrt{2}}{2}} = \frac{2 - 6}{-2 - 6} = \frac{1}{2}$$

(b) We set $f(\theta) = 4\sin 3\theta = 0$ and solve. This yields $\theta = 0$, $\theta = \pi/3$, $\theta = 2\pi/3$, $\theta = \pi$, $\theta = 4\pi/3$, and $\theta = 5\pi/3$.

(c) After noting that

$$\sin 3(\pi - \theta) = \sin(3\pi - 3\theta) = \sin 3\pi\cos 3\theta - \cos 3\pi\sin 3\theta$$
$$= \sin 3\theta$$

which implies symmetry with respect to the $y$-axis, we obtain a table of values and sketch the graph shown in Figure 7.

| $\theta$ | $r$ |
|---|---|
| 0 | 0 |
| $\pi/12$ | 2.8 |
| $\pi/6$ | 4 |
| $\pi/4$ | 2.8 |
| $\pi/3$ | 0 |
| $5\pi/12$ | −2.8 |
| $\pi/2$ | −4 |

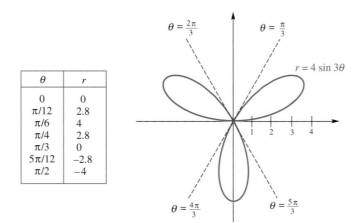

FIGURE 7

(d) $\quad A = \dfrac{1}{2}\displaystyle\int_0^{\pi/3} (4\sin 3\theta)^2\, d\theta = 8\int_0^{\pi/3} \sin^2 3\theta\, d\theta$

$\qquad = 4\displaystyle\int_0^{\pi/3}(1 - \cos 6\theta)\, d\theta = 4\int_0^{\pi/3} d\theta - \frac{4}{6}\int_0^{\pi/3}\cos 6\theta \cdot 6\, d\theta$

$\qquad = \left[4\theta - \dfrac{2}{3}\sin 6\theta\right]_0^{\pi/3} = \dfrac{4\pi}{3}$

## CONCEPTS REVIEW

**1.** The formula for the area $A$ of a sector of a circle of radius $r$ and angle $\theta$ (in radians) is $A =$ _____ .

**2.** The formula in Question 1 leads to the formula for the area $A$ of the region bounded by the curve $r = f(\theta)$ between $\theta = \alpha$ and $\theta = \beta$, namely, $A =$ _____ .

**3.** From the formula of Question 2, we conclude that the area $A$ of the region inside the cardioid $r = 2 + 2 \cos \theta$ can be expressed as $A =$ _____ .

**4.** The tangent lines to the polar curve $r = f(\theta)$ at the pole can be found by solving the equation _____ .

## PROBLEM SET 12.8

In each of the Problems 1–10, sketch the graph of the given equation and find the area of the region bounded by it.

**1.** $r = a$, $a > 0$     **2.** $r = 2a \cos \theta$, $a > 0$

**3.** $r = 2 + \cos \theta$     **4.** $r = 5 + 4 \cos \theta$

**5.** $r = 3 - 3 \sin \theta$     **6.** $r = 3 + 3 \sin \theta$

**7.** $r = a(1 + \cos \theta)$, $a > 0$   **8.** $r^2 = 6 \cos 2\theta$

**9.** $r^2 = 9 \sin 2\theta$     **10.** $r^2 = a \cos 2\theta$, $a > 0$

**11.** Sketch the limaçon $r = 3 - 4 \sin \theta$, and find the area of the region inside its small loop.

**12.** Sketch the limaçon $r = 2 - 4 \cos \theta$, and find the area of the region inside its small loop.

**13.** Sketch the limaçon $r = 2 - 3 \cos \theta$, and find the area of the region inside its large loop.

**14.** Sketch one leaf of the four-leaved rose $r = 3 \cos 2\theta$, and find the area of the region enclosed by it.

**15.** Sketch the three-leaved rose $r = 4 \cos 3\theta$, and find the area of the total region enclosed by it.

**16.** Sketch the three-leaved rose $r = 2 \sin 3\theta$, and find the area of the region bounded by it.

**17.** Find the area of the region between the two concentric circles $r = 7$ and $r = 10$.

**18.** Sketch the region that is inside the circle $r = 3 \sin \theta$ and outside the cardioid $r = 1 + \sin \theta$, and find its area.

**19.** Sketch the region that is outside the circle $r = 2$ and inside the lemniscate $r^2 = 8 \cos 2\theta$, and find its area.

**20.** Sketch the limaçon $r = 3 - 6 \sin \theta$ and find the area of the region that is inside its large loop and outside its small loop.

**21.** Sketch the region in the first quadrant that is inside the cardioid $r = 3 + 3 \cos \theta$ and outside the cardioid $r = 3 + 3 \sin \theta$, and find its area.

**22.** Sketch the region in the second quadrant that is inside the cardioid $r = 2 + 2 \sin \theta$ and outside the cardioid $r = 2 + 2 \cos \theta$, and find its area.

**23.** Find the slope of the tangent line to each of the following curves at $\theta = \pi/3$.

(a) $r = 2 \cos \theta$      (b) $r = 1 + \sin \theta$
(c) $r = \sin 2\theta$       (d) $r = 4 - 3 \cos \theta$

**24.** Find all points on the cardioid $r = a(1 + \cos \theta)$ where the tangent line is (a) horizontal, and (b) vertical.

**25.** Find all points on the limaçon $r = 1 - 2 \sin \theta$ where the tangent line is horizontal.

**26.** Let $r = f(\theta)$, where $f$ is continuous on the closed interval $[\alpha, \beta]$. Derive the following formula for the length $L$ of the corresponding polar curve from $\theta = \alpha$ to $\theta = \beta$.

$$L = \int_{\alpha}^{\beta} \sqrt{[f(\theta)]^2 + [f'(\theta)]^2} \, d\theta$$

**27.** Use the formula of Problem 26 to find the perimeter of the cardioid $r = a(1 + \cos \theta)$.

**28.** Find the length of the logarithmic spiral $r = e^{\theta/2}$ from $\theta = 0$ to $\theta = 2\pi$.

**29.** Find the total area of the rose $r = a \cos n\theta$, $n$ a positive integer.

**30.** Sketch the graph of the *strophoid* $r = \sec \theta - 2 \cos \theta$ and find the area of its loop.

**31.** Consider the two circles $r = 2a \sin \theta$ and $r = 2b \cos \theta$ with $a$ and $b$ positive.

(a) Find the area of the region inside both circles.
(b) Show that the two circles intersect at right angles.

**32.** Assume that a planet of mass $m$ is revolving around the sun (located at the pole) with constant angular momentum $mr^2 \, d\theta/dt$. Deduce Kepler's Second Law: The line from the sun to the planet sweeps out equal areas in equal times.

**33.** (First Old Goat Problem) A goat is tethered to the edge of a circular pond of radius $a$ by a rope of length $ka$ ($0 < k \leq 2$). Use the method of this section to find its grazing area (the shaded area in Figure 8). *Note*: We solved this problem once before (Problem 49 of Section 7.6); you should be able to make your answers agree.

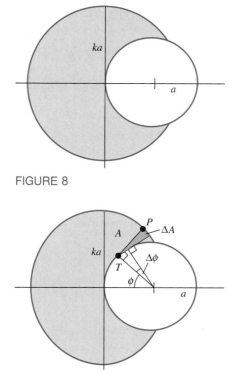

FIGURE 8

FIGURE 9

**34.** (Second Old Goat Problem) Do Problem 33 again but assume the pond has a fence around it so that in forming the wedge $A$, the rope wraps around the fence (Figure 9). *Hint*: If you are exceedingly ambitious, try the method of this section. Better, note that in the wedge $A$, $\Delta A \approx \left(\frac{1}{2}\right)|PT|^2 \Delta\phi$, which leads to a Riemann sum for an integral. The final answer is $a^2(\pi k^2/2 + k^3/3)$, a result needed in Problem 35.

C **35.** (Third Old Goat Problem) An untethered goat grazes inside a yard enclosed by a circular fence of radius $a$; another grazes outside the fence tethered as in Problem 34. Find the length of the rope if the two goats have the same grazing area.

PC Use a computer to do Problems 36–39. In each case be sure to make a mental estimate first. Note the length formula in Problem 26.

**36.** Find the total length of the limaçons $r = 2 + \cos t$ and $r = 2 + 4 \cos t$ (see Example 1 of this section and Example 1 of Section 12.7).

**37.** Find the total area and the total length of the three-leaved rose $r = 4 \sin 3t$ (see Example 4).

**38.** Find the total area and the total length of the lemniscate $r^2 = 8 \cos 2t$ (see Example 2 of Section 12.7).

**39.** Draw the curve $r = 4 \sin(3t/2)$, $0 \leq t \leq 4\pi$, and then find its total length.

**Answers to Concepts Review:** **1.** $\frac{1}{2}r^2\theta$ **2.** $\frac{1}{2}\int_\alpha^\beta [f(\theta)]^2 \, d\theta$ **3.** $\frac{1}{2}\int_0^{2\pi} (2 + 2\cos\theta)^2 \, d\theta$ **4.** $f(\theta) = 0$

## 12.9 CHAPTER REVIEW

### Concepts Test

Respond with true or false to each of the following assertions. Be prepared to justify your answer.

**1.** The graph of $y = ax^2 + bx + c$ is a parabola for all choices of $a$, $b$, and $c$.

**2.** The vertex of a parabola is midway between the focus and the directrix.

**3.** A vertex of an ellipse is closer to a directrix than to a focus.

**4.** The point on a parabola closest to its focus is the vertex.

**5.** The hyperbolas $x^2/a^2 - y^2/b^2 = 1$ and $y^2/b^2 - x^2/a^2 = 1$ have the same asymptotes.

**6.** The circumference $C$ of the ellipse $x^2/a^2 + y^2/b^2 = 1$ with $b < a$ satisfies $2\pi b < C < 2\pi a$.

**7.** The smaller the eccentricity $e$ of an ellipse, the more nearly circular the ellipse is.

**8.** The ellipse $6x^2 + 4y^2 = 24$ has its foci on the $x$-axis.

**9.** The equation $x^2 - y^2 = 0$ represents a hyperbola.

**10.** The equation $(y^2 - 4x + 1)^2 = 0$ represents a parabola.

**11.** If $k \neq 0$, $x^2/a^2 - y^2/b^2 = k$ is the equation of a hyperbola.

**12.** If $k \neq 0$, $x^2/a^2 + y^2/b^2 = k$ is the equation of an ellipse.

**13.** The distance between the foci of the graph of $x^2/a^2 + y^2/b^2 = 1$ is $2\sqrt{a^2 - b^2}$.

**14.** The graph of $x^2/9 - y^2/8 = -2$ does not intersect the $x$-axis.

**15.** Light emanating from a point between a focus and the nearest vertex of an elliptical mirror will be reflected beyond the other focus.

**16.** The set of points equidistant from the circle $x^2 + y^2 = 1$ and the line $x = 3$ is a parabola.

**17.** From Kepler's law on sweeping out areas, we conclude that a planet in its elliptical orbit around the sun reaches its maximum speed at the vertex nearest the sun.

**18.** An ellipse that is drawn using a string of length 8 units attached to foci 2 units apart will have minor diameter of length $\sqrt{60}$ units.

**19.** The graph of $x^2 + y^2 + Cx + Dy + F = 0$ is either a circle, a point, or the empty set.

**20.** The graph of $2x^2 + y^2 + Cx + Dy + F = 0$ cannot be a single point.

**21.** The graph of $Ax^2 + Bxy + Cy^2 + Dyx + Ey + F = 0$ is the intersection of a plane with a cone of two nappes for all choices of $A$, $B$, $C$, $D$, $E$, and $F$.

**22.** In an appropriate coordinate system, the inter-section of a plane with a cone of two nappes will have an equation of the form $Ax^2 + Cy^2 + Dx + Ey + F = 0$.

**23.** The graph of a hyperbola must enter all four quadrants.

**24.** If one of the conic sections passes through the four points $(1, 0)$, $(-1, 0)$, $(0, 1)$ and $(0, -1)$, it must be a circle.

**25.** The graph of the polar equation $r = 4\cos(\theta - \pi/3)$ is a circle.

**26.** Every point in the plane has infinitely many sets of polar coordinates.

**27.** All points of intersection of the graphs of the polar equations $r = f(\theta)$ and $r = g(\theta)$ can be found by solving these two equations simultaneously.

**28.** If $f$ is an odd function, then the graph of $r = f(\theta)$ is symmetric with respect to the $y$-axis (the line $\theta = \pi/2$).

**29.** If $f$ is an even function, then the graph of $r = f(\theta)$ is symmetric with respect to the $x$-axis (the line $\theta = 0$).

**30.** The graph of $r = 4\cos 3\theta$ is a rose of 3 leaves whose area is less than half that of the circle $r = 4$.

### Sample Test Problems

**1.** From the numbered list, pick the correct response to put in the blanks that follow.
(1) No graph.　　　　　(2) A single point.
(3) A single line.　　　 (4) Two parallel lines.
(5) Two intersecting lines.　(6) A circle.
(7) A parabola.　　　　(8) An ellipse.
(9) A hyperbola.　　　 (10) None of the above.

_____ (a) $x^2 - 4y^2 = 0$
_____ (b) $x^2 - 4y^2 = 0.01$
_____ (c) $x^2 - 4 = 0$
_____ (d) $x^2 - 4x + 4 = 0$
_____ (e) $x^2 + 4y^2 = 0$
_____ (f) $x^2 + 4y^2 = x$
_____ (g) $x^2 + 4y^2 = -x$
_____ (h) $x^2 + 4y^2 = -1$
_____ (i) $(x^2 + 4y - 1)^2 = 0$
_____ (j) $3x^2 + 4y^2 = -x^2 + 1$

In each of Problems 2–10, name the conic that has the given equation. Find its vertices and foci and sketch its graph.

**2.** $y^2 - 6x = 0$

**3.** $9x^2 + 4y^2 - 36 = 0$

**4.** $25x^2 - 36y^2 + 900 = 0$

**5.** $x^2 + 9y = 0$

**6.** $x^2 - 4y^2 - 16 = 0$

**7.** $9x^2 + 25y^2 - 225 = 0$

**8.** $9x^2 + 9y^2 - 225 = 0$

**9.** $r = \dfrac{5}{2 + 2\sin\theta}$

**10.** $r(2 + \cos\theta) = 3$

In each of Problems 11–18, find the Cartesian equation of the conic with the given properties.

**11.** Vertices $(\pm 4, 0)$ and eccentricity $\frac{1}{2}$.

**12.** Eccentricity 1, focus $(0, -3)$, and vertex $(0, 0)$.

**13.** Eccentricity 1, vertex $(0, 0)$, symmetric with respect to the $x$-axis, and passing through the point $(-1, 3)$.

**14.** Eccentricity $\frac{5}{3}$ and vertices $(0, \pm 3)$.

**15.** Vertices $(\pm 2, 0)$ and asymptotes $x \pm 2y = 0$.

**16.** Parabola with focus $(3, 2)$ and vertex $(3, 3)$.

**17.** Ellipse with center $(1, 2)$, and focus $(4, 2)$, and major diameter 10.

**18.** Hyperbola with vertices $(2, 0)$ and $(2, 6)$ and eccentricity $\frac{10}{3}$.

In Problems 19–22, use the process of completing the square to reduce the given equation to a standard form. Then name the corresponding curve and sketch its graph.

**19.** $4x^2 + 4y^2 - 24x + 36y + 81 = 0$

**20.** $4x^2 + 9y^2 - 24x - 36y + 36 = 0$

**21.** $x^2 + 8x + 6y + 28 = 0$

**22.** $3x^2 - 10y^2 + 36x - 20y + 68 = 0$

**23.** A rotation of axes through $\theta = 45°$ transforms $x^2 + 3xy + y^2 = 10$ into $ru^2 + sv^2 = 10$. Determine $r$ and $s$, name the corresponding conic, and find the distance between its foci.

**24.** Determine the rotation angle $\theta$ needed to eliminate the cross-product term in $7x^2 + 8xy + y^2 = 9$. Then obtain the corresponding $uv$-equation and identify the conic it represents.

In Problems 25–36, analyze the given polar equation and sketch its graph.

**25.** $r = 6 \cos \theta$

**26.** $r = \dfrac{5}{\sin \theta}$

**27.** $r = \cos 2\theta$

**28.** $r = \dfrac{3}{\cos \theta}$

**29.** $r = 4$

**30.** $r = 5 - 5 \cos \theta$

**31.** $r = 4 - 3 \cos \theta$

**32.** $r = 2 - 3 \cos \theta$

**33.** $\theta = \frac{2}{3}\pi$

**34.** $r = 4 \sin 3\theta$

**35.** $r^2 = 16 \sin 2\theta$

**36.** $r = -\theta, \theta \geq 0$

**37.** Find a Cartesian equation of the graph of

$$r^2 - 6r(\cos \theta + \sin \theta) + 9 = 0$$

and then sketch the graph.

**38.** Find a Cartesian equation of the graph of

$$r^2 \cos 2\theta = 9$$

and then sketch the graph.

**39.** Find the slope of the tangent to the graph of

$$r = 3 + 3 \cos \theta$$

at the point on the graph where $\theta = \frac{1}{6}\pi$.

**40.** Sketch the graphs of

$$r = 5 \sin \theta \quad \text{and} \quad r = 2 + \sin \theta$$

and find their points of intersection.

**41.** Find the area of the region bounded by the graph of

$$r = 5 - 5 \cos \theta$$

(see Problem 30).

**42.** Find the area of the region that is outside the limaçon $r = 2 + \sin \theta$ and inside the circle $r = 5 \sin \theta$ (see Problem 40).

**43.** A racing car on the elliptical race track $x^2/400 + y^2/100 = 1$ went out of control at the point $(16, 6)$ and thereafter continued on the tangent line till it hit a tree at $(14, k)$. Determine $k$.

**44.** Match up each of the following polar equations with the correct graph.

(a) $r = 1 - 2 \sin \theta$

(b) $r = 1 + \dfrac{\sin \theta}{2}$

(c) $r = 1 + 2 \cos \theta$

(d) $r = 1 + \dfrac{\cos \theta}{2}$

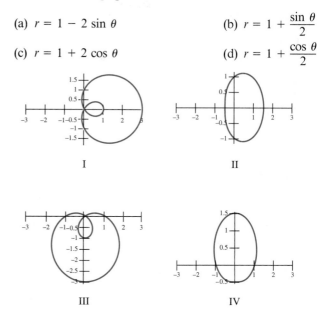

I  II

III  IV

**45.** Match up each of the following polar equations with the correct graph.

(a) $r = 4 \cos 2\theta$

(b) $r = 3 \cos 3\theta$

(c) $r = 5 \cos 5\theta$

(d) $r = 3 \sin 2\theta$

(e) $r = 4 \sin 4\theta$

I  II

III  IV

V

# TECHNOLOGY PROJECT

Most modern computing technologies can conveniently make polar plots; i.e., graph functions of the form $r = f(\theta)$. For example, in *Mathematica* one would plot $r = \sin 2\theta$ like this:

> PolarPlot [Sin[2 theta], {theta, 0, 2Pi}]

The result is shown in Figure 1 — it is called a *four-leaved rose*.

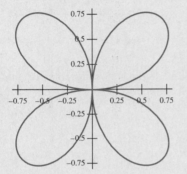

FIG. 1. $r = \sin 2\theta$ for $0 \le \theta \le 2\pi$

It is surprisingly difficult to predict the behavior of polar plots. For example, in the present example, since the period of $\sin 2\theta$ is $\pi$, we might think it sufficient to plot in the smaller range, 0 to $\pi$. But then we only get the right half of the graph. Try it on your computer technology. And if $\sin 2\theta$ produces a "four-leaved rose," then surely $\sin 3\theta$ produces 6 leaves? Nope. Three leaves and this time it is all done at $\pi$ instead of $2\pi$; see Figure 2.

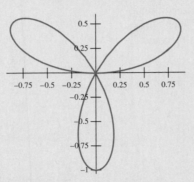

FIG. 2. $r = \sin 3\theta$ for $0 \le \theta \le \pi$

**Exercise 1** Experiment with plotting the equation $r = \sin k\theta$, $k = 1, 2, 3, \ldots$ and find out:

(a) How the plot range needed to get a complete graph depends on the integer $k$.

(b) How the number of "leaves" depends on the integer $k$.

Remark: Figures 1 and 2 give correct information; your results should be consistent with them. Also, don't forget about the case $k = 1$!

**Exercise 2** For $r = \sin\left(\dfrac{\theta}{k}\right)$, with $k$ still a positive integer, find out how the plot range needed to get a complete graph depends on $k$.

**Exercise 3 (Harder)** For $r = \sin\left(\dfrac{m\theta}{n}\right)$, with $m$ and $n$ positive integers, find out how the plot range needed to get a complete graph depends on $m$ and $n$.

*Hints*: You can assume that $m$ and $n$ have no common divisors. You already have some evidence from Exercises 1 and 2. Remark: The full polar plot of something like $\sin\left(\dfrac{6\theta}{7}\right)$ is quite attractive; we hope you enjoy seeing it.

**Exercise 4** Describe what happens if we use an irrational factor $b$ in $r = \sin b\theta$. In particular, for $b = \sqrt{2}$ plot this function on each of the intervals $[0, 12\pi]$, $[0, 24\pi]$, and $[0, 48\pi]$.
**Note:** In order to get a smooth graph some computing technologies allow you to increase the number of points in the plot. Look into this.

**Exercise 5** Investigate the polar plots of the family of functions given by $r(\theta) = 1 + p \sin \theta$ for various values of the *positive* parameter $p$. Begin by trying the values $p = 3/4$, 1, and 4/3. Describe how the family looks in terms of the $p$ value. In particular, tell what happens as $p \to 0$ and as $p \to +\infty$.

**Exercise 6** Hand in the most interesting polar plot you can create. Think about using exponentials, powers of trig functions, etc. Just to get your juices going, try $r(\theta) = 3 \cos^2 \theta - 1$ and $3 \cos^3 \theta - 1$, but we're sure you can do better!

# 13

# Geometry in the Plane, Vectors

**13.1**
**PLANE CURVES:**
**PARAMETRIC**
**REPRESENTATION**

We gave the general definition of a plane curve in Section 6.4 in connection with our derivation of the arc length formula. A **plane curve** is determined by a pair of parametric equations

$$x = f(t), \qquad y = g(t), \qquad t \text{ in } I$$

Not simple, not closed

Simple, not closed

Not simple, closed

Simple and closed

FIGURE 1

with $f$ and $g$ continuous on the interval $I$. Usually $I$ is a closed interval $[a, b]$. Think of $t$, called the **parameter**, as measuring time. As $t$ advances from $a$ to $b$, the point $(x, y)$ traces out the curve in the $xy$-plane. The points $P = (x(a), y(a))$ and $Q = (x(b), y(b))$ are the initial and final **endpoints**. If the endpoints coincide, the curve is **closed**. If distinct values of $t$ yield distinct points in the plane (except possibly for $t = a$ and $t = b$), we say the curve is a **simple** curve (Figure 1).

**Eliminating the Parameter**  To recognize a curve given by parametric equations, it may be desirable to eliminate the parameter. Sometimes this can be accomplished by solving one equation for $t$ and substituting in the other (Example 1). Often we can make use of a familiar identity, as in Example 2.

**EXAMPLE 1**  Eliminate the parameter in

$$x = t^2 + 2t, \qquad y = t - 3, \qquad -2 \le t \le 3$$

Then identify the corresponding curve and sketch its graph.

***Solution*** From the second equation, $t = y + 3$. Substituting this expression for $t$ in the first equation gives

$$x = (y + 3)^2 + 2(y + 3) = y^2 + 8y + 15$$

or

$$x + 1 = (y + 4)^2$$

This we recognize as a parabola with vertex at $(-1, -4)$ and opening to the right.

In graphing the given equation, we must be careful to display only that part of the parabola corresponding to $-2 \le t \le 3$. A table of values and the graph are shown in Figure 2. The arrow indicates the direction of increasing $t$.

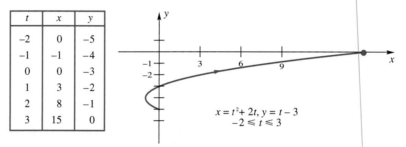

| $t$ | $x$ | $y$ |
|----|----|----|
| $-2$ | $0$ | $-5$ |
| $-1$ | $-1$ | $-4$ |
| $0$ | $0$ | $-3$ |
| $1$ | $3$ | $-2$ |
| $2$ | $8$ | $-1$ |
| $3$ | $15$ | $0$ |

$x = t^2 + 2t, \; y = t - 3$
$-2 \le t \le 3$

FIGURE 2

**EXAMPLE 2** Show that

$$x = a \cos t, \qquad y = b \sin t, \qquad 0 \le t \le 2\pi$$

represents the ellipse shown in Figure 3.

***Solution*** We solve the equations for $\cos t$ and $\sin t$, then square, and add.

$$\left(\frac{x}{a}\right)^2 + \left(\frac{y}{b}\right)^2 = \cos^2 t + \sin^2 t = 1$$

$$\frac{x^2}{a^2} + \frac{y^2}{b^2} = 1$$

A quick check of a few values for $t$ convinces us that we do get the complete ellipse. In particular, $t = 0$ and $t = 2\pi$ give the same point, namely, $(a, 0)$. If $a = b$, we get the circle $x^2 + y^2 = a^2$. ∎

Different pairs of parametric equations may have the same graph, as we now illustrate.

**EXAMPLE 3** Show that each of the following pairs of parametric equations have the same graph, namely, the semicircle shown in Figure 4.

(a) $x = \sqrt{1 - t^2}, \; y = t, \; -1 \le t \le 1$

(b) $x = \cos t, \; y = \sin t, \; -\dfrac{\pi}{2} \le t \le \dfrac{\pi}{2}$

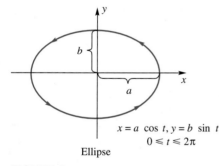

$x = a \cos t, y = b \sin t$
$0 \le t \le 2\pi$
Ellipse

FIGURE 3

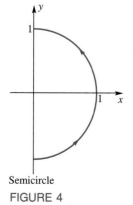

Semicircle
FIGURE 4

(c) $x = \dfrac{1 - t^2}{1 + t^2}, \; y = \dfrac{2t}{1 + t^2}, \; -1 \le t \le 1$

*Solution*  In each case, we discover that

$$x^2 + y^2 = 1$$

It is then just a matter of checking a few values of $t$ to make sure that the given intervals for $t$ yield the same section of the circle.  ∎

**EXAMPLE 4**  Show that each of the following pairs of parametric equations yields one branch of a hyperbola.

(a) $x = a \sec t, \; y = b \tan t, \; -\dfrac{\pi}{2} < t < \dfrac{\pi}{2}$

(b) $x = a \cosh t, \; y = b \sinh t, \; -\infty < t < \infty$

Assume in both cases that $a > 0$ and $b > 0$.

*Solution*
(a)  In the first case,

$$\left(\frac{x}{a}\right)^2 - \left(\frac{y}{b}\right)^2 = \sec^2 t - \tan^2 t = 1$$

(b)  In the second case,

$$\left(\frac{x}{a}\right)^2 - \left(\frac{y}{b}\right)^2 = \cosh^2 t - \sinh^2 t$$

$$= \left(\frac{e^t + e^{-t}}{2}\right)^2 - \left(\frac{e^t - e^{-t}}{2}\right)^2 = 1$$

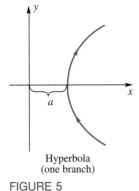

Hyperbola
(one branch)

FIGURE 5

Checking a few $t$-values shows that, in both cases, we obtain the branch of the hyperbola $x^2/a^2 - y^2/b^2 = 1$ shown in Figure 5.  ∎

**The Cycloid**  A cycloid is the curve traced by a point $P$ on the rim of a wheel as the wheel rolls along a straight line without slipping (Figure 6). The

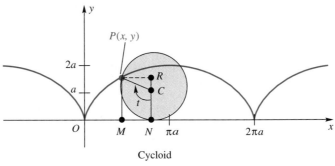

Cycloid

FIGURE 6

Cartesian equation of a cycloid is quite complicated, but simple parametric equations are readily found, as shown in the next example.

**EXAMPLE 5**  Find parametric equations for the cycloid.

*Solution*  Let the wheel roll along the *x*-axis with *P* initially at the origin. Denote the center of the wheel by *C* and let *a* be its radius. Choose for a parameter the radian measure *t* of the clockwise angle through which the line segment *CP* has turned from its vertical position when *P* was at the origin. All of this is shown in Figure 6.

Since $|ON| = \text{arc } PN = at,$

$$x = |OM| = |ON| - |MN| = at - a \sin t = a(t - \sin t)$$

and

$$y = |MP| = |NR| = |NC| + |CR| = a - a \cos t = a(1 - \cos t)$$

Thus the parametric equations for the cycloid are

$$x = a(t - \sin t), \qquad y = a(1 - \cos t) \qquad \blacksquare$$

The cycloid has a number of interesting applications, especially in mechanics. It is the "curve of fastest descent." If a particle, acted on only by gravity, is allowed to slide down some curve from a point *A* to a lower point *B* not on the same vertical line, it completes its journey in the shortest time when the curve is an inverted cycloid (Figure 7). Of course, the shortest distance is along the straight line segment *AB*, but the least time is used when the path is along a cycloid; this is because the acceleration when it is released depends on the steepness of descent, and along a cycloid it builds up velocity much more quickly than it does along a straight line.

Another interesting property is this: If *L* is the lowest point on an arch of an inverted cycloid, the time it takes a particle *P* to slide down the cycloid to *L* is the same no matter where *P* starts from on the inverted arch; thus, if several particles, $P_1$, $P_2$, and $P_3$, in different positions on the cycloid (Figure 8) start to slide at the same instant, all will reach the low point *L* at the same time.

In 1673, the Dutch astronomer Christian Huygens published a description of an ideal pendulum clock. Because the bob swings between cycloidal "cheeks," the path of the bob is a cycloid (Figure 9). This means that the period of the swing is independent of the amplitude, and so the period does not change as the clock's spring unwinds.

A surprising fact is that the three results just mentioned all date from the seventeenth century. To demonstrate them is a nontrivial task, as you may discover by looking at any book on the history of calculus.

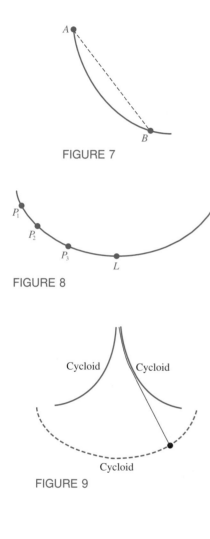

FIGURE 7

FIGURE 8

FIGURE 9

**Calculus For Curves Defined Parametrically**  Can we find the slope of the tangent line to a curve given parametrically without first eliminating the parameter? The answer is yes, according to the following theorem.

---

Theorem A

Let $f$ and $g$ be continuously differentiable with $f'(t) \neq 0$ on $\alpha < t < \beta$. Then the parametric equations

$$x = f(t), \; y = g(t)$$

define $y$ as a differentiable function of $x$ and

$$\frac{dy}{dx} = \frac{dy/dt}{dx/dt}$$

---

**Proof** Since $f'(t) \neq 0$ for $\alpha < t < \beta$, $f$ is strictly monotonic and so has a differentiable inverse $f^{-1}$ (see the Inverse Function Theorem (Theorem 7.2B)). Define $F$ by $F = g \circ f^{-1}$, so that

$$y = g(t) = g(f^{-1}(x)) = F(x) = F(f(t))$$

Then by the Chain Rule,

$$\frac{dy}{dt} = F'(f(t)) \cdot f'(t) = \frac{dy}{dx} \cdot \frac{dx}{dt}$$

Since $dx/dt \neq 0$, we have

$$\frac{dy}{dx} = \frac{dy}{dt} \div \frac{dx}{dt}$$ ∎

**EXAMPLE 6** Find the first two derivatives $dy/dx$ and $d^2y/dx^2$ for the function determined by

$$x = 5 \cos t, \qquad y = 4 \sin t, \qquad 0 < t < 3$$

and evaluate them at $t = \pi/6$ (see Example 2).

**Solution** Let $y'$ denote $dy/dx$. Then

$$\frac{dy}{dx} = \frac{dy}{dt} \div \frac{dx}{dt} = \frac{4 \cos t}{-5 \sin t} = -\frac{4}{5} \cot t$$

$$\frac{d^2y}{dx^2} = \frac{dy'}{dx} = \frac{dy'}{dt} \div \frac{dx}{dt} = \frac{\frac{4}{5} \csc^2 t}{-5 \sin t} = -\frac{4}{25} \csc^3 t$$

At $t = \pi/6$,

$$\frac{dy}{dx} = \frac{-4\sqrt{3}}{5}, \qquad \frac{d^2y}{dx^2} = \frac{-4}{25}(8) = \frac{-32}{25}$$

The first value is the slope of the tangent line to the ellipse $x^2/25 + y^2/16 = 1$ at the point $(5\sqrt{3}/2, 2)$. You can check that this is so by implicit differentiation. ∎

Sometimes a definite integral involves two variables, such as $x$ and $y$, in the integrand and differential, and $y$ may be defined as a function of $x$ by equations that give $x$ and $y$ in terms of a parameter such as $t$. In such cases, it is often convenient to evaluate the definite integral by expressing the integrand and the differential in terms of $t$ and $dt$ and adjusting the limits of integration before integrating with respect to $t$.

**EXAMPLE 7**   Evaluate (a) $\displaystyle\int_1^3 y\,dx$ and (b) $\displaystyle\int_1^3 xy^2\,dx$, using $x = 2t - 1$ and $y = t^2 + 2$.

*Solution*   From $x = 2t - 1$ we have $dx = 2\,dt$; when $x = 1$, $t = 1$; and when $x = 3$, $t = 2$.

(a) $\displaystyle\int_1^3 y\,dx = \int_1^2 (t^2 + 2)2\,dt = 2\left[\frac{t^3}{3} + 2t\right]_1^2 = \frac{26}{3}$

(b) $\displaystyle\int_1^3 xy^2\,dx = \int_1^2 (2t - 1)(t^2 + 2)^2 2\,dt$

$\displaystyle\qquad\qquad = 2\int_1^2 (2t^5 - t^4 + 8t^3 - 4t^2 + 8t - 4)\,dt = 86\frac{14}{15}$   ∎

**EXAMPLE 8**   Find the area $A$ under one arch of a cycloid (Figure 10). Also, find the length $L$ of this arch.

*Solution*   From Example 5, we know that we may represent one arch of the cycloid by

$$x = a(t - \sin t), \qquad y = a(1 - \cos t), \qquad 0 \le t \le 2\pi$$

Thus,

$$A = \int_0^{2\pi a} y\,dx = \int_0^{2\pi} a(1 - \cos t)\,d[a(t - \sin t)]$$

$$= a^2\int_0^{2\pi} (1 - \cos t)(1 - \cos t)\,dt$$

$$= a^2\int_0^{2\pi} (1 - 2\cos t + \cos^2 t)\,dt$$

$$= a^2\int_0^{2\pi} (1 - 2\cos t + \tfrac{1}{2} + \tfrac{1}{2}\cos 2t)\,dt$$

$$= a^2[\tfrac{3}{2}t - 2\sin t + \tfrac{1}{4}\sin 2t]_0^{2\pi} = 3\pi a^2$$

To calculate $L$, we recall the arc-length formula from Section 6.4,

$$L = \int_\alpha^\beta \sqrt{\left(\frac{dx}{dt}\right)^2 + \left(\frac{dy}{dt}\right)^2}\,dt$$

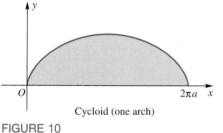

Cycloid (one arch)

**FIGURE 10**

which in our case reduces to

$$L = \int_0^{2\pi} \sqrt{a^2(1 - \cos t)^2 + a^2(\sin^2 t)} \, dt$$

$$= a \int_0^{2\pi} \sqrt{2(1 - \cos t)} \, dt$$

$$= a \int_0^{2\pi} \sqrt{4 \sin^2 \frac{t}{2}} \, dt$$

$$= 2a \int_0^{2\pi} \sin \frac{t}{2} \, dt$$

$$= \left[ -4a \cos \frac{t}{2} \right]_0^{2\pi} = 8a \qquad \blacksquare$$

---

## Two Fleas on a Trike

Two fleas are arguing about who will get the longest ride when Angie pedals her tricycle home from the park. A will ride between the treads of the front tire; B will ride between the treads of one of the rear tires. Settle the argument by showing that their paths will have equal lengths. Example 8 should help.

---

## CONCEPTS REVIEW

**1.** A circle is a premier example of a curve that is both _____ and _____; a figure eight is an example of a closed curve that is not _____.

**2.** We call two equations $x = f(t)$ and $y = g(t)$ a _____ representation of a curve, and $t$ is called a _____.

**3.** The path of a point on the rim of a rolling circle is called a _____.

**4.** The formula for $dy/dx$, given the representation $x = f(t)$ and $y = g(t)$, is $dy/dx = $ _____.

---

## PROBLEM SET 13.1

In each of the Problems 1–24, a parametric representation of a curve is given.

(a) Sketch the curve by assigning values to the parameter (see Example 1)
(b) Which of the following apply to the curve: simple or closed?
(c) Obtain the Cartesian equation of the curve by eliminating the parameter (see Examples 1–4).

1. $x = 3t, y = 2t; t$ in $\mathbb{R}$

2. $x = 2t, y = 3t; t$ in $\mathbb{R}$

3. $x = 3t - 1, y = t; 0 \le t \le 4$

4. $x = 4t - 2, y = 2t; 0 \le t \le 3$

5. $x = 4 - t, y = \sqrt{t}; 0 \le t \le 4$

6. $x = t - 3, y = \sqrt{2t}; 0 \le t \le 8$

7. $x = \frac{1}{s}, y = s; 1 \le s < 10$

8. $x = s, y = \frac{1}{s}; 1 \le s \le 10$

9. $x = r^2, y = r^3; -1 \le r \le 2$

10. $x = r^2, y = -r^5; -1 \le r \le 2$

11. $x = t^2 - 2, y = t^3; -3 \le t \le 3$

12. $x = t^3 - 2, y = t^3 - t; -3 \le t \le 3$

13. $x = t^3 - 4t, y = t^2 - 4; -3 \le t \le 3$

14. $x = t^3 - 2t, y = t^2 - 2t; -3 \le t \le 3$

15. $x = 2\sqrt{t - 2}, y = 3\sqrt{4 - t}; 2 \le t \le 4$

16. $x = 3\sqrt{t - 3}, y = 2\sqrt{4 - t}; 3 \le t \le 4$

17. $x = 2 \sin t, y = 3 \cos t; 0 \le t \le 2\pi$

18. $x = 3 \sin r, y = -2 \cos r; 0 \le r \le 2\pi$

19. $x = -2 \sin r, y = -3 \cos r; 0 \le r \le 4\pi$

20. $x = 2 \cos^2 r, y = 3 \sin^2 r; 0 \le r \le 2\pi$

21. $x = 9 \sin^2 \theta, y = 9 \cos^2 \theta; 0 \le \theta \le \pi$

**22.** $x = 9 \cos^2 \theta$, $y = 9 \sin^2 \theta$; $0 \le \theta \le \pi$

**23.** $x = \cos \theta$, $y = -2 \sin^2 2\theta$; $\theta$ in $\mathbb{R}$

**24.** $x = \sin \theta$, $y = 2 \cos^2 2\theta$; $\theta$ in $\mathbb{R}$

In Problems 25–36, find $dy/dx$ and $d^2y/dx^2$ without eliminating the parameter.

**25.** $x = 3\tau^2$, $y = 4\tau^3$; $\tau \neq 0$

**26.** $x = 6s^2$, $y = -2s^3$; $s \neq 0$

**27.** $x = 2\theta^2$, $y = \sqrt{5}\theta^3$; $\theta \neq 0$

**28.** $x = \sqrt{3}\theta^2$, $y = -\sqrt{3}\theta^3$; $\theta \neq 0$

**29.** $x = t - \dfrac{1}{t}$, $y = t + \dfrac{1}{t}$; $t \neq 0$

**30.** $x = 3t - \dfrac{4}{t}$, $y = 3t + \dfrac{2}{t}$; $t \neq 0$

**31.** $x = 1 - \cos t$, $y = 1 + \sin t$; $t \neq n\pi$

**32.** $x = 3 - 2\cos t$, $y = -1 + 5\sin t$; $t \neq n\pi$

**33.** $x = 3\tan t - 1$, $y = 5\sec t + 2$; $t \neq \dfrac{(2n+1)\pi}{2}$

**34.** $x = \cot t - 2$, $y = -2\csc t + 5$; $0 < t < \pi$

**35.** $x = \dfrac{1}{1+t^2}$, $y = \dfrac{1}{t(1-t)}$; $0 < t < 1$

**36.** $x = \dfrac{2}{1+t^2}$, $y = \dfrac{2}{t(1+t^2)}$; $t \neq 0$

In Problems 37–40, find the equation of the tangent to the given curve at the given point without eliminating the parameter. Make a sketch.

**37.** $x = t^2$, $y = t^3$; $t = 2$

**38.** $x = 3t$, $y = 8t^3$; $t = -\frac{1}{2}$

**39.** $x = 2\sec t$, $y = 2\tan t$; $t = -\dfrac{\pi}{6}$

**40.** $x = 2e^t$, $y = \frac{1}{3}e^{-t}$; $t = 0$

In Problems 41–56 find the length of the parametric curve defined over the given interval.

**41.** $x = 2t - 1$, $y = 3t - 4$; $0 \le t \le 3$

**42.** $x = 2 - t$, $y = 2t - 3$; $-3 \le t \le 3$

**43.** $x = t$, $y = t^{3/2}$; $0 \le t \le 3$

**44.** $x = 2\sin t$, $y = 2\cos t$; $0 \le t \le \pi$

**45.** $x = 3t^2$, $y = t^3$; $0 \le t \le 2$

**46.** $x = t + \dfrac{1}{t}$, $y = \ln t^2$; $1 \le t \le 4$

**47.** $x = 2e^t$, $y = 3e^{3t/2}$; $\ln 3 \le t \le 2\ln 3$

**48.** $x = \sqrt{1 - t^2}$, $y = 1 - t$; $0 \le t \le \dfrac{1}{4}$

**49.** $x = e^{2t}\sin t$, $y = e^{2t}\cos t$; $0 \le t \le \ln 2$

**50.** $x = \sin^3 t$, $y = \cos^3 t$; $0 \le t \le \pi$

**51.** $x = \cos t - t\sin t$, $y = \sin t + t\cos t$; $1 \le t \le 3$

**52.** $x = \ln\cos t$, $y = t - \tan t$; $0 \le t \le \dfrac{\pi}{6}$

**53.** $x = 4\sqrt{t}$, $y = t^2 + \dfrac{1}{2t}$; $\dfrac{1}{4} \le t \le 1$

**54.** $x = \tanh t$, $y = \ln(\cosh^2 t)$; $-3 \le t \le 3$

**55.** $x = \cos t$, $y = \ln(\sec t + \tan t) - \sin t$; $0 \le t \le \dfrac{\pi}{4}$

**56.** $x = \sin t - t\cos t$, $y = \cos t + t\sin t$; $\dfrac{\pi}{4} \le t \le \dfrac{\pi}{2}$

**57.** Find the length of the curve with the parametric equations

(a) $x = \sin \theta$, $y = \cos \theta$ for $0 \le \theta \le 2\pi$
(b) $x = \sin 3\theta$, $y = \cos 3\theta$ for $0 \le \theta \le 2\pi$
(c) Explain why the answers in part (a) and (b) are not equal.

You can generate surfaces by revolving smooth curves, given parametrically, about one of the coordinate axis. As $t$ increases from $a$ to $b$ a smooth curve $x = F(t)$ and $y = G(t)$ is traced out exactly once. Revolving this curve about the $x$-axis for $y \ge 0$ gives the surface of revolution with surface area

$$S = \int_a^b 2\pi y \sqrt{\left(\dfrac{dx}{dt}\right)^2 + \left(\dfrac{dy}{dt}\right)^2}\, dt$$

Problems 58–64 relate to such surfaces.

**58.** Derive a formula for the surface area generated by the rotation of the curve $x = F(t)$, $y = G(t)$ for $a \le t \le b$ about the $y$-axes for $x \ge 0$ and show that the result is given by

$$S = \int_a^b 2\pi x \sqrt{\left(\dfrac{dx}{dt}\right)^2 + \left(\dfrac{dy}{dt}\right)^2}\, dt$$

**59.** A parametrization of a circle of radius 1 about the point $(1, 0)$ in the $xy$-plane is given by $x = 1 + \cos t$, $y = \sin t$, for $0 \le t \le 2\pi$. Using this parametrization find the surface area when this curve is revolved about the $y$-axis.

**60.** Find the area of the surface generated by revolving the curve $x = \cos t$, $y = 3 + \sin t$, for $0 \le t \le 2\pi$ about the $x$-axis.

**61.** Find the area of the surface generated by revolving the curve $x = 2 + \cos t$, $y = 1 + \sin t$, for $0 \le t \le 2\pi$ about the $x$-axis.

**62.** Find the area of the surface generated by revolving the curve $x = (2/3)t^{3/2}$, $y = 2\sqrt{t}$, for $0 \le t \le 2\sqrt{3}$ about the y-axis.

**63.** Find the area of the surface generated by revolving the curve $x = t + \sqrt{7}$, $y = (t^2/2) + \sqrt{7}t$, for $-\sqrt{7} \le t \le \sqrt{7}$ about the y-axis.

**64.** Find the area of the surface generated by revolving the curve $x = (t^2/2) + at$, $y = t + a$, for $-\sqrt{a} \le t \le \sqrt{a}$ about the x-axis.

In Problems 65–66, evaluate the integrals.

**65.** $\displaystyle\int_0^1 (x^2 - 4y)\,dx$, where $x = t + 1$, $y = t^3 + 4$.

**66.** $\displaystyle\int_1^{\sqrt{3}} xy\,dy$, where $x = \sec t$, $y = \tan t$.

**67.** Find the area of the region between the curve $x = e^{2t}$, $y = e^{-t}$, and the x-axis from $t = 0$ to $t = \ln 5$. Make a sketch.

**68.** Find the area of the region bounded by the curve

$$x = t + \frac{1}{t}, \qquad y = t - \frac{1}{t}$$

and the line $3x - 10 = 0$, without eliminating the parameter. Make a sketch.

**69.** Modify the text discussion of the cycloid (and its accompanying diagram) to handle the case where the point P is $b < a$ units from the center of the wheel. Show that the corresponding parametric equations are

$$x = at - b \sin t \qquad y = a - b \cos t$$

Sketch the graph of these equations (called a **curtate cycloid**) when $a = 8$ and $b = 4$.

**70.** Follow the instructions of Problem 69 for the case $b > a$ (a flanged wheel, as on a train) showing that you get the same parametric equations. Sketch the graph of these equations (called a **prolate cycloid**) when $a = 6$ and $b = 8$.

**71.** The path of a projectile fired from level ground with a speed of $v_0$ feet per second at an angle $\alpha$ with the ground, is given by the parametric equations

$$x = (v_0 \cos \alpha)t \qquad y = -16t^2 + (v_0 \sin \alpha)t$$

(a) Show that the path is a parabola.
(b) Find the time of flight.
(c) Show that the range (horizontal distance traveled) is $(v_0^2/32)\sin 2\alpha$.
(d) For a given $v_0$, what value of $\alpha$ gives the largest possible range?

**72.** Each value of the parameter $t$ in the equations of a cycloid,

$$x = a(t - \sin t), \qquad y = a(1 - \cos t)$$

determines a unique point $P(x, y)$ on the cycloid and also determines a unique position of the rolling circle that generates the cycloid. Let $P_1(x_1, y_1)$ be the point on the cycloid determined by the value $t_1$ of the parameter $t$. Prove that the tangent to the cycloid at $P_1$ passes through the highest point on the rolling circle when the circle is in the position determined by $t_1$. Through what point on the circle does the normal pass?

**73.** Let a circle of radius $b$ roll, without slipping, inside a fixed circle of radius $a$, $a > b$. A point $P$ on the circumference of the rolling circle traces out a curve called a **hypocycloid**. Find parametric equations of the hypocycloid. *Hint:* Place the origin $O$ of Cartesian coordinates at the center of the fixed, larger circle and let the point $A(a, 0)$ be one position of the tracing point $P$. Denote by $B$ the moving point of tangency of the two circles and let $t$, the radian measure of the angle $AOB$, be the parameter (see Figure 11).

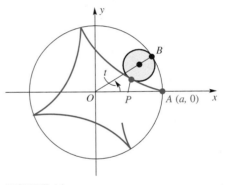

**FIGURE 11**

**74.** Show that if $b = a/4$ in Problem 73, the parametric equations of the hypocycloid may be simplified to

$$x = a \cos^3 t, \qquad y = a \sin^3 t$$

This is called a **hypocycloid of four cusps**. Sketch it carefully and show that its Cartesian equation is $x^{2/3} + y^{2/3} = a^{2/3}$.

**75.** The curve traced by a point on the circumference of a circle of radius $b$ as it rolls without slipping on the outside of a fixed circle of radius $a$ is called an **epicycloid**. Show that it has parametric equations

$$x = (a + b)\cos t - b \cos \frac{a + b}{b} t$$

$$y = (a + b)\sin t - b \sin \frac{a + b}{b} t$$

(See the hint in Problem 73.)

**76.** If $b = a$, the equations in Problem 75 are

$$x = 2a \cos t - a \cos 2t$$

$$y = 2a \sin t - a \sin 2t$$

Show that this special epicycloid is the cardioid $r = 2a(1 - \cos \theta)$, where the pole of the polar coordinate system is the point $(a, 0)$ in the Cartesian system and the polar axis has the direction of the positive $x$-axis. *Hint:* Find a Cartesian equation of the epicycloid by eliminating the parameter $t$ between the equations. Then show that the equations connecting the Cartesian and polar systems are

$$x = r \cos \theta + a \qquad y = r \sin \theta$$

and use these equations to transform the Cartesian equation into $r = 2a(1 - \cos \theta)$.

**77.** Consider a circle of radius $a$ centered at $(0, a)$, as in Figure 12. Let a line $OA$ intersect the line $y = 2a$ at $A$ and the circle $C$ at $B$. Finally, let $P$ be the point of intersection of a horizontal line through $B$ and a vertical line through $A$. As $\theta$, the angle $OA$ makes with the positive $x$-axis varies, $P$ traces out a curve, the **witch of Agnesi**.

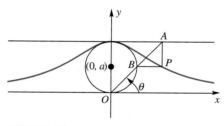

FIGURE 12

(a) Find parametric equations for the witch using $\theta$ as parameter.
(b) Find the Cartesian equation of the witch.

**78.** If $b = a/3$ in Problem 73, we obtain a hypocycloid of three cusps, called a **deltoid**, with parametric equations

$$x = \left(\frac{a}{3}\right)(2 \cos t + \cos 2t), \qquad y = \left(\frac{a}{3}\right)(2 \sin t - \sin 2t)$$

Find the length of the deltoid.

**79.** Find the length of the epicycloid of Problem 75 assuming $a/b$ is an integer and $0 \le t \le 2\pi$.

**80.** Consider the ellipse $x^2/a^2 + y^2/b^2 = 1$.

(a) Show that its perimeter is

$$P = 4a \int_0^{\pi/2} \sqrt{1 - e^2 \cos^2 t} \, dt,$$

where $e$ is the eccentricity.

⌐C⌐ (b) The integral in part a is called an *elliptic integral*. It has been studied at great length, and it is known that the integrand does not have an elementary antiderivative, so we must turn to approximate methods to evaluate $P$. Do so when $a = 1$ and $e = \frac{1}{4}$ using the Parabolic Rule with $n = 4$. (Your answer should be near $2\pi$. Why?)
⌐PC⌐ (c) Repeat part (b) using $n = 20$.

Personal computers and graphing calculators can draw parametric curves with ease. Use a computer to graph a selection of Problems 1–24. Then work the following problems.

**81.** The parametric curve given by $x = \cos 3t$ and $y = \sin 5t$ is difficult to sketch by hand. The result is known as a *Lissajous* figure. The $x$-coordinate oscillates three times between 1 and $-1$ as $t$ goes from 0 to $2\pi$, while the $y$-coordinate oscillates 5 times over the same $t$ interval. This behavior is repeated over every interval of length $2\pi$. The entire motion takes place in a unit square. Plot the following *Lissajous* figures for a range of $t$ which ensures that the resulting figure is a closed curve. In each case count the number of times the curve touches the horizontal and vertical borders of the unit square.

(a) $x = \sin t, y = \cos t$
(b) $x = \sin 3t, y = \cos 5t$
(c) $x = \cos 5t, y = \sin 15t$
(d) $x = \sin 2t, y = \cos 9t$

**82.** Sometimes the *Lissajous* figures do not appear to be closed. This is because the curve retraces itself. For example, the curves $x = \cos 2t$, $y = \sin 7t$, and $x = \cos(2t + 0.1), y = \sin 7t$ are plotted as Figures 13 and 14 below.

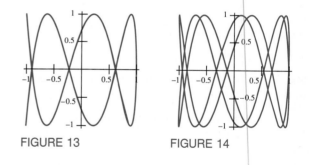

FIGURE 13          FIGURE 14

Using this knowledge, and the following *Lissajous* figures associated with $x = \cos at$ and $y = \sin bt$, match the appropriate figure with the correct ratio for $a/b$.

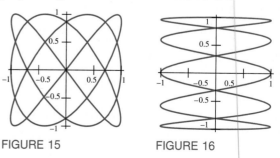

FIGURE 15          FIGURE 16

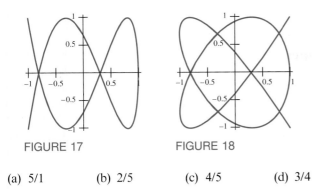

FIGURE 17 FIGURE 18

(a) 5/1 (b) 2/5 (c) 4/5 (d) 3/4

*Hint*: If a curve touches a corner of a square it counts as 1/2 a contact.

**83.** Using a computer algebra system plot the following parametric curves

(a) $x = \cos(t^2 - t)$, $y = \sin(t^2 - t)$
(b) $x = \cos(2t^2 + 3t + 1)$, $y = \sin(2t^2 + 3t + 1)$
(c) $x = \cos(-2 \ln t)$, $y = \sin(-2 \ln t)$
(d) $x = \cos(\sin t)$, $y = \sin(\sin t)$

After you have experimented and plotted these curves over various ranges of $t$, describe in words how the point moves around the curve in each of the cases.

**84.** Using a computer algebra system plot the following parametric curves for $0 \le t \le 2$. Describe the shape of the curve in each case and the similarities and differences among all the curves.

(a) $x = t$, $y = t^2$
(b) $x = t^3$, $y = t^6$
(c) $x = -t^4$, $y = -t^8$

**85.** It is easy to represent the motion along the $x$-axis, in terms of a parameter $t$ in terms of a simple equation, for a given interval in $t$. However, it is frequently important to visualize how the curve is actually traced out in time. One artifice for doing this is to introduce a new variable $y$ which is linear in $t$. This will smear out the motion in a two dimensional plane. Using a graphing calculator or a computer algebra system first graph the parametric equations given by

(a) $x = t^3 - t$, $y = 0$ over the range $-2 \le t \le 2$
(b) $x = \sin t$, $y = 0$ over the range $-2\pi \le t \le 2\pi$
(c) $x = 3t^2 - 2t + 6$, $y = 0$ over the range $-2 \le t \le 2$
(d) $x = 2t^4 - 10t^3 + 4t - 7$, $y = 0$ over the range $-2 \le t \le 2$.

Next, graph each of the following parametric equations

(a') $x = t^3 - t$, $y = t$ over the range $-2 \le t \le 2$
(b') $x = \sin t$, $y = t$ over the range $-2\pi \le t \le 2\pi$
(c') $x = 3t^2 - 2t + 6$, $y = t$ over the range $-2 \le t \le 2$
(d') $x = 2t^4 - 10t^3 + 4t - 7$, $y = t$ over the range $-2 \le t \le 2$.

What does the parametric path traced out in two dimensions tell you about the actual path of the point along the $x$-axis?

**86.** Draw the graph of each of the following for the interval $0 \le t \le 4\pi$.

(a) $x = 2(t - \sin t)$, $y = 2(1 - \cos t)$ (a cycloid)
(b) $x = 2t - \sin t$, $y = 2 - \cos t$ (a curtate cycloid; see Problem 69)
(c) $x = t - 2 \sin t$, $y = 1 - 2 \cos t$ (a prolate cycloid; see Problem 70)

**87.** Draw the graph of the hypocycloid (see Problem 73)

$$x = (a - b) \cos t + b \cos \frac{a - b}{b} t,$$

$$y = (a - b) \sin t - b \sin \frac{a - b}{b} t$$

for appropriate values of $t$ in each of the following cases.

(a) $a = 4$, $b = 1$       (b) $a = 3$, $b = 1$
(c) $a = 5$, $b = 2$       (d) $a = 7$, $b = 4$

Experiment with other positive integer values of $a$ and $b$ and then make conjectures about the length of the $t$-interval required for the curve to return to its starting point and about the number of cusps. What can you say if $a/b$ is irrational?

**88.** Draw the graph of the epicycloid (see Problem 75)

$$x = (a + b) \cos t - b \cos \frac{a + b}{b} t,$$

$$y = (a + b) \sin t - b \sin \frac{a + b}{b} t$$

for various values of $a$ and $b$. What conjectures can you make?

**89.** Draw the Folium of Descartes $x = 3t/(t^3 + 1)$, $y = 3t^2/(t^3 + 1)$. Then tell for what values of $t$ this graph is in each of the four quadrants.

---

**Answers to Concepts Review:** **1.** Simple; closed; simple **2.** Parametric; parameter **3.** Cycloid **4.** $(dy/dt)/(dx/dt)$

## 13.2
### VECTORS IN THE PLANE: GEOMETRIC APPROACH

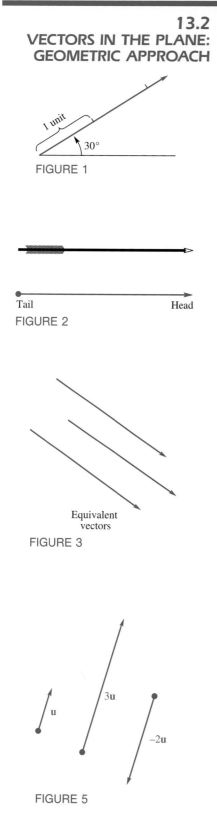

FIGURE 1

FIGURE 2

Tail          Head

Equivalent vectors

FIGURE 3

3u

u

−2u

FIGURE 5

Many quantities that occur in science (for example, length, mass, volume, and electric charge) can be specified by giving a single number. These quantities (and the numbers that measure them) are called **scalars**. Other quantities, such as velocity, force, torque, and displacement, require both a magnitude and a direction for complete specification. We call such quantities **vectors** and represent them by arrows (directed line segments). The length of the arrow is the magnitude of the vector; its direction is the direction of the vector. The vector in Figure 1 has length 2.3 units and direction 30° north of east (or 30° from the positive x-axis).

Arrows that we draw, like those shot from a bow, have two ends. There is the feather end (the initial point), called the **tail**, and the pointed end (the terminal point), called the **head**, or tip (Figure 2). Two vectors are considered to be **equivalent** if they have the same magnitude and direction (Figure 3). We shall symbolize vectors by boldface letters, such as **u** and **v**. Since this is hard to accomplish in normal writing, you might use $\vec{u}$ and $\vec{v}$. The magnitude, or length, of a vector **u** is symbolized by $|\mathbf{u}|$.

**Operations On Vectors**   To find the **sum**, or resultant, of **u** and **v**, move **v** without changing its magnitude or direction until its tail coincides with the head of **u**. Then **u** + **v** is the vector connecting the tail of **u** to the head of **v**. This method (called the *Triangle Law*) is illustrated in the left half of Figure 4.

As an alternative way to find **u** + **v**, move **v** so that its tail coincides with that of **u**. Then **u** + **v** is the vector with this common tail and coinciding with the diagonal of the parallelogram that has **u** and **v** as sides. This method (called the *Parallelogram Law*) is illustrated on the right in Figure 4.

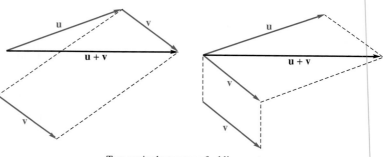

Two equivalent ways of adding vectors

FIGURE 4

You should convince yourself that vector addition is commutative and associative, that is,

$$\mathbf{u} + \mathbf{v} = \mathbf{v} + \mathbf{u}$$

$$(\mathbf{u} + \mathbf{v}) + \mathbf{w} = \mathbf{u} + (\mathbf{v} + \mathbf{w})$$

If **u** is a vector, then 3**u** is the vector with the same direction as **u** but three times as long; −2**u** is twice as long but oppositely directed (Figure 5). In general, c**u**, called a **scalar multiple** of **u**, has magnitude $|c|$ times that of **u**

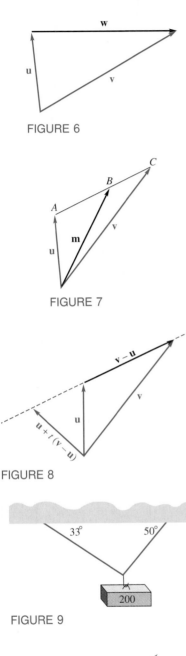

FIGURE 6

FIGURE 7

FIGURE 8

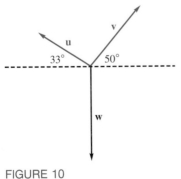

FIGURE 9

FIGURE 10

and is similarly or oppositely directed, depending on whether $c$ is positive or negative. In particular, $(-1)\mathbf{u}$ (usually written $-\mathbf{u}$) has the same length as $\mathbf{u}$ but opposite direction. It is called the **negative** of $\mathbf{u}$ because when we add it to $\mathbf{u}$, the result is a vector that has shriveled to a point. This latter vector (the only vector without a well-defined direction) is called the **zero vector** and is denoted by $\mathbf{0}$. It is the identity element for addition; that is, $\mathbf{u} + \mathbf{0} = \mathbf{0} + \mathbf{u} = \mathbf{u}$. Finally, subtraction is defined by

$$\mathbf{u} - \mathbf{v} = \mathbf{u} + (-\mathbf{v})$$

**EXAMPLE 1**   In Figure 6, express $\mathbf{w}$ in terms of $\mathbf{u}$ and $\mathbf{v}$.

**Solution**   Since $\mathbf{u} + \mathbf{w} = \mathbf{v}$, it follows that

$$\mathbf{w} = \mathbf{v} - \mathbf{u} \qquad \blacksquare$$

**EXAMPLE 2**   In Figure 7, $\overrightarrow{AB} = \frac{2}{3}\overrightarrow{AC}$. Express $\mathbf{m}$ in terms of $\mathbf{u}$ and $\mathbf{v}$.

**Solution**

$$\mathbf{m} = \mathbf{u} + \overrightarrow{AB} = \mathbf{u} + \tfrac{2}{3}\overrightarrow{AC}$$
$$= \mathbf{u} + \tfrac{2}{3}(\mathbf{v} - \mathbf{u})$$
$$= \tfrac{1}{3}\mathbf{u} + \tfrac{2}{3}\mathbf{v}$$

More generally, if $\overrightarrow{AB} = t\overrightarrow{AC}$ where $0 < t < 1$, then

$$\mathbf{m} = (1 - t)\mathbf{u} + t\mathbf{v} \qquad \blacksquare$$

The expression just obtained for $\mathbf{m}$ can also be written as

$$\mathbf{u} + t(\mathbf{v} - \mathbf{u})$$

If we allow $t$ to range over all scalars, we obtain the set of all vectors with heads on the line, shown in Figure 8. This fact will be important to us later in describing lines using vector language.

**Applications**   A force has both a magnitude and a direction. If two forces $\mathbf{u}$ and $\mathbf{v}$ act at a point, the resultant force at the point is the vector sum of the two forces.

**EXAMPLE 3**   A weight of 200 newtons is supported by two wires, as shown in Figure 9. Find the magnitude of the tension in each wire.

**Solution**   The weight $\mathbf{w}$ and the two tensions $\mathbf{u}$ and $\mathbf{v}$ are forces that behave as vectors (Figure 10). Each of these vectors can be expressed as a sum of a horizontal and a vertical component. For equilibrium, (1) the magnitude of the leftward force must equal the magnitude of the rightward force, and (2) the magnitude of the upward force must equal the magnitude of the downward force. Thus,

(1)
$$|\mathbf{u}|\cos 33° = |\mathbf{v}|\cos 50°$$

(2)
$$|\mathbf{u}|\sin 33° + |\mathbf{v}|\sin 50° = |\mathbf{w}| = 200$$

When we solve (1) for $|\mathbf{v}|$ and substitute in (2), we get

$$|\mathbf{u}|\sin 33° + \frac{|\mathbf{u}|\cos 33°}{\cos 50°}\sin 50° = 200$$

or

$$|\mathbf{u}| = \frac{200}{\sin 33° + \cos 33° \tan 50°} \approx 129.52 \text{ newtons}$$

Then,

$$|\mathbf{v}| = \frac{|\mathbf{u}|\cos 33°}{\cos 50°} \approx \frac{129.52 \cos 33°}{\cos 50°} \approx 168.99 \text{ newtons} \qquad ■$$

Velocities have both direction and magnitude and add together as vectors. The magnitude of a velocity vector is called **speed**.

**EXAMPLE 4**   A river 0.62 miles wide is flowing at 6 miles per hour. Karen's boat travels at 20 miles per hour in still water. In what direction should she point her boat if she wants to go straight across the river? How long will it take her to get across by traveling in this direction?

*Solution*   Our first job is to determine $\alpha$ in Figure 11.

$$\sin \alpha = \frac{6}{20}$$

$$\alpha \approx 17.46°$$

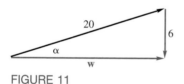

FIGURE 11

Next, we determine $|\mathbf{w}|$, the speed of the boat in the direction of $\mathbf{w}$.

$$|\mathbf{w}| \approx 20 \cos 17.46° \approx 19.08 \text{ miles per hour}$$

Finally, the time required to cross the river is

$$\frac{0.62}{19.08} \approx 0.0325 \text{ hours} = 1.95 \text{ minutes} \qquad ■$$

## CONCEPTS REVIEW

**1.** Vectors are distinguished from scalars in that vectors have both _____ and _____.

**2.** Two vectors are considered to be equivalent if _____.

**3.** If the tail of $\mathbf{v}$ coincides with the head of $\mathbf{u}$, then $\mathbf{u} + \mathbf{v}$ is the vector with tail at _____ and head at _____.

**4.** Two physical examples of vectors are _____ and _____.

## PROBLEM SET 13.2

In Problems 1–4, draw the vector **w**.

1. $\mathbf{w} = \mathbf{u} + \frac{3}{2}\mathbf{v}$

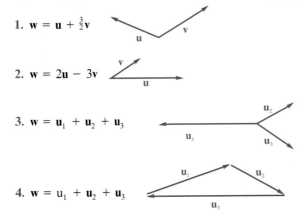

2. $\mathbf{w} = 2\mathbf{u} - 3\mathbf{v}$

3. $\mathbf{w} = \mathbf{u}_1 + \mathbf{u}_2 + \mathbf{u}_3$

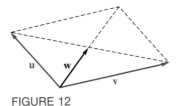

4. $\mathbf{w} = \mathbf{u}_1 + \mathbf{u}_2 + \mathbf{u}_3$

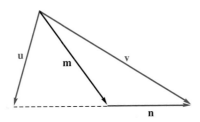

5. Figure 12 is a parallelogram. Express **w** in terms of **u** and **v**.

FIGURE 12

6. In the large triangle of Figure 13, **m** is a median (it bisects the side to which it is drawn). Express **m** and **n** in terms of **u** and **v**.

FIGURE 13

7. In Figure 14, $\mathbf{w} = -(\mathbf{u} + \mathbf{v})$ and $|\mathbf{u}| = |\mathbf{v}| = 1$. Find $|\mathbf{w}|$.

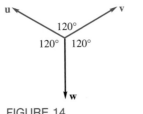

FIGURE 14

8. Do Problem 7 if the top angle is 90° and the two side angles are 135°.

[C] 9. In Figure 15, forces **u** and **v** each have magnitude 50 pounds. Find the magnitude and direction of the force **w** needed to counterbalance **u** and **v**.

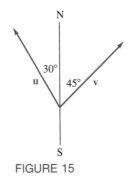

FIGURE 15

[≈] 10. John pushes on a post in the direction S 30° E (30° east of south) with a force of 60 pounds. Wayne pushes on the same post in the direction S 60° W with a force of 80 pounds. What are the magnitude and direction of the resultant force?

[≈] 11. A 300-newton weight rests on a smooth (friction negligible) inclined plane that makes an angle of 30° with the horizontal. What force parallel to the plane will just keep the weight from sliding down the plane? *Hint*: Consider the downward force of 300 newtons to be the sum of two forces, one parallel to the plane and one perpendicular to it.

[C] 12. A body weighing 258.5 pounds is held in equilibrium by two ropes that make angles of 27.34° and 39.22°, respectively, with the vertical. Find the magnitude of the force exerted on the body by each rope.

[C] 13. A wind with velocity 45 miles per hour is blowing in the direction N 20° W. An airplane that flies at 425 miles per hour in still air is supposed to fly straight north. How should the airplane be headed and how fast will it then be flying with respect to the ground?

14. A ship is sailing due south at 20 miles per hour. A man walks west (that is, at right angles to the side of the ship) across the deck at 3 miles per hour. What are the magnitude and direction of his velocity relative to the surface of the water?

15. A pilot, flying in a wind blowing 40 miles per hour due south, discovers that she is heading due east when she points her plane in the direction N 60° E. Find the air speed (speed in still air) of the plane.

[C] **16.** What heading and air speed are required for a plane to fly 837 miles per hour due north if a wind of 63 miles per hour is blowing in the direction S 11.5° E?

**17.** Prove, using vector methods, that the line segment joining the midpoints of two sides of a triangle is parallel to the third side.

**18.** Prove that the midpoints of the four sides of an arbitrary quadrilateral are the vertices of a parallelogram.

**19.** Let $\mathbf{v}_1, \mathbf{v}_2, \ldots, \mathbf{v}_n$ be the edges of a polygon arranged in cyclic order as shown in Figure 16. Show that $\mathbf{v}_1 + \mathbf{v}_2 + \cdots + \mathbf{v}_n = \mathbf{0}$.

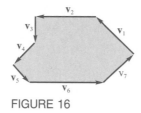

FIGURE 16

**20.** Let $n$ points be equally spaced on the rim of a circle and let $\mathbf{v}_1, \mathbf{v}_2, \ldots, \mathbf{v}_n$ be the vectors from the center of the circle to these $n$ points. Show that $\mathbf{v}_1 + \mathbf{v}_2 + \cdots + \mathbf{v}_n = \mathbf{0}$.

**21.** Consider a horizontal triangular table with each vertex angle less than 120°. At each vertex are frictionless pulleys over which pass strings knotted at $P$ and each with a weight $W$ attached as shown in Figure 17. Show that at equilibrium the three angles at $P$ are equal, that is, show that $\alpha + \beta = \alpha + \gamma = \beta + \gamma = 120°$.

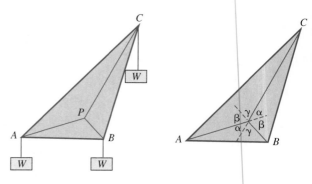

FIGURE 17

**22.** Show that the point $P$ of the triangle of Problem 21 that minimizes $|AP| + |BP| + |CP|$ is the point where the three angles at $P$ are equal. *Hint*: The system of Problem 21 is in equilibrium when the center of gravity of the three weights is lowest.

**23.** Let the weights at $A$, $B$, and $C$ of Problem 21 be $3w$, $4w$, and $5w$, respectively. Determine the three angles at $P$ at equilibrium. What geometric quantity (as in Problem 22) is now minimized?

**24.** A company will build a plant to manufacture refrigerators to be sold in cities A, B, and C in quantities $a$, $b$, and $c$, respectively, each year. Describe a physical experiment the company might use to determine the best location for the plant, that is, the location that will minimize delivery costs (see Problem 23).

---

**Answers to Concepts Review:**  **1.** Magnitude; direction
**2.** They have the same magnitude and direction
**3.** The tail of **u**; the head of **v**  **4.** Force; velocity

---

| 13.3 | From the geometric perspective of the previous section, a vector can be |

**13.3**

**VECTORS IN THE PLANE: ALGEBRAIC APPROACH**

From the geometric perspective of the previous section, a vector can be described as a family of arrows all having the same length and direction (Figure 1). Our aim now is to place vectors in an algebraic context. Here is how we do it.

We begin by imposing a Cartesian coordinate system on the plane. Then, for a given vector **u**, we pick as its representative the arrow that has its tail at the origin (Figure 2). This arrow is uniquely determined by the coordinates $u_1$ and $u_2$ of its head; that is, the vector **u** is completely described by the ordered pair $\langle u_1, u_2 \rangle$ (Figure 3). That being so, we consider $\langle u_1, u_2 \rangle$ to be the vector; it is the vector **u** in its algebraic clothes. Incidently, we use $\langle u_1, u_2 \rangle$ rather than $(u_1, u_2)$ because the latter symbol already has two meanings— open intervals and points in the plane.

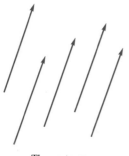

The vector **u**

FIGURE 1

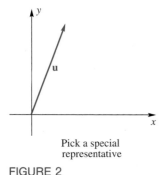

Pick a special
representative

FIGURE 2

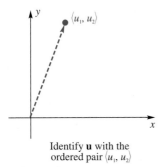

Identify **u** with the
ordered pair $\langle u_1, u_2 \rangle$

FIGURE 3

But why do we offer this new interpretation? There are two good answers. First the interplay between the geometric and algebraic aspects of vectors greatly enriches and clarifies the subject. Second, it is the algebraic viewpoint that most easily generalizes to higher dimensions. This is because it is almost as easy to talk about an ordered $n$-tuple $\langle u_1, u_2, \ldots, u_n \rangle$ as it is an ordered pair $\langle u_1, u_2 \rangle$.

**Operations on Vectors**   The numbers $u_1$ and $u_2$ are called **components** of $\mathbf{u} = \langle u_1, u_2 \rangle$. Two vectors $\mathbf{u} = \langle u_1, u_2 \rangle$ and $\mathbf{v} = \langle v_1, v_2 \rangle$ are equal if and only if $u_1 = v_1$ and $u_2 = v_2$. To add **u** and **v**, we add corresponding components—that is,

$$\mathbf{u} + \mathbf{v} = \langle u_1 + v_1, u_2 + v_2 \rangle$$

To multiply **u** by a scalar $c$, we multiply each component by $c$. Thus,

$$\mathbf{u}c = c\mathbf{u} = \langle cu_1, cu_2 \rangle$$

In particular,

$$-\mathbf{u} = \langle -u_1, -u_2 \rangle$$

and

$$\mathbf{0} = 0\mathbf{u} = \langle 0, 0 \rangle$$

Figure 4 shows that these definitions are equivalent to the earlier geometric ones.

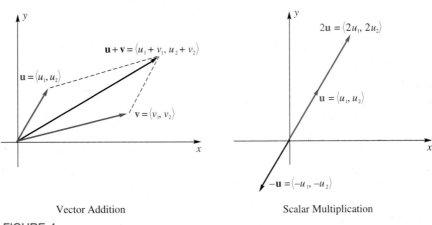

Vector Addition                   Scalar Multiplication

FIGURE 4

Using the algebraic interpretation of vectors, the following rules for operating with vectors are easily established.

Theorem A

For any vectors **u**, **v**, and **w** and any scalars $a$ and $b$, the following relationships hold.
1. $\mathbf{u} + \mathbf{v} = \mathbf{v} + \mathbf{u}$
2. $(\mathbf{u} + \mathbf{v}) + \mathbf{w} = \mathbf{u} + (\mathbf{v} + \mathbf{w})$
3. $\mathbf{u} + \mathbf{0} = \mathbf{0} + \mathbf{u} = \mathbf{u}$
4. $\mathbf{u} + (-\mathbf{u}) = \mathbf{0}$
5. $a(b\mathbf{u}) = (ab)\mathbf{u} = \mathbf{u}(ab)$
6. $a(\mathbf{u} + \mathbf{v}) = a\mathbf{u} + a\mathbf{v}$
7. $(a + b)\mathbf{u} = a\mathbf{u} + b\mathbf{u}$
8. $1\mathbf{u} = \mathbf{u}$

***Proof***  We illustrate the proof by demonstrating Rule 6. Be sure that you understand why each of the following steps is valid.

$$
\begin{aligned}
a(\mathbf{u} + \mathbf{v}) &= a(\langle u_1, u_2 \rangle + \langle v_1, v_2 \rangle) \\
&= a\langle u_1 + v_1, u_2 + v_2 \rangle \\
&= \langle a(u_1 + v_1), a(u_2 + v_2) \rangle \\
&= \langle au_1 + av_1, au_2 + av_2 \rangle \\
&= \langle au_1, au_2 \rangle + \langle av_1, av_2 \rangle \\
&= a\langle u_1, u_2 \rangle + a\langle v_1, v_2 \rangle \\
&= a\mathbf{u} + a\mathbf{v}
\end{aligned}
$$
■

**Length and the Dot Product**    The **length** (or magnitude) $|\mathbf{u}|$ of the vector $\mathbf{u} = \langle u_1, u_2 \rangle$ is given by

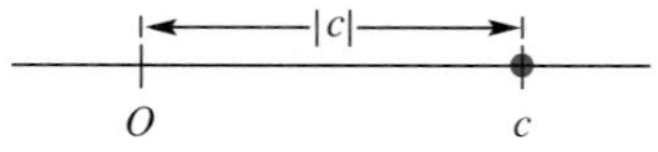

$$
\boxed{|\mathbf{u}| = \sqrt{u_1^2 + u_2^2}}
$$

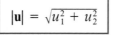

For example, if $\mathbf{u} = \langle 4, -2 \rangle$, then $|\mathbf{u}| = \sqrt{4^2 + (-2)^2} = 2\sqrt{5}$. If we multiply **u** by the scalar $c$, we multiply its length by $|c|$; that is,

$$
|c\mathbf{u}| = |c|\,|\mathbf{u}|
$$

Do not be confused by the apparent double usage of the symbol | |. The symbol $|c|$, called the *absolute value of c*, is the distance from the origin to $c$ on the real line (Figure 5). The symbol $|\mathbf{u}|$, called the *length* of **u**, is the distance from the origin to the head of **u** in the plane (Figure 6).

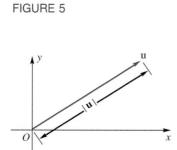

FIGURE 5

FIGURE 6

**EXAMPLE 1**    Let $\mathbf{u} = \langle 4, -3 \rangle$. Find $|\mathbf{u}|$ and $|-2\mathbf{u}|$. Also find a vector **v** with the same direction as **u**, but with length 1.

***Solution*** $|\mathbf{u}| = \sqrt{4^2 + (-3)^2} = 5$ and $|-2\mathbf{u}| = |-2|\,|\mathbf{u}| = 2 \cdot 5 = 10$. To find $\mathbf{v}$, simply divide $\mathbf{u}$ by its length $|\mathbf{u}|$; that is,

$$\mathbf{v} = \frac{\mathbf{u}}{|\mathbf{u}|} = \frac{\langle 4, -3 \rangle}{5} = \frac{1}{5}\langle 4, -3 \rangle = \left\langle \frac{4}{5}, \frac{-3}{5} \right\rangle \qquad \blacksquare$$

We have discussed scalar multiplication—that is, the multiplication of a vector $\mathbf{u}$ by a scalar $c$. The result $c\mathbf{u}$ is always a vector. Now we introduce a multiplication for two vectors $\mathbf{u}$ and $\mathbf{v}$. It is called the **dot product** and is symbolized by $\mathbf{u} \cdot \mathbf{v}$. We define it by the formula

$$\boxed{\mathbf{u} \cdot \mathbf{v} = u_1 v_1 + u_2 v_2}$$

For example, $\langle 4, -3 \rangle \cdot \langle 3, 2 \rangle = 12 + (-6) = 6$. Note that the dot product of two vectors is a scalar.

The properties of the dot product are easy to establish; we state them without proof.

---

### Theorem B

If $\mathbf{u}$, $\mathbf{v}$, and $\mathbf{w}$ are vectors and $c$ is a scalar, then these properties hold.
1. $\mathbf{u} \cdot \mathbf{v} = \mathbf{v} \cdot \mathbf{u}$
2. $\mathbf{u} \cdot (\mathbf{v} + \mathbf{w}) = \mathbf{u} \cdot \mathbf{v} + \mathbf{u} \cdot \mathbf{w}$
3. $c(\mathbf{u} \cdot \mathbf{v}) = (c\mathbf{u}) \cdot \mathbf{v} = \mathbf{u} \cdot (c\mathbf{v})$
4. $\mathbf{0} \cdot \mathbf{u} = 0$
5. $\mathbf{u} \cdot \mathbf{u} = |\mathbf{u}|^2$

---

To understand the significance of the dot product, we offer an alternative formula for it. If $\mathbf{u}$ and $\mathbf{v}$ are nonzero vectors, then

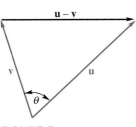

**FIGURE 7**

$$\boxed{\mathbf{u} \cdot \mathbf{v} = |\mathbf{u}|\,|\mathbf{v}|\cos \theta}$$

where $\theta$ is the angle between $\mathbf{u}$ and $\mathbf{v}$. By the **angle between $\mathbf{u}$ and $\mathbf{v}$**, we mean the smallest nonnegative angle between $\mathbf{u}$ and $\mathbf{v}$, so $0 \le \theta \le \pi$.

To derive this formula, apply the Law of Cosines to the triangle in Figure 7.

$$|\mathbf{u} - \mathbf{v}|^2 = |\mathbf{u}|^2 + |\mathbf{v}|^2 - 2|\mathbf{u}|\,|\mathbf{v}|\cos \theta$$

On the other hand, from the properties of the dot product stated in Theorem B,

$$|\mathbf{u} - \mathbf{v}|^2 = (\mathbf{u} - \mathbf{v}) \cdot (\mathbf{u} - \mathbf{v}) = \mathbf{u} \cdot (\mathbf{u} - \mathbf{v}) - \mathbf{v} \cdot (\mathbf{u} - \mathbf{v})$$

$$= \mathbf{u} \cdot \mathbf{u} - \mathbf{u} \cdot \mathbf{v} - \mathbf{v} \cdot \mathbf{u} + \mathbf{v} \cdot \mathbf{v}$$

$$= |\mathbf{u}|^2 + |\mathbf{v}|^2 - 2\mathbf{u} \cdot \mathbf{v}$$

The result follows by equating these two expressions for $|\mathbf{u} - \mathbf{v}|^2$.

An extremely important consequence of the formula just obtained is the following theorem.

> ### Theorem C
>
> **(Perpendicularity criterion).** Two vectors **u** and **v** are perpendicular (orthogonal) if and only if their dot product, **u** · **v**, is 0.

***Proof*** Two nonzero vectors are perpendicular if and only if the angle $\theta$ between them is $\pi/2$; that is, if and only if $\cos \theta = 0$. But $\cos \theta = 0$ if and only if **u** · **v** = 0. The result is valid for zero vectors, provided we agree that a zero vector is perpendicular to every vector. ∎

**EXAMPLE 2** Find $b$ so that **u** = $\langle 8, 6 \rangle$ and **v** = $\langle 3, b \rangle$ are perpendicular.

***Solution***

$$\mathbf{u} \cdot \mathbf{v} = (8)(3) + (6)(b) = 24 + 6b = 0$$

Thus, $b = -4$. ∎

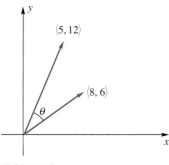

⟨5, 12⟩

⟨8, 6⟩

$\theta$

FIGURE 8

**EXAMPLE 3** Find the angle between **u** = $\langle 8, 6 \rangle$ and **v** = $\langle 5, 12 \rangle$ (Figure 8).

***Solution***

$$\cos \theta = \frac{\mathbf{u} \cdot \mathbf{v}}{|\mathbf{u}|\,|\mathbf{v}|} = \frac{(8)(5) + (6)(12)}{(10)(13)} = \frac{112}{130} \approx 0.862$$

Then,

$$\theta = \cos^{-1}(0.862) \approx 0.532 \text{ (or } 30.5°)$$ ∎

**Basis Vectors** Let **i** = $\langle 1, 0 \rangle$ and **j** = $\langle 0, 1 \rangle$ and note that these two vectors are perpendicular and of unit length. They are called **basis vectors**, because any vector **u** = $\langle u_1, u_2 \rangle$ can be represented in a unique way in terms of **i** and **j**. In fact,

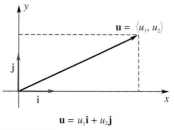

**u** = ⟨$u_1$, $u_2$⟩

**j**

**i**

**u** = $u_1$**i** + $u_2$**j**

FIGURE 9

$$\mathbf{u} = \langle u_1, u_2 \rangle = u_1 \langle 1, 0 \rangle + u_2 \langle 0, 1 \rangle = u_1\mathbf{i} + u_2\mathbf{j}$$

The geometric interpretation of this fact is shown in Figure 9.

**EXAMPLE 4** If **u** is the arrow from the point $P(2, -1)$ to $Q(-3, 7)$, write **u** in the form $u_1\mathbf{i} + u_2\mathbf{j}$.

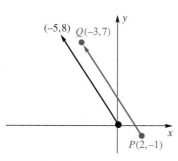

(−5,8)  $Q(-3,7)$

$P(2,-1)$

FIGURE 10

***Solution*** We first translate this arrow so it emanates from the origin (Figure 10). This can always be accomplished by subtracting the components of the initial point from those of the terminal point. Thus, the corresponding algebraic vector is $\langle -3 -2, 7 - (-1) \rangle = \langle -5, 8 \rangle$. We conclude that

$$\mathbf{u} = -5\mathbf{i} + 8\mathbf{j}$$ ∎

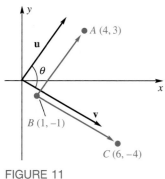

FIGURE 11

**EXAMPLE 5**    Find the measure of the angle $ABC$, where $A = (4, 3)$, $B = (1, -1)$, and $C = (6, -4)$, as in Figure 11.

*Solution*

$$\mathbf{u} = \overrightarrow{BA} = (4 - 1)\mathbf{i} + (3 + 1)\mathbf{j} = 3\mathbf{i} + 4\mathbf{j} = \langle 3, 4 \rangle$$
$$\mathbf{v} = \overrightarrow{BC} = (6 - 1)\mathbf{i} + (-4 + 1)\mathbf{j} = 5\mathbf{i} - 3\mathbf{j} = \langle 5, -3 \rangle$$
$$|\mathbf{u}| = \sqrt{3^2 + 4^2} = 5$$
$$|\mathbf{v}| = \sqrt{5^2 + (-3)^2} = \sqrt{34}$$
$$\mathbf{u} \cdot \mathbf{v} = (3)(5) + (4)(-3) = 3$$
$$\cos \theta = \frac{\mathbf{u} \cdot \mathbf{v}}{|\mathbf{u}|\,|\mathbf{v}|} = \frac{3}{5\sqrt{34}} \approx 0.1029$$
$$\theta \approx 1.468 \text{ (about } 84.09°)$$ ∎

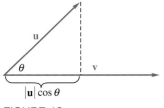

FIGURE 12

Let $\theta$ be the angle between $\mathbf{u}$ and $\mathbf{v}$. The scalar $|\mathbf{u}|\cos \theta$ is called the **scalar projection of $\mathbf{u}$ on $\mathbf{v}$** for reasons that should be clear from Figure 12. It is positive, zero, or negative, depending on whether $\theta$ is acute, right, or obtuse.

The work done by a constant force $\mathbf{F}$ in moving an object along the line from $P$ to $Q$ is the magnitude of the force in the direction of the motion times the distance moved. Thus, if $\mathbf{D}$ is the vector from $P$ to $Q$, the work done is

$$(\text{Scalar projection of } \mathbf{F} \text{ on } \mathbf{D})|\mathbf{D}| = |\mathbf{F}|\cos \theta |\mathbf{D}|$$

that is,

$$\boxed{\text{Work} = \mathbf{F} \cdot \mathbf{D}}$$

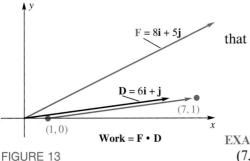

FIGURE 13

**EXAMPLE 6**    A force $\mathbf{F} = 8\mathbf{i} + 5\mathbf{j}$ in pounds moves an object from $(1, 0)$ to $(7, 1)$, distances measured in feet (Figure 13). How much work is done?

*Solution*    Let $\mathbf{D}$ be the vector from $(1, 0)$ to $(7, 1)$; that is, let $\mathbf{D} = 6\mathbf{i} + \mathbf{j}$. Then,

$$\text{Work} = \mathbf{F} \cdot \mathbf{D} = (8)(6) + (5)(1) = 53 \text{ foot-pounds}$$ ∎

## CONCEPTS REVIEW

**1.** If $\mathbf{u} = \langle u_1, u_2 \rangle$ and $\mathbf{v} = \langle v_1, v_2 \rangle$ are vectors, then $\mathbf{u} + \mathbf{v} =$ _____, $c\mathbf{u} =$ _____, and $|\mathbf{u}| =$ _____.

**2.** The dot product of $\mathbf{u} = \langle u_1, u_2 \rangle$ and $\mathbf{v} = \langle v_1, v_2 \rangle$ is defined by $\mathbf{u} \cdot \mathbf{v} =$ _____. The corresponding geometric formula for $\mathbf{u} \cdot \mathbf{v}$ is _____.

**3.** The vectors $\mathbf{i} = \langle 1, 0 \rangle$ and $\mathbf{j} = \langle 0, 1 \rangle$ are called _____ because any vector can be uniquely represented in the form $a\mathbf{i} + b\mathbf{j}$.

**4.** The work done by a force $\mathbf{F}$ in moving an object along the vector $\mathbf{D}$ is given by _____.

## PROBLEM SET 13.3

1. Let $\mathbf{a} = -2\mathbf{i} + 3\mathbf{j}$, $\mathbf{b} = 2\mathbf{i} - 3\mathbf{j}$, and $\mathbf{c} = -5\mathbf{j}$. Find each of the following.

(a) $2\mathbf{a} - 4\mathbf{b}$
(b) $\mathbf{a} \cdot \mathbf{b}$
(c) $\mathbf{a} \cdot (\mathbf{b} + \mathbf{c})$
(d) $(-2\mathbf{a} + 3\mathbf{b}) \cdot 5\mathbf{c}$
(e) $|\mathbf{a}|\mathbf{c} \cdot \mathbf{a}$
(f) $\mathbf{b} \cdot \mathbf{b} - |\mathbf{b}|$

2. Let $\mathbf{a} = \langle 3, -1 \rangle$, $\mathbf{b} = \langle 1, -1 \rangle$, and $\mathbf{c} = \langle 0, 5 \rangle$. Find each of the following.

(a) $-4\mathbf{a} + 3\mathbf{b}$
(b) $\mathbf{b} \cdot \mathbf{c}$
(c) $(\mathbf{a} + \mathbf{b}) \cdot \mathbf{c}$
(d) $2\mathbf{c} \cdot (3\mathbf{a} + 4\mathbf{b})$
(e) $|\mathbf{b}|\mathbf{b} \cdot \mathbf{a}$
(f) $|\mathbf{c}|^2 - \mathbf{c} \cdot \mathbf{c}$

3. Find the cosine of the angle between $\mathbf{a}$ and $\mathbf{b}$ and make a sketch.

(a) $\mathbf{a} = \langle 1, -3 \rangle$, $\mathbf{b} = \langle -1, 2 \rangle$
(b) $\mathbf{a} = \langle -1, -2 \rangle$, $\mathbf{b} = \langle 6, 0 \rangle$
(c) $\mathbf{a} = \langle 2, -1 \rangle$, $\mathbf{b} = \langle -2, -4 \rangle$
(d) $\mathbf{a} = \langle 4, -7 \rangle$, $\mathbf{b} = \langle -8, 10 \rangle$

4. Find the angle between $\mathbf{a}$ and $\mathbf{b}$ and make a sketch.

(a) $\mathbf{a} = 12\mathbf{i}$, $\mathbf{b} = -5\mathbf{i}$
(b) $\mathbf{a} = 4\mathbf{i} + 3\mathbf{j}$, $\mathbf{b} = -8\mathbf{i} - 6\mathbf{j}$
(c) $\mathbf{a} = -\mathbf{i} + 3\mathbf{j}$, $\mathbf{b} = 2\mathbf{i} - 6\mathbf{j}$
(d) $\mathbf{a} = \sqrt{3}\mathbf{i} + \mathbf{j}$, $\mathbf{b} = 3\mathbf{i} + \sqrt{3}\mathbf{j}$

5. Write the vector represented by $\overrightarrow{AB}$ in the form $\mathbf{a} = a_1\mathbf{i} + a_2\mathbf{j}$.

(a) $A(2, 2)$, $B(-3, 4)$
(b) $A(0, 4)$, $B(-6, 0)$
(c) $A(\sqrt{2}, -\pi)$, $B(0, 0)$
(d) $A(-7, e)$, $B(-4, \frac{1}{3})$

6. Find a unit vector $\mathbf{u}$ in the direction of $\mathbf{a}$ and express it in the form $\mathbf{u} = u_1\mathbf{i} + u_2\mathbf{j}$.

(a) $\mathbf{a} = \langle -3, -4 \rangle$
(b) $\mathbf{a} = \langle 1, -7 \rangle$
(c) $\mathbf{a} = \langle 0, -4 \rangle$
(d) $\mathbf{a} = \langle -5, -12 \rangle$

7. If $\mathbf{u} + \mathbf{v}$ is perpendicular to $\mathbf{u} - \mathbf{v}$, what can you say about the relative magnitudes of $\mathbf{u}$ and $\mathbf{v}$?

8. Show that $(\mathbf{u} + \mathbf{v}) \cdot (3\mathbf{u} - \mathbf{v}) = 3|\mathbf{u}|^2 - |\mathbf{v}|^2 + 2\mathbf{u} \cdot \mathbf{v}$.

In each of Problems 9–12, give a proof of the indicated property. Use $\mathbf{u} = \langle u_1, u_2 \rangle$, $\mathbf{v} = \langle v_1, v_2 \rangle$, and $\mathbf{w} = \langle w_1, w_2 \rangle$.

9. $(a + b)\mathbf{u} = a\mathbf{u} + b\mathbf{u}$

10. $\mathbf{u} \cdot (\mathbf{v} + \mathbf{w}) = \mathbf{u} \cdot \mathbf{v} + \mathbf{u} \cdot \mathbf{w}$

11. $c(\mathbf{u} \cdot \mathbf{v}) = (c\mathbf{u}) \cdot \mathbf{v}$

12. If $\mathbf{u} + \mathbf{v} = \mathbf{u}$, then $\mathbf{v} = \mathbf{0}$.

13. Find a vector that has the same direction as $6\mathbf{i} - 8\mathbf{j}$ and three times its length.

14. Find a vector that has the opposite direction to $-5\mathbf{i} + 12\mathbf{j}$ and unit length.

15. Show that the vectors $\langle 6, 3 \rangle$ and $\langle -1, 2 \rangle$ are orthogonal (perpendicular).

16. Show that the vectors $\langle -5, \sqrt{3} \rangle$ and $\langle \sqrt{27}, 15 \rangle$ are orthogonal.

17. For what numbers $c$ are $\langle c, 6 \rangle$ and $\langle c, -4 \rangle$ orthogonal?

18. For what numbers $c$ are $2c\mathbf{i} - 8\mathbf{j}$ and $3\mathbf{i} + c\mathbf{j}$ orthogonal?

19. Given $\mathbf{a} = 3\mathbf{i} - 2\mathbf{j}$ and $\mathbf{b} = -3\mathbf{i} + 4\mathbf{j}$, two noncollinear vectors (that is, vectors such that the angle $\theta$ between them satisfies $0 < \theta < \pi$), and another vector $\mathbf{r} = 7\mathbf{i} - 8\mathbf{j}$, find scalars $k$ and $m$ such that $\mathbf{r} = k\mathbf{a} + m\mathbf{b}$.

20. Given $\mathbf{a} = -4\mathbf{i} + 3\mathbf{j}$ and $\mathbf{b} = 2\mathbf{i} - \mathbf{j}$ (two noncollinear vectors) and another vector $\mathbf{r} = 6\mathbf{i} - 7\mathbf{j}$, find scalars $k$ and $m$ such that $\mathbf{r} = k\mathbf{a} + m\mathbf{b}$.

21. Let $\mathbf{a} = a_1\mathbf{i} + a_2\mathbf{j}$ and $\mathbf{b} = b_1\mathbf{i} + b_2\mathbf{j}$ be noncollinear vectors. If $\mathbf{r} = r_1\mathbf{i} + r_2\mathbf{j}$ is an arbitrarily chosen vector in the plane of $\mathbf{a}$ and $\mathbf{b}$, find scalars $k$ and $m$ such that $\mathbf{r} = k\mathbf{a} + m\mathbf{b}$.

22. Show that the vector $\mathbf{n} = a\mathbf{i} + b\mathbf{j}$ is perpendicular to the line with equation $ax + by = c$. *Hint*: Let $P_1(x_1, y_1)$ and $P_2(x_2, y_2)$ be two points on the line. Show that $\mathbf{n} \cdot \overrightarrow{P_1 P_2} = 0$.

23. Find the work done by the force $\mathbf{F} = 3\mathbf{i} + 10\mathbf{j}$ pounds in moving an object 10 feet north (that is, in the $\mathbf{j}$ direction).

24. Find the work done by a force of 100 dynes acting in the direction S 70° E in moving an object 30 centimeters straight east.

25. Find the work done by the force $\mathbf{F} = 6\mathbf{i} + 8\mathbf{j}$ pounds in moving an object from $(1, 0)$ to $(6, 8)$, distance in feet.

26. Find the work done by a force $\mathbf{F} = -5\mathbf{i} + 8\mathbf{j}$ newtons in moving an object 12 meters north.

27. Prove the **Cauchy-Schwarz Inequality**:

$$|\mathbf{u} \cdot \mathbf{v}| \leq |\mathbf{u}|\,|\mathbf{v}|$$

28. Prove the **Triangle Inequality** (see Figure 14):

$$|\mathbf{u} + \mathbf{v}| \leq |\mathbf{u}| + |\mathbf{v}|$$

*Hint*:

$$|\mathbf{u} + \mathbf{v}|^2 = (\mathbf{u} + \mathbf{v}) \cdot (\mathbf{u} + \mathbf{v}) = \mathbf{u} \cdot \mathbf{u} + 2\mathbf{u} \cdot \mathbf{v} + \mathbf{v} \cdot \mathbf{v}$$
$$= |\mathbf{u}|^2 + 2\mathbf{u} \cdot \mathbf{v} + |\mathbf{v}|^2$$

Now apply Problem 27.

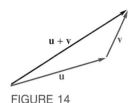

FIGURE 14

**29.** What can you say about the nonzero vectors **u** and **v** in each case?

(a) $|\mathbf{u}|^2 + |\mathbf{v}|^2 = 2\mathbf{u} \cdot \mathbf{v}$      (b) $|\mathbf{u}|^2 = |\mathbf{v}|^2 = 2\mathbf{u} \cdot \mathbf{v}$

**30.** Show that the diagonals of a rhombus (parallelogram with equal edges) are perpendicular. *Hint*: Let **u** and **v** denote adjacent edges. Calculate $(\mathbf{u} + \mathbf{v}) \cdot (\mathbf{u} - \mathbf{v})$.

**31.** Let $A$ and $B$ be the ends of the diameter of a circle and $C$ be any other point on the circle. Use vector methods to show that $AC$ is perpendicular to $BC$.

**32.** Prove that $\mathbf{w} = |\mathbf{v}|\mathbf{u} + |\mathbf{u}|\mathbf{v}$ bisects the angle between **u** and **v**.

**33.** Let $\mathbf{pr_v u}$ denote the **vector projection of u on v**; that is, $\mathbf{pr_v u}$ is the vector with the same direction as **v** having length $|\mathbf{u}|\cos\theta$ (see Figure 15). Show that

$$\mathbf{pr_v u} = \frac{\mathbf{u} \cdot \mathbf{v}}{|\mathbf{v}|^2} \mathbf{v}$$

and use this to calculate $\mathbf{pr_v u}$ in each case.

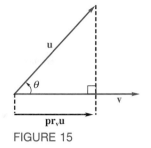

FIGURE 15

(a) $\mathbf{u} = \langle 0, 5\rangle$ and $\mathbf{v} = \langle 3, 4\rangle$
(b) $\mathbf{u} = \langle -3, 2\rangle$ and $\mathbf{v} = \langle 3, 4\rangle$

**34.** Consider the line $ax + by + c = 0$.

(a) Show that $\mathbf{n} = a\mathbf{i} + b\mathbf{j}$ is perpendicular to this line.
(b) Derive the formula $d = |ax_0 + by_0 + c|/\sqrt{a^2 + b^2}$ for the distance from $P(x_0, y_0)$ to the line. *Hint*: Let $Q(x_1, y_1)$ be a point on the line. Find the magnitude of the scalar projection of $\overrightarrow{QP}$ on **n**.

---

**Answers to Concepts Review:** **1.** $\langle u_1 + v_1, u_2 + v_2\rangle$; $\langle cu_1, cu_2\rangle$; $\sqrt{u_1^2 + u_2^2}$   **2.** $u_1v_1 + u_2v_2$; $|\mathbf{u}|\,|\mathbf{v}|\cos\theta$ **3.** Basis vectors **4.** $\mathbf{F} \cdot \mathbf{D}$

---

## 13.4
## VECTOR-VALUED FUNCTIONS AND CURVILINEAR MOTION

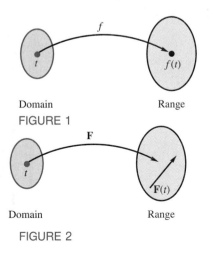

Domain        Range

FIGURE 1

Domain        Range

FIGURE 2

Recall that a **function** $f$ is a rule that associates with each member $t$ of one set (the **domain**) a unique value $f(t)$ from a second set (Figure 1). The set of values so obtained is the **range** of the function. So far in this book, our functions have been real-valued functions (scalar-valued functions) of a real variable; that is, both the domain and range have been sets of real numbers. A typical example is $f(t) = t^2$, which associates with each real number $t$ the real number $t^2$.

Now we offer the first of many generalizations (Figure 2). A **vector-valued function F** of a real variable $t$ associates with each real number $t$ a vector $\mathbf{F}(t)$. Thus,

$$\mathbf{F}(t) = f(t)\mathbf{i} + g(t)\mathbf{j} = \langle f(t), g(t)\rangle$$

where $f$ and $g$ are ordinary real-valued functions. A typical example is

$$\mathbf{F}(t) = t^2\mathbf{i} + e^t\mathbf{j} = \langle t^2, e^t\rangle$$

Note our use of a boldface letter; this helps us keep our vector functions and our scalar functions clearly distinguished.

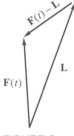

FIGURE 3

**Calculus for Vector Functions**   The most fundamental notion in calculus is that of limit. Intuitively, $\lim_{t \to c} \mathbf{F}(t) = \mathbf{L}$ means that the vector $\mathbf{F}(t)$ tends toward the vector $\mathbf{L}$ as $t$ tends toward $c$. Alternatively it means that the vector $\mathbf{F}(t) - \mathbf{L}$ approaches $\mathbf{0}$ as $t \to c$ (Figure 3). The precise $\varepsilon$, $\delta$ definition is identical with that given for real-valued functions in Section 2.5.

> **Definition**
>
> To say that $\lim_{t \to c} \mathbf{F}(t) = \mathbf{L}$ means that for each given $\varepsilon > 0$ (no matter how small), there is a corresponding $\delta > 0$ such that $|\mathbf{F}(t) - \mathbf{L}| < \varepsilon$, provided that $0 < |t - c| < \delta$; that is,
>
> $$0 < |t - c| < \delta \Rightarrow |\mathbf{F}(t) - \mathbf{L}| < \varepsilon$$

Only one word of explanation is necessary. The symbol $|\mathbf{F}(t) - \mathbf{L}|$ now stands for the length of the vector $\mathbf{F}(t) - \mathbf{L}$.

If we rephrase the definition of limit for $\mathbf{F}(t)$ in terms of its components $f(t)$ and $g(t)$, we obtain an important theorem, whose proof is given in the Appendix Section A.2, Theorem D.

> **Theorem A**
>
> Let $\mathbf{F}(t) = f(t)\mathbf{i} + g(t)\mathbf{j}$. Then $\mathbf{F}$ has a limit at $c$ if and only if $f$ and $g$ have limits at $c$. In that case,
>
> $$\lim_{t \to c} \mathbf{F}(t) = \left[ \lim_{t \to c} f(t) \right] \mathbf{i} + \left[ \lim_{t \to c} g(t) \right] \mathbf{j}$$

As you would expect, all the standard limit theorems hold. Also, continuity has its usual meaning; that is, $\mathbf{F}$ is **continuous** at $c$ if $\lim_{t \to c} \mathbf{F}(t) = \mathbf{F}(c)$.

From Theorem A, it is clear that $\mathbf{F}$ is continuous at $c$ if and only if both $f$ and $g$ are continuous there. Finally, the **derivative** $\mathbf{F}'(t)$ is defined just as for real-valued functions by

$$\mathbf{F}'(t) = \lim_{h \to 0} \frac{\mathbf{F}(t + h) - \mathbf{F}(t)}{h}$$

This also can be written in terms of components.

$$\mathbf{F}'(t) = \lim_{h \to 0} \frac{[f(t + h)\mathbf{i} + g(t + h)\mathbf{j}] - [f(t)\mathbf{i} + g(t)\mathbf{j}]}{h}$$

$$= \lim_{h \to 0} \frac{f(t + h) - f(t)}{h}\mathbf{i} + \lim_{h \to 0} \frac{g(t + h) - g(t)}{h}\mathbf{j}$$

$$= f'(t)\mathbf{i} + g'(t)\mathbf{j}$$

In summary, if $\mathbf{F}(t) = f(t)\mathbf{i} + g(t)\mathbf{j}$, then

$$\mathbf{F}'(t) = f'(t)\mathbf{i} + g'(t)\mathbf{j} = \langle f'(t), g'(t) \rangle$$

**EXAMPLE 1**  If $\mathbf{F}(t) = (t^2 + t)\mathbf{i} + e^t\mathbf{j}$, find $\mathbf{F}'(t)$, $\mathbf{F}''(t)$, and the angle $\theta$ between $\mathbf{F}'(0)$ and $\mathbf{F}''(0)$.

*Solution*  $\mathbf{F}'(t) = (2t + 1)\mathbf{i} + e^t\mathbf{j}$ and $\mathbf{F}''(t) = 2\mathbf{i} + e^t\mathbf{j}$. Thus, $\mathbf{F}'(0) = \mathbf{i} + \mathbf{j}$, $\mathbf{F}''(0) = 2\mathbf{i} + \mathbf{j}$, and

$$\cos\theta = \frac{\mathbf{F}'(0) \cdot \mathbf{F}''(0)}{|\mathbf{F}'(0)|\,|\mathbf{F}''(0)|} = \frac{(1)(2) + (1)(1)}{\sqrt{1^2 + 1^2}\sqrt{2^2 + 1^2}} = \frac{3}{\sqrt{2}\,\sqrt{5}}$$

$$\theta \approx 0.3218 \text{ (about } 18.43°)$$  ∎

Here are the rules for differentiation.

---
**Theorem B**

**(Differentiation formulas).** Let $\mathbf{F}$ and $\mathbf{G}$ be differentiable, vector-valued functions, $h$ a differentiable, real-valued function, and $c$ a scalar. Then:
1. $D_t[\mathbf{F}(t) + \mathbf{G}(t)] = \mathbf{F}'(t) + \mathbf{G}'(t)$;
2. $D_t[c\mathbf{F}(t)] = c\mathbf{F}'(t)$;
3. $D_t[h(t)\mathbf{F}(t)] = h(t)\mathbf{F}'(t) + h'(t)\mathbf{F}(t)$;
4. $D_t[\mathbf{F}(t) \cdot \mathbf{G}(t)] = \mathbf{F}(t) \cdot \mathbf{G}'(t) + \mathbf{G}(t) \cdot \mathbf{F}'(t)$;
5. $D_t[\mathbf{F}(h(t))] = \mathbf{F}'(h(t))h'(t)$ (Chain Rule).

---

*Proof*  We prove 4 and leave the other parts to the reader. Let

$$\mathbf{F}(t) = f_1(t)\mathbf{i} + f_2(t)\mathbf{j}, \qquad \mathbf{G}(t) = g_1(t)\mathbf{i} + g_2(t)\mathbf{j}$$

Then

$$\begin{aligned}
D_t[\mathbf{F}(t) \cdot \mathbf{G}(t)] &= D_t[f_1(t)g_1(t) + f_2(t)g_2(t)] \\
&= f_1(t)g_1'(t) + g_1(t)f_1'(t) + f_2(t)g_2'(t) + g_2(t)f_2'(t) \\
&= [f_1(t)g_1'(t) + f_2(t)g_2'(t)] + [g_1(t)f_1'(t) + g_2(t)f_2'(t)] \\
&= \mathbf{F}(t) \cdot \mathbf{G}'(t) + \mathbf{G}(t) \cdot \mathbf{F}'(t)
\end{aligned}$$  ∎

Since derivatives of vector-valued functions are found by differentiating components, it is natural to define integration in terms of components; that is, if $\mathbf{F}(t) = f(t)\mathbf{i} + g(t)\mathbf{j}$,

$$\int \mathbf{F}(t)\,dt = \left[\int f(t)\,dt\right]\mathbf{i} + \left[\int g(t)\,dt\right]\mathbf{j}$$

$$\int_a^b \mathbf{F}(t)\,dt = \left[\int_a^b f(t)\,dt\right]\mathbf{i} + \left[\int_a^b g(t)\,dt\right]\mathbf{j}$$

**EXAMPLE 2**   If $\mathbf{F}(t) = t^2\mathbf{i} + e^{-t}\mathbf{j}$, find (a) $D_t[t^3\mathbf{F}(t)]$, and (b) $\displaystyle\int_0^1 \mathbf{F}(t)\,dt$.

**Solution**

(a) $D_t[t^3\mathbf{F}(t)] = t^3(2t\mathbf{i} - e^{-t}\mathbf{j}) + 3t^2(t^2\mathbf{i} + e^{-t}\mathbf{j})$

$= 5t^4\mathbf{i} + (3t^2 - t^3)e^{-t}\mathbf{j}$

(b) $\displaystyle\int_0^1 \mathbf{F}(t)\,dt = \left(\int_0^1 t^2\,dt\right)\mathbf{i} + \left(\int_0^1 e^{-t}\,dt\right)\mathbf{j}$

$= \tfrac{1}{3}\mathbf{i} + (1 - e^{-1})\mathbf{j}$ ∎

**Curvilinear Motion**   We are going to use the theory developed above for vector-valued functions to study the motion of a point in the plane. Let $t$ measure time and suppose that the coordinates of a moving point $P$ are given by the parametric equations $x = f(t)$, $y = g(t)$. Then the vector

$$\mathbf{r}(t) = f(t)\mathbf{i} + g(t)\mathbf{j}$$

assumed to emanate from the origin, is called the **position vector** of the point. As $t$ varies, the head of $\mathbf{r}(t)$ traces the path of the moving point $P$ (Figure 4). This is a curve and we call the corresponding motion **curvilinear motion**.

In analogy with linear (straight line) motion, we define the **velocity $\mathbf{v}(t)$** and the **acceleration $\mathbf{a}(t)$** of the moving point $P$ by

$$\mathbf{v}(t) = \mathbf{r}'(t) = f'(t)\mathbf{i} + g'(t)\mathbf{j}$$
$$\mathbf{a}(t) = \mathbf{r}''(t) = f''(t)\mathbf{i} + g''(t)\mathbf{j}$$

Since

$$\mathbf{v}(t) = \lim_{h \to 0} \frac{\mathbf{r}(t + h) - \mathbf{r}(t)}{h}$$

it is clear (from Figure 5) that $\mathbf{v}(t)$ has the direction of the tangent line. The magnitude $|\mathbf{v}(t)|$ of the velocity is called **speed** of the moving point $P$. The acceleration vector $\mathbf{a}(t)$ points to the concave side of the curve (the side toward which the curve is bending).

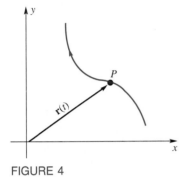

FIGURE 4

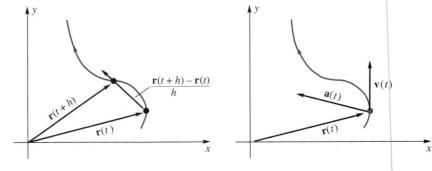

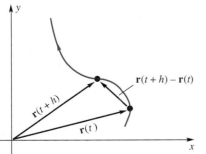

FIGURE 5

**EXAMPLE 3** **(Uniform circular motion).** Suppose that a point $P$ moves around a circle with center $(0, 0)$ and radius $r$ at a constant angular speed of $\omega$ radians per second. If its initial position is $(r, 0)$, find its acceleration.

*Solution*   The position vector at time $t$ is

$$\mathbf{r}(t) = r \cos \omega t \mathbf{i} + r \sin \omega t \mathbf{j}$$

Thus,

$$\mathbf{v}(t) = -r\omega \sin \omega t \mathbf{i} + r\omega \cos \omega t \mathbf{j}$$
$$\mathbf{a}(t) = -r\omega^2 \cos \omega t \mathbf{i} - r\omega^2 \sin \omega t \mathbf{j}$$

Note that

$$\mathbf{a}(t) = -\omega^2 \mathbf{r}(t)$$

Thus if we think of $\mathbf{a}$ as being based at $P$, it points to the origin and is perpendicular to $\mathbf{v}$ (Figure 6). ∎

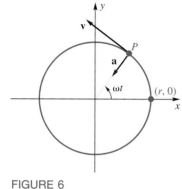

FIGURE 6

**EXAMPLE 4**   Parametric equations for a point $P$ moving in the plane are $x = 3 \cos t$ and $y = 2 \sin t$, where $t$ represents time.
(a) Graph the path of $P$.
(b) Find expressions for the velocity $\mathbf{v}(t)$, speed $|\mathbf{v}(t)|$, and acceleration $\mathbf{a}(t)$.
(c) Find the maximum and minimum values of the speed and where they occur.
(d) Show that the acceleration vector based at $P$ always points to the origin.

*Solution*
(a) Since $x^2/9 + y^2/4 = 1$, the path is the ellipse shown in Figure 7.
(b) The position vector is

$$\mathbf{r}(t) = 3 \cos t \mathbf{i} + 2 \sin t \mathbf{j}$$

and so

$$\mathbf{v}(t) = -3 \sin t \mathbf{i} + 2 \cos t \mathbf{j}$$
$$|\mathbf{v}(t)| = \sqrt{9 \sin^2 t + 4 \cos^2 t} = \sqrt{5 \sin^2 t + 4}$$
$$\mathbf{a}(t) = -3 \cos t \mathbf{i} - 2 \sin t \mathbf{j}$$

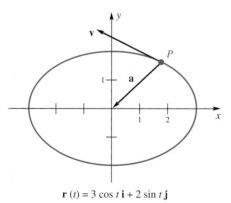

$\mathbf{r}(t) = 3 \cos t\, \mathbf{i} + 2 \sin t\, \mathbf{j}$

FIGURE 7

(c) Since the speed is given by $\sqrt{5 \sin^2 t + 4}$, the maximum speed of 3 occurs when $\sin t = \pm 1$,—that is, when $t = \pi/2$ or $3\pi/2$. This corresponds to the points $(0, \pm 2)$ on the ellipse. Similarly, the minimum speed of 2 occurs when $\sin t = 0$, which corresponds to the points $(\pm 3, 0)$.
(d) Note that $\mathbf{a}(t) = -\mathbf{r}(t)$. Thus, if we base $\mathbf{a}(t)$ at $P$, this vector will point to and exactly reach the origin. We conclude that $|\mathbf{a}(t)|$ is largest at $(\pm 3, 0)$ and smallest at $(0, \pm 2)$. ∎

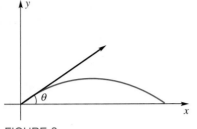

**FIGURE 8**

**EXAMPLE 5** A projectile is shot from the origin at an angle $\theta$ from the positive $x$-axis with an initial speed of $v_0$ feet per second (Figure 8). Neglecting friction, find expressions for the velocity $\mathbf{v}(t)$ and position $\mathbf{r}(t)$, and show that the path is a parabola.

*Solution* The acceleration due to gravity is $\mathbf{a}(t) = -32\mathbf{j}$ feet per second per second. The initial conditions are $\mathbf{r}(0) = \mathbf{0}$ and $\mathbf{v}(0) = v_0 \cos \theta \mathbf{i} + v_0 \sin \theta \mathbf{j}$. Starting with $\mathbf{a}(t) = -32\mathbf{j}$, we integrate twice.

$$\mathbf{v}(t) = \int \mathbf{a}(t)\,dt = \int -32\,dt\,\mathbf{j} = -32t\mathbf{j} + \mathbf{C}$$

The condition $\mathbf{v}(0) = v_0 \cos \theta \mathbf{i} + v_0 \sin \theta \mathbf{j}$ allows us to evaluate $\mathbf{C}$ and gives $\mathbf{C} = v_0 \cos \theta \mathbf{i} + v_0 \sin \theta \mathbf{j}$. Thus,

$$\mathbf{v}(t) = (v_0 \cos \theta)\mathbf{i} + (v_0 \sin \theta - 32t)\mathbf{j}$$

and

$$\mathbf{r}(t) = \int \mathbf{v}(t)\,dt = (tv_0 \cos \theta)\mathbf{i} + (tv_0 \sin \theta - 16t^2)\mathbf{j} + \mathbf{K}$$

The condition $\mathbf{r}(0) = \mathbf{0}$ implies $\mathbf{K} = \mathbf{0}$, and so

$$\mathbf{r}(t) = (tv_0 \cos \theta)\mathbf{i} + (tv_0 \sin \theta - 16t^2)\mathbf{j}$$

To find the equation of the path, we eliminate the parameter $t$ in the equations

$$x = (v_0 \cos \theta)t, \qquad y = (v_0 \sin \theta)t - 16t^2$$

Specifically, we solve the first equation for $t$ and substitute in the second. The result is

$$y = (\tan \theta)x - \left(\frac{4}{v_0 \cos \theta}\right)^2 x^2$$

the equation of a parabola. ∎

---

## CONCEPTS REVIEW

**1.** A function that associates with each real number $a$ single vector is called _____.

**2.** The function $\mathbf{F}(t) = f(t)\mathbf{i} + g(t)\mathbf{j}$ is continuous at $t = c$ if and only if _____. The derivative of $\mathbf{F}$ is given in terms of $f$ and $g$ by $\mathbf{F}'(t) =$ _____.

**3.** If a point moves along a curve so that it is at point

$P$ at time $t$, then the vector $\mathbf{r}(t)$ from the origin to $P$ is called the _____ vector of $P$.

**4.** In terms of $\mathbf{r}(t)$, the velocity of the point $P$ is _____ and its acceleration is _____. The velocity vector at $t$ is _____ to the curve, whereas the acceleration vector points to the _____ side of the curve.

## PROBLEM SET 13.4

In Problems 1–8, find the required limit or indicate that it does not exist.

1. $\lim_{t \to 1} [2t\mathbf{i} - t^2\mathbf{j}]$

2. $\lim_{t \to 3} [2(t - 3)^2\mathbf{i} - 7t^3\mathbf{j}]$

3. $\lim_{t \to 1} \left[ \frac{t - 1}{t^2 - 1}\mathbf{i} - \frac{t^2 + 2t - 3}{t - 1}\mathbf{j} \right]$

4. $\lim_{t \to -2} \left[ \frac{2t^2 - 10t - 28}{t + 2}\mathbf{i} - \frac{7t^3}{t - 3}\mathbf{j} \right]$

5. $\lim_{t \to 0} \left[ \frac{\sin t \cos t}{t}\mathbf{i} - \frac{7t^3}{e^t}\mathbf{j} \right]$

6. $\lim_{t \to \infty} \left[ \frac{t \sin t}{t^2}\mathbf{i} - \frac{7t^3}{t^3 - 3t}\mathbf{j} \right]$

7. $\lim_{t \to 0^+} \langle \ln(t^3), t^2\ln t \rangle$

8. $\lim_{t \to 0^-} \left\langle e^{-1/t^2}, \frac{t}{|t|} \right\rangle$

9. When no domain is given in the definition of a vector-valued function, it is to be understood that the domain is the set of all (real) scalars to which correspond real vectors (that is, vectors with real components). Find the domain of each of the following vector-valued functions.

(a) $\mathbf{r}(t) = \frac{2}{t - 4}\mathbf{i} + \sqrt{3 - t}\mathbf{j}$

(b) $\mathbf{r}(t) = [\![t^2]\!]\mathbf{i} - \sqrt{20 - t}\mathbf{j}$. ($[\![\ ]\!]$ denotes "greatest integer in.")

10. State the domain of each of the following vector-valued functions

(a) $\mathbf{r}(t) = \ln(t - 1)\mathbf{i} + \sqrt{20 - t}\mathbf{j}$
(b) $\mathbf{r}(t) = \ln(t^{-1})\mathbf{i} + \tan^{-1} t\mathbf{j}$

11. For what values of $t$ is each of the functions in Problem 9 continuous.

12. For what values of $t$ is each of the functions in Problem 10 continuous.

13. Find $D_t\mathbf{r}(t)$ and $D_t^2\mathbf{r}(t)$ for each of the following.

(a) $\mathbf{r}(t) = (3t + 4)^3\mathbf{i} + e^{t^2}\mathbf{j}$      (b) $\mathbf{r}(t) = \sin^2 t\mathbf{i} + \cos 3t\mathbf{j}$

14. Find $\mathbf{r}'(t)$ and $\mathbf{r}''(t)$ for each of the following.

(a) $\mathbf{r}(t) = (e^t + e^{-t^2})\mathbf{i} + 2^t\mathbf{j}$      (b) $\mathbf{r}(t) = \tan 2t\mathbf{i} + \arctan t\mathbf{j}$

15. If $\mathbf{r}(t) = e^{-t}\mathbf{i} - \ln(t^2)\mathbf{j}$ find $D_t[\mathbf{r}(t) \cdot \mathbf{r}'(t)]$.

16. If $\mathbf{r}(t) = \sin 3t\mathbf{i} - \cos 3t\mathbf{j}$ find $D_t[\mathbf{r}'(t) \cdot \mathbf{r}''(t)]$.

17. If $\mathbf{r}(t) = \sqrt{t - 1}\mathbf{i} + \ln(2t^2)\mathbf{j}$ and $h(t) = e^{-3t}$, find $D_t[h(t)\mathbf{r}(t)]$.

18. If $\mathbf{r}(t) = \sin 2t\mathbf{i} + \cosh t\mathbf{j}$ and $h(t) = \ln(3t - 2)$, find $D_t[h(t)\mathbf{r}(t)]$.

In Problems 19 and 20, $\mathbf{F}(t) = \mathbf{f}(g(t))$. Find $\mathbf{F}'(t)$ in terms of $t$.

19. $\mathbf{f}(u) = \cos u\mathbf{i} + e^{3u}\mathbf{j}$ and $g(t) = 3t^2 - 4$.

20. $\mathbf{f}(u) = u^2\mathbf{i} + \sin^2 u\mathbf{j}$ and $g(t) = \tan t$.

Evaluate the integrals in Problems 21 and 22.

21. $\int_0^1 (e^t\mathbf{i} + e^{-t}\mathbf{j})\, dt$

22. $\int_{-1}^1 [(1 + t)^{3/2}\mathbf{i} + (1 - t)^{3/2}\mathbf{j}]\, dt$

In each of Problems 23–30, the position of a moving particle at time $t$ is given by $\mathbf{r}(t)$. Find the velocity and acceleration vectors, $\mathbf{v}(t)$ and $\mathbf{a}(t)$, and the speed at the given time $t = t_1$. Sketch a portion of the graph of $\mathbf{r}(t)$ containing the position $P$ of the particle when $t = t_1$ and draw $\mathbf{v}(t_1)$ and $\mathbf{a}(t_1)$ with their initial points at $P$.

23. $\mathbf{r}(t) = e^{-t}\mathbf{i} + e^t\mathbf{j}; t_1 = 1$

24. $\mathbf{r}(t) = (3t^2 - 1)\mathbf{i} + t\mathbf{j}; t_1 = \frac{1}{2}$

25. $\mathbf{r}(t) = 2 \cos t\mathbf{i} - 3 \sin^2 t\mathbf{j}; t_1 = \frac{\pi}{3}$

26. $\mathbf{r}(t) = \tan t\mathbf{i} + \sin t\mathbf{j}; t_1 = \frac{\pi}{6}$

27. $\mathbf{r}(t) = 3t^2\mathbf{i} + t^3\mathbf{j}; t_1 = 2$

28. $\mathbf{r}(t) = a \sin t\mathbf{i} + 2a \cos t\mathbf{j}; t_1 = \frac{\pi}{4}, a > 0$

29. $\mathbf{r}(t) = \cos t\mathbf{i} - 2 \tan t\mathbf{j}; t_1 = -\frac{\pi}{4}$

30. $\mathbf{r}(t) = e^{t/2}\mathbf{i} + e^{-t}\mathbf{j}; t_1 = 2$

In Problems 31–34, use the given information to find the velocity vector $\mathbf{v}(t)$ and the position vector $\mathbf{r}(t)$ (see Example 5).

31. $\mathbf{a}(t) = -32\mathbf{j}, \mathbf{v}(0) = \mathbf{0}, \mathbf{r}(0) = \mathbf{0}$

32. $\mathbf{a}(t) = t\mathbf{j}, \mathbf{v}(0) = \mathbf{i} + 2\mathbf{j}, \mathbf{r}(0) = \mathbf{0}$

33. $\mathbf{a}(t) = \mathbf{i} + e^{-t}\mathbf{j}, \mathbf{v}(0) = 2\mathbf{i} + \mathbf{j}, \mathbf{r}(0) = \mathbf{i} + \mathbf{j}$

34. $\mathbf{a}(t) = -\cos t\mathbf{i} + \sin t\mathbf{j}, \mathbf{v}(0) = \mathbf{i}, \mathbf{r}(0) = \mathbf{i} + 3\mathbf{j}$

35. A point moves around the circle $x^2 + y^2 = 25$ at constant angular speed of 6 radians per second starting at (5, 0). Find expressions for $\mathbf{r}(t), \mathbf{v}(t), |\mathbf{v}(t)|$, and $\mathbf{a}(t)$ (see Example 3).

36. A point moves so that its speed is constant, that is, $\mathbf{v}(t) \cdot \mathbf{v}(t) = c$ (a constant). Show that the velocity and acceleration vectors are always perpendicular to each other.

**37.** In Example 5, suppose $\theta = 30°$ and $v_0 = 96$ feet per second. When does the projectile hit the ground, with what speed, and how far from the origin?

**38.** A point moves in the plane with constant acceleration vector $\mathbf{a} = a\mathbf{j}$. Show that its path is a parabola or a straight line.

**39.** A point moves on the hyperbola $x^2 - y^2 = 1$ with position vector

$$\mathbf{r}(t) = \cosh \omega t\,\mathbf{i} + \sinh \omega t\,\mathbf{j}$$

where $\omega$ is a constant. Show that $\mathbf{a}(t) = c\mathbf{r}(t)$, where $c$ is a positive constant.

**40.** A point moves on the ellipse $x^2/a^2 + y^2/b^2 = 1$ with position vector

$$\mathbf{r}(t) = a \cos \omega t\,\mathbf{i} + b \sin \omega t\,\mathbf{j}$$

where $\omega$ is a constant. Show that $\mathbf{a}(t) = c\mathbf{r}(t)$, where $c$ is a negative constant (see Example 4).

**41.** If a batted baseball leaves the bat at an angle of $45°$ with the horizontal and is caught by an outfielder 300 feet from home plate, what was the initial velocity of the ball? (See Example 5.)

**42.** A small ball is rolled off a horizontal table 4 feet high with a speed of 20 feet per second. At what angle and with what speed will it hit the floor?

**43.** An open can is attached to the rim of a vertical spinning wheel of radius 60 centimeters with its open end pointing to the center of the wheel. In the can is a marble. At least what angular speed must be maintained if the marble is not to fall out? (See Problem 44 for the value of $g$.)

**44.** A body of mass $m$ is revolving in a circular orbit of radius $r$ with constant speed $v$ around a body of mass $M$. From Example 3,

$$a = \left|\mathbf{a}(t)\right| = \omega^2 \left|\mathbf{r}(t)\right| = \frac{v^2}{r}$$

whereas from Newton's Inverse Square Law, $F = ma = GmM/r^2$ or

$$a = \frac{k}{r^2} \qquad (k \text{ a constant})$$

Let $T$ be the time for one revolution, $R \approx 3960$ miles (the radius of the earth), and $g \approx 32.17$ feet per second per second $\approx 980$ centimeters per second per second (the acceleration of gravity at the earth's surface). Use these results to show each of the following.

(a) $v^2 = \dfrac{k}{r}$

(b) $T^2 = \left(\dfrac{4\pi^2}{k}\right)r^3$ (Kepler's Third Law for circular orbits)

(c) $k = gR^2$

C **45.** Apply the results of Problem 44 to the problem of the earth and a satellite by determining each of the following.

(a) The speed of a satellite that orbits the earth 200 miles above its surface.

(b) The distance from the center of the earth to a communications satellite in *synchronous* orbit (revolving so as to stay directly over a point of the earth).

C **46.** The moon orbits the earth every 27.32 days. Determine the distance from the earth to the moon in miles (see Problem 44).

PC **47.** Draw the paths of the head of $\mathbf{r}(t) = (4 \cos t)\mathbf{i} + (3 \sin t)\mathbf{j}$, $\mathbf{r}'(t)$, and $\mathbf{r}''(t)$, $0 \le t \le \pi/2$. On these graphs, display the corresponding vectors for (a) $t = \pi/6$, (b) $t = \pi/4$, (c) $t = \pi/3$, and (d) $t = \pi/2$, thereby illustrating that $\mathbf{r}'(t)$ points in the direction of the tangent to the path and that $\mathbf{r}''(t)$ points to the concave side of the path.

PC **48.** Draw the paths of the head of $\mathbf{r}(t) = (a \cos t)\mathbf{i} + (b \sin t)\mathbf{j}$, $\mathbf{r}'(t)$, and $\mathbf{r}''(t)$, $0 \le t \le 2\pi$, for several values of $a$ and $b$. Make a conjecture about these paths and see if you can prove it.

PC **49.** Let $\mathbf{r}(t) = (3 \cos t - \cos 3t)\mathbf{i} + (3 \sin t - \sin 3t)\mathbf{j}$. Draw the paths of the head of $\mathbf{r}(t)$, $\mathbf{r}'(t)$, and $\mathbf{r}''(t)$ for $0 \le t \le \pi$. Then, using these graphs, display the corresponding vectors for (a) $t = \pi/6$, (b) $t = \pi/3$, (c) $t = \pi/2$, and (d) $t = 2\pi/3$.

---

**Answers to Concepts Review:** **1.** A vector-valued function of a real variable **2.** $f$ and $g$ are continuous at $c$; $f'(t)\mathbf{i} + g'(t)\mathbf{j}$ **3.** Position **4.** $\mathbf{r}'(t)$; $\mathbf{r}''(t)$; tangent; concave

---

## 13.5
## CURVATURE AND ACCELERATION

We want to introduce a number called *curvature*, which measures how sharply a curve bends. A line should have curvature 0, a sharply turning curve should have a large curvature (Figure 1). To lead up to an appropriate definition, we need to review some old ideas and introduce some new ones.

For $a \le t \le b$, let $\mathbf{r}(t) = f(t)\mathbf{i} + g(t)\mathbf{j} = \langle f(t), g(t)\rangle$ be the position vector for a point $P = P(t)$ in the plane. We suppose that $\mathbf{r}'(t)$ exists and is

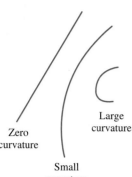

Zero
curvature

Small
curvature

Large
curvature

FIGURE 1

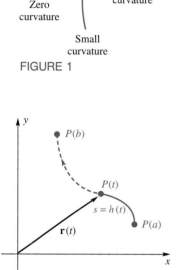

FIGURE 2

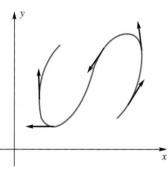

The unit tangent vector **T** $(t)$

FIGURE 3

continuous, and that $\mathbf{r}'(t) \neq \mathbf{0}$ on the interval $[a, b]$. Then (see Section 6.4) as $t$ increases, $P$ traces out a *smooth* curve (Figure 2) and the length $s = h(t)$ of the path from $P(a)$ to $P(t)$ is given by

$$s = h(t) = \int_a^t \sqrt{[f'(u)]^2 + [g'(u)]^2}\, du = \int_a^t |\mathbf{r}'(u)|\, du$$

The speed of the moving point is

$$\frac{ds}{dt} = |\mathbf{r}'(t)| = |\mathbf{v}(t)|$$

Since $\mathbf{r}'(t) \neq \mathbf{0}$, it follows that $|\mathbf{v}(t)| > 0$, and so $s$ increases as $t$ increases. By the Inverse Function Theorem (Theorem 7.2B), $s = h(t)$ has an inverse $t = h^{-1}(s)$ and

$$\frac{dt}{ds} = \frac{1}{ds/dt} = \frac{1}{|\mathbf{v}(t)|}$$

Let $\mathbf{T}(t)$, called the **unit tangent vector** at $P(t)$, be defined by

$$\mathbf{T}(t) = \frac{\mathbf{r}'(t)}{|\mathbf{r}'(t)|} = \frac{\mathbf{v}(t)}{|\mathbf{v}(t)|}$$

As $P(t)$ moves along the curve, the unit vector $\mathbf{T}(t)$ changes its direction (Figure 3). The rate of change of $\mathbf{T}$ with respect to arc length $s$, namely $d\mathbf{T}/ds$, is called the **curvature vector** at $P$. Finally, we define the **curvature** $\kappa$ (kappa) at $P$ to be the magnitude of $d\mathbf{T}/ds$; that is, $\kappa = |d\mathbf{T}/ds|$. Now, by the Chain Rule,

$$\frac{d\mathbf{T}}{ds} = \frac{d\mathbf{T}}{dt}\frac{dt}{ds} = \frac{\mathbf{T}'(t)}{|\mathbf{v}(t)|}$$

Thus,

$$\boxed{\kappa = \left|\frac{d\mathbf{T}}{ds}\right| = \frac{|\mathbf{T}'(t)|}{|\mathbf{v}(t)|} = \frac{|\mathbf{T}'(t)|}{|\mathbf{r}'(t)|}}$$

**Some Important Examples**    To convince you that our definition of curvature is sensible, we illustrate.

**EXAMPLE 1**    Show that the curvature of a line is identically zero.

*Solution*    This follows immediately from the fact that $\mathbf{T}$ is a constant vector. But to illustrate vector methods, we give an algebraic demonstration. Let $P$ and $Q$ be two fixed points on the line and let $\mathbf{a} = \overrightarrow{OP}$ and $\mathbf{b} = \overrightarrow{PQ}$. Then the vector form for the equation of the line may be written as (Figure 4)

$$\mathbf{r} = \mathbf{r}(t) = \mathbf{a} + t\mathbf{b}$$

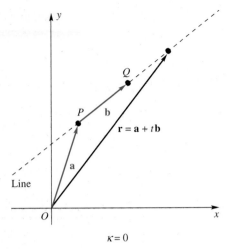

$\kappa = 0$

FIGURE 4

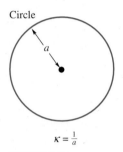

Circle

$\kappa = \frac{1}{a}$

FIGURE 5

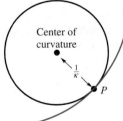

Circle of curvature

Center of
curvature

Curve

FIGURE 6

Thus,

$$\mathbf{v}(t) = \mathbf{r}'(t) = \mathbf{b}$$

$$\mathbf{T}(t) = \frac{\mathbf{b}}{|\mathbf{b}|}$$

$$\kappa = \frac{|\mathbf{T}'(t)|}{|\mathbf{v}(t)|} = \frac{0}{|\mathbf{b}|} = 0$$

$\blacksquare$

**EXAMPLE 2** Show that at each point the curvature of a circle of radius $a$ is $\kappa = 1/a$ (Figure 5).

*Solution* We may suppose the circle is centered at the origin. Then its vector equation may be written

$$\mathbf{r}(t) = a \cos t\mathbf{i} + a \sin t\mathbf{j}$$

Thus,

$$\mathbf{v}(t) = \mathbf{r}'(t) = -a \sin t\mathbf{i} + a \cos t\mathbf{j}$$

$$|\mathbf{v}(t)| = [a^2 \sin^2 t + a^2 \cos^2 t]^{1/2} = a$$

$$\mathbf{T}(t) = \frac{\mathbf{v}(t)}{|\mathbf{v}(t)|} = \frac{\mathbf{v}(t)}{a} = -\sin t\mathbf{i} + \cos t\mathbf{j}$$

$$\kappa = \frac{|\mathbf{T}'(t)|}{|\mathbf{v}(t)|} = \frac{|-\cos t\mathbf{i} - \sin t\mathbf{j}|}{a} = \frac{1}{a}$$

Since $\kappa$ is the reciprocal of the radius, the larger the circle, the smaller is its curvature. $\blacksquare$

The example of the circle leads to several new ideas. Let $P$ be a point on a curve where $\kappa \neq 0$. Consider the circle that is tangent to the curve at $P$ and has the same curvature there. Its center will lie on the concave side of the curve. This circle is called the **circle of curvature** (or osculating circle), its radius $R = 1/\kappa$ is the **radius of curvature**, and its center is the **center of curvature**. These notions are illustrated in Figure 6.

**EXAMPLE 3** Find the curvature and radius of curvature for the hypocycloid

$$\mathbf{r} = 8 \cos^3 t\mathbf{i} + 8 \sin^3 t\mathbf{j}$$

at the point $P$ where $t = \pi/12$. Then sketch the graph of this hypocycloid showing the circle of curvature at $P$.

*Solution* For $0 < t < \pi/2$,

$$\mathbf{v}(t) = \mathbf{r}'(t) = -24 \cos^2 t \sin t\mathbf{i} + 24 \sin^2 t \cos t\mathbf{j}$$

$$|\mathbf{v}(t)| = 24 \sin t \cos t$$

$$\mathbf{T}(t) = -\cos t\mathbf{i} + \sin t\mathbf{j}$$

$$\mathbf{T}'(t) = \sin t\mathbf{i} + \cos t\mathbf{j}$$

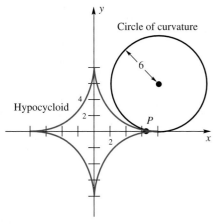

Circle of curvature

Hypocycloid

FIGURE 7

$$\kappa(t) = \frac{|\mathbf{T}'(t)|}{|\mathbf{v}(t)|} = \frac{|\sin t\mathbf{i} + \cos t\mathbf{j}|}{24 \sin t \cos t} = \frac{1}{24 \sin t \cos t}$$

$$= \frac{1}{12 \sin 2t}$$

$$\kappa\left(\frac{\pi}{12}\right) = \frac{1}{12 \cdot \frac{1}{2}} = \frac{1}{6}$$

$$R\left(\frac{\pi}{12}\right) = 6$$

The required graph is shown in Figure 7. Note that $P$ has the approximate coordinates $(7.21, 0.14)$. ∎

**Other Formulas for Curvature** Let $\phi$ denote the angle measured counterclockwise from $\mathbf{i}$ to $\mathbf{T}$ (Figure 8). Then, $\mathbf{T} = \cos \phi\mathbf{i} + \sin \phi\mathbf{j}$, and so

$$\frac{d\mathbf{T}}{d\phi} = -\sin \phi\mathbf{i} + \cos \phi\mathbf{j}$$

Now $d\mathbf{T}/d\phi$ is a unit vector (length 1) and $\mathbf{T} \cdot d\mathbf{T}/d\phi = 0$. Moreover,

$$\kappa = \left|\frac{d\mathbf{T}}{ds}\right| = \left|\frac{d\mathbf{T}}{d\phi}\frac{d\phi}{ds}\right| = \left|\frac{d\mathbf{T}}{d\phi}\right|\left|\frac{d\phi}{ds}\right|$$

and consequently

$$\kappa = \left|\frac{d\phi}{ds}\right|$$

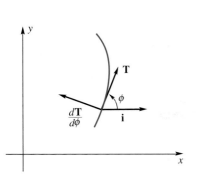

FIGURE 8

This formula for $\kappa$ helps our intuitive understanding of curvature (it measures the rate of change of $\phi$ with respect to $s$), and it also allows us to give a fairly simple proof of the following important theorem.

---

### Theorem A

Consider a curve with vector equation $\mathbf{r}(t) = f(t)\mathbf{i} + g(t)\mathbf{j}$, that is, with parametric equations $x = f(t)$ and $y = g(t)$. Then

$$\kappa = \frac{|x'y'' - y'x''|}{[x'^2 + y'^2]^{3/2}}$$

In particular, if the curve is the graph of $y = g(x)$, then

$$\kappa = \frac{|y''|}{[1 + y'^2]^{3/2}}$$

Primes indicate differentiation with respect to $t$ in the first formula and with respect to $x$ in the second formula.

***Proof***   We might calculate $\kappa$ directly from the formula $\kappa = |\mathbf{T}'(t)|/|\mathbf{r}'(t)|$, a task we propose in Problem 66. It is a good—but painful—exercise in differentiation and algebraic manipulation. Rather, we choose to use the formula $\kappa = |d\phi/ds|$ derived above. Refer to Figure 8, from which we see that

$$\tan \phi = \frac{dy}{dx} = \frac{dy/dt}{dx/dt} = \frac{y'}{x'}$$

Differentiate both sides of this equation with respect to $t$ to obtain

$$\sec^2\phi \frac{d\phi}{dt} = \frac{x'y'' - y'x''}{x'^2}$$

Then

$$\frac{d\phi}{dt} = \frac{x'y'' - y'x''}{x'^2 \sec^2\phi} = \frac{x'y'' - y'x''}{x'^2(1 + \tan^2\phi)}$$

$$= \frac{x'y'' - y'x''}{x'^2(1 + y'^2/x'^2)} = \frac{x'y'' - y'x''}{x'^2 + y'^2}$$

But

$$\kappa = \left|\frac{d\phi}{ds}\right| = \left|\frac{d\phi}{dt}\frac{dt}{ds}\right| = \left|\frac{d\phi}{dt} \div \frac{ds}{dt}\right| = \frac{|d\phi/dt|}{[x'^2 + y'^2]^{1/2}}$$

When we put these two results together, we obtain

$$\kappa = \frac{|x'y'' - y'x''|}{[x'^2 + y'^2]^{3/2}}$$

which is the first assertion of the theorem.

   To obtain the second assertion, simply regard $y = g(x)$ as corresponding to the parametric equations $x = t$, $y = g(t)$, so that $x' = 1$ and $x'' = 0$. The conclusion follows.   ∎

**EXAMPLE 4**   Find the curvature of the ellipse

$$x = 3 \cos t, \qquad y = 2 \sin t$$

at the points corresponding to $t = 0$ and $t = \pi/2$—that is, at $(3, 0)$ and $(0, 2)$. Sketch the ellipse showing the corresponding circles of curvature.

***Solution***   From the given equations,

$$x' = -3 \sin t \qquad y' = 2 \cos t$$
$$x'' = -3 \cos t \qquad y'' = -2 \sin t$$

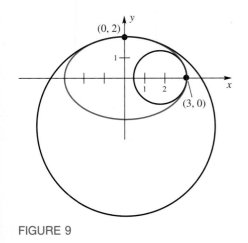

**FIGURE 9**

Thus,

$$\kappa = \kappa(t) = \frac{|x'y'' - y'x''|}{[x'^2 + y'^2]^{3/2}} = \frac{6 \sin^2 t + 6 \cos^2 t}{[9 \sin^2 t + 4 \cos^2 t]^{3/2}}$$

$$= \frac{6}{[5 \sin^2 t + 4]^{3/2}}$$

Consequently

$$\kappa(0) = \frac{6}{4^{3/2}} = \frac{3}{4}$$

$$\kappa\left(\frac{\pi}{2}\right) = \frac{6}{9^{3/2}} = \frac{2}{9}$$

Note that $\kappa(0)$ is larger than $\kappa(\pi/2)$, as it should be. Figure 9 shows the circle of curvature at $(3, 0)$, which has radius $\frac{4}{3}$, and the one at $(0, 2)$, which has radius $\frac{9}{2}$. ■

**EXAMPLE 5** Find the curvature of $y = \ln|\cos x|$ at $x = \pi/3$.

*Solution* We employ the second formula of Theorem A, noting that the primes now indicate differentiation with respect to $x$. Since $y' = \tan x$ and $y'' = \sec^2 x$,

$$\kappa = \frac{\sec^2 x}{(1 + \tan^2 x)^{3/2}} = \frac{\sec^2 x}{(\sec^2 x)^{3/2}} = |\cos x|$$

At $x = \pi/3$, $\kappa = \frac{1}{2}$. ■

**Normal and Tangential Components of Acceleration** Let $P = P(t)$ be a point on a smooth curve. Define a new vector $\mathbf{N} = \mathbf{N}(t)$, called the **unit normal vector** at $P$, by

$$\mathbf{N} = \frac{d\mathbf{T}/ds}{|d\mathbf{T}/ds|} = \frac{1}{\kappa} \frac{d\mathbf{T}}{ds}$$

so that

$$\boxed{\frac{d\mathbf{T}}{ds} = \kappa \mathbf{N}}$$

Now $\mathbf{T} = \cos \phi \mathbf{i} + \sin \phi \mathbf{j}$, and so

$$\frac{d\mathbf{T}}{ds} = \frac{d\mathbf{T}}{d\phi} \frac{d\phi}{ds} = (-\sin \phi \mathbf{i} + \cos \phi \mathbf{j}) \frac{d\phi}{ds}$$

from which it follows that $\mathbf{T} \cdot \mathbf{N} = 0$. Thus, $\mathbf{N}$ is a unit vector perpendicular to $\mathbf{T}$ and pointing (Problem 50) to the concave side of the curve.

We wish to express the acceleration vector $\mathbf{a}$ in terms of $\mathbf{T}$ and $\mathbf{N}$; that is, we want to decompose $\mathbf{a}$ into components tangent and normal to the curve. Since the velocity vector $\mathbf{v}$ satisfies

$$\mathbf{v} = |\mathbf{v}|\mathbf{T} = \frac{ds}{dt}\mathbf{T}$$

it follows that

$$\mathbf{a} = \frac{d\mathbf{v}}{dt} = \frac{d^2s}{dt^2}\mathbf{T} + \frac{ds}{dt}\frac{d\mathbf{T}}{dt}$$

$$= \frac{d^2s}{dt^2}\mathbf{T} + \frac{ds}{dt}\frac{d\mathbf{T}}{ds}\frac{ds}{dt}$$

Thus,

$$\mathbf{a} = \frac{d^2s}{dt^2}\mathbf{T} + \left(\frac{ds}{dt}\right)^2 \kappa\mathbf{N}$$

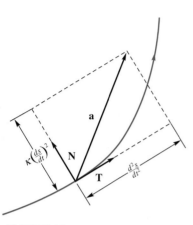

**FIGURE 10**

a result illustrated in Figure 10.

Let's interpret this result for the driver of a car who is pushing down on the accelerator as she travels along a straight road and then enters a curve. According to Newton's Law, force is mass times acceleration. Thus, the force on the car of mass $m$ is

$$\mathbf{F} = m\mathbf{a} = m\frac{d^2s}{dt^2}\mathbf{T} + m\kappa\left(\frac{ds}{dt}\right)^2\mathbf{N}$$

The driver of mass $m_0$ also experiences a force. On the straight road where $\kappa = 0$, she feels only a push from the seat against her body of magnitude $m_0 d^2s/dt^2$, this due to pushing down the accelerator. On entering the curve, she feels an additional inward push of magnitude $m_0\kappa(ds/dt)^2$. Note that this push is proportional both to the curvature and the square of the speed. The sharper the curve and the faster the driver is going, the stronger the push.

**EXAMPLE 6** A particle moves so that its position vector is $\mathbf{r}(t) = t^2\mathbf{i} + \frac{1}{3}t^3\mathbf{j}$, $t \geq 0$. Express $\mathbf{a}(t)$ in terms of $\mathbf{T}$ and $\mathbf{N}$, and evaluate when $t = 2$.

*Solution*

$$\mathbf{v}(t) = 2t\mathbf{i} + t^2\mathbf{j}$$

$$\frac{ds}{dt} = |\mathbf{v}(t)| = \sqrt{4t^2 + t^4} = t\sqrt{4 + t^2}$$

$$\frac{d^2s}{dt^2} = \frac{4 + 2t^2}{\sqrt{4 + t^2}}$$

$$\kappa \left(\frac{ds}{dt}\right)^2 = \frac{|x'y'' - y'x''|}{[x'^2 + y'^2]^{3/2}} \left(\frac{ds}{dt}\right)^2 = \frac{|x'y'' - y'x''|}{ds/dt}$$

$$= \frac{|(2t)(2t) - (t^2)(2)|}{t\sqrt{4 + t^2}} = \frac{2t^2}{t\sqrt{4 + t^2}} = \frac{2t}{\sqrt{4 + t^2}}$$

Thus,

$$\mathbf{a}(t) = \frac{d^2s}{dt^2}\mathbf{T} + \kappa \left(\frac{ds}{dt}\right)^2 \mathbf{N}$$

$$= \frac{4 + 2t^2}{\sqrt{4 + t^2}}\mathbf{T} + \frac{2t}{\sqrt{4 + t^2}}\mathbf{N}$$

and

$$\mathbf{a}(2) = \frac{12}{\sqrt{8}}\mathbf{T} + \frac{4}{\sqrt{8}}\mathbf{N} = 3\sqrt{2}\mathbf{T} + \sqrt{2}\mathbf{N} \qquad \blacksquare$$

We have expressed **a** in the form

$$\mathbf{a} = a_T\mathbf{T} + a_N\mathbf{N}$$

where the tangent and normal components are

$$a_T = \frac{d^2s}{dt^2} \quad \text{and} \quad a_N = \kappa \left(\frac{ds}{dt}\right)^2$$

To calculate $a_N$, it appears that we must calculate $\kappa$. However this can be avoided by noting that **T** and **N** are perpendicular and, therefore,

$$\boxed{|\mathbf{a}|^2 = a_T^2 + a_N^2}$$

We illustrate.

**EXAMPLE 7** Without finding $\kappa$, find $a_T$ and $a_N$ for the motion given by $\mathbf{r} = t^2\mathbf{i} + \frac{1}{3}t^3\mathbf{j}, t \geq 0$.

*Solution* The motion is the same as that of Example 6, and so our calculation of $a_T$ follows the lines of that example.

$$\mathbf{v} = 2t\mathbf{i} + t^2\mathbf{j}$$

$$\mathbf{a} = 2\mathbf{i} + 2t\mathbf{j}$$

$$\frac{ds}{dt} = |\mathbf{v}| = t\sqrt{4 + t^2}$$

$$a_T = \frac{d^2s}{dt^2} = \frac{4 + 2t^2}{\sqrt{4 + t^2}}$$

By the preceding boxed formula,

$$a_N^2 = |\mathbf{a}|^2 - a_T^2 = 4 + 4t^2 - \frac{(4 + 2t^2)^2}{4 + t^2} = \frac{4t^2}{4 + t^2}$$

$$a_N = \frac{2t}{\sqrt{4 + t^2}}$$

■

## CONCEPTS REVIEW

1. If the unit tangent vector $\mathbf{T}$ is considered to be a function of arc length $s$, then we may define the curvature $\kappa$ as $\kappa = $ _____.

2. The curvature of a circle of radius $a$ is constant and has value $\kappa = $ _____. The curvature of a line is _____.

3. The radius of curvature $R$ is related to the curvature $\kappa$ by _____.

4. The acceleration vector $\mathbf{a}$ can be expressed in terms of the unit tangent vector $\mathbf{T}$ and the unit normal vector $\mathbf{N}$ by $\mathbf{a} = $ _____.

## PROBLEM SET 13.5

In Problems 1–12, find the unit tangent vector $\mathbf{T}(t)$ and the curvature $\kappa(t)$ at the point where $t = t_1$. For calculating $\kappa$, we suggest using Theorem A, as in Example 4.

1. $\mathbf{r}(t) = 3t\mathbf{i} + 3t^2\mathbf{j}$; $t_1 = \frac{1}{3}$

2. $\mathbf{s}(t) = 3t^2\mathbf{i} + 3t\mathbf{j}$; $t_1 = \frac{1}{3}$

3. $\mathbf{u}(t) = 4t^2\mathbf{i} + 4t\mathbf{j}$; $t_1 = \frac{1}{2}$

4. $\mathbf{r}(t) = \frac{1}{3}t^3\mathbf{i} + \frac{1}{2}t^2\mathbf{j}$; $t_1 = 1$

5. $\mathbf{z}(t) = 3\cos t\mathbf{i} + 4\sin t\mathbf{j}$; $t_1 = \frac{\pi}{4}$

6. $\mathbf{r}(t) = e^t\mathbf{i} + e^t\mathbf{j}$; $t_1 = \ln 2$

7. $\mathbf{r}(t) = e^t\cos t\mathbf{i} + e^t\sin t\mathbf{j}$; $t_1 = \frac{\pi}{2}$

8. $\mathbf{r}(t) = e^{-3t}\mathbf{i} + e^t\mathbf{j}$; $t_1 = 0$

9. $x(t) = 1 - t^2$, $y(t) = 1 - t^3$; $t_1 = 1$

10. $x(t) = \sinh t$, $y(t) = \cosh t$; $t_1 = \ln 3$

11. $x(t) = e^{-t}\cos t$, $y(t) = e^{-t}\sin t$; $t_1 = 0$

12. $\mathbf{r}(t) = t\cos t\mathbf{i} + t\sin t\mathbf{j}$; $t_1 = 1$

In Problems 13–28, sketch the curve. Then for the given point, find the curvature and the radius of curvature. Finally draw the circle of curvature at the point. *Hint*: For the curvature, you will use the second formula in Theorem A, as in Example 5.

13. $y = 2x^2$, $(1, 2)$

14. $y = x(x - 4)^2$, $(4, 0)$

15. $y = \sin x$, $\left(\frac{\pi}{4}, \frac{\sqrt{2}}{2}\right)$

16. $y^2 = x - 1$, $(1, 0)$

17. $y = 2\ln x$, $(1, 0)$

18. $y = e^x - x$, $(0, 1)$

19. $y^2 - 4x^2 = 20$, $(2, 6)$    20. $y^2 - 4x^2 = 20$, $(2, -6)$

21. $y = \cos 2x$, $\left(\frac{1}{6}\pi, \frac{1}{2}\right)$    22. $y = \cosh \frac{1}{2}x$, $(0, 1)$

23. $y = \ln \sin x$, $\left(\frac{1}{4}\pi, -\ln\sqrt{2}\right)$

24. $y = e^{-x^2}$, $\left(1, \frac{1}{e}\right)$    25. $y = \tan x$, $(\pi/4, 1)$

26. $y = \sqrt{x}$, $(1, 1)$    27. $y = \sqrt[3]{x}$, $(1, 1)$

28. $y = \tanh x$, $\left(\ln 2, \frac{6}{10}\right)$

In Problems 29–34, find the point of the curve at which the curvature is a maximum.

29. $y = \ln x$    30. $y = \sin x$; $-\pi \le x \le \pi$

31. $y = \cosh x$    32. $y = \sinh x$

33. $y = e^x$

34. $y = \ln \cos x$ for $-\pi/2 < x < \pi/2$

In Problems 35–43, find the tangential and normal components ($a_T$ and $a_N$) of the acceleration vector at $t$. Then evaluate at $t = t_1$. See Examples 6 and 7.

35. $\mathbf{r}(t) = 3t\mathbf{i} + 3t^2\mathbf{j}$; $t_1 = \frac{1}{3}$

36. $\mathbf{r}(t) = t^2\mathbf{i} + t\mathbf{j}$; $t_1 = 1$

37. $\mathbf{r}(t) = (2t + 1)\mathbf{i} + (t^2 - 2)\mathbf{j}$; $t_1 = -1$

38. $\mathbf{r}(t) = a\cos t\mathbf{i} + a\sin t\mathbf{j}$; $t_1 = \frac{\pi}{6}$

39. $\mathbf{r}(t) = a\cosh t\mathbf{i} + a\sinh t\mathbf{j}$; $t_1 = \ln 3$

40. $\mathbf{r}(t) = \cos^3 t\mathbf{i} + \sin^3 t\mathbf{j}$; $t_1 = \frac{\pi}{3}$

41. $\mathbf{r}(t) = e^t\mathbf{i} + e^{-t}\mathbf{j}$; $t_1 = 0$

42. $x(t) = 1 + 3t$, $y(t) = 2 - 6t$; $t_1 = 2$

43. $x(t) = e^t\sin t$, $y(t) = e^t\cos t$; $t_1 = \frac{\pi}{3}$

**44.** Demonstrate that the second formula in Theorem A can also be written as $\kappa = |y'' \cos^3 \phi|$, where $\phi$ is the angle of inclination of the tangent line to the graph of $y = f(x)$.

**45.** Deduce from the definition

$$\mathbf{N} = \frac{d\mathbf{T}/ds}{|d\mathbf{T}/ds|}$$

the formula for $\mathbf{N}$ in terms of an arbitrary parameter $t$:

$$\mathbf{N} = \frac{d\mathbf{T}/dt}{|d\mathbf{T}/dt|}$$

**46.** Use the formula of Problem 45 to find $\mathbf{N}$ for the ellipse:

$$\mathbf{r}(t) = a \cos \omega t \mathbf{i} + b \sin \omega t \mathbf{j}$$

**47.** Sketch the path for a particle if its position vector is $\mathbf{r} = \sin t \mathbf{i} + \sin 2t \mathbf{j}$, $0 \le t \le 2\pi$ (you should get a figure eight). Where is the acceleration zero? Where does the acceleration vector point to the origin?

**48.** The position vector of a particle at time $t \ge 0$ is

$$\mathbf{r}(t) = (\cos t + t \sin t)\mathbf{i} + (\sin t - t \cos t)\mathbf{j}$$

(a) Show that the speed $ds/dt = t$.
(b) Show that $a_T = 1$ and $a_N = t$.

**49.** If for a particle, $a_T = 0$ for all $t$, what can you conclude about its speed? If $a_N = 0$ for all $t$, what can you conclude about its curvature?

**50.** Show that $\mathbf{N}$ points to the concave side of the curve. *Hint*: One method is to show

$$\mathbf{N} = (-\sin \phi \mathbf{i} + \cos \phi \mathbf{j}) \frac{d\phi/ds}{|d\phi/ds|}$$

Then consider the cases $d\phi/ds > 0$ (curve bends to the left) and $d\phi/ds < 0$ (curve bends to the right).

**51.** Show that $\mathbf{N}$ is perpendicular to $\mathbf{T}$ by differentiating $\mathbf{T} \cdot \mathbf{T} = 1$ with respect to $s$.

**52.** An object moves along the curve $y = \sin 2x$. Without doing any calculating, decide where $a_N = 0$.

**53.** A dog is running counterclockwise around the circle $x^2 + y^2 = 400$ (distances in feet). At the point $(-12, 16)$, it is running at 10 feet per second and is speeding up at 5 feet per second per second. Express its acceleration $\mathbf{a}$ at the point first in terms of $\mathbf{T}$ and $\mathbf{N}$ and then in terms of $\mathbf{i}$ and $\mathbf{j}$.

**54.** An object moves along the parabola $y = x^2$ with constant speed of 4. Express $\mathbf{a}$ at the point $(x, x^2)$ in terms of $\mathbf{T}$ and $\mathbf{N}$.

**55.** A car traveling at constant speed $v$ rounds a level curve, which we take to be a circle of radius $R$. If the car is to avoid sliding outward, the horizontal frictional force $F$ exerted by the road on the tires must at least balance the centrifugal force pulling outward. The force $F$ satisfies $F = \mu mg$, where $\mu$ is the *coefficient of friction*, $m$ is the mass of the car, and $g$ is the acceleration of gravity. Thus, $\mu mg \ge mv^2/R$. Show that $v_R$, the speed beyond which skidding will occur, satisfies

$$v_R = \sqrt{\mu g R}$$

and use this to determine $v_R$ for a curve with $R = 400$ feet and $\mu = 0.4$. Use $g = 32$ feet per second per second.

**56.** Consider again the car of Problem 55. Suppose the curve is icy at its worst spot ($\mu = 0$) but is banked at angle $\theta$ from the horizontal (Figure 11). Let $\mathbf{F}$ be the force exerted by the road on the car. Then at the critical speed $v_R$, $mg = |\mathbf{F}|\cos \theta$ and $mv_R^2/R = |\mathbf{F}|\sin \theta$.

(a) Show that $v_R = \sqrt{Rg \tan \theta}$.
(b) Find $v_R$ for a curve with $R = 400$ feet and $\theta = 10°$.

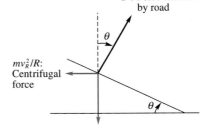

$mg$: Weight of car

FIGURE 11

**57.** Derive the polar coordinate curvature formula

$$\kappa = \frac{|r^2 + 2r'^2 - rr''|}{(r^2 + r'^2)^{3/2}}$$

where the derivatives are with respect to $\theta$.

In Problems 58–63, use the formula in Problem 57 to find the curvature $\kappa$

**58.** The circle $r = 4 \cos \theta$

**59.** The cardioid $r = 1 + \cos \theta$ at $\theta = 0$

**60.** $r = \theta$ at $\theta = 1$

**61.** $r = 4(1 + \cos\theta)$ at $\theta = \pi/2$

**62.** $r = e^{3\theta}$ at $\theta = 1$

**63.** $r = 4(1 + \sin\theta)$ at $\theta = \pi/2$

**64.** Show that the curvature of the polar curve $r = e^{6\theta}$ is proportional to $1/r$.

**65.** Show that the curvature of the polar curve $r^2 = \cos 2\theta$ is directly proportional to $r$ for $r > 0$.

**66.** Derive the first curvature formula in Theorem A by working directly with $\kappa = |\mathbf{T}'(t)|/|\mathbf{r}'(t)|$.

PC **67.** Draw the graph of $x = 4 \cos t$, $y = 3 \sin(t + 0.5)$, $0 \le t \le 2\pi$. Estimate its maximum and minimum curvature by looking at the graph (curvature is the reciprocal of the radius of curvature). Then use a max/min program to find these two numbers (accurate to four decimal places).

PC **68.** Draw the graph of $x = 2.5 \cos t + 0.5 \cos(t - 2\theta)$, $y = 2.5 \sin t - 0.5 \sin(t - 2\theta)$, $0 \le t \le 2\pi$, for $\theta = 0$, $\theta = \pi/6$, $\theta = \pi/3$, $\theta = \pi/2$, and $\theta = 3\pi/4$. Make a conjecture about the shape of these curves. Use this conjecture and Example 4 to determine the maximum and minimum curvature.

PC **69.** Generalize Problem 68 by drawing graphs for various values of $a$, $b$, and $\theta$ of

$$x = \frac{a + b}{2} \cos t + \frac{a - b}{2} \cos(t - 2\theta),$$

$$y = \frac{a + b}{2} \sin t - \frac{a - b}{2} \sin(t - 2\theta)$$

$0 \le t \le 2\pi$. Make a conjecture about the shape of these graphs. Can you prove your conjecture? (See Problem 40 of Section 12.2 for one idea.)

---

**Answers to Concepts Review:** **1.** $|d\mathbf{T}/ds|$ **2.** $1/a$; 0
**3.** $\kappa = 1/R$ **4.** $(d^2s/dt^2)\mathbf{T} + (ds/dt)^2\kappa\mathbf{N}$

---

## 13.6 CHAPTER REVIEW

### Concepts Test

Respond with true or false to each of the following assertions. Be prepared to justify your answer.

**1.** The parametric representation of a curve is unique.

**2.** The graph of $x = 2t^3$, $y = t^3$ is a line.

**3.** If $x = f(t)$ and $y = g(t)$, then we can find a function $h$ such that $y = h(x)$.

**4.** The curve with parametric representation $x = \ln t$ and $y = t^2 - 1$ passes through the origin.

**5.** If $x = f(t)$ and $y = g(t)$ and if both $f''$ and $g''$ exist, then $d^2y/dx^2 = g''(t)/f''(t)$ wherever $f''(t) \ne 0$.

**6.** A curve may have more than one tangent line at a point on the curve.

**7.** The vectors $2\mathbf{i} - 3\mathbf{j}$ and $6\mathbf{i} + 4\mathbf{j}$ are perpendicular.

**8.** If $\mathbf{u}$ and $\mathbf{v}$ are unit vectors, then the angle $\theta$ between them satisfies $\cos \theta = \mathbf{u} \cdot \mathbf{v}$.

**9.** The dot product for vectors satisfies the associative law.

**10.** If $\mathbf{u}$ and $\mathbf{v}$ are any two vectors, then $|\mathbf{u} \cdot \mathbf{v}| \le |\mathbf{u}| \, |\mathbf{v}|$.

**11.** $|\mathbf{u} \cdot \mathbf{v}| = |\mathbf{u}| \, |\mathbf{v}|$ for nonzero vectors $\mathbf{u}$ and $\mathbf{v}$ if and only if $\mathbf{u}$ is a scalar multiple of $\mathbf{v}$.

**12.** If $|\mathbf{u}| = |\mathbf{v}| = |\mathbf{u} + \mathbf{v}|$, then $\mathbf{u} = \mathbf{v} = \mathbf{0}$.

**13.** If $\mathbf{u} + \mathbf{v}$ and $\mathbf{u} - \mathbf{v}$ are perpendicular, then $|\mathbf{u}| = |\mathbf{v}|$.

**14.** For any two vectors $\mathbf{u}$ and $\mathbf{v}$, $|\mathbf{u} + \mathbf{v}|^2 = |\mathbf{u}|^2 + |\mathbf{v}|^2 + 2\mathbf{u} \cdot \mathbf{v}$.

**15.** The vector-valued function $\langle f(t), g(t) \rangle$ is continuous at $t = a$ if and only if both $f$ and $g$ are continuous at $t = a$.

**16.** $D_t[\mathbf{F}(t) \cdot \mathbf{F}(t)] = 2\mathbf{F}(t) \cdot \mathbf{F}'(t)$.

**17.** The curvature of the curve determined by $x = 3t + 4$ and $y = 2t - 1$ is zero for all $t$.

**18.** The curvature of the curve determined by $x = 2 \cos t$ and $y = 2 \sin t$ is 2 for all $t$.

**19.** If $\mathbf{T} = \mathbf{T}(t)$ is a unit vector tangent to a smooth curve, then $\mathbf{T}(t)$ and $\mathbf{T}'(t)$ are perpendicular.

**20.** If $v = |\mathbf{v}|$ is the speed of a particle moving along a smooth curve, then $|dv/dt|$ is the magnitude of the acceleration.

**21.** If $y = f(x)$ and $y'' = 0$ everywhere, the curvature of this curve is zero.

**22.** If $y = f(x)$ and $y''$ is a constant, then the curvature of this curve is a constant.

**23.** If $\mathbf{u} \cdot \mathbf{v} = 0$, then either $\mathbf{u} = \mathbf{0}$ or $\mathbf{v} = \mathbf{0}$, or both $\mathbf{u}$ and $\mathbf{v}$ are $\mathbf{0}$.

**24.** If $|\mathbf{r}(t)| = 1$ for all $t$, then $|\mathbf{r}'(t)| = $ constant.

**25.** If $\mathbf{v} \cdot \mathbf{v} = $ constant, then $\mathbf{v} \cdot \mathbf{v}' = 0$.

## Sample Test Problems

In Problems 1–4, a parametric representation of a curve is given. Eliminate the parameter to obtain the corresponding Cartesian equation. Sketch the given curve.

1. $x = 6t + 2$, $y = 2t$; $-\infty < t < \infty$

2. $x = 4t^2$, $y = 4t$; $-1 \le t \le 2$

3. $x = 4 \sin t - 2$, $y = 3 \cos t + 1$; $0 \le t \le 2\pi$

4. $x = 2 \sec t$, $y = \tan t$; $-\dfrac{\pi}{2} < t < \dfrac{\pi}{2}$

In Problems 5 and 6, find the equation of the tangent line and the normal line at $t = 0$.

5. $x = 2t^3 - 4t + 7$, $y = t + \ln(t + 1)$

6. $x = 3e^{-t}$, $y = \frac{1}{2}e^t$

7. Find the length of the curve
$$x = \cos t + t \sin t$$
$$y = \sin t - t \cos t$$
from 0 to $2\pi$. Make a sketch.

8. Find $\mathbf{u}$ and $|\mathbf{u}|$ if $\mathbf{u}$ is the vector from $P_1$ to $P_2$.

(a) $P_1 = (2, 4)$, $P_2 = (-1, 5)$
(b) $P_1 = (-3, 0)$, $P_2 = (-4, 5)$

9. Let $\mathbf{a} = \langle 2, -5 \rangle$, $\mathbf{b} = \langle 1, 1 \rangle$, and $\mathbf{c} = \langle -6, 0 \rangle$. Find each of the following.

(a) $3\mathbf{a} - 2\mathbf{b}$
(b) $\mathbf{a} \cdot \mathbf{b}$
(c) $\mathbf{a} \cdot (\mathbf{b} + \mathbf{c})$
(d) $(4\mathbf{a} + 5\mathbf{b}) \cdot 3\mathbf{c}$
(e) $|\mathbf{c}|\mathbf{c} \cdot \mathbf{b}$
(f) $\mathbf{c} \cdot \mathbf{c} - |\mathbf{c}|$

10. Find the cosine of the angle between $\mathbf{a}$ and $\mathbf{b}$ and make a sketch.

(a) $\mathbf{a} = 3\mathbf{i} + 2\mathbf{j}$, $\mathbf{b} = -\mathbf{i} + 4\mathbf{j}$
(b) $\mathbf{a} = -5\mathbf{i} - 3\mathbf{j}$, $\mathbf{b} = 2\mathbf{i} - \mathbf{j}$
(c) $\mathbf{a} = \langle 7, 0 \rangle$, $\mathbf{b} = \langle 5, 1 \rangle$

11. Given $\mathbf{a} = -2\mathbf{i}$ and $\mathbf{b} = 3\mathbf{i} - 2\mathbf{j}$ and another vector $\mathbf{r} = 5\mathbf{i} - 4\mathbf{j}$, find scalars $k$ and $m$ such that $\mathbf{r} = k\mathbf{a} + m\mathbf{b}$.

12. Find a vector of length 3 which is parallel to the tangent line to $y = x^2$ at $(-1, 1)$.

13. Find a vector of length 10 that makes an angle of $150°$ with the positive $x$-axis.

14. Two forces $\mathbf{F}_1 = 2\mathbf{i} - 3\mathbf{j}$ and $\mathbf{F}_2 = 3\mathbf{i} + 12\mathbf{j}$ are applied at a point. What force $\mathbf{F}$ must be applied at the point to counteract the resultant of these two forces?

15. What heading and air speed are required for a plane to fly 450 miles per hour due north if a wind of 100 miles per hour is blowing in the direction N 60° E?

16. If $\mathbf{r}(t) = \langle e^{2t}, e^{-t} \rangle$ find each of the following.

(a) $\displaystyle \lim_{t \to 0} \mathbf{r}(t)$
(b) $\displaystyle \lim_{h \to 0} \frac{\mathbf{r}(0 + h) - \mathbf{r}(0)}{h}$
(c) $\displaystyle \int_0^{\ln 2} \mathbf{r}(t)\, dt$
(d) $D_t[t\mathbf{r}(t)]$
(e) $D_t[\mathbf{r}(3t + 10)]$
(f) $D_t[\mathbf{r}(t) \cdot \mathbf{r}'(t)]$

17. Find $\mathbf{r}'(t)$ and $\mathbf{r}''(t)$ for each of the following.

(a) $\mathbf{r}(t) = (\ln t)\mathbf{i} - 3t^2\mathbf{j}$
(b) $\mathbf{r}(t) = \sin t\mathbf{i} + \cos 2t\mathbf{j}$
(c) $\mathbf{r}(t) = \tan t\mathbf{i} - t^4\mathbf{j}$

18. Find the length of the arc $\mathbf{r}(t) = 4t^{3/2}\mathbf{i} + 3t\mathbf{j}$ from $t = 0$ to $t = 2$.

In Problems 19 and 20, the position of a moving particle at time $t$ is given by $\mathbf{r}(t)$. Find the velocity and acceleration vectors, $\mathbf{v}(t)$ and $\mathbf{a}(t)$, and their values at the given time $t = t_1$. Also, find the speed of the particle for $t = t_1$.

19. $\mathbf{r}(t) = 2t^2\mathbf{i} + (4t + 2)\mathbf{j}$; $t_1 = -1$

20. $\mathbf{r}(t) = 4(1 - \sin t)\mathbf{i} + 4(t + \cos t)\mathbf{j}$; $t_1 = \frac{2}{3}\pi$.

21. Find the curvature $\kappa$ of the given curve at $P$.

(a) $y = x^2 - x$ at $P(1, 0)$.
(b) $\mathbf{r}(t) = (t + t^3)\mathbf{i} + (t + t^2)\mathbf{j}$ at $P(2, 2)$
(c) $y = a \cosh(x/a)$ at $P(a, a \cosh 1)$

22. Find the unit tangent vector $\mathbf{T}(t)$ for the curve $\mathbf{r}(t) = t\mathbf{i} + \frac{1}{3}t^3\mathbf{j}$. At the point $P(1)$ on the curve where $t = 1$, find $\mathbf{T}(1)$. Find the curvature $\kappa(1)$ of the curve at $P(1)$. Sketch the curve and draw the unit tangent vector $\mathbf{T}(1)$ with its initial point at $P(1)$.

23. If $\mathbf{r}(t) = (1 - t^2)\mathbf{i} + 2t\mathbf{j}$, find the tangential and the normal components, $a_T$ and $a_N$, of the acceleration $\mathbf{a}$ at $P(0, 2)$.

# TECHNOLOGY PROJECT

The equations $x = \cos t$, $y = \sin t$ with the parameter $t$ in the range $[0, 2\pi]$ give a familiar parametric representation of the unit circle (note that $x^2 + y^2 = 1$). Here is a *Maple* code to get a plot of a parametric form:

```
x := t → cos(t):
y := t → sin(t):
plot( [x(t), y(t), t = 0..2*Pi] );
```

The parametrization of the unit circle given above leads to a transversal of the circle with unit speed:

$$v(t) = \sqrt{(x'(t))^2 + (y'(t))^2}$$
$$= \sqrt{(-\sin t)^2 + (\cos t)^2}$$
$$= 1$$

**Exercise 1**   Show that the parametric equations $x(t) = (1 - t^2)/(1 + t^2)$, $y(t) = 2t/(1 + t^2)$ give an alternate representation of the unit circle (technically, the point $(-1, 0)$ is "missing"). Make a table showing the values of $x$, $y$, and the speed $v$ for the points $t = -5, -4, \ldots, 3, 4,$ and 5. Describe in words how the speed varies as the point moves around the unit circle. Discuss the "missing point" in terms of limits.

**The Hypocycloid.**   The *hypocycloid* is the curve generated by a circle of radius $b$ rolling on the inside of a larger circle of radius $a$, see Figure 1. The equations of

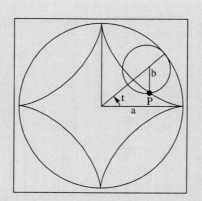

FIG. 1.  A hypocycloid with $b/a = 1/4$.

the hypocycloid are given parametrically as:

$$x(t) = (a - b)\cos t + b\cos\frac{a - b}{b}t$$
$$y(t) = (a - b)\sin t - b\sin\frac{a - b}{b}t, \quad a > b$$

We could factor out the scaling factor $a$ in each equation and study the family of hypocycloids in terms of the single parameter $b/a$. Alternately, we just set $a = 1$ but label our plots in terms of the ratio. For example, see Figure 2.

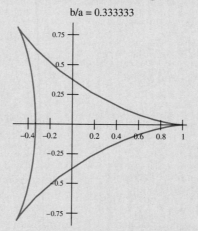

b/a = 0.333333

FIG. 2.  A hypocycloid with $b/a = 1/3$.

**Exercise 2**   Do graphical experiments with the above code to determine the period of the hypocycloid for the case when $a = 1$ and $b = 1/k$, with $k = 3, 4, 5, \ldots$ Write a sentence describing the nature of the graph.

**Exercise 3**   Continuing the previous problem, describe what happens for $a = 1$, $b = 1/k$, as $k \to \infty$.

**Exercise 4**   Determine the arclength of the limiting curve in the previous problem. **Hint**: Most approaches go better if you get the arclength for one "lobe" and multiply by the number of lobes.

**Exercise 5**   Do graphical experiments with the given hypocycloid code to determine its period when $a = 1$ and $b = j/16$, for $j = 1, 3, 5, \ldots, 15$. Write a paragraph describing what you observe about this sequence of plots.

**Exercise 6**   Perform experiments to determine the period of the hypocycloid for $a = 1$ and $b = j/k$ where $j$ and $k$ are integers with $j < k$.

**Exercise 7**   Hypocycloid pairs corresponding to $j/k$'s given by 1/3 and 2/3, or 1/4 and 3/4, or in general $1/k$ and $1 - 1/k$ have the same graph, but there is a difference. What is it? Be as quantitative as possible.

# Geometry in Space, Vectors

## 14.1
## CARTESIAN COORDINATES IN THREE-SPACE

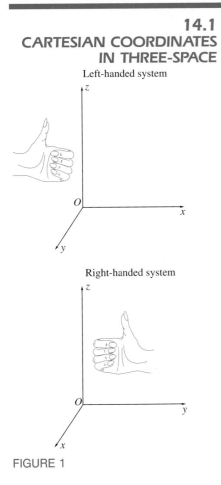

Left-handed system

Right-handed system

FIGURE 1

We have reached an important transition point in our study of calculus. Until now, we have been traveling across that broad flat expanse known as the Euclidean plane, or two-space. The concepts of calculus have been applied to functions of a single variable, functions whose graphs can be drawn in the plane.

The mountains lie ahead. Our charted course winds through three-space and occasionally into *n*-space. We are going to study *multiple variable calculus*, the calculus that applies to functions of two or more variables. All the familiar ideas (such as limit, derivative, integral) are to be explored again from a loftier perspective.

To begin, consider three mutually perpendicular coordinate lines (the *x*-, *y*-, and *z*-axes) with their zero points at a common point $O$, called the *origin*. Although these lines can be oriented in any way one pleases, we follow a standard custom in thinking of the *y*- and *z*-axes as lying in the plane of the paper with their positive directions to the right and upward, respectively. The *x*-axis is then perpendicular to the paper, and we suppose its positive end to point toward us, thus forming a **right-handed system.** We call it right-handed because if the fingers of the right hand are cupped so they curve from the positive *x*-axis toward the positive *y*-axis, the thumb points in the direction of the positive *z*-axis (Figure 1).

The three axes determine three planes, the *yz*-, *xz*-, and *xy*-planes, which divide space into eight octants (Figure 2). To each point $P$ in space corresponds an ordered triple of numbers $(x, y, z)$, its **Cartesian coordinates**, which measure its directed distances from the three planes (Figure 3).

Plotting points in the first octant (the octant where all three coordinates are positive) is relatively easy. In Figures 4 and 5, we illustrate something more difficult by plotting two points from other octants, the points $P(2, -3, 4)$ and $Q(-3, 2, -5)$.

667

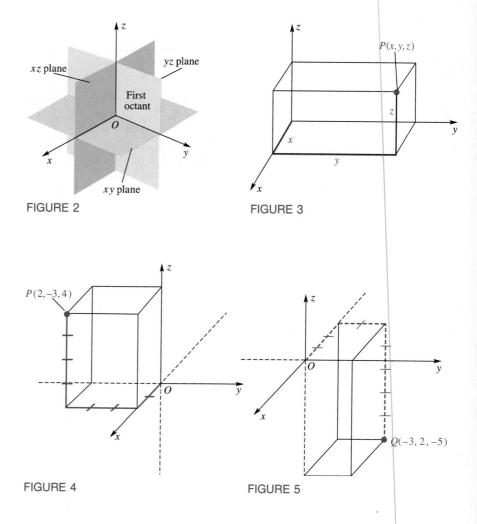

FIGURE 2

FIGURE 3

FIGURE 4

FIGURE 5

**The Distance Formula** Consider two points $P_1(x_1, y_1, z_1)$ and $P_2(x_2, y_2, z_2)$ in three-space ($x_1 \neq x_2$, $y_1 \neq y_2$, $z_1 \neq z_2$). They determine a rectangular box (parallelepiped), with $P_1$ and $P_2$ as opposite vertices and with edges parallel to the coordinate axes (Figure 6). The triangles $P_1QP_2$ and $P_1RQ$ are right triangles, and—by the Pythagorean Theorem—

$$|P_1P_2|^2 = |P_1Q|^2 + |QP_2|^2$$

and

$$|P_1Q|^2 = |P_1R|^2 + |RQ|^2$$

Thus,

$$|P_1P_2|^2 = |P_1R|^2 + |RQ|^2 + |QP_2|^2$$
$$= (x_2 - x_1)^2 + (y_2 - y_1)^2 + (z_1 - z_2)^2$$

FIGURE 6

This gives us the **Distance Formula** in three-space.

$$|P_1P_2| = \sqrt{(x_2 - x_1)^2 + (y_2 - y_1)^2 + (z_2 - z_1)^2}$$

The formula is correct even if some of the coordinates are identical.

**EXAMPLE 1**  Find the distance between the points $P(2, -3, 4)$ and $Q(-3, 2, -5)$, which were plotted in Figures 4 and 5.

*Solution*

$$|PQ| = \sqrt{(-3 - 2)^2 + (2 + 3)^2 + (-5 - 4)^2} = \sqrt{131} \approx 11.45 \quad \blacksquare$$

**Spheres and Their Equations**   It is a small step from the distance formula to the equation of a sphere. By a **sphere**, we mean the set of all points at constant distance (the radius) from a fixed point (the center). In fact, if $(x, y, z)$ is a point on the sphere of radius $r$ centered at $(h, k, l)$, then (see Figure 7)

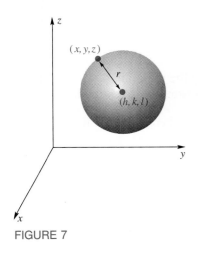

FIGURE 7

$$(x - h)^2 + (y - k)^2 + (z - l)^2 = r^2$$

We call this the **standard equation of a sphere**.

In expanded form, the boxed equation may be written as

$$x^2 + y^2 + z^2 + Gx + Hy + Iz + J = 0$$

Conversely, the graph of any equation of this form is either a sphere, a point (a degenerate sphere), or the empty set. To see why, consider the following example.

**EXAMPLE 2**  Find the center and radius of the sphere with equation

$$x^2 + y^2 + z^2 - 10x - 8y - 12z + 68 = 0$$

and sketch its graph.

*Solution*   We use the process of completing the square.

$$(x^2 - 10x +\quad) + (y^2 - 8y +\quad) + (z^2 - 12z +\quad) = -68$$
$$(x^2 - 10x + 25) + (y^2 - 8y + 16) + (z^2 - 12z + 36) =$$
$$-68 + 25 + 16 + 36$$
$$(x - 5)^2 + (y - 4)^2 + (z - 6)^2 = 9$$

Thus, the equation represents a sphere with center at $(5, 4, 6)$ and radius 3. Its graph is shown in Figure 8. $\quad \blacksquare$

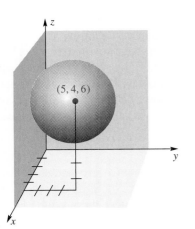

FIGURE 8

If, after completing the square in Example 2, the equation had been

$$(x - 5)^2 + (y - 4)^2 + (z - 6)^2 = 0$$

then the graph would be the single point (5, 4, 6); if the right side were negative, the graph would be the empty set.

Another simple result that follows from the Distance Formula is the **Midpoint Formula**. If $P_1(x_1, y_1, z_1)$ and $P_2(x_2, y_2, z_2)$ are endpoints of a line segment, then the midpoint $M(m_1, m_2, m_3)$ has coordinates

$$m_1 = \frac{x_1 + x_2}{2} \qquad m_2 = \frac{y_1 + y_2}{2} \qquad m_3 = \frac{z_1 + z_2}{2}$$

In other words, to find the coordinates of the midpoint of a segment, simply take the average of corresponding coordinates of the endpoints.

**EXAMPLE 3** Find the equation of the sphere that has the line segment joining $(-1, 2, 3)$ and $(5, -2, 7)$ as a diameter (Figure 9).

*Solution* The center of this sphere is at the midpoint of the segment, namely, at (2, 0, 5); the radius $r$ satisfies

$$r^2 = (5 - 2)^2 + (-2 - 0)^2 + (7 - 5)^2 = 17$$

We conclude that the equation of the sphere is

$$(x - 2)^2 + y^2 + (z - 5)^2 = 17 \qquad\blacksquare$$

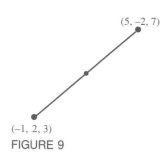

(5, –2, 7)

(–1, 2, 3)

FIGURE 9

**Graphs in Three-Space** It was natural to consider a quadratic equation first because of its relation to the distance formula. But presumably a **linear equation** in $x$, $y$, and $z$—that is, an equation of the form

$$Ax + By + Cz = D, \qquad A^2 + B^2 + C^2 \neq 0$$

should be even easier to analyze. As a matter of fact we will show in the next section that the graph of a linear equation is a plane. Taking this for granted for now, let's consider how we might graph such an equation.

If, as will often be the case, the plane intersects the three axes, we begin by finding these intersection points; that is, we find the $x$-, $y$-, and $z$-intercepts. These three points determine the plane and allow us to draw the (coordinate-plane) **traces**, which are the lines of intersection of that plane with the coordinate planes. Then with just a bit of artistry, we can shade in the plane.

**EXAMPLE 4** Sketch the graph of $3x + 4y + 2z = 12$.

*Solution* To find the $x$-intercept, set $y$ and $z$ equal to zero and solve for $x$, obtaining $x = 4$. The corresponding point is (4, 0, 0). Similarly, the $y$- and $z$-intercepts are (0, 3, 0) and (0, 0, 6). Next connect these points by line segments to get the traces. Then shade in (the first octant part of) the plane, thereby obtaining the result shown in Figure 10. $\blacksquare$

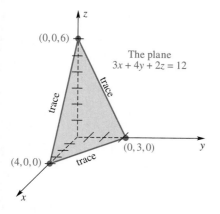

FIGURE 10

What if the plane does not intersect all three axes? This will happen, for example, if one of the variables in the equation of the plane is missing (that is, has a zero coefficient).

**EXAMPLE 5** Sketch the graph of the linear equation

$$2x + 3y = 6$$

in three-space.

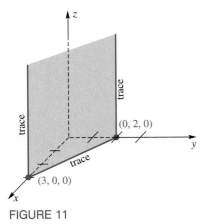

FIGURE 11

*Solution* The $x$- and $y$-intercepts are $(3, 0, 0)$ and $(0, 2, 0)$, respectively, and these points determine the trace in the $xy$-plane. The plane never crosses the $z$-axis ($x$ and $y$ cannot both be 0), and so the plane is parallel to the $z$-axis. We have sketched the graph in Figure 11. ■

Notice that in each of our examples, the graph of an equation in three-space was a surface. This contrasts with the two-space case, where the graph of an equation was usually a curve. We will have a good deal more to say about graphing equations and the corresponding surfaces in Section 14.6.

## CONCEPTS REVIEW

1. The numbers $x$, $y$, and $z$ in $(x, y, z)$ are called the _____ of a point in three-space.

2. The distance between the points $(-1, 3, 5)$ and $(x, y, z)$ is _____.

3. The equation $(x + 1)^2 + (y - 3)^2 + (z - 5)^2 = 16$ determines a sphere with center _____ and radius _____.

4. The graph of $3x - 2y + 4z = 12$ is a _____ with $x$-intercept _____, $y$-intercept _____, and $z$-intercept _____.

## PROBLEM SET 14.1

1. Plot the points whose coordinates are $(1, 2, 3)$, $(2, 0, 1)$, $(-2, 4, 5)$, $(0, 3, 0)$, and $(-1, -2, -3)$. If appropriate show the "box" as in Figures 4 and 5.

2. Follow the directions of Problem 1 for $(\sqrt{3}, -3, 3)$, $(0, \pi, -3)$, $(-2, \frac{1}{3}, 2)$, and $(0, 0, e)$.

3. What is peculiar to the coordinates of all points in the $yz$-plane? On the $z$-axis?

4. What is peculiar to the coordinates of all points in the $xz$-plane? On the $y$-axis?

5. Find the distance between the following pairs of points.

(a) $(6, -1, 0)$ and $(1, 2, 3)$
(b) $(-2, -2, 0)$ and $(2, -2, -3)$
(c) $(e, \pi, 0)$ and $(-\pi, -4, \sqrt{3})$

6. Show that $(4, 5, 3)$, $(1, 7, 4)$, and $(2, 4, 6)$ are vertices of an equilateral triangle.

7. Show that $(2, 1, 6)$, $(4, 7, 9)$, and $(8, 5, -6)$ are vertices of a right triangle. *Hint:* Only right triangles satisfy the Pythagorean Theorem.

8. Find the distance from $(2, 3, -1)$ to (a) the $xy$-plane, (b) the $y$-axis, and (c) the origin.

9. A rectangular box has its faces parallel to the coordinate planes and has $(2, 3, 4)$ and $(6, -1, 0)$ as the endpoints of a main diagonal. Sketch the box and find the coordinates of all eight vertices.

10. $P(x, 5, z)$ is on a line through $Q(2, -4, 3)$ that is parallel to one of the coordinate axes. Which axis must it be and what are $x$ and $z$?

11. Write the equation of the sphere with the given center and radius.

(a) $(1, 2, 3)$; 5          (b) $(-2, -3, -6)$; $\sqrt{5}$
(c) $(\pi, e, \sqrt{2})$; $\sqrt{\pi}$

**12.** Find the equation of the sphere whose center is $(2, 4, 5)$ and which is tangent to the $xy$-plane.

In Problems 13–16, complete the squares to find the center and radius of the sphere whose equation is given (see Example 2).

**13.** $x^2 + y^2 + z^2 - 12x + 14y - 8z + 1 = 0$

**14.** $x^2 + y^2 + z^2 + 2x - 6y - 10z + 34 = 0$

**15.** $4x^2 + 4y^2 + 4z^2 - 4x + 8y + 16z - 13 = 0$

**16.** $x^2 + y^2 + z^2 + 8x - 4y - 22z + 77 = 0$

In Problems 17–24, sketch the graphs of the given equations. Begin by sketching the traces in the coordinate planes (see Examples 4 and 5).

**17.** $2x + 6y + 3z = 12$    **18.** $3x - 4y + 2z = 24$

**19.** $x + 3y - z = 6$    **20.** $-3x + 2y + z = 6$

**21.** $x + 3y = 8$    **22.** $3x + 4z = 12$

**23.** $x^2 + y^2 + z^2 = 9$

**24.** $(x - 2)^2 + y^2 + z^2 = 4$

**25.** Find the equation of the sphere that has the line segment joining $(-2, 3, 6)$ and $(4, -1, 5)$ as a diameter (see Example 3).

**26.** Find the equations of the tangent spheres of equal radii whose centers are $(-3, 1, 2)$ and $(5, -3, 6)$.

**27.** Find the equation of the sphere that is tangent to the three coordinate planes, if its radius is 6 and its center is in the first octant.

**28.** Find the equation of the sphere with center $(1, 1, 4)$ that is tangent to the plane $x + y = 12$.

**29.** Describe the graph in 3-space of each of the following equations.

(a) $z = 2$    (b) $x = y$
(c) $xy = 0$    (d) $xyz = 0$
(e) $x^2 + y^2 = 4$    (f) $z = \sqrt{9 - x^2 - y^2}$

**30.** The sphere $(x - 1)^2 + (y + 2)^2 + (z + 1)^2 = 10$ intersects the plane $z = 2$ in a circle. Find its center and radius.

**31.** A point $P$ moves so that its distance from $(1, 2, -3)$ is twice its distance from $(1, 2, 3)$. Show that $P$ is on a sphere and find its center and radius.

**32.** A point $P$ moves so that its distance from $(1, 2, -3)$ equals its distance from $(2, 3, 2)$. Find the equation of the plane on which $P$ lies.

**33.** The balls $(x - 1)^2 + (y - 2)^2 + (z - 1)^2 \le 4$ and $(x - 2)^2 + (y - 4)^2 + (z - 3)^2 \le 4$ intersect in a solid. Find its volume.

**34.** Do Problem 33 assuming the second ball is $(x - 2)^2 + (y - 4)^2 + (z - 3)^2 \le 9$.

---

**Answers to Concepts Review:** **1.** Coordinates
**2.** $\sqrt{(x + 1)^2 + (y - 3)^2 + (z - 5)^2}$  **3.** $(-1, 3, 5)$; 4
**4.** Plane; 4; $-6$; 3

---

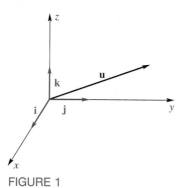

FIGURE 1

## 14.2
## VECTORS IN
## THREE-SPACE

The material from Sections 13.2 and 13.3 on vectors in the plane can be repeated almost word for word for vectors in space. About the only difference is that a vector $\mathbf{u}$ now has three components; that is,

$$\mathbf{u} = \langle u_1, u_2, u_3 \rangle = u_1\mathbf{i} + u_2\mathbf{j} + u_3\mathbf{k}$$

Here, $\mathbf{i}$, $\mathbf{j}$, and $\mathbf{k}$ are the standard unit vectors, called **basis vectors**, in the directions of the three positive coordinate axes (Figure 1). The **length** of $\mathbf{u}$, denoted by $|\mathbf{u}|$, is from the distance formula given by

$$\boxed{|\mathbf{u}| = \sqrt{u_1^2 + u_2^2 + u_3^2}}$$

Vectors in space are added, multiplied by scalars, and subtracted just as in the plane, and the algebraic laws that are satisfied agree with those studied earlier. The **dot product** of $\mathbf{u} = \langle u_1, u_2, u_3 \rangle$ and $\mathbf{v} = \langle v_1, v_2, v_3 \rangle$ is defined by

$$\mathbf{u} \cdot \mathbf{v} = u_1 v_1 + u_2 v_2 + u_3 v_3$$

and it has the geometric interpretation noted earlier, namely,

$$\mathbf{u} \cdot \mathbf{v} = |\mathbf{u}| \, |\mathbf{v}| \cos \theta$$

where $\theta$ is the angle between $\mathbf{u}$ and $\mathbf{v}$. Consequently, it continues to be true that two vectors are perpendicular if and only if their dot product is zero.

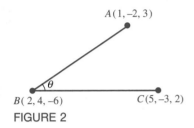

$A(1, -2, 3)$

$B(2, 4, -6)$      $C(5, -3, 2)$

**FIGURE 2**

**EXAMPLE 1** Find the angle $ABC$ if $A = (1, -2, 3)$, $B = (2, 4, -6)$, and $C = (5, -3, 2)$ (Figure 2).

**Solution** First we determine vectors $\mathbf{u}$ and $\mathbf{v}$ (emanating from the origin) equivalent to $\overrightarrow{BA}$ and $\overrightarrow{BC}$. This is done by subtracting the coordinates of the initial points from those of the terminal points—that is,

$$\mathbf{u} = \langle 1 - 2, -2 - 4, 3 + 6 \rangle = \langle -1, -6, 9 \rangle$$
$$\mathbf{v} = \langle 5 - 2, -3 - 4, 2 + 6 \rangle = \langle 3, -7, 8 \rangle$$

Thus,

$$\cos \theta = \frac{\mathbf{u} \cdot \mathbf{v}}{|\mathbf{u}| \, |\mathbf{v}|} = \frac{(-1)(3) + (-6)(-7) + (9)(8)}{\sqrt{1 + 36 + 81} \, \sqrt{9 + 49 + 64}} \approx 0.9251$$

$$\theta \approx 0.3894 \ (\text{about } 22.31°) \qquad \blacksquare$$

**EXAMPLE 2** Express $\mathbf{u} = \langle 2, 4, 5 \rangle$ as the sum of a vector $\mathbf{m}$ parallel to $\mathbf{v} = \langle 2, -1, -2 \rangle$ and a vector $\mathbf{n}$ perpendicular to $\mathbf{v}$.

**Solution** Figure 3 tells the story. First, we find $\mathbf{m} = \mathbf{pr_v u}$, the projection of $\mathbf{u}$ on $\mathbf{v}$ (see Problem 33 of Section 13.3).

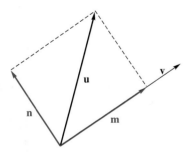

**FIGURE 3**

$$\mathbf{m} = |\mathbf{m}| \frac{\mathbf{v}}{|\mathbf{v}|} = |\mathbf{u}| \cos \theta \, \frac{\mathbf{v}}{|\mathbf{v}|}$$

$$= \frac{\mathbf{u} \cdot \mathbf{v}}{|\mathbf{v}|} \frac{\mathbf{v}}{|\mathbf{v}|} = \frac{\mathbf{u} \cdot \mathbf{v}}{|\mathbf{v}|^2} \mathbf{v}$$

$$= \frac{(2)(2) + (4)(-1) + (5)(-2)}{4 + 1 + 4} \langle 2, -1, -2 \rangle$$

$$= \left\langle \frac{-20}{9}, \frac{10}{9}, \frac{20}{9} \right\rangle$$

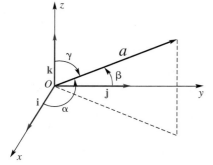

FIGURE 4

Then,

$$\mathbf{n} = \mathbf{u} - \mathbf{m} = \left\langle \frac{38}{9}, \frac{26}{9}, \frac{25}{9} \right\rangle$$

If you doubt that $\mathbf{m}$ and $\mathbf{n}$ are perpendicular, compute their dot product. You will get zero. ∎

**Direction Angles and Cosines** The (smallest nonnegative) angles between a nonzero vector $\mathbf{a}$ and the basis vectors $\mathbf{i}$, $\mathbf{j}$, and $\mathbf{k}$ are the **direction angles** of $\mathbf{a}$; they are designated by $\alpha$, $\beta$, and $\gamma$, respectively (Figure 4). It is generally more convenient to work with the **direction cosines** $\cos\alpha$, $\cos\beta$, and $\cos\gamma$. If $\mathbf{a} = a_1\mathbf{i} + a_2\mathbf{j} + a_3\mathbf{k}$, then

$$\cos\alpha = \frac{\mathbf{a}\cdot\mathbf{i}}{|\mathbf{a}|\,|\mathbf{i}|} = \frac{a_1}{|\mathbf{a}|}$$

and similarly

$$\cos\beta = \frac{a_2}{|\mathbf{a}|}, \qquad \cos\gamma = \frac{a_3}{|\mathbf{a}|}$$

Notice that

$$\cos^2\alpha + \cos^2\beta + \cos^2\gamma = 1$$

In fact, the vector $\langle\cos\alpha, \cos\beta, \cos\gamma\rangle$ is a unit vector with the same direction as the original vector $\mathbf{a}$.

**EXAMPLE 3** Find the direction angles for the vector $\mathbf{a} = 4\mathbf{i} - 5\mathbf{j} + 3\mathbf{k}$.

**Solution** Since $|\mathbf{a}| = \sqrt{4^2 + (-5)^2 + 3^2} = 5\sqrt{2}$,

$$\cos\alpha = \frac{4}{5\sqrt{2}} = \frac{2\sqrt{2}}{5}, \qquad \cos\beta = \frac{-\sqrt{2}}{2}, \qquad \cos\gamma = \frac{3\sqrt{2}}{10}$$

and

$$\alpha \approx 55.55°, \qquad \beta = 135°, \qquad \gamma \approx 64.90° \qquad ■$$

**EXAMPLE 4** Find a vector 5 units long that has $\alpha = 32°$ and $\beta = 100°$ as two of its direction angles.

**Solution** First we note that the third direction angle $\gamma$ must satisfy

$$\cos^2\gamma = 1 - \cos^2 32° - \cos^2 100° \approx 0.25066$$

Thus,

$$\cos \gamma \approx \pm 0.50066$$

Two vectors meet the requirements of the problem. They are

$$5\langle \cos \alpha, \cos \beta, \cos \gamma \rangle \approx 5\langle 0.84805, -0.17365, 0.50066 \rangle$$
$$= \langle 4.2403, -0.8683, 2.5033 \rangle$$

and $\langle 4.2403, -0.8683, -2.5033 \rangle$.  ∎

**Planes**    One fruitful way to describe a plane is by using vector language. Let $\mathbf{n} = \langle A, B, C \rangle$ be a fixed nonzero vector and $P_1(x_1, y_1, z_1)$ be a fixed point. The set of points $P(x, y, z)$ satisfying $\overrightarrow{P_1 P} \cdot \mathbf{n} = 0$ is the **plane** through $P_1$ perpendicular to $\mathbf{n}$. Since every plane contains a point and is perpendicular to some vector, a plane can be characterized in this way.

To get the Cartesian equation of the plane, write the vector $\overrightarrow{P_1 P}$ in component form, that is,

$$\overrightarrow{P_1 P} = \langle x - x_1, y - y_1, z - z_1 \rangle$$

Then, $\overrightarrow{P_1 P} \cdot \mathbf{n} = 0$ is equivalent to

$$\boxed{A(x - x_1) + B(y - y_1) + C(z - z_1) = 0}$$

This equation (in which at least one of $A$, $B$, $C$ is different from zero) is called the **standard form for the equation of a plane**.

If we remove the parentheses and simplify, the boxed equation takes the form of the general linear equation

$$Ax + By + Cz = D, \qquad A^2 + B^2 + C^2 \neq 0$$

*Thus, every plane has a linear equation. Conversely, the graph of a linear equation in three-space is always a plane.* To see the latter, let $(x_1, y_1, z_1)$ satisfy the equation, that is,

$$Ax_1 + By_1 + Cz_1 = D$$

When we subtract this equation from the one above, we have the boxed equation, which we know represents a plane.

**EXAMPLE 5**    Find the equation of the plane through $(5, 1, -2)$ perpendicular to $\mathbf{n} = \langle 2, 4, 3 \rangle$. Then find the angle between this plane and the one with equation $3x - 4y + 7z = 5$.

*Solution*    To perform the first task, simply apply the boxed result to the problem at hand. We obtain

$$2(x - 5) + 4(y - 1) + 3(z + 2) = 0$$

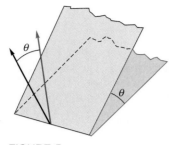

FIGURE 5

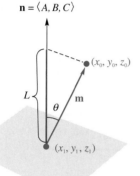

FIGURE 6

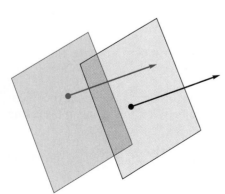

FIGURE 7

or, equivalently,

$$2x + 4y + 3z = 8$$

A vector **m** perpendicular to the second plane is $\mathbf{m} = \langle 3, -4, 7 \rangle$. The angle $\theta$ between two planes is the angle between their normals (Figure 5). Thus,

$$\cos \theta = \frac{\mathbf{m} \cdot \mathbf{n}}{|\mathbf{m}| \, |\mathbf{n}|} = \frac{(3)(2) + (-4)(4) + (7)(3)}{\sqrt{9 + 16 + 49} \, \sqrt{4 + 16 + 9}} \approx 0.2375$$

$$\theta \approx 76.26°$$

Actually there are two angles between two planes, but they are supplementary. The process just described will lead to one of them. The other, if desired, is obtained by subtracting the first value from 180°. In our case, it would be 103.74°. ∎

**EXAMPLE 6**   Show that the distance $L$ from the point $(x_0, y_0, z_0)$ to the plane $Ax + By + Cz = D$ is given by the formula

$$L = \frac{|Ax_0 + By_0 + Cz_0 - D|}{\sqrt{A^2 + B^2 + C^2}}$$

*Solution*   Let $(x_1, y_1, z_1)$ be a point on the plane and let $\mathbf{m} = \langle x_0 - x_1, y_0 - y_1, z_0 - z_1 \rangle$ be the vector from $(x_1, y_1, z_1)$ to $(x_0, y_0, z_0)$, as in Figure 6. Now $\mathbf{n} = \langle A, B, C \rangle$ is a vector perpendicular to the given plane, though it might point in the opposite direction of that in our figure. The number $L$ that we seek is the length of the projection of **m** on **n**. Thus,

$$L = \|\mathbf{m}\| \cos \theta| = \frac{|\mathbf{m} \cdot \mathbf{n}|}{|\mathbf{n}|}$$

$$= \frac{|A(x_0 - x_1) + B(y_0 - y_1) + C(z_0 - z_1)|}{\sqrt{A^2 + B^2 + C^2}}$$

$$= \frac{|Ax_0 + By_0 + Cz_0 - (Ax_1 + By_1 + Cz_1)|}{\sqrt{A^2 + B^2 + C^2}}$$

But $(x_1, y_1, z_1)$ is on the plane and so

$$Ax_1 + By_1 + Cz_1 = D$$

Substitution of this result in the expression for $L$ yields the desired formula. ∎

**EXAMPLE 7**   Find the distance between the parallel planes $3x - 4y + 5z = 9$ and $3x - 4y + 5z = 4$.

*Solution*   The planes are parallel, since the vector $\langle 3, -4, 5 \rangle$ is perpendicular to both of them (Figure 7). The point $(1, 1, 2)$ is easily seen to be on the first plane. We find the distance $L$ from $(1, 1, 2)$ to the second plane using the formula of Example 6.

$$L = \frac{|3(1) - 4(1) + 5(2) - 4|}{\sqrt{9 + 16 + 25}} = \frac{5}{5\sqrt{2}} \approx 0.7071 \qquad \blacksquare$$

## CONCEPTS REVIEW

**1.** Let $\mathbf{u} = \langle 2, -3, \sqrt{3} \rangle$ and $\mathbf{v} = \langle 3, 2, -2\sqrt{3} \rangle$ be two vectors. The length of $\mathbf{u}$ is $|\mathbf{u}| = $ _____ and the dot product of $\mathbf{u}$ and $\mathbf{v}$ is $\mathbf{u} \cdot \mathbf{v} = $ _____.

**2.** Two vectors are perpendicular if and only if their _____ is _____.

**3.** $3(x - 2) - 2(y + 1) + 4z = 0$ is the equation of a plane through the point _____ with _____ being a vector perpendicular to the plane.

**4.** The (smallest nonnegative) angle $\theta$ between the vectors $\mathbf{u}$ and $\mathbf{v}$ can be found from the geometric formula for the dot product, $\mathbf{u} \cdot \mathbf{v} = $ _____. This gives $\theta = $ _____.

## PROBLEM SET 14.2

**1.** For each pair of points $P_1$ and $P_2$ given below, sketch the directed line segment $\overrightarrow{P_1 P_2}$ and then write the corresponding vector in the form $a\mathbf{i} + b\mathbf{j} + c\mathbf{k}$.

(a) $P_1(1, 2, 4)$, $P_2(4, 5, 6)$
(b) $P_1(-1, -3, 204)$, $P_2(-14, 52, 26)$

**2.** Follow the directions of Problem 1.

(a) $P_1(-2, -2, -2)$, $P_2(-3, -4, 5)$
(b) $P_1(0, -1, e)$, $P_2(-\sqrt{14}, -5, \pi)$

**3.** Find the length of and direction cosines for each of the following vectors.

(a) $4\mathbf{i} + \mathbf{j} + 2\mathbf{k}$ $\qquad$ (b) $-2\mathbf{i} - 3\mathbf{j} + 7\mathbf{k}$

**4.** Follow the directions for Problem 3.

(a) $\langle 2, -1, -2 \rangle$ $\qquad$ (b) $\langle -1, 2, -2 \rangle$

**5.** Find the unit vector with the same direction as $\langle 3, -4, 5 \rangle$. Also, find a vector of length 5 oriented in the opposite direction.

**6.** Find a vector of length 10 with direction opposite to $-4\mathbf{i} + 3\mathbf{j} + -2\mathbf{k}$.

**7.** Find the angle between $\langle 4, -3, -1 \rangle$ and $\langle -2, -3, 5 \rangle$.

**8.** Find the angle between $-4\mathbf{i} + 2\mathbf{j} + 3\mathbf{k}$ and $2\mathbf{i} + \mathbf{j} + 5\mathbf{k}$.

**9.** Find two vectors of length 10, each of which is perpendicular to both $-4\mathbf{i} + 5\mathbf{j} + \mathbf{k}$ and $4\mathbf{i} + \mathbf{j}$.

**10.** Find all the vectors perpendicular to both $\langle 1, -2, -3 \rangle$ and $\langle -3, 2, 0 \rangle$.

**11.** Find the angle $ABC$ if $A = (1, 2, 3)$, $B = (-4, 5, 6)$, and $C = (1, 0, 1)$ (see Example 1).

**12.** Show that the triangle $ABC$ is a right triangle if $A = (6, 3, 3)$, $B = (3, 1, -1)$, and $C = (-1, 10, -2.5)$. *Hint*: Check the angle at $B$.

**13.** Find the *scalar projection* of $\mathbf{u} = -\mathbf{i} + 5\mathbf{j} + 3\mathbf{k}$ on $\mathbf{v} = -\mathbf{i} + \mathbf{j} - \mathbf{k}$. The scalar projection is the signed magnitude of the vector projection (Example 2), that is, it is $|\mathbf{u}| \cos \theta = \mathbf{u} \cdot \mathbf{v}/|\mathbf{v}|$.

**14.** Find the scalar projection of $\mathbf{u} = 5\mathbf{i} + 5\mathbf{j} + 2\mathbf{k}$ on $\mathbf{v} = -\sqrt{5}\mathbf{i} + \sqrt{5}\mathbf{j} + \mathbf{k}$.

**15.** If $\mathbf{u} = -3\mathbf{i} + 2\mathbf{j} + \mathbf{k}$ and $\mathbf{v} = -3\mathbf{i} + 5\mathbf{j} - 3\mathbf{k}$, express $\mathbf{u}$ as the sum of a vector $\mathbf{m}$ parallel to $\mathbf{v}$ and a vector $\mathbf{n}$ perpendicular to $\mathbf{v}$ (see Example 2).

**16.** Follow the directions of Problem 15 for $\mathbf{u} = e\mathbf{i} + \pi\mathbf{j} + \mathbf{k}$ and $\mathbf{v} = \mathbf{i} + \mathbf{j}$.

**17.** Find the direction angles for each vector.

(a) $\mathbf{u} = -3\mathbf{i} + 2\mathbf{j} + \mathbf{k}$ $\qquad$ (b) $\mathbf{u} = 3\mathbf{i} + 6\mathbf{j} - \mathbf{k}$

**18.** If $\alpha = 46°$ and $\beta = 108°$ are direction angles for a vector $\mathbf{u}$, find two possible values for the third angle (see Example 4).

**19.** A vector $\mathbf{u} = 2\mathbf{i} + 3\mathbf{j} + z\mathbf{k}$ emanating from the origin points into the first octant. If $|\mathbf{u}| = 5$, find $z$.

**20.** If $\mathbf{u} = 2\mathbf{i} + 3\mathbf{j} + z\mathbf{k}$ and $\mathbf{v} = 2\mathbf{i} + 6\mathbf{j} - 3\mathbf{k}$ are perpendicular, find $z$.

**21.** Find two perpendicular vectors $\mathbf{u}$ and $\mathbf{v}$ such that each is also perpendicular to $\mathbf{w} = \langle -4, 2, 5 \rangle$.

**22.** Find the vector emanating from the origin whose terminal point is the midpoint of the segment joining $(3, 2, -1)$ and $(5, -7, 2)$.

**23.** Which of the following do *not* make sense?

(a) $\mathbf{u} \cdot (\mathbf{v} \cdot \mathbf{w})$      (b) $(\mathbf{u} \cdot \mathbf{w}) + \mathbf{w}$
(c) $|\mathbf{u}| (\mathbf{v} \cdot \mathbf{w})$      (d) $(\mathbf{u} \cdot \mathbf{v})\mathbf{w}$
(e) $(|\mathbf{u}|\mathbf{v}) \cdot \mathbf{w}$      (f) $|\mathbf{u}| \cdot \mathbf{v}$

**24.** Let $\mathbf{a}$, $\mathbf{b}$, $\mathbf{c}$, and $\mathbf{d}$ be vectors emanating from the origin and terminating at $A$, $B$, $C$, and $D$, respectively. Use vector notation to express a necessary and sufficient condition that the figure $ABCD$ be a parallelogram.

**25.** Find the equation of the plane passing through $P$ and perpendicular to $\mathbf{n}$ (see Example 5).

(a) $P(1, 2, -3)$, $\mathbf{n} = 2\mathbf{i} - 4\mathbf{j} + 3\mathbf{k}$
(b) $P(-2, -3, 4)$, $\mathbf{n} = 3\mathbf{i} - 2\mathbf{j} - \mathbf{k}$

**26.** Find the smaller of the angles between the planes $3x - 2y + 5z = 7$ and $4x - 2y - 3z = 2$ (see Example 5).

**27.** Find the smaller of the angles between the two planes of Problem 25.

**28.** Find the equation of a plane through $(-1, 2, -3)$ and parallel to the plane $2x + 4y - z = 6$.

**29.** Find the equation of the plane through $(-4, -1, 2)$ and parallel (a) to the $xy$-plane, (b) to the plane $2x - 3y - 4z = 0$.

**30.** Find the distance from $(1, -1, 2)$ to the plane $x + 3y + z = 7$ (see Example 6).

**31.** Find the distance from $(2, 6, 3)$ to the plane $-3x + 2y + z = 9$.

**32.** Find the distance between the parallel planes $-3x + 2y + z = 9$ and $6x - 4y - 2z = 19$ (see Example 7).

**33.** Find the distance between the parallel planes $5x - 3y - 2z = 5$ and $-5x + 3y + 2z = 7$.

**34.** Find the equation of the plane, each of whose points is equidistant from $(-2, 1, 4)$ and $(6, 1, -2)$.

**35.** Prove $|\mathbf{u} + \mathbf{v}|^2 + |\mathbf{u} - \mathbf{v}|^2 = 2|\mathbf{u}|^2 + 2|\mathbf{v}|^2$. *Hint:* $|\mathbf{w}|^2 = \mathbf{w} \cdot \mathbf{w}$.

**36.** Prove $\mathbf{u} \cdot \mathbf{v} = \frac{1}{4}|\mathbf{u} + \mathbf{v}|^2 - \frac{1}{4}|\mathbf{u} - \mathbf{v}|^2$.

**37.** Find the angle between a main diagonal of a cube and one of its faces.

**38.** Find a unit vector whose direction angles are equal.

**39.** Find the smallest angle between the main diagonals of a rectangular box 4 feet by 6 feet by 10 feet.

**40.** A constant force of $\mathbf{F} = 3\mathbf{i} - 6\mathbf{j} + 7\mathbf{k}$ pounds is applied to an object in moving it from $(2, 1, 3)$ to $(9, 4, 6)$, coordinates given in feet. Find the work done. Recall that $W = \mathbf{F} \cdot \mathbf{D}$ (see Section 13.3).

**41.** How much work is done by a force of 5 newtons acting in the direction $2\mathbf{i} + 2\mathbf{j} - \mathbf{k}$ in moving an object from $(0, 1, 2)$ to $(3, 5, 7)$, distances being measured in meters? (See Problem 40.)

**42.** A weight of 30 pounds is suspended by 3 wires with resulting tensions $3\mathbf{i} + 4\mathbf{j} + 15\mathbf{k}$, $-8\mathbf{i} - 2\mathbf{j} + 10\mathbf{k}$, and $a\mathbf{i} + b\mathbf{j} + c\mathbf{k}$. Determine $a$, $b$, and $c$, assuming $\mathbf{k}$ points straight up.

**43.** Find the point one-fifth of the way from $(2, 3, -1)$ to $(7, -2, 9)$.

**44.** Suppose the three coordinate planes bounding the first octant are mirrors. A light ray with direction $a\mathbf{i} + b\mathbf{j} + c\mathbf{k}$ is reflected successively from the $xy$-plane, the $xz$-plane, and the $yz$-plane. Determine the direction of the ray after each reflection and state a nice conclusion concerning the final reflected ray.

**45.** Find the distance from the sphere $x^2 + y^2 + z^2 + 2x + 6y - 8z = 0$ to the plane $3x + 4y + z = 15$.

**46.** Refine the method of Example 7 by showing that the distance $L$ between the parallel planes $Ax + By + Cz = D$ and $Ax + By + Cz = E$ is

$$L = \frac{|D - E|}{\sqrt{A^2 + B^2 + C^2}}$$

**47.** Let $\mathbf{a} = \langle a_1, a_2, a_3 \rangle$ and $\mathbf{b} = \langle b_1, b_2, b_3 \rangle$ be fixed vectors. Show that $(\mathbf{x} - \mathbf{a}) \cdot (\mathbf{x} - \mathbf{b}) = 0$ is the equation of a sphere and find its center and radius.

**48.** Show that the work done by a constant force $\mathbf{F}$ on an object that moves completely around a closed polygonal path is 0.

**49.** The medians of a triangle meet in a point $P$ (the centroid by Problem 24 of Section 6.8) that is two-thirds of the way from a vertex to the midpoint of the opposite edge. Show that $P$ is the head of the position vector $(\mathbf{a} + \mathbf{b} + \mathbf{c})/3$, where $\mathbf{a}$, $\mathbf{b}$, and $\mathbf{c}$ are the position vectors of the vertices and use this to find $P$ if the vertices are $(2, 6, 5)$, $(4, -1, 2)$ and $(6, 1, 2)$.

**50.** Let $\mathbf{a}$, $\mathbf{b}$, $\mathbf{c}$, and $\mathbf{d}$ be the position vectors of the vertices of a tetrahedron. Show that the lines joining the vertices to the centroids of the opposite faces meet in a point $P$ and give a nice vector formula for it, thus generalizing Problem 49.

---

**Answers to Concepts Review:**    **1.** $4; -6$    **2.** Dot product; 0
**3.** $(2, -1, 0)$; $\langle 3, -2, 4 \rangle$
**4.** $|\mathbf{u}| |\mathbf{v}| \cos \theta$; $\cos^{-1}(\mathbf{u} \cdot \mathbf{v}/|\mathbf{u}| |\mathbf{v}|)$

## 14.3
## THE CROSS PRODUCT

The dot product of two vectors is a scalar. We have explored some of its uses in earlier sections. Now we introduce the **cross product** (or vector product); it will also have many uses. The cross product $\mathbf{u} \times \mathbf{v}$ of $\mathbf{u} = \langle u_1, u_2, u_3 \rangle$ and $\mathbf{v} = \langle v_1, v_2, v_3 \rangle$ is defined by

$$\mathbf{u} \times \mathbf{v} = \langle u_2 v_3 - u_3 v_2, \, u_3 v_1 - u_1 v_3, \, u_1 v_2 - u_2 v_1 \rangle$$

In this form, the formula is hard to remember and its significance is not obvious. Note the one thing that is obvious. The cross product of two vectors is a vector.

To help us remember the formula for the cross product, we recall a subject from an earlier mathematics course, namely, *determinants*. First, the value of a $2 \times 2$ determinant is

$$\begin{vmatrix} a & b \\ c & d \end{vmatrix} = ad - bc$$

Then the value of a $3 \times 3$ determinant is (expanding according to the top row)

$$\begin{vmatrix} a_1 & a_2 & a_3 \\ b_1 & b_2 & b_3 \\ c_1 & c_2 & c_3 \end{vmatrix} = a_1 \begin{vmatrix} a_1 & a_2 & a_3 \\ b_1 & b_2 & b_3 \\ c_1 & c_2 & c_3 \end{vmatrix} - a_2 \begin{vmatrix} a_1 & a_2 & a_3 \\ b_1 & b_2 & b_3 \\ c_1 & c_2 & c_3 \end{vmatrix} + a_3 \begin{vmatrix} a_1 & a_2 & a_3 \\ b_1 & b_2 & b_3 \\ c_1 & c_2 & c_3 \end{vmatrix}$$

$$= a_1 \begin{vmatrix} b_2 & b_3 \\ c_2 & c_3 \end{vmatrix} - a_2 \begin{vmatrix} b_1 & b_3 \\ c_1 & c_3 \end{vmatrix} + a_3 \begin{vmatrix} b_1 & b_2 \\ c_1 & c_2 \end{vmatrix}$$

Using determinants, we may write the definition of $\mathbf{u} \times \mathbf{v}$ as

$$\mathbf{u} \times \mathbf{v} = \begin{vmatrix} \mathbf{i} & \mathbf{j} & \mathbf{k} \\ u_1 & u_2 & u_3 \\ v_1 & v_2 & v_3 \end{vmatrix} = \begin{vmatrix} u_2 & u_3 \\ v_2 & v_3 \end{vmatrix} \mathbf{i} - \begin{vmatrix} u_1 & u_3 \\ v_1 & v_3 \end{vmatrix} \mathbf{j} + \begin{vmatrix} u_1 & u_2 \\ v_1 & v_2 \end{vmatrix} \mathbf{k}$$

Note that the components of the left vector $\mathbf{u}$ go in the second row, and those of the right vector $\mathbf{v}$ go in the third row. This is important because if we interchange the positions of $\mathbf{u}$ and $\mathbf{v}$, we interchange the second and third rows of the determinant, and this changes the sign of the determinant's value, as you may check. Thus,

$$\mathbf{u} \times \mathbf{v} = -(\mathbf{v} \times \mathbf{u})$$

which is sometimes called the *anticommutative law*.

**EXAMPLE 1** Let $\mathbf{u} = \langle 1, -2, -1 \rangle$ and $\mathbf{v} = \langle -2, 4, 1 \rangle$. Calculate $\mathbf{u} \times \mathbf{v}$ and $\mathbf{v} \times \mathbf{u}$ using the determinant definition.

---

### Torque

The cross product plays an important role in mechanics. Let $O$ be a fixed point in a body and suppose a force $\mathbf{F}$ is applied at another point $P$ of the body. Then, $\mathbf{F}$ tends to rotate the body about an axis through $O$ and perpendicular to the plane of $OP$ and $\mathbf{F}$. The vector

$$\tau = \overrightarrow{OP} \times \mathbf{F}$$

is called the **torque**. It points in the direction of the axis and has magnitude $|\overrightarrow{OP}|\,|\mathbf{F}|\sin \theta$, which is just the moment of force about the axis due to $\mathbf{F}$.

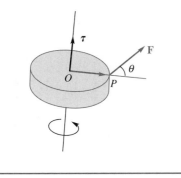

*Solution*

$$\mathbf{u} \times \mathbf{v} = \begin{vmatrix} \mathbf{i} & \mathbf{j} & \mathbf{k} \\ 1 & -2 & -1 \\ -2 & 4 & 1 \end{vmatrix} = \mathbf{i} \begin{vmatrix} -2 & -1 \\ 4 & 1 \end{vmatrix} - \mathbf{j} \begin{vmatrix} 1 & -1 \\ -2 & 1 \end{vmatrix} + \mathbf{k} \begin{vmatrix} 1 & -2 \\ -2 & 4 \end{vmatrix}$$

$$= 2\mathbf{i} + \mathbf{j} + 0\mathbf{k}$$

$$\mathbf{v} \times \mathbf{u} = \begin{vmatrix} \mathbf{i} & \mathbf{j} & \mathbf{k} \\ -2 & 4 & 1 \\ 1 & -2 & -1 \end{vmatrix} = \mathbf{i} \begin{vmatrix} 4 & 1 \\ -2 & -1 \end{vmatrix} - \mathbf{j} \begin{vmatrix} -2 & 1 \\ 1 & -1 \end{vmatrix} + \mathbf{k} \begin{vmatrix} -2 & 4 \\ 1 & -2 \end{vmatrix}$$

$$= -2\mathbf{i} - \mathbf{j} + 0\mathbf{k} \qquad \blacksquare$$

**Geometric Interpretation of u × v** Like the dot product, the cross product gains significance from its geometric interpretation.

---

### Theorem A

Let **u** and **v** be vectors in three-space and $\theta$ be the angle between them. Then:

1. $\mathbf{u} \cdot (\mathbf{u} \times \mathbf{v}) = 0 = \mathbf{v} \cdot (\mathbf{u} \times \mathbf{v})$—that is, $\mathbf{u} \times \mathbf{v}$ is perpendicular to both **u** and **v**;
2. **u**, **v**, and $\mathbf{u} \times \mathbf{v}$ form a right-handed triple;
3. $|\mathbf{u} \times \mathbf{v}| = |\mathbf{u}| \, |\mathbf{v}| \sin \theta$.

---

*Proof* Let $\mathbf{u} = \langle u_1, u_2, u_3 \rangle$ and $\mathbf{v} = \langle v_1, v_2, v_3 \rangle$.

1. $\mathbf{u} \cdot (\mathbf{u} \times \mathbf{v}) = u_1(u_2 v_3 - u_3 v_2) + u_2(u_3 v_1 - u_1 v_3) + u_3(u_1 v_2 - u_2 v_1)$. When we remove parentheses, the six terms cancel in pairs. A similar event occurs when we expand $\mathbf{v} \cdot (\mathbf{u} \times \mathbf{v})$.

2. The meaning of right-handedness for the triple **u**, **v**, $\mathbf{u} \times \mathbf{v}$ is illustrated in Figure 1. There $\theta$ is the angle between **u** and **v**, and the hand is cupped in the direction of the rotation through $\theta$ that makes **u** coincide with **v**. It seems difficult to establish analytically that the indicated triple is right-handed, but you might check it with a few examples. Note in particular that since $\mathbf{i} \times \mathbf{j} = \mathbf{k}$, the triple **i**, **j**, $\mathbf{i} \times \mathbf{j}$ is right-handed.

3. We need Lagrange's identity

$$|\mathbf{u} \times \mathbf{v}|^2 = |\mathbf{u}|^2 |\mathbf{v}|^2 - (\mathbf{u} \cdot \mathbf{v})^2$$

whose proof is a simple algebraic exercise (Problem 23). Taking it for granted, we may write

$$|\mathbf{u} \times \mathbf{v}|^2 = |\mathbf{u}|^2 |\mathbf{v}|^2 - (|\mathbf{u}| \, |\mathbf{v}| \cos \theta)^2$$

$$= |\mathbf{u}|^2 |\mathbf{v}|^2 (1 - \cos^2 \theta)$$

$$= |\mathbf{u}|^2 |\mathbf{v}|^2 \sin^2 \theta$$

Since $0 \le \theta \le \pi$, $\sin \theta \ge 0$. Thus, taking principal square roots yields

$$|\mathbf{u} \times \mathbf{v}| = |\mathbf{u}| \, |\mathbf{v}| \sin \theta \qquad \blacksquare$$

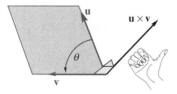

**FIGURE 1**

It is important that we have geometric interpretations of both $\mathbf{u} \cdot \mathbf{v}$ and $\mathbf{u} \times \mathbf{v}$. It means that, while both products were originally defined in terms of components that depend on a choice of coordinate system, they are actually independent of coordinate systems. They are intrinsic geometric quantities, and you will get the same answers for $\mathbf{u} \cdot \mathbf{v}$ and $\mathbf{u} \times \mathbf{v}$ no matter how you introduce the coordinates used to compute them.

Here is a simple consequence of Theorem A (part 3) and the fact that vectors are parallel if and only if the angle $\theta$ between them is either $0°$ or $180°$.

---

### Theorem B

Two vectors $\mathbf{u}$ and $\mathbf{v}$ in three-space are parallel if and only if $\mathbf{u} \times \mathbf{v} = \mathbf{0}$.

---

**Applications**   Our first application is to find the equation of the plane through three noncollinear points.

**EXAMPLE 2**   Find the equation of the plane (Figure 2) through the three points $P_1(1, -2, 3)$, $P_2(4, 1, -2)$, and $P_3(-2, -3, 0)$.

**Solution**   Let $\mathbf{u} = \overrightarrow{P_2 P_1} = \langle -3, -3, 5 \rangle$ and $\mathbf{v} = \overrightarrow{P_2 P_3} = \langle -6, -4, 2 \rangle$. Then,

$$\mathbf{u} \times \mathbf{v} = \begin{vmatrix} \mathbf{i} & \mathbf{j} & \mathbf{k} \\ -3 & -3 & 5 \\ -6 & -4 & 2 \end{vmatrix} = 14\mathbf{i} - 24\mathbf{j} - 6\mathbf{k}$$

The plane through $(4, 1, -2)$ with normal $14\mathbf{i} - 24\mathbf{j} - 6\mathbf{k}$ has equation (see Section 14.2)

$$14(x - 4) - 24(y - 1) - 6(z + 2) = 0$$

or

$$14x - 24y - 6z = 44 \qquad \blacksquare$$

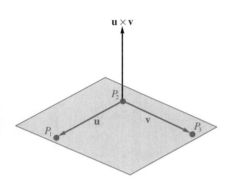

FIGURE 2

**EXAMPLE 3**   Show that the area of a parallelogram with $\mathbf{a}$ and $\mathbf{b}$ as adjacent sides is $|\mathbf{a} \times \mathbf{b}|$.

**Solution**   Recall that the area of a parallelogram is the product of the base times the height. Now look at Figure 3 and use the fact that $|\mathbf{a} \times \mathbf{b}| = |\mathbf{a}|\,|\mathbf{b}|\sin \theta$. $\qquad \blacksquare$

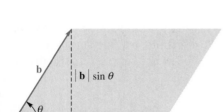

FIGURE 3

**EXAMPLE 4**   Let $\mathbf{a}$, $\mathbf{b}$, and $\mathbf{c}$ be vectors not all lying in the same plane. Show that the volume of the parallelepiped determined by $\mathbf{a}$, $\mathbf{b}$, and $\mathbf{c}$ is

$$V = |\mathbf{a} \cdot (\mathbf{b} \times \mathbf{c})| = \left|\begin{vmatrix} a_1 & a_2 & a_3 \\ b_1 & b_2 & b_3 \\ c_1 & c_2 & c_3 \end{vmatrix}\right|$$

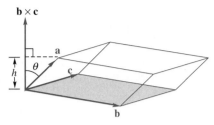

FIGURE 4

*Solution* Refer to Figure 4 and regard the parallelogram determined by **b** and **c** as the base of the parallelepiped. The area of this base is $|\mathbf{b} \times \mathbf{c}|$ by Example 3; the height $h$ of the parallelepiped is the absolute value of the scalar projection of **a** on $\mathbf{b} \times \mathbf{c}$. Thus,

$$h = |\mathbf{a}| \, |\cos \theta| = \frac{|\mathbf{a}| \, |\mathbf{a} \cdot (\mathbf{b} \times \mathbf{c})|}{|\mathbf{a}| \, |\mathbf{b} \times \mathbf{c}|} = \frac{|\mathbf{a} \cdot (\mathbf{b} \times \mathbf{c})|}{|\mathbf{b} \times \mathbf{c}|}$$

and

$$V = h|\mathbf{b} \times \mathbf{c}| = |\mathbf{a} \cdot (\mathbf{b} \times \mathbf{c})|$$

That $V$ can also be expressed as a determinant is established by expanding $|\mathbf{a} \cdot (\mathbf{b} \times \mathbf{c})|$ in terms of components and then comparing it with the value of the indicated determinant. ∎

**Algebraic Properties** The rules for calculating with cross products are summarized in the following theorem. Proving this theorem is a matter of writing everything out in terms of components and will be left as an exercise.

---

**Theorem C**

If **u**, **v**, and **w** are vectors in three-space and $k$ is a scalar, then:
1. $\mathbf{u} \times \mathbf{v} = -(\mathbf{v} \times \mathbf{u})$   (anticommutative law);
2. $\mathbf{u} \times (\mathbf{v} + \mathbf{w}) = (\mathbf{u} \times \mathbf{v}) + (\mathbf{u} \times \mathbf{w})$   (left distributive law);
3. $k(\mathbf{u} \times \mathbf{v}) = (k\mathbf{u}) \times \mathbf{v} = \mathbf{u} \times (k\mathbf{v})$;
4. $\mathbf{u} \times \mathbf{0} = \mathbf{0} \times \mathbf{u} = \mathbf{0}, \mathbf{u} \times \mathbf{u} = \mathbf{0}$;
5. $(\mathbf{u} \times \mathbf{v}) \cdot \mathbf{w} = \mathbf{u} \cdot (\mathbf{v} \times \mathbf{w})$;
6. $\mathbf{u} \times (\mathbf{v} \times \mathbf{w}) = (\mathbf{u} \cdot \mathbf{w})\mathbf{v} - (\mathbf{u} \cdot \mathbf{v})\mathbf{w}$.

---

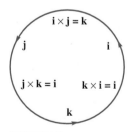

FIGURE 5

Once the rules in Theorem C are mastered, complicated calculations with vectors can be done with ease. We illustrate by calculating a cross product in a new way. We will need the following simple but important products.

$$\boxed{\mathbf{i} \times \mathbf{j} = \mathbf{k} \qquad \mathbf{j} \times \mathbf{k} = \mathbf{i} \qquad \mathbf{k} \times \mathbf{i} = \mathbf{j}}$$

These results have a cyclic order, which can be remembered by appealing to Figure 5.

**EXAMPLE 5** Calculate $\mathbf{u} \times \mathbf{v}$ if $\mathbf{u} = 3\mathbf{i} - 2\mathbf{j} + \mathbf{k}$ and $\mathbf{v} = 4\mathbf{i} + 2\mathbf{j} - 3\mathbf{k}$.

*Solution* We appeal to Theorem C, especially the distributive law and the anticommutative law.

$$\mathbf{u} \times \mathbf{v} = (3\mathbf{i} - 2\mathbf{j} + \mathbf{k}) \times (4\mathbf{i} + 2\mathbf{j} - 3\mathbf{k})$$

$$= 12(\mathbf{i} \times \mathbf{i}) + 6(\mathbf{i} \times \mathbf{j}) - 9(\mathbf{i} \times \mathbf{k}) - 8(\mathbf{j} \times \mathbf{i}) - 4(\mathbf{j} \times \mathbf{j})$$
$$+ 6(\mathbf{j} \times \mathbf{k}) + 4(\mathbf{k} \times \mathbf{i}) + 2(\mathbf{k} \times \mathbf{j}) - 3(\mathbf{k} \times \mathbf{k})$$

$$= 12(\mathbf{0}) + 6(\mathbf{k}) - 9(-\mathbf{j}) - 8(-\mathbf{k}) - 4(\mathbf{0})$$
$$+ 6(\mathbf{i}) + 4(\mathbf{j}) + 2(-\mathbf{i}) - 3(\mathbf{0})$$

$$= 4\mathbf{i} + 13\mathbf{j} + 14\mathbf{k}$$

Experts would do most of this in their heads; novices might find the determinant method easier. ∎

## CONCEPTS REVIEW

**1.** The cross product of $\mathbf{u} = \langle -1, 2, 1 \rangle$ and $\mathbf{v} = \langle 3, 1, -1 \rangle$ is given by a specific determinant; evaluation of this determinant gives $\mathbf{u} \times \mathbf{v} =$ _____.

**2.** Geometrically, $\mathbf{u} \times \mathbf{v}$ is a vector perpendicular to the plane of $\mathbf{u}$ and $\mathbf{v}$ and has length $|\mathbf{u} \times \mathbf{v}| =$ _____.

**3.** The cross product is anticommutative; that is, $\mathbf{u} \times \mathbf{v} =$ _____.

**4.** Two vectors are _____ if and only if their cross product is $\mathbf{0}$.

## PROBLEM SET 14.3

**1.** Let $\mathbf{a} = -3\mathbf{i} + 2\mathbf{j} - 2\mathbf{k}$, $\mathbf{b} = -\mathbf{i} + 2\mathbf{j} - 4\mathbf{k}$, and $\mathbf{c} = 7\mathbf{i} + 3\mathbf{j} - 4\mathbf{k}$. Find each of the following.

(a) $\mathbf{a} \times \mathbf{b}$      (b) $\mathbf{a} \times (\mathbf{b} + \mathbf{c})$
(c) $\mathbf{a} \cdot (\mathbf{b} + \mathbf{c})$      (d) $\mathbf{a} \times (\mathbf{b} \times \mathbf{c})$

**2.** If $\mathbf{a} = \langle 3, 3, 1 \rangle$, $\mathbf{b} = \langle -2, -1, 0 \rangle$, and $\mathbf{c} = \langle -2, -3, -1 \rangle$, find each of the following.

(a) $\mathbf{a} \times \mathbf{b}$      (b) $\mathbf{a} \times (\mathbf{b} + \mathbf{c})$
(c) $\mathbf{a} \cdot (\mathbf{b} \times \mathbf{c})$      (d) $\mathbf{a} \times (\mathbf{b} \times \mathbf{c})$

**3.** Find all vectors perpendicular to both of the vectors $\mathbf{a} = \mathbf{i} + 2\mathbf{j} + 3\mathbf{k}$ and $\mathbf{b} = -2\mathbf{i} + 2\mathbf{j} - 4\mathbf{k}$.

**4.** Find all vectors perpendicular to both of the vectors $\mathbf{a} = -2\mathbf{i} + 5\mathbf{j} - 2\mathbf{k}$ and $\mathbf{b} = 3\mathbf{i} - 2\mathbf{j} + 4\mathbf{k}$.

**5.** Find the unit vectors perpendicular to the plane determined by the three points $(1, 3, 5)$, $(3, -1, 2)$, and $(4, 0, 1)$.

**6.** Find the unit vectors perpendicular to the plane determined by the three points $(-1, 3, 0)$, $(5, 1, 2)$, and $(4, -3, -1)$.

**7.** Find the area of the parallelogram with $\mathbf{a} = -\mathbf{i} + \mathbf{j} - 3\mathbf{k}$ and $\mathbf{b} = 4\mathbf{i} + 2\mathbf{j} - 4\mathbf{k}$ as the adjacent sides.

**8.** Find the area of the parallelogram with $\mathbf{a} = 2\mathbf{i} + 2\mathbf{j} - \mathbf{k}$ and $\mathbf{b} = -\mathbf{i} + \mathbf{j} - 4\mathbf{k}$ as the adjacent sides.

**9.** Find the area of the triangle with $(3, 2, 1)$, $(2, 4, 6)$, and $(-1, 2, 5)$ as vertices.

**10.** Find the area of the triangle with $(1, 2, 3)$, $(3, 1, 5)$, and $(4, 5, 6)$ as vertices.

**11.** Find the equation of the plane through $(1, 3, 2)$, $(0, 3, 0)$, and $(2, 4, 3)$ (see Example 2).

**12.** Find the equation of the plane through $(1, 1, 2)$, $(0, 0, 1)$, and $(-2, -3, 0)$.

**13.** Find the equation of the plane through $(-1, -2, 3)$ and perpendicular to both the planes $x - 3y + 2z = 7$ and $2x - 2y - z = -3$.

**14.** Find the equation of the plane through $(2, -3, 2)$ and parallel to the plane of the vectors $4\mathbf{i} + 3\mathbf{j} - \mathbf{k}$ and $2\mathbf{i} - 5\mathbf{j} + 6\mathbf{k}$.

**15.** Find the equation of the plane through $(6, 2, -1)$ and perpendicular to the line of intersection of the planes $4x - 3y + 2z + 5 = 0$ and $3x + 2y - z + 11 = 0$.

**16.** Let **a** and **b** be nonzero vectors with the same initial point and let **c** be any nonzero vector. Show that $(\mathbf{a} \times \mathbf{b}) \times \mathbf{c}$ is a vector parallel to the plane of **a** and **b**.

**17.** Find the volume of the parallelepiped with edges $\langle 2, 3, 4 \rangle$, $\langle 0, 4, -1 \rangle$, and $\langle 5, 1, 3 \rangle$ (see Example 4).

**18.** Find the volume of the parallelepiped with edges $3\mathbf{i} - 4\mathbf{j} + 2\mathbf{k}$, $-\mathbf{i} + 2\mathbf{j} + \mathbf{k}$, and $3\mathbf{i} - 2\mathbf{j} + 5\mathbf{k}$.

**19.** Let $K$ be the parallelepiped determined by $\mathbf{u} = \langle 3, 2, 1 \rangle$, $\mathbf{v} = \langle 1, 1, 2 \rangle$, and $\mathbf{w} = \langle 1, 3, 3 \rangle$.

(a) Find the volume of $K$.
(b) Find the area of the face determined by **u** and **v**.
(c) Find the angle between **u** and the plane containing the face determined by **v** and **w**.

**20.** Show that if **a**, **b**, **c**, and **d** all lie in the same plane, then

$$(\mathbf{a} \times \mathbf{b}) \times (\mathbf{c} \times \mathbf{d}) = \mathbf{0}$$

**21.** The volume of a tetrahedron is known to be $\frac{1}{3}$ (area of base)(height). From this, show that the volume of the tetrahedron with edges **a**, **b**, and **c** is $\frac{1}{6} |\mathbf{a} \cdot (\mathbf{b} \times \mathbf{c})|$.

**22.** Find the volume of the tetrahedron with vertices $(-1, 2, 3)$, $(4, -1, 2)$, $(5, 6, 3)$, and $(1, 1, -2)$ (see Problem 21).

**23.** Prove **Lagrange's Identity**

$$|\mathbf{u} \times \mathbf{v}|^2 = |\mathbf{u}|^2|\mathbf{v}|^2 - (\mathbf{u} \cdot \mathbf{v})^2$$

Do this without using Theorem A.

**24.** Prove the left distributive law

$$\mathbf{u} \times (\mathbf{v} + \mathbf{w}) = (\mathbf{u} \times \mathbf{v}) + (\mathbf{u} \times \mathbf{w})$$

**25.** Use Problem 24 and the anticommutative law to prove the right distributive law.

**26.** If both $\mathbf{u} \times \mathbf{v} = \mathbf{0}$ and $\mathbf{u} \cdot \mathbf{v} = 0$, what can you conclude about **u** or **v**?

**27.** Use Example 3 to develop a formula for the area of the triangle with vertices $P(a, 0, 0)$, $Q(0, b, 0)$, and $R(0, 0, c)$ shown in the top half of Figure 6.

**28.** Show that the triangle in the plane with vertices $(x_1, y_1)$, $(x_2, y_2)$, and $(x_3, y_3)$ has area equal to one-half the absolute value of the determinant

$$\begin{vmatrix} x_1 & y_1 & 1 \\ x_2 & y_2 & 1 \\ x_3 & y_3 & 1 \end{vmatrix}$$

**29.** (A Pythagorean Theorem in Three-Space) As in Figure 6, let $P$, $Q$, $R$, and $O$ be the vertices of a (right-angled) tetrahedron and let $A$, $B$, $C$, and $D$ be the areas of the opposite faces, respectively. Show that $A^2 + B^2 + C^2 = D^2$.

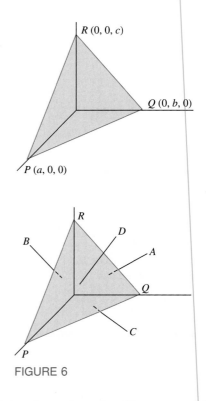

**FIGURE 6**

**30.** Let vectors **a**, **b**, and **c** with common initial point determine a tetrahedron and let **m**, **n**, **p**, and **q** be vectors perpendicular to the four faces, pointing outward, and having length equal to the area of the corresponding face. Show that $\mathbf{m} + \mathbf{n} + \mathbf{p} + \mathbf{q} = \mathbf{0}$.

**31.** Let **a**, **b**, and $\mathbf{a} - \mathbf{b}$ denote the three edges of a triangle with lengths $a$, $b$, and $c$, respectively. Use Lagrange's Identity together with $2\mathbf{a} \cdot \mathbf{b} = |\mathbf{a}|^2 + |\mathbf{b}|^2 - |\mathbf{a} - \mathbf{b}|^2$ to prove **Heron's Formula** for the area $A$ of a triangle, namely,

$$A = \sqrt{s(s - a)(s - b)(s - c)}$$

where $s$ is the semiperimeter $(a + b + c)/2$.

**32.** Use the method of Example 5 to show directly that if $\mathbf{u} = u_1\mathbf{i} + u_2\mathbf{j} + u_3\mathbf{k}$ and $\mathbf{v} = v_1\mathbf{i} + v_2\mathbf{j} + v_3\mathbf{k}$, then $\mathbf{u} \times \mathbf{v} = (u_2v_3 - u_3v_2)\mathbf{i} + (u_3v_1 - u_1v_3)\mathbf{j} + (u_1v_2 - u_2v_1)\mathbf{k}$

**Answers to Concepts Review:** **1.** $-3\mathbf{i} + 2\mathbf{j} - 7\mathbf{k}$
**2.** $|\mathbf{u}| |\mathbf{v}|\sin\theta$ **3.** $-(\mathbf{v} \times \mathbf{u})$ **4.** Parallel

## 14.4
## LINES AND CURVES
## IN THREE-SPACE

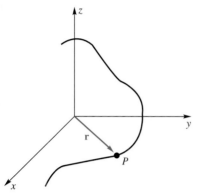

FIGURE 1

Our study of lines and curves in the plane extends easily to three-space. A **space curve** is determined by a triple of parametric equations

$$x = f(t), \qquad y = g(t), \qquad z = h(t), \qquad t \in I$$

with $f$, $g$, and $h$ continuous on the interval $I$. In vector language a curve is specified by giving the position vector $\mathbf{r} = \mathbf{r}(t)$ of a point $P = P(t)$, namely,

$$\mathbf{r} = \mathbf{r}(t) = \langle f(t), g(t), h(t) \rangle = f(t)\mathbf{i} + g(t)\mathbf{j} + h(t)\mathbf{k}$$

The tip of $\mathbf{r}$ traces out the curve as $t$ ranges over the interval $I$, as we see in Figure 1.

**Lines**  The simplest of all curves is a line. A line is determined by a fixed point $P_0$ and a fixed vector $\mathbf{v} = a\mathbf{i} + b\mathbf{j} + c\mathbf{k}$. It is the set of all points $P$ such that $\overrightarrow{P_0P}$ is parallel to $\mathbf{v}$—that is, that satisfy

$$\overrightarrow{P_0P} = t\mathbf{v}$$

for some real number $t$ (Figure 2). If $\mathbf{r} = \overrightarrow{OP}$ and $\mathbf{r}_0 = \overrightarrow{OP}_0$ are the position vectors of $P$ and $P_0$, respectively, then $\overrightarrow{P_0P} = \mathbf{r} - \mathbf{r}_0$ and the equation of the line can thus be written

$$\mathbf{r} = \mathbf{r}_0 + t\mathbf{v}$$

If we write $\mathbf{r} = \langle x, y, z \rangle$ and $\mathbf{r}_0 = \langle x_0, y_0, z_0 \rangle$ and equate components in the last equation above, we obtain

$$x = x_0 + at, \qquad y = y_0 + bt, \qquad z = z_0 + ct$$

These are **parametric equations** of the line through $(x_0, y_0, z_0)$ and parallel to $\mathbf{v} = \langle a, b, c \rangle$. The numbers $a$, $b$, and $c$ are called **direction numbers** for the line. They are not unique; any nonzero constant multiples $ka$, $kb$, and $kc$ are also direction numbers.

**EXAMPLE 1**  Find parametric equations for the line through $(3, -2, 4)$ and $(5, 6, -2)$ (see Figure 3).

*Solution*  A vector parallel to the given line is

$$\mathbf{v} = \langle 5 - 3, 6 + 2, -2 - 4 \rangle = \langle 2, 8, -6 \rangle$$

If we choose $(x_0, y_0, z_0)$ as $(3, -2, 4)$, we obtain the parametric equations

$$x = 3 + 2t, \qquad y = -2 + 8t, \qquad z = 4 - 6t$$

FIGURE 2

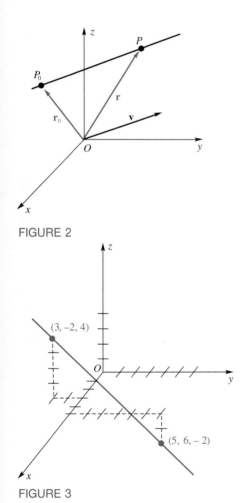

FIGURE 3

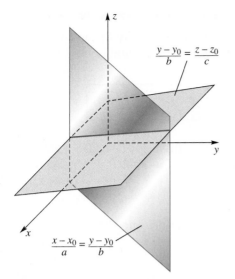

$$\frac{y - y_0}{b} = \frac{z - z_0}{c}$$

$$\frac{x - x_0}{a} = \frac{y - y_0}{b}$$

FIGURE 4

Note that $t = 0$ determines the point $(3, -2, 4)$, whereas $t = 1$ gives $(5, 6, -2)$. In fact, $0 \leq t \leq 1$ corresponds to the segment joining these two points. ∎

If we solve each of the parametric equations for $t$ (assuming $a$, $b$, and $c$ are all different from zero) and equate the results, we obtain the **symmetric equations** for the line through $(x_0, y_0, z_0)$ with direction numbers $a$, $b$, $c$—namely

$$\boxed{\frac{x - x_0}{a} = \frac{y - y_0}{b} = \frac{z - z_0}{c}}$$

This is the conjunction of the two equations

$$\frac{x - x_0}{a} = \frac{y - y_0}{b} \quad \text{and} \quad \frac{y - y_0}{b} = \frac{z - z_0}{c}$$

both of which are the equations of planes (Figure 4); and of course, the intersection of two planes is a line.

**EXAMPLE 2** Find the symmetric equations of the line that is parallel to the vector $\langle 4, -3, 2 \rangle$ and goes through $(2, 5, -1)$.

*Solution*

$$\frac{x - 2}{4} = \frac{y - 5}{-3} = \frac{z + 1}{2}$$

∎

**EXAMPLE 3** Find the symmetric equations of the line of intersection of the planes

$$2x - y - 5z = -14 \quad \text{and} \quad 4x + 5y + 4z = 28$$

*Solution* We begin by finding two points on the line. Any two points would do, but we choose to find the points where the line pierces the $yz$-plane and the $xz$-plane (Figure 5). The former is obtained by setting $x = 0$ and solving the resulting equations $-y - 5z = -14$ and $5y + 4z = 28$ simultaneously. This yields the point $(0, 4, 2)$. A similar procedure with $y = 0$ gives the point $(3, 0, 4)$. Consequently, a vector parallel to the required line is

$$\langle 3 - 0, 0 - 4, 4 - 2 \rangle = \langle 3, -4, 2 \rangle$$

Using $(3, 0, 4)$ for $(x_0, y_0, z_0)$, we get

$$\frac{x - 3}{3} = \frac{y - 0}{-4} = \frac{z - 4}{2}$$

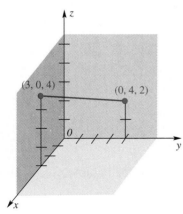

FIGURE 5

An alternative solution is based on the fact that the line of intersection of two planes is perpendicular to both of their normals. $\mathbf{u} = \langle 2, -1, -5 \rangle$ is a normal to the first plane; $\mathbf{v} = \langle 4, 5, 4 \rangle$ is a normal to the second. Since

$$\mathbf{u} \times \mathbf{v} = \begin{vmatrix} \mathbf{i} & \mathbf{j} & \mathbf{k} \\ 2 & -1 & -5 \\ 4 & 5 & 4 \end{vmatrix} = 21\mathbf{i} - 28\mathbf{j} + 14\mathbf{k}$$

the vector $\mathbf{w} = \langle 21, -28, 14 \rangle$ is parallel to the required line. This implies that $\frac{1}{7}\mathbf{w} = \langle 3, -4, 2 \rangle$ also has this property. Next, find any point on the line of intersection—for example, $(3, 0, 4)$—and proceed as in the earlier solution. ∎

**EXAMPLE 4**   Find symmetric equations or parametric equations of the line through $(1, -2, 3)$ that is perpendicular to both the $x$-axis and the line

$$\frac{x - 4}{2} = \frac{y - 3}{-1} = \frac{z}{5}$$

*Solution*   The $x$-axis and the given line have directions $\mathbf{u} = \langle 1, 0, 0 \rangle$ and $\mathbf{v} = \langle 2, -1, 5 \rangle$, respectively. A vector perpendicular to both $\mathbf{u}$ and $\mathbf{v}$ is

$$\mathbf{u} \times \mathbf{v} = \begin{vmatrix} \mathbf{i} & \mathbf{j} & \mathbf{k} \\ 1 & 0 & 0 \\ 2 & -1 & 5 \end{vmatrix} = 0\mathbf{i} - 5\mathbf{j} - \mathbf{k}$$

The required line is parallel to $\langle 0, -5, -1 \rangle$ and so also to $\langle 0, 5, 1 \rangle$. Since the first direction number is zero, the line does not have symmetric equations. Its parametric equations are

$$x = 1, \qquad y = -2 + 5t, \qquad z = 3 + t$$  ∎

**The Tangent Line to a Curve**   Let

$$\mathbf{r} = \mathbf{r}(t) = f(t)\mathbf{i} + g(t)\mathbf{j} + h(t)\mathbf{k}$$

be the position vector determining a curve in three-space (Figure 6). In complete analogy with what we did in the plane (Section 13.4), we define $\mathbf{r}'(t)$ by

$$\mathbf{r}'(t) = \lim_{h \to 0} \frac{\mathbf{r}(t + h) - \mathbf{r}(t)}{h}$$

It follows that $\mathbf{r}'(t)$, if it exists, has the direction of the tangent line to the curve at the point $P(t)$ corresponding to $t$. Moreover, $\mathbf{r}'(t)$ exists if and only if $f'(t)$, $g'(t)$, and $h'(t)$ exist and, in this case,

$$\mathbf{r}'(t) = f'(t)\mathbf{i} + g'(t)\mathbf{j} + h'(t)\mathbf{k}$$

Thus, $f'(t)$, $g'(t)$, and $h'(t)$ are direction numbers for the tangent line at $P$.

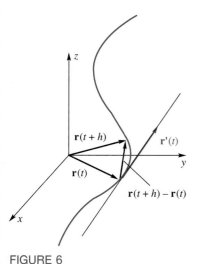

FIGURE 6

**EXAMPLE 5** Find the symmetric equations for the tangent line to the curve determined by

$$\mathbf{r}(t) = t\mathbf{i} + \tfrac{1}{2}t^2\mathbf{j} + \tfrac{1}{3}t^3\mathbf{k}$$

at $P(2) = (2, 2, \tfrac{8}{3})$.

*Solution*

$$\mathbf{r}'(t) = \mathbf{i} + t\mathbf{j} + t^2\mathbf{k}$$

and

$$\mathbf{r}'(2) = \mathbf{i} + 2\mathbf{j} + 4\mathbf{k}$$

so the tangent line has direction $\langle 1, 2, 4 \rangle$. Its symmetric equations are

$$\frac{x - 2}{1} = \frac{y - 2}{2} = \frac{z - \tfrac{8}{3}}{4}$$

■

## CONCEPTS REVIEW

**1.** The parametric equations for a line through $(1, -3, 2)$ parallel to the vector $\langle 4, -2, -1 \rangle$ are $x =$ _____, $y =$ _____, $z =$ _____.

**2.** The symmetric equations for the line of Question 1 are _____.

**3.** If $\mathbf{r}(t) = t^2\mathbf{i} - 3t\mathbf{j} + t^3\mathbf{k}$, then $\mathbf{r}'(t) =$ _____.

**4.** A vector parallel to the tangent line at $t = 1$ of the curve determined by the position vector $\mathbf{r}(t)$ of Question 3 is _____. This tangent line has symmetric equations _____.

## PROBLEM SET 14.4

In Problems 1–4, find the parametric equations of the line through the given pair of points.

**1.** $(1, -2, 3), (4, 5, 6)$     **2.** $(2, -1, -5), (7, -2, 3)$

**3.** $(4, 2, 3), (6, 2, -1)$     **4.** $(5, -3, -3), (5, 4, 2)$

In Problems 5–8, write both the parametric equations and the symmetric equations for the line through the given point parallel to the given vector.

**5.** $(4, 5, 6), \langle 3, 2, 1 \rangle$

**6.** $(-1, 3, -6), \langle -2, 0, 5 \rangle$

**7.** $(1, 1, 1), \langle -10, -100, -1000 \rangle$

**8.** $(-2, 2, -2), \langle 7, -6, 3 \rangle$

In Problems 9–12, find the symmetric equations of the line of intersection of the given pair of planes.

**9.** $4x + 3y - 7z = 1, 10x + 6y - 5z = 10$

**10.** $x + y - z = 2, 3x - 2y + z = 3$

**11.** $x + 4y - 2z = 13, 2x - y - 2z = 5$

**12.** $x - 3y + z = -1, 6x - 5y + 4z = 9$

**13.** Find the symmetric equations of the line through $(4, 0, 6)$ and perpendicular to the plane $x - 5y + 2z = 10$.

**14.** Find the symmetric equations of the line through $(-5, 7, -2)$ and perpendicular to both $\langle 2, 1, -3 \rangle$ and $\langle 5, 4, -1 \rangle$.

**15.** Find the parametric equations of the line through $(5, -3, 4)$ that intersects the $z$-axis at right angles.

**16.** Find the symmetric equations of the line through $(2, -4, 5)$ that is parallel to the plane $3x + y - 2z = 5$ and perpendicular to the line

$$\frac{x + 8}{2} = \frac{y - 5}{3} = \frac{z - 1}{-1}$$

**17.** Find the equation of the plane that contains the parallel lines

$$\begin{cases} x = -2 + 2t \\ y = 1 + 4t \\ z = 2 - t \end{cases} \quad \text{and} \quad \begin{cases} x = 2 - 2t \\ y = 3 - 4t \\ z = 1 + t \end{cases}$$

**18.** Show that the lines

$$\frac{x-1}{-4} = \frac{y-2}{3} = \frac{z-4}{-2}$$

and

$$\frac{x-2}{-1} = \frac{y-1}{1} = \frac{z+2}{6}$$

intersect and find the equation of the plane they determine.

**19.** Find the equation of the plane containing the line $x = 1 + 2t, y = -1 + 3t, z = 4 + t$ and the point $(1, -1, 5)$.

**20.** Find the equation of the plane containing the line $x = 3t, y = 1 + t, z = 2t$ and parallel to the intersection of the planes $2x - y + z = 0$ and $y + z + 1 = 0$.

**21.** Find the distance between the skew (nonintersecting) lines $x = 2 - t, y = 3 + 4t, z = 2t$ and $x = -1 + t, y = 2, z = -1 + 2t$ by using the following steps.

(a) Note by putting $t = 0$ that $(2, 3, 0)$ is on the first line.
(b) Find the equation of the plane $\pi$ through $(2, 3, 0)$ parallel to both given lines (that is with normal perpendicular to both).
(c) Find a point $Q$ on the second line.
(d) Find the distance from $Q$ to the plane $\pi$. (See Example 6 of Section 14.2.)
See Problem 30 for another way to do this problem.

**22.** Find the distance between the skew lines $x = 1 + 2t, y = -3 + 4t, z = -1 - t$ and $x = 4 - 2t, y = 1 + 3t, z = 2t$ (see Problem 21).

**23.** Find the symmetric equations of the tangent line to the curve with equation

$$\mathbf{r}(t) = 2 \cos t\mathbf{i} + 6 \sin t\mathbf{j} + t\mathbf{k}$$

at $t = \pi/3$.

**24.** Find the parametric equations of the tangent line to the curve $x = 2t^2, y = 4t, z = t^3$ at $t = 1$.

**25.** Find the equation of the plane perpendicular to the curve $x = 3t, y = 2t^2, z = t^5$ at $t = -1$.

**26.** Find the equation of the plane perpendicular to the curve

$$\mathbf{r}(t) = t \sin t\mathbf{i} + 3t\mathbf{j} + 2t \cos t\mathbf{k}$$

at $t = \pi/2$.

**27.** Consider the curve $\mathbf{r}(t) = \sin t \cos t\mathbf{i} + \sin^2 t\mathbf{j} + \cos t\mathbf{k}$. $0 \le t \le 2\pi$.

(a) Show that the curve lies on a sphere centered at the origin.
(b) Where does the tangent line at $t = \pi/6$ intersect the $xy$-plane?

**28.** (Point to Plane) Let $P$ be a point on a plane with normal $\mathbf{n}$ and $Q$ be a point off the plane (Figure 7). Show the distance $d$ from $Q$ to the plane is given by

$$d = \frac{|\overrightarrow{PQ} \cdot \mathbf{n}|}{|\mathbf{n}|}$$

and use this result to find the distance from $(4, -2, 3)$ to the plane $4x - 4y + 2z = 2$. Check with Example 6 of Section 14.2.

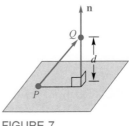

FIGURE 7

**29.** (Point to Line) Let $P$ be a point on a line with direction $\mathbf{n}$ and $Q$ a point off the line (Figure 8). Show that the distance $d$ from $Q$ to the line is given by

$$d = \frac{|\overrightarrow{PQ} \times \mathbf{n}|}{|\mathbf{n}|}$$

and use this result to find each distance in (a) and (b).

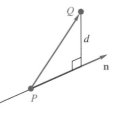

FIGURE 8

(a) From $Q(1, 0, -4)$ to the line $\frac{x-3}{2} = \frac{y+2}{-2} = \frac{z-1}{1}$.
(b) From $Q(2, -1, 3)$ to the line $x = 1 + 2t, y = -1 + 3t, z = -6t$.

**30.** (Line to Line) Let $P$ and $Q$ be points on nonintersecting skew lines with directions $\mathbf{n}_1$ and $\mathbf{n}_2$ and let $\mathbf{n} = \mathbf{n}_1 \times \mathbf{n}_2$ (Figure 9). Show that the distance $d$ between these lines is given by

$$d = \frac{|\overrightarrow{PQ} \cdot \mathbf{n}|}{|\mathbf{n}|}$$

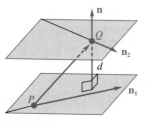

FIGURE 9

and use this result to find the distance between each pair of lines in (a) and (b).

(a) $\dfrac{x - 3}{1} = \dfrac{y + 2}{1} = \dfrac{z - 1}{2}$ and $\dfrac{x + 4}{3} = \dfrac{y + 5}{4} = \dfrac{z}{5}$

(b) $x = 1 + 2t, y = -2 + 3t, z = -4t$ and $x = 3t, y = 1 + t,$ $z = -5t$

---

**Answers to Concepts Review:** **1.** $1 + 4t; -3 - 2t; 2 - t$

**2.** $\dfrac{x - 1}{4} = \dfrac{y + 3}{-2} = \dfrac{z - 2}{-1}$ **3.** $2t\mathbf{i} - 3\mathbf{j} + 3t^2\mathbf{k}$

**4.** $<2, -3, 3>; \dfrac{x - 1}{2} = \dfrac{y + 3}{-3} = \dfrac{z - 1}{3}$

---

## 14.5
## VELOCITY, ACCELERATION, AND CURVATURE

All that we did with curvilinear motion in the plane (Sections 13.4 and 13.5) generalizes in a natural way to three-space. Let

$$\mathbf{r}(t) = f(t)\mathbf{i} + g(t)\mathbf{j} + h(t)\mathbf{k}, \qquad a \le t \le b$$

be the position vector for a point $P = P(t)$ that is tracing out a curve as $t$ increases (Figure 1). We suppose that $\mathbf{r}'(t)$ exists and is continuous and $\mathbf{r}'(t) \ne \mathbf{0}$, in which case the curve is said to be **smooth**. The length $s$ of the arc from $P(a)$ to $P(t)$ is given by

$$s = \int_a^t |\mathbf{r}'(u)|\, du = \int_a^t \sqrt{[f'(u)]^2 + [g'(u)]^2 + [h'(u)]^2}\, du$$

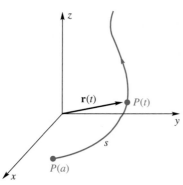

FIGURE 1

If $t$ measures time, we may define velocity, speed, and acceleration of the moving point $P$ by

Velocity:  $\mathbf{v}(t) = \mathbf{r}'(t)$

Speed:  $\dfrac{ds}{dt} = |\mathbf{r}'(t)| = |\mathbf{v}(t)|$

Acceleration: $\mathbf{a}(t) = \mathbf{r}''(t)$

**An Example: The Circular Helix**  Suppose that a point $P$ moves so that its position vector at time $t$ is

$$\mathbf{r}(t) = a \cos t\,\mathbf{i} + a \sin t\,\mathbf{j} + ct\mathbf{k}$$

where $a$ and $c$ are positive constants. Then $P$ traces a curve that winds around the right circular cylinder with parametric equations $x = a \cos t$, $y = a \sin t$ but spirals higher and higher because $z = ct$ increases with $t$. The curve is called a **circular helix**; part of it is shown in Figure 2.

**EXAMPLE 1**  Find the arc length of the circular helix for $0 \le t \le 2\pi$.

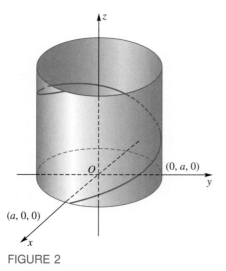

FIGURE 2

*Solution*

$$s = \int_0^{2\pi} \sqrt{(-a \sin t)^2 + (a \cos t)^2 + c^2} \, dt$$

$$= \int_0^{2\pi} \sqrt{a^2 + c^2} \, dt = 2\pi\sqrt{a^2 + c^2}$$

Note that when $c = 0$, this reduces to $2\pi a$ (the circumference of a circle of radius $a$), as it should.    ∎

**EXAMPLE 2**    For the motion $\mathbf{r}(t) = a \cos t\mathbf{i} + a \sin t\mathbf{j} + ct\mathbf{k}$ described above, calculate the acceleration $\mathbf{a}$ at $t = 2\pi$.

*Solution*

$$\mathbf{v}(t) = \mathbf{r}'(t) = -a \sin t\mathbf{i} + a \cos t\mathbf{j} + c\mathbf{k}$$

$$\mathbf{a}(t) = \mathbf{r}''(t) = -a \cos t\mathbf{i} - a \sin t\mathbf{j}$$

$$\mathbf{a}(2\pi) = -a\mathbf{i}$$    ∎

**EXAMPLE 3**    Beginning at $t = 0$, a bee flew so its position vector was $\mathbf{r} = t \cos t\mathbf{i} + t \sin t\mathbf{j} + t\mathbf{k}$ until $t = 4\pi$, at which time it went off on a tangent at the speed then attained. What was the total length of its flight path on the interval $0 \le t \le 4\pi + 3$?

*Solution*    The path consists of a spiral part and a straight-line part with lengths $L_1$ and $L_2$, respectively. On the spiral part,

$$\mathbf{r}'(t) = (\cos t - t \sin t)\mathbf{i} + (\sin t + t \cos t)\mathbf{j} + \mathbf{k}$$

and

$$|\mathbf{r}'(t)| = [(\cos t - t \sin t)^2 + (\sin t + t \cos t)^2 + 1]^{1/2} = \sqrt{2 + t^2}$$

Upon using Formula 44 at the end of the book, we find

$$L_1 = \int_0^{4\pi} \sqrt{2 + t^2} \, dt \approx 82.336$$

Also

$$L_2 = 3|\mathbf{r}'(4\pi)| = 3\sqrt{2 + 16\pi^2} \approx 37.937$$

We conclude that $L_1 + L_2 \approx 120.273$.    ∎

**Curvature**    As we noted in Section 14.4, $\mathbf{v}(t) = \mathbf{r}'(t)$ is a vector with the same direction as the tangent to the curve at $P(t)$. Thus,

$$\mathbf{T} = \mathbf{T}(t) = \frac{\mathbf{v}(t)}{|\mathbf{v}(t)|}$$

is a **unit tangent vector** at $P(t)$. Since $s$ denotes the arc length measured from some fixed point in the direction of increasing $t$, $d\mathbf{T}/ds$ measures the rate of change of direction of the tangent with respect to the distance along the curve. By the Chain Rule,

$$\frac{d\mathbf{T}}{ds} = \frac{d\mathbf{T}}{dt}\frac{dt}{ds} = \frac{\mathbf{T}'(t)}{|\mathbf{v}(t)|}$$

Thus, just as for plane curves, we may define the **curvature** $\kappa$ of a space curve by

$$\kappa = \kappa(t) = \left|\frac{d\mathbf{T}}{ds}\right| = \frac{|\mathbf{T}'(t)|}{|\mathbf{v}(t)|}$$

**EXAMPLE 4**  Find the curvature of the circular helix

$$\mathbf{r}(t) = a\cos t\mathbf{i} + a\sin t\mathbf{j} + ct\mathbf{k}, \qquad a > 0$$

*Solution*

$$\mathbf{v}(t) = -a\sin t\mathbf{i} + a\cos t\mathbf{j} + c\mathbf{k}$$

$$\mathbf{T}(t) = \frac{\mathbf{v}(t)}{|\mathbf{v}(t)|} = \frac{1}{\sqrt{a^2 + c^2}}(-a\sin t\mathbf{i} + a\cos t\mathbf{j} + c\mathbf{k})$$

$$\kappa(t) = \frac{|\mathbf{T}'(t)|}{|\mathbf{v}(t)|} = \frac{1}{a^2 + c^2}|-a\cos t\mathbf{i} - a\sin t\mathbf{j}|$$

$$= \frac{a}{a^2 + c^2}$$

Thus, $\kappa$ is a constant for the circular helix.  ∎

The radius of curvature $R$ is the reciprocal of $\kappa$. In the example above, $R = (a^2 + c^2)/a$. This reduces to $R = a$ when $c = 0$, which corresponds to the fact that the motion is then along the circle $\mathbf{r} = a\cos t\mathbf{i} + a\sin t\mathbf{j}$ in the $xy$-plane. When $c$ is large, $R$ is large as we should expect.

**Components of Acceleration**  Just as in the plane case, we define the **principal unit normal vector** $\mathbf{N}$ at $P$ by

$$\mathbf{N} = \frac{d\mathbf{T}/ds}{|d\mathbf{T}/ds|} = \frac{1}{\kappa}\frac{d\mathbf{T}}{ds}$$

It is obvious that $\mathbf{N}$ is a unit vector. That $\mathbf{N}$ is normal (perpendicular) to the curve follows by differentiating $\mathbf{T} \cdot \mathbf{T} = 1$ with respect to $s$. This gives

$$2\mathbf{T} \cdot \frac{d\mathbf{T}}{ds} = 0$$

which implies that $d\mathbf{T}/ds$ is perpendicular to $\mathbf{T}$. With these facts in hand, we can mimic a derivation given in the plane case (Section 13.5) to obtain (Figure 3)

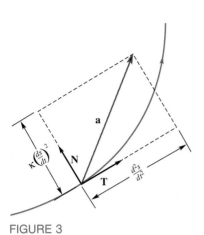

$$\mathbf{a} = \frac{d^2s}{dt^2}\,\mathbf{T} + \left(\frac{ds}{dt}\right)^2 \kappa\mathbf{N} = a_T\mathbf{T} + a_N\mathbf{N}$$

If we take the dot product of this last result with $\mathbf{T}$, we get

$$\mathbf{T} \cdot \mathbf{a} = a_T\mathbf{T} \cdot \mathbf{T} + a_N\mathbf{T} \cdot \mathbf{N} = a_T$$

**FIGURE 3**

or

$$a_T = \mathbf{T} \cdot \mathbf{a} = \frac{\mathbf{r}' \cdot \mathbf{r}''}{|\mathbf{r}'|}$$

If we take the cross product of $\mathbf{T}$ with $\mathbf{a}$, we get

$$\mathbf{T} \times \mathbf{a} = a_T(\mathbf{T} \times \mathbf{T}) + a_N(\mathbf{T} \times \mathbf{N}) = a_N(\mathbf{T} \times \mathbf{N})$$

and so

$$|\mathbf{T} \times \mathbf{a}| = a_N|\mathbf{T} \times \mathbf{N}| = a_N|\mathbf{T}|\,|\mathbf{N}| \sin\theta = a_N$$

or

$$a_N = |\mathbf{T} \times \mathbf{a}| = \frac{|\mathbf{r}' \times \mathbf{r}''|}{|\mathbf{r}'|}$$

Since $a_N = (ds/dt)^2\kappa = |\mathbf{r}'|^2\kappa$, we conclude that

$$\kappa = \frac{|\mathbf{r}' \times \mathbf{r}''|}{|\mathbf{r}'|^3}$$

**EXAMPLE 5**    At the point $(1, 1, \frac{1}{3})$, find $\mathbf{T}$, $\mathbf{N}$, $a_T$, $a_N$, and $\kappa$ for the curvilinear motion

$$\mathbf{r}(t) = t\mathbf{i} + t^2\mathbf{j} + \tfrac{1}{3}t^3\mathbf{k}$$

***Solution***

$$\mathbf{r}'(t) = \mathbf{i} + 2t\mathbf{j} + t^2\mathbf{k}$$
$$\mathbf{r}''(t) = 2\mathbf{j} + 2t\mathbf{k}$$

At $t = 1$ (which gives the point $(1, 1, \frac{1}{3})$), we have

$$\mathbf{r}' = \mathbf{i} + 2\mathbf{j} + \mathbf{k}$$

$$\mathbf{r}'' = 2\mathbf{j} + 2\mathbf{k}$$

$$\mathbf{T} = \frac{\mathbf{r}'}{|\mathbf{r}'|} = \frac{\mathbf{i} + 2\mathbf{j} + \mathbf{k}}{\sqrt{6}}$$

$$a_T = \frac{\mathbf{r}' \cdot \mathbf{r}''}{|\mathbf{r}'|} = \frac{6}{\sqrt{6}}$$

$$a_N = \frac{|\mathbf{r}' \times \mathbf{r}''|}{|\mathbf{r}'|} = \frac{1}{\sqrt{6}} \begin{vmatrix} \mathbf{i} & \mathbf{j} & \mathbf{k} \\ 1 & 2 & 1 \\ 0 & 2 & 2 \end{vmatrix} = \frac{1}{\sqrt{6}} |2\mathbf{i} - 2\mathbf{j} + 2\mathbf{k}| = \sqrt{2}$$

$$\mathbf{N} = \frac{\mathbf{a} - a_T\mathbf{T}}{a_N} = \frac{(2\mathbf{j} + 2\mathbf{k}) - (\mathbf{i} + 2\mathbf{j} + \mathbf{k})}{\sqrt{2}} = \frac{-\mathbf{i} + \mathbf{k}}{\sqrt{2}}$$

$$\kappa = \frac{|\mathbf{r}' \times \mathbf{r}''|}{|\mathbf{r}'|^3} = \frac{a_N}{|\mathbf{r}'|^2} = \frac{\sqrt{2}}{6}$$   ∎

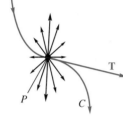

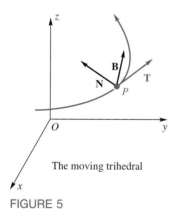

FIGURE 4

**The Binormal at P** Given a curve $C$ and the unit tangent vector $\mathbf{T}$ at $P$, there are, of course, infinitely many unit vectors perpendicular to $\mathbf{T}$ at $P$ (Figure 4). We picked one of them,

$$\mathbf{N} = \frac{d\mathbf{T}/ds}{|d\mathbf{T}/ds|}$$

and called it the principal normal. The vector

$$\mathbf{B} = \mathbf{T} \times \mathbf{N}$$

is called the **binormal**. It, too, is a unit vector and it is perpendicular to both $\mathbf{T}$ and $\mathbf{N}$. (Why?)

If the unit tangent vector $\mathbf{T}$, the principal normal $\mathbf{N}$, and the binormal $\mathbf{B}$ have their initial points at $P$, they form a right-handed, mutually perpendicular triple of unit vectors known as the **trihedral** at $P$ (Figure 5). This moving trihedral plays a crucial role in a subject called differential geometry. The plane of $\mathbf{T}$ and $\mathbf{N}$ is called the **osculating plane** at $P$.

The moving trihedral

FIGURE 5

## CONCEPTS REVIEW

**1.** If $\mathbf{r}(t)$ is the position vector for a point $P = P(t)$ moving along a curve, then the velocity of this point is the vector _____ and the acceleration is _____.

**2.** The arc length of the curve in Question 1 between $P(a)$ and $P(b)$ is given by _____.

**3.** Just as in the two-space case, $\mathbf{r}'(t)$ is _____ to

the tangent vector at $P(t)$ and $\mathbf{r}''(t)$ points to the _____ side of the curve.

**4.** If $\mathbf{T} = \mathbf{T}(s)$ denotes the unit tangent vector expressed in terms of arc length, then the curvature $\kappa$ is defined by $\kappa = $ _____.

## PROBLEM SET 14.5

In Problems 1–12, find the velocity $\mathbf{v}$, acceleration $\mathbf{a}$, and speed $s$ at the indicated time $t = t_1$.

1. $\mathbf{r}(t) = 4t\mathbf{i} + 5(t^2 - 1)\mathbf{j} + 2t\mathbf{k}; \ t_1 = 1$

2. $\mathbf{r}(t) = t\mathbf{i} + (t - 1)^2\mathbf{j} + (t - 3)^3\mathbf{k}; \ t_1 = 0$

3. $\mathbf{r}(t) = (1/t)\mathbf{i} + (t^2 - 1)^{-1}\mathbf{j} + t^5\mathbf{k}; \ t_1 = 2$

4. $\mathbf{r}(t) = t^6\mathbf{i} + (6t^2 - 5)^6\mathbf{j} + t\mathbf{k}; \ t_1 = 1$

5. $\mathbf{r}(t) = \mathbf{i} + \left(\int_0^t x^2 \, dx\right)\mathbf{j} + t^{\frac{2}{3}}\mathbf{k}; \ t_1 = 2$

6. $\mathbf{r}(t) = \int_1^t \left[x^2\mathbf{i} + 5(x - 1)^3\mathbf{j} + (\sin \pi x)\mathbf{k}\right] dx; \ t_1 = 2$

7. $\mathbf{r}(t) = \cos t\mathbf{i} + \sin t\mathbf{j} + t\mathbf{k}; \ t_1 = \pi$

8. $\mathbf{r}(t) = \sin 2t\mathbf{i} + \cos 3t\mathbf{j} + \cos 4t\mathbf{k}; \ t_1 = \dfrac{\pi}{2}$

9. $\mathbf{r}(t) = \tan t\mathbf{i} + 3e^t\mathbf{j} + \cos 4t\mathbf{k}; \ t_1 = \dfrac{\pi}{4}$

10. $\mathbf{r}(t) = \left(\int_t^1 e^x \, dx\right)\mathbf{i} + \left(\int_t^\pi \sin \pi\theta d\theta\right)\mathbf{j} + t^{\frac{2}{3}}\mathbf{k}; \ t_1 = 2$

11. $\mathbf{r}(t) = t \sin \pi t\mathbf{i} + t \cos \pi t\mathbf{j} + e^{-t}\mathbf{k}; \ t_1 = 2$

12. $\mathbf{r}(t) = \ln t\mathbf{i} + \ln t^2\mathbf{j} + \ln t^3\mathbf{k}; \ t_1 = 2$

13. Show that if the speed of a moving particle is constant, its acceleration vector is always perpendicular to its velocity vector.

14. Prove that $|\mathbf{r}(t)|$ is constant if and only if $\mathbf{r}(t) \cdot \mathbf{r}'(t) = 0$.

In Problems 15–26, find the length of the curve with the given vector equation.

15. $\mathbf{r}(t) = t\mathbf{i} \sin + t\mathbf{j} + \cos t\mathbf{k}; 0 \le t \le 2$

16. $\mathbf{r}(t) = t \cos t\mathbf{i} + t \sin t\mathbf{j} + \sqrt{2}t\mathbf{k}; 0 \le t \le 2$

17. $\mathbf{r}(t) = \sqrt{6}t^2\mathbf{i} + \frac{2}{3}t^3\mathbf{j} + 6t\mathbf{k}; 3 \le t \le 6$

18. $\mathbf{r}(t) = t^2\mathbf{i} - 2t^3\mathbf{j} + 6t^3\mathbf{k}; 0 \le t \le 1$

19. $\mathbf{r}(t) = t^3\mathbf{i} - 2t^3\mathbf{j} + 6t^3\mathbf{k}; 0 \le t \le 1$

20. $\mathbf{r}(t) = \sqrt{7}t^7\mathbf{i} - \sqrt{2}t^7\mathbf{j} + 6t^7\mathbf{k}; 0 \le t \le 1$

21. $\mathbf{r}(t) = \cos^3 t\mathbf{i} + \sin^3 t\mathbf{j}; 0 \le t \le \pi/2$

22. $\mathbf{r}(t) = e^{2t} \cos t\mathbf{i} + e^{2t} \sin t\mathbf{j} + e^{2t}\mathbf{k}; 0 \le t \le \pi$

23. $\mathbf{r}(t) = \sinh t\mathbf{i} + \cosh t\mathbf{j} + t\mathbf{k}; 0 \le t \le \pi$

24. $\mathbf{r}(t) = \sinh 3t\mathbf{i} + \cosh 3t\mathbf{j} + 3t\mathbf{k}; 0 \le t \le \pi/3$

25. $\mathbf{r}(t) = (t \sin t)\mathbf{i} + (2\sqrt{2}/3)t^{3/2}\mathbf{j} + (t \cos t)\mathbf{k}; 0 \le t \le \pi$

26. $\mathbf{r}(t) = \frac{1}{3}t^3\mathbf{i} + \frac{2}{9}(t^3 + 4)^{3/2}\mathbf{j} + \frac{1}{3}t^3\mathbf{k}; 0 \le t \le 3$

In Problems 27–39, find the curvature, the unit tangent vector, the principal normal, and the binormal at $t = t_1$.

27. $\mathbf{r}(t) = (t^2 - 1)\mathbf{i} + (2t + 3)\mathbf{j} + (t^2 - 4t)\mathbf{k}; \ t_1 = 2$

28. $\mathbf{r}(t) = \frac{1}{3}t^3\mathbf{i} + \frac{1}{2}t^2\mathbf{k}; \ t_1 = 1$

29. $\mathbf{r}(t) = \frac{1}{2}t^2\mathbf{i} + t\mathbf{j} + \frac{1}{3}t^3\mathbf{k}; \ t_1 = 2$

30. $x = \sin 3t, \ y = \cos 3t, \ z = t, \ t_1 = \pi/9$

31. $x = 7 \sin 3t, \ y = 7 \cos 3t, \ z = 14t, \ t_1 = \pi/3$

32. $\mathbf{r}(t) = \cos^3 t\mathbf{i} + \sin^3 t\mathbf{k}; \ t_1 = \pi/2$

33. $\mathbf{r}(t) = 3 \cosh(t/3)\mathbf{i} + t\mathbf{j}; \ t_1 = 1$

34. $\mathbf{r}(t) = (\sin t - t \cos t)\mathbf{i} + 5\mathbf{j} + (\cos t + t \sin t)\mathbf{k}; \ t_1 = \pi$

35. $\mathbf{r}(t) = -\sinh t\mathbf{i} + \cosh t\mathbf{j}; \ t_1 = 1$

36. $\mathbf{r}(t) = e^{7t} \cos 2t\mathbf{i} + e^{7t} \sin 2t\mathbf{j} + e^{7t}\mathbf{k}; \ t_1 = \pi/3$

37. $\mathbf{r}(t) = e^{-2t}\mathbf{i} + e^{2t}\mathbf{j} + 2\sqrt{2}t\mathbf{k}; \ t_1 = 0$

38. $x = \ln t, \ y = 3t, \ z = t^2; \ t_1 = 2$

39. $x = \sqrt{3} \sin t, \ y = \cos t, \ z = \cos t; \ t_1 = \pi/2$

In Problems 40–47, find the tangential and the normal vector components $a_T$ and $a_N$ of the acceleration vector at any time $t$.

40. $x = t, \ y = t^2, \ z = t^3$

41. $\mathbf{r}(t) = (t + 1)\mathbf{i} + 3t\mathbf{j} + t^2\mathbf{k}$

42. $\mathbf{r}(t) = (t - 2)^2\mathbf{i} - t^2\mathbf{j} + t\mathbf{k}$

43. $\mathbf{r}(t) = (t - \frac{1}{3}t^3)\mathbf{i} - (t + \frac{1}{3}t^3)\mathbf{j} + t\mathbf{k}$

44. $\mathbf{r}(t) = t\mathbf{i} + \frac{1}{3}t^3\mathbf{j} + t^{-1}\mathbf{k}, \ t > 0$

45. $\mathbf{r}(t) = (\ln \sin t)\mathbf{i} + (\ln \cos t)\mathbf{j} + t\mathbf{k}, \ 0 < t < \pi/2$

46. $\mathbf{r}(t) = t \sin t\mathbf{i} + t \cos t\mathbf{j} + t^2\mathbf{k}$

47. $x = e^{-t}, \ y = 2t, \ z = e^t$

In Problems 48–57, find the vectors $\mathbf{T}$, $\mathbf{N}$ and $\mathbf{B}$ at the indicated point.

48. $\mathbf{r}(t) = \cos^3 t\mathbf{i} + \sin^3 t\mathbf{k}, \ t = \pi/6$

49. $\mathbf{r}(t) = c \sinh(t/c)\mathbf{i} + t\mathbf{k}, \ t = \pi/6$

50. $\mathbf{r}(t) = t \sin t\mathbf{i} + t \cos t\mathbf{j} + t^2\mathbf{k}, \ t = \pi/2$

51. $\mathbf{r}(t) = t \cos t\mathbf{i} + t \sin t\mathbf{j} + t^2\mathbf{k}, \ t = \pi$

52. $\mathbf{r}(t) = t\mathbf{i} + \frac{1}{3}t^3\mathbf{j} + t^{-1}\mathbf{k}, \ t = 1$

53. $\mathbf{r}(t) = e^t\mathbf{i} + e^t \cos t\mathbf{j} + e^t \sin t\mathbf{k}, \ t = \pi/3$

54. $\mathbf{r}(t) = e^t\mathbf{i} + e^t \sin t\mathbf{j} + e^t \cos t\mathbf{k}, \ t = \pi/3$

55. $x = t, \ y = t^2, \ z = \frac{2}{3}t^3, \ t = 1$

56. $\mathbf{r}(t) = 5\mathbf{i} + (\cos t + t \sin t)\mathbf{j} + (\sin t - t \cos t)\mathbf{k}, \ t = \pi$

57. $\mathbf{r}(t) = \sinh t\mathbf{i} + t\mathbf{j} + \cosh t\mathbf{k}, \ t = \pi/3$

58. Consider the motion of a particle along a helix given by $\mathbf{r}(t) = \sin t\mathbf{i} + \cos t\mathbf{j} + (t^2 - 3t + 2)\mathbf{k}$, where the $\mathbf{k}$ component measures the height above the ground and $t \ge 0$.

(a) Does the particle ever move downwards?

(b) Does the particle ever stop moving?

(c) At what times does it reach a position 12 units above the ground?

(d) What is the velocity of the particle when it is 12 units above the ground?

(e) If the particle leaves the helix and moves along the line tangent to the helix when it is 12 units above the ground, give a vector describing its path.

**59.** In many places in the universe moons orbit planets which in turn orbit a star. These orbits are very close to conic sections and in some cases are very close to circular. We will assume that these orbits are circular with the star at the center of the planet's orbit and the planet at the center of the moon's orbit. We will further assume that all motion is in a single $xy$-plane. Suppose that in the time the planet orbits the star once, the moon orbits the planet 10 times.

(a) If the radius of the moon's orbit is $R_m$ and the radius of the planet's orbit about the star is $R_p$ show that the motion of the moon with respect to the star at the origin could be given by

$$x = R_p \cos t + R_m \cos 10t, \qquad y = R_p \sin t + R_m \sin 10t$$

(b) Find values for $R_p$, $R_m$ and $t$ so that at that instant the moon is motionless with respect to the star.

**60.** Assuming that the orbits of the earth about the sun and the moon about the earth lie in the same plane and are circular we can represent the motion of the moon by

$$\mathbf{r}(t) = [93 \cos(2\pi t) + 0.24 \cos(26\pi t)]\mathbf{i} \\ + [93 \sin(2\pi t) + 0.24 \sin(26\pi t)]\mathbf{j}$$

where $\mathbf{r}(t)$ is measured in millions of miles.

(a) What are the proper units for $t$?

(b) What is the period of each of the two motions?

(c) What is the maximum distance that the moon is from the sun?

(d) What is the minimum distance that the moon is from the sun?

(e) Is there ever a time that the moon is stationary with respect to the sun?

(f) What is the velocity, speed and acceleration of the moon when $t = 1/2$?

**61.** Describe in general terms the following "helical" type motions

(a) $\mathbf{r}(t) = \sin t\mathbf{i} + \cos t\mathbf{j} + t\mathbf{k}$

(b) $\mathbf{r}(t) = \sin t^3\mathbf{i} + \cos t^3\mathbf{j} + t^3\mathbf{k}$

(c) $\mathbf{r}(t) = \sin(t^3 + \pi)\mathbf{i} + t^3\mathbf{j} + \cos(t^3 + \pi)\mathbf{k}$

(d) $\mathbf{r}(t) = t \sin t\mathbf{i} + t \cos t\mathbf{j} + t\mathbf{k}$

(e) $\mathbf{r}(t) = t^{-2} \sin t\mathbf{i} + t^{-2} \cos t\mathbf{j} + t\mathbf{k}, t > 0$

(f) $\mathbf{r}(t) = t^2 \sin (\ln t)\,\mathbf{i} + \ln t\mathbf{j} + t^2 \cos (\ln t)\,\mathbf{k}, t > 1$

**62.** The aesthetics of form place a premium on smooth transitions. It is generally acknowledged that the eye can see abrupt changes (discontinuities) in the curvature. Most products are designed so that any cross section of the surface will have a continuous second derivative. As a first step, use

a graphing device to display the curves $y = 0$, for $x \le 0$ together with $y = x^2$ and $y = x^3$ for $x > 0$. Show that the curve

$$y = \begin{cases} 0 & \text{if } x \le 0 \\ x^3 & \text{if } x > 0 \end{cases}$$

has continuous slope, first derivatives and curvature at all points.

**63.** Find a curve given by a polynominal $P_5(x)$ which provides a smooth transition between two horizontal lines. That is, assume a function of the form $P_5(x) = a_0 + a_1 x + a_2 x^2 + a_3 x^3 + a_4 x^4 + a_5 x^5$ which provides a smooth transition between $y = 0$ for $x \le 0$ and $y = 1$ for $x \ge 1$ in such a way that the function, its slope and curvature are all continuous for all values of $x$.

$$y = \begin{cases} 0 & \text{if } x \le 0 \\ P_5(x) & \text{if } 0 < x < 1 \\ 1 & \text{if } x \ge 1 \end{cases}$$

*Hint*: $P_5(x)$ must satisfy the six conditions $P_5(0) = 0$, $P_5'(0) = 0$, $P_5''(0) = 0$, $P_5(1) = 1$, $P_5'(1) = 0$, and $P_5''(1) = 0$. Use these five conditions to determine $a_0, \ldots, a_5$ uniquely and thus find $P_5(x)$.

**64.** It is possible to find a curve which matches a straight line section so that all derivatives are continuous. Consider the function given by

$$F(x) = \begin{cases} 0 & \text{if } x \le 0 \\ e^{-1/x} & \text{if } x > 0 \end{cases}$$

Show that $F(x)$ has continuous derivatives of all orders for all $x$. *Hint*: Use l'Hôpital's rule. See Problem 82 for $\dfrac{d}{ds}(u \times v)$.

**65.** In Section 13.5 the curvature for a plane curve was defined as $\kappa = \left| \dfrac{d\phi}{ds} \right|$, where $\phi$ is the angle measured clockwise from $\mathbf{i}$ to $\mathbf{T}$. Show that this is consistent with the definition of curvature given in three dimensional space.

**66.** Show that the unit binormal vector $\mathbf{B} = \mathbf{T} \times \mathbf{N}$ has the property that $\dfrac{d\mathbf{B}}{ds}$ is perpendicular to $\mathbf{B}$.

**67.** Show that the unit binormal vector $\mathbf{B} = \mathbf{T} \times \mathbf{N}$ has the property that $\dfrac{d\mathbf{B}}{ds}$ is perpendicular to $\mathbf{T}$.

**68.** Using the result obtained in Problems 66–67 show that $\dfrac{d\mathbf{B}}{ds}$ must be parallel to $\mathbf{N}$, and consequently there must be a number $\tau$ depending on $s$ such that $\dfrac{d\mathbf{B}}{ds} = -\tau(s)\mathbf{N}$. The function $\tau(s)$ is called the torsion of curve and measures the twist of the curve from the plane determined by $\mathbf{T}$ and $\mathbf{N}$.

**69.** Show that for a plane curve the torsion $\tau(s) = 0$.

**70.** Show that for a straight line $\mathbf{r}(t) = \mathbf{r}_0 + a_0 t\mathbf{i} + b_0 t\mathbf{j} + c_0 t\mathbf{k}$ that both $\kappa$ and $\tau$ are zero.

**71.** There are a set of formulas called the *Frenet-Serret formulas* which connect $\mathbf{T}$, $\mathbf{B}$ and $\mathbf{N}$ with the torsion $\tau$ and curvature $\kappa$. These formulas play a fundamental role in understanding the moving trihedral and in classical differential geometry. The first formula follows from the definition of the principal unit normal $\mathbf{N}$ and is given by $\dfrac{d\mathbf{T}}{ds} = \kappa(s)\mathbf{N}$. Another formula comes from Problem 68 and is given by $\dfrac{d\mathbf{B}}{ds} = -\tau(s)\mathbf{N}$. The last formula is given by $\dfrac{d\mathbf{N}}{ds} = \kappa(s)\mathbf{T} + \tau(s)\mathbf{B}$. Derive this formula using the two previous formulas and the relation $\mathbf{N} = \mathbf{B} \times \mathbf{T}$.

**72.** Using the Frenet-Serret formulas

$$\frac{d\mathbf{T}}{ds} = \kappa(s)\,\mathbf{N}$$

$$\frac{d\mathbf{N}}{ds} = \kappa(s)\mathbf{T} + \tau(s)\mathbf{B}$$

$$\frac{d\mathbf{B}}{ds} = -\kappa(s)\mathbf{N}$$

derive each of the following formulas where primes denote derivatives with respect to $t$. *Hint*: You will have to use the chain rule repeatedly to change derivatives with respect to $s$ to ones with respect to $t$ and remember that $\kappa$ and $\tau$ are both functions of $s$.

(a) $\mathbf{r}'' = s''\mathbf{T} + \kappa \cdot (s')^2\mathbf{N}$

(b) $\mathbf{r}' \times \mathbf{r}'' = \kappa \cdot (s')^3\mathbf{B}$

(c) $\mathbf{r}''' = [s''' - \kappa^2 \cdot (s')^3]\mathbf{T}$
$+ [3\kappa \cdot s's'' + k' \cdot (s')^2]\mathbf{N}$
$+ \kappa \cdot \tau \cdot (s')^3\mathbf{B}$

(d) $\tau = \dfrac{(\mathbf{r}' \times \mathbf{r}'') \cdot \mathbf{r}'''}{|\mathbf{r}' \times \mathbf{r}''|^2}$

(e) $\tau = -\dfrac{1}{s'}\,(\mathbf{B}' \cdot \mathbf{N})$

*Hint*: Start from $\tau = -\left(\dfrac{d\mathbf{B}}{ds}\right) \cdot \mathbf{N}$ and use the chain rule to change the derivatives to be with respect to $t$.

**73.** Use the formulas $\tau = \dfrac{(\mathbf{r}' \times \mathbf{r}'') \cdot \mathbf{r}'''}{|\mathbf{r}' \times \mathbf{r}''|^2}$ and $\kappa = \dfrac{|\mathbf{r}' \times \mathbf{r}''|}{|\mathbf{r}'|^3}$ for the torsion and curvature to find $\tau$ and $\kappa$ for the following curves

(a) $\mathbf{r}(t) = \cos t\mathbf{i} + \sin t\mathbf{j} + t\mathbf{k}$

(b) $\mathbf{r}(t) = at\mathbf{i} + c\sin t\mathbf{j} + c\cos t\mathbf{k}$

(c) $\mathbf{r}(t) = \sinh t\mathbf{i} + \cosh t\mathbf{j} + t\mathbf{k}$ at $t = \ln 2$

(d) $\mathbf{r}(t) = t\mathbf{i} + \frac{1}{2}t^2\mathbf{j} + \frac{1}{3}t^3\mathbf{k}$

**74.** For the helix $\mathbf{r}(t) = at\mathbf{i} + c\sin t\mathbf{j} + c\cos t\mathbf{k}$ with $a$ and $c \geq 0$, and $a$ fixed, maximize the $\kappa$ by choosing $c$. Justify your answer.

**75.** For the helix $\mathbf{r}(t) = at\mathbf{i} + c\sin t\mathbf{j} + c\cos t\mathbf{k}$ with $a$ and $c \geq 0$, and $a$ fixed, maximize the $\tau$ by choosing $a$. Justify your answer.

$\boxed{\text{C}}$ **76.** The DNA molecule in humans is a double helix, each with about $2.9 \times 10^8$ complete turns. Each helix has radius about 10 angstroms and rises about 34 angstroms on each complete turn (an angstrom is $10^{-8}$ centimeter). What is the total length of such a helix (see Example 1)?

**77.** A fly is crawling along a wire helix so that its position vector is $\mathbf{r} = 6\cos \pi t\mathbf{i} + 6\sin \pi t\mathbf{j} + 2t\mathbf{k}$, $t \geq 0$. At what point will the fly hit the sphere $x^2 + y^2 + z^2 = 100$ and how far did it travel in getting there (assuming it started when $t = 0$)?

**78.** A bee flew the spiral path

$$\mathbf{r} = 100e^{-t}(\cos t\mathbf{i} + \sin t\mathbf{j} + \mathbf{k})$$

Where was its final resting place and how far did it travel during its lifetime $0 \leq t \leq \infty$?

**79.** Solve the following vector differential equations ($c$ and $\mathbf{c}$ are constants), subject to $\mathbf{r} = \mathbf{r}_0$ at $t = 0$.

(a) $\dfrac{d\mathbf{r}}{dt} = \mathbf{0}$        (b) $\dfrac{d\mathbf{r}}{dt} = \mathbf{c}$

(c) $\dfrac{d^2\mathbf{r}}{dt^2} = \mathbf{c}$ subject also to $\mathbf{r}' = \mathbf{v}_0$ at $t = 0$

(d) $\dfrac{d\mathbf{r}}{dt} = c\mathbf{r}$

**80.** A bee was flying along a helical path so that its position vector at time $t$ was $\mathbf{r} = \cos t\mathbf{i} + \sin t\mathbf{j} + 16t\mathbf{k}$. At $t = 12$, it had a heart attack and died instantly. Where did it land (that is, hit the $xy$-plane)? Assume distance is in feet, time is in seconds, and $g = 32$ feet per second per second. *Hint*: Measure time from the instant of the heart attack and use the result of Problem 79(c).

**81.** Where will the bee of Example 3 hit the plane $x + y = 30$?

**82.** Prove that for differentiable vector-valued functions $\mathbf{F}(t)$ and $\mathbf{G}(t)$,

$$\frac{d}{dt}\,[\mathbf{F}(t) \times \mathbf{G}(t)] = \mathbf{F}(t) \times \mathbf{G}'(t) + \mathbf{F}'(t) \times \mathbf{G}(t)$$

and use this to show that

$$\frac{d}{dt}\,[\mathbf{r}(t) \times \mathbf{r}'(t)] = \mathbf{r}(t) \times \mathbf{r}''(t)$$

**83.** Show that if an object moves subject only to a central force (that is, $\mathbf{r}''(t) = c\mathbf{r}(t)$), then the object moves in a plane. *Hint*: Show that $\mathbf{r}(t) \times \mathbf{r}'(t)$ is a constant.

**84.** The angular momentum $\mathbf{L}(t)$ and torque $\tau(t)$ of a moving particle of mass $m$ and position vector $\mathbf{r}(t)$ are

$$\mathbf{L}(t) = m\mathbf{r}(t) \times \mathbf{v}(t) \qquad \tau(t) = m\mathbf{r}(t) \times \mathbf{a}(t)$$

Show that (a) $\mathbf{L}'(t) = \tau(t)$, (b) if $\tau(t) = \mathbf{0}$ for all $t$, then $\mathbf{L}(t)$ is constant (the law of conservation of angular momentum), and (c) angular momentum is conserved for a particle moving under a central force.

**85.** Let $\mathbf{r} = r\cos\theta\mathbf{i} + r\sin\theta\mathbf{j}$, where $r$ and $\theta$ are polar coordinates of an object of mass $m$ moving in the $xy$-plane. Show that the angular momentum $\mathbf{L}(t)$ is $(mr^2\, d\theta/dt)\mathbf{k}$. *Note*: This result together with Problems 83 and 84 and Problem 32 of Section 12.8 establish Kepler's Second Law.

**86.** Find the equation of the osculating plane at $P(1, 1, \frac{1}{3})$ for the motion of Example 5.

**87.** Prove that $\mathbf{N} = \mathbf{B} \times \mathbf{T}$, where $\mathbf{T}$, $\mathbf{N}$, and $\mathbf{B}$ form a trihedral.

---

**Answers to Concepts Review:** **1.** $\mathbf{r}'(t)$; $\mathbf{r}''(t)$ **2.** $\int_a^b |\mathbf{r}'(t)|\, dt$
**3.** Parallel; concave **4.** $|d\mathbf{T}/ds|$

---

## 14.6 SURFACES IN THREE-SPACE

The graph of an equation in three variables is normally a surface. We have met two examples already. The graph of $Ax + By + Cz = D$ is a plane; the graph of $(x - h)^2 + (y - k)^2 + (z - l)^2 = r^2$ is a sphere. Graphing surfaces can be very complicated. It is best accomplished by finding the intersections of the surface with well-chosen planes. These intersections are called **cross sections** (Figure 1); those with the coordinate planes are also called **traces**.

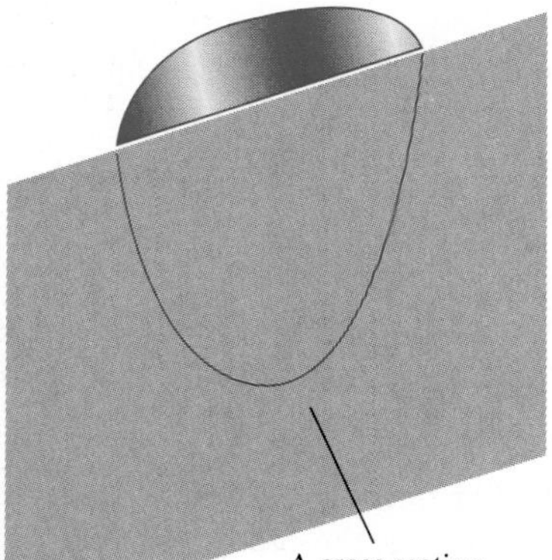

**EXAMPLE 1** Sketch the graph of

$$\frac{x^2}{16} + \frac{y^2}{25} + \frac{z^2}{9} = 1$$

*Solution* To find the trace in the $xy$-plane, we set $z = 0$ in the given equation. The graph of the resulting equation

$$\frac{x^2}{16} + \frac{y^2}{25} = 1$$

is an ellipse. The traces in the $xz$-plane and the $yz$-plane (obtained by setting $y = 0$ and $x = 0$, respectively) are also ellipses. These three traces are shown in Figure 2 and help provide a good visual image of the required surface (called an ellipsoid). ∎

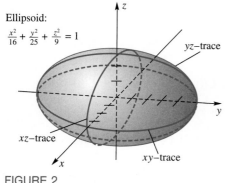

A cross section

FIGURE 1

Ellipsoid:
$\frac{x^2}{16} + \frac{y^2}{25} + \frac{z^2}{9} = 1$

yz–trace

xz–trace

xy–trace

FIGURE 2

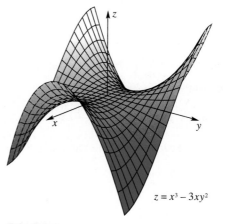

$z = x^3 - 3xy^2$

FIGURE 3

If the surface is very complicated, it may be useful to show the cross sections with many planes parallel to the coordinate planes. Here a computer with graphics capability can be very helpful. In Figure 3 on the previous page, we have shown a typical computer-generated graph, the graph of the so-called monkey saddle: $z = x^3 - 3xy^2$. We have more to say about computer-generated graphs in the next chapter.

**Cylinders**    You should be familiar with right circular cylinders from high school geometry. Here the word *cylinder* will denote a much more extensive class of surfaces.

Let $C$ be a plane curve and let $l$ be a line intersecting $C$ that is not in the plane of $C$. The set of all points on lines that are parallel to $l$ and that intersect $C$ is called a **cylinder** (Figure 4).

Cylinders occur naturally when we graph an equation in three-space that involves just two variables. Consider as a first example,

$$\frac{y^2}{a^2} - \frac{x^2}{b^2} = 1$$

in which the variable $z$ is missing. This equation determines a curve $C$ in the $xy$-plane, namely, a hyperbola. Moreover, if $(x_1, y_1, 0)$ satisfies the equation, so does $(x_1, y_1, z)$. As $z$ runs through all real values, the point $(x_1, y_1, z)$ traces out a line parallel to the $z$-axis. We conclude that the graph of the given equation is a cylinder, a hyperbolic cylinder (Figure 5).

A second example is the graph of $z = \sin y$ (Figure 6).

**Quadric Surfaces**    If a surface is the graph in three-space of an equation of second degree, it is called a **quadric surface**. Plane sections of a quadric surface are conics.

The general second-degree equation has the form

$$Ax^2 + By^2 + Cz^2 + Dxy + Exz + Fyz + Gx + Hy + Iz + J = 0$$

It can be shown that any such equation can be reduced, by rotation and translation of coordinate axes, to one of the two forms

$$Ax^2 + By^2 + Cz^2 + J = 0$$

or

$$Ax^2 + By^2 + Iz = 0$$

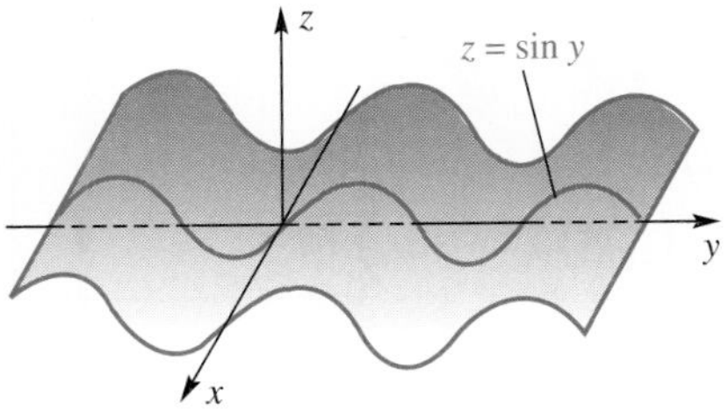

The quadric surfaces represented by the first of these equations are symmetric with respect to the coordinate planes and the origin. They are called **central quadrics**.

In Figures 7 to 12 (pages 700–701), we show six general types of quadric surfaces. Study them carefully. The graphs were drawn by an artist; we do not expect that most of our readers will be able to duplicate them in doing the problems. A more reasonable drawing for most people to make is one like that is shown in Figure 13 with our next example on page 702.

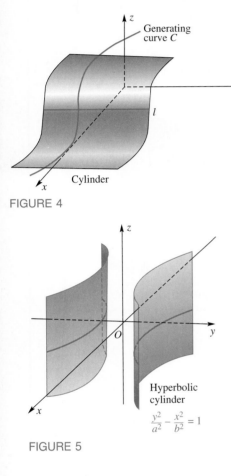

Generating curve $C$

$l$

Cylinder

FIGURE 4

Hyperbolic cylinder

$\dfrac{y^2}{a^2} - \dfrac{x^2}{b^2} = 1$

FIGURE 5

$z = \sin y$

FIGURE 6

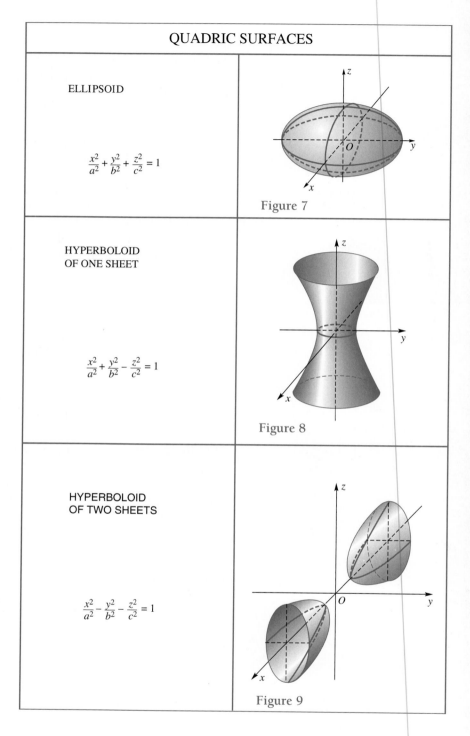

## QUADRIC SURFACES

ELLIPSOID

$$\frac{x^2}{a^2} + \frac{y^2}{b^2} + \frac{z^2}{c^2} = 1$$

Figure 7

HYPERBOLOID OF ONE SHEET

$$\frac{x^2}{a^2} + \frac{y^2}{b^2} - \frac{z^2}{c^2} = 1$$

Figure 8

HYPERBOLOID OF TWO SHEETS

$$\frac{x^2}{a^2} - \frac{y^2}{b^2} - \frac{z^2}{c^2} = 1$$

Figure 9

## QUADRIC SURFACES

| | |
|---|---|
| ELLIPTIC PARABOLOID<br><br>$z = \dfrac{x^2}{a^2} + \dfrac{y^2}{b^2}$ | 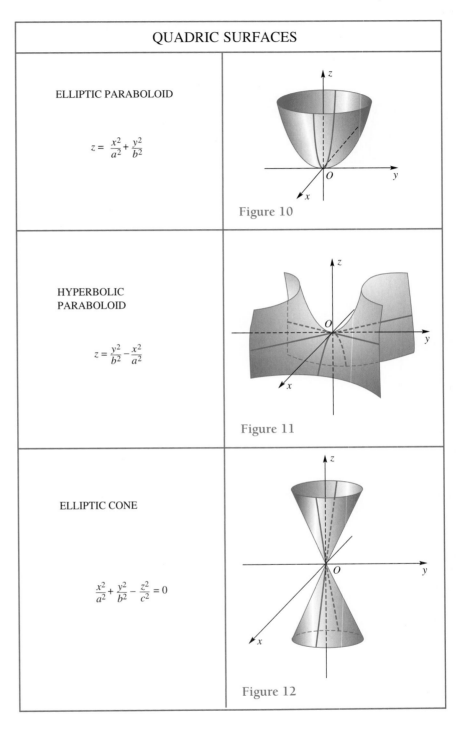<br>Figure 10 |
| HYPERBOLIC PARABOLOID<br><br>$z = \dfrac{y^2}{b^2} - \dfrac{x^2}{a^2}$ | Figure 11 |
| ELLIPTIC CONE<br><br>$\dfrac{x^2}{a^2} + \dfrac{y^2}{b^2} - \dfrac{z^2}{c^2} = 0$ | Figure 12 |

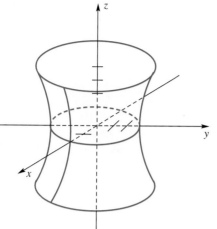

**FIGURE 13**

**EXAMPLE 2**   Analyze the equation

$$\frac{x^2}{4} + \frac{y^2}{9} - \frac{z^2}{16} = 1$$

and sketch its graph.

*Solution*   The traces in the three coordinate planes are obtained by setting $z = 0$, $y = 0$, and $x = 0$, respectively.

$$xy\text{-plane:}\qquad \frac{x^2}{4} + \frac{y^2}{9} = 1,\qquad \text{an ellipse}$$

$$xz\text{-plane:}\qquad \frac{x^2}{4} - \frac{z^2}{16} = 1,\qquad \text{a hyperbola}$$

$$yz\text{-plane:}\qquad \frac{y^2}{9} - \frac{z^2}{16} = 1,\qquad \text{a hyperbola}$$

These traces are graphed in Figure 13. We have also shown the cross sections in the planes $z = 4$ and $z = -4$. Note that when we substitute $z = \pm 4$ in the original equation, we get

$$\frac{x^2}{4} + \frac{y^2}{9} - \frac{16}{16} = 1$$

which is equivalent to

$$\frac{x^2}{8} + \frac{y^2}{18} = 1$$

an ellipse.   ∎

**EXAMPLE 3**   By inspection, name the graph of each of the following equations.
(a) $4x^2 + 4y^2 - 25z^2 + 100 = 0$
(b) $y^2 + z^2 - 12y = 0$
(c) $x^2 - z^2 = 0$
(d) $9x^2 + 4z^2 - 36y = 0$

*Solution*
(a) After dividing this equation by $-100$, it takes the form

$$-\frac{x^2}{25} - \frac{y^2}{25} + \frac{z^2}{4} = 1$$

Its graph is a hyperboloid of two sheets. It does not intersect the $xy$-plane but cross sections in planes parallel to this plane (and at least 2 units away) are circles.
(b) The variable $x$ does not appear, so the graph is a cylinder parallel to the $x$-axis. Since the equation can be written in the form $(y - 6)^2 + z^2 = 36$, its graph is a circular cylinder.

(c) Since the variable $y$ is missing, the graph is a cylinder. The given equation can be written $(x - z)(x + z) = 0$; so its graph consists of the two planes $x = z$ and $x = -z$.

(d) The equation can be rewritten as

$$\frac{x^2}{4} + \frac{z^2}{9} = y$$

which has an elliptic paraboloid as its graph. It is symmetric with respect to the $y$-axis. ∎

## CONCEPTS REVIEW

**1.** The intersections of a surface with the coordinate planes are called _____. More generally, intersections with any plane are called _____.

**2.** Equations involving just two variables when graphed in three-space generate surfaces called _____. In particular, the graph of $x^2 + y^2 = 1$ is an ordinary right circular cylinder whose axis is the _____.

**3.** The graph of $3x^2 + 2y^2 + 4z^2 = 12$ is a surface called an _____.

**4.** The graph of $4z = x^2 + 2y^2$ is a surface called an _____.

## PROBLEM SET 14.6

Name and sketch each of the following equations in three-space.

**1.** $4x^2 + 36y^2 = 144$     **2.** $y^2 + z^2 = 15$

**3.** $3x + 2z = 10$     **4.** $z^2 = 3y$

**5.** $x^2 + y^2 - 8x + 4y + 13 = 0$

**6.** $2x^2 - 16z^2 = 0$

**7.** $4x^2 + 9y^2 + 49z^2 = 1764$

**8.** $9x^2 - y^2 + 9z^2 - 9 = 0$

**9.** $4x^2 + 16y^2 - 32z = 0$   **10.** $-x^2 + y^2 + z^2 = 0$

**11.** $y = e^{2z}$          **12.** $6x - 3y = \pi$

**13.** $x^2 - z^2 + y = 0$

**14.** $x^2 + y^2 - 4z^2 + 4 = 0$

**15.** $9x^2 + 4z^2 - 36y = 0$

**16.** $9x^2 + 25y^2 + 9z^2 = 225$

**17.** $5x + 8y - 2z = 10$     **18.** $y = \cos x$

**19.** $z = \sqrt{16 - x^2 - y^2}$     **20.** $z = \sqrt{x^2 + y^2 + 1}$

**21.** The graph of an equation in $x$, $y$, and $z$ is symmetric with respect to the $xy$-plane if replacing $z$ by $-z$ results in an equivalent equation. What condition leads to a graph that is symmetric with respect to each of the following?

(a) The $yz$-plane.     (b) The $z$-axis.
(c) The $x$-axis.     (d) The origin.

**22.** Which of the equations in Problems 1–20 has a graph which is symmetric with respect to each of the following?

(a) The $xy$-plane.     (b) The $z$-axis.

**23.** If the curve $z = x^2$ is revolved about the $z$-axis, the resulting surface has equation $z = x^2 + y^2$, obtained as a result of replacing $x$ by $\sqrt{x^2 + y^2}$. If $y = 2x^2$ is revolved about the $y$-axis, what is the equation of the resulting surface?

**24.** Find the equation of the surface that results when the curve $z = 2y$ is revolved about the $z$-axis.

**25.** Find the equation of the surface that results when the curve $4x^2 + 3y^2 = 12$ is revolved about the $y$-axis.

**26.** Find the equation of the surface that results when the curve $4x^2 - 3y^2 = 12$ is revolved about the $x$-axis.

**27.** Find the coordinates of the foci of the ellipse which is the intersection of $z = x^2/4 + y^2/9$ with the plane $z = 4$.

**28.** Find the coordinates of the focus of the parabola which is the intersection of $z = x^2/4 + y^2/9$ with $x = 4$.

**29.** Find the area of the elliptical cross section cut from the surface $x^2/a^2 + y^2/b^2 + z^2/c^2 = 1$ by the plane $z = h$, $-c < h < c$. *Recall*: The area of the ellipse $x^2/A^2 + y^2/B^2 = 1$ is $\pi AB$.

**30.** Show that the volume of the solid bounded by the elliptic paraboloid $x^2/a^2 + y^2/b^2 = h - z$, $h > 0$, and the $xy$-plane is $\pi abh^2/2$, that is, the volume is one half the area of the base times the height. *Hint*: Use the method of slabs of Section 6.2.

**31.** Show that the projection in the $xz$-plane of the curve that is the intersection of the surfaces $y = 4 - x^2$ and $y = x^2 + z^2$ is an ellipse, and find its major and minor diameters.

**32.** Sketch the triangle in the plane $y = x$ that is above the plane $z = y/2$, below the plane $z = 2y$, and inside the cylinder $x^2 + y^2 = 8$. Then find the area of this triangle.

**33.** Show that the spiral $\mathbf{r} = t \cos t\mathbf{i} + t \sin t\mathbf{j} + t\mathbf{k}$ of Example 3 in Section 14.5 lies on the circular cone $x^2 + y^2 - z^2 = 0$. On what surface does the spiral $\mathbf{r} = 3t \cos t\mathbf{i} + t \sin t\mathbf{j} + t\mathbf{k}$ lie?

**34.** Show that the curve determined by $\mathbf{r} = t\mathbf{i} + t\mathbf{j} + t^2\mathbf{k}$ is a parabola and find the coordinates of its focus.

---

**Answers to Concepts Review:** **1.** Traces; cross sections **2.** Cylinders; $z$-axis **3.** Ellipsoid **4.** Elliptic paraboloid

---

## 14.7
## CYLINDRICAL AND
## SPHERICAL COORDINATES

Giving the Cartesian (rectangular) coordinates $(x, y, z)$ is just one of many ways of specifying the position of a point in three-space. Two other kinds of coordinates that play a significant role in calculus are cylindrical coordinates $(r, \theta, z)$ and spherical coordinates $(\rho, \theta, \phi)$. The meaning of the three kinds of coordinates is illustrated for the same point $P$ in Figure 1.

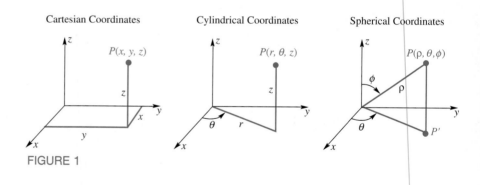

FIGURE 1

The **cylindrical coordinate system** uses the polar coordinates $r$ and $\theta$ (Section 12.6) in place of Cartesian coordinates $x$ and $y$ in the plane. The $z$-coordinate is the same as in Cartesian coordinates. We will usually require that $r \geq 0$ and we will restrict $\theta$ so that $0 \leq \theta < 2\pi$.

A point $P$ has **spherical coordinates** $(\rho, \theta, \phi)$ if $\rho$ (the Greek letter rho) is the distance $|OP|$ from the origin to $P$, $\theta$ is the polar angle associated with the projection $P'$ of $P$ onto the $xy$-plane, and $\phi$ is the angle between the positive $z$-axis and the line segment $OP$. We require that

$$\rho \geq 0, \qquad 0 \leq \theta < 2\pi, \qquad 0 \leq \phi \leq \pi$$

Cylindrical and spherical coordinates are designed to help us with certain kinds of problems. We consider some of them now. Further uses will appear in Chapter 16.

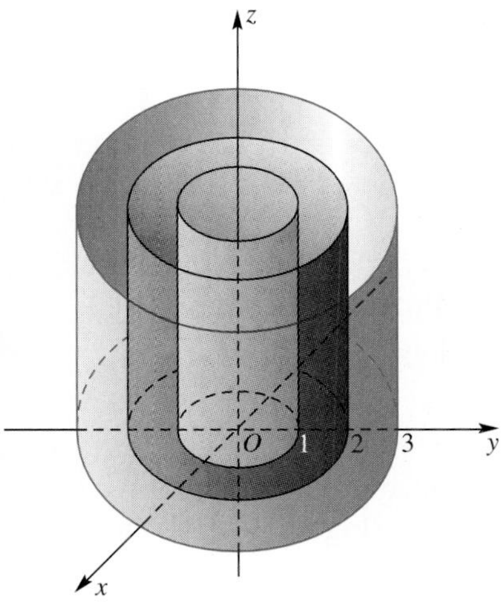

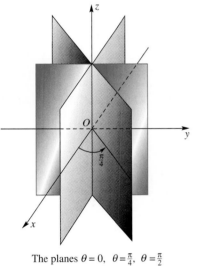

The cylinders $r = 1, r = 2, r = 3$

FIGURE 2

**Cylindrical Coordinates**    If a solid or a surface has an axis of symmetry, it is often wise to orient it so this axis is the $z$-axis and then use cylindrical coordinates. Note in particular the simplicity of the equation of a circular cylinder with $z$-axis symmetry (Figure 2) and also of a plane containing the $z$-axis (Figure 3). In Figure 3, we have allowed $r < 0$.

Cylindrical and Cartesian coordinates are related by the equations

$$x = r \cos \theta, \qquad\qquad y = r \sin \theta, \qquad z = z$$

$$r^2 = x^2 + y^2, \qquad \tan \theta = \frac{y}{x}$$

With these relationships, we can go back and forth between the two coordinate systems.

**EXAMPLE 1**    Find (a) the Cartesian coordinates of the point with cylindrical coordinates $(4, 2\pi/3, 5)$ and (b) the cylindrical coordinates of the point with Cartesian coordinates $(-5, -5, 2)$.

*Solution*

(a)    $x = 4 \cos \dfrac{2\pi}{3} = 4 \cdot \left(-\dfrac{1}{2}\right) = -2$

$y = 4 \sin \dfrac{2\pi}{3} = 4 \cdot \left(\dfrac{\sqrt{3}}{2}\right) = 2\sqrt{3}$

Thus, the Cartesian coordinates of $(4, 2\pi/3, 5)$ are $(-2, 2\sqrt{3}, 5)$.

(b)    $r = \sqrt{(-5)^2 + (-5)^2} = 5\sqrt{2}$

$\tan \theta = \dfrac{-5}{-5} = 1, \qquad$ so $\theta = \dfrac{5\pi}{4}$

The cylindrical coordinates of $(-5, -5, 2)$ are $(5\sqrt{2}, 5\pi/4, 2)$.    ■

**EXAMPLE 2**    Find the equations in cylindrical coordinates of the paraboloid and cylinder whose Cartesian equations are $x^2 + y^2 = 4 - z$ and $x^2 + y^2 = 2x$.

*Solution*

Paraboloid:    $r^2 = 4 - z$

Cylinder:    $r^2 = 2r \cos \theta \qquad$ or (equivalently) $\qquad r = 2 \cos \theta$

Division of an equation by a variable creates the potential for losing a solution. For example, dividing $x^2 = x$ by $x$ gives $x = 1$ and loses the solution $x = 0$. Similarly, dividing $r^2 = 2r \cos \theta$ by $r$ gives $r = 2 \cos \theta$ and appears to lose the solution $r = 0$ (the origin). However, the origin satisfies the equation $r = 2 \cos \theta$ with coordinates $(0, \pi/2)$. Thus, $r^2 = 2r \cos \theta$ and $r = 2 \cos \theta$ have identical polar graphs (see CAUTION in Section 12.6).    ■

The planes $\theta = 0, \ \theta = \frac{\pi}{4}, \ \theta = \frac{\pi}{2}$

FIGURE 3

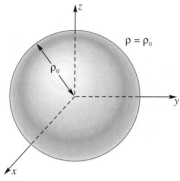

FIGURE 4

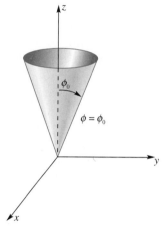

FIGURE 5

**EXAMPLE 3**    Find the Cartesian equations of the surfaces whose equations in cylindrical coordinates are $r^2 + 4z^2 = 16$ and $r^2 \cos 2\theta = z$.

*Solution*    Since $r^2 = x^2 + y^2$, the surface $r^2 + 4z^2 = 16$ has the Cartesian equation $x^2 + y^2 + 4z^2 = 16$, or $x^2/16 + y^2/16 + z^2/4 = 1$. Its graph is an ellipsoid.

Since $\cos 2\theta = \cos^2\theta - \sin^2\theta$, the second equation can be written $r^2 \cos^2\theta - r^2 \sin^2\theta = z$. In Cartesian coordinates it becomes $x^2 - y^2 = z$, the graph of which is a hyperbolic paraboloid.    ∎

**Spherical Coordinates**    When a solid or a surface is symmetric with respect to a point, spherical coordinates are likely to play a simplifying role. In particular, a sphere centered at the origin (Figure 4) has the simple equation $\rho = \rho_0$. Also note the equation of a cone with axis along the $z$-axis and vertex at the origin (Figure 5), namely, $\phi = \phi_0$.

It is easy to determine the relationship between spherical and cylindrical coordinates and then between spherical and Cartesian coordinates.

$$
\begin{array}{lll}
r = \rho \sin \phi, & \theta = \theta, & z = \rho \cos \phi \\
x = \rho \sin \phi \cos \theta, & y = \rho \sin \phi \sin \theta, & z = \rho \cos \phi \\
\end{array}
$$
$$
\rho = \sqrt{x^2 + y^2 + z^2}
$$

**EXAMPLE 4**    Find the Cartesian coordinates of the point $P$ with spherical coordinates $(8, \pi/3, 2\pi/3)$.

*Solution*    We have plotted the point $P$ in Figure 6.

$$x = 8 \sin \frac{2\pi}{3} \cos \frac{\pi}{3} = 8 \frac{\sqrt{3}}{2} \frac{1}{2} = 2\sqrt{3}$$

$$y = 8 \sin \frac{2\pi}{3} \sin \frac{\pi}{3} = 8 \frac{\sqrt{3}}{2} \frac{\sqrt{3}}{2} = 6$$

$$z = 8 \cos \frac{2\pi}{3} = 8 \left( -\frac{1}{2} \right) = -4$$

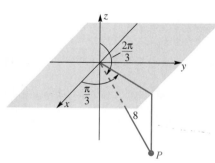

FIGURE 6

Thus, $P$ has Cartesian coordinates $(2\sqrt{3}, 6, -4)$.    ∎

**EXAMPLE 5**    Describe the graph of $\rho = 2 \cos \phi$.

*Solution*    We change to Cartesian coordinates. Multiply both sides by $\rho$ to obtain

$$\rho^2 = 2\rho \cos \phi$$
$$x^2 + y^2 + z^2 = 2z$$
$$x^2 + y^2 + (z - 1)^2 = 1$$

The graph is a sphere of radius 1 centered at the point with Cartesian coordinates $(0, 0, 1)$.   ∎

**EXAMPLE 6**   Find the equation of the paraboloid $z = x^2 + y^2$ in spherical coordinates.

*Solution*   Substituting for $x$, $y$, and $z$ yields

$$\rho \cos \phi = \rho^2 \sin^2\phi \cos^2\theta + \rho^2 \sin^2\phi \sin^2\theta$$

$$\rho \cos \phi = \rho^2 \sin^2\phi(\cos^2\theta + \sin^2\theta)$$

$$\rho \cos \phi = \rho^2 \sin^2\phi$$

$$\cos \phi = \rho \sin^2\phi$$

$$\rho = \cos \phi \csc^2\phi$$

Note that $\phi = \pi/2$ yields $\rho = 0$, which shows that we did not lose the origin when we canceled $\rho$ at the fourth step.   ∎

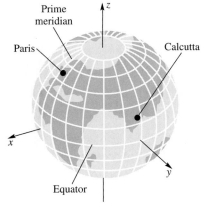

Prime meridian

Paris

Calcutta

Equator

**FIGURE 7**

**Spherical Coordinates in Geography**   Geographers and navigators use a coordinate system very closely related to spherical coordinates, the longitude-latitude system. Suppose that the earth is a sphere with center at the origin, that the positive $z$-axis passes through the North Pole, and that the positive $x$-axis passes through the prime meridian (Figure 7). By convention, longitudes are specified in degrees east or west of the prime meridian and latitudes in degrees north or south of the equator. It is a simple matter to determine spherical coordinates from such data.

**EXAMPLE 7**   Assuming the earth to be a sphere of radius 3960 miles, find the great-circle distance from Paris (longitude 2.2° E, latitude 48.4° N) to Calcutta (longitude 88.2° E, latitude 22.3° N).

*Solution*   We first calculate the spherical angles $\theta$ and $\phi$ for the two cities.

| | |
|---|---|
| Paris: | $\theta = 2.2° \approx 0.0384$ radians |
| | $\phi = 90° - 48.4° = 41.6° \approx 0.7261$ radians |
| Calcutta: | $\theta = 88.2° \approx 1.5394$ radians |
| | $\phi = 90° - 22.3° = 67.7° \approx 1.1816$ radians |

From these data and $\rho = 3960$ miles, we determine the Cartesian coordinates, as illustrated in Example 4.

| | |
|---|---|
| Paris: | $P_1(2627.2, 100.9, 2961.3)$ |
| Calcutta: | $P_2(115.1, 3662.0, 1502.6)$ |

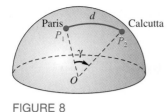

FIGURE 8

Next, referring to Figure 8, we determine $\gamma$, the angle between $\overrightarrow{OP}_1$ and $\overrightarrow{OP}_2$.

$$\cos \gamma = \frac{\overrightarrow{OP}_1 \cdot \overrightarrow{OP}_2}{|\overrightarrow{OP}_1| \, |\overrightarrow{OP}_2|} \approx \frac{(2627.2)(115.1) + (100.9)(3662) + (2961.3)(1502.6)}{(3960)(3960)}$$

$$\approx 0.3266$$

Thus, $\gamma \approx 1.2381$ radians and the great-circle distance $d$ is

$$d = \rho\gamma \approx (3960)(1.2381) \approx 4903 \text{ miles} \qquad \blacksquare$$

## CONCEPTS REVIEW

1. In cylindrical coordinates, the graph of $r = 6$ is a _____; in spherical coordinates, the graph of $\rho = 6$ is a _____.

2. In cylindrical coordinates, the graph of $\theta = \pi/6$ is a _____; in spherical coordinates, the graph of $\phi = \pi/6$ is a _____.

3. The equation _____ connects $\rho$ with $r$ and $z$.

4. The equation $\rho^2 = 4\rho \cos \phi$ in spherical coordinates becomes the equation _____ when written in rectangular coordinates.

## PROBLEM SET 14.7

1. Change the following from cylindrical to Cartesian (rectangular) coordinates.

(a) $(6, \pi/6, -2)$        (b) $(4, 4\pi/3, -8)$

2. Change the following from spherical to Cartesian coordinates.

(a) $(8, \pi/4, \pi/6)$        (b) $(4, \pi/3, 3\pi/4)$

3. Change the following from Cartesian to spherical coordinates.

(a) $(2, -2\sqrt{3}, 4)$        (b) $(-\sqrt{2}, \sqrt{2}, 2\sqrt{3})$

4. Change the following from Cartesian to cylindrical coordinates.

(a) $(2, 2, 3)$        (b) $(4\sqrt{3}, -4, 6)$

In Problems 5–14, sketch the graph of the given cylindrical or spherical equation.

5. $r = 5$        6. $\rho = 5$

7. $\phi = \pi/6$        8. $\theta = \pi/6$

9. $r = 3 \cos \theta$        10. $r = 2 \sin 2\theta$

11. $\rho = 3 \cos \phi$        12. $\rho = \sec \phi$

13. $r^2 + z^2 = 9$        14. $r^2 \cos^2\theta + z^2 = 4$

In Problems 15–28, make the required change in the given equation.

15. $x^2 + y^2 = 9$ to cylindrical coordinates.

16. $x^2 - y^2 = 25$ to cylindrical coordinates.

17. $x^2 + y^2 + 4z^2 = 10$ to cylindrical coordinates.

18. $x^2 + y^2 + 4z^2 = 10$ to spherical coordinates.

19. $2x^2 + 2y^2 - 4z^2 = 0$ to spherical coordinates.

20. $x^2 - y^2 - z^2 = 1$ to spherical coordinates.

21. $r^2 + 2z^2 = 4$ to spherical coordinates.

22. $\rho = 2 \cos \phi$ to cylindrical coordinates.

23. $x + y = 4$ to cylindrical coordinates.

24. $x + y + z = 1$ to spherical coordinates.

25. $x^2 + y^2 = 9$ to spherical coordinates.

26. $r = 2 \sin \theta$ to Cartesian coordinates.

27. $r^2 \cos 2\theta = z$ to Cartesian coordinates.

28. $\rho \sin \phi = 1$ to Cartesian coordinates.

29. The parabola $z = 2x^2$ is revolved about the $z$-axis. Write the equation of the resulting surface in cylindrical coordinates.

30. The hyperbola $2x^2 - z^2 = 2$ is revolved about the $z$-axis. Write the equation of the resulting surface in cylindrical coordinates.

C 31. Find the great-circle distance from St. Paul (longitude 93.1° W, latitude 45° N) to Oslo (longitude 10.5° E, latitude 59.6° N). See Example 7.

**C** 32. Find the great-circle distance from New York (longitude 74° W, latitude 40.4° N) to Greenwich (longitude 0°, latitude 51.3° N).

**C** 33. Find the great-circle distance from St. Paul (longitude 93.1° W, latitude 45° N) to Turin, Italy (longitude 7.4° E, latitude 45° N).

**C** 34. What is the distance along the 45° parallel between St. Paul and Turin? See Problem 33.

**C** 35. How close does the great-circle route from St. Paul to Turin get to the North Pole? See Problem 33.

36. Let $(\rho_1, \theta_1, \phi_1)$ and $(\rho_2, \theta_2, \phi_2)$ be the spherical coordinates of two points and let $d$ be the straight-line distance between them. Show that

$$d = \{(\rho_1 - \rho_2)^2 + 2\rho_1\rho_2[1 - \cos(\theta_1 - \theta_2)\sin \phi_1 \sin \phi_2$$
$$- \cos \phi_1 \cos \phi_2]\}^{1/2}$$

37. Let $(a, \theta_1, \phi_1)$ and $(a, \theta_2, \phi_2)$ be two points on the sphere $\rho = a$. Show (using Problem 36) that the great-circle distance between these points is $a\gamma$, where $0 \le \gamma \le \pi$ and

$$\cos \gamma = \cos(\theta_1 - \theta_2)\sin \phi_1 \sin \phi_2 + \cos \phi_1 \cos \phi_2$$

38. As you may have guessed, there is a simple formula for expressing great-circle distance directly in terms of longitude and latitude. Let $(\alpha_1, \beta_1)$ and $(\alpha_2, \beta_2)$ be the longitude-latitude coordinates of two points on the surface of the earth, where we interpret N and E as positive and S and W as negative. Show that the great-circle distance between these points is $3960\gamma$ miles, where $0 \le \gamma \le \pi$ and

$$\cos \gamma = \cos(\alpha_1 - \alpha_2)\cos \beta_1 \cos \beta_2 + \sin \beta_1 \sin \beta_2$$

**C** 39. Use Problem 38 to find the great-circle distance between each pair of places.

(a) New York and Greenwich (see Problem 32).
(b) St. Paul and Turin (see Problem 33).
(c) Turin and the South Pole (use $\alpha_1 = \alpha_2$).
(d) New York and Cape Town (18.4° E, 33.9° S).
(e) Two points on the equator with longitudes 100° E and 80° W, respectively.

40. It is easy to see that the graph of $\rho = 2a \cos \phi$ is a sphere of radius $a$ sitting on the $xy$-plane at the origin. But what is the graph of $\rho = 2a \sin \phi$?

---

**Answers to Concepts Review:** 1. Circular cylinder; sphere 2. Plane; cone 3. $\rho^2 = r^2 + z^2$ 4. $x^2 + y^2 + (z - 2)^2 = 4$

---

# 14.8 CHAPTER REVIEW

## Concepts Test

Respond with true or false to each of the following assertions. Be prepared to justify your answer.

1. Each point in three-space has a unique set of Cartesian coordinates.

2. The equation $x^2 + y^2 + z^2 - 4x + 9 = 0$ represents a sphere.

3. The linear equation $Ax + By + Cz = D$ represents a plane in three-space provided $A$, $B$, and $C$ are not all zero.

4. In three-space, the equation $Ax + By = C$ represents a line.

5. The planes $3x - 2y + 4z = 12$ and $3x - 2y + 4z = -12$ are parallel and 24 units apart.

6. The vector $\langle 1, -2, 3 \rangle$ is parallel to the plane $2x - 4y + 6z = 5$.

7. The line $x = 2t - 1$, $y = 4t + 2$, $z = 6t - 5$ goes through the point $(0, 4, -2)$.

8. If $\mathbf{u} = a\mathbf{i} + b\mathbf{j} + c\mathbf{k}$ is a unit vector then $a$, $b$, and $c$ are direction cosines for $\mathbf{u}$.

9. For any vector $\mathbf{u}$, $\|\mathbf{u}|\mathbf{u}|\|^2 = |\mathbf{u}|^2$.

10. For any vector $\mathbf{u}$, $|\mathbf{u}| \cdot \mathbf{u} = \mathbf{u} \cdot |\mathbf{u}|$.

11. For any vectors $\mathbf{u}$ and $\mathbf{v}$, $|\mathbf{u} \times \mathbf{v}| = |\mathbf{v} \times \mathbf{u}|$.

12. If $\mathbf{u}$ is a scalar multiple of $\mathbf{v}$, then $\mathbf{u} \times \mathbf{v} = \mathbf{0}$.

13. The cross product of two unit vectors is a unit vector.

14. Multiplying each component of a vector $\mathbf{v}$ by the scalar $a$ multiplies the length of $\mathbf{v}$ by $a$.

15. For any nonzero and nonperpendicular vectors $\mathbf{u}$ and $\mathbf{v}$ with angle $\theta$ between them, $|\mathbf{u} \times \mathbf{v}| \div (\mathbf{u} \cdot \mathbf{v}) = \tan \theta$.

16. If $\mathbf{u} \cdot \mathbf{v} = 0$ and $\mathbf{u} \times \mathbf{v} = \mathbf{0}$, then $\mathbf{u}$ or $\mathbf{v}$ is $\mathbf{0}$.

17. The volume of the parallelepiped determined by $2\mathbf{i}$, $2\mathbf{j}$, and $\mathbf{j} \times \mathbf{i}$ is 4.

18. For all vectors $\mathbf{u}$, $\mathbf{v}$, and $\mathbf{w}$, $\mathbf{u} \times (\mathbf{v} \times \mathbf{w}) = (\mathbf{u} \times \mathbf{v}) \times \mathbf{w}$.

19. If $a_1\mathbf{i} + a_2\mathbf{j} + a_3\mathbf{k}$ is a vector in the plane $b_1x + b_2y + b_3z = 0$, then $a_1b_1 + a_2b_2 + a_3b_3 = 0$.

20. Any line can be represented by both parametric equations and by symmetric equations.

21. When $\kappa = 0$ the path is a straight line.

**22.** When $\mathbf{r}'(t)$ and $\mathbf{r}''(t)$ are parallel then curvature must be zero.

**23.** The curvature depends on the shape of the curve and the speed with which you move along the curve.

**24.** If the velocity of the motion along the curve is of constant magnitude then there can be no acceleration.

**25.** $\mathbf{T}$, $\mathbf{N}$, and $\mathbf{B}$ depend only on the shape of the curve and not on the speed of motion along the curve.

**26.** If $\mathbf{v}$ is perpendicular to $\mathbf{a}$ then the speed of motion along the curve must be a constant.

**27.** If $\mathbf{v}$ is perpendicular to $\mathbf{a}$ then the path of motion must be a circle.

**28.** The only curves with constant curvature are straight lines and circles.

**29.** An ellipse has its maximum curvature at points on the major axis.

**30.** If an object moves subject only to a central force then it must move in a plane.

### Sample Test Problems

**1.** Find the equation of the sphere with $(-2, 3, 3)$ and $(4, 1, 5)$ as endpoints of a diameter.

**2.** Find the center and radius of the sphere with equation $x^2 + y^2 + z^2 - 6x + 2y - 8z = 0$.

**3.** Sketch the two position vectors $\mathbf{a} = 2\mathbf{i} - \mathbf{j} + 2\mathbf{k}$ and $\mathbf{b} = 5\mathbf{i} + \mathbf{j} - 3\mathbf{k}$. Then find each of the following.

(a) Their lengths.
(b) Their direction cosines.
(c) The unit vector with the same direction as $\mathbf{a}$.
(d) The angle $\theta$ between $\mathbf{a}$ and $\mathbf{b}$.

**4.** Let $\mathbf{a} = 2\mathbf{i} - \mathbf{j} + \mathbf{k}$, $\mathbf{b} = -\mathbf{i} + 3\mathbf{j} + 2\mathbf{k}$, and $\mathbf{c} = \mathbf{i} + 2\mathbf{j} - \mathbf{k}$. Find each of the following.
(a) $\mathbf{a} \times \mathbf{b}$       (b) $\mathbf{a} \times (\mathbf{b} + \mathbf{c})$
(c) $\mathbf{a} \cdot (\mathbf{b} \times \mathbf{c})$     (d) $\mathbf{a} \times (\mathbf{b} \times \mathbf{c})$

**5.** Find all vectors that are perpendicular to both of the vectors $3\mathbf{i} + 3\mathbf{j} - \mathbf{k}$ and $-\mathbf{i} - 2\mathbf{j} + 4\mathbf{k}$.

**6.** Find the unit vectors that are perpendicular to the plane determined by the three points $(3, -6, 4)$, $(2, 1, 1)$, and $(5, 0, -2)$.

**7.** Write the equation of the plane through the point $(-5, 7, -2)$ that satisfies each condition.

(a) Parallel to the $xz$-plane.
(b) Perpendicular to the $x$-axis.
(c) Parallel to both the $x$- and $y$-axes.
(d) Parallel to the plane $3x - 4y + z = 7$.

**31.** The curves given by $r_1(t) = \sin t\mathbf{i} + \cos t\mathbf{j} + t^3\mathbf{k}$ and $\mathbf{r}_2(t) = \sin t^3\mathbf{i} + \cos t^3\mathbf{j} + t^9\mathbf{k}$ for $0 \le t \le 1$ are identical.

**32.** The motions along the curves given by $r_1(t) = \sin t\mathbf{i} + \cos t\mathbf{j} + t^3\mathbf{k}$ and $\mathbf{r}_2(t) = \sin t^3\mathbf{i} = \cos t^3\mathbf{j} + t^9\mathbf{k}$ for $0 \le t \le 1$ are identical.

**33.** The length of a given curve is independent of the parameterization used to describe a given motion along the curve.

**34.** If a curve lies in a plane then the binormal vector $\mathbf{B}$ must be a constant.

**35.** If $|\mathbf{r}(t)| = \text{constant}$ then $\mathbf{r}'(t) = \mathbf{0}$.

**36.** The curve which is the intersection of the sphere $x^2 + y^2 + z^2 = 1$ and the plane $ax + by + cz = 0$ has constant curvature 1.

**37.** The graph of the equation $z = \rho$ is the $z$-axis (here $\rho$ is a spherical coordinate).

**38.** The graph of $y = x^2$ in three-space is a paraboloid.

**39.** If we restrict $\rho$, $\theta$, and $\phi$ by $\rho \ge 0$, $0 \le \theta < 2\pi$, and $0 \le \phi \le \pi$, then each point in three-space has a unique set of spherical coordinates.

**8.** A plane through the point $(2, -4, -5)$ is perpendicular to the line joining the points $(-1, 5, -7)$ and $(4, 1, 1)$.
(a) Write a vector equation of the plane.
(b) Find a Cartesian equation of the plane.
(c) Sketch the plane by drawing its traces.

**9.** Find the value of $C$ if the plane $x + 5y + Cz + 6 = 0$ is perpendicular to the plane $4x - y + z - 17 = 0$.

**10.** Find a Cartesian equation of the plane through the three points $(2, 3, -1)$, $(-1, 5, 2)$, and $(-4, -2, 2)$.

**11.** Find parametric equations for the line through $(-2, 1, 5)$ and $(6, 2, -3)$.

**12.** Find the points where the line of intersection of the planes $x - 2y + 4z - 14 = 0$ and $-x + 2y - 5z + 30 = 0$ pierces the $yz$- and $xz$-planes.

**13.** Write the equation of the line in Problem 12 in parametric form.

**14.** Find symmetric equations of the line through $(4, 5, 8)$ and perpendicular to the plane $3x + 5y + 2z = 30$. Sketch the plane and the line.

**15.** Write a vector equation of the line through $(2, -2, 1)$ and $(-3, 2, 4)$.

**16.** Sketch the curve whose vector equation is $\mathbf{r}(t) = t\mathbf{i} + \frac{1}{2}t^2\mathbf{j} + \frac{1}{3}t^3\mathbf{k}$, $-2 \le t \le 3$.

**17.** Find the symmetric equations for the tangent line to the curve of Problem 16 at the point where $t = 2$. Also find the equation of the normal plane at this point.

**18.** Find $\mathbf{r}'(\pi/2)$, $\mathbf{T}(\pi/2)$, and $\mathbf{r}''(\pi/2)$ if

$$\mathbf{r}(t) = \langle t \cos t, t \sin t, 2t \rangle.$$

**19.** Find the length of the curve

$$\mathbf{r}(t) = e^t \sin t \mathbf{i} + e^t \cos t \mathbf{j} + e^t \mathbf{k}, \qquad 1 \le t \le 5$$

**20.** Suppose an object is moving so that its position vector at time $t$ is

$$\mathbf{r}(t) = e^t \mathbf{i} + e^{-t} \mathbf{j} + 2t \mathbf{k}$$

Find $\mathbf{v}(t)$, $\mathbf{a}(t)$, and $\kappa(t)$, at $t = \ln 2$.

**21.** If $\mathbf{r}(t) = t\mathbf{i} + t^2\mathbf{j} + t^3\mathbf{k}$ is the position vector for a moving particle at time $t$, find the tangential and normal components, $a_T$ and $a_N$, of the acceleration vector at $t = 1$.

For each of the equations in Problems 22–30, name and sketch the graph in three-space.

**22.** $x^2 + y^2 = 81$

**23.** $x^2 + y^2 + z^2 = 81$

**24.** $z^2 = 4y$

**25.** $x^2 + z^2 = 4y$

**26.** $3y - 6z - 12 = 0$

**27.** $3x + 3y - 6z - 12 = 0$

**28.** $x^2 + y^2 - z^2 - 1 = 0$

**29.** $3x^2 + 4y^2 + 9z^2 - 36 = 0$

**30.** $3x^2 + 4y^2 + 9z^2 + 36 = 0$

**31.** Write the following Cartesian equations in cylindrical coordinate form.

(a) $x^2 + y^2 = 9$  (b) $x^2 + 4y^2 = 16$
(c) $x^2 + y^2 = 9z$  (d) $x^2 + y^2 + 4z^2 = 10$

**32.** Find the Cartesian equation corresponding to each of the following cylindrical coordinate equations.

(a) $r^2 + z^2 = 9$  (b) $r^2 \cos^2\theta + z^2 = 4$
(c) $r^2 \cos 2\theta + z^2 = 1$

**33.** Write the following equations in spherical coordinate form.

(a) $x^2 + y^2 + z^2 = 4$  (b) $2x^2 + 2y^2 - 2z^2 = 0$
(c) $x^2 - y^2 - z^2 = 1$  (d) $x^2 + y^2 = z$

**34.** Find the (straight-line) distance between the points whose spherical coordinates are $(8, \pi/4, \pi/6)$ and $(4, \pi/3, 3\pi/4)$.

**35.** Find the distance between the parallel planes $2x - 3y + \sqrt{3}z = 4$ and $2x - 3y + \sqrt{3}z = 9$.

**36.** Find the acute angle between the planes $2x - 4y + z = 7$ and $3x + 2y - 5z = 9$.

**37.** Show that if the speed of a moving particle is constant, then its velocity and acceleration vectors are orthogonal.

# TECHNOLOGY PROJECT

The standard parametric representation of a sphere with radius $a$, and its interior, is:

$$x = r \cos\theta \sin\phi, \qquad 0 \le r \le a,$$
$$y = r \sin\theta \sin\phi, \text{ with } 0 \le \phi \le \pi, \quad (1)$$
$$z = r \cos\phi, \qquad 0 \le \theta \le 2\pi.$$

**Exercise 1**   Verify that Equations 1 satisfy the equation of the solid sphere,

$$x^2 + y^2 + z^2 = r^2, \qquad 0 \le r \le a.$$

To get a space curve out of Equations 1, we first set $r = 1$ to obtain the parametric representation of the *surface* of the unit sphere. Then to obtain a parametric representation of a *curve* on the surface of the unit sphere, we specify $\phi$ and $\theta$ in terms of a single, common parameter $t$. For a curve that starts at the North pole of the sphere, winds around the sphere $n$ times and then ends up at the South pole, we set $\phi = t$ and give $t$ the range $[0, \pi]$. That gets us from the North pole to the South pole. To get the $n$ twists around the sphere we let $\theta = 2nt$. Thus, we have the following parametric representation of our twisting sphere curves:

$$x = \cos 2nt \sin t,$$
$$y = \sin 2nt \sin t, \qquad (2)$$
$$z = \cos t,$$

with $0 \le t \le \pi$ and $n = 0, 1, 2, \ldots$, governing the number of twists (or orbits) in our space curve.

If you have a 3-dim plotting capability, you can generate a plot something like Figure 1. Try it.

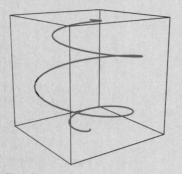

**Fig. 1.**
Three twists down the sphere.

**Exercise 2**   Verify that for any value of $n$, the curve specified by Equations (2) lies on the unit sphere.

**Exercise 3**   Verify that the parametric equations

$$x = \cos 2nt \sin t,$$
$$y = \sin 2nt \sin t, \qquad (3)$$
$$z = 3 \cos t,$$

with $0 \le t \le \pi$, correspond to curves twisting down the ellipsoid $x^2 + y^2 + (\frac{z}{3})^2 = 1$.

We now try our hand at computing the *arclengths* of some curves. As is typical for arclength integrals, the exact answer cannot be obtained. For example, in *Mathematica* the integrand for the arclength for the curves given by Equations (2) can be computed:

```
dsdt = Sqrt [x'[t]^2 + y'[t]^2 +
        z'[t]^2];

              2      2
Sqrt[1 + 2 n  - 2 n  Cos[2 t]]
```

Although this is a relatively simple looking integrand, it cannot be integrated exactly. So here is a loop in *Mathematica* to numerically compute the arclength for several values of $n$:

```
Do [ length = NIntegrate[dsdt,
        {t, 0, Pi}];
      Print [n, " ", length],

  {n, 3, 6}]
```

```
3  12.6116
4  16.4951
5  20.4185
6  24.364
```

**Exercise 4**   It looks like the arclengths of the twisting sphere curves are approaching $4n$ where $n$ is the number of twists. Investigate this for larger $n$ and see if this holds up.

**Exercise 5**   Repeat Exercise 4 for the curves defined by Equations (3).

**Exercise 6**   The formula obtained for the differential of arclength in Exercise 4 was

$$ds = \sqrt{1 + 2n^2(1 - \cos 2t)}\, dt$$

Use the fact that 1 is negligible compared to $n^2$, for $n$ large, to explain the numerical results of that Exercise. *Hint*: After dropping the "1", exercise those good old trig identities to write $1 - \cos 2t = 2 \sin^2 t$. Then you can integrate exactly.

# The Derivative in *n*-Space

## 15.1
## FUNCTIONS OF TWO OR MORE VARIABLES

Two kinds of functions have so far made their appearance. The first, typified by $f(x) = x^2$, associates with the real number $x$ another real number $f(x)$. We call it a real-valued function of a real variable. The second type of function, illustrated by $\mathbf{f}(x) = \langle x^3, e^x \rangle$, associates with the real number $x$ a vector $\mathbf{f}(x)$. We call it a vector-valued function of a real variable.

Our interest shifts now to a **real-valued function of two real variables**; that is, a function $f$ (Figure 1) that assigns to each ordered pair $(x, y)$ in some set $D$ of the plane a (unique) real number $f(x, y)$. Examples are

(1) $$f(x, y) = x^2 + 3y^2$$

(2) $$g(x, y) = 2x\sqrt{y}$$

Note that $f(-1, 4) = (-1)^2 + 3(4)^2 = 49$ and $g(-1, 4) = 2(-1)\sqrt{4} = -4$.

The set $D$ is called the **domain** of the function. If it is not specified, we take $D$ to be the natural domain, that is, the set of all points $(x, y)$ in the plane for which the function rule makes sense and gives a real number value. For $f(x, y) = x^2 + 3y^2$, the natural domain is the whole plane; for $g(x, y) = 2x\sqrt{y}$, it is $\{(x, y): x \in \mathbb{R}, y \geq 0\}$. The **range** of a function is its set of values. If $z = f(x, y)$, we call $x$ and $y$ the **independent variables** and $z$ the **dependent variable**.

All that we have said extends in a natural way to real-valued functions of three real variables (or even $n$ real variables). We will feel free to use such functions without further comment.

**EXAMPLE 1** In the *xy*-plane, sketch the natural domain for

$$f(x, y) = \frac{\sqrt{y - x^2}}{x^2 + (y - 1)^2}$$

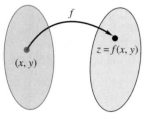

Domain    Range

FIGURE 1

713

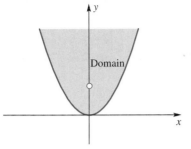

FIGURE 2

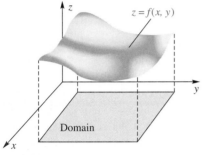

FIGURE 3

*Solution* For this rule to make sense, we must exclude $\{(x, y): y < x^2\}$ and the point $(0, 1)$. The resulting domain is shown in Figure 2. ∎

**Graphs** By the **graph** of a function $f$ of two variables, we mean the graph of the equation $z = f(x, y)$. This graph will normally be a surface (Figure 3) and, since to each $(x, y)$ in the domain there corresponds just one value $z$, each line perpendicular to the $xy$-plane intersects the surface in at most one point.

**EXAMPLE 2** Sketch the graph of $f(x, y) = \frac{1}{3}\sqrt{36 - 9x^2 - 4y^2}$.

*Solution* Let $z = \frac{1}{3}\sqrt{36 - 9x^2 - 4y^2}$ and note that $z \geq 0$. If we square both sides and simplify, we obtain the equation

$$9x^2 + 4y^2 + 9z^2 = 36$$

which we recognize as the equation of an ellipsoid (see Section 14.6). The graph of the given function is the upper half of this ellipsoid; it is shown in Figure 4. ∎

**EXAMPLE 3** Sketch the graph of $z = f(x, y) = y^2 - x^2$.

*Solution* The graph is a hyperbolic paraboloid (see Section 14.6); it is sketched in Figure 5. ∎

**Computer Graphs** A computer with a software package such as *True BASIC MacFunction* allows us to produce complicated three-dimensional graphs with ease. In Figures 6–9, we show four such graphs. To enhance the visual presentation, we are able to show these graphs in various orientations. Often, as in these four examples, we choose to show the graph with the $y$-axis pointing partially toward the viewer rather than keeping it in the plane of the paper, as we usually do for hand-drawn graphs.

FIGURE 4

FIGURE 5

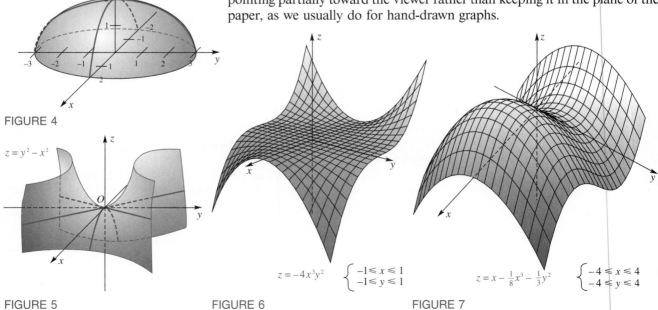

$z = -4x^3y^2 \quad \begin{cases} -1 \leq x \leq 1 \\ -1 \leq y \leq 1 \end{cases}$

FIGURE 6

$z = x - \frac{1}{8}x^3 - \frac{1}{3}y^2 \quad \begin{cases} -4 \leq x \leq 4 \\ -4 \leq y \leq 4 \end{cases}$

FIGURE 7

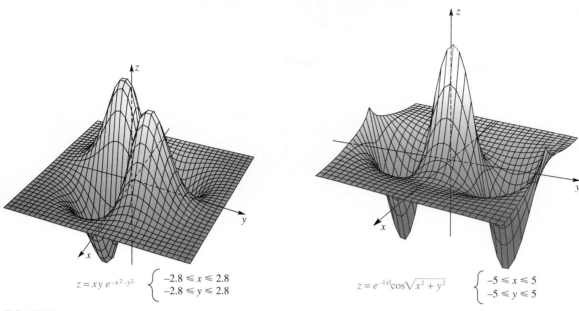

$z = xy \, e^{-x^2-y^2}$ $\begin{cases} -2.8 \leq x \leq 2.8 \\ -2.8 \leq y \leq 2.8 \end{cases}$

**FIGURE 8**

$z = e^{-|x|}\cos\sqrt{x^2 + y^2}$ $\begin{cases} -5 \leq x \leq 5 \\ -5 \leq y \leq 5 \end{cases}$

**FIGURE 9**

**Level Curves** To sketch the surface corresponding to the graph of a function $z = f(x, y)$ of two variables is often very difficult. Map makers have given us another and usually simpler way to picture a surface, the so-called contour map. Each horizontal plane $z = c$ intersects the surface in a curve. The projection of this curve on the $xy$-plane is called a **level curve** (Figure 10), and a collection of such curves is a **contour map**. We have shown a contour map for a hill-shaped surface in Figure 11.

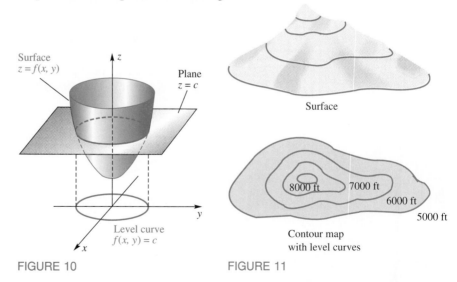

**FIGURE 10**

**FIGURE 11**

**EXAMPLE 4** Draw contour maps for the surfaces corresponding to $z = \frac{1}{3}\sqrt{36 - 9x^2 - 4y^2}$ and $z = y^2 - x^2$ (see Examples 2 and 3).

***Solution*** The level curves of $z = \frac{1}{3}\sqrt{36 - 9x^2 - 4y^2}$ corresponding to $z = 0, 1, 1.5, 1.75, 2$ are shown in Figure 12. They are ellipses. Similarly, in Figure 13, we show the level curves of $z = y^2 - x^2$ for $z = -5, -4, -3, \ldots, 2, 3, 4$. These curves are hyperbolas. ∎

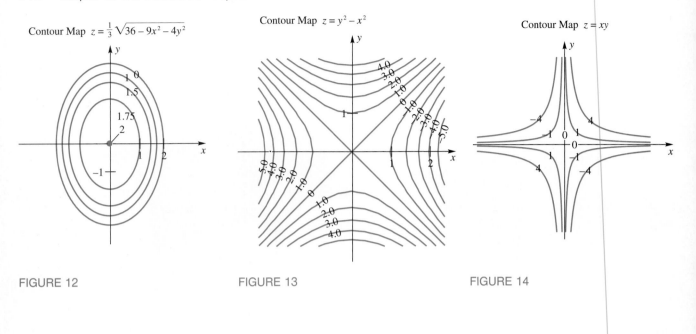

Contour Map $z = \frac{1}{3}\sqrt{36 - 9x^2 - 4y^2}$

FIGURE 12

Contour Map $z = y^2 - x^2$

FIGURE 13

Contour Map $z = xy$

FIGURE 14

**EXAMPLE 5** Sketch a contour map for $z = f(x, y) = xy$.

*Solution* The level curves corresponding to $z = -4, -1, 0, 1, 4$ are shown in Figure 14. It can be shown that they are hyperbolas. Comparing the contour map of Figure 14 with that of Figure 13 suggests that the graph of $z = xy$ might be a hyperbolic paraboloid but with axes rotated through 45°. The suggestion is correct. ∎

**Computer Graphs and Level Curves** In Figures 15–18, we have drawn four more surfaces, but we now also show the corresponding level curves. Note that we have rotated the *xy*-plane so the *x*-axis points to the right, making it easier to relate the surface and the level curves.

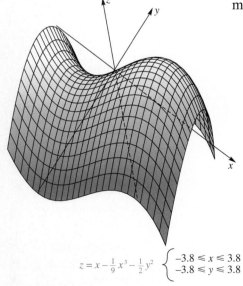

$z = x - \frac{1}{9}x^3 - \frac{1}{2}y^2$ $\begin{cases} -3.8 \le x \le 3.8 \\ -3.8 \le y \le 3.8 \end{cases}$

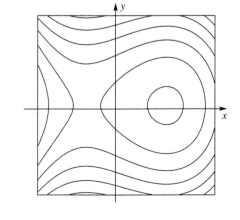

FIGURE 15

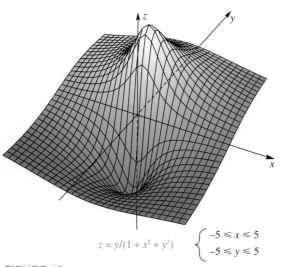
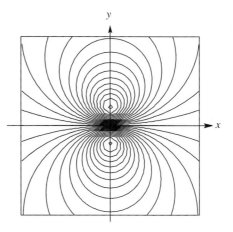

$$z = y/(1 + x^2 + y^2) \qquad \begin{cases} -5 \leq x \leq 5 \\ -5 \leq y \leq 5 \end{cases}$$

FIGURE 16

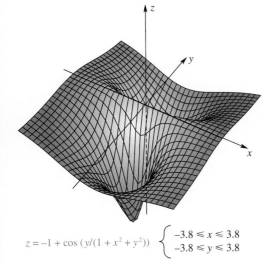
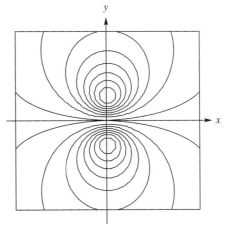

$$z = -1 + \cos(y/(1 + x^2 + y^2)) \qquad \begin{cases} -3.8 \leq x \leq 3.8 \\ -3.8 \leq y \leq 3.8 \end{cases}$$

FIGURE 17

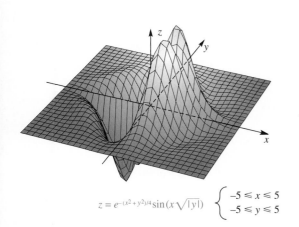
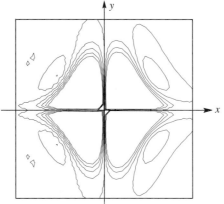

$$z = e^{-(x^2 + y^2)/4} \sin(x\sqrt{|y|}) \qquad \begin{cases} -5 \leq x \leq 5 \\ -5 \leq y \leq 5 \end{cases}$$

FIGURE 18

## CONCEPTS REVIEW

**1.** A function $f$ determined by $z = f(x, y)$ is called a _____.

**2.** The projection of the curve $z = f(x, y) = c$ to the $xy$-plane is called a _____ and a collection of such curves is called a _____.

**3.** The contour map for $z = x^2 + y^2$ consists of _____.

**4.** The contour map for $z = x^2$ consists of _____.

## PROBLEM SET 15.1

**1.** Let $f(x, y) = x^2y + \sqrt{y}$. Find each value.

(a) $f(2, 1)$        (b) $f(3, 0)$
(c) $f(1, 4)$        (d) $f(a, a^4)$
(e) $f(1/x, x^4)$    (f) $f(2, -4)$

What is the natural domain for this function?

**2.** Let $f(x, y) = y/x + xy$. Find each value.

(a) $f(1, 2)$        (b) $f(\frac{1}{4}, 4)$
(c) $f(4, \frac{1}{4})$      (d) $f(a, a)$
(e) $f(1/x, x^2)$    (f) $f(0, 0)$

What is the natural domain for this function?

**3.** Let $g(x, y, z) = x^2 \sin yz$. Find each value.

(a) $g(1, \pi, 2)$        (b) $g(2, 1, \pi/6)$
(c) $g(4, 2, \pi/4)$    [C] (d) $g(\pi, \pi, \pi)$
[C] (e) $g(1.2, 3.1, 4.2)$

**4.** Let $g(x, y, z) = \sqrt{x} \cos y + z^2$. Find each value.

(a) $g(4, 0, 2)$        (b) $g(-9, \pi, 3)$
(c) $g(2, \pi/3, -1)$    [C] (d) $g(3, 6, 1.2)$
[C] (e) $g(-2, 2, 3)$

**5.** Find $F(f(t), g(t))$ if $F(x, y) = x^2y$ and $f(t) = t \cos t$, $g(t) = \sec^2 t$.

**6.** Find $F(f(t), g(t))$ if $F(x, y) = e^x + y^2$ and $f(t) = \ln t^2$, $g(t) = e^{t/2}$.

In Problems 7–16, sketch the graph of $f$.

**7.** $f(x, y) = 6$          **8.** $f(x, y) = 6 - x$

**9.** $f(x, y) = 6 - x - 2y$    **10.** $f(x, y) = 6 - x^2$

**11.** $f(x, y) = \sqrt{16 - x^2 - y^2}$

**12.** $f(x, y) = \sqrt{16 - 4x^2 - y^2}$

**13.** $f(x, y) = 3 - x^2 - y^2$    **14.** $f(x, y) = 2 - x - y^2$

**15.** $f(x, y) = e^{-(x^2 + y^2)}$      **16.** $f(x, y) = x^2/y, \, y > 0$

In Problems 17–22, sketch the level curve $z = k$ for the indicated values of $k$.

**17.** $z = \frac{1}{2}(x^2 + y^2), \, k = 0, 2, 4, 6, 8$

**18.** $z = \dfrac{x}{y}, \, k = -2, -1, 0, 1, 2$

**19.** $z = \dfrac{x^2}{y}, \, k = -4, -1, 0, 1, 4$

**20.** $z = x^2 + y, \, k = -4, -1, 0, 1, 4$

**21.** $z = \dfrac{x^2 + y}{x + y^2}, \, k = 0, 1, 2, 4$

**22.** $z = y - \sin x, \, k = -2, -1, 0, 1, 2$

**23.** If $T(x, y)$ is the temperature at a point $(x, y)$ in the plane, the level curves of $T$ are called **isothermal curves**. Draw the isothermal curves for

$$T(x, y) = \frac{x^2}{x^2 + y^2}$$

corresponding to $T = \frac{1}{10}, \frac{1}{5}, \frac{1}{2}, 0$.

**24.** If $V(x, y)$ is the voltage at a point $(x, y)$ in the plane, the level curves of $V$ are called **equipotential curves**. Draw the equipotential curves for

$$V(x, y) = \frac{4}{\sqrt{(x - 2)^2 + (y + 3)^2}}$$

corresponding to $V = \frac{1}{2}, 1, 2, 4$.

In Problems 25–28, describe geometrically the domain of each of the indicated functions of three variables.

**25.** $f(x, y, z) = \sqrt{x^2 + y^2 + z^2 - 16}$

**26.** $f(x, y, z) = \sqrt{x^2 + y^2 - z^2 - 9}$

**27.** $f(x, y, z) = \sqrt{144 - 16x^2 - 9y^2 - 144z^2}$

**28.** $f(x, y, z) = \dfrac{(144 - 16x^2 - 16y^2 + 9z^2)^{3/2}}{xyz}$

A **level surface** for a function $f$ of three variables is the graph of the set of points in three-dimensional space whose coordinates satisfy an equation $f(x, y, z) = k$, where $k$ is a constant. Describe geometrically the level surfaces for the functions defined in Problems 29–33.

**29.** $f(x, y, z) = x^2 + y^2 + z^2; k > 0$

**30.** $f(x, y, z) = 100x^2 + 16y^2 + 25z^2; k > 0$

**31.** $f(x, y, z) = 16x^2 + 16y^2 - 9z^2; k \varepsilon \mathbb{R}$

**32.** $f(x, y, z) = 9x^2 - 4y^2 - z^2; k \varepsilon \mathbb{R}$

**33.** $f(x, y, z) = 4x^2 - 9y^2; k \varepsilon \mathbb{R}$

**34.** Sketch (as best you can) the graph of the monkey saddle $z = x(x^2 - 3y^2)$. Begin by noting where $z = 0$.

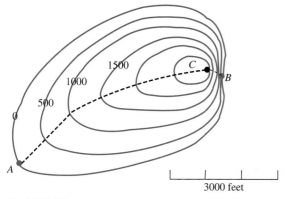

FIGURE 19

**35.** The contour map in Figure 19 shows level curves for a mountain 3000 feet high.

(a) What is special about the path to the top labeled $AC$? What is special about $BC$?

(b) Make good estimates of the total lengths of path $AC$ and path $BC$.

**36.** Identify the graph of $f(x, y) = x^2 - x + 3y^2 + 12y - 13$, state where it attains its minimum value, and find this minimum value.

PC For the each of the functions in Problems 37–40, draw the graph and the corresponding contour map.

**37.** $f(x, y) = \sin\sqrt{2x^2 + y^2}; -2 \le x \le 2, -2 \le y \le 2$

**38.** $f(x, y) = \sin(x^2 + y^2)/(x^2 + y^2), f(0, 0) = 1;$
$\quad -2 \le x \le 2, -2 \le y \le 2$

**39.** $f(x, y) = (2x - y^2)\exp(-x^2 - y^2);$
$\quad -2 \le x \le 2, -2 \le y \le 2$

**40.** $f(x, y) = (\sin x \sin y)/(1 + x^2 + y^2);$
$\quad -2.9 \le x \le 2.9, -2.9 \le y \le 2.9$

**Answers to Concepts Review:** **1.** Real-valued function of two real variables **2.** Level curve; contour map **3.** Concentric circles **4.** Parallel lines

---

## 15.2
## PARTIAL DERIVATIVES

Suppose that $f$ is a function of two variables $x$ and $y$. If $y$ is held constant, say $y = y_0$, then $f(x, y_0)$ is a function of the single variable $x$. Its derivative at $x = x_0$ is called the **partial derivative of $f$ with respect to $x$** at $(x_0, y_0)$ and is denoted by $f_x(x_0, y_0)$. Thus,

$$f_x(x_0, y_0) = \lim_{\Delta x \to 0} \frac{f(x_0 + \Delta x, y_0) - f(x_0, y_0)}{\Delta x}$$

Similarly, the partial derivative of $f$ with respect to $y$ at $(x_0, y_0)$ is denoted by $f_y(x_0, y_0)$ and is given by

$$f_y(x_0, y_0) = \lim_{\Delta y \to 0} \frac{f(x_0, y_0 + \Delta y) - f(x_0, y_0)}{\Delta y}$$

Rather than calculate $f_x(x_0, y_0)$ and $f_y(x_0, y_0)$ directly from the boxed definitions, we typically find $f_x(x, y)$ and $f_y(x, y)$ using the standard rules for derivatives; then we substitute $x = x_0$ and $y = y_0$.

**EXAMPLE 1**  Find $f_x(1, 2)$ and $f_y(1, 2)$ if $f(x, y) = x^2y + 3y^3$.

**Solution**  To find $f_x(x, y)$, we treat $y$ as a constant and differentiate with respect to $x$, obtaining

$$f_x(x, y) = 2xy + 0$$

Thus,

$$f_x(1, 2) = 2 \cdot 1 \cdot 2 = 4$$

Similarly, we treat $x$ as a constant and differentiate with respect to $y$, obtaining

$$f_y(x, y) = x^2 + 9y^2$$

and so

$$f_y(1, 2) = 1^2 + 9 \cdot 2^2 = 37$$  ■

If $z = f(x, y)$, we use the following alternative notations.

$$f_x(x, y) = \frac{\partial z}{\partial x} = \frac{\partial f(x, y)}{\partial x} \qquad f_y(x, y) = \frac{\partial z}{\partial y} = \frac{\partial f(x, y)}{\partial y}$$

$$f_x(x_0, y_0) = \frac{\partial z}{\partial x}\bigg|_{(x_0, y_0)} \qquad f_y(x_0, y_0) = \frac{\partial z}{\partial y}\bigg|_{(x_0, y_0)}$$

The symbol $\partial$ is special to mathematics and is called the partial derivative sign.

**EXAMPLE 2**  If $z = x^2 \sin(xy^2)$, find $\partial z/\partial x$ and $\partial z/\partial y$.

**Solution**

$$\frac{\partial z}{\partial x} = x^2 \frac{\partial}{\partial x} [\sin(xy^2)] + \sin(xy^2) \frac{\partial}{\partial x} (x^2)$$

$$= x^2 \cos(xy^2) \frac{\partial}{\partial x} (xy^2) + \sin(xy^2) \cdot 2x$$

$$= x^2 \cos(xy^2) \cdot y^2 + 2x \sin(xy^2)$$

$$= x^2y^2 \cos(xy^2) + 2x \sin(xy^2)$$

$$\frac{\partial z}{\partial y} = x^2 \cos(xy^2) \cdot 2xy = 2x^3y \cos(xy^2)$$  ■

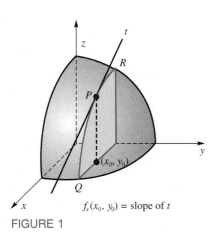

$f_x(x_0, y_0)$ = slope of $t$

FIGURE 1

**Geometric and Physical Interpretations**  Consider the surface whose equation is $z = f(x, y)$. The plane $y = y_0$ intersects this surface in the plane curve $QPR$ (Figure 1) and the value of $f_x(x_0, y_0)$ is the slope of the

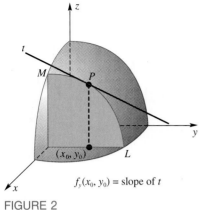

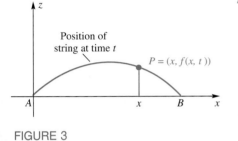

$f_y(x_0, y_0) = $ slope of $t$

FIGURE 2

Position of
string at time $t$

$P = (x, f(x, t))$

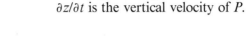

FIGURE 3

tangent line to this curve at $P(x_0, y_0, f(x_0, y_0))$. Similarly, the plane $x = x_0$ intersects the surface in the plane curve $LPM$ (Figure 2) and $f_y(x_0, y_0)$ is the slope of the tangent line to this curve at $P$.

Partial derivatives may also be interpreted as (instantaneous) rates of change. Suppose that a violin string is fixed at points $A$ and $B$ and vibrates in the $xz$-plane. Figure 3 shows the position of the string at a typical time $t$. If $z = f(x, t)$ denotes the height of the string at the point $P$ with abscissa $x$ at time $t$, then $\partial z/\partial x$ is the slope of the string at $P$ and $\partial z/\partial t$ is the time rate of change of height of $P$ along the indicated vertical line. In other words, $\partial z/\partial t$ is the vertical velocity of $P$.

**EXAMPLE 3**   The surface $z = f(x, y) = \sqrt{9 - 2x^2 - y^2}$ and the plane $y = 1$ intersect in a curve as in Figure 1. Find parametric equations for the tangent line at $(\sqrt{2}, 1, 2)$.

*Solution*

$$f_x(x, y) = \tfrac{1}{2}(9 - 2x^2 - y^2)^{-1/2}(-4x)$$

and so $f_x(\sqrt{2}, 1) = -\sqrt{2}$. This number is the slope of the tangent line to the curve at $(\sqrt{2}, 1, 2)$, that is, $-\sqrt{2}/1$ is the ratio of rise to run along the tangent line. It follows that this line has direction numbers $(1, 0, -\sqrt{2})$ and, since it goes through $(\sqrt{2}, 1, 2)$.

$$x = \sqrt{2} + t, \qquad y = 1, \qquad z = 2 - \sqrt{2}t$$

provide the required parametric equations.   ∎

**EXAMPLE 4**   The volume of a certain gas is related to its temperature $T$ and its pressure $P$ by the gas law $PV = 10T$, where $V$ is measured in cubic inches, $P$ in pounds per square inch, and $T$ in degrees Celsius. If $T$ is kept constant at 200, what is the rate of change of pressure with respect to volume at $V = 50$?

*Solution*   Since $P = 10T/V$,

$$\frac{\partial P}{\partial V} = \frac{-10T}{V^2}$$

Thus,

$$\left.\frac{\partial P}{\partial V}\right|_{T = 200, \, V = 50} = \frac{(-10)(200)}{(50)^2} = -\frac{4}{5}$$   ∎

**Higher Partial Derivatives**   Since a partial derivative of a function of $x$ and $y$ is, in general, another function of these same two variables, it may be differentiated partially with respect to either $x$ or $y$, resulting in four **second**

**partial derivatives** of $f$:

$$f_{xx} = \frac{\partial}{\partial x}\left(\frac{\partial f}{\partial x}\right) = \frac{\partial^2 f}{\partial x^2} \qquad\qquad f_{yy} = \frac{\partial}{\partial y}\left(\frac{\partial f}{\partial y}\right) = \frac{\partial^2 f}{\partial y^2}$$

$$f_{xy} = (f_x)_y = \frac{\partial}{\partial y}\left(\frac{\partial f}{\partial x}\right) = \frac{\partial^2 f}{\partial y \partial x} \qquad f_{yx} = (f_y)_x = \frac{\partial}{\partial x}\left(\frac{\partial f}{\partial y}\right) = \frac{\partial^2 f}{\partial x \partial y}$$

**EXAMPLE 5**  Find the four second partial derivatives of

$$f(x, y) = xe^y - \sin(x/y) + x^3 y^2.$$

*Solution*

$$f_x(x, y) = e^y - \frac{1}{y}\cos\left(\frac{x}{y}\right) + 3x^2 y^2$$

$$f_y(x, y) = xe^y + \frac{x}{y^2}\cos\left(\frac{x}{y}\right) + 2x^3 y$$

$$f_{xx}(x, y) = \frac{1}{y^2}\sin\left(\frac{x}{y}\right) + 6xy^2$$

$$f_{yy}(x, y) = xe^y + \frac{x^2}{y^4}\sin\left(\frac{x}{y}\right) - \frac{2x}{y^3}\cos\left(\frac{x}{y}\right) + 2x^3$$

$$f_{xy}(x, y) = e^y - \frac{x}{y^3}\sin\left(\frac{x}{y}\right) + \frac{1}{y^2}\cos\left(\frac{x}{y}\right) + 6x^2 y$$

$$f_{yx}(x, y) = e^y - \frac{x}{y^3}\sin\left(\frac{x}{y}\right) + \frac{1}{y^2}\cos\left(\frac{x}{y}\right) + 6x^2 y$$

Notice that in Example 5, $f_{xy} = f_{yx}$, which is usually the case for the functions of two variables encountered in a first course. A criterion for this equality will be given in Section 15.3 (Theorem B).

Partial derivatives of the third and higher orders are defined analogously, and the notation for them is similar. Thus, if $f$ is a function of the two variables $x$ and $y$, the third partial derivative of $f$ obtained by differentiating $f$ partially, first with respect to $x$ and then twice with respect to $y$, will be indicated by

$$\frac{\partial}{\partial y}\left(\frac{\partial^2 f}{\partial y \partial x}\right) = \frac{\partial^3 f}{\partial y^2 \partial x} = f_{xyy}$$

**More Than Two Variables**  Let $f$ be a function of three variables, $x$, $y$, and $z$. The **partial derivative of $f$ with respect to $x$** at $(x, y, z)$ is denoted by $f_x(x, y, z)$ or $\partial f(x, y, z)/\partial x$ and is defined by

$$f_x(x, y, z) = \lim_{\Delta x \to 0} \frac{f(x + \Delta x, y, z) - f(x, y, z)}{\Delta x}$$

Thus, $f_x(x, y, z)$ may be obtained by treating $y$ and $z$ as constants and differentiating with respect to $x$.

The partial derivatives with respect to $y$ and $z$ are defined in an analogous way.

**EXAMPLE 6**   If $f(x, y, z) = xy + 2yz + 3zx$, find $f_x$, $f_y$, and $f_z$.

*Solution*   To get $f_x$, we think of $y$ and $z$ as constants and differentiate with respect to the variable $x$. Thus,

$$f_x(x, y, z) = y + 3z$$

To find $f_y$, we treat $x$ and $z$ as constants and differentiate with respect to $y$:

$$f_y(x, y, z) = x + 2z$$

Similarly,

$$f_z(x, y, z) = 2y + 3x$$ ∎

**EXAMPLE 7**   If $f(x, y, z) = x \cos(y - z)$, find $\partial f/\partial x$, $\partial f/\partial y$, and $\partial f/\partial z$.

*Solution*

$$\frac{\partial}{\partial x}[x \cos(y - z)] = \cos(y - z)$$

$$\frac{\partial}{\partial y}[x \cos(y - z)] = -x \sin(y - z)$$

$$\frac{\partial}{\partial z}[x \cos(y - z)] = x \sin(y - z)$$ ∎

## CONCEPTS REVIEW

**1.** $f_x(x_0, y_0)$ is defined as a limit by _____ and is called the _____ at $(x_0, y_0)$.

**2.** If $f(x, y) = x^3 + xy$, then $f_x(1, 2) =$ _____ and $f_y(1, 2) =$ _____ .

**3.** Another notation for $f_{xy}(x, y)$ is _____ .

**4.** If $f(x, y) = g(x) + h(y)$, then $f_{xy}(x, y) =$ _____ .

## PROBLEM SET 15.2

In Problems 1–16, find the first partial derivatives of the given functions with respect to each independent variable.

**1.** $f(x, y) = (2x - y)^4$

**2.** $f(x, y) = (4x - y^2)^{3/2}$

**3.** $f(x, y) = \dfrac{x^2 - y^2}{xy}$

**4.** $f(x, y) = e^x \cos y$

**5.** $f(x, y) = e^y \sin x$

**6.** $f(x, y) = (3x^2 + y^2)^{-1/3}$

**7.** $f(x, y) = \sqrt{x^2 - y^2}$

**8.** $f(u, v) = e^{uv}$

**9.** $g(x, y) = e^{-xy}$

**10.** $f(s, t) = \ln(s^2 - t^2)$

**11.** $f(x, y) = \tan^{-1}(4x - 7y)$

**12.** $F(w, z) = w \sin^{-1}\left(\dfrac{w}{z}\right)$

13. $f(x, y) = y \cos(x^2 + y^2)$  14. $f(s, t) = e^{t^2 - s^2}$

15. $F(x, y) = 2 \sin x \cos y$   16. $f(r, \theta) = 3r^3 \cos 2\theta$

In Problems 17–20, verify that

$$\frac{\partial^2 f}{\partial y \partial x} = \frac{\partial^2 f}{\partial x \partial y}$$

17. $f(x, y) = 2x^2y^3 - x^3y^5$  18. $f(x, y) = (x^3 + y^2)^5$

19. $f(x, y) = 3e^{2x} \cos y$   20. $f(x, y) = \tan^{-1} xy$

21. If $F(x, y) = \dfrac{2x - y}{xy}$, find $F_x(3, -2)$ and $F_y(3, -2)$.

22. If $F(x, y) = \ln(x^2 + xy + y^2)$, find $F_x(-1, 4)$ and $F_y(-1, 4)$.

23. If $f(x, y) = \tan^{-1}(y^2/x)$, find $f_x\left(\sqrt{5}, -2\right)$ and $f_y\left(\sqrt{5}, -2\right)$.

24. If $f(x, y) = e^y \cosh x$, find $f_x(-1, 1)$ and $f_y(-1, 1)$.

25. Find the slope of the tangent to the curve of intersection of the surface $36z = 4x^2 + 9y^2$ and the plane $x = 3$ at the point $(3, 2, 2)$.

26. Find the slope of the tangent to the curve of intersection of the surface $3z = \sqrt{36 - 9x^2 - 4y^2}$ and the plane $x = 1$ at the point $\left(1, -2, \sqrt{11/3}\right)$.

27. Find the slope of the tangent to the curve of intersection of the surface $2z = \sqrt{9x^2 + 9y^2 - 36}$ and the plane $y = 1$ at the point $\left(2, 1, \frac{3}{2}\right)$.

28. Find the slope of the tangent to the curve of intersection of the cylinder $4z = 5\sqrt{16 - x^2}$ and the plane $y = 3$ at the point $\left(2, 3, 5\sqrt{3}/2\right)$.

29. The volume $V$ of a right circular cylinder is given by $V = \pi r^2 h$, where $r$ is the radius and $h$ is the height. If $h$ is held fixed at $h = 10$ inches, find the rate of change of $V$ with respect to $r$ when $r = 6$ inches.

30. The temperature in degrees Celsius on a metal plate in the $xy$-plane is given by $T(x, y) = 4 + 2x^2 + y^3$. What is the rate of change of temperature with respect to distance (measured in feet) if we start moving from $(3, 2)$ in the direction of the positive $y$-axis?

31. According to the ideal gas law, the pressure, temperature, and volume of a gas are related by $PV = kT$, where $k$ is a constant. Find the rate of change of pressure (pounds per square inch) with respect to temperature when the temperature is $40°C$ if the volume is kept fixed at 100 cubic inches.

32. Show that for the gas law of Problem 31,

$$V \frac{\partial P}{\partial V} + T \frac{\partial P}{\partial T} = 0 \quad \text{and} \quad \frac{\partial P}{\partial V} \frac{\partial V}{\partial T} \frac{\partial T}{\partial P} = -1$$

A function of two variables that satisfies **Laplace's Equation**,

$$\frac{\partial^2 f}{\partial x^2} + \frac{\partial^2 f}{\partial y^2} = 0$$

is said to be **harmonic**. (Pierre-Simon de Laplace (1749–1827) was a gifted French mathematician.) Show that the functions defined in Problems 33 and 34 are harmonic functions.

33. $f(x, y) = x^3y - xy^3$

34. $f(x, y) = \ln(4x^2 + 4y^2)$

35. If $F(x, y) = 3x^4y^5 - 2x^2y^3$, find $\partial^3 F(x, y)/\partial y^3$.

36. If $f(x, y) = \cos(2x^2 - y^2)$, find $\partial^3 f(x, y)/\partial y \, \partial x^2$.

37. Express the following in $\partial$ notation.

(a) $f_{yyy}$         (b) $f_{xxy}$         (c) $f_{xyyy}$

38. Express the following in subscript notation.

(a) $\dfrac{\partial^3 f}{\partial x^2 \, \partial y}$   (b) $\dfrac{\partial^4 f}{\partial x^2 \, \partial y^2}$   (c) $\dfrac{\partial^5 f}{\partial x^3 \, \partial y^2}$

39. If $f(x, y, z) = 3x^2y - xyz + y^2z^2$, find each of the following.

(a) $f_x(x, y, z)$   (b) $f_y(0, 1, 2)$   (c) $f_{xy}(x, y, z)$

40. If $f(x, y, z) = (x^3 + y^2 + z)^4$, find each of the following.

(a) $f_x(x, y, z)$   (b) $f_y(0, 1, 1)$   (c) $f_{zz}(x, y, z)$

41. If $f(x, y, z) = e^{-xyz} - \ln(xy - z^2)$, find $f_x(x, y, z)$.

42. If $f(x, y, z) = (xy/z)^{1/2}$, find $f_x(-2, -1, 8)$.

43. A bee was flying upward along the curve that is the intersection of $z = x^4 + xy^3 + 12$ with the plane $x = 1$. At the point $(1, -2, 5)$, it went off on the tangent line. Where did the bee hit the $xz$-plane? (See Example 3.)

44. Let $A(x, y)$ be the area of a nondegenerate rectangle of dimensions $x$ and $y$, the rectangle being inside a circle of radius 10. Determine the domain and range for this function.

45. The interval $[0, 1]$ is to be separated into three pieces by making cuts at $x$ and $y$. Let $A(x, y)$ be the area of any nondegenerate triangle that can be formed from these three pieces. Determine the domain and range for this function.

46. The **wave equation** $c^2 \, \partial^2 u/\partial x^2 = \partial^2 u/\partial t^2$ and the **heat equation** $c \, \partial^2 u/\partial x^2 = \partial u/\partial t$ are two of the most important equations in physics ($c$ is a constant). Show each of the following.

(a) $u = \cos x \cos ct$ and $u = e^x \cosh ct$ satisfy the wave equation.

(b) $u = e^{-ct} \sin x$ and $u = t^{-1/2} e^{-x^2/(4ct)}$ satisfy the heat equation.

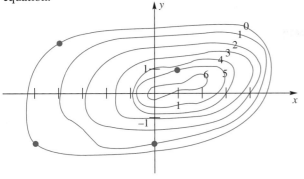

**FIGURE 4**

**47.** For the contour map for $z = f(x, y)$ shown in Figure 4, estimate each value.

(a) $f_y(1, 1)$          (b) $f_x(-4, 2)$
(c) $f_x(-5, -2)$      (d) $f_y(0, -2)$

PC **48.** *True BASIC MacFunction* allows us to calculate and graph partial derivatives. Draw the graphs of each of the following.

(a) $\sin(x + y^2)$           (b) $D_x \sin(x + y^2)$
(c) $D_y \sin(x + y^2)$      (d) $D_x(D_y \sin(x + y^2))$

---

**Answers to Concepts Review:**   **1.** $\lim\limits_{h \to 0}[f(x_0 + h, y_0) - f(x_0, y_0)]/h$; partial derivative of $f$ with respect to $x$
**2.** 5; 1   **3.** $\partial^2 f/\partial y \, \partial x$   **4.** 0

---

**15.3**
**LIMITS AND CONTINUITY**

Although partial derivative is a concept for functions of several variables, the theory needed so far is only that for functions of one variable (since all but one of the variables are held fixed). In particular, the notion of limit in Section 15.2 was the familiar one treated in Section 2.5. To go further, we will need a more general notion of limit.

Our aim is to give meaning to the symbol

$$\lim_{(x, y) \to (a, b)} f(x, y) = L$$

It has the usual intuitive meaning: namely, the values of $f(x, y)$ get closer and closer to the number $L$ as $(x, y)$ approaches $(a, b)$. The problem is that $(x, y)$ can approach $(a, b)$ in infinitely many different ways (Figure 1). We want a definition that gives the same $L$ no matter what path $(x, y)$ takes in approaching $(a, b)$. Fortunately, the formal definition given first for real-valued functions of one variable (Section 2.5) and then for vector-valued functions of one variable (Section 13.4) is still appropriate, provided we interpret it correctly (see Figure 2).

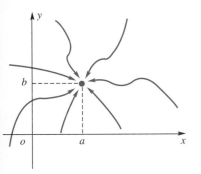

**FIGURE 1**

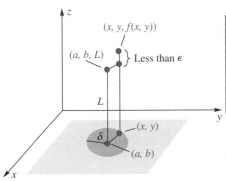

**FIGURE 2**

> **Definition**
>
> To say that $\lim\limits_{(x, y) \to (a, b)} f(x, y) = L$ means that for each $\varepsilon > 0$ (no matter how small) there is a corresponding $\delta > 0$ such that $|f(x, y) - L| < \varepsilon$ provided that $0 < |(x, y) - (a, b)| < \delta$.

Only the briefest explanation is necessary. To interpret $|(x, y) - (a, b)|$, think of $(x, y)$ and $(a, b)$ as vectors. Then

$$|(x, y) - (a, b)| = \sqrt{(x - a)^2 + (y - b)^2}$$

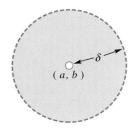

FIGURE 3

and the points satisfying $0 < |(x, y) - (a, b)| < \delta$ are those points inside a circle of radius $\delta$ excluding the center $(a, b)$ (see Figure 3).

Note several aspects of this definition.

1. The path of approach to $(a, b)$ is irrelevant. This means that if different paths of approach lead to different $L$-values, then the limit does not exist.
2. The behavior of $f(x, y)$ at $(a, b)$ is irrelevant; the function does not even have to be defined at $(a, b)$. This follows from the restriction $0 < |(x, y) - (a, b)|$.
3. The definition is phrased so that it immediately extends to functions of three (or more) variables. Simply replace $(x, y)$ and $(a, b)$ by $(x, y, z)$ and $(a, b, c)$ wherever they occur.

What you expect to be true is generally true. For example,

$$\lim_{(x, y) \to (1, 2)} [x^2 y + 3y] = 1^2 \cdot 2 + 3 \cdot 2 = 8$$

$$\lim_{(x, y) \to (3, 4)} \sqrt{x^2 + y^2} = \sqrt{3^2 + 4^2} = 5$$

In fact, all the usual limit theorems are also valid for this new kind of limit. But the following example illustrates that we must be careful.

**EXAMPLE 1**  Show that the function $f$ defined by

$$f(x, y) = \frac{x^2 - y^2}{x^2 + y^2}$$

has no limit at the origin (Figure 4).

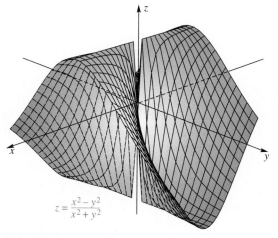

$$z = \frac{x^2 - y^2}{x^2 + y^2}$$

FIGURE 4

*Solution*  The function $f$ is defined everywhere in the $xy$-plane except at the origin.

At all points on the $x$-axis, different from the origin, the value of $f$ is

$$f(x, 0) = \frac{x^2 - 0}{x^2 + 0} = 1$$

Thus, the limit of $f(x, y)$ as $(x, y)$ approaches $(0, 0)$ along the $x$-axis is

$$\lim_{(x, 0) \to (0, 0)} f(x, 0) = \lim_{(x, 0) \to (0, 0)} \frac{x^2 - 0}{x^2 + 0} = +1$$

Similarly, the limit of $f(x, y)$ as $(x, y)$ approaches $(0, 0)$ along the $y$-axis is

$$\lim_{(0, y) \to (0, 0)} f(0, y) = \lim_{(0, y) \to (0, 0)} \frac{0 - y^2}{0 + y^2} = -1$$

Thus, we get different values depending on how $(x, y) \to (0, 0)$. In fact, there are points arbitrarily close to $(0, 0)$ at which the value of $f$ is 1 and other points equally close at which the value of $f$ is $-1$. The limit cannot exist. ∎

See Problem 17 for an even more unusual example where a limit fails to exist and Problem 27 for another approach to Example 1.

**Continuity at a Point** To say that $f(x, y)$ is **continuous** at the point $(a, b)$, we require that (1) $f$ has a value at $(a, b)$, (2) $f$ has a limit at $(a, b)$, and (3) the value of $f$ at $(a, b)$ is equal to the limit there. In summary, we require that

$$\lim_{(x, y) \to (a, b)} f(x, y) = f(a, b)$$

Intuitively this means that $f$ has no jumps or wild fluctuations at $(a, b)$.

As with functions of a single variable, sums, products, and quotients of continuous functions are continuous (provided, in the last case, that we avoid division by zero). It follows that *polynomial functions* of two variables are continuous everywhere, since they are sums and products of the continuous functions $ax$, $by$, and $c$, where $a$, $b$, and $c$ are constants. For example, the function $f(x, y) = 5x^4y^2 - 2xy^3 + 4$ is continuous at all points in the $xy$-plane.

*Rational functions* of two variables are quotients of polynomial functions and thus are continuous wherever the denominator is not zero. To illustrate, $f(x, y) = (2x + 3y)/(y^2 - 4x)$ is continuous everywhere in the $xy$-plane except at points on the parabola $y^2 = 4x$.

As with functions of one variable, a continuous function of a continuous function is continuous.

---

**Theorem A**

**(Composition of functions).** If a function $g$ of two variables is continuous at $(a, b)$ and a function $f$ of one variable is continuous at $g(a, b)$, then the composite function $f \circ g$, defined by $(f \circ g)(x, y) = f(g(x, y))$, is continuous at $(a, b)$.

---

Proof of this theorem is similar to the proof of Theorem 2.7D.

**EXAMPLE 2** Show that $F(x, y) = \cos(x^3 - 4xy + y^2)$ is continuous at every point of the plane.

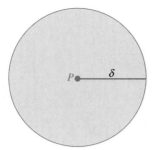

A neighborhood in 2-space

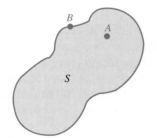

A neighborhood in 3-space

**FIGURE 5**

**FIGURE 6**

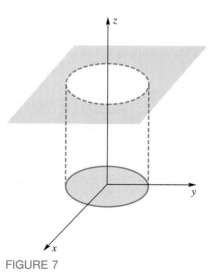

**FIGURE 7**

*Solution* The function $g(x, y) = x^3 - 4xy + y^2$, being a polynomial, is continuous everywhere. Also $f(t) = \cos t$ is continuous at each number $t$ in $\mathbb{R}$. We conclude from Theorem A, that $F(x, y) = f(g(x, y))$ is continuous at all $(x, y)$ in the plane. ∎

**Continuity on a Set** To say that $f(x, y)$ is continuous on a set $S$ ought to mean that $f(x, y)$ is continuous at every point of the set. It does mean that, but there are some subtleties connected with this statement that need to be cleared up.

First we need to introduce some language relative to sets in the plane (and higher-dimensional spaces). By a **neighborhood** of radius $\delta$ of a point $P$, we mean the set of all points $Q$ satisfying $|Q - P| < \delta$. In two-space, a neighborhood is the "inside" of a circle; in three-space, it is the inside of a sphere (Figure 5). A point $P$ is an **interior point** of a set $S$ if there is a neighborhood of $P$ contained in $S$. The set of all interior points of $S$ is the **interior** of $S$. On the other hand, $P$ is a **boundary point** of $S$ if every neighborhood of $P$ contains points that are in $S$ and points that are not in $S$. The set of all boundary points of $S$ is called the **boundary** of $S$. In Figure 6, $A$ is an interior point and $B$ is a boundary point of $S$. Finally, a set is **open** if all its points are interior points and it is **closed** if it contains all its boundary points. The analogy with open and closed intervals in one-space should occur to you.

If $S$ is an open set, to say that $f$ is continuous on $S$ means precisely that $f$ is continuous at every point of $S$. On the other hand, if $S$ contains some or all of its boundary points, we must be careful to give the right interpretation of continuity at such points (recall that in one-space, we had to talk about left and right continuity at the endpoints of an interval). To say that $f$ is continuous at a boundary point $P$ of $S$ means that $f(Q)$ must approach $f(P)$ as $Q$ approaches $P$ through points of $S$.

Here is an example that will help clarify what we have said (see Figure 7). Let

$$f(x, y) = \begin{cases} 0 & \text{if } x^2 + y^2 \le 1 \\ 4 & \text{otherwise} \end{cases}$$

If $S$ is the set $\{(x, y): x^2 + y^2 \le 1\}$, it is correct to say that $f(x, y)$ is continuous on $S$. On the other hand, it would be incorrect to say that $f(x, y)$ is continuous on the whole plane.

We said in Section 15.2 that for most functions of two variables studied in a first course, $f_{xy} = f_{yx}$; that is, the order of differentiation in mixed partial derivatives is immaterial. Now that continuity is defined, conditions for this to be true can be simply stated.

---

**Theorem B**

**(Equality of mixed partials).** Let $f_{xy}$ and $f_{yx}$ be continuous on an open set $S$. Then $f_{xy} = f_{yx}$ at each point of $S$.

---

A proof of this theorem is given in books on advanced calculus. A counterexample where continuity of $f_{xy}$ is lacking is given in Problem 32.

Our discussion of continuity has dealt mainly with functions of two variables. We believe you can make the simple changes that are required to describe continuity for functions of three or more variables.

## CONCEPTS REVIEW

**1.** In intuitive language, to say $\lim_{(x, y) \to (1, 2)} f(x, y) = 3$ means that $f(x, y)$ gets close to _____ when _____.

**2.** For $f(x, y)$ to be continuous at $(1, 2)$ means that _____.

**3.** The point $P$ is an interior point of set $S$ if there is a neighborhood of $P$ that is _____.

**4.** The set $S$ is open if every point of $S$ is _____; $S$ is closed if $S$ contains all its _____.

## PROBLEM SET 15.3

In Problems 1–8, find the indicated limit or state that it does not exist.

**1.** $\lim_{(x, y) \to (1, 3)} (3x^2y - xy^3)$

**2.** $\lim_{(x, y) \to (-2, 1)} (xy^3 - xy + 3y^2)$

**3.** $\lim_{(x, y) \to (2, \pi)} [x \cos^2 xy - \sin(xy/3)]$

**4.** $\lim_{(x, y) \to (-1, 2)} \dfrac{xy - y^3}{(x + y + 1)^2}$

**5.** $\lim_{(x, y) \to (0, 0)} \dfrac{\sin(x^2 + y^2)}{3x^2 + 3y^2}$

**6.** $\lim_{(x, y) \to (0, 0)} \dfrac{\tan(x^2 + y^2)}{x^2 + y^2}$

**7.** $\lim_{(x, y) \to (0, 0)} \dfrac{x^2 + y^2}{x^4 - y^4}$

**8.** $\lim_{(x, y) \to (0, 0)} \dfrac{x^4 - y^4}{x^2 + y^2}$

In Problems 9–14, describe the largest set $S$ on which it is correct to say that $f$ is continuous.

**9.** $f(x, y) = \dfrac{x^3 + xy - 5}{x^2 + y^2 + 1}$

**10.** $f(x, y) = \ln(1 - x^2 - y^2)$

**11.** $f(x, y) = \dfrac{x^2 + 3xy + y^2}{y - x^2}$

**12.** $f(x, y) = \begin{cases} \dfrac{\sin(xy)}{xy} & xy \neq 0 \\ 1 & xy = 0 \end{cases}$

**13.** $f(x, y) = \sqrt{x - y + 1}$

**14.** $f(x, y) = (4 - x^2 - y^2)^{-1/2}$

**15.** Show that
$$\lim_{(x, y) \to (0, 0)} \dfrac{xy}{x^2 + y^2}$$
does not exist by considering the $x$-axis as one path to the origin and the line $y = x$ as another.

**16.** Show that
$$\lim_{(x, y) \to (0, 0)} \dfrac{xy + y^3}{x^2 + y^2}$$
does not exist.

**17.** Let $f(x, y) = x^2y/(x^4 + y^2)$.

(a) Show that $f(x, y) \to 0$ as $(x, y) \to (0, 0)$ along any straight line $y = mx$.
(b) Show that $f(x, y) \to \frac{1}{2}$ as $(x, y) \to (0, 0)$ along the parabola $y = x^2$.
(c) What conclusion do you draw?

**18.** Show that $\lim_{(x, y) \to (0, 0)} [xy^2/(x^2 + y^2)] = 0$.

$Hint:$ $\left| \dfrac{xy^2}{x^2 + y^2} \right| \leq \dfrac{(\sqrt{x^2 + y^2})(x^2 + y^2)}{x^2 + y^2} = \sqrt{x^2 + y^2}.$

In Problems 19–24, sketch the indicated set. Describe the boundary of the set. Finally state whether the set is open, closed, or neither.

**19.** $\{(x, y): 2 \leq x \leq 4, 1 \leq y \leq 5\}$

**20.** $\{(x, y): x^2 + y^2 < 4\}$

**21.** $\{(x, y): 0 < x^2 + y^2 \leq 4\}$

**22.** $\{(x, y): 1 < x \leq 4\}$

**23.** $\{(x, y): x > 0, y < \sin(1/x)\}$

**24.** $\{(x, y): x = 0, y = 1/n, n \text{ a positive integer}\}$

**25.** Let

$$f(x, y) = \begin{cases} \dfrac{x^2 - 4y^2}{x - 2y} & x \neq 2y \\ g(x) & x = 2y \end{cases}$$

If $f$ is continuous in the whole plane, find a formula for $g(x)$.

**26.** Prove that

$$\lim_{(x, y) \to (a, b)} [f(x, y) + g(x, y)]$$

$$= \lim_{(x, y) \to (a, b)} f(x, y) + \lim_{(x, y) \to (a, b)} g(x, y)$$

provided the latter two limits exist.

**27.** Occasionally, it is easier to analyze the continuity of $f(x, y)$ by changing to polar coordinates. Let $f(x, y) = (x^2 - y^2)/(x^2 + y^2)$ for $(x, y) \neq (0, 0)$ and $f(0, 0) = 0$ (see Example 1). In polar coordinates,

$$f(x, y) = \frac{r^2 \cos^2\theta - r^2 \sin^2\theta}{r^2} = \cos 2\theta$$

which takes all values between $-1$ and $1$ in every neighborhood of the origin. We conclude that $\lim_{(x, y) \to (0, 0)} f(x, y)$ does not exist and $f$ is not continuous at the origin. Let each of the following functions have the value 0 at $(0, 0)$. Which of them are continuous at $(0, 0)$ and which are discontinuous there?

(a) $f(x, y) = \dfrac{xy}{\sqrt{x^2 + y^2}}$

(b) $f(x, y) = \dfrac{xy}{x^2 + y^2}$

(c) $f(x, y) = \dfrac{x^{7/3}}{x^2 + y^2}$

(d) $f(x, y) = xy\dfrac{x^2 - y^2}{x^2 + y^2}$

(e) $f(x, y) = \dfrac{x^2 y^2}{x^2 + y^4}$

(f) $f(x, y) = \dfrac{xy^2}{x^2 + y^4}$

**28.** Let $f(x, y)$ be the shortest distance that a raindrop landing at latitude $x$ and longitude $y$ in the state of Colorado must travel to reach the ocean. Where in Colorado is this function discontinuous?

**29.** Let $H$ be the hemispherical shell $x^2 + y^2 + (z - 1)^2 = 1$, $0 \leq z < 1$, shown in Figure 8 and let $D = \{(x, y, z): 1 \leq z \leq 2\}$. For each of the functions defined below, determine its set of discontinuities within $D$.

(a) $f(x, y, z)$ is the time required for a particle dropped from $(x, y, z)$ to reach the level $z = 0$.
(b) $f(x, y, z)$ is the area of the inside of $H$ (assumed opaque) that can be seen from $(x, y, z)$.
(c) $f(x, y, z)$ is the area of the shadow of $H$ on the $xy$-plane due to a point light source at $(x, y, z)$.
(d) $f(x, y, z)$ is the distance along the shortest path from $(x, y, z)$ to $(0, 0, 0)$ that does not penetrate $H$.

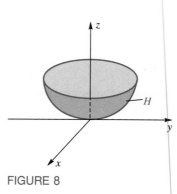

**FIGURE 8**

**30.** Let $f$, a function of $n$ variables, be continuous on an open set $D$ and suppose that $P_0$ is in $D$ with $f(P_0) > 0$. Prove that there is a $\delta > 0$ such that $f(P) > 0$ in a neighborhood of $P_0$ with radius $\delta$.

**31.** (The French Railroad) Paris is located at the origin of the $xy$-plane. Rail lines emanate from Paris along all rays, and these are the only rail lines. Determine the set of discontinuities of the following functions.

(a) $f(x, y)$ is the distance from $(x, y)$ to $(1, 0)$ on the French railroad.
(b) $g(u, v, x, y)$ is the distance from $(u, v)$ to $(x, y)$ on the French railroad.

**32.** Let $f(x, y) = xy\dfrac{x^2 - y^2}{x^2 + y^2}$ if $(x, y) \neq 0$ and $f(0, 0) = 0$. Show that $f_{xy}(0, 0) \neq f_{yx}(0, 0)$ by completing the following steps.

(a) Show $f_x(0, y) = \lim_{h \to 0} \dfrac{f(0 + h, y) - f(0, y)}{h} = -y$ for all $y$.

(b) Similarly, show that $f_y(x, 0) = x$ for all $x$.

(c) Show that $f_{yx}(0, 0) = \lim_{h \to 0} \dfrac{f_y(0 + h, 0) - f_y(0, 0)}{h} = 1$.

(d) Similarly, show that $f_{xy}(0, 0) = -1$.

PC **33.** Draw the graph of the function mentioned in Problem 32. Do you see why this surface is sometimes called the dog saddle?

PC **34.** Draw the graphs of each of the following functions on $-2 \leq x \leq 2$, $-2 \leq y \leq 2$, and determine where on this set they are discontinuous.

(a) $f(x, y) = x^2/(x^2 + y^2)$, $f(0, 0) = 0$
(b) $f(x, y) = \tan(x^2 + y^2)/(x^2 + y^2)$, $f(0, 0) = 0$

PC **35.** Draw the graph of $f(x, y) = x^2 y/(x^4 + y^2)$ in an orientation that illustrates its unusual characteristics (see Problem 17).

---

**Answers to Concepts Review:** **1.** 3; $(x, y)$ approaches $(1, 2)$ **2.** $\lim_{(x, y) \to (1, 2)} f(x, y) = f(1, 2)$ **3.** Contained in $S$
**4.** An interior point of $S$; boundary points

**15.4**
**DIFFERENTIABILITY**

For a function of a single variable, differentiability of $f$ at $x$ meant the existence of the derivative $f'(x)$. This, in turn, was equivalent to the graph of $f$ having a nonvertical tangent line at $x$.

Now we ask: What is the right concept of differentiability for a function of two variables? Surely it must correspond in a natural way to the existence of a tangent plane and clearly this requires more than the mere existence of the partial derivatives of $f$, for they reflect the behavior of $f$ in only two directions. To emphasize this point, consider

$$f(x, y) = -10 \sqrt{|xy|}$$

for which we have shown a computer-generated graph (Figure 1). Note that $f_x(0, 0)$ and $f_y(0, 0)$ both exist and equal 0; yet no one would claim that the graph has a tangent plane at the origin. The reason is, of course, that the graph of $f$ is not well approximated there by any plane (in particular, the $xy$-plane) except in two directions. A tangent plane ought to approximate the graph very well in all directions.

Consider a second question. What plays the role of *the* derivative for a function of two variables? Again the partial derivatives fall short, if for no other reason than because there are two of them.

To answer these two questions, we start by downplaying the distinction between the point $(x, y)$ and the vector $\langle x, y \rangle$. Thus, we write $\mathbf{p} = (x, y) = \langle x, y \rangle$ and $f(\mathbf{p}) = f(x, y)$. Recall that

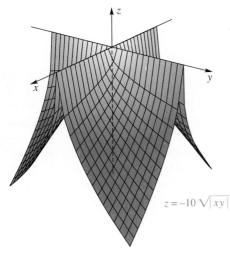

$z = -10 \sqrt{|xy|}$

FIGURE 1

$$(1) \qquad f'(x_0) = \lim_{x \to x_0} \frac{f(x) - f(x_0)}{x - x_0} = \lim_{h \to 0} \frac{f(x_0 + h) - f(x_0)}{h}$$

The analogue would seem to be

$$(2) \qquad f'(\mathbf{p}_0) = \lim_{\mathbf{p} \to \mathbf{p}_0} \frac{f(\mathbf{p}) - f(\mathbf{p}_0)}{\mathbf{p} - \mathbf{p}_0} = \lim_{h \to 0} \frac{f(\mathbf{p}_0 + \mathbf{h}) - f(\mathbf{p}_0)}{\mathbf{h}}$$

but, unfortunately, the division by a vector makes no sense.

But let us not give up too quickly. The existence of the limit (1) means that there is a number $m = f'(x_0)$ such that

$$\left| \frac{f(x) - f(x_0)}{x - x_0} - m \right| = \varepsilon_1$$

where $\varepsilon_1 \to 0$ as $x \to x_0$. We may rewrite this as

$$|f(x) - f(x_0) - m(x - x_0)| = \varepsilon_1 |x - x_0|$$

or (after replacing $\pm \varepsilon_1$ by $\varepsilon$) as

$$f(x) - f(x_0) - m(x - x_0) = \varepsilon |x - x_0|$$

Thus, to say that $f$ is differentiable at $x_0$ means that there is a number $m = f'(x_0)$ such that

$$(3) \qquad f(x) = f(x_0) + m(x - x_0) + |x - x_0| \varepsilon$$

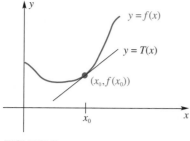

FIGURE 2

where $\varepsilon \to 0$ as $x \to x_0$. Now, look at Figure 2 and note that $y = T(x) = f(x_0) + m(x - x_0)$ is the equation of the line through $(x_0, f(x_0))$ with slope $m$. Consequently, we can reformulate (3) to say that $f$ is differentiable at $x_0$ provided there is a line $y = T(x)$ through $(x_0, f(x_0))$ that approximates $y = f(x)$ so well near $x_0$ that $|f(x) - T(x)|/|x - x_0| \to 0$ when $x \to x_0$. This line is what we have called the tangent line.

The way is now clear for an appropriate generalization to higher dimensions. To say that $f(\mathbf{p})$ is differentiable at $\mathbf{p}_0$ will mean that there is a vector $\mathbf{m}$ such that

$$(4) \qquad f(\mathbf{p}) = f(\mathbf{p}_0) + \mathbf{m} \cdot (\mathbf{p} - \mathbf{p}_0) + |\mathbf{p} - \mathbf{p}_0|\varepsilon$$

where $\varepsilon \to 0$ as $\mathbf{p} \to \mathbf{p}_0$. Here,

$$(5) \qquad z = T(\mathbf{p}) = f(\mathbf{p}_0) + \mathbf{m} \cdot (\mathbf{p} - \mathbf{p}_0)$$

is the equation of a plane through $(\mathbf{p}_0, f(\mathbf{p}_0))$, and so differentiability of $f(\mathbf{p})$ at $\mathbf{p}_0$ is equivalent to the existence of a plane $z = T(\mathbf{p})$ that approximates $z = f(\mathbf{p})$ in the sense indicated. Naturally, this plane is called the **tangent plane** (Figure 3).

We are finally ready to give a formal definition. We use the h-notation, since this allows us to drop the subscript 0. In (4), replace $\mathbf{p}_0$ by $\mathbf{p}$ and replace $\mathbf{p}$ by $\mathbf{p} + \mathbf{h}$. Note also our use of the symbol $\varepsilon(\mathbf{h})$ to indicate that $\varepsilon$ does depend on $\mathbf{h}$.

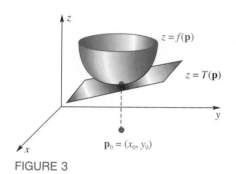

FIGURE 3

---

**Definition**

We say that $f$ is **differentiable** at $\mathbf{p}$ if there exists a vector $\mathbf{m}$ such that

$$f(\mathbf{p} + \mathbf{h}) = f(\mathbf{p}) + \mathbf{m} \cdot \mathbf{h} + |\mathbf{h}|\varepsilon(\mathbf{h})$$

where $\varepsilon(\mathbf{h}) \to 0$ as $\mathbf{h} \to \mathbf{0}$.

---

It can be shown that if the vector $\mathbf{m}$ exists, it is unique (Problem 19). We call this vector the **gradient** of $f$ at $\mathbf{p}$ and denote it by $\nabla f(\mathbf{p})$. Thus, if $f$ is differentiable at $\mathbf{p}$, it has a gradient $\nabla f(\mathbf{p})$ and

$$\boxed{f(\mathbf{p} + \mathbf{h}) = f(\mathbf{p}) + \nabla f(\mathbf{p}) \cdot \mathbf{h} + |\mathbf{h}|\varepsilon(\mathbf{h})}$$

where $\varepsilon(\mathbf{h}) \to 0$ as $\mathbf{h} \to \mathbf{0}$.

We call your attention to several aspects of this definition.

1. The derivative $f'(x)$ is a number, but the gradient $\nabla f(\mathbf{p})$ is a vector.
2. The dot in $\nabla f(\mathbf{p}) \cdot \mathbf{h}$ indicates the dot product of two vectors.
3. The definition makes sense in any number of dimensions.

**Computing Gradients**    To find a gradient from the definition is burdensome (as it was for derivatives). We need rules for calculation. Here is a very useful result, which we state without proof (but see Problem 20).

---

Theorem A

If $f$, a function of two variables, is differentiable at $\mathbf{p} = (x, y)$, then the first partial derivatives of $f$ exist at $\mathbf{p}$ and

$$\nabla f(\mathbf{p}) = \frac{\partial f}{\partial x}(\mathbf{p})\mathbf{i} + \frac{\partial f}{\partial y}(\mathbf{p})\,\mathbf{j}$$

Similarly, if $g$, a function of three variables, is differentiable at $\mathbf{p} = (x, y, z)$, then

$$\nabla g(\mathbf{p}) = \frac{\partial g}{\partial x}(\mathbf{p})\mathbf{i} + \frac{\partial g}{\partial y}(\mathbf{p})\mathbf{j} + \frac{\partial g}{\partial z}(\mathbf{p})\mathbf{k}$$

---

To use Theorem A, we still need to know that $f$ and $g$ are differentiable, which is troublesome. Fortunately, there is a simple criterion, proved in most books on advanced calculus.

---

Theorem B

If $f$ has first partial derivatives in a neighborhood of $\mathbf{p}$ and if these partial derivatives are continuous at $\mathbf{p}$, then $f$ is differentiable at $\mathbf{p}$.

---

Theorems A and B together provide just what we need to handle most problems.

**EXAMPLE 1**    Show that $f(x, y) = xe^y + x^2y$ is differentiable everywhere and calculate its gradient. Then find the equation $z = T(x, y)$ of the tangent plane at $(2, 0)$.

*Solution*    We note first that

$$\frac{\partial f}{\partial x} = e^y + 2xy, \qquad \frac{\partial f}{\partial y} = xe^y + x^2$$

Both of these functions are continuous everywhere, and so by Theorem B, $f$ is differentiable everywhere. Moreover, by Theorem A,

$$\nabla f(x, y) = (e^y + 2xy)\mathbf{i} + (xe^y + x^2)\mathbf{j}$$

Thus,

$$\nabla f(2, 0) = \mathbf{i} + 6\mathbf{j} = \langle 1, 6 \rangle$$

and, as in (5), the equation of the tangent plane is

$$z = f(2, 0) + \nabla f(2, 0) \cdot \langle x - 2, y \rangle$$
$$= 2 + \langle 1, 6 \rangle \cdot \langle x - 2, y \rangle$$
$$= 2 + x - 2 + 6y = x + 6y \qquad \blacksquare$$

**EXAMPLE 2**   For $f(x, y, z) = x \sin z + x^2 y$, find $\nabla f(1, 2, 0)$.

*Solution*   Since the partial derivatives

$$\frac{\partial f}{\partial x} = \sin z + 2xy, \qquad \frac{\partial f}{\partial y} = x^2, \qquad \frac{\partial f}{\partial z} = x \cos z$$

are all continuous, the gradient exists. Moreover at (1, 2, 0), these partials have the values 4, 1, and 1, respectively. Thus,

$$\nabla f(1, 2, 0) = 4\mathbf{i} + \mathbf{j} + \mathbf{k} \qquad \blacksquare$$

**Rules for Gradients**   In many respects, gradients behave like derivatives. Recall that $D$ considered as an operator is linear. So is the operator $\nabla$, often called the *del* operator.

---

### Theorem C

$\nabla$ is a linear operator; that is:

**(i)**   $\nabla[f(\mathbf{p}) + g(\mathbf{p})] = \nabla f(\mathbf{p}) + \nabla g(\mathbf{p})$;
**(ii)**  $\nabla[\alpha f(\mathbf{p})] = \alpha \nabla f(\mathbf{p})$.

Also, we have the product rule.

**(iii)** $\nabla[f(\mathbf{p})g(\mathbf{p})] = f(\mathbf{p})\nabla g(\mathbf{p}) + g(\mathbf{p})\nabla f(\mathbf{p})$

---

*Proof*   All three results follow from the corresponding facts for partial derivatives. We prove (iii) in the two-variable case, suppressing the point $\mathbf{p}$ for brevity.

$$\nabla fg = \frac{\partial(fg)}{\partial x}\mathbf{i} + \frac{\partial(fg)}{\partial y}\mathbf{j}$$

$$= \left( f\frac{\partial g}{\partial x} + g\frac{\partial f}{\partial x} \right)\mathbf{i} + \left( f\frac{\partial g}{\partial y} + g\frac{\partial f}{\partial y} \right)\mathbf{j}$$

$$= f\left( \frac{\partial g}{\partial x}\mathbf{i} + \frac{\partial g}{\partial y}\mathbf{j} \right) + g\left( \frac{\partial f}{\partial x}\mathbf{i} + \frac{\partial f}{\partial y}\mathbf{j} \right)$$

$$= f\nabla g + g\nabla f \qquad \blacksquare$$

**Continuity versus Differentiability**   Recall for functions of one variable that differentiability implies continuity, but not vice versa. The same is true here.

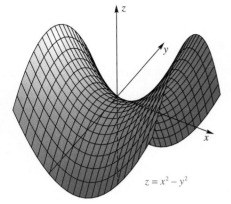

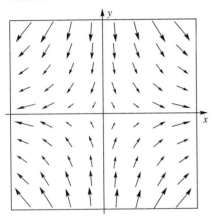

FIGURE 4

$z = x^2 - y^2$

FIGURE 5

**Theorem D**

If $f$ is differentiable at $\mathbf{p}$, then $f$ is continuous at $\mathbf{p}$.

**Proof** Since $f$ is differentiable at $\mathbf{p}$,

$$f(\mathbf{p} + \mathbf{h}) - f(\mathbf{p}) = \nabla f(\mathbf{p}) \cdot \mathbf{h} + |\mathbf{h}|\varepsilon(\mathbf{h})$$

Thus,

$$|f(\mathbf{p} + \mathbf{h}) - f(\mathbf{p})| \le |\nabla f(\mathbf{p}) \cdot \mathbf{h}| + |\mathbf{h}| \, |\varepsilon(\mathbf{h})|$$
$$= |\nabla f(\mathbf{p})| \, |\mathbf{h}| \, |\cos \theta| + |\mathbf{h}| \, |\varepsilon(\mathbf{h})|$$

Both of the latter terms approach 0 as $\mathbf{h} \to \mathbf{0}$, and so

$$\lim_{\mathbf{h} \to \mathbf{0}} f(\mathbf{p} + \mathbf{h}) = f(\mathbf{p})$$

This last equality is one way of formulating the continuity of $f$ at $\mathbf{p}$. ∎

**The Gradient Field** The gradient $\nabla f$ associates with each point $\mathbf{p}$ in the domain of $f$ a vector $\nabla f(\mathbf{p})$. The set of all these vectors is called the **gradient field** for $f$. In Figures 4 and 5, we show computer-drawn graphs of the surface $z = x^2 - y^2$ and the corresponding gradient field. Do these figures suggest something about the direction in which the gradient vectors point? We explore this subject in the next section.

## CONCEPTS REVIEW

**1.** The analog of the derivative $f'(x)$ for a function of more than one variable is the _____ denoted by $\nabla f(\mathbf{p})$.

**2.** $\nabla f(\mathbf{p})$ satisfies the relation $f(\mathbf{p} + \mathbf{h}) - f(\mathbf{p}) =$ _____, where $\varepsilon(\mathbf{h}) \to 0$ as $\mathbf{h} \to \mathbf{0}$.

**3.** We can calculate $\nabla f$ for a function of two variables from the formula $\nabla f(\mathbf{p}) =$ _____. Thus, if $f(x, y) = xy^2$, $\nabla f(x, y) =$ _____.

**4.** $f(x, y)$ being differentiable at $(x_0, y_0)$ is equivalent to the existence of a _____ to the graph at this point.

## PROBLEM SET 15.4

In Problems 1–10, find the gradient $\nabla f$.

**1.** $f(x, y) = x^2y + 3xy$    **2.** $f(x, y) = x^3y - y^3$

**3.** $f(x, y) = xe^{xy}$    **4.** $f(x, y) = x^2y \cos y$

**5.** $f(x, y) = x^2y/(x + y)$    **6.** $f(x, y) = \sin^3(x^2y)$

**7.** $f(x, y, z) = \sqrt{x^2 + y^2 + z^2}$

**8.** $f(x, y, z) = x^2y + y^2z + z^2x$

**9.** $f(x, y, z) = x^2ye^{x - z}$

**10.** $f(x, y, z) = xz \ln(x + y + z)$

In Problems 11–14, find the gradient vector of the given function at the given point $\mathbf{p}$. Then find the equation of the tangent plane at $\mathbf{p}$ (see Example 1).

**11.** $f(x, y) = x^2y - xy^2$, $\mathbf{p} = (-2, 3)$

**12.** $f(x, y) = x^3y + 3xy^2$, $\mathbf{p} = (2, -2)$

**13.** $f(x, y) = \cos \pi x \sin \pi y + \sin 2\pi y$, $\mathbf{p} = \left(-1, \tfrac{1}{2}\right)$

**14.** $f(x, y) = \dfrac{x^2}{y}$, $\mathbf{p} = (2, -1)$

In Problems 15 and 16, find the equation $w = T(x, y, z)$ of the tangent "hyperplane" at **p**.

**15.** $f(x, y, z) = 3x^2 - 2y^2 + xz^2$, $\mathbf{p} = (1, 2, -1)$

**16.** $f(x, y, z) = xyz + x^2$, $\mathbf{p} = (2, 0, -3)$

**17.** Show that

$$\nabla\left(\frac{f}{g}\right) = \frac{g\nabla f - f\nabla g}{g^2}$$

**18.** Show that

$$\nabla(f^r) = rf^{r-1}\,\nabla f$$

**19.** Show that if

$$f(\mathbf{p} + \mathbf{h}) - f(\mathbf{p}) = \mathbf{q}_1 \cdot \mathbf{h} + |\mathbf{h}|\varepsilon_1(\mathbf{h})$$

and

$$f(\mathbf{p} + \mathbf{h}) - f(\mathbf{p}) = \mathbf{q}_2 \cdot \mathbf{h} + |\mathbf{h}|\varepsilon_2(\mathbf{h})$$

then $\mathbf{q}_1 = \mathbf{q}_2$. *Hint*: Calculate $|\mathbf{q}_1 - \mathbf{q}_2|$ by using

$$\mathbf{q}_1 \cdot \mathbf{h} - \mathbf{q}_2 \cdot \mathbf{h} = (\mathbf{q}_1 - \mathbf{q}_2) \cdot \mathbf{h} = |\mathbf{q}_1 - \mathbf{q}_2| \, |\mathbf{h}|\cos\theta$$

**20.** By letting $\mathbf{h} = (h, 0)$ and then $\mathbf{h} = (0, k)$ in the definition of the gradient, show that $\nabla f(x, y)$ must have the form asserted in Theorem A.

**21.** Refer to Figure 1. Find the equation of the tangent plane to $z = -10\sqrt{|xy|}$ at $(1, -1)$. *Recall*: $d|x|/dx = |x|/x$ for $x \neq 0$.

**22. Mean Value Theorem (Several Variables)** If $f$ is differentiable at each point of the line segment from **a** to **b**, then there exists on that line segment a point **c** between **a** and **b** such that

$$f(\mathbf{b}) - f(\mathbf{a}) = \nabla f(\mathbf{c}) \cdot (\mathbf{b} - \mathbf{a})$$

Taking this result for granted, show that if $f$ is differentiable on a convex set $S$ and if $\nabla f(\mathbf{p}) = \mathbf{0}$ on $S$, then $f$ is constant on $S$. *Note*: A set $S$ is convex if each pair of points in $S$ can be connected by a line segment in $S$.

**23.** Use the result of Problem 22 to show that if $\nabla f(\mathbf{p}) = \nabla g(\mathbf{p})$ for all **p** in a convex set $S$, then $f$ and $g$ differ by a constant on $S$.

**24.** Find the most general function $f(\mathbf{p})$ satisfying $\nabla f(\mathbf{p}) = \mathbf{p}$.

**25.** Let

$$f(x, y) = \begin{cases} \dfrac{\sin(x^2 + y^2)}{x^2 + y^2} & \text{if } (x, y) \neq (0, 0) \\ 1 & \text{if } (x, y) = (0, 0) \end{cases}$$

Show that $f(x, y)$ is differentiable at $(0, 0)$ with $\nabla f(0, 0) = \langle 0, 0\rangle$. *Hint*: Let $\mathbf{h} = (h, k)$ and note that we can write

$$f(0 + \mathbf{h}) = f(h, k)$$
$$= 1 + 0 + \frac{\sin(h^2 + k^2) - (h^2 + k^2)}{(h^2 + k^2)^{3/2}} \sqrt{h^2 + k^2}$$

What must be shown for the multiplier of $\sqrt{h^2 + k^2}$?

**26.** Is $f(x, y) = \sqrt{x^2 + y^2}$ differentiable at $(0, 0)$? Justify your answer.

PC **27.** Draw the graph of $f(x, y) = -|xy|$ together with its gradient field.

(a) Based on this and Figures 4 and 5, make a conjecture about the direction in which a gradient vector points.
(b) Is $f$ differentiable at the origin? Justify your answer.

PC **28.** Draw the graph of $f(x, y) = \sin x + \sin y - \sin(x + y)$ on $0 \leq x \leq 2\pi$, $0 \leq y \leq 2\pi$. Also draw the gradient field to see if your conjecture in Problem 27 is confirmed.

---

**Answers to Concepts Review:** **1.** Gradient
**2.** $\nabla f(\mathbf{p}) \cdot \mathbf{h} + |\mathbf{h}|\varepsilon(\mathbf{h})$ **3.** $\dfrac{\partial f}{\partial x}(\mathbf{p})\mathbf{i} + \dfrac{\partial f}{\partial y}(\mathbf{p})\mathbf{j}$; $y^2\mathbf{i} + 2xy\mathbf{j}$
**4.** Tangent plane

---

## 15.5 DIRECTIONAL DERIVATIVES AND GRADIENTS

Consider again a function $f(x, y)$ of two variables. The partial derivatives $f_x(x, y)$ and $f_y(x, y)$ measure the rate of change (and the slope of the tangent line) in directions parallel to the $x$- and $y$-axes. Our goal now is to study the rate of change of $f$ in an arbitrary direction. This leads to the concept of the directional derivative, which in turn is related to the gradient.

It will be convenient to use vector notation. Let $\mathbf{p} = (x, y)$ and let **i** and **j** be the unit vectors in the positive $x$- and $y$-directions. Then the two partial derivatives at **p** may be written as follows.

$$f_x(\mathbf{p}) = \lim_{h \to 0} \frac{f(\mathbf{p} + h\mathbf{i}) - f(\mathbf{p})}{h}$$

$$f_y(\mathbf{p}) = \lim_{h \to 0} \frac{f(\mathbf{p} + h\mathbf{j}) - f(\mathbf{p})}{h}$$

To get the concept we are after, all we have to do is replace $\mathbf{i}$ or $\mathbf{j}$ by an arbitrary unit vector $\mathbf{u}$.

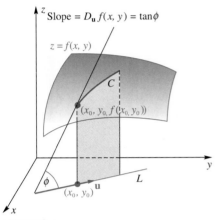

Slope $= D_{\mathbf{u}} f(x, y) = \tan\phi$

$z = f(x, y)$

$C$

$(x_0, y_0, f(x_0, y_0))$

$L$

$(x_0, y_0)$  $\mathbf{u}$

$\phi$

**FIGURE 1**

> **Definition**
>
> For each unit vector $\mathbf{u}$, let
>
> $$D_{\mathbf{u}} f(\mathbf{p}) = \lim_{h \to 0} \frac{f(\mathbf{p} + h\mathbf{u}) - f(\mathbf{p})}{h}$$
>
> This limit, if it exists, is called the **directional derivative** of $f$ at $\mathbf{p}$ in the direction $\mathbf{u}$.

Thus, $D_{\mathbf{i}} f(\mathbf{p}) = f_x(\mathbf{p})$ and $D_{\mathbf{j}} f(\mathbf{p}) = f_y(\mathbf{p})$. Since $\mathbf{p} = (x, y)$, we also use the notation $D_{\mathbf{u}} f(x, y)$. Figure 1 gives the geometric interpretation of $D_{\mathbf{u}} f(x_0, y_0)$. The vector $\mathbf{u}$ determines a line $L$ in the $xy$-plane through $(x_0, y_0)$. The plane through $L$ perpendicular to the $xy$-plane intersects the surface $z = f(x, y)$ in a curve $C$. Its tangent at the point $(x_0, y_0, f(x_0, y_0))$ has slope $D_{\mathbf{u}} f(x_0, y_0)$. Another useful interpretation is that $D_{\mathbf{u}} f(x_0, y_0)$ measures the rate of change of $f$ with respect to distance in the direction $\mathbf{u}$.

**Connection with the Gradient**  Recall from Section 15.4 that $\nabla f(\mathbf{p})$ is given by

$$\nabla f(\mathbf{p}) = f_x(\mathbf{p})\mathbf{i} + f_y(\mathbf{p})\mathbf{j}$$

> **Theorem A**
>
> Let $f$ be differentiable at $\mathbf{p}$. Then $f$ has a directional derivative at $\mathbf{p}$ in the direction of the unit vector $\mathbf{u} = u_1\mathbf{i} + u_2\mathbf{j}$ and
>
> $$D_{\mathbf{u}} f(\mathbf{p}) = \mathbf{u} \cdot \nabla f(\mathbf{p})$$
>
> that is,
>
> $$D_{\mathbf{u}} f(x, y) = u_1 f_x(x, y) + u_2 f_y(x, y)$$

**_Proof_**  Since $f$ is differentiable at $\mathbf{p}$,

$$f(\mathbf{p} + h\mathbf{u}) - f(\mathbf{p}) = h\mathbf{u} \cdot \nabla f(\mathbf{p}) + |h\mathbf{u}|\varepsilon(h\mathbf{u})$$

where $\varepsilon(h\mathbf{u}) \to 0$ as $h \to 0$. Thus,

$$\frac{f(\mathbf{p} + h\mathbf{u}) - f(\mathbf{p})}{h} = \mathbf{u} \cdot \nabla f(\mathbf{p}) \pm \varepsilon(h\mathbf{u})$$

The conclusion follows by taking limits as $h \to 0$. ∎

**EXAMPLE 1** If $f(x, y) = 4x^2 - xy + 3y^2$, find the directional derivative of $f$ at $(2, -1)$ in the direction of the vector $\mathbf{a} = 4\mathbf{i} + 3\mathbf{j}$.

*Solution* The unit vector $\mathbf{u}$ in the direction of $\mathbf{a}$ is $(\frac{4}{5})\mathbf{i} + (\frac{3}{5})\mathbf{j}$. Also, $f_x(x, y) = 8x - y$ and $f_y(x, y) = -x + 6y$; thus, $f_x(2, -1) = 17$ and $f_y(2, -1) = -8$. Consequently, by Theorem A,

$$D_\mathbf{u} f(2, -1) = \tfrac{4}{5}(17) + \tfrac{3}{5}(-8) = \tfrac{44}{5}$$ ∎

Although we will not go through the details, we assert that what we have done is valid for functions of three or more variables, with obvious modifications. We illustrate.

**EXAMPLE 2** Find the directional derivative of the function $f(x, y, z) = xy \sin z$ at the point $(1, 2, \pi/2)$ in the direction of the vector $\mathbf{a} = \mathbf{i} + 2\mathbf{j} + 2\mathbf{k}$.

*Solution* The unit vector $\mathbf{u}$ in the direction of $\mathbf{a}$ is $\frac{1}{3}\mathbf{i} + \frac{2}{3}\mathbf{j} + \frac{2}{3}\mathbf{k}$. Also, $f_x(x, y, z) = y \sin z$, $f_y(x, y, z) = x \sin z$, and $f_z(x, y, z) = xy \cos z$, and so $f_x(1, 2, \pi/2) = 2$, $f_y(1, 2, \pi/2) = 1$, and $f_z(1, 2, \pi/2) = 0$. We conclude that

$$D_\mathbf{u} f\left(1, 2, \frac{\pi}{2}\right) = \frac{1}{3}(2) + \frac{2}{3}(1) + \frac{2}{3}(0) = \frac{4}{3}$$ ∎

**Maximum Rate of Change** For a given function $f$ at a given point $\mathbf{p}$, it is natural to ask in what direction the function is changing most rapidly—that is, in what direction $D_\mathbf{u} f(\mathbf{p})$ is the largest. From the geometric formula for the dot product (Section 14.2), we may write

$$D_\mathbf{u} f(\mathbf{p}) = \mathbf{u} \cdot \nabla f(\mathbf{p}) = |\mathbf{u}| \, |\nabla f(\mathbf{p})| \cos \theta = |\nabla f(\mathbf{p})| \cos \theta$$

where $\theta$ is the angle between $\mathbf{u}$ and $\nabla f(\mathbf{p})$. Thus, $D_\mathbf{u} f(\mathbf{p})$ is maximized when $\theta = 0$ and minimized when $\theta = \pi$. We summarize as follows.

> **Theorem B**
>
> A function increases most rapidly at $\mathbf{p}$ in the direction of the gradient (with rate $|\nabla f(\mathbf{p})|$) and decreases most rapidly in the opposite direction (with rate $-|\nabla f(\mathbf{p})|$).

**EXAMPLE 3** Suppose a bug is located on the hyperbolic paraboloid $z = y^2 - x^2$ at the point $(1, 1, 0)$, as in Figure 2. In what direction should it move for the steepest climb and what is the slope as it starts out?

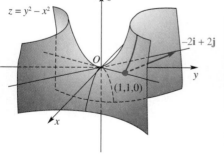

FIGURE 2

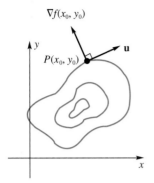

FIGURE 3

*Solution*  Let $f(x, y) = y^2 - x^2$. Since $f_x(x, y) = -2x$ and $f_y(x, y) = 2y$,

$$\nabla f(1, 1) = f_x(1, 1)\mathbf{i} + f_y(1, 1)\mathbf{j} = -2\mathbf{i} + 2\mathbf{j}$$

Thus the bug should move from $(1, 1, 0)$ in the direction $-2\mathbf{i} + 2\mathbf{j}$, where the slope will be $|-2\mathbf{i} + 2\mathbf{j}| = \sqrt{8}$.  ■

**Level Curves and Gradients**  Recall from Section 15.1 that the *level curves* of a surface $z = f(x, y)$ are the projections onto the $xy$-plane of the curves of intersection of the surface with planes $z = k$ that are parallel to the $xy$-plane. The value of the function at all points on the same level curve is constant (Figure 3).

Denote by $L$ the level curve of $f(x, y)$ that passes through an arbitrarily chosen point $P(x_0, y_0)$ in the domain of $f$ and let the unit vector $\mathbf{u}$ be tangent to $L$ at $P$. Since the value of $f$ is the same at all points on the level curve $L$, its directional derivative $D_\mathbf{u} f(x_0, y_0)$, which is the rate of change of $f(x, y)$ in the direction $\mathbf{u}$, is zero when $\mathbf{u}$ is tangent to $L$. (This statement—which seems very clear intuitively—requires justification, a justification we omit since the result we want also follows from an argument to be given in Section 15.7.) Since

$$0 = D_\mathbf{u} f(x_0, y_0) = \nabla f(x_0, y_0) \cdot \mathbf{u}$$

we conclude that $\nabla f$ and $\mathbf{u}$ are perpendicular, a result worthy of theorem status.

### Theorem C

The gradient of $f$ at a point $P$ is perpendicular to the level curve of $f$ that goes through $P$.

**EXAMPLE 4**  For the paraboloid $z = x^2/4 + y^2$, find the equation of its level curve that passes through the point $P(2, 1)$ and sketch it. Find the gradient vector of the paraboloid at $P$ and draw the gradient with its initial point at $P$.

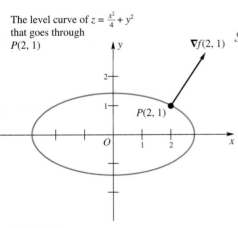

The level curve of $z = \frac{x^2}{4} + y^2$ that goes through $P(2, 1)$

FIGURE 4

*Solution*  The level curve of the paraboloid that corresponds to the plane $z = k$ has the equation $x^2/4 + y^2 = k$. To find the value of $k$ belonging to the level curve through $P$, we substitute $(2, 1)$ for $(x, y)$ and obtain $k = 2$. Thus, the equation of the level curve that goes through $P$ is the ellipse

$$\frac{x^2}{8} + \frac{y^2}{2} = 1$$

Next let $f(x, y) = x^2/4 + y^2$. Since $f_x(x, y) = x/2$ and $f_y(x, y) = 2y$, the gradient of the paraboloid at $P(2, 1)$ is

$$\nabla f(2, 1) = f_x(2, 1)\mathbf{i} + f_y(2, 1)\mathbf{j} = \mathbf{i} + 2\mathbf{j}$$

The level curve and the gradient at $P$ are shown in Figure 4.  ■

To provide additional illustration of Theorems B and C, we asked our computer to draw the surface $z = |xy|$, together with its contour map and gradient field. The results are shown in Figure 5. Note that the gradient vectors are perpendicular to the level curves and that they do point in the direction of greatest increase of $z$.

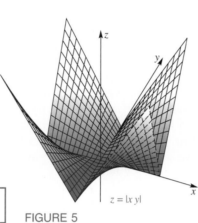

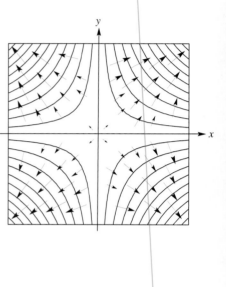

$z = |x\,y|$

**FIGURE 5**

| *From 2 to 3 Variables* | |
|---|---|
| $z = f(x, y)$ | $w = f(x, y, z)$ |
| 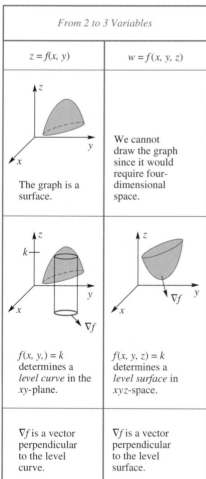 The graph is a surface. | We cannot draw the graph since it would require four-dimensional space. |
| $f(x, y,) = k$ determines a *level curve* in the $xy$-plane. | $f(x, y, z) = k$ determines a *level surface* in $xyz$-space. |
| $\nabla f$ is a vector perpendicular to the level curve. | $\nabla f$ is a vector perpendicular to the level surface. |

**Higher Dimensions** The concept of level curves for functions of two variables generalizes to level surfaces for functions of three variables. If $f$ is a function of three variables, the surface $f(x, y, z) = k$, where $k$ is a constant, is called a **level surface** for $f$. At all points on a level surface, the value of the function is the same, and the gradient vector of $f(x, y, z)$ at a point $P(x, y, z)$ in its domain is normal to the level surface of $f$ that goes through $P$.

In problems of heat conduction in a homogeneous body where $w = f(x, y, z)$ gives the temperature at the point $(x, y, z)$, the level surface $f(x, y, z) = k$ is called an *isothermal surface* because all points on it have the same temperature $k$. At any given point of the body, heat flows in the direction opposite to the gradient (that is, in the direction of the greatest decrease in temperature) and therefore perpendicular to the isothermal surface through the point. Again, if $w = f(x, y, z)$ gives the electrostatic potential (voltage) at any point in an electric potential field, the level surfaces of the function are called *equipotential surfaces*. All points on an equipotential surface have the same electrostatic potential, and the direction of flow of electricity is along the negative gradient—that is, in the direction of greatest drop in potential.

**EXAMPLE 5** If the temperature at any point in a homogeneous body is given by $T = e^{xy} - xy^2 - x^2yz$, what is the direction of the greatest drop in temperature at the point $(1, -1, 2)$?

*Solution* The greatest decrease in temperature at $(1, -1, 2)$ is in the direction of the negative gradient at that point.

Since $\nabla T = (ye^{xy} - y^2 - 2xyz)\mathbf{i} + (xe^{xy} - 2xy - x^2z)\mathbf{j} + (-x^2y)\mathbf{k}$, $-\nabla T$ at $(1, -1, 2)$ is

$$(e^{-1} - 3)\mathbf{i} - e^{-1}\mathbf{j} - \mathbf{k} \qquad \blacksquare$$

## CONCEPTS REVIEW

**1.** The directional derivative of $f$ at $\mathbf{p}$ in the direction of the unit vector $\mathbf{u}$ is denoted by $D_\mathbf{u} f(\mathbf{p})$ and is defined as $\lim_{h \to 0} \underline{\hspace{2cm}}$.

**2.** If $\mathbf{u} = (u_1, u_2)$ is a unit vector, then we may calculate $D_\mathbf{u} f(x, y)$ from the formula $D_\mathbf{u} f(x, y) = \underline{\hspace{2cm}}$.

**3.** The gradient vector $\nabla f$ always points in the direction of $\underline{\hspace{2cm}}$ of $f$.

**4.** The gradient vector of $f$ at $P$ is always perpendicular to the $\underline{\hspace{2cm}}$ of $f$ through $P$.

## PROBLEM SET 15.5

In Problems 1–8, find the directional derivative of $f$ at the point $\mathbf{p}$ in the direction of $\mathbf{a}$.

**1.** $f(x, y) = x^2y$; $\mathbf{p} = (1, 2)$; $\mathbf{a} = 3\mathbf{i} - 4\mathbf{j}$

**2.** $f(x, y) = y^2 \ln x$; $\mathbf{p} = (1, 4)$; $\mathbf{a} = \mathbf{i} - \mathbf{j}$

**3.** $f(x, y) = 2x^2 + xy - y^2$; $\mathbf{p} = (3, -2)$; $\mathbf{a} = \mathbf{i} - \mathbf{j}$

**4.** $f(x, y) = x^2 - 3xy + 2y^2$; $\mathbf{p} = (-1, 2)$; $\mathbf{a} = 2\mathbf{i} - \mathbf{j}$

**5.** $f(x, y) = e^x \sin y$; $\mathbf{p} = (0, \pi/4)$; $\mathbf{a} = \mathbf{i} + \sqrt{3}\mathbf{j}$

**6.** $f(x, y) = e^{-xy}$; $\mathbf{p} = (1, -1)$; $\mathbf{a} = -\mathbf{i} + \sqrt{3}\mathbf{j}$

**7.** $f(x, y, z) = x^3y - y^2z^2$;
$\mathbf{p} = (-2, 1, 3)$; $\mathbf{a} = \mathbf{i} - 2\mathbf{j} + 2\mathbf{k}$

**8.** $f(x, y, z) = x^2 + y^2 + z^2$;
$\mathbf{p} = (1, -1, 2)$; $\mathbf{a} = \sqrt{2}\mathbf{i} - \mathbf{j} - \mathbf{k}$

In Problems 9–12, find a unit vector in the direction in which $f$ increases most rapidly at $\mathbf{p}$. What is the rate of change in this direction?

**9.** $f(x, y) = x^3 - y^5$; $\mathbf{p} = (2, -1)$

**10.** $f(x, y) = e^y \sin x$; $\mathbf{p} = (5\pi/6, 0)$

**11.** $f(x, y, z) = x^2yz$; $\mathbf{p} = (1, -1, 2)$

**12.** $f(x, y, z) = xe^{yz}$; $\mathbf{p} = (2, 0, -4)$

**13.** In what direction $\mathbf{u}$ does $f(x, y) = 1 - x^2 - y^2$ decrease most rapidly at $\mathbf{p} = (-1, 2)$?

**14.** In what direction $\mathbf{u}$ does $f(x, y) = \sin(3x - y)$ decrease most rapidly at $\mathbf{p} = (\pi/6, \pi/4)$?

**15.** Sketch the level curve of $f(x, y) = y/x^2$ that goes through $\mathbf{p} = (1, 2)$. Calculate the gradient vector $\nabla f(\mathbf{p})$ and

draw this vector, placing its initial point at $\mathbf{p}$. What should be true about $\nabla f(\mathbf{p})$?

**16.** Follow the instructions of Problem 15 for $f(x, y) = x^2 + 4y^2$ and $\mathbf{p} = (2, 1)$.

**17.** Find the directional derivative of $f(x, y, z) = xy + z^2$ at $(1, 1, 1)$ in the direction toward $(5, -3, 3)$.

**18.** Find the directional derivative of $f(x, y) = e^{-x} \cos y$ at $(0, \pi/3)$ in the direction toward the origin.

**19.** The temperature at $(x, y, z)$ of a ball centered at the origin is given by

$$T(x, y, z) = \frac{200}{5 + x^2 + y^2 + z^2}$$

(a) By inspection, decide where the ball is hottest.
(b) Find a vector pointing in the direction of greatest increase of temperature at $(1, -1, 1)$.
(c) Does the vector of (b) point toward the origin?

**20.** The temperature at $(x, y, z)$ of a ball centered at the origin is $T(x, y, z) = 100e^{-(x^2 + y^2 + z^2)}$. Note that this ball is hottest at the origin. Show that the direction of greatest decrease in temperature is always a vector pointing away from the origin.

**21.** The elevation of a mountain above sea level at the point $(x, y)$ is $f(x, y)$. A mountain climber at $\mathbf{p}$ notes that the slope in the easterly direction is $-\frac{1}{2}$ and the slope in the northerly direction is $-\frac{1}{4}$. In what direction should he move for fastest descent?

**22.** Given that $f_x(2, 4) = -3$ and $f_y(2, 4) = 8$, find the directional derivative of $f$ at $(2, 4)$ in the direction toward $(5, 0)$.

**23.** The elevation of a mountain above sea level at $(x, y)$ is $3000e^{-(x^2 + 2y^2)/100}$ meters. The positive $x$-axis points east and the positive $y$-axis points north. A climber is directly above (10, 10). If the climber moves northwest, will she ascend or descend and at what slope?

**24.** If the temperature of a plate at the point $(x, y)$ is $T(x, y) = 10 + x^2 - y^2$, find the path a heat-seeking particle (which always moves in the direction of greatest increase in temperature) would follow if it starts at $(-2, 1)$. *Hint:* The particle moves in the direction of the gradient

$$\nabla T = 2x\mathbf{i} - 2y\mathbf{j}$$

We may write the path in parametric form as

$$\mathbf{r}(t) = x(t)\mathbf{i} + y(t)\mathbf{j}$$

and we want $x(0) = -2$ and $y(0) = 1$. To move in the required direction means that $\mathbf{r}'(t)$ should be parallel to $\nabla T$. This will be satisfied if

$$\frac{x'(t)}{2x(t)} = -\frac{y'(t)}{2y(t)}$$

together with the conditions $x(0) = -2$ and $y(0) = 1$. Now solve this differential equation and evaluate the arbitrary constant of integration.

**25.** Do Problem 24 assuming $T(x, y) = 20 - 2x^2 - y^2$.

**26.** The point $P(1, -1, -10)$ is on the surface $z = -10\sqrt{|xy|}$ (see Figure 1 of Section 15.4). Starting at $P$, in what direction $\mathbf{u} = u_1\mathbf{i} + u_2\mathbf{j}$ should one move in each case?

(a) To climb most rapidly.
(b) To stay at the same level.
(c) To climb at slope 1.

**27.** The temperature $T$ in degrees Celsius at $(x, y, z)$ is given by $T = 10/(x^2 + y^2 + z^2)$, where distances are in meters. A bee is flying away from the hot spot at the origin on a spiral path so that its position vector at time $t$ seconds is $\mathbf{r} = t \cos \pi t\mathbf{i} + t \sin \pi t\mathbf{j} + t\mathbf{k}$. Determine the rate of change of $T$ in each case.

(a) With respect to distance traveled at $t = 1$.
(b) With respect to time at $t = 1$. (Think of two ways to do this.)

**28.** Let $\mathbf{u} = (3\mathbf{i} - 4\mathbf{j})/5$ and $\mathbf{v} = (4\mathbf{i} + 3\mathbf{j})/5$ and suppose that at some point $P$, $D_\mathbf{u} f = -6$ and $D_\mathbf{v} f = 17$.

(a) Find $\nabla f$ at $P$.
(b) Note that $|\nabla f|^2 = (D_\mathbf{u} f)^2 + (D_\mathbf{v} f)^2$ in part (a). Show that this relation always holds if $\mathbf{u}$ and $\mathbf{v}$ are perpendicular.

**29.** Figure 6 shows the contour map for a hill 60 feet high, which we assume has equation $z = f(x, y)$.

(a) A raindrop landing on the hill above point $A$ will reach the $xy$-plane at $A'$, by following the path of steepest descent from $A$. Draw this path and use it to estimate $A'$.
(b) Do the same for point $B$.
(c) Estimate $f_x$ at $C$, $f_y$ at $D$, and $D_\mathbf{u} f$ at $E$, where $\mathbf{u} = (\mathbf{i} + \mathbf{j})/\sqrt{2}$.

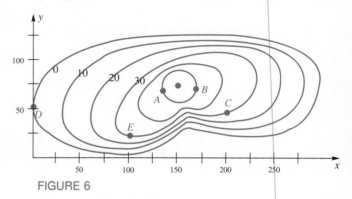

FIGURE 6

**30.** According to Theorem A, the differentiability of $f$ at $\mathbf{p}$ implies the existence of $D_\mathbf{u} f(\mathbf{p})$ in all directions. Show that the converse is false by considering

$$f(x, y) = \begin{cases} 1 & \text{if } 0 < y < x^2 \\ 0 & \text{otherwise} \end{cases}$$

at the origin.

PC **31.** Draw the graph of $z = x^2 - y^2$ on $-5 \le x \le 5$, $-5 \le y \le 5$; also draw its contour map and gradient field, thus illustrating Theorems B and C. Then estimate the $xy$-coordinates of the point where a raindrop landing above the point $(-5, -0.1)$ will leave this surface. (*True BASIC MacFunction* allows one to draw flow lines and to find coordinates of points.)

PC **32.** Follow the directions of Problem 31 for $z = x - x^3/9 - y^2$.

PC **33.** For the monkey saddle $z = x^3 - 3xy^2$ on $-5 \le x \le 5$, $-5 \le y \le 5$, estimate the $xy$-coordinates of the point where a raindrop landing above the point $(5, -0.2)$ will leave the surface.

PC **34.** Where will a raindrop landing above the point $(4, 1)$ on the surface $z = \sin x + \sin y - \sin(x + y)$, $0 \le x \le 2\pi$, $0 \le y \le 2\pi$, come to rest?

**Answers to Concepts Review:** **1.** $[f(\mathbf{p} + h\mathbf{u}) - f(\mathbf{p})]/h$
**2.** $u_1 f_x(x, y) + u_2 f_y(x, y)$   **3.** Greatest increase
**4.** Level curve

**15.6**
**THE CHAIN RULE**

The Chain Rule for composite functions of one variable is by now familiar to all our readers. If $y = f(x(t))$, where both $f$ and $x$ are differentiable functions, then

$$\frac{dy}{dt} = \frac{dy}{dx}\frac{dx}{dt}$$

Our goal is to obtain generalizations for functions of several variables.

**First Version**    If $z = f(x, y)$, where $x$ and $y$ are functions of $t$, then it makes sense to ask for $dz/dt$, and there ought to be a formula for it.

---

**Theorem A**

**(Chain Rule).** Let $x = x(t)$ and $y = y(t)$ be differentiable at $t$ and let $z = f(x, y)$ be differentiable at $(x(t), y(t))$. Then, $z = f(x(t), y(t))$ is differentiable at $t$ and

$$\frac{dz}{dt} = \frac{\partial z}{\partial x}\frac{dx}{dt} + \frac{\partial z}{\partial y}\frac{dy}{dt}$$

---

**Beauty and Generality**

Does the general analogue of the one variable Chain Rule (Theorem A, Section 3.5) hold? Yes, and here is a particularly elegant statement of it. Let $\mathbb{R}^n$ denote Euclidean $n$-space, let $g$ be a function from $\mathbb{R}$ to $\mathbb{R}^n$, and let $f$ be a function from $\mathbb{R}^n$ to $\mathbb{R}$. If $g$ is differentiable at $t$ and if $f$ is differentiable at $g(t)$, then the composite function $f \circ g$ is differentiable at $g(t)$ and

$$(f \circ g)'(t) = \nabla f(g(t)) \cdot g'(t)$$

All the machinery needed to demonstrate this is available; see if you can give the proof.

**Proof**    We mimic the one-variable proof of Appendix A.2, Theorem B. To simplify notation, let $\mathbf{p} = (x, y)$, $\Delta\mathbf{p} = (\Delta x, \Delta y)$, and $\Delta z = f(\mathbf{p} + \Delta\mathbf{p}) - f(\mathbf{p})$. Then, since $f$ is differentiable,

$$\Delta z = f(\mathbf{p} + \Delta\mathbf{p}) - f(\mathbf{p}) = \nabla f(\mathbf{p}) \cdot \Delta\mathbf{p} + |\Delta\mathbf{p}|\varepsilon(\Delta\mathbf{p})$$
$$= f_x(\mathbf{p})\Delta x + f_y(\mathbf{p})\Delta y + |\Delta\mathbf{p}|\varepsilon(\Delta\mathbf{p})$$

with $\varepsilon(\Delta\mathbf{p}) \to 0$ as $\Delta\mathbf{p} \to \mathbf{0}$.

When we divide both sides by $\Delta t$, we obtain

$$(1) \qquad \frac{\Delta z}{\Delta t} = f_x(\mathbf{p})\frac{\Delta x}{\Delta t} + f_y(\mathbf{p})\frac{\Delta y}{\Delta t} + \frac{|\Delta\mathbf{p}|}{\Delta t}\varepsilon(\Delta\mathbf{p})$$

Now

$$\left|\frac{|\Delta\mathbf{p}|}{\Delta t}\right| = \frac{\sqrt{(\Delta x)^2 + (\Delta y)^2}}{|\Delta t|} = \sqrt{\left(\frac{\Delta x}{\Delta t}\right)^2 + \left(\frac{\Delta y}{\Delta t}\right)^2}$$

and the latter approaches

$$\sqrt{\left(\frac{dx}{dt}\right)^2 + \left(\frac{dy}{dt}\right)^2}$$

as $\Delta t \to 0$. Also, when $\Delta t \to 0$, both $\Delta x$ and $\Delta y$ approach 0 (remember that $x(t)$ and $y(t)$ are continuous, being differentiable). It follows that $\Delta\mathbf{p} \to \mathbf{0}$.

Consequently, when we let $\Delta t \to 0$ in (1), we get

$$\frac{dz}{dt} = f_x(\mathbf{p}) \frac{dx}{dt} + f_y(\mathbf{p}) \frac{dy}{dt}$$

a result equivalent to the claimed assertion. ∎

**EXAMPLE 1** Suppose $z = x^3y$ where $x = 2t$ and $y = t^2$. Find $dz/dt$.

**Solution**

$$\frac{dz}{dt} = \frac{\partial z}{\partial x}\frac{dx}{dt} + \frac{\partial z}{\partial y}\frac{dy}{dt}$$
$$= (3x^2y)(2) + (x^3)(2t)$$
$$= 6(2t)^2(t^2) + 2(2t)^3(t)$$
$$= 40t^4$$

∎

We could have done Example 1 without use of the Chain Rule. By direct substitution,

$$z = x^3y = (2t)^3t^2 = 8t^5$$

and so $dz/dt = 40t^4$. However, the direct substitution method is often not available or not convenient—witness the next example.

**EXAMPLE 2** As a solid right circular cylinder is heated, its radius $r$ and height $h$ increase; hence, so does its surface area $S$. Suppose that at the instant when $r = 10$ centimeters and $h = 100$ centimeters, $r$ is increasing at 0.2 centimeters per hour and $h$ is increasing at 0.5 centimeters per hour. How fast is $S$ increasing at this instant?

**Solution** The formula for the total surface area of a cylinder (Figure 1) is

$$S = 2\pi rh + 2\pi r^2$$

Thus,

$$\frac{dS}{dt} = \frac{\partial S}{\partial r}\frac{dr}{dt} + \frac{\partial S}{\partial h}\frac{dh}{dt}$$
$$= (2\pi h + 4\pi r)(0.2) + (2\pi r)(0.5)$$

At $r = 10$ and $h = 100$,

$$\frac{dS}{dt} = (2\pi \cdot 100 + 4\pi \cdot 10)(0.2) + (2\pi \cdot 10)(0.5)$$

$$= 58\pi \text{ square centimeters per hour}$$

∎

The result in Theorem A extends readily to a function of three variables, as we now illustrate.

---

**The Chain Rule:**

**Two Variable Case**

Here is a device that may help you to remember the Chain Rule.

$z = f(x, y)$ — Dependent variable

$\dfrac{\partial z}{\partial x}$　　$\dfrac{\partial z}{\partial y}$

$x$　　　$y$ — Middle variables

$\dfrac{dx}{dt}$　　$\dfrac{dy}{dt}$

$t$ — Independent variable

$$\frac{dz}{dt} = \frac{\partial z}{\partial x}\frac{dx}{dt} + \frac{\partial z}{\partial y}\frac{dy}{dt}$$

---

**FIGURE 1**

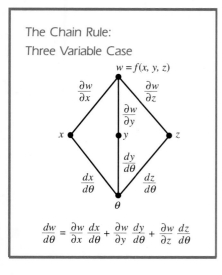

The Chain Rule:
Three Variable Case

$$\frac{dw}{d\theta} = \frac{\partial w}{\partial x}\frac{dx}{d\theta} + \frac{\partial w}{\partial y}\frac{dy}{d\theta} + \frac{\partial w}{\partial z}\frac{dz}{d\theta}$$

**EXAMPLE 3**  Suppose $w = x^2 y + y + xz$, where $x = \cos \theta$, $y = \sin \theta$, and $z = \theta^2$. Find $dw/d\theta$ and evaluate it at $\theta = \pi/3$.

**Solution**

$$\frac{dw}{d\theta} = \frac{\partial w}{\partial x}\frac{dx}{d\theta} + \frac{\partial w}{\partial y}\frac{dy}{d\theta} + \frac{\partial w}{\partial z}\frac{dz}{d\theta}$$

$$= (2xy + z)(-\sin \theta) + (x^2 + 1)(\cos \theta) + (x)(2\theta)$$

$$= -2\cos \theta \sin^2\theta - \theta^2 \sin \theta + \cos^3\theta + \cos \theta + 2\theta \cos \theta$$

At $\theta = \pi/3$,

$$\frac{dw}{d\theta} = -2 \cdot \frac{1}{2} \cdot \frac{3}{4} - \frac{\pi^2}{9} \cdot \frac{\sqrt{3}}{2} + \left(\frac{1}{4} + 1\right)\frac{1}{2} + \frac{2\pi}{3} \cdot \frac{1}{2}$$

$$= -\frac{1}{8} - \frac{\pi^2 \sqrt{3}}{18} + \frac{\pi}{3}$$  ■

**Second Version**  Suppose next that $z = f(x, y)$ where $x = x(s, t)$ and $y = y(s, t)$. Then it makes sense to ask for $\partial z/\partial s$ and $\partial z/\partial t$.

---

**Theorem B**

**(Chain Rule).** Let $x = x(s, t)$ and $y = y(s, t)$ have first partial derivatives at $(s, t)$ and let $z = f(x, y)$ be differentiable at $(x(s, t), y(s, t))$. Then $z = f(x(s, t), y(s, t))$ has first partial derivatives given by:

**(i)** $\dfrac{\partial z}{\partial s} = \dfrac{\partial z}{\partial x}\dfrac{\partial x}{\partial s} + \dfrac{\partial z}{\partial y}\dfrac{\partial y}{\partial s};$

**(ii)** $\dfrac{\partial z}{\partial t} = \dfrac{\partial z}{\partial x}\dfrac{\partial x}{\partial t} + \dfrac{\partial z}{\partial y}\dfrac{\partial y}{\partial t}.$

---

**Proof**  If $s$ is held fixed, then $x(s, t)$ and $y(s, t)$ become functions of $t$ alone, which means that Theorem A applies. When we use this theorem with $\partial$ replacing $d$ to indicate that $s$ is fixed, we obtain the formula in (ii) for $\partial z/\partial t$. The formula for $\partial z/\partial s$ is obtained in a similar way by holding $t$ fixed.  ■

**EXAMPLE 4**  If $z = 3x^2 - y^2$ where $x = 2s + 7t$ and $y = 5st$, find $\partial z/\partial t$ and express it in terms of $s$ and $t$.

**Solution**

$$\frac{\partial z}{\partial t} = \frac{\partial z}{\partial x}\frac{\partial x}{\partial t} + \frac{\partial z}{\partial y}\frac{\partial y}{\partial t}$$

$$= (6x)(7) + (-2y)(5s)$$

$$= 42(2s + 7t) - 10st(5s)$$

$$= 84s + 294t - 50s^2t$$  ■

Here is the corresponding result for three intermediate variables illustrated in an example.

**EXAMPLE 5**   If $w = x^2 + y^2 + z^2 + xy$, where $x = st$, $y = s - t$, and $z = s + 2t$, find $\partial w/\partial t$.

*Solution*

$$\frac{\partial w}{\partial t} = \frac{\partial w}{\partial x}\frac{\partial x}{\partial t} + \frac{\partial w}{\partial y}\frac{\partial y}{\partial t} + \frac{\partial w}{\partial z}\frac{\partial z}{\partial t}$$

$$= (2x + y)(s) + (2y + x)(-1) + (2z)(2)$$

$$= (2st + s - t)(s) + (2s - 2t + st)(-1) + (2s + 4t)2$$

$$= 2s^2t + s^2 - 2st + 2s + 10t \qquad \blacksquare$$

**Implicit Functions**   Suppose that $F(x, y) = 0$ defines $y$ implicitly as a function of $x$—for example, $y = g(x)$—but that the function $g$ is difficult or impossible to determine. We can still find $dy/dx$. One method for doing this, implicit differentiation, was discussed in Section 3.8. Here is another method.

Let's differentiate both sides of $F(x, y) = 0$ with respect to $x$ using the Chain Rule. We obtain

$$\frac{\partial F}{\partial x}\frac{dx}{dx} + \frac{\partial F}{\partial y}\frac{dy}{dx} = 0$$

Solving for $dy/dx$ yields the formula

$$\boxed{\frac{dy}{dx} = -\frac{\partial F/\partial x}{\partial F/\partial y}}$$

**EXAMPLE 6**   Find $dy/dx$ if $x^3 + x^2y - 10y^4 = 0$.

*Solution*   Let $F(x, y) = x^3 + x^2y - 10y^4$. Then

$$\frac{dy}{dx} = -\frac{\partial F/\partial x}{\partial F/\partial y} = -\frac{3x^2 + 2xy}{x^2 - 40y^3}$$

This may be compared with Example 3 of Section 3.8, where the same problem was done by implicit differentiation. $\qquad \blacksquare$

If $z$ is an implicit function of $x$ and $y$ defined by the equation $F(x, y, z) = 0$, then differentiation of both sides with respect to $x$ holding $y$ fixed yields

$$\frac{\partial F}{\partial x}\frac{\partial x}{\partial x} + \frac{\partial F}{\partial y}\frac{\partial y}{\partial x} + \frac{\partial F}{\partial z}\frac{\partial z}{\partial x} = 0$$

If we solve for $\partial z/\partial x$ and note that $\partial y/\partial x = 0$, we get the first of the formulas below. A similar calculation holding $x$ fixed and differentiating with respect to $y$ produces the second formula.

$$\frac{\partial z}{\partial x} = -\frac{\partial F/\partial x}{\partial F/\partial z}, \qquad \frac{\partial z}{\partial y} = -\frac{\partial F/\partial y}{\partial F/\partial z}$$

**EXAMPLE 7**   If $F(x, y, z) = x^3 e^{y+z} - y\sin(x-z) = 0$ defines $z$ implicitly as a function of $x$ and $y$, find $\partial z/\partial x$.

*Solution*

$$\frac{\partial z}{\partial x} = -\frac{\partial F/\partial x}{\partial F/\partial z} = -\frac{3x^2 e^{y+z} - y\cos(x-z)}{x^3 e^{y+z} + y\cos(x-z)}$$    ∎

## CONCEPTS REVIEW

**1.** If $z = f(x, y)$, where $x = g(t)$ and $y = h(t)$, then the Chain Rule says that $dz/dt = $ _____.

**2.** Thus, if $z = xy^2$, where $x = \sin t$ and $y = \cos t$, then $dz/dt = $ _____.

**3.** If $z = f(x, y)$, where $x = g(s, t)$ and $y = h(s, t)$, then the Chain Rule says that $\partial z/\partial t = $ _____.

**4.** Thus, if $z = xy^2$, where $x = st$ and $y = s^2 + t^2$, then $\partial z/\partial t$ at $s = 1$ and $t = 1$ has the value _____.

## PROBLEM SET 15.6

In Problems 1–6, find $dw/dt$ by using the Chain Rule. Express your final answer in terms of $t$.

**1.** $w = x^2 y^3; x = t^3, y = t^2$

**2.** $w = x^2 y - y^2 x; x = \cos t, y = \sin t$

**3.** $w = e^x \sin y + e^y \sin x; x = 3t, y = 2t$

**4.** $w = \ln(x/y); x = \tan t, y = \sec^2 t$

**5.** $w = \sin(xyz^2); x = t^3, y = t^2, z = t$

**6.** $w = xy + yz + xz; x = t^2, y = 1 - t^2, z = 1 - t$

In Problems 7–12, find $\partial w/\partial t$ by using the Chain Rule. Express your final answer in terms of $s$ and $t$.

**7.** $w = x^2 y; x = st, y = s - t$

**8.** $w = x^2 - y \ln x; x = s/t, y = s^2 t$

**9.** $w = e^{x^2 + y^2}; x = s \sin t, y = t \sin s$

**10.** $w = \ln(x + y) - \ln(x - y); x = te^s, y = e^{st}$

**11.** $w = \sqrt{x^2 + y^2 + z^2}; x = \cos st, y = \sin st, z = s^2 t$

**12.** $w = e^{xy+z}; x = s + t, y = s - t, z = t^2$

**13.** If $z = x^2 y; x = 2t + s, y = 1 - st^2$, find

$$\left.\frac{\partial z}{\partial t}\right|_{s = 1, t = -2}$$

**14.** If $z = xy + x + y, x = r + s + t$, and $y = rst$, find

$$\left.\frac{\partial z}{\partial s}\right|_{r = 1, s = -1, t = 2}$$

**15.** If $w = u^2 - u \tan v, u = x$, and $v = \pi x$, find

$$\left.\frac{dw}{dx}\right|_{x = 1/4}$$

**16.** If $w = x^2 y + z^2, x = \rho \cos\theta\sin\phi, y = \rho\sin\theta\sin\phi$, and $z = \rho\cos\phi$, find

$$\left.\frac{\partial w}{\partial\theta}\right|_{\rho = 2, \theta = \pi, \phi = \pi/2}$$

**17.** The part of a tree normally sawed into lumber is the trunk, a solid shaped approximately like a right circular cylinder. If the radius of the trunk of a certain tree is growing $\frac{1}{2}$ inch per year and the height is increasing 8 inches per year, how fast is the volume increasing when the radius is 20 inches and the height is 400 inches? Express your answer in board feet per year (1 board foot = 1 inch by 12 inches by 12 inches).

**18.** The temperature of a metal plate at $(x, y)$ is $e^{-x-3y}$ degrees. A bug is walking northeast at a rate of $\sqrt{8}$ feet per minute (that is, $dx/dt = dy/dt = 2$). From the bug's point of view, how is the temperature changing with time as it crosses the origin?

**19.** A boy's toy boat slips from his grasp at the edge of a straight river. The stream carries it along at 5 feet per second. A crosswind blows it toward the opposite shore at 4 feet per second. If the boy runs along the shore at 3 feet per second following his boat, how fast is the boat moving away from him when $t = 3$ seconds?

**20.** Sand is pouring onto a conical pile in such a way that at a certain instant the height is 100 inches and increasing at 3 inches per minute and the radius is 40 inches and increasing at 2 inches per minute. How fast is the volume increasing at that instant?

In Problems 21–24, use the method of Example 6 to find $dy/dx$.

**21.** $x^3 + 2x^2y - y^3 = 0$

**22.** $ye^{-x} + 5x - 17 = 0$

**23.** $x \sin y + y \cos x = 0$

**24.** $x^2 \cos y - y^2 \sin x = 0$

**25.** If $3x^2z + y^3 - xyz^3 = 0$, find $\partial z/\partial x$ (Example 7).

**26.** If $ye^{-x} + z \sin x = 0$, find $\partial x/\partial z$ (Example 7).

**27.** If $T = f(x, y, z, w)$ and $x, y, z,$ and $w$ are each functions of $s$ and $t$, write a chain rule for $\partial T/\partial s$.

**28.** Let $z = f(x, y)$, where $x = r \cos \theta$ and $y = r \sin \theta$. Show that

$$\left(\frac{\partial x}{\partial x}\right)^2 + \left(\frac{\partial z}{\partial y}\right)^2 = \left(\frac{\partial z}{\partial r}\right)^2 + \frac{1}{r^2}\left(\frac{\partial z}{\partial \theta}\right)^2$$

**29.** The wave equation of physics is the partial differential equation

$$\frac{\partial^2 y}{\partial t^2} = c^2 \frac{\partial^2 y}{\partial x^2}$$

where $c$ is a constant. Show that if $f$ is any twice differentiable function, then

$$y(x, t) = \tfrac{1}{2}[f(x - ct) + f(x + ct)]$$

satisfies this equation.

**30.** Show that if $w = f(r - s, s - t, t - r)$, then $\dfrac{\partial w}{\partial r} + \dfrac{\partial w}{\partial s} + \dfrac{\partial w}{\partial t} = 0$.

**31.** Let $F(t) = \displaystyle\int_{g(t)}^{h(t)} f(u)\, du$, where $f$ is continuous and $g$ and $h$ are differentiable. Show that

$$F'(t) = f(h(t))h'(t) - f(g(t))g'(t)$$

and use this result to find $F'(\sqrt{2})$, where

$$F(t) = \int_{\sin \sqrt{2}\pi t}^{t^2} \sqrt{9 + u^4}\, du$$

**32.** Call a function $f(x, y)$ *homogeneous of degree 1* if $f(tx, ty) = tf(x, y)$ for all $t > 0$. For example, $f(x, y) = x + ye^{y/x}$ satisfies this criterion. Prove Euler's Theorem that such a function satisfies

$$f(x, y) = x\frac{\partial f}{\partial x} + y\frac{\partial f}{\partial y}$$

*Note*: Let $f(x, y)$ denote the value of production from $x$ units of capital and $y$ units of labor, a homogeneous function (for example, doubling capital and labor doubles production). Euler's Theorem then asserts an important law of economics that may be phrased as follows: The value of production $f(x, y)$ equals the cost of capital plus the cost of labor provided they are paid for at their respective marginal rates $\partial f/\partial x$ and $\partial f/\partial y$.

C **33.** Leaving from the same point $P$, plane A flies due east while plane B flies N 50° E. At a certain instant, A is 200 miles from $P$ flying at 450 miles per hour and B is 150 miles from $P$ flying at 400 miles per hour. How fast are they separating at that instant?

**34.** Recall Newton's Law of Gravitation, which asserts that the magnitude $F$ of the force of attraction between objects of mass $M$ and $m$ is $F = GMm/r^2$, where $r$ is the distance between them and $G$ is a universal constant. Let an object of mass $M$ be located at the origin and suppose a second object of changing mass $m$ (say from fuel consumption) is moving away from the origin so that its position vector is $\mathbf{r} = x\mathbf{i} + y\mathbf{j} + z\mathbf{k}$. Obtain a formula for $dF/dt$ in terms of the time derivatives of $m, x, y,$ and $z$.

---

**Answers to Concepts Review:** 1. $\dfrac{\partial z}{\partial x}\dfrac{dx}{dt} + \dfrac{\partial z}{\partial y}\dfrac{dy}{dt}$
2. $y^2 \cos t + 2xy(-\sin t) = \cos^3 t - 2\sin^2 t \cos t$
3. $\dfrac{\partial z}{\partial x}\dfrac{\partial x}{\partial t} + \dfrac{\partial z}{\partial y}\dfrac{\partial y}{\partial t}$  4. 12

**15.7**
**TANGENT PLANES,**
**APPROXIMATIONS**

We introduced the notion of a tangent plane to a surface in Section 15.4 but dealt only with surfaces determined by equations of the form $z = f(x, y)$ (Figure 1). Now we want to consider the more general situation of a surface determined by $F(x, y, z) = k$. (Note that $z = f(x, y)$ can be written as $F(x, y, z) = f(x, y) - z = 0$.) Consider a curve on this surface passing through the point $(x_0, y_0, z_0)$. If $x = x(t)$, $y = y(t)$, and $z = z(t)$ are parametric equations for this curve, then for all $t$,

$$F(x(t), y(t), z(t)) = k$$

By the Chain Rule,

$$\frac{dF}{dt} = \frac{\partial F}{\partial x}\frac{dx}{dt} + \frac{\partial F}{\partial y}\frac{dy}{dt} + \frac{\partial F}{\partial z}\frac{dz}{dt} = \frac{dk}{dt} = 0$$

We can express this in terms of the gradient of $F$ and the derivative of the vector expression for the curve $\mathbf{r}(t) = x(t)\mathbf{i} + y(t)\mathbf{j} + z(t)\mathbf{k}$ as

$$\nabla F \cdot \frac{d\mathbf{r}}{dt} = 0$$

As we learned earlier (Section 14.4), $d\mathbf{r}/dt$ is tangent to the curve. In summary, the gradient at $(x_0, y_0, z_0)$ is perpendicular to the tangent line at this point.

The argument just given is valid for any curve through $(x_0, y_0, z_0)$ that lies in the surface $F(x, y, z) = k$ (Figure 2). This suggests the following general definition.

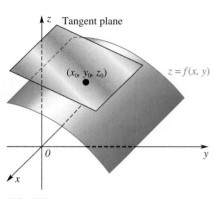

Tangent plane

$z = f(x, y)$

$(x_0, y_0, z_0)$

FIGURE 1

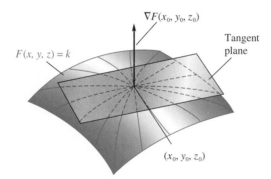

$\nabla F(x_0, y_0, z_0)$

Tangent plane

$F(x, y, z) = k$

$(x_0, y_0, z_0)$

FIGURE 2

---

**Definition**

Let $F(x, y, z) = k$ determine a surface and suppose that $F$ is differentiable at a point $P(x_0, y_0, z_0)$ of this surface with $\nabla F(x_0, y_0, z_0) \neq \mathbf{0}$. Then the plane through $P$ perpendicular to $\nabla F(x_0, y_0, z_0)$ is called the **tangent plane** to the surface at $P$.

---

As a consequence of this definition and Section 14.2, we can write the equation of the tangent plane.

### Theorem A

**(Tangent planes).** For the surface $F(x, y, z) = k$, the equation of the tangent plane at $(x_0, y_0, z_0)$ is $\nabla F(x_0, y_0, z_0) \cdot \langle x - x_0, y - y_0, z - z_0 \rangle = 0$, that is,

$$F_x(x_0, y_0, z_0)(x - x_0) + F_y(x_0, y_0, z_0)(y - y_0)$$
$$+ F_z(x_0, y_0, z_0)(z - z_0) = 0$$

In particular, for the surface $z = f(x, y)$, the equation of the tangent plane at $(x_0, y_0, f(x_0, y_0))$ is

$$z - z_0 = f_x(x_0, y_0)(x - x_0) + f_y(x_0, y_0)(y - y_0)$$

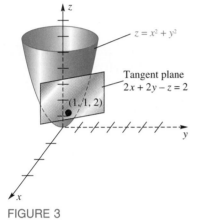

FIGURE 3

**Proof** The first statement is immediate and the second follows from it by considering $F(x, y, z) = f(x, y) - z$. ∎

**EXAMPLE 1** Find the equation of the tangent plane (Figure 3) to $z = x^2 + y^2$ at the point $(1, 1, 2)$.

**Solution** Let $f(x, y) = x^2 + y^2$ and note that $\nabla f(x, y) = 2x\mathbf{i} + 2y\mathbf{j}$. Thus, $\nabla f(1, 1) = 2\mathbf{i} + 2\mathbf{j}$ and, from Theorem A, the required equation is

$$z - 2 = 2(x - 1) + 2(y - 1)$$

or

$$2x + 2y - z = 2$$ ∎

**EXAMPLE 2** Find the equation of the tangent plane and the normal line to the surface $x^2 + y^2 + 2z^2 = 23$ at $(1, 2, 3)$.

**Solution** Let $F(x, y, z) = x^2 + y^2 + 2z^2$, so that $\nabla F(x, y, z) = 2x\mathbf{i} + 2y\mathbf{j} + 4z\mathbf{k}$ and $\nabla F(1, 2, 3) = 2\mathbf{i} + 4\mathbf{j} + 12\mathbf{k}$. According to Theorem A, the equation of the tangent plane at $(1, 2, 3)$ is

$$2(x - 1) + 4(y - 2) + 12(z - 3) = 0$$

Similarly the symmetric equations of the normal line through $(1, 2, 3)$ are

$$\frac{x - 1}{2} = \frac{y - 2}{4} = \frac{z - 3}{12}$$ ∎

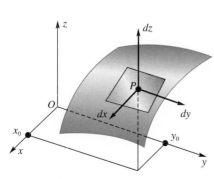

FIGURE 4

**Differentials and Approximations** We suggest that you review Section 3.10, where the topic of differentials and approximations is treated for functions of one variable.

Let $z = f(x, y)$ and let $P(x_0, y_0, z_0)$ be a fixed point on the corresponding surface. Introduce new coordinate axes (the $dx$-, $dy$-, and $dz$-axes), parallel to the old axes, with $P$ as origin (Figure 4). In the old system, the tangent plane at $P$ has equation

$$z - z_0 = f_x(x_0, y_0)(x - x_0) + f_y(x_0, y_0)(y - y_0)$$

but in the new system this takes the simple form

$$dz = f_x(x_0, y_0)\, dx + f_y(x_0, y_0)\, dy$$

This suggests a definition.

---

### Definition

Let $z = f(x, y)$, where $f$ is a differentiable function, and let $dx$ and $dy$ (called the differentials of $x$ and $y$) be variables. The **differential of the dependent variable**, $dz$, also called the total **differential of $f$** and written $df(x, y)$, is defined by

$$dz = df(x, y) = f_x(x, y)\, dx + f_y(x, y)\, dy$$

---

The significance of $dz$ arises from the fact that if $dx = \Delta x$ and $dy = \Delta y$ represent small changes in $x$ and $y$, respectively, then $dz$ will be a good approximation to $\Delta z$, the corresponding change in $z$. This is illustrated in Figure 5 and, while $dz$ does not appear to be a very good approximation to $\Delta z$, you can see that it will get better and better as $\Delta x$ and $\Delta y$ get smaller.

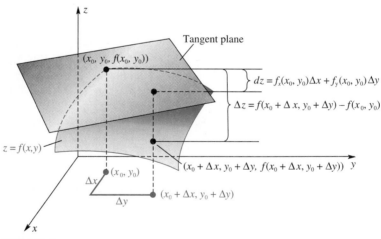

FIGURE 5

**EXAMPLE 3**  Let $z = f(x, y) = 2x^3 + xy - y^3$. Compute $\Delta z$ and $dz$ as $(x, y)$ changes from $(2, 1)$ to $(2.03, 0.98)$.

**Solution**

$$
\begin{aligned}
\Delta z &= f(2.03, 0.98) - f(2, 1) \\
&= 2(2.03)^3 + (2.03)(0.98) - (0.98)^3 - [2(2)^3 + 2(1) - 1^3] \\
&= 0.779062
\end{aligned}
$$

$$dz = f_x(x, y)\,\Delta x + f_y(x, y)\,\Delta y$$
$$= (6x^2 + y)\,\Delta x + (x - 3y^2)\,\Delta y$$

At $(2, 1)$ with $\Delta x = 0.03$ and $\Delta y = -0.02$,

$$dz = (25)(0.03) + (-1)(-0.02) = 0.77 \qquad \blacksquare$$

**EXAMPLE 4** The formula $P = k(T/V)$, where $k$ is a constant, gives the pressure $P$ of a confined gas of volume $V$ and temperature $T$. Find, approximately, the maximum percentage error in $P$ introduced by an error of $\pm 0.4\%$ in measuring the temperature and an error of $\pm 0.9\%$ in measuring the volume.

*Solution* The error in $P$ is $\Delta P$, which we will approximate by $dP$. Thus,

$$|\Delta P| \approx |dP| = \left| \frac{\partial P}{\partial T}\,\Delta T + \frac{\partial P}{\partial V}\,\Delta V \right|$$
$$\leq \left| \frac{k}{V}\,(\pm 0.004T) \right| + \left| -\frac{kT}{V^2}\,(\pm 0.009V) \right|$$
$$= \frac{kT}{V}\,(0.004 + 0.009) = 0.013\,\frac{kT}{V} = 0.013P$$

The maximum relative error, $|\Delta P|/P$, is approximately 0.013, and the maximum percentage error is approximately 1.3%. $\qquad \blacksquare$

The subject of approximation goes far beyond what we have done here. Recall that, for functions of one variable, the first-order approximation

$$f(x) \approx f(x_0) + f'(x_0)(x - x_0)$$

was extended to the second-order Taylor approximation

$$f(x) \approx f(x_0) + f'(x_0)(x - x_0) + \tfrac{1}{2}f''(x_0)(x - x_0)^2$$

and then to *n*th-order Taylor approximations (see Section 10.1). The analogues for functions of two variables are

$$f(x, y) \approx f(x_0, y_0) + f_x(x_0, y_0)(x - x_0) + f_y(x_0, y_0)(y - y_0)$$

and

$$f(x, y) \approx f(x_0, y_0) + f_x(x_0, y_0)(x - x_0) + f_y(x_0, y_0)(y - y_0)$$
$$+ \tfrac{1}{2}[f_{xx}(x_0, y_0)(x - x_0)^2 + 2f_{xy}(x_0, y_0)(x - x_0)(y - y_0)$$
$$+ f_{yy}(x_0, y)(y - y_0)^2]$$

Moreover, these extend both to the *n*th-order case and to functions of more than two variables. The details are best left to higher-level books.

# CONCEPTS REVIEW

**1.** Let $F(x, y, z) = k$ determine a surface. The direction of the gradient vector $\nabla F$ is _____ to the surface.

**2.** Let $z = x^2 + xy$ determine a surface. A vector at $(1, 1, 2)$ perpendicular to this surface is _____.

**3.** Let $xy^2z^3 = 2$ determine a surface. An equation for the tangent plane at $(2, 1, 1)$ is _____.

**4.** We define the total differential of $f(x, y)$ by $df(x, y) =$ _____.

# PROBLEM SET 15.7

In Problems 1–8, find the equation of the tangent plane to the given surface at the indicated point.

**1.** $x^2 + y^2 + z^2 = 16$; $(2, 3, \sqrt{3})$

**2.** $8x^2 + y^2 + 8z^2 = 16$; $(1, 2, \sqrt{2}/2)$

**3.** $x^2 - y^2 + z^2 + 1 = 0$; $(1, 3, \sqrt{7})$

**4.** $x^2 + y^2 - z^2 = 4$; $(2, 1, 1)$

**5.** $z = \dfrac{x^2}{4} + \dfrac{y^2}{4}$; $(2, 2, 2)$

**6.** $z = xe^{-2y}$; $(1, 0, 1)$

**7.** $z = 2e^{3y} \cos 2x$; $(\pi/3, 0, -1)$

**8.** $z = x^{1/2} + y^{1/2}$; $(1, 4, 3)$

In Problems 9–12, use the total differential $dz$ to approximate the change in $z$ as $(x, y)$ moves from $P$ to $Q$. Then use a calculator to find the corresponding exact change $\Delta z$ (to the accuracy of your calculator). See Example 3.

C **9.** $z = 2x^2y^3$; $P(1, 1)$, $Q(0.99, 1.02)$

C **10.** $z = x^2 - 5xy + y$; $P(2, 3)$, $Q(2.03, 2.98)$

C **11.** $z = \ln(x^2y)$; $P(-2, 4)$, $Q(-1.98, 3.96)$

C **12.** $z = \tan^{-1} xy$; $P(-2, -0.5)$, $Q(-2.03, -0.51)$

**13.** Find all points on the surface $z = x^2 - 2xy - y^2 - 8x + 4y$, where the tangent plane is horizontal.

**14.** Find a point on the surface $z = 2x^2 + 3y^2$ where the tangent plane is parallel to the plane $8x - 3y - z = 0$.

**15.** Show that the surface $x^2 + 4y + z^2 = 0$ and $x^2 + y^2 + z^2 - 6z + 7 = 0$ are tangent to each other at $(0, -1, 2)$; that is, show that they have the same tangent plane at $(0, -1, 2)$.

**16.** Show that the surfaces $z = x^2y$ and $y = \frac{1}{4}x^2 + \frac{3}{4}$ intersect at $(1, 1, 1)$ and have perpendicular tangent planes there.

**17.** Find a point on the surface $x^2 + 2y^2 + 3z^2 = 12$ where the tangent plane is perpendicular to the line with parametric equations: $x = 1 + 2t$, $y = 3 + 8t$, $z = 2 - 6t$.

**18.** Show that the equation of the tangent plane to the ellipsoid

$$\frac{x^2}{a^2} + \frac{y^2}{b^2} + \frac{z^2}{c^2} = 1$$

at $(x_0, y_0, z_0)$ can be written in the form

$$\frac{x_0 x}{a^2} + \frac{y_0 y}{b^2} + \frac{z_0 z}{c^2} = 1$$

**19.** Find the parametric equations of the line that is tangent to the curve of intersection of the surfaces

$$f(x, y, z) = 9x^2 + 4y^2 + 4z^2 - 41 = 0$$

and

$$g(x, y, z) = 2x^2 - y^2 + 3z^2 - 10 = 0$$

at the point $(1, 2, 2)$. *Hint*: This line is perpendicular to $\nabla f(1, 2, 2)$ and $\nabla g(1, 2, 2)$.

**20.** Find the parametric equations of the line that is tangent to the curve of intersection of the surfaces $x = z^2$ and $y = z^3$ at $(1, 1, 1)$ (see Problem 19).

**21.** In determining the specific gravity of an object, its weight in air is found to be $A = 36$ pounds and its weight in water is $W = 20$ pounds, with a possible error in each measurement of 0.02 pound. Find, approximately, the maximum possible error in calculating its specific gravity $S$, where $S = A/(A - W)$.

**22.** Use differentials to find the approximate amount of copper in the four sides and bottom of a rectangular copper tank that is 6 feet long, 4 feet wide, and 3 feet deep *inside*, if the sheet copper is $\frac{1}{4}$ inch thick. *Hint*: Make a sketch.

**23.** The radius and height of a right circular cone are measured with errors of at most 2% and 3%, respectively. Use differentials to estimate the maximum percentage error in the calculated volume (see Example 4).

**24.** The period $T$ of a pendulum of length $L$ is given by $T = 2\pi\sqrt{L/g}$, where $g$ is the acceleration of gravity. Show that $dT/T = \frac{1}{2}[dL/L - dg/g]$ and use this result to estimate the maximum percentage error in $T$ due to an error of 0.5% in measuring $L$ and 0.3% in measuring $g$.

**25.** The formula $1/R = 1/R_1 + 1/R_2$ determines the combined resistance $R$ when resistors of resistance $R_1$ and $R_2$ are connected in parallel. Suppose that $R_1$ and $R_2$ were measured at 25 and 100 ohms, respectively, with possible errors in each measurement of 0.5 ohms. Calculate $R$ and give an estimate for the maximum error in this value.

**26.** A bee sat at the point $(1, 2, 1)$ on the ellipsoid $x^2 + y^2 + 2z^2 = 6$ (distances in feet). At $t = 0$, it took off along the normal line at a speed of 4 feet per second. Where and when did it hit the plane $2x + 3y + z = 49$?

**27.** Show that the tangent plane to the surface $xyz = k$ at any point forms with the coordinate planes a tetrahedron of fixed volume and find this volume.

**28.** Find and simplify the equation of the tangent plane at $(x_0, y_0, z_0)$ to the surface $\sqrt{x} + \sqrt{y} + \sqrt{z} = a$. Then show that the sum of the intercepts of this plane with the coordinate axes is $a^2$.

**29.** For the function $f(x, y) = \sqrt{x^2 + y^2}$, find the second-order Taylor approximation based at $(x_0, y_0) = (3, 4)$. Then estimate $f(3.1, 3.9)$ using (a) the first-order approximation, (b) the second-order approximation, and (c) your calculator directly.

**Answers to Concepts Review:** **1.** Perpendicular **2.** $\langle 3, 1, -1 \rangle$ **3.** $x - 2 + 4(y - 1) + 6(z - 1) = 0$ **4.** $\frac{\partial f}{\partial x} dx + \frac{\partial f}{\partial y} dy$

## 15.8
## MAXIMA AND MINIMA

Our goal is to extend the notions of Chapter 4 to functions of several variables; a quick review of that chapter, especially Sections 4.1 and 4.3, will be helpful. The definitions given there extend almost without change, but for clarity we repeat them. In what follows, let $\mathbf{p} = (x, y)$ and $\mathbf{p}_0 = (x_0, y_0)$ be a variable point and a fixed point, respectively, in two-space (they could just as well be points in *n*-space).

---

**Definition**

Let $\mathbf{p}_0$ be a point in $S$, the domain of $f$.
**(i)** $f(\mathbf{p}_0)$ is a (global) **maximum value** of $f$ on $S$ if $f(\mathbf{p}_0) \geq f(\mathbf{p})$ for all $\mathbf{p}$ in $S$.
**(ii)** $f(\mathbf{p}_0)$ is a (global) **minimum value** of $f$ on $S$ if $f(\mathbf{p}_0) \leq f(\mathbf{p})$ for all $\mathbf{p}$ in $S$.
**(iii)** $f(\mathbf{p}_0)$ is a (global) **extreme value** of $f$ on $S$ if it is either a (global) maximum value or a (global) minimum value.
The same definitions hold with the word *global* replaced by *local* if, in (i) and (ii), we require only that the inequalities hold on $N \cap S$, where $N$ is a neighborhood of $\mathbf{p}_0$.

---

Figure 1 gives a geometric interpretation of the concepts we have defined. Note that a global maximum (or minimum) is automatically a local maximum (or minimum).

Our first theorem is a big one—difficult to prove, but intuitively clear.

---

**Theorem A**

**(Max-Min Existence Theorem).** If $f$ is continuous on a closed bounded set $S$, then $f$ attains both a (global) maximum value and a (global) minimum value there.

---

The proof may be found in most books on advanced calculus.

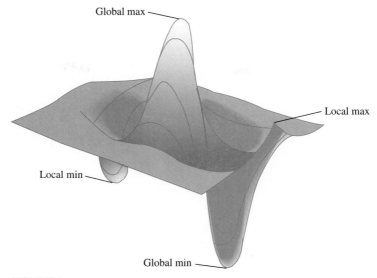

FIGURE 1

**Where Do Extreme Values Occur?**   The situation is analogous to the one-variable case. The **critical points** of $f$ on $S$ are of three types.

1. **Boundary points**. See Section 15.3.
2. **Stationary points**. We call $\mathbf{p}_0$ a stationary point if $\mathbf{p}_0$ is an interior point of $S$ where $f$ is differentiable and $\nabla f(\mathbf{p}_0) = \mathbf{0}$. At such a point, the tangent plane is horizontal.
3. **Singular points**. We call $\mathbf{p}_0$ a singular point if $\mathbf{p}_0$ is an interior point of $S$ where $f$ is not differentiable—for example, a point where the graph of $f$ has a sharp corner.

Now we can state another big theorem; we can actually prove this one.

---

Theorem B

**(Critical Point Theorem).** Let $f$ be defined on a set $S$ containing $\mathbf{p}_0$. If $f(\mathbf{p}_0)$ is an extreme value, then $\mathbf{p}_0$ must be a critical point; that is, either $\mathbf{p}_0$ is:

**(i)**   a boundary point of $S$; or

**(ii)**   a stationary point of $f$; or

**(iii)**   a singular point of $f$.

---

***Proof***   Suppose that $\mathbf{p}_0$ is neither a boundary point nor a singular point (so that $\mathbf{p}_0$ is an interior point where $\nabla f$ exists). We will be done if we can show that $\nabla f(\mathbf{p}_0) = \mathbf{0}$. For simplicity, set $\mathbf{p}_0 = (x_0, y_0)$; the higher-dimensional cases will follow in a similar fashion.

Since $f$ has an extreme value at $(x_0, y_0)$, the function $g(x) = f(x, y_0)$ has an extreme value at $x_0$. Moreover $g$ is differentiable at $x_0$ since $f$ is differentiable at $(x_0, y_0)$ and therefore, by the Critical Point Theorem for functions of one variable (Theorem 4.1B),

$$g'(x_0) = f_x(x_0, y_0) = 0$$

Similarly, the function $h(y) = f(x_0, y)$ has an extreme value at $y_0$ and satisfies

$$h'(y_0) = f_y(x_0, y_0) = 0$$

The gradient is **0** since both partials are 0. ∎

The theorem and its proof are valid whether the extreme values are global or local extreme values.

**EXAMPLE 1**   Find the local maximum or minimum values of $f(x, y) = x^2 - 2x + y^2/4$.

***Solution***   The given function is differentiable throughout its domain, the *xy*-plane. Thus, the only possible critical points are the stationary points obtained by setting $f_x(x, y)$ and $f_y(x, y)$ equal to zero. But $f_x(x, y) = 2x - 2$ and $f_y(x, y) = y/2$ are zero only when $x = 1$ and $y = 0$. It remains to decide whether $(1, 0)$ gives a maximum or a minimum or neither. We will develop a simple tool for this soon, but for now we must use a little ingenuity. Note that $f(1, 0) = -1$ and

$$f(x, y) = x^2 - 2x + \frac{y^2}{4} = x^2 - 2x + 1 + \frac{y^2}{4} - 1$$

$$= (x - 1)^2 + \frac{y^2}{4} - 1 \geq -1$$

Thus, $f(1, 0)$ is actually a global minimum for $f$. There are no local maximum values. ∎

**EXAMPLE 2**   Find the local minimum or maximum values of $f(x, y) = -x^2/a^2 + y^2/b^2$.

***Solution***   The only critical points are obtained by setting $f_x(x, y) = -2x/a^2$ and $f_y(x, y) = 2y/b^2$ equal to zero. This yields the point $(0, 0)$, which gives neither a maximum nor minimum (see Figure 2). It is called a **saddle point**. The given function has no local extrema. ∎

Example 2 illustrates the troublesome fact that $\nabla f(x_0, y_0) = \mathbf{0}$ does not guarantee that there is a local extremum at $(x_0, y_0)$. Fortunately, there is a nice criterion for deciding what is happening at a stationary point—our next topic.

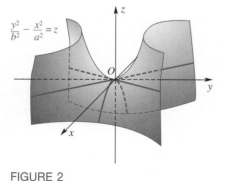

$\dfrac{y^2}{b^2} - \dfrac{x^2}{a^2} = z$

FIGURE 2

**Sufficient Conditions for Extrema**   You should think of the next theorem as being analogous to the Second Derivative Test for functions of one variable (Theorem 4.3B). The proof may be found in books on advanced calculus.

> ## Theorem C
>
> **(Second-Partials Test).** Suppose that $f(x, y)$ has continuous second partial derivatives in a neighborhood of $(x_0, y_0)$ and that $\nabla f(x_0, y_0) = \mathbf{0}$. Let
>
> $$D = D(x_0, y_0) = f_{xx}(x_0, y_0)f_{yy}(x_0, y_0) - f_{xy}^2(x_0, y_0)$$
>
> Then:
> **(i)**   if $D > 0$ and $f_{xx}(x_0, y_0) < 0$, $f(x_0, y_0)$ is a local maximum value;
> **(ii)**  if $D > 0$ and $f_{xx}(x_0, y_0) > 0$, $f(x_0, y_0)$ is a local minimum value;
> **(iii)** if $D < 0$, $f(x_0, y_0)$ is not an extreme value ($(x_0, y_0)$ is a saddle point);
> **(iv)**  if $D = 0$, the test is inconclusive.

**EXAMPLE 3**   Find the extrema, if any, of the function $F$ defined by $F(x, y) = 3x^3 + y^2 - 9x + 4y$.

***Solution***   Since $F_x(x, y) = 9x^2 - 9$ and $F_y(x, y) = 2y + 4$, the critical points, obtained by solving the simultaneous equations $F_x(x, y) = F_y(x, y) = 0$, are $(1, -2)$ and $(-1, -2)$.

Now $F_{xx}(x, y) = 18x$, $F_{yy}(x, y) = 2$, and $F_{xy} = 0$. Thus at the critical point $(1, -2)$,

$$D = F_{xx}(1, -2) \cdot F_{yy}(1, -2) - F_{xy}^2(1, -2) = 18(2) - 0 = 36 > 0$$

Furthermore, $F_{xx}(1, -2) = 18 > 0$, and so by (ii), $F(1, -2) = -10$ is a local minimum value of $F$.

In testing the given function at the other critical point, $(-1, -2)$, we find $F_{xx}(-1, -2) = -18$, $F_{yy}(-1, -2) = 2$, and $F_{xy}(-1, -2) = 0$, which makes $D = -36 < 0$. Thus, by (iii), $(-1, -2)$ is a saddle point and $F(-1, -2)$ is not an extremum.   ∎

**EXAMPLE 4**   Find the minimum distance between the origin and the surface $z^2 = x^2y + 4$.

***Solution***   Let $P(x, y, z)$ be any point on the surface. The square of the distance between the origin and $P$ is $d^2 = x^2 + y^2 + z^2$. We seek the coordinates of $P$ that make $d^2$ (and hence $d$) a minimum.

Since $P$ is on the surface, its coordinates satisfy the equation of the surface. Substituting $z^2 = x^2y + 4$ in $d^2 = x^2 + y^2 + z^2$, we obtain $d^2$ as a function of two variables $x$ and $y$:

$$d^2 = f(x, y) = x^2 + y^2 + x^2y + 4$$

To find the critical points, we set $f_x(x, y) = 0$ and $f_y(x, y) = 0$, obtaining

$$2x + 2xy = 0 \qquad \text{and} \qquad 2y + x^2 = 0$$

By eliminating $y$ between these equations, we get

$$2x - x^3 = 0$$

Thus, $x = 0$ or $x = \pm\sqrt{2}$. Substituting these values in the second of the equations, we obtain $y = 0$ and $y = -1$. Therefore, the critical points are $(0, 0)$, $(\sqrt{2}, -1)$, and $(-\sqrt{2}, -1)$.

To test each of these, we need $f_{xx}(x, y) = 2 + 2y$, $f_{yy}(x, y) = 2$, $f_{xy}(x, y) = 2x$, and

$$D(x, y) = f_{xx}f_{yy} - f_{xy}^2 = 4 + 4y - 4x^2$$

Since $D(\pm\sqrt{2}, -1) = -8 < 0$, neither $(\sqrt{2}, -1)$ nor $(-\sqrt{2}, -1)$ yields an extremum. However, $D(0, 0) = 4 > 0$ and $f_{xx}(0, 0) = 2 > 0$; so $(0, 0)$ yields the minimum distance. Substituting $x = 0$ and $y = 0$ in the expression for $d^2$, we find $d^2 = 4$.

The minimum distance between the origin and the given surface is 2. ∎

None of your examples so far has involved boundary points. Here is one that does.

**EXAMPLE 5** Find the maximum and minimum values of $f(x, y) = 2x^2 + y^2 - 4x - 2y + 5$ on the closed set $S = \{(x, y): x^2 + \frac{1}{2}y^2 \leq 1\}$; see Figure 3.

*Solution* Since $f_x(x, y) = 4x - 4$ and $f_y(x, y) = 2y - 2$, the only possible interior critical point is $(1, 1)$. However, this point lies outside $S$, so it can be ignored.

The boundary of $S$ is the ellipse $x^2 + \frac{1}{2}y^2 = 1$, which can be described parametrically by

$$x = \cos t, \qquad y = \sqrt{2} \sin t, \qquad 0 \leq t \leq 2\pi$$

We wish to maximize and minimize the function of one variable

$$g(t) = f(\cos t, \sqrt{2} \sin t), \qquad 0 \leq t \leq 2\pi$$

By the Chain Rule,

$$
\begin{aligned}
g'(t) &= \frac{\partial f}{\partial x}\frac{dx}{dt} + \frac{\partial f}{\partial y}\frac{dy}{dt} \\
&= (4x - 4)(-\sin t) + (2y - 2)(\sqrt{2}\cos t) \\
&= (4\cos t - 4)(-\sin t) + (2\sqrt{2}\sin t - 2)(\sqrt{2}\cos t) \\
&= 4\sin t - 2\sqrt{2}\cos t
\end{aligned}
$$

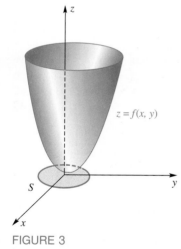

$z = f(x, y)$

$S$

FIGURE 3

Setting $g'(t) = 0$ yields $\tan t = \sqrt{2}/2$ with two solutions $t_1 = \tan^{-1}(\sqrt{2}/2)$ and $t_2 = \pi + t_1$. Thus, $g$ has the four critical points $0$, $t_1$, $t_2$, and $2\pi$ on the interval $[0, 2\pi]$. These, in turn, determine the three points $(1, 0)$, $(2/\sqrt{6}, 2/\sqrt{6})$, and $(-2/\sqrt{6}, -2/\sqrt{6})$ on the boundary of $S$. The corresponding values of $f$ are

$$f(1, 0) = 3$$

$$f\left(\frac{2}{\sqrt{6}}, \frac{2}{\sqrt{6}}\right) \approx 2.101$$

$$f\left(\frac{-2}{\sqrt{6}}, \frac{-2}{\sqrt{6}}\right) \approx 11.899$$

We conclude that the minimum value of $f$ on $S$ is 2.101 and the maximum value is 11.899.  ∎

## CONCEPTS REVIEW

**1.** If $f(x, y)$ is continuous on a _____ set $S$, then $f$ attains both a maximum value and a minimum value on $S$.

**2.** If $f(x, y)$ attains a maximum value at a point $(x_0, y_0)$, then $(x_0, y_0)$ is either a _____ point or a _____ point or a _____ point.

**3.** If $(x_0, y_0)$ is a stationary point for $f$, then $f$ is differentiable there and _____.

**4.** In the Second Derivative Test for a function $f$ of two variables, the number $D = $ _____ plays a crucial role.

## PROBLEM SET 15.8

In Problems 1–10, find all critical points. Indicate whether each such point gives a local maximum, a local minimum, or whether it is a saddle point. *Hint*: Use Theorem C.

**1.** $f(x, y) = x^2 + 4y^2 - 4x$

**2.** $f(x, y) = x^2 + 4y^2 - 2x + 8y - 1$

**3.** $f(x, y) = 2x^4 - x^2 + 3y^2$

**4.** $f(x, y) = xy^2 - 6x^2 - 3y^2$

**5.** $f(x, y) = xy$

**6.** $f(x, y) = x^3 + y^3 - 6xy$

**7.** $f(x, y) = xy + \dfrac{2}{x} + \dfrac{4}{y}$

**8.** $f(x, y) = e^{-(x^2 + y^2 - 4y)}$

**9.** $f(x, y) = \cos x + \cos y + \cos(x + y)$; $0 < x < \pi/2, 0 < y < \pi/2$

**10.** $f(x, y) = x^2 + a^2 - 2ax \cos y$; $-\pi < y < \pi$

In Problems 11–14, find the global maximum value and global minimum value of $f$ on $S$ and indicate where they occur.

**11.** $f(x, y) = 3x + 4y$;
$S = \{(x, y): 0 \le x \le 1, -1 \le y \le 1\}$

**12.** $f(x, y) = x^2 + y^2$;
$S = \{(x, y): -1 \le x \le 3, -1 \le y \le 4\}$

**13.** $f(x, y) = x^2 - y^2 + 1$; $S = \{(x, y): x^2 + y^2 \le 1\}$ (see Example 5).

**14.** $f(x, y) = x^2 - 6x + y^2 - 8y + 7$;
$S = \{(x, y): x^2 + y^2 \le 1\}$

**15.** Express a positive number $N$ as a sum of three positive numbers such that the product of these three numbers is a maximum.

**16.** Use the methods of this section to find the shortest distance from the origin to the plane $x + 2y + 3z = 12$.

**17.** Find the shape of the closed rectangular box of volume $V_0$ with minimum surface area.

**18.** Find the shape of the rectangular box of volume $V_0$ for which the sum of the edge lengths is least.

**19.** A rectangular metal tank with open top is to hold 256 cubic feet of liquid. What are the dimensions of the tank that require the least material to build?

**20.** A rectangular box, whose edges are parallel to the coordinate axes, is inscribed in the ellipsoid $96x^2 + 4y^2 + 4z^2 = 36$. What is the greatest possible volume for such a box?

**21.** Find the three-dimensional vector with length 9, the sum of whose components is a maximum.

**22.** Find the minimum distance between the point $(1, 2, 0)$ and the quadric cone $z^2 = x^2 + y^2$.

**23.** An open gutter with cross section in the form of a trapezoid with equal base angles is to be made by bending up equal strips along both sides of a long piece of metal 12 inches wide. Find the base angles and the width of the sides for maximum carrying capacity.

**24.** Find the minimum distance between the lines having parametric equations $x = t - 1, y = 2t, z = t + 3$ and $x = 3s, y = s + 2, z = 2s - 1$.

**25.** Convince yourself that the maximum and minimum values of a linear function $f(x, y) = ax + by + c$ on a closed polygonal set will always occur at a vertex. Then use this fact to find each value.

(a) Maximum value of $2x + 3y + 4$ on the closed polygon with vertices $(-1, 2), (0, 1), (1, 0), (-3, 0)$, and $(0, -4)$.
(b) Minimum value of $-3x + 2y + 1$ on the closed polygon with vertices $(-3, 0), (0, 5), (2, 3), (4, 0)$, and $(1, -4)$.

**26.** Use the result of Problem 25 to maximize $2x + y$ subject to the constraints $4x + y \leq 8, 2x + 3y \leq 14, x \geq 0$, and $y \geq 0$. *Hint*: Begin by graphing the set determined by the constraints.

**27.** Find the maximum and minimum values of $z = y^2 - x^2$ (Figure 2) on the closed triangle with vertices $(0, 0)$, $(1, 2)$, and $(2, -2)$.

**28.** (Least Squares) Given the $n$ points $P_1(x_1, y_1)$, $P_2(x_2, y_2), \ldots, P_n(x_n, y_n)$ in the plane, we wish to find the line $y = mx + b$ such that the sum of the squares of the vertical distances from the points to the line is a minimum; that is, we wish to minimize

$$f(m, b) = \sum_{i=1}^{n} (y_i - mx_i - b)^2$$

Show that this minimum will occur where

$$m \sum x_i^2 + b \sum x_i = \sum x_i y_i$$
$$m \sum x_i + nb = \sum y_i$$

**29.** Find the least-squares line (Problem 28) for the data $(3, 2), (4, 3), (5, 4), (6, 4)$, and $(7, 5)$.

**30.** Find the maximum and minimum values of $z = 2x^2 + y^2 - 4x - 2y + 5$ (Figure 3) on the set bounded by the closed triangle with vertices $(0, 0), (4, 0)$, and $(0, 1)$.

**31.** Suppose the temperature $T$ on the circular plate $\{(x, y): x^2 + y^2 \leq 1\}$ is given by $T = 2x^2 + y^2 - y$. Find the hottest and coldest spots on the plate.

**32.** A wire of length $k$ is to be cut into (at most) 3 pieces to form a circle and two squares, any of which may be degenerate. How should this be done to maximize and minimize the area thus enclosed? *Hint*: Reduce the problem to that of optimizing $x^2 + y^2 + z^2$ on the part of the plane $2\sqrt{\pi}x + 4y + 4z = k$ in the first octant. Then reason geometrically.

**33.** Find the shape of the triangle of largest area that can be inscribed in a circle of radius $r$. *Hint*: Let $\alpha, \beta$, and $\gamma$ be the central angles that subtend the three sides of the triangle. Show that the area of the triangle is $\frac{1}{2}r^2[\sin \alpha + \sin \beta - \sin(\alpha + \beta)]$. Maximize.

**34.** Let $(a, b, c)$ be a fixed point in the first octant. Find the plane through this point that cuts off from the first octant the tetrahedron of minimum volume, and determine the resulting volume.

PC Sometimes the finding of extrema for a function of two variables can best be handled by commonsense methods using a computer. To illustrate, look at the pictures of the surfaces and the corresponding contour maps for the four functions shown near the end of Section 15.1. Note that these pictures allow us to locate the extrema geometrically. With the additional ability (provided by *True BASIC Mac-Function* in the *Level Curve* section) to quickly evaluate the function at points, we can experimentally find maxima and minima to rather good accuracy. In each of the following, use this method to find the point where the indicated maximum or minimum occurs and give the functional value at this point. Note that Problems 35–38 refer to the four functions from Section 15.1.

**35.** $f(x, y) = x - x^3/9 - y^2/2; -3.8 \leq x \leq 3.8, -3.8 \leq y \leq 3.8$; local maximum point near $(2, 0)$; also global maximum point. Check with calculus.

**36.** $f(x, y) = y/(1 + x^2 + y^2); -5 \leq x \leq 5, -5 \leq y \leq 5$; global maximum point and global minimum point. Check with calculus.

**37.** $f(x, y) = -1 + \cos(y/(1 + x^2 + y^2))$; $-3.8 \le x \le 3.8$, $-3.8 \le y \le 3.8$; global minimum points.

**38.** $f(x, y) = \exp(-(x^2 + y^2)/4)\sin\left(x\sqrt{|y|}\right) - 5 \le x \le 5$, $-5 \le y \le 5$; global maximum and global minimum points.

**39.** $f(x, y) = -x/(x^2 + y^2)$, $f(0, 0) = 0$; $-1 \le x \le 1$, $-1 \le y \le 1$; global maximum and global minimum points. Be careful.

**40.** $f(x, y) = (\sin x)/(6 + x + |y|)$; $-3 \le x \le 3$, $-3 \le y \le 3$; global maximum and global minimum points.

**41.** $f(x, y) = \cos(|x| + y^2) + 10x \exp(-x^2 - y^2)$; $-2 \le x \le 2$, $-2 \le y \le 2$; global maximum and global minimum points.

**42.** $f(x, y) = (x^2 - x - 5)(1 - 9y) \sin x \sin y$; $-6 \le x \le 6$, $-6 \le y \le 6$; global maximum and global minimum points.

**43.** $f(x, y) = 2 \sin x + \sin y - \sin(x + y)$; $0 \le x \le 2\pi$, $0 \le y \le 2\pi$; global maximum and global minimum points.

**44.** Let three arms of lengths 6, 8, and 10 emanate from $N$, as shown in Figure 4. Let $K(\alpha, \beta)$ and $L(\alpha, \beta)$ denote the area and perimeter, respectively, of the triangle $ABC$ determined by these arms.

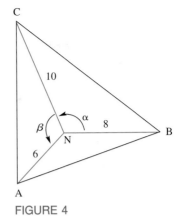

FIGURE 4

(a) Find formulas for $K(\alpha, \beta)$ and $L(\alpha, \beta)$.
(b) Determine $(\alpha, \beta)$ in $D = \{(\alpha, \beta): 0 \le \alpha \le \pi, 0 \le \beta \le \pi\}$ that maximizes $K(\alpha, \beta)$.
(c) Determine $(\alpha, \beta)$ in $D$ that maximizes $L(\alpha, \beta)$

---

**Answers to Concepts Review:** **1.** Closed bounded **2.** Boundary; stationary; singular **3.** $\nabla f(x_0, y_0) = \mathbf{0}$ **4.** $f_{xx}(x_0, y_0)f_{yy}(x_0, y_0) - f_{xy}^2(x_0, y_0)$

---

## 15.9
## LAGRANGE'S METHOD

We begin by distinguishing between two kinds of problems. To find the minimum value of $x^2 + 2y^2 + z^4 + 4$ is a *free extremum* problem. To find the minimum of $x^2 + 2y^2 + z^4 + 4$ subject to the condition that $x + 3y - z = 7$ is a *constrained extremum* problem. Many of the problems of the real world, especially those in economics, are of the latter type. For example, a manufacturer may wish to maximize profits, but is likely to be constrained by the amount of raw materials available, the size of its labor force, and so on.

Example 4 of the previous section was a constrained extremum problem. We were asked to find the minimum distance from the surface $z^2 = x^2y + 4$ to the origin. We formulated the problem as that of minimizing $d^2 = x^2 + y^2 + z^2$ subject to the constraint $z^2 = x^2y + 4$. We handled that problem by substituting the value for $z^2$ from the constraint in the expression for $d^2$ and then solving the resulting free extremum problem. However, it often happens that the constraint equation is not easily solved for one of the variables and, even if this can be done, another method may be more practical. This is the method of **Lagrange multipliers**, named after Joseph-Louis Lagrange.

**Geometric Interpretation of the Method** Let us consider first the case where we wish to maximize or minimize $f(x, y)$ subject to the constraint $g(x, y) = 0$. Figure 1 suggests a geometric interpretation of the problem.

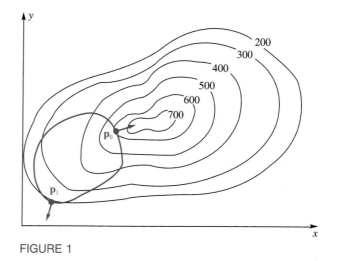

FIGURE 1

The level curves of $f$ are the curves $f(x, y) = k$, where $k$ is a constant. They are shown as black curves in Figure 1 for $k = 200, 300, \ldots, 700$. The graph of the constraint $g(x, y) = 0$ is also a curve; it is shown in blue in Figure 1.

To maximize $f$ subject to the constraint $g(x, y) = 0$ is to find the level curve with the greatest possible $k$ that intersects the constraint curve. It is geometrically evident from Figure 1 that such a level curve is tangent to the constraint curve at a point $P_0(x_0, y_0)$ and therefore that the maximum value of $f$ subject to the constraint $g(x, y) = 0$ is $f(x_0, y_0)$. The other point of tangency $P_1(x_1, y_1)$ gives the minimum value $f(x_1, y_1)$ of $f$ subject to the constraint $g(x, y) = 0$.

Lagrange's method provides an algebraic procedure for finding the points $P_0$ and $P_1$. Since at such a point, the level curve and the constraint curve are tangent (that is, have a common tangent line), the two curves have a common perpendicular line. But at any point of a level curve, the gradient vector $\nabla f$ is perpendicular to the level curve (Section 15.5), and similarly $\nabla g$ is perpendicular to the constraint curve. Thus, $\nabla f$ and $\nabla g$ are parallel at $P_0$ and also at $P_1$; that is,

$$\nabla f(P_0) = \lambda_0 \, \nabla g(P_0) \qquad \text{and} \qquad \nabla f(P_1) = \lambda_1 \, \nabla g(P_1)$$

for some nonzero numbers $\lambda_0$ and $\lambda_1$.

The argument just given is admittedly an intuitive one but it can be made completely rigorous under appropriate hypotheses. Moreover, this argument works just as well for the problem of maximizing or minimizing $f(x, y, z)$ subject to the constraint $g(x, y, z) = 0$. We simply consider level surfaces rather than level curves. In fact, the result is valid in any number of variables.

All of this suggests the following formulation of the method of Lagrange multipliers.

Theorem A

**(Lagrange's Method).** To maximize or minimize $f(\mathbf{p})$ subject to the constraint $g(\mathbf{p}) = 0$, solve the system of equations

$$\nabla f(\mathbf{p}) = \lambda \, \nabla g(\mathbf{p}) \qquad \text{and} \qquad g(\mathbf{p}) = 0$$

for $\mathbf{p}$ and $\lambda$. Each such point $\mathbf{p}$ is a critical point for the constrained extremum problem and the corresponding $\lambda$ is called a Lagrange multiplier.

**Applications** We illustrate the method with several examples.

**EXAMPLE 1** What is the greatest area that a rectangle can have if the length of its diagonal is 2?

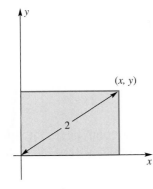

**FIGURE 2**

*Solution* Place the rectangle in the first quadrant with two of its sides along the coordinate axes; then the vertex opposite the origin has coordinates $(x, y)$, with $x$ and $y$ positive (Figure 2). The length of its diagonal is $\sqrt{x^2 + y^2} = 2$ and its area is $xy$.

   Thus, we may formulate the problem to be that of maximizing $f(x, y) = xy$ subject to the constraint $g(x, y) = x^2 + y^2 - 4 = 0$. The corresponding gradients are

$$\nabla f(x, y) = f_x(x, y)\mathbf{i} + f_y(x, y)\mathbf{j} = y\mathbf{i} + x\mathbf{j}$$
$$\nabla g(x, y) = g_x(x, y)\mathbf{i} + g_y(x, y)\mathbf{j} = 2x\mathbf{i} + 2y\mathbf{j}$$

Lagrange's equations thus become

(1) $$y = \lambda(2x)$$

(2) $$x = \lambda(2y)$$

(3) $$x^2 + y^2 = 4$$

which we must solve simultaneously. If we multiply the first equation by $y$ and the second by $x$, we get $y^2 = 2\lambda xy$ and $x^2 = 2\lambda xy$, from which

(4) $$y^2 = x^2$$

From (3) and (4), we find $x = \sqrt{2}$ and $y = \sqrt{2}$; and by substituting values in (1), we obtain $\lambda = \frac{1}{2}$. Thus the solution to equations (1)–(3), keeping $x$ and $y$ positive, is $x = \sqrt{2}$, $y = \sqrt{2}$, $\lambda = \frac{1}{2}$.

   We conclude that the rectangle of greatest area with diagonal 2 is the square of side length $\sqrt{2}$. Its area is 2. A geometric interpretation of this problem is shown in Figure 3. ∎

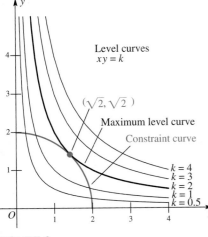

**FIGURE 3**

**EXAMPLE 2**    Use Lagrange's method to find the maximum and minimum values of

$$f(x, y) = y^2 - x^2$$

on the ellipse $x^2/4 + y^2 = 1$.

**Solution**    Refer to Figure 2 of Section 15.8 for a graph of the hyperbolic paraboloid $z = f(x, y) = y^2 - x^2$. From this figure, we would certainly guess that the minimum value occurs at $(\pm 2, 0)$ and the maximum value at $(0, \pm 1)$. But let us justify this conjecture.

We may write the constraint as $g(x, y) = x^2 + 4y^2 - 4 = 0$. Now

$$\nabla f = -2x\mathbf{i} + 2y\mathbf{j}$$

and

$$\nabla g = 2x\mathbf{i} + 8y\mathbf{j}$$

The Lagrange equations are

(1)                                      $-2x = \lambda 2x$

(2)                                       $2y = \lambda 8y$

(3)                          $x^2 + 4y^2 = 4$

Note from the third equation that $x$ and $y$ cannot both be 0. If $x \neq 0$, the first equation implies that $\lambda = -1$, and the second equation then requires that $y = 0$. We conclude from the third equation that $x = \pm 2$. We have thus obtained the critical points $(\pm 2, 0)$.

Exactly the same argument with $y \neq 0$ yields $\lambda = \frac{1}{4}$ from the second equation, then $x = 0$ from the first equation, and finally $y = \pm 1$ from the third equation. We conclude that $(0, \pm 1)$ are also critical points.

Now for $f(x, y) = y^2 - x^2$,

$$f(2, 0) = -4$$
$$f(-2, 0) = -4$$
$$f(0, 1) = 1$$
$$f(0, -1) = 1$$

The minimum value of $f(x, y)$ on the given ellipse is $-4$; the maximum value is 1.    ∎

**EXAMPLE 3**    Find the minimum of $f(x, y, z) = 3x + 2y + z + 5$, subject to the constraint $g(x, y, z) = 9x^2 + 4y^2 - z = 0$.

**Solution**    The gradients of $f$ and $g$ are $\nabla f(x, y, z) = 3\mathbf{i} + 2\mathbf{j} + \mathbf{k}$ and $\nabla g(x, y, z) = 18x\mathbf{i} + 8y\mathbf{j} - \mathbf{k}$. To find the critical points, we solve the equations

$$\nabla f(x, y, z) = \lambda \nabla g(x, y, z) \qquad \text{and} \qquad g(x, y, z) = 0$$

for $(x, y, z, \lambda)$, in which $\lambda$ is a Lagrange multiplier. This is equivalent, in the present problem, to solving the following system of four simultaneous equations in the four variables $x$, $y$, $z$, and $\lambda$.

(1) $\qquad\qquad\qquad\qquad\qquad\qquad\qquad\qquad 3 = 18x\lambda$

(2) $\qquad\qquad\qquad\qquad\qquad\qquad\qquad\qquad 2 = 8y\lambda$

(3) $\qquad\qquad\qquad\qquad\qquad\qquad\qquad\qquad 1 = -\lambda$

(4) $\qquad\qquad\qquad\qquad\qquad\qquad 9x^2 + 4y^2 - z = 0$

From (3), $\lambda = -1$. Substituting this result in equations (1) and (2), we get $x = -\frac{1}{6}$ and $y = -\frac{1}{4}$. By putting these values for $x$ and $y$ in equation 4, we obtain $z = \frac{1}{2}$. Thus the solution of the foregoing system of four simultaneous equations is $(-\frac{1}{6}, -\frac{1}{4}, \frac{1}{2}, -1)$, and the only critical point is $(-\frac{1}{6}, -\frac{1}{4}, \frac{1}{2})$. Therefore, the minimum of $f(x, y, z)$, subject to the constraint $g(x, y, z) = 0$, is $f(-\frac{1}{6}, -\frac{1}{4}, \frac{1}{2}) = 4\frac{1}{2}$. (How do we know that this value is a minimum rather than a maximum?) ∎

When more than one constraint is imposed on the variables of a function that is to be maximized or minimized, additional Lagrange multipliers are used (one for each constraint). For example, if we seek the extrema of a function $f$ of three variables, subject to the two constraints $g(x, y, z) = 0$ and $h(x, y, z) = 0$, we solve the equations

$$\nabla f(x, y, z) = \lambda \nabla g(x, y, z) + \mu \nabla h(x, y, z), \quad g(x, y, z) = 0, \quad h(x, y, z) = 0$$

for $x$, $y$, $z$, $\lambda$, and $\mu$ where $\lambda$ and $\mu$ are Lagrange multipliers. This is equivalent to finding the solutions of the system of five simultaneous equations in the variables $x$, $y$, $z$, $\lambda$, and $\mu$.

(1) $\qquad\qquad\qquad f_x(x, y, z) = \lambda g_x(x, y, z) + \mu h_x(x, y, z)$

(2) $\qquad\qquad\qquad f_y(x, y, z) = \lambda g_y(x, y, z) + \mu h_y(x, y, z)$

(3) $\qquad\qquad\qquad f_z(x, y, z) = \lambda g_z(x, y, z) + \mu h_z(x, y, z)$

(4) $\qquad\qquad\qquad g(x, y, z) = 0$

(5) $\qquad\qquad\qquad h(x, y, z) = 0$

From the solutions of this system we obtain the critical points.

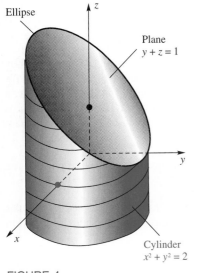

Ellipse

Plane $y + z = 1$

Cylinder $x^2 + y^2 = 2$

FIGURE 4

**EXAMPLE 4**  Find the maximum and minimum values of $f(x, y, z) = x + 2y + 3z$ on the ellipse that is the intersection of the cylinder $x^2 + y^2 = 2$ and the plane $y + z = 1$ (see Figure 4).

*Solution*   We want to maximize and minimize $f(x, y, z)$ subject to $g(x, y, z) = x^2 + y^2 - 2 = 0$ and $h(x, y, z) = y + z - 1 = 0$. The corresponding Lagrange equations are

(1) $$1 = 2\lambda x$$

(2) $$2 = 2\lambda y + \mu$$

(3) $$3 = \mu$$

(4) $$x^2 + y^2 - 2 = 0$$

(5) $$y + z - 1 = 0$$

From (1), $x = 1/2\lambda$; from (2) and (3), $y = -1/2\lambda$. Thus, from (4), $(1/2\lambda)^2 + (-1/2\lambda)^2 = 2$, which implies that $\lambda = \pm\frac{1}{2}$. The solution $\lambda = \frac{1}{2}$ yields the critical point $(x, y, z) = (1, -1, 2)$ and $\lambda = -\frac{1}{2}$ yields the critical point $(x, y, z) = (-1, 1, 0)$. We conclude that $f(1, -1, 2) = 5$ is the maximum value and $f(-1, 1, 0) = 1$ is the minimum value.  ■

## CONCEPTS REVIEW

**1.** To maximize $f(x, y)$ is a _____ extremum problem; to maximize $f(x, y)$ subject to $g(x, y) = 0$ is a _____ extremum problem.

**2.** The method of Lagrange multipliers depends on the fact that at an extreme point, the vectors $\nabla f$ and $\nabla g$ are _____.

**3.** Thus, to use the method of Lagrange, we attempt to solve the equations $\nabla f(x, y) = \lambda \nabla g(x, y)$ and _____ simultaneously.

**4.** Sometimes simple geometric reasoning yields a solution. The maximum value of $f(x, y) = x^4 + y^4$ on the circle $(x - 1)^2 + (y - 1)^2 = 2$ clearly occurs at _____.

## PROBLEM SET 15.9

**1.** Find the minimum of $f(x, y) = x^2 + y^2$, subject to the constraint $g(x, y) = xy - 3 = 0$.

**2.** Find the maximum of $f(x, y) = xy$, subject to the constraint $g(x, y) = 4x^2 + 9y^2 - 36 = 0$.

**3.** Find the maximum of $f(x, y) = 4x^2 - 4xy + y^2$, subject to the constraint $x^2 + y^2 = 1$.

**4.** Find the minimum of $f(x, y) = x^2 + 4xy + y^2$, subject to the constraint $x - y - 6 = 0$.

**5.** Find the minimum of $f(x, y, z) = x^2 + y^2 + z^2$, subject to the constraint $x + 3y - 2z = 12$.

**6.** Find the minimum of $f(x, y, z) = 4x - 2y + 3z$, subject to the constraint $2x^2 + y^2 - 3z = 0$.

**7.** What are the dimensions of the rectangular box, open at the top, which has maximum volume when the surface area is 48?

**8.** Find the least distance between the origin and the plane $x + 3y - 2z = 4$.

**9.** The material for the bottom of a rectangular box costs three times as much per square foot as the material for the sides and top. Find the greatest capacity such a box can have if the total amount of money available for material is $12 and the material for the bottom costs $0.60 per square foot.

**10.** Find the least distance between the origin and the surface $x^2 y - z^2 + 9 = 0$.

**11.** Find the maximum volume of a closed rectangular box with faces parallel to the coordinate planes inscribed in the ellipsoid

$$\frac{x^2}{a^2} + \frac{y^2}{b^2} + \frac{z^2}{c^2} = 1$$

**12.** Find the maximum volume of the first octant rectangular box with faces parallel to the coordinate planes, one vertex at $(0, 0, 0)$, and diagonally opposite vertex on the plane

$$\frac{x}{a} + \frac{y}{b} + \frac{z}{c} = 1$$

**13.** Redo Problem 34 of Section 15.8 the easy way using Lagrange multipliers. *Hint*: Let the plane be $x/A + y/B + z/C = 1$.

**14.** Find the shape of the triangle of maximum perimeter that can be inscribed in a circle of radius $r$. *Hint*: Let $\alpha$, $\beta$, and $\gamma$ be as in Figure 5 and reduce the problem to maximizing $P = 2r(\sin \alpha/2 + \sin \beta/2 + \sin \gamma/2)$ subject to $\alpha + \beta + \gamma = 2\pi$.

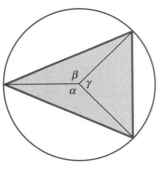

FIGURE 5

**15.** Consider the Cobb-Douglas production model for a manufacturing process depending on three inputs $x$, $y$, and $z$ with unit costs $a$, $b$, and $c$, respectively, given by

$$P = kx^\alpha y^\beta z^\gamma, \qquad \alpha > 0, \beta > 0, \gamma > 0, \alpha + \beta + \gamma = 1$$

subject to the cost constraint $ax + by + cz = d$. Determine $x$, $y$, and $z$ to maximize the production $P$.

**16.** Find the minimum distance from the origin to the line of intersection of the two planes

$$x + y + z = 8 \quad \text{and} \quad 2x - y + 3z = 28$$

**17.** Find the maximum and minimum of $f(x, y, z) = -x + 2y + 2z$ on the ellipse $x^2 + y^2 = 2$, $y + 2z = 1$ (see Example 4).

**18.** Let $w = x_1 x_2 \cdots x_n$.

(a) Maximize $w$ subject to $x_1 + x_2 + \cdots + x_n = 1$ and all $x_i > 0$.

(b) Use part (a) to deduce the famous **Geometric Mean-Arithmetic Mean Inequality** for positive numbers $a_1, a_2, \ldots, a_n$, namely,

$$\sqrt[n]{a_1 a_2 \cdots a_n} \le \frac{(a_1 + a_2 + \cdots + a_n)}{n}$$

**19.** Maximize $w = a_1 x_1 + a_2 x_2 + \cdots + a_n x_n$, all $a_i > 0$, subject to $x_1^2 + x_2^2 + \cdots + x_n^2 = 1$.

[PC] Drawing surfaces and level curves plus using a little common sense can allow us to solve some constrained extremum problems. Solve the following, which are based on the functions of Figures 6–9 of Section 15.1.

**20.** Maximize $z = -4x^3y^2$ subject to $x^2 + y^2 = 1$.

**21.** Minimize $z = x - x^3/8 - y^2/3$ subject to $x^2/16 + y^2 = 1$.

**22.** Maximize $z = xy \exp(-x^2 - y^2)$ subject to $xy = 2$.

**23.** Minimize $z = \exp(-|x|) \cos\sqrt{x^2 + y^2}$ subject to $x^2 + y^2/9 = 1$.

**Answers to Concepts Review:** **1.** Free; constrained **2.** Parallel **3.** $g(x, y) = 0$ **4.** $(2, 2)$

# 15.10 CHAPTER REVIEW

## Concepts Test

Respond with true or false to each of the following assertions. Be prepared to justify your answer.

**1.** The level curves of $z = 2x^2 + 3y^2$ are ellipses.

**2.** If $f_x(0, 0) = f_y(0, 0)$, then $f(x, y)$ is continuous at the origin.

**3.** If $f_x(0, 0)$ exists, then $g(x) = f(x, 0)$ is continuous at $x = 0$.

**4.** If $\lim\limits_{(x, y) \to (0, 0)} f(x, y) = L$, then $\lim\limits_{y \to 0} f(y, y) = L$.

**5.** If $f(x, y) = g(x)h(y)$ where $g$ and $h$ are continuous for all $x$ and $y$, respectively, then $f$ is continuous on the whole $xy$-plane.

**6.** If $f(x, y) = g(x)h(y)$ where both $g$ and $h$ are twice differentiable, then

$$\frac{\partial^2 f}{\partial x^2} + \frac{\partial^2 f}{\partial y^2} = g''(x)h(y) + g(x)h''(y)$$

**7.** If $f(x, y)$ and $g(x, y)$ have the same gradient, then they are identical functions.

**8.** The gradient of $f$ is perpendicular to the graph of $z = f(x, y)$.

**9.** If $f$ is differentiable and $\nabla f(a, b) = \mathbf{0}$, then the graph of $z = f(x, y)$ has a horizontal tangent plane at $(a, b)$.

**10.** If $\nabla f(\mathbf{p}_0) = \mathbf{0}$, then $f$ has an extreme value at $\mathbf{p}_0$.

**11.** If $T = e^y \sin x$ gives the temperature at a point $(x, y)$ in the plane, then a heat-seeking object would move away from the origin in the direction $\mathbf{i}$.

**12.** The function $f(x, y) = \sqrt[3]{x^2 + y^4}$ has a global minimum value at the origin.

**13.** The function $f(x, y) = \sqrt[3]{x + y^4}$ has neither a global minimum nor a global maximum value.

### Sample Test Problems

**1.** Find and sketch the domain of each indicated function of two variables, showing clearly any points on the boundary of the domain that belong to the domain.

(a) $z = \sqrt{x^2 + 4y^2 - 100}$  (b) $z = -\sqrt{2x - y - 1}$

**2.** Sketch the level curves of $f(x, y) = (x + y^2)$ for $k = 0, 1, 2, 4$.

In Problems 3–6, find $\partial f/\partial x$, $\partial^2 f/\partial x^2$, and $\partial^2 f/\partial y\, \partial x$.

**3.** $f(x, y) = 3x^4 y^2 + 7x^2 y^7$

**4.** $f(x, y) = \cos^2 x - \sin^2 y$

**5.** $f(x, y) = e^{-y} \tan x$

**6.** $f(x, y) = e^{-x} \sin y$

**7.** If $F(x, y) = 5x^3 y^6 - xy^7$, find $\partial^3 F(x, y)/\partial x\, \partial y^2$.

**8.** If $f$ is the function of three variables defined by $f(x, y, z) = xy^3 - 5x^2 yz^4$, find $f_x(2, -1, 1)$, $f_y(2, -1, 1)$, and $f_z(2, -1, 1)$.

**9.** Find the slope of the tangent to the curve of intersection of the surface $z = x^2 + y^2/4$ and the plane $x = 2$ at the point $(2, 2, 5)$.

**10.** For what points is the function defined by $f(x, y) = xy/(x^2 - y)$ continuous?

**11.** Does

$$\lim_{(x, y) \to (0, 0)} \frac{x - y}{x + y}$$

exist? Explain.

**12.** In each case, find the indicated limit or state that it does not exist.

(a) $\displaystyle \lim_{(x, y) \to (2, 2)} \frac{x^2 - 2y}{x^2 + 2y}$  (b) $\displaystyle \lim_{(x, y) \to (2, 2)} \frac{x^2 + 2y}{x^2 - 2y}$

**14.** If $f(x, y) = 4x + 4y$, then $|D_{\mathbf{u}} f(x, y)| \le 4$.

**15.** If $D_{\mathbf{u}} f(x, y)$ exists, then $D_{-\mathbf{u}} f(x, y) = -D_{\mathbf{u}} f(x, y)$.

**16.** The set $\{(x, y): y = x, 0 \le x \le 1\}$ is a closed set in the plane.

**17.** If $f(x, y)$ is continuous on a closed bounded set $S$, then $f$ attains a maximum value on $S$.

**18.** If $f(x, y)$ attains its maximum value at an interior point $(x_0, y_0)$ of $S$, then $\nabla f(x_0, y_0) = \mathbf{0}$.

**19.** The function $f(x, y) = \sin(xy)$ does not attain a maximum value on the set $\{(x, y): x^2 + y^2 < 4\}$.

**20.** If $f_x(x_0, y_0)$ and $f_y(x_0, y_0)$ both exist, then $f$ is differentiable at $(x_0, y_0)$.

(c) $\displaystyle \lim_{(x, y) \to (0, 0)} \frac{x^4 - 4y^4}{x^2 + 2y^2}$

**13.** Find $\nabla f(1, 2, -1)$.

(a) $f(x, y, z) = x^2 yz^3$  (b) $f(x, y, z) = y^2 \sin xz$

**14.** Find the directional derivative of $f(x, y) = \tan^{-1}(3xy)$. What is its value at the point $(4, 2)$ in the direction $\mathbf{u} = (\sqrt{3}/2)\mathbf{i} - (1/2)\mathbf{j}$?

**15.** Find the slope of the tangent line to the curve of intersection of the vertical plane $x - \sqrt{3}y + 2\sqrt{3} - 1 = 0$ and the surface $z = x^2 + y^2$ at the point $(1, 2, 5)$.

**16.** In what direction is $f(x, y) = 9x^4 + 4y^2$ increasing most rapidly at $(1, 2)$?

**17.** For $f(x, y) = x^2/2 + y^2$,

(a) find the equation of its level curve that goes through the point $(4, 1)$ in its domain;
(b) find the gradient vector $\nabla f$ at $(4, 1)$;
(c) draw the level curve and draw the gradient vector with its initial point at $(4, 1)$.

**18.** If $F(u, v) = \tan^{-1}(uv)$, $u = \sqrt{xy}$, $v = \sqrt{x} - \sqrt{y}$, find $\partial F/\partial x$ and $\partial F/\partial y$ in terms of $u$, $v$, $x$, and $y$.

**19.** If $f(u, v) = u/v$, $u = x^2 - 3y + 4z$, and $v = xyz$, find $f_x$, $f_y$, and $f_z$ in terms of $x$, $y$, and $z$.

**20.** If $F(x, y) = x^3 - xy^2 - y^4$, $x = 2 \cos 3t$, and $y = 3 \sin t$, find $dF/dt$ at $t = 0$.

**21.** If $F(x, y, z) = (5x^2 y/z^3)$, $x = t^{3/2} + 2$, $y = \ln 4t$, and $z = e^{3t}$, find $dF/dt$ in terms of $x$, $y$, $z$, and $t$.

**22.** A triangle has vertices $A$, $B$, and $C$. The length of the side $c = AB$ is increasing at the rate of 3 inches per second, the side $b = AC$ is decreasing at 1 inch per second, and the included angle $\alpha$ is increasing at 0.1 radian per second. If

$c = 10$ inches and $b = 8$ inches when $\alpha = \pi/6$, how fast is the area changing then?

**23.** Find the gradient vector of the ellipsoid $9x^2 + 4y^2 + 9z^2 = 34$ at the point $P(1, 2, -1)$. Write the equation of the tangent plane to this surface at the given point.

**24.** A right circular cylinder is measured to have a radius of $10 \pm 0.02$ inches and a height of $6 \pm 0.01$ inches. Calculate its volume and use differentials to give an estimate of the possible error.

**25.** If $f(x, y, z) = xy^2/(1 + z^2)$, use differentials to estimate $f(1.01, 1.98, 2.03)$.

**26.** Find the extrema of $f(x, y) = x^2y - 6y^2 - 3x^2$.

**27.** A rectangular box whose edges are parallel to the coordinate axes is inscribed in the ellipsoid $36x^2 + 4y^2 + 9z^2 = 36$. What is the greatest possible volume for such a box?

**28.** Use Lagrange multipliers to find the maximum and the minimum of $f(x, y) = xy$, subject to the constraint $x^2 + y^2 = 1$.

**29.** Use Lagrange multipliers to find the dimensions of the right circular cylinder with maximum volume if its surface area is $24\pi$.

# TECHNOLOGY PROJECT

Recall that Newton's method for solving $f(x) = 0$ provides the iteration method:

$$x_{n+1} = x_n - f(x_n)/f'(x_n)$$

We know that if there really is a solution near a good starting value, Newton's method usually (but not always) converges rapidly to that solution.

Now consider the problem in *two* dimensions. Suppose we have the two equations

$$z = f(x, y) \text{ and } z = g(x, y).$$

and a starting value $(x_0, y_0)$ near a simultaneous root of the equations (that is, both $f(x_0, y_0) \approx 0$ and $g(x_0, y_0) \approx 0$). Again, we seek a better value by using the linear approximation to the equations:

$$z \approx f + f_x(x - x_0) + f_y(y - y_0) \tag{1}$$
$$z \approx g + g_x(x - x_0) + g_y(y - y_0),$$

where all the functions are evaluated at $(x_0, y_0)$. Put $z = 0$ in each of these approximate equations and then solve for $(x, y)$. If we denote the solution by $(x_1, y_1)$, the first Newton iterate, we obtain:

$$x_1 = x_0 - \frac{fg_y - gf_y}{f_x g_y - g_x f_y}$$
$$\tag{2}$$
$$y_1 = y_0 - \frac{gf_x - fg_y}{f_x g_y - g_x f_y}$$

We want the minimum distance between a surface defined by $z = h(x, y)$ and a point,

$\mathbf{P} = (p, q, r)$, off the surface. A good approach is to minimize the function, $F(x, y) =$ "square of the distance between $\mathbf{P}$ and a point $(x, y, h(x, y))$ on the surface".

**Exercise 1** a) Explain how the minimization of the above function $F$ leads to the equations:

$$x - p + (h(x, y) - r)h_x = 0,$$
$$y - q + (h(x, y) - r)h_y = 0.$$

b) If $h(x, y) = x^2 + 4y^2$, so the surface is a paraboloid, derive the equations:

$$f(x, y) = 2x^3 + 8xy^2 + (1 - 2r)x - p = 0 \tag{3}$$
$$g(x, y) = 32y^3 + 8x^2y + (1 - 8r)y - q = 0.$$

The task now is to solve Equations (3). Getting good starting values for $x$ and $y$ for the upcoming Newton iterations is much harder now than for the $f(x) = 0$ case. Some computing technologies have an 'implicit plot' capability which can plot the two functional relations between $x$ and $y$ defined by Equations (3). Figure 1 shows the results of an implicit plot for the case $(p, q, r) = (4, 1, 0)$. The intersection of the two level curves denotes the desired solution to $f = g = 0$.

Thus, we get the starting values, roughly $x = 1.1$, $y = 0.1$.

Here is some *Maple* code to show the Newton iteration. Assume functions $f$ and $g$ and the four partial derivatives $dfx = \partial f/\partial x$, etc. have been defined. For brevity, we don't show the arguments; e.g. $f$ is an abbreviation for $f(xn, yn)$, etc.

```
xn := 1.1:      yn := .1:
for n to 8 do
  xn := xn - (f*dgy - g*dfy)/
       (dfx*dgy - dgx*dfy):
  yn := yn - (g*dfx - f*dgx)/
       (dfx*dgy - dgx*dfy):
  lprint( xn, yn )
od;
```

The results of these 8 Newton iterations is $xn = 1.119976715$ and $yn = .8860526308$. Try to get this result with your computing technology.

**Exercise 2** Find the minimum distance from the point $(2, 2, 0)$ to the paraboloid in Exercise 1. **Note**: If you cannot do implicit plots, use the starting value: $(xn, yn) = (.8, .3)$.

**Exercise 3** Repeat the last exercise for the given point $(4, 1, 4)$.
*Warning*: An implicit plot shows three candidates for the minimum, approximately: $(2, .2)$, $(-.6, 0)$ and $(-1.5, 0)$. Which of these leads to the actual minimum distance?

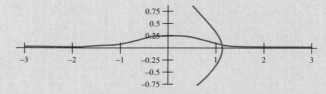

Fig. 1. An implicit plot for the Exercise 1.

# The Integral
# in *n*-Space

---

**16.1**
**DOUBLE INTEGRALS OVER RECTANGLES**

Differentiation and integration are the major processes of calculus. We have studied differentiation in *n*-space (Chapter 15); it is time to consider integration in *n*-space. The theory and the applications of single (Riemann) integrals are to be generalized to multiple integrals. In Chapter 6 we used single integrals to calculate the area of curved planar regions, to find the length of planar curves, and to determine the center of mass of straight wires of variable density. In this chapter we use multiple integrals to find the volume of general solids, the area of general surfaces, and the center of mass of laminas and solids of variable density.

The intimate connection between integration and differentiation was enunciated in the Fundamental Theorem of Calculus; this theorem provided the principal theoretical tool for evaluating single integrals. Here we reduce multiple integration to a succession of single integrations where again the Fundamental Theorem will play a crucial role. The integration skills you learned in Chapters 5–8 will be severely tested.

The Riemann integral for a function of one variable was defined in Section 5.5, a section worth reviewing. Recall that we formed a partition $P$ of the interval $[a, b]$ into subintervals of length $\Delta x_k$, $k = 1, 2, \ldots, n$, picked a sample point $\overline{x}_k$ from the $k$th subinterval, and then wrote

$$\int_a^b f(x)\, dx = \lim_{|P| \to 0} \sum_{k=1}^n f(\overline{x}_k)\Delta x_k$$

We proceed in a very similar fashion to define the integral for a function of two variables.

Let $R$ be a rectangle with sides parallel to the coordinate axes; that is, let

$$R = \{(x, y): a \leq x \leq b, c \leq y \leq d\}$$

771

Form a partition $P$ of $R$ by means of lines parallel to the $x$- and $y$-axes, as in Figure 1. This divides $R$ into subrectangles, say $n$ of them, which we denote by $R_k$, $k = 1, 2, \ldots, n$. Let $\Delta x_k$ and $\Delta y_k$ be the lengths of the sides of $R_k$ and let $\Delta A_k = \Delta x_k \Delta y_k$ be its area. On $R_k$, pick a sample point $(\overline{x}_k, \overline{y}_k)$ and form the Riemann sum

$$\sum_{k=1}^{n} f(\overline{x}_k, \overline{y}_k)\, \Delta A_k$$

which corresponds (if $f(x, y) \geq 0$) to the sum of the volumes of $n$ boxes (Figures 2 and 3). Letting the partition get finer and finer in such a way that all the $R_k$'s get smaller will lead to the concept that we want.

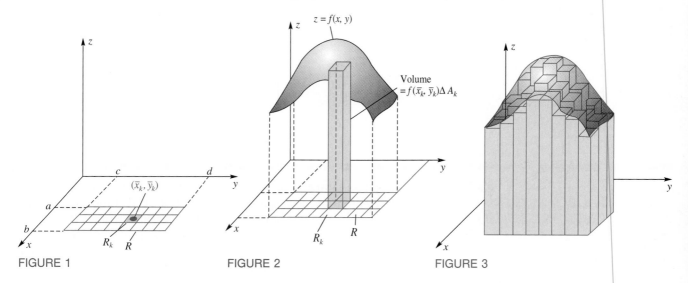

FIGURE 1　　　　FIGURE 2　　　　FIGURE 3

We are ready for a formal definition. We use the notation introduced above, with the additional proviso that the norm of the partition $P$—denoted by $|P|$—is the length of the longest diagonal of any subrectangle in the partition.

### Definition

**(The double integral).** Let $f$ be a function of two variables that is defined on a closed rectangle $R$. If

$$\lim_{|P| \to 0} \sum_{k=1}^{n} f(\overline{x}_k, \overline{y}_k)\, \Delta A_k$$

exists, we say $f$ is integrable on $R$. Moreover, $\displaystyle\iint_R f(x, y)\, dA$, called the **double integral** of $f$ over $R$, is then given by

$$\iint_R f(x, y)\, dA = \lim_{|P| \to 0} \sum_{k=1}^{n} f(\overline{x}_k, \overline{y}_k)\, \Delta A_k$$

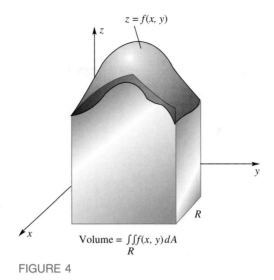

$$\text{Volume} = \iint_R f(x, y)\,dA$$

FIGURE 4

Recall that if $f(x) \geq 0$, $\int_a^b f(x)\,dx$ represents the area of the region under the curve, $y = f(x)$ between $a$ and $b$. In a similar manner, if $f(x, y) \geq 0$, $\iint_R f(x, y)\,dA$ represents the **volume** of the solid under the surface $z = f(x, y)$ and above the rectangle $R$ (Figure 4). In fact, we take this integral as the definition of the volume of such a solid.

**The Existence Question**  Not every function of two variables is integrable on a given rectangle $R$. The reasons are the same as in the one-variable case (Section 5.5). In particular, a function which is unbounded on $R$ will always fail to be integrable. Fortunately, there is a natural generalization of Theorem 5.5A. Its proof is beyond the level of a first course.

> ### Theorem A
>
> **(Integrability Theorem).** If $f$ is bounded on the closed rectangle $R$ and if it is continuous there except on a finite number of smooth curves, then $f$ is integrable on $R$. In particular, if $f$ is continuous on all of $R$, then $f$ is integrable there.

As a consequence, most of the common functions (provided they are bounded) are integrable on every rectangle. For example,

$$f(x, y) = e^{\sin(xy)} - y^3 \cos(x^2 y)$$

is integrable on every rectangle. On the other hand,

$$g(x, y) = \frac{x^2 y - 2x}{y - x^2}$$

would fail to be integrable on any rectangle that intersected the parabola $y = x^2$. The "staircase" function of Figure 5 is integrable on $R$ because its discontinuities occur along two line segments.

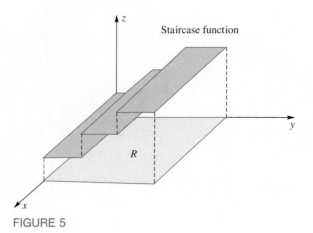

FIGURE 5

**Properties of the Double Integral** The double integral inherits most of the properties of the single integral.

1. The double integral is linear; that is:

   **(a)** $\iint\limits_R kf(x, y)\, dA = k \iint\limits_R f(x, y)\, dA;$

   **(b)** $\iint\limits_R [f(x, y) + g(x, y)]\, dA = \iint\limits_R f(x, y)\, dA + \iint\limits_R g(x, y)\, dA.$

2. The double integral is additive on rectangles (Figure 6) that overlap only on a line segment.

$$\iint\limits_R f(x, y)\, dA = \iint\limits_{R_1} f(x, y)\, dA + \iint\limits_{R_2} f(x, y)\, dA$$

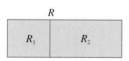

FIGURE 6

3. The comparison property holds. If $f(x, y) \le g(x, y)$ for all $(x, y)$ in $R$, then

$$\iint\limits_R f(x, y)\, dA \le \iint\limits_R g(x, y)\, dA$$

All of these properties hold on more general sets than rectangles, but that is a matter we take up in Section 16.3.

**Evaluation of Double Integrals** This topic will receive major attention in the next section, where we will develop a powerful tool for evaluating double integrals. However, we can already evaluate a few integrals and we can approximate others.

Note first of all that if $f(x, y) = 1$ on $R$, then the double integral is the area of $R$, and from this it follows that

$$\iint\limits_{R} k\, dA = k \iint\limits_{R} 1\, dA = kA(R)$$

**EXAMPLE 1**   Let $f$ be the staircase function of Figure 5—that is, let

$$f(x, y) = \begin{cases} 1 & 0 \le x \le 3, 0 \le y < 1 \\ 2 & 0 \le x \le 3, 1 \le y < 2 \\ 3 & 0 \le x \le 3, 2 \le y \le 3 \end{cases}$$

Calculate $\iint\limits_{R} f(x, y)\, dA$ where $R = \{(x, y): 0 \le x \le 3, 0 \le y \le 3\}$.

*Solution*   Introduce rectangles $R_1$, $R_2$, and $R_3$ as follows.

$$R_1 = \{(x, y): 0 \le x \le 3, 0 \le y \le 1\}$$
$$R_2 = \{(x, y): 0 \le x \le 3, 1 \le y \le 2\}$$
$$R_3 = \{(x, y): 0 \le x \le 3, 2 \le y \le 3\}$$

Then, using the additivity property of the double integral, we obtain

$$\iint\limits_{R} f(x, y)\, dA = \iint\limits_{R_1} f(x, y)\, dA + \iint\limits_{R_2} f(x, y)\, dA + \iint\limits_{R_3} f(x, y)\, dA$$

$$= 1\, A(R_1) + 2A(R_2) + 3A(R_3)$$
$$= 1 \cdot 3 + 2 \cdot 3 + 3 \cdot 3 = 18$$

In this derivation, we also used the fact that the value of $f$ on the boundary of a rectangle does not affect the value of the integral.   ■

Example 1 was a minor accomplishment, and to be honest we cannot do much more without more tools. However, we can always approximate a double integral by calculating a Riemann sum. In general, we can expect the approximation to be better the finer the partition we use.

**EXAMPLE 2**   Approximate $\iint\limits_{R} f(x, y)\, dA$, where

$$f(x, y) = \frac{64 - 8x + y^2}{16}$$

and

$$R = \{(x, y): 0 \le x \le 4, 0 \le y \le 8\}$$

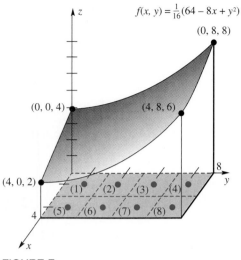

$$f(x, y) = \tfrac{1}{16}(64 - 8x + y^2)$$

FIGURE 7

Do this by calculating the Riemann sum obtained by dividing $R$ into eight equal squares and using the center of each square as the sample point (Figure 7).

*Solution*   The required sample points and the corresponding values of the function are as follows.

$$(\overline{x}_1, \overline{y}_1) = (1, 1), \qquad f(\overline{x}_1, \overline{y}_1) = \tfrac{57}{16}$$

$$(\overline{x}_2, \overline{y}_2) = (1, 3), \qquad f(\overline{x}_2, \overline{y}_2) = \tfrac{65}{16}$$

$$(\overline{x}_3, \overline{y}_3) = (1, 5), \qquad f(\overline{x}_3, \overline{y}_3) = \tfrac{81}{16}$$

$$(\overline{x}_4, \overline{y}_4) = (1, 7), \qquad f(\overline{x}_4, \overline{y}_4) = \tfrac{105}{16}$$

$$(\overline{x}_5, \overline{y}_5) = (3, 1), \qquad f(\overline{x}_5, \overline{y}_5) = \tfrac{41}{16}$$

$$(\overline{x}_6, \overline{y}_6) = (3, 3), \qquad f(\overline{x}_6, \overline{y}_6) = \tfrac{49}{16}$$

$$(\overline{x}_7, \overline{y}_7) = (3, 5), \qquad f(\overline{x}_7, \overline{y}_7) = \tfrac{65}{16}$$

$$(\overline{x}_8, \overline{y}_8) = (3, 7), \qquad f(\overline{x}_8, \overline{y}_8) = \tfrac{89}{16}$$

Thus, since $\Delta A_k = 4$,

$$\iint\limits_{R} f(x, y)\, dA \approx \sum_{k=1}^{8} f(\overline{x}_k, \overline{y}_k)\, \Delta A_k$$

$$= 4 \sum_{k=1}^{8} f(\overline{x}_k, \overline{y}_k)$$

$$= \frac{4(57 + 65 + 81 + 105 + 41 + 49 + 65 + 89)}{16} = 138$$

In Section 16.2, we shall learn how to find the exact value of this integral. It is $138\tfrac{2}{3}$.  ∎

# CONCEPTS REVIEW

**1.** Assume the rectangle $R$ has been partitioned into $n$ subrectangles of area $\Delta A_k$ with sample points $(\bar{x}_k, \bar{y}_k)$, $k = 1, 2, \ldots, n$. Then $\iint\limits_R f(x, y)\, dA = \lim\limits_{|P| \to 0}$ _____.

**2.** If $f(x, y) \geq 0$ on $R$, then $\iint\limits_R f(x, y)\, dA$ can be interpreted geometrically as _____.

**3.** If $f$ is _____ on $R$, then $f$ is integrable there.

**4.** If $f(x, y) = 6$ on the rectangle $R = \{(x, y): 1 \leq x \leq 2, 0 \leq y \leq 2\}$, then $\iint\limits_R f(x, y)\, dA$ has the value _____.

# PROBLEM SET 16.1

In Problems 1–4, let $R = \{(x, y): 1 \leq x \leq 4, 0 \leq y \leq 2\}$. Evaluate $\iint\limits_R f(x, y)\, dA$, where $f$ is the given function (see Example 1).

**1.** $f(x, y) = \begin{cases} 2 & 1 \leq x < 3, 0 \leq y \leq 2 \\ 3 & 3 \leq x \leq 4, 0 \leq y \leq 2 \end{cases}$

**2.** $f(x, y) = \begin{cases} -1 & 1 \leq x \leq 4, 0 \leq y < 1 \\ 2 & 1 \leq x \leq 4, 1 \leq y \leq 2 \end{cases}$

**3.** $f(x, y) = \begin{cases} 2 & 1 \leq x < 3, 0 \leq y < 1 \\ 1 & 1 \leq x < 3, 1 \leq y \leq 2 \\ 3 & 3 \leq x \leq 4, 0 \leq y \leq 2 \end{cases}$

**4.** $f(x, y) = \begin{cases} 2 & 1 \leq x \leq 4, 0 \leq y < 1 \\ 3 & 1 \leq x < 3, 1 \leq y \leq 2 \\ 1 & 3 \leq x \leq 4, 1 \leq y \leq 2 \end{cases}$

Suppose that $R = \{(x, y): 0 \leq x \leq 2, 0 \leq y \leq 2\}$, $R_1 = \{(x, y): 0 \leq x \leq 2, 0 \leq y \leq 1\}$, and $R_2 = \{(x, y): 0 \leq x \leq 2, 1 \leq y \leq 2\}$. Suppose, in addition, that $\iint\limits_R f(x, y)\, dA = 3$, $\iint\limits_R g(x, y)\, dA = 5$, and $\iint\limits_{R_1} g(x, y)\, dA = 2$. Use the properties of integrals to evaluate each of the following.

**5.** $\iint\limits_R [3f(x, y) - g(x, y)]\, dA$

**6.** $\iint\limits_R [2f(x, y) + 5g(x, y)]\, dA$

**7.** $\iint\limits_{R_2} g(x, y)\, dA$

**8.** $\iint\limits_{R_1} [2g(x, y) + 3]\, dA$

In Problems 9–14, $R = \{(x, y): 0 \leq x \leq 6, 0 \leq y \leq 4\}$ and $P$ is the partition of $R$ into six equal squares by the lines $x = 2$, $x = 4$, and $y = 2$. Approximate $\iint\limits_R f(x, y)\, dA$ by calculating the corresponding Riemann sum $\sum\limits_{k=1}^{6} f(\bar{x}_k, \bar{y}_k)\, \Delta A_k$, assuming that $(\bar{x}_k, \bar{y}_k)$ are the centers of the six squares (see Example 2).

**9.** $f(x, y) = 12 - x - y$

**10.** $f(x, y) = 10 - y^2$

**11.** $f(x, y) = x^2 + 2y^2$

**12.** $f(x, y) = \frac{1}{6}(48 - 4x - 3y)$

C **13.** $f(x, y) = \sqrt{x + y}$

C **14.** $f(x, y) = e^{xy}$

**15.** Calculate $\iint\limits_R (6 - y)\, dA$, where $R = \{(x, y): 0 \leq x \leq 1, 0 \leq y \leq 1\}$. *Hint*: This integral represents the volume of a certain solid. Sketch this solid and calculate its volume from elementary principles.

**16.** Calculate $\iint\limits_R (1 + x)\, dA$, where $R = \{(x, y): 0 \leq x \leq 2, 0 \leq y \leq 1\}$. See the hint of Problem 15.

**17.** Use the comparison property of double integrals to show that if $f(x, y) \geq 0$ on $R$, then $\iint\limits_R f(x, y)\, dA \geq 0$.

**18.** Suppose $m \leq f(x, y) \leq M$ on $R$. Show that

$$m A(R) \leq \iint\limits_R f(x, y)\, dA \leq M A(R)$$

C **19.** Let $R$ be the rectangle shown in Figure 8. For the indicated partition into 12 equal squares, calculate the smallest and largest Riemann sums for $\iint\limits_R \sqrt{x^2 + y^2}\, dA$ and thereby obtain numbers $c$ and $C$ such that

$$c \leq \iint\limits_R \sqrt{x^2 + y^2}\, dA \leq C$$

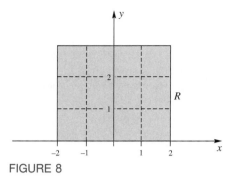

FIGURE 8

**20.** Evaluate $\iint\limits_R x \cos^2(xy)\,dA$, where $R$ is the rectangle of Figure 8. *Hint*: Does the graph of the integrand have any kind of symmetry?

**21.** Recall that $[\![x]\!]$ is the greatest integer in $x$. For $R$ of Figure 8, evaluate:

(a) $\displaystyle\iint\limits_R [\![x]\!][\![y]\!]\,dA$   (b) $\displaystyle\iint\limits_R ([\![x]\!] + [\![y]\!])\,dA$

**22.** Suppose the rectangle of Figure 8 represents a thin plate (lamina) whose mass density at $(x, y)$ is $\delta(x, y)$, say in grams per square centimeter. What does $\iint\limits_R \delta(x, y)\,dA$ represent?

**23.** Colorado is a rectangular state (assuming we ignore the curvature of the earth). Let $f(x, y)$ be the number of inches of rainfall during 1980 at the point $(x, y)$ in that state. What does $\iint\limits_{colorado} f(x, y)\,dA$ represent? What does this number divided by the area of Colorado represent?

**24.** Let $f(x, y) = 1$ if both $x$ and $y$ are rational numbers and let $f(x, y) = 0$ otherwise. Show that $f(x, y)$ is not integrable over the rectangle $R$ in Figure 8.

---

**Answers to Concepts Review:**  **1.** $\displaystyle\sum_{k=1}^{n} f(\bar{x}_k, \bar{y}_k)\Delta A_k$

**2.** The volume of the solid under $z = f(x, y)$ and above $R$
**3.** Continuous  **4.** 12

---

**16.2**
**ITERATED INTEGRALS**

Now we face in earnest the problem of evaluating $\iint\limits_R f(x, y)\,dA$, where $R$ is the rectangle

$$R = \{(x, y): a \le x \le b, c \le y \le d\}$$

Suppose for the time being that $f(x, y) \ge 0$ on $R$ so that we may interpret the double integral as the volume $V$ of the solid under the surface of Figure 1.

(1)
$$V = \iint\limits_R f(x, y)\,dA$$

There is another way to calculate the volume of this solid, which at least intuitively seems just as valid. Slice the solid into thin slabs by means of planes parallel to the $xz$-plane. A typical such slab is shown in Figure 2(a). The area of the face of this slab depends on how far it is from the $xz$-plane, that is, it depends on $y$; therefore, we denote this area by $A(y)$ (see Figure 2(b)).

The volume $\Delta V$ of the slab is given approximately by

$$\Delta V \approx A(y)\,\Delta y$$

and, recalling our old motto (*slice, approximate, integrate*), we may write

$$V = \int_c^d A(y)\,dy$$

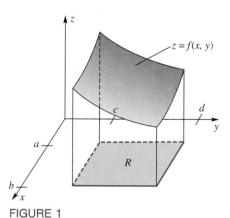

FIGURE 1

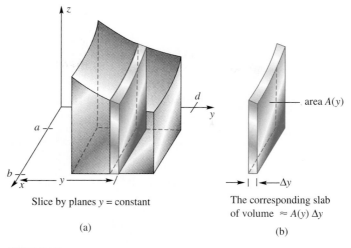

Slice by planes $y$ = constant

(a)

area $A(y)$

$\Delta y$

The corresponding slab
of volume $\approx A(y)\,\Delta y$

(b)

FIGURE 2

On the other hand, for fixed $y$ we may calculate $A(y)$ by means of an ordinary single integral; in fact,

$$A(y) = \int_a^b f(x, y)\,dx$$

We conclude that

(2)
$$V = \int_c^d \left[ \int_a^b f(x, y)\,dx \right] dy$$

an expression called an **iterated integral**.

When we equate the expressions for $V$ from (1) and (2), we obtain the result we want.

$$\iint_R f(x, y)\,dA = \int_c^d \left[ \int_a^b f(x, y)\,dx \right] dy$$

If we had begun the process above by slicing the solid with planes parallel to the $yz$-plane, we would have obtained another iterated integral with the integrations occurring in the opposite order.

$$\iint_R f(x, y)\,dA = \int_a^b \left[ \int_c^d f(x, y)\,dy \right] dx$$

Two remarks are in order. First, while the two boxed results were derived under the assumption that $f$ was nonnegative, they are valid in general. Second, the whole exercise was rather pointless unless iterated integrals can be evaluated. Fortunately, iterated integrals are often easy to evaluate as we demonstrate next.

---

What if *f* Is Negative?

If $f(x, y)$ is negative on part of $R$, then $\iint_R f(x, y)\,dA$ gives the *signed* volume of the solid between the surface $z = f(x, y)$ and the rectangle $R$ of the $xy$-plane.

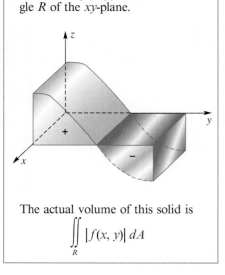

The actual volume of this solid is

$$\iint_R |f(x, y)|\,dA$$

**Evaluating Iterated Integrals**    We begin with a simple example.

**EXAMPLE 1**    Evaluate $\int_0^3 \left[ \int_1^2 (2x + 3y)\,dx \right] dy.$

*Solution*    In the inner integration $y$ is a constant, so

$$\int_1^2 (2x + 3y)\,dx = \left[ x^2 + 3yx \right]_1^2 = 4 + 6y - (1 + 3y) = 3 + 3y$$

Consequently,

$$\int_0^3 \left[ \int_1^2 (2x + 3y)\,dx \right] dy = \int_0^3 [3 + 3y]\,dy = \left[ 3y + \tfrac{3}{2}y^2 \right]_0^3$$

$$= 9 + \tfrac{27}{2} = \tfrac{45}{2} \qquad \blacksquare$$

**EXAMPLE 2**    Evaluate $\int_1^2 \left[ \int_0^3 (2x + 3y)\,dy \right] dx.$

*Solution*    Note that we have simply reversed the order of integration from Example 1; we expect the same answer as in that example.

$$\int_0^3 (2x + 3y)\,dy = \left[ 2xy + \tfrac{3}{2}y^2 \right]_0^3$$

$$= 6x + \tfrac{27}{2}$$

Thus,

$$\int_1^2 \left[ \int_0^3 (2x + 3y)\,dy \right] dx = \int_1^2 [6x + \tfrac{27}{2}]\,dx = \left[ 3x^2 + \tfrac{27}{2}x \right]_1^2$$

$$= 12 + 27 - (3 + \tfrac{27}{2}) = \tfrac{45}{2} \qquad \blacksquare$$

From now on, we shall usually omit the brackets in the iterated integral, allowing the order $dx\,dy$ or $dy\,dx$ to specify which integration is to be done first. Of course, the limits on the inner integral sign refer to the first variable to be integrated.

**EXAMPLE 3**    Evaluate $\int_0^8 \int_0^4 \tfrac{1}{16}[64 - 8x + y^2]\,dx\,dy.$

*Solution*    Note that this iterated integral corresponds to the double integral of Example 2 of Section 16.1 for which we claimed the answer $138\tfrac{2}{3}$. Now

$$\tfrac{1}{16} \int_0^4 (64 - 8x + y^2)\,dx = \tfrac{1}{16} \left[ 64x - 4x^2 + y^2x \right]_0^4$$

$$= \tfrac{1}{16}[256 - 64 + 4y^2] = 12 + \tfrac{1}{4}y^2$$

The required iterated integral has the value

$$\int_0^8 \left( 12 + \frac{1}{4}y^2 \right) dy = \left[ 12y + \frac{y^3}{12} \right]_0^8 = 96 + \frac{512}{12}$$

$$= 96 + \frac{128}{3} = 138\frac{2}{3}$$ ∎

**Calculating Volumes**  Now we can calculate volumes for a wide variety of solids.

**EXAMPLE 4**  Find the volume $V$ of the solid under the surface $z = 4 - x^2 - y$ and over the rectangle $R = \{(x, y): 0 \le x \le 1, 0 \le y \le 2\}$ (see Figure 3).

≈ *Solution*  Let's estimate this volume by assuming the solid has constant height 2.5, giving it a volume of $(2.5)(2) = 5$. If the following calculation gives an answer that is not close to 5, we will know we have made a mistake.

$$V = \iint_R (4 - x^2 - y) \, dA = \int_0^2 \int_0^1 (4 - x^2 - y) \, dx \, dy$$

$$= \int_0^2 \left[ 4x - \frac{x^3}{3} - yx \right]_0^1 dy = \int_0^2 \left[ 4 - \frac{1}{3} - y \right] dy$$

$$= \left[ 4y - \frac{1}{3}y - \frac{1}{2}y^2 \right]_0^2 = 8 - \frac{2}{3} - 2 = \frac{16}{3}$$ ∎

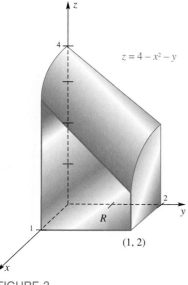

$z = 4 - x^2 - y$

$R$

$(1, 2)$

**FIGURE 3**

---

## CONCEPTS REVIEW

**1.** The expression $\int_a^b \left[ \int_c^a f(x, y) \, dy \right] dx$ is called an _____ integral.

**2.** Let $R = \{(x, y): -1 \le x \le 2, 0 \le y \le 2\}$. Then $\iint_R f(x, y) \, dA$ can be expressed as an interated integral either as _____ or as _____.

**3.** For a general function $f$ defined on $R$, $\iint_R f(x, y) \, dA$ can be interpreted as the _____ volume of the solid between the surface $z = f(x, y)$ and the $xy$-plane; the part above this plane gets a _____ sign; the part below, a _____ sign.

**4.** Thus if a double integral turns out to have a negative value, we know that more than half of the solid _____.

---

## PROBLEM SET 16.2

In Problems 1–12, evaluate each of the iterated integrals.

**1.** $\int_0^2 \int_1^3 x^2 y \, dy \, dx$

**2.** $\int_{-1}^4 \int_1^2 (x + y^2) \, dy \, dx$

**3.** $\int_1^2 \int_0^3 (xy + y^2) \, dx \, dy$

**4.** $\int_{-1}^1 \int_1^2 (x^2 + y^2) \, dx \, dy$

5. $\displaystyle\int_0^\pi \int_0^1 x \sin y \, dx \, dy$    6. $\displaystyle\int_0^{\ln 3} \int_0^{\ln 2} e^{x+y} \, dy \, dx$

7. $\displaystyle\int_0^{\pi/2} \int_0^1 x \sin xy \, dy \, dx$    8. $\displaystyle\int_0^1 \int_0^1 xe^{xy} \, dy \, dx$

9. $\displaystyle\int_0^3 \int_0^1 2x\sqrt{x^2 + y} \, dx \, dy$

10. $\displaystyle\int_0^1 \int_0^1 \frac{y}{(xy + 1)^2} \, dx \, dy$

11. $\displaystyle\int_0^{\ln 3} \int_0^1 xye^{xy^2} \, dy \, dx$

12. $\displaystyle\int_0^1 \int_0^2 \frac{y}{1 + x^2} \, dy \, dx$

Evaluate the indicated double integral over *R*.

13. $\displaystyle\iint_R xy^3 \, dA; \; R = \{(x, y): 0 \le x \le 1, -1 \le y \le 1\}$

14. $\displaystyle\iint_R (x^2 + y^2) \, dA;$

$R = \{(x, y): -1 \le x \le 1, 0 \le y \le 2\}$

15. $\displaystyle\iint_R \sin(x + y) \, dA;$

$R = \{(x, y): 0 \le x \le \pi/2, 0 \le y \le \pi/2\}$

16. $\displaystyle\iint_R xy\sqrt{1 + x^2} \, dA;$

$R = \{(x, y): 0 \le x \le \sqrt{3}, 1 \le y \le 2\}$

In Problems 17–20, sketch the solid whose volume is the indicated iterated integral.

17. $\displaystyle\int_0^1 \int_0^2 \frac{x}{2} \, dx \, dy$

18. $\displaystyle\int_0^1 \int_0^1 (2 - x - y) \, dy \, dx$

19. $\displaystyle\int_0^2 \int_0^2 (x^2 + y^2) \, dy \, dx$

20. $\displaystyle\int_0^2 \int_0^2 (4 - y^2) \, dy \, dx$

≈ In Problems 21–24, find the volume of the given solid. First, sketch the solid; then estimate its volume; finally determine its exact volume.

21. The solid under the plane $z = x + y + 1$ over $R = \{(x, y): 0 \le x \le 1, 1 \le y \le 3\}$.

22. The solid under the plane $z = 2x + 3y$ and over $R = \{(x, y): 1 \le x \le 2, 0 \le y \le 4\}$.

23. The solid between $z = x^2 + y^2 + 2$ and $z = 1$ and lying above $R = \{(x, y): -1 \le x \le 1, 0 \le y \le 1\}$.

24. The solid in the first octant enclosed by $z = 4 - x^2$ and $y = 2$.

25. Show that if $f(x, y) = g(x)h(y)$, then

$$\int_a^b \int_c^d f(x, y) \, dy \, dx = \left[\int_a^b g(x) \, dx\right] \left[\int_c^d h(y) \, dy\right]$$

26. Use Problem 25 to evaluate

$$\int_0^{\sqrt{\ln 2}} \int_0^1 \frac{xye^{x^2}}{1 + y^2} \, dy \, dx$$

27. Evaluate

$$\int_0^1 \int_0^1 xye^{x^2 + y^2} \, dy \, dx$$

28. Find the volume of the solid trapped between the surface $z = \cos x \cos y$ and the *xy*-plane, where $-\pi \le x \le \pi$, $-\pi \le y \le \pi$.

29. Evaluate each of the following.

(a) $\displaystyle\int_{-2}^2 \int_{-1}^1 |x^2 y^3| \, dy \, dx$    (b) $\displaystyle\int_{-2}^2 \int_{-1}^1 [\![x^2]\!] y^3 \, dy \, dx$

(c) $\displaystyle\int_{-2}^2 \int_{-1}^1 [\![x^2]\!] |y^3| \, dy \, dx$

30. Evaluate $\displaystyle\int_0^{\sqrt{3}} \int_0^1 \frac{8x}{(x^2 + y^2 + 1)^2} \, dy \, dx$. *Hint*: Reverse the order of integration.

31. Prove the Cauchy-Schwarz Inequality for Integrals

$$\left[\int_a^b f(x)g(x) \, dx\right]^2 \le \int_a^b f^2(x) \, dx \int_a^b g^2(x) \, dx$$

*Hint*: Consider the double integral of

$$F(x, y) = [f(x)g(y) - f(y)g(x)]^2$$

over the rectangle $R = \{(x, y): a \le x \le b, a \le y \le b\}$.

32. Suppose $f$ is increasing on $[a, b]$ and $\displaystyle\int_a^b f(x) \, dx > 0$. Prove that $\displaystyle\int_a^b xf(x) \, dx \Big/ \int_a^b f(x) \, dx > (a + b)/2$ and give a physical interpretation of this result. *Hint*: Let $F(x, y) = [y - x][f(y) - f(x)]$ and use the hint of Problem 31.

**Answers to Concepts Review:**  **1.** Iterated
**2.** $\displaystyle\int_{-1}^2 \left[\int_0^2 f(x, y) \, dy\right] dx; \int_0^2 \left[\int_{-1}^2 f(x, y) \, dx\right] dy$
**3.** Signed; plus; minus  **4.** Is below the *xy*-plane

## 16.3
## DOUBLE INTEGRALS OVER NONRECTANGULAR REGIONS

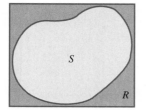

FIGURE 1

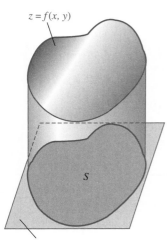

$z = f(x, y)$

$f(x, y) = 0$

FIGURE 2

Consider an arbitrary closed bounded set $S$ in the plane. Surround $S$ by a rectangle $R$ with sides parallel to the coordinate axes (Figure 1). Suppose $f(x, y)$ is defined on $S$ and define (or redefine, if necessary) $f(x, y) = 0$ on the part of $R$ outside of $S$ (Figure 2). We say that $f$ is integrable on $S$ if it is integrable on $R$ and write

$$\iint_S f(x, y)\, dA = \iint_R f(x, y)\, dA$$

We assert that the double integral on general sets $S$ is (1) linear, (2) additive on sets that overlap only on smooth curves, and (3) satisfies the comparison property (see Section 16.1).

**Evaluation of Double Integrals Over General Sets**    Sets with curved boundaries can be very complicated. For our purposes, it will be sufficient to consider so-called $x$-simple sets and $y$-simple sets (and finite unions of such sets). A set $S$ is $y$-simple if it is simple in the $y$-direction, meaning that a line in this direction intersects $S$ in a single interval (or point or not at all). Thus, a set $S$ is **$y$-simple** (Figure 3) if there are functions $\phi_1$ and $\phi_2$ on $[a, b]$ such that

$$S = \{(x, y):\ \phi_1(x) \le y \le \phi_2(x),\ a \le x \le b\}$$

A set $S$ is **$x$-simple** (Figure 4) if there are functions $\psi_1$ and $\psi_2$ on $[c, d]$ such that

$$S = \{(x, y):\ \psi_1(y) \le x \le \psi_2(y),\ c \le y \le d\}$$

Figure 5 exhibits a set that is neither $x$-simple nor $y$-simple.

Now suppose we wish to evaluate the double integral of a function $f(x, y)$ over a $y$-simple set $S$. We enclose $S$ in a rectangle $R$ (Figure 6) and

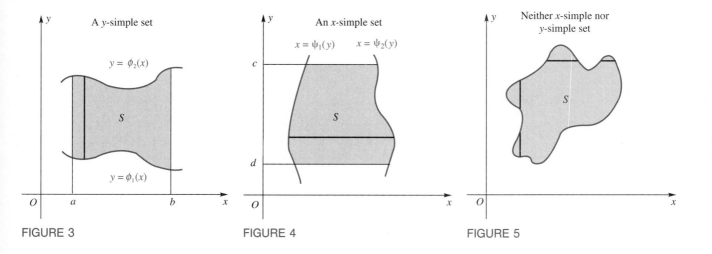

FIGURE 3

FIGURE 4

FIGURE 5

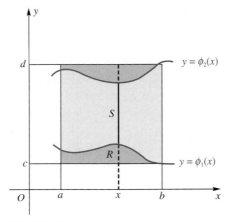

FIGURE 6

make $f(x, y) = 0$ outside of $S$. Then

$$\iint\limits_{S} f(x, y)\,dA = \iint\limits_{R} f(x, y)\,dA = \int_{a}^{b} \left[ \int_{c}^{d} f(x, y)\,dy \right] dx$$

$$= \int_{a}^{b} \left[ \int_{\phi_1(x)}^{\phi_2(x)} f(x, y)\,dy \right] dx$$

In summary,

$$\iint\limits_{S} f(x, y)\,dA = \int_{a}^{b} \int_{\phi_1(x)}^{\phi_2(x)} f(x, y)\,dy\,dx$$

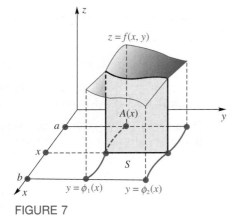

FIGURE 7

In the inner integration, $x$ is held fixed; thus, that integration is along the heavy line of Figure 6. This integration yields the area $A(x)$ of the cross section shown in Figure 7. Finally, $A(x)$ is integrated from $a$ to $b$.

If the set $S$ is $x$-simple (Figure 4), similar reasoning leads to the formula

$$\iint\limits_{S} f(x, y)\,dA = \int_{c}^{d} \int_{\psi_1(y)}^{\psi_2(y)} f(x, y)\,dx\,dy$$

If the set $S$ is neither $x$-simple nor $y$-simple (Figure 5), it can usually be considered as a union of pieces which have one or the other of these properties. For example the annulus of Figure 8 is not simple in either direction, but it is the union of the two $y$-simple sets $S_1$ and $S_2$. The integrals on these pieces can be calculated and added together to obtain the integral over $S$.

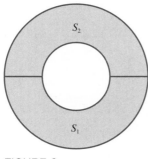

FIGURE 8

**EXAMPLES** For a little preliminary practice, we evaluate two iterated integrals, where the limits on the inner integral sign are variables.

**EXAMPLE 1**   Evaluate the iterated integral

$$\int_3^5 \int_{-x}^{x^2} (4x + 10y)\, dy\, dx$$

*Solution*   We first perform the inner integration with respect to $y$, temporarily thinking of $x$ as constant, and obtain

$$\int_3^5 \int_{-x}^{x^2} (4x + 10y)\, dy\, dx = \int_3^5 \left[ 4xy + 5y^2 \right]_{-x}^{x^2} dx$$

$$= \int_3^5 \left[ (4x^3 + 5x^4) - (-4x^2 + 5x^2) \right] dx$$

$$= \int_3^5 (5x^4 + 4x^3 - x^2)\, dx = \left[ x^5 + x^4 - \frac{x^3}{3} \right]_3^5$$

$$= 3393\frac{1}{3}$$  ■

Notice in iterated integrals that the outer integral always has constant limits.

**EXAMPLE 2**   Evaluate the iterated integral

$$\int_0^1 \int_0^{y^2} 2ye^x\, dx\, dy$$

*Solution*

$$\int_0^1 \int_0^{y^2} 2ye^x\, dx\, dy = \int_0^1 \left[ \int_0^{y^2} 2ye^x\, dx \right] dy$$

$$= \int_0^1 \left[ 2ye^x \right]_0^{y^2} dy = \int_0^1 (2ye^{y^2} - 2ye^0)\, dy$$

$$= \int_0^1 e^{y^2} d(y^2) - 2 \int_0^1 y\, dy$$

$$= \left[ e^{y^2} \right]_0^1 - 2 \left[ \frac{y^2}{2} \right]_0^1 = e - 1 - 2\left( \frac{1}{2} \right) = e - 2$$  ■

We turn to the problem of calculating volumes by means of iterated integrals.

**EXAMPLE 3**   Use double integration to find the volume of the tetrahedron bounded by the coordinate planes and the plane $3x + 6y + 4z - 12 = 0$.

*Solution*   Denote by $S$ the triangular region in the $xy$-plane that forms the base of the tetrahedron (Figure 9). We seek the volume of the solid under the surface $z = \frac{3}{4}(4 - x - 2y)$ and above the region $S$.

The given plane intersects the $xy$-plane in the line $x + 2y - 4 = 0$, a segment of which belongs to the boundary of $S$. Since this equation can

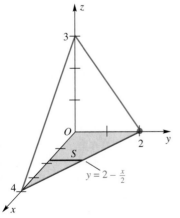

FIGURE 9

be written $y = 2 - x/2$ and $x = 4 - 2y$, $S$ can be thought of as the $y$-simple set

$$S = \left\{ (x, y): 0 \le x \le 4, 0 \le y \le 2 - \frac{x}{2} \right\}$$

or as the $x$-simple set

$$S = \{(x, y): 0 \le x \le 4 - 2y, 0 \le y \le 2\}$$

We will treat $S$ as a $y$-simple set; the final result would be the same either way, as you should verify.

The volume $V$ of the solid is

$$V = \iint_S \frac{3}{4}(4 - x - 2y)\, dA$$

In writing this as an iterated integral, we fix $x$ and integrate along a line (Figure 9) from $y = 0$ to $y = 2 - x/2$ and then integrate the result from $x = 0$ to $x = 4$. Thus,

$$
\begin{aligned}
V &= \int_0^4 \int_0^{2 - x/2} \frac{3}{4} (4 - x - 2y)\, dy\, dx \\
&= \int_0^4 \left[ \frac{3}{4} \int_0^{2 - x/2} (4 - x - 2y)\, dy \right] dx \\
&= \int_0^4 \frac{3}{4} \left[ 4y - xy - y^2 \right]_0^{2 - x/2} dx \\
&= \frac{3}{16} \int_0^4 (16 - 8x + x^2)\, dx \\
&= \frac{3}{16} \left[ 16x - 4x^2 + \frac{x^3}{3} \right]_0^4 = 4
\end{aligned}
$$

You may recall that the volume of a tetrahedron is one-third the area of the base times, the height. In the case at hand, $V = \frac{1}{3}(4)(3) = 4$. This confirms our answer. ∎

≈ **EXAMPLE 4** Find the volume of the solid in the first octant ($x \ge 0$, $y \ge 0$, $z \ge 0$) bounded by the circular paraboloid $z = x^2 + y^2$, the cylinder $x^2 + y^2 = 4$, and the coordinate planes (Figure 10).

*Solution* The region $S$ in the first quadrant of the $xy$-plane is bounded by a quarter of the circle $x^2 + y^2 = 4$ and the lines $x = 0$ and $y = 0$. Although $S$ can be thought of as either a $y$-simple or an $x$-simple region, we shall treat $S$ as the latter and write its boundary curves as $x = \sqrt{4 - y^2}$, $x = 0$, and $y = 0$. Thus,

$$S = \{(x, y): 0 \le x \le \sqrt{4 - y^2}, 0 \le y \le 2\}$$

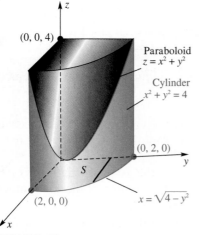

$z$

$(0, 0, 4)$

Paraboloid $z = x^2 + y^2$

Cylinder $x^2 + y^2 = 4$

$(0, 2, 0)$

$y$

$S$

$(2, 0, 0)$

$x = \sqrt{4 - y^2}$

$x$

FIGURE 10

Our goal is to calculate

$$V = \iint_S (x^2 + y^2)\, dA$$

by means of an iterated integral. This time we first fix $y$ and integrate along a line (Figure 10) from $x = 0$ to $x = \sqrt{4 - y^2}$, and then integrate the result from $y = 0$ to $y = 2$.

$$V = \iint_S (x^2 + y^2)\, dA = \int_0^2 \int_0^{\sqrt{4 - y^2}} (x^2 + y^2)\, dx\, dy$$

$$= \int_0^2 \left[ \frac{1}{3}(4 - y^2)^{3/2} + y^2 \sqrt{4 - y^2} \right] dy$$

By the trigonometric substitution $y = 2 \sin \theta$, the latter integral can be rewritten as

$$\int_0^{\pi/2} \left[ \frac{8}{3} \cos^3 \theta + 8 \sin^2 \theta \cos \theta \right] 2 \cos \theta\, d\theta$$

$$= \int_0^{\pi/2} \left[ \frac{16}{3} \cos^4 \theta + 16 \sin^2 \theta \cos^2 \theta \right] d\theta$$

$$= \frac{16}{3} \int_0^{\pi/2} \cos^2 \theta (1 - \sin^2 \theta + 3 \sin^2 \theta)\, d\theta$$

$$= \frac{16}{3} \int_0^{\pi/2} (\cos^2 \theta + 2 \sin^2 \theta \cos^2 \theta)\, d\theta$$

$$= \frac{16}{3} \int_0^{\pi/2} \left( \cos^2 \theta + \frac{1}{2} \sin^2 2\theta \right) d\theta$$

$$= \frac{16}{3} \int_0^{\pi/2} \left( \frac{1 + \cos 2\theta}{2} + \frac{1 - \cos 4\theta}{4} \right) d\theta = 2\pi$$

Is this answer reasonable? Note the volume of the complete quarter cylinder in Figure 10 is $\frac{1}{4} \pi r^2 h = \frac{1}{4} \pi (2^2)(4) = 4\pi$. One-half this number is certainly a reasonable value for the required volume.    ■

**EXAMPLE 5**  By changing the order of integration, evaluate

$$\int_0^4 \int_{x/2}^2 e^{y^2}\, dy\, dx$$

*Solution*  The inner integral cannot be evaluated as it stands because $e^{y^2}$ does not have an elementary antiderivative. However, we recognize that the given iterated integral is equal to

$$\iint_S e^{y^2}\, dA$$

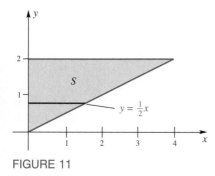

**FIGURE 11**

where $S = \{(x, y): x/2 \le y \le 2, 0 \le x \le 4\}$ (see Figure 11). If we write this double integral as an iterated integral with the $x$-integration performed first, we get

$$\int_0^2 \int_0^{2y} e^{y^2} \, dx \, dy = \int_0^2 \left[ xe^{y^2} \right]_0^{2y} dy$$

$$= \int_0^2 2ye^{y^2} \, dy = \left[ e^{y^2} \right]_0^2 = e^4 - 1 \quad \blacksquare$$

## CONCEPTS REVIEW

**1.** For an arbitrary set $S$, we define $\iint_S f(x, y) \, dA$ as $\iint_R f(x, y) \, dA$, where $R$ is _____ and $f(x, y) =$ _____ outside of $S$.

**2.** A set $S$ is called $y$-simple if there are functions $\phi_1$ and $\phi_2$ on $[a, b]$ such that $S = \{(x, y):$ _____$, a \le x \le b\}$.

**3.** If $S$ is a $y$-simple set, as in Question 2, then the double integral over $S$ can be written as an iterated integral, namely, $\iint_S f(x, y) \, dA =$ _____.

**4.** If $S$ is the triangle in the first quadrant bounded by $x + y = 1$, then $\iint_S 2x \, dA$ can be written as the iterated integral _____, which has value _____.

## PROBLEM SET 16.3

Evaluate the iterated integrals in Problems 1–12.

**1.** $\int_0^1 \int_0^{3x} x^2 \, dy \, dx$

**2.** $\int_1^2 \int_0^{x-1} y \, dy \, dx$

**3.** $\int_{-1}^3 \int_0^{3y} (x^2 + y^2) \, dx \, dy$

**4.** $\int_{-3}^1 \int_0^x (x^2 - y^3) \, dy \, dx$

**5.** $\int_1^3 \int_{-y}^{2y} xe^{y^3} \, dx \, dy$

**6.** $\int_1^5 \int_0^x \frac{3}{x^2 + y^2} \, dy \, dx$

**7.** $\int_{1/2}^1 \int_0^{2x} \cos(\pi x^2) \, dy \, dx$

**8.** $\int_0^{\pi/4} \int_{\sqrt{2}}^{\sqrt{2}\cos\theta} r \, dr \, d\theta$

**9.** $\int_0^{\pi/9} \int_{\pi/4}^{3r} \sec^2\theta \, d\theta \, dr$

**10.** $\int_0^{\pi/2} \int_0^{\sin y} e^x \cos y \, dx \, dy$

**11.** $\int_0^2 \int_0^{\sqrt{4 - x^2}} (x + y) \, dy \, dx$

**12.** $\int_{\pi/6}^{\pi/2} \int_0^{\sin\theta} 6r \cos\theta \, dr \, d\theta$

In Problems 13–18, evaluate the given double integral by changing it to an iterated integral.

**13.** $\iint_S xy \, dA$; $S$ is the region bounded by $y = x^2$ and $y = 1$.

**14.** $\iint_S (x + y) \, dA$; $S$ is the triangular region with vertices $(0, 0)$, $(0, 4)$, and $(1, 4)$.

**15.** $\iint_S (x^2 + 2y) \, dA$; $S$ is the region between $y = x^2$ and $y = \sqrt{x}$.

**16.** $\iint_S (x^2 - xy) \, dA$; $S$ is the region between $y = x$ and $y = 3x - x^2$.

**17.** $\iint_S \frac{2}{1 + x^2} \, dA$; $S$ is the triangular region with vertices at $(0, 0)$, $(2, 2)$, and $(0, 2)$.

**18.** $\iint_S x \, dA$; $S$ is the region between $y = x$ and $y = x^3$. (Note that $S$ has two parts.)

In Problems 19–30, sketch the indicated solid. Then find its volume by an iterated integration.

**19.** The tetrahedron bounded by the coordinate planes and the plane $z = 6 - 2x - 3y$.

**20.** The tetrahedron bounded by the coordinate planes and the plane $3x + 4y + z - 12 = 0$.

**21.** The wedge bounded by the coordinate planes and the planes $x = 5$ and $y + 2z - 4 = 0$.

**22.** The solid in the first octant bounded by the coordinate planes and the planes $2x + y - 4 = 0$ and $8x + y - 4z = 0$.

**23.** The solid in the first octant bounded by the surface $9x^2 + 4y^2 = 36$ and the plane $9x + 4y - 6z = 0$.

**24.** The solid in the first octant bounded by the surface $z = 9 - x^2 - y^2$ and the coordinate planes.

**25.** The solid in the first octant bounded by the cylinder $y = x^2$ and the planes $x = 0$, $z = 0$, and $y + z = 1$.

**26.** The solid bounded by the parabolic cylinder $x^2 = 4y$ and the planes $z = 0$ and $5y + 9z - 45 = 0$.

**27.** The solid in the first octant bounded by the cylinder $z = \tan x^2$ and the planes $x = y$, $x = 1$, and $y = 0$.

**28.** The solid in the first octant bounded by the surface $z = e^{x-y}$, the plane $x + y = 1$, and the coordinate planes.

**29.** The solid in the first octant bounded by the surface $9z = 36 - 9x^2 - 4y^2$ and the coordinate planes.

**30.** The solid in the first octant bounded by the circular cylinders $x^2 + z^2 = 16$ and $y^2 + z^2 = 16$, and the coordinate planes.

In Problems 31–36, write the given iterated integral as an iterated integral with the order of integration interchanged. *Hint*: Begin by sketching a region $S$, as in Example 5.

**31.** $\displaystyle\int_0^1 \int_0^x f(x, y)\, dy\, dx$ **32.** $\displaystyle\int_0^2 \int_{y^2}^{2y} f(x, y)\, dx\, dy$

**33.** $\displaystyle\int_0^1 \int_{x^2}^{x^{1/4}} f(x, y)\, dy\, dx$ **34.** $\displaystyle\int_{1/2}^1 \int_{x^3}^x f(x, y)\, dy\, dx$

**35.** $\displaystyle\int_0^1 \int_{-y}^y f(x, y)\, dx\, dy$ **36.** $\displaystyle\int_{-1}^0 \int_{-\sqrt{y+1}}^{\sqrt{y+1}} f(x, y)\, dx\, dy$

**37.** Evaluate $\iint_S xy^2\, dA$, where $S$ is the region shown in Figure 12.

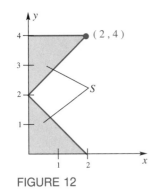

FIGURE 12

**38.** Evaluate $\iint_S xy\, dA$, where $S$ is the region in Figure 13.

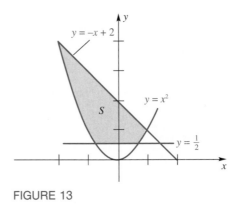

FIGURE 13

**39.** Evaluate $\iint_S \sin(y^3)\, dA$, where $S$ is the region bounded by $y = \sqrt{x}$, $y = 2$, and $x = 0$. *Hint*: If one order of integration does not work, try the other.

**40.** Evaluate $\iint_S \sin(xy^2)\, dA$, where $S$ is the annulus $\{(x, y): 1 \le x^2 + y^2 \le 4\}$. *Hint*: Done without thinking, this problem is hard; using symmetry it is trivial.

**41.** Evaluate $\iint_S (x^2 + x^4y)\, dA$, where $S = \{(x, y): 1 \le x^2 + y^2 \le 4\}$. *Hint*: Use symmetry to reduce the problem to evaluating $4[\iint_{S_1} x^2\, dA + \iint_{S_2} x^2\, dA]$, where $S_1$ and $S_2$ are as in Figure 14.

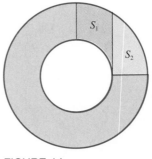

FIGURE 14

**42.** Evaluate $\iint_S x^2\, dA$, where $S$ is the region between the ellipse $x^2 + 2y^2 = 4$ and the circle $x^2 + y^2 = 4$.

---

**Answers to Concepts Review:** **1.** A rectangle containing $S$; 0

**2.** $\phi_1(x) \le y \le \phi_2(x)$ **3.** $\displaystyle\int_a^b \int_{\phi_1(x)}^{\phi_2(x)} f(x, y)\, dy\, dx$

**4.** $\displaystyle\int_0^1 \int_0^{1-x} 2x\, dy\, dx; \frac{1}{3}$

## 16.4
## DOUBLE INTEGRALS IN POLAR COORDINATES

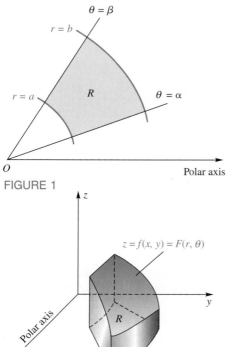

FIGURE 1

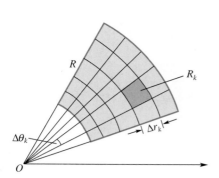

FIGURE 2

Certain curves in the plane, such as circles, cardioids, and roses, are easier to describe in terms of polar coordinates than in Cartesian (rectangular) coordinates. Thus, we can expect that double integrals over regions enclosed by such curves are more easily evaluated using polar coordinates.

Let $R$ have the shape shown in Figure 1, which we call a *polar rectangle* and will describe analytically in a moment. Let $z = f(x, y)$ determine a surface over $R$ and suppose $f$ is continuous and nonnegative. Then the volume $V$ of the solid under this surface and above $R$ (Figure 2) is given by

$$(1) \qquad V = \iint\limits_{R} f(x, y)\, dA$$

In polar coordinates, a polar rectangle $R$ has the form

$$R = \{(r, \theta): a \le r \le b, \alpha \le \theta \le \beta\}$$

where $a \ge 0$ and $\beta - \alpha \le 2\pi$. Also, the equation of the surface can be written as

$$z = f(x, y) = f(r \cos \theta, r \sin \theta) = F(r, \theta)$$

We are going to calculate the volume $V$ in a new way using polar coordinates.

Partition $R$ into smaller polar rectangles $R_1, R_2, \ldots, R_n$ by means of a polar grid and let $\Delta r_k$ and $\Delta \theta_k$ denote the dimensions of the typical piece $R_k$, as shown in Figure 3. The area $A(R_k)$ is given by (see Problem 30)

$$A(R_k) = \bar{r}_k\, \Delta r_k\, \Delta \theta_k$$

where $\bar{r}_k$ is the average radius of $R_k$. Thus,

$$V \approx \sum_{k=1}^{n} F(\bar{r}_k, \bar{\theta}_k) \bar{r}_k\, \Delta r_k\, \Delta \theta_k$$

When we take the limit as the norm of the partition approaches zero, we ought to get the actual volume. This limit is a double integral.

$$(2) \qquad V = \iint\limits_{R} F(r, \theta) r\, dr\, d\theta = \iint\limits_{R} f(r \cos \theta, r \sin \theta) r\, dr\, d\theta$$

Now we have two expressions for $V$, namely, (1) and (2). Equating them yields

$$\boxed{\iint\limits_{R} f(x, y)\, dA = \iint\limits_{R} f(r \cos \theta, r \sin \theta) r\, dr\, d\theta}$$

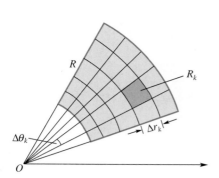

FIGURE 3

The boxed result was derived under the assumption that $f$ was nonnegative, but it is valid for very general functions, in particular for continuous functions of arbitrary sign.

**Iterated Integrals**    The result announced above becomes useful when we write the polar double integral as an iterated integral, a statement we now illustrate.

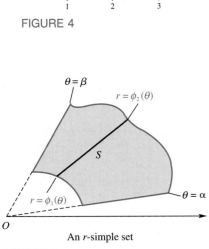

FIGURE 4

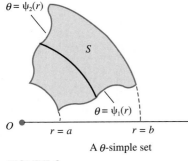

An r-simple set

FIGURE 5

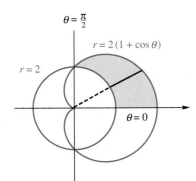

A θ-simple set

FIGURE 6

**EXAMPLE 1**  Find the volume $V$ of the solid above the polar rectangle (Figure 4).

$$R = \left\{ (r, \theta): 1 \le r \le 3, 0 \le \theta \le \frac{\pi}{4} \right\}$$

and under the surface $z = e^{x^2 + y^2}$.

*Solution*  Since $x^2 + y^2 = r^2$,

$$V = \iint\limits_R e^{x^2 + y^2} \, dA$$

$$= \int_0^{\pi/4} \left[ \int_1^3 e^{r^2} r \, dr \right] d\theta$$

$$= \int_0^{\pi/4} \left[ \frac{1}{2} e^{r^2} \right]_1^3 d\theta$$

$$= \int_0^{\pi/4} \frac{1}{2} (e^9 - e) \, d\theta = \frac{\pi}{8} (e^9 - e) \approx 3181$$

Without the help of polar coordinates, we could not have done this problem. Note how the extra factor of $r$ was just what we needed in order to antidifferentiate $e^{r^2}$.  ∎

**General Regions**  Recall how we extended the double integral over an ordinary rectangle $R$ to the integral over a general set $S$. We simply enclosed $S$ in a rectangle and gave the function to be integrated the value zero outside of $S$. We can do the same thing for double polar integrals, except that we use polar rectangles rather than ordinary rectangles. Omitting the details, we simply assert that the boxed result stated earlier holds for general sets $S$.

Of special interest for polar integration are what we shall call *r*-simple and *θ*-simple sets. Call a set $S$ an **r-simple** set if it has the form (Figure 5)

$$S = \{(r, \theta): \phi_1(\theta) \le r \le \phi_2(\theta), \alpha \le \theta \le \beta\}$$

and call it **θ-simple** if it has the form (Figure 6)

$$S = \{(r, \theta): a \le r \le b, \psi_1(r) \le \theta \le \psi_2(r)\}$$

**EXAMPLE 2**  Evaluate

$$\iint\limits_R y \, dA$$

where $S$ is the region in the first quadrant that is outside the circle $r = 2$ and inside the cardioid $r = 2(1 + \cos \theta)$ (see Figure 7).

*Solution*  Since $S$ is an *r*-simple set, we write the given integral as an iterated polar integral, with $r$ as the first variable of integration. In this inner integration, $\theta$ is held fixed; the integration is along the heavy line of Figure 7 from $r = 2$ to $r = 2(1 + \cos \theta)$.

FIGURE 7

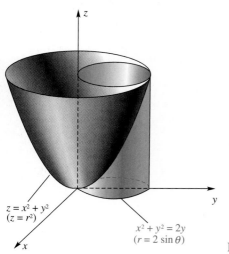

$z = x^2 + y^2$
$(z = r^2)$

$x^2 + y^2 = 2y$
$(r = 2 \sin \theta)$

**FIGURE 8**

$$\iint\limits_{S} y \, dA = \int_0^{\pi/2} \int_2^{2(1 + \cos \theta)} (r \sin \theta) r \, dr \, d\theta$$

$$= \int_0^{\pi/2} \left[ \frac{r^3}{3} \sin \theta \right]_2^{2(1 + \cos \theta)} d\theta$$

$$= \frac{8}{3} \int_0^{\pi/2} [(1 + \cos \theta)^3 \sin \theta - \sin \theta] \, d\theta$$

$$= \frac{8}{3} \left[ -\frac{1}{4} (1 + \cos \theta)^4 + \cos \theta \right]_0^{\pi/2}$$

$$= \frac{8}{3} \left[ -\frac{1}{4} + 0 - (-4 + 1) \right] = \frac{22}{3} \qquad \blacksquare$$

**EXAMPLE 3**   Find the volume of the solid under the surface $z = x^2 + y^2$, above the *xy*-plane, and inside the cylinder $x^2 + y^2 = 2y$ (Figure 8).

*Solution*   From symmetry, we can double the volume in the first octant. When we use $x = r \cos \theta$ and $y = r \sin \theta$, the equation of the surface becomes $z = r^2$ and that of the cylinder $r = 2 \sin \theta$. Let $S$ denote the region shown in Figure 9. The required volume $V$ is given by

$$V = 2 \iint\limits_{S} (x^2 + y^2) \, dA = 2 \int_0^{\pi/2} \int_0^{2 \sin \theta} r^2 r \, dr \, d\theta$$

$$= 2 \int_0^{\pi/2} \left[ \frac{r^4}{4} \right]_0^{2 \sin \theta} d\theta = 8 \int_0^{\pi/2} \sin^4 \theta \, d\theta$$

$$= 8 \left( \frac{3}{8} \cdot \frac{\pi}{2} \right) = \frac{3\pi}{2}$$

The last integral was evaluated by means of Formula 113 in the table of integrals at the end of the book. $\qquad \blacksquare$

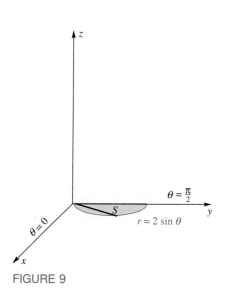

$\theta = \frac{\pi}{2}$

$\theta = 0$

$r = 2 \sin \theta$

**FIGURE 9**

**A Probability Integral**   In the theory of probability and statistics, it is essential to be able to calculate the improper integral

$$\int_{-\infty}^{\infty} e^{-x^2} \, dx = 2 \int_0^{\infty} e^{-x^2} \, dx$$

**EXAMPLE 4**   Show that $I = \displaystyle\int_0^{\infty} e^{-x^2} \, dx = \sqrt{\pi}/2$.

*Solution*   We are going to sneak up on this problem in a roundabout, but decidedly ingenious, way. First recall that

$$I = \int_0^{\infty} e^{-x^2} \, dx = \lim_{b \to \infty} \int_0^{b} e^{-x^2} \, dx$$

 **Common Sense**

To estimate the volume in Example 3, note that the height of the cylinder displayed in Figure 8 is 4 (let $x = 0$ and $y = 2$ in $z = x^2 + y^2$). Thus, the desired volume is somewhat less than half the volume of a cylinder of radius 1 and height 4—that is, it is less than $(\frac{1}{2})\pi(1^2)4 = 2\pi$. The answer we got—namely, $3\pi/2$—is reasonable.

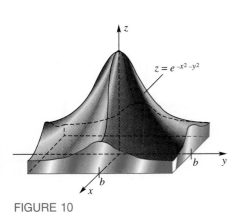

FIGURE 10

Now let $V_b$ be the volume of the solid (Figure 10) that lies under the surface $z = e^{-x^2 - y^2}$ and above the square with vertices $(\pm b, \pm b)$. Then,

$$V_b = \int_{-b}^{b} \int_{-b}^{b} e^{-x^2 - y^2} \, dy \, dx = \int_{-b}^{b} e^{-x^2} \left[ \int_{-b}^{b} e^{-y^2} \, dy \right] dx$$

$$= \int_{-b}^{b} e^{-x^2} \, dx \int_{-b}^{b} e^{-y^2} \, dy = \left[ \int_{-b}^{b} e^{-x^2} \, dx \right]^2 = 4 \left[ \int_{0}^{b} e^{-x^2} \, dx \right]^2$$

It follows that the volume of the region under $z = e^{-x^2 - y^2}$ above the whole $xy$-plane is

$$(1) \qquad V = \lim_{b \to \infty} V_b = \lim_{b \to \infty} 4 \left[ \int_0^b e^{-x^2} \, dx \right]^2$$

$$= 4 \left[ \int_0^\infty e^{-x^2} \, dx \right]^2 = 4I^2$$

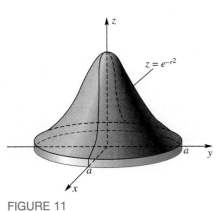

FIGURE 11

On the other hand, we can also calculate $V$ using polar coordinates. Here, $V$ is the limit as $a \to \infty$ of $V_a$, the volume of the solid under the surface $z = e^{-x^2 - y^2} = e^{-r^2}$ above the circular region of radius $a$ centered at the origin (Figure 11).

$$(2) \quad V = \lim_{a \to \infty} V_a = \lim_{a \to \infty} \int_0^{2\pi} \int_0^a e^{-r^2} r \, dr \, d\theta = \lim_{a \to \infty} \int_0^{2\pi} \left[ -\tfrac{1}{2} e^{-r^2} \right]_0^a d\theta$$

$$= \lim_{a \to \infty} \tfrac{1}{2} \int_0^{2\pi} [1 - e^{-a^2}] \, d\theta = \lim_{a \to \infty} \pi [1 - e^{-a^2}] = \pi$$

Equating the two values obtained for $V$ in (1) and (2) yields $4I^2 = \pi$, or $I = \tfrac{1}{2}\sqrt{\pi}$, as desired. ∎

## CONCEPTS REVIEW

**1.** A polar rectangle $R$ has the form $R = \{(r, \theta): \underline{\hspace{2cm}}\}$.

**2.** The $dy \, dx$ of integrals in Cartesian (rectangular) coordinates transforms to $\underline{\hspace{2cm}}$ for integrals in polar coordinates.

**3.** The integral $\iint_S (x^2 + y^2) \, dA$, where $S$ is the semicircle bounded by $y = \sqrt{4 - x^2}$ and $y = 0$, becomes the iterated integral $\underline{\hspace{2cm}}$ in polar coordinates.

**4.** The value of the integral in Question 3 is $\underline{\hspace{2cm}}$.

## PROBLEM SET 16.4

Evaluate the following iterated integrals.

**1.** $\displaystyle\int_0^{\pi/2} \int_0^{\cos \theta} r^2 \sin \theta \, dr \, d\theta$

**2.** $\displaystyle\int_0^{\pi/2} \int_0^{\sin \theta} r \, dr \, d\theta$

**3.** $\displaystyle\int_0^{\pi} \int_0^{\sin \theta} r^2 \, dr \, d\theta$

**4.** $\displaystyle\int_0^{\pi} \int_0^{1 - \cos \theta} r \sin \theta \, dr \, d\theta$

In Problems 5–10, find the area of the given region $S$ by calculating $\iint_S r\, dr\, d\theta$. Be sure to make a sketch of the region first.

**5.** $S$ is the region inside the circle $r = 4\cos\theta$ and outside the circle $r = 2$.

**6.** $S$ is the smaller region bounded by $\theta = \pi/6$ and $r = 4\sin\theta$.

**7.** $S$ is one leaf of the four-leaved rose $r = a\sin 2\theta$.

**8.** $S$ is the region inside the cardioid $r = 6 - 6\sin\theta$.

**9.** $S$ is the region inside the larger loop of the limaçon $r = 2 - 4\sin\theta$.

**10.** $S$ is the region outside the circle $r = 2$ and inside the lemniscate $r^2 = 9\cos 2\theta$.

In Problems 11–18, evaluate by using polar coordinates. Sketch the region of integration first.

**11.** $\iint_S e^{x^2 + y^2}\, dA$, where $S$ is the region enclosed by $x^2 + y^2 = 4$.

**12.** $\iint_S \sqrt{4 - x^2 - y^2}\, dA$, where $S$ is the first quadrant sector of the circle $x^2 + y^2 = 4$ between $y = 0$ and $y = x$.

**13.** $\iint_S \dfrac{1}{4 + x^2 + y^2}\, dA$, where $S$ is as in Problem 12.

**14.** $\iint_S y\, dA$, where $S$ is the first quadrant polar rectangle inside $x^2 + y^2 = 4$ and outside $x^2 + y^2 = 1$.

**15.** $\displaystyle\int_0^1 \int_0^{\sqrt{1 - x^2}} (4 - x^2 - y^2)^{-1/2}\, dy\, dx$

**16.** $\displaystyle\int_0^1 \int_0^{\sqrt{1 - y^2}} \sin(x^2 + y^2)\, dx\, dy$

**17.** $\displaystyle\int_0^1 \int_x^1 x^2\, dy\, dx$

**18.** $\displaystyle\int_1^2 \int_0^{\sqrt{2x - x^2}} (x^2 + y^2)^{-1/2}\, dy\, dx$

≈ **19.** Find the volume of the solid in the first octant under the paraboloid $z = x^2 + y^2$ and inside the cylinder $x^2 + y^2 = 9$ by using polar coordinates.

≈ **20.** Find the volume of the solid bounded above by $2x^2 + 2y^2 + z^2 = 18$, below by $z = 0$, and laterally by $x^2 + y^2 = 4$ using polar coordinates.

**21.** Switch to rectangular coordinates and then evaluate.

$$\int_{3\pi/4}^{4\pi/3} \int_0^{-5\sec\theta} r^3 \sin^2\theta\, dr\, d\theta$$

**22.** Let $V = \iint_S \sin\sqrt{x^2 + y^2}\, dA$ and
$W = \iint_S \left|\sin\sqrt{x^2 + y^2}\right| dA$, where $S$ is the region inside the circle $x^2 + y^2 = 4\pi^2$.

(a) Without calculation, determine the sign of $V$.
(b) Evaluate $V$.
(c) Evaluate $W$.

**23.** The centers of two balls of radius $a$ are $2b$ units apart with $b \le a$. Find the volume of their intersection in terms of $d = a - b$

**24.** The depth (in feet) of water distributed by a rotating lawn sprinkler in an hour is $ke^{-r/10}$, $0 \le r \le 10$, where $r$ is the distance from the sprinkler and $k$ is a constant. Determine $k$ if 100 cubic feet of water are distributed in 1 hour.

**25.** Find the volume of the solid cut from the ball $r^2 + z^2 \le a^2$ by the cylinder $r = a\sin\theta$.

**26.** Find the volume of the wedge cut from a tall right cylinder of radius $a$ by a plane through a diameter of its base and making an angle $\alpha$ ($0 < \alpha < \pi/2$) with the base (compare Problem 33, Section 6.2).

**27.** Consider the ring $A$ of height $2b$ obtained from a ball of radius $a$ when a hole of radius $c$ ($c < a$) is bored through the center of the ball (left part of Figure 12). Show that the volume of $A$ is $4\pi b^3/3$, which is remarkable for two reasons. It is independent of the radius $a$ and it is the same as the volume of a sphere of radius $b$ (right part of Figure 12).

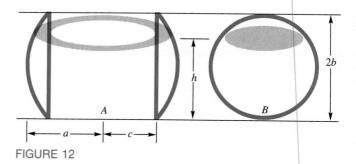

FIGURE 12

**28.** There is a simple explanation for the remarkable result in Problem 27. Show that the cross-sectional areas of the two solids (ring $A$ and ball $B$) in Figure 12 determined by a horizontal plane at height $h$ are the same size, implying that the solids below this plane have the same volume. Find a nice formula for this volume in terms of $h$ and $b$.

**29.** Show that

$$\int_0^\infty \int_0^\infty \frac{1}{(1 + x^2 + y^2)^2}\, dy\, dx = \frac{\pi}{4}$$

**30.** Recall the formula $A = \frac{1}{2}r^2\theta$ for the area of the sector of a circle of radius $r$ and central angle $\theta$ radians (Section 12.8). Use this to obtain the formula

$$A = \frac{r_1 + r_2}{2}(r_2 - r_1)(\theta_2 - \theta_1)$$

for the area of the polar rectangle $\{(r, \theta): r_1 \le r \le r_2, \theta_1 \le \theta \le \theta_2\}$.

---

**Answers to Concepts Review: 1.** $a \le r \le b; \alpha \le \theta \le \beta$
**2.** $r\,dr\,d\theta$ **3.** $\int_0^\pi \int_0^2 r^3\,dr\,d\theta$ **4.** $4\pi$

---

## 16.5
## APPLICATIONS OF
## DOUBLE INTEGRALS

The most obvious application of double integrals is in calculating volumes of solids. That use of double integrals has been amply illustrated, so now we turn to other applications (mass, center of mass, moment of inertia, and radius of gyration).

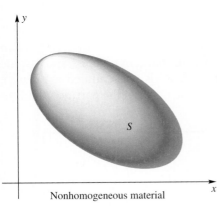

FIGURE 1

Nonhomogeneous material

Consider a flat sheet that is so thin that we may consider it to be two-dimensional. In Section 6.6, we called such a sheet a lamina, but there we considered only laminas of constant density. Here we wish to study laminas of variable density—that is, laminas made of nonhomogeneous material (Figure 1).

Suppose that a lamina covers a region $S$ in the $xy$-plane and let the density (mass per unit area) at $(x, y)$ be denoted by $\delta(x, y)$. Partition $S$ into small rectangles $R_1, R_2, \ldots, R_k$ as shown in Figure 2. Pick a point $(\overline{x}_k, \overline{y}_k)$ on $R_k$. Then the mass of $R_k$ is approximately $\delta(\overline{x}_k, \overline{y}_k)A(R_k)$ and the total mass of the lamina is approximately

$$m \approx \sum_{k=1}^n \delta(\overline{x}_k, \overline{y}_k)A(R_k)$$

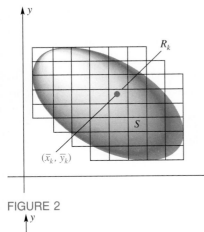

FIGURE 2

The actual mass $m$ is obtained by taking the limit of the above expression as the norm of the partition approaches zero which is, of course, a double integral.

$$m = \iint_S \delta(x, y)\,dA$$

**EXAMPLE 1** A lamina with density $\delta(x, y) = xy$ is bounded by the $x$-axis, the line $x = 8$, and the curve $y = x^{2/3}$ (Figure 3). Find its total mass.

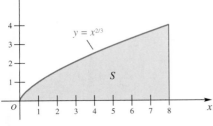

FIGURE 3

**Solution**

$$m = \iint_S xy\,dA = \int_0^8 \int_0^{x^{2/3}} xy\,dy\,dx$$

$$= \int_0^8 \left[\frac{xy^2}{2}\right]_0^{x^{2/3}}\,dx = \frac{1}{2}\int_0^8 x^{7/3}\,dx$$

$$= \frac{1}{2}\left[\frac{3}{10}x^{10/3}\right]_0^8 = \frac{768}{5} = 153.6 \qquad \blacksquare$$

**Center of Mass**  We suggest that you review the concept of center of mass from Section 6.6. There we learned that if $m_1, m_2, \ldots, m_n$ is a collection of point masses situated at $(x_1, y_1), (x_2, y_2), \ldots, (x_n, y_n)$, respectively, in the plane, then the total moments with respect to the $y$-axis and the $x$-axis are given by

$$M_y = \sum_{k=1}^{n} x_k m_k \qquad M_x = \sum_{k=1}^{n} y_k m_k$$

Moreover, the coordinates $(\bar{x}, \bar{y})$ of the center of mass (balance point) are

$$\bar{x} = \frac{M_y}{m} = \frac{\sum_{k=1}^{n} x_k m_k}{\sum_{k=1}^{n} m_k} \qquad \bar{y} = \frac{M_x}{m} = \frac{\sum_{k=1}^{n} y_k m_k}{\sum_{k=1}^{n} m_k}$$

Consider now a lamina of variable density $\delta(x, y)$ covering a region $S$ in the $xy$-plane, as in Figure 1. Partition this lamina as in Figure 2 and assume as an approximation that the mass of each $R_k$ is concentrated at $(\bar{x}_k, \bar{y}_k)$, $k = 1, 2, \ldots, n$. Finally, take the limit as the norm of the partition tends to zero. This leads to the formulas

$$\bar{x} = \frac{M_y}{m} = \frac{\iint\limits_{S} x\delta(x, y)\,dA}{\iint\limits_{S} \delta(x, y)\,dA} \qquad \bar{y} = \frac{M_x}{m} = \frac{\iint\limits_{S} y\delta(x, y)\,dA}{\iint\limits_{S} \delta(x, y)\,dA}$$

**EXAMPLE 2**  Find the center of mass of the lamina of Example 1.

**Solution**  In Example 1, we showed that the mass $m$ of this lamina is $\frac{768}{5}$. The moments $M_y$ and $M_x$ with respect to the $y$-axis and $x$-axis, respectively, are

$$M_y = \iint\limits_{S} x\delta(x, y)\,dA = \int_0^8 \int_0^{x^{2/3}} x^2 y \, dy \, dx$$

$$= \frac{1}{2}\int_0^8 x^{10/3}\,dx = \frac{12{,}288}{13} \approx 945.23$$

$$M_x = \iint\limits_{S} y\delta(x, y)\,dA = \int_0^8 \int_0^{x^{2/3}} xy^2 \, dy \, dx$$

$$= \frac{1}{3}\int_0^8 x^3\,dx = \frac{1024}{3} \approx 341.33$$

We conclude that

$$\bar{x} = \frac{M_y}{m} = 6\tfrac{2}{13} \approx 6.15, \qquad \bar{y} = \frac{M_x}{m} = 2\tfrac{2}{9} \approx 2.22$$

Note the $(\bar{x}, \bar{y})$ is in the upper right portion of $S$; but that is to be expected since a lamina with density $\delta(x, y) = xy$ gets heavier as the distance from the $x$- and $y$-axes increases. ∎

**EXAMPLE 3** Find the center of mass of a lamina in the shape of a quarter-circle of radius $a$ whose density is proportional to the distance from the center of the circle (Figure 4).

*Solution* By hypothesis, $\delta(x, y) = k\sqrt{x^2 + y^2}$, where $k$ is a constant. The shape of $S$ suggests the use of polar coordinates.

$$m = \iint_S k\sqrt{x^2 + y^2}\, dA = k \int_0^{\pi/2} \int_0^a rr\, dr\, d\theta$$

$$= k \int_0^{\pi/2} \frac{a^3}{3}\, d\theta = \frac{k\pi a^3}{6}$$

Also,

$$M_y = \iint_S xk\sqrt{x^2 + y^2}\, dA = k \int_0^{\pi/2} \int_0^a (r \cos\theta)r^2\, dr\, d\theta$$

$$= k \int_0^{\pi/2} \frac{a^4}{4} \cos\theta\, d\theta = \left[ \frac{ka^4}{4} \sin\theta \right]_0^{\pi/2} = \frac{ka^4}{4}$$

We conclude that

$$\bar{x} = \frac{M_y}{m} = \frac{ka^4/4}{k\pi a^3/6} = \frac{3a}{2\pi}$$

Because of the symmetry of the lamina, we recognize that $\bar{y} = \bar{x}$, so no further calculation is needed. ∎

A perceptive reader might well ask a question at this point. What if a lamina is homogeneous; that is, what if $\delta(x, y) = k$, a constant? Will the formulas derived in this section, which involve double integrals, agree with those of Section 6.6, which involved only single integrals? The answer is yes. To give a partial justification, consider calculating $M_y$ for a $y$-simple region $S$ (Figure 5)

$$M_y = \iint_S xk\, dA = k \int_a^b \int_{\phi_1(x)}^{\phi_2(x)} x\, dy\, dx = k \int_a^b x[\phi_2(x) - \phi_1(x)]\, dx$$

The single integral on the right is the one given in Section 6.6.

**Moment of Inertia** From physics, we learn that the kinetic energy, $KE$, of a particle of mass $m$ and velocity $v$, moving in a straight line, is

$$(1) \qquad\qquad KE = \tfrac{1}{2}mv^2$$

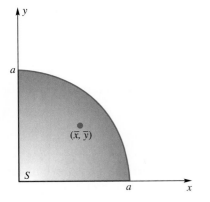

FIGURE 4

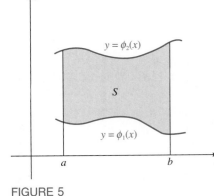

FIGURE 5

If instead of moving in a straight line, the particle rotates about an axis with an angular velocity of $\omega$ radians per unit of time, its linear velocity is $v = r\omega$, where $r$ is the radius of its circular path. When we substitute this in (1), we obtain

$$KE = \tfrac{1}{2}(r^2m)\omega^2$$

The expression $r^2m$ is called the **moment of inertia** of the particle and is denoted by $I$. Thus, for a rotating particle,

(2)
$$KE = \tfrac{1}{2}I\omega^2$$

We conclude from (1) and (2) that the moment of inertia of a body in circular motion plays a role similar to the mass of a body in linear motion.

For a system of $n$ particles in a plane with masses $m_1, m_2, \ldots, m_n$ and at distances $r_1, r_2, \ldots, r_n$ from line $L$, the moment of inertia of the system about $L$ is defined to be

$$I = m_1 r_1^2 + m_2 r_2^2 + \cdots + m_n r_n^2 = \sum_{k=1}^{n} m_k r_k^2$$

In other words, we add the moments of inertia of the individual particles.

Now consider a lamina with density $\delta(x, y)$ covering a region $S$ of the $xy$-plane (Figure 1). If we partition $S$ as in Figure 2, approximate the moments of inertia of each piece $R_k$, add up, and take the limit, we are led to the following formulas. The **moments of inertia** (also called the second moments) of the lamina about the $x$-, $y$-, and $z$-axes are given by

$$I_x = \iint\limits_{S} y^2\delta(x, y)\,dA \qquad I_y = \iint\limits_{S} x^2\delta(x, y)\,dA$$

$$I_z = \iint\limits_{S} (x^2 + y^2)\delta(x, y)\,dA = I_x + I_y$$

**EXAMPLE 4**  Find the moments of inertia about the $x$-, $y$-, and $z$-axes of the lamina of Example 1.

*Solution*

$$I_x = \iint\limits_{S} xy^3\,dA = \int_0^8 \int_0^{x^{2/3}} xy^3\,dy\,dx = \frac{1}{4}\int_0^8 x^{11/3}\,dx = \frac{6144}{7} \approx 877.71$$

$$I_y = \iint\limits_{S} x^3y\,dA = \int_0^8 \int_0^{x^{2/3}} x^3y\,dy\,dx = \frac{1}{2}\int_0^8 x^{13/3}\,dx = 6144$$

$$I_z = I_x + I_y = \frac{49{,}152}{7} \approx 7021.71$$

■

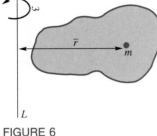

FIGURE 6

Consider the problem of replacing a general mass system of total mass $m$ by a single point mass $m$ with the same moment of inertia $I$ with respect to a line $L$ (Figure 6). How far should this point be from $L$? The answer is $\bar{r}$ where $m\bar{r}^2 = I$. The number

$$\bar{r} = \sqrt{\frac{I}{m}}$$

is called the **radius of gyration** of the system. Thus the kinetic energy of the system rotating about $L$ with angular velocity $\omega$ is

$$KE = \tfrac{1}{2}m\bar{r}^2\omega^2$$

## CONCEPTS REVIEW

**1.** If the density at $(x, y)$ is $x^2y^4$, then the mass $m$ of the lamina $S$ is given by $m =$ _____.

**2.** The $y$-coordinate of the center of mass of the lamina of Question 1 is given by $\bar{y} =$ _____.

**3.** The moment of inertia with respect to the $y$-axis of the lamina $S$ of Question 1 is given by $I_y =$ _____.

**4.** If $S = \{(x, y): 0 \le x \le 1, 0 \le y \le 1\}$, then geometric reasoning says that both $\bar{x}$ and $\bar{y}$ are _____ than $\tfrac{1}{2}$.

## PROBLEM SET 16.5

In Problems 1–8, find the mass $m$ and center of mass $(\bar{x}, \bar{y})$ of the lamina bounded by the given curves and with the indicated density.

**1.** $x = 0, x = 4, y = 0, y = 3; \delta(x, y) = y + 1$

**2.** $y = 0, y = \sqrt{4 - x^2}; \delta(x, y) = y$

**3.** $y = 0, y = \sin x, 0 \le x \le \pi; \delta(x, y) = y$

**4.** $y = 1/x, y = x, y = 0, x = 2; \delta(x, y) = x$

**5.** $y = e^{-x}, y = 0, x = 0, x = 1; \delta(x, y) = y^2$

**6.** $y = e^x, y = 0, x = 0, x = 1; \delta(x, y) = 2 - x + y$

**7.** $r = 2 \sin \theta; \delta(r, \theta) = r$

**8.** $r = 1 + \cos \theta; \delta(r, \theta) = r$

In Problems 9–12, find the moments of inertia $I_x, I_y,$ and $I_z$ for the lamina bounded by the given curves and with the indicated density $\delta$.

**9.** $y = \sqrt{x}, x = 9, y = 0; \delta(x, y) = x + y$

**10.** $y = x^2, y = 4; \delta(x, y) = y$

**11.** Square with vertices $(0, 0), (0, a), (a, a), (a, 0)$; $\delta(x, y) = x + y$

**12.** Triangle with vertices $(0, 0), (0, a), (a, 0); \delta(x, y) = x^2 + y^2$

**13.** Find the radius of gyration of the lamina of Problem 11 with respect to the $x$-axis.

**14.** Find the radius of gyration of the lamina of Problem 12 with respect to the $y$-axis.

**15.** Find the moment of inertia and radius of gyration of a homogeneous ($\delta$ a constant) circular lamina of radius $a$ with respect to a diameter.

**16.** Show that the moment of inertia of a homogeneous rectangular lamina with sides of length $a$ and $b$ about a perpendicular axis through its center of mass is

$$I = \tfrac{1}{12}(a^3b + ab^3).$$

Here $k$ is the constant density.

**17.** Find the moment of inertia of the lamina of Problem 15 about a line tangent to its boundary. *Hint*: Let the circle be $r = 2a \sin \theta$; then the tangent line is the $x$-axis. Formula 113 at the back of the book may help with the integration.

**18.** Consider the lamina $S$ of constant density $k$ bounded by the cardioid $r = a(1 + \sin \theta)$, as shown in Figure 7. Find its center of mass and moment of inertia with respect to the $x$-axis. *Hint*: Problem 7 of Section 12.8 gives the useful fact that $S$ has area $3\pi a^2/2$; also, Formula 113 at the back of the book may prove helpful.

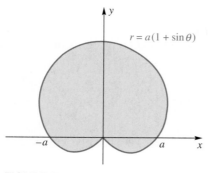

$r = a(1 + \sin\theta)$

FIGURE 7

**19.** Find the center of mass of the part of the cardioid of Problem 18, that is, outside the circle $r = a$.

**20.** (Parallel Axis Theorem)   Consider a lamina $S$ of mass $m$ together with parallel lines $L$ and $L'$ in the plane of $S$, the first line $L$ passing through the center of mass of $S$. Show that if $I$ and $I'$ are the moments of inertia of $S$ about $L$ and $L'$, then $I' = I + d^2m$, where $d$ is the distance between $L$ and $L'$. *Hint*: Assume $S$ lies in the $xy$-plane, $L$ is the $y$-axis, and $L'$ is the line $x = -d$.

**21.** Refer to the lamina of Problem 11, for which we found $I_y = 5a^5/12$. Find (a) $m$, (b) $\bar{x}$, and (c) $I_L$, where $L$ is a line through $(\bar{x}, \bar{y})$ parallel to the $y$-axis (see Problem 20).

**22.** Use the Parallel Axis Theorem together with Problem 15 to solve Problem 17 another way.

**23.** Find $I_x$, $I_y$, and $I_z$ for the two-piece lamina of constant density $k$ shown in Figure 8 (see Problems 15 and 20).

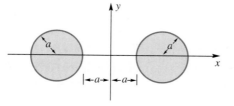

FIGURE 8

**24.** The Parallel Axis Theorem also holds for lines that are perpendicular to a lamina. Use this fact to find the moment of inertia of the rectangular lamina of Problem 16 about an axis perpendicular to the lamina and through a corner.

**25.** Let $S_1$ and $S_2$ be disjoint laminas in the $xy$-plane of mass $m_1$ and $m_2$ with centers of mass $(\bar{x}_1, \bar{y}_1)$ and $(\bar{x}_2, \bar{y}_2)$. Show that the center of mass $(\bar{x}, \bar{y})$ of the combined lamina $S_1 \cup S_2$ satisfies

$$\bar{x} = \bar{x}_1 \frac{m_1}{m_1 + m_2} + \bar{x}_2 \frac{m_2}{m_1 + m_2}$$

with a similar formula for $\bar{y}$. Conclude that in finding $(\bar{x}, \bar{y})$, the two laminas can be treated as if they were point masses at $(\bar{x}_1, \bar{y}_1)$ and $(\bar{x}_2, \bar{y}_2)$.

**26.** Let $S_1$ and $S_2$ be the homogeneous circular laminas of radius $a$ and $ta$ ($t > 0$), centered at $(-a, a)$ and $(ta, 0)$, respectively. Use Problem 25 to find the center of mass of $S_1 \cup S_2$.

**27.** Let $S$ be a lamina in the $xy$-plane with center of mass at the origin and let $L$ be the line $ax + by = 0$, which goes through the origin. Show that the (signed) distance $d$ of a point $(x, y)$ from $L$ is $d = (ax + by)/\sqrt{a^2 + b^2}$ and use this to conclude that the moment of $S$ with respect to $L$ is 0. *Note*: This shows that a lamina will balance on any line through its center of mass.

**28.** For the lamina of Example 3, find the equation of the balance line that makes an angle of 135° with the positive $x$-axis (see Problem 27). Write your answer in the form $Ax + By = C$.

---

**Answers to Concepts Review:    1.** $\iint_S x^2y^4 \, dA$

**2.** $\iint_S x^2y^5 \, dA/m$   **3.** $\iint_S x^4y^4 \, dA$   **4.** Greater

---

**16.6**
**SURFACE AREA**

In this section, we develop the formula for the area of a surface. We begin by considering a rectangle $T$, whose projection in the $xy$-plane is another rectangle $R$ (see Figure 1). Clearly, their areas $A(T)$ and $A(R)$ are related by

(1)                                    $A(T) = A(R)\sec\gamma$

where $\gamma$ is the acute angle between the two rectangles.

Next consider any (nonvertical) plane making an acute angle $\gamma$ with the $xy$-plane; that is, $\gamma$ is the angle between the corresponding unit normal

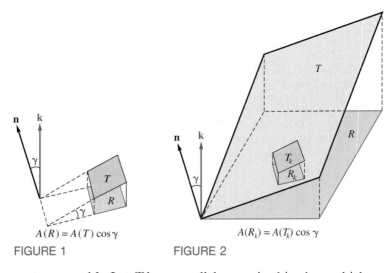

$A(R) = A(T) \cos \gamma$

FIGURE 1

$A(R_k) = A(T_k) \cos \gamma$

FIGURE 2

vectors **n** and **k**. Let $T$ be a parallelogram in this plane which projects onto a rectangle $R$ in the $xy$-plane (Figure 2). If we partition $R$ by means of lines parallel to and perpendicular to $\mathbf{n} \times \mathbf{k}$ into small rectangles $R_k$, the corresponding figures on $T$ are rectangles $T_k$ for which the formula (1) holds. It follows (by a limiting argument) that the formula holds for $R$ and $T$ as well.

**General Surfaces**   Let $F$ be a function of three variables having continuous first partial derivatives $F_x$, $F_y$, and $F_z$, with $F_z \neq 0$. Consider the surface $F(x, y, z) = k$ and denote by $G$ a part of this surface that projects onto a closed bounded region $S$ in the $xy$-plane (Figure 3). We wish to give meaning to the concept *area of G* and find a formula for calculating it.

Form a partition $P$ of $S$ by means of lines parallel to the $x$- and $y$-axes. Let $R_k$, $k = 1, 2, \ldots, n$, denote the resulting rectangles that lie completely within $S$. For each $k$, let $G_k$ be the part of the surface $G$ that projects onto $R_k$ and let $P_k$ be the point of $G_k$ that projects onto the corner of $R_k$ with the smallest $x$- and $y$-coordinates (Figure 4(a)). Finally, let $T_k$ denote the parallelogram from the tangent plane at $P_k$ that projects onto $R_k$ (Figure 4b) and $\gamma_k$ denote the acute angle that the upward normal at $P_k$ makes with **k**.

If the norm $|P|$ of the partition of $S$ is small, the set of tangent parallelograms $T_k$ will approximately conform to the surface $G$, and the smaller $|P|$ is taken, the better the conformation. This suggests the following definition of the area of $G$.

The area of the surface $G$ is defined to be

$$(2) \quad A(G) = \lim_{|P| \to 0} \sum_{k=1}^{n} A(T_k) = \lim_{|P| \to 0} \sum_{k=1}^{n} \sec \gamma_k A(R_k) = \iint_S \sec \gamma \, dA$$

where $\gamma$ is the acute angle that the upward normal to the surface makes with the vertical vector **k**.

To turn (2) into a usable formula, note first that the gradient vector $\nabla F$ is normal to the surface $G$, though it might point down rather than up. It follows (Section 14.2) that

$$\cos \gamma = \frac{|\nabla F \cdot \mathbf{k}|}{|\nabla F| \, |\mathbf{k}|}$$

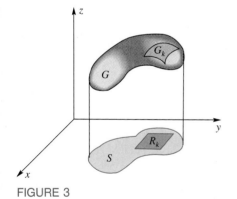

FIGURE 3

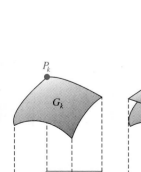

(a)

(b)

FIGURE 4

### A Quickie Problem

John's house has a rectangular base and a gable roof. Peter's house has the same base with a pyramidal-type roof. The slopes of all parts of both roofs are the same. Peter and John are arguing about whose roof has the smaller area. Settle the argument.

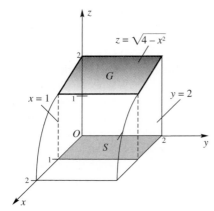

But $\nabla F = F_x\mathbf{i} + F_y\mathbf{j} + F_z\mathbf{k}$, and so

$$\sec \gamma = \frac{\sqrt{F_x^2 + F_y^2 + F_z^2}}{|F_z|}$$

In the special case where the surface is given by $z = f(x, y)$, we may write $F(x, y, z) = f(x, y) - z$ to obtain

$$\sec \gamma = \sqrt{f_x^2 + f_y^2 + 1}$$

This yields the important formula

$$A(G) = \iint_S \sqrt{f_x^2 + f_y^2 + 1} \, dA$$

**EXAMPLES** We illustrate the boxed formula with two examples.

**EXAMPLE 1** If $S$ is the rectangular region in the $xy$-plane that is bounded by the lines $x = 0$, $x = 1$, $y = 0$, and $y = 2$, find the area of the part of the semicylindrical surface $z = \sqrt{4 - x^2}$ that projects onto $S$ (Figure 5).

*Solution* Let $f(x, y) = \sqrt{4 - x^2}$. Then $f_x = -x/\sqrt{4 - x^2}$, $f_y = 0$, and

$$A(G) = \iint_S \sqrt{f_x^2 + f_y^2 + 1} \, dA = \iint_S \sqrt{\frac{x^2}{4 - x^2} + 1} \, dA = \iint_S \frac{2}{\sqrt{4 - x^2}} \, dA$$

$$= \int_0^1 \int_0^2 \frac{2}{\sqrt{4 - x^2}} \, dy \, dx = 4 \int_0^1 \frac{1}{\sqrt{4 - x^2}} \, dx = 4\left[ \sin^{-1} \frac{x}{2} \right]_0^1 = \frac{2\pi}{3} \quad \blacksquare$$

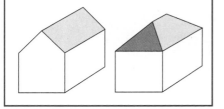

**FIGURE 5**

**EXAMPLE 2** Find the area of the surface $z = x^2 + y^2$ below the plane $z = 9$.

*Solution* The designated part $G$ of the surface projects onto the circular region $S$ inside the circle $x^2 + y^2 = 9$ (Figure 6). Let $f(x, y) = x^2 + y^2$. Then $f_x = 2x$, $f_y = 2y$, and

$$A(G) = \iint_S \sqrt{4x^2 + 4y^2 + 1} \, dA$$

The shape of $S$ suggests use of polar coordinates.

$$A(G) = \int_0^{2\pi} \int_0^3 \sqrt{4r^2 + 1} \, r \, dr \, d\theta$$

$$= \int_0^{2\pi} \frac{1}{8} \left[ \frac{2}{3} (4r^2 + 1)^{3/2} \right]_0^3 d\theta$$

$$= \int_0^{2\pi} \tfrac{1}{12}(37^{3/2} - 1) \, d\theta = \frac{\pi}{6}(37^{3/2} - 1) \approx 117.32 \quad \blacksquare$$

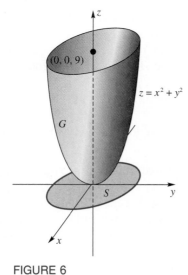

**FIGURE 6**

Our next example states the remarkable result that two planes perpendicular to the axis of a circular cylinder cut off equal areas on the cylinder and an inscribed sphere (Figure 7). Both areas are equal to $2\pi ah$, where $a$ is the radius of the cylinder and $h$ is the distance between the planes.

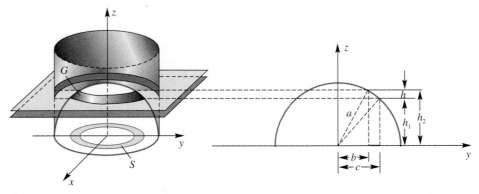

FIGURE 7

**EXAMPLE 3**   Show that the area of the surface $G$ cut off from the sphere $x^2 + y^2 + z^2 = a^2$ by the planes $z = h_1$ and $z = h_2$ $(-a \le h_1 \le h_2 \le a)$ is

$$A(G) = 2\pi a(h_2 - h_1) = 2\pi ah$$

**Solution**   We may assume $h_1 \ge 0$. Let $F(x, y, z) = x^2 + y^2 + z^2$. Then

$$\sec \gamma = \frac{\sqrt{F_x^2 + F_y^2 + F_z^2}}{|F_z|} = \frac{2\sqrt{x^2 + y^2 + z^2}}{2|z|} = \frac{a}{\sqrt{a^2 - x^2 - y^2}}$$

Referring to Figure 7, we obtain

$$A(G) = \iint\limits_{S} \sec \gamma \, dA = \int_0^{2\pi} \int_b^c \frac{a}{\sqrt{a^2 - r^2}} \, r \, dr \, d\theta = \left[ -2\pi a\sqrt{a^2 - r^2} \right]_b^c$$

$$= 2\pi a[\sqrt{a^2 - b^2} - \sqrt{a^2 - c^2}] = 2\pi a(h_2 - h_1) \qquad \blacksquare$$

## CONCEPTS REVIEW

**1.** Let plane $T$ make angle $\gamma$ with horizontal plane $R$. If region $G$ in plane $T$ projects onto the region $S$ in plane $R$, then the area $A(G)$ relates to the area $A(S)$ by the formula _____.

**2.** More generally, if $z = f(x, y)$ determines a surface $G$ that projects onto the region $S$ in the $xy$-plane, then $A(G)$ is given by the formula $A(G) =$ _____.

**3.** Applying Question 2 with $z = (a^2 - x^2 - y^2)^{1/2}$ leads to the integral formula $A =$ _____ for the area of a hemisphere of radius $a$. When this integral is evaluated, we obtain the familiar formula $A =$ _____.

**4.** Consider a sphere inscribed in a cylindrical can of radius $a$. Two planes, both perpendicular to the axis of the cylinder and separated by distance $h$, will cut off regions on both the cylinder and the sphere of area _____.

## PROBLEM SET 16.6

In Problems 1–14, find the area of the indicated surface. Make a sketch in each case.

**1.** The part of the plane $3x + 4y + 6z = 12$ that is above the rectangle in the $xy$-plane with vertices $(0, 0)$, $(2, 0)$, $(2, 1)$, and $(0, 1)$.

**2.** The part of the plane $3x - 2y + 6z = 12$ that is bounded by the planes $x = 0$, $y = 0$, and $3x + 2y = 12$.

**3.** The part of the surface $z = \sqrt{4 - y^2}$ that is directly above the square in the $xy$-plane with vertices $(1, 0)$, $(2, 0)$, $(2, 1)$, and $(1, 1)$.

**4.** The part of the surface $z = \sqrt{4 - y^2}$ in the first octant that is directly above the circle $x^2 + y^2 = 4$ in the $xy$-plane.

**5.** The part of the cylinder $x^2 + z^2 = 9$ that is directly over the rectangle in the $xy$-plane with vertices $(0, 0)$, $(2, 0)$, $(2, 3)$, and $(0, 3)$.

**6.** The part of the paraboloid $z = x^2 + y^2$ that is cut off by the plane $z = 4$.

**7.** The part of the conical surface $x^2 + y^2 = z^2$ that is directly over the triangle in the $xy$-plane with vertices $(0, 0)$, $(4, 0)$, and $(0, 4)$.

**8.** The part of the surface $z = x^2/4 + 4$ that is cut off by the planes $x = 0$, $x = 1$, $y = 0$, and $y = 2$.

**9.** The part of the sphere $x^2 + y^2 + z^2 = a^2$ inside the circular cylinder $x^2 + y^2 = b^2$, where $0 < b \le a$.

**10.** The part of the sphere $x^2 + y^2 + z^2 = a^2$ inside the elliptic cylinder $b^2x^2 + a^2y^2 = a^2b^2$, where $0 < b \le a$.

**11.** The part of the sphere $x^2 + y^2 + z^2 = a^2$ inside the circular cylinder $x^2 + y^2 = ay$ ($r = a \sin \theta$ in polar coordinates), $a > 0$.

**12.** The part of the cylinder $x^2 + y^2 = ay$ inside the sphere $x^2 + y^2 + z^2 = a^2$, $a > 0$, *Hint*: Project to the $yz$-plane to get the region of integration.

**13.** The part of the saddle $az = x^2 - y^2$ inside the cylinder $x^2 + y^2 = a^2$, $a > 0$.

**14.** The surface of the solid that is the intersection of the two solid cylinders $x^2 + z^2 \le a^2$ and $x^2 + y^2 \le a^2$. *Hint*: You may need the integration formula $\int (1 + \sin \theta)^{-1} \, d\theta = -\tan[(\pi - 2\theta)/4] + C$.

Problems 15–17 are related to Example 3.

**15.** Find the centroid (center of mass) of the part of the homogeneous sphere $x^2 + y^2 + z^2 = a^2$ between the planes $z = h_1$ and $z = h_2$, $-a \le h_1 < h_2 \le a$.

**16.** Show that the polar cap (Figure 8) on a sphere of radius $a$ determined by the spherical angle $\phi$ has area $2\pi a^2(1 - \cos \phi)$.

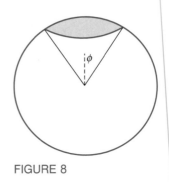

**FIGURE 8**

**17.** (Another Goat Problem) Four goats have grazing areas $A$, $B$, $C$, and $D$, respectively. The first three goats are each tethered by ropes of length $b$, the first on a flat plane, the second on the outside of a sphere of radius $a$, and the third on the inside of a sphere of radius $a$. The fourth goat must stay inside a ring of radius $b$ that has been dropped over a sphere of radius $a$. Determine formulas for $A$, $B$, $C$, and $D$ and arrange them in order of size. Assume $b < a$.

**18.** Let $S$ be a planar region in three-space and let $S_{xy}$, $S_{xz}$, and $S_{yz}$ be the projections on the three coordinate planes (Figure 9). Show that

$$[A(S)]^2 = [A(S_{xy})]^2 + [A(S_{xz})]^2 + [A(S_{yz})]^2$$

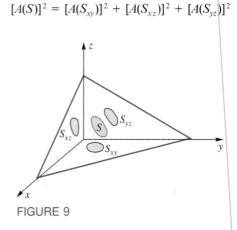

**FIGURE 9**

**19.** Assume the region $S$ of Figure 9 lies in the plane $z = f(x, y) = ax + by + c$ and that $S$ is above the $xy$-plane. Show that the volume of the solid cylinder under $S$ is $A(S_{xy})f(\bar{x}, \bar{y})$, where $(\bar{x}, \bar{y})$ is the centroid of $S_{xy}$.

**20.** Let $z = f(x, y)$ and let $T_k$ be the parallelogram in the tangent plane at $P_k$ that projects onto a rectangle $R_k$ in the $xy$-plane, as shown in Figure 10. Then the vectors $\mathbf{u}$ and $\mathbf{v}$ from $P_k$, which form the sides of $T_k$, are

$$\mathbf{u} = \Delta x_k \mathbf{i} + f_x(x_k, y_k)\Delta x_k \mathbf{k}$$

$$\mathbf{v} = \Delta y_k \mathbf{j} + f_y(x_k, y_k)\Delta y_k \mathbf{k}$$

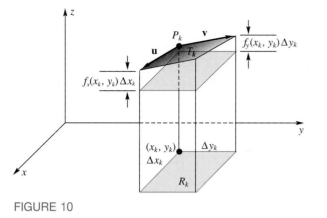

FIGURE 10

From Section 14.3, the area $A(T_k)$ is given by $|\mathbf{u} \times \mathbf{v}|$. Use this to show that

$$A(T_k) = \sqrt{[f_x(x_k, y_k)]^2 + [f_y(x_k, y_k)]^2 + 1}\,\Delta x_k \Delta y_k$$

and thereby give another derivation of the surface area formula in the text.

---

**Answers to Concepts Review:** **1.** $A(G) = A(S)\sec \gamma$

**2.** $\iint\limits_{S} \sqrt{f_x^2 + f_y^2 + 1}\, dA$

**3.** $\displaystyle\int_{-a}^{a} \int_{-\sqrt{a^2 - x^2}}^{\sqrt{a^2 - x^2}} \left(a/\sqrt{a^2 - x^2 - y^2}\right) dy\, dx =$
$$\int_{0}^{2\pi} \int_{0}^{a} \left(ar/\sqrt{a^2 - r^2}\right) dr\, d\theta;\ 2\pi a^2$$

**4.** $2\pi ah$

---

**16.7**<br>**TRIPLE INTEGRALS**<br>**(CARTESIAN**<br>**COORDINATES)**

The concept embodied in single and double integrals extends in a natural way to triple, and even $n$-fold, integrals.

Consider a function $f$ of three variables defined over a box-shaped region $B$ with faces parallel to the coordinate planes. We can no longer graph $f$ (four dimensions would be required), but we can picture $B$ (Figure 1). Form a partition $P$ of $B$ by passing planes through $B$ parallel to the coordinate planes, thus cutting $B$ into small sub-boxes $B_1, B_2, \ldots, B_n$; a typical one—$B_k$—is shown in Figure 1. On $B_k$, pick a sample point $(\overline{x}_k, \overline{y}_k, \overline{z}_k)$ and consider the Riemann sum

$$\sum_{k=1}^{n} f(\overline{x}_k, \overline{y}_k, \overline{z}_k)\, \Delta V_k$$

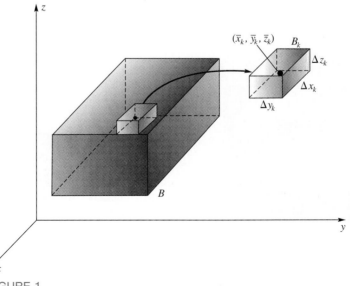

FIGURE 1

where $\Delta V_k = \Delta x_k \Delta y_k \Delta z_k$ is the volume of $B_k$. Let the norm of the partition $|P|$ be the length of the longest diagonal of all the sub-boxes. Then we define the **triple integral** by

$$\iiint_B f(x, y, z) \, dV = \lim_{|P| \to 0} \sum_{k=1}^{n} f(\bar{x}_k, \bar{y}_k, \bar{z}_k) \Delta V_k$$

provided this limit exists.

The question of what kind of functions are integrable arises here, as it did for single and double integrals. It is certainly sufficient that $f$ be continuous on $B$. Actually we can allow some discontinuities, for example, on a finite number of smooth surfaces. We do not prove this (a very difficult task), but we assert that it is true.

As you would expect, the triple integral has the standard properties: linearity, additivity on sets that overlap only on a boundary surface, and the comparison property. Finally, triple integrals can be written as triple iterated integrals, as we now illustrate.

**EXAMPLE 1**   Evaluate $\iiint_B x^2 yz \, dV$, where $B$ is the box

$$B = \{(x, y, z): 1 \le x \le 2, 0 \le y \le 1, 1 \le z \le 2\}$$

*Solution*

$$\iiint_B x^2 yz \, dV = \int_0^2 \int_0^1 \int_1^2 x^2 yz \, dx \, dy \, dz$$

$$= \int_0^2 \int_0^1 \left[ \frac{1}{3} x^3 yz \right]_1^2 dy \, dz = \int_0^2 \int_0^1 \frac{7}{3} yz \, dy \, dz$$

$$= \frac{7}{3} \int_0^2 \left[ \frac{1}{2} y^2 z \right]_0^1 dz = \frac{7}{3} \int_0^2 \frac{1}{2} z \, dz$$

$$= \frac{7}{6} \left[ \frac{z^2}{2} \right]_0^2 = \frac{7}{3}$$

There are six possible orders of integration. Any one of them will yield the answer $\frac{7}{3}$.  ∎

**General Regions**   Consider a closed bounded set $S$ in three-space and enclose it in a box $B$, as shown in Figure 2. Let $f(x, y, z)$ be defined on $S$ and give $f$ the value zero outside of $S$. Then we define

$$\iiint_S f(x, y, z) \, dV = \iiint_B f(x, y, z) \, dV$$

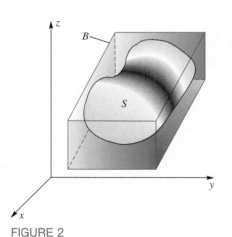

FIGURE 2

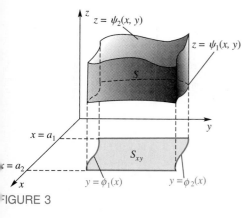

**FIGURE 3**

The integral on the right was defined in our opening remarks, but that does not mean that it is easy to evaluate. In fact, if the set $S$ is sufficiently complicated, we may not be able to make the evaluation.

Let $S$ be a $z$-simple set (vertical lines intersect $S$ in a single line segment) and let $S_{xy}$ be its projection in the $xy$-plane (Figure 3). Then

$$\iiint_S f(x, y, z)\,dV = \iint_{S_{xy}} \left[ \int_{\psi_1(x, y)}^{\psi_2(x, y)} f(x, y, z)\,dz \right] dA$$

If, in addition, $S_{xy}$ is a $y$-simple set (as shown in Figure 3), we can rewrite the outer double integral as an iterated integral.

$$\iiint_S f(x, y, z)\,dV = \int_{a_1}^{a_2} \int_{\phi_1(x)}^{\phi_2(x)} \int_{\psi_1(x, y)}^{\psi_2(x, y)} f(x, y, z)\,dz\,dy\,dx$$

Other orders of integration may be possible, depending on the shape of $S$, but in each case we should expect the limits on the inner integral to be functions of two variables, those on the middle integral to be functions of one variable, and those on the outer integral to be constants.

We give several examples. The first simply illustrates evaluation of a triple iterated integral.

**EXAMPLE 1**   Evaluate the iterated integral

$$\int_{-2}^{5} \int_{0}^{3x} \int_{y}^{x+2} 4\,dz\,dy\,dx$$

**Solution**

$$\int_{-2}^{5} \int_{0}^{3x} \int_{y}^{x+2} 4\,dz\,dy\,dx = \int_{-2}^{5} \int_{0}^{3x} \left( \int_{y}^{x+2} 4\,dz \right) dy\,dx$$

$$= \int_{-2}^{5} \int_{0}^{3x} \left[ 4z \right]_{y}^{x+2} dy\,dx$$

$$= \int_{-2}^{5} \int_{0}^{3x} (4x - 4y + 8)\,dy\,dx$$

$$= \int_{-2}^{5} \left[ 4xy - 2y^2 + 8y \right]_{0}^{3x} dx$$

$$= \int_{-2}^{5} (-6x^2 + 24x)\,dx = -14 \qquad \blacksquare$$

**EXAMPLE 2**   Evaluate the triple integral of $f(x, y, z) = 2xyz$ over the solid region $S$ that is bounded by the parabolic cylinder $z = 2 - \frac{1}{2}x^2$ and the planes $z = 0$, $y = x$, and $y = 0$.

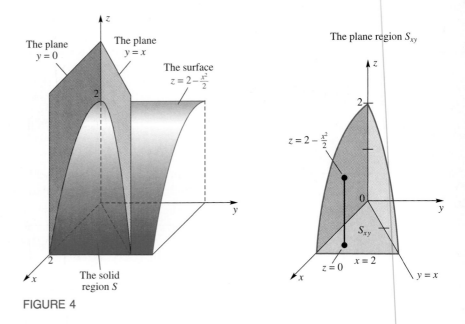

FIGURE 4

**Solution** The solid region $S$ is shown in Figure 4. The triple integral

$$\iiint\limits_{S} 2xyz \, dV$$

can be evaluated by an iterated integral.

Note first that $S$ is a $z$-simple set and that its projection $S_{xy}$ in the $xy$-plane is $y$-simple (also $x$-simple). In the first integration, $x$ and $y$ are fixed; we integrate along a vertical line from $z = 0$ to $z = 2 - x^2/2$. The result is then integrated over the set $S_{xy}$.

$$\iiint\limits_{S} 2xyz \, dV = \int_0^2 \int_0^x \int_0^{2 - x^2/2} 2xyz \, dz \, dy \, dx$$

$$= \int_0^2 \int_0^x \left[ xyz^2 \right]_0^{2 - x^2/2} dy \, dx$$

$$= \int_0^2 \int_0^x (4xy - 2x^3y + \tfrac{1}{4}x^5y) \, dy \, dx$$

$$= \int_0^2 (2x^3 - x^5 + \tfrac{1}{8}x^7) \, dx = \tfrac{4}{3}$$ ∎

Many different orders of integration are possible in Example 2. We illustrate another way to do this problem.

**EXAMPLE 3** Evaluate the integral of Example 2 by doing the integration in the order $dy \, dx \, dz$.

The plane region $S_{xz}$

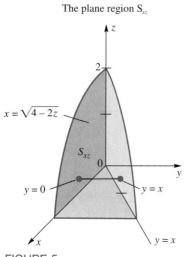

**FIGURE 5**

**Solution** Note that the solid $S$ is $y$-simple and that it projects onto the plane set $S_{xz}$ shown in Figure 5. We first integrate along a horizontal line from $y = 0$ to $y = x$; then we integrate the result over $S_{xz}$.

$$\iiint\limits_{S} 2xyz\, dV = \int_0^2 \int_0^{\sqrt{4-2z}} \int_0^x 2xyz\, dy\, dx\, dz$$

$$= \int_0^2 \int_0^{\sqrt{4-2z}} x^3 z\, dx\, dz = \tfrac{1}{4} \int_0^2 (\sqrt{4-2z})^4 z\, dz$$

$$= \tfrac{1}{4} \int_0^2 (16z - 16z^2 + 4z^3)\, dz = \tfrac{4}{3}$$

■

**Mass and Center of Mass** The concepts of mass and center of mass generalize easily to solid regions. By now, the process that leads to the right formula is very familiar and can be summarized in our motto: *slice, approximate, integrate.* Figure 6 gives away the whole idea. The symbol $\delta(x, y, z)$ denotes the density (mass per unit volume) at $(x, y, z)$.

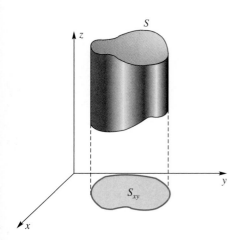

**FIGURE 6**

Mass of $B_k \approx \delta(\bar{x}_k, \bar{y}_k, \bar{z}_k)\Delta V_k$
Moment of $B_k$
about $xy$-plane $\approx \bar{z}_k\, \delta(\bar{x}_k, \bar{y}_k, \bar{z}_k)\, \Delta V_k$

The corresponding integral formulas for the mass $m$ of the solid $S$, moment $M_{xy}$ of $S$ with respect to the $xy$-plane, and $z$-coordinate $\bar{z}$ of the center of mass are

$$m = \iiint\limits_{S} \delta(x, y, z)\, dV$$

$$M_{xy} = \iiint\limits_{S} z\delta(x, y, z)\, dV$$

$$\bar{z} = \frac{M_{xy}}{m}$$

There are similar formulas for $M_{yz}$, $M_{xz}$, $\bar{x}$, and $\bar{y}$.

**EXAMPLE 4**   Find the mass and center of mass of the solid $S$ of Example 2, assuming that its density is proportional to the distance from its base in the $xy$-plane.

*Solution*   By hypothesis, $\delta(x, y, z) = kz$, where $k$ is a constant. Thus,

$$m = \iiint\limits_{S} kz \, dV = \int_0^2 \int_0^x \int_0^{2 - x^2/2} kz \, dz \, dy \, dx$$

$$= k \int_0^2 \int_0^x \frac{1}{2} \left( 2 - \frac{x^2}{2} \right)^2 dy \, dx = k \int_0^2 \int_0^x \left( 2 - x^2 + \frac{1}{8} x^4 \right) dy \, dx$$

$$= k \int_0^2 \left( 2x - x^3 + \frac{1}{8} x^5 \right) dx = k \left[ x^2 - \frac{x^4}{4} + \frac{x^6}{48} \right]_0^2 = \frac{4}{3} k$$

$$M_{xy} = \iiint\limits_{S} kz^2 \, dV = \int_0^2 \int_0^x \int_0^{2 - x^2/2} kz^2 \, dz \, dy \, dx$$

$$= \frac{k}{3} \int_0^2 \int_0^x \left( 2 - \frac{x^2}{2} \right)^3 dy \, dx$$

$$= \frac{k}{3} \int_0^2 \int_0^x \left( 8 - 6x^2 + \frac{3}{2} x^4 - \frac{1}{8} x^6 \right) dy \, dx$$

$$= \frac{k}{3} \int_0^2 \left( 8x - 6x^3 + \frac{3}{2} x^5 - \frac{1}{8} x^7 \right) dx$$

$$= \frac{k}{3} \left[ 4x^2 - \frac{3}{2} x^4 + \frac{1}{4} x^6 - \frac{1}{64} x^8 \right]_0^2 = \frac{4}{3} k$$

$$M_{xz} = \iiint\limits_{S} kyz \, dV = \int_0^2 \int_0^x \int_0^{2 - x^2/2} kyz \, dz \, dy \, dx$$

$$= k \int_0^2 \int_0^x \frac{1}{2} y \left( 2 - \frac{x^2}{2} \right)^2 dy \, dx = k \int_0^2 \frac{1}{4} x^2 \left( 2 - \frac{x^2}{2} \right)^2 dx$$

$$= k \int_0^2 \left( x^2 - \frac{1}{2} x^4 + \frac{1}{16} x^6 \right) dx = \frac{64}{105} k$$

$$M_{yz} = \iiint\limits_{S} kxz \, dV = \int_0^2 \int_0^x \int_0^{2 - x^2/2} kxz \, dz \, dy \, dx = \frac{128}{105} k$$

$$\bar{z} = \frac{M_{xy}}{m} = \frac{4k/3}{4k/3} = 1$$

$$\bar{y} = \frac{M_{xz}}{m} = \frac{64k/105}{4k/3} = \frac{16}{35}$$

$$\bar{x} = \frac{M_{yz}}{m} = \frac{128k/105}{4k/3} = \frac{32}{35}$$

---

## CONCEPTS REVIEW

**1.** $\iiint_S 1\, dV$ gives the _____ of the solid $S$.

**2.** If the density at $(x, y, z)$ is $|xyz|$, then the mass of $S$ is _____.

**3.** $\displaystyle\int_0^1 \int_0^1 \int_{x^2}^x f(x, y, z)\, dy\, dx\, dz =$

$$\int_0^1 \int_0^1 \int_{g(y)}^{h(y)} f(x, y, z)\, dx\, dy\, dx,$$

where $g(y) =$ _____ and $h(y) =$ _____.

**4.** Let $S$ be the unit ball centered at the origin. Then, from symmetry, we conclude that $\iiint_S (x + y + z)\, dV =$ _____.

---

## PROBLEM SET 16.7

In Problems 1–8, evaluate the iterated integrals.

**1.** $\displaystyle\int_{-3}^7 \int_0^{2x} \int_y^{x-1} dz\, dy\, dx$

**2.** $\displaystyle\int_0^2 \int_{-1}^4 \int_0^{3y+x} dz\, dy\, dx$

**3.** $\displaystyle\int_1^4 \int_{z-1}^{2z} \int_0^{y+2z} dx\, dy\, dz$

**4.** $\displaystyle\int_0^5 \int_{-2}^4 \int_1^2 6xy^2 z^3\, dx\, dy\, dz$

**5.** $\displaystyle\int_0^2 \int_1^z \int_0^{\sqrt{x/z}} 2xyz\, dy\, dx\, dz$

**6.** $\displaystyle\int_0^{\pi/2} \int_0^z \int_0^y \sin(x + y + z)\, dx\, dy\, dz$

**7.** $\displaystyle\int_{-2}^4 \int_{x-1}^{x+1} \int_0^{\sqrt{2y/x}} 3xyz\, dz\, dy\, dx$

**8.** $\displaystyle\int_0^{\pi/2} \int_{\sin 2z}^0 \int_0^{2yz} \sin\left(\frac{x}{y}\right) dx\, dy\, dz$

In Problems 9–18, sketch the solid $S$. Then write an iterated integral for

$$\iiint_S f(x, y, z)\, dV$$

**9.** $S = \{(x, y, z): 0 \le x \le 1, 0 \le y \le 3, 0 \le z \le \frac{1}{6}(12 - 3x - 2y)\}$

**10.** $S = \{(x, y, z): 0 \le x \le \sqrt{4 - y^2}, 0 \le y \le 2, 0 \le z \le 3\}$

**11.** $S = \{(x, y, z): 0 \le x \le \frac{1}{2}y, 0 \le y \le 4, 0 \le z \le 2\}$

**12.** $S = \{(x, y, z): 0 \le x \le \sqrt{y}, 0 \le y \le 4, 0 \le z \le \frac{3}{2}x\}$

**13.** $S = \{(x, y, z): 0 \le x \le 3z, 0 \le y \le 4 - x - 2z, 0 \le z \le 2\}$

**14.** $S = \{(x, y, z): 0 \le x \le y^2, 0 \le y \le \sqrt{z}, 0 \le z \le 1\}$

**15.** $S$ is the tetrahedron with vertices $(0, 0, 0)$, $(3, 2, 0)$, $(0, 3, 0)$, and $(0, 0, 2)$.

**16.** $S$ is the region in the first octant bounded by the surface $z = 9 - x^2 - y^2$ and the coordinate planes.

**17.** $S$ is the region in the first octant bounded by the cylinder $y^2 + z^2 = 1$ and the planes $x = 1$ and $x = 4$.

**18.** $S$ is the smaller region bounded by the cylinder $x^2 + y^2 - 2y = 0$ and the planes $x - y = 0$, $z = 0$, and $z = 3$.

In Problems 19–26, use triple iterated integrals to find the indicated quantities.

**19.** The volume of the solid in the first octant bounded by $y = 2x^2$ and $y + 4z = 8$.

**20.** The volume of the solid in the first octant bounded by the elliptic cylinder $y^2 + 64z^2 = 4$ and the plane $y = x$.

**21.** The volume of the solid bounded by the cylinders $x^2 = y$ and $z^2 = y$, and the plane $y = 1$.

**22.** The volume of the solid bounded by the cylinder $y = x^2 + 2$ and the planes $y = 4$, $z = 0$, and $3y - 4z = 0$.

**23.** The center of mass of the tetrahedron bounded by the planes $x + y + z = 1$, $x = 0$, $y = 0$, and $z = 0$ if the density is proportional to the sum of the coordinates of the point.

**24.** The center of mass of the solid bounded by the cylinder $x^2 + y^2 = 9$ and the planes $z = 0$ and $z = 4$ if the density is proportional to the square of the distance from the origin.

**25.** The center of mass of that part of the ball $\{(x, y, z):$ $x^2 + y^2 + z^2 \le a^2\}$ that lies in the first octant, assuming it has constant density.

**26.** The moment of inertia $I_x$ about the $x$-axis of the solid bounded by the cylinder $y^2 + z^2 = 4$ and the planes $x - y = 0$, $x = 0$, and $z = 0$ if the density $\delta(x, y, z) = z$. *Hint*: You will need to develop your own formula; *slice, approximate, integrate.*

In Problems 27–30, write the given iterated integral as an iterated integral with the indicated order of integration.

**27.** $\int_0^1 \int_0^{\sqrt{1-y^2}} \int_0^{\sqrt{1-y^2-z^2}} f(x, y, z)\, dx\, dz\, dy;\ dz\, dy\, dx$

**28.** $\int_0^2 \int_0^{4-2y} \int_0^{4-2y-z} f(x, y, z)\, dx\, dz\, dy;\ dz\, dy\, dx$

**29.** $\int_0^2 \int_0^{9-x^2} \int_0^{2-x} f(x, y, z)\, dz\, dy\, dx;\ dy\, dx\, dz$

**30.** $\int_0^2 \int_0^{9-x^2} \int_0^{2-x} f(x, y, z)\, dz\, dy\, dx;\ dz\, dx\, dy$

**31.** Consider the solid (Figure 7) in the first octant cut off from the square cylinder with sides $x = 0$, $x = 1$, $z = 0$, and $z = 1$ by the plane $2x + y + 2z = 6$. Find its volume in three ways.

(a) Hard way: by a $dz\, dy\, dx$ integration.
(b) Easier way: by a $dy\, dx\, dz$ integration.
(c) Easiest way: by Problem 19 of Section 16.6.

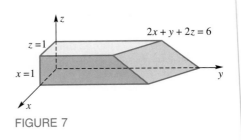

**FIGURE 7**

**32.** Assuming the density of the solid of Figure 7 is a constant $k$, find the moment of inertia of the solid with respect to the $y$-axis.

**33.** If the temperature at $(x, y, z)$ is $T = 30 - z$ degrees, find the average temperature of the solid of Figure 7. Also find the $z$-coordinate of its centroid.

**34.** (Soda Can Problem) A full soda can of height $h$ stands on the $xy$-plane. Punch a hole in the base and watch $\bar{z}$ (the $z$-coordinate of the center of mass) as the soda leaks away. Starting at $h/2$, $\bar{z}$ gradually drops to a minimum and then rises back to $h/2$ when the can is empty. Show that $\bar{z}$ is least when it coincides with the height of the soda. Would the same conclusion hold for a soda bottle? *Hint*: Don't calculate; think geometrically.

**35.** Let $S = \{(x, y, z): x^2/a^2 + y^2/b^2 + z^2/c^2 \le 1\}$. Evaluate $\iiint_S (xy + xz + yz)\, dV$.

---

**Answers to Concepts Review:** **1.** Volume **2.** $\iiint_S |xyz|\, dV$
**3.** $y;\ \sqrt{y}$ **4.** 0

---

## 16.8
### TRIPLE INTEGRALS (CYLINDRICAL AND SPHERICAL COORDINATES)

When a solid region $S$ in three-space has an axis of symmetry, the evaluation of triple integrals over $S$ is often facilitated by using cylindrical coordinates. Similarly if $S$ is symmetric with respect to a point, spherical coordinates may be helpful. Cylindrical and spherical coordinates were introduced in Section 14.7, a section you may wish to review before going on.

**Cylindrical Coordinates** Figure 1 serves to remind us of the meaning of cylindrical coordinates and displays the symbols we will use. Cylindrical and Cartesian (rectangular) coordinates are related by the equations

$$x = r \cos \theta, \qquad y = r \sin \theta, \qquad x^2 + y^2 = r^2$$

As a result the function $f(x, y, z)$ transforms to

$$f(x, y, z) = f(r \cos \theta, r \sin \theta, z) = F(r, \theta, z)$$

when written in cylindrical coordinates.

Suppose now that we wish to evaluate $\iiint_S f(x, y, z)\, dV$, where $S$ is a solid region. Consider partitioning $S$ by means of a cylindrical grid, where

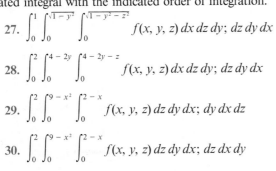

**FIGURE 1**

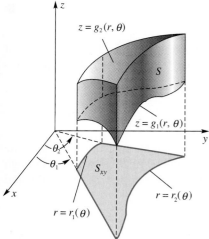

FIGURE 2

the typical volume element has the shape shown in Figure 2. Since this piece (called a *cylindrical wedge*) has volume $\Delta V_k = \bar{r}_k \Delta r_k \Delta \theta_k \Delta z_k$, the sum that approximates the integral has the form

$$\sum_{k=1}^{n} F(\bar{r}_k, \bar{\theta}_k, \bar{z}_k) \bar{r}_k \Delta z_k \Delta r_k \Delta \theta_k$$

Taking the limit as the norm of the partition tends to zero leads to a new integral and suggests an important formula for changing from Cartesian to cylindrical coordinates in a triple integral.

Let $S$ be a $z$-simple solid and suppose its projection $S_{xy}$ in the $xy$-plane is $r$-simple, as shown in Figure 3. If $f$ is continuous on $S$, then

$$\iiint_S f(x, y, z)\, dV = \int_{\theta_1}^{\theta_2} \int_{r_1(\theta)}^{r_2(\theta)} \int_{g_1(r, \theta)}^{g_2(r, \theta)} f(r \cos \theta, r \sin \theta, z) r\, dz\, dr\, d\theta$$

The key fact to note is that the $dz\, dy\, dx$ of Cartesian coordinates becomes $r\, dz\, dr\, d\theta$ in cylindrical coordinates.

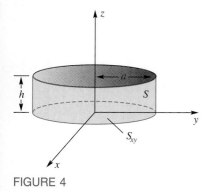

FIGURE 3

**EXAMPLE 1**   Find the mass and center of mass of a solid cylinder $S$, assuming the density is proportional to the distance from the base.

**Solution**   With $S$ oriented as shown in Figure 4, we can write the density function as $\delta(x, y, z) = kz$, where $k$ is a constant. Then,

$$m = \iiint_S \delta(x, y, z)\, dV = k \int_0^{2\pi} \int_0^a \int_0^h zr\, dz\, dr\, d\theta$$

$$= k \int_0^{2\pi} \int_0^a \frac{1}{2} h^2 r\, dr\, d\theta = \frac{1}{2} kh^2 \int_0^{2\pi} \int_0^a r\, dr\, d\theta$$

$$= \frac{1}{2} kh^2 \int_0^{2\pi} \frac{1}{2} a^2\, d\theta = \frac{1}{2} kh^2 \pi a^2$$

$$M_{xy} = \iiint_S z\delta(x, y, z)\, dV = k \int_0^{2\pi} \int_0^a \int_0^h z^2 r\, dz\, dr\, d\theta$$

$$= k \int_0^{2\pi} \int_0^a \frac{1}{3} h^3 r\, dr\, d\theta = \frac{1}{3} kh^3 \int_0^{2\pi} \int_0^a r\, dr\, d\theta$$

$$= \frac{1}{3} kh^3 \pi a^2$$

$$\bar{z} = \frac{M_{xy}}{m} = \frac{\frac{1}{3} kh^3 \pi a^2}{\frac{1}{2} kh^2 \pi a^2} = \frac{2}{3} h$$

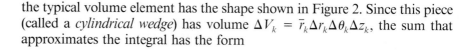

FIGURE 4

By symmetry, $\bar{x} = \bar{y} = 0$.

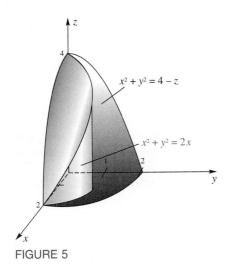

**FIGURE 5**

**EXAMPLE 2** Find the volume of the solid region $S$ bounded above by the paraboloid $z = 4 - x^2 - y^2$, below by $z = 0$, and laterally by $y = 0$ and the cylinder $x^2 + y^2 = 2x$, as shown in Figure 5.

**Solution** In cylindrical coordinates, the paraboloid is $z = 4 - r^2$ and the cylinder is $r = 2 \cos \theta$. Thus,

$$V = \iiint_S 1 \, dV = \int_0^{\pi/2} \int_0^{2\cos\theta} \int_0^{4-r^2} r \, dz \, dr \, d\theta$$

$$= \int_0^{\pi/2} \int_0^{2\cos\theta} r(4 - r^2) \, dr \, d\theta = \int_0^{\pi/2} \left[ 2r^2 - \frac{1}{4}r^4 \right]_0^{2\cos\theta} d\theta$$

$$= \int_0^{\pi/2} (8 \cos^2\theta - 4 \cos^4\theta) \, d\theta$$

$$= 8 \cdot \frac{1}{2} \cdot \frac{\pi}{2} - 4 \cdot \frac{3}{8} \cdot \frac{\pi}{2} = \frac{5\pi}{4}$$

We used Formula 113 from the table of integrals at the end of the book to make the last calculation. ∎

**Spherical Coordinates** Figure 6 serves to remind us of the meaning of spherical coordinates, which were introduced in Section 14.7. There we learned that the equations

$$x = \rho \sin \phi \cos \theta, \qquad y = \rho \sin \phi \sin \theta, \qquad z = \rho \cos \phi$$

relate spherical coordinates and Cartesian coordinates. Figure 7 exhibits the volume element in spherical coordinates (called a spherical wedge). Though we omit the details, it can be shown that the volume of the indicated spherical wedge is

$$\Delta V = \bar{\rho}^2 \sin \bar{\phi} \, \Delta\rho \, \Delta\theta \, \Delta\phi$$

where $(\bar{\rho}, \bar{\theta}, \bar{\phi})$ is an appropriately chosen point in the wedge.

Partitioning a solid $S$ by means of a spherical grid, forming the appropriate sum, and taking the limit leads to an integral in which $dz \, dy \, dx$ is replaced by $\rho^2 \sin \phi \, d\rho \, d\theta \, d\phi$.

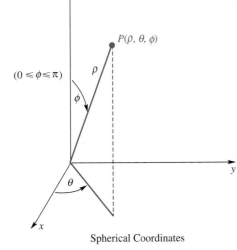

Spherical Coordinates

**FIGURE 6**

$$\iiint_S f(x, y, z) \, dV$$

$$= \iiint_{\substack{\text{appropriate} \\ \text{limits}}} f(\rho \sin \phi \cos \theta, \rho \sin \phi \sin \theta, \rho \cos \phi)\rho^2 \sin \phi \, d\rho \, d\theta \, d\phi$$

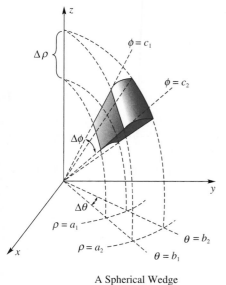

A Spherical Wedge

**FIGURE 7**

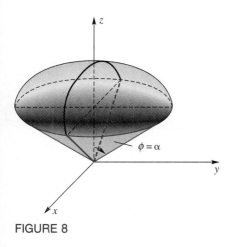

$\phi = \alpha$

**FIGURE 8**

**EXAMPLE 3** Find the mass of a solid ball $S$ if its density $\delta$ is proportional to the distance from the center.

**Solution** Center the ball at the origin and let its radius be $a$. The density $\delta$ is given by $\delta = k\sqrt{x^2 + y^2 + z^2} = k\rho$. Thus, the mass $m$ is given by

$$m = \iiint_S \delta\, dV = k \int_0^\pi \int_0^{2\pi} \int_0^a \rho \rho^2 \sin\phi\, d\rho\, d\theta\, d\phi$$

$$= k\frac{a^4}{4} \int_0^\pi \int_0^{2\pi} \sin\phi\, d\theta\, d\phi = \frac{1}{2}\, k\pi a^4 \int_0^\pi \sin\phi\, d\phi$$

$$= k\pi a^4 \qquad\qquad \blacksquare$$

**EXAMPLE 4** Find the volume and center of mass of the homogeneous solid $S$ that is bounded above by the sphere $\rho = a$ and below by the cone $\phi = \alpha$, where $a$ and $\alpha$ are constants (Figure 8).

**Solution** The volume $V$ is given by

$$V = \int_0^\alpha \int_0^{2\pi} \int_0^a \rho^2 \sin\phi\, d\rho\, d\theta\, d\phi$$

$$= \int_0^\alpha \int_0^{2\pi} \left(\frac{a^3}{3}\right) \sin\phi\, d\theta\, d\phi$$

$$= \frac{2\pi a^3}{3} \int_0^\alpha \sin\phi\, d\phi = \frac{2\pi a^3}{3}(1 - \cos\alpha)$$

It follows that the mass $m$ of the solid is

$$m = kV = \frac{2\pi a^3 k}{3}(1 - \cos\alpha)$$

where $k$ is the constant density.

From symmetry, the center of mass is on the $z$-axis; that is, $\bar{x} = \bar{y} = 0$. To find $\bar{z}$, we first calculate $M_{xy}$.

$$M_{xy} = \iiint_S kz\, dV = \int_0^\alpha \int_0^{2\pi} \int_0^a k(\rho\cos\phi)\rho^2 \sin\phi\, d\rho\, d\theta\, d\phi$$

$$= \int_0^\alpha \int_0^{2\pi} \int_0^a k\rho^3 \sin\phi\cos\phi\, d\rho\, d\theta\, d\phi$$

$$= \int_0^\alpha \int_0^{2\pi} \frac{1}{4} ka^4 \sin\phi\cos\phi\, d\theta\, d\phi$$

$$= \int_0^\alpha \frac{1}{2} \pi ka^4 \sin\phi\cos\phi\, d\phi = \frac{1}{4} \pi a^4 k \sin^2\alpha$$

Thus,

$$\bar{z} = \frac{\frac{1}{4}\pi a^4 k \sin^2\alpha}{\frac{2}{3}\pi a^3 k(1 - \cos\alpha)} = \frac{3a\sin^2\alpha}{8(1 - \cos\alpha)}$$

$$= \frac{3}{8}\,a(1 + \cos\alpha)$$    ∎

**Change of Variables in Multiple Integrals**    The two boxed formulas of this section and the one in Section 16.4 are examples of change of variable formulas. They illustrate a very general result, which we wish to discuss briefly.

Suppose that

$$x = m(u, v) \qquad y = n(u, v)$$

relate the old variables $x$ and $y$ to new variables $u$ and $v$. Define a function $J(u, v)$, called the **Jacobian**, by

$$J(u, v) = \begin{vmatrix} \dfrac{\partial x}{\partial u} & \dfrac{\partial x}{\partial v} \\ \dfrac{\partial y}{\partial u} & \dfrac{\partial y}{\partial v} \end{vmatrix}$$

Then, under suitable conditions on the functions $m$ and $n$ and with appropriate limits on the integral signs,

$$\iint f(x, y)\,dx\,dy = \iint f[m(u, v), n(u, v)]\,|J(u, v)|\,du\,dv$$

Note, for example, that if

$$x = r\cos\theta, \qquad y = r\sin\theta$$

then

$$J(r, \theta) = \begin{vmatrix} \dfrac{\partial x}{\partial r} & \dfrac{\partial x}{\partial \theta} \\ \dfrac{\partial y}{\partial r} & \dfrac{\partial y}{\partial \theta} \end{vmatrix} = \begin{vmatrix} \cos\theta & -r\sin\theta \\ \sin\theta & r\cos\theta \end{vmatrix} = r$$

This is exactly the extra factor that occurred when we changed from rectangular to polar coordinates in Section 16.4.

In the three-variable case, where

$$x = m(u, v, w), \qquad y = n(u, v, w), \qquad z = p(u, v, w)$$

the **Jacobian** $J(u, v, w)$ is defined by

$$J(u, v, w) = \begin{vmatrix} \dfrac{\partial x}{\partial u} & \dfrac{\partial x}{\partial v} & \dfrac{\partial x}{\partial w} \\[2ex] \dfrac{\partial y}{\partial u} & \dfrac{\partial y}{\partial v} & \dfrac{\partial y}{\partial w} \\[2ex] \dfrac{\partial z}{\partial u} & \dfrac{\partial z}{\partial v} & \dfrac{\partial z}{\partial w} \end{vmatrix}$$

Again, $|J(u, v, w)|$ is the extra factor that appears in the transformed integral. In Problems 22 and 23, we ask you to show that

$$J(r, \theta, z) = r, \qquad J(\rho, \theta, \phi) = \rho^2 \sin \phi$$

are the Jacobians corresponding to changing from rectangular to cylindrical and spherical coordinates, respectively. For more details about the change of variable problem, see any book on advanced calculus.

## CONCEPTS REVIEW

**1.** $dz\,dy\,dx$ takes the form _____ in cylindrical coordinates and the form _____ in spherical coordinates.

**2.** $\displaystyle\int_0^1 \int_0^{\sqrt{1-x^2}} \int_0^3 xy\,dz\,dy\,dx$ becomes _____ in cylindrical coordinates.

**3.** If $S$ is the unit ball centered at the origin, then $\iiint\limits_S z^2\,dV$, when written as an iterated integral in spherical coordinates, becomes _____.

**4.** The value of the integral in Question 3 is _____.

## PROBLEM SET 16.8

In Problems 1–6, use cylindrical coordinates to find the indicated quantity.

**1.** The volume of the solid bounded by the paraboloid $z = x^2 + y^2$ and the plane $z = 4$.

**2.** The volume of the solid bounded above by the sphere $x^2 + y^2 + z^2 = 9$, below by the plane $z = 0$, and laterally by the cylinder $x^2 + y^2 = 4$.

**3.** The volume of the solid bounded above by the sphere $r^2 + z^2 = 5$ and below by the paraboloid $r^2 = 4z$.

**4.** The volume of the solid under the surface $z = xy$, above the $xy$-plane, and within the cylinder $x^2 + y^2 = 2x$.

**5.** The center of mass of the homogeneous solid bounded above by $z = 12 - 2x^2 - 2y^2$ and below by $z = x^2 + y^2$.

**6.** The center of mass of the homogeneous solid inside $x^2 + y^2 = 4$, outside $x^2 + y^2 = 1$, below $z = 12 - x^2 - y^2$, and above $z = 0$.

In Problems 7–14, use spherical coordinates to find the indicated quantity.

**7.** The mass of the solid inside the sphere $\rho = b$ and outside the sphere $\rho = a\ (a < b)$ if the density is proportional to the distance from the origin.

**8.** The mass of a solid inside a sphere of radius $2a$ and outside a circular cylinder of radius $a$ whose axis is a diameter of the sphere, if the density is proportional to the square of the distance from the center of the sphere.

**9.** The center of mass of a solid hemisphere of radius $a$, if the density is proportional to the distance from the center of the sphere.

**10.** The center of mass of a solid hemisphere of radius $a$, if the density is proportional to the distance from the axis of symmetry.

**11.** The moment of inertia of the solid of Problem 10 with respect to its axis of symmetry.

**12.** The volume of the solid within the sphere $x^2 + y^2 + z^2 = 16$, outside the cone $z = \sqrt{x^2 + y^2}$, and above the $xy$-plane.

**13.** The volume of the smaller wedge cut from the unit sphere by two planes which meet at a diameter at an angle of $30°$.

**14.** $\int_{-3}^{3} \int_{-\sqrt{9-x^2}}^{\sqrt{9-x^2}} \int_{-\sqrt{9-x^2-z^2}}^{\sqrt{9-x^2-z^2}} (x^2 + y^2 + z^2)^{3/2} \, dy \, dz \, dx$

**15.** Find the volume of the solid bounded above by the plane $z = y$ and below by the paraboloid $z = x^2 + y^2$. *Hint*: In cylindrical coordinates the plane has equation $z = r \sin \theta$ and the paraboloid has equation $z = r^2$. Solve simultaneously to get the projection in the $xy$-plane.

**16.** Find the volume of the solid inside both of the spheres $\rho = 2\sqrt{2} \cos \phi$ and $\rho = 2$.

**17.** For a ball of radius $a$, find each average distance.

(a) From its center.
(b) From a diameter.
(c) From a point on its boundary (consider $\rho = 2a \cos \phi$).

**18.** For any homogeneous solid $S$, show that the average value of the linear function $f(x, y, z) = ax + by + cz + d$ on $S$ is $f(\bar{x}, \bar{y}, \bar{z})$ where $(\bar{x}, \bar{y}, \bar{z})$ is the center of mass.

**19.** A homogeneous ball of radius $a$ is centered at the origin. For the section $S$ bounded by the half-planes $\theta = -\alpha$ and $\theta = \alpha$ (like a section of an orange), find each value.

(a) The $x$-coordinate of the center of mass.
(b) The average distance from the $z$-axis.

**20.** All balls in this problem have radius $a$, constant density $k$, and mass $m$. Find in terms of $a$ and $m$ the moment of inertia of each of the following.

(a) A ball about a diameter.

(b) A ball about a tangent line to its boundary (the Parallel Axis Theorem holds also for solids—see Problem 20 of Section 16.5).
(c) The two-ball solid of Figure 9 about the $z$-axis.

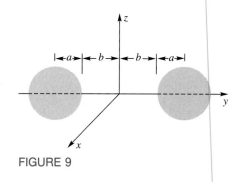

**FIGURE 9**

**21.** Suppose that the left ball in Figure 9 has density $k$ and the right ball density $ck$. Find the $y$-coordinate of the center of mass of this two-ball solid (convince yourself that the analogue of Problem 25 of Section 16.5 is valid).

**22.** Show that the Jacobian for changing from Cartesian to cylindrical coordinates has value $r$.

**23.** Show that the Jacobian for changing from Cartesian to spherical coordinates has value $\rho^2 \sin \phi$.

**24.** Find the volume of the ellipsoid $x^2/a^2 + y^2/b^2 + z^2/c^2 \le 1$ by making the change of variables $x = ua$, $y = vb$, and $z = cw$. Also find the moment of inertia of this solid about the $z$-axis assuming it has constant density $k$.

---

**Answers to Concepts Review:** **1.** $r \, dz \, dr \, d\theta$;

$\rho^2 \sin \phi \, d\rho \, d\theta \, d\phi$ **2.** $\int_0^{\pi/2} \int_0^1 \int_0^3 r^3 \cos \theta \sin \theta \, dz \, dr \, d\theta$

**3.** $\int_0^\pi \int_0^{2\pi} \int_0^1 \rho^4 \cos^2\phi \sin \phi \, d\rho \, d\theta \, d\phi$ **4.** $4\pi/15$

---

## 16.9 CHAPTER REVIEW

### Concepts Test

Respond with true or false to each of the following assertions. Be prepared to defend your answer.

**1.** $\int_a^b \int_a^b f(x)f(y) \, dy \, dx = \left[ \int_a^b f(x) \, dx \right]^2$

**2.** $\int_0^1 \int_0^x f(x, y) \, dy \, dx = \int_0^1 \int_0^y f(x, y) \, dx \, dy$

**3.** $\int_0^2 \int_{-1}^1 \sin(x^3 y^3) \, dx \, dy = 0$

**4.** $\int_{-1}^1 \int_{-1}^1 e^{x^2 + 2y^2} \, dy \, dx = 4 \int_0^1 \int_0^1 e^{x^2 + 2y^2} \, dy \, dx$

**5.** $\int_1^2 \int_0^2 \sin^2(x/y) \, dx \, dy \le 2$.

**6.** If $f$ is continuous and nonnegative on $R$ and $f(x_0, y_0) > 0$ where $(x_0, y_0)$ is an interior point of $R$, then $\iint_R f(x, y) \, dA > 0$.

7. $\iint_R f(x, y)\, dA \le \iint_R g(x, y)\, dA$, then $f(x, y) \le g(x, y)$ on $R$.

8. If $f(x, y) \ge 0$ on $R$ and $\iint_R f(x, y)\, dA = 0$, then $f(x, y) = 0$ for all $(x, y)$ in $R$.

9. If $\delta(x, y) = k$ gives the density of a lamina at $(x, y)$, the coordinates of the center of mass of the lamina do not involve $k$.

10. If $\delta(x, y) = y^2/(1 + x^2)$ gives the density of the lamina $\{(x, y): 0 \le x \le 1, 0 \le y \le 1\}$, we know without calculating that $\bar{x} < \frac{1}{2}$ and $\bar{y} > \frac{1}{2}$.

11. If $S = \{(x, y, z): 1 \le x^2 + y^2 + z^2 \le 16\}$, then $\iiint_S dV = 84\pi$.

## Sample Test Problems

In Problems 1–4, evaluate the integrals.

1. $\displaystyle\int_0^1 \int_x^{\sqrt{x}} xy\, dy\, dx$

2. $\displaystyle\int_{-2}^2 \int_{-\sqrt{4 - y^2}}^{\sqrt{4 - y^2}} 2xy^2\, dx\, dy$

3. $\displaystyle\int_0^{\pi/2} \int_0^{2 \sin \theta} r \cos \theta\, dr\, d\theta$

4. $\displaystyle\int_1^2 \int_3^x \int_0^{\sqrt{3y}} \frac{y}{y^2 + z^2}\, dz\, dy\, dx$

In Problems 5–8, rewrite the iterated integrals with the indicated order of integration. Make a sketch first.

5. $\displaystyle\int_0^1 \int_x^1 f(x, y)\, dy\, dx; \; dx\, dy$

6. $\displaystyle\int_0^1 \int_0^{\cos^{-1} y} f(x, y)\, dx\, dy; \; dy\, dx$

7. $\displaystyle\int_0^1 \int_0^{(1 - x)/2} \int_0^{1 - x - 2y} f(x, y, z)\, dz\, dy\, dx; \; dx\, dz\, dy$

8. $\displaystyle\int_0^2 \int_{x^2}^4 \int_0^{4 - y} f(x, y, z)\, dz\, dy\, dx; \; dx\, dy\, dz$

9. Write the triple iterated integrals for the volume of a sphere of radius $a$ in each case.

(a) Cartesian coordinates.  (b) Cylindrical coordinates.
(c) Spherical coordinates.

10. Evaluate $\iint_S (x + y)\, dA$, where $S$ is the region bounded by $y = \sin x$ and $y = 0$ between $x = 0$ and $x = \pi$.

11. Evaluate $\iiint_S z^2\, dV$, where $S$ is the region bounded by $x^2 + z = 1$ and $y^2 + z = 1$ and the $xy$-plane.

12. If the top of a right circular cylinder of radius 1 is sliced off by a plane that makes an angle of 30° with the base of the cylinder, the area of the resulting slanted top is $2\sqrt{3}\pi/3$.

13. There are eight possible orders of integration for a triple iterated integral.

14. $\displaystyle\int_0^2 \int_0^{2\pi} \int_0^1 dr\, d\theta\, dz$ represents the volume of a right circular cylinder of radius 1 and height 2.

15. If $|f_x| \le 2$ and $|f_y| \le 2$, then the surface $G$ determined by $z = f(x, y)$, $0 \le x \le 1$, $0 \le y \le 1$, has area at most 3.

12. Evaluate $\displaystyle\iint_S \frac{1}{x^2 + y^2}\, dA$ where $S$ is the region between the circles $x^2 + y^2 = 4$ and $x^2 + y^2 = 9$.

13. Find the center of mass of the rectangular lamina bounded by $x = 1$, $x = 3$, $y = 0$, and $y = 2$ if the density is $\delta(x, y) = xy^2$.

14. Find the moment of inertia of the lamina of Problem 13 with respect to the $x$-axis.

15. Find the area of the surface of the cylinder $z^2 + y^2 = 9$ lying in the first octant between the planes $y = x$ and $y = 3x$.

16. Evaluate by changing to cylindrical or spherical coordinates

(a) $\displaystyle\int_0^3 \int_0^{\sqrt{9 - x^2}} \int_0^2 \sqrt{x^2 + y^2}\, dz\, dy\, dx$

(b) $\displaystyle\int_0^2 \int_0^{\sqrt{4 - x^2}} \int_0^{\sqrt{4 - x^2 - y^2}} z\sqrt{4 - x^2 - y^2}\, dz\, dy\, dx$

17. Find the mass of the solid between the spheres $x^2 + y^2 + z^2 = 1$ and $x^2 + y^2 + z^2 = 9$ if the density is proportional to the distance from the origin.

18. Find the center of mass of the homogeneous lamina bounded by the cardioid $r = 4(1 + \sin \theta)$.

19. Find the mass of the solid in the first octant under the plane $x/a + y/b + z/c = 1$ ($a, b, c$ positive) if the density is $\delta(x, y, z) = kx$.

20. Compute the volume of the solid bounded by $z = x^2 + y^2$, $z = 0$, and $x^2 + (y - 1)^2 = 1$.

# TECHNOLOGY PROJECT

**Work in lifting a load.** Consider the work done in lifting a load from the "floor", $z = 0$, to a space above the floor. See Figure 1. The load could be a solid or "fluid" (e.g., water, sand, bricks).

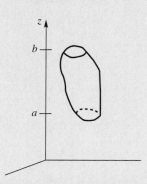

Fig. 1. The space occupied by the lifted material

Let $\rho$ denote the density of the material and $g$ the gravitation constant. One can derive this formula for the work in raising the load to its new space.

$$W = \rho g \iiint_V z \, dx \, dy \, dz,$$

where the subscript $V$ denotes the space now occupied. This is a familiar integral related to the $z$-component of the center of mass. Thus $W = \rho g V \bar{z}$ or $\boxed{W = Mg\bar{z}.}$ Here, $M$ denotes the total mass of the material lifted. The boxed result has important implications. If we know the height $\bar{z}$ for the space in question, and we know the total mass $M$ of the material, then computing the work is an easy calculation.

**Exercise 1**   As an example of the above observation, consider the work done in filling a spherical tank of radius $R$ whose lowest point is at $z = h$. Show that one gets $W = 4\pi\rho gR^3(h + R)/3$ without doing any integrals!

**The inverse square law.** The force law, $F = -mg$, is just an approximation to the inverse square law that is valid near the surface of the earth. It is natural to wonder if the mass center trick also works for the full fledged inverse square law; i.e. *Is the force between two bodies the same as the force between two point masses separated by a distance equal to the distance of the mass centers of the original bodies?* (Note: Newton showed this *is* true for spheres.)

The inverse square law force between two mass particles with respective masses $M$ and $m$ and separated by a distance $q$ has magnitude

$$F_{point} = -\frac{GMm}{q^2} \qquad (1)$$

and acts along the line connecting the two particles.

**Exercise 2**   You are to investigate the force between a particle of mass $m$ located at the point $(0, 0, q)$ on the $z$-axis and a thin square plate of negligible thickness $h$ and of density $\rho$ filling the region $-a/2 \le x \le a/2$, $-a/2 \le y \le a/2$, $-h/2 \le z \le h/2$. Clearly the resultant force is wholly in the $z$-direction. This vertical force between the particle and a volume element of the box is given by

$$\Delta F = \frac{Gm \, \Delta M \cos \alpha}{R^2},$$

where $R = \sqrt{x^2 + y^2 + q^2}$, $\Delta M = \rho h \, \Delta x \, \Delta y$, and $\cos \alpha = q/R$. (Note that if the plate were collapsed onto its mass center (the origin), then the force between the true particle and the fictitious one at the mass center of the plate would be given by Equation (1)).

a) Draw a sketch and justify the formulas given above for $R$, $\Delta M$, and $\cos \alpha$.

b) Assuming that the vertical variation in force is negligible over the small thickness $h$, show that

$$F = -Gmh\rho q \int_{-a/2}^{a/2} \int_{-a/2}^{a/2} \frac{1}{R^3} \, dx \, dy.$$

c) To make our formula for the force look more like the point force formula, eliminate the density $\rho$ and show that the resulting expression is

$$F = -\frac{GMmq}{a^2} \int_{-a/2}^{a/2} \int_{-a/2}^{a/2} \frac{1}{R^3} \, dx \, dy.$$

d) Evaluate the integral in the previous part using your technology. (Set $a = q = 1$ if necessary.) Is $F = F_{point}$?

e) Take $a = 1$ and table the $F/F_{point}$ ratio for $q = 2^k$, $k = 0$, $1, \ldots, 8$. Does $F \to F_{point}$ as $q$ gets large? (Large $q$ results are referred to as "far field" results in the scientific literature.)

# 17

# Vector Calculus

## 17.1
## VECTOR FIELDS

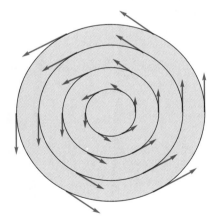

FIGURE 1

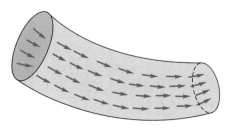

FIGURE 2

All that we have done relates in some way to the concept of a function. This concept and the associated calculus have been steadily generalized as we have moved along.

1. Real-valued functions of one real variable (Chapters 1–12).
2. Vector-valued functions of one real variable (Chapters 13–14).
3. Real-valued functions of several real variables (Chapters 15–16).

You will correctly guess that the next step is to study vector-valued functions of several real variables, which is in fact the principal task of the present chapter. This is the final step in a first calculus course; more advanced courses will carry the process of generalization further—to complex-valued functions of complex variables, to real-valued functions of infinitely many variables, and so on.

Consider then a function $\mathbf{F}$ which associates with each point $\mathbf{p}$ in $n$-space a vector $\mathbf{F(p)}$. A typical example in 2-space is

$$\mathbf{F(p)} = \mathbf{F}(x, y) = -\tfrac{1}{2}y\mathbf{i} + \tfrac{1}{2}x\mathbf{j}$$

For historical reasons, we refer to such a function as a **vector field**, a name arising from a visual image that we now describe. Imagine that to each point $\mathbf{p}$ in a region of space is attached a vector $\mathbf{F(p)}$ emanating from $\mathbf{p}$. We cannot draw all of these vectors, but a representative sample can give us a good intuitive picture of a field. Figure 1 is just such a picture for the vector field $\mathbf{F}(x, y) = -\tfrac{1}{2}y\mathbf{i} + \tfrac{1}{2}x\mathbf{j}$ mentioned earlier. It is the velocity field of a wheel spinning at a constant rate of $\tfrac{1}{2}$ radian per unit of time (see Example 2). Figure 2 might represent the velocity field for water flowing in a curved pipe.

821

Other vector fields that arise naturally in science are electric fields, magnetic fields, force fields, and gravitational fields. We consider only the case in which these fields are independent of time, which we call **steady vector fields**. In contrast to a vector field, a function $F$ that attaches a number to each point in space is called a **scalar field**. The function that gives the temperature at each point would be a good physical example of a scalar field.

**EXAMPLE 1**  Sketch a representative sample of vectors from the vector field

$$\mathbf{F}(x, y) = \frac{x\mathbf{i} + y\mathbf{j}}{\sqrt{x^2 + y^2}}$$

*Solution*  $\mathbf{F}(x, y)$ is a unit vector pointing in the same direction as $x\mathbf{i} + y\mathbf{j}$—that is, away from the origin. Several such vectors are shown in Figure 3. ∎

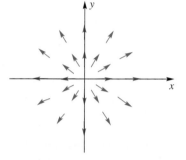

**FIGURE 3**

**EXAMPLE 2**  Show that each vector from the vector field

$$\mathbf{F}(x, y) = -\tfrac{1}{2}y\mathbf{i} + \tfrac{1}{2}x\mathbf{j}$$

is tangent to a circle centered at the origin and has length equal to one-half the radius of that circle (see Figure 1).

*Solution*  If $\mathbf{r} = x\mathbf{i} + y\mathbf{j}$ is the position vector of the point $(x, y)$, then

$$\mathbf{r} \cdot \mathbf{F}(x, y) = -\tfrac{1}{2}xy + \tfrac{1}{2}xy = 0$$

Thus, $\mathbf{F}(x, y)$ is perpendicular to $\mathbf{r}$ and is therefore tangent to the circle of radius $|\mathbf{r}|$. Finally,

$$|\mathbf{F}(x, y)| = \sqrt{(-\tfrac{1}{2}y)^2 + (\tfrac{1}{2}x)^2} = \tfrac{1}{2}\,|\mathbf{r}|$$  ∎

According to Isaac Newton, the magnitude of the force of attraction between objects of mass $M$ and $m$, respectively, is given by $GMm/d^2$, where $d$ is the distance between the objects and $G$ is a universal constant. This is the famous inverse square law of gravitational attraction. It supplies us with an important example of a force field.

**EXAMPLE 3**  Suppose that a spherical object of mass $M$ (for example, the earth) is centered at the origin. Derive the formula for the gravitational field of force $\mathbf{F}(x, y, z)$ exerted by this mass on an object of mass $m$ located at a point $(x, y, z)$ in space. Then sketch this field.

*Solution*  We assume that we may treat the object of mass $M$ as a point mass located at the origin. Let $\mathbf{r} = x\mathbf{i} + y\mathbf{j} + z\mathbf{k}$. Then $\mathbf{F}$ has magnitude

$$|\mathbf{F}| = \frac{GMm}{|\mathbf{r}|^2}$$

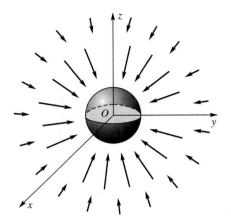

FIGURE 4

The direction of $\mathbf{F}$ is toward the origin; that is, $\mathbf{F}$ has the direction of the unit vector $-\mathbf{r}/|\mathbf{r}|$. We conclude that

$$\mathbf{F}(x, y, z) = \frac{GMm}{|\mathbf{r}|^2} \left( \frac{-\mathbf{r}}{|\mathbf{r}|} \right) = -GMm \frac{\mathbf{r}}{|\mathbf{r}|^3}$$

This field is sketched in Figure 4.  ■

**The Gradient of a Scalar Field**  Let $f(x, y, z)$ determine a scalar field and suppose that $f$ is differentiable. Then the gradient of $f$, denoted by $\nabla f$, is the vector field given by

$$\mathbf{F}(x, y, z) = \nabla f(x, y, z) = \frac{\partial f}{\partial x}\mathbf{i} + \frac{\partial f}{\partial y}\mathbf{j} + \frac{\partial f}{\partial z}\mathbf{k}$$

We first met gradient fields in Sections 15.4 and 15.5. There we learned that $\nabla f(x, y, z)$ points in the direction of greatest increase of $f(x, y, z)$. A vector field $\mathbf{F}$ that is the gradient of a scalar field $f$ is called a **conservative vector field**, and $f$ is its **potential function** (the origin of these names will be clarified in Section 17.3). Such fields and their potential functions are important in physics. In particular, fields that obey the inverse square law (for example, electric fields and gravitational fields) are conservative, as we now show.

**EXAMPLE 4**  Let $\mathbf{F}$ result from an inverse square law; that is, let

$$\mathbf{F}(x, y, z) = -c\frac{\mathbf{r}}{|\mathbf{r}|^3} = -c\frac{x\mathbf{i} + y\mathbf{j} + z\mathbf{k}}{(x^2 + y^2 + z^2)^{3/2}}$$

where $c$ is a constant (see Example 3). Show that

$$f(x, y, z) = \frac{c}{(x^2 + y^2 + z^2)^{1/2}} = c(x^2 + y^2 + z^2)^{-1/2}$$

is a potential function for $\mathbf{F}$, and therefore that $\mathbf{F}$ is conservative (for $\mathbf{r} \neq \mathbf{0}$).

*Solution*

$$\nabla f(x, y, z) = \frac{\partial f}{\partial x}\mathbf{i} + \frac{\partial f}{\partial y}\mathbf{j} + \frac{\partial f}{\partial z}\mathbf{k}$$

$$= \frac{-c}{2}(x^2 + y^2 + c^2)^{-3/2}(2x\mathbf{i} + 2y\mathbf{j} + 2z\mathbf{k})$$

$$= \mathbf{F}(x, y, z) \qquad ■$$

Example 4 was really too easy since we gave the function $f$. A much harder and more significant problem is this. Given a vector field $\mathbf{F}$, decide whether it is conservative and, if so, find its potential function. We discuss this problem in Section 17.3.

**The Divergence and Curl of a Vector Field** With a given vector field

$$\mathbf{F}(x, y, z) = M(x, y, z)\mathbf{i} + N(x, y, z)\mathbf{j} + P(x, y, z)\mathbf{k}$$

are associated two other important fields. The first, called the **divergence** of **F**, is a scalar field; the second, called the **curl** of **F**, is a vector field.

---

**Definition**

**(Div and curl).** Let $\mathbf{F} = M\mathbf{i} + N\mathbf{j} + P\mathbf{k}$ be a vector field for which $\partial M/\partial x$, $\partial N/\partial y$, and $\partial P/\partial z$ exist. Then

$$\operatorname{div} \mathbf{F} = \frac{\partial M}{\partial x} + \frac{\partial N}{\partial y} + \frac{\partial P}{\partial z}$$

$$\operatorname{curl} \mathbf{F} = \left(\frac{\partial P}{\partial y} - \frac{\partial N}{\partial z}\right)\mathbf{i} + \left(\frac{\partial M}{\partial z} - \frac{\partial P}{\partial x}\right)\mathbf{j} + \left(\frac{\partial N}{\partial x} - \frac{\partial M}{\partial y}\right)\mathbf{k}$$

---

**What Do They Mean?**

To help you visualize the divergence and curl, we offer this physical interpretation. If **F** denotes the velocity field for a fluid, then div **F** at a point **p** measures the tendency of that fluid to diverge away from **p** (div **F** > 0) or accumulate toward **p** (div **F** < 0). On the other hand, curl **F** picks out the direction of the axis about which the fluid rotates (curls) most rapidly and |curl **F**| is a measure of the speed of this rotation. The direction of rotation is according to the right hand rule. We will expand on this discussion later in the chapter.

At this point it is hard to see the significance of these fields; that will be apparent later. Our interest now is in learning to calculate divergence and curl easily and in relating them to the gradient operator $\nabla$. Recall that $\nabla$ is the operator

$$\nabla = \frac{\partial}{\partial x}\mathbf{i} + \frac{\partial}{\partial y}\mathbf{j} + \frac{\partial}{\partial z}\mathbf{k}$$

When $\nabla$ operates on a function $f$, it produces the gradient; that is,

$$\nabla f = \operatorname{grad} f$$

Note that by a slight (but very helpful) abuse of notation, we can write

$$\nabla \cdot \mathbf{F} = \left(\frac{\partial}{\partial x}\mathbf{i} + \frac{\partial}{\partial y}\mathbf{j} + \frac{\partial}{\partial z}\mathbf{k}\right) \cdot (M\mathbf{i} + N\mathbf{j} + P\mathbf{k})$$

$$= \frac{\partial M}{\partial x} + \frac{\partial N}{\partial y} + \frac{\partial P}{\partial z} = \operatorname{div} \mathbf{F}$$

$$\nabla \times \mathbf{F} = \begin{vmatrix} \mathbf{i} & \mathbf{j} & \mathbf{k} \\ \dfrac{\partial}{\partial x} & \dfrac{\partial}{\partial y} & \dfrac{\partial}{\partial z} \\ M & N & P \end{vmatrix}$$

$$= \mathbf{i}\left(\frac{\partial P}{\partial y} - \frac{\partial N}{\partial z}\right) - \mathbf{j}\left(\frac{\partial P}{\partial x} - \frac{\partial M}{\partial z}\right) + \mathbf{k}\left(\frac{\partial N}{\partial x} - \frac{\partial M}{\partial y}\right) = \operatorname{curl} \mathbf{F}$$

Thus, grad $f$, div **F**, and curl **F** can all be written in terms of the operator $\nabla$; this is, in fact, the way to remember how these fields are defined.

**EXAMPLE 5**   Let

$$\mathbf{F}(x, y, z) = x^2yz\mathbf{i} + 3xyz^3\mathbf{j} + (x^2 - z^2)\mathbf{k}$$

Find div $\mathbf{F}$ and curl $\mathbf{F}$.

*Solution*

$$\text{div } \mathbf{F} = \nabla \cdot \mathbf{F} = 2xyz + 3xz^3 - 2z$$

$$\text{curl } \mathbf{F} = \nabla \times \mathbf{F} = \begin{vmatrix} \mathbf{i} & \mathbf{j} & \mathbf{k} \\ \dfrac{\partial}{\partial x} & \dfrac{\partial}{\partial y} & \dfrac{\partial}{\partial z} \\ x^2yz & 3xyz^3 & x^2 - z^2 \end{vmatrix}$$

$$= -\mathbf{i}(9xyz^2) - \mathbf{j}(2x - x^2y) + \mathbf{k}(3yz^3 - x^2z) \qquad \blacksquare$$

## CONCEPTS REVIEW

**1.** A function that associates with each point $(x, y, z)$ in space a vector $\mathbf{F}(x, y, z)$ is called a _____ .

**2.** In particular, the function that associates with the scalar function $f(x, y, z)$ the vector $\nabla f(x, y, z)$ is called a _____ .

**3.** Two important examples of vector fields in physics

that arise as gradients of scalar fields are _____ and _____ .

**4.** Given a vector field $\mathbf{F} = M\mathbf{i} + N\mathbf{j} + P\mathbf{k}$, we introduce a corresponding scalar field div $\mathbf{F}$ and a vector field curl $\mathbf{F}$. They can be defined symbolically by div $\mathbf{F}$ = _____ and curl $\mathbf{F}$ = _____ .

## PROBLEM SET 17.1

In Problems 1–6, sketch a sample of vectors for the given vector field $\mathbf{F}$.

**1.** $\mathbf{F}(x, y) = x\mathbf{i} + y\mathbf{j}$       **2.** $\mathbf{F}(x, y) = x\mathbf{i} - y\mathbf{j}$

**3.** $\mathbf{F}(x, y) = -x\mathbf{i} + 2y\mathbf{j}$     **4.** $\mathbf{F}(x, y) = 3x\mathbf{i} + y\mathbf{j}$

**5.** $\mathbf{F}(x, y, z) = x\mathbf{i} + 0\mathbf{j} + \mathbf{k}$

**6.** $\mathbf{F}(x, y, z) = -z\mathbf{k}$

In Problems 7–12, find $\nabla f$.

**7.** $f(x, y, z) = x^2 - 3xy + 2z$

**8.** $f(x, y, z) = \sin(xyz)$

**9.** $f(x, y, z) = \ln|xyz|$

**10.** $f(x, y, z) = \frac{1}{2}(x^2 + y^2 + z^2)$

**11.** $f(x, y, z) = xe^y \cos z$

**12.** $f(x, y, z) = y^2e^{-2z}$

In Problems 13–18, find div $\mathbf{F}$ and curl $\mathbf{F}$.

**13.** $\mathbf{F}(x, y, z) = x^2\mathbf{i} - 2xy\mathbf{j} + yz^2\mathbf{k}$

**14.** $\mathbf{F}(x, y, z) = x^2\mathbf{i} + y^2\mathbf{j} + z^2\mathbf{k}$

**15.** $\mathbf{F}(x, y, z) = yz\mathbf{i} + xz\mathbf{j} + xy\mathbf{k}$

**16.** $\mathbf{F}(x, y, z) = \cos x\mathbf{i} + \sin y\mathbf{j} + 3\mathbf{k}$

**17.** $\mathbf{F}(x, y, z) = e^x \cos y\mathbf{i} + e^x \sin y\mathbf{j} + z\mathbf{k}$

**18.** $\mathbf{F}(x, y, z) = (y + z)\mathbf{i} + (x + z)\mathbf{j} + (x + y)\mathbf{k}$

**19.** Let $f$ be a scalar field and $\mathbf{F}$ a vector field. Indicate which of the following are scalar fields, vector fields, or meaningless.

(a)  div $f$             (b)  grad $f$
(c)  curl $\mathbf{F}$          (d)  div (grad $f$)
(e)  curl (grad $f$)    (f)  grad (div $\mathbf{F}$)
(g)  curl (curl $\mathbf{F}$)    (h)  div (div $\mathbf{F}$)
(i)  grad (grad $f$)    (j)  div (curl (grad $f$ ))
(k)  curl (div(grad $f$ ))

**20.** Assuming that the required partial derivatives exist and are continuous, show that

(a)  div (curl $\mathbf{F}$) = 0;
(b)  curl (grad $f$) = $\mathbf{0}$;
(c)  div ($f\mathbf{F}$) = ($f$)(div $\mathbf{F}$) + (grad $f$) $\cdot$ $\mathbf{F}$;
(d)  curl ($f\mathbf{F}$) = ($f$)(curl $\mathbf{F}$) + (grad $f$) $\times$ $\mathbf{F}$.

**21.** Let $\mathbf{F}(x, y, z) = c\mathbf{r}/|\mathbf{r}|^3$ be an inverse square law field (see Examples 3 and 4). Show that curl $\mathbf{F} = \mathbf{0}$ and div $\mathbf{F} = 0$. *Hint*: Use Problem 20 with $f = -c/|\mathbf{r}|^3$.

**22.** Let $\mathbf{F}(x, y, z) = c\mathbf{r}/|\mathbf{r}|^m$, $c \neq 0$, $m \neq 3$. Show in contrast to Problem 21 that div $\mathbf{F} \neq 0$, though curl $\mathbf{F} = \mathbf{0}$.

**23.** Let $\mathbf{F}(x, y, z) = f(r)\mathbf{r}$, where $r = |\mathbf{r}| = \sqrt{x^2 + y^2 + z^2}$ and $f$ is a differentiable scalar function (except possibly at $r = 0$). Show that curl $\mathbf{F} = \mathbf{0}$ (except at $r = 0$). *Hint*: First show that grad $f = f'(r)\mathbf{r}/r$ and then apply Problem 20d.

**24.** Let $\mathbf{F}(x, y, z)$ be as in Problem 23. Show that if div $\mathbf{F} = 0$, then $f(r) = cr^{-3}$, where $c$ is a constant.

**25.** This problem relates to the interpretation of div and curl given in the margin on page 824. Consider the four velocity fields $\mathbf{F}$, $\mathbf{G}$, $\mathbf{H}$, and $\mathbf{L}$, which have for every $z$ the configuration illustrated in Figure 5. Decide each of the following by geometric reasoning.

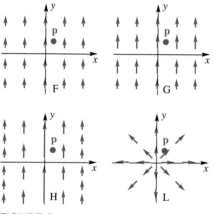

**FIGURE 5**

(a) Is the divergence at $\mathbf{p}$ positive, negative, or zero?
(b) Will a paddle wheel at $\mathbf{p}$ with vertical axis (Figure 4, Section 17.7) rotate clockwise, counterclockwise, or not at all?

Now let $\mathbf{F} = c\mathbf{j}$, $\mathbf{G} = e^{-y^2}\mathbf{j}$, $\mathbf{H} = e^{-x^2}\mathbf{j}$, and $\mathbf{L} = (x\mathbf{i} + y\mathbf{j})/\sqrt{x^2 + y^2}$ which might be modeled as in Figure 5.

(c) Calculate the divergence and curl for each of these fields and thereby confirm your answers in parts (a) and (b).

**26.** Consider the velocity field $\mathbf{v}(x, y, z) = \omega y\mathbf{i} + \omega x\mathbf{j}$, $\omega > 0$, (see Example 2 and Figure 1). Note that $\mathbf{v}$ is perpendicular to $x\mathbf{i} + y\mathbf{j}$ and that $|\mathbf{v}| = \omega\sqrt{x^2 + y^2}$. Thus, $\mathbf{v}$ describes a fluid that is rotating (like a solid) about the $z$-axis with constant angular velocity $\omega$. Show that div $\mathbf{v} = 0$ and curl $\mathbf{v} = 2\omega\mathbf{k}$.

**27.** An object of mass $m$, which is revolving in a circular orbit with constant angular velocity $\omega$, is subject to the centrifugal force given by

$$\mathbf{F}(x, y, z) = m\omega^2\mathbf{r} = m\omega^2(x\mathbf{i} + y\mathbf{j} + z\mathbf{k})$$

Show that

$$f(x, y, z) = \tfrac{1}{2}m\omega^2(x^2 + y^2 + z^2)$$

is a potential function for $\mathbf{F}$.

**28.** The scalar function div(grad $f$) $= \nabla \cdot \nabla f$ (also written $\nabla^2 f$) is called the *Laplacian* and a function $f$ satisfying $\nabla^2 f = 0$ is said to be *harmonic*, concepts important in physics. Show that $\nabla^2 f = f_{xx} + f_{yy} + f_{zz}$. Then find $\nabla^2 f$ for each of the following and decide which are harmonic.

(a) $f(x, y, z) = 2x^2 - y^2 - z^2$
(b) $f(x, y, z) = xyz$
(c) $f(x, y, z) = x^3 - 3xy^2 + 3z$
(d) $f(x, y, z) = (x^2 + y^2 + z^2)^{-1/2}$

**29.** Show that:

(a) div$(\mathbf{F} \times \mathbf{G}) = \mathbf{G} \cdot$ curl $\mathbf{F} - \mathbf{F} \cdot$ curl $\mathbf{G}$
(b) div$(\nabla f \times \nabla g) = 0$

**30.** By analogy with earlier definitions, define each of the following.

(a) $\displaystyle \lim_{(x, y, z) \to (a, b, c)} \mathbf{F}(x, y, z) = \mathbf{L}$
(b) $\mathbf{F}(x, y, z)$ is continuous at $(a, b, c)$.

---

**Answers to Concepts Review:** **1.** Vector-valued function of three real variables or a vector field **2.** Gradient field **3.** Gravitational fields; electric fields **4.** $\nabla \cdot \mathbf{F}$, $\nabla \times \mathbf{F}$

---

**17.2**
**LINE INTEGRALS**

One kind of generalization of the definite integral $\int_a^b f(x)\, dx$ is obtained by replacing the set $[a, b]$ over which we integrate by two- and three-dimensional sets. This led us to the double and triple integrals of Chapter 16. A very different generalization is obtained by replacing $[a, b]$ with a curve $C$. The resulting integral $\int_C f(x, y)\, ds$ is called a *line integral* but would more properly be called a *curve integral*.

Let $C$ be a smooth plane curve; that is, let $C$ be given parametrically by

$$x = x(t), \qquad y = y(t), \qquad a \le t \le b$$

where $x'$ and $y'$ are continuous and not simultaneously zero on $[a, b]$. We suppose that $C$ is **positively oriented** (that is, that its positive direction corresponds to increasing values of $t$) and that $C$ is traced only once as $t$ varies from $a$ to $b$. Thus, $C$ has the initial point $A = (x(a), y(a))$ and the terminal point $B = (x(b), y(b))$. Consider the partition $P$ of the parameter interval $[a, b]$ obtained by inserting the points

$$a = t_0 < t_1 < t_2 < \cdots < t_n = b$$

This partition of $[a, b]$ results in a division of the curve $C$ into $n$ subarcs $P_{i-1}P_i$ in which the point $P_i$ corresponds to $t_i$. Let $\Delta s_i$ denote the length of the arc $P_{i-1}P_i$ and let $|P|$ be the norm of the partition $P$; that is, let $|P|$ be the largest $\Delta t_i = t_i - t_{i-1}$. Finally, choose a sample point $Q_i(\bar{x}_i, \bar{y}_i)$ on the subarc $P_{i-1}P_i$.

Now consider the Riemann sum

$$\sum_{i=1}^{n} f(\bar{x}_i, \bar{y}_i)\, \Delta s_i$$

If $f$ is nonnegative, this sum approximates the area of the curved vertical curtain shown in Figure 1. If $f$ is continuous on a region $D$ containing the curve $C$, then this Riemann sum has a limit as $|P| \to 0$. This limit is called the **line integral of $f$ along $C$ from $A$ to $B$ with respect to arc length**; that is,

$$\int_C f(x, y)\, ds = \lim_{|P| \to 0} \sum_{i=1}^{n} f(\bar{x}_i, \bar{y}_i)\, \Delta s_i$$

It represents (for $f(x, y) \ge 0$) the exact area of the curved curtain of Figure 1.

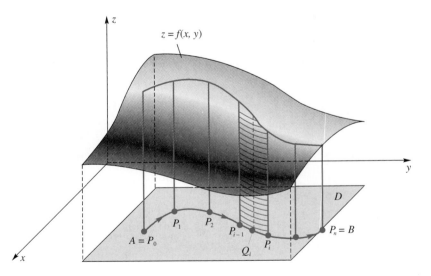

FIGURE 1

The definition does not provide a very good way of evaluating $\int_C f(x, y)\,ds$. That is best accomplished by expressing everything in terms of the parameter $t$ and leads to an ordinary definite integral (recall Section 6.4).

$$\int_C f(x, y)\,ds = \int_a^b f(x(t), y(t))\sqrt{[x'(t)]^2 + [y'(t)]^2}\,dt$$

Of course, a curve can be parameterized in many different ways; fortunately, it can be proved that any parameterization results in the same value for $\int_C f(x, y)\,ds$.

The definition of a line integral can be extended to the case where $C$, though not smooth itself, is piecewise smooth—that is, consists of several smooth curves $C_1, C_2, \ldots, C_k$ joined together, as shown in Figure 2. We simply define the integral over $C$ to be the sum of the integrals over the individual curves.

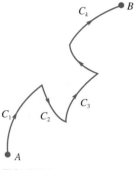

FIGURE 2

**Examples and Applications**   We begin with two examples where $C$ is part of a circle.

**EXAMPLE 1**   Evaluate $\int_C x^2 y\,ds$, where $C$ is determined by the parametric equations $x = 3\cos t$, $y = 3\sin t$, $0 \le t \le \pi/2$. Also show that the parameterization $x = \sqrt{9 - y^2}$, $y = y$, $0 \le y \le 3$, gives the same value.

*Solution*   Formulas for $ds$ were given in Section 6.4. Using the first parameterization, we obtain

$$\int_C x^2 y\,ds = \int_0^{\pi/2} (3\cos t)^2 (3\sin t)\sqrt{(-3\sin t)^2 + (3\cos t)^2}\,dt$$

$$= 81 \int_0^{\pi/2} \cos^2 t \sin t\,dt$$

$$= \left[-\frac{81}{3}\cos^3 t\right]_0^{\pi/2} = 27$$

For the second parameterization,

$$ds = \sqrt{1 + \left(\frac{dx}{dy}\right)^2}\,dy = \sqrt{1 + \frac{y^2}{9 - y^2}}\,dy = \frac{3}{\sqrt{9 - y^2}}\,dy$$

and

$$\int_C x^2 y\,ds = \int_0^3 (9 - y^2)\,y\,\frac{3}{\sqrt{9 - y^2}}\,dy = 3\int_0^3 \sqrt{9 - y^2}\,y\,dy$$

$$= -\left[(9 - y^2)^{3/2}\right]_0^3 = 27 \qquad \blacksquare$$

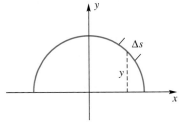

**FIGURE 3**

**EXAMPLE 2** A thin wire is bent in the shape of the semicircle

$$x = a \cos t, \qquad y = a \sin t, \qquad 0 \le t \le \pi, \qquad a > 0$$

If the density of the wire at a point is proportional to its distance from the $x$-axis, find the mass and center of mass of the wire.

**Solution** Our old motto—*slice, approximate, integrate*—is still appropriate. The mass of a small piece of wire of length $\Delta s$ (Figure 3) is approximately $\delta(x, y)\,\Delta s$ where $\delta(x, y) = ky$ is the density at $(x, y)$ ($k$ is a constant). Thus, the mass $m$ of the whole wire is

$$m = \int_C ky\,ds = \int_0^\pi ka \sin t\sqrt{a^2 \sin^2 t + a^2 \cos^2 t}\,dt$$

$$= ka^2 \int_0^\pi \sin t\,dt$$

$$= \left[ -ka^2 \cos t \right]_0^\pi = 2ka^2$$

The moment of the wire with respect to the $x$-axis is given by

$$M_x = \int_C y\,ky\,ds = \int_0^\pi ka^3 \sin^2 t\,dt$$

$$= \frac{ka^3}{2} \int_0^\pi (1 - \cos 2t)\,dt$$

$$= \frac{ka^3}{2} \left[ t - \tfrac{1}{2} \sin 2t \right]_0^\pi = \frac{ka^3 \pi}{2}$$

Thus,

$$\bar{y} = \frac{M_x}{m} = \frac{\tfrac{1}{2}ka^3 \pi}{2ka^2} = \tfrac{1}{4}\pi a$$

From symmetry, $\bar{x} = 0$, so the center of mass is at $(0, \pi a/4)$. ■

All that we have done extends easily to a smooth curve $C$ in three-space. In particular, if $C$ is given parametrically by

$$x = x(t), \qquad y = y(t), \qquad z = z(t), \qquad a \le t \le b$$

then

$$\boxed{\int_C f(x, y, z)\,ds = \int_a^b f(x(t), y(t), z(t))\sqrt{[x'(t)]^2 + [y'(t)]^2 + [z'(t)]^2}\,dt}$$

**EXAMPLE 3** Find the mass of a wire of density $\delta(x, y, z) = kz$ if it has the shape of the helix $C$ with parameterization

$$x = 3 \cos t, \qquad y = 3 \sin t, \qquad z = 4t, \qquad 0 \le t \le \pi$$

*Solution*

$$m = \int_C kz\,ds = k\int_0^\pi (4t)\sqrt{9\sin^2 t + 9\cos^2 t + 16}\,dt$$

$$= 20k\int_0^\pi t\,dt = \left[20k\frac{t^2}{2}\right]_0^\pi = 10k\pi^2$$

The units for $m$ depend on those for length and density.  ■

**Work**  Suppose that the force acting at a point $(x, y, z)$ in space is given by the vector field

$$\mathbf{F}(x, y, z) = M(x, y, z)\mathbf{i} + N(x, y, z)\mathbf{j} + P(x, y, z)\mathbf{k}$$

where $M$, $N$, and $P$ are continuous. We want to find the work $W$ done by $\mathbf{F}$ in moving a particle along a smooth oriented curve $C$. Let $\mathbf{r} = x\mathbf{i} + y\mathbf{j} + z\mathbf{k}$ be the position vector for a point $Q(x, y, z)$ on the curve (Figure 4). If $\mathbf{T}$ is the unit tangent vector $d\mathbf{r}/ds$ at $Q$, then $\mathbf{F} \cdot \mathbf{T}$ is the tangential component of $\mathbf{F}$ at $Q$. The work done by $\mathbf{F}$ in moving the particle from $Q$ a short distance $\Delta s$ along the curve is approximately $\mathbf{F} \cdot \mathbf{T}\,\Delta s$, and consequently the work done in moving the particle from $A$ to $B$ along $C$ is defined to be $\int_C \mathbf{F} \cdot \mathbf{T}\,ds$. From Section 13.5, we know that $\mathbf{T} = (d\mathbf{r}/dt)(dt/ds)$, and so we have the following alternative formulas for work.

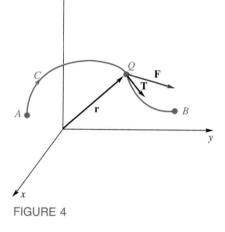

FIGURE 4

$$W = \int_C \mathbf{F} \cdot \mathbf{T}\,ds = \int_C \mathbf{F} \cdot \frac{d\mathbf{r}}{dt}\,dt = \int_C \mathbf{F} \cdot d\mathbf{r}$$

To interpret the last expression, think of $\mathbf{F} \cdot d\mathbf{r}$ as representing the work done by $\mathbf{F}$ in moving a particle along the "infinitesimal" tangent vector $d\mathbf{r}$, a formulation preferred by physicists and many applied mathematicians.

There is still another expression for work that is often useful in calculations. If we agree to write $d\mathbf{r} = dx\mathbf{i} + dy\mathbf{j} + dz\mathbf{k}$, then

$$\mathbf{F} \cdot d\mathbf{r} = (M\mathbf{i} + N\mathbf{j} + P\mathbf{k}) \cdot (dx\mathbf{i} + dy\mathbf{j} + dz\mathbf{k}) = M\,dx + N\,dy + P\,dz$$

and

$$W = \int_C \mathbf{F} \cdot d\mathbf{r} = \int_C M\,dx + N\,dy + P\,dz$$

The integrals $\int_C M\,dx$, $\int_C N\,dy$, and $\int_C P\,dz$ are a special kind of line integral. They are defined just as $\int_C f\,ds$ was defined at the beginning of the section, except that $\Delta s_i$ is replaced by $\Delta x_i$, $\Delta y_i$, and $\Delta z_i$, respectively. However, we point out that while $\Delta s_i$ is always taken to be positive, $\Delta x_i$, $\Delta y_i$, and $\Delta z_i$ may well be negative on a path $C$. The result of this is that a change in the orientation of $C$ switches the sign of $\int_C M\,dx$, $\int_C N\,dy$, and $\int_C P\,dz$ while leaving that of $\int_C f\,ds$ unchanged (see Problem 31).

**EXAMPLE 4**   Find the work done by the inverse-square law force field

$$\mathbf{F}(x, y, z) = \frac{-c\mathbf{r}}{|\mathbf{r}|^3} = \frac{-c(x\mathbf{i} + y\mathbf{j} + z\mathbf{k})}{(x^2 + y^2 + z^2)^{3/2}} = M\mathbf{i} + N\mathbf{j} + P\mathbf{k}$$

in moving a particle along the straight-line curve $C$ from $(0, 3, 0)$ to $(4, 3, 0)$ shown in Figure 5.

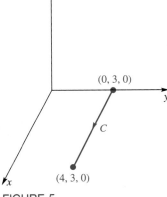

**Solution**   Along $C$, $y = 3$ and $z = 0$, so $dy = dz = 0$. Using $x$ as the parameter, we obtain

$$W = \int_C M\,dx + N\,dy + P\,dz = -c \int_C \frac{x\,dx + y\,dy + z\,dz}{(x^2 + y^2 + z^2)^{3/2}}$$

$$= -c \int_0^4 \frac{x}{(x^2 + 9)^{3/2}}\,dx = \left[\frac{c}{(x^2 + 9)^{1/2}}\right]_0^4 = \frac{-2c}{15}$$

FIGURE 5

Of course, appropriate units must be assigned, depending on those for length and force.   ■

Here is a planar version of the line integral introduced above.

**EXAMPLE 5**   Evaluate the line integral

$$\int_C (x^2 - y^2)\,dx + 2xy\,dy$$

along the curve $C$ whose parametric equations are $x = t^2$, $y = t^3$, $0 \le t \le \frac{3}{2}$.

**Solution**   Since $dx = 2t\,dt$ and $dy = 3t^2\,dt$,

$$\int_C (x^2 - y^2)\,dx + 2xy\,dy = \int_0^{3/2} [(t^4 - t^6)2t + 2t^5(3t^2)]\,dt$$

$$= \int_0^{3/2} (2t^5 + 4t^7)\,dt = \frac{8505}{512} \approx 16.61$$   ■

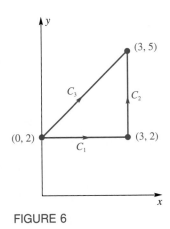

FIGURE 6

**EXAMPLE 6**   Evaluate $\int_C xy^2\,dx + xy^2\,dy$ along the path $C = C_1 \cup C_2$ shown in Figure 6. Also evaluate this integral along the straight path $C_3$ from $(0, 2)$ to $(3, 5)$.

***Solution*** On $C_1$, $y = 2$, $dy = 0$, and

$$\int_{C_1} xy^2\, dx + xy^2\, dy = \int_0^3 4x\, dx = \left[2x^2\right]_0^3 = 18$$

On $C_2$, $x = 3$, $dx = 0$, and

$$\int_{C_2} xy^2\, dx + xy^2\, dy = \int_2^5 3y^2\, dy = \left[y^3\right]_2^5 = 117$$

We conclude that

$$\int_C xy^2\, dx + xy^2\, dy = 18 + 117 = 135$$

On $C_3$, $y = x + 2$, $dy = dx$, and so

$$\int_{C_3} xy^2\, dx + xy^2\, dy = 2 \int_0^3 x(x + 2)^2\, dx$$

$$= 2 \int_0^3 (x^3 + 4x^2 + 4x)\, dx = 2\left[\frac{x^4}{4} + \frac{4x^3}{3} + 2x^2\right]_0^3$$

$$= \frac{297}{2}$$

Note that the two paths from $(0, 2)$ to $(3, 5)$ give different values for the integral. ∎

## CONCEPTS REVIEW

**1.** A curve $C$ given parametrically by $x = x(t)$, $y = y(t)$, $a \le t \le b$, is said to be positively oriented if its positive direction corresponds to _____.

**2.** The line integral $\int_C f(x, y)\, ds$, where $C$ is the positively oriented curve of Question 1, is defined as $\lim\limits_{|P| \to 0}$ _____.

**3.** The line integral in Question 2 transforms to the ordinary integral $\int_a^b$ _____ $dt$.

**4.** If $\mathbf{r} = x(t)\mathbf{i} + y(t)\mathbf{j}$ is the position vector of a point on the curve $C$ of Question 1 and if $\mathbf{F} = M(x, y)\mathbf{i} + N(x, y)\mathbf{j}$ is a force field in the plane, then the work $W$ done by $\mathbf{F}$ in moving an object along $C$ is given by $\int_C$ _____ $dt$.

## PROBLEM SET 17.2

In Problems 1–16, evaluate the line integrals.

**1.** $\int_C (x^3 + y)\, ds$; $C$ is the curve $x = 3t$, $y = t^3$, $0 \le t \le 1$.

**2.** $\int_C xy^{2/5}\, ds$; $C$ is the curve $x = \frac{1}{2}t$, $y = t^{5/2}$, $0 \le t \le 1$.

**3.** $\int_C (\sin x + \cos y)\, ds$; $C$ is the line segment from $(0, 0)$ to $(\pi, 2\pi)$.

**4.** $\int_C xe^y\, ds$; $C$ is the line segment from $(-1, 2)$ to $(1, 1)$.

**5.** $\int_C (2x + 9z)\, ds$; $C$ is the curve $x = t$, $y = t^2$, $z = t^3$, $0 \le t \le 1$.

**6.** $\int_C (x^2 + y^2 + z^2)\, ds$; $C$ is the curve $x = 4\cos t$, $y = 4\sin t$, $z = 3t$, $0 \le t \le 2\pi$.

7. $\int_C y\,dx + x^2\,dy$; $C$ is the curve $x = 2t$, $y = t^2 - 1$, $0 \le t \le 2$.

8. $\int_C y\,dx + x^2\,dy$; $C$ is the right-angle curve from $(0, -1)$ to $(4, -1)$ to $(4, 3)$.

9. $\int_C y^3\,dx + x^3\,dy$; $C$ is the right-angle curve from $(-4, 1)$ to $(-4, -2)$ to $(2, -2)$.

10. $\int_C y^3\,dx + x^3\,dy$; $C$ is the curve $x = 2t$, $y = t^2 - 3$, $-2 \le t \le 1$.

11. $\int_C (x + 2y)\,dx + (x - 2y)\,dy$; $C$ is the line segment from $(1, 1)$ to $(3, -1)$.

12. $\int_C y\,dx + x\,dy$; $C$ is the curve $y = x^2$, $0 \le x \le 1$.

13. $\int_C (x + y + z)\,dx + x\,dy - yz\,dz$; $C$ is the line segment from $(1, 2, 1)$ to $(2, 1, 0)$.

14. $\int_C xz\,dx + (y + z)\,dy + x\,dz$; $C$ is the curve $x = e^t$, $y = e^{-t}$, $z = e^{2t}$, $0 \le t \le 1$.

15. $\int_C (x + y + z)\,dx + (x - 2y + 3z)\,dy + (2x + y - z)\,dz$; $C$ is the line-segment path from $(0, 0, 0)$ to $(2, 0, 0)$ to $(2, 3, 0)$ to $(2, 3, 4)$.

16. Same integral as in Problem 15; $C$ is the line segment from $(0, 0, 0)$ to $(2, 3, 4)$.

17. Find the mass of a wire with the shape of the curve $y = x^2$ between $(-2, 4)$ and $(2, 4)$ if the density is given by $\delta(x, y) = k|x|$.

18. A wire of constant density has the shape of the helix $x = a \cos t$, $y = a \sin t$, $z = bt$, $0 \le t \le 3\pi$. Find its mass and center of mass.

In Problems 19–24, find the work done by the force field **F** in moving a particle along the curve $C$.

19. $\mathbf{F}(x, y) = (x^3 - y^3)\mathbf{i} + xy^2\mathbf{j}$; $C$ is the curve $x = t^2$, $y = t^3$, $-1 \le t \le 0$.

20. $\mathbf{F}(x, y) = e^x\mathbf{i} - e^{-y}\mathbf{j}$; $C$ is the curve $x = 3 \ln t$, $y = \ln 2t$, $1 \le t \le 5$.

21. $\mathbf{F}(x, y) = (x + y)\mathbf{i} + (x - y)\mathbf{j}$; $C$ is the quarter-ellipse $x = a \cos t$, $y = b \sin t$, $0 \le t \le \pi/2$.

22. $\mathbf{F}(x, y, z) = (2x - y)\mathbf{i} + 2z\mathbf{j} + (y - z)\mathbf{k}$; $C$ is the line segment from $(0, 0, 0)$ to $(1, 1, 1)$.

23. Same **F** as in Problem 22; $C$ is the curve $x = \sin(\pi t/2)$, $y = \sin(\pi t/2)$, $z = t$, $0 \le t \le 1$.

24. $\mathbf{F}(x, y, z) = y\mathbf{i} + z\mathbf{j} + x\mathbf{k}$; $C$ is the curve $x = t$, $y = t^2$, $z = t^3$, $0 \le t \le 2$.

25. Karen plans to paint both sides of a fence whose base is in the $xy$-plane with shape $x = 30 \cos^3 t$, $y = 30 \sin^3 t$, $0 \le t \le \pi/2$ and whose height at $(x, y)$ is $1 + \frac{1}{3}y$, all measured in feet. Sketch a picture of the fence and decide how much paint she will need if a gallon covers 200 square feet.

26. A squirrel weighing 1.2 pounds climbed a cylindrical tree by following the helical path $x = \cos t$, $y = \sin t$, $z = 4t$, $0 \le t \le 8\pi$ (distance measured in feet). How much work did it do? Use a line integral, but then think of a trivial way to answer this question.

27. Use a line integral to find the area of the part cut out of the vertical square cylinder $|x| + |y| = a$ by the sphere $x^2 + y^2 + z^2 = a^2$. Check your answer by finding a trivial way to do this problem.

28. A wire of constant density $k$ has the shape $|x| + |y| = a$. Find its moment of inertia with respect to the $y$-axis and with respect to the $z$-axis.

29. Use a line integral to find the area of that part of the cylinder $x^2 + y^2 = ay$ inside the sphere $x^2 + y^2 + z^2 = a^2$ (compare with Problem 12, Section 16.6). *Hint*: Use polar coordinates where $ds = [r^2 + (dr/d\theta)^2]^{1/2}\,d\theta$; see Problem 26, Section 12.8.

30. Two circular cylinders of radius $a$ intersect so their axes meet at right angles. Use a line integral to find the area of the part from one cut off by the other (compare with Problem 14, Section 16.6).

31. Evaluate (a) $\int_C x^2 y\,ds$ using the parameterization $x = 3 \sin t$, $y = 3 \cos t$, $0 \le t \le \pi/2$, which reverses the orientation of $C$ in Example 1 and (b) $\int_{C_4} xy^2\,dx + xy^2\,dy$ using the parameterization $x = 3 - t$, $y = 5 - t$, $0 \le t \le 3$, and note that $C_4$ has the reverse orientation of $C_3$ in Example 6. Orientation-reversing parameterizations do not change the sign of $\int_C f\,ds$ but do change the sign of the other types of line integrals considered in this section.

---

**Answers to Concepts Review:** 1. Increasing values of $t$

2. $\sum_{i=1}^{n} f(\bar{x}_i, \bar{y}_i)\,\Delta s_i$   3. $f(x(t), y(t))\sqrt{[x'(t)]^2 + [y'(t)]^2}$

4. $\mathbf{F} \cdot d\mathbf{r}/dt$

**17.3**
**INDEPENDENCE**
**OF PATH**

The basic tool in evaluating ordinary definite integrals is the Fundamental Theorem of Calculus. In symbols, it says that

$$\int_a^b f'(x)\, dx = f(b) - f(a)$$

Now we ask the question: Is there an analogous theorem for line integrals? The answer is yes.

In what follows, interpret $\mathbf{r}(t)$ as $x(t)\mathbf{i} + y(t)\mathbf{j}$ if the context is two-space and as $x(t)\mathbf{i} + y(t)\mathbf{j} + z(t)\mathbf{k}$ if it is three-space. Correspondingly, $f(\mathbf{r})$ will mean $f(x, y)$ in the first case and $f(x, y, z)$ in the second.

### Theorem A

**(Fundamental Theorem for Line Integrals).** Let $C$ be a piecewise smooth curve given parametrically by $\mathbf{r} = \mathbf{r}(t)$, $a \le t \le b$, which begins at $\mathbf{a} = \mathbf{r}(a)$ and ends at $\mathbf{b} = \mathbf{r}(b)$. If $f$ is continuously differentiable on an open set containing $C$, then

$$\int_C \nabla f(\mathbf{r}) \cdot d\mathbf{r} = f(\mathbf{b}) - f(\mathbf{a})$$

**Proof** We suppose first that $C$ is smooth. Then

$$\int_C \nabla f(\mathbf{r}) \cdot d\mathbf{r} = \int_a^b [\nabla f(\mathbf{r}(t)) \cdot \mathbf{r}'(t)]\, dt$$

$$= \int_a^b \frac{d}{dt}\, [f(\mathbf{r}(t))]\, dt = f(\mathbf{r}(b)) - f(\mathbf{r}(a))$$

$$= f(\mathbf{b}) - f(\mathbf{a})$$

Note how we first wrote the line integral as an ordinary definite integral, then applied the Chain Rule, and finally used the Fundamental Theorem of Calculus.

If $C$ is not smooth but only piecewise smooth, we simply apply the above result to the individual pieces. We leave the details to you. ∎

**EXAMPLE 1** Recall from Example 4 of Section 17.1 that

$$f(x, y, z) = f(\mathbf{r}) = \frac{c}{|\mathbf{r}|} = \frac{c}{\sqrt{x^2 + y^2 + z^2}}$$

is a potential function for the inverse square law field $\mathbf{F}(\mathbf{r}) = -c\mathbf{r}/|\mathbf{r}|^3$. Calculate $\displaystyle\int_C \mathbf{F}(\mathbf{r}) \cdot d\mathbf{r}$ where $C$ is any simple piecewise smooth curve from $(0, 3, 0)$ to $(4, 3, 0)$ that misses the origin

***Solution*** Since $\mathbf{F}(\mathbf{r}) = \nabla f(\mathbf{r})$,

$$\int_C \mathbf{F}(\mathbf{r}) \cdot d\mathbf{r} = \int_C \nabla f(\mathbf{r}) \cdot d\mathbf{r} = f(4, 3, 0) - f(0, 3, 0)$$

$$= \frac{c}{\sqrt{16 + 9}} - \frac{c}{\sqrt{9}} = \frac{-2c}{15} \qquad \blacksquare$$

Now compare Example 1 with Example 4 of the previous section. There we calculated the same integral, but for a specific curve $C$, namely, the line segment from $(0, 3, 0)$ to $(4, 3, 0)$. Surprisingly, we will get the same answer no matter what curve we take from $(0, 3, 0)$ to $(4, 3, 0)$. We say that the given line integral is independent of path.

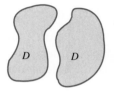

A Connected Set

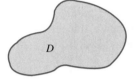

A Disconnected Set

**FIGURE 1**

**Criteria for Independence of Path** Call a set $D$ **connected** if any two points in $D$ can be joined by a piecewise smooth curve lying entirely in $D$ (Figure 1). Then call $\int_C \mathbf{F}(\mathbf{r}) \cdot d\mathbf{r}$ **independent of path in $D$** if for any two points $A$ and $B$ in $D$, the line integral has the same value for every path $C$ in $D$ that is positively oriented from $A$ to $B$.

One consequence of Theorem A is that if $\mathbf{F}$ is the gradient of another function $f$, then $\int_C \mathbf{F}(\mathbf{r}) \cdot d\mathbf{r}$ is independent of path. The converse is also true.

---

**Theorem B**

Let $\mathbf{F}(\mathbf{r})$ be continuous on an open connected set $D$. Then the line integral $\int_C \mathbf{F}(\mathbf{r}) \cdot d\mathbf{r}$ is independent of path if and only if $\mathbf{F}(\mathbf{r}) = \nabla f(\mathbf{r})$ for some scalar function $f$—that is, if and only if $\mathbf{F}$ is a conservative vector field on $D$.

---

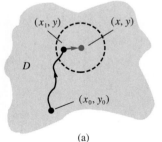

(a)

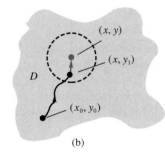

(b)

**FIGURE 2**

***Proof*** Theorem $A$ takes care of the "if" statement. Suppose then that $\int_C \mathbf{F}(\mathbf{r}) \cdot d\mathbf{r}$ is independent of path in $D$. Our task is to construct a function $f$ satisfying $\nabla f = \mathbf{F}$; that is, we must find a potential for the vector field $\mathbf{F}$. For simplicity, we restrict ourselves to the two-dimensional case where $D$ is a plane set and $\mathbf{F}(\mathbf{r}) = M(x, y)\mathbf{i} + N(x, y)\mathbf{j}$.

Let $(x_0, y_0)$ be a fixed point of $D$ and let $(x, y)$ be any other point of $D$. Choose a third point $(x_1, y)$ in $D$ and slightly to the left of $(x, y)$ and join it to $(x, y)$ by a horizontal segment in $D$. Then join $(x_0, y_0)$ to $(x_1, y)$ by a curve in $D$. (All this is possible because $D$ is both open and connected; see Figure 2(a)). Finally, let $C$ denote the path from $(x_0, y_0)$ to $(x, y)$ composed of these two pieces and define $f$ by

$$f(x, y) = \int_C \mathbf{F}(\mathbf{r}) \cdot d\mathbf{r} = \int_{(x_0, y_0)}^{(x_1, y)} \mathbf{F}(\mathbf{r}) \cdot d\mathbf{r} + \int_{(x_1, y)}^{(x, y)} \mathbf{F}(\mathbf{r}) \cdot d\mathbf{r}$$

That we get a unique value is clear from the assumed independence of path.

The first integral on the right above does not depend on $x$; the second, which has $y$ fixed, can be written as an ordinary definite integral using—for example—$t$ as a parameter. It follows that

$$\frac{\partial f}{\partial x} = 0 + \frac{\partial}{\partial x} \int_{x_1}^{x} M(t, y)\, dt = M(x, y)$$

The last equality is a consequence of the theorem on differentiating an integral with respect to its upper limit (Theorem 5.7D).

A similar argument using Figure 2(b) shows that $\partial f/\partial y = N(x, y)$. We conclude that $\nabla f = M(x, y)\mathbf{i} + N(x, y)\mathbf{j} = \mathbf{F}$ as desired. ∎

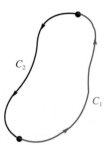

**FIGURE 3**

The condition that $\int_{C} \mathbf{F}(\mathbf{r}) \cdot d\mathbf{r}$ is independent of path in $D$ implies that if $C$ is any closed, oriented curve in $D$, then $\int_{C} \mathbf{F}(\mathbf{r}) \cdot d\mathbf{r} = 0$. To see this, consider $C$ to be composed of two oriented curves $C_1$ and $C_2$, as shown in Figure 3. Let $-C_2$ denote the curve $C_2$ with opposite orientation. Since $C_1$ and $-C_2$ have the same initial and terminal points, the independence of path guarantees that

$$\int_{C} \mathbf{F}(\mathbf{r}) \cdot d\mathbf{r} = \int_{C_1} \mathbf{F}(\mathbf{r}) \cdot d\mathbf{r} + \int_{C_2} \mathbf{F}(\mathbf{r}) \cdot d\mathbf{r}$$

$$= \int_{C_1} \mathbf{F}(\mathbf{r}) \cdot d\mathbf{r} - \int_{-C_2} \mathbf{F}(\mathbf{r}) \cdot d\mathbf{r}$$

$$= \int_{C_1} \mathbf{F}(\mathbf{r}) \cdot d\mathbf{r} - \int_{C_1} \mathbf{F}(\mathbf{r}) \cdot d\mathbf{r} = 0$$

The argument just given is reversible. Hence we have three equivalent conditions.

1. $\mathbf{F} = \nabla f$ for some function $f$ ($\mathbf{F}$ is conservative).
2. $\int_{C} \mathbf{F}(\mathbf{r}) \cdot d\mathbf{r}$ is independent of path.
3. $\int_{C} \mathbf{F}(\mathbf{r}) \cdot d\mathbf{r} = 0$ for every closed path.

There is an interesting physical interpretation of Condition 3. The work done by a conservative force field as it moves a particle around a closed path is zero. In particular, this is true of both gravitational fields and electric fields, since they are conservative.

While Conditions 2 and 3 each imply that $\mathbf{F}$ is the gradient of a scalar function $f$, they are not particularly useful in this connection. A more useful criterion is given in the following theorem. We need, however, to impose an additional condition on $D$, namely, that it is **simply connected**. In two-space, this means that $D$ has no "holes" and in three-space that it has no "tunnels" all the way through $D$. (For the technical definition, see any advanced calculus book.)

---

### Theorem C

Let $\mathbf{F} = M\mathbf{i} + N\mathbf{j} + P\mathbf{k}$, where $M$, $N$, and $P$ are continuous together with their first-order partial derivatives in an open connected set $D$, which is also simply connected. Then $\mathbf{F}$ is conservative ($\mathbf{F} = \nabla f$) if and only if curl $\mathbf{F} = \mathbf{0}$, that is, if and only if

$$\frac{\partial M}{\partial y} = \frac{\partial N}{\partial x}, \qquad \frac{\partial M}{\partial z} = \frac{\partial P}{\partial x}, \qquad \frac{\partial N}{\partial z} = \frac{\partial P}{\partial y}$$

In particular, in the two-variable case where $\mathbf{F} = M\mathbf{i} + N\mathbf{j}$, $\mathbf{F}$ is conservative if and only if

$$\frac{\partial M}{\partial y} = \frac{\partial N}{\partial x}$$

---

The "only if" statement is easy to prove (Problem 17). The "if" statement follows from Green's Theorem (Theorem 17.4A) in the two-variable case and from Stokes's Theorem in the three-variable case (see Example 4 of Section 17.7). Problem 25 shows the need for simple connectedness.

**Recovering a Function from Its Gradient**    Suppose that we are given a vector field $\mathbf{F}$ satisfying the conditions of Theorem C. Then we know there is a function $f$ satisfying $\nabla f = \mathbf{F}$. But how can we find $f$? We illustrate the answer first for a two-dimensional vector field.

**EXAMPLE 2**    Determine whether $\mathbf{F} = (4x^3 + 9x^2y^2)\mathbf{i} + (6x^3y + 6y^5)\mathbf{j}$ is conservative and, if so, find the function $f$ of which it is the gradient.

*Solution*    $M(x, y) = 4x^3 + 9x^2y^2$ and $N(x, y) = 6x^3y + 6y^5$. In this, the two-variable case, the conditions of Theorem C reduce to showing

$$\frac{\partial M}{\partial y} = \frac{\partial N}{\partial x}$$

Now

$$\frac{\partial M}{\partial y} = 18x^2y, \qquad \frac{\partial N}{\partial x} = 18x^2y$$

so the condition is satisfied and $f$ must exist.

To find $f$, we first note that

$$\nabla f = \frac{\partial f}{\partial x}\mathbf{i} + \frac{\partial f}{\partial y}\mathbf{j} = M\mathbf{i} + N\mathbf{j}$$

Thus,

(1)
$$\frac{\partial f}{\partial x} = 4x^3 + 9x^2y^2, \qquad \frac{\partial f}{\partial y} = 6x^3y + 6y^5$$

If we antidifferentiate the left equation with respect to $x$, we obtain

(2)
$$f(x, y) = x^4 + 3x^3y^2 + C_1(y)$$

in which the "constant" of integration $C_1$ may depend on $y$. But the partial with respect to $y$ of the expression in (2) must equal $6x^3y + 6y^5$.

$$\frac{\partial f}{\partial y} = 6x^3y + C_1'(y) = 6x^3y + 6y^5$$

We conclude that $C_1'(y) = 6y^5$. Another antidifferentiation gives

$$C_1(y) = y^6 + C$$

where $C$ is a constant (independent of both $x$ and $y$). Substitution of this result in (1) yields

$$f(x, y) = x^4 + 3x^3y^2 + y^6 + C \qquad \blacksquare$$

Next we use the result of Example 2 to calculate a line integral.

**EXAMPLE 3**   Let $\mathbf{F}(\mathbf{r}) = \mathbf{F}(x, y) = (4x^3 + 9x^2y^2)\mathbf{i} + (6x^3y + 6y^5)\mathbf{j}$. Calculate $\int_C \mathbf{F}(\mathbf{r}) \cdot d\mathbf{r} = \int_C (4x^3 + 9x^2y^2)\, dx + (6x^3y + 6y^5)\, dy$, where $C$ is any path from $(0, 0)$ to $(1, 2)$.

*Solution*   Example 1 shows that $\mathbf{F} = \nabla f$, where

$$f(x, y) = x^4 + 3x^3y^2 + y^6 + C$$

and thus the given line integral is independent of path. In fact, by Theorem A,

$$\int_C \mathbf{F}(\mathbf{r}) \cdot d\mathbf{r} = \left[ x^4 + 3x^3y^2 + y^6 + C \right]_{(0, 0)}^{(1, 2)}$$

$$= 1 + 12 + 64 = 77 \qquad \blacksquare$$

**EXAMPLE 4**   Show that $\mathbf{F} = (e^x \cos y + yz)\mathbf{i} + (xz - e^x \sin y)\mathbf{j} + xy\mathbf{k}$ is conservative and find $f$ such that $\mathbf{F} = \nabla f$.

*Solution*

$$M = e^x \cos y + yz, \qquad N = xz - e^x \sin y, \qquad P = xy$$

and so

$$\frac{\partial M}{\partial y} = -e^x \sin y + z = \frac{\partial N}{\partial x}, \qquad \frac{\partial M}{\partial z} = y = \frac{\partial P}{\partial x}, \qquad \frac{\partial N}{\partial z} = x = \frac{\partial P}{\partial y}$$

which are the conditions of Theorem C. Now

(3)
$$\frac{\partial f}{\partial x} = e^x \cos y + yz$$

$$\frac{\partial f}{\partial y} = xz - e^x \sin y$$

$$\frac{\partial f}{\partial z} = xy$$

When we antidifferentiate the first of these with respect to $x$, we get

(4)
$$f(x, y, z) = e^x \cos y + xyz + C_1(y, z)$$

Now differentiate (4) with respect to $y$ and set the result equal to the second expression in (3).

$$-e^x \sin y + xz + \frac{\partial C_1}{\partial y} = xz - e^x \sin y$$

or

$$\frac{\partial C_1(y, z)}{\partial y} = 0$$

Antidifferentiating the latter with respect to $y$ gives

$$C_1(y, z) = C_2(z)$$

which we in turn substitute in (4).

(5)
$$f(x, y, z) = e^x \cos y + xyz + C_2(z)$$

When we differentiate (5) with respect to $z$ and equate the result to the third expression in (3), we get

$$\frac{\partial f}{\partial z} = xy + C_2'(z) = xy$$

or $C_2'(z) = 0$ and $C_2(z) = C$. We conclude that

$$f(x, y, z) = e^x \cos y + xyz + C \qquad \blacksquare$$

**Conservation of Energy**   Let us make an application to physics and at the same time offer a reason for the name *conservative* force field. We will establish the Law of Conservation of Energy, which says that the sum of the kinetic energy and the potential energy of an object due to a conservative force is constant.

Suppose that an object of mass $m$ is moving along a smooth curve $C$ given by

$$\mathbf{r} = \mathbf{r}(t) = x(t)\mathbf{i} + y(t)\mathbf{j} + z(t)\mathbf{k}, \qquad a \le t \le b$$

under the influence of a conservative force $\mathbf{F}(\mathbf{r}) = \nabla f(\mathbf{r})$. From physics, we learn three facts about the object at time $t$.

1. $\mathbf{F}(\mathbf{r}(t)) = m\mathbf{a}(t) = m\mathbf{r}''(t)$   (Newton's Second Law).
2. $\text{KE} = \frac{1}{2}m|\mathbf{r}'(t)|^2$   (KE = kinetic energy).
3. $\text{PE} = -f(\mathbf{r})$   (PE = potential energy).

Thus,

$$\frac{d}{dt}(\text{KE} + \text{PE}) = \frac{d}{dt}\left[\frac{1}{2}m|\mathbf{r}'(t)|^2 - f(\mathbf{r})\right]$$

$$= \frac{m}{2}\frac{d}{dt}[\mathbf{r}'(t) \cdot \mathbf{r}'(t)] - \left[\frac{\partial f}{\partial x}\frac{dx}{dt} + \frac{\partial f}{\partial y}\frac{dy}{dt} + \frac{\partial f}{\partial z}\frac{dz}{dt}\right]$$

$$= m\mathbf{r}''(t) \cdot \mathbf{r}'(t) - \nabla f(\mathbf{r}) \cdot \mathbf{r}'(t)$$

$$= [m\mathbf{r}''(t) - \nabla f(\mathbf{r})] \cdot \mathbf{r}'(t)$$

$$= [\mathbf{F}(\mathbf{r}) - \mathbf{F}(\mathbf{r})] \cdot \mathbf{r}'(t) = 0$$

We conclude that KE + PE is constant.

## CONCEPTS REVIEW

**1.** Let $C$ be determined by $\mathbf{r} = \mathbf{r}(t)$, $a \le t \le b$, and let $\mathbf{a} = \mathbf{r}(a)$ and $\mathbf{b} = \mathbf{r}(b)$. Then, by the Fundamental Theorem for Line Integrals, $\int_C \nabla f(\mathbf{r}) \, d\mathbf{r} = $ _____.

**2.** $\int_C \mathbf{F}(\mathbf{r}) \, d\mathbf{r}$ is independent of path if and only if

$\mathbf{F}$ is a _____ vector field—that is, if and only if $\mathbf{F}(\mathbf{r}) = $ _____ for some scalar function $f$.

**3.** If curl $\mathbf{F} = $ _____ in an open connected and simply connected set D, then $\mathbf{F} = \nabla f$ for some $f$ defined on D. Conversely, curl$(\nabla f) = $ _____.

**4.** Let $\mathbf{F} = f(x)\mathbf{i} + g(y)\mathbf{j}$ be a two-dimensional vector field. Since $\partial f/\partial y = \partial g/\partial x$, we conclude that _____.

## PROBLEM SET 17.3

In Problems 1–10, determine whether the given field $\mathbf{F}$ is conservative. If so, find $f$ so that $\mathbf{F} = \nabla f$; if not, state that $\mathbf{F}$ is not conservative. See Examples 2 and 4.

**1.** $\mathbf{F}(x, y) = (10x - 7y)\mathbf{i} - (7x - 2y)\mathbf{j}$

**2.** $\mathbf{F}(x, y) = (12x^2 + 3y^2 + 5y)\mathbf{i} + (6xy - 3y^2 + 5x)\mathbf{j}$

**3.** $\mathbf{F}(x, y) = (45x^4y^2 - 6y^6 + 3)\mathbf{i} + (18x^5y - 12xy^5 + 7)\mathbf{j}$

**4.** $\mathbf{F}(x, y) = (35x^4 - 3x^2y^4 + y^9)\mathbf{i} - (4x^3y^3 - 9xy^8)\mathbf{j}$

**5.** $\mathbf{F}(x, y) = \left(\frac{6x^2}{5y^2}\right)\mathbf{i} - \left(\frac{4x^3}{5y^3}\right)\mathbf{j}$

**6.** $\mathbf{F}(x, y) = 4y^2 \cos(xy^2)\mathbf{i} + 8x \cos(xy^2)\mathbf{j}$

**7.** $\mathbf{F}(x, y) = (2e^y - ye^x)\mathbf{i} + (2xe^y - e^x)\mathbf{j}$

8. $\mathbf{F}(x, y) = -e^{-x} \ln y \mathbf{i} + e^{-x} y^{-1} \mathbf{j}$

9. $\mathbf{F}(x, y, z) = 3x^2 \mathbf{i} + 6y^2 \mathbf{j} + 9z^2 \mathbf{k}$

10. $\mathbf{F}(x, y, z) = (2xy + z^2) \mathbf{i} + x^2 \mathbf{j} + (2xz + \pi \cos \pi z) \mathbf{k}$

In Problems 11–16, show that the given line integral is independent of path (use Theorem C) and then evaluate the integral (either by choosing a convenient path or, if you prefer, by finding a potential function $f$ and applying Theorem A).

11. $\displaystyle\int_{(-1, 2)}^{(3, 1)} (y^2 + 2xy) \, dx + (x^2 + 2xy) \, dy$

12. $\displaystyle\int_{(0, 0)}^{(1, \pi/2)} e^x \sin y \, dx + e^x \cos y \, dy$

13. $\displaystyle\int_{(0, 0, 0)}^{(1, 1, 1)} (6xy^3 + 2z^2) \, dx + 9x^2 y^2 \, dy + (4xz + 1) \, dz$

*Hint*: Try the path consisting of line segments from $(0, 0, 0)$ to $(1, 0, 0)$ to $(1, 1, 0)$ to $(1, 1, 1)$.

14. $\displaystyle\int_{(0, 1, 0)}^{(1, 1, 1)} (yz + 1) \, dx + (xz + 1) \, dy + (xy + 1) \, dz$

15. $\displaystyle\int_{(0, 0, 0)}^{(-1, 0, \pi)} (y + z) \, dx + (x + z) \, dy + (x + y) \, dz$

16. $\displaystyle\int_{(0, 0, 0)}^{(\pi, \pi, 0)} (\cos x + 2yz) \, dx + (\sin y + 2xz) \, dy$
$$+ (z + 2xy) \, dz$$

17. Suppose $\nabla f(x, y, z) = M(x, y, z) \mathbf{i} + N(x, y, z) \mathbf{j} + P(x, y, z) \mathbf{k}$ where $M$, $N$, and $P$ have continuous first-order partial derivatives in an open set $D$. Prove that

$$\frac{\partial M}{\partial y} = \frac{\partial N}{\partial x}, \qquad \frac{\partial M}{\partial z} = \frac{\partial P}{\partial x}, \qquad \frac{\partial N}{\partial z} = \frac{\partial P}{\partial y}$$

in $D$. *Hint*: Use Theorem 15.3B on $f$.

18. For each $(x, y, z)$, let $\mathbf{F}(x, y, z)$ be a vector pointed toward the origin with magnitude inversely proportional to the distance from the origin; that is, let

$$\mathbf{F}(x, y, z) = \frac{-k(x\mathbf{i} + y\mathbf{i} + z\mathbf{k})}{(x^2 + y^2 + z^2)}$$

Show that $\mathbf{F}$ is conservative by finding a potential function for $\mathbf{F}$. *Hint*: If this looks like hard work, see Problem 20.

19. Follow the directions of Problem 18 for $\mathbf{F}(x, y, z)$ directed away from the origin with magnitude that is proportional to the distance from the origin.

20. Generalize Problems 18 and 19 by showing that if

$$\mathbf{F}(x, y, z) = [g(x^2 + y^2 + z^2)](x\mathbf{i} + y\mathbf{i} + z\mathbf{k})$$

where $g$ is a continuous function of one variable, then $\mathbf{F}$ is

conservative. *Hint*: Show that $\mathbf{F} = \nabla f$, where $f(x, y, z) = \frac{1}{2} h(x^2 + y^2 + z^2)$ and $h(u) = \int g(u) \, du$.

21. Suppose that an object of mass $m$ is moved along a smooth curve $C$ described by

$$\mathbf{r} = \mathbf{r}(t) = x(t) \mathbf{i} + y(t) \mathbf{j} + z(t) \mathbf{k}, \qquad a \le t \le b$$

while subject only to the continuous force $\mathbf{F}$. Show that the work done is equal to the change in the kinetic energy of the object, that is, show that

$$\int_C \mathbf{F} \cdot d\mathbf{r} = \frac{m}{2} [|\mathbf{r}'(b)|^2 - |\mathbf{r}'(a)|^2]$$

*Hint*: $\mathbf{F}(\mathbf{r}(t)) = m\mathbf{r}''(t)$.

22. Roberto moved a heavy object along the ground from $A$ to $B$. The object was at rest at the beginning and at the end. Does Problem 21 imply that Roberto did no work? Explain.

23. We normally consider the gravitational force of the earth on an object of mass $m$ to be given by the constant $\mathbf{F} = -gm\mathbf{k}$, but, of course, this is valid only in small regions near the earth's surface. Find the potential function $f$ for $\mathbf{F}$ and use it to show that the work done by $\mathbf{F}$ when an object is moved from $(x_1, y_1, z_1)$ to a nearby point $(x_2, y_2, z_2)$ is $mg(z_1 - z_2)$.

C 24. The distance from the earth (mass $m$) to the sun (mass $M$) varies from a maximum (aphelion) of 152.1 million kilometers to a minimum (perihelion) of 147.1 million kilometers. Assume that Newton's Inverse Square Law $\mathbf{F} = -GMm\mathbf{r}/|\mathbf{r}|^3$ holds with $G = 6.67 \times 10^{-11}$ newton-meters$^2$/kilogram$^2$, $M = 1.99 \times 10^{30}$ kilograms, and $m = 5.97 \times 10^{24}$ kilograms. How much work does $\mathbf{F}$ do in moving the earth in each case?

(a) From aphelion to perihelion.
(b) Around a complete orbit.

25. This problem shows the need for simple connectedness in the "if" statement of Theorem C. Let $\mathbf{F} = (y\mathbf{i} - x\mathbf{j})/(x^2 + y^2)$ on the set $D = \{(x, y): x^2 + y^2 \ne 0\}$. Show each of the following.

(a) The condition $\partial M/\partial y = \partial N/\partial x$ holds on $D$.
(b) $\mathbf{F}$ is not conservative on $D$.

*Hint*: To establish (b), show that $\displaystyle\int_C \mathbf{F} \cdot d\mathbf{r} = -2\pi$, where $C$ is the circle with parametric equations $x = \cos t$, $y = \sin t$, $0 \le t \le 2\pi$.

26. Let $f(x, y) = \tan^{-1}(y/x)$. Show that $\nabla f = (y\mathbf{i} - x\mathbf{j})/(x^2 + y^2)$, which is the vector function of Problem 25. Why doesn't the hint to that problem violate Theorem A?

**Answers to Concepts Review:** 1. $f(\mathbf{b}) - f(\mathbf{a})$
2. Gradient; $\nabla f(\mathbf{r})$ 3. 0; 0 4. $\mathbf{F}$ is conservative

## 17.4
### GREEN'S THEOREM IN THE PLANE

We begin with another look at the Fundamental Theorem of Calculus,

$$\int_a^b f'(x)\, dx = f(b) - f(a)$$

It says that the integral of a function over a set $S = [a, b]$ is equal to a related function (the antiderivative) evaluated in a certain way on the boundary of $S$, which in this case consists of just the two points $a$ and $b$. In the remainder of this chapter, we are going to give three generalizations of this result: the theorems of Green, Gauss, and Stokes. These theorems are applied in physics, particularly in the study of heat, electricity, magnetism, and fluid flow. The first of these theorems is due to George Green (1793–1841), a self-taught English mathematical physicist.

We suppose that $C$ is a simple closed curve (Section 13.1), which forms the boundary of a region $S$ in the $xy$-plane. Let $C$ be oriented so that traversing $C$ in its positive direction keeps $S$ to the left (the counterclockwise orientation). The corresponding line integral of $\mathbf{F}(x, y) = M(x, y)\mathbf{i} + N(x, y)\mathbf{j}$ around $C$ is denoted by

$$\oint_C M\, dx + N\, dy$$

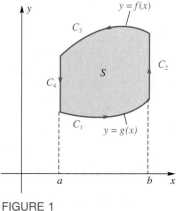

**FIGURE 1**

---

**Theorem A**

**(Green's Theorem).** Let $C$ be a piecewise smooth, simple closed curve that forms the boundary of a region $S$ in the $xy$-plane. If $M(x, y)$ and $N(x, y)$ are continuous and have continuous partial derivatives on $S$ and its boundary $C$, then

$$\iint_S \left( \frac{\partial N}{\partial x} - \frac{\partial M}{\partial y} \right) dA = \oint_C M\, dx + N\, dy$$

---

**A Promised Result**

Suppose that $\partial N/\partial x = \partial M/\partial y$. Then Green's Theorem tells us that

$$\oint_C M\, dx + N\, dy = 0$$

This in turn implies that the field $\mathbf{F} = M\mathbf{i} + N\mathbf{j}$ is conservative. This is part of what we claimed in Theorem 17.3C for the two-variable case

**Proof** We prove the theorem for the case where $S$ is both an $x$-simple and a $y$-simple set and then discuss extensions to the general case.

Since $S$ is $y$-simple, it has the shape of Figure 1; that is,

$$S = \{(x, y): g(x) \le y \le f(x), a \le x \le b\}$$

Its boundary $C$ consists of four arcs $C_1$, $C_2$, $C_3$, and $C_4$ ($C_2$ or $C_4$ could be degenerate) and

$$\oint_C M\, dx = \int_{C_1} M\, dx + \int_{C_2} M\, dx + \int_{C_3} M\, dx + \int_{C_4} M\, dx$$

The integrals over $C_2$ and $C_4$ are zero, since on these curves $x$ is constant, so that $dx = 0$. Thus,

$$\oint_C M \, dx = \int_a^b M(x, g(x)) \, dx + \int_b^a M(x, f(x)) \, dx$$

$$= -\int_a^b [M(x, f(x)) - M(x, g(x))] \, dx$$

$$= -\int_a^b \int_{g(x)}^{f(x)} \frac{\partial M(x, y)}{\partial y} \, dy \, dx$$

$$= -\iint_S \frac{\partial M}{\partial y} \, dA$$

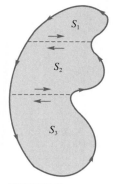

FIGURE 2

Similarly, by treating $S$ as an $x$-simple set, we obtain

$$\oint_C N \, dy = \iint_S \frac{\partial N}{\partial x} \, dA$$

We conclude that Green's Theorem holds on a set that is both $x$-simple and $y$-simple.

The result extends easily to a region $S$ that decomposes into a union of regions $S_1, S_2, \ldots, S_k$, which are both $x$- and $y$-simple (Figure 2). We simply apply the theorem in the form we have proved to each of these sets and then add the results. Note that the contributions of the line integrals cancel on boundaries shared by adjoining regions, since these boundaries are traversed twice but in opposite directions. ■

FIGURE 3

Green's Theorem even holds for a region $S$ with one or more holes (Figure 3), provided that each part of the boundary is oriented so $S$ is always on the left as one traverses the curve in its positive direction. We simply decompose it into ordinary regions in the manner shown in Figure 4.

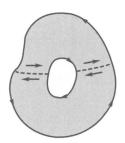

FIGURE 4

**Examples and Applications**   Sometimes, Green's Theorem provides the simplest way of evaluating a line integral.

**EXAMPLE 1**   Let $C$ be the boundary of the triangle with vertices $(0, 0)$, $(1, 2)$, and $(0, 2)$ (Figure 5). Calculate

$$\oint_C 4x^2y \, dx + 2y \, dy$$

(a) by the direct method, (b) by Green's Theorem.

**Solution**   (a) On $C_1$, $y = 2x$ and $dy = 2 \, dx$, so

$$\int_{C_1} 4x^2y \, dx + 2y \, dy = \int_0^1 8x^3 \, dx + 8x \, dx = \left[ 2x^4 + 4x^2 \right]_0^1 = 6$$

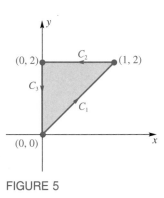

FIGURE 5

Also,

$$\int_{C_2} 4x^2y\,dx + 2y\,dy = \int_1^0 8x^2\,dx = \left[\frac{8x^3}{3}\right]_1^0 = -\frac{8}{3}$$

$$\int_{C_3} 4x^2y\,dx + 2y\,dy = \int_2^0 2y\,dy = \left[y^2\right]_2^0 = -4$$

Thus,

$$\oint_C 4x^2y\,dx + 2y\,dy = 6 - \tfrac{8}{3} - 4 = -\tfrac{2}{3}$$

(b) By Green's Theorem,

$$\oint_C 4x^2y\,dx + 2y\,dy = \int_0^1 \int_{2x}^2 (0 - 4x^2)\,dy\,dx$$

$$= \int_0^1 \left[-4x^2y\right]_{2x}^2\,dx = \int_0^1 (-8x^2 + 8x^3)\,dx$$

$$= \left[\frac{-8x^3}{3} + 2x^4\right]_0^1 = -\frac{2}{3}$$ ∎

**EXAMPLE 2**  Show that if a region $S$ in the plane has boundary $C$, where $C$ is a piecewise smooth, simple closed curve, then the area of $S$ is given by

$$A(S) = \tfrac{1}{2} \oint_C x\,dy - y\,dx$$

*Solution*  Let $M(x, y) = -y/2$ and $N(x, y) = x/2$ and apply Green's Theorem.

$$\oint_C \left(-\frac{y}{2}\,dx + \frac{x}{2}\,dy\right) = \iint_S \left(\frac{1}{2} + \frac{1}{2}\right)dA = A(S)$$ ∎

**EXAMPLE 3**  Use the result of Example 2 to find the area enclosed by the ellipse $b^2x^2 + a^2y^2 = a^2b^2$.

*Solution*  The given ellipse has parametric equations

$$x = a \cos t, \qquad y = b \sin t, \qquad 0 \le t \le 2t$$

Thus,

$$A(S) = \tfrac{1}{2}\oint_C x\,dy - y\,dx$$

$$= \tfrac{1}{2} \int_0^{2\pi} (a \cos t)(b \cos t\,dt) - (b \sin t)(-a \sin t\,dt)$$

$$= \tfrac{1}{2} \int_0^{2\pi} ab(\cos^2 t + \sin^2 t)\,dt$$

$$= \tfrac{1}{2}ab \int_0^{2\pi} dt = \pi ab$$ ∎

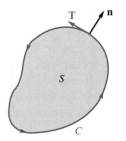

FIGURE 6

**EXAMPLE 4** Use Green's Theorem to evaluate the line integral

$$\oint_C (x^3 + 2y)\,dx + (4x - 3y^2)\,dy$$

where $C$ is the ellipse $b^2x^2 + a^2y^2 = a^2b^2$.

***Solution*** Let $M(x, y) = x^3 + 2y$, $N(x, y) = 4x - 3y^2$ so that $\partial M/\partial y = 2$ and $\partial N/\partial x = 4$. By Green's Theorem and Example 3,

$$\oint_C (x^3 + 2y)\,dx + (4x - 3y^2)\,dy = \iint_S (4 - 2)\,dA$$

$$= 2A(S) = 2\pi ab \qquad \blacksquare$$

**Vector Forms of Green's Theorem**  Our next goal is to restate Green's Theorem for the plane in its vector form and in two different ways. It is these forms that we will generalize later to two important theorems in three-space.

 We suppose that $C$ is a smooth, simple closed curve in the $xy$-plane and that it has been given a counterclockwise orientation by means of its arc length parameterization $x = x(s)$ and $y = y(s)$. Then

$$\mathbf{T} = \frac{dx}{ds}\mathbf{i} + \frac{dy}{ds}\mathbf{j}$$

is a unit tangent vector and

$$\mathbf{n} = \frac{dy}{ds}\mathbf{i} - \frac{dx}{ds}\mathbf{j}$$

is a unit normal vector pointing out of the region $S$ bounded by $C$ (Figure 6). (Note that $\mathbf{T} \cdot \mathbf{n} = 0$.) If $\mathbf{F}(x, y) = M(x, y)\mathbf{i} + N(x, y)\mathbf{j}$ is a vector field, then

$$\oint_C \mathbf{F} \cdot \mathbf{n}\,ds = \oint_C (M\mathbf{i} + N\mathbf{j}) \cdot \left(\frac{dy}{ds}\mathbf{i} - \frac{dx}{ds}\mathbf{j}\right) ds = \oint_C - N\,dx + M\,dy$$

$$= \iint_S \left(\frac{\partial M}{\partial x} + \frac{\partial N}{\partial y}\right) dA$$

The last equality comes from Green's Theorem. On the other hand,

$$\text{div } \mathbf{F} = \nabla \cdot \mathbf{F} = \frac{\partial M}{\partial x} + \frac{\partial N}{\partial y}$$

We conclude that

$$\boxed{\oint_C \mathbf{F} \cdot \mathbf{n}\,ds = \iint_S \text{div } \mathbf{F}\,dA = \iint_S \nabla \cdot \mathbf{F}\,dA}$$

a result sometimes called Gauss's Divergence Theorem in the plane.

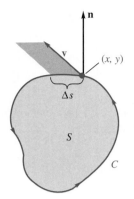

**FIGURE 7**

We give a physical interpretation to this last formula and thereby also understand the origin of the term *divergence*. Imagine a uniform layer of a fluid of constant density moving across the $xy$-plane, a layer so thin that we may consider it to be two-dimensional. We wish to compute the rate at which the fluid in a region $S$ crosses its boundary curve $C$ (Figure 7).

Let $\mathbf{F}(x, y) = \mathbf{v}(x, y)$ denote the velocity vector of the fluid at $(x, y)$ and let $\Delta s$ be the length of a short segment of the curve with initial point $(x, y)$. The amount of fluid crossing this segment per unit of time is approximately the area of the parallelogram of Figure 7, namely, $\mathbf{v} \cdot \mathbf{n} \Delta s$. The (net) amount of fluid leaving $S$ per unit of time, called the **flux** of the vector field $\mathbf{F}$ across the curve $C$ in the outward direction, is therefore

$$\text{flux of } \mathbf{F} \text{ across } C = \oint_C \mathbf{F} \cdot \mathbf{n} \, ds$$

Now consider a fixed point $(x_0, y_0)$ in $S$ and a small circle $C_r$, of radius $r$ around it. On $S_r$, the circular region with boundary $C_r$, div $\mathbf{F}$ will be approximately equal to its value div $\mathbf{F}(x_0, y_0)$ at the center (we are assuming that div $\mathbf{F}$ is continuous); so by Green's Theorem,

$$\text{flux of } \mathbf{F} \text{ across } C_r = \oint_C \mathbf{F} \cdot \mathbf{n} \, ds$$

$$= \iint_{S_r} \text{div } \mathbf{F} \, dA \approx \text{div } \mathbf{F}(x_0, y_0)(\pi r^2)$$

We conclude that div $\mathbf{F}(x_0, y_0)$ measures the rate at which the fluid is "diverging away" from $(x_0, y_0)$. If div $\mathbf{F}(x_0, y_0) > 0$, there is a *source* of fluid at $(x_0, y_0)$; if div $\mathbf{F}(x_0, y_0) < 0$, there is a *sink* for the fluid at $(x_0, y_0)$. If the flux across the boundary of a region is zero, then the sources and sinks in the region must balance each other. On the other hand, if there are no sources or sinks in a region $S$, then div $\mathbf{F} = 0$ and, by Green's Theorem, there is a net flow of zero across the boundary of $S$.

There is another vector form for Green's Theorem. We redraw Figure 6 but now as a subset of three-space (Figure 8). If $\mathbf{F} = M\mathbf{i} + N\mathbf{j} + 0\mathbf{k}$, then Green's Theorem says

$$\oint_C \mathbf{F} \cdot \mathbf{T} \, ds = \oint_C M \, dx + N \, dy = \iint_S \left( \frac{\partial N}{\partial x} - \frac{\partial M}{\partial y} \right) dA$$

On the other hand,

$$\text{curl } \mathbf{F} = \nabla \times \mathbf{F} = \begin{vmatrix} \mathbf{i} & \mathbf{j} & \mathbf{k} \\ \dfrac{\partial}{\partial x} & \dfrac{\partial}{\partial y} & \dfrac{\partial}{\partial z} \\ M & N & 0 \end{vmatrix} = \left( \frac{\partial N}{\partial x} - \frac{\partial M}{\partial y} \right) \mathbf{k}$$

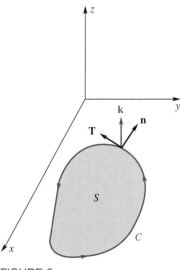

**FIGURE 8**

so that

$$(\text{curl } \mathbf{F}) \cdot \mathbf{k} = \left( \frac{\partial N}{\partial x} - \frac{\partial M}{\partial y} \right)$$

Green's Theorem thus takes the form

$$\oint_C \mathbf{F} \cdot \mathbf{T} \, ds = \iint_S (\text{curl } \mathbf{F}) \cdot \mathbf{k} \, dA$$

which is sometimes called Stoke's Theorem in the plane.

If we apply this result to a small circle $C_r$ centered at $(x_0, y_0)$, we obtain

$$\oint_{C_r} \mathbf{F} \cdot \mathbf{T} \, ds \approx (\text{curl } \mathbf{F}(x_0, y_0)) \cdot \mathbf{k} \, (\pi r^2)$$

This says that the flow in the direction of the tangent to $C_r$ (the so-called *circulation* of $\mathbf{F}$ around $C_r$) is measured by the curl of $\mathbf{F}$. In other words, curl $\mathbf{F}$ measures the tendency of the fluid to rotate about $(x_0, y_0)$. If curl $\mathbf{F} = \mathbf{0}$ in a region $S$, the corresponding fluid flow is said to be *irrotational*.

**EXAMPLE 5** The vector field $\mathbf{F}(x, y) = -\frac{1}{2}y\mathbf{i} + \frac{1}{2}x\mathbf{j} = M\mathbf{i} + N\mathbf{j}$ is the velocity field of a steady counterclockwise rotation of a wheel about the z-axis (see Example 2 of Section 17.1). Calculate $\oint_C \mathbf{F} \cdot \mathbf{n} \, ds$ and $\oint_C \mathbf{F} \cdot \mathbf{T} \, ds$ for any closed curve $C$ in the $xy$-plane.

**Solution** If $S$ is the region enclosed by $C$,

$$\oint_C \mathbf{F} \cdot \mathbf{n} \, ds = \iint_S \text{div } \mathbf{F} \, dA = \iint_S \left( \frac{\partial M}{\partial x} + \frac{\partial N}{\partial y} \right) dA = 0$$

$$\oint_C \mathbf{F} \cdot \mathbf{T} \, ds = \iint_S (\text{curl } \mathbf{F}) \cdot \mathbf{k} \, dA = \iint_S \left( \frac{\partial N}{\partial x} - \frac{\partial M}{\partial y} \right) dA$$

$$= \iint_S \left( \frac{1}{2} + \frac{1}{2} \right) dA = \text{area } S \qquad \blacksquare$$

## CONCEPTS REVIEW

**1.** Let $C$ be a simple closed curve bounding a region $S$ in the $xy$-plane. Then by Green's Theorem,

$$\oint_C M \, dx + N \, dy = \iint_S \underline{\hspace{1cm}} dA.$$

**2.** Thus, if $C$ is the boundary of the square $S = \{(x, y): 0 \le x \le 1, 0 \le y \le 1\}$,

$$\oint_C y \, dx - x \, dy = \iint_S \underline{\hspace{1cm}} dA = \underline{\hspace{1cm}}.$$

**3.** The div $\mathbf{F}(x, y)$ measures the rate at which a homogeneous fluid flow with velocity field $\mathbf{F}$ diverges away from $(x, y)$. If div $\mathbf{F}(x, y) > 0$, there is a _____ of fluid at $(x, y)$; if div $\mathbf{F}(x, y) < 0$, there is a _____ at $(x, y)$.

**4.** On the other hand, curl $\mathbf{F}(x, y)$ measures the tendency of the fluid to _____ about $(x, y)$. If curl $\mathbf{F}(x, y) = \mathbf{0}$ in a region, the flow is _____.

## PROBLEM SET 17.4

In Problems 1–6, use Green's Theorem to evaluate the given line integrals. Begin by sketching the region $S$.

**1.** $\oint_C 2xy\, dx + y^2\, dy$, where $C$ is the closed curve formed by $y = x/2$ and $y = \sqrt{x}$ between $(0, 0)$ and $(4, 2)$.

**2.** $\oint_C \sqrt{y}\, dx + \sqrt{x}\, dy$, where $C$ is the closed curve formed by $y = 0$, $x = 2$, and $y = x^2/2$.

**3.** $\oint_C (2x + y^2)\, dx + (x^2 + 2y)\, dy$, where $C$ is the closed curve formed by $y = 0$, $x = 2$, and $y = x^3/4$.

**4.** $\oint_C xy\, dx + (x + y)\, dy$, where $C$ is the triangle with vertices $(0, 0)$, $(2, 0)$, and $(0, 1)$.

**5.** $\oint_C (x^2 + 4xy)\, dx + (2x^2 + 3y)\, dy$, where $C$ is the ellipse $9x^2 + 16y^2 = 144$.

**6.** $\oint_C (e^{3x} + 2y)\, dx + (x^2 + \sin y)\, dy$, where $C$ is the rectangle with vertices $(2, 1)$, $(6, 1)$, $(6, 4)$, and $(2, 4)$.

In Problems 7–8, use the result of Example 2 to find the area of the indicated region $S$. Make a sketch.

**7.** $S$ is bounded by the curves $y = 4x$ and $y = 2x^2$.

**8.** $S$ is bounded by the curves $y = \frac{1}{2}x^3$ and $y = x^2$.

In Problems 9–12, use the vector forms of Green's Theorem to calculate (a) $\oint_C \mathbf{F} \cdot \mathbf{n}\, ds$, (b) $\oint_C \mathbf{F} \cdot \mathbf{T}\, ds$.

**9.** $\mathbf{F} = y^2\mathbf{i} + x^2\mathbf{j}$; $C$ is the boundary of unit square with vertices $(0, 0)$, $(1, 0)$, $(1, 1)$, and $(0, 1)$.

**10.** $\mathbf{F} = ay\mathbf{i} + bx\mathbf{j}$; $C$ as in Problem 9.

**11.** $\mathbf{F} = y^3\mathbf{i} + x^3\mathbf{j}$; $C$ is the unit circle.

**12.** $\mathbf{F} = x\mathbf{i} + y\mathbf{j}$; $C$ is the unit circle.

**13.** Suppose the integrals $\oint \mathbf{F} \cdot \mathbf{T}\, ds$ taken counterclockwise around the circles $x^2 + y^2 = 36$ and $x^2 + y^2 = 1$ are 30 and $-20$, respectively. Calculate $\iint_S (\text{curl } \mathbf{F}) \cdot \mathbf{k}\, dA$, where $S$ is the region between the circles.

**14.** If $\mathbf{F} = (x^2 + y^2)\mathbf{i} + 2xy\mathbf{j}$, find the flux of $\mathbf{F}$ across the boundary $C$ of the unit square with vertices $(0, 0)$, $(1, 0)$, $(1, 1)$, and $(0, 1)$; that is, calculate

$$\oint_C \mathbf{F} \cdot \mathbf{n}\, ds$$

**15.** Find the work done by $\mathbf{F} = (x^2 + y^2)\mathbf{i} - 2xy\mathbf{j}$ in moving a body counterclockwise around the curve $C$ of Problem 14.

**16.** If $\mathbf{F} = (x^2 + y^2)\mathbf{i} + 2xy\mathbf{j}$, calculate the circulation of $\mathbf{F}$ around $C$ of Problem 14; that is, calculate

$$\oint_C \mathbf{F} \cdot \mathbf{T}\, ds$$

**17.** Show that the work done by a constant force $\mathbf{F}$ in moving a body around a simple closed curve is 0.

**18.** Use Green's Theorem to prove the plane case of Theorem 17.3C; that is, show that $\partial N/\partial x = \partial M/\partial y$ implies $\oint_C M\, dx + N\, dy = 0$, which implies that $\mathbf{F} = M\mathbf{i} + N\mathbf{j}$ is conservative.

**19.** Let

$$\mathbf{F} = \frac{y}{x^2 + y^2}\mathbf{i} - \frac{x}{x^2 + y^2}\mathbf{j} = M\mathbf{i} + N\mathbf{j}$$

(a) Show that $\partial N/\partial x = \partial M/\partial y$.
(b) Show, by using the parameterization $x = \cos t$, $y = \sin t$, that $\oint_C M\, dx + N\, dy = -2\pi$, where $C$ is the unit circle.
(c) Why doesn't this contradict Green's Theorem?

**20.** Let $\mathbf{F}$ be as in Problem 19. Calculate $\oint_C M\, dx + N\, dy$ where (a) $C$ is the ellipse $x^2/9 + y^2/4 = 1$, (b) $C$ is the square with vertices $(1, -1)$, $(1, 1)$, $(-1, 1)$, and $(-1, -1)$, and (c) $C$ is the triangle with vertices $(1, 0)$, $(2, 0)$, and $(1, 1)$.

**21.** Let the piecewise smooth, simple closed curve $C$ be the boundary of a region $S$ in the $xy$-plane. Modify the argument in Example 2 to show that

$$A(S) = \oint_C (-y)\, dx = \oint_C x\, dy$$

**22.** Let $S$ and $C$ be as in Problem 21. Show that the moments $M_x$ and $M_y$ about the $x$- and $y$-axes are given by

$$M_x = -\frac{1}{2}\oint_C y^2\,dx, \qquad M_y = \frac{1}{2}\oint_C x^2\,dy$$

**23.** Calculate the area of the asteroid $x^{2/3} + y^{2/3} = a^{2/3}$. *Hint*: Parameterize by $x = a\cos^3 t$, $y = a\sin^3 t$, $0 \le t \le 2\pi$.

**24.** Calculate the work done by $\mathbf{F} = 2y\mathbf{i} - 3x\mathbf{j}$ in moving an object around the astroid of Problem 23.

**25.** Let $\mathbf{F}(\mathbf{r}) = \mathbf{r}/|\mathbf{r}|^2 = (x\mathbf{i} + y\mathbf{j})/(x^2 + y^2)$.

(a) Show $\displaystyle\int_C \mathbf{F}\cdot\mathbf{n}\,ds = 2\pi$, where $C$ is the circle centered at the origin of radius $a$ and $\mathbf{n} = (x\mathbf{i} + y\mathbf{j})/\sqrt{x^2 + y^2}$ is the exterior unit normal to $C$.
(b) Show that div $\mathbf{F} = 0$.
(c) Explain why the results of (a) and (b) do not contradict the vector form of Green's Theorem on page 845.
(d) Show that if $C$ is a smooth simple closed curve, then $\displaystyle\int_C \mathbf{F}\cdot\mathbf{n}\,ds$ equals $2\pi$ or $0$ accordingly as the origin is inside or outside $C$.

**26.** (Area of a Polygon) Let $V_0(x_0, y_0)$, $V_1(x_1, y_1)$, ..., $V_n(x_n, y_n)$ be the vertices of a simple polygon $P$, labeled counterclockwise and with $V_0 = V_n$. Show each of the following.

(a) $\displaystyle\int_C x\,dy = \frac{1}{2}(x_1 + x_0)(y_1 - y_0)$, where $C$ is the edge $V_0 V_1$.

(b) Area $(P) = \displaystyle\sum_{i=1}^{n} \frac{x_i + x_{i-1}}{2}(y_i - y_{i-1})$.

(c) The area of a polygon with vertices having integral coordinates is always a multiple of $\frac{1}{2}$.

(d) The formula in (b) gives the correct answer for the polygon with vertices $(2, 0)$, $(2, -2)$, $(6, -2)$, $(6, 0)$, $(10, 4)$, and $(-2, 4)$.

PC In each of the following problems, draw the graph of $f(x, y)$ and the corresponding gradient field $\mathbf{F} = \nabla f$ on $S = \{(x, y): -3 \le x \le 3, -3 \le y \le 3\}$. Note that, in each case, curl $\mathbf{F} = \mathbf{0}$ (Theorem C of Section 17.3) and so there is no tendency for rotation around any point.

**27.** Let $f(x, y) = x^2 + y^2$.

(a) By visually examining the field $\mathbf{F}$, convince yourself that div $\mathbf{F} > 0$ on $S$. Then calculate div $\mathbf{F}$.
(b) Calculate the flux of $\mathbf{F}$ across the boundary of $S$.

**28.** Let $f(x, y) = \ln(\cos(x/3)) - \ln(\cos(y/3))$.

(a) Guess whether div $\mathbf{F}$ is positive or negative at a few points and then calculate div $\mathbf{F}$ to check on your guesses.
(b) Calculate the flux of $\mathbf{F}$ across the boundary of $S$.

**29.** Let $f(x, y) = \sin x \sin y$.

(a) By visually examining the field $\mathbf{F}$, guess where div $\mathbf{F}$ is positive and where it is negative. Then calculate div $\mathbf{F}$ to check on your guesses.
(b) Calculate the flux of $\mathbf{F}$ across the boundary of $S$; then calculate it across the boundary of $T = \{(x, y): 0 \le x \le 3, 0 \le y \le 3\}$.

**30.** Let $f(x, y) = \exp(-(x^2 + y^2)/4)$. Guess at where div $\mathbf{F}$ is positive and where it is negative. Then determine this analytically.

---

**Answers to Concepts Review:** **1.** $\dfrac{\partial N}{\partial x} - \dfrac{\partial M}{\partial y}$  **2.** $-2; -2$  **3.** Source; sink  **4.** Rotate; irrotational

---

## 17.5
## SURFACE INTEGRALS

A line integral generalizes the ordinary definite integral; in a similar way, a surface integral generalizes a double integral.

Let the surface $G$ be the graph of $z = f(x, y)$ where $(x, y)$ ranges over a rectangle $R$ in the $xy$-plane. Let $P$ be a partition of $R$ into $n$ subrectangles $R_i$; this results in a corresponding partition of the surface $G$ into $n$ pieces $G_i$ (Figure 1). Choose a sample point $(\bar{x}_i, \bar{y}_i)$ in $R_i$ and let $(\bar{x}_i, \bar{y}_i, \bar{z}_i) = (\bar{x}_i, \bar{y}_i, f(\bar{x}_i, \bar{y}_i))$ be the corresponding point on $G_i$. Then define the **surface integral** of $g$ over $G$ by

$$\iint_G g(x, y, z)\,dS = \lim_{|P|\to 0} \sum_{i=1}^{n} g(\bar{x}_i, \bar{y}_i, \bar{z}_i)\,\Delta S_i$$

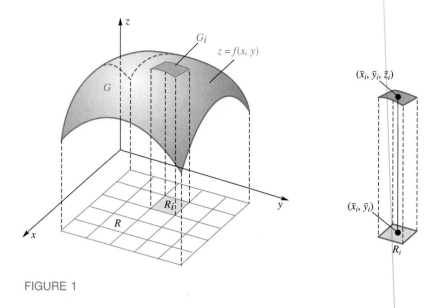

FIGURE 1

where $\Delta S_i$ is the area of $G_i$. Finally, extend the definition to the case where $R$ is a general closed, bounded set in the $xy$-plane in the usual way (by giving $g$ the value 0 outside of $R$).

**Evaluating Surface Integrals**    A definition is not enough; we need a practical way to evaluate a surface integral. The development in Section 16.6 suggests the correct result. There we showed that under appropriate hypotheses, the area of a small patch $G_i$ of the surface satisfies

$$A(G_i) \approx \sec \gamma_i \, A(R_i) = \sec \gamma_i \, \Delta y_i \, \Delta x_i$$

where $\gamma_i$ is the angle between **k** and the upward unit normal **n** (Figure 2). This leads to the following theorem.

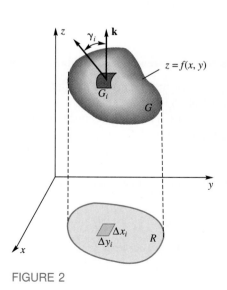

FIGURE 2

---

**Theorem A**

Let $G$ be a surface given by $z = f(x, y)$, where $(x, y)$ is in $R$. If $f$ has continuous first-order partial derivatives and $g(x, y, z) = g(x, y, f(x, y))$ is continuous on $R$, then

$$\iint_G g(x, y, z) \, dS = \iint_R g(x, y, f(x, y)) \sec \gamma \, dA$$

$$= \iint_R g(x, y, f(x, y)) \sqrt{f_x^2 + f_y^2 + 1} \, dy \, dx$$

where $\gamma$ is the angle between the upward unit normal **n** at $(x, y, f(x, y))$ and the positive $z$-axis.

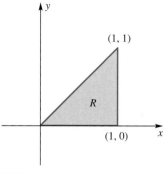

FIGURE 3

**EXAMPLE 1** Evaluate $\iint_G (xy + z) \, dS$, where $G$ is the part of the plane $2x - y + z = 3$ above the triangle $R$ sketched in Figure 3.

**Solution** In this case, $z = 3 + y - 2x = f(x, y)$, $f_x = -2$, $f_y = 1$, and $g(x, y, z) = xy + 3 + y - 2x$. Thus,

$$\iint_G (xy + z) \, dS = \int_0^1 \int_0^x (xy + 3 + y - 2x)\sqrt{(-2)^2 + 1^2 + 1} \, dy \, dx$$

$$= \sqrt{6} \int_0^1 \left[ \frac{xy^2}{2} + 3y + \frac{y^2}{2} - 2xy \right]_0^x dx$$

$$= \sqrt{6} \int_0^1 \left[ \frac{x^3}{2} + 3x - \frac{3x^2}{2} \right] dx = \frac{9\sqrt{6}}{8} \qquad \blacksquare$$

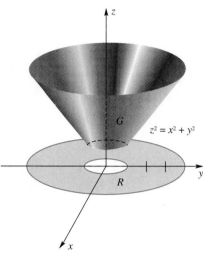

FIGURE 4

**EXAMPLE 2** Evaluate $\iint_G xyz \, dS$, where $G$ is the portion of the cone $z^2 = x^2 + y^2$ between the planes $z = 1$ and $z = 4$ (Figure 4).

**Solution** We may write

$$z = (x^2 + y^2)^{1/2} = f(x, y)$$

from which

$$f_x^2 + f_y^2 + 1 = \frac{x^2}{x^2 + y^2} + \frac{y^2}{x^2 + y^2} + 1 = 2$$

Thus,

$$\iint_G xyz \, dS = \iint_R xy\sqrt{x^2 + y^2} \, \sqrt{2} \, dy \, dx$$

After a change to polar coordinates, this becomes

$$\sqrt{2} \int_0^{2\pi} \int_1^4 (r \cos \theta)(r \sin \theta)r^2 \, dr \, d\theta = \sqrt{2} \int_0^{2\pi} \left[ \sin \theta \cos \theta \frac{r^5}{5} \right]_1^4 d\theta$$

$$= \frac{1023\sqrt{2}}{5} \left[ \frac{\sin^2\theta}{2} \right]_0^{2\pi} = 0 \qquad \blacksquare$$

**EXAMPLE 3** The partial spherical surface $G$ with equation

$$z = f(x, y) = \sqrt{9 - x^2 - y^2}, \qquad 0 \le x^2 + y^2 \le 4$$

has a thin metal covering whose density at $(x, y, z)$ is $\delta(x, y, z) = z$. Find the mass of this covering.

***Solution*** Let $R$ be the projection of $G$ in the $xy$-plane; that is, let $R = \{(x, y): 0 \le x^2 + y^2 \le 4\}$. Then

$$m = \iint_G \delta(x, y, z)\, dS = \iint_R z\sqrt{f_x^2 + f_y^2 + 1}\, dA$$

$$= \iint_R z\,\sqrt{\frac{x^2}{9 - x^2 - y^2} + \frac{y^2}{9 - x^2 - y^2} + 1}\, dA$$

$$= \iint_R z\,\frac{3}{z}\, dA = 3(\pi 2^2) = 12\pi \qquad \blacksquare$$

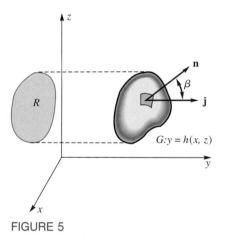

FIGURE 5

Let the surface $G$ be given by an equation of the form $y = h(x, z)$ and let $R$ be its projection in the $xz$-plane (Figure 5). Then the appropriate formula for the surface integral is

$$\iint_G g(x, y, z)\, dS = \iint_R g(x, h(x, z), z)\sec\beta\, dA$$

$$= \iint_R g(x, h(x, z), z)\sqrt{h_x^2 + h_z^2 + 1}\, dx\, dz$$

There is a corresponding formula when the surface $G$ is given by $x = k(y, z)$. It involves the projection of $G$ on the $yz$-plane and the angle $\alpha$ that the outward normal $\mathbf{n}$ makes with $\mathbf{i}$.

**EXAMPLE 4** Evaluate $\iint_G (x^2 + z^2)\, dS$, where $G$ is the part of the paraboloid $y = 1 - x^2 - z^2$ that projects onto $R = \{(x, z): 0 \le x^2 + z^2 \le 1\}$.

***Solution***

$$\iint_G (x^2 + z^2)\, dS = \iint_R (x^2 + z^2)\sqrt{4x^2 + 4z^2 + 1}\, dA$$

If we use polar coordinates, this becomes

$$\int_0^{2\pi} \int_0^1 r^2\sqrt{4r^2 + 1}\, r\, dr\, d\theta$$

In the inner integral, let $u = \sqrt{4r^2 + 1}$, so $u^2 = 4r^2 + 1$ and $u\, du = 4r\, dr$. We obtain

$$\frac{1}{16}\int_0^{2\pi}\int_1^{\sqrt{5}} (u^2 - 1)u^2\, du\, d\theta = \frac{(25\sqrt{5} + 1)\pi}{60} \approx 2.979 \qquad \blacksquare$$

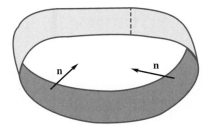

FIGURE 6

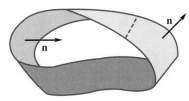

FIGURE 7

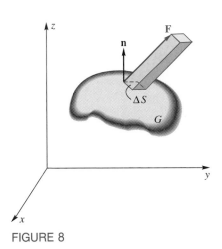

FIGURE 8

**The Flux of a Vector Field Through a Surface**   For what we discuss now and for later applications, we need to limit further the kind of surfaces we consider. Most surfaces that arise in practice have two sides. However, it is surprisingly easy to construct a surface with just one side. Take a paper band (Figure 6), slit it at the dotted line, give one end a half twist, and paste it back together (Figure 7). You obtain a famous one-sided surface, called a Möbius band.

From now on, we consider only two-sided surfaces, so it will make sense to talk about a fluid flowing through the surface from one side to the other. We also suppose that the surface is *smooth*, meaning that it has a continuously varying unit normal **n**. Let $G$ be such a smooth, two-sided surface and assume that it is submerged in a fluid with a continuous velocity field $\mathbf{F}(x, y, z)$. If $\Delta S$ is the area of a small piece of $G$, then **F** is almost constant there, and the volume $\Delta V$ of fluid crossing this piece in the direction of the unit normal **n** (Figure 8) is

$$\Delta V \approx \mathbf{F} \cdot \mathbf{n}\Delta S$$

We conclude that

$$\text{Flux of } \mathbf{F} \text{ across } G = \iint_G \mathbf{F} \cdot \mathbf{n} \, dS$$

**EXAMPLE 5**   Find the upward flux of $\mathbf{F} = -y\mathbf{i} + x\mathbf{j} + 9\mathbf{k}$ across the part of the spherical surface $G$ determined by

$$z = f(x, y) = \sqrt{9 - x^2 - y^2}, \qquad 0 \le x^2 + y^2 \le 4$$

*Solution*   Note that the field **F** is a rotating stream flowing in the direction of the positive $z$-axis.

The equation of the surface may be written

$$H(x, y, z) = z - \sqrt{9 - x^2 - y^2} = z - f(x, y) = 0$$

and thus

$$\mathbf{n} = \frac{\nabla H}{|\nabla H|} = \frac{-f_x\mathbf{i} - f_y\mathbf{j} + \mathbf{k}}{\sqrt{f_x^2 + f_y^2 + 1}} = \frac{(x/z)\mathbf{i} + (y/z)\mathbf{j} + \mathbf{k}}{\sqrt{(x/z)^2 + (y/z)^2 + 1}}$$

is a unit vector normal to the surface. The vector $-\mathbf{n}$ is also normal to the surface, but, since we desire the normal unit vector that points upward, **n** is the right choice. A straightforward computation using the fact that $x^2 + y^2 + z^2 = 9$ gives

$$\mathbf{n} = \frac{(x/z)\mathbf{i} + (y/z)\mathbf{j} + \mathbf{k}}{3/z} = \frac{x}{3}\mathbf{i} + \frac{y}{3}\mathbf{j} + \frac{z}{3}\mathbf{k}$$

(A simple geometric argument will also give this result; the normal must point directly away from the origin.)

The flux of $\mathbf{F}$ across $G$ is given by

$$\text{flux} = \iint\limits_{G} \mathbf{F} \cdot \mathbf{n}\, dS$$

$$= \iint\limits_{G} (-y\mathbf{i} + x\mathbf{j} + 9\mathbf{k}) \cdot \left(\frac{x}{3}\mathbf{i} + \frac{y}{3}\mathbf{j} + \frac{z}{3}\mathbf{k}\right) dS$$

$$= \iint\limits_{G} 3z\, dS$$

Finally we write this surface integral as a double integral, using the fact that $R$ is a circle of radius 2 and that $\sec \gamma = \sqrt{f_x^2 + f_y^2 + 1} = 3/z$.

$$\text{flux} = \iint\limits_{G} 3z\, dS = \iint\limits_{R} 3z\, \frac{3}{z}\, dA = 9(\pi \cdot 2^2) = 36\pi$$

The total flux across $G$ in one unit of time is $36\pi$ cubic units. ∎

An observant reader, after noting the cancellation that occurred in Example 5, will suspect that a theorem is lurking near.

### Theorem B

Let $G$ be a smooth, two-sided surface given by $z = f(x, y)$, where $(x, y)$ is in $R$, and let $\mathbf{n}$ denote the upward unit normal on $G$. If $f$ has continuous first-order partial derivatives and $\mathbf{F} = M\mathbf{i} + N\mathbf{j} + P\mathbf{k}$ is a continuous vector field, then the flux of $\mathbf{F}$ across $G$ is given by

$$\text{flux } \mathbf{F} = \iint\limits_{G} \mathbf{F} \cdot \mathbf{n}\, dS = \iint\limits_{R} [-Mf_x - Nf_y + P]\, dx\, dy$$

***Proof*** If we write $H(x, y, z) = z - f(x, y)$, we obtain

$$\mathbf{n} = \frac{\nabla H}{|\nabla H|} = \frac{-f_x\mathbf{i} - f_y\mathbf{j} + \mathbf{k}}{\sqrt{f_x^2 + f_y^2 + 1}} = (-f_x\mathbf{i} - f_y\mathbf{j} + \mathbf{k})\cos \gamma$$

where $\gamma$ is the angle between the upward normal and the positive $z$-axis. It follows from Theorem A that

$$\iint\limits_{G} \mathbf{F} \cdot \mathbf{n}\, dS = \iint\limits_{R} (M\mathbf{i} + N\mathbf{j} + P\mathbf{k}) \cdot (-f_x\mathbf{i} - f_y\mathbf{j} + \mathbf{k})\cos \gamma \sec \gamma\, dA$$

$$= \iint\limits_{R} (-Mf_x - Nf_y + P)\, dx\, dy$$

∎

You might try reworking Example 5 using Theorem B. We offer a different example.

**EXAMPLE 6** Evaluate the flux for the vector field $\mathbf{F} = x\mathbf{i} + y\mathbf{j} + z\mathbf{k}$ across the part $G$ of the paraboloid $z = 1 - x^2 - y^2$ that lies above the $xy$-plane, taking $\mathbf{n}$ to be the upward normal.

*Solution*

$$f(x, y) = 1 - x^2 - y^2, \qquad f_x = -2x, \qquad f_y = -2y$$

$$-Mf_x - Nf_y + P = 2x^2 + 2y^2 + z$$

$$= 2x^2 + 2y^2 + 1 - x^2 - y^2 = 1 + x^2 + y^2$$

$$\iint\limits_{G} \mathbf{F} \cdot \mathbf{n}\, dS = \iint\limits_{R} (1 + x^2 + y^2)\, dx\, dy$$

$$= \int_0^{2\pi} \int_0^1 (1 + r^2) r\, dr\, d\theta = \frac{3}{2}\pi \qquad \blacksquare$$

## CONCEPTS REVIEW

**1.** A _____ generalizes the ordinary double integral similar to the way a line integral generalizes the definite integral.

**2.** If $G$ is a surface, $\iint\limits_{G} g(x, y, z)\, dS = \lim\limits_{|P| \to 0}$ _____.

**3.** Let $G$ be a surface given by $z = f(x, y)$, where $(x, y)$ is in $R$. Then $\iint\limits_{G} g(x, y, z)\, dS = \iint\limits_{R} g(x, y, f(x, y))$ _____ $dy\, dx$.

**4.** Consider the cone with axis along the $z$-axis, with vertex at the origin, and making an angle of $30°$ with the $z$-axis. If $G$ is the portion of this cone above the set $R = \{(x, y): x^2 + y^2 \le 9\}$, then $\iint\limits_{G} dS = \iint\limits_{R}$ _____ $dy\, dx =$ _____.

## PROBLEM SET 17.5

In Problems 1–8, evaluate $\iint\limits_{G} g(x, y, z)\, dS$.

**1.** $g(x, y, z) = x^2 + y^2 + z$; $G:z = x + y + 1$, $0 \le x \le 1, 0 \le y \le 1$.

**2.** $g(x, y, z) = x$; $G:x + y + 2z = 4, 0 \le x \le 1$, $0 \le y \le 1$.

**3.** $g(x, y, z) = x + y$; $G:z = \sqrt{4 - x^2}, 0 \le x \le \sqrt{3}$, $0 \le y \le 1$.

**4.** $g(x, y, z) = 2y^2 + z$; $G:z = x^2 - y^2$, $0 \le x^2 + y^2 \le 1$.

**5.** $g(x, y, z) = \sqrt{4x^2 + 4y^2 + 1}$; $G$ is the part of $z = x^2 + y^2$ below $y = z$.

**6.** $g(x, y, z) = y$; $G:z = 4 - y^2, 0 \le x \le 3, 0 \le y \le 2$.

**7.** $g(x, y, z) = x + y$; $G$ is the surface of the cube $0 \le x \le 1, 0 \le y \le 1, 0 \le z \le 1$.

**8.** $g(x, y, z) = z$; $G$ is the tetrahedron bounded by the coordinate planes and the plane $4x + 8y + 2z = 16$.

Use Theorem B to calculate the flux of $\mathbf{F}$ across $G$ in Problems 9–12.

**9.** $\mathbf{F}(x, y, z) = -y\mathbf{i} + x\mathbf{j}$; $G$ is the part of the plane $z = 8x - 4y - 5$ above the triangle with vertices $(0, 0, 0)$, $(0, 1, 0)$, and $(1, 0, 0)$.

**10.** $\mathbf{F}(x, y, z) = (9 - x^2)\mathbf{j}$; $G$ is the part of the plane $2x + 3y + 6z = 6$ in the first octant.

**11.** $\mathbf{F}(x, y, z) = y\mathbf{i} - x\mathbf{j} + 2\mathbf{k}$; $G$ is the surface determined by $z = \sqrt{1 - y^2}, 0 \le x \le 5$.

**12.** $\mathbf{F}(x, y, z) = 2\mathbf{i} + 5\mathbf{j} + 3\mathbf{k}$; $G$ is the part of the cone $z = (x^2 + y^2)^{1/2}$ that is inside the cylinder $x^2 + y^2 = 1$.

**13.** Find the mass of the triangle with vertices $(a, 0, 0)$, $(0, a, 0)$, and $(0, 0, a)$ if its density $\delta$ satisfies $\delta(x, y, z) = kx^2$.

**14.** Find the mass of the surface $z = 1 - \frac{1}{2}(x^2 + y^2)$ over $0 \le x \le 1$, $0 \le y \le 1$, $z = 0$, if $\delta(x, y, z) = kxy$.

**15.** Find the center of mass of the homogeneous triangle with vertices $(a, 0, 0)$, $(0, a, 0)$, and $(0, 0, a)$.

**16.** Refer to Example 3. The hemispherical surface $z = f(x, y) = \sqrt{9 - x^2 - y^2}$ has a thin metal covering with density $\delta(x, y, z) = z$. Find the mass of this covering. Note that Theorem A does not apply directly, since $f_x$ and $f_y$ are undefined on the boundary $x^2 + y^2 = 9$ of $R$. Therefore, proceed by letting $R_\varepsilon$ be the region $0 \le x^2 + y^2 \le (3 - \varepsilon)^2$, make the calculation, and then let $\varepsilon \to 0$. Discover that you get the same answer as you would if you ignored this subtle point.

**17.** Let $G$ be the sphere $x^2 + y^2 + z^2 = a^2$. Evaluate each of the following.

(a) $\displaystyle\iint_G z\, dS$

(b) $\displaystyle\iint_G \frac{x + y^3 + \sin z}{1 + z^4}\, dS$

(c) $\displaystyle\iint_G (x^2 + y^2 + z^2)\, dS$

(d) $\displaystyle\iint_G x^2\, dS$

(e) $\displaystyle\iint_G (x^2 + y^2)\, dS$

*Hint*: Use symmetry properties to make this a trivial problem.

**18.** The sphere $x^2 + y^2 + z^2 = a^2$ has constant area density $k$. Find each moment of inertia.

(a) About a diameter.
(b) About a tangent line (assume the Parallel Axis Theorem).

**19.** Find the total force against the surface of a tank full of a liquid of weight density $k$ for each tank shape.

(a) A sphere of radius $a$.
(b) A hemisphere of radius $a$ with a flat base.
(c) A vertical cylinder of radius $a$ and height $h$.

*Hint*: The force against a small patch of area $\Delta G$ is approximately $kd\, \Delta G$, where $d$ is the depth of the water at the patch.

**20.** Find the center of mass of the part of the sphere $x^2 + y^2 + z^2 = a^2$ between the planes $z = h_1$ and $z = h_2$, where $0 \le h_1 \le h_2 \le a$. Do this by the methods of this section and then compare with Problem 15 of Section 16.6.

---

**Answers to Concepts Review:**   **1.** Surface integral
**2.** $\displaystyle\sum_{i=1}^{n} g(\bar{x}_i, \bar{y}_i, \bar{z}_i)\, \Delta S_i$   **3.** $\sqrt{f_x^2 + f_y^2 + 1}$   **4.** $2; 18\pi$

---

## 17.6
## GAUSS'S DIVERGENCE THEOREM

The theorems of Green, Gauss, and Stokes all relate an integral over a set $S$ to another integral over the boundary of $S$. To emphasize the similarity in these theorems, we introduce the notation $\partial S$ to stand for the boundary of $S$. Thus, one form of Green's Theorem (Section 17.4) can be written

$$\oint_{\partial S} \mathbf{F} \cdot \mathbf{n}\, ds = \iint_S \operatorname{div} \mathbf{F}\, dA$$

It says that the flux of $\mathbf{F}$ across the boundary $\partial S$ of a closed bounded plane region $S$ is equal to the double integral of div $\mathbf{F}$ over that region. Gauss's Theorem lifts this result up a dimension.

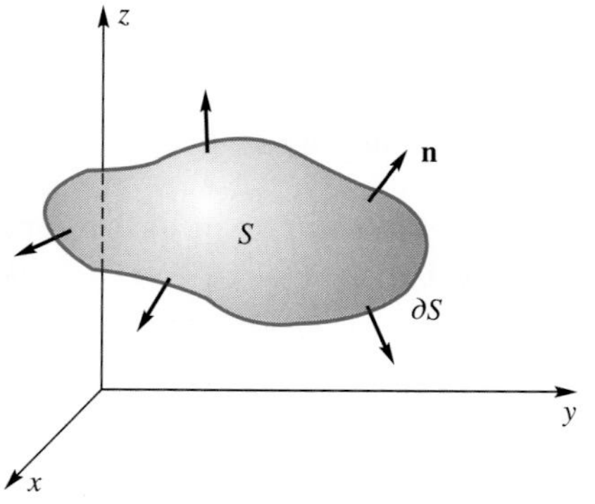

FIGURE 1

**Gauss's Theorem**   Let $S$ be a closed bounded solid in three-space that is completely enclosed by a piecewise smooth surface $\partial S$ (Figure 1).

### Theorem A

**(Gauss's Theorem).**    Let $\mathbf{F} = M\mathbf{i} + N\mathbf{j} + P\mathbf{k}$ be a vector field such that $M$, $N$, and $P$ have continuous first-order partial derivatives on a solid $S$ with boundary $\partial S$. If $\mathbf{n}$ denotes the outer unit normal to $\partial S$, then

$$\iint\limits_{\partial S} \mathbf{F} \cdot \mathbf{n} \, dS = \iiint\limits_{S} \operatorname{div} \mathbf{F} \, dV$$

In other words, the flux of $\mathbf{F}$ across the boundary of a closed region in three-space is the triple integral of its divergence over the region.

It is useful both for some applications and for the proof to state the conclusion to Gauss's Theorem in its Cartesian (nonvector) form. We may write

$$\mathbf{n} = \cos \alpha \, \mathbf{i} + \cos \beta \, \mathbf{j} + \cos \gamma \, \mathbf{k}$$

where $\alpha$, $\beta$, and $\gamma$ are the direction angles for $\mathbf{n}$. Thus, $\mathbf{F} \cdot \mathbf{n} = M \cos \alpha + N \cos \beta + P \cos \gamma$, and so Gauss's formula becomes

$$\iint\limits_{\partial S} (M \cos \alpha + N \cos \beta + P \cos \gamma) \, dS = \iiint\limits_{S} \left( \frac{\partial M}{\partial x} + \frac{\partial N}{\partial y} + \frac{\partial P}{\partial z} \right) dV$$

***Proof of Gauss's Theorem***    We first consider the case where the region $S$ is $x$-simple, $y$-simple, and $z$-simple. It will be sufficient to show that

$$\iint\limits_{\partial S} M \cos \alpha \, dS = \iiint\limits_{S} \frac{\partial M}{\partial x} \, dV$$

$$\iint\limits_{\partial S} N \cos \beta \, dS = \iiint\limits_{S} \frac{\partial N}{\partial y} \, dV$$

$$\iint\limits_{\partial S} P \cos \gamma \, dS = \iiint\limits_{S} \frac{\partial P}{\partial z} \, dV$$

Since these demonstrations are similar, we show only the third.

Since $S$ is $z$-simple, it can be described by the inequalities $f_1(x, y) \leq z \leq f_2(x, y)$. As in Figure 2, $\partial S$ consists of three parts: $S_1$, corresponding to $z = f_1(x, y)$; $S_2$, corresponding to $z = f_2(x, y)$; and the lateral surface $S_3$, which may be empty. On $S_3$, $\cos \gamma = \cos 90° = 0$, so we can ignore its contribution. Also, from Theorem 17.5A,

$$\iint\limits_{S_2} P \cos \gamma \, dS = \iint\limits_{R} P(x, y, f_2(x, y)) \, dx \, dy$$

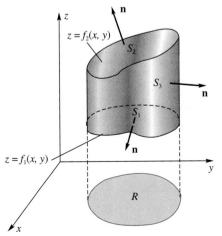

FIGURE 2

The result to which we just referred assumes that the normal **n** points upward. Hence, when we apply it to $S_1$, where **n** is a lower normal (Figure 2), we must reverse the sign.

$$\iint\limits_{S_1} P \cos \gamma \, dS = -\iint\limits_{R} P(x, y, f_1(x, y)) \, dx \, dy$$

It follows that

$$\iint\limits_{\partial S} P \cos \gamma \, dS = \iint\limits_{R} [P(x, y, f_2(x, y)) - P(x, y, f_1(x, y))] \, dx \, dy$$

$$= \iint\limits_{R} \left[ \int_{f_1(x, y)}^{f_2(x, y)} \frac{\partial P}{\partial z} \, dz \right] dx \, dy$$

$$= \iiint\limits_{S} \frac{\partial P}{\partial z} \, dV$$

The result just proved extends easily to regions that are finite unions of the type considered. We omit the details. ■

**EXAMPLE 1** Verify Gauss's Theorem for $\mathbf{F} = x\mathbf{i} + y\mathbf{j} + z\mathbf{k}$ and $S = \{(x, y, z): x^2 + y^2 + z^2 \le a^2\}$ by independently calculating (a) $\iint\limits_{\partial S} \mathbf{F} \cdot \mathbf{n} \, dS$ and (b) $\iiint\limits_{S} \operatorname{div} \mathbf{F} \, dV$.

*Solution* (a) On $\partial S$, $\mathbf{n} = (x\mathbf{i} + y\mathbf{j} + z\mathbf{k})/a$, and so $\mathbf{F} \cdot \mathbf{n} = (x^2 + y^2 + z^2)/a = a$. Thus,

$$\iint\limits_{\partial S} \mathbf{F} \cdot \mathbf{n} \, dS = a \iint\limits_{\partial S} dS = a(4\pi a^2) = 4\pi a^3$$

(b) Since div $\mathbf{F} = 3$,

$$\iiint\limits_{S} \operatorname{div} \mathbf{F} \, dV = 3 \iiint\limits_{S} dV = 3\left(\frac{4\pi a^3}{3}\right) = 4\pi a^3 \qquad ■$$

**EXAMPLE 2** Compute the flux of the vector field $\mathbf{F} = x^2 y\mathbf{i} + 2xz\mathbf{j} + yz^3\mathbf{k}$ across the surface of the rectangular solid $S$ determined by (Figure 3)

$$0 \le x \le 1, \qquad 0 \le y \le 2, \qquad 0 \le z \le 3$$

(a) by a direct method and (b) by Gauss's Theorem.

*Solution* (a) To calculate $\iint\limits_{\partial S} \mathbf{F} \cdot \mathbf{n} \, dS$ directly, we calculate this integral over the six faces and add the results. On the face $x = 1$, $\mathbf{n} = \mathbf{i}$, and $\mathbf{F} \cdot \mathbf{n} =$

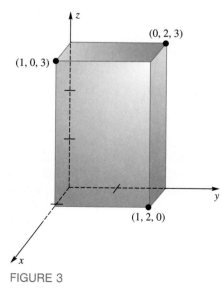

FIGURE 3

$x^2y = 1^2y = y$, so $\iint\limits_{x=1} \mathbf{F} \cdot \mathbf{n}\, dS = \int_0^3 \int_0^2 y\, dy\, dz = 6$. By similar calculations, we can construct the following table.

| Face | $\mathbf{n}$ | $\mathbf{F} \cdot \mathbf{n}$ | $\iint\limits_{\text{face}} \mathbf{F} \cdot \mathbf{n}\, dS$ |
|------|------|------|------|
| $x = 1$ | $\mathbf{i}$ | $y$ | 6 |
| $x = 0$ | $-\mathbf{i}$ | 0 | 0 |
| $y = 2$ | $\mathbf{j}$ | $2xz$ | 9/2 |
| $y = 0$ | $-\mathbf{j}$ | $-2xz$ | $-9/2$ |
| $z = 3$ | $\mathbf{k}$ | $27y$ | 54 |
| $z = 0$ | $-\mathbf{k}$ | 0 | 0 |

Thus,

$$\iint\limits_{\partial S} \mathbf{F} \cdot \mathbf{n}\, dS = 6 + 0 + \frac{9}{2} - \frac{9}{2} + 54 + 0 = 60$$

(b) By Gauss's Theorem,

$$\iint\limits_{\partial S} \mathbf{F} \cdot \mathbf{n}\, dS = \iiint\limits_{S} (2xy + 0 + 3yz^2)\, dV$$

$$= \int_0^1 \int_0^2 \int_0^3 (2xy + 3yz^2)\, dz\, dy\, dx = \int_0^1 \int_0^2 (6xy + 27y)\, dy\, dx$$

$$= \int_0^1 (12x + 54)\, dx = \left[ 6x^2 + 54x \right]_0^1 = 60 \qquad \blacksquare$$

**EXAMPLE 3**   Let $S$ be the solid cylinder bounded by $x^2 + y^2 = 4$, $z = 0$, and $z = 3$, and let $\mathbf{n}$ be the outer unit normal to the boundary $\partial S$ (Figure 4). If $\mathbf{F} = (x^3 + \tan yz)\mathbf{i} + (y^3 - e^{xz})\mathbf{j} + (3z + x^3)\mathbf{k}$, find the flux of $\mathbf{F}$ across $\partial S$.

*Solution*   Imagine the difficulty in trying to evaluate $\iint\limits_{\partial S} \mathbf{F} \cdot \mathbf{n}\, dS$ directly. However,

$$\text{div } \mathbf{F} = 3x^2 + 3y^2 + 3 = 3(x^2 + y^2 + 1)$$

and so by Gauss's Theorem and a shift to cylindrical coordinates,

$$\iint\limits_{\partial S} \mathbf{F} \cdot \mathbf{n}\, dS = 3 \iiint\limits_{S} (x^2 + y^2 + 1)\, dV$$

$$= 3 \int_0^{2\pi} \int_0^2 \int_0^3 (r^2 + 1)r\, dz\, dr\, d\theta$$

$$= 9 \int_0^{2\pi} \int_0^2 (r^3 + r)\, dr\, d\theta$$

$$= 9 \int_0^{2\pi} 6\, d\theta = 108\pi \qquad \blacksquare$$

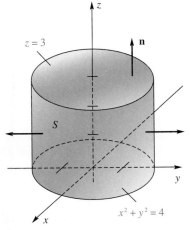

$z = 3$

$\mathbf{n}$

$S$

$x^2 + y^2 = 4$

FIGURE 4

**Extensions and Applications** So far we have implicitly assumed that the solid $S$ has no holes in its interior and that its boundary $\partial S$ consists of one connected surface. In fact, Gauss's Theorem holds for a solid with holes, like a chunk of Swiss cheese, provided we always require $\mathbf{n}$ to point away from the interior of the solid. For example, let $S$ be the solid shell between two concentric spheres centered at the origin. Gauss's Theorem applies, provided we recognize that $\partial S$ now consists of two surfaces (an outer surface where $\mathbf{n}$ points away from the origin and an inner surface where $\mathbf{n}$ points toward the origin).

**EXAMPLE 4** Let $S$ be the solid determined by

$$1 \le x^2 + y^2 + z^2 \le 4$$

and let $\mathbf{F} = x\mathbf{i} + (2y + z)\mathbf{j} + (z + x^2)\mathbf{k}$. Evaluate

$$\iint_{\partial S} \mathbf{F} \cdot \mathbf{n} \, dS$$

*Solution*

$$\iint_{\partial S} \mathbf{F} \cdot \mathbf{n} \, dS = \iiint_{S} \operatorname{div} \mathbf{F} \, dV$$

$$= \iiint_{S} (1 + 2 + 1) \, dV$$

$$= 4 \left[ \frac{4}{3} \pi(2^3) - \frac{4}{3} \pi(1^3) \right] = \frac{112\pi}{3} \quad \blacksquare$$

Recall from Section 17.1 that the gravitational field $\mathbf{F}$ due to a point mass $M$ at the origin has the form

$$\mathbf{F}(x, y, z) = -cM \frac{\mathbf{r}}{|\mathbf{r}|^3}$$

where $\mathbf{r} = x\mathbf{i} + y\mathbf{j} + z\mathbf{k}$ and $c$ is a constant.

**EXAMPLE 5** Let $S$ be a solid region containing a point mass $M$ at the origin in its interior and with the corresponding field $\mathbf{F} = -cM\mathbf{r}/|\mathbf{r}|^3$. Show that the flux of $\mathbf{F}$ across $\partial S$ is $-4\pi cM$, regardless of the shape of $S$.

*Solution* Since $\mathbf{F}$ is discontinuous at the origin, Gauss's Theorem does not apply directly. However, let us imagine that a small solid sphere $S_a$ centered at the origin and of radius $a$ has been removed from $S$, leaving a solid $W$

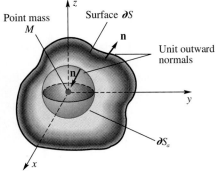

Point mass $M$

Surface $\partial S$

$\mathbf{n}$

Unit outward normals

$\mathbf{n}$

$\partial S_a$

FIGURE 5

with outer boundary $\partial S$ and inner boundary $\partial S_a$ (Figure 5). When we apply Gauss's Theorem to $W$, we get

$$\iint_{\partial S} \mathbf{F} \cdot \mathbf{n} \, dS + \iint_{\partial S_a} \mathbf{F} \cdot \mathbf{n} \, dS = \iint_{\partial W} \mathbf{F} \cdot \mathbf{n} \, dS$$

$$= \iiint_{W} \operatorname{div} \mathbf{F} \, dV$$

But div $\mathbf{F} = 0$, which is easy to check (Problem 21 of Section 17.1), and so

$$\iint_{\partial S} \mathbf{F} \cdot \mathbf{n} \, dS = -\iint_{\partial S_a} \mathbf{F} \cdot \mathbf{n} \, dS$$

On the surface $\partial S_a$, $\mathbf{n} = -\mathbf{r}/|\mathbf{r}|$ and $|\mathbf{r}| = a$. Consequently,

$$-\iint_{\partial S_a} \mathbf{F} \cdot \mathbf{n} \, dS = -\iint_{\partial S_a} \left( -cM \frac{\mathbf{r}}{|\mathbf{r}|^3} \right) \cdot \left( -\frac{\mathbf{r}}{|\mathbf{r}|} \right) dS$$

$$= -cM \iint_{\partial S_a} \frac{\mathbf{r} \cdot \mathbf{r}}{a^4} \, dS$$

$$= -cM \iint_{\partial S_a} \frac{1}{a^2} \, dS$$

$$= \frac{-cM}{a^2} (4\pi a^2) = -4\pi cM \qquad \blacksquare$$

We can extend the result of Example 5 to the case where a solid $S$ contains $k$ point masses $M_1, M_2, \ldots, M_k$ in its interior. The result, known as *Gauss's Law*, gives the flux of $\mathbf{F}$ across $\partial S$ as

$$\iint_{\partial S} \mathbf{F} \cdot \mathbf{n} \, dS = -4\pi c(M_1 + M_2 + \cdots + M_k)$$

Finally, Gauss's Law can be extended to a body $B$ with a continuously distributed mass of size $M$ by subdividing it into small pieces and approximating these pieces by point masses. The result is

$$\iint_{\partial S} \mathbf{F} \cdot \mathbf{n} \, dS = -4\pi cM$$

for any region $S$ containing $B$.

# CONCEPTS REVIEW

**1.** The theorems of Green, Gauss, and Stokes all relate an integral over $S$ to another integral over the _____ of $S$, which is denoted by _____.

**2.** In particular, the theorem of Gauss says that $\iiint_S \operatorname{div} \mathbf{F} \, dV = \iint_{\partial S}$ _____ $dS$.

**3.** Another way to state Gauss's Theorem is to say

that the flux of $\mathbf{F}$ across the boundary of $S$ equals $\iiint_S$ _____ $dV$.

**4.** A consequence of Gauss's Theorem is that the _____ of the gravitational field due to a mass $M$ across the boundary of any solid $S$ containing $M$ is $-4\pi cM$; that is, it is independent of _____ of $S$.

# PROBLEM SET 17.6

In Problems 1–10, use Gauss's Divergence Theorem to calculate $\iint_{\partial S} \mathbf{F} \cdot \mathbf{n} \, dS$.

**1.** $\mathbf{F}(x, y, z) = z\mathbf{i} + x\mathbf{j} + y\mathbf{k}$; $S$ is the hemisphere $0 \le z \le \sqrt{9 - x^2 - y^2}$.

**2.** $\mathbf{F}(x, y, z) = x\mathbf{i} + 2y\mathbf{j} + 3z\mathbf{k}$; $S$ is the cube $0 \le x \le 1$, $0 \le y \le 1$, $0 \le z \le 1$.

**3.** $\mathbf{F}(x, y, z) = x^2yz\mathbf{i} + xy^2z\mathbf{j} + xyz^2\mathbf{k}$; $S$ is the box $0 \le x \le a$, $0 \le y \le b$, $0 \le z \le c$.

**4.** $\mathbf{F}(x, y, z) = 3x\mathbf{i} - 2y\mathbf{j} + 4z\mathbf{k}$; $S$ is the ball $x^2 + y^2 + z^2 \le 9$.

**5.** $\mathbf{F}(x, y, z) = x^2\mathbf{i} + y^2\mathbf{j} + z^2\mathbf{k}$; $S$ is the parabolic solid $0 \le z \le 4 - x^2 - y^2$.

**6.** $\mathbf{F}(x, y, z) = (x^2 + \cos yz)\mathbf{i} + (y - e^z)\mathbf{j} + (z^2 + x^2)\mathbf{k}$; $S$ is the solid bounded by $x^2 + y^2 = 4$, $x + z = 2$, $z = 0$.

**7.** $\mathbf{F}(x, y, z) = (x + z^2)\mathbf{i} + (y - z^2)\mathbf{j} + x\mathbf{k}$; $S$ is the solid $0 \le y^2 + z^2 \le 1$, $0 \le x \le 2$.

**8.** $\mathbf{F}(x, y, z) = x^2\mathbf{i} + y^2\mathbf{j} + z^2\mathbf{k}$; $S$ is the solid enclosed by $x + y + z = 4$, $x = 0$, $y = 0$, $z = 0$.

**9.** $\mathbf{F}(x, y, z) = 2x\mathbf{i} + 3y\mathbf{j} + 4z\mathbf{k}$; $S$ is the solid spherical shell $9 \le x^2 + y^2 + z^2 \le 25$.

**10.** $\mathbf{F}(x, y, z) = 2z\mathbf{i} + x\mathbf{j} + z^2\mathbf{k}$; $S$ is the solid cylindrical shell $1 \le x^2 + y^2 \le 4$, $0 \le z \le 2$.

**11.** Let $\mathbf{F}(x, y, z) = x\mathbf{i} + y\mathbf{j} + z\mathbf{k}$ and let $S$ be a solid to which Gauss's Divergence Theorem applies. Show that the volume of $S$ is given by

$$V(S) = \tfrac{1}{3} \iint_{\partial S} \mathbf{F} \cdot \mathbf{n} \, dS$$

**12.** Use the result of Problem 11 to verify the formula for the volume of a right circular cone of height $h$ and radius $a$.

**13.** Consider the plane $ax + by + cz = d$, where $a$, $b$, $c$, and $d$ are all positive. Use Problem 11 to show that the volume of the tetrahedron cut from the first octant by this plane is $dD/(3\sqrt{a^2 + b^2 + c^2})$, where $D$ is the area of the part of the plane in the first octant.

**14.** Let $\mathbf{F}$ be a constant vector field. Show that

$$\iint_{\partial S} \mathbf{F} \cdot \mathbf{n} \, dS = 0$$

for any "nice" solid $S$. What should we mean by "nice"?

**15.** Calculate $\iint_{\partial S} \mathbf{F} \cdot \mathbf{n} \, dS$ for each of the following. Looked at the right way, all are quite easy and some are even trivial.

(a) $\mathbf{F} = (2x + yz)\mathbf{i} + 3y\mathbf{j} + z^2\mathbf{k}$; $S$ is the ball $x^2 + y^2 + z^2 \le 1$.
(b) $\mathbf{F} = (x^2 + y^2 + z^2)^{5/3}(x\mathbf{i} + y\mathbf{j} + z\mathbf{k})$; $S$ as in (a).
(c) $\mathbf{F} = x^2\mathbf{i} + y^2\mathbf{j} + z^2\mathbf{k}$; $S$ is the ball $(x - 2)^2 + y^2 + z^2 \le 1$.
(d) $\mathbf{F} = x^2\mathbf{i}$; $S$ is the cube $0 \le x \le 1$, $0 \le y \le 1$, $0 \le z \le 1$.
(e) $\mathbf{F} = (x + z)\mathbf{i} + (y + x)\mathbf{j} + (z + y)\mathbf{k}$; $S$ is the tetrahedron cut from the first octant by the plane $3x + 4y + 2z = 12$.
(f) $\mathbf{F} = x^3\mathbf{i} + y^3\mathbf{j} + z^3\mathbf{k}$; $S$ as in (a).
(g) $\mathbf{F} = (x\mathbf{i} + y\mathbf{j})\ln(x^2 + y^2)$; $S$ is the solid cylinder $x^2 + y^2 \le 4$, $0 \le z \le 2$.

**16.** Calculate $\iint_{\partial S} \mathbf{F} \cdot \mathbf{n} \, dS$. In each case, $\mathbf{r} = x\mathbf{i} + y\mathbf{j} + z\mathbf{k}$.

(a) $\mathbf{F} = \mathbf{r}/|\mathbf{r}|^3$; $S$ is the ball $(x - 2)^2 + y^2 + z^2 \le 1$.
(b) $\mathbf{F} = \mathbf{r}/|\mathbf{r}|^3$; $S$ is the ball $x^2 + y^2 + z^2 \le a^2$.
(c) $\mathbf{F} = \mathbf{r}/|\mathbf{r}|^2$; $S$ as in (b).
(d) $\mathbf{F} = f(|\mathbf{r}|)\mathbf{r}$, $f$ any any scalar function; $S$ as in (b).
(e) $\mathbf{F} = |\mathbf{r}|^n\mathbf{r}$, $n \ge 0$; $S$ is the ball $x^2 + y^2 + z^2 \le az$ ($\rho \le a \cos \phi$ in spherical coordinates).

**17.** We have defined the laplacian of a scalar field by

$$\operatorname{lap} f = \nabla^2 f = \frac{\partial^2 f}{\partial x^2} + \frac{\partial^2 f}{\partial y^2} + \frac{\partial^2 f}{\partial z^2}$$

Show that

$$\iint_{\partial S} D_{\mathbf{n}} f \, dS = \iiint_{S} \nabla^2 f \, dV$$

Here $D_{\mathbf{n}} f$ is the directional derivative (Section 15.5).

**18.** Suppose $\nabla^2 f$ is identically zero in a region $S$. Show that

$$\iint_{\partial S} f D_{\mathbf{n}} f \, dS = \iiint_{S} |\nabla f|^2 \, dV$$

**19.** Establish Green's First Identity

$$\iint_{\partial S} f D_{\mathbf{n}} g \, dS = \iiint_{S} (f \nabla^2 g + \nabla f \cdot \nabla g) \, dV$$

by applying Gauss's Divergence Theorem to $\mathbf{F} = f \nabla g$.

**20.** Establish Green's Second Identity

$$\iint_{\partial S} (f D_{\mathbf{n}} g - g D_{\mathbf{n}} f) \, dS = \iiint_{S} (f \nabla^2 g - g \nabla^2 f) \, dV$$

---

**Answers to Concepts Review:** **1.** Boundary; $\partial S$ **2.** $\mathbf{F} \cdot \mathbf{n}$ **3.** div $\mathbf{F}$ **4.** Flux; the shape

---

**17.7**
**STOKES'S THEOREM**

We showed in Section 17.4 that the conclusion to Green's Theorem could be written as

$$\oint_{\partial S} \mathbf{F} \cdot \mathbf{T} \, ds = \iint_{S} (\text{curl } \mathbf{F}) \cdot \mathbf{k} \, dA$$

As stated, it was a theorem for a plane set $S$ bounded by a simple closed curve $\partial S$. We are going to generalize this result to the case where $S$ is a curved surface in 3-space. In this form, the theorem is due to the Irish scientist George Gabriel Stokes (1819–1903).

We will need to put some restrictions on the surface $S$. First, we suppose that $S$ is two-sided with a continuously varying unit normal $\mathbf{n}$ (the one-sided Möbius band of Section 17.5 is thereby eliminated from our discussion). Second, we require that the boundary $\partial S$ be a piecewise smooth, simple closed curve, oriented consistently with $\mathbf{n}$. This means that if you stand near the edge of the surface with your head in the direction $\mathbf{n}$ and your eyes looking in the direction of the curve, the surface is to your left (Figure 1).

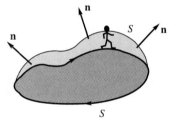

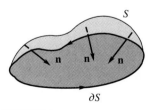

FIGURE 1

**Theorem A**

**(Stokes's Theorem).** Let $S$, $\partial S$, and $\mathbf{n}$ be as indicated above and suppose $\mathbf{F} = M\mathbf{i} + N\mathbf{j} + P\mathbf{k}$ is a vector field with $M$, $N$, and $P$ having continuous first-order partial derivatives on $S$ and its boundary $\partial S$. If $\mathbf{T}$ denotes the unit tangent vector to $\partial S$, then

$$\oint_{\partial S} \mathbf{F} \cdot \mathbf{T} \, ds = \iint_{S} (\text{curl } \mathbf{F}) \cdot \mathbf{n} \, dS$$

**Examples and Applications** The proof of Stokes's Theorem is more appropriate for a course in advanced calculus. However, we can at least verify the theorem in an example.

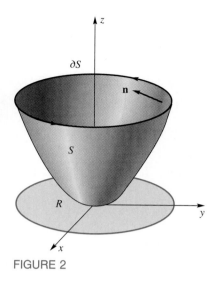

FIGURE 2

**EXAMPLE 1**    Verify Stokes's Theorem for $\mathbf{F} = y\mathbf{i} - x\mathbf{j} + yz\mathbf{k}$ if $S$ is the paraboloid $z = x^2 + y^2$ with the circle $x^2 + y^2 = 1$, $z = 1$ as its boundary (Figure 2).

*Solution*    We may describe $\partial S$ by the parametric equations

$$x = \cos t, \qquad y = \sin t, \qquad z = 1$$

Then $dz = 0$ and (see Section 17.2)

$$\oint_{\partial S} \mathbf{F} \cdot \mathbf{T} \, ds = \oint_{\partial S} y \, dx - x \, dy = \int_0^{2\pi} [\sin t(-\sin t) \, dt - \cos t \cos t \, dt]$$

$$= -\int_0^{2\pi} [\sin^2 t + \cos^2 t] \, dt = -2\pi$$

On the other hand, to calculate $\iint_S (\text{curl } \mathbf{F}) \cdot \mathbf{n} \, dS$, we first obtain

$$\text{curl } \mathbf{F} = \nabla \times \mathbf{F} = \begin{vmatrix} \mathbf{i} & \mathbf{j} & \mathbf{k} \\ \dfrac{\partial}{\partial x} & \dfrac{\partial}{\partial y} & \dfrac{\partial}{\partial z} \\ y & -x & yz \end{vmatrix} = z\mathbf{i} + 0\mathbf{j} - 2\mathbf{k}$$

Then, by Theorem 17.5B,

$$\iint_S (\text{curl } \mathbf{F}) \cdot \mathbf{n} \, dS = \iint_R [-z(2x) - 0(2y) - 2] \, dx \, dy$$

$$= -2 \iint_R [xz + 1] \, dx \, dy$$

$$= -2 \iint_R [x(x^2 + y^2) + 1] \, dx \, dy$$

$$= -2 \int_0^{2\pi} \int_0^1 [r^3 \cos \theta + 1] r \, dr \, d\theta$$

$$= -2 \int_0^{2\pi} [\tfrac{1}{5} \cos \theta + \tfrac{1}{2}] \, d\theta = -2\pi \qquad \blacksquare$$

**EXAMPLE 2**    Let $S$ be the part of the spherical surface $x^2 + y^2 + (z - 4)^2 = 10$ below the plane $z = 1$, and let $\mathbf{F} = y\mathbf{i} - x\mathbf{j} + yz\mathbf{k}$. Use Stokes's Theorem to calculate

$$\iint_S (\text{curl } \mathbf{F}) \cdot \mathbf{n} \, dS$$

where $\mathbf{n}$ is the upward unit normal.

***Solution*** Note that the field **F** is the same as that of Example 1 and also that $S$ has the same circle as its boundary curve. We conclude that

$$\iint_S (\text{curl } \mathbf{F}) \cdot \mathbf{n}\, dS = \oint_{\partial S} \mathbf{F} \cdot \mathbf{n}\, ds = -2\pi$$

In fact, we conclude that the flux of curl **F** is $-2\pi$ for all surfaces $S$ that have the circle $\partial S$ of Figure 2 as their oriented boundary. ∎

**EXAMPLE 3** Use Stokes's Theorem to evaluate $\oint_C \mathbf{F} \cdot \mathbf{T}\, ds$, where $\mathbf{F} = 2z\mathbf{i} + (8x - 3y)\mathbf{j} + (3x + y)\mathbf{k}$ and $C$ is the triangular curve of Figure 3.

***Solution*** We could let $S$ be any surface with $C$ as its oriented boundary, but it is to our advantage to choose the simplest such surface—namely, the flat planar triangle $T$. To determine **n** for this surface, we note that the vectors

$$\mathbf{A} = (0 - 1)\mathbf{i} + (0 - 0)\mathbf{j} + (2 - 0)\mathbf{k} = -\mathbf{i} + 2\mathbf{k}$$
$$\mathbf{B} = (0 - 1)\mathbf{i} + (1 - 0)\mathbf{j} + (0 - 0)\mathbf{k} = -\mathbf{i} + \mathbf{j}$$

lie on this surface and hence

$$\mathbf{N} = \mathbf{A} \times \mathbf{B} = \begin{vmatrix} \mathbf{i} & \mathbf{j} & \mathbf{k} \\ -1 & 0 & 2 \\ -1 & 1 & 0 \end{vmatrix} = -2\mathbf{i} - 2\mathbf{j} - \mathbf{k}$$

is perpendicular to it. The upward unit normal **n** is therefore

$$\mathbf{n} = \frac{2\mathbf{i} + 2\mathbf{j} + \mathbf{k}}{\sqrt{4 + 4 + 1}} = \frac{2}{3}\mathbf{i} + \frac{2}{3}\mathbf{j} + \frac{1}{3}\mathbf{k}$$

Also

$$\text{curl } \mathbf{F} = \begin{vmatrix} \mathbf{i} & \mathbf{j} & \mathbf{k} \\ \frac{\partial}{\partial x} & \frac{\partial}{\partial y} & \frac{\partial}{\partial z} \\ 2z & 8x - 3y & 3x + y \end{vmatrix} = \mathbf{i} - \mathbf{j} + 8\mathbf{k}$$

and curl $\mathbf{F} \cdot \mathbf{n} = \frac{8}{3}$. We conclude that

$$\oint_C \mathbf{F} \cdot \mathbf{T}\, ds = \iint_T (\text{curl } \mathbf{F}) \cdot \mathbf{n}\, dS = \tfrac{8}{3}(\text{area of } T) = \tfrac{8}{3}\left(\tfrac{3}{2}\right) = 4 \qquad ∎$$

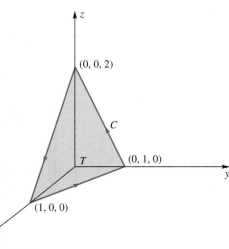

(0, 0, 2)

$C$

$T$    (0, 1, 0)

$y$

(1, 0, 0)

FIGURE 3

**EXAMPLE 4** Let the vector field **F** and the region $D$ satisfy the hypotheses of Theorem 17.3C. Show that if curl **F** = **0** in $D$, then **F** is conservative there.

*Solution* From the discussion in Section 17.3, we conclude that it is enough to show that $\oint_C \mathbf{F} \cdot d\mathbf{r} = 0$ for any simple closed path $C$ in $D$. Let $S$ be a surface having $C$ as its boundary and oriented consistently with $C$ (the simple connectedness of $D$ can be shown to guarantee the existence of such a surface). Then from Stokes's Theorem,

$$\oint_C \mathbf{F} \cdot d\mathbf{r} = \oint_C \mathbf{F} \cdot \mathbf{T} \, ds = \iint_S (\text{curl } \mathbf{F}) \cdot \mathbf{n} \, dS = 0$$

∎

**Physical Interpretation of the Curl** We offered an interpretation of the curl in Section 17.4. Now we can amplify that discussion. Let $C$ be a circle of radius $a$ centered at the point $P$. Then

$$\oint_C \mathbf{F} \cdot \mathbf{T} \, ds$$

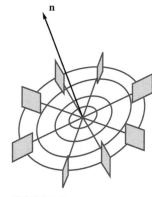

**n**

FIGURE 4

is called the *circulation* of **F** around $C$ and measures the tendency of a fluid with velocity field **F** to circulate around $C$. Now if **F** is continuous and $C$ is very small, Stokes's Theorem gives

$$\oint_C \mathbf{F} \cdot \mathbf{T} \, ds = \iint_S (\text{curl } \mathbf{F}) \cdot \mathbf{n} \, dS \approx [\text{curl } \mathbf{F}(P)] \cdot \mathbf{n} \, (\pi a^2)$$

The expression on the right will have the largest magnitude if **n** has the same direction as curl **F**($P$).

Suppose a small paddle wheel is placed in the fluid with center at $P$ and axis having direction **n** (Figure 4). This wheel will rotate most rapidly if **n** has the direction of curl **F**. The direction of rotation will be that determined by the right-hand rule.

## CONCEPTS REVIEW

**1.** Stokes's Theorem in three-space says that under appropriate hypotheses, $\int_{\partial S} \mathbf{F} \cdot \mathbf{T} \, dS = \iint_S$ _____ $dS$. Here $S$ is a surface and $\partial S$ is its boundary.

**2.** One of these hypotheses is that $S$ be two-sided. An important example of a one-sided surface is the _____  obtained by cutting an ordinary cylindrical band, giving it a half twist, and pasting it back together.

**3.** It follows from Stokes's Theorem that all two-sided surfaces with the same boundary $\partial S$ give the same value for _____.

**4.** A paddle wheel centered at $P$ and immersed in a fluid with velocity field **F** will rotate most rapidly about $P$ if **n** has the direction of _____.

## PROBLEM SET 17.7

In Problems 1–6, use Stokes's Theorem to calculate

$$\iint_S (\text{curl } \mathbf{F}) \cdot \mathbf{n}\, dS.$$

**1.** $\mathbf{F} = x^2\mathbf{i} + y^2\mathbf{j} + z^2\mathbf{k}$; $S$ is the hemisphere $z = \sqrt{1 - x^2 - y^2}$ and $\mathbf{n}$ is the upper normal.

**2.** $\mathbf{F} = xy\mathbf{i} + yz\mathbf{j} + xz\mathbf{k}$; $S$ is the triangular surface with vertices $(0, 0, 0)$, $(1, 0, 0)$, and $(0, 2, 1)$ and $\mathbf{n}$ is the upper normal.

**3.** $\mathbf{F} = (y + z)\mathbf{i} + (x^2 + z^2)\mathbf{j} + y\mathbf{k}$; $S$ is the half-cylinder $z = \sqrt{1 - x^2}$ between $y = 0$ and $y = 1$ and $\mathbf{n}$ is the upper normal.

**4.** $\mathbf{F} = xz^2\mathbf{i} + x^3\mathbf{j} + \cos xz\,\mathbf{k}$; $S$ is the part of the ellipsoid $x^2 + y^2 + 3z^2 = 1$ below the $xy$-plane and $\mathbf{n}$ is the lower normal.

**5.** $\mathbf{F} = yz\mathbf{i} + 3xz\mathbf{j} + z^2\mathbf{k}$; $S$ is the part of the sphere $x^2 + y^2 + z^2 = 16$ below the plane $z = 2$ and $\mathbf{n}$ is the outward normal.

**6.** $\mathbf{F} = (z - y)\mathbf{i} + (z + x)\mathbf{j} - (x + y)\mathbf{k}$; $S$ is the part of the paraboloid $z = 1 - x^2 - y^2$ above the $xy$-plane and $\mathbf{n}$ is the upward normal.

In Problems 7–12, use Stokes's Theorem to calculate $\oint_C \mathbf{F} \cdot \mathbf{T}\, ds$.

**7.** $\mathbf{F} = 2z\mathbf{i} + x\mathbf{j} + 3y\mathbf{k}$; $C$ is the ellipse that is the intersection of the plane $z = x$ and the cylinder $x^2 + y^2 = 4$, oriented clockwise as viewed from above.

**8.** $\mathbf{F} = y\mathbf{i} + z\mathbf{j} + x\mathbf{k}$; $C$ is the triangular curve with vertices $(0, 0, 0)$, $(2, 0, 0)$, and $(0, 2, 2)$, oriented counterclockwise as viewed from above.

**9.** $\mathbf{F} = (y - x)\mathbf{i} + (x - z)\mathbf{j} + (x - y)\mathbf{k}$; $C$ is the boundary of the plane $x + 2y + z = 2$ in the first octant, oriented clockwise as viewed from above.

**10.** $\mathbf{F} = y(x^2 + y^2)\mathbf{i} - x(x^2 + y^2)\mathbf{j}$; $C$ is the rectangular path from $(0, 0, 0)$ to $(1, 0, 0)$ to $(1, 1, 1)$ to $(0, 1, 1)$ to $(0, 0, 0)$.

**11.** $\mathbf{F} = (z - y)\mathbf{i} + y\mathbf{j} + x\mathbf{k}$; $C$ is the intersection of the cylinder $x^2 + y^2 = x$ with the sphere $x^2 + y^2 + z^2 = 1$, oriented counterclockwise as viewed from above.

**12.** $\mathbf{F} = (y - z)\mathbf{i} + (z - x)\mathbf{j} + (x - y)\mathbf{k}$; $C$ is the ellipse which is the intersection of the plane $x + z = 1$ and the cylinder $x^2 + y^2 = 1$, oriented clockwise as viewed from above.

**13.** Suppose the surface $S$ is determined by the formula $z = g(x, y)$. Show that the surface integral in Stokes's

Theorem can be written as a double integral in the following way.

$$\iint_S (\text{curl } \mathbf{F}) \cdot \mathbf{n}\, dS = \iint_{S_{xy}} (\text{curl } \mathbf{F}) \cdot (-g_x\mathbf{i} - g_y\mathbf{j} + \mathbf{k})\, dA$$

where $\mathbf{n}$ is the upward normal to $S$ and $S_{xy}$ is the projection of $S$ in the $xy$-plane.

**14.** Let $\mathbf{F} = x^2\mathbf{i} - 2xy\mathbf{j} + yz^2\mathbf{k}$ and $\partial S$ be the boundary of the surface $z = xy$, $0 \le x \le 1$, $0 \le y \le 1$, oriented counterclockwise as viewed from above. Use Stokes's Theorem and Problem 13 to evaluate $\oint_{\partial S} \mathbf{F} \cdot \mathbf{T}\, ds$.

**15.** Let $\mathbf{F} = 2\mathbf{i} + xz\mathbf{j} + z^3\mathbf{k}$ and $\partial S$ be the boundary of the surface $z = xy^2$, $0 \le x \le 1$, $0 \le y \le 1$, oriented counterclockwise as viewed from above. Evaluate $\oint_{\partial S} \mathbf{F} \cdot \mathbf{T}\, ds$.

**16.** Let $\mathbf{F} = 2\mathbf{i} + xz\mathbf{j} + z^3\mathbf{k}$ and $\partial S$ be the boundary of the surface $z = x^2y^2$, $x^2 + y^2 \le a^2$, oriented counterclockwise as viewed from above. Evaluate $\oint_{\partial S} \mathbf{F} \cdot \mathbf{T}\, ds$.

**17.** Let $\mathbf{F} = 2z\mathbf{i} + 2y\mathbf{k}$ and let $\partial S$ be the intersection of the cylinder $x^2 + y^2 = ay$ with the hemisphere $z = \sqrt{a^2 - x^2 - y^2}$, $a > 0$. Assuming distances in meters and force in newtons, find the work done by the force $\mathbf{F}$ in moving an object around $\partial S$ in the counterclockwise direction as viewed from above.

**18.** A central force is one of the form $\mathbf{F} = f(|\mathbf{r}|)\mathbf{r}$, where $f$ has a continuous derivative (except possibly at $|\mathbf{r}| = 0$). Show that the work done by such a force in moving an object around a closed path that misses the origin is 0.

**19.** Let $S$ be a solid sphere (or any solid enclosed by a "nice" surface $\partial S$). Show that

$$\iint_{\partial S} (\text{curl } \mathbf{F}) \cdot \mathbf{n}\, dS = 0$$

(a) by using Stokes's Theorem;
(b) by using Gauss's Theorem. *Hint:* Show div(curl $\mathbf{F}$) = 0.

**20.** Show

$$\oint_{\partial S} (f \nabla g) \cdot \mathbf{T}\, ds = \iint_S (\nabla f \times \nabla g) \cdot \mathbf{n}\, dS.$$

**Answers to Concept Review:** **1.** (curl $\mathbf{F}$) $\cdot$ $\mathbf{n}$
**2.** Möbius band  **3.** $\iint_S$ (curl $\mathbf{F}$) $\cdot$ $\mathbf{n}\, dS$  **4.** curl $\mathbf{F}$

## 17.8 CHAPTER REVIEW

### Concepts Test

Respond with true or false to each of the following assertions. Be prepared to justify your answer.

1. Inverse square law fields are conservative.

2. The divergence of a vector field is another vector field.

3. A physicist might be interested in both curl (grad $f$) and grad(curl $\mathbf{F}$).

4. If $f$ has continuous second partial derivatives, then curl (grad $f$) = $\mathbf{0}$.

5. The work done by a conservative force field as it moves an object around a closed path is zero.

6. If $\int_C \mathbf{F}(\mathbf{r}) \cdot d\mathbf{r} = 0$ for every closed path in an open connected set $D$, then there is a function $f$ such that $\nabla f = \mathbf{F}$ in $D$.

7. The field $\mathbf{F}(x, y, z) = (2x + 2y)\mathbf{i} + 2x\mathbf{j} + yz^2\mathbf{k}$ is conservative.

8. Green's theorem holds for a region $S$ with a hole provided the complete boundary of $S$ is oriented correctly.

9. The double integral is a special case of a surface integral.

10. A surface always has two sides.

11. If there are no sources or sinks in a region, then the net flow across the boundary of the region is zero.

12. If $S$ is a sphere with outward normal $\mathbf{n}$ and $\mathbf{F}$ is a constant vector field, then

$$\iint_S (\mathbf{F} \cdot \mathbf{n})\, dS = 0$$

### Sample Test Problems

1. Sketch a sample of vectors from the vector field $\mathbf{F}(x, y) = x\mathbf{i} + 2y\mathbf{j}$.

2. Find div $\mathbf{F}$, curl $\mathbf{F}$, grad(div $\mathbf{F}$), and div(curl $\mathbf{F}$) if $\mathbf{F}(x, y, z) = 2xyz\mathbf{i} - 3y^2\mathbf{j} + 2y^2z\mathbf{k}$.

3. We showed in Problem 20, Section 17.1, that

$$\text{curl} (f\mathbf{F}) = f(\text{curl } \mathbf{F}) + \nabla f \times \mathbf{F}$$

and in Problem 17, Section 17.3, that

$$\text{curl}(\nabla f) = \mathbf{0}$$

Use these facts to show that

$$\text{curl}(f\nabla f) = \mathbf{0}$$

4. Find a function $f$ satisfying

(a) $\nabla f = (2xy + y)\mathbf{i} + (x^2 + x + \cos y)\mathbf{j}$;
(b) $\nabla f = (yz - e^{-x})\mathbf{i} + (xz + e^y)\mathbf{j} + xy\mathbf{k}$.

5. Calculate.

(a) $\int_C (1 - y^2)\, ds$; $C$ is the quarter circle from $(0, -1)$ to $(1, 0)$, centered at the origin.

(b) $\int_C xy\, dx + z \cos x\, dy + z\, dz$; $C$ is the curve $x = t$, $y = \cos t$, $z = \sin t$, $0 \le t \le \pi/2$.

6. Show that $\int_C y^2\, dx + 2xy\, dy$ is independent of path and use this to calculate the integral on any path from $(0, 0)$ to $(1, 2)$.

7. Find the work done by $\mathbf{F} = y^2\mathbf{i} + 2xy\mathbf{j}$ in moving an object from $(1, 1)$ to $(3, 4)$ (see Problem 6).

8. Evaluate (see Problem 4b).

$$\int_{(0, 0, 0)}^{(1, 1, 4)} (yz - e^{-x})\, dx + (xy + e^y)\, dy + xy\, dz$$

9. Evaluate $\oint_C xy\, dx + (x^2 + y^2)\, dy$ if

(a) $C$ is the square path $(0, 0)$ to $(1, 0)$ to $(1, 1)$ to $(0, 1)$ to $(0, 0)$;
(b) $C$ is the triangular path $(0, 0)$ to $(2, 0)$ to $(2, 1)$ to $(0, 0)$;
(c) $C$ is the circle $x^2 + y^2 = 1$ traversed in the clockwise direction.

10. Calculate the flux of $\mathbf{F} = x\mathbf{i} + y\mathbf{j}$ across the square curve $C$ with vertices $(1, 1)$, $(-1, 1)$, $(-1, -1)$, and $(1, -1)$; that is, calculate $\oint_C \mathbf{F} \cdot \mathbf{n}\, ds$.

11. Calculate the flux of $\mathbf{F} = x\mathbf{i} + y\mathbf{j} + 3\mathbf{k}$ across the sphere $x^2 + y^2 + z^2 = 1$.

12. Evaluate $\iint_G xyz\, dS$, where $G$ is the part of the plane $z = x + y$ above the triangular region with vertices $(0, 0, 0)$, $(1, 0, 0)$, and $(0, 2, 0)$.

13. Evaluate $\iint\limits_{G} (\text{curl } \mathbf{F}) \cdot \mathbf{n} \, dS$, where

$$\mathbf{F} = x^3 y \mathbf{i} + e^y \mathbf{j} + z \tan\left(\frac{xyz}{4}\right) \mathbf{k}$$

and $G$ is the part of the sphere $x^2 + y^2 + z^2 = 2$ above the plane $z = 1$ and $\mathbf{n}$ is the upward unit normal.

14. Evaluate $\iint\limits_{G} \mathbf{F} \cdot \mathbf{n} \, dS$, where

$$\mathbf{F} = \sin x \mathbf{i} + (1 - \cos x) y \mathbf{j} + 4z \mathbf{k}$$

and $G$ is the closed surface bounded by $z = \sqrt{9 - x^2 - y^2}$, and $z = 0$ with outward unit normal $\mathbf{n}$.

15. Let $C$ be the circle that is the intersection of the plane $ax + by + z = 0$ ($a \geq 0$, $b \geq 0$) and the sphere $x^2 + y^2 + z^2 = 9$. For $\mathbf{F} = y\mathbf{i} - x\mathbf{j} + 3y\mathbf{k}$, evaluate

$$\oint_{C} \mathbf{F} \cdot \mathbf{T} \, ds$$

*Hint*: Use Stokes's Theorem.

## TECHNOLOGY PROJECT

An important notion associated with line integrals is "independence of path"; that is, in going from point $A$ to point $B$ (in the plane or in $R^3$) does the path taken matter? If the application is arc length, for example, the obvious answer is "yes"; and it comes as a bit of a surprise that in many line integral problems the answer is "no". The path independence issue is especially important in applications involving work, as one often wishes to minimize the amount of work, energy, etc. associated with a given task.

**Exercise 1** Suppose you are to power a boat from point $A = (0, 0)$ to point $B = (2, 1)$, and the primary consideration is the force of the wind which generally opposes you. You are to investigate the effect of taking different paths from $A$ to $B$. Note: The wind forces considered have negative components, indicating that the wind opposes the direction of the path. But think of the work we do as that required to *counteract* the wind, thus producing a positive value.

a) Suppose the wind force is $F = (-a, -b)$, for $a$ and $b$ positive. Compute the work done along the straight line path between $A$ and $B$. Also compute the work done along a second path, of your choice. Does the path matter here? Why?

b) Now, due to the effect of the harbor you are entering, the wind force is $(-a, -ae^{-y})$. Again, compute the work along two paths as in part (a).

Does the path matter here? Why?

c) Now let $F = (-a, -ae^{-y+cx})$ for $c > 0$. Compute the work along the straight line path and show that work $= 2 + (1 - \exp(2c - 1))/(1 - 2c)$. (Note: For $c = 0$ you should agree with the work in part (b)).

d) Now for a bit of a surprise. For the force of part (c), compute the work along the broken line path: up to $(0, 1)$ and over to $(2, 1)$. Explain why one does *not* have independence of path for this force. Comment on the fact that this path completely avoids the resistive part of the force represented by $e^{cx}$.

An important type of problem in several fields of application is: Can we find the "optimal path" (e.g., to minimize the work)? You are to explore this issue in regard to the wind force in part (c) of the previous exercise, $F = (-a, -ae^{-y+cx})$. Along with the paths you used in Exercise 1, you are to consider some new paths. In particular, consider a family of parabolas described for $0 \le t \le 1$, by:

$$x = 2t, \quad y = 2t(\beta - 2\alpha t)$$

where $\alpha > 0$.

**Exercise 2** For the parabolic paths, and for the force $F = (-a, -ae^{-y+cx})$ with $c = 0.1$:

a) To go from $(0, 0)$ to $(2, 1)$, show that:

$$\beta = (1 + 4\alpha)/2.$$

b) Graphically show that for $\alpha$ "large" the path swings "north", i.e. it goes above the

line $y = 1$ before ending at $y = 1$. (This could be bad). What's the situation for $\alpha = 0$? Moreover, find $\alpha$ such that the path comes into the final destination horizontally. (This could be good. You may want to find $\alpha$ analytically).

c) Now compute the work in going from $(0, 0)$ to $(2, 1)$ along various parabolic paths. What is the situation regarding the minimizing of work, when there are no restrictions on the parabolic path? Is this surprising in any way?

d) Same problem as in part (c), except now you have the restriction that the path cannot, due to the coast line, swing "north." There *is* an optimal path in this case; what is it?

e) We have so far assumed $\alpha > 0$. What happens for $\alpha < 0$, both graphically and regarding the work?

**Exercise 3** Now consider a circuitous path from $(0, 0)$ to $(1, 2)$ described by:

$$x = 2t + \sin 2\pi t,$$
$$y = t^2 + 1 - \cos 2\pi t.$$

where $0 \le t \le 1$.

a) Compute the arc length of this path.

b) Compute the work on this path for the current force. How does it compare with the work along the parabolic paths?

c) Compute the work on this path for the force $F = (-a, -ae^{-y})$. Compare with the results of Exercise 1b), and comment.

# Differential Equations

We call an equation involving one or more derivatives of an unknown function a **differential equation**. In particular, an equation of the form

$$F(x, y, y^{(1)}, y^{(2)}, \ldots, y^{(n)}) = 0$$

in which $y^{(k)}$ denotes the $k$th derivative of $y$ with respect to $x$, is called an **ordinary differential equation of order $n$**. Examples of order 1, 2, and 3 are

$$y' + 2 \sin x = 0$$

$$\frac{d^2y}{dx^2} + 3x\frac{dy}{dx} - 2y = 0$$

$$\frac{d^3y}{dx^3} + \left(\frac{dy}{dx}\right)^2 - e^x = 0$$

If, when $f(x)$ is substituted for $y$ in the differential equation, the resulting equation is an identity for all $x$ in some interval, then $f(x)$ is called a **solution** of the differential equation. Thus, $f(x) = 2 \cos x + 10$ is a solution to $y' + 2 \sin x = 0$ since

$$f'(x) + 2 \sin x = -2 \sin x + 2 \sin x = 0$$

for all $x$. We call $2 \cos x + C$ the **general solution** of the given equation, since it can be shown that every solution can be written in this form. In contrast, $2 \cos x + 10$ is called a **particular solution** of the equation.

Differential equations appeared earlier in this book, principally in two sections. In Section 5.2, we introduced the technique called *separation of variables* and used it to solve a wide variety of first-order equations. In

Section 7.5, we solved the differential equation $y' = ky$ of exponential growth and decay.

In this chapter we consider only **linear** differential equations—that is, equations of the form

$$y^{(n)} + a_1(x)y^{(n-1)} + \cdots + a_{n-1}(x)y' + a_n(x)y = k(x)$$

(Note that $y$ and all its derivatives occur to the first power.) This is called a linear equation because if it is written in operator notation,

$$[D_x^n + a_1(x)D_x^{n-1} + \cdots + a_{n-1}(x)D_x + a_n(x)]y = k(x)$$

the operator in brackets is a linear operator. Thus, if $L$ denotes this operator and if $f$ and $g$ are functions and $c$ is constant,

$$L(f + g) = L(f) + L(g)$$
$$L(cf) = cL(f)$$

That $L$ has these properties follows readily from the corresponding properties for the derivative operators $D, D^2, \ldots, D^n$.

Of course, not all operators are linear; in fact, many important differential equations are nonlinear. A simple example is

$$\frac{dy}{dx} + y^2 = 0$$

The presence of the exponent 2 on $y$ is enough to spoil the linearity, as you may check. The theory of nonlinear differential equations is both complicated and fascinating, but best left for more advanced courses.

**The General First-Order Linear Equation**  The equations we now consider can be put in the form

$$\frac{dy}{dx} + P(x)y = Q(x)$$

An equation of this type can always be solved in principle. We first multiply both sides by the **integrating factor**

$$e^{\int P(x)\, dx}$$

obtaining

$$e^{\int P(x)\, dx}\frac{dy}{dx} + e^{\int P(x)\, dx} P(x)y = Q(x)e^{\int P(x)\, dx}$$

Then we recognize the left side as the derivative of $ye^{\int P(x)\, dx}$, so that the equation takes the form

$$\frac{d}{dx}\left(ye^{\int P(x)\, dx}\right) = Q(x)e^{\int P(x)\, dx}$$

Integration of both sides yields

$$ye^{\int P(x)\,dx} = \int (Q(x)e^{\int P(x)\,dx})\,dx$$

and thus

$$y = e^{-\int P(x)\,dx} \int (Q(x)e^{\int P(x)\,dx})\,dx$$

It is not worth memorizing this final result; the process of getting there is easily recalled, and that is what we illustrate.

**EXAMPLE 1**  Solve

$$\frac{dy}{dx} + \frac{2}{x}y = \frac{\sin 3x}{x^2}$$

*Solution*  Our integrating factor is

$$e^{\int P(x)\,dx} = e^{\int (2/x)\,dx} = e^{2\ln|x|} = e^{\ln x^2} = x^2$$

(We have taken the arbitrary constant from the integration $\int P(x)\,dx$ to be 0, since it would cancel out in the end anyway.) Multiplying both sides of the original equation by $x^2$, we obtain

$$x^2\frac{dy}{dx} + 2xy = \sin 3x$$

or, equivalently,

$$\frac{d}{dx}(x^2y) = \sin 3x$$

Integration of both members yields

$$x^2y = \int \sin 3x\,dx = -\tfrac{1}{3}\cos 3x + C$$

or

$$y = (-\tfrac{1}{3}\cos 3x + C)x^{-2} \qquad\blacksquare$$

**EXAMPLE 2**  Find the particular solution of

$$\frac{dy}{dx} - 3y = xe^{3x}$$

that satisfies $y = 4$ when $x = 0$.

*Solution* The appropriate integrating factor is

$$e^{\int -3\,dx} = e^{-3x}$$

Upon multiplication by this factor, our equation takes the form

$$\frac{d}{dx}(e^{-3x}y) = x$$

or

$$e^{-3x}y = \int x\,dx = \frac{1}{2}x^2 + C$$

Thus, the general solution is

$$y = \frac{1}{2}x^2e^{3x} + Ce^{3x}$$

Substitution of $y = 4$ when $x = 0$ makes $C = 4$. The desired particular solution is

$$y = \frac{1}{2}x^2e^{3x} + 4e^{3x}$$

■

**Applications**  We begin with a mixture problem, typical of many problems that arise in chemistry.

EXAMPLE 3  A tank initially contains 120 gallons of brine, holding 75 pounds of dissolved salt in solution. Salt water containing 1.2 pounds of salt per gallon is entering the tank at the rate of 2 gallons per minute and brine flows out at the same rate (Figure 1). If the mixture is kept uniform by constant stirring, find the amount of salt in the tank at the end of 1 hour.

*Solution*  Let $y$ be the number of pounds of salt in the tank at the end of $t$ minutes. From the brine flowing in, the tank gains 2.4 pounds of salt per minute; from that flowing out it loses $\frac{2}{120}y$ pounds per minute. Thus,

$$\frac{dy}{dt} = 2.4 - \frac{1}{60}y$$

subject to the condition $y = 75$ when $t = 0$. The equivalent equation

$$\frac{dy}{dt} + \frac{1}{60}y = 2.4$$

has the integrating factor $e^{t/60}$, and so

$$\frac{d}{dt}[ye^{t/60}] = 2.4e^{t/60}$$

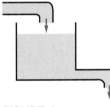

FIGURE 1

A General Principle

In flow problems such as Example 3, we apply a general principle. Let $y$ measure the quantity of interest that is in the tank at time $t$. Then the rate of change of $y$ with respect to time is the input rate minus the output rate; that is

$$\frac{dy}{dt} = \text{rate in} - \text{rate out}$$

We conclude that

$$ye^{t/60} = \int 2.4e^{t/60}\, dt = (60)(2.4)e^{t/60} + C$$

Substituting $y = 75$ when $t = 0$ yields $C = -69$, and so

$$y = e^{-t/60}[144e^{t/60} - 69] = 144 - 69e^{-t/60}$$

At the end of 1 hour ($t = 60$),

$$y = 144 - 69e^{-1} \approx 118.62 \text{ pounds}$$

Note that the limiting value for $y$ as $t \to \infty$ is 144. This corresponds to the fact that the tank will ultimately take on the complexion of the brine entering the tank. One hundred twenty gallons of brine with a concentration of 1.2 pounds of salt per gallon will contain 144 pounds of salt. ∎

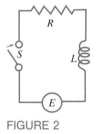

**FIGURE 2**

We turn next to an example from electricity. According to Kirchhoff's Law, a simple electrical circuit (Figure 2) containing a resistor with a resistance of $R$ ohms and an inductor with an inductance of $L$ henrys in series with a source of electromotive force (a battery or generator) that supplies a voltage of $E(t)$ volts at time $t$ satisfies

$$L\frac{dI}{dt} + RI = E(t)$$

where $I$ is the current measured in amperes. This is a linear equation, easily solved by the method of this section.

**EXAMPLE 4**  Consider a circuit (Figure 2) with $L = 2$ henrys, $R = 6$ ohms, and a battery supplying a constant voltage of 12 volts. If $I = 0$ at $t = 0$ (when the switch $S$ is closed), find $I$ at time $t$.

*Solution*  The differential equation is

$$2\frac{dI}{dt} + 6I = 12, \quad \text{or} \quad \frac{dI}{dt} + 3I = 6$$

Following our standard procedure (multiply by the integrating factor $e^{3t}$, integrate, and multiply by $e^{-3t}$), we obtain

$$I = e^{-3t}(2e^{3t} + C) = 2 + Ce^{-3t}$$

The initial condition, $I = 0$ at $t = 0$, gives $C = -2$; hence

$$I = 2 - 2e^{-3t}$$

As $t$ increases, the current tends toward a current of 2 amps. ∎

**EXAMPLE 5**  If the battery of Example 4 is replaced by an alternating current generator that supplies a voltage of $E(t) = 12 \sin 9t$ volts, find $I$ at time $t$ and note its behavior for large $t$.

*Solution*  The differential equation may be written as

$$\frac{dI}{dt} + 3I = 6 \sin 9t$$

Multiplying by $e^{3t}$ gives

$$\frac{d}{dt}(Ie^{3t}) = 6e^{3t} \sin 9t$$

By Formula 67 from the integral tables at the end of the book,

$$Ie^{3t} = 6 \int e^{3t} \sin 9t \, dt$$

$$= \frac{6e^{3t}}{9 + 81}(3 \sin 9t - 9 \cos 9t) + C$$

Thus,

$$I = \tfrac{1}{5} \sin 9t - \tfrac{3}{5} \cos 9t + Ce^{-3t}$$

Next, the initial condition $I = 0$ at $t = 0$ gives $C = \tfrac{3}{5}$. Finally, we obtain

$$I = \tfrac{1}{5} \sin 9t - \tfrac{3}{5} \cos 9t + \tfrac{3}{5}e^{-3t}$$

For large $t$, $e^{-3t}$ is negligible, and

$$I \approx \tfrac{1}{5} \sin 9t - \tfrac{3}{5} \cos 9t$$

which is an alternating current with the same frequency as the imposed voltage.  ∎

## CONCEPTS REVIEW

**1.** The general first-order linear differential equation has the form $dy/dx + P(x)y = Q(x)$. An integrating factor for this equation is _____.

**2.** Multiplying both sides of the first-order linear differential equation in Question 1 by its integrating factor makes the left side $\dfrac{d}{dx}$ (_____).

**3.** The integrating factor for $dy/dx - (1/x)y = x$ is _____. When we multiply both sides by this factor, the equation takes the form _____. The general solution to this equation is $y =$ _____.

**4.** The solution to the differential equation in Question 1 satisfying $y(a) = b$ is called a _____ solution.

## PROBLEM SET 18.1

In Problems 1–14, solve the differential equations.

1. $\dfrac{dy}{dx} + y = e^{-x}$

2. $(x + 1)\dfrac{dy}{dx} + y = x^2 - 1$

3. $(1 - x^2)\dfrac{dy}{dx} + xy = ax,\ |x| < 1$

4. $y' + y \tan x = \sec x$    5. $\dfrac{dy}{dx} - \dfrac{y}{x} = xe^x$

6. $y' - ay = f(x)$         7. $\dfrac{dy}{dx} + \dfrac{y}{x} = \dfrac{1}{x}$

8. $y' + \dfrac{2y}{x + 1} = (x + 1)^3$

9. $y' + yf(x) = f(x)$

10. $\dfrac{dy}{dx} + 2y = x$

11. $\dfrac{dy}{dx} - \dfrac{y}{x} = 3x^3;\ y = 3$ when $x = 1$.

12. $y' = e^{2x} - 3y;\ y = 1$ when $x = 0$.

13. $xy' + (1 + x)y = e^{-x};\ y = 0$ when $x = 1$.

14. $\sin x\dfrac{dy}{dx} + 2y \cos x = \sin 2x;\ y = 2$ when $x = \dfrac{\pi}{6}$.

15. A tank contains 20 gallons of a solution, with 10 pounds of chemical A in the solution. At a certain instant, we begin pouring in a solution containing the same chemical in a concentration of 2 pounds per gallon. We pour at a rate of 3 gallons per minute while simultaneously draining off the resulting (well-stirred) solution at the same rate. Find the amount of chemical A in the tank after 20 minutes.

16. A tank initially contains 200 gallons of brine, with 50 pounds of salt in solution. Brine containing 2 pounds of salt per gallon is entering the tank at the rate of 4 gallons per minute and is flowing out at the same rate. If the mixture in the tank is kept uniform by constant stirring, find the amount of salt in the tank at the end of 40 minutes.

17. A tank initially contains 120 gallons of pure water. Brine with 1 pound of salt per gallon flows into the tank at 4 gallons per minute, and the well-stirred solution runs out at 6 gallons per minute. How much salt is in the tank after $t$ minutes, $0 \le t \le 60$?

18. A tank initially contains 50 gallons of brine, with 30 pounds of salt in solution. Water runs into the tank at 3 gallons per minute and the well-stirred solution runs out at 2 gallons per minute. How long will it be until there are 25 pounds of salt in the tank?

19. Find the current $I$ as a function of time in the circuit of Figure 3 if the switch $S$ is closed and $I = 0$ at $t = 0$.

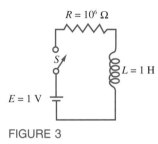

$R = 10^6\ \Omega$

$L = 1$ H

$E = 1$ V

**FIGURE 3**

20. Find $I$ as a function of time in the circuits of Figure 4, assuming the switch is closed and $I = 0$ at $t = 0$.

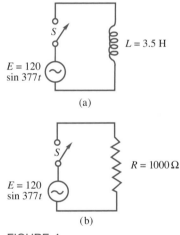

$L = 3.5$ H

$E = 120$ $\sin 377t$

(a)

$R = 1000\ \Omega$

$E = 120$ $\sin 377t$

(b)

**FIGURE 4**

21. Using Figure 5, find (a) $I$ as a function of time $t$, assuming the switch is closed and $I = 0$ at $t = 0$, (b) the steady-state current as a function of time (let $t \to \infty$).

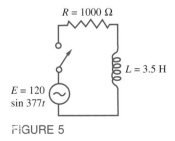

$R = 1000\ \Omega$

$L = 3.5$ H

$E = 120$ $\sin 377t$

**FIGURE 5**

**22.** Suppose that tank 1 initially contains 100 gallons of solution, with 50 pounds of dissolved salt, and tank 2 contains 200 gallons, with 150 pounds of dissolved salt. Pure water flows into tank 1 at 2 gallons per minute, the well-mixed solution flows out and into tank 2 at the same rate, and—finally—the solution in tank 2 drains away also at the same rate. Let $x(t)$ and $y(t)$ denote the amounts of salt in tanks 1 and 2, respectively, at time $t$. Find $y(t)$. *Hint*: First find $x(t)$ and use it in setting up the differential equation for tank 2.

**23.** A tank of capacity 100 gallons is initially full of pure alcohol. The flow rate of the drain pipe is 5 gallons per minute; the flow rate of the filler pipe can be adjusted to $c$ gallons per minute. An unlimited amount of 25% alcohol solution can be brought in through the filler pipe. Our goal is to reduce the amount of alcohol in the tank so that it will contain 100 gallons of 50% solution. Let $T$ be the number of minutes required to accomplish the desired change.

(a) Evaluate $T$ if $c = 5$ and both pipes are opened.
(b) Evaluate $T$ if $c = 5$ and we first drain away a sufficient amount of the pure alcohol and then close the drain and open the filler pipe.
(c) For what values of $c$ (if any) would strategy (b) give a faster time than (a)?
(d) Suppose $c = 4$. Determine the equation for $T$ if we initially open both pipes and then close the drain.
(e) Solve the equation in (d), perhaps by Newton's Method.

**24.** The differential equation for a body falling under the influence of gravity near the earth's surface with air resistance proportional to the velocity $v$ is $dv/dt = -g - av$, where $g = 32$ feet per second per second is the acceleration of gravity and $a > 0$ is the *drag coefficient*. Show each of the following.

(a) $v(t) = (v_0 - v_\infty)e^{-at} + v_\infty$, where $v_0 = v(0)$ and
$$v_\infty = -g/a = \lim_{t \to \infty} v(t)$$
the so-called terminal velocity.

(b) If $y(t)$ denotes the altitude, then
$$y(t) = y_0 + tv_\infty + (1/a)(v_0 - v_\infty)(1 - e^{-at})$$

**25.** A ball was thrown straight up from ground level with an initial velocity $v_0 = 120$ feet per second. Assuming a drag coefficient of $a = 0.05$, determine each of the following.

(a) The maximum altitude.
(b) The equation for $T$, the time when the ball hit the ground.
(c) The value of $T$ (perhaps by Newton's Method).

**26.** Andrea bailed out of her plane at an altitude of 8000 feet, fell freely for 15 seconds, and then opened her parachute. Assume the drag coefficients are $a = 0.10$ for free fall and $a = 1.6$ with parachute. When did she hit the ground?

**Answers to Concepts Review:** **1.** $\exp(\int P(x)\,dx)$
**2.** $y\exp(\int P(x)\,dx)$  **3.** $1/x$; $\dfrac{d}{dx}\left(\dfrac{y}{x}\right) = 1$; $x^2 + Cx$
**4.** Particular

---

## 18.2
## SECOND-ORDER HOMOGENEOUS EQUATIONS

A second-order linear differential equation has the form

$$y'' + a_1(x)y' + a_2(x)y = k(x)$$

In this section, we make two simplifying assumptions: (1) $a_1(x)$ and $a_2(x)$ are constants, and (2) $k(x)$ is identically zero (the homogeneous case). Thus, our initial task is to solve

$$y'' + a_1y' + a_2y = 0$$

To solve a first-order equation required one integration and led to a general solution with one arbitrary constant. By analogy, we might expect solving a second-order equation to involve two integrations and thus that the general solution would have two arbitrary constants. Our expectations are correct. In fact, a second-order homogeneous linear differential equation always has two fundamental solutions $u_1(x)$ and $u_2(x)$, which are **independent** of each other (that is, neither function is a constant multiple of the other). By the linearity of the operator $D^2 + a_1 D + a_2$,

$$C_1 u_1(x) + C_2 u_2(x)$$

is also a solution. Moreover, it can be shown that every solution has this form.

**The Auxiliary Equation**    Because $D_x(e^{rx}) = re^{rx}$, it seems likely that $e^{rx}$ will be a solution to our differential equation for an appropriate choice of $r$. To test this possibility, we first write the equation in the operator form

(1) $$(D^2 + a_1 D + a_2)y = 0$$

Now

$$\begin{aligned}
(D^2 + a_1 D + a_2)e^{rx} &= D^2(e^{rx}) + a_1 D(e^{rx}) + a_2 e^{rx} \\
&= r^2 e^{rx} + a_1 r e^{rx} + a_2 e^{rx} \\
&= e^{rx}(r^2 + a_1 r + a_2)
\end{aligned}$$

The latter expression is zero, provided

(2) $$r^2 + a_1 r + a_2 = 0$$

Equation (2) is called the **auxiliary equation** for (1) (note the similarity in form). It is an ordinary quadratic equation and can be solved by factoring or, if necessary, by the Quadratic Formula. There are three cases to consider, corresponding to whether the auxiliary equation has two distinct, real roots, a single, repeated root, or two complex conjugate roots.

---

Theorem A

**(Distinct real roots).**   If $r_1$ and $r_2$ are distinct, real roots of the auxiliary equation, then the general solution to $y'' + a_1 y' + a_2 y = 0$ is

$$y = C_1 e^{r_1 x} + C_2 e^{r_2 x}$$

---

EXAMPLE 1    Find the general solution to $y'' + 7y' + 12y = 0$.

*Solution*   The auxiliary equation

$$r^2 + 7r + 12 = (r + 3)(r + 4) = 0$$

has the two roots $-3$ and $-4$. Since $e^{-3x}$ and $e^{-4x}$ are independent solutions, the general solution to the differential equation is

$$y = C_1 e^{-3x} + C_2 e^{-4x}$$    ■

**EXAMPLE 2**    Find the solution of $y'' - 2y' - y = 0$ that satisfies $y(0) = 0$ and $y'(0) = \sqrt{2}$.

*Solution*    The auxiliary equation $r^2 - 2r - 1 = 0$ is best solved by the Quadratic Formula.

$$r = \frac{-b \pm \sqrt{b^2 - 4ac}}{2a} = \frac{2 \pm \sqrt{4 + 4}}{2} = 1 \pm \sqrt{2}$$

The general solution to the differential equation is, therefore,

$$y = C_1 e^{(1 + \sqrt{2})x} + C_2 e^{(1 - \sqrt{2})x}$$

The condition $y(0) = 0$ implies that $C_2 = -C_1$. Then

$$y' = C_1 (1 + \sqrt{2})e^{(1 + \sqrt{2})x} - C_1 (1 - \sqrt{2})e^{(1 - \sqrt{2})x}$$

and so

$$\sqrt{2} = y'(0) = C_1 (1 + \sqrt{2}) - C_1 (1 - \sqrt{2}) = 2C_1\sqrt{2}$$

We conclude that $C_1 = \frac{1}{2}$ and

$$y = \tfrac{1}{2}e^{(1 + \sqrt{2})x} - \tfrac{1}{2}e^{(1 - \sqrt{2})x} \qquad \blacksquare$$

This is all fine if the auxiliary equation has distinct real roots. But what if it has the form

$$r^2 - 2r_1 r + r_1^2 = (r - r_1)^2 = 0$$

Then our method produces the single fundamental solution $e^{r_1 x}$ and we must find another solution independent of this one. Such a solution is $xe^{r_1 x}$, as we now demonstrate.

$$(D^2 - 2r_1 D + r_1^2)xe^{r_1 x} = D^2(xe^{r_1 x}) - 2r_1 D(xe^{r_1 x}) + r_1^2 xe^{r_1 x}$$
$$= (xr_1^2 e^{r_1 x} + 2r_1 e^{r_1 x}) - 2r_1(xr_1 e^{r_1 x} + e^{r_1 x}) + r_1^2 xe^{r_1 x}$$
$$= 0$$

---

### Theorem B

**(A repeated root).**    If the auxiliary equation has the single repeated root $r_1$, then the general solution to $y'' + a_1 y' + a_2 y = 0$ is

$$y = C_1 e^{r_1 x} + C_2 xe^{r_1 x}$$

---

**EXAMPLE 3**    Solve $y'' - 6y' + 9y = 0$.

*Solution*    The auxiliary equation has 3 as a repeated root. Thus,

$$y = C_1 e^{3x} + C_2 xe^{3x} \qquad \blacksquare$$

Summary

Consider the second order differential equation
$$y'' + a_1 y' + a_2 = 0$$
with auxiliary equation
$$r^2 + a_1 r + a_2 = 0$$
The latter equation may have two real roots $r_1$ and $r_2$ or two complex roots $\alpha \pm \beta i$.

| Roots | Solution to Differential Equation |
|-------|-----------------------------------|
| $r_1 \neq r_2$ | $y = C_1 e^{r_1 x} + C_2 e^{r_2 x}$ |
| $r_1 = r_2$ | $y = C_1 e^{r_1 x} + C_2 x e^{r_1 x}$ |
| $\alpha \pm \beta i$ | $y = C_1 e^{\alpha x} \cos \beta$ $\quad + C_2 e^{\alpha x} \sin \beta x$ |

Finally, we consider the case where the auxiliary equation has conjugate complex roots. The simple equation

$$(D^2 + \beta^2)y = 0$$

with auxiliary equation $r^2 + \beta^2 = 0$ and roots $\pm \beta i$ offers a hint. Its fundamental solutions are easily seen to be $\sin \beta x$ and $\cos \beta x$. You can check by direct differentiation that the general situation is as follows.

**Theorem C**

**(Complex conjugate roots).** If the auxiliary equation has complex conjugate roots $\alpha \pm \beta i$, then the general solution to $y'' + a_1 y' + a_2 y = 0$ is

$$y = C_1 e^{\alpha x} \cos \beta x + C_2 e^{\alpha x} \sin \beta x$$

**EXAMPLE 4**   Solve $y'' - 4y' + 13y = 0$.

*Solution*   The roots of the auxiliary equation $r^2 - 4r + 13 = 0$ are $2 \pm 3i$. Hence the general solution is

$$y = C_1 e^{2x} \cos 3x + C_2 e^{2x} \sin 3x \qquad \blacksquare$$

**Higher-Order Equations**   All of what we have done extends to higher-order linear homogeneous equations with constant coefficients. To solve

$$y^{(n)} + a_1 y^{(n-1)} + \cdots + a_{n-1} y' + a_n y = 0$$

find the roots of the auxiliary equation

$$r^n + a_1 r^{n-1} + \cdots + a_{n-1} r + a_n = 0$$

and make the obvious generalizations of the second-order case. For example, if the auxiliary equation is

$$(r - r_1)(r - r_2)^3 \, [r - (\alpha + \beta i)] \, [r - (\alpha - \beta i)] = 0$$

the general solution to the differential equation is

$$y = C_1 e^{r_1 x} + (C_2 + C_3 x + C_4 x^2) e^{r_2 x} + e^{\alpha x}[C_5 \cos \beta x + C_6 \sin \beta x]$$

**EXAMPLE 5**   Solve $\dfrac{d^4 y}{dx^4} - \dfrac{d^3 y}{dx^3} - 20 \dfrac{d^2 y}{dx^2} = 0$.

*Solution*   The auxiliary equation is

$$r^4 - r^3 - 20r^2 = r^2(r - 5)(r + 4) = 0$$

with roots $-4$, $5$, and a double root of $0$. Hence, the general solution is

$$y = C_1 + C_2 x + C_3 e^{5x} + C_4 e^{-4x} \qquad \blacksquare$$

# CONCEPTS REVIEW

1. The auxiliary equation corresponding to the differential equation $(D^2 + a_1 D + a_2)y = 0$ is _____ . This equation may have two real roots, a single repeated root or _____ .

2. The general solution to $(D^2 - 1)y = 0$ is $y =$ _____ .

3. The general solution to $(D^2 - 2D + 1)y = 0$ is $y =$ _____ .

4. The general solution to $(D^2 + 1)y = 0$ is $y =$ _____ .

# PROBLEM SET 18.2

In Problems 1–16, solve the differential equations.

1. $y'' - 5y' + 6y = 0$     2. $y'' + 5y' - 6y = 0$

3. $y'' + 6y' - 7y = 0$; $y = 0$, $y' = 4$ at $x = 0$.

4. $y'' - 3y' - 10y = 0$; $y = 1$, $y' = 10$ at $x = 0$.

5. $y'' - 4y' + 4y = 0$

6. $y'' + 10y' + 25y = 0$

7. $y'' - 4y' + y = 0$

8. $y'' + 6y' - 2y = 0$

9. $y'' + 4y = 0$; $y = 2$ at $x = 0$, $y = 3$ at $x = \pi/4$.

10. $y'' + 9y = 0$; $y = 3$, $y' = 3$ at $x = \pi/3$.

11. $y'' + 2y' + 2y = 0$

12. $y'' + y' + y = 0$

13. $y^{(4)} + 3y''' - 4y'' = 0$

14. $y^{(4)} - y = 0$

15. $(D^4 + 3D^2 - 4)y = 0$

16. $[(D^2 + 1)(D^2 - D - 6)]y = 0$

17. Solve $y'' - 4y = 0$ and express your answer in terms of the hyperbolic functions cosh and sinh.

18. Show that the solution of

$$\frac{d^2y}{dx^2} - 2b\frac{dy}{dx} - c^2y = 0$$

can be written

$$y = e^{bx}\left(D_1 \cosh\sqrt{b^2 + c^2}x + D_2 \sinh\sqrt{b^2 + c^2}x\right)$$

19. Solve $y^{(4)} + 2y^{(3)} + 3y'' + 2y' + y = 0$. *Hint*: First show that the auxiliary equation is $(r^2 + r + 1)^2 = 0$.

20. Solve $y'' - 2y' + 2y = 0$ and express your answer in the form $ce^{\alpha x} \sin(\beta x + \gamma)$. *Hint*: Let $\sin \gamma = C_1/c$ and $\cos \gamma = C_2/c$, where $c = \sqrt{C_1^2 + C_2^2}$.

21. Solve $x^2y'' + 5xy' + 4y = 0$ by first making the substitution $x = e^z$.

22. Show that the substitution $x = e^z$ transforms the Euler equation $ax^2y'' + bxy' + cy = 0$ to a homogeneous linear equation with constant coefficients.

23. Recall that complex numbers have the form $a + bi$, where $a$ and $b$ are real. These numbers behave much like the real numbers with the proviso that $i^2 = -1$. Show each of the following.

(a) $e^{bi} = \cos b + i \sin b$
(b) $e^{a + bi} = e^a(\cos b + i \sin b)$
(c) $D_x e^{(\alpha + \beta i)x} = (\alpha + \beta i)e^{(\alpha + \beta i)x}$

*Hint*: For (a), use the Maclaurin series for $e^u$.

24. Let the roots of the auxiliary equation $r^2 + a_1 r + a_2 = 0$ be $\alpha \pm \beta i$. From Problem 23c, it follows just as in the real case that $y = c_1 e^{(\alpha + \beta i)x} + c_2 e^{(\alpha - \beta i)x}$ satisfies $(D^2 + a_1 D + a_2)y = 0$. Show that this solution can be rewritten in the form

$$y = C_1 e^{\alpha x} \cos \beta x + C_2 e^{\alpha x} \sin \beta x$$

giving another approach to Theorem C.

PC *True BASIC Calculus* has a section for solving second-order linear differential equations. Use it to redo Problems 1–12. Then use it to solve each of the following equations.

25. $y'' - 4y' - 6y = 0$; $y(0) = 1$, $y'(0) = 2$

26. $y'' + 5y' + 6.25y = 0$; $y(0) = 2$, $y'(0) = -1.5$

27. $2y'' + y' + 2y = 0$; $y(0) = 0$, $y'(0) = 1.25$

28. $3y'' - 2y + y = 0$; $y(0) = 2.5$, $y'(0) = -1.5$

---

**Answers to Concepts Review:**   1. $r^2 + a_1 r + a_2 = 0$; complex conjugate roots   2. $C_1 e^{-x} + C_2 e^x$
3. $(C_1 + C_2 x)e^x$   4. $C_1 \cos x + C_2 \sin x$

## 18.3
## THE NONHOMOGENEOUS EQUATION

Consider the general nonhomogeneous linear equation with constant coefficients

$$y^{(n)} + a_1 y^{(n-1)} + \cdots + a_{n-1} y' + a_n y = k(x)$$

Solving this equation can be reduced to three steps.

**1.** Find the general solution

$$y_h = C_1 u_1(x) + C_2 u_2(x) + \cdots + C_n u_n(x)$$

to the homogeneous equation as described in the previous section.

**2.** Find a particular solution $y_p$ to the nonhomogeneous equation.

**3.** Add the solutions from Steps 1 and 2. We state the result as a formal theorem.

---

### Theorem A

If $y_p$ is any particular solution to the nonhomogeneous equation

$$(*) \qquad L(y) = (D^n + a_1 D^{n-1} + \cdots + a_{n-1} D + a_n) y = k(x)$$

and if $y_h$ is the general solution to the corresponding homogeneous equation, then

$$y = y_p + y_h$$

is the general solution of $(*)$.

---

**Proof** The linearity of the operator $L$ is the key element in the proof. Let $y_p$ and $y_h$ be as described. Then

$$L(y_p + y_h) = L(y_p) + L(y_h) = k(x) + 0$$

and so $y = y_p + y_h$ is a solution to $(*)$.

Conversely, let $y$ be any solution to $(*)$. Then

$$L(y - y_p) = L(y) - L(y_p) = k(x) - k(x) = 0$$

and so $y - y_p$ is a solution to the homogeneous equation. Consequently, $y = y_p + (y - y_p)$ can be written as $y_p$ plus a solution to the homogeneous equation, as we wished to show. ∎

Now we apply this result to second-order equations.

**The Method of Undetermined Coefficients**   Consider the equation

$$y'' + a_1 y' + a_2 y = k(x)$$

It turns out that the functions $k(x)$ most apt to occur in applications are polynomials, exponentials, sines, and cosines. For these functions, we offer a procedure for finding $y_p$ based on trial solutions.

| If $k(x) =$ | Try $y_p =$ |
|---|---|
| $b_m x^m + \cdots + b_1 x + b_0$ | $B_m x^m + \cdots + B_1 x + B_0$ |
| $be^{\alpha x}$ | $Be^{\alpha x}$ |
| $b \cos \beta x + c \sin \beta x$ | $B \cos \beta x + C \sin \beta x$ |
| Modification. If a term of $k(x)$ is a solution to the homogeneous equation, multiply the trial solution by $x$ (or perhaps by a higher power of $x$). | |

To illustrate the table, we suggest the appropriate trial solution $y_p$ in six cases. The first three are straightforward; the last three are modified because a term on the right side of the differential equation is present in the solution to the homogeneous equation.

1. $y'' - 3y' - 4y = 3x^2 + 2$ $\qquad y_p = B_2 x^2 + B_1 x + B_0$
2. $y'' - 3y' - 4y = e^{2x}$ $\qquad\qquad y_p = Be^{2x}$
3. $y'' + 4y = 2 \sin x$ $\qquad\qquad\, y_p = B \cos x + C \sin x$
4. $y'' + 2y' = 3x^2 + 2$ $\qquad\qquad y_p = B_2 x^3 + B_1 x^2 + B_0 x$
   (2 is a solution to the homogeneous equation)
5. $y'' - 3y' - 4y = e^{4x}$ $\qquad\qquad y_p = Bxe^{4x}$
   ($e^{4x}$ is a solution to the homogeneous equation)
6. $y'' + 4y = \sin 2x$ $\qquad\qquad\; y_p = Bx \cos 2x + Cx \sin 2x$
   ($\sin 2x$ is a solution to the homogeneous equation)

Next we carry out the details in four specific examples.

**EXAMPLE 1**   Solve $y'' + y' - 2y = 2x^2 - 10x + 3$.

*Solution*   The auxiliary equation $r^2 + r - 2 = 0$ has roots $-2$ and $1$, and so

$$y_h = C_1 e^{-2x} + C_2 e^x$$

To find a particular solution to the nonhomogeneous equation, we try

$$y_p = Ax^2 + Bx + C$$

Substitution of this expression in the differential equation gives

$$2A + (2Ax + B) - 2(Ax^2 + Bx + C) = 2x^2 - 10x + 3$$

Equating coefficients of $x^2$, $x$, and 1, we find

$$-2A = 2, \qquad 2A - 2B = -10, \qquad 2A + B - 2C = 3$$

or $A = -1$, $B = 4$, and $C = -\frac{1}{2}$. Hence

$$y_p = -x^2 + 4x - \tfrac{1}{2}$$

and

$$y = -x^2 + 4x - \tfrac{1}{2} + C_1 e^{-2x} + C_2 e^x$$ ∎

**EXAMPLE 2**   Solve $y'' - 2y' - 3y = 8e^{3x}$.

*Solution*   Since the auxiliary equation $r^2 - 2r - 3 = 0$ has roots $-1$ and $3$, we have

$$y_h = C_1 e^{-x} + C_2 e^{3x}$$

Note that $k(x) = 8e^{3x}$ is a solution to the homogeneous equation. Thus, we use the *modified* trial solution

$$y_p = Bxe^{3x}$$

Substituting $y_p$ in the differential equation gives

$$(Bxe^{3x})'' - 2(Bxe^{3x})' - 3Bxe^{3x} = 8e^{3x}$$

or

$$(9xBe^{3x} + 6Be^{3x}) - 2(3Bxe^{3x} + Be^{3x}) - 3Bxe^{3x} = 8e^{3x}$$

or finally

$$4Be^{3x} = 8e^{3x}$$

We conclude that $B = 2$ and

$$y = 2xe^{3x} + C_1 e^{-x} + C_2 e^{3x}$$ ∎

If 3 had been a double root of the auxiliary equation in Example 2, we would have used $Bx^2 e^{3x}$ as our trial solution.

**EXAMPLE 3**   Solve $y'' - 2y' - 3y = \cos 2x$.

*Solution*   The homogeneous equation agrees with that of Example 2, and so

$$y_h = C_1 e^{-x} + C_2 e^{3x}$$

For the trial solution $y_p$, we use

$$y_p = B \cos 2x + C \sin 2x$$

Now

$$Dy_p = -2B \sin 2x + 2C \cos 2x$$
$$D^2 y_p = -4B \cos 2x - 4C \sin 2x$$

Hence substitution of $y_p$ in the differential equation gives (after collecting terms)

$$(-7B - 4C)\cos 2x + (4B - 7C)\sin 2x = \cos 2x$$

Thus, $-7B - 4C = 1$ and $4B - 7C = 0$, which imply that $B = -\frac{7}{65}$ and $C = -\frac{4}{65}$. We conclude that

$$y = -\tfrac{7}{65} \cos 2x - \tfrac{4}{65} \sin 2x + C_1 e^{-x} + C_2 e^{3x}$$   ∎

**EXAMPLE 4**   Solve $y'' - 2y' - 3y = 8e^{3x} + \cos 2x$.

*Solution*   Combining the results of Examples 2 and 3 and using the linearity of the operator $D^2 - 2D - 3$, we obtain

$$y = 2xe^{3x} - \tfrac{7}{65} \cos 2x - \tfrac{4}{65} \sin 2x + C_1 e^{-x} + C_2 e^{3x}$$   ∎

**The Method of Variation of Parameters**   A more general method than that of undetermined coefficients is the method of variation of parameters. If $u_1(x)$ and $u_2(x)$ are independent solutions to the homogeneous equation, then it can be shown (see Problem 23) that there is a particular solution to the nonhomogeneous equation of the form

$$y_p = v_1(x)u_1(x) + v_2(x)u_2(x)$$

in which

$$v'_1 u_1 + v'_2 u_2 = 0$$
$$v'_1 u'_1 + v'_2 u'_2 = k(x)$$

We show how this method works in an example.

**EXAMPLE 5**   Find the general solution of $y'' + y = \sec x$.

*Solution*   The general solution to the homogeneous equation is

$$y_h = C_1 \cos x + C_2 \sin x$$

To find a particular solution to the nonhomogeneous equation, we set

$$y_p = v_1(x)\cos x + v_2(x)\sin x$$

and impose the conditions

$$v'_1 \cos x + v'_2 \sin x = 0$$
$$-v'_1 \sin x + v'_2 \cos x = \sec x$$

When we solve this system of equations for $v'_1$ and $v'_2$, we obtain $v'_1 = -\tan x$ and $v'_2 = 1$. Thus,

$$v_1(x) = \int -\tan x \, dx = \ln|\cos x|$$

$$v_2(x) = \int dx = x$$

(We can omit the arbitrary constants in the above integrations, since any solutions $v_1$ and $v_2$ will do.) A particular solution is therefore

$$y_p = (\ln|\cos x|)\cos x + x \sin x$$

a result which is easy to check by direct substitution in the original differential equation. We conclude that

$$y = (\ln|\cos x|)\cos x + x \sin x + C_1 \cos x + C_2 \sin x \qquad \blacksquare$$

## CONCEPTS REVIEW

**1.** The general solution to a nonhomogeneous equation has the form $y = y_p + y_h$ where $y_p$ is a _____ and $y_h$ is the general solution to the _____.

**2.** Thus, after noting that $y'' - y' - 6y = 6$ has the particular solution $y = -6$, we conclude that the general solution is $y = $ _____.

**3.** The method of undetermined coefficients suggests trying a particular solution of the form $y = $ _____ for $y'' - y' - 6y = x^2$.

**4.** The method of undetermined coefficients suggests trying a particular solution of the form $y = $ _____ for $y'' - y' - 6y = e^{3x}$.

## PROBLEM SET 18.3

In Problems 1–16, use the method of undetermined coefficients to solve the differential equations.

1. $y'' - 9y = x$

2. $y'' + y' - 6y = 2x^2$

3. $y'' - 2y' + y = x^2 + x$

4. $y'' + y' = 4x$

5. $y'' - 5y' + 6y = e^x$

6. $y'' + 6y' + 9y = 2e^{-x}$

7. $y'' + 4y' + 3y = e^{-3x}$

8. $y'' + 2y' + 2y = 3e^{-2x}$

9. $y'' - y' - 2y = 2 \sin x$

10. $y'' + 4y' = \cos x$

11. $y'' + 4y = 2 \cos 2x$

12. $y'' + 9y = \sin 3x$

13. $y'' + 9y = \sin x + e^{2x}$

14. $y'' + y' = e^x + 3x$

15. $y'' - 5y' + 6y = 2e^x$; $y = 1$, $y' = 0$ when $x = 0$.

16. $y'' - 4y = 4 \sin x$; $y = 4$, $y' = 0$ when $x = 0$.

In Problems 17–22, solve the differential equations by variation of parameters.

17. $y'' - 3y' + 2y = 5x + 2$

18. $y'' - 4y = e^{2x}$

19. $y'' + y = \csc x \cot x$

20. $y'' + y = \cot x$

21. $y'' - 3y' + 2y = \dfrac{e^x}{e^x + 1}$

22. $y'' - 5y' + 6y = 2e^x$.

23. Let $L(y) = y'' + by' + cy = 0$ have solutions $u_1$ and $u_2$ and let $y_p = v_1 u_1 + v_2 u_2$. Show that

$$L(y_p) = v_1(u''_1 + bu'_1 + cu_1) + v_2(u''_2 + bu'_2 + cu_2)$$
$$+ b(v'_1 u_1 + v'_2 u_2) + (v'_1 u_1 + v'_2 u_2)' + (v'_1 u'_1 + v'_2 u'_2)$$

Thus, if the conditions of the method of variation of param-

eters hold,

$$L(y_p) = (v_1)(0) + (v_2)(0) + (b)(0) + 0 + k(x) = k(x)$$

24. Solve $y'' + 4y = \sin^3 x$

---

**Answers to Concepts Review:** **1.** Particular solution to the nonhomogeneous equation; homogeneous equation **2.** $-6 + C_1 e^{-2x} + C_2 e^{3x}$ **3.** $y = Ax^2 + Bx + C$ **4.** $y = Bxe^{3x}$

---

## 18.4
## APPLICATIONS OF
## SECOND-ORDER
## EQUATIONS

Many problems in physics lead to second-order linear differential equations. We first consider the problem of a vibrating spring, under various assumptions. Then we return to and generalize an earlier application to electric circuits.

**A Vibrating Spring (Simple Harmonic Motion)** Consider a coiled spring weighted by an object A and hanging vertically from a support, as in Figure 1(a). We wish to consider the motion of the point $P$ if the spring is pulled $y_0$ units below its equilibrium position (Figure 1(b)) and released. We assume friction to be negligible.

According to Hooke's Law, the force $F$ tending to restore $P$ to its equilibrium position at $y = 0$ satisfies $F = -ky$, where $k$ is a constant depending on the characteristics of the spring and $y$ is the $y$-coordinate of $P$. But by Newton's Second Law, $F = ma = (w/g)a$, where $w$ is the weight of the object A, $a$ is the acceleration of $P$, and $g$ is the constant acceleration due to gravity ($g = 32$ feet per second per second). Thus,

$$\frac{w}{g}\frac{d^2y}{dt^2} = -ky, \qquad k > 0$$

is the differential equation of the motion.

If we let $kg/w = B^2$, this equation takes the form

$$\frac{d^2y}{dt^2} + B^2 y = 0$$

and has the general solution

$$y = C_1 \cos Bt + C_2 \sin Bt$$

The conditions $y = y_0$ and $y' = 0$ at $t = 0$ determine the constants $C_1$ and $C_2$ to be $y_0$ and 0, respectively. Thus,

$$y = y_0 \cos Bt$$

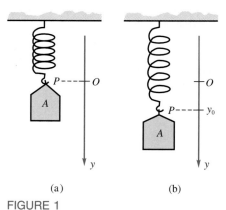

(a)    (b)

FIGURE 1

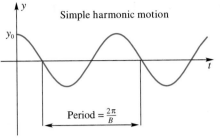

Simple harmonic motion

Period $= \frac{2\pi}{B}$

**FIGURE 2**

We say that the spring is executing **simple harmonic motion** with amplitude $y_0$ and period $2\pi/B$ (Figure 2).

**EXAMPLE 1**   When an object weighing 5 pounds is attached to the lowest point $P$ of a spring that hangs vertically, the spring is extended 6 inches. The 5-pound weight is replaced by a 20-pound weight, and the system is allowed to come to equilibrium. If the 20-pound weight is now pulled downward another 2 feet and then released, describe the motion of the lowest point $P$ of the spring.

*Solution*   The first sentence of the example allows us to determine the spring constant. By Hooke's Law, $|F| = ks$, where $s$ is the amount in feet the spring is stretched, and so $5 = k\left(\frac{1}{2}\right)$, or $k = 10$. Now put the origin at the equilibrium point after the 20-pound weight has been attached. From the derivation just before the example, we know that $y = y_0 \cos Bt$. In the present case, $y_0 = 2$ and $B^2 = kg/w = (10)(32)/20 = 16$. We conclude that

$$y = 2 \cos 4t$$

The motion of $P$ is simple harmonic motion, with period $\frac{1}{2}\pi$ and amplitude 2 feet. That is, $P$ oscillates up and down from 2 feet below 0 to 2 feet above 0 and then back to 2 feet below 0 every $\frac{1}{2}\pi \approx 1.57$ seconds. ∎

**Damped Vibrations**   So far, we have assumed a simplified situation, in which there is no friction either within the spring or resulting from the resistance of the air. We can take friction into account by assuming a retarding force that is proportional to the velocity $dy/dt$. The differential equation describing the motion then takes the form

$$\frac{w}{g}\frac{d^2y}{dt^2} = -ky - q\frac{dy}{dt}, \qquad k > 0, q > 0$$

By letting $E = qg/w$ and $B^2 = kg/w$, this equation can be written as

$$\frac{d^2y}{dt^2} + E\frac{dy}{dt} + B^2y = 0$$

an equation to which the methods of Section 18.2 apply. There are three cases to consider.

**Case 1 ($E^2 - 4B^2 < 0$).**   The roots of the auxiliary equation are complex conjugates; we will denote them by $-\alpha \pm \beta i$, where $\alpha$ and $\beta$ are positive. The general solution is

$$y = e^{-\alpha t}(C_1 \cos \beta t + C_2 \sin \beta t)$$

which can be rewritten in the form (see Problem 15)

$$y = Ce^{-\alpha t} \sin(\beta t + \gamma)$$

The factor $e^{-\alpha t}$, called the **damping factor**, causes the amplitude of the motion to approach zero as $t \to \infty$ (Figure 3(a)).

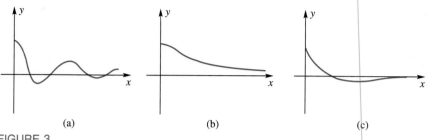

**FIGURE 3**

*Case 2* $(E^2 - 4B^2 = 0)$. In this case, the auxiliary equation has a double root $-\alpha$ and the general solution of the differential equation is

$$y = C_1 e^{-\alpha t} + C_2 t e^{-\alpha t}$$

The motion described by this equation is said to be **critically damped**.

*Case 3* $(E^2 - 4B^2 > 0)$. The auxiliary equation has roots $-\alpha_1$ and $-\alpha_2$, and the general solution of the differential equation is

$$y = C_1 e^{-\alpha_1 t} + C_2 e^{-\alpha_2 t}$$

It describes a motion that is said to be **overdamped**.

The graphs in the critically damped and overdamped cases cross the $t$-axis at most once and may look something like Figure 3(b) or 3(c).

**EXAMPLE 2** If a damping force with $q = 0.2$ is imposed on the system of Example 1, find the equation of motion.

*Solution* $E = qg/w = (0.2)(32)/20 = 0.32$ and $B^2 = (10)(32)/20 = 16$, and so we must solve

$$\frac{d^2y}{dt^2} + 0.32\frac{dy}{dt} + 16y = 0$$

The auxiliary equation $r^2 + 0.32r + 16 = 0$ has roots $r = -0.16 \pm \sqrt{15.9744}i \approx -0.16 \pm 4i$, and thus

$$y = e^{-0.16t}(C_1 \cos 4t + C_2 \sin 4t)$$

When we impose the conditions $y = 2$ and $y' = 0$ at $t = 0$, we find $C_1 = 2$ and $C_2 = 0.08$. Consequently,

$$y = e^{-0.16t}(2 \cos 4t + 0.08 \sin 4t) \qquad\blacksquare$$

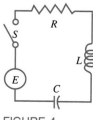

**FIGURE 4**

**Electric Circuits** Consider a circuit (Figure 4) with a resistor ($R$ ohms), an inductor ($L$ henrys), and a capacitor ($C$ farads) in series with a source of electromotive force supplying $E(t)$ volts. The new feature in comparison to the circuits of Section 18.1 is the presence of a capacitor (also called a condenser). Kirchhoff's Law in this situation says that the charge $Q$ on the capacitor, measured in coulombs, satisfies

(1)
$$L\frac{d^2Q}{dt^2} + R\frac{dQ}{dt} + \frac{1}{C}Q = E(t)$$

The current $I = dQ/dt$, measured in amperes, satisfies the equation obtained by differentiating equation (1) with respect to $t$; that is,

(2)
$$L\frac{d^2I}{dt^2} + R\frac{dI}{dt} + \frac{1}{C}I = E'(t)$$

Either of these equations can be solved by the methods of Sections 18.2 and 18.3.

**EXAMPLE 3**   Find the charge $Q$ and the current $I$ as functions of time $t$ in an RCL circuit (Figure 4) if $R = 16$, $L = 0.02$, $C = 2 \times 10^{-4}$, and $E = 12$. Assume that $Q = 0$ and $I = 0$ at $t = 0$ (when the switch is closed).

*Solution*   By Kirchhoff's Law as expressed in equation (1)

$$\frac{d^2Q}{dt^2} + 800\frac{dQ}{dt} + 250{,}000Q = 600$$

The auxiliary equation has roots

$$\frac{-800 \pm \sqrt{640{,}000 - 1{,}000{,}000}}{2} = -400 \pm 300i$$

so

$$Q_h = e^{-400t}(C_1 \cos 300t + C_2 \sin 300t)$$

By inspection, a particular solution is $Q_p = 2.4 \times 10^{-3}$. Therefore, the general solution is

$$Q = 2.4 \times 10^{-3} + e^{-400t}(C_1 \cos 300t + C_2 \sin 300t)$$

When we impose the given initial conditions, we find $C_1 = -2.4 \times 10^{-3}$ and $C_2 = -3.2 \times 10^{-3}$. We conclude that

$$Q = 10^{-3}[2.4 - e^{-400t}(2.4 \cos 300t + 3.2 \sin 300t)]$$

and by differentiation that

$$I = \frac{dQ}{dt} = 2e^{-400t} \sin 300t$$

■

## CONCEPTS REVIEW

**1.** A spring that vibrates without friction might obey a law of motion such as $y = 3 \cos 2t$. We say it is executing simple harmonic motion with amplitude _____ and period _____.

**2.** A spring vibrating in the presence of friction might obey a law of motion such as $y = 3e^{-0.1t}\cos 2t$, called damped harmonic motion. The "period" is still _____ but now the amplitude _____ as time increases.

**3.** If the friction is very great, the law of motion might take the form $y = 3e^{-0.1t} + te^{-0.1t}$, the critically damped case, in which $y$ slowly fades to _____ as time increases.

**4.** Kirchhoff's Law says that an _____ satisfies a second-order linear differential equation.

## PROBLEM SET 18.4

**1.** A spring with a spring constant $k$ of 20 pounds per foot is loaded with a 10-pound weight and allowed to reach equilibrium. It is then raised 1 foot and released. Find the equation of motion and the period. Neglect friction.

**2.** A spring with a spring constant $k$ of 100 pounds per foot is loaded with a 1-pound weight and brought to equilibrium. It is then stretched an additional 1 inch and released. Find the equation of motion, the amplitude, and the period. Neglect friction.

**3.** In Problem 1, what is the absolute value of the velocity of the moving weight as it passes through the equilibrium position?

**4.** A 10-pound weight stretches a spring 4 inches. This weight is removed and replaced with a 20-pound weight, which is then allowed to reach equilibrium. The weight is next raised 1 foot and released with an initial velocity of 2 feet per second downward. What is the equation of motion? Neglect friction.

**5.** A spring with a spring constant $k$ of 20 pounds per foot is loaded with a 10-pound weight and allowed to reach equilibrium. It is then displaced 1 foot downward and released. If the weight experiences a retarding force in pounds equal to one-tenth the velocity, find the equation of motion.

**6.** Determine the motion in Problem 5 if the retarding force equals four times the velocity at every point.

**7.** In Problem 5, how long will it take the oscillations to diminish to one-tenth their original amplitude?

**8.** In Problem 5, what will be the equation of motion if the weight is given an upward velocity of 1 foot per second at the moment of release?

**9.** Using Figure 5, find the charge $Q$ on the capacitor as a function of time if $S$ is closed at $t = 0$. Assume that the capacitor is initially uncharged.

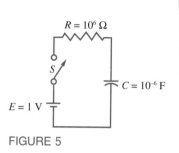

$R = 10^6\ \Omega$

$C = 10^{-6}$ F

$E = 1$ V

**FIGURE 5**

**10.** Find the current $I$ as a function of time in Problem 9 if the capacitor has an initial charge of 4 coulombs.

**11.** Use Figure 6.

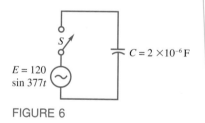

$C = 2 \times 10^{-6}$ F

$E = 120 \sin 377t$

**FIGURE 6**

(a) Find $Q$ as a function of time. Assume the capacitor initially uncharged.
(b) Find $I$ as a function of time.

**12.** Using Figure 7, find the current as a function of time if the capacitor is initially uncharged and $S$ is closed at $t = 0$. *Hint*: The current at $t = 0$ will equal 0, since the current through an inductance cannot change instantaneously.

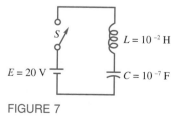

$L = 10^{-2}$ H

$E = 20$ V

$C = 10^{-7}$ F

**FIGURE 7**

**13.** Using Figure 8, find the steady-state current as a function of time; that is, find a formula for $I$ that is valid when $t$ is very large ($t \to \infty$).

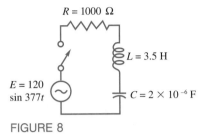

FIGURE 8

**14.** Suppose that an undamped spring is subjected to an external periodic force so that its differential equation has the form

$$\frac{d^2y}{dt^2} + B^2y = c \sin At, \qquad c > 0$$

(a) Show that the equation of motion for $A \neq B$ is

$$y = C_1 \cos Bt + C_2 \sin Bt + \frac{c}{B^2 - A^2}\sin At$$

(b) Solve the differential equation when $A = B$ (the resonance case).
(c) What happens to the amplitude of the motion in part b when $t \to \infty$?

**15.** Show that $C_1 \cos \beta t + C_2 \sin \beta t$ can be written in the form $C \sin(\beta t + \gamma)$. *Hint*: Let $C = \sqrt{C_1^2 + C_2^2}$, $\sin \gamma = C_1/C$, and $\cos \gamma = C_2/C$.

**16.** Show that the motion of part a of Problem 14 is periodic if $B/A$ is rational.

**17.** Refer to Figure 9, which shows a pendulum bob of mass $m$ supported by a weightless wire of length $L$. Derive the equation of motion, that is, derive the differential equation satisfied by $\theta$. *Suggestion*: Use the fact from Section

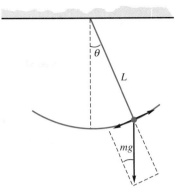

FIGURE 9

13.5 that the scalar tangential component of the acceleration is $d^2s/dt^2$, where $s$ measures arc length in the counterclockwise direction.

**18.** The equation derived in Problem 17 is nonlinear, but—for small $\theta$—it is customary to approximate it by the equation

$$\frac{d^2\theta}{dt^2} + \frac{g}{L}\theta = 0$$

Here $g = GM/R^2$, where $G$ is a universal constant, $M$ is the mass of the earth, and $R$ is the distance from the pendulum to the center of the earth. Two clocks, with pendulums of length $L_1$ and $L_2$ and located at distances $R_1$ and $R_2$ from the center of the earth, have periods $p_1$ and $p_2$, respectively.

(a) Show that $\dfrac{p_1}{p_2} = \dfrac{R_1\sqrt{L_1}}{R_2\sqrt{L_2}}$.

(b) Find the height of a mountain if a clock which kept perfect time at sea level ($R = 3960$ miles) with $L = 81$ inches had to have its pendulum shortened to $L = 80.85$ inches to keep perfect time at the top of the mountain.

---

**Answers to Concepts Review:** **1.** 3; $\pi$ **2.** $\pi$; decreases **3.** 0 **4.** Electric circuit

---

## 18.5 CHAPTER REVIEW

### Concepts Test

Respond with true or false to each of the following assertions. Be prepared to justify your answer.

**1.** $y'' + y^2 = 0$ is a linear differential equation.

**2.** $y'' + x^2y = 0$ is a linear differential equation.

**3.** $y = \tan x + \sec x$ is a solution of $2y' - y^2 = 1$.

**4.** The general solution to $[D^2 + aD + b]^3y = 0$ should involve 8 arbitrary constants.

**5.** An integrating factor for $y' + \dfrac{4}{x}y = e^x$ is $x^4$.

**6.** If $u_1(x)$ and $u_2(x)$ are two solutions to $y'' + a_1y' + a_2y = f(x)$, then $C_1u_1(x) + C_2u_2(x)$ is also a solution.

**7.** The general solution to $y''' + 3y'' + 3y' + y = 0$ is $y = C_1 e^{-x} + C_2 xe^{-x} + C_3 x^2 e^{-x}$.

**8.** If $u_1(x)$ and $u_2(x)$ are solutions to the linear differential equation $L(y) = f(x)$, then $u_1(x) - u_2(x)$ is a solution to $L(y) = 0$.

**9.** The equation $y'' + 9y = 2 \sin 3x$ has a particular solution of the form $y_p = B \sin 3x + C \cos 3x$.

**10.** An expression of the form $C_1 \cos \beta t + C_2 \sin \beta t$ can always be written in the form $C \sin(\beta t + \gamma)$.

## Sample Test Problems

In Problems 1–19, solve the differential equations.

**1.** $\dfrac{dy}{dx} + \dfrac{y}{x} = 0$

**2.** $\dfrac{dy}{dx} - \dfrac{x^2 - 2y}{x} = 0$

**3.** $\dfrac{dy}{dx} + 2x(y - 1) = 0$; $y = 3$ when $x = 0$.

**4.** $\dfrac{dy}{dx} - ay = e^{ax}$

**5.** $\dfrac{dy}{dx} - 2y = e^x$

**6.** $\dfrac{dy}{dx} + y \tan x = 2 \sec x$

**7.** $\dfrac{d^2y}{dx^2} + 3\dfrac{dy}{dx} = e^x$

*Suggestion*: Let $u = dy/dx$.

**8.** $y'' - y = 0$

**9.** $y'' - 3y' + 2y = 0$; $y = 0$, $y' = 3$ when $x = 0$.

**10.** $4y'' + 12y' + 9y = 0$

**11.** $y'' - y = 1$

**12.** $y'' + 4y' + 4y = 3e^x$

**13.** $y'' + 4y' + 4y = e^{-2x}$

**14.** $y'' + 4y = 0$; $y = 0$, $y' = 2$ when $x = 0$.

**15.** $y'' + 6y' + 25y = 0$

**16.** $y'' + y = \sec x \tan x$

**17.** $y''' + 2y'' - 8y' = 0$

**18.** $y^{(4)} - 3y'' - 10y = 0$

**19.** $y^{(4)} - 4y'' + 4y = 0$

**20.** Suppose that glucose is infused into the bloodstream of a patient at the rate of 3 grams per minute, but that the patient's body converts and removes glucose from its blood at a rate proportional to the amount present (with constant of proportionality 0.02). If $Q(t)$ is the amount present at time $t$ and $Q(0) = 120$,

(a) write the differential equation for $Q$;
(b) solve this differential equation;
(c) determine what happens to $Q$ in the long run.

**21.** A spring with a spring constant $k$ of 5 pounds per foot is loaded with a 10-pound weight and allowed to reach equilibrium. It is then raised 1 foot and released. What are the equation of motion, the amplitude, and the period? Neglect friction.

**22.** In Problem 21, what is the absolute value of the velocity of the moving weight as it passes through the equilibrium position?

**23.** Suppose the switch of the circuit in Figure 1 is closed at $t = 0$. Find $I$ as a function of time if $C$ is initially uncharged. (The current at $t = 0$ will equal zero, since current through an inductance cannot change instantaneously.)

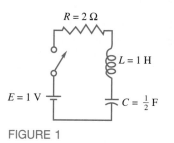

FIGURE 1

# TECHNOLOGY PROJECT

In this project we will seek solutions to differential equations, concentrating on the *nonhomogeneous* or *forced* equation. For example, in the mass-spring problem an external force $f$ is applied to the mass resulting in $my'' + Ky = f$.

We will concentrate on determining approximate solutions via Taylor series, $y(x) = \sum_{k=0}^{\infty} a_k x^k$, valid for some interval about $x = 0$. In practice one usually truncates this infinite series to a finite number of terms; then ask: "how many terms should we take?" and, "Can we be confident of the resulting approximation?" One approach is to expand the forcing term in a similar series,

$f(x) = \sum_{k=0}^{\infty} b_k x^k$ where the $b_k$ are assumed to be known. Then we somehow "integrate" the $f$-series to get the desired series for $y$.

**Exercise 1**   Solve the following differential equations with the initial conditions $y(0) = \alpha$, and $y'(0) = \beta$.

a) $y'' = 3x - 5x^3$. Hint: Just integrate twice.

b) $y'' = \sin x^2$. Since you can't integrate $\sin x^2$, seek a series for $y$. Get the first four non-zero terms of a series for $\sin x^2$ by replacing $u$ by $x^2$ in the series for $\sin u$. Integrate the resulting approximate differential equation twice to get $y$.

**Exercise 2**   Consider the forced mass-spring problem, $y'' + Cy = f$, where $C$ is a constant and $f(x) = \sum_{k=0}^{\infty} b_k x^k$ with the

$b_k$ known. Seek the solution in the form $y(x) = \sum_{k=0}^{\infty} a_k x^k$.

a) Derive the following *recurrence formula* for the $a_k$:

$$a_{k+2} = \frac{b_k - Ca_k}{(k+1)(k+2)}, \quad k \geq 0.$$

b) For $f(x) = \sin x^2$, find the first five non-zero terms for $y$. Again use the initial conditions, $y(0) = \alpha$, and $y'(0) = \beta$.

**Illustrative Exercise.**

We find the first ten terms of the series solution to $y'' + y = 3 \cos 2x$, with initial conditions, $y(0) = 2$, and $y'(0) = 0$. We first compute the $b_k$ by expanding out $f$ by a "taylor" command (or whatever command works on your technology). We get the following approximating polynomial $P_n$:

$P_n =$

$$3 - 6x^2 + 2x^4 - \frac{4}{15}x^6 + \frac{2}{105}x^8.$$

Then we compute the $a_k$ from the recurrence formula of Exercise 2 and form the approximation $y_n$:

$$y_n = 2 + \frac{1}{2}x^2 - \frac{13}{24}x^4 + \frac{61}{720}x^6$$

$$- \frac{253}{40320}x^8 + \frac{1021}{3628800}x^{10}.$$

If we were to graphically compare $y_n$ with the exact solution $(y = 3 \cos x - \cos 2x)$ we would see that $y_n$ is quite accurate on, say, $[-2, 2]$. However, typical of series solutions, it starts breaking down on larger intervals. Try it on your technology.

**Exercise 3**   Referring to the *Illustrative Exercise*:

a) Verify that $y = 3 \cos x - \cos 2x$ is the exact solution to this initial value problem.

b) Note that the series for $y_n$ has two more terms than the series for $f$. Why is that?

c) While the above approximate solution seems adequate on the interval $[-\pi/2, \pi/2]$, it is not so for $[-\pi, \pi]$. Take enough terms so that you get good graphical agreement on $[-\pi, \pi]$.

**Exercise 4**   For the differential equation, $y'' + y = f(x) = \ln(1 + x) \cos 2x$, with initial conditions, $y(0) = 0 = y'(0)$, a solution in terms of elementary functions is no longer available. Suppose we need an approximate solution that is graphically good on the interval $[0, \pi/4]$.

a) Compute enough coefficients $b_k$ so that, graphically, your Taylor polynomial approximation of $f(x)$ is adequate. Record the maximum numerical error in your approximation.

b) Compute the coefficients $a_k$ of the desired solution corresponding to your $b_k$. Then form the approximate solution, say, $y_n$. Graph $y_n$ on $[0, \pi/4]$. Comment on your confidence in $y_n$ at this point.

c) To check the accuracy of the above $y_n$, either do a numerical solution of this problem with your technology, or use the *variation of parameters* solution (introduced briefly in the text):

$$y(x) = \int_0^x f(t) \sin(x - t)\, dt.$$

Since you will not be able to integrate this integral exactly, do numerical evaluations for several $x$ values on $[0, \pi/4]$ in order to check your results of part b).

# Appendix

## A.1 MATHEMATICAL INDUCTION

Often in mathematics, we are faced with the task of wanting to establish that a certain proposition $P_n$ is true for every integer $n \geq 1$ (or perhaps every integer $n \geq N$). Here are three examples.

1. $P_n$:   $1^2 + 2^2 + 3^2 + \cdots + n^2 = \dfrac{n(n + 1)(2n + 1)}{6}$

2. $Q_n$:   $2^n > n + 20$

3. $R_n$:   $n^2 - n + 41$ is prime.

Proposition $P_n$ is true for every positive integer and $Q_n$ is true for every integer greater than or equal to 5 (as we will show soon). The third proposition, $R_n$, is interesting. Note that for $n = 1, 2, 3, \ldots$, the values of $n^2 - n + 41$ are 41, 43, 47, 53, 61, ... (prime numbers so far). In fact, we will get a prime number for all $n$'s through 40; but at $n = 41$, the formula yields the composite number $1681 = (41)(41)$. Showing the truth of a proposition for 40 (or 40 million) individual cases may make a proposition plausible, but it most certainly does not prove it is true for all $n$. The chasm between any finite number of cases and *all* cases is infinitely wide.

What is to be done? Is there a procedure for establishing that a proposition $P_n$ is true for *all n*? An affirmative answer is provided by the **Principle of Mathematical Induction**, a favorite of all mathematics students.

**(Principle of Mathematical Induction).** Let $\{P_n\}$ be a sequence of propositions (statements) satisfying these two conditions.

**(i)** $P_N$ is true (usually $N$ will be 1).

**(ii)** The truth of $P_i$ implies the truth of $P_{i+1}$, $i \geq N$.
Then, $P_n$ is true for every integer $n \geq N$.

We do not prove this principle; it is often taken as an axiom, and we hope it seems obvious. After all, if the first domino falls and if each domino knocks over the next one, then the whole row of dominoes will fall. Our efforts will be directed toward illustrating how we use mathematical induction.

**EXAMPLE 1** Prove that

$$P_n: \quad 1^2 + 2^2 + 3^2 + \cdots + n^2 = \frac{n(n + 1)(2n + 1)}{6}$$

is true for all $n \geq 1$.

*Solution* First, we note that

$$P_1: \quad 1^2 = \frac{1(1 + 1)(2 + 1)}{6}$$

is a true statement.

Second, we demonstrate Implication (ii). We begin by writing the statements $P_i$ and $P_{i+1}$.

$$P_i: \quad 1^2 + 2^2 + \cdots + i^2 = \frac{i(i + 1)(2i + 1)}{6}$$

$$P_{i+1}: \quad 1^2 + 2^2 + \cdots + i^2 + (i + 1)^2 = \frac{(i + 1)(i + 2)(2i + 3)}{6}$$

We must show that $P_i$ implies $P_{i+1}$, so we assume $P_i$ is true. Then the left side of $P_{i+1}$ can be written as follows (* indicates where $P_i$ is used).

$$[1^2 + 2^2 + \cdots + i^2] + (i + 1)^2 \overset{*}{=} \frac{i(i + 1)(2i + 1)}{6} + (i + 1)^2$$

$$= (i + 1)\frac{2i^2 + i + 6i + 6}{6}$$

$$= (i + 1)\frac{2i^2 + 7i + 6}{6}$$

$$= \frac{(i + 1)(i + 2)(2i + 3)}{6}$$

Read from first expression to last expression, this chain of equalities is the statement $P_{i+1}$. Thus the truth of $P_i$ does imply the truth of $P_{i+1}$. By the Principle of Mathematical Induction, $P_n$ is true for each positive integer $n$. ∎

**EXAMPLE 2** Prove $P_n: \quad 2^n > n + 20$ is true for each integer $n \geq 5$.

*Solution* First, we note that $P_5: \quad 2^5 > 5 + 20$ is true. Second, we suppose that $P_i: \quad 2^i > i + 20$ is true and attempt to deduce from this that $P_{i+1}:$

$2^{i+1} > i + 1 + 20$ is true. But

$$2^{i+1} = 2 \cdot 2^i \overset{*}{>} 2(i + 20) = 2i + 40 > i + 21$$

Read from left to right, this is proposition $P_{i+1}$. We conclude that $P_n$ is true for $n \geq 5$. ∎

**EXAMPLE 3**  Prove that

$$P_n: \quad x - y \text{ is a factor of } x^n - y^n$$

is true for each integer $n \geq 1$.

*Solution*  Trivially, $x - y$ is a factor of $x - y$, so $P_1$ is true. Suppose that $x - y$ is a factor of $x^i - y^i$, that is, suppose that

$$x^i - y^i = Q(x, y)(x - y)$$

for some polynomial $Q(x, y)$. Then

$$
\begin{aligned}
x^{i+1} - y^{i+1} &= x^{i+1} - x^i y + x^i y - y^{i+1} \\
&= x^i(x - y) + y(x^i - y^i) \\
&\overset{*}{=} x^i(x - y) + yQ(x, y)(x - y) \\
&= [x^i + yQ(x, y)](x - y)
\end{aligned}
$$

which displays $x - y$ as a factor. Thus, the truth of $P_i$ does imply the truth of $P_{i+1}$. We conclude by the Principle of Mathematical Induction that $P_n$ is true for all $n \geq 1$. ∎

## PROBLEM SET A.1

In Problems 1–8, use the Principle of Mathematical Induction to prove the given proposition is true for each integer $n \geq 1$.

1. $1 + 2 + 3 + \cdots + n = \dfrac{n(n + 1)}{2}$

2. $1 + 3 + 5 + \cdots + (2n - 1) = n^2$

3. $1 \cdot 2 + 2 \cdot 3 + 3 \cdot 4 + \cdots + n(n + 1) = \dfrac{n(n + 1)(n + 2)}{3}$

4. $1^2 + 3^2 + 5^2 + \cdots + (2n - 1)^2 = \dfrac{n(2n - 1)(2n + 1)}{3}$

5. $1^3 + 2^3 + 3^3 + \cdots + n^3 = \left[\dfrac{n(n + 1)}{2}\right]^2$

6. $1^4 + 2^4 + 3^4 + \cdots + n^4 = \dfrac{n(n + 1)(6n^3 + 9n^2 + n - 1)}{30}$

7. $n^3 - n$ is divisible by 6.

8. $n^3 + (n + 1)^3 + (n + 2)^3$ is divisible by 9.

In Problems 9–12, determine the first integer $N$ for which the proposition is true for all $n \geq N$ and then prove the proposition for all $n \geq N$.

9. $3n + 25 < 3^n$      10. $n - 100 > \log_{10} n$

11. $n^2 \leq 2^n$

12. $|\sin nx| \leq n|\sin x|$ for all $x$

In Problems 13–20, indicate what conclusion about $P_n$ can be drawn from the given information.

13. $P_5$ is true and $P_i$ true implies $P_{i+2}$ true.

14. $P_1$ and $P_2$ are true and $P_i$ true implies $P_{i+2}$ true.

15. $P_{30}$ is true and $P_i$ true implies $P_{i-1}$ true.

16. $P_{30}$ is true and $P_i$ true implies both $P_{i+1}$ and $P_{i-1}$ true.

**17.** $P_1$ is true and $P_i$ true implies both $P_{4i}$ and $P_{i-1}$ true.

**18.** $P_1$ is true and $P_{2i}$ true implies $P_{2i+1}$ true.

**19.** $P_1$ and $P_2$ are true, $P_i$ and $P_{i+1}$ true imply $P_{i+2}$ true.

**20.** $P_1$ is true and $P_j$ true for $j \leq i$ imply $P_{i+1}$ true.

In Problems 21–27, decide for what $n$'s the given proposition is true and then use mathematical induction (perhaps in one of the alternative forms you may have discovered in Problems 13–20) to prove each of the following.

**21.** $x + y$ is a factor of $x^n + y^n$.

**22.** The sum of the measures of the interior angles of an $n$-sided convex (no holes or dents) polygon is $(n - 2)\pi$.

**23.** The number of diagonals of an $n$-sided convex polygon is $\dfrac{n(n - 3)}{2}$.

**24.** $\dfrac{1}{n + 1} + \dfrac{1}{n + 2} + \dfrac{1}{n + 3} + \cdots + \dfrac{1}{2n} > \dfrac{3}{5}$

**25.** $\left(1 - \dfrac{1}{4}\right)\left(1 - \dfrac{1}{9}\right)\left(1 - \dfrac{1}{16}\right) \cdots \left(1 - \dfrac{1}{n^2}\right) = \dfrac{n + 1}{2n}$

**26.** Let $f_0 = 0$, $f_1 = 1$, and $f_{n+2} = f_{n+1} + f_n$ for $n \geq 0$ (this is the Fibonacci sequence). Then,

$$f_n = \frac{1}{\sqrt{5}}\left[\left(\frac{1 + \sqrt{5}}{2}\right)^n - \left(\frac{1 - \sqrt{5}}{2}\right)^n\right]$$

**27.** Let $a_0 = 0$, $a_1 = a$, and $a_{n+2} = (a_{n+1} + a_n)/2$ for $n \geq 0$. Then,

$$a_n = \tfrac{2}{3}\left[1 - \left(-\tfrac{1}{2}\right)^n\right]$$

**28.** What is wrong with the following argument, which purports to show that all people in any set of $n$ people are the same age? The statement is certainly true for a set consisting of one person. Suppose it is true for any set of $i$ people and consider a set $W$ of $i + 1$ people. We may think of $W$ as the union of sets $X$ and $Y$, each consisting of $i$ people (draw a picture, for example, when $W$ has 6 people). By supposition, each of these sets consists of identically aged people. But $X$ and $Y$ overlap (in $X \cap Y$) and so all members of $W = X \cup Y$ also are the same age.

---

## A.2
## PROOFS OF
## SEVERAL THEOREMS

Theorem A

**(Main Limit Theorem).**   Let $n$ be a positive integer, $k$ be a constant, and $f$ and $g$ be functions which have limits at $c$. Then

**1.** $\lim\limits_{x \to c} k = k$;

**2.** $\lim\limits_{x \to c} x = c$;

**3.** $\lim\limits_{x \to c} kf(x) = k \lim\limits_{x \to c} f(x)$;

**4.** $\lim\limits_{x \to c}[f(x) + g(x)] = \lim\limits_{x \to c} f(x) + \lim\limits_{x \to c} g(x)$;

**5.** $\lim\limits_{x \to c}[f(x) - g(x)] = \lim\limits_{x \to c} f(x) - \lim\limits_{x \to c} g(x)$;

**6.** $\lim\limits_{x \to c}[f(x) \cdot g(x)] = \lim\limits_{x \to c} f(x) \cdot \lim\limits_{x \to c} g(x)$;

**7.** $\lim\limits_{x \to c} \dfrac{f(x)}{g(x)} = \dfrac{\lim\limits_{x \to c} f(x)}{\lim\limits_{x \to c} g(x)}$, provided $\lim\limits_{x \to c} g(x) \neq 0$;

**8.** $\lim\limits_{x \to c}[f(x)]^n = \left[\lim\limits_{x \to c} f(x)\right]^n$;

**9.** $\lim\limits_{x \to c} \sqrt[n]{f(x)} = \sqrt[n]{\lim\limits_{x \to c} f(x)}$, provided $\lim\limits_{x \to c} f(x) > 0$ when $n$ is even.

***Proof***   We proved parts 1–5 near the end of Section 2.6, so we should start with part 6. However, we choose first to prove a special case of part 8, namely,

$$\lim_{x \to c}[g(x)]^2 = \left[\lim_{x \to c} g(x)\right]^2$$

To see this, recall that we have proved that $\lim_{x \to c} x^2 = c^2$ (Example 6 of Section 2.5) and so $f(x) = x^2$ is continuous everywhere. Thus, by the Composite Limit Theorem (Theorem 2.7D),

$$\lim_{x \to c}[g(x)]^2 = \lim_{x \to c} f(g(x)) = f\left[\lim_{x \to c} g(x)\right] = \left[\lim_{x \to c} g(x)\right]^2$$

Next, write

$$f(x)g(x) = \tfrac{1}{4}\{[f(x) + g(x)]^2 - [f(x) - g(x)]^2\}$$

and apply parts 3, 4, and 5, plus what we have just proved. Part 6 falls out.

To prove part 7, apply the Composite Limit Theorem with $f(x) = 1/x$ and use Example 7 of Section 2.5. Then

$$\lim_{x \to c} \frac{1}{g(x)} = \lim_{x \to c} f(g(x)) = f\left(\lim_{x \to c} g(x)\right) = \frac{1}{\lim_{x \to c} g(x)}$$

Finally, by part 6,

$$\lim_{x \to c} \frac{f(x)}{g(x)} = \lim_{x \to c} \left[f(x) \cdot \frac{1}{g(x)}\right] = \lim_{x \to c} f(x) \cdot \lim_{x \to c} \frac{1}{g(x)}$$

from which the result follows.

Part 8 follows from repeated use of part 6 (technically, by mathematical induction).

We prove part 9 only for square roots. Let $f(x) = \sqrt{x}$, which is continuous for positive numbers by Example 4 of Section 2.5. By the Composite Limit Theorem,

$$\lim_{x \to c} \sqrt{g(x)} = \lim_{x \to c} f(g(x)) = f\left(\lim_{x \to c} g(x)\right) = \sqrt{\lim_{x \to c} g(x)}$$

which is equivalent to the desired result.    ∎

---

### Theorem B

**(Chain Rule).**   If $g$ is differentiable at $a$ and $f$ is differentiable at $g(a)$, then $f \circ g$ is differentiable at $a$ and

$$(f \circ g)'(a) = f'(g(a))g'(a)$$

***Proof*** We offer a proof that generalizes easily to higher dimensions (see Section 15.6).

By hypothesis, $f$ is differentiable at $b = g(a)$; that is, there is a number $f'(b)$ such that

$$(1) \qquad \lim_{\Delta u \to 0} \frac{f(b + \Delta u) - f(b)}{\Delta u} = f'(b)$$

Define a function $\varepsilon$ depending on $\Delta u$ by

$$\varepsilon(\Delta u) = \frac{f(b + \Delta u) - f(b)}{\Delta u} - f'(b)$$

and multiply both sides by $\Delta u$ to obtain

$$(2) \qquad f(b + \Delta u) - f(b) = f'(b)\,\Delta u + \Delta u\,\varepsilon(\Delta u)$$

The existence of the limit in (1) is equivalent to $\varepsilon(\Delta u) \to 0$ as $\Delta u \to 0$ in (2). If in (2), we replace $\Delta u$ by $g(a + \Delta x) - g(a)$ and $b$ by $g(a)$, we get

$$f(g(a + \Delta x)) - f(g(a)) = f'(g(a))[g(a + \Delta x) - g(a)]$$
$$+\ [g(a + \Delta x) - g(a)]\varepsilon(\Delta u)$$

or upon dividing both sides by $\Delta x$,

$$(3) \qquad \frac{f(g(a + \Delta x)) - f(g(a))}{\Delta x} = f'(g(a))\frac{g(a + \Delta x) - g(a)}{\Delta x}$$
$$+\ \frac{g(a + \Delta x) - g(a)}{\Delta x}\varepsilon(\Delta u)$$

In (3), let $\Delta x \to 0$. Since $g$ is differentiable at $a$, it is continuous there, so $\Delta x \to 0$ forces $\Delta u \to 0$; this, in turn, makes $\varepsilon(\Delta u) \to 0$. We conclude that

$$\lim_{\Delta x \to 0} \frac{f(g(a + \Delta x)) - f(g(a))}{\Delta x} = f'(g(a)) \lim_{\Delta x \to 0} \frac{g(a + \Delta x) - g(a)}{\Delta x} + 0$$

that is, $f \circ g$ is differentiable at $a$ and

$$(f \circ g)'(a) = f'(g(a))g'(a) \qquad \blacksquare$$

---

## Theorem C

**(Power Rule).** If $r$ is rational, then $x^r$ is differentiable at any $x$ that is in an open interval on which $x^{r-1}$ is real and

$$D_x(x^r) = rx^{r-1}$$

***Proof***   Consider first the case, where $r = 1/q$, $q$ a positive integer. Recall that $a^q - b^q$ factors as

$$a^q - b^q = (a - b)(a^{q-1} + a^{q-2}b + \cdots + ab^{q-2} + b^{q-1})$$

so

$$\frac{a - b}{a^q - b^q} = \frac{1}{a^{q-1} + a^{q-2}b + \cdots + ab^{q-2} + b^{q-1}}$$

Thus, if $f(t) = t^{1/q}$,

$$f'(x) = \lim_{t \to x} \frac{t^{1/q} - x^{1/q}}{t - x} = \lim_{t \to x} \frac{t^{1/q} - x^{1/q}}{(t^{1/q})^q - (x^{1/q})^q}$$

$$= \lim_{t \to x} \frac{1}{t^{(q-1)/q} + t^{(q-2)/q}x^{1/q} + \cdots + x^{(q-1)/q}}$$

$$= \frac{1}{qx^{(q-1)/q}} = \frac{1}{q} x^{1/q - 1}$$

Now by the Chain Rule, with $p$ an integer,

$$D_x(x^{p/q}) = D_x[(x^{1/q})^p] = p(x^{1/q})^{p-1} D_x(x^{1/q})$$

$$= px^{p/q - 1/q} \frac{1}{q} x^{1/q - 1} = \frac{p}{q} x^{p/q - 1}$$   ∎

---

### Theorem D

**(Vector limits).**   Let $\mathbf{F}(t) = f(t)\mathbf{i} + g(t)\mathbf{j}$. Then $\mathbf{F}$ has a limit at $c$ if and only if $f$ and $g$ have limits at $c$. In that case,

$$\lim_{t \to c} \mathbf{F}(t) = \left[\lim_{t \to c} f(t)\right]\mathbf{i} + \left[\lim_{t \to c} g(t)\right]\mathbf{j}$$

---

***Proof***   First, note that for any vector $\mathbf{u} = u_1\mathbf{i} + u_2\mathbf{j}$,

$$\boxed{|u_1| \leq |\mathbf{u}| \leq |u_1| + |u_2|}$$

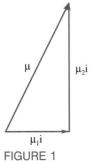

μ

$\mu_2\mathbf{i}$

$\mu_1\mathbf{i}$

FIGURE 1

This fact is readily seen from Figure 1.

Now suppose $\lim_{t \to c} \mathbf{F}(t) = \mathbf{L} = a\mathbf{i} + b\mathbf{j}$. This means that for any $\varepsilon > 0$ there is a corresponding $\delta > 0$ such that

$$0 < |t - c| < \delta \quad \Rightarrow \quad |\mathbf{F}(t) - \mathbf{L}| < \varepsilon$$

But by the left part of the boxed inequality,

$$|f(t) - a| \leq |\mathbf{F}(t) - \mathbf{L}|$$

and so

$$0 < |t - c| < \delta \quad \Rightarrow \quad |f(t) - a| < \varepsilon$$

This shows that $\lim_{t \to c} f(t) = a$. A similar argument establishes that $\lim_{t \to c} g(t) = b$. The first half of our theorem is complete.

Conversely, suppose that

$$\lim_{t \to c} f(t) = a \qquad \lim_{t \to c} g(t) = b$$

and let $\mathbf{L} = a\mathbf{i} + b\mathbf{j}$. For any given $\varepsilon > 0$, there is a corresponding $\delta > 0$ such that $0 < |t - c| < \delta$ implies both

$$|f(t) - a| < \frac{\varepsilon}{2} \quad \text{and} \quad |g(t) - b| < \frac{\varepsilon}{2}$$

Hence, by the right part of the boxed inequality,

$$0 < |t - c| < \delta \quad \Rightarrow \quad |\mathbf{F}(t) - \mathbf{L}| \leq \frac{\varepsilon}{2} + \frac{\varepsilon}{2} = \varepsilon$$

Thus,

$$\lim_{t \to c} \mathbf{F}(t) = \mathbf{L} = a\mathbf{i} + b\mathbf{j} = \lim_{t \to c} f(t)\mathbf{i} + \lim_{t \to c} g(t)\mathbf{j} \qquad \blacksquare$$

---

## A.3
## A BACKWARD LOOK

Those of you who have persisted to the end of this book are to be complemented. Now it is time to turn around, to survey the road we have traveled, to develop a bit of perspective. Just what have we been studying? What is calculus, this many-faceted subject that some have called the greatest invention of the human mind?

Suppose we divide school mathematics into three broad areas: geometry, algebra, and calculus. Then, one author has said that geometry is the study of *shape*, algebra is the study of *quantity*, and calculus is the study of *change*. It is true that the part of our subject known as differential calculus (including differential equations) is largely the study of rates of change. But what about integral calculus? It hardly fits under this rubric.

Another writer has suggested that we can distinguish between algebra and calculus by saying that algebra deals with *finite processes* (addition, multiplication, exponentiation, and so on), whereas calculus deals with *infinite processes* (differentiation, integration, summation of series, and so on).

We prefer to define calculus as the study of *limits*. The notion of limit pervades all of calculus; every major idea is determined by a certain kind of limit. The accompanying chart has the word limit at the center; the principal concepts (in ovals) emanate from it.

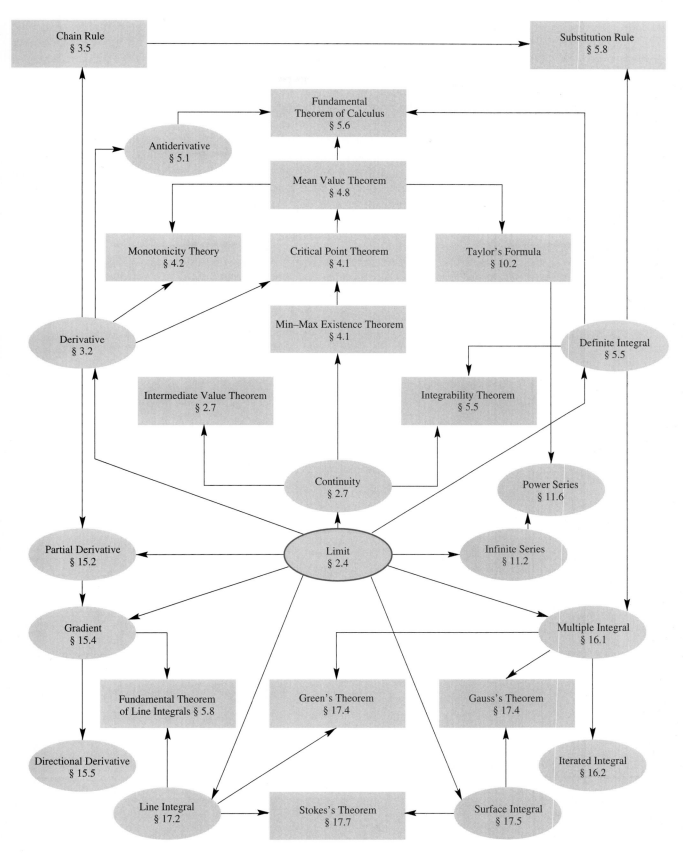

The machinery of calculus is embodied in its theorems. If we include the many that are stated in problems, the theorems in this book must number in the hundreds. Undoubtedly, you have managed to forget most of them—a practice we actually encourage. One should rather master the most important ones and know how they fit together. To help you with this task, our chart lists the principal theorems of calculus (in rectangles) and suggests how they relate to the concepts and to each other.

A student who understands what our chart represents is well prepared for those subjects that build on elementary calculus. And what are they? Suffice it to mention advanced calculus, ordinary and partial differential equations, integral equations, power series, Fourier series, real analysis, differential geometry, complex analysis, probability, Lebesgue integration, measure theory, calculus of variations, and abstract analysis. There is plenty to challenge the best of us.

## A.4
## NUMERICAL TABLES

TABLE I
Trigonometric Functions, Degree Measure

| Angle | sin | tan | cot | cos | — | | Angle | sin | tan | cot | cos | — |
|---|---|---|---|---|---|---|---|---|---|---|---|---|
| 0.0° | 0.0000 | 0.0000 | — | 1.0000 | 90.0° | | 22.5° | 0.3827 | 0.4142 | 2.4142 | 0.9239 | 67.5° |
| 0.5° | 0.0087 | 0.0087 | 114.59 | 1.0000 | 89.5° | | 23.0° | 0.3907 | 0.4245 | 2.3559 | 0.9205 | 67·0° |
| 1.0° | 0.0175 | 0.0175 | 57.290 | 0.9998 | 89.0° | | 23.5° | 0.3987 | 0.4348 | 2.2998 | 0.9171 | 66.5° |
| 1.5° | 0.0262 | 0.0262 | 38.188 | 0.9997 | 88.5° | | 24.0° | 0.4067 | 0.4452 | 2.2460 | 0.9135 | 66.0° |
| 2.0° | 0.0349 | 0.0349 | 28.636 | 0.9994 | 88.0° | | 24.5° | 0.4147 | 0.4557 | 2.1943 | 0.9100 | 65.5° |
| 2.5° | 0.0436 | 0.0437 | 22.904 | 0.9990 | 87.5° | | 25.0° | 0.4226 | 0.4663 | 2.1445 | 0.9063 | 65.0° |
| 3.0° | 0.0523 | 0.0524 | 19.081 | 0.9986 | 87.0° | | 25.5° | 0.4305 | 0.4770 | 2.0965 | 0.9026 | 64.5° |
| 3.5° | 0.0610 | 0.0612 | 16.350 | 0.9981 | 86.5° | | 26.0° | 0.4384 | 0.4877 | 2.0503 | 0.8988 | 64.0° |
| 4.0° | 0.0698 | 0.0699 | 14.301 | 0.9976 | 86.0° | | 26.5° | 0.4462 | 0.4986 | 2.0057 | 0.8949 | 63.5° |
| 4.5° | 0.0785 | 0.0787 | 12.706 | 0.9969 | 85.5° | | 27.0° | 0.4540 | 0.5095 | 1.9626 | 0.8910 | 63.0° |
| 5.0° | 0.0872 | 0.0875 | 11.430 | 0.9962 | 85.0° | | 27.5° | 0.4617 | 0.5206 | 1.9210 | 0.8870 | 62.5° |
| 5.5° | 0.0958 | 0.0963 | 10.385 | 0.9954 | 84.5° | | 28.0° | 0.4695 | 0.5317 | 1.8807 | 0.8829 | 62.0° |
| 6.0° | 0.1045 | 0.1051 | 9.5144 | 0.9945 | 84.0° | | 28.5° | 0.4772 | 0.5430 | 1.8418 | 0.8788 | 61.5° |
| 6.5° | 0.1132 | 0.1139 | 8.7769 | 0.9936 | 83.5° | | 29.0° | 0.4848 | 0.5543 | 1.8040 | 0.8746 | 61.0° |
| 7.0° | 0.1219 | 0.1228 | 8.1443 | 0.9925 | 83.0° | | 29.5° | 0.4924 | 0.5658 | 1.7675 | 0.8704 | 60.5° |
| 7.5° | 0.1305 | 0.1317 | 7.5958 | 0.9914 | 82.5° | | 30.0° | 0.5000 | 0.5774 | 1.7321 | 0.8660 | 60.0° |
| 8.0° | 0.1392 | 0.1405 | 7.1154 | 0.9903 | 82.0° | | 30.5° | 0.5075 | 0.5890 | 1.6977 | 0.8616 | 59.5° |
| 8.5° | 0.1478 | 0.1495 | 6.6912 | 0.9890 | 81.5° | | 31.0° | 0.5150 | 0.6009 | 1.6643 | 0.8572 | 59.0° |
| 9.0° | 0.1564 | 0.1584 | 6.3138 | 0.9877 | 81.0° | | 31.5° | 0.5225 | 0.6128 | 1.6319 | 0.8526 | 58.5° |
| 9.5° | 0.1650 | 0.1673 | 5.9758 | 0.9863 | 80.5° | | 32.0° | 0.5299 | 0.6249 | 1.6003 | 0.8480 | 58.0° |
| 10.0° | 0.1736 | 0.1763 | 5.6713 | 0.9848 | 80.0° | | 32.5° | 0.5373 | 0.6371 | 1.5697 | 0.8434 | 57.5° |
| 10.5° | 0.1822 | 0.1853 | 5.3955 | 0.9833 | 79.5° | | 33.0° | 0.5446 | 0.6494 | 1.5399 | 0.8387 | 57.0° |
| 11.0° | 0.1908 | 0.1944 | 5.1446 | 0.9816 | 79.0° | | 33.5° | 0.5519 | 0.6619 | 1.5108 | 0.8339 | 56.5° |
| 11.5° | 0.1994 | 0.2035 | 4.9152 | 0.9799 | 78.5° | | 34.0° | 0.5592 | 0.6745 | 1.4826 | 0.8290 | 56.0° |
| 12.0° | 0.2079 | 0.2126 | 4.7046 | 0.9781 | 78.0° | | 34.5° | 0.5664 | 0.6873 | 1.4550 | 0.8241 | 55.5° |
| 12.5° | 0.2164 | 0.2217 | 4.5107 | 0.9763 | 77.5° | | 35.0° | 0.5736 | 0.7002 | 1.4281 | 0.8192 | 55.0° |
| 13.0° | 0.2250 | 0.2309 | 4.3315 | 0.9744 | 77.0° | | 35.5° | 0.5807 | 0.7133 | 1.4019 | 0.8141 | 54.5° |
| 13.5° | 0.2334 | 0.2401 | 4.1653 | 0.9724 | 76.5° | | 36.0° | 0.5878 | 0.7265 | 1.3764 | 0.8090 | 54.0° |
| 14.0° | 0.2419 | 0.2493 | 4.0108 | 0.9703 | 76.0° | | 36.5° | 0.5948 | 0.7400 | 1.3514 | 0.8039 | 53.5° |
| 14.5° | 0.2504 | 0.2586 | 3.8667 | 0.9681 | 75.5° | | 37.0° | 0.6018 | 0.7536 | 1.3270 | 0.7986 | 53.0° |
| 15.0° | 0.2588 | 0.2679 | 3.7321 | 0.9659 | 75.0° | | 37.5° | 0.6088 | 0.7673 | 1.3032 | 0.7934 | 52.5° |
| 15.5° | 0.2672 | 0.2773 | 3.6059 | 0.9636 | 74.5° | | 38.0° | 0.6157 | 0.7813 | 1.2799 | 0.7880 | 52.0° |
| 16.0° | 0.2756 | 0.2867 | 3.4874 | 0.9613 | 74.0° | | 38.5° | 0.6225 | 0.7954 | 1.2572 | 0.7826 | 51.5° |
| 16.5° | 0.2840 | 0.2962 | 3.3759 | 0.9588 | 73.5° | | 39.0° | 0.6293 | 0.8098 | 1.2349 | 0.7771 | 51.0° |
| 17.0° | 0.2924 | 0.3057 | 3.2709 | 0.9563 | 73.0° | | 39.5° | 0.6361 | 0.8243 | 1.2131 | 0.7716 | 50.5° |
| 17.5° | 0.3007 | 0.3153 | 3.1716 | 0.9537 | 72.5° | | 40.0° | 0.6428 | 0.8391 | 1.1918 | 0.7660 | 50.0° |
| 18.0° | 0.3090 | 0.3249 | 3.0777 | 0.9511 | 72.0° | | 40.5° | 0.6494 | 0.8541 | 1.1708 | 0.7604 | 49.5° |
| 18.5° | 0.3173 | 0.3346 | 2.9887 | 0.9483 | 71.5° | | 41.0° | 0.6561 | 0.8693 | 1.1504 | 0.7547 | 49.0° |
| 19.0° | 0.3256 | 0.3443 | 2.9042 | 0.9455 | 71.0° | | 41.5° | 0.6626 | 0.8847 | 1.1303 | 0.7490 | 48.5° |
| 19.5° | 0.3338 | 0.3541 | 2.8239 | 0.9426 | 70.5° | | 42.0° | 0.6691 | 0.9004 | 1.1106 | 0.7431 | 48.0° |
| 20.0° | 0.3420 | 0.3640 | 2.7475 | 0.9397 | 70.0° | | 42.5° | 0.6756 | 0.9163 | 1.0913 | 0.7373 | 47.5° |
| 20.5° | 0.3502 | 0.3739 | 2.6746 | 0.9367 | 69.5° | | 43.0° | 0.6820 | 0.9325 | 1.0724 | 0.7314 | 47.0° |
| 21.0° | 0.3584 | 0.3839 | 2.6051 | 0.9336 | 69.0° | | 43.5° | 0.6884 | 0.9490 | 1.0538 | 0.7254 | 46.5° |
| 21.5° | 0.3665 | 0.3939 | 2.5386 | 0.9304 | 68.5° | | 44.0° | 0.6947 | 0.9657 | 1.0355 | 0.7193 | 46.0° |
| 22.0° | 0.3746 | 0.4040 | 2.4751 | 0.9272 | 68.0° | | 44.5° | 0.7009 | 0.9827 | 1.0176 | 0.7133 | 45.5° |
| 22.5° | 0.3827 | 0.4142 | 2.4142 | 0.9239 | 67.5° | | 45.0° | 0.7071 | 1.0000 | 1.0000 | 0.7071 | 45.0° |
| — | cos | cot | tan | sin | Angle | | — | cos | cot | tan | sin | Angle |

TABLE II
Trigonometric Functions, Radian Measure

| Radians | sin | cos | tan | Radians | sin | cos | tan |
|---|---|---|---|---|---|---|---|
| 0.00 | 0.0000 | 1.0000 | 0.0000 | 0.40 | 0.3894 | 0.9211 | 0.4228 |
| 0.01 | 0.0100 | 1.0000 | 0.0100 | 0.41 | 0.3986 | 0.9171 | 0.4346 |
| 0.02 | 0.0200 | 0.9998 | 0.0200 | 0.42 | 0.4078 | 0.9131 | 0.4466 |
| 0.03 | 0.0300 | 0.9996 | 0.0300 | 0.43 | 0.4169 | 0.9090 | 0.4586 |
| 0.04 | 0.0400 | 0.9992 | 0.0400 | 0.44 | 0.4259 | 0.9048 | 0.4708 |
| 0.05 | 0.0500 | 0.9988 | 0.0500 | 0.45 | 0.4350 | 0.9004 | 0.4831 |
| 0.06 | 0.0600 | 0.9982 | 0.0601 | 0.46 | 0.4439 | 0.8961 | 0.4954 |
| 0.07 | 0.0699 | 0.9976 | 0.0701 | 0.47 | 0.4529 | 0.8916 | 0.5080 |
| 0.08 | 0.0799 | 0.9968 | 0.0802 | 0.48 | 0.4618 | 0.8870 | 0.5206 |
| 0.09 | 0.0899 | 0.9960 | 0.0902 | 0.49 | 0.4706 | 0.8823 | 0.5334 |
| 0.10 | 0.0998 | 0.9950 | 0.1003 | 0.50 | 0.4794 | 0.8776 | 0.5463 |
| 0.11 | 0.1098 | 0.9940 | 0.1104 | 0.51 | 0.4882 | 0.8727 | 0.5594 |
| 0.12 | 0.1197 | 0.9928 | 0.1206 | 0.52 | 0.4969 | 0.8678 | 0.5726 |
| 0.13 | 0.1296 | 0.9916 | 0.1307 | 0.53 | 0.5055 | 0.8628 | 0.5859 |
| 0.14 | 0.1395 | 0.9902 | 0.1409 | 0.54 | 0.5141 | 0.8577 | 0.5994 |
| 0.15 | 0.1494 | 0.9888 | 0.1511 | 0.55 | 0.5227 | 0.8525 | 0.6131 |
| 0.16 | 0.1593 | 0.9872 | 0.1614 | 0.56 | 0.5312 | 0.8473 | 0.6269 |
| 0.17 | 0.1692 | 0.9856 | 0.1717 | 0.57 | 0.5396 | 0.8419 | 0.6410 |
| 0.18 | 0.1790 | 0.9838 | 0.1820 | 0.58 | 0.5480 | 0.8365 | 0.6552 |
| 0.19 | 0.1889 | 0.9820 | 0.1923 | 0.59 | 0.5564 | 0.8309 | 0.6696 |
| 0.20 | 0.1987 | 0.9801 | 0.2027 | 0.60 | 0.5646 | 0.8253 | 0.6841 |
| 0.21 | 0.2085 | 0.9780 | 0.2131 | 0.61 | 0.5729 | 0.8196 | 0.6989 |
| 0.22 | 0.2182 | 0.9759 | 0.2236 | 0.62 | 0.5810 | 0.8139 | 0.7139 |
| 0.23 | 0.2280 | 0.9737 | 0.2341 | 0.63 | 0.5891 | 0.8080 | 0.7291 |
| 0.24 | 0.2377 | 0.9713 | 0.2447 | 0.64 | 0.5972 | 0.8021 | 0.7445 |
| 0.25 | 0.2474 | 0.9689 | 0.2553 | 0.65 | 0.6052 | 0.7961 | 0.7602 |
| 0.26 | 0.2571 | 0.9664 | 0.2660 | 0.66 | 0.6131 | 0.7900 | 0.7761 |
| 0.27 | 0.2667 | 0.9638 | 0.2768 | 0.67 | 0.6210 | 0.7838 | 0.7923 |
| 0.28 | 0.2764 | 0.9611 | 0.2876 | 0.68 | 0.6288 | 0.7776 | 0.8087 |
| 0.29 | 0.2860 | 0.9582 | 0.2984 | 0.69 | 0.6365 | 0.7712 | 0.8253 |
| 0.30 | 0.2955 | 0.9553 | 0.3093 | 0.70 | 0.6442 | 0.7648 | 0.8423 |
| 0.31 | 0.3051 | 0.9523 | 0.3203 | 0.71 | 0.6518 | 0.7584 | 0.8595 |
| 0.32 | 0.3146 | 0.9492 | 0.3314 | 0.72 | 0.6594 | 0.7518 | 0.8771 |
| 0.33 | 0.3240 | 0.9460 | 0.3425 | 0.73 | 0.6669 | 0.7452 | 0.8949 |
| 0.34 | 0.3335 | 0.9428 | 0.3537 | 0.74 | 0.6743 | 0.7385 | 0.9131 |
| 0.35 | 0.3429 | 0.9394 | 0.3650 | 0.75 | 0.6816 | 0.7317 | 0.9316 |
| 0.36 | 0.3523 | 0.9359 | 0.3764 | 0.76 | 0.6889 | 0.7248 | 0.9505 |
| 0.37 | 0.3616 | 0.9323 | 0.3879 | 0.77 | 0.6961 | 0.7179 | 0.9697 |
| 0.38 | 0.3709 | 0.9287 | 0.3994 | 0.78 | 0.7033 | 0.7109 | 0.9893 |
| 0.39 | 0.3802 | 0.9249 | 0.4111 | 0.79 | 0.7104 | 0.7038 | 1.009 |

TABLE II
Trigonometric Functions, Radian Measure (cont.)

| Radians | sin | cos | tan | Radians | sin | cos | tan |
|---|---|---|---|---|---|---|---|
| 0.80 | 0.7174 | 0.6967 | 1.030 | 1.20 | 0.9320 | 0.3624 | 2.572 |
| 0.81 | 0.7243 | 0.6895 | 1.050 | 1.21 | 0.9356 | 0.3530 | 2.650 |
| 0.82 | 0.7311 | 0.6822 | 1.072 | 1.22 | 0.9391 | 0.3436 | 2.733 |
| 0.83 | 0.7379 | 0.6749 | 1.093 | 1.23 | 0.9425 | 0.3342 | 2.820 |
| 0.84 | 0.7446 | 0.6675 | 1.116 | 1.24 | 0.9458 | 0.3248 | 2.912 |
| 0.85 | 0.7513 | 0.6600 | 1.138 | 1.25 | 0.9490 | 0.3153 | 3.010 |
| 0.86 | 0.7578 | 0.6524 | 1.162 | 1.26 | 0.9521 | 0.3058 | 3.113 |
| 0.87 | 0.7643 | 0.6448 | 1.185 | 1.27 | 0.9551 | 0.2963 | 3.224 |
| 0.88 | 0.7707 | 0.6372 | 1.210 | 1.28 | 0.9580 | 0.2867 | 3.341 |
| 0.89 | 0.7771 | 0.6294 | 1.235 | 1.29 | 0.9608 | 0.2771 | 3.467 |
| 0.90 | 0.7833 | 0.6216 | 1.260 | 1.30 | 0.9636 | 0.2675 | 3.602 |
| 0.91 | 0.7895 | 0.6137 | 1.286 | 1.31 | 0.9662 | 0.2579 | 3.747 |
| 0.92 | 0.7956 | 0.6058 | 1.313 | 1.32 | 0.9687 | 0.2482 | 3.903 |
| 0.93 | 0.8016 | 0.5978 | 1.341 | 1.33 | 0.9711 | 0.2385 | 4.072 |
| 0.94 | 0.8076 | 0.5898 | 1.369 | 1.34 | 0.9735 | 0.2288 | 4.256 |
| 0.95 | 0.8134 | 0.5817 | 1.398 | 1.35 | 0.9757 | 0.2190 | 4.455 |
| 0.96 | 0.8192 | 0.5735 | 1.428 | 1.36 | 0.9779 | 0.2092 | 4.673 |
| 0.97 | 0.8249 | 0.5653 | 1.459 | 1.37 | 0.9799 | 0.1994 | 4.913 |
| 0.98 | 0.8305 | 0.5570 | 1.491 | 1.38 | 0.9819 | 0.1896 | 5.177 |
| 0.99 | 0.8360 | 0.5487 | 1.524 | 1.39 | 0.9837 | 0.1798 | 5.471 |
| 1.00 | 0.8415 | 0.5403 | 1.557 | 1.40 | 0.9854 | 0.1700 | 5.798 |
| 1.01 | 0.8468 | 0.5319 | 1.592 | 1.41 | 0.9871 | 0.1601 | 6.165 |
| 1.02 | 0.8521 | 0.5234 | 1.628 | 1.42 | 0.9887 | 0.1502 | 6.581 |
| 1.03 | 0.8573 | 0.5148 | 1.665 | 1.43 | 0.9901 | 0.1403 | 7.055 |
| 1.04 | 0.8624 | 0.5062 | 1.704 | 1.44 | 0.9915 | 0.1304 | 7.602 |
| 1.05 | 0.8674 | 0.4976 | 1.743 | 1.45 | 0.9927 | 0.1205 | 8.238 |
| 1.06 | 0.8724 | 0.4889 | 1.784 | 1.46 | 0.9939 | 0.1106 | 8.989 |
| 1.07 | 0.8772 | 0.4801 | 1.827 | 1.47 | 0.9949 | 0.1006 | 9.887 |
| 1.08 | 0.8820 | 0.4713 | 1.871 | 1.48 | 0.9959 | 0.0907 | 10.98 |
| 1.09 | 0.8866 | 0.4625 | 1.917 | 1.49 | 0.9967 | 0.0807 | 12.35 |
| 1.10 | 0.8912 | 0.4536 | 1.965 | 1.50 | 0.9975 | 0.0707 | 14.10 |
| 1.11 | 0.8957 | 0.4447 | 2.014 | 1.51 | 0.9982 | 0.0608 | 16.43 |
| 1.12 | 0.9001 | 0.4357 | 2.066 | 1.52 | 0.9987 | 0.0508 | 19.67 |
| 1.13 | 0.9044 | 0.4267 | 2.120 | 1.53 | 0.9992 | 0.0408 | 24.50 |
| 1.14 | 0.9086 | 0.4176 | 2.176 | 1.54 | 0.9995 | 0.0308 | 32.46 |
| 1.15 | 0.9128 | 0.4085 | 2.234 | 1.55 | 0.9998 | 0.0208 | 48.08 |
| 1.16 | 0.9168 | 0.3993 | 2.296 | 1.56 | 0.9999 | 0.0108 | 92.62 |
| 1.17 | 0.9208 | 0.3902 | 2.360 | 1.57 | 1.0000 | 0.0008 | 1256. |
| 1.18 | 0.9246 | 0.3809 | 2.427 | | | | |
| 1.19 | 0.9284 | 0.3717 | 2.498 | | | | |

TABLE III
Natural Logarithms

| | 0.00 | 0.01 | 0.02 | 0.03 | 0.04 | 0.05 | 0.06 | 0.07 | 0.08 | 0.09 |
|---|---|---|---|---|---|---|---|---|---|---|
| 1.0 | 0.0000 | 0.0100 | 0.0198 | 0.0296 | 0.0392 | 0.0488 | 0.0583 | 0.0677 | 0.0770 | 0.0862 |
| 1.1 | 0.0953 | 0.1044 | 0.1133 | 0.1222 | 0.1310 | 0.1398 | 0.1484 | 0.1570 | 0.1655 | 0.1740 |
| 1.2 | 0.1823 | 0.1906 | 0.1989 | 0.2070 | 0.2151 | 0.2231 | 0.2311 | 0.2390 | 0.2469 | 0.2546 |
| 1.3 | 0.2624 | 0.2700 | 0.2776 | 0.2852 | 0.2927 | 0.3001 | 0.3075 | 0.3148 | 0.3221 | 0.3293 |
| 1.4 | 0.3365 | 0.3436 | 0.3507 | 0.3577 | 0.3646 | 0.3716 | 0.3784 | 0.3853 | 0.3920 | 0.3988 |
| 1.5 | 0.4055 | 0.4121 | 0.4187 | 0.4253 | 0.4318 | 0.4383 | 0.4447 | 0.4511 | 0.4574 | 0.4637 |
| 1.6 | 0.4700 | 0.4762 | 0.4824 | 0.4886 | 0.4947 | 0.5008 | 0.5068 | 0.5128 | 0.5188 | 0.5247 |
| 1.7 | 0.5306 | 0.5365 | 0.5423 | 0.5481 | 0.5539 | 0.5596 | 0.5653 | 0.5710 | 0.5766 | 0.5822 |
| 1.8 | 0.5878 | 0.5933 | 0.5988 | 0.6043 | 0.6098 | 0.6152 | 0.6206 | 0.6259 | 0.6313 | 0.6366 |
| 1.9 | 0.6419 | 0.6471 | 0.6523 | 0.6575 | 0.6627 | 0.6678 | 0.6729 | 0.6780 | 0.6831 | 0.6881 |
| 2.0 | 0.6931 | 0.6981 | 0.7031 | 0.7080 | 0.7130 | 0.7178 | 0.7227 | 0.7275 | 0.7324 | 0.7372 |
| 2.1 | 0.7419 | 0.7467 | 0.7514 | 0.7561 | 0.7608 | 0.7655 | 0.7701 | 0.7747 | 0.7793 | 0.7839 |
| 2.2 | 0.7885 | 0.7930 | 0.7975 | 0.8020 | 0.8065 | 0.8109 | 0.8154 | 0.8198 | 0.8242 | 0.8286 |
| 2.3 | 0.8329 | 0.8372 | 0.8416 | 0.8459 | 0.8502 | 0.8544 | 0.8587 | 0.8629 | 0.8671 | 0.8713 |
| 2.4 | 0.8755 | 0.8796 | 0.8838 | 0.8879 | 0.8920 | 0.8961 | 0.9002 | 0.9042 | 0.9083 | 0.9123 |
| 2.5 | 0.9163 | 0.9203 | 0.9243 | 0.9282 | 0.9322 | 0.9361 | 0.9400 | 0.9439 | 0.9478 | 0.9517 |
| 2.6 | 0.9555 | 0.9594 | 0.9632 | 0.9670 | 0.9708 | 0.9746 | 0.9783 | 0.9821 | 0.9858 | 0.9895 |
| 2.7 | 0.9933 | 0.9969 | 1.0006 | 1.0043 | 1.0080 | 1.0116 | 1.0152 | 1.0188 | 1.0225 | 1.0260 |
| 2.8 | 1.0296 | 1.0332 | 1.0367 | 1.0403 | 1.0438 | 1.0473 | 1.0508 | 1.0543 | 1.0578 | 1.0613 |
| 2.9 | 1.0647 | 1.0682 | 1.0716 | 1.0750 | 1.0784 | 1.0818 | 1.0852 | 1.0886 | 1.0919 | 1.0953 |
| 3.0 | 1.0986 | 1.1019 | 1.1053 | 1.1086 | 1.1119 | 1.1151 | 1.1184 | 1.1217 | 1.1249 | 1.1282 |
| 3.1 | 1.1314 | 1.1346 | 1.1378 | 1.1410 | 1.1442 | 1.1474 | 1.1506 | 1.1537 | 1.1569 | 1.1600 |
| 3.2 | 1.1632 | 1.1663 | 1.1694 | 1.1725 | 1.1756 | 1.1787 | 1.1817 | 1.1848 | 1.1878 | 1.1909 |
| 3.3 | 1.1939 | 1.1970 | 1.2000 | 1.2030 | 1.2060 | 1.2090 | 1.2119 | 1.2149 | 1.2179 | 1.2208 |
| 3.4 | 1.2238 | 1.2267 | 1.2296 | 1.2326 | 1.2355 | 1.2384 | 1.2413 | 1.2442 | 1.2470 | 1.2499 |
| 3.5 | 1.2528 | 1.2556 | 1.2585 | 1.2613 | 1.2641 | 1.2669 | 1.2698 | 1.2726 | 1.2754 | 1.2782 |
| 3.6 | 1.2809 | 1.2837 | 1.2865 | 1.2892 | 1.2920 | 1.2947 | 1.2975 | 1.3002 | 1.3029 | 1.3056 |
| 3.7 | 1.3083 | 1.3110 | 1.3137 | 1.3164 | 1.3191 | 1.3218 | 1.3244 | 1.3271 | 1.3297 | 1.3324 |
| 3.8 | 1.3350 | 1.3376 | 1.3403 | 1.3429 | 1.3455 | 1.3481 | 1.3507 | 1.3533 | 1.3558 | 1.3584 |
| 3.9 | 1.3610 | 1.3635 | 1.3661 | 1.3686 | 1.3712 | 1.3737 | 1.3762 | 1.3788 | 1.3813 | 1.3838 |
| 4.0 | 1.3863 | 1.3888 | 1.3913 | 1.3938 | 1.3962 | 1.3987 | 1.4012 | 1.4036 | 1.4061 | 1.4085 |
| 4.1 | 1.4110 | 1.4134 | 1.4159 | 1.4183 | 1.4207 | 1.4231 | 1.4255 | 1.4279 | 1.4303 | 1.4327 |
| 4.2 | 1.4351 | 1.4375 | 1.4398 | 1.4422 | 1.4446 | 1.4469 | 1.4493 | 1.4516 | 1.4540 | 1.4563 |
| 4.3 | 1.4586 | 1.4609 | 1.4633 | 1.4656 | 1.4679 | 1.4702 | 1.4725 | 1.4748 | 1.4770 | 1.4793 |
| 4.4 | 1.4816 | 1.4839 | 1.4861 | 1.4884 | 1.4907 | 1.4929 | 1.4952 | 1.4974 | 1.4996 | 1.5019 |
| 4.5 | 1.5041 | 1.5063 | 1.5085 | 1.5107 | 1.5129 | 1.5151 | 1.5173 | 1.5195 | 1.5217 | 1.5239 |
| 4.6 | 1.5261 | 1.5282 | 1.5304 | 1.5326 | 1.5347 | 1.5369 | 1.5390 | 1.5412 | 1.5433 | 1.5454 |
| 4.7 | 1.5476 | 1.5497 | 1.5518 | 1.5539 | 1.5560 | 1.5581 | 1.5602 | 1.5623 | 1.5644 | 1.5665 |
| 4.8 | 1.5686 | 1.5707 | 1.5728 | 1.5748 | 1.5769 | 1.5790 | 1.5810 | 1.5831 | 1.5851 | 1.5872 |
| 4.9 | 1.5892 | 1.5913 | 1.5933 | 1.5953 | 1.5974 | 1.5994 | 1.6014 | 1.6034 | 1.6054 | 1.6074 |
| 5.0 | 1.6094 | 1.6114 | 1.6134 | 1.6154 | 1.6174 | 1.6194 | 1.6214 | 1.6233 | 1.6253 | 1.6273 |
| 5.1 | 1.6292 | 1.6312 | 1.6332 | 1.6351 | 1.6371 | 1.6390 | 1.6409 | 1.6429 | 1.6448 | 1.6467 |
| 5.2 | 1.6487 | 1.6506 | 1.6525 | 1.6544 | 1.6563 | 1.6582 | 1.6601 | 1.6620 | 1.6639 | 1.6658 |
| 5.3 | 1.6677 | 1.6696 | 1.6715 | 1.6734 | 1.6752 | 1.6771 | 1.6790 | 1.6808 | 1.6827 | 1.6845 |
| 5.4 | 1.6864 | 1.6882 | 1.6901 | 1.6919 | 1.6938 | 1.6956 | 1.6974 | 1.6993 | 1.7011 | 1.7029 |

$$\ln (N \cdot 10^m) = \ln N + m \ln 10, \quad \ln 10 = 2.3026$$

TABLE III
Natural Logarithms (cont.)

|      | 0.00   | 0.01   | 0.02   | 0.03   | 0.04   | 0.05   | 0.06   | 0.07   | 0.08   | 0.09   |
|------|--------|--------|--------|--------|--------|--------|--------|--------|--------|--------|
| 5.5  | 1.7047 | 1.7066 | 1.7084 | 1.7102 | 1.7120 | 1.7138 | 1.7156 | 1.7174 | 1.7192 | 1.7210 |
| 5.6  | 1.7228 | 1.7246 | 1.7263 | 1.7281 | 1.7299 | 1.7317 | 1.7334 | 1.7352 | 1.7370 | 1.7387 |
| 5.7  | 1.7405 | 1.7422 | 1.7440 | 1.7457 | 1.7475 | 1.7492 | 1.7509 | 1.7527 | 1.7544 | 1.7561 |
| 5.8  | 1.7579 | 1.7596 | 1.7613 | 1.7630 | 1.7647 | 1.7664 | 1.7682 | 1.7699 | 1.7716 | 1.7733 |
| 5.9  | 1.7750 | 1.7766 | 1.7783 | 1.7800 | 1.7817 | 1.7834 | 1.7851 | 1.7867 | 1.7884 | 1.7901 |
| 6.0  | 1.7918 | 1.7934 | 1.7951 | 1.7967 | 1.7984 | 1.8001 | 1.8017 | 1.8034 | 1.8050 | 1.8066 |
| 6.1  | 1.8083 | 1.8099 | 1.8116 | 1.8132 | 1.8148 | 1.8165 | 1.8181 | 1.8197 | 1.8213 | 1.8229 |
| 6.2  | 1.8245 | 1.8262 | 1.8278 | 1.8294 | 1.8310 | 1.8326 | 1.8342 | 1.8358 | 1.8374 | 1.8390 |
| 6.3  | 1.8406 | 1.8421 | 1.8437 | 1.8453 | 1.8469 | 1.8485 | 1.8500 | 1.8516 | 1.8532 | 1.8547 |
| 6.4  | 1.8563 | 1.8579 | 1.8594 | 1.8610 | 1.8625 | 1.8641 | 1.8656 | 1.8672 | 1.8687 | 1.8703 |
| 6.5  | 1.8718 | 1.8733 | 1.8749 | 1.8764 | 1.8779 | 1.8795 | 1.8810 | 1.8825 | 1.8840 | 1.8856 |
| 6.6  | 1.8871 | 1.8886 | 1.8901 | 1.8916 | 1.8931 | 1.8946 | 1.8961 | 1.8976 | 1.8991 | 1.9006 |
| 6.7  | 1.9021 | 1.9036 | 1.9051 | 1.9066 | 1.9081 | 1.9095 | 1.9110 | 1.9125 | 1.9140 | 1.9155 |
| 6.8  | 1.9169 | 1.9184 | 1.9199 | 1.9213 | 1.9228 | 1.9242 | 1.9257 | 1.9272 | 1.9286 | 1.9301 |
| 6.9  | 1.9315 | 1.9330 | 1.9344 | 1.9359 | 1.9373 | 1.9387 | 1.9402 | 1.9416 | 1.9430 | 1.9445 |
| 7.0  | 1.9459 | 1.9473 | 1.9488 | 1.9502 | 1.9516 | 1.9530 | 1.9544 | 1.9559 | 1.9573 | 1.9587 |
| 7.1  | 1.9601 | 1.9615 | 1.9629 | 1.9643 | 1.9657 | 1.9671 | 1.9685 | 1.9699 | 1.9713 | 1.9727 |
| 7.2  | 1.9741 | 1.9755 | 1.9769 | 1.9782 | 1.9796 | 1.9810 | 1.9824 | 1.9838 | 1.9851 | 1.9865 |
| 7.3  | 1.9879 | 1.9892 | 1.9906 | 1.9920 | 1.9933 | 1.9947 | 1.9961 | 1.9974 | 1.9988 | 2.0001 |
| 7.4  | 2.0015 | 2.0028 | 2.0042 | 2.0055 | 2.0069 | 2.0082 | 2.0096 | 2.0109 | 2.0122 | 2.0136 |
| 7.5  | 2.0149 | 2.0162 | 2.0176 | 2.0189 | 2.0202 | 2.0215 | 2.0229 | 2.0242 | 2.0255 | 2.0268 |
| 7.6  | 2.0282 | 2.0295 | 2.0308 | 2.0321 | 2.0334 | 2.0347 | 2.0360 | 2.0373 | 2.0386 | 2.0399 |
| 7.7  | 2.0412 | 2.0425 | 2.0438 | 2.0451 | 2.0464 | 2.0477 | 2.0490 | 2.0503 | 2.0516 | 2.0528 |
| 7.8  | 2.0541 | 2.0554 | 2.0567 | 2.0580 | 2.0592 | 2.0605 | 2.0618 | 2.0631 | 2.0643 | 2.0656 |
| 7.9  | 2.0669 | 2.0681 | 2.0694 | 2.0707 | 2.0719 | 2.0732 | 2.0744 | 2.0757 | 2.0769 | 2.0782 |
| 8.0  | 2.0794 | 2.0807 | 2.0819 | 2.0832 | 2.0844 | 2.0857 | 2.0869 | 2.0882 | 2.0894 | 2.0906 |
| 8.1  | 2.0919 | 2.0931 | 2.0943 | 2.0956 | 2.0968 | 2.0980 | 2.0992 | 2.1005 | 2.1017 | 2.1029 |
| 8.2  | 2.1041 | 2.1054 | 2.1066 | 2.1078 | 2.1090 | 2.1102 | 2.1114 | 2.1126 | 2.1138 | 2.1150 |
| 8.3  | 2.1163 | 2.1175 | 2.1187 | 2.1190 | 2.1211 | 2.1223 | 2.1235 | 2.1247 | 2.1258 | 2.1270 |
| 8.4  | 2.1282 | 2.1294 | 2.1306 | 2.1318 | 2.1330 | 2.1342 | 2.1353 | 2.1365 | 2.1377 | 2.1389 |
| 8.5  | 2.1401 | 2.1412 | 2.1424 | 2.1436 | 2.1448 | 2.1459 | 2.1471 | 2.1483 | 2.1494 | 2.1506 |
| 8.6  | 2.1518 | 2.1529 | 2.1541 | 2.1552 | 2.1564 | 2.1576 | 2.1587 | 2.1599 | 2.1610 | 2.1622 |
| 8.7  | 2.1633 | 2.1645 | 2.1656 | 2.1668 | 2.1679 | 2.1691 | 2.1702 | 2.1713 | 2.1725 | 2.1736 |
| 8.8  | 2.1748 | 2.1759 | 2.1770 | 2.1782 | 2.1793 | 2.1804 | 2.1815 | 2.1827 | 2.1838 | 2.1849 |
| 8.9  | 2.1861 | 2.1872 | 2.1883 | 2.1894 | 2.1905 | 2.1917 | 2.1928 | 2.1939 | 2.1950 | 2.1961 |
| 9.0  | 2.1972 | 2.1983 | 2.1994 | 2.2006 | 2.2017 | 2.2028 | 2.2039 | 2.2050 | 2.2061 | 2.2072 |
| 9.1  | 2.2083 | 2.2094 | 2.2105 | 2.2116 | 2.2127 | 2.2138 | 2.2148 | 2.2159 | 2.2170 | 2.2181 |
| 9.2  | 2.2192 | 2.2203 | 2.2214 | 2.2225 | 2.2235 | 2.2246 | 2.2257 | 2.2268 | 2.2279 | 2.2289 |
| 9.3  | 2.2300 | 2.2311 | 2.2322 | 2.2332 | 2.2343 | 2.2354 | 2.2364 | 2.2375 | 2.2386 | 2.2396 |
| 9.4  | 2.2407 | 2.2418 | 2.2428 | 2.2439 | 2.2450 | 2.2460 | 2.2471 | 2.2481 | 2.2492 | 2.2502 |
| 9.5  | 2.2513 | 2.2523 | 2.2534 | 2.2544 | 2.2555 | 2.2565 | 2.2576 | 2.2586 | 2.2597 | 2.2607 |
| 9.6  | 2.2618 | 2.2628 | 2.2638 | 2.2649 | 2.2659 | 2.2670 | 2.2680 | 2.2690 | 2.2701 | 2.2711 |
| 9.7  | 2.2721 | 2.2732 | 2.2742 | 2.2752 | 2.2762 | 2.2773 | 2.2783 | 2.2793 | 2.2803 | 2.2814 |
| 9.8  | 2.2824 | 2.2834 | 2.2844 | 2.2854 | 2.2865 | 2.2875 | 2.2885 | 2.2895 | 2.2905 | 2.2915 |
| 9.9  | 2.2925 | 2.2935 | 2.2946 | 2.2956 | 2.2966 | 2.2976 | 2.2986 | 2.2996 | 2.3006 | 2.3016 |

TABLE IV
Exponential and Hyperbolic Functions

| $x$ | $e^x$ | $e^{-x}$ | sinh $x$ | cosh $x$ | tanh $x$ |
|------|--------|--------|--------|--------|--------|
| 0.00 | 1.0000 | 1.0000 | 0.0000 | 1.0000 | 0.0000 |
| 0.01 | 1.0101 | 0.9900 | 0.0100 | 1.0001 | 0.0100 |
| 0.02 | 1.0202 | 0.9802 | 0.0200 | 1.0002 | 0.0200 |
| 0.03 | 1.0305 | 0.9704 | 0.0300 | 1.0005 | 0.0300 |
| 0.04 | 1.0408 | 0.9608 | 0.0400 | 1.0008 | 0.0400 |
| 0.05 | 1.0513 | 0.9512 | 0.0500 | 1.0013 | 0.0500 |
| 0.06 | 1.0618 | 0.9418 | 0.0600 | 1.0018 | 0.0599 |
| 0.07 | 1.0725 | 0.9324 | 0.0701 | 1.0025 | 0.0699 |
| 0.08 | 1.0833 | 0.9231 | 0.0801 | 1.0032 | 0.0798 |
| 0.09 | 1.0942 | 0.9139 | 0.0901 | 1.0041 | 0.0898 |
| 0.10 | 1.1052 | 0.9048 | 0.1002 | 1.0050 | 0.0997 |
| 0.11 | 1.1163 | 0.8958 | 0.1102 | 1.0061 | 0.1096 |
| 0.12 | 1.1275 | 0.8869 | 0.1203 | 1.0072 | 0.1194 |
| 0.13 | 1.1388 | 0.8781 | 0.1304 | 1.0085 | 0.1293 |
| 0.14 | 1.1503 | 0.9694 | 0.1405 | 1.0098 | 0.1391 |
| 0.15 | 1.1618 | 0.8607 | 0.1506 | 1.0113 | 0.1489 |
| 0.16 | 1.1735 | 0.8521 | 0.1607 | 1.0128 | 0.1586 |
| 0.17 | 1.1853 | 0.8437 | 0.1708 | 1.0145 | 0.1684 |
| 0.18 | 1.1972 | 0.8353 | 0.1810 | 1.0162 | 0.1781 |
| 0.19 | 1.2092 | 0.8270 | 0.1911 | 1.0181 | 0.1877 |
| 0.20 | 1.2214 | 0.8187 | 0.2013 | 1.0201 | 0.1974 |
| 0.21 | 1.2337 | 0.8106 | 0.2115 | 1.0221 | 0.2070 |
| 0.22 | 1.2461 | 0.8025 | 0.2218 | 1.0243 | 0.2165 |
| 0.23 | 1.2586 | 0.7945 | 0.2320 | 1.0266 | 0.2260 |
| 0.24 | 1.2712 | 0.7866 | 0.2423 | 1.0289 | 0.2355 |
| 0.25 | 1.2840 | 0.7788 | 0.2526 | 1.0314 | 0.2449 |
| 0.26 | 1.2969 | 0.7711 | 0.2629 | 1.0340 | 0.2543 |
| 0.27 | 1.3100 | 0.7634 | 0.2733 | 1.0367 | 0.2636 |
| 0.28 | 1.3231 | 0.7558 | 0.2837 | 1.0395 | 0.2729 |
| 0.29 | 1.3364 | 0.7483 | 0.2941 | 1.0423 | 0.2821 |
| 0.30 | 1.3499 | 0.7408 | 0.3045 | 1.0453 | 0.2913 |
| 0.31 | 1.3634 | 0.7334 | 0.3150 | 1.0484 | 0.3004 |
| 0.32 | 1.3771 | 0.7261 | 0.3255 | 1.0516 | 0.3095 |
| 0.33 | 1.3910 | 0.7189 | 0.3360 | 1.0549 | 0.3185 |
| 0.34 | 1.4049 | 0.7118 | 0.3466 | 1.0584 | 0.3275 |
| 0.35 | 1.4191 | 0.7047 | 0.3572 | 1.0619 | 0.3364 |
| 0.36 | 1.4333 | 0.6977 | 0.3678 | 1.0655 | 0.3452 |
| 0.37 | 1.4477 | 0.6907 | 0.3785 | 1.0692 | 0.3540 |
| 0.38 | 1.4623 | 0.6839 | 0.3892 | 1.0731 | 0.3627 |
| 0.39 | 1.4770 | 0.6771 | 0.4000 | 1.0770 | 0.3714 |
| 0.40 | 1.4918 | 0.6703 | 0.4108 | 1.0811 | 0.3799 |
| 0.41 | 1.5068 | 0.6637 | 0.4216 | 1.0852 | 0.3885 |
| 0.42 | 1.5220 | 0.6570 | 0.4325 | 1.0895 | 0.3969 |
| 0.43 | 1.5373 | 0.6505 | 0.4434 | 1.0939 | 0.4053 |
| 0.44 | 1.5527 | 0.6440 | 0.4543 | 1.0984 | 0.4136 |

TABLE IV
Exponential and Hyperbolic Functions (cont.)

| $x$ | $e^x$ | $e^{-x}$ | $\sinh x$ | $\cosh x$ | $\tanh x$ |
|------|--------|---------|-----------|-----------|-----------|
| 0.45 | 1.5683 | 0.6376 | 0.4653 | 1.1030 | 0.4219 |
| 0.46 | 1.5841 | 0.6313 | 0.4764 | 1.1077 | 0.4301 |
| 0.47 | 1.6000 | 0.6250 | 0.4875 | 1.1125 | 0.4382 |
| 0.48 | 1.6161 | 0.6188 | 0.4986 | 1.1174 | 0.4462 |
| 0.49 | 1.6323 | 0.6126 | 0.5098 | 1.1225 | 0.4542 |
| 0.50 | 1.6487 | 0.6065 | 0.5211 | 1.1276 | 0.4621 |
| 0.51 | 1.6653 | 0.6005 | 0.5324 | 1.1329 | 0.4699 |
| 0.52 | 1.6820 | 0.5945 | 0.5438 | 1.1383 | 0.4777 |
| 0.53 | 1.6989 | 0.5886 | 0.5552 | 1.1438 | 0.4854 |
| 0.54 | 1.7160 | 0.5827 | 0.5666 | 1.1494 | 0.4930 |
| 0.55 | 1.7333 | 0.5769 | 0.5782 | 1.1551 | 0.5005 |
| 0.56 | 1.7507 | 0.5712 | 0.5897 | 1.1609 | 0.5080 |
| 0.57 | 1.7683 | 0.5655 | 0.6014 | 1.1669 | 0.5154 |
| 0.58 | 1.7860 | 0.5599 | 0.6131 | 1.1730 | 0.5227 |
| 0.59 | 1.8040 | 0.5543 | 0.6248 | 1.1792 | 0.5299 |
| 0.60 | 1.8221 | 0.5488 | 0.6367 | 1.1855 | 0.5370 |
| 0.61 | 1.8044 | 0.5434 | 0.6485 | 1.1919 | 0.5441 |
| 0.62 | 1.8589 | 0.5379 | 0.6605 | 1.1984 | 0.5511 |
| 0.63 | 1.8776 | 0.5326 | 0.6725 | 1.2051 | 0.5581 |
| 0.64 | 1.8965 | 0.5273 | 0.6846 | 1.2119 | 0.5649 |
| 0.65 | 1.9155 | 0.5220 | 0.6967 | 1.2188 | 0.5717 |
| 0.66 | 1.9348 | 0.5169 | 0.7090 | 1.2258 | 0.5784 |
| 0.67 | 1.9542 | 0.5117 | 0.7213 | 1.2330 | 0.5850 |
| 0.68 | 1.9739 | 0.5066 | 0.7336 | 1.2402 | 0.5915 |
| 0.69 | 1.9937 | 0.5016 | 0.7461 | 1.2476 | 0.5980 |
| 0.70 | 2.0138 | 0.4966 | 0.7586 | 1.2552 | 0.6044 |
| 0.71 | 2.0340 | 0.4916 | 0.7712 | 1.2628 | 0.6107 |
| 0.72 | 2.0544 | 0.4868 | 0.7838 | 1.2706 | 0.6169 |
| 0.73 | 2.0751 | 0.4819 | 0.7966 | 1.2785 | 0.6231 |
| 0.74 | 2.0959 | 0.4771 | 0.8094 | 1.2865 | 0.6291 |
| 0.75 | 2.1170 | 0.4724 | 0.8223 | 1.2947 | 0.6351 |
| 0.76 | 2.1383 | 0.4677 | 0.8353 | 1.3030 | 0.6411 |
| 0.77 | 2.1598 | 0.4630 | 0.8484 | 1.3114 | 0.6469 |
| 0.78 | 2.1815 | 0.4584 | 0.8615 | 1.3199 | 0.6527 |
| 0.79 | 2.2034 | 0.4538 | 0.8748 | 1.3286 | 0.6584 |
| 0.80 | 2.2255 | 0.4493 | 0.8881 | 1.3374 | 0.6640 |
| 0.81 | 2.2479 | 0.4449 | 0.9015 | 1.3464 | 0.6696 |
| 0.82 | 2.2705 | 0.4404 | 0.9150 | 1.3555 | 0.6751 |
| 0.83 | 2.2933 | 0.4360 | 0.9286 | 1.3647 | 0.6805 |
| 0.84 | 2.3164 | 0.4317 | 0.9423 | 1.3740 | 0.6858 |
| 0.85 | 2.3396 | 0.4274 | 0.9561 | 1.3835 | 0.6911 |
| 0.86 | 2.3632 | 0.4232 | 0.9700 | 1.3932 | 0.6963 |
| 0.87 | 2.3869 | 0.4190 | 0.9840 | 1.4029 | 0.7014 |
| 0.88 | 2.4109 | 0.4148 | 0.9981 | 1.4128 | 0.7064 |
| 0.89 | 2.4351 | 0.4107 | 1.0122 | 1.4229 | 0.7114 |

TABLE IV
Exponential and Hyperbolic Functions (cont.)

| $x$ | $e^x$ | $e^{-x}$ | $\sinh x$ | $\cosh x$ | $\tanh x$ |
|------|--------|----------|-----------|-----------|-----------|
| 0.90 | 2.4596 | 0.4066 | 1.0265 | 1.4331 | 0.7163 |
| 0.91 | 2.4843 | 0.4025 | 1.0409 | 1.4434 | 0.7211 |
| 0.92 | 2.5093 | 0.3985 | 1.0554 | 1.4539 | 0.7259 |
| 0.93 | 2.5345 | 0.3946 | 1.0700 | 1.4645 | 0.7306 |
| 0.94 | 2.5600 | 0.3906 | 1.0847 | 1.4753 | 0.7352 |
| 0.95 | 2.5857 | 0.3867 | 1.0995 | 1.4862 | 0.7398 |
| 0.96 | 2.6117 | 0.3829 | 1.1144 | 1.4973 | 0.7443 |
| 0.97 | 2.6379 | 0.3791 | 1.1294 | 1.5085 | 0.7487 |
| 0.98 | 2.6645 | 0.3753 | 1.1446 | 1.5199 | 0.7531 |
| 0.99 | 2.6912 | 0.3716 | 1.1598 | 1.5314 | 0.7574 |
| 1.00 | 2.7183 | 0.3679 | 1.1752 | 1.5431 | 0.7616 |
| 1.05 | 2.8577 | 0.3499 | 1.2539 | 1.6038 | 0.7818 |
| 1.10 | 3.0042 | 0.3329 | 1.3356 | 1.6685 | 0.8005 |
| 1.15 | 3.1582 | 0.3166 | 1.4208 | 1.7374 | 0.8178 |
| 1.20 | 3.3201 | 0.3012 | 1.5085 | 1.8107 | 0.8337 |
| 1.25 | 3.4903 | 0.2865 | 1.6019 | 1.8884 | 0.8483 |
| 1.30 | 3.6693 | 0.2725 | 1.6984 | 1.9709 | 0.8617 |
| 1.35 | 3.8574 | 0.2592 | 1.7991 | 2.0583 | 0.8741 |
| 1.40 | 4.0552 | 0.2466 | 1.9043 | 2.1509 | 0.8854 |
| 1.45 | 4.2631 | 0.2346 | 2.0143 | 2.2488 | 0.8957 |
| 1.50 | 4.4817 | 0.2231 | 2.1293 | 2.3524 | 0.9051 |
| 1.55 | 4.7115 | 0.2122 | 2.2496 | 2.4619 | 0.9138 |
| 1.60 | 4.9530 | 0.2019 | 2.3756 | 2.5775 | 0.9217 |
| 1.65 | 5.2070 | 0.1920 | 2.5075 | 2.6995 | 0.9289 |
| 1.70 | 5.4739 | 0.1827 | 2.6456 | 2.8283 | 0.9354 |
| 1.75 | 5.7546 | 0.1738 | 2.7904 | 2.9642 | 0.9414 |
| 1.80 | 6.0496 | 0.1653 | 2.9422 | 3.1075 | 0.9468 |
| 1.85 | 6.3598 | 0.1572 | 3.1013 | 3.2585 | 0.9517 |
| 1.90 | 6.6859 | 0.1496 | 3.2682 | 3.4177 | 0.9562 |
| 1.95 | 7.0287 | 0.1423 | 3.4432 | 3.5855 | 0.9603 |
| 2.00 | 7.3891 | 0.1353 | 3.6269 | 3.7622 | 0.9640 |
| 2.05 | 7.7679 | 0.1287 | 3.8196 | 3.9483 | 0.9674 |
| 2.10 | 8.1662 | 0.1225 | 4.0219 | 4.1443 | 0.9705 |
| 2.15 | 8.5849 | 0.1165 | 4.2342 | 4.3507 | 0.9732 |
| 2.20 | 9.0250 | 0.1108 | 4.4571 | 4.5679 | 0.9757 |
| 2.25 | 9.4877 | 0.1054 | 4.6912 | 4.7966 | 0.9780 |
| 2.30 | 9.9742 | 0.1003 | 4.9370 | 5.0372 | 0.9801 |
| 2.35 | 10.486 | 0.0954 | 5.1951 | 5.2905 | 0.9820 |
| 2.40 | 11.023 | 0.0907 | 5.4662 | 5.5569 | 0.9837 |
| 2.45 | 11.588 | 0.0863 | 5.7510 | 5.8373 | 0.9852 |
| 2.50 | 12.182 | 0.0821 | 6.0502 | 6.1323 | 0.9866 |
| 2.55 | 12.807 | 0.0781 | 6.3645 | 6.4426 | 0.9879 |
| 2.60 | 13.464 | 0.0743 | 6.6947 | 6.7690 | 0.9890 |
| 2.65 | 14.154 | 0.0707 | 7.0417 | 7.1123 | 0.9901 |
| 2.70 | 14.880 | 0.0672 | 7.4063 | 7.4735 | 0.9910 |

TABLE IV
Exponential and Hyperbolic Functions (cont.)

| $x$ | $e^x$ | $e^{-x}$ | $\sinh x$ | $\cosh x$ | $\tanh x$ |
|---|---|---|---|---|---|
| 2.75 | 15.643 | 0.0639 | 7.7894 | 7.8533 | 0.9919 |
| 2.80 | 16.445 | 0.0608 | 8.1919 | 8.2527 | 0.9926 |
| 2.85 | 17.288 | 0.0578 | 8.6150 | 8.6728 | 0.9933 |
| 2.90 | 18.174 | 0.0550 | 9.0596 | 9.1146 | 0.9940 |
| 2.95 | 19.106 | 0.0523 | 9.5268 | 9.5791 | 0.9945 |
| 3.00 | 20.086 | 0.0498 | 10.018 | 10.068 | 0.9951 |
| 3.05 | 21.115 | 0.0474 | 10.534 | 10.581 | 0.9955 |
| 3.10 | 22.198 | 0.0450 | 11.076 | 11.122 | 0.9959 |
| 3.15 | 23.336 | 0.0429 | 11.647 | 11.689 | 0.9963 |
| 3.20 | 24.533 | 0.0408 | 12.246 | 12.287 | 0.9967 |
| 3.25 | 25.790 | 0.0388 | 12.876 | 12.915 | 0.9970 |
| 3.30 | 27.113 | 0.0369 | 13.538 | 13.575 | 0.9973 |
| 3.35 | 28.503 | 0.0351 | 14.234 | 14.269 | 0.9975 |
| 3.40 | 29.964 | 0.0334 | 14.965 | 14.999 | 0.9978 |
| 3.45 | 31.500 | 0.0317 | 15.734 | 15.766 | 0.9980 |
| 3.50 | 33.115 | 0.0302 | 16.543 | 16.573 | 0.9982 |
| 3.55 | 34.813 | 0.0287 | 17.392 | 17.421 | 0.9983 |
| 3.60 | 36.598 | 0.0273 | 18.286 | 18.313 | 0.9985 |
| 3.65 | 38.475 | 0.0260 | 19.224 | 19.250 | 0.9986 |
| 3.70 | 40.447 | 0.0247 | 20.211 | 20.236 | 0.9988 |
| 3.75 | 42.521 | 0.0235 | 21.249 | 21.272 | 0.9989 |
| 3.80 | 44.701 | 0.0224 | 22.339 | 22.362 | 0.9990 |
| 3.85 | 46.993 | 0.0213 | 23.486 | 23.507 | 0.9991 |
| 3.90 | 49.402 | 0.0202 | 24.691 | 24.711 | 0.9992 |
| 3.95 | 51.935 | 0.0193 | 25.958 | 25.977 | 0.9993 |
| 4.00 | 54.598 | 0.0183 | 27.290 | 27.308 | 0.9993 |
| 4.10 | 60.340 | 0.0166 | 30.162 | 30.178 | 0.9995 |
| 4.20 | 66.686 | 0.0150 | 33.336 | 33.351 | 0.9996 |
| 4.30 | 73.700 | 0.0136 | 36.843 | 36.857 | 0.9996 |
| 4.40 | 81.451 | 0.0123 | 40.719 | 40.732 | 0.9997 |
| 4.50 | 90.017 | 0.0111 | 45.003 | 45.014 | 0.9998 |
| 4.60 | 99.484 | 0.0101 | 49.737 | 49.747 | 0.9998 |
| 4.70 | 109.95 | 0.0091 | 54.969 | 54.978 | 0.9998 |
| 4.80 | 121.51 | 0.0082 | 60.751 | 60.759 | 0.9999 |
| 4.90 | 134.29 | 0.0074 | 67.141 | 67.149 | 0.9999 |
| 5.00 | 148.41 | 0.0067 | 74.203 | 74.210 | 0.9999 |
| 5.20 | 181.27 | 0.0055 | 90.633 | 90.639 | 0.9999 |
| 5.40 | 221.41 | 0.0045 | 110.70 | 110.71 | 1.0000 |
| 5.60 | 270.43 | 0.0037 | 135.21 | 135.22 | 1.0000 |
| 5.80 | 330.30 | 0.0030 | 165.15 | 165.15 | 1.0000 |
| 6.00 | 403.43 | 0.0025 | 201.71 | 201.72 | 1.0000 |
| 7.00 | 1096.6 | 0.0009 | 548.32 | 548.32 | 1.0000 |
| 8.00 | 2981.0 | 0.0003 | 1490.5 | 1490.5 | 1.0000 |
| 9.00 | 8103.1 | 0.0001 | 4051.5 | 4051.5 | 1.0000 |
| 10.00 | 22026. | 0.00005 | 11013. | 11013. | 1.0000 |

# Answers to Odd-Numbered Problems

**PROBLEM SET 1.1**

**1.** 16    **3.** −148    **5.** $\frac{58}{91}$    **7.** $\frac{1}{24}$    **9.** $\frac{6}{49}$

**11.** $\frac{7}{15}$    **13.** $\frac{1}{3}$    **15.** 2    **17.** −6    **19.** $\frac{16}{81}$

**21.** $3x^2 - x - 4$    **23.** $6x^2 - 15x - 9$

**25.** $9t^4 - 6t^3 + 7t^2 - 2t + 1$    **27.** $x + 2$    **29.** $t - 7$

**31.** $\dfrac{2(3x+10)}{x(x+2)}$    **33.** $\dfrac{-(t+4)(x-7)}{(x-1)(t-5)}$

**35.** **(a)** 0;   **(b)** Undefined;   **(c)** 0;   **(d)** Undefined;
     **(e)** 0;   **(f)** 1

**37.** **(a)** False;   **(b)** True;   **(c)** False;   **(d)** True;
     **(e)** True;   **(f)** False

**41.** **(a)** $3 \cdot 3 \cdot 3 \cdot 3 \cdot 3$ or $3^5$;   **(b)** $1 \cdot 127$;
     **(c)** $2 \cdot 2 \cdot 3 \cdot 5 \cdot 5 \cdot 17$ or $2^2 \cdot 3 \cdot 5^5 \cdot 17$;   **(d)** $2 \cdot 173$

**47.** **(a)** Rational;   **(b)** Rational;   **(c)** Irrational;
     **(d)** Irrational;   **(e)** Rational;   **(f)** Irrational

**PROBLEM SET 1.2**

**1.** 0.08333...    **3.** 0.142857...    **5.** 3.6666....

**7.** $\frac{41}{333}$    **9.** $\frac{254}{99}$    **11.** $\frac{1}{5}$

**13.** Those that can be expressed by a terminating decimal followed by zeros

**19.** No    **21.** Irrational    **23.** 20.39230485

**25.** 0.000283073883    **27.** 12.43322783

**29.** 78,760.57007    **31.** 0.000691744752

**35.** 132,700,874 ft    **37.** 651,441 board ft

**39.** **(a)** 286.866542;   **(b)** 9.16925;
     **(c)** 16.34874967;   **(d)** 4.292

**41.** **(a)** −2;   **(b)** −2;   **(c)** $\frac{22}{9}$;   **(d)** 1;   **(e)** $\frac{3}{2}$;   **(f)** $\sqrt{2}$

**PROBLEM SET 1.3**

**1.** **(a)**
     **(b)**
     **(c)**
     **(d)**
     **(e)**
     **(f)**

**3.** $(-2, \infty)$;

**5.** $\left[-\frac{5}{2}, \infty\right)$;

**7.** $(2, \infty)$;

**9.** $(-2, 1)$;

**11.** $\left[-\frac{1}{2}, \frac{2}{3}\right)$;

**13.** $\left(\frac{1}{2}, 3\right)$;

**15.** $\left(-1-\sqrt{13}, -1+\sqrt{13}\right)$;

**17.** $\left[-\frac{1}{3}, 4\right]$;

**19.** $(-\infty, -3) \cup \left(\frac{1}{2}, \infty\right)$;

**21.** $[-4, 3)$;

**23.** $(-\infty, 0) \cup \left(\frac{2}{5}, \infty\right)$;

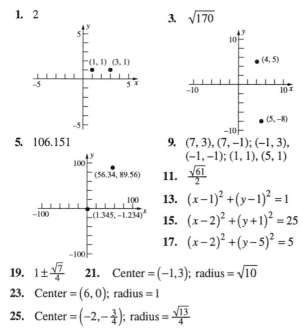

**25.** $\left(-\infty, \frac{2}{3}\right) \cup \left[\frac{3}{4}, \infty\right)$;

**27.** $(-\infty, -10) \cup (-4, \infty)$;

**29.** $(-2, 1) \cup (3, \infty)$;

**31.** $\left(-\infty, \frac{3}{2}\right] \cup [3, \infty)$;

**33.** $(-\infty, -1) \cup (0, 6)$;

**35. (a)** $(-2, 1)$;    **(b)** $(-2, \infty)$;    **(c)** No values

**37. (a)** $[-3, -1] \cup [2, \infty)$;    **(b)** $(-\infty, -2] \cup [2, \infty)$;
**(c)** $(-2, -1) \cup (1, 2)$

**39.** $\frac{60}{11} \le R \le \frac{120}{3}$

## PROBLEM SET 1.4

**1.** $(-3, -1)$    **3.** $\left(-\infty, -\frac{1}{2}\right) \cup \left(\frac{3}{2}, \infty\right)$    **5.** $(-8, 0)$

**7.** $(-\infty, 2) \cup (5, \infty)$    **9.** $(-\infty, -3] \cup [2, \infty)$

**11.** $(-\infty, -5) \cup \left(-\frac{5}{3}, 0\right) \cup (0, \infty)$    **13.** $(-\infty, -1] \cup [4, \infty)$

**15.** $(-\infty, -6) \cup \left(\frac{1}{3}, \infty\right)$    **21.** $\frac{\varepsilon}{3}$    **23.** $\frac{\varepsilon}{6}$

**25.** 0.006 in.    **27.** $\left(-\infty, \frac{7}{3}\right) \cup (5, \infty)$    **29.** $\left(-\frac{4}{5}, \frac{16}{3}\right)$

## PROBLEM SET 1.5

**1.** 2

**3.** $\sqrt{170}$

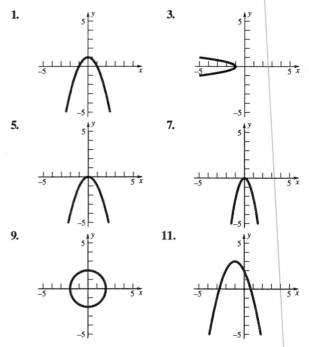

**5.** 106.151

**9.** $(7, 3), (7, -1); (-1, 3),$
$(-1, -1); (1, 1), (5, 1)$

**11.** $\frac{\sqrt{61}}{2}$

**13.** $(x-1)^2 + (y-1)^2 = 1$

**15.** $(x-2)^2 + (y+1)^2 = 25$

**17.** $(x-2)^2 + (y-5)^2 = 5$

**19.** $1 \pm \frac{\sqrt{7}}{4}$    **21.** Center $= (-1, 3)$; radius $= \sqrt{10}$

**23.** Center $= (6, 0)$; radius $= 1$

**25.** Center $= \left(-2, -\frac{3}{4}\right)$; radius $= \frac{\sqrt{13}}{4}$

**27.** Inscribed: $(x-4)^2 + (y-1)^2 = 4$;
circumscribed: $(x-4)^2 + (y-1)^2 = 8$

**29.** Cheaper by plane; \$1,344.75.    **35.** 42.6

**39.** $18 + 2\sqrt{17} + 4\pi \approx 38.8$

## PROBLEM SET 1.6

**1.** 1    **3.** $\frac{9}{7}$    **5.** $-\frac{5}{3}$    **7.** $-1.03$

**9.** $y = -x + 4$; $x + y - 4 = 0$    **11.** $y = 2x + 3$; $-2x + y - 3 = 0$

**13.** $y = \frac{5}{2}x - 2$; $-\frac{5}{2}x + y + 2 = 0$    **15.** $x = 2$; $x + 0y - 2 = 0$

**17.** Slope $= -\frac{2}{3}$; $y$-intercept $= \frac{1}{3}$

**19.** Slope $= -5$; $y$-intercept $= 4$

**21. (a)** $y = 2x - 9$;    **(b)** $y = -\frac{1}{2}x - \frac{3}{2}$;    **(c)** $y = -\frac{2}{3}x - 1$;
**(d)** $y = \frac{3}{2}x - \frac{15}{2}$;    **(e)** $y = -\frac{3}{4}x - \frac{3}{4}$;    **(f)** $x = 3$;
**(g)** $y = -3$

**23.** $y = \frac{3}{2}x + 2$    **25.** It lies above the line.

**27.** $(-1, 2)$; $y = \frac{3}{2}x + \frac{7}{2}$    **29.** $(3, 1)$; $y = -\frac{4}{3}x + 5$

**31.** $\frac{7}{5}$    **33.** $\frac{18}{13}$    **35.** $\frac{1}{\sqrt{5}}$    **37.** $V = 120,000 - 9600t$

**39.** $N = 700,000 + 12,000n$; $1,180,000$

**41. (a)** $P$ is negative, which indicates that money has been lost.
**(b)** 450; it is the amount of money added to the profit per
item sold.

**47.** $(3, 3)$    **49.** $x + \sqrt{3}y = 12$ and $x - \sqrt{3}y = 12$    **53.** 8

## PROBLEM SET 1.7

**1.**

**3.**

**5.**

**7.**

**9.**

**11.**

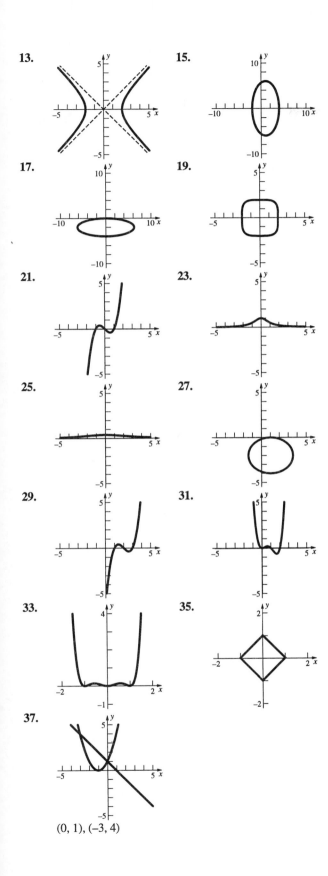

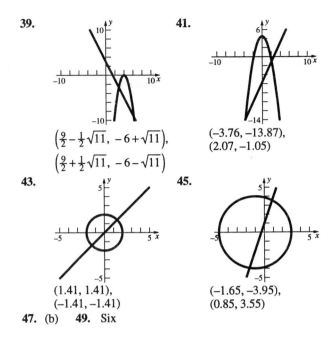

**13.**

**15.**

**17.**

**19.**

**21.**

**23.**

**25.**

**27.**

**29.**

**31.**

**33.**

**35.**

**37.**

$(0, 1), (-3, 4)$

**39.**

**41.**

$\left(\frac{9}{2} - \frac{1}{2}\sqrt{11}, -6 + \sqrt{11}\right),$

$\left(\frac{9}{2} + \frac{1}{2}\sqrt{11}, -6 - \sqrt{11}\right)$

$(-3.76, -13.87),$
$(2.07, -1.05)$

**43.**

**45.**

$(1.41, 1.41),$
$(-1.41, -1.41)$

$(-1.65, -3.95),$
$(0.85, 3.55)$

**47.** (b)    **49.** Six

## CHAPTER REVIEW 1.8

### Concepts Test

**1.** True    **3.** False    **5.** False    **7.** False    **9.** True
**11.** True    **13.** True    **15.** True    **17.** True    **19.** True
**21.** True    **23.** True    **25.** True    **27.** True    **29.** True
**31.** True    **33.** True    **35.** False    **37.** False

### Sample Test Problems

**1.** (a)  $2, \frac{25}{4}, \frac{4}{25}$;    (b)  $1, 9, 49$;    (c)  $64, 8, \frac{1}{8}$;
   (d)  $1, \frac{\sqrt{2}}{2}, \sqrt{2}$

**7.** 2.66

**9.** $\{x : x < \frac{1}{3}\}$; $\left(-\infty, \frac{1}{3}\right)$;

**11.** $\{x : \frac{1}{3} \le x \le 3\}$; $\left[\frac{1}{3}, 3\right]$;

**13.** $\{t : \frac{3}{7} \le t \le \frac{5}{3}\}$; $\left[\frac{3}{7}, \frac{5}{3}\right]$;

**15.** $\{x : -4 \le x \le 3\}$; $[-4, 3]$;

**17.** $\{x : x \le -\frac{1}{2} \text{ or } x > 1\}$; $\left(-\infty, -\frac{1}{2}\right] \cup (1, \infty)$;

**19.** $\{x : x \le \frac{8}{5} \text{ or } x \ge 8\}$; $\left(-\infty, \frac{8}{5}\right] \cup [8, \infty)$ ;

**21.** Any negative number    **23.** $t \le 5$

**27.**

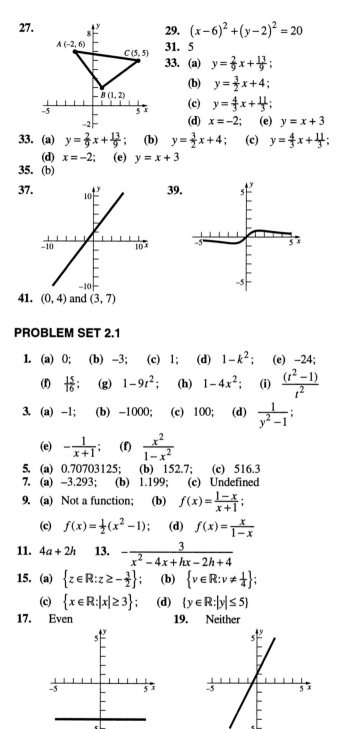

**29.** $(x-6)^2 + (y-2)^2 = 20$

**31.** 5

**33.** (a) $y = \frac{2}{9}x + \frac{13}{9}$;

(b) $y = \frac{3}{2}x + 4$;

(c) $y = \frac{4}{3}x + \frac{11}{3}$;

(d) $x = -2$;    (e) $y = x + 3$

**33.** (a) $y = \frac{2}{9}x + \frac{13}{9}$;    (b) $y = \frac{3}{2}x + 4$;    (c) $y = \frac{4}{3}x + \frac{11}{3}$;

(d) $x = -2$;    (e) $y = x + 3$

**35.** (b)

**37.**

**39.**

**41.** $(0, 4)$ and $(3, 7)$

## PROBLEM SET 2.1

**1.** (a) 0;    (b) –3;    (c) 1;    (d) $1 - k^2$;    (e) –24;

(f) $\frac{15}{16}$;    (g) $1 - 9t^2$;    (h) $1 - 4x^2$;    (i) $\frac{(t^2 - 1)}{t^2}$

**3.** (a) –1;    (b) –1000;    (c) 100;    (d) $\frac{1}{y^2 - 1}$;

(e) $-\frac{1}{x+1}$;    (f) $\frac{x^2}{1 - x^2}$

**5.** (a) 0.70703125;    (b) 152.7;    (c) 516.3

**7.** (a) –3.293;    (b) 1.199;    (c) Undefined

**9.** (a) Not a function;    (b) $f(x) = \frac{1-x}{x+1}$;

(c) $f(x) = \frac{1}{2}(x^2 - 1)$;    (d) $f(x) = \frac{x}{1-x}$

**11.** $4a + 2h$    **13.** $-\frac{3}{x^2 - 4x + hx - 2h + 4}$

**15.** (a) $\left\{z \in \mathbb{R} : z \geq -\frac{3}{2}\right\}$;    (b) $\left\{v \in \mathbb{R} : v \neq \frac{1}{4}\right\}$;

(c) $\left\{x \in \mathbb{R} : |x| \geq 3\right\}$;    (d) $\{y \in \mathbb{R} : |y| \leq 5\}$

**17.** Even    **19.** Neither

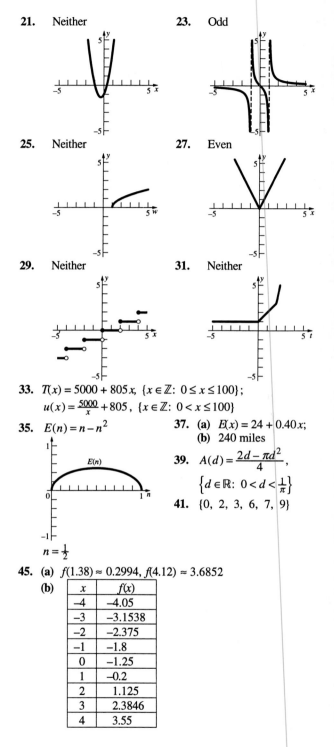

**21.** Neither    **23.** Odd

**25.** Neither    **27.** Even

**29.** Neither    **31.** Neither

**33.** $T(x) = 5000 + 805x$, $\{x \in \mathbb{Z} : 0 \leq x \leq 100\}$;

$u(x) = \frac{5000}{x} + 805$, $\{x \in \mathbb{Z} : 0 < x \leq 100\}$

**35.** $E(n) = n - n^2$

$n = \frac{1}{2}$

**37.** (a) $E(x) = 24 + 0.40x$;

(b) 240 miles

**39.** $A(d) = \frac{2d - \pi d^2}{4}$,

$\left\{d \in \mathbb{R} : 0 < d < \frac{1}{\pi}\right\}$

**41.** $\{0, 2, 3, 6, 7, 9\}$

**45.** (a) $f(1.38) \approx 0.2994$, $f(4.12) \approx 3.6852$

(b)

| $x$ | $f(x)$ |
|----|--------|
| –4 | –4.05 |
| –3 | –3.1538 |
| –2 | –2.375 |
| –1 | –1.8 |
| 0 | –1.25 |
| 1 | –0.2 |
| 2 | 1.125 |
| 3 | 2.3846 |
| 4 | 3.55 |

**47.**

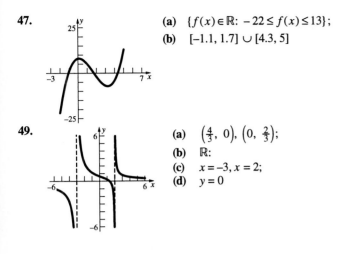

(a)  $\{f(x) \in \mathbb{R}: -22 \leq f(x) \leq 13\}$;

(b)  $[-1.1, 1.7] \cup [4.3, 5]$

**49.**

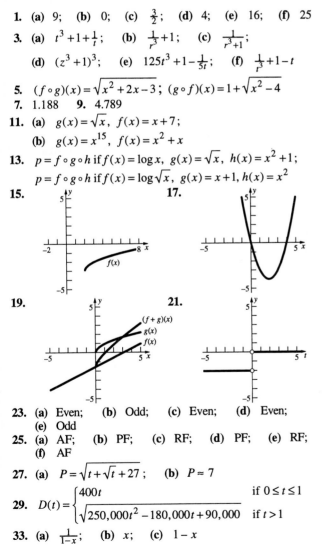

(a)  $\left(\frac{4}{3}, 0\right), \left(0, \frac{2}{3}\right)$;

(b)  $\mathbb{R}$:

(c)  $x = -3, x = 2$;

(d)  $y = 0$

## PROBLEM SET 2.2

**1.**  (a)  9;    (b)  0;    (c)  $\frac{3}{2}$;    (d)  4;    (e)  16;    (f)  25

**3.**  (a)  $t^3 + 1 + \frac{1}{t}$;    (b)  $\frac{1}{r^3} + 1$;    (c)  $\frac{1}{r^3 + 1}$;

(d)  $(z^3 + 1)^3$;    (e)  $125t^3 + 1 - \frac{1}{5t}$;    (f)  $\frac{1}{t^3} + 1 - t$

**5.**  $(f \circ g)(x) = \sqrt{x^2 + 2x - 3}$;  $(g \circ f)(x) = 1 + \sqrt{x^2 - 4}$

**7.**  1.188    **9.**  4.789

**11.**  (a)  $g(x) = \sqrt{x}$,  $f(x) = x + 7$;

(b)  $g(x) = x^{15}$,  $f(x) = x^2 + x$

**13.**  $p = f \circ g \circ h$ if $f(x) = \log x$, $g(x) = \sqrt{x}$, $h(x) = x^2 + 1$;

$p = f \circ g \circ h$ if $f(x) = \log \sqrt{x}$, $g(x) = x + 1$, $h(x) = x^2$

**15.**    **17.**

**19.**    **21.**

**23.**  (a)  Even;    (b)  Odd;    (c)  Even;    (d)  Even;

(e)  Odd

**25.**  (a)  AF;    (b)  PF;    (c)  RF;    (d)  PF;    (e)  RF;

(f)  AF

**27.**  (a)  $P = \sqrt{t + \sqrt{t + 27}}$;    (b)  $P \approx 7$

**29.**  $D(t) = \begin{cases} 400t & \text{if } 0 \leq t \leq 1 \\ \sqrt{250,000t^2 - 180,000t + 90,000} & \text{if } t > 1 \end{cases}$

**33.**  (a)  $\frac{1}{1-x}$;    (b)  $x$;    (c)  $1 - x$

**35.**

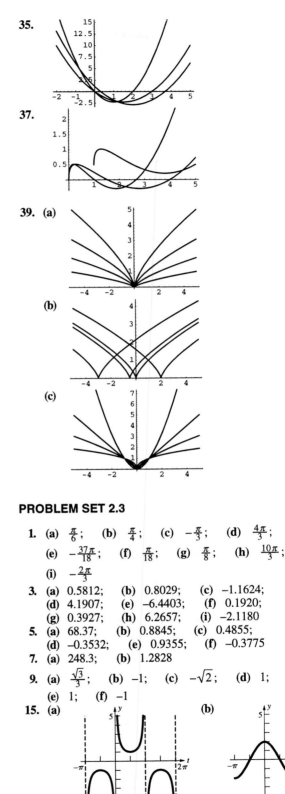

**37.**

**39.**  (a)

(b)

(c)

## PROBLEM SET 2.3

**1.**  (a)  $\frac{\pi}{6}$;    (b)  $\frac{\pi}{4}$;    (c)  $-\frac{\pi}{3}$;    (d)  $\frac{4\pi}{3}$;

(e)  $-\frac{37\pi}{18}$;    (f)  $\frac{\pi}{18}$;    (g)  $\frac{\pi}{8}$;    (h)  $\frac{10\pi}{3}$;

(i)  $-\frac{2\pi}{3}$

**3.**  (a)  0.5812;    (b)  0.8029;    (c)  −1.1624;

(d)  4.1907;    (e)  −6.4403;    (f)  0.1920;

(g)  0.3927;    (h)  6.2657;    (i)  −2.1180

**5.**  (a)  68.37;    (b)  0.8845;    (c)  0.4855;

(d)  −0.3532;    (e)  0.9355;    (f)  −0.3775

**7.**  (a)  248.3;    (b)  1.2828

**9.**  (a)  $\frac{\sqrt{3}}{3}$;    (b)  −1;    (c)  $-\sqrt{2}$;    (d)  1;

(e)  1;    (f)  −1

**15.**  (a)    (b)

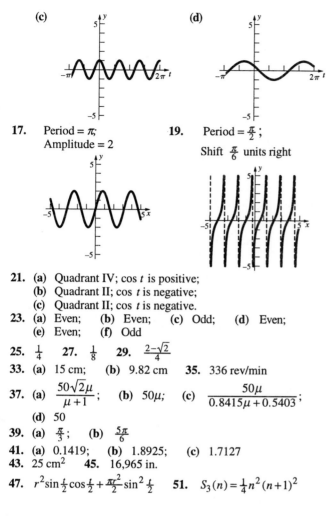

**(c)**

**(d)**

**17.** Period $= \pi$; Amplitude $= 2$

**19.** Period $= \frac{\pi}{2}$; Shift $\frac{\pi}{6}$ units right

**21.** (a) Quadrant IV; $\cos t$ is positive;
(b) Quadrant II; $\cos t$ is negative;
(c) Quadrant II; $\cos t$ is negative.

**23.** (a) Even;  (b) Even;  (c) Odd;  (d) Even;
(e) Even;  (f) Odd

**25.** $\frac{1}{4}$   **27.** $\frac{1}{8}$   **29.** $\frac{2-\sqrt{2}}{4}$

**33.** (a) 15 cm;  (b) 9.82 cm   **35.** 336 rev/min

**37.** (a) $\frac{50\sqrt{2}\mu}{\mu+1}$;  (b) $50\mu$;  (c) $\frac{50\mu}{0.8415\mu + 0.5403}$;
(d) 50

**39.** (a) $\frac{\pi}{3}$;  (b) $\frac{5\pi}{6}$

**41.** (a) 0.1419;  (b) 1.8925;  (c) 1.7127

**43.** 25 cm$^2$   **45.** 16,965 in.

**47.** $r^2 \sin \frac{t}{2} \cos \frac{t}{2} + \frac{\pi r^2}{2} \sin^2 \frac{t}{2}$   **51.** $S_3(n) = \frac{1}{4}n^2(n+1)^2$

## PROBLEM SET 2.4

**1.** $-2$   **3.** $-1$   **5.** $\frac{\sqrt{2}}{2}$   **7.** 4   **9.** 12   **11.** $-2t$

**13.** $\frac{\sqrt{6}}{9}$   **15.** 36   **17.** 0.5   **19.** 0   **21.** 2   **23.** 0

**25.** 0.25

**27.** (a) 2;  (b) 1;  (c) Does not exist;  (d) $\frac{5}{2}$;
(e) 2;  (f) Does not exist;  (g) 2;  (h) 1

**29.**
(a) 0;
(b) Does not exist;
(c) 2;
(d) 2

**31.**
(a) 0;  (b) Does not exist;  (c) 1;  (d) $\frac{1}{2}$

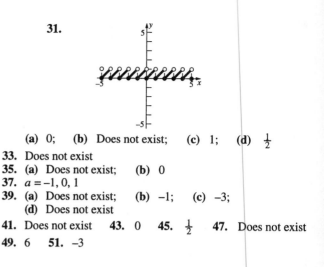

**33.** Does not exist

**35.** (a) Does not exist;  (b) 0

**37.** $a = -1, 0, 1$

**39.** (a) Does not exist;  (b) $-1$;  (c) $-3$;
(d) Does not exist

**41.** Does not exist   **43.** 0   **45.** $\frac{1}{2}$   **47.** Does not exist

**49.** 6   **51.** $-3$

## PROBLEM SET 2.5

**1.** $0 < |t-a| < \delta \Rightarrow |f(t) - M| < \varepsilon$

**3.** $0 < |z-d| < \delta \Rightarrow |h(z) - P| < \varepsilon$

**5.** $0 < c - x < \delta \Rightarrow |f(x) - L| < \varepsilon$

**27.** (b), (c)

**29.** (a) $\dfrac{x^3 - x^2 - 2x - 4}{x^4 - 4x^3 + x^2 + x + 6}$;  (b) No;  (c) 3

## PROBLEM SET 2.6

**1.** 3   **3.** $-3$   **5.** $-5$   **7.** 2   **9.** $-1$   **11.** 2

**13.** $-\frac{2}{3}$   **15.** Does not exist   **17.** $\frac{3}{2}$   **19.** $\frac{x+2}{5}$

**21.** $-1$   **23.** $\sqrt{10}$   **25.** $-6$   **27.** 6   **29.** 12

**31.** $-\frac{1}{4}$   **41.** 0   **43.** 0   **45.** $\frac{2}{5}$   **47.** $-1$

**51.** (a) 1;  (b) 0

## PROBLEM SET 2.7

**1.** Continuous

**3.** Not continuous; $\lim\limits_{x \to 3} h(x)$ and $h(3)$ do not exist.

**5.** Not continuous; $\lim\limits_{t \to 3} h(t)$ and $h(3)$ do not exist.

**7.** Continuous   **9.** Not continuous; $h(3)$ does not exist.

**11.** Continuous   **13.** Continuous   **15.** Continuous

**17.** $(-\infty, -5) \cup (-5, -2) \cup (-2, 4) \cup (4, 6) \cup (6, 8) \cup (8, \infty)$

**19.** Define $f(3) = -12$.   **21.** Define $H(1) = \frac{1}{2}$.

**23.** Define $F(-1) = -\sin 2$.   **25.** 3, $\pi$

**27.** $n\pi + \frac{\pi}{2}$ where $n$ is an integer.   **29.** $-1$

**31.** $(-\infty, -2) \cup (2, \infty)$   **33.** 1

**35.** Every $n + \frac{1}{2}$ where $n$ is an integer.

**37.**

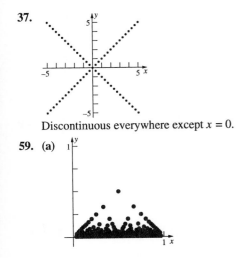

Discontinuous everywhere except $x = 0$.

**59. (a)**

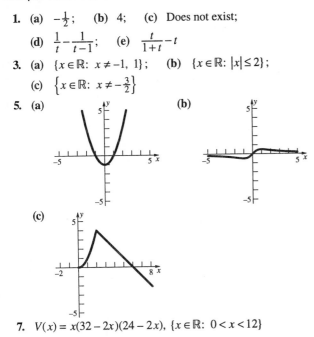

## CHAPTER REVIEW 2.8

### Concepts Test

| | | | | |
|---|---|---|---|---|
| **1.** True | **3.** True | **5.** False | **7.** True | **9.** True |
| **11.** False | **13.** True | **15.** True | **17.** False | |
| **19.** True | **21.** False | **23.** True | **25.** False | |
| **27.** True | **29.** False | **31.** False | **33.** True | |
| **35.** True | **37.** True | **39.** True | **41.** False | **43.** True |

### Sample Test Problems

**1. (a)** $-\frac{1}{2}$;  **(b)** 4;  **(c)** Does not exist;

**(d)** $\frac{1}{t} - \frac{1}{t-1}$;  **(e)** $\frac{t}{1+t} - t$

**3. (a)** $\{x \in \mathbb{R}: x \neq -1, 1\}$;  **(b)** $\{x \in \mathbb{R}: |x| \leq 2\}$;

**(c)** $\left\{x \in \mathbb{R}: x \neq -\frac{3}{2}\right\}$

**5. (a)**                              **(b)**

**(c)**

**7.** $V(x) = x(32 - 2x)(24 - 2x)$, $\{x \in \mathbb{R}: 0 < x < 12\}$

**9. (a)**                    **(b)**

**(c)**

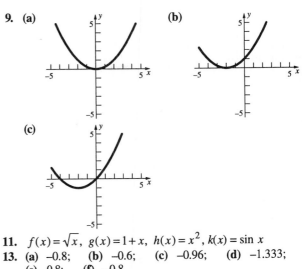

**11.** $f(x) = \sqrt{x}$, $g(x) = 1 + x$, $h(x) = x^2$, $k(x) = \sin x$

**13. (a)** −0.8;  **(b)** −0.6;  **(c)** −0.96;  **(d)** −1.333;
**(e)** 0.8;  **(f)** −0.8

**15.** 18.85 in.  **17.** 2  **19.** $\frac{1}{8}$  **21.** $\frac{1}{2}$  **23.** 4

**25.** −1  **27.** −1  **29. (a)** $x = -1, 1$;  **(b)** $f(-1) = -1$

**31. (a)** 14;  **(b)** −12;  **(c)** −2;  **(d)** −2;  **(e)** 5;
**(f)** 0

**33.** $a = 2, b = -1$

## PROBLEM SET 3.1

**1.** 4

**3.**                              **5.**

$\frac{5}{2}$

**7. (a), (b)**                    **(c)** 2;
**(d)** 2.01;
**(e)** 2

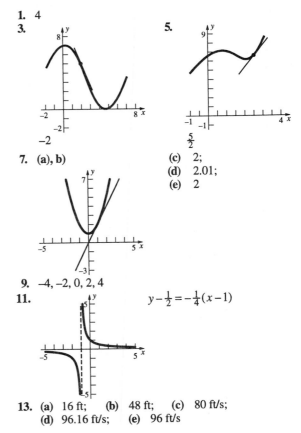

**9.** −4, −2, 0, 2, 4

**11.**                    $y - \frac{1}{2} = -\frac{1}{4}(x - 1)$

**13. (a)** 16 ft;  **(b)** 48 ft;  **(c)** 80 ft/s;
**(d)** 96.16 ft/s;  **(e)** 96 ft/s

**15.** (a) $\dfrac{1}{\sqrt{2\alpha+1}}$ ft/s;    (b)  1.5 sec

**17.** (a)  0.02005 g;    (b)  2.005 g/h;    (c)  2 g/h

**19.** (a)  49 g/cm;    (b)  27 g/cm

**21.** 4    **23.** 29,167 gal/h, 75,000 gal/h    **25.** $24\pi\,\text{km}^2/\text{day}$

**27.**

(a)  7;
(b)  0;
(c)  –1;
(d)  17.92

**29.** 2.818

## PROBLEM SET 3.2

**1.** 2    **3.** 5    **5.** 2    **7.** $6x$    **9.** $2ax+b$

**11.** $3x^2+4x$    **13.** $-\dfrac{2}{x^2}$    **15.** $-\dfrac{12}{(x^2+1)^2}$

**17.** $-\dfrac{7}{(x-4)^2}$    **19.** $\dfrac{3}{2\sqrt{3x}}$    **21.** $-\dfrac{3}{2(x-2)^{3/2}}$

**23.** $2x-3$    **25.** $-\dfrac{5}{(x-5)^2}$    **27.** $f(x)=2x^3$ at $x=5$

**29.** $f(x)=x^2$ at $x=2$    **31.** $f(x)=x^2$ at $x$

**33.** $f(t)=\dfrac{2}{t}$ at $t$    **35.** $f(x)=\cos x$ at $x$    **37.** $-\dfrac{1}{2},\,1,\,\dfrac{2}{3},\,-3$

**39.**

**41.** (a)  $\dfrac{5}{2},\dfrac{3}{2},1.8,-0.6$;
(b)  3.6;
(c)  5;
(d)  3, 5;
(e)  1, 3, 5;
(f)  0;
(g)  –0.7, 1.5, $5<x<7$

**43.** $m=4$, $b=-4$    **45.** (a)  $m$;    (b)  $-m$

**47.**

(a)  $\left(0,\dfrac{8}{3}\right)$;    (b)  $\left(0,\dfrac{8}{3}\right)$;
(c)  $f(x)$ decreases as $x$ increases when $f'(x)<0$.

## PROBLEM SET 3.3

**1.** $4x$    **3.** $\pi$    **5.** $-4x^{-3}$    **7.** $-\dfrac{\pi}{x^2}$    **9.** $-\dfrac{500}{x^6}$

**11.** $2x+2$    **13.** $4x^3+3x^2+2x+1$

**15.** $7\pi x^6-10x^4+10x^{-3}$    **17.** $-\dfrac{9}{x^4}-4x^{-5}$

**19.** $-\dfrac{2}{x^2}+\dfrac{2}{x^3}$    **21.** $-\dfrac{1}{2x^2}+2$    **23.** $3x^2+1$

**25.** $8x+4$    **27.** $5x^4+6x^2+2x$

**29.** $5x^4+42x^2+2x-51$    **31.** $60x^3-30x^2-32x+14$

**33.** $-\dfrac{6x}{(3x^2+1)^2}$    **35.** $\dfrac{-8x+3}{(4x^2-3x+9)^2}$    **37.** $\dfrac{2}{(x+1)^2}$

**39.** $\dfrac{6x^2+20x+3}{(3x+5)^2}$    **41.** $\dfrac{4x^2+4x-5}{(2x+1)^2}$    **43.** $\dfrac{x^2-1}{(x^2+1)^2}$

**45.** (a)  23;    (b)  4;    (c)  $-\dfrac{17}{9}$

**49.** $y=1$    **51.** $(0,0)$ and $\left(\dfrac{2}{3},-\dfrac{4}{27}\right)$

**53.** (a)  –24 ft/s;    (b)  1.25 s    **55.** $y=2x+1$, $y=-2x+9$

**57.** $3\sqrt{5}$    **59.** $681\ \text{cm}^3$ per week

## PROBLEM SET 3.4

**1.** $2\cos x-3\sin x$    **3.** 0    **5.** $\sec x\tan x$    **7.** $\sec^2 x$

**9.** $\sec^2 x$    **11.** $-x^2\sin x+2x\cos x$

**13.** $y-0.5403=-0.8415(x-1)$    **15.** $-60$ ft/sec    **17.** $\dfrac{1}{2}$

**19.** 3    **21.** $\dfrac{1}{2\pi}$    **27.** $x_0=\dfrac{\pi}{3}$, $\tan^{-1}3$

**29.** (b)  $\dfrac{\sin t(1-\cos t)}{t-\sin t\cos t}$    (c)  0.75

**33.** (a)

(b)  6; 5;    (c)  $f(x)=x\sin x$ is a counterexample;
(d)  24.93

## PROBLEM SET 3.5

**1.** $15(1+x)^{14}$    **3.** $-10(3-2x)^4$

**5.** $11(x^3-2x^2+3x+1)^{10}(3x^2-4x+3)$

**7.** $111(x^3-2x^2+3x+1)^{110}(3x^2-4x+3)$

**9.** $-\dfrac{5}{(x+3)^6}$    **11.** $(2x+1)\cos(x^2+x)$

**13.** $-3\sin x\cos^2 x$    **15.** $-\dfrac{6(x+1)^2}{(x-1)^4}$

**17.** $\dfrac{3x^2(3x+1)^2(2x^2+6x-6\pi x-\pi+1)}{(1-2x^2-\pi)^4}$

**19.** $-\dfrac{3(2x-x^2)}{(1-x)^2}\cos^2\left(\dfrac{x^2}{1-x}\right)\sin\left(\dfrac{x^2}{1-x}\right)$

**21.** $2(3x-2)(3-x^2)(9+4x-9x^2)$    **23.** $\dfrac{(x+1)(3x-11)}{(3x-4)^2}$

**25.** $\dfrac{4x(3x^2+2)(3x^2-17)}{(2x^5-5)^2}$    **27.** $\dfrac{51(3t-2)^2}{(t+5)^4}$

**29.** $\dfrac{(6t+47)(3t-2)^2}{(t+5)^2}$    **31.** $3\sin^2\theta\cos\theta$

**33.** $\dfrac{3\sin^2 x(\cos x\cos 2x+2\sin x\sin 2x)}{\cos^4 2x}$    **35.** $\dfrac{48}{5}$

**37.** 1.4183    **39.** $4(2x+3)\sin^3(x^2+3x)\cos(x^2+3x)$

**41.** $-3\sin t\sin^2(\cos t)\cos(\cos t)$

**43.** $-8\theta\cos^3(\sin\theta^2)\sin(\sin\theta^2)(\cos\theta^2)$

**45.** $-2\cos[\cos(\sin 2x)]\sin(\sin 2x)(\cos 2x)$

**47.** $y-32=224(x-1)$

**49.** (a) $(10\cos 8\pi t,\ 10\sin 8\pi t)$; (b) $80\pi$ cm/s

**51.** (a) $(\cos 2\pi t,\ \sin 2\pi t)$; (b) $\sin 2\pi t+\sqrt{25-\cos^2 2\pi t}$;

 (c) $2\pi\cos 2\pi t\left(1+\dfrac{\sin 2\pi t}{\sqrt{25-\cos^2 2\pi t}}\right)$

**53.** (a) $\dfrac{2x\left|x^2-1\right|}{x^2-1}$; (b) $\cot x\left|\sin x\right|$

**55.** (a) $2\sin^3 2x$; (b) $\sin^4 2x$ **57.** 16

**59.** (a)

Odd

 (b)

Even

 (c) 0.678; (d) 1

## PROBLEM SET 3.6

**1.** 1.5 **3.** 0.0081 **5.** $2x$ **7.** $-\dfrac{1}{(x+1)^2}$ **9.** $\sin 2x$

**11.** $2x\sec^2(x^2)$ **13.** $\dfrac{4(x^2+1)^3(2x\cos x+\sin x+x^2\sin x)}{\cos^5 x}$

**15.** $\sin 2x\cos(x^2)-2x\sin^2 x\sin(x^2)$

**17.** $8x\sin^3(x^2+3)\cos(x^2+3)$

**19.** $\dfrac{16x}{(x^2-2)^2}\cos\left(\dfrac{x^2+2}{x^2-2}\right)\sin\left(\dfrac{x^2+2}{x^2-2}\right)$

**21.** $3(\sin^2 t\cos t-\sin t\cos^2 t)$ **23.** $-9\pi r^2-22\pi r-6\pi$

**25.** $\frac{375}{8}$ **27.** (a) $-5$; (b) $-11$; (c) $\frac{5}{9}$; (d) 4

**29.** (a) 2; (b) $-1$

**31.** (a) 19,200 cm$^3$/min; (b) 2880 cm$^2$/min

**33.** $\dfrac{\sqrt{2\pi}}{4}$ **37.** 0.38 in./min

## PROBLEM SET 3.7

**1.** 6 **3.** 162 **5.** $-343\cos(7x)$ **7.** $-\dfrac{6}{(x-1)^4}$ **9.** 2

**11.** $\frac{1}{2}$ **13.** $2\pi^2$ **15.** $-900$

**19.** (a) 0; (b) 0; (c) 0

**21.** $f''(-5)=-24$; $f''(3)=24$

**23.** (a) $v(t)=12-4t$; $a(t)=-4$ (b) $t<3$; (c) $t>3$;
 (d) All $t$; (e)

**25.** (a) $v(t)=3t^2-18t+24$; $a(t)=6t-18$;
 (b) $t<2$ or $t>4$; (c) $2<t<4$; (d) $t<3$
 (e)

**27.** (a) $v(t)=2t-\dfrac{16}{t^2}$; $a(t)=2+\dfrac{32}{t^3}$; (b) $t>2$;
 (c) $0<t<2$; (d) No $t$;
 (e)

**29.** $v(1)=11$; $v(4)=-16$

**31.** (a) $\frac{3}{4}$ s; (b) $\frac{1}{2}$ s, $\frac{3}{4}$ s; (c) 0 s, $\frac{3}{2}$ s

**33.** (a) 48 ft/s; (b) $\frac{3}{2}$ s; (c) 292 ft; (d) 5.77 s;
 (e) 137 ft/s

**35.** 581 ft/s **37.** $t<-2,\ 1<t<4$

**39.** Let $s=f(t)$ denote distance.

 (a) $\dfrac{ds}{dt}=ks$; (b) $\dfrac{d^2s}{dt^2}>0$;

 (c) $\dfrac{d^3s}{dt^3}<0,\ \dfrac{d^2s}{dt^2}>0$; (d) $\dfrac{d^2s}{dt^2}=10$ mph/min;

 (e) $\dfrac{d^2s}{dt^2}<0,\ \left|\dfrac{d^2s}{dt^2}\right|$ very small; (f) $s=kt$

**41.** (a) Let $c=f(t)$ denote the cost at time $t$.

 $\dfrac{dc}{dt}>0,\ \dfrac{d^2c}{dt^2}>0$;

 (b) Let $u=f(t)$ denote oil consumption at time $t$.

 $\dfrac{du}{dt}<0,\ \dfrac{d^2u}{dt^2}>0$;

 (c) Let $P=f(t)$ denote world population at time $t$.

 $\dfrac{dP}{dt}>0,\ \dfrac{d^2P}{dt^2}<0$;

 (d) Let $s=f(t)$ denote the distance at time $t$.

 $\dfrac{d^2s}{dt^2}=k>0$;

 (e) Let $\theta=f(t)$ denote the angle at time $t$.

 $\dfrac{d\theta}{dt}>0,\ \dfrac{d^2\theta}{dt^2}>0$;

 (f) Let $P=f(t)$ denote profit at time $t$.

 $\dfrac{dP}{dt}>0,\ \dfrac{d^2P}{dt^2}<0$;

 (g) Let $u=f(t)$ denote the amount of money at time $t$.

 $\dfrac{du}{dt}<0,\ \dfrac{d^2u}{dt^2}>0$

**43.** $D_x^n(uv) = \sum_{k=0}^{n} \binom{n}{k} D_x^{n-k}(u) D_x^k(v)$ where $\binom{n}{k}$ is the binomial coefficient $\dfrac{n!}{(n-k)!k!}$.

**45. (a)** 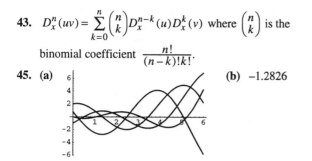 **(b)** $-1.2826$

## PROBLEM SET 3.8

**1.** $\dfrac{x}{y}$ **3.** $-\dfrac{y}{x}$ **5.** $\dfrac{1-y^2}{2xy}$ **7.** $\dfrac{12x^2+7y^2}{6y^2-14xy}$

**9.** $\dfrac{y^3-\dfrac{5y}{2\sqrt{5xy}}}{\dfrac{5x}{2\sqrt{5xy}}+2-2y-3xy^2}$ **11.** $-\dfrac{y}{x}$

**13.** $y-3=-\frac{9}{7}(x-1)$ **15.** $y=1$

**17.** $y+1=\frac{1}{2}(x-1)$ **19.** $5x^{2/3}+\dfrac{1}{2\sqrt{x}}$

**21.** $\dfrac{1}{3\sqrt[3]{x^2}}-\dfrac{1}{3\sqrt[3]{x^4}}$ **23.** $\dfrac{3x-2}{2\sqrt[4]{(3x^2-4x)^3}}$

**25.** $-\dfrac{6x^2+4}{3\sqrt[3]{(x^3+2x)^5}}$ **27.** $\dfrac{2x+\cos x}{2\sqrt{x^2+\sin x}}$

**29.** $-\dfrac{x^2\cos x+2x\sin x}{3\sqrt[3]{(x^2\sin x)^4}}$ **31.** $-\dfrac{(x+1)\sin(x^2+2x)}{2\sqrt[4]{[1+\cos(x^2+2x)]^3}}$

**33.** $\dfrac{ds}{dt}=-\dfrac{s^2+3t^2}{2st}$; $\dfrac{dt}{ds}=-\dfrac{2st}{s^2+3t^2}$

**35.** $\sqrt{3}y+x=0$, $\sqrt{3}y-x=0$

**37. (a)** $y'=-\dfrac{y}{x+3y^2}$; **(b)** $y''=\dfrac{2xy}{(x+3y^2)^3}$ **39.** $-15$;

**45.** $\theta\approx 2.0344$ **47.** $y=2(x+4);\ y=2(x-4)$ **49.** $\frac{13}{3}$

## PROBLEM SET 3.9

**1.** $1296$ in.$^3$/s **3.** $392$ mi/h **5.** $471$ mi/h
**7.** $0.258$ ft/s **9.** $0.0796$ ft/s **11.** $\frac{1}{12}$ ft/s
**13.** $1.018$ in.$^2$/s **15.** $15.71$ km/min
**17. (a)** $\frac{1}{2}$ ft/s; **(b)** $\frac{5}{2}$ ft/s **(c)** $\frac{1}{24}$ rad/s
**19.** $110$ ft/s **21.** $-0.016$ ft/h **23.** $134$ in.$^3$/min
**25. (a)** $-1.125$ ft/s; **(b)** $-0.08$ ft/s$^2$
**27. (a)** Proof omitted; **(b)** $3$ hours
**29.** $\frac{16}{3}$ ft/s when the girl is at least $30$ ft from the light pole and $\frac{80}{17}$ ft/s when she is less than $30$ ft from the pole.

## PROBLEM SET 3.10

**1.** $dy=(2x+1)dx$ **3.** $dy=-8(2x+3)^{-5}dx$
**5.** $dy=3(\sin x+\cos x)^2(\cos x-\sin x)dx$
**7.** $dy=-\frac{3}{2}(7x^2+3x-1)^{-5/2}(14x+3)dx$
**9.** $dF=\dfrac{2}{5\sqrt[5]{(t-2)^3}}dt$
**11. (a)** $0.75$; **(b)** $2.25$

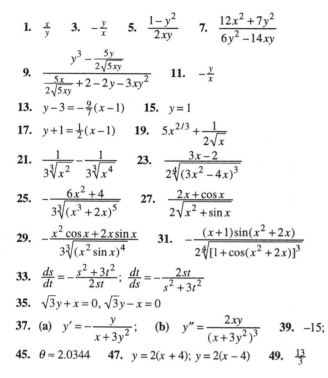

**13. (a)** $3.25$; **(b)** $0.984375$
**15. (a)** $\Delta y=2.25,\ dy=2$;
  **(b)** $\Delta y=-0.7056,\ dy=-0.72$
**17.** $20.05$ **19.** $5.9917$ **21.** $39.27$ cm$^3$ **23.** $893$ ft$^3$
**25.** $12.6$ ft **27.** $4189\pm 63$ cm$^3$ **29.** $79.097\pm 0.729$ cm
**31.** $dy=0.01;\ |\Delta y-dy|\le 0.000003$ **33.** $754$ cm$^3$
**35.** $9.5\%$

## CHAPTER REVIEW 3.11

### Concepts Test

**1.** False **3.** True **5.** True **7.** True **9.** True
**11.** True **13.** False **15.** True **17.** False
**19.** True **21.** True **23.** True **25.** True
**27.** False **29.** True **31.** True **33.** True **35.** True

### Sample Test Problems

**1. (a)** $9x^2$; **(b)** $10x^4+3$; **(c)** $-\dfrac{1}{3x^2}$;
  **(d)** $-\dfrac{6x}{(3x^2+2)^2}$; **(e)** $\dfrac{3}{2\sqrt{3x}}$; **(f)** $3\cos 3x$;
  **(g)** $\dfrac{x}{\sqrt{x^2+5}}$; **(h)** $-\pi\sin\pi x$

**3. (a)** $f(x)=3x$ at $x=1$; **(b)** $f(x)=4x^3$ at $x=2$;
  **(c)** $f(x)=\sqrt{x^3}$ at $x=1$; **(d)** $f(x)=\sin x$ at $x=\pi$;
  **(e)** $f(x)=\frac{4}{x}$ at $x$; **(f)** $f(x)=-\sin 3x$ at $x$;
  **(g)** $f(x)=\tan x$ at $x=\frac{\pi}{4}$; **(h)** $f(x)=\dfrac{1}{\sqrt{x}}$ at $x=5$

**5.** $15x^4$ **7.** $3z^2+8z+2$ **9.** $\dfrac{-24t^2+60t+10}{(6t^2+2t)^2}$

**11.** $\dfrac{-4x^4+10x^2+2}{(x^3+x)^2}$ **13.** $-\dfrac{x}{\sqrt{(x^2+4)^3}}$

**15.** $-\sin\theta+6\sin^2\theta\cos\theta-3\cos^3\theta$ **17.** $2\theta\cos(\theta^2)$

**19.** $2\pi \sin(\sin(\pi\theta))\cos(\sin(\pi\theta))\cos(\pi\theta)$    **21.** $3\sec^2 3\theta$

**23.** 672    **25.** $\dfrac{-\csc^2 x - 2x\cot x\tan x^2}{\sec x^2}$    **27.** $16 - 4\pi$

**29.** 458.8

**31.** $F'(r(x)+s(x))(r''(x)+s''(x))$
$+(r'(x)+s'(x))^2 F''(r(x)+s(x))+s''(x)$

**33.** $27z^2 \cos(9z^3)$

**35.** $\left(\dfrac{7}{4}, \dfrac{1}{16}\right)$    **37.** $31.4 \text{ m}^3$    **39.** $0.167 \text{ ft/min}$

**41.** (a) $1 < t < 3$;    (b) $a(1) = -6, a(3) = 6$;    (c) $t > 2$

**43.** (a) $\dfrac{1-x}{y}$;    (b) $-\dfrac{y^2 + 2xy}{x^2 + 2xy}$;    (c) $\dfrac{x^2 y^3 - x^2}{y^2 - x^3 y^2}$;

(d) $\dfrac{2x - \sin(xy) - xy\cos(xy)}{x^2 \cos(xy)}$;

(e) $-\dfrac{\tan(xy) + xy\sec^2(xy)}{x^2 \sec^2(xy)}$

**45.** 0.0714    **47.** (a) 84;    (b) 23;    (c) 20;    (d) 26

**49.** 104 mi/h    **51.** (a) $\cot\theta|\sin\theta|$;    (b) $-\tan\theta|\cos\theta|$

## ADDITIONAL PROBLEMS 3.12

**5.** $\dfrac{m}{n} x^{\frac{m}{n}-1}$

**7.** (a)

| $x$ | 0.01 | 0.02 | 0.03 | 0.04 | 0.05 |
|-----|------|------|------|------|------|
| $\sqrt{1+x}$ | 1.005 | 1.01 | 1.0149 | 1.0198 | 1.024 |
| $1+\frac{x}{2}$ | 1.005 | 1.01 | 1.015 | 1.025 | 1.025 |

(b)

| $dx$ | 0.01 | 0.02 | 0.03 | 0.04 | 0.05 |
|------|------|------|------|------|------|
| $dy$ | 0.005 | 0.01 | 0.015 | 0.02 | 0.025 |

(c) Positive; negative;

(d) $(1+x)^{1/2} \approx \sqrt{5} + \dfrac{1}{2\sqrt{5}}(x-4)$ for $x \approx 4$

**9.** (a) 15, 18, 18;    (b) 2;    (c) $y = 17x - 51$;    (d) 69.7

**13.** (a) How quickly water is flowing into or out of the tank;
(b) How quickly the water is flowing through the pipe.

**15.** (b)        (c)

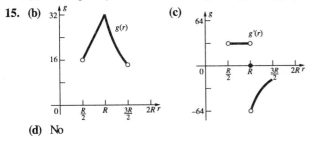

(d) No

## PROBLEM SET 4.1

**1.** Critical points: $-4, -2, 0$;
maximum value 4, minimum value 0

**3.** Critical points: $-2, -\dfrac{3}{2}, 1$;
maximum value 4, minimum value $-\dfrac{9}{4}$

**5.** Critical points: $-1, 1$;
no maximum value, minimum value $-1$

**7.** Critical points: $-1, 3$;
No maximum value, no minimum value

**9.** Critical point: 0; maximum value 1, no minimum value

**11.** Critical points: $-\dfrac{\pi}{4}, \dfrac{\pi}{6}$;
Maximum value $\dfrac{\sqrt{3}}{2}$, minimum value $-\dfrac{1}{\sqrt{2}}$

**13.** Critical points: 0, 1, 3; maximum value 2, minimum value 0

**15.** Critical points: $-1, 0, 27$;
maximum value 3, minimum value $-1$

**17.** 5 and 5    **19.** 50 ft by 50 ft    **21.** 1024 in.$^3$

**23.** 20 ft by 40 ft    **25.** 40 ft by 100 ft    **27.** 4 by 8

**29.** $\dfrac{\pi}{3}$    **31.** 55 mi/h    **33.** $P(2\sqrt{2}, 2), \ Q(0, 0)$

**35.** $x = 1, y = 3, z = 3$

**37.** (a) Maximum value 3.57, minimum value $-2.71$
(b) Maximum value 3.57, minimum value 0

## PROBLEM SET 4.2

**1.** Increasing on $(-\infty, \infty)$

**3.** Increasing on $[-1, \infty)$, decreasing on $(-\infty, -1]$

**5.** Increasing on $(-\infty, 1] \cup [2, \infty)$, decreasing on $[1, 2]$

**7.** Increasing on $[2, \infty)$, decreasing on $(-\infty, 2]$

**9.** Increasing on $\left[0, \dfrac{\pi}{2}\right] \cup \left[\dfrac{3\pi}{2}, 2\pi\right]$, decreasing on $\left[\dfrac{\pi}{2}, \dfrac{3\pi}{2}\right]$

**11.** Concave up for all $x$; no inflection points

**13.** Concave up on $(0, \infty)$, concave down on $(-\infty, 0)$;
inflection point $(0, 0)$

**15.** Concave up on $(-\infty, -1) \cup (4, \infty)$, concave down on $(-1, 4)$; inflection points $(-1, -19)$ and $(4, -499)$

**17.** Concave up for all $x$; no inflection points

**19.** Increasing on $(-\infty, -2] \cup [2, \infty)$, decreasing on $[-2, 2]$;
concave up on $(0, \infty)$, concave down on $(-\infty, 0)$

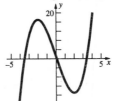

**21.** Increasing on $[1, \infty)$, decreasing on $(-\infty, 1]$; concave up on $(-\infty, 0) \cup \left(\dfrac{2}{3}, \infty\right)$, concave down on $\left(0, \dfrac{2}{3}\right)$

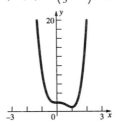

**23.** Increasing on $(-\infty, -1] \cup [1, \infty)$, decreasing on $[-1, 1]$; concave up on $\left(-\frac{1}{\sqrt{2}}, 0\right) \cup \left(\frac{1}{\sqrt{2}}, \infty\right)$, concave down on $\left(-\infty, -\frac{1}{\sqrt{2}}\right) \cup \left(0, \frac{1}{\sqrt{2}}\right)$

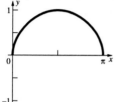

**25.** Increasing on $\left[0, \frac{\pi}{2}\right]$, decreasing on $\left[\frac{\pi}{2}, \pi\right]$; concave down on $(0, \pi)$

**27.** Increasing on $\left[0, \frac{2}{5}\right]$, decreasing on $(-\infty, 0] \cup \left[\frac{2}{5}, \infty\right)$; concave up on $\left(-\infty, -\frac{1}{5}\right)$, concave down on $\left(-\frac{1}{5}, 0\right) \cup (0, \infty)$.

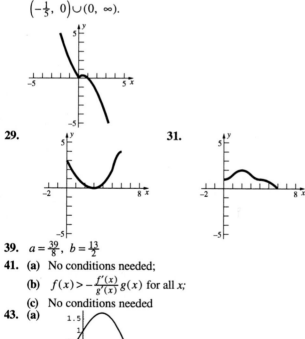

**29.**    **31.**

**39.** $a = \frac{39}{8}$, $b = \frac{13}{2}$

**41.** (a) No conditions needed;
   (b) $f(x) > -\frac{f'(x)}{g'(x)} g(x)$ for all $x$;
   (c) No conditions needed

**43.** (a)

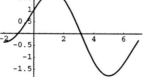

   (b) $(1.3, 5)$;    (c) $(-0.25, 3.1) \cup (6.5, 7]$

(d)

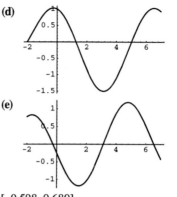

(e)

**45.** $[-0.598, 0.680]$

## PROBLEM SET 4.3

**1.** Critical points: 0, 4; local minimum at $x = 4$; local maximum at $x = 0$

**3.** No critical points; no local minima or maxima on $\left(0, \frac{\pi}{4}\right)$

**5.** Critical points: 0, $\frac{\pi}{2}$, $\pi$; local minima at $\theta = 0$, $\pi$; local maximum at $\theta = \frac{\pi}{2}$

**7.** Critical points: –1, 1; local minimum value $f(1) = -2$; local maximum value $f(-1) = 2$

**9.** Critical points 0, $\frac{3}{2}$; local minimum value $H\left(\frac{3}{2}\right) = -\frac{27}{16}$; no local maximum

**11.** Critical point: 2; no local minimum values; local maximum value $g(2) = \pi$

**13.** No critical points
No local minimum or maximum values

**15.** No critical points
No local minimum or maximum values

**17.** Minimum value –4; maximum value $\frac{9}{4}$

**19.** Minimum value 125; no maximum values    **21.** $\approx 277$

**23.** Local minimum at $x = 3$; local maximum at $x = -2$

**25.** (a) Increasing on $(-\infty, -3] \cup [-1, 0)$; decreasing on $[-3, -1] \cup (0, \infty)$;
   (b) Concave up on $(-2, 0) \cup (0, 2)$; concave down on $(-\infty, -2) \cup (2, \infty)$;
   (c) Local maximum at $x = -3$; local minimum at $x = -1$;
   (d) $x = -2, 2$

**27.** Local minima $f(-2) = 12$, $f(\sqrt{3}) \approx -6.4$; local maxima $f(-\sqrt{3}) \approx 14.4$, $f(2.5) = 23.53125$

**29.** Local minimum at $x \approx 2.1$; local maximum at $x = 1$

## PROBLEM SET 4.4

**1.** –4 and 4    **3.** $\frac{1}{16}$    **5.** $\left(-\frac{3}{\sqrt{2}}, \frac{9}{2}\right)$, $\left(\frac{3}{\sqrt{2}}, \frac{9}{2}\right)$

**7.** $x = 15\sqrt{3}$ ft, $y = 20\sqrt{3}$ ft

**9.** $x = \dfrac{10\sqrt{5}}{\sqrt{3}}$ ft, $y = 6\sqrt{15}$ ft

**11.** width $= \dfrac{4\sqrt{2}}{\sqrt{3}}$, depth $= 2\sqrt{6}$

**13.** 2.27 miles down the shore from $P$   **15.** At the town

**17.** 8:09 A.M.   **19.** $\dfrac{4\pi\sqrt{3}}{9}r^3$

**21.** $h = \sqrt{2}r$, $x = \dfrac{r}{\sqrt{2}}$ where $h =$ height of the cylinder, $x =$ radius of the cylinder, $r =$ radius of the sphere

**23.** **(a)** 43.48 cm from one end; shorter length bent to form square
   **(b)** No cut, wire bent to form square

**25.** height $= \left(\dfrac{3V}{\pi}\right)^{1/3}$, radius $= \tfrac{1}{2}\left(\dfrac{3V}{\pi}\right)^{1/3}$

**27.** $r = \sqrt{A}$, $\theta = 2$   **31.** 11.18 ft

**33.** **(a)** $\tfrac{2a}{3}$;   **(b)** $\tfrac{2a}{3}$;   **(c)** $\tfrac{3a}{4}$

**35.** **(a)** $L' = 3$, $L = 4$, $\phi = 90°$;
   **(b)** $L' = 5$, $L = 12$, $\phi = 90°$;
   **(c)** $\phi = 90°$, $L = \sqrt{m^2 - h^2}$, $L' = h$

## PROBLEM SET 4.5

**1.** $55   **3.** $p(n) = 300 - \dfrac{n}{2}$; $R(n) = 300n - \dfrac{n^2}{2}$

**5.** $n = 200$

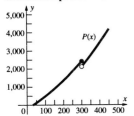

**7.** $1.92 per unit; $1.33

**9.** **(a)** $R(x) = 20x + 4x^2 - \dfrac{x^3}{3}$; $\dfrac{dR}{dx} = 20 + 8x - x^2$
   **(b)** $0 \le x \le 10$   **(c)** 4

**11.** $x_1 = 25$, $\dfrac{dR}{dx} = 0$ at $x_1$

**13.** $p(x) = 8.4 - 0.0006x$; $4.20 per yd

**15.** **(a)** $C(x) = \begin{cases} 6000 + 1.40x & \text{if } 0 \le x \le 4500 \\ 6000 + 1.60x & \text{if } 4500 < x \end{cases}$
   **(b)** $p(x) = 11 - 0.001x$   **(c)** 4500

**17.** Maximum profit of $4172.50 at $x = 450$

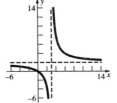

**19.** $1713.14

## PROBLEM SET 4.6

**1.** 1   **3.** $-1$   **5.** $-1$   **7.** $\tfrac{1}{2}$   **9.** $\tfrac{3}{\pi}$   **11.** $\dfrac{3}{\sqrt{2}}$

**13.** 2   **15.** 2   **17.** 0   **19.** $-\infty$   **21.** $\infty$   **23.** $\infty$

**25.** $\infty$   **27.** $-\infty$   **29.** 5   **31.** 0   **33.** $-1$   **35.** $-\infty$

**37.** Horizontal asymptote: $y = 0$; vertical asymptote: $x = -1$

**39.** Horizontal asymptote: $y = 2$; vertical asymptote: $x = 3$

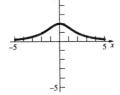

**41.** Horizontal asymptote: $y = 0$; no vertical asymptotes

**43.** $y = 2x + 3$

**45.** **(a)** We say that $\lim\limits_{x \to c^+} f(x) = -\infty$ if for each negative number $M$, there corresponds a $\delta > 0$ such that $0 < x - c < \delta \Rightarrow f(x) < M.$;
   **(b)** We say that $\lim\limits_{x \to c^-} f(x) = \infty$ if for each positive number $M$, there corresponds a $\delta > 0$ such that $0 < c - x < \delta \Rightarrow f(x) > M.$

**49.** **(a)** Does not exist;   **(b)** 0;   **(c)** 1;   **(d)** $\infty$;
   **(e)** 0;   **(f)** $\tfrac{1}{2}$;   **(g)** Does not exist;   **(h)** 0

**51.** $\tfrac{3}{2}$   **53.** $-\dfrac{3}{2\sqrt{2}}$   **55.** 1   **57.** $\infty$   **59.** $-1$

**61.** $-\infty$   **63.** $e \approx 2.718$   **65.** 1

**PROBLEM SET 4.7**

1.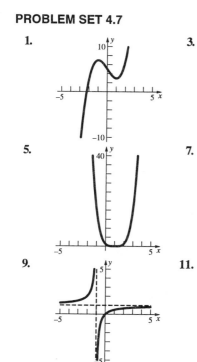

3.

5.

7.

9.

11.

13.

15.

17.

19.

21.

23.

25.

27.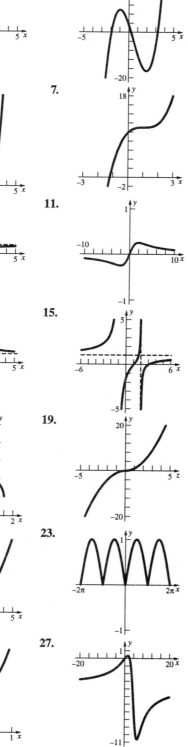

29.

31.

33.

35.

37.

39.

41. $a \neq 0$

43.

45.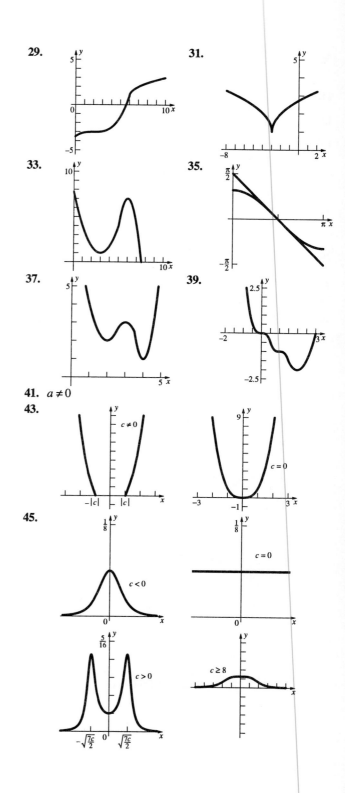

**47.**

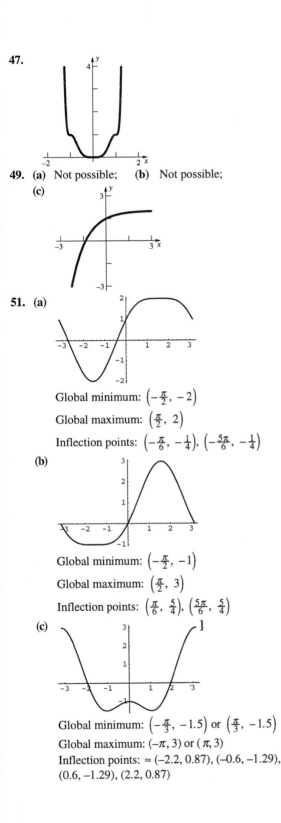

**49. (a)** Not possible; **(b)** Not possible;
**(c)**

**51. (a)**

Global minimum: $\left(-\frac{\pi}{2}, -2\right)$

Global maximum: $\left(\frac{\pi}{2}, 2\right)$

Inflection points: $\left(-\frac{\pi}{6}, -\frac{1}{4}\right)$, $\left(-\frac{5\pi}{6}, -\frac{1}{4}\right)$

**(b)**

Global minimum: $\left(-\frac{\pi}{2}, -1\right)$

Global maximum: $\left(\frac{\pi}{2}, 3\right)$

Inflection points: $\left(\frac{\pi}{6}, \frac{5}{4}\right)$, $\left(\frac{5\pi}{6}, \frac{5}{4}\right)$

**(c)**

Global minimum: $\left(-\frac{\pi}{3}, -1.5\right)$ or $\left(\frac{\pi}{3}, -1.5\right)$

Global maximum: $(-\pi, 3)$ or $(\pi, 3)$

Inflection points: $\approx (-2.2, 0.87)$, $(-0.6, -1.29)$,
$(0.6, -1.29)$, $(2.2, 0.87)$

**(d)**

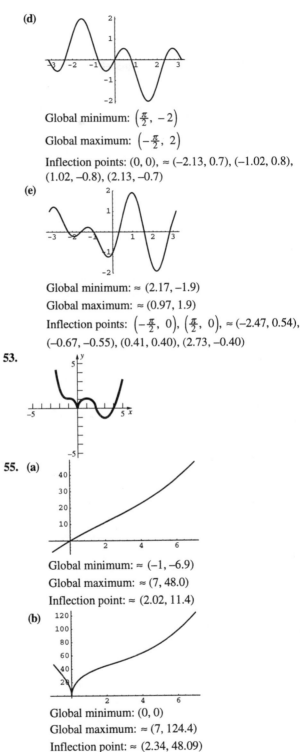

Global minimum: $\left(\frac{\pi}{2}, -2\right)$

Global maximum: $\left(-\frac{\pi}{2}, 2\right)$

Inflection points: $(0, 0)$, $\approx (-2.13, 0.7)$, $(-1.02, 0.8)$,
$(1.02, -0.8)$, $(2.13, -0.7)$

**(e)**

Global minimum: $\approx (2.17, -1.9)$

Global maximum: $\approx (0.97, 1.9)$

Inflection points: $\left(-\frac{\pi}{2}, 0\right)$, $\left(\frac{\pi}{2}, 0\right)$, $\approx (-2.47, 0.54)$,
$(-0.67, -0.55)$, $(0.41, 0.40)$, $(2.73, -0.40)$

**53.**

**55. (a)**

Global minimum: $\approx (-1, -6.9)$

Global maximum: $\approx (7, 48.0)$

Inflection point: $\approx (2.02, 11.4)$

**(b)**

Global minimum: $(0, 0)$

Global maximum: $\approx (7, 124.4)$

Inflection point: $\approx (2.34, 48.09)$

(c)

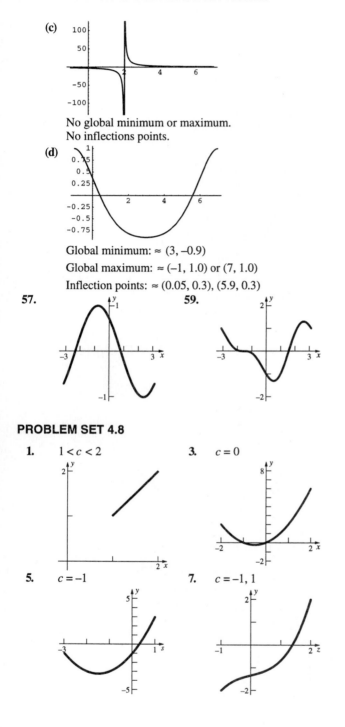

No global minimum or maximum.
No inflections points.

(d)

Global minimum: $\approx (3, -0.9)$
Global maximum: $\approx (-1, 1.0)$ or $(7, 1.0)$
Inflection points: $\approx (0.05, 0.3), (5.9, 0.3)$

**57.**

**59.**

**PROBLEM SET 4.8**

**1.** $1 < c < 2$

**3.** $c = 0$

**5.** $c = -1$

**7.** $c = -1, 1$

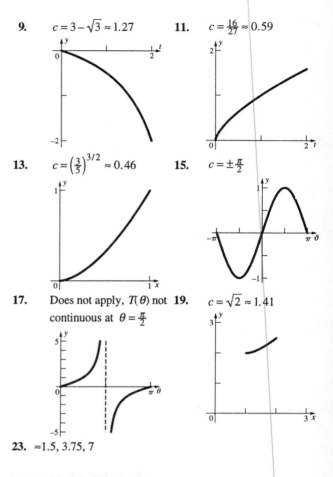

**9.** $c = 3 - \sqrt{3} \approx 1.27$

**11.** $c = \frac{16}{27} \approx 0.59$

**13.** $c = \left(\frac{3}{5}\right)^{3/2} \approx 0.46$

**15.** $c = \pm\frac{\pi}{2}$

**17.** Does not apply, $T(\theta)$ not continuous at $\theta = \frac{\pi}{2}$

**19.** $c = \sqrt{2} \approx 1.41$

**23.** $\approx 1.5, 3.75, 7$

**CHAPTER REVIEW 4.9**

Concepts Test

**1.** True  **3.** True  **5.** True  **7.** True  **9.** True
**11.** False  **13.** True  **15.** True  **17.** True
**19.** False  **21.** False  **23.** True  **25.** True
**27.** True  **29.** True  **31.** False  **33.** True

Sample Test

**1.** Critical points: 0, 1, 4; minimum value $f(1) = -1$; maximum value $f(4) = 8$

**3.** Critical points: $-2, -\frac{1}{2}$; minimum value $f(-2) = \frac{1}{4}$; maximum value $f\left(-\frac{1}{2}\right) = 4$

**5.** Critical points: $-\frac{1}{2}, 0, 1$; minimum value $f(0) = 0$; maximum value $f(1) = 1$

**7.** Critical points: $-2, 0, 1, 3$; minimum value $f(1) = -1$; maximum value $f(3) = 135$

**9.** Critical points: $-1, 0, 2, 3$; minimum value $f(2) = -9$; maximum value $f(3) = 88$

**11.** Critical points: $\frac{\pi}{4}, \frac{\pi}{2}, \frac{3\pi}{4}$; minimum value $f\left(\frac{4\pi}{3}\right) \approx -0.87$; maximum value $f\left(\frac{\pi}{2}\right) = 1$

**13.** Increasing: $\left(-\infty, \frac{3}{2}\right)$; concave down: $(-\infty, \infty)$

**15.** Increasing: $(-\infty, -1] \cup [1, \infty)$; concave down: $(-\infty, 0)$

**17.** Increasing: $\left[0, \frac{1}{5}\right]$; concave down: $\left(\frac{3}{20}, \infty\right)$

**19.** Increasing: $\left(-\infty, \frac{3}{4}\right]$; concave down: $(-\infty, 0) \cup \left(\frac{1}{2}, \infty\right)$

**21.** Increasing: $(-\infty, 0] \cup \left[\frac{8}{3}, \infty\right)$; decreasing: $\left[0, \frac{8}{3}\right]$;

local minimum value $f\left(\frac{8}{3}\right) = -\frac{256}{27}$

Local maximum value $f(0) = 0$

Inflection point: $\left(\frac{4}{3}, -\frac{128}{27}\right)$

**23.**    **25.**

**27.**    **29.**

**31.**    **33.**

**35.**    **37.**

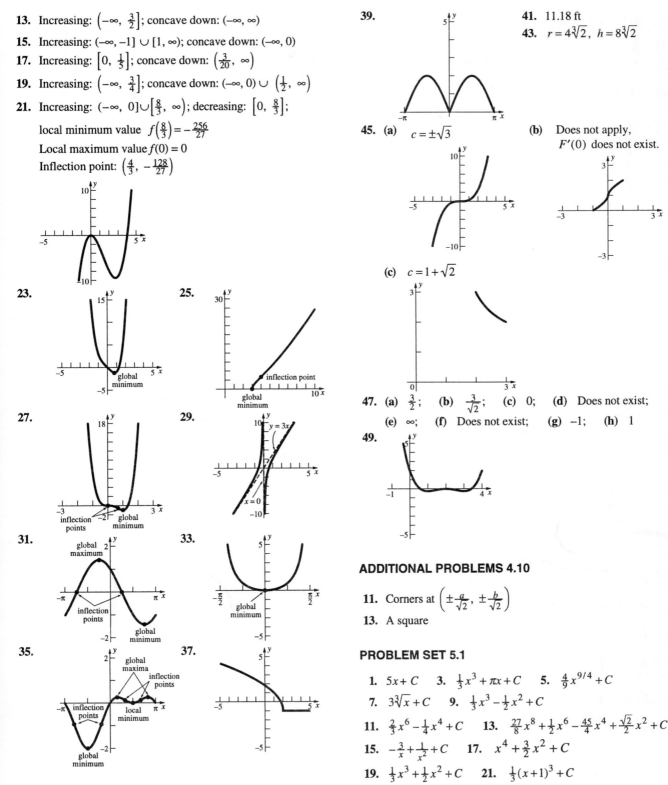

**39.**

**41.** 11.18 ft

**43.** $r = 4\sqrt[3]{2}, \ h = 8\sqrt[3]{2}$

**45.** **(a)** $c = \pm\sqrt{3}$    **(b)** Does not apply, $F'(0)$ does not exist.

**(c)** $c = 1 + \sqrt{2}$

**47.** **(a)** $\frac{3}{2}$;    **(b)** $\frac{3}{\sqrt{2}}$;    **(c)** 0;    **(d)** Does not exist;

**(e)** $\infty$;    **(f)** Does not exist;    **(g)** $-1$;    **(h)** 1

**49.**

## ADDITIONAL PROBLEMS 4.10

**11.** Corners at $\left(\pm\frac{a}{\sqrt{2}}, \pm\frac{b}{\sqrt{2}}\right)$

**13.** A square

## PROBLEM SET 5.1

**1.** $5x + C$    **3.** $\frac{1}{3}x^3 + \pi x + C$    **5.** $\frac{4}{9}x^{9/4} + C$

**7.** $3\sqrt[3]{x} + C$    **9.** $\frac{1}{3}x^3 - \frac{1}{2}x^2 + C$

**11.** $\frac{2}{3}x^6 - \frac{1}{4}x^4 + C$    **13.** $\frac{27}{8}x^8 + \frac{1}{2}x^6 - \frac{45}{4}x^4 + \frac{\sqrt{2}}{2}x^2 + C$

**15.** $-\frac{3}{x} + \frac{1}{x^2} + C$    **17.** $x^4 + \frac{3}{2}x^2 + C$

**19.** $\frac{1}{3}x^3 + \frac{1}{2}x^2 + C$    **21.** $\frac{1}{3}(x+1)^3 + C$

**23.** $\frac{2}{9}z^{9/2} + \frac{4}{5}z^{5/2} + 2z^{1/2}$    **25.** $-\cos\theta - \sin\theta + C$

**27.** $\frac{1}{4}(\sqrt{2}x+1)^4+C$   **29.** $\frac{1}{21}(5x^3+3x-8)^7+C$

**31.** $\frac{9}{16}\sqrt[3]{(2t^2-11)^4}+C$   **33.** $\frac{1}{2}x^3+\frac{1}{2}x^2+C_1x+C_2$

**35.** $\frac{4}{15}x^{5/2}+C_1x+C_2$   **37.** $\frac{1}{6}x^3+\frac{1}{2x}+C_1x+C_2$

**41.** $x^2\sqrt{x-1}+C$   **45.** $\dfrac{5x^3+2}{2\sqrt{x^3+1}}+C$

**49.** $\frac{1}{2}x^2+C$ if $x\ge0$, $-\frac{1}{2}x^2+C$ if $x<0$

## PROBLEM SET 5.2

**5.** $y=\frac{1}{3}x^3+x+C$; $y=\frac{1}{3}x^3+x-\frac{1}{3}$

**7.** $y=\pm\sqrt{x^2+C}$; $y=\sqrt{x^2}$

**9.** $z=\dfrac{3}{C-t^3}$; $z=\dfrac{3}{10-t^3}$

**11.** $s=\frac{16}{3}t^3+2t^2-t+C$

$s=\frac{16}{3}t^3+2t^2-t+100$

**13.** $y=\frac{1}{10}(2x+1)^5+C$

$y=\frac{1}{10}(2x+1)^5+\frac{59}{10}$

**15.** $y=\frac{3}{2}x^2+\frac{1}{2}$   **17.** $v=5$ cm/s; $s=\frac{22}{3}$ cm

**19.** $v\approx2.83$ cm/s; $s\approx11.48$ cm

**21.** 144 ft   **23.** $v=32.24$ ft/s; $s=1198.54$ ft

**27.** Moon: $\approx1.470$ mi/s; Venus: $\approx6.257$ mi/s

Jupiter: $\approx36.812$ mi/s; Sun: $\approx382.908$ mi/s

**29.** 2.2 ft/s$^2$   **31.** 5500 m

**33. (a)**   **(b)** 36 mi/h;

**(c)** 0.9 mi/min$^2$

**35. (a)** $\dfrac{dV}{dt}=C_1\dfrac{\sqrt{V}}{10}$, $V(0)=1600$, $V(40)=0$;

**(b)** $V=\frac{1}{400}(-20t+800)^2$;   **(c)** 900 cm$^3$

**37. (a)** $v(t)=\begin{cases}-32t & \text{for } 0\le t\le1 \\ -32(t-1)+24 & \text{for } 1<t\le2.5\end{cases}$

**(b)** $t\approx0.66$, 1.75 s

## PROBLEM SET 5.3

**1.** 15   **3.** $\frac{481}{280}$   **5.** $\frac{85}{2}$   **7.** 3   **9.** $\sum\limits_{i=1}^{41}i$

**11.** $\sum\limits_{i=1}^{100}\frac{1}{i}$   **13.** $\sum\limits_{i=1}^{50}a_{2i-1}$   **15.** $\sum\limits_{i=1}^{n}f(c_i)$   **17.** 90

**19.** $-10$   **21.** $\frac{40}{41}$   **23.** $-\frac{48}{441}$   **25.** 14,950

**27.** 2640   **29.** $\dfrac{4n^3-3n^2-n}{6}$   **31.** $\sum\limits_{k=1}^{17}(k+2)k$

**33.** $\sum\limits_{i=1}^{11}\frac{i-1}{i}$   **35.** $\frac{33}{5}$

**37. (a)** $1-\left(\frac{1}{2}\right)^{10}$;   **(b)** $2^{11}-2$

**39.** $S=\dfrac{(n+1)(2a+nd)}{2}$;   **41.** $\bar{x}\approx7.86$; $s^2\approx12.41$

**49.** 715; 55,675; $S=\dfrac{m(m+1)(3n-m+1)}{6}$

## PROBLEM SET 5.4

**1.** $\frac{7}{2}$   **3.** $\frac{9}{2}$   **5.** $\frac{23}{8}$

**7.** $A=6$   **9.** $A=\frac{1243}{216}$

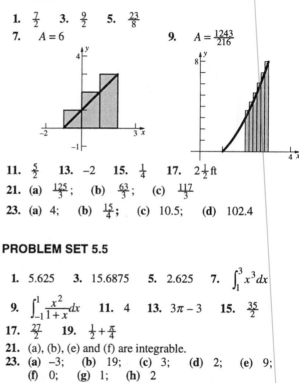

**11.** $\frac{5}{2}$   **13.** $-2$   **15.** $\frac{1}{4}$   **17.** $2\frac{1}{2}$ ft

**21. (a)** $\frac{125}{3}$;   **(b)** $\frac{63}{3}$;   **(c)** $\frac{117}{3}$

**23. (a)** 4;   **(b)** $\frac{15}{4}$;   **(c)** 10.5;   **(d)** 102.4

## PROBLEM SET 5.5

**1.** 5.625   **3.** 15.6875   **5.** 2.625   **7.** $\int_1^3 x^3\,dx$

**9.** $\int_{-1}^1\frac{x^2}{1+x}dx$   **11.** 4   **13.** $3\pi-3$   **15.** $\frac{35}{2}$

**17.** $\frac{27}{2}$   **19.** $\frac{1}{2}+\frac{\pi}{4}$

**21.** (a), (b), (e) and (f) are integrable.

**23. (a)** $-3$;   **(b)** 19;   **(c)** 3;   **(d)** 2;   **(e)** 9;

**(f)** 0;   **(g)** 1;   **(h)** 2

**27.** Left: 5.24; Right: 6.84; Midpoint: 5.98

**29.** Left: 0.8638; Right: 0.8178; Midpoint: 0.8418

**31.** 4   **33.** 0.6   **35.** Not integrable

## PROBLEM SET 5.6

**1.** 4   **3.** 15   **5.** $\frac{3}{4}$   **7.** $\frac{16}{3}$   **9.** $\frac{1783}{96}$   **11.** 1

**13.** $\frac{22}{5}$   **15.** $\frac{2047}{11}$   **17.** $\frac{4}{5}$   **19.** $\frac{122}{9}$   **21.** 0

**23.** $\frac{1}{3}$   **25.** $\frac{\pi^2}{4}+1$   **27.** 14   **29.** $\frac{38}{15}$   **31.** 9

**33.** 2   **35.** $\frac{77}{200}=0.385$; $\frac{1}{3}=0.\overline{333}$   **37.** 6   **39.** $-16$

**41.** $-24$   **43.** $\dfrac{([\![b]\!]-1)([\![b]\!]-2)}{2}+[\![b]\!](b-[\![b]\!])$

## PROBLEM SET 5.7

**1.** 6   **3.** 14   **5.** –31   **7.** 23   **9.** $3+\frac{\pi}{2}$   **11.** $2x$

**13.** $2x^2+\sqrt{x}$   **15.** $-(x-2)\cot(2x)$   **17.** $2x\sin(x^2)$

**19.** $\exp x^2+2x\exp x^4$   **23.** $\frac{26}{3}$   **25.** 4   **29.** 40   **31.** $\frac{13}{4}$

**33.** 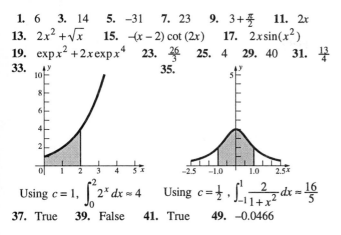   **35.**

Using $c=1$, $\int_0^2 2^x\,dx\approx 4$   Using $c=\frac{1}{2}$, $\int_{-1}^1\frac{2}{1+x^2}\,dx\approx\frac{16}{5}$

**37.** True   **39.** False   **41.** True   **49.** –0.0466

## PROBLEM SET 5.8

**1.** $\frac{2}{9}(3x+2)^{3/2}+C$   **3.** $\frac{4}{27}(6x-7)^{9/8}+C$

**5.** $\frac{1}{3}\sin(3x+2)+C$   **7.** $-\frac{1}{6}\cos(6x-7)+C$

**9.** $\frac{1}{3}(x^2+4)^{3/2}+C$   **11.** $-\frac{7}{10}(x^2+3)^{-5/7}+C$

**13.** $-\frac{1}{2}\cos(x^2+4)+C$   **15.** $-\frac{1}{18}\cos(6x^3-7)+C$

**17.** $-\cos\sqrt{x^2+4}+C$   **19.** $\frac{1}{27}\sin[(x^3+5)^9]+C$

**21.** $\frac{1}{3}[\sin(x^2+4)]^{3/2}+C$   **23.** $-\frac{1}{30}\cos^{10}(x^3+5)+C$

**25.** $\sec^{1/2}(x^2+2x)+C$   **27.** $\frac{1}{2}(\sqrt{t}+4)^4+C$

**29.** $\frac{5}{16}(\sqrt[5]{32z}+\pi)^8+C$   **31.** $\frac{85}{4}$   **33.** $\frac{2}{117}$   **35.** $\frac{5}{12}$

**37.** $\frac{1}{64}$   **39.** $\frac{\sin 3}{3}$   **41.** $\frac{1}{\pi}$   **43.** 1   **45.** $1-\cos 1$

**47.** $\frac{1-\cos^4 1}{8}$   **49.** $\frac{5}{36}$   **51.** 0   **53.** 0   **55.** $2\pi$

**57.** $\frac{8}{3}$   **59.** $\frac{1}{2}$   **61.** 4

**63.** Even: $\int_{-b}^{-a}f(x)dx=\int_a^b f(x)dx$; Odd: $\int_{-b}^{-a}f(x)dx=-\int_a^b f(x)dx$

**65.** 8   **67.** $\int_1^{1+\pi}|\sin x|dx=2$   **69.** 883.2   **71.** $\frac{\pi^2}{4}-2$

**73.** (a) Even;   (b) $2\pi$;
  (c)

| Interval | Value of Integral |
| --- | --- |
| $\left[0,\frac{\pi}{2}\right]$ | 0.46 |
| $\left[-\frac{\pi}{2},\frac{\pi}{2}\right]$ | 0.92 |
| $\left[0,\frac{3\pi}{2}\right]$ | –0.46 |
| $\left[-\frac{3\pi}{2},\frac{3\pi}{2}\right]$ | –0.92 |
| $[0,2\pi]$ | 0 |
| $\left[\frac{\pi}{6},\frac{13\pi}{6}\right]$ | 0 |
| $\left[\frac{\pi}{6},\frac{4\pi}{3}\right]$ | –0.44 |
| $\left[\frac{13\pi}{6},\frac{10\pi}{3}\right]$ | –0.44 |

## 5.9 CHAPTER REVIEW

### Concepts Test

**1.** True   **3.** True   **5.** True   **7.** False   **9.** True
**11.** True   **13.** True   **15.** True   **17.** True   **19.** True
**21.** False   **23.** True   **25.** True   **27.** False
**29.** False   **31.** False   **33.** True   **35.** False
**37.** True   **39.** False   **41.** True   **43.** True

### Sample Test Problems

**1.** $\frac{5}{4}$   **3.** $\frac{1}{3}y^3+9\cos y-\frac{26}{y}+C$

**5.** $\frac{3}{16}(2z^2-3)^{4/3}+C$   **7.** $\frac{1}{18}\tan^3(3\pi^2)$   **9.** 46.9

**11.** $\frac{5}{24}(2y^3+3y^2+6y)^{4/5}+C$   **13.** $y=2\sqrt{x+1}+14$

**15.** $y=\frac{1}{3}(2t-1)^{3/2}-1$   **17.** $y=\sqrt{3x^2-\frac{1}{4}x^4+9}$

**19.** $y=\frac{x^3}{6}+1$   **21.** 7 s; –176 ft/s

**23.** $\frac{7}{4}$

**25.** $\frac{5}{6}$   **27.** $\frac{39}{4}$   **29.** 1870

**31.** (a) $\sum_{n=2}^{78}\frac{1}{n}$;   (b) $\sum_{n=1}^{50}nx^{2n}$

**33.** (a) –2;   (b) –4;   (c) 6;   (d) –12;   (e) 2
**35.** (a) –8;   (b) 8;   (c) 0;   (d) –16;   (e) –2;
  (f) –5

**37.** $c=-\sqrt{7}$

**39.** (a) $\sin^2 x$;   (b) $f(x+1)-f(x)$;
  (c) $-\frac{1}{x^2}\int_0^x f(z)d(z)+\frac{1}{x}f(x)$;   (d) $\int_0^x f(t)dt$;
  (e) $g'(g(x))g'(x)$;   (f) $-f(x)$

## ADDITIONAL PROBLEMS 5.10

**1.** Figure 1
**3.** (a) 11,010 ft;   (b) 8370 ft;   (c) 9690 ft;
  (d) (a) uses the largest and (b) the smallest value of each
  pair. (c) averages each pair so the total value is the
  average of (a) and (b).

**7.** (a) 0;   (b) $\frac{2\hat{V}}{\pi}$;   (d) 169.7 Volts
**9.** (a) Local minima at 0, ≈ 1.8, ≈ 3.8, ≈ 5.8,≈ 7.8;
  local maxima at ≈ 1, ≈ 3.1, ≈ 5.2, ≈ 7.3;
  (b) Absolute minimum at 0; absolute maximum at 10;
  (c) $(0.5,1.5)\cup(2.5,3.5)\cup(4.5,5.5)\cup(6.5,7.5)\cup(8.5,9.5)$;

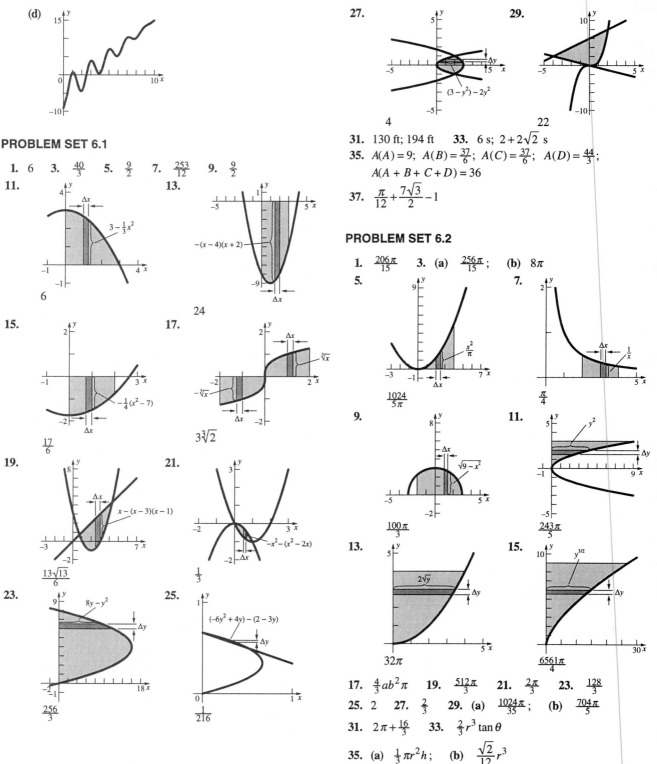

**(d)**

## PROBLEM SET 6.1

**1.** 6   **3.** $\frac{40}{3}$   **5.** $\frac{9}{2}$   **7.** $\frac{253}{12}$   **9.** $\frac{9}{2}$

**11.**

6

**13.**

24

$-(x-4)(x+2)$

**15.**

$\frac{17}{6}$

$-\frac{1}{4}(x^2-7)$

**17.**

$3\sqrt[3]{2}$

$\sqrt[3]{x}$ , $-\sqrt[3]{x}$

**19.**

$\frac{13\sqrt{13}}{6}$

$x-(x-3)(x-1)$

**21.**

$\frac{1}{3}$

$-x^2-(x^2-2x)$

**23.**

$\frac{256}{3}$

$8y-y^2$

**25.**

$216$

$(-6y^2+4y)-(2-3y)$

**27.**

4

$(3-y^2)-2y^2$

**29.**

22

**31.** 130 ft; 194 ft   **33.** 6 s; $2+2\sqrt{2}$ s

**35.** $A(A)=9$; $A(B)=\frac{37}{6}$; $A(C)=\frac{37}{6}$; $A(D)=\frac{44}{3}$; $A(A+B+C+D)=36$

**37.** $\frac{\pi}{12}+\frac{7\sqrt{3}}{2}-1$

## PROBLEM SET 6.2

**1.** $\frac{206\pi}{15}$   **3. (a)** $\frac{256\pi}{15}$;   **(b)** $8\pi$

**5.**

$\frac{1024}{5\pi}$

$\frac{x^2}{\pi}$

**7.**

$\frac{\pi}{4}$

$\frac{1}{x}$

**9.**

$\frac{100\pi}{3}$

$\sqrt{9-x^2}$

**11.**

$\frac{243\pi}{5}$

$y^2$

**13.**

$32\pi$

$2\sqrt{y}$

**15.**

$\frac{6561\pi}{4}$

$y^{3/2}$

**17.** $\frac{4}{3}ab^2\pi$   **19.** $\frac{512\pi}{3}$   **21.** $\frac{2\pi}{3}$   **23.** $\frac{128}{3}$

**25.** 2   **27.** $\frac{2}{3}$   **29. (a)** $\frac{1024\pi}{35}$;   **(b)** $\frac{704\pi}{5}$

**31.** $2\pi+\frac{16}{3}$   **33.** $\frac{2}{3}r^3\tan\theta$

**35. (a)** $\frac{1}{3}\pi r^2 h$;   **(b)** $\frac{\sqrt{2}}{12}r^3$

## PROBLEM SET 6.3

**1. (a), (b)**

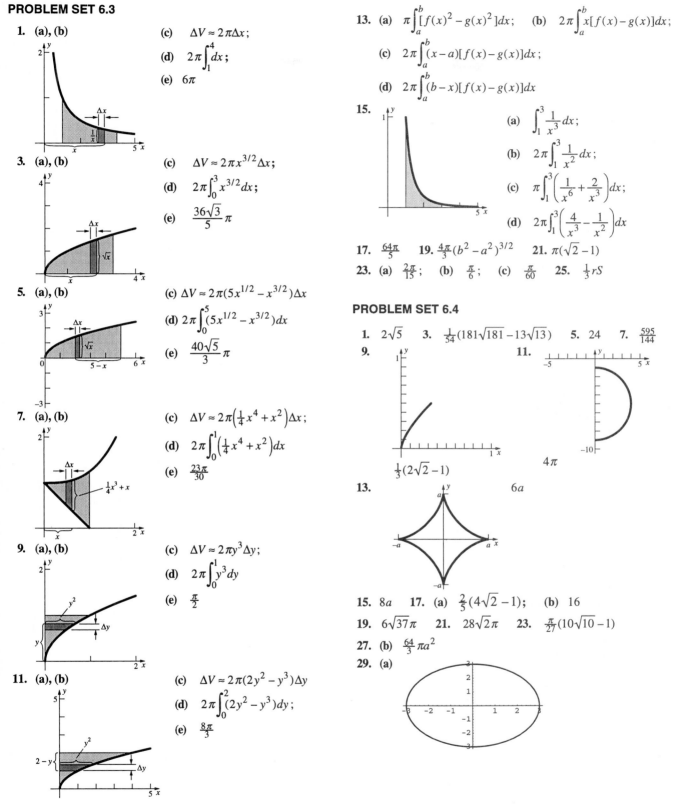

(c) $\Delta V \approx 2\pi\Delta x$;

(d) $2\pi\displaystyle\int_1^4 dx$;

(e) $6\pi$

**3. (a), (b)**

(c) $\Delta V \approx 2\pi x^{3/2}\Delta x$;

(d) $2\pi\displaystyle\int_0^3 x^{3/2}dx$;

(e) $\dfrac{36\sqrt{3}}{5}\pi$

**5. (a), (b)**

(c) $\Delta V \approx 2\pi(5x^{1/2} - x^{3/2})\Delta x$

(d) $2\pi\displaystyle\int_0^5 (5x^{1/2} - x^{3/2})dx$

(e) $\dfrac{40\sqrt{5}}{3}\pi$

**7. (a), (b)**

(c) $\Delta V \approx 2\pi\left(\frac{1}{4}x^4 + x^2\right)\Delta x$;

(d) $2\pi\displaystyle\int_0^1 \left(\frac{1}{4}x^4 + x^2\right)dx$

(e) $\dfrac{23\pi}{30}$

**9. (a), (b)**

(c) $\Delta V \approx 2\pi y^3\Delta y$;

(d) $2\pi\displaystyle\int_0^1 y^3 dy$

(e) $\dfrac{\pi}{2}$

**11. (a), (b)**

(c) $\Delta V \approx 2\pi(2y^2 - y^3)\Delta y$

(d) $2\pi\displaystyle\int_0^2 (2y^2 - y^3)dy$;

(e) $\dfrac{8\pi}{3}$

**13. (a)** $\pi\displaystyle\int_a^b [f(x)^2 - g(x)^2]dx$;   **(b)** $2\pi\displaystyle\int_a^b x[f(x) - g(x)]dx$;

**(c)** $2\pi\displaystyle\int_a^b (x-a)[f(x) - g(x)]dx$;

**(d)** $2\pi\displaystyle\int_a^b (b-x)[f(x) - g(x)]dx$

**15.**

**(a)** $\displaystyle\int_1^3 \frac{1}{x^3}dx$;

**(b)** $2\pi\displaystyle\int_1^3 \frac{1}{x^2}dx$;

**(c)** $\pi\displaystyle\int_1^3 \left(\frac{1}{x^6} + \frac{2}{x^3}\right)dx$;

**(d)** $2\pi\displaystyle\int_1^3 \left(\frac{4}{x^3} - \frac{1}{x^2}\right)dx$

**17.** $\dfrac{64\pi}{5}$   **19.** $\dfrac{4\pi}{3}(b^2 - a^2)^{3/2}$   **21.** $\pi(\sqrt{2} - 1)$

**23. (a)** $\dfrac{2\pi}{15}$;   **(b)** $\dfrac{\pi}{6}$;   **(c)** $\dfrac{\pi}{60}$   **25.** $\dfrac{1}{3}rS$

## PROBLEM SET 6.4

**1.** $2\sqrt{5}$   **3.** $\dfrac{1}{54}(181\sqrt{181} - 13\sqrt{13})$   **5.** $24$   **7.** $\dfrac{595}{144}$

**9.**

$\dfrac{1}{3}(2\sqrt{2} - 1)$

**11.**

$4\pi$

**13.**

$6a$

**15.** $8a$   **17. (a)** $\dfrac{2}{5}(4\sqrt{2} - 1)$;   **(b)** $16$

**19.** $6\sqrt{37}\pi$   **21.** $28\sqrt{2}\pi$   **23.** $\dfrac{\pi}{27}(10\sqrt{10} - 1)$

**27. (b)** $\dfrac{64}{3}\pi a^2$

**29. (a)**

**(b)**

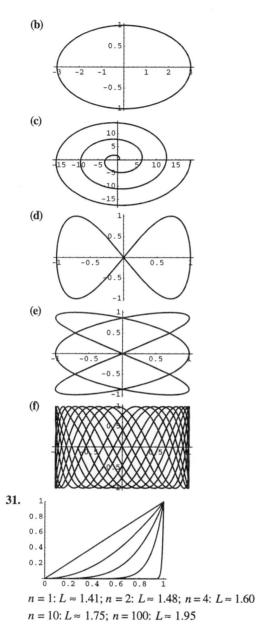

**(c)**

**(d)**

**(e)**

**(f)**

**31.**

$n = 1: L \approx 1.41$; $n = 2: L \approx 1.48$; $n = 4: L \approx 1.60$
$n = 10: L \approx 1.75$; $n = 100: L \approx 1.95$

**PROBLEM SET 6.5**

**1.** 1.5 ft-lb    **3.** 200 ergs    **7.** 18 ft-lb    **9.** 52,000 ft-lb
**11.** 76,128 ft-lb    **13.** 125,664 ft-lb    **17.** 2075.83 in.-lb
**19.** 350,000 ft-lb    **21.** 952,381 mi-lb    **23.** 43,200 ft-lb
**25.** 10,477,274 ft-lb    **27.** 264.84 ft-lb

**PROBLEM SET 6.6**

**1.** $\frac{5}{21}$    **3.** $\frac{21}{5}$    **5.** $M_y = 17$, $M_x = -3$; $\bar{x} = 1$, $\bar{y} = -\frac{3}{17}$
**7.** $\bar{x} = -\frac{3}{14}$, $\bar{y} = \frac{1}{14}$    **9.** $\bar{x} = \frac{9}{16}$, $\bar{y} = \frac{31}{16}$

**11.**

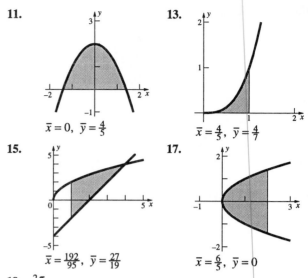

$\bar{x} = 0$, $\bar{y} = \frac{4}{5}$

**13.**

$\bar{x} = \frac{4}{5}$, $\bar{y} = \frac{4}{7}$

**15.**

$\bar{x} = \frac{192}{95}$, $\bar{y} = \frac{27}{19}$

**17.**

$\bar{x} = \frac{6}{5}$, $\bar{y} = 0$

**19.** $\frac{2\pi}{5}$

**21.** The centroid is $\frac{4a}{3\pi}$ units perpendicular from the center of the diameter.

**23.** (a) $V = 2\pi \int_c^d (e - y)w(y)dy$

**25.** (a) $4\pi r^3 n \sin\frac{\pi}{2n}\cos^2\frac{\pi}{2n}$

**CHAPTER REVIEW 6.7**

**Concepts Test**

**1.** False    **3.** False    **5.** True    **7.** False    **9.** False
**11.** False    **13.** True    **15.** True    **17.** True

**Sample Test**

**1.** $\frac{1}{6}$    **3.** $\frac{\pi}{6}$    **5.** $\frac{5\pi}{6}$
**7.** $V(S_1) = \frac{\pi}{30}$; $V(S_2) = \frac{\pi}{6}$; $V(S_3) = \frac{7\pi}{10}$; $V(S_4) = \frac{5\pi}{6}$
**9.** 205,837 ft-lb    **11.** (a), (b). $\frac{32}{3}$    **13.** $\frac{2048\pi}{15}$

**15.** $\frac{53}{6}$    **17.** 36    **19.** $\pi\int_a^b [f^2(x) - g^2(x)]dx$

**21.** $M_x = \frac{\delta}{2}\int_a^b [f^2(x) - g^2(x)]dx$

$M_y = \delta\int_a^b x[f(x) - g(x)]dx$

**23.** $2\pi\int_a^b f(x)\sqrt{1 + [f'(x)]^2}\,dx$

$+ 2\pi\int_a^b g(x)\sqrt{1 + [g'(x)]^2}\,dx$

$+ \pi[f^2(a) - g^2(a)] + \pi[f^2(b) - g^2(b)]$

## ADDITIONAL PROBLEMS 6.8

**1.** (a) ft/sec; $F(T)$ is the change in velocity from 6 seconds to $T$ seconds.;
(b) ft-lb; $F(s)$ is the change in work from 3 feet to $s$ feet.;
(c) in.; $F(r)$ is the center of mass of an object $r$ inches long whose mass at $x$ is $f(x)$.

**3.** (a) $\frac{1}{3}$;   (b) $\frac{6}{343}$;   (c) $\frac{1}{27}$

**7.** (a) $\frac{1}{21}$;

(b) $P(t)=\begin{cases} \frac{2}{4375}t^5 - \frac{6}{875}t^4 + \frac{24}{875}t^3 & \text{if } 0\le t\le 5 \\ -\frac{1}{21}t^2 + \frac{16}{21}t - \frac{43}{21} & \text{if } 5<t<8 \\ 1 & \text{otherwise} \end{cases}$

**11.** 4602.96 lb   **13.** 4.54 ft
**15.** (a) 19 hp;   (b) 88 hp;   (c) 1,898,424 ft-lb

## PROBLEM SET 7.1

**1.** (a) 1.792;   (b) 0.406;   (c) 4.396;   (d) 0.3465;
(e) −3.584;   (f) 3.871

**3.** $\frac{2x+3}{x^2+3x+\pi}$   **5.** $\frac{3}{x-4}$   **7.** $\frac{3}{x}$

**9.** $2x+4x\ln x+\frac{3}{x}(\ln x)^2$   **11.** $\frac{1}{\sqrt{x^2+1}}$   **13.** $\frac{1}{243}$

**15.** $\frac{1}{2}\ln|2x+1|+C$   **17.** $\ln|3v^2+9v|+C$

**19.** $(\ln x)^2+C$   **21.** $\frac{1}{10}[\ln(486+\pi)-\ln\pi]$

**23.** $\ln\frac{(x+1)^2}{x}$   **25.** $\ln\frac{x^2(x-2)}{x+2}$

**27.** $\frac{-(x^3+33x^2+8)}{2(x^3-4)^{3/2}}$   **29.** $\frac{-(10x^2+219x-118)}{6(x-4)^2(x+13)^{1/2}(2x+1)^{4/3}}$

**31.**   **33.**

**35.**

**37.** Minimum $f(1)=-1$   **39.** $\lim\limits_{x\to\infty}\ln x=\infty$   **41.** $x=3$
**43.** $\ln 2$   **45.** (a) 1   (b) 3   **47.** $\pi\ln 4\approx 4.355$
**51.** (a) Maxima: $\left(\frac{\pi}{2},0.916\right)$, $\left(\frac{5\pi}{2},0.916\right)$;
minimum: $\left(\frac{3\pi}{2},-0.693\right)$;
(b) (3.871, −0.182), (5.553, −0.182);   (c) 4.042

**53.**
(a) 0.139;   (b) 0.260

## PROBLEM SET 7.2

**1.** $f^{-1}(2)=4$   **3.** No inverse   **5.** $f^{-1}(2)=-1$
**15.** $f^{-1}(x)=x-1$   **17.** $f^{-1}(x)=x^2-1, x\ge 0$
**19.** $f^{-1}(x)=3-\frac{1}{x}$   **21.** $f^{-1}(x)=-\frac{\sqrt{x}}{2}$
**23.** $f^{-1}(x)=1+\sqrt[3]{x}$   **25.** $f^{-1}(x)=\frac{1+x}{1-x}$
**27.** $f^{-1}(x)=\left(\frac{2-x}{x-1}\right)^{1/3}$   **29.** $(-\infty,-0.25]$ or $[-0.25,\infty)$
**31.**   **33.**
$(f^{-1})'(3)\approx\frac{1}{10}$   $(f^{-1})'(3)\approx-\frac{1}{3}$
**35.** $\frac{1}{16}$   **37.** $\frac{1}{4}$   **41.** (a) 1;   (b) $\frac{2}{\sqrt{7}}$;   (c) $\frac{1}{\sqrt{2}}$
**43.** $\frac{3}{5}$

## PROBLEM SET 7.3

**1.** (a) 20.086;   (b) 0.0183;   (c) 8.1662;   (d) 544.6;
(e) 4.1;   (f) 0.47;   (g) 1.20
**3.** $x^3$   **5.** $\cos x$   **7.** $3\ln x-3x$   **9.** $3x^2$
**11.** $e^{x+2}$   **13.** $\frac{e^{\sqrt{x+2}}}{2\sqrt{x+2}}$   **15.** $2x$
**17.** $x^2e^x(x+3)$   **19.** $x\sqrt{e^{x^2}}+\frac{x}{|x|}e^{\sqrt{x^2}}$   **21.** $-\frac{y}{x}$
**23.** (a)   (b)

**25.**
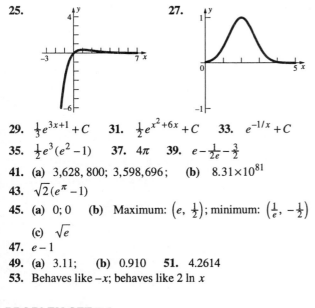

**27.**

**25.**
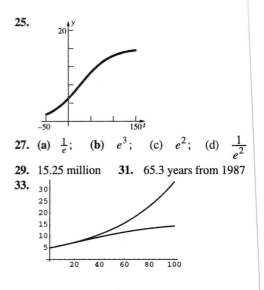

**29.** $\frac{1}{3}e^{3x+1}+C$   **31.** $\frac{1}{2}e^{x^2+6x}+C$   **33.** $e^{-1/x}+C$

**35.** $\frac{1}{2}e^3(e^2-1)$   **37.** $4\pi$   **39.** $e-\frac{1}{2e}-\frac{3}{2}$

**41.** (a) $3{,}628{,}800$; $3{,}598{,}696$;   (b) $8.31\times10^{81}$

**43.** $\sqrt{2}(e^\pi-1)$

**45.** (a) $0$; $0$   (b) Maximum: $\left(e, \frac{1}{2}\right)$; minimum: $\left(\frac{1}{e}, -\frac{1}{2}\right)$

   (c) $\sqrt{e}$

**47.** $e-1$

**49.** (a) $3.11$;   (b) $0.910$   **51.** $4.2614$

**53.** Behaves like $-x$; behaves like $2\ln x$

## PROBLEM SET 7.4

**1.** $3$   **3.** $8$   **5.** $9$   **7.** $1$   **9.** $1.540$   **11.** $0.1747$

**13.** $4.08746$   **15.** $1.9307$   **17.** $2\cdot6^{2x}\ln6$   **19.** $\dfrac{1}{\ln3}$

**21.** $3^z\left[\dfrac{1}{z+5}+\ln(z+5)\ln3\right]$   **23.** $\dfrac{2^{x^2}}{2\ln2}+C$   **25.** $\dfrac{10}{\ln5}$

**27.** $10^{x^2}2x\ln10+20x^{19}$   **29.** $(\pi+1)x^\pi+(\pi+1)^x\ln(\pi+1)$

**31.** $(x^2+1)^{\ln x}\left(\dfrac{\ln(x^2+1)}{x}+\dfrac{2x\ln x}{x^2+1}\right)$   **33.** $\sin1$

**35.** $\log_{1/2}x=-\log_2 x$

**37.** $E\approx5.017\times10^8$ kW-h for magnitude 7;
   $E\approx1.560\times10^{10}$ kW-h for magnitude 8

**39.** $r=2^{1/12}\approx1.0595$; frequency of $\overline{C}=440\sqrt[4]{2}\approx523.25$

**47.** $\lim\limits_{x\to0^+}x^x=1$; minimum: $(e, e^e)$   **49.** $20.2259$

## PROBLEM SET 7.5

**1.** $y=4e^{6t}$   **3.** $y=0.005t+1.95$   **5.** $56{,}125$

**7.** $15.9$ days

**9.** $4.64$ million; $4.79$ million; $6.17$ million; $105$ million

**11.** $7.43$ g   **13.** $2950$ years ago   **15.** $81.6°$

**17.** (a) \$449.63;   (b) \$453.13;   (c) \$453.46;
   (d) \$453.47

**19.** (a) $5.805$ years (5 years 10 months);
   (b) $5.776$ years (5 years, 9 months, 9 days)

**21.** \$82.7 billion

**27.** (a) $\frac{1}{e}$;   (b) $e^3$;   (c) $e^2$;   (d) $\dfrac{1}{e^2}$

**29.** $15.25$ million   **31.** $65.3$ years from 1987

**33.**

## PROBLEM SET 7.6

**1.** $\frac{\pi}{4}$   **3.** $-\frac{\pi}{3}$   **5.** $\frac{\pi}{3}$   **7.** $-\frac{\pi}{6}$   **9.** $-\frac{\pi}{6}$

**11.** $0.4567$   **13.** $0.1115$   **15.** $-1.412$   **17.** $0.9548$

**19.** $2.038$   **21.** $0.6259$   **23.** $-1.160$   **25.** $\theta=\sin^{-1}\frac{x}{8}$

**27.** $\theta=\sin^{-1}\frac{5}{x}$   **29.** $\theta=\tan^{-1}\frac{3}{x}-\tan^{-1}\frac{1}{x}$   **31.** $\frac{1}{9}$

**33.** $\frac{56}{65}$   **41.** (a) $\frac{\pi}{2}$;   (b) $-\frac{\pi}{2}$

**43.**

**45.** (a) $\frac{1}{2}\arccos\frac{x}{3}$, $0\le x\le\frac{\pi}{2}$;

   (b) $\frac{1}{3}\arcsin\frac{x}{2}$, $-\frac{\pi}{6}\le x\le\frac{\pi}{6}$;

   (c) $\arctan 2x$, $-\frac{\pi}{2}\le x\le\frac{\pi}{2}$;

   (d) $\dfrac{1}{\arcsin x}$, $-\infty<x<-\frac{2}{\pi}$ or $\frac{2}{\pi}<x<\infty$

**49.** $\pi b^2-b^2\cos^{-1}\dfrac{b}{2a}-2a^2\sin^{-1}\dfrac{b}{2a}+\dfrac{b\sqrt{4a^2-b^2}}{2}$

**51.** $\frac{\pi}{2}-\arcsin x=\arccos x$

## PROBLEM SET 7.7

**1.** $2\sin(x-1)\cos(x-1)$   **3.** $-\csc x\cot x$

**5.** $e^{\tan x}\sec^2 x$   **7.** $-(\sin x)e^{\cos x}(1+\cos x)$   **9.** $\sec x$

**11.** $\dfrac{4x}{\sqrt{1-4x^4}}$   **13.** $x^2\left[\dfrac{xe^x}{1+e^{2x}}+3\tan^{-1}(e^x)\right]$

**15.** $\text{arccot}\,x-\dfrac{x}{1+x^2}$   **17.** $-\dfrac{x}{(1+x^2)^{3/2}}$

**19.** $\dfrac{3}{|x|\sqrt{x^6-1}}$   **21.** $\dfrac{1}{1+x^2}$   **23.** $-\tfrac{1}{2}\cos(x^2)+C$

**25.** $-\ln|\cos x|+C$   **27.** $\ln|\tan x|+C$   **29.** $\dfrac{\sin e^2-\sin 1}{2}$

**31.** $\tfrac{\pi}{4}$   **33.** $\tfrac{\pi}{2}$   **35.** $\tfrac{1}{2}\arctan 2x+C$   **43.** $\pi$

**45.** 4.9 ft   **47.** $\tfrac{1}{13}$ rad/s   **49.** 1 rev/min

**51.** $3.96\times10^{-4}$ rad/s   **53.** $\approx[0.2513,\,0.2585]$

## PROBLEM SET 7.8

**13.** $2\sinh x\cosh x$   **15.** $10\sinh x\cosh x$

**17.** $3\sinh(3x+1)$   **19.** $\coth x$   **21.** $x^2\sinh x+2x\cosh x$

**23.** $\cosh 3x\cosh x+3\sinh 3x\sinh x$

**25.** $2\tanh x\cosh 2x+\sinh 2x\,\mathrm{sech}^2 x$   **27.** $\dfrac{2x}{\sqrt{x^4+1}}$

**29.** $-\dfrac{1}{2(x^2-3x+2)}$   **31.** $\dfrac{3x}{\sqrt{9x^2-1}}+\cosh^{-1}3x$

**33.** $\dfrac{1}{\sqrt{x^2-1}\cosh^{-1}x}$   **35.** $\cos x\cosh(\sin x)$

**37.** $-\dfrac{\sin x}{\sqrt{\cos^2 x+1}}$   **39.** $-\csc^2 x\,\mathrm{sech}^2(\cot x)$   **41.** $\tfrac{20}{9}$

**43.** $\tfrac{1}{2\pi}\sinh(\pi x^2+5)+C$   **45.** $2\cosh(2z^{1/4})+C$

**47.** $\cosh(\sin x)+C$   **49.** $\tfrac{1}{4}[\ln(\sinh x^2)]^2+C$   **51.** $\tfrac{1}{4}$

**53.** $\tfrac{\pi}{2}+\tfrac{\pi}{4}\sinh 2$   **55.** $\pi+\tfrac{\pi}{2}\sinh 2$

**59. (a)**

**(b)** 42,200 ft$^3$;   **(c)** 5640 ft$^2$

**65.**

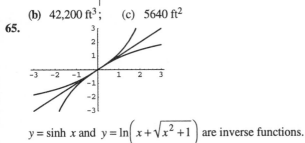

$y=\sinh x$ and $y=\ln\!\left(x+\sqrt{x^2+1}\right)$ are inverse functions.

## CHAPTER REVIEW 7.9

Concepts Test

**1.** False   **3.** True   **5.** True   **7.** False   **9.** True
**11.** True   **13.** True   **15.** True   **17.** False   **19.** True
**21.** False   **23.** False   **25.** False   **27.** True
**29.** False   **31.** True   **33.** True   **35.** True   **37.** True
**39.** True   **41.** False   **43.** True

## Sample Test

**1.** $\tfrac{4}{x}$   **3.** $(2x-4)e^{x^2-4x}$   **5.** $\sec^2 x$   **7.** $\dfrac{\mathrm{sech}^2\sqrt{x}}{\sqrt{x}}$

**9.** $|\sec x|$   **11.** $\dfrac{1}{\sqrt{e^{2x}-1}}$   **13.** $\dfrac{15e^{5x}}{e^{5x}+1}$

**15.** $-\dfrac{e^{\sqrt{x}}\sin e^{\sqrt{x}}}{2\sqrt{x}}$   **17.** $-\dfrac{1}{\sqrt{x-x^2}}$

**19.** $-\dfrac{\csc\sqrt{x}\cot\sqrt{x}}{\sqrt{x}}$   **21.** $20\sec 5x(2\sec^2 5x-1)$

**23.** $x^{1+x}\!\left(\ln x+1+\tfrac{1}{x}\right)$   **25.** $\tfrac{1}{3}e^{3x-1}+C$

**27.** $-\cos e^x+C$   **29.** $\dfrac{\ln(e^{x+3}+1)}{e}+C$

**31.** $2\sin^{-1}2x+C$   **33.** $-\tan^{-1}(\ln x)+C$

**35.** Increasing: $\left[-\tfrac{\pi}{2},\tfrac{\pi}{4}\right]$; decreasing: $\left[\tfrac{\pi}{4},\tfrac{\pi}{2}\right]$;

concave up: $\left(-\tfrac{\pi}{2},-\tfrac{\pi}{4}\right)$; concave down: $\left(-\tfrac{\pi}{4},\tfrac{\pi}{2}\right)$

inflection point: $\left(-\tfrac{\pi}{4},0\right)$; global minimum: $\left(-\tfrac{\pi}{2},-1\right)$

global maximum: $\left(\tfrac{\pi}{4},\sqrt{2}\right)$

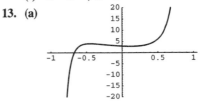

**37. (b)** 1   **(c)** $\tfrac{1}{15}$
**39. (a)** \$112;   **(b)** \$112.68;   **(c)** \$112.75;
   **(d)** \$112.75
**41.** $y=1$

## ADDITIONAL PROBLEMS 7.10

**1.** \$1051.27
**3. (a)** \$1232.61;

   **(b)** $\dfrac{100e^{0.05\cdot(30/365)}-100e^{0.05\cdot(390/365)}}{1-e^{0.05\cdot(30/365)}}$

**5. (c)** 10 years
**11. (a)** In order of increasing slope: $y=2^x$; $y=3^x$, $y=4^x$;
   **(b)** $\ln y$ is linear with respect to $x$;
   **(c)** $b=2^{5/2}$; $c=2^{3/2}$
**13. (a)**

(b)

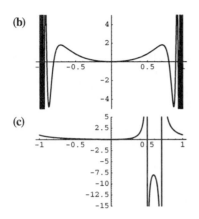

(c)

## PROBLEM SET 8.1

**1.** $\frac{1}{6}(x-2)^6 + C$   **3.** $\frac{1}{12}(x^2+1)^6 + C$

**5.** $\frac{1}{12}\tan^{-1}\left(\frac{x}{2}\right) + C$   **7.** $\frac{1}{2}\ln(x^2+4) + C$

**9.** $2(4+z^2)^{3/2} + C$   **11.** $\frac{1}{2}\tan^2 z + C$

**13.** $-2\cos\sqrt{t} + C$   **15.** $\tan^{-1}\dfrac{\sqrt{2}}{2}$

**17.** $\frac{3}{2}x^2 - x + \ln|x+1| + C$   **19.** $\frac{4}{3}(\tan x)^{3/2} + C$

**21.** $-\frac{1}{2}\cos(\ln 4x^2) + C$   **23.** $6\sin^{-1}(e^x) + C$

**25.** $-3\sqrt{1-e^{2x}} + C$   **27.** $\frac{1}{\ln 3}$   **29.** $x - \ln|\sin x| + C$

**31.** $\frac{1}{2}\ln\left|\sec(z^2+4z-3)\right| + C$   **33.** $\ln\left|\sec e^x + \tan e^x\right| + C$

**35.** $\tan x + e^{\sin x} + C$   **37.** $-\dfrac{1}{3\sin(t^3-2)} + C$

**39.** $-\frac{1}{3}[\cot(t^3-2) + t^3] + C$   **41.** $\frac{1}{2}e^{\tan^{-1}2t} + C$

**43.** $\frac{1}{6}\sin^{-1}\left(\dfrac{3y^2}{4}\right) + C$   **45.** $\tan^{-1}(\sec x) + C$

**47.** $\frac{1}{3}\sin^{-1}\left(\dfrac{e^{3t}}{2}\right) + C$   **49.** $\frac{1}{4}\tan^{-1}\left(\frac{1}{4}\right)$

**51.** $\frac{1}{2}\tan^{-1}\left(\dfrac{x+1}{2}\right) + C$   **53.** $\frac{1}{3}\tan^{-1}(3x+3) + C$

**55.** $\frac{1}{18}\ln\left|9x^2 + 18x + 10\right| + C$   **57.** $\frac{1}{3}\sec^{-1}\left(\dfrac{|\sqrt{2}t|}{3}\right) + C$

**59.** $\frac{2}{135}(9x-4)(3x+2)^{3/2} + C$   **61.** $\frac{1}{24}\ln\left|\dfrac{4x+3}{4x-3}\right| + C$

**63.** $\frac{x}{16}(4x^2-9)\sqrt{9-2x^2} + \frac{81\sqrt{2}}{23}\sin^{-1}\left(\dfrac{\sqrt{2}x}{3}\right) + C$

**65.** $\frac{1}{\sqrt{3}}\ln\left|\sqrt{3}x + \sqrt{3x^2+5}\right| + C$

**67.** $\ln\left|t+1+\sqrt{t^2+2t-3}\right| + C$

**69.** $\frac{2}{27}(3\sin t - 10)\sqrt{3\sin t + 5} + C$   **71.** $\ln\left|\sqrt{2}+1\right|$   **73.** $\pi^2$

## PROBLEM SET 8.2

**1.** $\frac{1}{2}x - \frac{1}{4}\sin 2x + C$   **3.** $-\cos x + \frac{1}{3}\cos^3 x + C$

**5.** $\frac{8}{15}$   **7.** $\frac{1}{4}\tan^2 2z + \frac{1}{2}\ln|\cos 2z| + C$

**9.** $-\frac{1}{12}\cos^3 4x + \frac{1}{10}\cos^5 4x - \frac{1}{28}\cos^7 4x + C$

**11.** $-\frac{1}{3}\csc 3\theta - \frac{1}{3}\sin 3\theta + C$

**13.** $\frac{3}{128}t - \frac{1}{384}\sin 12t + \frac{1}{3072}\sin 24t + C$

**15.** $\frac{1}{20}\sec^5 4y - \frac{1}{12}\sec^3 4y + C$

**17.** $-\frac{1}{2}\cot^2 x - \frac{1}{4}\cot^4 x + C$   **19.** $-\frac{1}{3}\tan^{-3}q + C$

**21.** $\frac{1}{2}\cos y - \frac{1}{18}\cos 9y + C$

**23.** $-\frac{1}{6}\cot^3 2x + \frac{1}{2}\cot 2x + x + C$

**25.** $\frac{1}{7}\tan 7x + \frac{1}{21}\tan^3 7x + C$   **27.** $-\frac{1}{2}\cot^2 x - \ln|\sin x| + C$

**29.** $\tan x - \cot x + C$   **31.** $\dfrac{5\pi^2}{2} + \dfrac{\pi^4}{3}$

## PROBLEM SET 8.3

**1.** $\frac{2}{5}(x+1)^{5/2} - \frac{2}{3}(x+1)^{3/2} + C$

**3.** $\frac{2}{27}(3t+4)^{3/2} - \frac{8}{9}(3t+4)^{1/2} + C$

**5.** $2\sqrt{2} - 2 - 2e\ln\left(\dfrac{\sqrt{2}+e}{1+e}\right)$

**7.** $\frac{2}{63}(3t+2)^{7/2} - \frac{4}{45}(3t+2)^{5/2} + C$

**9.** $2\ln\left|\dfrac{2-\sqrt{4-x^2}}{x}\right| + \sqrt{4-x^2} + C$   **11.** $\pm\frac{1}{3}\sec^{-1}\left(\frac{x}{3}\right) + C$

**13.** $\dfrac{2\sqrt{2}}{3} - \dfrac{\sqrt{3}}{2}$   **15.** $-\sqrt{1-t^2} + C$

**17.** $-2\sqrt{1-z^2} - 3\sin^{-1}z + C$   **19.** $\sqrt{y^2+9} + \dfrac{9}{\sqrt{y^2+9}} + C$

**21.** $\ln\left|\sqrt{x^2+2x+5} + x+1\right| + C$

**23.** $3\sqrt{x^2+2x+5} - 3\ln\left|\sqrt{x^2+2x+5} + x+1\right| + C$

**25.** $\frac{9}{2}\sin^{-1}\left(\dfrac{x+2}{3}\right) + \dfrac{x+2}{2}\sqrt{5-4x-x^2} + C$

**27.** $\sin^{-1}\left(\frac{x-2}{2}\right) + C$

**29.** $\ln\left|x^2+2x+2\right| - \tan^{-1}(x+1) + C$

**31.** $\frac{\pi}{16}\left(\frac{1}{10} + \frac{\pi}{4} - \tan^{-1}\frac{1}{2}\right)$   **33.** $\frac{1}{2}\ln\left|x^2+9\right| + C$

**35.** $2\ln\left|\dfrac{2-\sqrt{4-x^2}}{x}\right| + \sqrt{4-x^2} + C$

**39.** $y = -\sqrt{a^2 - x^2} - a\ln\left|\dfrac{a - \sqrt{a^2 - x^2}}{x}\right|$

## PROBLEM SET 8.4

**1.** $xe^x - e^x + C$    **3.** $\frac{1}{5}te^{5t+\pi} - \frac{1}{25}e^{5t+\pi} + C$

**5.** $x\sin x + \cos x + C$

**7.** $(t-3)\sin(t-3) + \cos(t-3) + C$

**9.** $\frac{2}{3}t(t+1)^{3/2} - \frac{4}{15}(t+1)^{5/2} + C$    **11.** $x\ln 3x - x + C$

**13.** $x\arctan x - \frac{1}{2}\ln(1+x^2) + C$    **15.** $-\frac{\ln x}{x} - \frac{1}{x} + C$

**17.** $\frac{2}{3}t^{3/2}\ln t - \frac{4}{9}t^{3/2} + C$    **19.** $\frac{1}{4}z^4\ln z - \frac{1}{16}z^4 + C$

**21.** $t\arctan\left(\frac{1}{t}\right) + \frac{1}{2}\ln(1+t^2) + C$

**23.** $-\frac{x}{3}\cos^3 x + \frac{1}{3}\sin x - \frac{1}{9}\sin^3 x + C$    **25.** $\dfrac{\pi}{2\sqrt{3}} + \ln 2$

**27.** $\dfrac{\pi}{4} - \dfrac{\pi}{6\sqrt{3}} + \dfrac{1}{2}\ln\dfrac{2}{3}$

**29.** $\frac{2}{9}x^3(x^3+4)^{3/2} - \frac{4}{45}(x^3+4)^{5/2} + C$

**31.** $\dfrac{t^4}{6(7-3t^4)^{1/2}} + \frac{1}{9}(7-3t^4)^{1/2} + C$

**33.** $\dfrac{z^4}{4(4-z^4)} + \frac{1}{4}\ln\left|4 - z^4\right| + C$

**35.** $x\cosh x - \sinh x + C$

**37.** $\frac{1}{2}\sec x\tan x + \frac{1}{2}\ln\left|\sec x + \tan x\right| + C$

**39.** $2\sqrt{x}\ln x - 4\sqrt{x} + C$    **41.** $\dfrac{x}{\ln 2}2^x - \dfrac{1}{(\ln 2)^2}2^x + C$

**43.** $x^2 e^x - 2xe^x + 2e^x + C$    **45.** $z\ln^2 z - 2z\ln z + 2z + C$

**47.** $\frac{1}{2}e^t(\sin t + \cos t) + C$

**49.** $x^2\sin x + 2x\cos x - 2\sin x + C$

**51.** $\frac{x}{2}[\sin(\ln x) - \cos(\ln x)] + C$

**53.** $x\ln^3 x - 3x\ln^2 x + 6x\ln x - 6x + C$    **73.** $1$

**75.** $9 - \dfrac{9}{e^3}$

**77.** $\dfrac{\sqrt{2}\pi}{4} - 1$    **79.** $\bar{x} = \dfrac{e^2+1}{4}$, $\bar{y} = \dfrac{e-2}{4}$

**81. (a)** $(x^3 - 2x)e^x - (3x^2 - 2)e^x + 6xe^x - 6e^x + C$

**(b)** $(x^2 - 3x + 1)(-\cos x) - (2x - 3)(-\sin x) + 2\cos x + C$

**83.** $\begin{cases} 0 & \text{if } n \text{ is odd} \\ \dfrac{1\cdot 3\cdot 5\cdots(n-1)}{2\cdot 4\cdot 6\cdots n}2\pi & \text{if } n \text{ is even} \end{cases}$

**97.** $e^x(3x^4 - 12x^3 + 38x^2 - 76x + 76)$

## PROBLEM SET 8.5

**1.** $\ln|x| - \ln|x+1| + C$

**3.** $-\frac{3}{2}\ln|x+1| + \frac{3}{2}\ln|x-1| + C$

**5.** $3\ln|x+4| - 2\ln|x-1| + C$

**7.** $4\ln|x+5| - \ln|x-2| + C$

**9.** $2\ln|2x-1| - \ln|x+5| + C$

**11.** $\frac{5}{3}\ln|3x-2| + 4\ln|x+1| + C$

**13.** $2\ln|x| - \ln|x+1| + \ln|x-2| + C$

**15.** $\ln|2x-1| - \ln|x+3| + 3\ln|x-2| + C$

**17.** $\frac{1}{2}x^2 - x + \frac{8}{3}\ln|x+2| + \frac{1}{3}\ln|x-1| + C$

**19.** $\frac{1}{2}x^2 - 2\ln|x| + 7\ln|x+2| + 7\ln|x-2| + C$

**21.** $\ln|x-3| - \dfrac{4}{x-3} + C$    **23.** $-\dfrac{3}{x+1} + \dfrac{1}{2(x+1)^2} + C$

**25.** $2\ln|x| + \ln|x-4| + \dfrac{1}{x-4} + C$

**27.** $-2\ln|x| + \frac{1}{2}\tan^{-1}\left(\frac{x}{2}\right) + 2\ln\left|x^2+4\right| + C$

**29.** $-2\ln|2x-1| + \frac{3}{2}\ln\left|x^2+9\right| + C$

**31.** $-\frac{2}{125}\ln|x-1| - \dfrac{1}{25(x-1)} + \frac{2}{125}\ln|x+4| - \dfrac{1}{25(x+4)} + C$

**33.**

$\sin t - \frac{50}{13}\ln|\sin t + 3| - \frac{68}{13}\tan^{-1}(\sin t - 2) - \frac{41}{26}\ln\left|\sin^2 t - 4\sin t + 5\right| + C$

**35.** $\frac{1}{2}\ln\left|x^2+1\right| + \dfrac{5}{2(x^2+1)} + C$

**37.** $\frac{3}{2}\tan^{-1}\dfrac{x}{2} + \dfrac{2x-5}{2(x^2+4)} + C$

**39.** $\frac{1}{8}\ln\left|\dfrac{\sqrt{2}+1}{\sqrt{2}-1}\right| + \frac{1}{2}\tan^{-1}\dfrac{1}{2} + \dfrac{1}{6\sqrt{2}}$

**43. (a)** $y = \dfrac{16}{1 + 7e^{-\left(\frac{1}{50}\ln\frac{7}{3}\right)t}}$;    **(b)** 6.34 billion;    **(c)** In 2055

**45.** $\displaystyle\lim_{t\to\infty} y = M$

## CHAPTER REVIEW 8.6

### Concepts Test

**1.** True    **3.** False    **5.** True    **7.** True    **9.** True
**11.** False    **13.** True    **15.** True    **17.** False
**19.** True    **21.** False    **23.** False    **25.** True

### Sample Test

**1.** 2    **3.** $e-1$

**5.** $\frac{1}{3}y^3 - \frac{1}{2}y^2 + 2y - 2\ln|1+y| + C$

**7.** $\frac{1}{2}\ln\left|y^2 - 4y + 2\right| + C$    **9.** $e^t + 2\ln\left|e^t - 2\right| + C$

**11.** $\dfrac{1}{\sqrt{2}}\sin^{-1}\left(\dfrac{x-1}{3}\right) + C$    **13.** $\dfrac{1}{\sqrt{3}}\ln\left|\sqrt{y^2 + \frac{2}{3}} + y\right| + C$

**15.** $-\ln|\ln|\cos x|| + C$    **17.** $\cosh x + C$

**19.** $-x\cot x - \tfrac{1}{2}x^2 + \ln|\sin x| + C$    **21.** $\tfrac{1}{4}\left[\ln(t^2)\right]^2 + C$

**23.** $-\tfrac{3}{82}e^{t/3}(9\cos 3t - \sin 3t) + C$    **25.**
$-\tfrac{1}{2}\cos x - \tfrac{1}{4}\cos 2x + C$

**27.** $\tfrac{1}{6}\sec^3(2x) - \tfrac{1}{2}\sec(2x) + C$

**29.** $\tfrac{2}{5}\tan^{5/2}x + \tfrac{2}{9}\tan^{9/2}x + C$    **31.** $-\sqrt{9 - e^{2y}} + C$

**33.** $3\sin x + C$    **35.** $\tfrac{1}{4}\tan^{-1}(e^{4x}) + C$

**37.** $\tfrac{2}{3}(w+5)^{3/2} - 10(w+5)^{1/2} + C$

**39.** $-\tfrac{1}{6}\tan^{-1}\left(\dfrac{\cos^2 y}{3}\right) + C$

**41.** $\ln|x| - \dfrac{2}{x} - \tfrac{1}{2}\ln|x^2+3| + \dfrac{2}{\sqrt{3}}\tan^{-1}\left(\dfrac{x}{\sqrt{3}}\right) + C$

**43. (a)** $\dfrac{A}{2x+1} + \dfrac{B}{(2x+1)^2} + \dfrac{C}{(2x+1)^3}$

**(b)** $\dfrac{A}{x-1} + \dfrac{B}{(x-1)^2} + \dfrac{C}{2-x} + \dfrac{D}{(2-x)^2} + \dfrac{E}{(2-x)^3}$

**(c)** $\dfrac{Ax+B}{x^2+x+10} + \dfrac{Cx+D}{(x^2+x+10)^2}$

**(c)** $\dfrac{A}{1-x} + \dfrac{B}{(1-x)^2} + \dfrac{C}{1+x} + \dfrac{D}{(1+x)^2}$
$+ \dfrac{Ex+F}{x^2-x+10} + \dfrac{Gx+H}{(x^2-x+10)^2}$

**(e)** $\dfrac{A}{x+3} + \dfrac{B}{(x+3)^2} + \dfrac{C}{(x+3)^3} + \dfrac{D}{(x+3)^4}$
$+ \dfrac{Ex+F}{x^2+2x+10} + \dfrac{Gx+H}{(x^2+2x+10)^2}$

**(f)** $\dfrac{Ax+B}{2x^2+x+10} + \dfrac{Cx+D}{(2x^2+x+10)^2} + \dfrac{Ex+F}{(2x^2+x+10)^3}$

**45.** $\sqrt{5} + 4\ln\left(\dfrac{1+\sqrt{5}}{2}\right)$    **47.** $2\pi\ln\tfrac{32}{25}$

**49.** $4\pi\left[2 - \ln 3 - \tfrac{1}{2}(\ln 3)^2\right]$    **51.** $\ln 7 - \tfrac{6}{7}$

**53.** $\ln\left(\dfrac{2\sqrt{3}+3}{3}\right)$

## PROBLEM SET 9.1

**1.** 1    **3.** $-1$    **5.** $-\tfrac{2}{7}$    **7.** $-\infty$    **9.** 0

**11.** $-\tfrac{3}{2}$    **13.** $-\tfrac{2}{7}$    **15.** $-\tfrac{1}{4}$    **17.** $\infty$    **19.** $-\tfrac{1}{24}$

**21.** $-\infty$    **23.** 1    **27. (a)** $\tfrac{3}{4}$;    **(b)** $\tfrac{1}{2}$    **29.** $4\pi b^2$

**33. (a)** $\tfrac{1}{2}$;    **(b)** 2

## PROBLEM SET 9.2

**1.** 0    **3.** 0    **5.** 3    **7.** 0    **9.** $\infty$    **11.** 0    **13.** 1

**15.** 1    **17.** 0    **19.** $e^4$    **21.** 1    **23.** 1    **25.** 0

**27.** 1    **29.** 0    **31.** $\infty$    **33.** 1

**35.** Limit does not exist.    **37.** 0    **39.** 1

**41. (a)** 1;    **(b)** 1;    **(c)** $\ln a$;    **(d)** $\infty$

**43.**

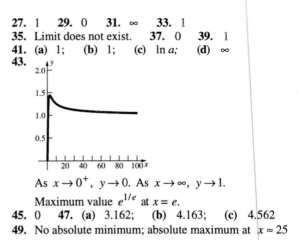

As $x \to 0^+$, $y \to 0$. As $x \to \infty$, $y \to 1$.
Maximum value $e^{1/e}$ at $x = e$.

**45.** 0    **47. (a)** 3.162;    **(b)** 4.163;    **(c)** 4.562

**49.** No absolute minimum; absolute maximum at $x \approx 25$

## PROBLEM SET 9.3

**1.** Diverges    **3.** $\tfrac{1}{e}$    **5.** Diverges    **7.** 100,000

**9.** Diverges    **11.** Diverges    **13.** $\tfrac{1}{2}(\ln 2 + 1)$

**15.** $-\tfrac{1}{4}$    **17.** Diverges    **19.** $\tfrac{\pi}{3}$    **21.** $\pi$    **23.** $\tfrac{1}{2}$

**25.** $\tfrac{1}{2}\ln 3$    **29.** \$1,250,000

**37.** $\displaystyle\int_1^{100} \dfrac{1}{x^2}\,dx = 0.99$;   $\displaystyle\int_1^{100} \dfrac{1}{x^{1.1}}\,dx \approx 3.69$
$\displaystyle\int_1^{100} \dfrac{1}{x^{1.01}}\,dx \approx 4.50$;   $\displaystyle\int_1^{100} \dfrac{1}{x}\,dx = \ln 100 \approx 4.61$;
$\displaystyle\int_1^{100} \dfrac{1}{x^{0.99}}\,dx \approx 4.71$

**39.** $\displaystyle\int_{-1}^1 \dfrac{1}{\sqrt{2\pi}}\exp(-0.5x^2)\,dx \approx 0.6827$;
$\displaystyle\int_{-2}^2 \dfrac{1}{\sqrt{2\pi}}\exp(-0.5x^2)\,dx \approx 0.9545$;
$\displaystyle\int_{-3}^3 \dfrac{1}{\sqrt{2\pi}}\exp(-0.5x^2)\,dx \approx 0.9973$;
$\displaystyle\int_{-4}^4 \dfrac{1}{\sqrt{2\pi}}\exp(-0.5x^2)\,dx \approx 0.9999$

## PROBLEM SET 9.4

**1.** $\dfrac{3}{\sqrt[3]{2}}$    **3.** $2\sqrt{7}$    **5.** $\tfrac{\pi}{2}$    **7.** Diverges    **9.** $\tfrac{21}{2}$

**11.** $\tfrac{1}{2}(2^{2/3} - 10^{2/3})$    **13.** Diverges    **15.** Diverges

**17.** Diverges    **19.** Diverges    **21.** Diverges

**23.** Diverges    **25.** Diverges    **27.** $2\sqrt{2}$    **29.** Diverges

**31.** $\ln(2 + \sqrt{3})$    **35.** 0    **37.** Diverges    **41.** 6

**43. (a)** 3    **45.** No    **53. (a)** $\tfrac{\pi}{2}$;    **(b)** $\pi$

## 9.5 CHAPTER REVIEW 9.5

### Concepts Test

**1.** True    **3.** False    **5.** False    **7.** True    **9.** True
**11.** False    **13.** True    **15.** True    **17.** False
**19.** True    **21.** True    **23.** True    **25.** False

### Sample Test Problems

**1.** 4    **3.** 0    **5.** 2    **7.** 0    **9.** 0    **11.** 1    **13.** 0
**15.** 0    **17.** 1    **19.** 1    **21.** $\frac{1}{2}e^2$    **23.** Diverges
**25.** $1-\frac{\pi}{4}$    **27.** Diverges    **29.** $\frac{1}{\ln 2}$    **31.** 6
**33.** Diverges    **35.** $\frac{\pi}{4}$    **37.** 0
**39.** Converges: $p > 1$; diverges: $p \le 1$
**41.** Converges    **43.** Diverges

## 9.6 ADDITIONAL PROBLEMS

**1.** (b) $\frac{\pi}{4}$    **3.** (a) $C = k$;    (b) 0

**5.** (a) $C = K$;    (b) $\mu = \frac{1}{K}$, $\sigma = \frac{1}{K}$;    (c) $e^{-K}$

**7.** (a) $C = k$;    (b) $\mu = \frac{kM}{k-1}$, finite when $k > 1$;

    (c) $\sigma^2 = \dfrac{kM^2}{(k-1)^2(k-2)}$ when $k > 2$

**9.** (a)

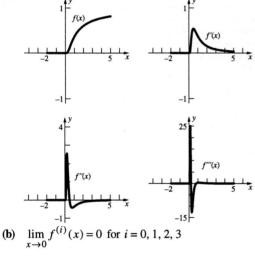

   (b) $\displaystyle\lim_{x \to 0} f^{(i)}(x) = 0$ for $i = 0, 1, 2, 3$

## PROBLEM SET 10.1

**1.** 33.820022    **3.** 106.562271    **5.** 812.616954
**7.** $1 + 2x + 2x^2 + \frac{4}{3}x^3 + \frac{2}{3}x^4$; 1.2712
**9.** $2x - \frac{4}{3}x^3$; 0.2377
**11.** $x - \frac{1}{2}x^2 + \frac{1}{3}x^3 - \frac{1}{4}x^4$; 0.1133
**13.** $x - \frac{1}{3}x^3$; 0.1194
**15.** $e + e(x-1) + \frac{e}{2}(x-1)^2 + \frac{e}{6}(x-1)^3$
**17.** $\frac{\sqrt{3}}{3} + \frac{4}{3}\left(x - \frac{\pi}{6}\right) + \frac{4\sqrt{3}}{9}\left(x - \frac{\pi}{6}\right)^2 + \frac{8}{9}\left(x - \frac{\pi}{6}\right)^3$

**19.** $\frac{\pi}{4} - \frac{1}{2}(x-1) + \frac{1}{4}(x-1)^2 - \frac{1}{12}(x-1)^3$
**21.** $7 + 2(x-1) + (x-1)^2 + (x-1)^3$
**23.** (a) 1.1111;    (b) 1.9375;    (c) 4.0951;    (d) 31
**27.** (c)

| $r$ | $n$ (exact) | $n$ (approx.) | $n$ (rule 70) |
|---|---|---|---|
| 0.05 | 13.892 | 13.889 | 14 |
| 0.10 | 6.960 | 6.959 | 7 |
| 0.15 | 4.650 | 4.649 | 4.667 |
| 0.20 | 3.495 | 3.494 | 3.5 |

**29.** (a)              (b)

**31.** (a) $x - 3x^2 + 2x^3$;    (b) $x - 3x^2 + 2x^3 - x^4$;
    (c) $x - 3x^2 + 2x^3 - x^4$;    (d) $x - 3x^2 + 2x^3 - x^4$;
    (e) $1 + x + x^2 + x^3 + x^4$;    (f) $x - \frac{1}{3}x^3$;
    (g) $1 + x^2 + x^4$;    (h) $1 + 2x + x^2 + \frac{2}{3}x^3 + x^4$

## PROBLEM SET 10.2

**1.** $e^6 + 1$    **3.** $2\sqrt{2}\pi$    **5.** $\frac{e^4}{3}$    **7.** $\frac{17}{10\ln 2}$

**9.** $R_6(x) = \dfrac{x^7}{7(2+c)^7}$; $8.72 \times 10^{-6}$

**11.** $R_6(x) = -\dfrac{\cos c}{5040}\left(x - \frac{\pi}{4}\right)^7$; $2.69 \times 10^{-8}$

**13.** $3.77 \times 10^{-6} \le R_3 \le 4.17 \times 10^{-6}$; actual error: $4.08 \times 10^{-6}$

**15.** $n \ge 9$    **17.** $1 - \frac{1}{2}x + \frac{3}{8}x^2 - \frac{5}{16}x^3$; $|R_3(x)| \le 2.15 \times 10^{-6}$

**19.** 0.1224; $|\text{Error}| \le 0.00013025$

**21.** $-1 - (x-1)^2 + (x-1)^3 + (x-1)^4$

**23.** 0.681998; $|R_3| \le 6.19 \times 10^{-8}$

**29.** (b) $L_{52} = \dfrac{(x-x_1)(x-x_3)(x-x_4)(x-x_5)}{(x_2-x_1)(x_2-x_3)(x_2-x_4)(x_2-x_5)}$;

      $L_{53} = \dfrac{(x-x_1)(x-x_2)(x-x_4)(x-x_5)}{(x_3-x_1)(x_3-x_2)(x_3-x_4)(x_3-x_5)}$;

      $L_{54} = \dfrac{(x-x_1)(x-x_2)(x-x_3)(x-x_5)}{(x_4-x_1)(x_4-x_2)(x_4-x_3)(x_4-x_5)}$;

      $L_{55} = \dfrac{(x-x_1)(x-x_2)(x-x_3)(x-x_4)}{(x_5-x_1)(x_5-x_2)(x_5-x_3)(x_5-x_4)}$

   (d) $-0.75x^2 + 2.75x$

**35.** (a) $(a+b+c) + (2a+b)(x-1) + a(x-1)^2$;
   (b) $c + (a+b)(x) + a(x)(x-1)$

**37.** $P_2(x) = -0.769125 + 0.8425x - 0.073375x^2$;
   $P_2(2) = 0.622375$; $|R_3(2)| \le 1$; $|\text{Error}| \approx 0.0708$

**39.** Maclaurin: $P_2(x) = 1 + x + \frac{1}{2}x^2$, $|R_2(x)| < 0.00607$,

$|e^{0.1} - P_2(0.1)| \approx 0.0001709$; Interpolation:

$P_2(x) = 1 + \frac{2.965}{3}x + \frac{175}{3}x^2$, $|R_2(x)| < 0.00056$,

$|e^{0.1} - P_2(0.1)| \approx 0.0005043$

## PROBLEM SET 10.3

**1.** 0.6766; 0.6671; $\frac{2}{3}$    **3.** 1.8615; 1.8755; $\frac{4\sqrt{2}}{3}$
**5.** 1.5708; 1.9541; 1.9886    **7.** 3.1416    **9.** $n = 5$; 0.74
**11.** $n = 5$; 0.27    **13.** $n = 16$; −7.219    **17.** $n \geq 40{,}825$
**21.** 4570 ft$^2$    **23.** 1,074,585,600 ft$^3$
**25.** (c) 0.99997; $|E_{10}| \leq 0.0235$
**29.** −0.9580
**31.** (c) $\frac{h}{3}(y_1 + 4y_2 + y_3)$;    (e) Yes, the result is valid.
**33.** (a) 1.41615;    (b) 1.42506;    (c) 1.41016
**35.** (a) $|f''(x)| < 25.5$;

  (b) $n \geq 47$;    (c) $|f^{(4)}(x)| < 295$;    (d) $n \geq 7$;

  (e) 1.60024; $|E_{10}| < 0.000164$

## PROBLEM SET 10.4

**1.** 1.46    **3.** 1.45    **5.** −0.12061    **7.** 1.37015
**9.** 0.45018    **11.** 2, 0.58579, 3.41421    **13.** 0.52658
**15.** 1.81712    **17.** 4.49341; −0.21723    **19.** $1.08 \times 10^{-19}$
**21.** If $x_1 = 1.2$, the algorithm fails to converge. If $x_1 = 0.5$,
  $r \approx 0.36788$.
**23.** (c) $i = 0.0151308$; $r = 18.157\%$
**25.** (a) The algorithm computes the root of $\frac{1}{x} - a = 0$ for $x_1$
    close to $\frac{1}{a}$.
**27.** −1.87939, 0.34730, 1.53209
**29.** −2.08204, 0.09251, 0.91314, 1.62015, 1.85411

## PROBLEM SET 10.5

**1.** 0.09237    **3.** 2.21756
**5.** (a)

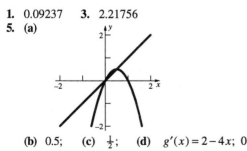

  (b) 0.5;    (c) $\frac{1}{2}$;    (d) $g'(x) = 2 - 4x$; 0

**7.** (a)

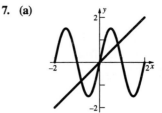

  (b) The algorithm does not yield a convergent sequence.;
  (c) $g'(x) = \frac{3\pi}{2}\cos \pi x$, $|g'(x)| > 1$ near the fixed points
**9.** (b) 0.81673, −0.81673    **11.** 1.839    **13.** 1.7725
**15.** (a) 293.75    (b) 0.01595
**19.** 0, 0.6    **21.** 0.55801, 0.76457
**23.** 0.875, 0.38282, 0.82694, 0.50088

## CHAPTER REVIEW 10.6

### Concepts Test

**1.** True    **3.** True    **5.** True    **7.** True    **9.** False
**11.** True    **13.** True    **15.** True    **17.** False    **19.** False

### Sample Test Problems

**1.** 47.258608    **3.** $3 + 9(x - 2) + 4(x - 2)^2 + (x - 2)^3$
**5.** $\frac{1}{2} - \frac{1}{4}(x - 1) + \frac{1}{8}(x - 1)^2 - \frac{1}{16}(x - 1)^3 + \frac{1}{32}(x - 1)^4$
**7.** $x^2 - \frac{1}{3}x^4$; $|R_4(x)| < 2.85 \times 10^{-6}$
**9.** −0.00269867; $|\text{Error}| \leq 1.63 \times 10^{-5}$
**11.** −0.00269939; $|E_8| \leq 2.04 \times 10^{-7}$    **13.** 0.281785
**15.** 4.49341    **17.** $x \approx -3.18306$

## ADDITIONAL PROBLEMS 10.7

**1.**

| $i$ | $(x_i, y_i)$ |
|---|---|
| 0 | (3.1416, 1) |
| 1 | (3.4558, 0.9511) |
| 2 | (3.7699, 0.8090) |
| 3 | (4.0841, 0.5878) |
| 4 | (4.3982, 0.3090) |
| 5 | (4.7124, 0) |
| 6 | (5.0265, −0.3090) |
| 7 | (5.3407, −0.5878) |
| 8 | (5.6549, −0.8090) |
| 9 | (5.969, −0.9511) |
| 10 | (6.2832, −1) |

**3.** (a) 2.69486; 2.714081;    (b) 3.30660; 3.42816

## PROBLEM SET 11.1

**1.** $\frac{1}{3}$    **3.** 4    **5.** 1    **7.** Diverges    **9.** 0
**11.** Diverges    **13.** 0    **15.** 2    **17.** 0    **19.** $e$
**21.** $a_n = \frac{n}{n+1}$; 1    **23.** $a_n = (-1)^n \frac{n}{2n-1}$; diverges
**25.** $a_n = \frac{n}{2n-1}$; $\frac{1}{2}$    **27.** $a_n = n\sin\frac{1}{n}$; 1

**29.** $a_n = \dfrac{2^n}{n^2}$; diverges   **31.** $\frac{1}{2}, \frac{5}{4}, \frac{9}{8}, \frac{13}{16}$

**33.** $\frac{3}{4}, \frac{2}{3}, \frac{5}{8}, \frac{3}{5}$   **35.** $1, \frac{3}{2}, \frac{7}{4}, \frac{15}{8}$   **37.** 2.3028

**39.** $\frac{1}{2}(1+\sqrt{13})$   **41.** 1.1118   **43.** $1 - \cos 1$   **51.** No

**53.** $\dfrac{\pi}{2\sqrt{3}}$   **55.** $e^{1/2}$   **57.** $e^{-2}$   **59.** $e^{-1}$

## PROBLEM SET 11.2

**1.** $\frac{1}{6}$   **3.** $\frac{31}{6}$   **5.** Diverges   **7.** $-1$   **9.** Diverges

**11.** $\dfrac{e^2}{\pi(\pi - e)}$   **13.** 3   **15.** $\frac{2}{9}$   **17.** $\frac{13}{999}$   **19.** $\frac{1}{2}$

**21.** 1   **25.** 500 ft   **27.** \$4 billion   **29.** $\frac{1}{4}$   **31.** $\frac{4}{5}$

**33.** $111\frac{1}{9}$ yd   **37.** (b) Indefinitely

**41.** (a) 2;   (b) 1   **3.** (a) $\dfrac{Ce^{kt}}{e^{kt}-1}$;   (b) $\frac{8}{3}$ mg

**45.** 1

## PROBLEM SET 11.3

**1.** Diverges   **3.** Diverges   **5.** Diverges
**7.** Diverges   **9.** Converges   **11.** Converges
**13.** Diverges   **15.** Diverges   **17.** Diverges
**19.** Converges   **21.** Converges   **23.** 0.0404
**25.** 0.1974   **27.** $p > 1$
**31.** The rationality of $\gamma$ is an unsolved problem.
**33.** 272,404,866

## PROBLEM SET 11.4

**1.** Diverges   **3.** Converges   **5.** Converges
**7.** Diverges   **9.** Converges
**11.** Diverges; $n$th-Term Test
**13.** Converges; Limit Comparison Test
**15.** Converges; Ratio Test
**17.** Converges; Limit Comparison Test
**19.** Converges; Limit Comparison Test
**21.** Converges; Limit Comparison Test
**23.** Converges; Ratio Test
**25.** Converges; Integral Test
**27.** Diverges; $n$th Term Test
**29.** Converges; Comparison Test
**31.** Converges; Ratio Test
**33.** Converges; Ratio Test
**43.** (a) Diverges;   (b) Converges;   (c) Converges;
(d) Converges;   (e) Converges   (f) Converge
**45.** Converges for $p > 1$, diverges for $p \leq 1$.

## PROBLEM SET 11.5

**1.** $|S - S_9| \leq 0.065$   **3.** $|S - S_9| \leq 0.417$

**5.** $|S - S_9| \leq 0.230$   **13.** Conditionally convergent

**15.** Divergent   **17.** Conditionally convergent
**19.** Absolutely convergent

**21.** Conditionally convergent   **23.** Conditionally convergent
**25.** Absolutely convergent   **27.** Conditionally convergent
**29.** Divergent
**35.** (a) $1 + \frac{1}{3} \approx 1.33$;   (b) $1 + \frac{1}{3} - \frac{1}{2} \approx 0.833$   **45.** ln 2

## PROBLEM SET 11.6

**1.** $-1 \leq x \leq 1$   **3.** All $x$   **5.** $-1 < x < 1$   **7.** $-1 < x \leq 1$
**9.** $-1 \leq x \leq 1$   **11.** $-2 < x < 2$   **13.** All $x$
**15.** $0 \leq x < 2$   **17.** $-3 < x < 1$   **19.** $-6 \leq x \leq -4$

**21.** If $\displaystyle\lim_{n\to\infty} \dfrac{x_0^{\,n}}{n!} \neq 0$, then $\displaystyle\sum \dfrac{x_0^{\,n}}{n!}$ will not converge.

**23.** $\sqrt{2}$   **25.** $\dfrac{1}{4-x}$; $2 < x < 4$

**27.** (a) $-1 \leq x < \frac{1}{3}$;   (b) $-\frac{1}{2} < x \leq \frac{7}{2}$

**29.** $S(x) = \dfrac{a_0 + a_1 x + a_2 x^3}{1 - x^3}$, $|x| < 1$

## PROBLEM SET 11.7

**1.** $1 - x + x^2 - x^3 + x^4 - x^5 + \ldots$; 1

**3.** $1 + 3x + 6x^2 + 10x^3 + \ldots$; 1

**5.** $\dfrac{1}{2} + \dfrac{3x}{4} + \dfrac{9x^2}{8} + \dfrac{27x^3}{16} + \ldots$; $\frac{2}{3}$

**7.** $x^2 + x^6 + x^{10} + x^{14} + \ldots$; 1

**9.** $\dfrac{x^2}{2} - \dfrac{x^3}{6} + \dfrac{x^4}{12} - \dfrac{x^5}{20} + \ldots$; 1

**11.** $2x + \dfrac{2x^3}{3} + \dfrac{2x^5}{5} + \ldots$; 1

**13.** $1 - x + \dfrac{x^2}{2!} - \dfrac{x^3}{3!} + \dfrac{x^4}{4!} - \dfrac{x^5}{5!} + \ldots$

**15.** $2 + \dfrac{2x^2}{2!} + \dfrac{2x^4}{4!} + \dfrac{2x^6}{6!} + \ldots$

**17.** $1 + \dfrac{x^2}{2} + \dfrac{x^3}{3} + \dfrac{3x^4}{8} + \dfrac{11x^5}{30} + \ldots$

**19.** $x - x^2 + \dfrac{x^3}{6} + \dfrac{x^4}{6} + \dfrac{3x^5}{40} + \ldots$

**21.** $x + \dfrac{2x^3}{3} + \dfrac{13x^5}{15} - \dfrac{29x^7}{105} + \ldots$

**23.** $x + \dfrac{x^3}{6} - \dfrac{x^4}{12} + \dfrac{3x^5}{40} - \ldots$

**25.** (a) $\dfrac{x}{1+x}$;   (b) $\dfrac{e^x - (1+x)}{x^2}$;   (c) $-\ln(1 - 2x)$

**27.** $\dfrac{x}{(1-x)^2}$, $-1 < x < 1$

**29.** (a) $x + \dfrac{x^2}{2} - \dfrac{x^3}{6} - \ldots$;   (b) $1 + x + x^2 + \dfrac{5x^3}{6} + \ldots$

**31.** $\dfrac{x}{2} + \dfrac{3x^2}{4} + \dfrac{7x^3}{8} + \ldots$   **33.** $\dfrac{x}{1 - x - x^2}$   **35.** 3.14159

## PROBLEM SET 11.8

**1.** $x + \dfrac{x^3}{3} + \dfrac{2x^5}{15}$    **3.** $x + x^2 + \dfrac{x^3}{3} - \dfrac{x^5}{30}$

**5.** $x - \dfrac{x^2}{2} - \dfrac{x^3}{6} + \dfrac{3x^5}{40}$    **7.** $1 + 3x + \dfrac{x^2}{2} + \dfrac{x^4}{24} + \dfrac{x^5}{60}$

**9.** $1 + x + \dfrac{3x^2}{2} + \dfrac{3x^3}{3} + \dfrac{37x^4}{24} + \dfrac{37x^5}{24}$

**11.** $1 - x + x^3 - x^4$    **13.** $x^3 - \dfrac{x^5}{2}$

**15.** $2x - \dfrac{x^3}{6} + \dfrac{61x^5}{120}$

**17.** $1 + \dfrac{3x}{2} + \dfrac{3x^2}{8} - \dfrac{x^3}{16} + \dfrac{3x^4}{128} - \dfrac{3x^5}{256}$

**19.** $e + e(x-1) + \dfrac{e}{2}(x-1)^2 + \dfrac{e}{6}(x-1)^3$

**21.** $\dfrac{1}{2} - \dfrac{\sqrt{3}}{2}\left(x - \dfrac{\pi}{3}\right) - \dfrac{1}{4}\left(x - \dfrac{\pi}{3}\right)^2 + \dfrac{\sqrt{3}}{12}\left(x - \dfrac{\pi}{3}\right)^3$

**23.** $3 + 5(x-1) + 4(x-1)^2 + (x-1)^3$

**27.** $x + \dfrac{x^3}{6} + \dfrac{3x^5}{40} + \dfrac{5x^7}{112}$    **29.** 0.9045

**31.** $1 - (x-1) + (x-1)^2 - (x-1)^3 + \ldots$

**33.** **(a)** 25;    **(b)** −3;    **(c)** 0;    **(d)** 4e;    **(e)** −4

**35.** $x - \dfrac{x^3}{3} + \dfrac{2x^5}{15}$    **39.** $x - \dfrac{x^3}{6} + \dfrac{x^5}{120} - \dfrac{x^7}{5040}$

**41.** $-2 + x - x^2 - \dfrac{5x^3}{6}$    **43.** $x + \dfrac{x^2}{2} - \dfrac{5x^4}{24} - \dfrac{23x^5}{120}$

**45.** $x + x^2 + \dfrac{x^3}{3} - \dfrac{x^5}{30}$

## CHAPTER REVIEW 11.9

### Concepts Test

**1.** False    **3.** True    **5.** False    **7.** False    **9.** True
**11.** True    **13.** False    **15.** True    **17.** False
**19.** True    **21.** True    **23.** True    **25.** True
**27.** True    **29.**    True    **31.** True    **33.** False
**35.** True

### Sample Test Problems

**1.** 3    **3.** $e^4$    **5.** 1    **7.** 0    **9.** 1    **11.** Diverges

**13.** $\dfrac{e^2}{e^2 - 1}$    **15.** $\dfrac{91}{99}$    **17.** cos 2    **19.** Diverges

**21.** Converges    **23.** Converges    **25.** Diverges
**27.** Converges    **29.** Diverges    **31.** Converges
**33.** Conditionally convergent    **35.** Divergent
**37.** $-1 \le x \le 1$    **39.** $3 < x \le 5$    **41.** $1 < x < 5$
**43.** $1 - 2x + 3x^2 - 4x^3 + \ldots$; $-1 < x < 1$

**45.** $x^2 - \dfrac{x^4}{3} + \dfrac{2x^6}{45} - \dfrac{x^8}{315} + \ldots$; all $x$

**47.** $1 + x - \dfrac{x^2}{2!} - \dfrac{x^3}{3!} + \dfrac{x^4}{4!} + \dfrac{x^5}{5!} - \ldots$; all $x$

**49.** 0.21046    **51.** 0.000004167

## PROBLEM SET 12.1

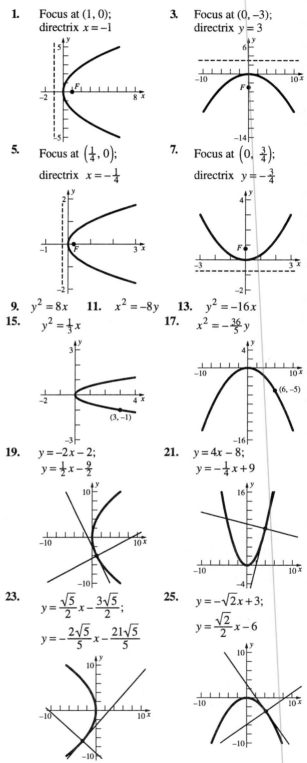

**1.** Focus at $(1, 0)$; directrix $x = -1$

**3.** Focus at $(0, -3)$; directrix $y = 3$

**5.** Focus at $\left(\frac{1}{4}, 0\right)$; directrix $x = -\frac{1}{4}$

**7.** Focus at $\left(0, \frac{3}{4}\right)$; directrix $y = -\frac{3}{4}$

**9.** $y^2 = 8x$    **11.** $x^2 = -8y$    **13.** $y^2 = -16x$

**15.** $y^2 = \frac{1}{3}x$    **17.** $x^2 = -\frac{36}{5}y$

**19.** $y = -2x - 2$; $y = \frac{1}{2}x - \frac{9}{2}$

**21.** $y = 4x - 8$; $y = -\frac{1}{4}x + 9$

**23.** $y = \dfrac{\sqrt{5}}{2}x - \dfrac{3\sqrt{5}}{2}$; $y = -\dfrac{2\sqrt{5}}{5}x - \dfrac{21\sqrt{5}}{5}$

**25.** $y = -\sqrt{2}x + 3$; $y = \dfrac{\sqrt{2}}{2}x - 6$

**27.**

$(4,\ 2\sqrt{5})$

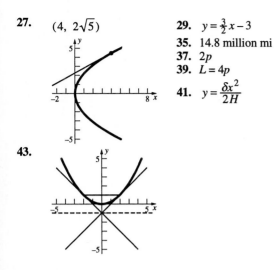

**29.** $y = \frac{3}{2}x - 3$

**35.** 14.8 million mi

**37.** $2p$

**39.** $L = 4p$

**41.** $y = \frac{\delta x^2}{2H}$

**43.**

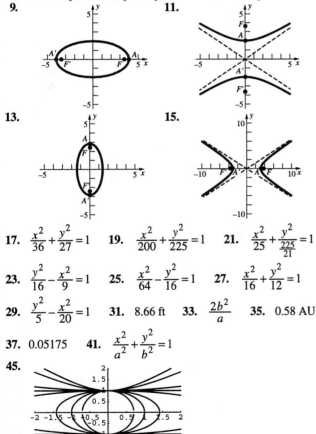

## PROBLEM SET 12.2

**1.** Horizontal ellipse **3.** Vertical hyperbola

**5.** Vertical parabola (opens up) **7.** Vertical ellipse

**9.** **11.**

**13.** **15.**

**17.** $\frac{x^2}{36} + \frac{y^2}{27} = 1$ **19.** $\frac{x^2}{200} + \frac{y^2}{225} = 1$ **21.** $\frac{x^2}{25} + \frac{y^2}{\frac{225}{21}} = 1$

**23.** $\frac{y^2}{16} - \frac{x^2}{9} = 1$ **25.** $\frac{x^2}{64} - \frac{y^2}{16} = 1$ **27.** $\frac{x^2}{16} + \frac{y^2}{12} = 1$

**29.** $\frac{y^2}{5} - \frac{x^2}{20} = 1$ **31.** 8.66 ft **33.** $\frac{2b^2}{a}$ **35.** 0.58 AU

**37.** 0.05175 **41.** $\frac{x^2}{a^2} + \frac{y^2}{b^2} = 1$

**45.**

## PROBLEM SET 12.3

**1.** $\frac{x^2}{88} + \frac{y^2}{169} = 1$ **3.** $\frac{x^2}{36} - \frac{y^2}{13} = 1$ **5.** $x + \sqrt{6}y = 9$

**7.** $x - \sqrt{6}y = 9$ **9.** $5x + 12y = 169$ **11.** $y = 13$

**13.** $\left(-\sqrt{3},\ \frac{3}{2}\right),\ \left(\sqrt{3},\ \frac{3}{2}\right)$ **15.** $(-7, 3), (7, -3)$

**17.** $\pi ab$ **19.** $\frac{\pi b^2}{3a^2}\left[(a^2 + b^2)^{3/2} - 3a^2\sqrt{a^2 + b^2} + 2a^3\right]$

**21.** $a\sqrt{2}$ by $b\sqrt{2}$ **23.** $\left(6,\ 5\sqrt{3}\right)$

**25.** 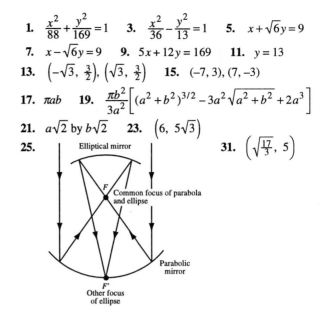 **31.** $\left(\sqrt{\frac{17}{3}},\ 5\right)$

## PROBLEM SET 12.4

**1.** Circle **3.** Ellipse **5.** Point **7.** Parabola

**9.** Empty set **11.** Intersecting lines **13.** Line

**15.** Ellipse

**17.** **19.**

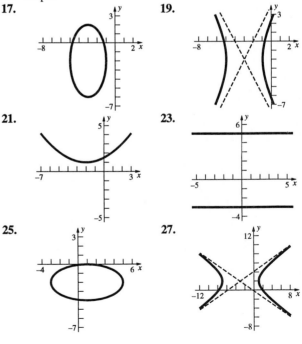

**21.** **23.**

**25.** **27.**

**29.**

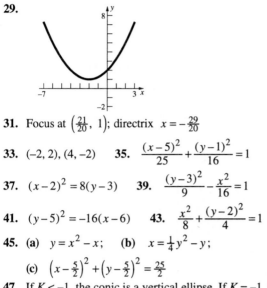

**31.** Focus at $\left(\frac{21}{20}, 1\right)$; directrix $x = -\frac{29}{20}$

**33.** $(-2, 2), (4, -2)$  **35.** $\dfrac{(x-5)^2}{25} + \dfrac{(y-1)^2}{16} = 1$

**37.** $(x-2)^2 = 8(y-3)$  **39.** $\dfrac{(y-3)^2}{9} - \dfrac{x^2}{16} = 1$

**41.** $(y-5)^2 = -16(x-6)$  **43.** $\dfrac{x^2}{8} + \dfrac{(y-2)^2}{4} = 1$

**45.** (a) $y = x^2 - x$;  (b) $x = \frac{1}{4}y^2 - y$;

   (c) $\left(x - \frac{5}{2}\right)^2 + \left(y - \frac{5}{2}\right)^2 = \frac{25}{2}$

**47.** If $K < -1$, the conic is a vertical ellipse. If $K = -1$, the conic is a circle. If $-1 < K < 0$, the conic is a horizontal ellipse. If $K = 0$, the conic is a horizontal parabola. If $K > 0$, the conic is a horizontal hyperbola.

## PROBLEM SET 12.5

**1.** $\dfrac{u^2}{4} + \dfrac{v^2}{12} = 1$  **3.** $\dfrac{u^2}{\frac{112}{9}} + \dfrac{v^2}{16} = 1$

**5.** $\dfrac{v^2}{4} - \dfrac{u^2}{36} = 1$  **7.** $\dfrac{(u-2)^2}{4} - \dfrac{v^2}{3} = 1$

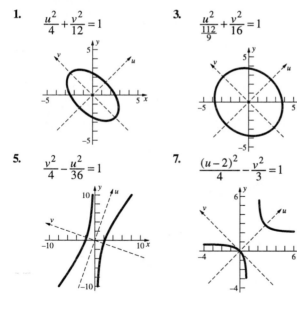

**9.** $\dfrac{(u+2)^2}{2} + \dfrac{(v+3)^2}{4} = 1$  **11.** $\dfrac{\left(u - \sqrt{10}\right)^2}{44} + \dfrac{\left(v - 2\sqrt{10}\right)^2}{4} = 1$

**15.** $u = x\cos\theta + y\sin\theta, v = -x\sin\theta + y\cos\theta$

**17.** $\left(-\frac{1}{5}, -\frac{7}{5}\right), \left(\frac{1}{5}, \frac{7}{5}\right)$

**23.** (a) $-2 < B < 2$;  (b) $B = 0$;  (c) $B < -2$ or $B > 2$;
   (d) $B = \pm 2$

## PROBLEM SET 12.6

**1.** $\left(1, \frac{1}{2}\pi\right) \left(\frac{5}{3}, \frac{1}{2}\pi\right)$

$\left(4, \frac{1}{3}\pi\right)$
$\left(3, \frac{1}{3}\pi\right)$
$(4, 0)$
$(1, 4\pi)$
$(0, \pi)$
$\left(3, \frac{11}{7}\pi\right)$

**3.** $\left(-2, -\frac{1}{4}\pi\right)$  $\left(-1, -\frac{1}{2}\pi\right)$
$\left(\sqrt{3}, -\frac{7}{6}\pi\right)$
$(3, 2\pi)$
$(-1, 1)$
$(1, -4\pi)$
$\left(-2, \frac{1}{4}\pi\right)$
$\left(-2, \frac{1}{3}\pi\right)$

**5.**

(a) $\left(1, -\frac{3}{2}\pi\right), \left(1, \frac{5}{2}\pi\right), \left(-1, -\frac{1}{2}\pi\right), \left(-1, \frac{3}{2}\pi\right)$

(b) $\left(1, -\frac{3}{4}\pi\right), \left(1, \frac{5}{4}\pi\right), \left(-1, -\frac{7}{4}\pi\right), \left(-1, \frac{9}{4}\pi\right)$

(c) $\left(\sqrt{2}, -\frac{7}{3}\pi\right), \left(\sqrt{2}, \frac{5}{3}\pi\right), \left(-\sqrt{2}, -\frac{4}{3}\pi\right), \left(-\sqrt{2}, \frac{2}{3}\pi\right)$

(d) $\left(\sqrt{2}, -\frac{1}{2}\pi\right), \left(\sqrt{2}, \frac{3}{2}\pi\right), \left(-\sqrt{2}, -\frac{3}{2}\pi\right), \left(-\sqrt{2}, \frac{1}{2}\pi\right)$

**7.** (a) $(0, 1)$;  (b) $\left(-\dfrac{\sqrt{2}}{2}, -\dfrac{\sqrt{2}}{2}\right)$;  (c) $\left(\dfrac{\sqrt{2}}{2}, -\dfrac{\sqrt{6}}{2}\right)$;

   (d) $(0, -\sqrt{2})$

**9.** (a) $\left(6, \frac{1}{6}\pi\right)$; (b) $\left(4, \frac{5}{6}\pi\right)$; (c) $\left(2, \frac{5}{4}\pi\right)$;
(d) $(0, 0)$

**11.** $r = \dfrac{2}{3\sin\theta - \cos\theta}$

**13.** $r = -2\csc\theta$

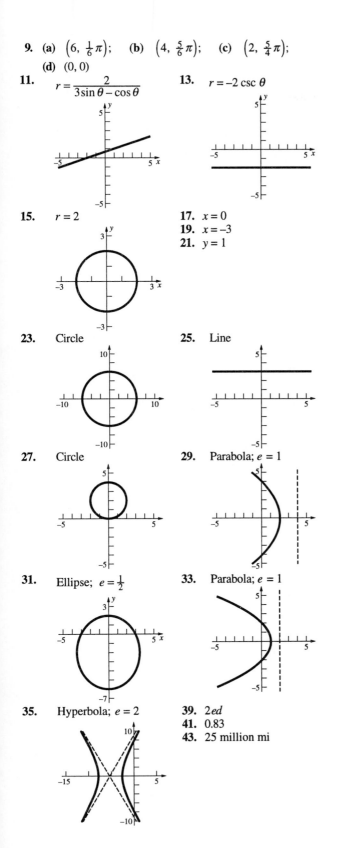

**15.** $r = 2$

**17.** $x = 0$
**19.** $x = -3$
**21.** $y = 1$

**23.** Circle

**25.** Line

**27.** Circle

**29.** Parabola; $e = 1$

**31.** Ellipse; $e = \frac{1}{2}$

**33.** Parabola; $e = 1$

**35.** Hyperbola; $e = 2$

**39.** $2ed$
**41.** 0.83
**43.** 25 million mi

**45.** $e = 0.1$

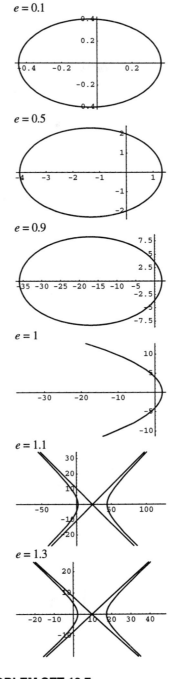

$e = 0.5$

$e = 0.9$

$e = 1$

$e = 1.1$

$e = 1.3$

**PROBLEM SET 12.7**

**1.**

**3.**

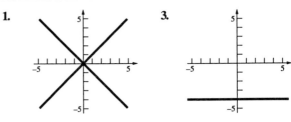

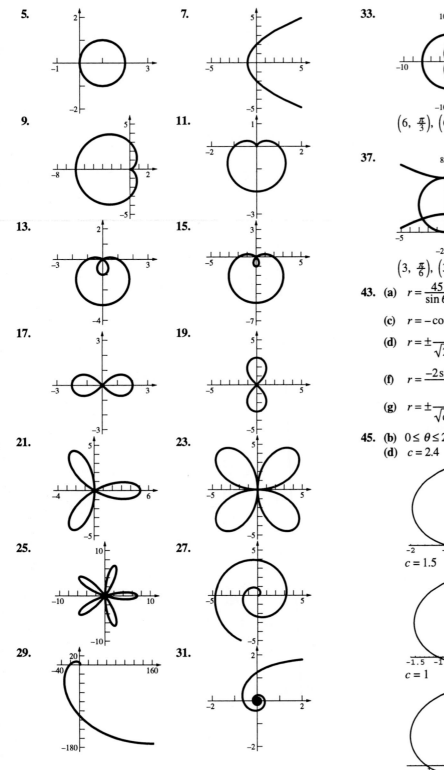

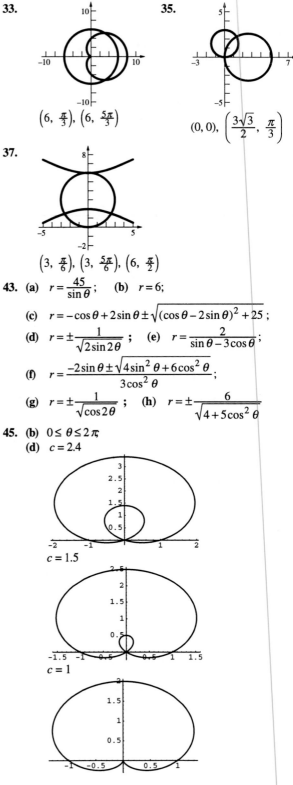

**33.** $\left(6, \frac{\pi}{3}\right), \left(6, \frac{5\pi}{3}\right)$

**35.** $(0, 0), \left(\frac{3\sqrt{3}}{2}, \frac{\pi}{3}\right)$

**37.** $\left(3, \frac{\pi}{6}\right), \left(3, \frac{5\pi}{6}\right), \left(6, \frac{\pi}{2}\right)$

**43.** (a) $r = \dfrac{45}{\sin\theta}$; (b) $r = 6$;

(c) $r = -\cos\theta + 2\sin\theta \pm \sqrt{(\cos\theta - 2\sin\theta)^2 + 25}$;

(d) $r = \pm\dfrac{1}{\sqrt{2\sin 2\theta}}$; (e) $r = \dfrac{2}{\sin\theta - 3\cos\theta}$;

(f) $r = \dfrac{-2\sin\theta \pm \sqrt{4\sin^2\theta + 6\cos^2\theta}}{3\cos^2\theta}$;

(g) $r = \pm\dfrac{1}{\sqrt{\cos 2\theta}}$; (h) $r = \pm\dfrac{6}{\sqrt{4 + 5\cos^2\theta}}$

**45.** (b) $0 \le \theta \le 2\pi$;

(d) $c = 2.4$

$c = 1.5$

$c = 1$

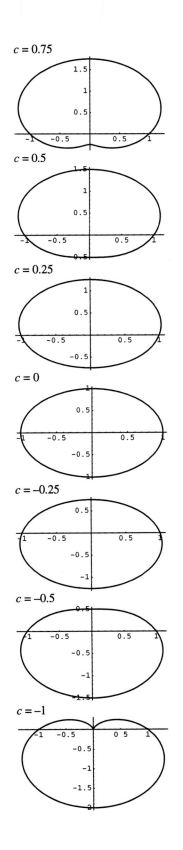

$c = 0.75$

$c = 0.5$

$c = 0.25$

$c = 0$

$c = -0.25$

$c = -0.5$

$c = -1$

$c = -2$

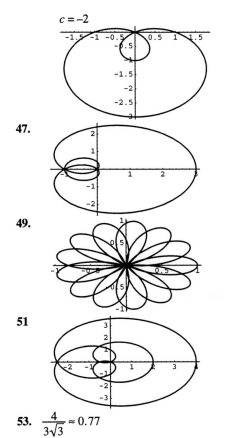

**47.**

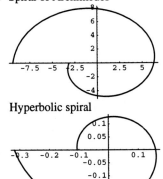

**49.**

**51**

**53.** $\dfrac{4}{3\sqrt{3}} \approx 0.77$

**55.** If $n$ is even, the number of loops is $2n$.
If $n$ is odd, the number of loops is $n$.

**57.** (a) The graph for $\phi = 0$ is the graph for $\phi \neq 0$ rotated by $\phi$ counterclockwise about the pole.

(b) As $n$ increases, the number of "leaves" increases.

**59.** The spiral will unwind clockwise for $c < 0$. The spiral will unwind counter-clockwise for $c > 0$.

**61.** Spiral of Archimedes

Hyperbolic spiral

Parabolic spiral

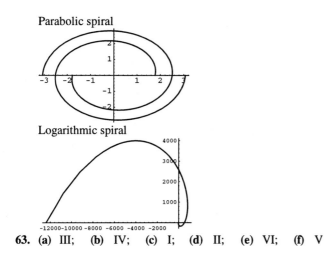

Logarithmic spiral

**63.** (a) III;   (b) IV;   (c) I;   (d) II;   (e) VI;   (f) V

## PROBLEM SET 12.8

**1.**

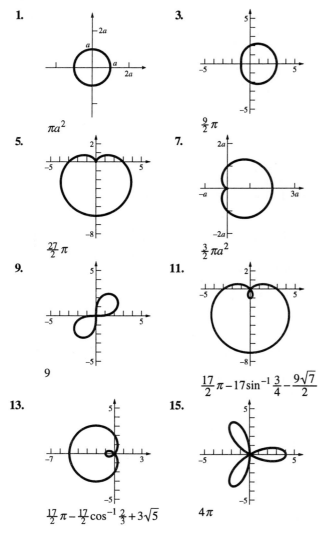

$\pi a^2$

**3.**

$\frac{9}{2}\pi$

**5.**

$\frac{27}{2}\pi$

**7.**

$\frac{3}{2}\pi a^2$

**9.**

9

**11.**

$\frac{17}{2}\pi - 17\sin^{-1}\frac{3}{4} - \frac{9\sqrt{7}}{2}$

**13.**

$\frac{17}{2}\pi - \frac{17}{2}\cos^{-1}\frac{2}{3} + 3\sqrt{5}$

**15.**

$4\pi$

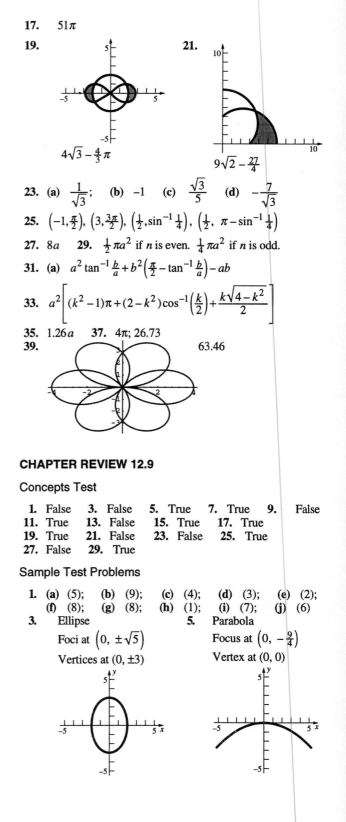

**17.**   $51\pi$

**19.**

$4\sqrt{3} - \frac{4}{3}\pi$

**21.**

$9\sqrt{2} - \frac{27}{4}$

**23.** (a) $\dfrac{1}{\sqrt{3}}$;   (b) $-1$   (c) $\dfrac{\sqrt{3}}{5}$   (d) $-\dfrac{7}{\sqrt{3}}$

**25.** $\left(-1, \frac{\pi}{2}\right)$, $\left(3, \frac{3\pi}{2}\right)$, $\left(\frac{1}{2}, \sin^{-1}\frac{1}{4}\right)$, $\left(\frac{1}{2},\ \pi - \sin^{-1}\frac{1}{4}\right)$

**27.** $8a$   **29.** $\frac{1}{2}\pi a^2$ if $n$ is even. $\frac{1}{4}\pi a^2$ if $n$ is odd.

**31.** (a) $a^2\tan^{-1}\frac{b}{a} + b^2\left(\frac{\pi}{2} - \tan^{-1}\frac{b}{a}\right) - ab$

**33.** $a^2\left[(k^2-1)\pi + (2-k^2)\cos^{-1}\left(\frac{k}{2}\right) + \frac{k\sqrt{4-k^2}}{2}\right]$

**35.** $1.26a$   **37.** $4\pi$; $26.73$

**39.**                                                    $63.46$

## CHAPTER REVIEW 12.9

### Concepts Test

**1.** False   **3.** False   **5.** True   **7.** True   **9.**   False
**11.** True   **13.** False   **15.** True   **17.** True
**19.** True   **21.** False   **23.** False   **25.** True
**27.** False   **29.** True

### Sample Test Problems

**1.** (a) (5);   (b) (9);   (c) (4);   (d) (3);   (e) (2);
(f) (8);   (g) (8);   (h) (1);   (i) (7);   (j) (6)

**3.**   Ellipse                                   **5.**   Parabola
Foci at $\left(0,\ \pm\sqrt{5}\right)$                Focus at $\left(0, -\frac{9}{4}\right)$
Vertices at $(0, \pm 3)$                    Vertex at $(0, 0)$

**7.** Ellipse
Foci at $(\pm 4, 0)$
Vertices at $(\pm 5, 0)$

**9.** Parabola
Foci at $(0, 0)$
Vertices at $\left(0, \frac{5}{4}\right)$

**37.** $(x-3)^2 + (y-3)^2 = 9$   **39.** $-1$   **41.** $\frac{75}{2}\pi$

**43.** $\frac{22}{3}$

**45.** (a) I;   (b) IV;
(c) III;   (d) II;
(e) V

**11.** $\frac{x^2}{16} + \frac{y^2}{12} = 1$   **13.** $y^2 = -9x$   **15.** $\frac{x^2}{4} - y^2 = 1$

**17.** $\frac{(x-1)^2}{25} + \frac{(y-2)^2}{16} = 1$

**19.** Circle   **21.** Parabola

**23.** $r = \frac{5}{2}$;  $s = -\frac{1}{2}$;  hyperbola;  $4\sqrt{6}$

**25.**   **27.**

**29.**   **31.**

**33.**   **35.**

## PROBLEM SET 13.1

**1.** (a)
(b) Simple; not closed
(c) $y = \frac{2}{3}x$

**3.** (a)
(b) Simple, not closed
(c) $y = \frac{1}{3}(x+1)$

**5.** (a)
(b) Simple; not closed
(c)

**7.** (a)
(b) Simple; not closed
(c) $y = \frac{1}{x}$

**9.** (a)
(b) Simple; not closed
(c) $y^2 = x^3$

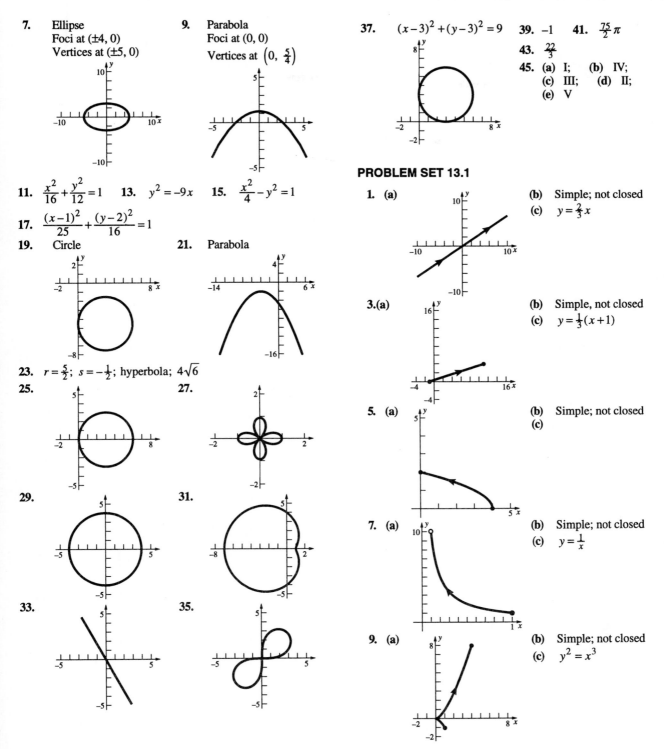

**11. (a)**

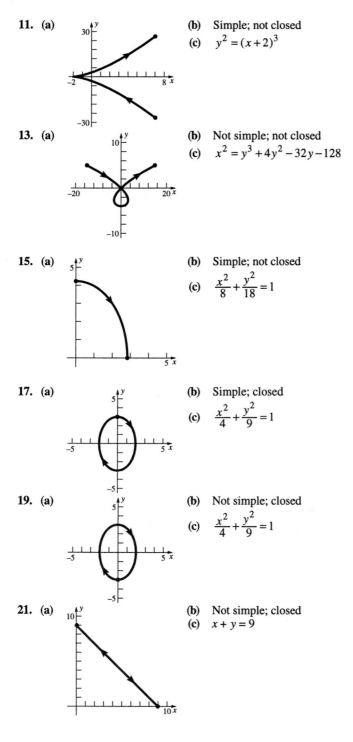

**(b)** Simple; not closed
**(c)** $y^2 = (x+2)^3$

**13. (a)**

**(b)** Not simple; not closed
**(c)** $x^2 = y^3 + 4y^2 - 32y - 128$

**15. (a)**

**(b)** Simple; not closed
**(c)** $\dfrac{x^2}{8} + \dfrac{y^2}{18} = 1$

**17. (a)**

**(b)** Simple; closed
**(c)** $\dfrac{x^2}{4} + \dfrac{y^2}{9} = 1$

**19. (a)**

**(b)** Not simple; closed
**(c)** $\dfrac{x^2}{4} + \dfrac{y^2}{9} = 1$

**21. (a)**

**(b)** Not simple; closed
**(c)** $x + y = 9$

**23. (a)**

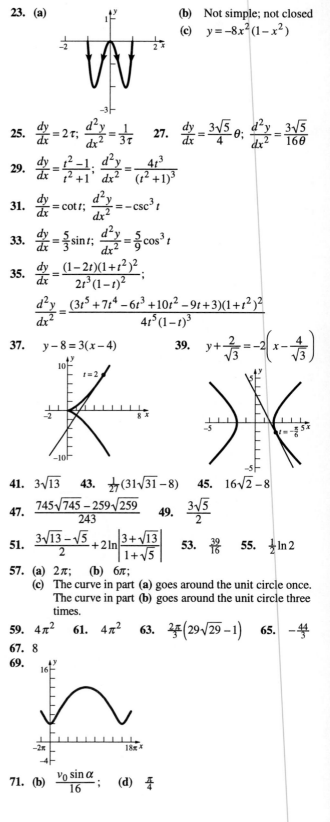

**(b)** Not simple; not closed
**(c)** $y = -8x^2(1 - x^2)$

**25.** $\dfrac{dy}{dx} = 2\tau;\ \dfrac{d^2y}{dx^2} = \dfrac{1}{3\tau}$    **27.** $\dfrac{dy}{dx} = \dfrac{3\sqrt{5}}{4}\theta;\ \dfrac{d^2y}{dx^2} = \dfrac{3\sqrt{5}}{16\theta}$

**29.** $\dfrac{dy}{dx} = \dfrac{t^2 - 1}{t^2 + 1};\ \dfrac{d^2y}{dx^2} = \dfrac{4t^3}{(t^2 + 1)^3}$

**31.** $\dfrac{dy}{dx} = \cot t;\ \dfrac{d^2y}{dx^2} = -\csc^3 t$

**33.** $\dfrac{dy}{dx} = \dfrac{5}{3}\sin t;\ \dfrac{d^2y}{dx^2} = \dfrac{5}{9}\cos^3 t$

**35.** $\dfrac{dy}{dx} = \dfrac{(1 - 2t)(1 + t^2)^2}{2t^3(1 - t)^2};$

$\dfrac{d^2y}{dx^2} = \dfrac{(3t^5 + 7t^4 - 6t^3 + 10t^2 - 9t + 3)(1 + t^2)^2}{4t^5(1 - t)^3}$

**37.** $y - 8 = 3(x - 4)$    **39.** $y + \dfrac{2}{\sqrt{3}} = -2\left(x - \dfrac{4}{\sqrt{3}}\right)$

**41.** $3\sqrt{13}$    **43.** $\frac{1}{27}(31\sqrt{31} - 8)$    **45.** $16\sqrt{2} - 8$

**47.** $\dfrac{745\sqrt{745} - 259\sqrt{259}}{243}$    **49.** $\dfrac{3\sqrt{5}}{2}$

**51.** $\dfrac{3\sqrt{13} - \sqrt{5}}{2} + 2\ln\left|\dfrac{3 + \sqrt{13}}{1 + \sqrt{5}}\right|$    **53.** $\frac{39}{16}$    **55.** $\frac{1}{2}\ln 2$

**57. (a)** $2\pi$;    **(b)** $6\pi$;
**(c)** The curve in part **(a)** goes around the unit circle once. The curve in part **(b)** goes around the unit circle three times.

**59.** $4\pi^2$    **61.** $4\pi^2$    **63.** $\frac{2\pi}{3}(29\sqrt{29} - 1)$    **65.** $-\frac{44}{3}$

**67.** 8

**69.**

**71. (b)** $\dfrac{v_0 \sin\alpha}{16}$;    **(d)** $\frac{\pi}{4}$

**73.**  $x = (a-b)\cos t + b\cos\left(\dfrac{a-b}{b}t\right),$

$y = (a-b)\sin t - b\sin\left(\dfrac{a-b}{a}t\right)$

**77.**  **(a)**  $x = 2a\cot\theta;\ y = 2a\sin^2\theta$    **(b)**  $y = \dfrac{8a^3}{x^2 + 4a^2}$

**79.**  $8(a+b)$

**81.**  **(a)**

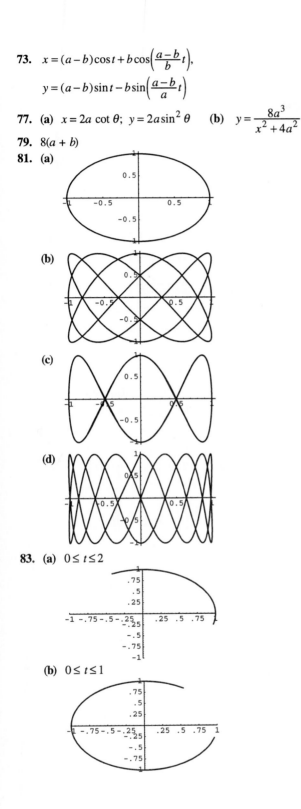

**(b)**

**(c)**

**(d)**

**83.**  **(a)**  $0 \le t \le 2$

**(b)**  $0 \le t \le 1$

**(c)**  $0.25 \le t \le 2$

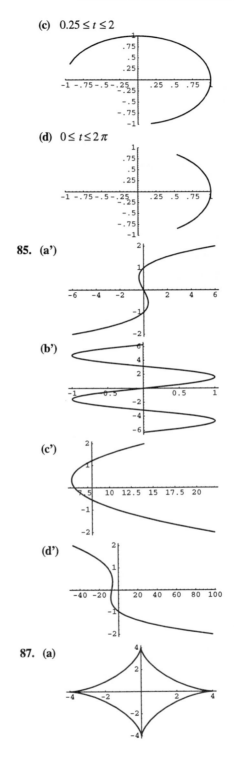

**(d)**  $0 \le t \le 2\pi$

**85.**  **(a')**

**(b')**

**(c')**

**(d')**

**87.**  **(a)**

**(b)**

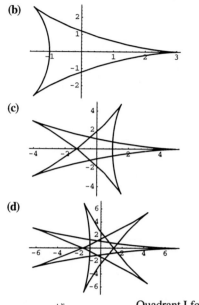

**(c)**

**(d)**

**89.**

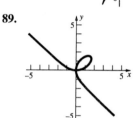

Quadrant I for $t > 0$, quadrant II for $-1 < t < 0$, quadrant III for no $t$, quadrant IV for $t < -1$.

## PROBLEM SET 13.2

**1.**

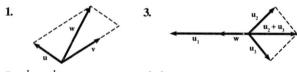

**3.**

**5.** $\frac{1}{2}\mathbf{u} - \frac{1}{2}\mathbf{v}$   **7.** 1   **9.** $|\mathbf{w}| \approx 79.34$; $\theta = 7.5°$
**11.** 150 N   **13.** N 2.08° E; 467 mi/h   **15.** 80 mi/h
**23.** $\alpha + \beta = 143.13°$, $\beta + \gamma = 126.87°$, $\alpha + \gamma = 90°$

## PROBLEM SET 13.3

**1.** **(a)** $-12\mathbf{i} + 18\mathbf{j}$;   **(b)** $-13$;   **(c)** $-28$;   **(d)** $375$;
**(e)** $-15\sqrt{13}$   **(f)** $13 - \sqrt{13}$
**3.** **(a)** $-\dfrac{7}{5\sqrt{2}}$   **(b)** $-\dfrac{1}{\sqrt{5}}$

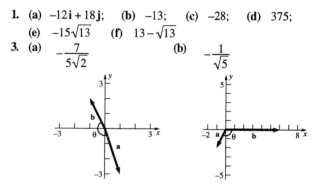

**(c)** 0

**(d)** $-\dfrac{51}{\sqrt{2665}}$

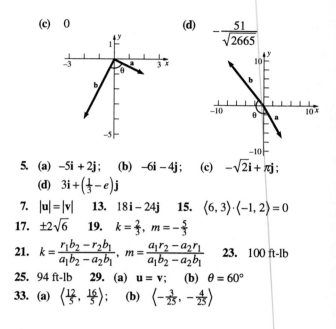

**5.** **(a)** $-5\mathbf{i} + 2\mathbf{j}$;   **(b)** $-6\mathbf{i} - 4\mathbf{j}$;   **(c)** $-\sqrt{2}\mathbf{i} + \pi\mathbf{j}$;
**(d)** $3\mathbf{i} + \left(\frac{1}{3} - e\right)\mathbf{j}$

**7.** $|\mathbf{u}| = |\mathbf{v}|$   **13.** $18\mathbf{i} - 24\mathbf{j}$   **15.** $\langle 6, 3\rangle \cdot \langle -1, 2\rangle = 0$

**17.** $\pm 2\sqrt{6}$   **19.** $k = \frac{2}{3}$, $m = -\frac{5}{3}$

**21.** $k = \dfrac{r_1 b_2 - r_2 b_1}{a_1 b_2 - a_2 b_1}$, $m = \dfrac{a_1 r_2 - a_2 r_1}{a_1 b_2 - a_2 b_1}$   **23.** 100 ft-lb

**25.** 94 ft-lb   **29.** **(a)** $\mathbf{u} = \mathbf{v}$;   **(b)** $\theta = 60°$
**33.** **(a)** $\langle \frac{12}{5}, \frac{16}{5}\rangle$;   **(b)** $\langle -\frac{3}{25}, -\frac{4}{25}\rangle$

## PROBLEM SET 13.4

**1.** $2\mathbf{i} - \mathbf{j}$   **3.** $\frac{1}{2}\mathbf{i} - 4\mathbf{j}$   **5.** $\mathbf{i}$   **7.** Does not exist
**9.** **(a)** $\{t \in \mathbb{R}: t \le 3\}$;   **(b)** $\{t \in \mathbb{R}: t \le 20\}$
**11.** **(a)** $\{t \in \mathbb{R}: t < 3\}$;   **(b)** $\{t \in \mathbb{R}: t < 20, t^2 \text{ not an integer}\}$
**13.** **(a)** $9(3t + 4)^2\mathbf{i} + 2te^{t^2}\mathbf{j}$; $54(3t + 4)\mathbf{i} + 2(2t^2 + 1)e^{t^2}\mathbf{j}$;
**(b)** $\sin 2t\,\mathbf{i} - 3\sin 3t\,\mathbf{j}$; $2\cos 2t\,\mathbf{i} - 9\cos 3t\,\mathbf{j}$
**15.** $2e^{-2t} + \dfrac{4}{t^2} - \dfrac{2}{t^2}\ln(t^2)$

**17.** $-\dfrac{e^{-3t}}{2}\left(\dfrac{6t - 7}{\sqrt{t - 1}}\right)\mathbf{i} + e^{-3t}\left(\dfrac{2}{t} - 3\ln(2t^2)\right)\mathbf{j}$

**19.** $-6t\sin(3t^2 - 4)\mathbf{i} + 18te^{9t^2 - 12}\mathbf{j}$
**21.** $(e - 1)\mathbf{i} + (1 - e^{-1})\mathbf{j}$

**23.** $\mathbf{v} = -e^{-1}\mathbf{i} + e\mathbf{j}$; $\mathbf{a} = e^{-1}\mathbf{i} + e\mathbf{j}$; $|\mathbf{v}| = \sqrt{e^{-2} + e^2}$

**25.** $v = -\sqrt{3}i - \dfrac{3\sqrt{3}}{2}j;\ a = -i + 3j;\ |v| = \dfrac{\sqrt{39}}{2}$

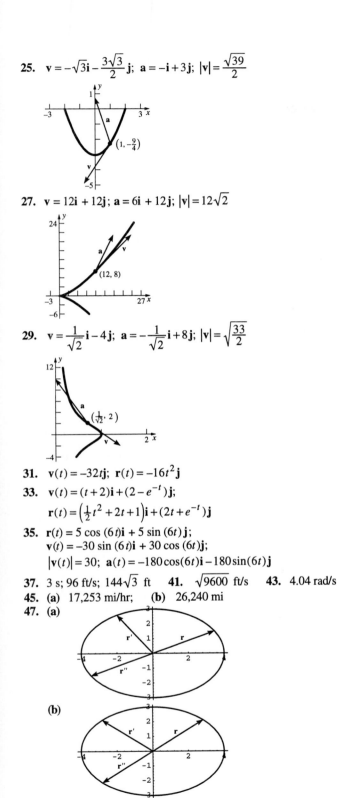

**27.** $v = 12i + 12j;\ a = 6i + 12j;\ |v| = 12\sqrt{2}$

**29.** $v = \dfrac{1}{\sqrt{2}}i - 4j;\ a = -\dfrac{1}{\sqrt{2}}i + 8j;\ |v| = \sqrt{\dfrac{33}{2}}$

**31.** $v(t) = -32tj;\ r(t) = -16t^2 j$

**33.** $v(t) = (t+2)i + (2 - e^{-t})j;$
$r(t) = \left(\frac{1}{2}t^2 + 2t + 1\right)i + (2t + e^{-t})j$

**35.** $r(t) = 5\cos(6t)i + 5\sin(6t)j;$
$v(t) = -30\sin(6t)i + 30\cos(6t)j;$
$|v(t)| = 30;\ a(t) = -180\cos(6t)i - 180\sin(6t)j$

**37.** 3 s; 96 ft/s; $144\sqrt{3}$ ft  **41.** $\sqrt{9600}$ ft/s  **43.** 4.04 rad/s
**45. (a)** 17,253 mi/hr;  **(b)** 26,240 mi
**47. (a)**

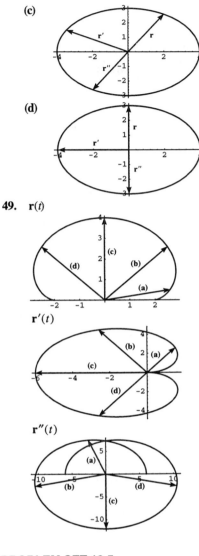

**(c)**

**(d)**

**49.** $r(t)$

$r'(t)$

$r''(t)$

## PROBLEM SET 13.5

**1.** $\dfrac{3}{\sqrt{13}}i + \dfrac{2}{\sqrt{13}}j;\ \dfrac{18}{13\sqrt{13}}$  **3.** $\dfrac{1}{\sqrt{2}}i + \dfrac{1}{\sqrt{2}}j;\ \dfrac{1}{4\sqrt{2}}$

**5.** $-\dfrac{3}{5}i + \dfrac{4}{5}j;\ \dfrac{24\sqrt{2}}{125}$  **7.** $-\dfrac{1}{\sqrt{2}}i + \dfrac{1}{\sqrt{2}}j;\ \dfrac{1}{\sqrt{2}e^{\pi/2}}$

**9.** $-\dfrac{2}{\sqrt{13}}i - \dfrac{3}{\sqrt{13}}j;\ \dfrac{6}{13\sqrt{13}}$  **11.** $-\dfrac{1}{\sqrt{2}}i + \dfrac{1}{\sqrt{2}}j;\ \dfrac{1}{\sqrt{2}}$

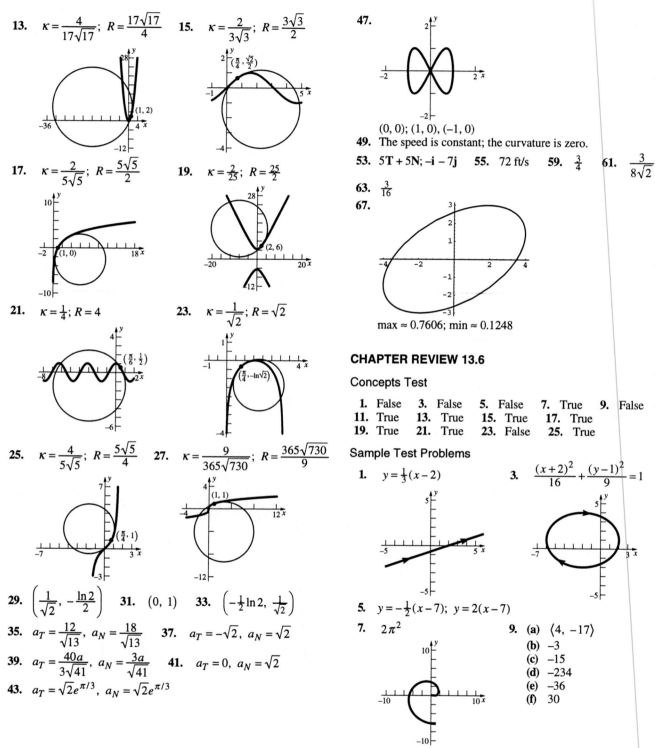

**13.** $\kappa = \dfrac{4}{17\sqrt{17}}$; $R = \dfrac{17\sqrt{17}}{4}$  **15.** $\kappa = \dfrac{2}{3\sqrt{3}}$; $R = \dfrac{3\sqrt{3}}{2}$

**17.** $\kappa = \dfrac{2}{5\sqrt{5}}$; $R = \dfrac{5\sqrt{5}}{2}$  **19.** $\kappa = \dfrac{2}{25}$; $R = \dfrac{25}{2}$

**21.** $\kappa = \frac{1}{4}$; $R = 4$  **23.** $\kappa = \dfrac{1}{\sqrt{2}}$; $R = \sqrt{2}$

**25.** $\kappa = \dfrac{4}{5\sqrt{5}}$; $R = \dfrac{5\sqrt{5}}{4}$  **27.** $\kappa = \dfrac{9}{365\sqrt{730}}$; $R = \dfrac{365\sqrt{730}}{9}$

**29.** $\left(\dfrac{1}{\sqrt{2}}, -\dfrac{\ln 2}{2}\right)$  **31.** $(0, 1)$  **33.** $\left(-\frac{1}{2}\ln 2, \dfrac{1}{\sqrt{2}}\right)$

**35.** $a_T = \dfrac{12}{\sqrt{13}}$, $a_N = \dfrac{18}{\sqrt{13}}$  **37.** $a_T = -\sqrt{2}$, $a_N = \sqrt{2}$

**39.** $a_T = \dfrac{40a}{3\sqrt{41}}$, $a_N = \dfrac{3a}{\sqrt{41}}$  **41.** $a_T = 0$, $a_N = \sqrt{2}$

**43.** $a_T = \sqrt{2}e^{\pi/3}$, $a_N = \sqrt{2}e^{\pi/3}$

**47.**

$(0, 0)$; $(1, 0)$, $(-1, 0)$

**49.** The speed is constant; the curvature is zero.

**53.** $5\mathbf{T} + 5\mathbf{N}$; $-\mathbf{i} - 7\mathbf{j}$  **55.** 72 ft/s  **59.** $\frac{3}{4}$  **61.** $\dfrac{3}{8\sqrt{2}}$

**63.** $\frac{3}{16}$

**67.**

max $\approx 0.7606$; min $\approx 0.1248$

## CHAPTER REVIEW 13.6

### Concepts Test

**1.** False  **3.** False  **5.** False  **7.** True  **9.** False
**11.** True  **13.** True  **15.** True  **17.** True
**19.** True  **21.** True  **23.** False  **25.** True

### Sample Test Problems

**1.** $y = \frac{1}{3}(x - 2)$  **3.** $\dfrac{(x+2)^2}{16} + \dfrac{(y-1)^2}{9} = 1$

**5.** $y = -\frac{1}{2}(x - 7)$; $y = 2(x - 7)$

**7.** $2\pi^2$

**9.** (a) $\langle 4, -17\rangle$
(b) $-3$
(c) $-15$
(d) $-234$
(e) $-36$
(f) $30$

**11.** $k = \frac{1}{2}$, $m = 2$  **13.** $\langle -5\sqrt{3}, 5\rangle$
**15.** N 12.22° W; 409.27 mi/h

**17. (a)** $\frac{1}{t}\mathbf{i} - 6t\mathbf{j}$; $-\frac{1}{t^2}\mathbf{i} - 6\mathbf{j}$;

**(b)** $\cos t\,\mathbf{i} - 2\sin 2t\mathbf{j}$; $-\sin t\,\mathbf{i} - 4\cos 2t\,\mathbf{j}$;

**(c)** $\sec^2 t\mathbf{i} - 4t^3\mathbf{j}$; $2\sec^2 t\tan t\mathbf{i} - 12t^2\mathbf{j}$

**19.** $\mathbf{v} = -4\mathbf{i} + 4\mathbf{j}$; $\mathbf{a} = 4\mathbf{i}$; $|\mathbf{v}| = 4\sqrt{2}$

**21. (a)** $\frac{1}{\sqrt{2}}$; **(b)** $\frac{2}{25}$; **(c)** $\frac{1}{|a|}\operatorname{sech}^2 1$

**23.** $a_T = \sqrt{2}$, $a_N = \sqrt{2}$

## PROBLEM SET 14.1

**1.** $A(1,2,3)$, $B(2,0,1)$, $C(-2,4,5)$, $D(0,3,0)$, $E(-1,-2,-3)$

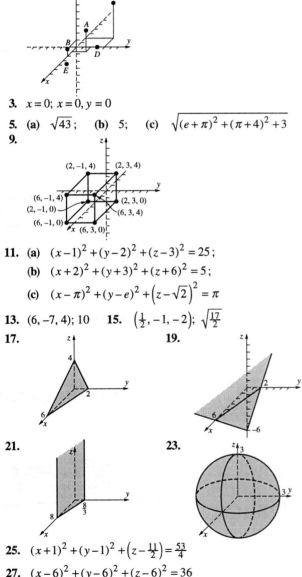

**3.** $x = 0$; $x = 0, y = 0$

**5. (a)** $\sqrt{43}$; **(b)** 5; **(c)** $\sqrt{(e+\pi)^2 + (\pi+4)^2 + 3}$

**9.**

**11. (a)** $(x-1)^2 + (y-2)^2 + (z-3)^2 = 25$;

**(b)** $(x+2)^2 + (y+3)^2 + (z+6)^2 = 5$;

**(c)** $(x-\pi)^2 + (y-e)^2 + \left(z-\sqrt{2}\right)^2 = \pi$

**13.** $(6, -7, 4)$; 10    **15.** $\left(\frac{1}{2}, -1, -2\right)$; $\sqrt{\frac{17}{2}}$

**17.**    **19.**

**21.**    **23.**

**25.** $(x+1)^2 + (y-1)^2 + \left(z - \frac{11}{2}\right) = \frac{53}{4}$

**27.** $(x-6)^2 + (y-6)^2 + (z-6)^2 = 36$

**29. (a)** Plane parallel to and 2 units above the $xy$-plane;

**(b)** Plane perpendicular to the $xy$-plane, whose trace in the $xy$-plane is the line $x = y$;

**(c)** Union of the $yz$-plane ($x = 0$) and the $xz$-plane ($y = 0$);

**(d)** Union of the three coordinate planes;

**(e)** Cylinder of radius 2, parallel to the $z$-axis;

**(f)** Top half of the sphere with center $(0, 0, 0)$ and radius 3

**33.** $\frac{11\pi}{12}$

## PROBLEM SET 14.2

**1. (a)**    **(b)**

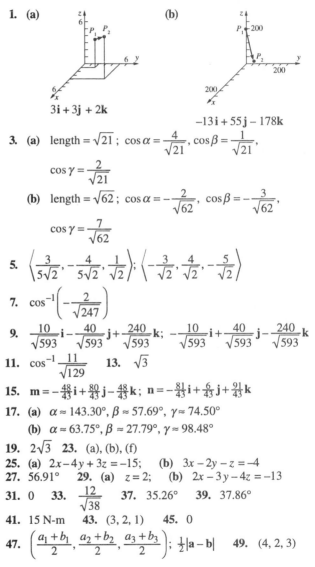

$3\mathbf{i} + 3\mathbf{j} + 2\mathbf{k}$

$-13\mathbf{i} + 55\mathbf{j} - 178\mathbf{k}$

**3. (a)** length $= \sqrt{21}$; $\cos\alpha = \frac{4}{\sqrt{21}}$, $\cos\beta = \frac{1}{\sqrt{21}}$,

$\cos\gamma = \frac{2}{\sqrt{21}}$

**(b)** length $= \sqrt{62}$; $\cos\alpha = -\frac{2}{\sqrt{62}}$, $\cos\beta = -\frac{3}{\sqrt{62}}$,

$\cos\gamma = \frac{7}{\sqrt{62}}$

**5.** $\left\langle \frac{3}{5\sqrt{2}}, -\frac{4}{5\sqrt{2}}, \frac{1}{\sqrt{2}} \right\rangle$; $\left\langle -\frac{3}{\sqrt{2}}, \frac{4}{\sqrt{2}}, -\frac{5}{\sqrt{2}} \right\rangle$

**7.** $\cos^{-1}\left(-\frac{2}{\sqrt{247}}\right)$

**9.** $\frac{10}{\sqrt{593}}\mathbf{i} - \frac{40}{\sqrt{593}}\mathbf{j} + \frac{240}{\sqrt{593}}\mathbf{k}$; $-\frac{10}{\sqrt{593}}\mathbf{i} + \frac{40}{\sqrt{593}}\mathbf{j} - \frac{240}{\sqrt{593}}\mathbf{k}$

**11.** $\cos^{-1}\frac{11}{\sqrt{129}}$    **13.** $\sqrt{3}$

**15.** $\mathbf{m} = -\frac{48}{43}\mathbf{i} + \frac{80}{43}\mathbf{j} - \frac{48}{43}\mathbf{k}$; $\mathbf{n} = -\frac{81}{43}\mathbf{i} + \frac{6}{43}\mathbf{j} + \frac{91}{43}\mathbf{k}$

**17. (a)** $\alpha \approx 143.30°$, $\beta \approx 57.69°$, $\gamma \approx 74.50°$

**(b)** $\alpha \approx 63.75°$, $\beta \approx 27.79°$, $\gamma \approx 98.48°$

**19.** $2\sqrt{3}$    **23.** (a), (b), (f)

**25. (a)** $2x - 4y + 3z = -15$;    **(b)** $3x - 2y - z = -4$

**27.** $56.91°$    **29. (a)** $z = 2$;    **(b)** $2x - 3y - 4z = -13$

**31.** 0    **33.** $\frac{12}{\sqrt{38}}$    **37.** $35.26°$    **39.** $37.86°$

**41.** 15 N-m    **43.** $(3, 2, 1)$    **45.** 0

**47.** $\left( \frac{a_1 + b_1}{2}, \frac{a_2 + b_2}{2}, \frac{a_3 + b_3}{2} \right)$; $\frac{1}{2}|\mathbf{a} - \mathbf{b}|$    **49.** $(4, 2, 3)$

## PROBLEM SET 14.3

**1. (a)** $-4\mathbf{i} - 10\mathbf{j} - 4\mathbf{k}$;    **(b)** $-6\mathbf{i} - 36\mathbf{j} - 27\mathbf{k}$;    **(c)** 8;

**(d)** $-98\mathbf{i} - 59\mathbf{j} + 88\mathbf{k}$

**3.** $c(-14\mathbf{i} - 2\mathbf{j} + 6\mathbf{k}), c$ in $\mathbb{R}$    **5.** $\pm\left\langle \dfrac{7}{\sqrt{86}}, -\dfrac{1}{\sqrt{86}}, \dfrac{6}{\sqrt{86}} \right\rangle$

**7.** $2\sqrt{74}$    **9.** $4\sqrt{6}$    **11.** $2x - y - z = -3$

**13.** $7x + 5y + 4z = -5$    **15.** $-x + 10y + 17z = -3$    **17.** $69$

**19.** (a) $9$;    (b) $\sqrt{35}$;    (c) $40.01°$

**27.** $\frac{1}{2}\sqrt{a^2 b^2 + a^2 c^2 + b^2 c^2}$

## PROBLEM SET 14.4

**1.** $x = 1 + 3t, y = -2 + 7t, z = 3 + 3t$

**3.** $x = 4 + t, y = 2, z = 3 - 2t$

**5.** $x = 4 + 3t, y = 5 + 2t, z = 6 + t$; $\dfrac{x-4}{3} = \dfrac{y-5}{2} = \dfrac{z-6}{1}$

**7.** $x = 1 + t, y = 1 + 10t, z = 1 + 100t$; $\dfrac{x-1}{1} = \dfrac{y-1}{10} = \dfrac{z-1}{100}$

**9.** $\dfrac{x-4}{27} = \dfrac{y+5}{-50} = \dfrac{z}{-6}$    **11.** $\dfrac{x+8}{10} = \dfrac{y}{2} = \dfrac{z+\frac{21}{2}}{9}$

**13.** $\dfrac{x-4}{1} = \dfrac{y}{-5} = \dfrac{z-6}{-2}$    **15.** $x = 5t, y = -3t, z = 4$

**17.** $x + y + 6z = 11$    **19.** $3x - 2y = 5$

**21.** (b) $2x + y - z = 7$;    (c) $(-1, 2, -1)$;    (d) $\sqrt{6}$

**23.** $\dfrac{x-1}{-\sqrt{3}} = \dfrac{y - 3\sqrt{3}}{3} = \dfrac{z - \frac{\pi}{3}}{1}$    **25.** $3x - 4y + 5z = -22$

**27.** (b) $\left( \dfrac{3\sqrt{3}}{4}, \dfrac{7}{4}, 0 \right)$    **29.** (a) $\dfrac{8\sqrt{2}}{3}$;    (b) $\dfrac{3\sqrt{26}}{7}$

## SECTION 14.5

**1.** $\mathbf{v}(1) = 4\mathbf{i} + 10\mathbf{j} + 2\mathbf{k}$; $\mathbf{a}(1) = 10\mathbf{j}$; $s(1) = 2\sqrt{30}$

**3.** $\mathbf{v}(2) = -\frac{1}{4}\mathbf{i} - \frac{4}{9}\mathbf{j} + 80\mathbf{k}$; $\mathbf{a}(2) = \frac{1}{4}\mathbf{i} + \frac{26}{27}\mathbf{j} + 160\mathbf{k}$;

$s(2) = \dfrac{\sqrt{8,294,737}}{36}$

**5.** $\mathbf{v}(2) = 4\mathbf{j} + \dfrac{2^{2/3}}{3}\mathbf{k}$; $\mathbf{a}(2) = 4\mathbf{j} - \dfrac{1}{9\sqrt[3]{2}}\mathbf{k}$;

$s(2) = \sqrt{16 + \dfrac{2^{4/3}}{9}}$

**7.** $\mathbf{v}(\pi) = -\mathbf{j} + \mathbf{k}$; $\mathbf{a}(\pi) = \mathbf{i}$, $s(\pi) = \sqrt{2}$

**9.** $\mathbf{v}\left(\frac{\pi}{4}\right) = 2\mathbf{i} + 3e^{\pi/4}\mathbf{j}$; $\mathbf{a}\left(\frac{\pi}{4}\right) = 4\mathbf{i} + 3e^{\pi/4}\mathbf{j} + 16\mathbf{k}$;

$s\left(\frac{\pi}{4}\right) = \sqrt{4 + 9e^{\pi/2}}$

**11.** $\mathbf{v}(2) = 2\pi\mathbf{i} + \mathbf{j} - e^{-2}\mathbf{k}$; $\mathbf{a}(2) = 2\pi\mathbf{i} - 2\pi^2\mathbf{j} + e^{-2}\mathbf{k}$;

$s(2) = \sqrt{4\pi^2 + 1 + e^{-4}}$

**15.** $2\sqrt{2}$    **17.** $144$    **19.** $\sqrt{41}$    **21.** $\frac{3}{2}$

**23.** $\sqrt{2}\sinh\pi$    **25.** $\frac{1}{2}\pi^2 + \pi$

**27.** $\kappa = \dfrac{\sqrt{6}}{10\sqrt{5}}$; $\mathbf{T} = \dfrac{2}{\sqrt{5}}\mathbf{i} + \dfrac{1}{\sqrt{5}}\mathbf{j}$; $\mathbf{N} = \dfrac{1}{\sqrt{30}}\mathbf{i} - \dfrac{2}{\sqrt{30}}\mathbf{j} + \dfrac{5}{\sqrt{30}}\mathbf{k}$;

$\mathbf{B} = \dfrac{1}{\sqrt{6}}\mathbf{i} - \dfrac{2}{\sqrt{6}}\mathbf{j} - \dfrac{1}{\sqrt{6}}\mathbf{k}$

**29.** $\kappa = \dfrac{\sqrt{11}}{21\sqrt{7}}$; $\mathbf{T} = \dfrac{2}{\sqrt{21}}\mathbf{i} + \dfrac{1}{\sqrt{21}}\mathbf{j} + \dfrac{4}{\sqrt{21}}\mathbf{k}$;

$\mathbf{N} = -\dfrac{5}{\sqrt{77}}\mathbf{i} - \dfrac{6}{\sqrt{77}}\mathbf{j} + \dfrac{4}{\sqrt{77}}\mathbf{k}$;

$\mathbf{B} = \dfrac{4}{\sqrt{33}}\mathbf{i} - \dfrac{4}{\sqrt{33}}\mathbf{j} - \dfrac{1}{\sqrt{33}}\mathbf{k}$

**31.** $\kappa = \frac{9}{91}$; $\mathbf{T} = \left\langle -\dfrac{3}{\sqrt{13}}, 0, \dfrac{2}{\sqrt{13}} \right\rangle$; $\mathbf{N} = \langle 0, 1, 0 \rangle$;

$\mathbf{B} = \left\langle -\dfrac{2}{\sqrt{13}}, 0, -\dfrac{3}{\sqrt{13}} \right\rangle$

**33.** $\kappa = \frac{1}{3}\operatorname{sech}^2\frac{1}{3}$; $\mathbf{T} = \tanh\frac{1}{3}\mathbf{i} + \operatorname{sech}\frac{1}{3}\mathbf{j}$;

$\mathbf{N} = \operatorname{sech}\frac{1}{3}\mathbf{i} - \tanh\frac{1}{3}\mathbf{j}$; $\mathbf{B} = \left( -\operatorname{sech}^2\frac{1}{3} - \tanh^2\frac{1}{3} \right)\mathbf{k}$

**35.** $\kappa = \dfrac{1}{(\cosh^2 1 + \sinh^2 1)^{3/2}}$; $\mathbf{T} = \dfrac{-\cosh 1\mathbf{i} + \sinh 1\mathbf{j}}{\sqrt{\cosh^2 1 + \sinh^2 1}}$;

$\mathbf{N} = \dfrac{\sinh 1\mathbf{i} + \cosh 1\mathbf{j}}{\sqrt{\cosh^2 1 + \sinh^2 1}}$; $\mathbf{B} = -\mathbf{k}$

**37.** $\kappa = \dfrac{1}{2\sqrt{2}}$; $\mathbf{T} = -\frac{1}{2}\mathbf{i} + \frac{1}{2}\mathbf{j} + \dfrac{1}{\sqrt{2}}\mathbf{k}$;

$\mathbf{N} = \dfrac{1}{\sqrt{2}}\mathbf{i} + \dfrac{1}{\sqrt{2}}\mathbf{j}$; $\mathbf{B} = -\frac{1}{2}\mathbf{i} + \frac{1}{2}\mathbf{j} - \dfrac{1}{\sqrt{2}}\mathbf{k}$

**39.** $\kappa = \dfrac{\sqrt{3}}{2}$; $\mathbf{T} = \left\langle 0, -\dfrac{1}{\sqrt{2}}, -\dfrac{1}{\sqrt{2}} \right\rangle$;

$\mathbf{N} = \langle -1, 0, 0 \rangle$; $\mathbf{B} = \left\langle 0, \dfrac{1}{\sqrt{2}}, -\dfrac{1}{\sqrt{2}} \right\rangle$

**41.** $a_T(t) = \dfrac{4t}{\sqrt{10 + 4t^2}}$; $a_N(t) = 2\sqrt{\dfrac{5}{5 + 2t^2}}$

**43.** $a_T(t) = \dfrac{4t^3}{\sqrt{2t^4 + 3}}$; $a_N(t) = \dfrac{2\sqrt{6}t}{\sqrt{2t^4 + 3}}$

**45.** $a_T(t) = \dfrac{\tan t\sec^2 t - \cot t\csc^2 t}{\sqrt{1 + \cot^2 t + \tan^2 t}}$;

$a_N(t) = \dfrac{\sqrt{\csc^4 t + 4\csc^2 t\sec^2 t + \sec^4 t}}{\sqrt{1 + \cot^2 t + \tan^2 t}}$

**47.** $a_T(t) = \dfrac{e^{2t} - e^{-2t}}{\sqrt{e^{2t} + 4 + e^{-2t}}}$; $a_N(t) = 2\sqrt{\dfrac{e^{2t} + 1 + e^{-2t}}{e^{2t} + 4 + e^{-2t}}}$

**49.** $\mathbf{T} = \dfrac{1}{\sqrt{\cosh^2\frac{\pi}{6c} + 1}}\left( \cosh\frac{\pi}{6c}\mathbf{i} + \mathbf{k} \right)$;

$\mathbf{N} = \dfrac{1}{\sqrt{\cosh^2\frac{\pi}{6c} + 1}}\left( \mathbf{i} - \cos\frac{\pi}{6c}\mathbf{k} \right)$;

$\mathbf{B} = \mathbf{j}$

**51.** $T = \dfrac{1}{\sqrt{1+5\pi^2}}(-\pi i - j + 2\pi k)$ ;

$N = \dfrac{[(-2-5\pi^2)i + (5\pi^3 + 6\pi)j + 2k]}{\sqrt{(8+16\pi^2 + 5\pi^4)(1+5\pi^2)}}$ ;

$B = \dfrac{1}{\sqrt{8+16\pi^2 + 5\pi^4}}\left[(-2-2\pi^2)i - 2\pi j - (2+\pi^2)k\right]$

**53.** $T = \dfrac{1}{\sqrt{3}}i + \dfrac{1}{2}\left(\dfrac{1}{\sqrt{3}}-1\right)j + \dfrac{1}{2}\left(\dfrac{1}{\sqrt{3}}+1\right)k$ ;

$N = -\dfrac{1+\sqrt{3}}{2\sqrt{2}}j + \dfrac{1-\sqrt{3}}{2\sqrt{2}}k$ ;

$B = \sqrt{\dfrac{2}{3}}i + \dfrac{1}{12}\left(3\sqrt{2}-\sqrt{6}\right)j - \dfrac{1}{12}\left(3\sqrt{2}+\sqrt{6}\right)k$

**55.** $T = \left\langle \frac{1}{3}, \frac{2}{3}, \frac{2}{3}\right\rangle$ ; $N = \left\langle -\frac{2}{3}, -\frac{1}{3}, \frac{2}{3}\right\rangle$ ; $B = \left\langle \frac{2}{3}, -\frac{2}{3}, \frac{1}{3}\right\rangle$

**57.** $T = \dfrac{1}{\sqrt{2}}i + \dfrac{1}{\sqrt{2}}\operatorname{sech}\dfrac{\pi}{3}j + \dfrac{1}{\sqrt{2}}\tanh\dfrac{\pi}{3}k$ ;

$N = -\tanh\dfrac{\pi}{3}j + \operatorname{sech}\dfrac{\pi}{3}k$ ;

$B = \dfrac{1}{\sqrt{2}}i - \dfrac{1}{\sqrt{2}}\operatorname{sech}\dfrac{\pi}{3}j - \dfrac{1}{\sqrt{2}}\tanh\dfrac{\pi}{3}k$

**59.** (b) $R_p = 10R_m$ ; $t = \frac{\pi}{9}$

**61.** (a) Winding upward around the right circular cylinder $x = \sin t, y = \cos t$, as $t$ increases.

(b) Same as part (a), but winding much faster by a factor of $3t^2$.

(c) With standard orientation of the axes, the motion is winding to the right around the right circular cylinder $x = \sin t, z = \cos t$.

(d) Spiraling upward, with increasing radius, along the spiral $x = t\sin t, y = t\cos t$.

(e) Spiraling upward, with decreasing radius, along the spiral $x = \dfrac{1}{t^2}\sin t, \; y = \dfrac{1}{t^2}\cos t$.

(f) Spiraling to the right, with increasing radius along the spiral $x = t^2\sin(\ln t), \; z = t^2\cos(\ln t)$.

**63.** $P_5(x) = 10x^3 - 15x^4 + 6x^5$

**73.** (a) $\tau = \frac{1}{2}$ ; $\kappa = \frac{1}{2}$ ;

(b) $\tau = -\dfrac{a}{a^2+c^2}$ ; $\kappa = \dfrac{|c|}{a^2+c^2}$ ;

(c) $\tau = -\frac{8}{25}$ ; $\kappa = \frac{8}{25}$ ;

(d) $\tau = \dfrac{2}{1+4t^2+t^4}$ ; $\kappa = \sqrt{\dfrac{t^4+4t^2+1}{(1+t^2+t^4)^3}}$

**75.** $a = 0$    **77.** $(6,0,8)$ ; $8\sqrt{9\pi^2+1}$

**79.** (a) $r(t) = r_0$ ;    (b) $r(t) = ct + r_0$ ;

(c) $r(t) = \frac{1}{2}ct^2 + v_0 t + r_0$ ;    (d) $r(t) = r_0 e^{ct}$

**81.** $\left(\dfrac{16\pi^2+30}{4\pi+1}, \dfrac{120\pi-16\pi^2}{4\pi+1}, \dfrac{16\pi^2+30}{4\pi+1}\right)$

## PROBLEM SET 14.6

**1.** Elliptic cylinder    **3.** Plane

**5.** Circular cylinder    **7.** Ellipsoid

**11.** Cylinder    **13** Hyperbolic paraboloid

**15.** Elliptic paraboloid    **17** Plane

**19.** Hemisphere

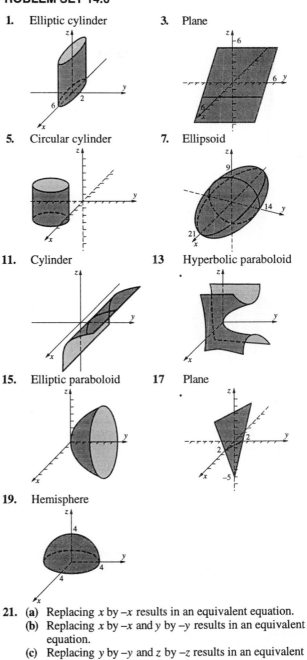

**21.** (a) Replacing $x$ by $-x$ results in an equivalent equation.

(b) Replacing $x$ by $-x$ and $y$ by $-y$ results in an equivalent equation.

(c) Replacing $y$ by $-y$ and $z$ by $-z$ results in an equivalent equation.

(d) Replacing $x$ by $-x, y$ by $-y$, and $z$ by $-z$ results in an equivalent equation.

**23.** $y = 2x^2 + 2z^2$    **25.** $4x^2 + 3y^2 + 4z^2 = 12$

**27.** $\left(0, \pm 2\sqrt{5}, 4\right)$    **29.** $\dfrac{\pi ab(c^2-h^2)}{c^2}$

**31.** Major diameter 4; minor diameter $2\sqrt{2}$

**33.** $x^2 + 9y^2 - 9z^2 = 0$

## PROBLEM SET 14.7

**1. (a)** $\left(3\sqrt{3}, 3, -2\right)$;   **(b)** $\left(-2, -2\sqrt{3}, -8\right)$

**3. (a)** $\left(4\sqrt{2}, \frac{5\pi}{3}, \frac{\pi}{4}\right)$;   **(b)** $\left(4, \frac{3\pi}{4}, \frac{\pi}{6}\right)$

**5.**

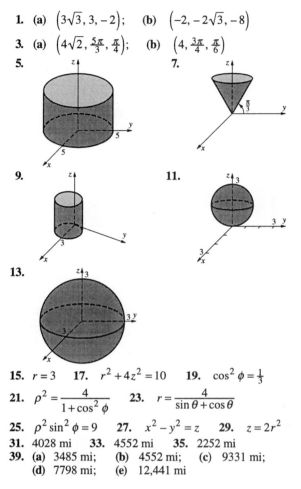

**7.**

**9.**

**11.**

**13.**

**15.** $r = 3$   **17.** $r^2 + 4z^2 = 10$   **19.** $\cos^2 \phi = \frac{1}{3}$

**21.** $\rho^2 = \frac{4}{1 + \cos^2 \phi}$   **23.** $r = \frac{4}{\sin\theta + \cos\theta}$

**25.** $\rho^2 \sin^2 \phi = 9$   **27.** $x^2 - y^2 = z$   **29.** $z = 2r^2$

**31.** 4028 mi   **33.** 4552 mi   **35.** 2252 mi

**39. (a)** 3485 mi;   **(b)** 4552 mi;   **(c)** 9331 mi;
   **(d)** 7798 mi;   **(e)** 12,441 mi

## 14.8 CHAPTER REVIEW

### Concepts Test

**1.** True   **3.** True   **5.** False   **7.** True   **9.** True
**11.** True   **13.** False   **5.** True   **17.** True   **19.** True
**21.** True   **23.** False   **25.** True   **27.** False
**29.** True   **31.** True   **33.** True   **35.** False
**37.** False   **39.** False

### Sample Test Problems

**1.** $(x-1)^2 + (y-2)^2 + (z-4)^2 = 11$

**3.**

**(a)** $3; \sqrt{35}$ ;

**(b)** $\frac{2}{3}, -\frac{1}{3}, \frac{2}{3}$ ;
   $\frac{5}{\sqrt{35}}, \frac{1}{\sqrt{35}}, -\frac{3}{\sqrt{35}}$ ;

**(c)** $\frac{2}{3}\mathbf{i} - \frac{1}{3}\mathbf{j} + \frac{2}{3}\mathbf{k}$ ;

**(d)** $\cos^{-1}\frac{1}{\sqrt{35}}$

**5.** $c\langle 10, -11, -3 \rangle$, $c$ in $\mathbb{R}$

**7. (a)** $y = 7$;   **(b)** $x = -5$;   **(c)** $z = -2$;
   **(d)** $3x - 4y + z = -45$

**9.** 1   **11.** $x = -2 + 8t, y = 1 + t, z = 5 - 8t$

**13.** $x = 2t, y = 25 + t, z = 16$   **15.** $\langle 2, -2, 1 \rangle + t\langle 5, -4, -3 \rangle$

**17.** Tangent line: $\dfrac{x-2}{1} = \dfrac{y-2}{2} = \dfrac{z - \frac{8}{3}}{4}$

Normal Plane: $3x + 6y + 12z = 50$

**19.** $\sqrt{3}(e^5 - e)$   **21.** $a_T = \dfrac{22}{\sqrt{14}}$; $a_N = \dfrac{2\sqrt{19}}{\sqrt{14}}$

**23.** Sphere   **25.** Circular paraboloid

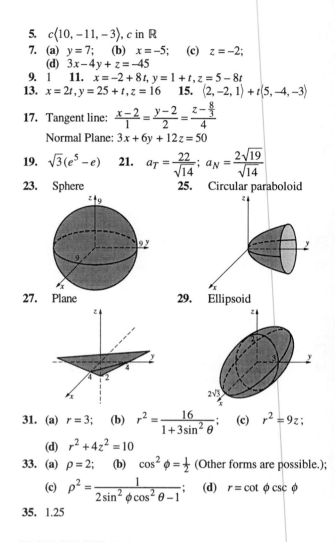

**27.** Plane   **29.** Ellipsoid

**31. (a)** $r = 3$;   **(b)** $r^2 = \dfrac{16}{1 + 3\sin^2\theta}$;   **(c)** $r^2 = 9z$;
   **(d)** $r^2 + 4z^2 = 10$

**33. (a)** $\rho = 2$;   **(b)** $\cos^2 \phi = \frac{1}{2}$ (Other forms are possible.);
   **(c)** $\rho^2 = \dfrac{1}{2\sin^2 \phi \cos^2 \theta - 1}$;   **(d)** $r = \cot\phi \csc\phi$

**35.** 1.25

## PROBLEM SET 15.1

**1. (a)** 5;   **(b)** 0;   **(c)** 6;   **(d)** $a^6 + a^2$;   **(e)** $2x^2$;
   **(f)** $(2, -4)$ is not in the domain of $f$.

**3. (a)** 0;   **(b)** 2;   **(c)** 16;   **(d)** $-4.2469$;   **(e)** 0.6311

**5.** $t^2$

**7.**       **9.**

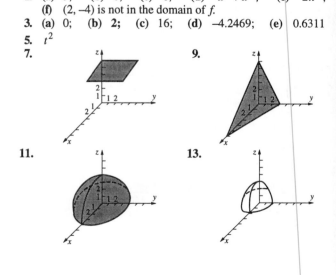

**11.**       **13.**

**15.**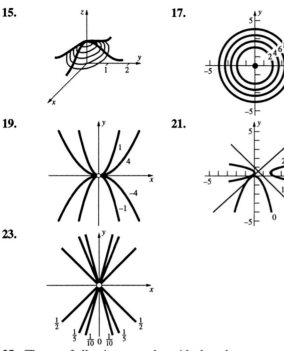

**17.**

**19.**

**21.**

**23.**

**25.** The set of all points on and outside the sphere
$x^2 + y^2 + z^2 = 16$.

**27.** The set of all points on and inside the ellipsoid
$x^2/9 + y^2/16 + z^2/1 = 1$.

**29.** The set of all spheres with centers at the origin.

**31.** A set of hyperboloids of revolution about the $z$-axis.

**33.** A set of hyperbolic cylinders parallel to the $z$-axis.

**35.** (a) gentle climb, steep climb;   (b) 6700 ft, 3040 ft

**37.** Elliptic paraboloid, $(\frac{1}{2}, -2)$, $-25.25$

## PROBLEM SET 15.2

**1.** $f_x(x,y) = 8(2x-y)^3$; $f_y(x,y) = -4(2x-y)^3$

**3.** $f_x(x,y) = (x^2+y^2)/(x^2y)$; $f_y(x,y) = -(x^2+y^2)/(xy^2)$

**5.** $f_x(x,y) = e^y \cos x$; $f_y(x,y) = e^y \sin x$

**7.** $f_x(x,y) = x(x^2-y^2)^{-1/2}$; $f_y(x,y) = -y(x^2-y^2)^{-1/2}$

**9.** $g_x(x,y) = -ye^{-xy}$; $g_y(x,y) = -xe^{-xy}$

**11.** $f_x(x,y) = 4/[1+(4x-7y)^2]$;
$f_y(x,y) = -7/[1+(4x-7y)^2]$

**13.** $f_x(x,y) = -2xy\sin(x^2+y^2)$;
$f_y(x,y) = -2y^2\sin(x^2+y^2) + \cos(x^2+y^2)$

**15.** $F_x(x,y) = 2\cos x \cos y$; $F_y(x,y) = -2\sin x \sin y$

**17.** $f_{xy}(x,y) = 12xy^2 - 15x^2y^4 = f_{yx}(x,y)$

**19.** $f_{xy}(x,y) = -6e^{2x}\sin y = f_{yx}(x,y)$

**21.** $F_x(3,-2) = \frac{1}{9}$; $F_y(3,-2) = -\frac{1}{2}$

**23.** $f_x(\sqrt5,-2) = -\frac{4}{21}$; $f_y(\sqrt5,-2) = -4\sqrt5/21$

**25.** 1   **27.** 3   **29.** $120\pi$   **31.** $k/100$

**33.** $\partial^2 f/\partial x^2 = 6xy$; $\partial^2 f/\partial y^2 = -6xy$

**35.** $180x^4y^2 - 12x^2$

**37.** (a) $\partial^3 f/\partial y^3$;   (b) $\partial^3 f/\partial y \partial x^2$;   (c) $\partial^4 f/\partial y^3 \partial x$

**39.** (a) $6xy - yz$;   (b) 8;   (c) $6x - z$

**41.** $-yze^{-xyz} - y(xy - z^2)^{-1}$   **43.** $(1, 0, 29)$

**45.** $\{(x,y): x < \frac{1}{2}, y > \frac{1}{2}, y < x + \frac{1}{2}\}$
$\cup \{(x,y): x > \frac{1}{2}, y < \frac{1}{2}, x < y + \frac{1}{2}\}$, $\{z: 0 < z \le \sqrt3/36\}$

**47.** (a) $-4$;   (b) $\frac{2}{3}$;   (c) $\frac{2}{5}$;   (d) $\frac{8}{3}$

## PROBLEM SET 15.3

**1.** $-18$   **3.** $2 - \frac{1}{2}\sqrt3$   **5.** $\frac{1}{3}$   **7.** Does not exist.

**9.** Entire plane.   **11.** $\{(x,y): y \ne x^2\}$

**13.** $\{(x,y): y \le x+1\}$

**15.** $\lim_{x\to0} f(x,0) = \lim_{x\to0}[0/(x^2+0)] = 0$;
$\lim_{x\to0} f(x,x) = \lim_{x\to0}[x^2/(x^2+x^2)] = \frac{1}{2}$

**17.** (a) $\lim_{x\to0} f(x,mx) = \lim_{x\to0} mx^3/(x^4 + m^2x^2)$
$= \lim_{x\to0} mx/(x^2 + m^2) = 0$;

(b) $\lim_{x\to0} f(x,x^2) = \lim_{x\to0} x^4/(x^4 + x^4) = \frac{1}{2}$;

(c) $\lim_{(x,y)\to(0,0)} f(x,y)$ does not exist.

**19.** The boundary consists of the line segments that form the outer edges of the given rectangle; the set is closed.

**21.** Boundary: $\{(x,y): x^2 + y^2 = 4\} \cup \{(0,0)\}$;
the set is neither open nor closed.

**23.** Boundary: $\{(x,y): y = \sin(1/x), x > 0\}$
$\cup \{(x,y): x = 0, y \le 1\}$; the set is open.

**25.** $g(x) = 2x$

**27.** (a) Continuous;   (b) discontinuous;   (c) continuous;
(d) continuous;   (e) continuous;   (f) discontinuous

**29.** (a) $\{(x,y,z): x^2 + y^2 = 1, \ 1 \le z \le 2\}$;
(b) $\{(x,y,z): x^2 + y^2 = 1, \ z = 1\}$;   (c) $\{(x,y,z): z = 1\}$;
(d) empty set.

**31.** (a) $\{(x, y): x > 0\}$;
(b) $\{(u,v,x,y): \langle x,y \rangle = k\langle u,v \rangle, \ k > 0, \ \langle u,v \rangle \ne \langle 0,0 \rangle\}$

## PROBLEM SET 15.4

**1.** $(2xy + 3y)\mathbf{i} + (x^2 + 3x)\mathbf{j}$   **3.** $e^{xy}(1 + xy)\mathbf{i} + x^2e^{xy}\mathbf{j}$

**5.** $(x+y)^{-2}[(x^2y + 2xy^2)\mathbf{i} + x^3\mathbf{j}]$

**7.** $(x^2 + y^2 + z^2)^{-1/2}(x\mathbf{i} + y\mathbf{j} + z\mathbf{k})$

**9.** $xe^{x-z}[(yx + 2y)\mathbf{i} + x\mathbf{j} - xy\mathbf{k}]$

**11.** $\langle -21, 16 \rangle$, $z = -21x + 16y - 60$

**13.** $\langle 0, -2\pi \rangle$, $z = -2\pi y + \pi - 1$

**15.** $w = 7x - 8y - 2z + 3$ **21.** $z = -5x + 5y$

## PROBLEM SET 15.5

**1.** $\frac{8}{5}$ **3.** $3\sqrt{2}/2$ **5.** $(\sqrt{2} + \sqrt{6})/4$ **7.** $\frac{52}{3}$

**9.** 13 **11.** $\sqrt{21}$ **13.** $(1/\sqrt{5})(-\mathbf{i} + 2\mathbf{j})$

**15.** $\nabla f(\mathbf{p}) = -4\mathbf{i} + \mathbf{j}$ is perpendicular to the tangent line at $\mathbf{p}$.

**17.** $\frac{2}{3}$ **19.** (a) $(0, 0, 0)$; (b) $-\mathbf{i} + \mathbf{j} - \mathbf{k}$; (c) yes.

**21.** N 63.43° E **23.** Descend; $-300\sqrt{2}e^{-3}$

**25.** $x = -2y^2$

**27.** (a) $-10/\sqrt{2 + \pi^2}$ deg/m; (b) $-10$ deg/s

**29.** (a) $(90, 125)$; (b) $(190, 35)$; (c) $-\frac{1}{3}$, $0$, $\frac{2}{5}$

**31.** Leave at about $(-0.1, -5)$. **33.** Leave at about $(3, 5)$.

## PROBLEM SET 15.6

**1.** $12t^{11}$ **3.** $e^{3t}(3\sin 2t + 2\cos 2t) + e^{2t}(3\cos 3t + 2\sin 3t)$

**5.** $7t^6 \cos(t^7)$ **7.** $2s^3 t - 3s^2 t^2$

**9.** $(2s^2 \sin t \cos t + 2t \sin^2 s)\exp(s^2 \sin^2 t + t^2 \sin^2 s)$

**11.** $s^4 t(1 + s^4 t^2)^{-1/2}$ **13.** 72 **15.** $-\frac{1}{2}(\pi + 1)$

**17.** 244.35 board ft per year **19.** $\sqrt{20}$ ft/s

**21.** $(3x^2 + 4xy)/(3y^2 - 2x^2)$

**23.** $(y\sin x - \sin y)/(x\cos y + \cos x)$

**25.** $(yz^3 - 6xz)/(3x^3 - 3xyz^2)$

**27.** $\partial T/\partial s = (\partial f/\partial x)(\partial x/\partial s) + (\partial f/\partial y)(\partial y/\partial s)$
$+ (\partial f/\partial z)(\partial z/\partial s) + (\partial f/\partial w)(\partial w/\partial s)$

**31.** $10\sqrt{2} - 3\pi\sqrt{2}$ **33.** 288 mi/h

## PROBLEM SET 15.7

**1.** $2(x - 2) + 3(y - 3) + \sqrt{3}(z - \sqrt{3}) = 0$

**3.** $(x - 1) - 3(y - 3) + \sqrt{7}(z - \sqrt{7}) = 0$

**5.** $x + y - z = 2$ **7.** $z + 1 = -2\sqrt{3}(x - \frac{1}{3}\pi) - 3y$

**9.** 0.08; 0.08017992 **11.** $-0.03$; $-0.03015101$

**13.** $(3, -1, -14)$

**15.** The common tangent plane is $y + z = 1$.

**17.** $(1, 2, -1)$ and $(-1, -2, 1)$

**19.** $x = 1 + 32t$; $y = 2 - 19t$; $z = 2 - 17t$ **21.** 0.004375 lb

**23.** 7% **25.** $20 \pm 0.34$ **27.** $V = 9|k|/2$

**29.** (a) 4.98; (b) 4.98196; (c) 4.9819675

## PROBLEM SET 15.8

**1.** $(2, 0)$: local minimum point.

**3.** $(0, 0)$: saddle point; $(\pm\frac{1}{2}, 0)$; local minimum points.

**5.** $(0, 0)$: saddle point (special argument needed).

**7.** $(1, 2)$: local minimum point. **9.** No critical points.

**11.** Global maximum of 7 at $(1, 1)$; global minimum of $-4$ at $(0, -1)$.

**13.** Global maximum of 2 at $(\pm 1, 0)$; global minimum of 0 at $(0, \pm 1)$.

**15.** Each of the three numbers is $N/3$. **17.** A cube.

**19.** Base 8 ft by 8 ft; depth 4 ft. **21.** $3\sqrt{3}(\mathbf{i} + \mathbf{j} + \mathbf{k})$

**23.** $2\pi/3$; 4 in. **25.** (a) 8; (b) $-11$

**27.** Maximum of 3 at $(1, 2)$; minimum of $-\frac{12}{5}$ at $\left(\frac{8}{5}, -\frac{2}{5}\right)$.

**29.** $y = \frac{7}{10}x + \frac{1}{10}$

**31.** $\left(\pm\sqrt{3}/2, -\frac{1}{2}\right)$ where $T = 9/4$; $(0, 1/2)$, where $T = 9/4$

**33.** Equilateral triangle.

**35.** Local maximum: $f(1.75, 0) = 1.15$;
global maximum: $f(-3.8, 0) = 2.30$

**37.** Global minimum: $f(0, 1) = f(0, -1) = -0.12$.

**39.** No global maximum or global minimum.

**41.** Global maximum: $f(0.67, 0) = 5.06$;
global minimum: $f(-0.75, 0) = -3.54$.

**43.** Global maximum: $f(2.1, 2.1) = 3.5$;
global minimum: $f(4.2, 4.2) = -3.5$

## PROBLEM SET 15.9

**1.** $f(\sqrt{3}, \sqrt{3}) = f(-\sqrt{3}, -\sqrt{3}) = 6$

**3.** $f(2/\sqrt{5}, -1/\sqrt{5}) = f(-2/\sqrt{5}, 1/\sqrt{5}) = 5$

**5.** $f\left(\frac{6}{7}, \frac{18}{7}, -\frac{12}{7}\right) = \frac{72}{7}$ **7.** Base is 4 by 4; depth is 2.

**9.** $10\sqrt{5}$ ft$^3$ **11.** $8abc/3\sqrt{3}$

**13.** $x/a + y/b + z/c = 3$, $V = \frac{9}{2}abc$

**15.** $x = \alpha d/a$, $y = \beta d/b$. $z = \gamma d/c$

**17.** $f(-1, 1, 0) = 3, f(1, -1, 1) = -1$

**19.** $x_i = a_i/\sqrt{a_1^2 + ... + a_n^2}$, $w = \sqrt{a_1^2 + ... + a_n^2}$

**21.** $f(4, 0) = -4$. **23.** $f(0, 3) = f(0, -3) = -0.99$

## CHAPTER REVIEW 15.10

### Concepts Test

**1.** True **3.** True **5.** True **7.** False **9.** True
**11.** True **13.** True **15.** True **17.** True **19.** False

### Sample Test Problems

**1.** (a) $\{(x, y): x^2 + 4y^2 \geq 100\}$

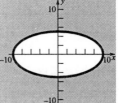

**(b)** $\{(x, y): 2x - y \geq 1\}$

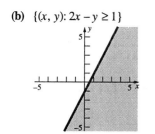

**3.** $12x^3y^2 + 14xy^7$; $36x^2y^2 + 14y^7$; $24x^3y + 98xy^6$

**5.** $e^{-y}\sec^2 x$; $2e^{-y}\sec^2 x \tan x$; $-e^{-y}\sec^2 x$

**7.** $450x^2y^4 - 42y^5$  **9.** 1  **11.** Does not exist.

**13.** **(a)** $-4\mathbf{i} - \mathbf{j} + 6\mathbf{k}$; $-4(\cos 1\mathbf{i} + \sin 1\mathbf{j} - \cos 1\mathbf{k})$

**15.** $\sqrt{3} + 2$  **17.** **(a)** $x^2 + 2y^2 = 18$;  **(b)** $4\mathbf{i} + 2\mathbf{j}$

**19.** $(x^2 + 3y - 4z)/x^2yz$; $(-x^2 - 4x)/xy^2z$; $(3y - x^2)/xyz^2$

**21.** $15xy\sqrt{t}/z^3 + 5x^2/tz^3 - 45x^2ye^{3t}/z^4$

**23.** $18\mathbf{i} + 16\mathbf{j} - 18\mathbf{k}$; $9x + 8y - 9z = 34$

**25.** $0.7728$  **27.** $16\sqrt{3}/3$  **29.** Radius 2; height 4.

## PROBLEM SET 16.1

**1.** 14  **3.** 12  **5.** 4  **7.** 3  **9.** 168  **11.** 520
**13.** 52.57  **15.** 5.5  **19.** $c = 15.30$, $C = 30.97$
**21.** **(a)** $-6$;  **(b)** 6
**23.** Number of cubic inches of rain that fell on all of Colorado in 1980; average rainfall in Colorado during 1980.

## PROBLEM SET 16.2

**1.** $\frac{32}{3}$  **3.** $\frac{55}{4}$  **5.** 1  **7.** $\pi/2 - 1$

**9.** $\frac{4}{15}[31 - 9\sqrt{3}] \approx 4.110$  **11.** $1 - \frac{1}{2}\ln 3 \approx 0.4507$

**13.** 0  **15.** 2

**17.**  **19.**

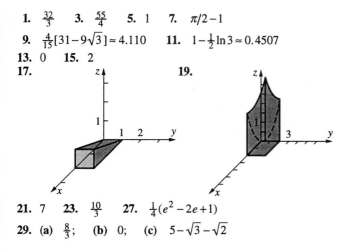

**21.** 7  **23.** $\frac{10}{3}$  **27.** $\frac{1}{4}(e^2 - 2e + 1)$

**29.** **(a)** $\frac{8}{3}$;  **(b)** 0;  **(c)** $5 - \sqrt{3} - \sqrt{2}$

## PROBLEM SET 16.3

**1.** $\frac{3}{4}$  **3.** 240  **5.** $\frac{1}{2}(e^{27} - e)$  **7.** $-\sqrt{2}/(2\pi)$

**9.** $(3\ln 2 - \pi)/9$  **11.** $\frac{16}{3}$  **13.** 0  **15.** $\frac{27}{70}$

**17.** $4\tan^{-1} 2 - \ln 5$  **19.** 6  **21.** 20  **23.** 10

**25.** $\frac{4}{15}$  **27.** $\frac{1}{2}\ln \sec 1$  **29.** $3\pi$

**31.** $\int_0^1 \int_y^1 f(x, y)\, dy\, dx$  **33.** $\int_0^1 \int_{y^4}^{\sqrt{y}} f(x, y)\, dx\, dy$

**35.** $\int_{-1}^0 \int_{-x}^1 f(x, y)\, dy\, dx + \int_0^1 \int_x^1 f(x, y)\, dy\, dx$  **37.** $\frac{256}{15}$

**39.** $\frac{1}{3}(1 - \cos 8)$  **41.** $15\pi/4$

## PROBLEM SET 16.4

**1.** $\frac{1}{12}$  **3.** $\frac{4}{9}$  **5.** $2\sqrt{3} + \frac{4}{3}\pi \approx 7.653$  **7.** $\pi a^2/8$

**9.** $8\pi + 6\sqrt{3} \approx 35.525$  **11.** $\pi(e^4 - 1) \approx 168.384$

**13.** $(\pi \ln 2)/8 \approx 0.272$  **15.** $\pi(2 - \sqrt{3})/2 \approx 0.421$

**17.** $\frac{1}{12}$  **19.** $81\pi/8 \approx 31.809$

**21.** $625(3\sqrt{3} + 1)/12 \approx 322.716$  **23.** $\frac{2}{3}\pi d^2(3a - d)$

**25.** $\frac{1}{9}a^3(6\pi - 8)$

## PROBLEM SET 16.5

**1.** $m = 30$; $\bar{x} = 2$; $\bar{y} = 1.8$

**3.** $m = \pi/4$; $\bar{x} = \pi/2$; $\bar{y} = 16/(9\pi)$

**5.** $m \approx 0.1056$; $\bar{x} \approx 0.281$; $\bar{y} \approx 0.581$

**7.** $m = 32/9$; $\bar{x} = 0$; $\bar{y} = 6/5$

**9.** $I_x \approx 269$; $I_y \approx 5194$; $I_z \approx 5463$

**11.** $I_x = I_y = 5a^5/12$; $I_z \approx 5a^5/6$

**13.** $\bar{r} = \sqrt{5/12}a \approx 0.6455a$  **15.** $I_x = \pi\delta a^4/4$; $\bar{r} = a/2$

**17.** $5\pi\delta a^4/4$  **19.** $\bar{x} = 0$, $\bar{y} = (15\pi + 32)a/(6\pi + 48)$

**21.** **(a)** $a^3$;  **(b)** $7a/12$;  **(c)** $11a^5/144$

**23.** $I_x = \pi ka^4/2$, $I_y = 17k\pi a^4/2$, $I_z = 9\pi ka^4$

## PROBLEM SET 16.6

**1.** $\sqrt{16}/3$  **3.** $\pi/3$  **5.** $9\sin^{-1}\left(\frac{2}{3}\right)$  **7.** $8\sqrt{2}$

**9.** $4\pi a(a - \sqrt{a^2 - b^2})$  **11.** $2a^2(\pi - 2)$

**13.** $\frac{1}{6}\pi a^2(5\sqrt{5} - 1)$  **15.** $\bar{x} = \bar{y} = 0$, $\bar{z} = \frac{1}{2}(h_1 + h_2)$

**17.** $A = \pi b^2$, $B = 2\pi a^2[1 - \cos(b/a)]$, $C = \pi b^2$,

$D = \pi b^2[2a/(a + \sqrt{a^2 + b^2})]$, $B < A = C < D$

## PROBLEM SET 16.7

**1.** $-40$  **3.** $\frac{189}{2}$  **5.** $\frac{2}{3}$  **7.** 156

**9.** $\int_0^1 \int_0^3 \int_0^{(1/6)(12-3x-2y)} f(x,y,z)\,dz\,dy\,dx$

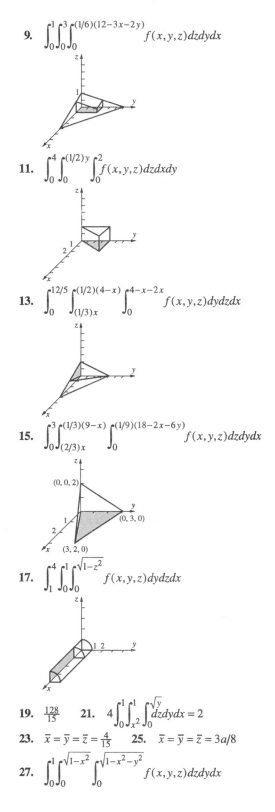

**11.** $\int_0^4 \int_0^{(1/2)y} \int_0^2 f(x,y,z)\,dz\,dx\,dy$

**13.** $\int_0^{12/5} \int_{(1/3)x}^{(1/2)(4-x)} \int_0^{4-x-2x} f(x,y,z)\,dy\,dz\,dx$

**15.** $\int_0^3 \int_{(2/3)x}^{(1/3)(9-x)} \int_0^{(1/9)(18-2x-6y)} f(x,y,z)\,dz\,dy\,dx$

**17.** $\int_1^4 \int_0^1 \int_0^{\sqrt{1-z^2}} f(x,y,z)\,dy\,dz\,dx$

**19.** $\frac{128}{15}$   **21.** $4\int_0^1 \int_{x^2}^1 \int_0^{\sqrt{y}} dz\,dy\,dx = 2$

**23.** $\bar{x} = \bar{y} = \bar{z} = \frac{4}{15}$   **25.** $\bar{x} = \bar{y} = \bar{z} = 3a/8$

**27.** $\int_0^1 \int_0^{\sqrt{1-x^2}} \int_0^{\sqrt{1-x^2-y^2}} f(x,y,z)\,dz\,dy\,dx$

**29.** $\int_0^2 \int_0^{2-z} \int_0^{9-x^2} f(x,y,z)\,dy\,dx\,dz$   **31.** 4

**33.** Ave $T = 29.54$, $\bar{z} = \frac{11}{24}$

**PROBLEM SET 16.8**

**1.** $8\pi$   **3.** $2\pi(5\sqrt{5}-4)/3 \approx 15.038$

**5.** $\bar{x} = \bar{y} = 0$; $\bar{z} = \frac{16}{3}$   **7.** $k\pi(b^4-a^4)$

**9.** $\bar{x} = \bar{y} = 0$; $\bar{z} = 2a/5$   **11.** $k\pi^2 a^6/16$   **13.** $\pi/9$

**15.** $\pi/32$   **17.** (a) $3a/4$;   (b) $3\pi a/16$;   (c) $6a/5$

**19.** (a) $3\pi a \sin\alpha/16\alpha$;   (b) $3\pi a/16$

**21.** $(a+b)(c-1)/(c+1)$

**CHAPTER REVIEW 16.9**

Concepts Test

**1.** True   **3.** True   **5.** True   **7.** False
**9.** True   **11.** True   **13.** False   **15.** True

Sample Test Problems

**1.** $\frac{1}{24}$   **3.** $\frac{2}{3}$   **5.** $\int_0^1 \int_0^y f(x,y)\,dx\,dy$

**7.** $\int_0^{1/2} \int_0^{1-2y} \int_0^{1-2y-z} f(x,y,z)\,dx\,dz\,dy$

**9.** (a) $8\int_0^a \int_0^{\sqrt{a^2-x^2}} \int_0^{\sqrt{a^2-x^2-y^2}} dz\,dy\,dx$;

(b) $8\int_0^{\pi/2} \int_0^a \int_0^{\sqrt{a^2-r^2}} r\,dz\,dr\,d\theta$;

(c) $8\int_0^{\pi/2} \int_0^{\pi/2} \int_0^a \rho^2 \sin\phi\,d\rho\,d\phi\,d\theta$

**11.** $0.8857$   **13.** $\bar{x} = \frac{13}{6}$; $\bar{y} = \frac{3}{2}$   **15.** 6

**17.** $80\pi k$   **19.** $ka^2 bc/24$

**PROBLEM SET 17.1**

**1.**   **3.**

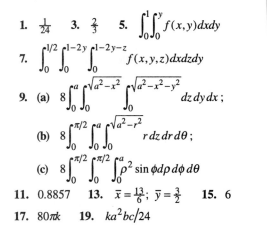

**5.**

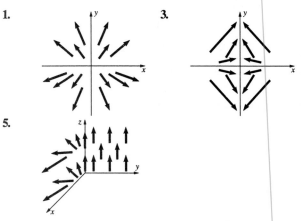

**7.** $(2x - 3y)\mathbf{i} - 3x\mathbf{j} + 2\mathbf{k}$   **9.** $x^{-1}\mathbf{i} + y^{-1}\mathbf{j} + z^{-1}\mathbf{k}$

**11.** $e^y \cos z\mathbf{i} + xe^y \cos z\mathbf{j} - xe^y \sin z\mathbf{k}$   **13.** $2yz;\ z^2\mathbf{i} - 2y\mathbf{k}$

**15.** $0; 0$   **17.** $2e^x \cos y + 1;\ 2e^x \sin y\mathbf{k}$

**19. (a)** Meaningless;   **(b)** vector field;   **(c)** meaningless;
**(d)** scalar field;   **(e)** vector field;   **(f)** vector field;
**(g)** vector field;   **(h)** meaningless;   **(i)** meaningless;
**(j)** scalar field;   **(k)** meaningless.

**25. (a)** div $\mathbf{F} = 0$, div $\mathbf{G} < 0$, div $\mathbf{H} = 0$, div $\mathbf{L} > 0$;
**(b)** clockwise for $\mathbf{H}$, not at all for others.

**(c)** div $\mathbf{F} = 0$, curl $\mathbf{F} = 0$, div $\mathbf{G} = -2ye^{-y^2}$, curl $\mathbf{G} = 0$,

div $\mathbf{H} = 0$, curl $\mathbf{H} = -2xe^{-x^2}\mathbf{k}$, div $\mathbf{L} = 1/\sqrt{x^2 + y^2}$,

curl $\mathbf{L} = 0$

## PROBLEM SET 17.2

**1.** $14(2\sqrt{2} - 1)$   **3.** $2\sqrt{5}$   **5.** $\frac{1}{16}(14\sqrt{14} - 1)$

**7.** $\frac{100}{3}$   **9.** 144   **11.** 0   **13.** $\frac{17}{6}$   **15.** 19

**17.** $k(17\sqrt{17} - 1)/6$   **19.** $-\frac{7}{44}$   **21.** $-\frac{1}{2}(a^2 + b^2)$

**23.** $2 - 2/\pi$   **25.** 2.25 gal   **27.** $2\pi a^2$   **29.** $4a^2$

**31. (a)** 27;   **(b)** $-297/2$

## PROBLEM SET 17.3

**1.** $f(x, y) = 5x^2 - 7xy + y^2 + C$   **3.** Not conservative.

**5.** $f(x, y) = \frac{2}{5}x^3 y^{-2} + C$   **7.** $f(x, y) = 2xe^y - ye^x + C$

**9.** $f(x, y, z) = x^3 + 2y^3 + 3z^3 + C$   **11.** 14   **13.** 6

**15.** $-\pi$   **19.** $f(x, y, z) = \frac{1}{2}k(x^2 + y^2 + z^2)$

**21.** $\int_C \mathbf{F} \cdot d\mathbf{r} = \int_a^b m\mathbf{r}''(t) \cdot \mathbf{r}'(t)\,dt$

$= \frac{1}{2}m\int_a^b (d/dt)[\mathbf{r}'(t) \cdot \mathbf{r}'(t)]\,dt = \frac{1}{2}m\int_a^b (d/dt)|\mathbf{r}'(t)|^2\,dt$

$= \left[\frac{1}{2}m|\mathbf{r}'(t)|^2\right]_a^b = \frac{1}{2}m\left[|\mathbf{r}'(b)|^2 - |\mathbf{r}'(a)|^2\right]$

**23.** $f(x, y, z) = -gmz$

## PROBLEM SET 17.4

**1.** $-\frac{64}{15}$   **3.** $\frac{72}{35}$   **5.** 0   **7.** $\frac{8}{3}$

**9. (a)** 0;   **(b)** 0   **11. (a)** 0;   **(b)** 0   **13.** 50

**15.** $-2$   **19. (c)** $M$ and $N$ have a discontinuity at $(0, 0)$.

**23.** $3\pi a^2/8$   **27. (a)** div $\mathbf{F} = 4$;   **(b)** 144

**29. (a)** div $\mathbf{F} < 0$ in quadrants I and III;
div $\mathbf{F} > 0$ in quadrants II and IV;
**(b)** 0; $-2(1 - \cos 3)^2$

## PROBLEM SET 17.5

**1.** $8\sqrt{3}/3$   **3.** $2 + \pi/3$   **5.** $5\pi/8$   **7.** 6   **9.** 2

**11.** 20   **13.** $\sqrt{3}ka^4/12$   **15.** $\bar{x} = \bar{y} = \bar{z} = a/3$

**17. (a)** 0;   **(b)** 0;   **(c)** $4\pi a^4$;   **(d)** $4\pi a^4/3$;
**(e)** $8\pi a^4/3$

**19. (a)** $4k\pi a^3$;   **(b)** $2k\pi a^3$;   **(c)** $hk\pi a(a + b)$

## PROBLEM SET 17.6

**1.** 0   **3.** $3a^2 b^2 c^2/4$   **5.** $64\pi/3$   **7.** $4\pi$   **9.** $1176\pi$

**11.** $\nabla \cdot \mathbf{F} = 3$ and so $\iint_{\delta S} \mathbf{F} \cdot \mathbf{n}\,dS = \iiint_S 3\,dV = 3V(S)$.

**15. (a)** $20\pi/3$;   **(b)** $4\pi$;   **(c)** $16\pi/3$;   **(d)** 1;
**(e)** 36;   **(f)** $12\pi/5$;   **(g)** $32\pi \ln 2$

## PROBLEM SET 17.7

**1.** 0   **3.** $-2$   **5.** $-48\pi$   **7.** $8\pi$   **9.** 2   **11.** $\pi/4$

**15.** $1/3$   **17.** $\frac{4}{3}a^2 J$

## CHAPTER REVIEW 17.8

### Concepts Test

**1.** True.   **3.** False   **5.** True   **7.** False
**9.** True   **11.** True.

### Sample Test Problems

**3.** $\text{curl}(f\nabla f) = f\,\text{curl}(\nabla f) + \nabla f \times \nabla f = 0 + 0 = 0$

**5. (a)** $\pi/4$;   **(b)** $(3\pi - 5)/6$   **7.** 47

**9. (a)** $\frac{1}{2}$;   **(b)** $\frac{4}{3}$;   **(c)** 0   **11.** $\frac{8}{3}\pi$   **13.** 0

**15.** $9\pi(3a - 2)/\sqrt{a^2 + b^2 + 1}$

## PROBLEM SET 18.1

**1.** $y = xe^{-x} + Ce^{-x}$   **3.** $y = a + C\sqrt{1 - x^2}$

**5.** $y = xe^x + Cx$   **7.** $y = 1 + Cx^{-1}$

**9.** $y = 1 + Ce^{-\int f(x)dx}$   **11.** $y = x^4 + 2x$

**13.** $y = e^{-x}(1 - x^{-1})$   **15.** 38.506 lb

**17.** $2(60 - t) - \frac{1}{1800}(60 - t)^3$

**19.** $I = 10^{-6}[1 - \exp(-10^6 t)]$

**21. (a)** $I = 0.0438 \sin 377t - 0.0578 \cos 377t + 0.0578e^{-285.7t}$;
**(b)** $I = 0.0438 \sin 377t - 0.0578 \cos 377t$

**23. (a)** 21.97;   **(b)** 26.67;   **(c)** $c > 7.72$;
**(d)** $20 + 80e^{-0.05T} + 0.25T = 50$;   **(e)** 24.10

**25. (a)** 200.32 ft;   **(b)** $15200(1 - e^{-0.05T}) - 640T = 0$;
**(c)** 7.08

## PROBLEM SET 18.2

**1.** $y = C_1 e^{2x} + C_2 e^{3x}$   **3.** $y = \frac{1}{2}e^x - \frac{1}{2}e^{-7x}$

**5.** $y = (C_1 + C_2 x)e^{2x}$   **7.** $y = e^{2x}(C_1 e^{\sqrt{3}x} + C_2 e^{-\sqrt{3}x})$

**9.** $y = 3\sin 2x + 2\cos 2x$

**11.** $y = e^{-x}(C_1 \cos x + C_2 \sin x)$

**13.** $y = C_1 + C_2 x + C_3 e^{-4x} + C_4 e^x$

**15.** $y = C_1 e^x + C_2 e^{-x} + C_3 \cos 2x + C_4 \sin 2x$

**17.** $y = D_1 \cosh 2x + D_2 \sinh 2x$

**19.** $y = e^{-x/2}[(C_1 + C_2 x)\cos(\sqrt{3}/2)x$
$\quad + (C_3 + C_4 x)\sin(\sqrt{3}/2)x]$

**21.** $y = (C_1 + C_2 \ln x)x^{-2}$

**25.** $y = 0.5 e^{5.16228x} + 0.5 e^{-5.16228x}$

**27.** $y = 1.29099 e^{-0.25x}\sin(0.968246x)$

**7.** $y = \frac{1}{4}e^x + C_1 e^{-3x} + C_2$   **9.** $y = 3e^{2x} - 3e^x$

**11.** $y = C_1 e^x + C_2 e^{-x} - 1$   **13.** $y = (C_1 + C_2 x + \frac{1}{2}x^2)e^{-2x}$

**15.** $y = e^{-3x}(C_1 \cos 4x + C_2 \sin 4x)$

**17.** $y = C_1 + C_2 e^{-4x} + C_3 e^{2x}$

**19.** $y = (C_1 + C_2 x)e^{\sqrt{2}x} + (C_3 + C_4 x)e^{-\sqrt{2}x}$

**21.** $y = -\cos 4t$; 1; $\pi/2$   **23.** $I = e^{-t}\sin t$

## PROBLEM SET 18.3

**1.** $y = C_1 e^{3x} + C_2 e^{-3x} - \frac{1}{9}x$

**3.** $y = (C_1 + C_2 x)e^x + x^2 + 5x + 8$

**5.** $y = C_1 e^{2x} + C_2 e^{3x} + \frac{1}{2}e^x$

**7.** $y = C_1 e^{-3x} + C_2 e^{-x} - \frac{1}{2}xe^{-3x}$

**9.** $y = C_1 e^{2x} + C_2 e^{-x} - \frac{3}{5}\sin x + \frac{1}{5}\cos x$

**11.** $y = C_1 \cos 2x + C_2 \sin 2x + \frac{1}{2}x\sin 2x$

**13.** $y = C_1 \cos 3x + C_2 \sin 3x + \frac{1}{8}\sin x + \frac{1}{13}e^{2x}$

**15.** $y = e^{2x} - e^{3x} + e^x$   **17.** $y = C_1 e^{2x} + C_2 e^x + \frac{5}{2}x + \frac{19}{4}$

**19.** $y = C_1 \sin x + C_2 \cos x - x\sin x - \cos x \ln|\sin x|$

**21.** $y = C_1 e^x + C_2 e^{2x} + (e^x + e^{2x})\ln(1 + e^{-x})$

## PROBLEM SET 18.4

**1.** $y = -\cos 8t$; $\pi/4$ s   **3.** 8 ft/s

**5.** $y \approx e^{-0.16t}(\cos 8t + 0.02\sin 8t)$   **7.** 14.4 s

**9.** $Q = 10^{-6}(1 - e^{-t})$

**11. (a)** $Q = 2.4 \times 10^{-4}\sin 377t$;
   **(b)** $I = 9.05 \times 10^{-2}\cos 377t$

**13.** $I \approx 12 \times 10^{-2}\sin 377t$   **17.** $d^2\theta/dt^2 = -(g/L)\sin\theta$

## CHAPTER REVIEW 18.5

Concepts Test

**1.** False   **3.** True   **5.** True   **7.** True   **9.** False

Sample Test Problems

**1.** $y = Cx^{-1}$   **3.** $y = 2e^{-x^2} + 1$   **5.** $y = -e^x + Ce^{2x}$

# Index

3.14159

FORMULA CARD
for
CALCULUS

**FORMULA CARD
to accompany
CALCULUS, 7/E**
*Varberg and Purcell*

## DERIVATIVES

$$D_x\, x^r = rx^{r-1} \qquad\qquad D_x\, |x| = \frac{|x|}{x}$$

$$D_x \sin x = \cos x \qquad\qquad D_x \cos x = -\sin x$$

$$D_x \tan x = \sec^2 x \qquad\qquad D_x \cot x = -\csc^2 x$$

$$D_x \sec x = \sec x \tan x \qquad\qquad D_x \csc x = -\csc x \cot x$$

$$D_x \sinh x = \cosh x \qquad\qquad D_x \coth x = -\operatorname{csch}^2 x$$

$$D_x \cosh x = \sinh x \qquad\qquad D_x \operatorname{sech} x = -\operatorname{sech} x \tanh x$$

$$D_x \tanh x = \operatorname{sech}^2 x \qquad\qquad D_x \operatorname{csch} x = -\operatorname{csch} x \coth x$$

$$D_x \ln x = \frac{1}{x} \qquad\qquad D_x \log_a x = \frac{1}{x \ln a}$$

$$D_x\, e^x = e^x \qquad\qquad D_x\, a^x = a^x \ln a$$

$$D_x \sin^{-1}x = \frac{1}{\sqrt{1-x^2}} \qquad D_x \cos^{-1}x = \frac{-1}{\sqrt{1-x^2}}$$

$$D_x \tan^{-1}x = \frac{1}{1+x^2} \qquad D_x \sec^{-1}x = \frac{1}{|x|\sqrt{x^2-1}}$$

## INTEGRALS

1. $\int u\, dv = uv - \int v\, du$

2. $\int u^n\, du = \dfrac{1}{n+1} u^{n+1} + C,\ n \neq -1$

3. $\int \dfrac{1}{u}\, du = \ln|u| + C$

4. $\int e^u\, du = e^u + C$

5. $\int a^u\, du = \dfrac{a^u}{\ln a} + C$

6. $\int \sin u\, du = -\cos u + C$

7. $\int \cos u\, du = \sin u + C$

8. $\int \sec^2 u\, du = \tan u + C$

9. $\int \csc^2 u\, du = -\cot u + C$

10. $\int \sec u \tan u\, du = \sec u + C$

11. $\int \csc u \cot u\, du = -\csc u + C$

12. $\int \tan u\, du = \ln|\sec u| + C$

13. $\int \cot u\, du = \ln|\sin u| + C$

14. $\int \sec u\, du = \ln|\sec u + \tan u| + C$

15. $\int \csc u\, du = \ln|\csc u - \cot u| + C$

16. $\int \dfrac{1}{\sqrt{a^2 - u^2}}\, du = \sin^{-1}\dfrac{u}{a} + C$

17. $\int \dfrac{1}{a^2 + u^2}\, du = \dfrac{1}{a}\tan^{-1}\dfrac{u}{a} + C$

18. $\int \dfrac{1}{a^2 - u^2}\, du = \dfrac{1}{2a}\ln\left|\dfrac{u+a}{u-a}\right| + C$

19. $\int \dfrac{1}{u\sqrt{u^2 - a^2}}\, du = \dfrac{1}{a}\sec^{-1}\left|\dfrac{u}{a}\right| + C$

## GEOMETRY

### Triangles

Pythagorean Theorem

$$a^2 + b^2 = c^2$$

Right triangle

Angles $\alpha + \beta + \gamma = 180°$
Area $A = \frac{1}{2}bh$

Any triangle

### Circles

Circumference $\quad C = 2\pi r$
Area $\qquad\qquad A = \pi r^2$

### Cylinders

Surface area $\quad S = 2\pi r^2 + 2\pi rh$
Volume $\qquad\quad V = \pi r^2 h$

### Cones

Surface area $\quad S = \pi r^2 + \pi r\sqrt{r^2 + h^2}$
Volume $\qquad\quad V = \frac{1}{3}\pi r^2 h$

### Spheres

Surface area $\qquad S = 4\pi r^2$
Volume $\qquad\qquad V = \frac{4}{3}\pi r^3$

### Conversions

1 inch = 2.54 centimeters
1 liter = 1000 cubic centimeters
1 kilogram = 2.20 pounds
1 kilometer = .62 miles
1 liter = 1.057 quarts
1 pound = 453.6 grams
$\pi$ radians = 180 degrees

# TRIGONOMETRY

## Basic Identities

$$\tan t = \frac{\sin t}{\cos t} \qquad \cot t = \frac{\cos t}{\sin t} \qquad \cot t = \frac{1}{\tan t}$$

$$\sec t = \frac{1}{\cos t} \qquad \csc t = \frac{1}{\sin t}$$

$$1 + \tan^2 t = \sec^2 t \qquad \sin^2 t + \cos^2 t = 1 \qquad 1 + \cot^2 t = \csc^2 t$$

## Cofunction Identities

$$\sin\left(\frac{\pi}{2} - t\right) = \cos t \qquad \cos\left(\frac{\pi}{2} - t\right) = \sin t \qquad \tan\left(\frac{\pi}{2} - t\right) = \cot t$$

## Odd-even Identities

$$\sin(-t) = -\sin t \qquad \cos(-t) = \cos t \qquad \tan(-t) = -\tan t$$

## Addition Formulas

$$\sin(s + t) = \sin s \cos t + \cos s \sin t$$
$$\cos(s + t) = \cos s \cos t - \sin s \sin t$$
$$\tan(s + t) = \frac{\tan s + \tan t}{1 - \tan s \tan t}$$

$$\sin(s - t) = \sin s \cos t - \cos s \sin t$$
$$\cos(s - t) = \cos s \cos t + \sin s \sin t$$
$$\tan(s - t) = \frac{\tan s - \tan t}{1 + \tan s \tan t}$$

## Double Angle Formulas

$$\sin 2t = 2 \sin t \cos t$$

$$\cos 2t = \cos^2 t - \sin^2 t = 1 - 2 \sin^2 t = 2 \cos^2 t - 1$$

$$\tan 2t = \frac{2 \tan t}{1 - \tan^2 t}$$

## Half Angle Formulas

$$\sin\frac{t}{2} = \pm\sqrt{\frac{1 - \cos t}{2}} \qquad \cos\frac{t}{2} = \pm\sqrt{\frac{1 + \cos t}{2}}$$

$$\tan\frac{t}{2} = \frac{1 - \cos t}{\sin t} \qquad \tan\frac{t}{2} = \frac{1 - \cos t}{\sin t}$$

## Product Formulas

$$2 \sin s \cos t = \sin(s + t) + \sin(s - t)$$
$$2 \cos s \cos t = \cos(s + t) + \cos(s - t)$$
$$2 \cos s \sin t = \sin(s + t) - \sin(s - t)$$
$$2 \sin s \sin t = \cos(s - t) - \cos(s + t)$$

## Factoring Formulas

$$\sin s + \sin t = 2 \cos\frac{s - t}{2} \sin\frac{s + t}{2}$$
$$\cos s + \cos t = 2 \cos\frac{s - t}{2} \cos\frac{s + t}{2}$$
$$\sin s - \sin t = 2 \cos\frac{s + t}{2} \sin\frac{s - t}{2}$$
$$\cos s - \cos t = -2 \sin\frac{s + t}{2} \sin\frac{s - t}{2}$$

## Laws of Sines and Cosines

$$\frac{\sin \alpha}{a} = \frac{\sin \beta}{b} = \frac{\sin \gamma}{c}$$

$$a^2 = b^2 + c^2 - 2bc \cos \alpha$$

## Graphs

$$\sin t = \sin \theta = y = \frac{b}{r} \qquad \cos t = \cos \theta = x = \frac{a}{r}$$

$$\tan t = \tan \theta = \frac{y}{x} = \frac{b}{a} \qquad \cot t = \cot \theta = \frac{x}{y} = \frac{a}{b}$$

$$y = \sin t \qquad y = \cos t$$
$$y = \tan t \qquad y = \cot t$$
$$y = \csc t \qquad y = \sec t$$

## Inverse Trigonometric Functions

$$y = \sin^{-1} x \Longleftrightarrow x = \sin y, \ -\pi/2 \leq y \leq \pi/2$$
$$y = \cos^{-1} x \Longleftrightarrow x = \cos y, \ 0 \leq y \leq \pi$$
$$y = \tan^{-1} x \Longleftrightarrow x = \tan y, \ -\pi/2 < y < \pi/2$$
$$y = \sec^{-1} x \Longleftrightarrow x = \sec y, \ 0 \leq y \leq \pi, y \neq \pi/2$$

$$\sec^{-1} x = \cos^{-1}(1/x)$$

## Hyperbolic Functions

$$\sinh x = \frac{1}{2}(e^x - e^{-x}) \qquad \cosh x = \frac{1}{2}(e^x + e^{-x})$$

$$\tanh x = \frac{\sinh x}{\cosh x} \qquad \coth x = \frac{\cosh x}{\sinh x}$$

$$\text{sech } x = \frac{1}{\cosh x} \qquad \text{csch } x = \frac{1}{\sinh x}$$

## Series

$$\tan^{-1} x = x - \frac{x^3}{3} + \frac{x^5}{5} - \frac{x^7}{7} + \cdots - 1 \leq x \leq 1$$

$$\ln(1 + x) = x - \frac{x^2}{2} + \frac{x^3}{3} - \frac{x^4}{4} + \cdots - 1 < x \leq 1$$

$$\frac{1}{1 - x} = 1 + x + x^2 + x^3 + \cdots - 1 < x < 1$$

$$e^x = 1 + x + \frac{x^2}{2!} + \frac{x^3}{3!} + \cdots$$

$$\sin x = x - \frac{x^3}{3!} + \frac{x^5}{5!} - \frac{x^7}{7!} + \cdots$$

$$\cos x = 1 - \frac{x^2}{2!} + \frac{x^4}{4!} - \frac{x^6}{6!} + \cdots$$

$$\sinh x = x + \frac{x^3}{3!} + \frac{x^5}{5!} + \frac{x^7}{7!} + \cdots$$

$$\cosh x = 1 + \frac{x^2}{2!} + \frac{x^4}{4!} + \frac{x^6}{6!} + \cdots$$

$$(1 + x)^p = 1 + \binom{p}{1}x + \binom{p}{2}x^2 + \binom{p}{3}x^3 + \cdots - 1 < x < 1$$

$$\binom{p}{k} = \frac{p(p - 1)(p - 2)\cdots(p - k + 1)}{k!}$$

# Table of Integrals

ELEMENTARY FORMS

$1 \quad \int u\,dv = uv - \int v\,du$

$2 \quad \int u^n\,du = \dfrac{1}{n+1}\,u^{n+1} + C \quad \text{if } n \neq -1$

$3 \quad \int \dfrac{du}{u} = \ln|u| + C$

$4 \quad \int e^u\,du = e^u + C$

$5 \quad \int a^u\,du = \dfrac{a^u}{\ln a} + C$

$6 \quad \int \sin u\,du = -\cos u + C$

$7 \quad \int \cos u\,du = \sin u + C$

$8 \quad \int \sec^2 u\,du = \tan u + C$

$9 \quad \int \csc^2 u\,du = -\cot u + C$

$10 \quad \int \sec u \tan u\,du = \sec u + C$

$11 \quad \int \csc u \cot u\,du = -\csc u + C$

$12 \quad \int \tan u\,du = \ln|\sec u| + C$

$13 \quad \int \cot u\,du = \ln|\sin u| + C$

$14 \quad \int \sec u\,du = \ln|\sec u + \tan u| + C$

$15 \quad \int \csc u\,du = \ln|\csc u - \cot u| + C$

$16 \quad \int \dfrac{du}{\sqrt{a^2 - u^2}} = \sin^{-1} \dfrac{u}{a} + C$

$17 \quad \int \dfrac{du}{a^2 + u^2} = \dfrac{1}{a}\tan^{-1}\dfrac{u}{a} + C$

$18 \quad \int \dfrac{du}{a^2 - u^2} = \dfrac{1}{2a}\ln\left|\dfrac{u+a}{u-a}\right| + C$

$19 \quad \int \dfrac{du}{u\sqrt{u^2 - a^2}} = \dfrac{1}{a}\sec^{-1}\left|\dfrac{u}{a}\right| + C$

TRIGONOMETRIC FORMS

$20 \quad \int \sin^2 u\,du = \dfrac{1}{2}u - \dfrac{1}{4}\sin 2u + C$

$21 \quad \int \cos^2 u\,du = \dfrac{1}{2}u + \dfrac{1}{4}\sin 2u + C$

$22 \quad \int \tan^2 u\,du = \tan u - u + C$

$23 \quad \int \cot^2 u\,du = -\cot u - u + C$

$24 \quad \int \sin^3 u\,du = -\dfrac{1}{3}(2 + \sin^2 u)\cos u + C$

$25 \quad \int \cos^3 u\,du = \dfrac{1}{3}(2 + \cos^2 u)\sin u + C$

$26 \quad \int \tan^3 u\,du = \dfrac{1}{2}\tan^2 u + \ln|\cos u| + C$

$27 \quad \int \cot^3 u\,du = -\dfrac{1}{2}\cot^2 u - \ln|\sin u| + C$

$28 \quad \int \sec^3 u\,du = \dfrac{1}{2}\sec u \tan u + \dfrac{1}{2}\ln|\sec u + \tan u| + C$

$29 \quad \int \csc^3 u\,du = -\dfrac{1}{2}\csc u \cot u + \dfrac{1}{2}\ln|\csc u - \cot u| + C$

$30 \quad \int \sin au \sin bu\,du = \dfrac{\sin(a-b)u}{2(a-b)} - \dfrac{\sin(a+b)u}{2(a+b)} + C \quad \text{if } a^2 \neq b^2$

$31 \quad \int \cos au \cos bu\,du = \dfrac{\sin(a-b)u}{2(a-b)} + \dfrac{\sin(a+b)u}{2(a+b)} + C \quad \text{if } a^2 \neq b^2$

$32 \quad \int \sin au \cos bu\,du = -\dfrac{\cos(a-b)u}{2(a-b)} - \dfrac{\cos(a+b)u}{2(a+b)} + C \quad \text{if } a^2 \neq b^2$

$33 \quad \int \sin^n u\,du = -\dfrac{1}{n}\sin^{n-1} u \cos u + \dfrac{n-1}{n}\int \sin^{n-2} u\,du$

$34 \quad \int \cos^n u\,du = \dfrac{1}{n}\cos^{n-1} u \sin u + \dfrac{n-1}{n}\int \cos^{n-2} u\,du$

$35 \quad \int \tan^n u\,du = \dfrac{1}{n-1}\tan^{n-1} u - \int \tan^{n-2} u\,du \quad \text{if } n \neq 1$

$36 \quad \int \cot^n u\,du = \dfrac{-1}{n-1}\cot^{n-1} u - \int \cot^{n-2} u\,du \quad \text{if } n \neq 1$

$37 \quad \int \sec^n u\,du = \dfrac{1}{n-1}\sec^{n-2} u \tan u + \dfrac{n-2}{n-1}\int \sec^{n-2} u\,du \quad \text{if } n \neq 1$

$38 \quad \int \csc^n u\,du = \dfrac{-1}{n-1}\csc^{n-2} u \cot u + \dfrac{n-2}{n-1}\int \csc^{n-2} u\,du \quad \text{if } n \neq 1$

$39a \quad \int \sin^n u \cos^m u\,du = -\dfrac{\sin^{n-1} u \cos^{m+1} u}{n+m} + \dfrac{n-1}{n+m}\int \sin^{n-2} u \cos^m u\,du \quad \text{if } n \neq -m$

$39b \quad \int \sin^n u \cos^m u\,du = \dfrac{\sin^{n+1} u \cos^{m-1} u}{n+m} + \dfrac{m-1}{n+m}\int \sin^n u \cos^{m-2} u\,du \quad \text{if } m \neq -n$

$40 \quad \int u \sin u\,du = \sin u - u \cos u + C$

$41 \quad \int u \cos u\,du = \cos u + u \sin u + C$

$42 \quad \int u^n \sin u\,du = -u^n \cos u + n\int u^{n-1} \cos u\,du$

$43 \quad \int u^n \cos u\,du = u^n \sin u - n\int u^{n-1} \sin u\,du$

## FORMS INVOLVING $\sqrt{u^2 \pm a^2}$

44  $\displaystyle\int \sqrt{u^2 \pm a^2}\ du = \frac{u}{2}\sqrt{u^2 \pm a^2} \pm \frac{a^2}{2}\ln|u + \sqrt{u^2 \pm a^2}| + C$

45  $\displaystyle\int \frac{du}{\sqrt{u^2 \pm a^2}} = \ln|u + \sqrt{u^2 \pm a^2}| + C$

46  $\displaystyle\int \frac{\sqrt{u^2 + a^2}}{u}\ du = \sqrt{u^2 + a^2} - a\ln\left(\frac{a + \sqrt{u^2 + a^2}}{u}\right) + C$

47  $\displaystyle\int \frac{\sqrt{u^2 - a^2}}{u}\ du = \sqrt{u^2 - a^2} - a\sec^{-1}\frac{u}{a} + C$

48  $\displaystyle\int u^2 \sqrt{u^2 \pm a^2}\ du = \frac{u}{8}(2u^2 \pm a^2)\sqrt{u^2 \pm a^2} - \frac{a^4}{8}\ln|u + \sqrt{u^2 \pm a^2}| + C$

49  $\displaystyle\int \frac{u^2\ du}{\sqrt{u^2 \pm a^2}} = \frac{u}{2}\sqrt{u^2 \pm a^2} \mp \frac{a^2}{2}\ln|u + \sqrt{u^2 \pm a^2}| + C$

50  $\displaystyle\int \frac{du}{u^2\sqrt{u^2 \pm a^2}} = \mp \frac{\sqrt{u^2 \pm a^2}}{a^2 u} + C$

51  $\displaystyle\int \frac{\sqrt{u^2 \pm a^2}}{u^2}\ du = -\frac{\sqrt{u^2 \pm a^2}}{u} + \ln|u + \sqrt{u^2 \pm a^2}| + C$

52  $\displaystyle\int \frac{du}{(u^2 \pm a^2)^{3/2}} = \frac{\pm u}{a^2\sqrt{u^2 \pm a^2}} + C$

53  $\displaystyle\int (u^2 \pm a^2)^{3/2}\ du = \frac{u}{8}(2u^2 \pm 5a^2)\sqrt{u^2 \pm a^2} + \frac{3a^4}{8}\ln|u + \sqrt{u^2 \pm a^2}| + C$


## FORMS INVOLVING $\sqrt{a^2 - u^2}$

54  $\displaystyle\int \sqrt{a^2 - u^2}\ du = \frac{u}{2}\sqrt{a^2 - u^2} + \frac{a^2}{2}\sin^{-1}\frac{u}{a} + C$

55  $\displaystyle\int \frac{\sqrt{a^2 - u^2}}{u}\ du = \sqrt{a^2 - u^2} - a\ln\left|\frac{a + \sqrt{a^2 - u^2}}{u}\right| + C$

56  $\displaystyle\int \frac{u^2\ du}{\sqrt{a^2 - u^2}} = -\frac{u}{2}\sqrt{a^2 - u^2} + \frac{a^2}{2}\sin^{-1}\frac{u}{a} + C$

57  $\displaystyle\int u^2 \sqrt{a^2 - u^2}\ du = \frac{u}{8}(2u^2 - a^2)\sqrt{a^2 - u^2} + \frac{a^4}{8}\sin^{-1}\frac{u}{a} + C$

58  $\displaystyle\int \frac{du}{u^2\sqrt{a^2 - u^2}} = -\frac{\sqrt{a^2 - u^2}}{a^2 u} + C$

59  $\displaystyle\int \frac{\sqrt{a^2 - u^2}}{u^2}\ du = -\frac{\sqrt{a^2 - u^2}}{u} - \sin^{-1}\frac{u}{a} + C$

60  $\displaystyle\int \frac{du}{u\sqrt{a^2 - u^2}} = -\frac{1}{a}\ln\left|\frac{a + \sqrt{a^2 - u^2}}{u}\right| + C$

61  $\displaystyle\int \frac{du}{(a^2 - u^2)^{3/2}} = \frac{u}{a^2\sqrt{a^2 - u^2}} + C$

62  $\displaystyle\int (a^2 - u^2)^{3/2}\ du = \frac{u}{8}(5a^2 - 2u^2)\sqrt{a^2 - u^2} + \frac{3a^4}{8}\sin^{-1}\frac{u}{a} + C$


## EXPONENTIAL AND LOGARITHMIC FORMS

63  $\displaystyle\int ue^u\ du = (u - 1)e^u + C$

64  $\displaystyle\int u^n e^u\ du = u^n e^u - n\int u^{n-1} e^u\ du$

65  $\displaystyle\int \ln u\ du = u\ln u - u + C$

66  $\displaystyle\int u^n \ln u\ du = \frac{u^{n+1}}{n+1}\ln u - \frac{u^{n+1}}{(n+1)^2} + C$

67  $\displaystyle\int e^{au}\sin bu\ du = \frac{e^{au}}{a^2 + b^2}(a\sin bu - b\cos bu) + C$

68  $\displaystyle\int e^{au}\cos bu\ du = \frac{e^{au}}{a^2 + b^2}(a\cos bu + b\sin bu) + C$


## INVERSE TRIGONOMETRIC FORMS

69  $\displaystyle\int \sin^{-1} u\ du = u\sin^{-1} u + \sqrt{1 - u^2} + C$

70  $\displaystyle\int \tan^{-1} u\ du = u\tan^{-1} u - \frac{1}{2}\ln(1 + u^2) + C$

71  $\displaystyle\int \sec^{-1} u\ du = u\sec^{-1} u - \ln|u + \sqrt{u^2 - 1}| + C$

72  $\displaystyle\int u\sin^{-1} u\ du = \frac{1}{4}(2u^2 - 1)\sin^{-1} u + \frac{u}{4}\sqrt{1 - u^2} + C$

73  $\displaystyle\int u\tan^{-1} u\ du = \frac{1}{2}(u^2 + 1)\tan^{-1} u - \frac{u}{2} + C$

74  $\displaystyle\int u\sec^{-1} u\ du = \frac{u^2}{2}\sec^{-1} u - \frac{1}{2}\sqrt{u^2 - 1} + C$

75  $\displaystyle\int u^n \sin^{-1} u\ du = \frac{u^{n+1}}{n+1}\sin^{-1} u - \frac{1}{n+1}\int \frac{u^{n+1}}{\sqrt{1 - u^2}}\ du + C$  if $n \neq -1$

76  $\displaystyle\int u^n \tan^{-1} u\ du = \frac{u^{n+1}}{n+1}\tan^{-1} u - \frac{1}{n+1}\int \frac{u^{n+1}}{1 + u^2}\ du + C$  if $n \neq -1$

77  $\displaystyle\int u^n \sec^{-1} u\ du = \frac{u^{n+1}}{n+1}\sec^{-1} u - \frac{1}{n+1}\int \frac{u^n}{\sqrt{u^2 - 1}}\ du + C$  if $n \neq -1$